Organic Reactions

Organic Reactions

VOLUME 74

WILEY

A JOHN WILEY & SONS, INC., PUBLICATION

Published by John Wiley & Sons, Inc., Hoboken, New Jersey

Published simultaneously in Canada.

Library of Congress Catalog Card Number: 42-20265
ISBN 978-0-470-53048-1

Printed in the United States of America

10 9 8 7 6 5 4 3 2 1

INTRODUCTION TO THE SERIES
ROGER ADAMS, 1942

In the course of nearly every program of research in organic chemistry, the investigator finds it necessary to use several of the better-known synthetic reactions. To discover the optimum conditions for the application of even the most familiar one to a compound not previously subjected to the reaction often requires an extensive search of the literature; even then a series of experiments may be necessary. When the results of the investigation are published, the synthesis, which may have required months of work, is usually described without comment. The background of knowledge and experience gained in the literature search and experimentation is thus lost to those who subsequently have occasion to apply the general method. The student of preparative organic chemistry faces similar difficulties. The textbooks and laboratory manuals furnish numerous examples of the application of various syntheses, but only rarely do they convey an accurate conception of the scope and usefulness of the processes.

For many years American organic chemists have discussed these problems. The plan of compiling critical discussions of the more important reactions thus was evolved. The volumes of *Organic Reactions* are collections of chapters each devoted to a single reaction, or a definite phase of a reaction, of wide applicability. The authors have had experience with the processes surveyed. The subjects are presented from the preparative viewpoint, and particular attention is given to limitations, interfering influences, effects of structure, and the selection of experimental techniques. Each chapter includes several detailed procedures illustrating the significant modifications of the method. Most of these procedures have been found satisfactory by the author or one of the editors, but unlike those in *Organic Syntheses*, they have not been subjected to careful testing in two or more laboratories. Each chapter contains tables that include all the examples of the reaction under consideration that the author has been able to find. It is inevitable, however, that in the search of the literature some examples will be missed, especially when the reaction is used as one step in an extended synthesis. Nevertheless, the investigator will be able to use the tables and their accompanying bibliographies in place of most or all of the literature search so often required. Because of the systematic arrangement of the material in the chapters and the entries in the tables, users of the books will be able to find information desired by reference to the table of contents of the appropriate chapter. In the interest of economy, the entries in the indices have been kept to a minimum, and, in particular, the compounds listed in the tables are not repeated in the indices.

The success of this publication, which will appear periodically, depends upon the cooperation of organic chemists and their willingness to devote time and effort to the preparation of the chapters. They have manifested their interest already by the almost unanimous acceptance of invitations to contribute to the work. The editors will welcome their continued interest and their suggestions for improvements in *Organic Reactions*.

INTRODUCTION TO THE SERIES
SCOTT E. DENMARK, 2008

In the intervening years since "The Chief" wrote this introduction to the second of his publishing creations, much in the world of chemistry has changed. In particular, the last decade has witnessed a revolution in the generation, dissemination, and availability of the chemical literature with the advent of electronic publication and abstracting services. Although the exponential growth in the chemical literature was one of the motivations for the creation of *Organic Reactions*, Adams could never have anticipated the impact of electronic access to the literature. Yet, as often happens with visionary advances, the value of this critical resource is now even greater than at its inception.

From 1942 to the 1980's the challenge that *Organic Reactions* successfully addressed was the difficulty in compiling an authoritative summary of a preparatively useful organic reaction from the primary literature. Practitioners interested in executing such a reaction (or simply learning about the features, advantages, and limitations of this process) would have a valuable resource to guide their experimentation. As abstracting services, in particular *Chemical Abstracts* and later *Beilstein*, entered the electronic age, the challenge for the practitioner was no longer to locate all of the literature on the subject. However, *Organic Reactions* chapters are much more than a surfeit of primary references; they constitute a distillation of this avalanche of information into the knowledge needed to correctly implement a reaction. It is in this capacity, namely to provide focused, scholarly, and comprehensive overviews of a given transformation, that *Organic Reactions* takes on even greater significance for the practice of chemical experimentation in the 21st century.

Adams' description of the content of the intended chapters is still remarkably relevant today. The development of new chemical reactions over the past decades has greatly accelerated and has embraced more sophisticated reagents derived from elements representing all reaches of the Periodic Table. Accordingly, the successful implementation of these transformations requires more stringent adherence to important experimental details and conditions. The suitability of a given reaction for an unknown application is best judged from the informed vantage point provided by precedent and guidelines offered by a knowledgeable author.

As Adams clearly understood, the ultimate success of the enterprise depends on the willingness of organic chemists to devote their time and efforts to the preparation of chapters. The fact that, at the dawn of the 21st century, the series continues to thrive is fitting testimony to those chemists whose contributions serve as the foundation of this edifice. Chemists who are considering the preparation of a manuscript for submission to *Organic Reactions* are urged to contact the Editor-in-Chief.

PREFACE TO VOLUME 74

Although not by design, Volume 74 comprises three chapters that represent two of the most fundamental processes in organic chemistry, namely, oxidation and reduction. In fact, 31 of the 241 chapters published in *Organic Reactions* are dedicated to the oxidation or reduction of organic compounds in some form. Thus, our three newest chapter family members are in good company.

The first chapter by Hans-Ulrich Blaser and Felix Spindler covers the catalytic, enantioselective hydrogenation of the azomethine function including imines, activated imines, oximes, hydrazones, and azacycles to form enantiomerically enriched amines. Organic compounds containing nitrogen atoms on one or more stereogenic centers are ubiquitous in nature and are of unparalleled importance in pharmaceutical substances. From amino acids and amino sugars to alkaloids and penicillins, chiral nitrogen-containing compounds are of significant synthetic and biological interest. Because chiral amines were traditionally available from classical resolution, the development of general methods for the enantioselective construction of stereogenic carbons bearing nitrogen substituents lagged behind that for the synthesis of chiral alcohols. Whereas catalytic, asymmetric hydrogenation of alkenes and carbonyl compounds is now a well-established process, the development of catalysts for production of enantiomerically enriched amines by hydrogenation of azomethines has only recently reached a level of practicality and selectivity that rivals those for carbonyl compounds. Blaser and Spindler clearly outline the unique challenges associated with the formation and reduction of imines and the currently available solutions. As is often the case, the selection of the appropriate transition metal, ligand and reaction conditions is crucial for success and these authors provide expert guidance from their extensive practical knowledge.

The second chapter by James M. Bobbitt, Christian Brückner, and Nabyl Merbouh covers the use of oxoammonium salts and nitroxides as catalysts for the oxidation of alcohols. The discovery of stable nitroxide-based free radicals opened many opportunities for their use as ESR spin labels in medicine. However, their impact in synthetic organic chemistry has been primarily as mild and selective catalysts for the oxidation of alcohols in the presence of a stoichiometric amount of a secondary oxidant. Bobbitt, Brückner, and Merbouh have critically distilled the vast literature on the most important practical aspects of these oxidations such as the choice of nitroxide (often the familiar TEMPO), the choice of stoichiometric oxidant such as hypochlorite, chlorite, peroxides, iodine(III) agents, and oxygen (in the presence of metal co-catalysts), and the importance of the medium (acidic or basic). The use of metal-free oxidations (particularly on large scale) benefits from a more selective oxidation process and a less toxic waste stream. Of special note is the ability of this method to prepare enantiomerically enriched aldehydes bearing configurationally labile stereogenic centers without racemization. The authors thoroughly cover the practical advances available for this method including the use of polymer-supported reagents, reactions in ionic liquids, and applications to sensitive carbohydrates and nucleosides.

The third chapter by Michael J. Porter and John Skidmore covers the asymmetric epoxidation of alkenes bearing electron-withdrawing groups. The synthetic utility of epoxides cannot be overstated and indeed, no fewer than five *Organic Reactions* chapters are concerned with the synthesis or transformations of oxiranes. However, the special challenges associated with oxidizing electron-deficient alkenes requires alternative solutions compared to normal, electron-rich alkenes. Whereas the asymmetric epoxidations of allylic alcohols and isolated alkenes have general solutions, the currently available methods for, inter alia, unsaturated ketones, aldehydes, esters, nitriles, and sulfones are much more varied. Accordingly, the thorough and focused presentation provided by Porter and Skidmore represents a welcome addition to this rapidly expanding field. The authors present a scholarly overview of the wide range of methods available and the various mechanistic and stereochemical attributes of each (ranging from the use of polypeptides, to transition metal catalysts to phase transfer catalysts to dioxiranes). The authors then describe the enantioselective oxidations of all the classes of substrates organized by functional group. This presentation will be of tremendous utility to the experimentalist hoping to apply this method to a given target. In view of the diversity of reagents used for these oxidations, the experimental conditions and representative procedures are particularly valuable. Together with the chapter on the Katsuki-Sharpless oxidation by Tsutomu Katsuki and Victor S. Martin (Volume 48) and the chapter on dioxirane epoxidations by Waldemar Adam, Chantu R. Saha-Möller, and Gong-Gui Zhao (Volume 61), this chapter provides coverage of all three major classes of alkenes that can be converted to enantiomerically enriched epoxides.

It is appropriate here to acknowledge the expert assistance of the entire editorial board, in particular Vittorio Farina, Michael Martinelli and Peter Wipf who shepherded the three chapters in this volume, respectively. The contributions of the authors, editors, and the publisher were expertly coordinated by the responsible secretary, Jeffery B. Press. In addition, the *Organic Reactions* enterprise could not maintain the quality of production without the dedicated efforts of its editorial staff, Dr. Linda S. Press and Dr. Danielle Soenen. Insofar as the essence of *Organic Reactions* chapters resides in the massive tables of examples, the authors' and editorial coordinators' painstaking efforts are highly prized.

SCOTT E. DENMARK
Urbana, Illinois

RALPH F. HIRSCHMANN
May 6, 1922 – June 20, 2009

Ralph F. Hirschmann, one of our most distinguished scientists, passed away peacefully on June 20, 2009. Born on May 6, 1922, in Fürth, Barvaria, Germany, the youngest son of Carl and Alice Hirschmann, he attended the Gymnasium in Fürth until December 1936 when, due to the rise of Nazi Germany, he convinced his parents to leave Germany for the United States. Settling in Kansas City, Missouri, Ralph completed high school and in 1939 entered Oberlin College in Ohio, graduating in 1943 with a degree in Chemistry. He immediately enlisted in the U.S. Army, became a naturalized citizen in 1944, and then served for three years in the Pacific theater, stationed primarily in Guadalcanal and Okinawa.

After World War II, Ralph was admitted to the University of Wisconsin, where in 1950 he completed his Ph.D. degree in Chemistry under the mentorship of Professor W. S. Johnson. During a visit to the University by Dr. Max Tishler of Merck and Co., Ralph was persuaded to join the Merck Process Laboratories in Rahway, New Jersey. There he initiated what was to become a stellar career in both medicinal and bioorganic chemistry, and management in the pharmaceutical industry. Shortly after joining Merck, Ralph met and married fellow Merck chemist, Lucy Aliminosa.

The early days at Merck were principally concerned with steroid and vitamin K synthesis, the chemical frontiers of that time. Ralph's interests, however, were not confined to organic chemistry! He quickly recognized the power of organic synthesis, in the best traditions of W.S. Johnson and Max Tishler, to explore relevant biomedical issues. This interest led to the early introduction of the principal of "drug latentiation," now widely recognized as the prodrug approach.

With the steroid field maturing in the early 1960s, the powers at Merck asked Ralph to initiate a program in peptide/protein synthesis. This proved to be a brilliant change in direction, and in 1969, in collaboration with Dr. Robert Denkewalter, Ralph led the Merck team to achieve the first total synthesis of an enzyme, ribonuclease A, simultaneously with the Merrifield synthesis at Rockefeller University. The Hirschmann approach entailed the tactic of fragment coupling, while the Merrifield approach exploited their recently introduced stepwise solid-phase synthetic method. Nearly every major newspaper in the western world reported their achievements as front-page news.

Although Ralph maintained a strong interest in basic biomedical research at Merck during the 1970s and 1980s, with a particular focus on the mechanism of action and bioactive conformation of somatostatin, a neuropeptide hormone, his administrative duties increasingly removed him from the laboratory. Indeed, at the time of his retirement in 1987 as Senior Vice-President for Basic Research, Ralph had responsibilities for much of the company's basic research, including the sites at Rahway, New Jersey, West Point, Pennsylvania, Terlings Park, UK, and Merck Frosst, Canada. During this time, under Ralph's leadership, Merck developed Vasotec, Lisinopril, Primaxin, Mevacor, Proscar and Ivomec. It was also during this period, that Ralph championed, and was a major benefactor to, many young academic chemists at the outset of their academic careers.

Upon retirement from Merck in 1987, Ralph was invited to join the faculty of the University of Pennsylvania, first as a Research Professor and later as the Rao Makineni Professor of Bioorganic Chemistry. He also held a concurrent appointment at the Medical University of South Carolina between 1987 and 1999. At Penn he presented a yearly graduate course on Bioorganic and Medicinal Chemistry and he served as research mentor to 56 graduate students and postdoctoral fellows. His research at Penn focused initially on two areas. Working with the Nicolaou group he focused on the design and synthesis of small mimics of the neuropeptide hormone somatostatin, introducing β-D-glucose as a privileged scaffold, and with the Smith group, the design, synthesis and evaluation of what are now known as nonpeptide foldamers and inhibitors of the aspartic acid proteases, specifically renin and HIV-1. In 1989, a third major program evolved that joined forces with Professor Stephen Benkovic, a biochemist at The Pennsylvania State University, and the synthetic chemistry of the Smith laboratory to develop a transition state analog that would hold the promise of eliciting catalytic antibodies capable of peptide and protein ligation. Ralph knew only too well the potential that such an antibody would hold for protein synthesis, having achieved the total synthesis of ribonuclease. Each of these programs proved highly successful. Some 59 joint publications emanated from these efforts. Moreover, during this period, Ralph presented more than 100 plenary lectures both in this country and abroad about the Penn research.

Ralph served as a member of the Board of Editors for Organic Reactions from 1976–1985 (Volumes 24–33), and was a member of the Organic Reaction Board of Directors from 1982–1987.

For his many scholarly achievements both in medicinal and bioorganic chemistry, Ralph received numerous prestigious honors, including three honorary Doctor of Science Degrees: Oberlin College (1969), University of Wisconsin, Madison (1996), and the Medical College of South Carolina (1997). He was elected a member of the American Academy of Arts and Sciences, the National Academy of Sciences, and a Senior Fellow of the Institute of Medicine. His awards included the Intra-Science Research Foundation Award, the Merck Board of Director's Award, the Nichols Medal, the Chemical Pioneer Award of the American Institute of Chemists, the Max Bergmann Kreis Award, the Carothers Award, the Guthikonda Memorial Award from Columbia University, the Joseph Rudiger Award of the European Peptide Society, and the Edward E. Smissman-Bristol Myers Award, and he was the only person to receive both the Alfred Burger Award in Medicinal Chemistry and the ACS Award in Medicinal Chemistry. In addition, he received the Williard Gibbs Medal, National Academy of Sciences Award for Industrial Application of Sciences, Arthur C. Cope Award, one of the highest awards of the American Chemical Society, and the National Medal of Science from President Clinton. Finally, the ACS-administered Award in Peptide Research and the Hirschmann-Makineni Professor in Chemistry at the University of Pennsylvania were established in his honor.

Ralph is survived by his wife, Lucy Hirschmann; his son, Ralph Frederick Hirschmann of Pasadena, California, and his daughter, Carla Hummel, MD, of Las Vegas, Nevada; his daughter-in-law, Karen Montle Hirschmann, and his son-in-law, Paul Hummel, MD; and his grandchildren, Brian, Lisa, Brendan, and Lauren Hirschmann, and Patrick and Christina Hummel.

Amos B. Smith, III

University of Pennsylvania
August 11, 2009

CONTENTS

CHAPTER 1

CATALYTIC ASYMMETRIC HYDROGENATION OF C=N FUNCTIONS

HANS-ULRICH BLASER and FELIX SPINDLER

Solvias AG, P.O. Box, CH-4002 Basel, Switzerland

CONTENTS

hans-ulrich.blaser@solvias.com

Organic Reactions, Vol. 74, Edited by Scott E. Denmark et al.

INTRODUCTION

Chiral amines are important targets for synthetic chemists and attempts to prepare such compounds via enantioselective hydrogenation of an appropriate C=N function date back to 1941.[1] Originally, only heterogeneous hydrogenation catalysts such as Pt black, Pd/C, or Raney nickel were employed. These classical hydrogenation catalysts were modified with chiral additives in the hope that some asymmetric induction in the delivery of dihydrogen to the reactant might occur. Only very few substrates were studied and not surprisingly, enantioselectivities were low and results could not always be reproduced.[2] The first reports on the use of homogeneous ruthenium[3] and rhodium[4,5] diphosphine complexes appeared in 1975, but useful enantioselectivities were not reported until 1984.[6] Remarkable progress has been made since the 1990's and a variety of very selective catalysts are now available for the enantioselective reduction of different types of C=N functions.[7−15] Moreover, the first industrial application was announced in 1996.[16] Despite this progress, the enantioselective hydrogenation of prochiral C=N groups such as imines, oximes, or hydrazones to the corresponding chiral amines still represents a major challenge.

Whereas many highly enantioselective catalysts have been developed for the asymmetric hydrogenation of alkenes and ketones bearing various functional groups, fewer catalysts are effective for the hydrogenation of substrates with a C=N function. Several reasons can be cited for this situation. The enantioselective hydrogenation of enamides and other C=C groups, and later of C=O compounds, has been so successful that most attention has been directed to these substrates. In addition, C=N compounds have some chemical peculiarities that make their stereoselective reduction more complex than that of C=O and C=C compounds. Even though the preparation of imines starting from the corresponding amine derivative and carbonyl compound is relatively simple, complete conversion is not always possible and formation of trimers or oligomers can occur. In addition, the resulting C=N compounds are often sensitive to hydrolysis, and, because many of the homogeneous catalysts can complex with both the starting material and the amine product, catalytic activity is often low. Further problems arise from the fact that imines can be in equilibrium with their corresponding enamines, which can also be reduced but with different stereoselectivities. Another problem is the potential coexistence of syn/anti imine isomers. These different forms may be reduced with different selectivities, as has been shown for the reduction of an oxime.[17]

Generally, the C=N substrates are prepared from the corresponding ketone and amino derivative and are hydrogenated as isolated (and purified) compounds. However, reductive amination where the C=N function is prepared in situ is attractive from an industrial point of view and indeed a number of successful examples have been reported.[18–20]

This chapter provides a comprehensive overview of the enantioselective hydrogenation (Eq. 1) and transfer hydrogenation (Eq. 2) of various C=N functions using chiral catalysts. Because the net transformation is the same for both types of reduction, the results for the various substrate types are summarized together. However, it should be noted that in many cases different catalyst types are required. For example, if dihydrogen is used, the metal must be able to activate the very strong H–H bond. In contrast, the activation seems to be more facile for the transfer of hydrogen from donor molecules such as formic acid and even more so from Hantzsch esters, which form aromatic products. In all equations, the specific hydrogen donor is shown in the equations, whereas only the pressure is given when dihydrogen is used. This review covers the literature up to September 2007. Several reviews on the asymmetric reduction of C=N functions have been published,[7,9–11,15,21,22] and the topic has been covered as part of various overviews on asymmetric hydrogenation.[8,12–14,23,24] Alternative reduction methods for C=N functions such as hydride reductions,[25,26] hydrosilylation,[26–28] and biocatalysis[29] are addressed briefly.

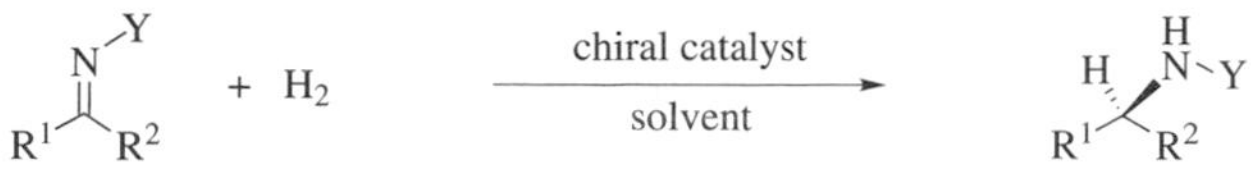

(Eq. 1)

$Y = R^3$, OH, NHR^3, SO_2R^3, $POPh_2$
$R^{1\text{-}3}$ = alkyl, (het)aryl

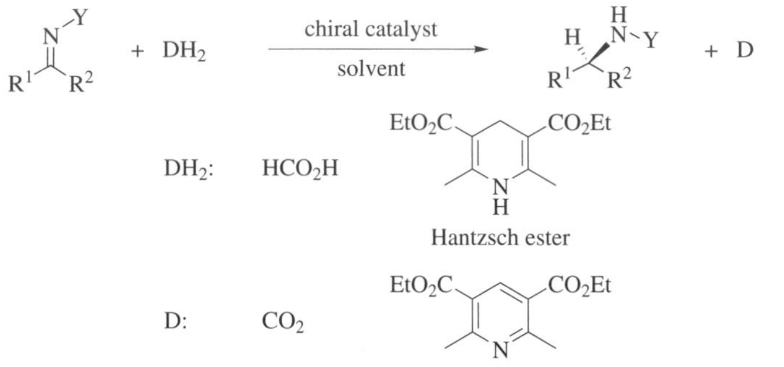

(Eq. 2)

SCOPE AND LIMITATIONS

Ligands and Catalysts

Most catalysts effective for the enantioselective (transfer) hydrogenation of C=N bonds are homogeneous complexes consisting of a central metal ion, one or more (chiral) ligands, and anions. There is always an interdependence between the nature of the C=N function and the most suitable catalyst. To identify an effective catalyst for any specific substrate, not only the optimum metal, but also the optimum ligand and, with somewhat lower priority, the optimum anion have to be chosen. Other reaction parameters are solvent, temperature, hydrogen pressure, and sometimes additives. Experience has shown that low-valent ruthenium, rhodium, and iridium complexes stabilized by tertiary (chiral) phosphorus-based ligands are the most active and the most versatile hydrogenation catalysts. As a result, the majority of research has focused on these types of complexes. Complexes that are able to directly activate dihydrogen have somewhat different ligand requirements than transfer hydrogenation catalysts, which formally transfer a hydrogen molecule from a suitable donor. For particular applications, cyclopentadienyl titanium and zirconium complexes and recently some Pd-diphosphine complexes show very good enantioselectivities in conjunction with dihydrogen; organocatalysts based on phosphoric acid are also promising.

Chiral Ligands. A plethora of chiral ligands have been developed for enantioselective hydrogenation. However, relatively few have proven effective as well as practical, and have actually been applied to the catalytic hydrogenation.[30,31] For the reduction of C=N functions, the most effective and most frequently used ligands are depicted in Fig. 1. For hydrogenations with dihydrogen, diphosphines such as binap (sometimes in combination with a diamine), biphep (and analogs), duphos, josiphos, and phosphine oxazoline ligands are most effective. For transfer hydrogenations, tosylated diphenylethylendiamine ligands (dpenTs) are most useful.

Recently, substituted binol esters of phosphoric acid (binol-P(O)OH) have been shown to be effective catalysts for transfer hydrogenation with Hantzsch

For a glossary of ligand structures, refer to Chart 1 immediately preceding the Tabular Survey.

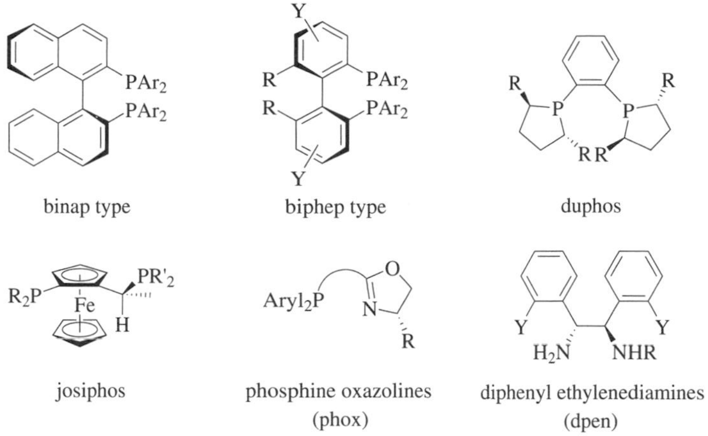

Figure 1. Structures and names of privileged ligands.

esters as donors. The naming of new ligands does not follow any rules. In this review we will use the name given by the creator of the ligand but will use small letters only, except when a chemical group is specified, as in MeO-biphep. For other ligands that have not been named, a bold number will be used with a short descriptor, for example: amino alcohol **14**. The reader is referred to Chart 1 preceding the Tabular Survey for structures of all ligands and catalysts referred to in this text. Those catalysts referred to by bold numbers only are reproduced in the text for convenience.

Metal Complexes. *Rhodium Catalysts.* Rh-diphosphine catalysts can be easily prepared from a rhodium precursor and a chiral ligand. The catalysts are either prepared in situ or applied as preformed and isolated complexes. In most cases, the in situ method is preferred because it offers greater flexibility, but there are cases where preformed complexes are required, either because the ligand is not stable, the complex formation is too slow, or the performance is superior. The most common and commercially available rhodium precursors are $[Rh(cod)Cl]_2$ and $[Rh(nbd)Cl]_2$ complexes for catalysts with a covalently bound anion, and $[Rh(nbd)_2]BF_4$ for cationic catalysts. Preformed complexes are of the type [Rh(nbd)(diphosphine)]Y (Y = BF_4, OTf, SbF_6, ClO_4). The catalytically active species are obtained by hydrogenation of the diene (cod or nbd), a process which can take some time.[32] Pentamethylcyclopentadienyl rhodium (cp*Rh) complexes having a tosylated diamine or an amino alcohol ligand are active transfer hydrogenation catalysts.

Iridium Catalysts. Among the various catalyst types investigated in recent years for the hydrogenation of imines, Ir-diphosphine and Ir-phosphine oxazoline complexes have proven to be most versatile. Ir-diphosphine catalysts are usually generated in situ from commercially available $[Ir(cod)Cl]_2$ and a chiral diphosphine and used in the presence of an iodide source. Also common are preformed complexes of the type $[Ir(diphosphine)Cl]_2$ or $[Ir(diphosphine)(cod)]BF_4$ and some Ir(III) complexes like HBrIr(diphosphine)OAc. The Ir-phox catalysts, [Ir(phox)(cod)]Y, are prepared starting from $[Ir(cod)Cl]_2$ and the PN ligand. The chloride is then exchanged for a non-coordinating anion Y, preferably BARF or PF_6. As in the case for rhodium, the diene must bc hydrogenated to obtain the active iridium catalyst. Cp*Ir complexes with a tosylated diamine or an amino alcohol ligand are active transfer hydrogenation catalysts.

Ruthenium Catalysts. In contrast to the wide scope of Ru-diphosphine complexes for the hydrogenation of ketones, their use for C=N reduction is still somewhat limited due to the tendency of these catalysts to deactivate in the presence of bases. For hydrogenation with dihydrogen, ruthenium catalysts are usually applied as preformed complexes of the type $Ru(diphosphine)Y_2$ (Y = Cl, OAc) or $Ru(diphosphine)(diamine)Cl_2$.

For transfer hydrogenations with formic acid–triethylamine, (arene)Ru (dpenTs)Cl (arene = benzene or *p*-cymene) complexes (Noyori transfer hydrogenation catalysts described later in the text) are the catalysts of choice. It is also possible to prepare these catalysts in situ from $Ru(cod)Cl_2$ or $[Ru(cymene)_2Cl]_2$ and the required sulfonylated diamine.

Palladium Catalysts. A very recent development is the use of Pd-diphosphine catalysts, especially for C=N–Y functions and imines of α-keto esters. Either complexes formed in situ from $Pd(CF_3CO_2)_2$ and a diphosphine or preformed $Pd(diphosphine)(CF_3CO_2)_2$ complexes can be used.

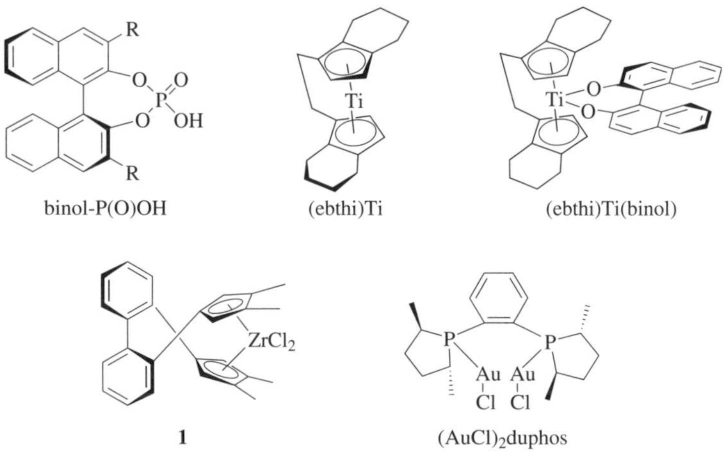

Figure 2. Structures of miscellaneous catalysts.

For a glossary of ligand structures, refer to Chart 1 immediately preceding the Tabular Survey.

Miscellaneous Catalysts (Fig. 2). Very recently, sterically hindered binol-derived phosphoric acids (binol-P(O)OH) have shown potential for metal free, "organocatalytic" transfer hydrogenations with Hantzsch esters as hydrogen donors. Bis(cyclopentadienyl) complexes (ebthi)Ti catalysts and Zr complex **1** can achieve remarkable enantioselectivities for the hydrogenation of cyclic imines with molecular hydrogen. However, their synthetic potential may be rather low, because the ligands and complexes are difficult to prepare, the activation of the catalyst precursor is tricky, and high catalyst loadings are needed. An unusual $(AuCl)_2$(Me-duphos) complex has been described for the hydrogenation of an *N*-benzyl imine where duphos is postulated to be coordinated to two gold atoms.

Substrates

The electronic and steric nature of the substituent directly attached to the nitrogen atom affects the properties of the C=N function (basicity, reduction potential, size, etc.) more than the substituents on the carbon atom. As a consequence, catalyst specificity can be highly substrate dependent. For example, Ir-diphosphine catalysts that are very active for *N*-aryl imines were found to deactivate rapidly when used with imines possessing aliphatic N-substituents;[33] titanocene-based catalysts are active for *N*-alkyl imines but not for *N*-aryl imines.[34,35] Oximes and other C=N–Y compounds show even more pronounced differences in reactivity. Because quite different catalysts and/or reaction conditions are optimal for a particular type of substrate, and to facilitate the search for the optimal catalyst for the reduction of a particular type of C=N compound, the following classes of substrates are distinguished: *N*-aryl imines, *N*-alkyl imines, endocyclic imines (including iminium derivatives), *N*-heteroarenes (including pyridinium ylides), and C=N–Y functions (including nitrones) (Fig. 3). Furthermore, the hydrogenation of α- and β-carboxy imine derivatives is discussed and compiled separately. The reductive amination of ketones where the imine is formed in situ is also considered separately.

When assessing the results compiled in this review, one has to keep in mind that most new ligands have only been tested under standard conditions with selected model substrates. The results are usually optimized for enantioselectivity, whereas catalyst productivity [substrate to catalyst ratio (s/c) or turnover number (TON, measured as mol product per mol catalyst)] and catalyst activity [turnover frequency (TOF, measured as TON per reaction time) at high conversions] are only a preliminary indication of the performance of a ligand. The decisive test, namely the application of a new ligand to "real world problems" are often yet to come and will eventually tell about the scope and limitations of a given ligand, or family of ligands, vs. changes in the substrate structure and/or the presence of functional groups. Indeed, relatively few catalyst systems have been optimized to date for complex synthetic or industrial applications.

For a glossary of ligand structures, refer to Chart 1 immediately preceding the Tabular Survey.

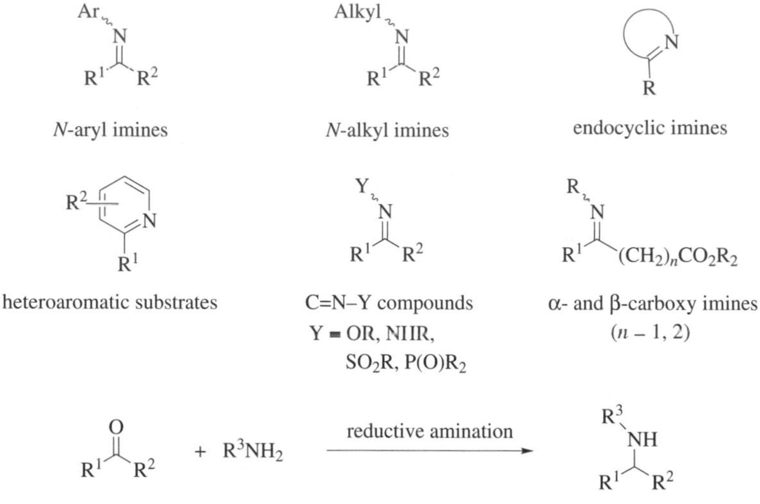

Figure 3. Substrate classes.

***N*-Aryl Imines.** The switch from the racemic form of metolachlor (see below), one of the major herbicides, to its *S*-enantiomer was undoubtedly the driving force for the development of suitable ligands and catalysts for the enantioselective hydrogenation of *N*-aryl imines.[16,36] Accordingly, much effort has been devoted to finding catalysts able to hydrogenate hindered *N*-aryl imines (Eq. 3). Interestingly, many iridium catalysts give very high enantioselectivities for 2,6-disubstituted *N*-aryl groups[33,37,38] and in some cases ee values of >99% are achieved with an f-binaphane ligand.[39]

(Eq. 3)

R^1	Ar	Catalyst		Ref.
$MeOCH_2$	2,6-$Me_2C_6H_3$	Ir-bdpp	96% conv., 90% ee	38
Ph	2,6-$Me_2C_6H_3$	Ir-josiphos	100% conv., 96% ee	37
Ph	2,6-$Me_2C_6H_3$	Ir-f-binaphane	77% conv., >99% ee	39
4-$CF_3C_6H_4$	2,6-$Me_2C_6H_3$	Ir-f-binaphane	80% conv., 99% ee	39

A number of Ir-diphosphine and Ir-phosphine oxazoline catalysts can achieve medium to very high enantioselectivities for model substrates derived from (substituted) acetophenones and (substituted) anilines (Eq. 4). Enantioselectivities of 96 to >99% and moderate catalytic activities are observed for iridium complexes of f-binaphane,[39] *P,N*-ferrocene **2**,[40] phosphino oxazoline **3**,[41] phosphino sulfoxime **4**,[42] *t*-Bu-bisP*,[43] and josiphos.[37] Enantioselectivities of 90–94% can be obtained with iridium complexes of phosphino oxazolines **5**[44] and **6**[45], with ddppm,[46] as well as with a Ru-duphos-dach catalyst[47] and the binol-P(O)OH transfer hydrogenation catalyst **7c**.[48] Enantiomeric excesses of 73–87% have been described for iridium complexes with phox2,[49,50] phosphine olefin **8**,[51] phosphino oxazolines **9**[52] and **10**,[53] phosphoramidite **11**,[54] and binol-P(O)OH **7b**.[55] In addition, a number of less selective iridium catalysts with P,N ligands (ee <52%) have been described.[56–58] Several systematic studies have shown that the effect of the substituents R^1 and R^2 on enantioselectivity is often significant, but no clear correlation between enantioselectivity and type and position of the R groups has been established.

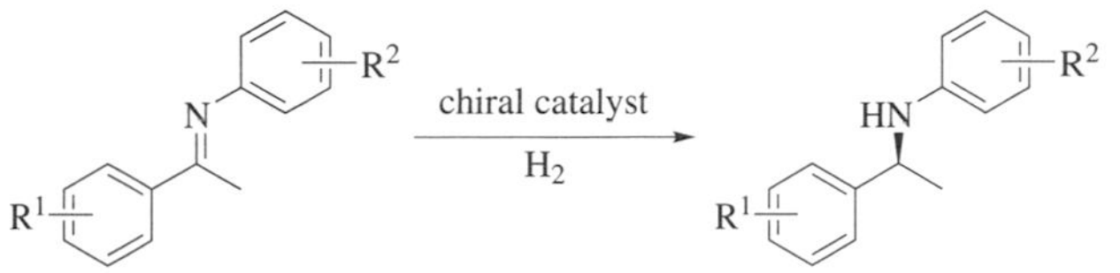

(Eq. 4)

R^1, R^2 = H, Me, MeO, CF_3, halogen

Chiral catalyst, reaction conditions		Ref.
f-binaphane, $[Ir(cod)Cl]_2$, s/c 100, DCM, 70 bar, 24–44 h	72–100% conv., 81 to >99% ee	39
[Ir(**2**)(cod)]BARF, s/c 100, toluene/MeOH, 10 bar, rt, 2–6 h	>99.5% conv., 84–99% ee	40
[Ir(**3**)(cod)]BARF, s/c 100, TBME, 4 Å MS, 1 bar, 10°, 20 h	>99.5% conv., 90–97% ee	41
4, $[Ir(cod)Cl]_2$, s/c 100, toluene, I_2, 20 bar, rt, 4–6 h	>99% conv., 90–96% ee	42
[Ir(*t*-Bu-bisP*)(cod)]BARF, s/c 200, DCM, 1 bar, rt, 2–12 h	(91–100%) 69–99% ee	43
josiphos (Ph/4-$CF_3C_6H_4$), $[Ir(cod)Cl]_2$, s/c 200, toluene, AcOH, TBAI, 30 bar, rt	100% conv., 96% ee	37
$[Ir(ddppm)(cod)]PF_6$, s/c 100, DCM, 1 bar, rt, 24 h	99–100% conv., 81–94% ee	46
[Ir(**5**)(cod)]BARF, s/c 200, DCM, 20 bar, rt, 2 h	53–99% conv., 83–90% ee	44
Hantzsch ester,7**c**, s/c 100, toluene, 35°, 42–72 h	84–98% conv., 80–93% ee	48
[Ir(**6**)(cod)]BARF, s/c 50, DCM, 20 bar, rt, 12 h	100% conv., 90% ee	45
Ru((*R,R*)-Et-duphos)((*R,R*)-dach)Cl_2, s/c 100, *t*-BuOH, *t*-BuOK, 15 bar, 65°, 20 h	92% conv., 92% ee	47

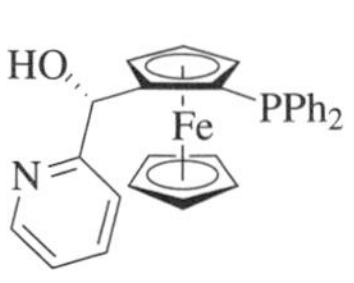

2

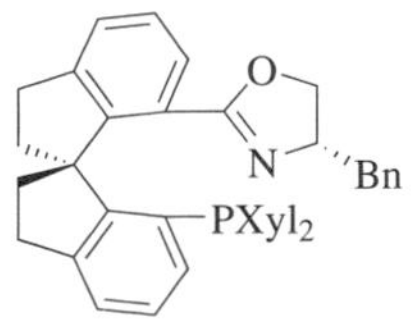

3

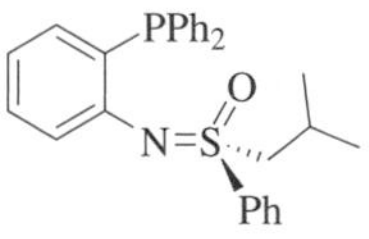

(*R*)-**4**

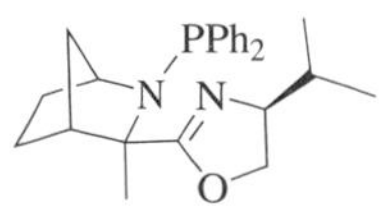

5

Naphthyl methyl ketone imines (Eq. 5) can be hydrogenated with similar enantioselectivities using iridium complexes with phosphino oxazoline **5**[44] or phosphino sulfoxime **4**.[42] Interestingly, the 1-naphthyl and 2-naphthyl derivatives

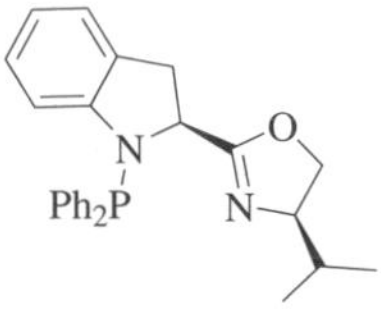

6

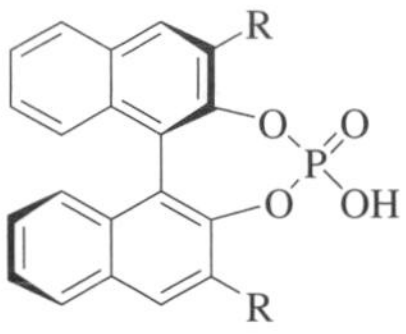

(*R*)-**7**

	R
7a	$Si(Ph)_3$
7b	3,5-$(CF_3)_2C_6H_3$
7c	2,4,6-$(i\text{-}Pr)_3C_6H_2$
7d	9-phenanthryl
7e	9-anthryl

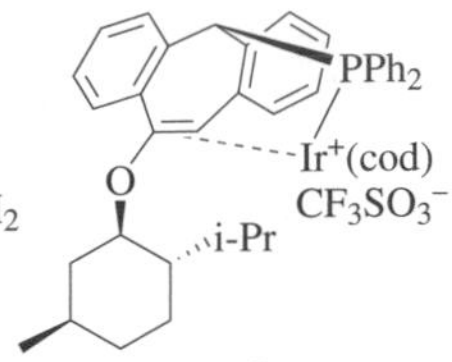

8

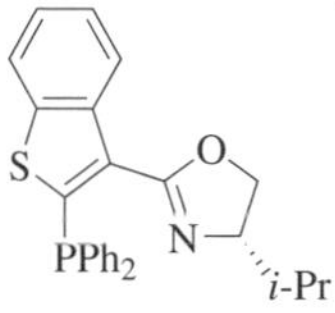

9

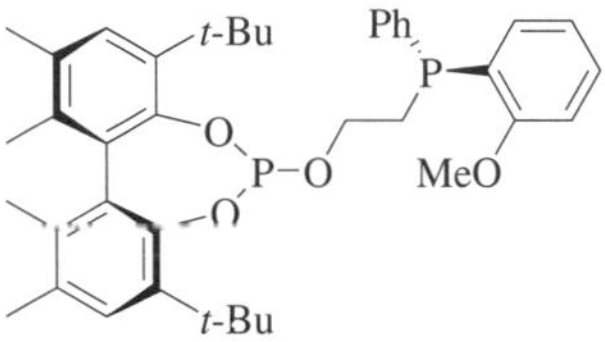

10

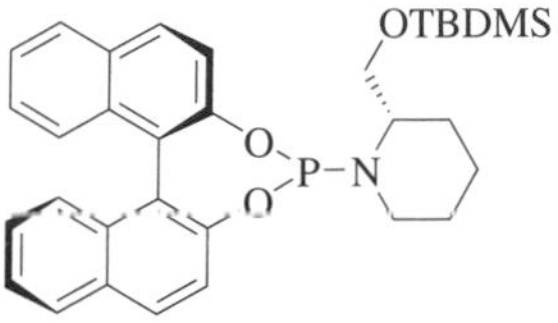

11

lead to opposite enantioselectivities with the catalyst system featuring ligand **4**. The iridium/phosphino sulfoxime **4** complex has also been shown to hydrogenate the imine from 4-methoxyaniline and tetralone in 91% ee.[42]

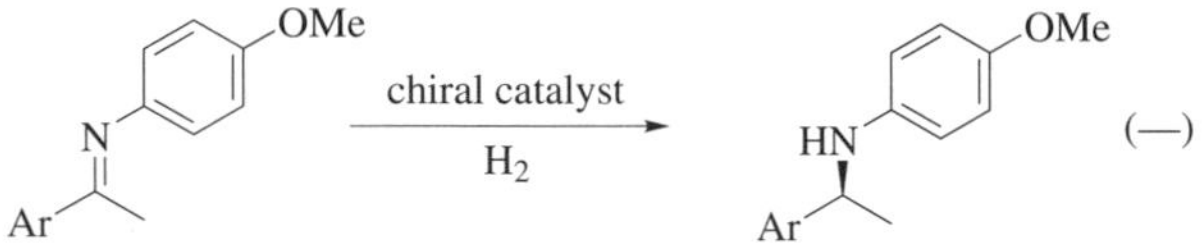

(Eq. 5)

Chiral catalyst, reaction conditions	Ar	% ee	Ref.
4, $[Ir(cod)Cl]_2$, s/c 100, toluene, I_2, 20 bar, rt, 4–6 h	1-Np	98 (+)	42
	2-Np	69 (–)	42
[Ir(**5**)(cod)]BARF, s/c 200, DCM, 20 bar, rt, 1.5 h	Ph	89 (+)	44

***N*-Alkyl Imines.** To date, few reductions of acyclic *N*-alkyl imines to the corresponding amines are of synthetic or industrial importance. Most studies reported in this area were carried out with simple model substrates, especially with the *N*-benzyl imine of acetophenone, related substituted derivatives, and some analogs thereof. One reason for this substrate choice could be the easy preparation of a pure crystalline starting material. Another is that synthetically useful chiral phenethylamines can be obtained by hydrogenolysis of the benzyl group. In comparison with the reduction of *N*-aryl imines, ee values obtained with *N*-alkyl imines are generally modest. Enantioselectivities of >90% can be achieved with rhodium complexes of bddp, either sulfated[59,60] or in reversed micelles,[61] with a Rh(cycphos) complex,[62] with a Ru(dppach)(dach)HCl complex,[63] and with Ru-dpenTs in water[64] (Eq. 6). Enantioselectivities of 70–83% have been described for (substituted) *N*-benzyl imines of acetophenone, 2-furyl, and 2-naphthyl methyl ketones with iridium complexes of phosphino oxazoline **6**,[45] phox2,[50,65] binol-POH **12**,[66] tol-binap,[67] and phosphine oxide **13**;[68] with a Rh-bdpch complex,[69] with (ebthi)Ti(binol),[34,70] and, somewhat surprisingly, with an Au/Me-duphos complex.[71] Several papers have described results with <70% ee for a number of Ir-P,N[57,72–75] and Ru[76] catalysts.

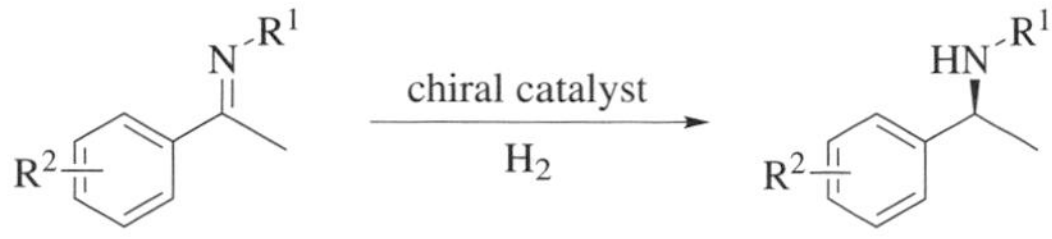

(Eq. 6)

R^1 = Bn, *n*-Bu; R^2 = H, MeO

Chiral catalyst, reaction conditions		Ref.
$bdpp_{sulf}$, $[Rh(cod)Cl]_2$, s/c 100, H_2O/AcOEt, 70 bar, rt, 16 h	93–96% conv., 86–96% ee	59, 60
12, $[Ir(cod)Cl]_2$, PPh_3, s/c 100, DCM, 50 bar, rt, 48 h	99–100% conv., 88–92% ee	66
Ru(dppach)(dach)HCl, s/c 1500, *i*-PrOK, 3 bar, 20°, 60 h	91% conv., 92% ee	63
$[Rh(bdpp)(nbd)]ClO_4$, s/c 100, C_6H_6/reverse micelles, 70 bar, 4–8°, 21–73 h	(95%) 92% ee	61
cycphos, $[Rh(cod)Cl]_2$, s/c 100, C_6H_6/MeOH, KI, 70 bar, rt, 90–144 h	(>99%) 91% ee	62
HCO_2Na, $[(C_6Me_6)Ru(dachTs)H_2O]BF_4$, s/c 100, H_2O, pH 9, 60°, 2–5 h	100% conv., 91% ee	64

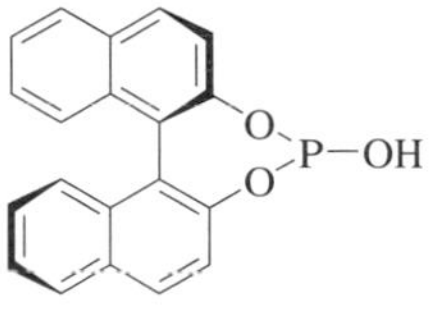

12

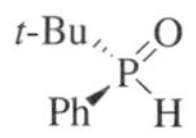

(*R*)-**13**

Only (ebthi)Ti(binol) catalysts have been described to hydrogenate *N*-alkyl imines of aliphatic ketones (Eq. 7), and very high pressures are required for good results.[34,70]

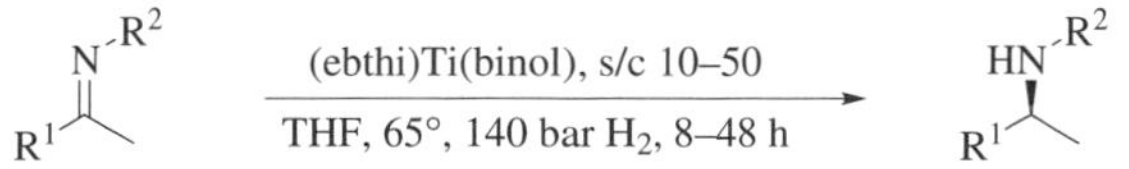

(Eq. 7)

R^1 = *n*-, *i*- and cycloalkyl
R^2 = Me, *n*-Pr, Bn

(64–93%) 53–92% ee

The ruthenium-catalyzed transfer hydrogenation of α-substituted exocyclic imines occurs with excellent cis-diastereoselectivity by a dynamic kinetic asymmetric transformation, with up to 97% ee for 5-membered rings but only 50% for 6-membered rings (Eq. 8).[77]

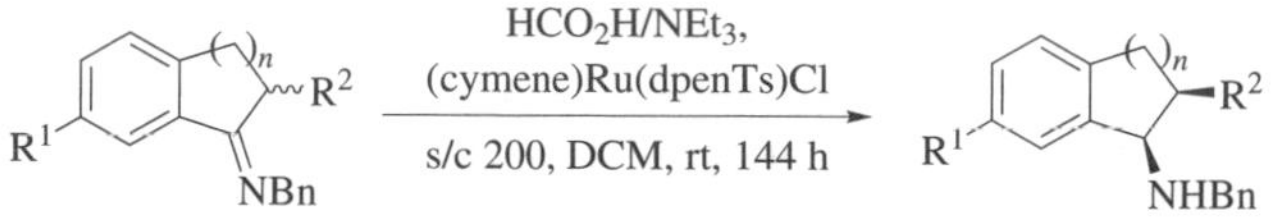
HCO2H/NEt3,
(cymene)Ru(dpenTs)Cl
s/c 200, DCM, rt, 144 h
R1
R2
NBn
NHBn

(Eq. 8)

n	R^1	R^2	
1	H	Me	(70%) 96% ee
1	Me	Me	(82%) 97% ee
1	Me	$CH_2{=}CHCH_2$	(67%) 92% ee
2	H	Me	(45%) 50% ee

Endocyclic Imines. Because cyclic imines do not have the problem of syn/anti isomerism, in principle higher enantioselectivities might be expected in their reduction. While this expectation is not generally met, in conjunction with several cyclic model substrates the (ebthi)Ti catalyst achieves up to 99% ee (Eq. 9). With this catalyst, enantioselectivities for acyclic imines are ≤92%, as described above.[34,78]

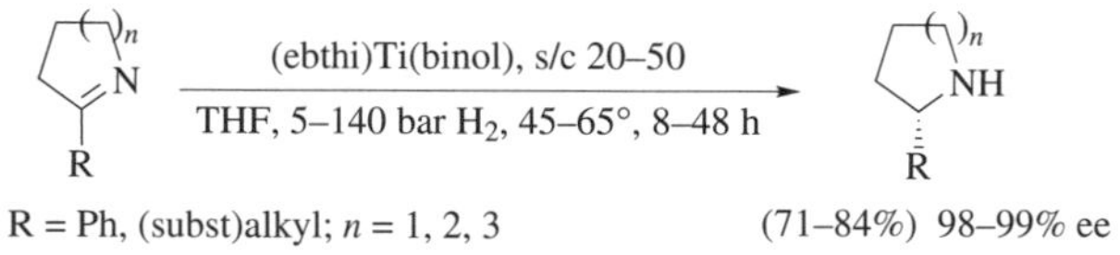
(ebthi)Ti(binol), s/c 20–50
THF, 5–140 bar H2, 45–65°, 8–48 h
N
NH
R
R = Ph, (subst)alkyl; n = 1, 2, 3
(71–84%) 98–99% ee

(Eq. 9)

One enantiomer of racemic disubstituted pyrrolines can be reduced with very high selectivities (Eq. 10a, kinetic resolution).[79] Unfortunately, these highly selective catalysts operate at rather low s/c ratios, exhibit TOFs of $<3\ h^{-1}$ and, in addition, functional groups like esters, carboxylic acids, or nitriles are not tolerated. Zirconium complex **1** exhibits similar properties as the (ebthi)Ti catalyst (ee 96% for R = Ph, $n = 1$ in Eq. 9) but is more effective (TON up to 1000).[80] Simple 5- and 6-membered endocyclic imines are hydrogenated using Ir-binap catalysts[67,81] with enantioselectivities of 89–91% ee. Moderate enantioselectivities of 50–78% ee, but quite good catalytic activities (TOF 100–1000 h^{-1}) are obtained for the hydrogen transfer hydrogenation of various azirines using a catalyst prepared from $[RuCl_2(cymene)]_2$ and amino alcohol **14** (Eq. 10b).[82]

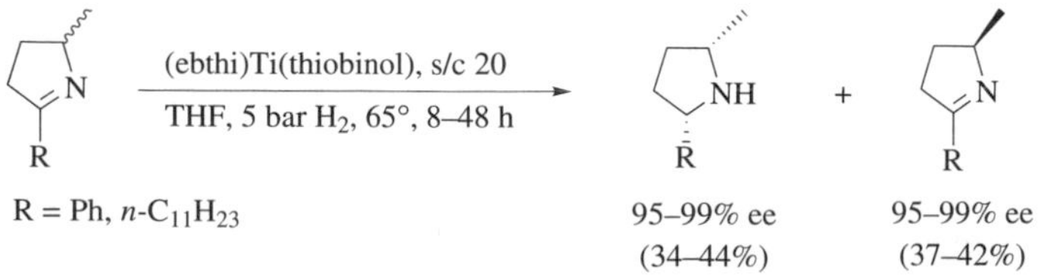
(ebthi)Ti(thiobinol), s/c 20
THF, 5 bar H2, 65°, 8–48 h
N
NH
+
R
R = Ph, n-C11H23
95–99% ee
(34–44%)
95–99% ee
(37–42%)

(Eq. 10a)

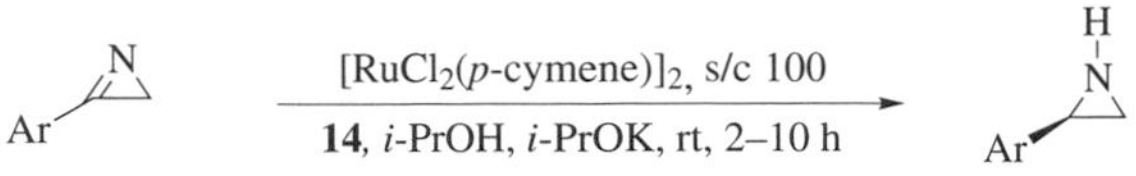

(Eq. 10b)

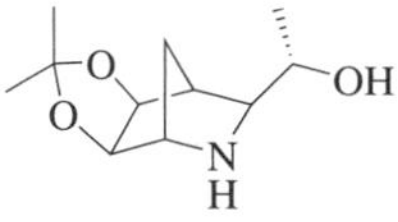

14

Ar		
Ph	(80%)	70% ee
4-BrC_6H_4	(92%)	50% ee
2-Np	(92%)	50% ee

A number of bi- and tricyclic imines have been investigated extensively. 2,3,3-Trimethyl-3*H*-indole (TMI) can be reduced in up to 94% ee using a variety of iridium complexes with diphosphine ligands such as bdpp,[83] bicp,[84] ferrocenyl based ligands,[37,85,86] bcpm,[87] the diop analog **15**,[88] as well as monophos[89] (Eq. 11, absolute configuration not reported). Only moderate TONs (s/c 100–250) and rather low activities (TOF 1–10 h^{-1}) are achieved, most likely because of steric hindrance. Interestingly, the best enantioselectivities are observed in the presence of a variety of additives such as phthalimides, iodine, or iodide/acid. The role of these additives is not clear. The reaction with an Ir-josiphos catalyst can also be carried out in ionic liquids with slightly lower enantioselectivities but similar catalyst activities.[85] Whereas no successful transfer hydrogenation is reported for TMI, the *N*-benzylated iminium derivative is reduced with formic acid–triethylamine in the presence of cp*Rh(dpenTs)Cl with 76% ee.[90]

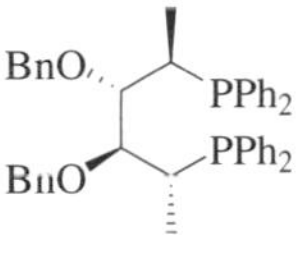

15

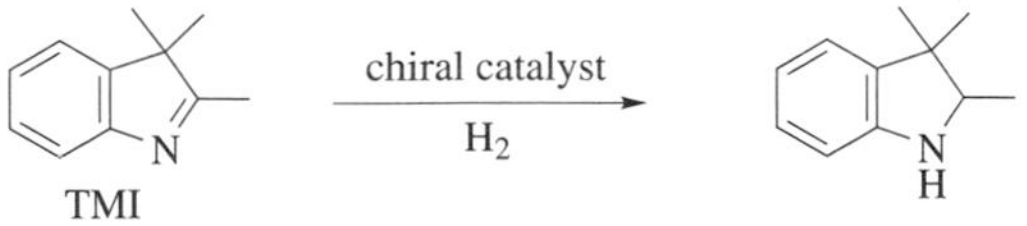

(Eq. 11)

TMI

Chiral catalyst, reaction conditions		Ref.
bicp, $[Ir(cod)Cl]_2$, s/c 100, DCM, phthalimide, 70 bar, 0°, 100 h	100% conv., 95% ee	84
josiphos (Xyl/Xyl), $[Ir(cod)Cl]_2$, s/c 250, toluene, TFA/TBAI, 40 bar, 30°, 47 h	100% conv., 95% ee	37
bcpm, $[Ir(cod)Cl]_2$, s/c 100, C_6H_6/MeOH, BiI_3, 100 bar, –30°, 90 h	92% conv., 91% ee	87

Several bi- and tricyclic imines have been investigated as intermediates or model substrates for biologically active compounds (Eqs. 12 and 13; see also Applications to Synthesis). These compounds are reduced with good to very good enantioselectivities using a number of different catalytic systems. Interestingly, most reactions were not reported by catalyst specialists, but rather by synthetic organic chemists. This might explain why transfer hydrogenation was the preferred experimental method in these studies. Enantioselectivities up to 99% are described for variants of the Noyori type catalysis, i.e., transfer hydrogenation using formic acid–triethylamine as the reducing agent in the presence of an arene ruthenium complex with dpenTs as the chiral ligand.[91–98] With water-soluble ligands, the reaction can be carried out with comparable enantioselectivities in water with sodium formate as the reducing agent.[64,99,100] Similar results are also obtained with cp*Rh(dpenTs) complexes.[101] The s/c ratios vary widely between 20 for some hindered substrates[101] and a respectable 1000 for more active catalysts.[91] (Ebthi)Ti[34], Ir-bcpm, or Ir-binap catalysts in the presence of additives[102,103] can achieve 86–98% ee with s/c ratios of 20–100. An Ir-P,N catalyst was less stereoselective (34% ee).[58]

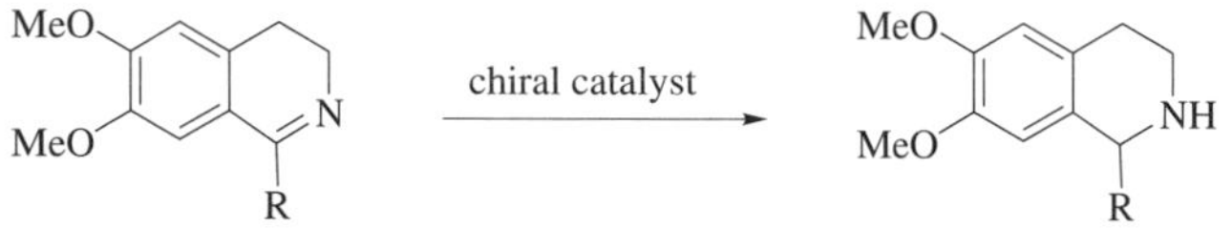

(Eq. 12)

R = alkyl, cycloalkyl

Chiral catalyst, reaction conditions		Ref.
(ebthi)Ti(binol), s/c 20, THF, 135 bar H_2, 65°, 8–48 h	(82%) 98% ee	34
HCO_2H/NEt_3, cp*Rh(dpenTs)Cl, s/c 200, DCM, 20°, 0.15 h	(93–96%) 83–99% ee	101
HCO_2H/NEt_3, (cymene)Ru(dpenTs)Cl, s/c 100–1000, MeCN, 28°, 12 h	(90–>99%) 84–95% ee	91
HCO_2Na, $dpenTs_{sulf}$, $[(cymene)RuCl_2]_2$, s/c 100, H_2O, CTAB, 28°, 10–25 h	(85–98%) 90–98% ee	99
HCO_2Na, $dpenTs_{amin}$, $[Cp^*RhCl_2]_2$, s/c 100, H_2O, 28°, 8 h	(95%) 93% ee	100
Bcpm, $[Ir(cod)Cl]_2$, s/c 100, toluene/MeOH, var. additives, 100 bar H_2, rt, 24–72 h	(84–99%) 86–89% ee	102, 103
HCO_2Na, $[(cymene)Ru(dachTs)H_2O]BF_4$, s/c 100, H_2O, pH 9, 60°, 2–5 h	100% conv., 88% ee	64

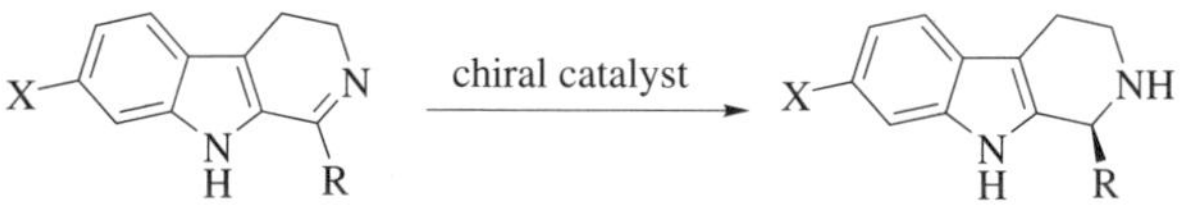

(Eq. 13)

X = H, Br; R = alkyl, (het)aryl

Chiral catalyst, reaction conditions		Ref.
HCO_2Na, $dpenTs_{sulf}$, $[(cymene)RuCl_2]_2$, s/c 100, s/c 500, H_2O, CTAB, 28°, 4–30 h	(83–99%) 98–99% ee	99
HCO_2H/NEt_3, (cymene)Ru(dpenTs)Cl, s/c 230, MeCN, rt, 12 h	(70–85%) >98% ee	92
HCO_2H/NEt_3, (cymene)Ru(dpenTs)Cl, s/c 200, DMF, 28°, 5 h	(83–86%) 96–97% ee	91
HCO_2Na, $dpenTs_{amin}$, $[cp^*RhCl_2]_2$, s/c 100, H_2O, 20°, 10 h	(94%) 93% ee	100
HCO_2H/NEt_3, (cymene)Ru((S,S)-dpenTs)Cl, s/c 25, DMF, 20°, 12 h	(89–96%) 93–96% ee	97

Various bicyclic imines can be reduced with high enantioselectivities but moderate to low catalytic activity via an organocatalytic hydrogen transfer reaction with sterically hindered binol phosphoric acid catalysts **7a** and **7d** using a Hantzsch ester as the reducing agent (Eq. 14).[104,105] In some cases, s/c ratios as high as 1000 are reported, albeit with very long reaction times.

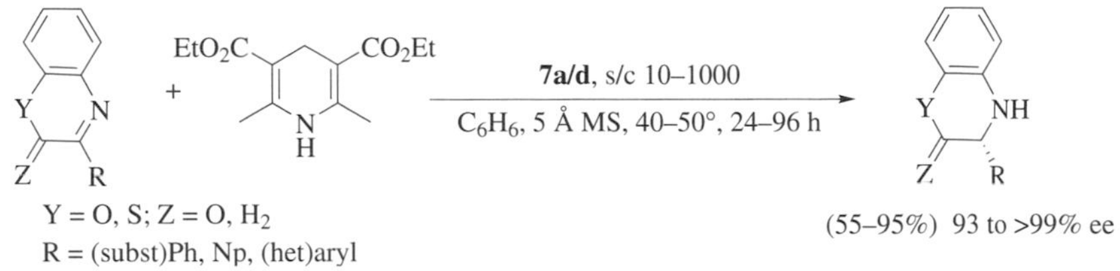
EtO2C
CO2Et
+
Y
N
Z
R
N
H
7a/d, s/c 10–1000
C6H6, 5 Å MS, 40–50°, 24–96 h
Y
NH
Z
R
Y = O, S; Z = O, H2
R = (subst)Ph, Np, (het)aryl
(55–95%) 93 to >99% ee

(Eq. 14)

Tri- and tetracyclic iminium compounds (intermediates in the synthesis of alkaloids) are amenable to the Ru-dpenTs-catalyzed transfer-hydrogenation with 79–92% ee but modest s/c ratios (Eq. 15).[106,107] Similar cyclic amines can be obtained in 70% yield and 50–70% ee via tricyclic iminium compounds that are formed in situ.[108]

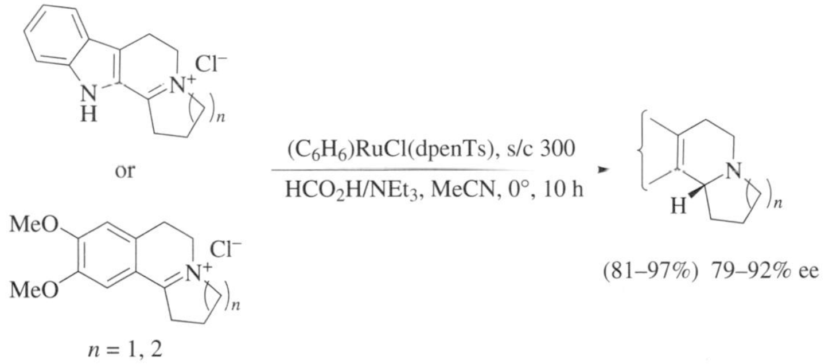
Cl⁻
N
H
N+
n
or
MeO
MeO
Cl⁻
N+
n
n = 1, 2
(C6H6)RuCl(dpenTs), s/c 300
HCO2H/NEt3, MeCN, 0°, 10 h
N
H
n
(81–97%) 79–92% ee

(Eq. 15)

Heteroaromatic Substrates. Until very recently, the hydrogenation of heteroaromatic substrates with homogeneous catalysts was considered to be very difficult. In the last few years, a number of catalytic systems with reasonable activities have been developed for the partial hydrogenation of substituted quinolines, giving access to a variety of cyclic amines in fair to very good enantioselectivities. However, up to now, results for pyridines or pyrazines are disappointing (ee <30%), probably due to their more aromatic character.[109] Exceptions are the Ir-josiphos-catalyzed hydrogenation of a pyrazine ester (Eq. 16) with up to 78% ee but very low catalyst activity[110] and the recently reported organocatalytic transfer hydrogenation of pyridines with electron-withdrawing substituents in the 3-position catalyzed by binol-P(O)OH **7e** (Eqs. 17a and 17b).[111] Furthermore, *N*-iminium pyridine ylides can be hydrogenated with up to 90% ee using Ir-phox complexes (Eq. 18).[112] Dimethyl derivatives give preferentially cis-substituted piperidines (>95% for the 2,3- and 57% for the 2,5-isomer, respectively).

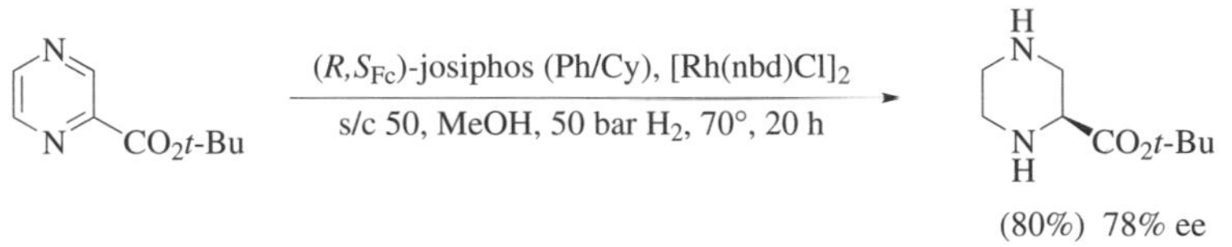

(Eq. 16)

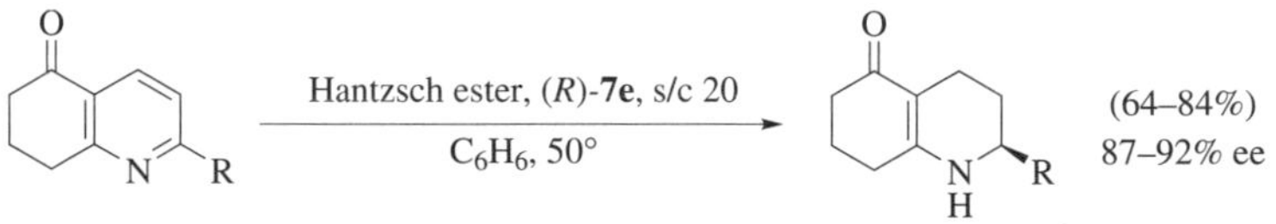

(Eq. 17a)

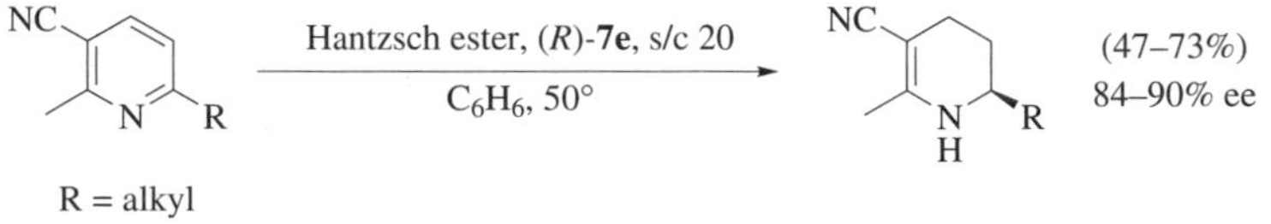

(Eq. 17b)

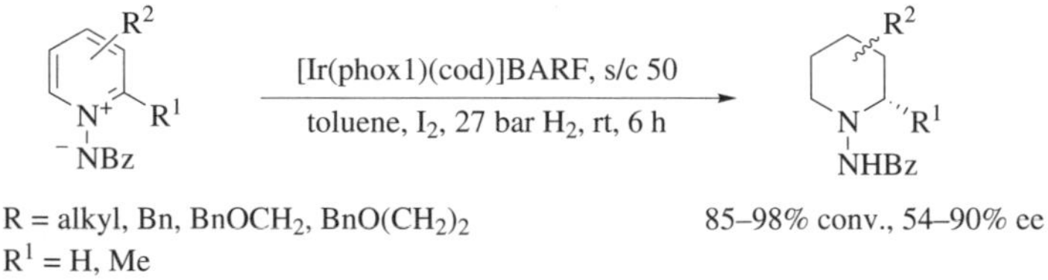

(Eq. 18)

Many reports describe the partial hydrogenation of a variety of substituted quinoline derivatives to the corresponding tetrahydro derivatives (Eq. 19). With the exception of the transfer hydrogenation in the presence of binol-P(O)OH **7d**,[113] all effective catalysts are iridium phosphine complexes. Enantioselectivities range from modest to >99%, depending mainly on the catalysts used and the nature of the substituent R^1 on the heteroaromatic ring. Yields are good to quantitative at s/c ratios that are usually around 100 but can go up to 1000. Enantioselectivities of 87–96% have been reported for atropisomeric diphosphines such as MeO-biphep,[114] *P*-phos,[115] dendrimeric binap **16**,[116] or segphos (transfer hydrogenation with Hantzsch ester).[117] Similar results are achieved using diphosphinites H8-binapo[118] or **17**[119] and ligand combination **18** featuring an achiral ligand together with a diphosphonite.[120] The best catalyst activities are obtained with the ferrocene-based phosphino oxazoline **19**[121] (TOF >80 h^{-1}) and with the dendrimeric binap **16**[116] (up to 43,100 turnovers in 48 hours).

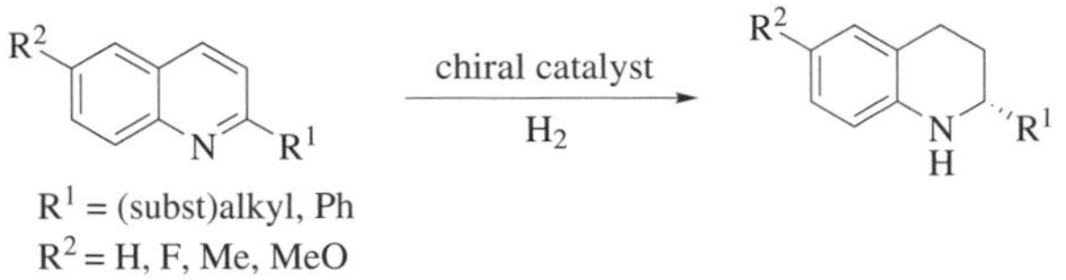

(Eq. 19)

Chiral catalyst, reaction conditions		Ref.
MeO-biphep, $[Ir(cod)Cl]_2$, s/c 100, toluene, I_2, 50 bar, rt, 18 h	(83–94%) 75–96% ee	114
P-phos, $[Ir(cod)Cl]_2$, s/c 100, THF, I_2, 50 bar, rt, 20 h	(97–99%) 90–92% ee	115
16, $[Ir(cod)Cl]_2$, s/c 400, THF, I_2, 45 bar, rt, 1.5 h	77–95% conv., 76–92% ee	116
H8-binapo; $[Ir(cod)Cl]_2$, s/c 100, $DMPEG_{500}$/hexane, I_2, 50 bar, rt, 20 h	(90–99%) 87–97% ee	118
17, $[Ir(cod)Cl]_2$, s/c 100, THF or $DMPEG_{500}$/hexane, I_2, 50 bar, rt, 18 h	60–100% conv., 65–93% ee	119
18, $[Ir(cod)Cl]_2$, s/c 200, toluene, I_2, 60 bar, rt, 20 h	>96% conv., 80–96% ee	120
19, $[Ir(cod)Cl]_2$, s/c 1000, toluene, I_2, 40 bar, rt, 12 h	(82 to >95%) 79–92% ee	121
Hantzsch ester, **7d**, s/c 50, C_6H_6, 69°, 12–60 h	(54–95%) 87 to >99% ee	113
Hantzsch ester, (*S*)-segphos, $[Ir(cod)Cl]_2$, s/c 100, toluene/dioxane, I_2, 40 bar, rt, 42–79 h	(43–98%) 68–88% ee	117

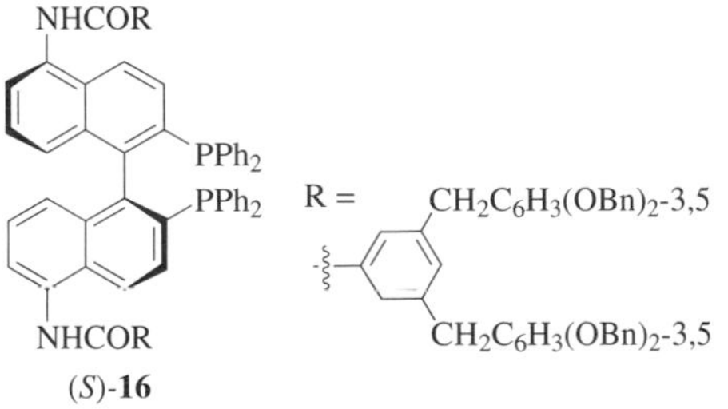

(*S*)-**16**

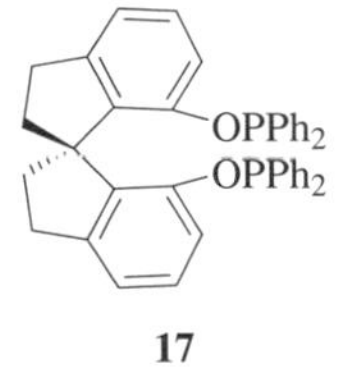

17

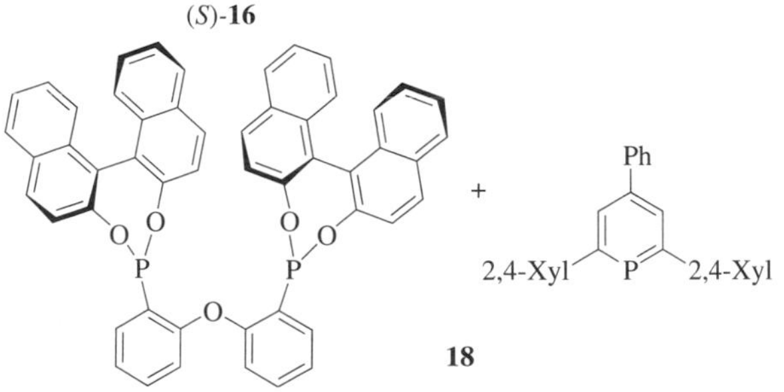

18

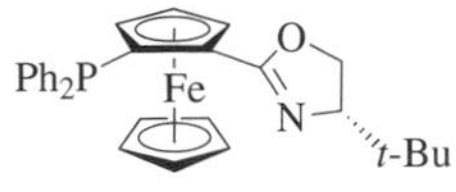

(*S*,S_{Fc})-**19**

The reduction of quinolines can also be carried out in the presence of chloroformates with an Ir-segphos catalyst[122] leading to the corresponding protected tetrahydroquinolines with moderate to good enantioselectivities and yields (Eq. 20). The hydrogenation of isoquinolines has been investigated less. Interestingly, the Ir-segphos catalyst[122] under the same conditions in the presence of chloroformates does not lead to the expected tetrahydroisoquinolines but rather to the corresponding N-protected dihydroisoquinolines, in moderate enantioselectivities and yields (Eq. 21).

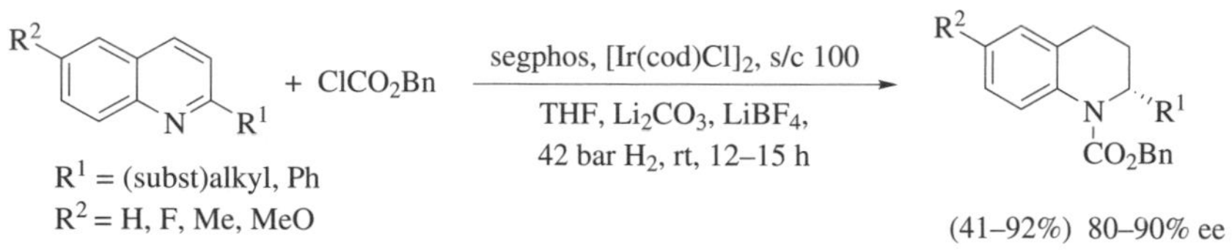
R2
+ ClCO2Bn
segphos, [Ir(cod)Cl]2, s/c 100
THF, Li2CO3, LiBF4,
42 bar H2, rt, 12–15 h
R1 = (subst)alkyl, Ph
R2 = H, F, Me, MeO
CO2Bn
(41–92%) 80–90% ee

(Eq. 20)

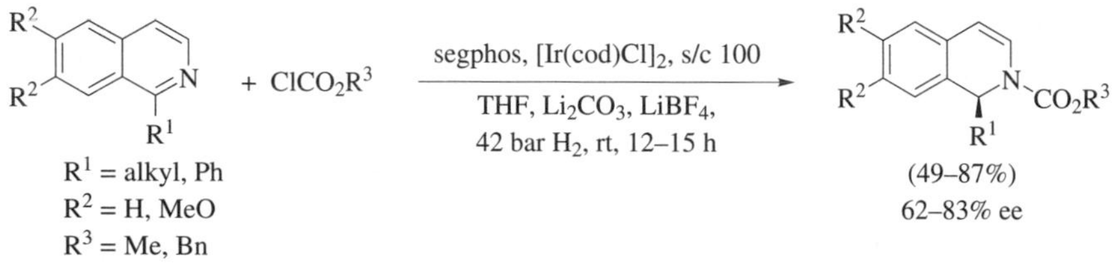
+ ClCO2R3
segphos, [Ir(cod)Cl]2, s/c 100
THF, Li2CO3, LiBF4,
42 bar H2, rt, 12–15 h
R1 = alkyl, Ph
R2 = H, MeO
R3 = Me, Bn
CO2R3
(49–87%)
62–83% ee

(Eq. 21)

Quinoxalines can be considered to be model substrates for the reduction of folic acid (see Applications to Synthesis). Only two successful catalysts have been described. The first success (actually one of the first hydrogenations of an aromatic substrate) was achieved with the uncommon tetradentate iridium complex **20** in good ee but low yield.[123] Ru(hexaphemp)(dach)Cl_2 gives better yields of product but only 69% ee (Eq. 22).[47]

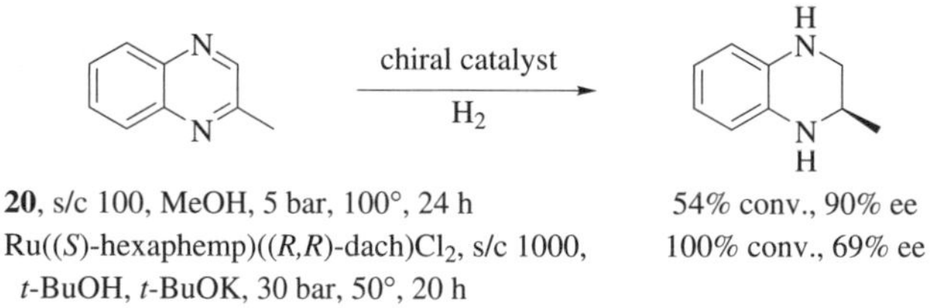
chiral catalyst
H2
20, s/c 100, MeOH, 5 bar, 100°, 24 h
54% conv., 90% ee
Ru((S)-hexaphemp)((R,R)-dach)Cl2, s/c 1000,
t-BuOH, t-BuOK, 30 bar, 50°, 20 h
100% conv., 69% ee

(Eq. 22)

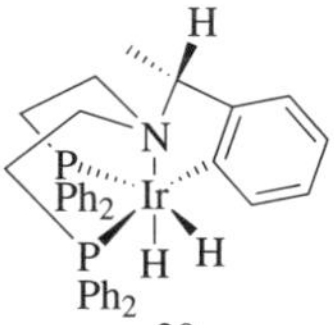
Ph2
Ir
Ph2

20

C=N–Y Functions (Y = OR, NHCOAr, Ts, POAr$_2$). Oxime derivatives were among the first C=N functions to be tried as substrates for enantioselective reduction. However, with ee values of <30% both with modified heterogeneous catalysts[2] as well as homogeneous catalysts,[3] the results were disappointing, especially for α-keto acid derivatives that provide access to α-amino acids. A few examples of oxime hydrogenation with Rh-binap (30–66% ee)[17] and Ir-dpampp (93% ee)[124] are known, but high pressures and/or temperatures are required to give reasonable catalyst activities. An interesting variant is the hydrogenation of nitrones (Eq. 23), which can be carried out with an Ir-binap catalyst with moderate enantioselectivity and often low chemical yields to provide the hydroxylamines.[125]

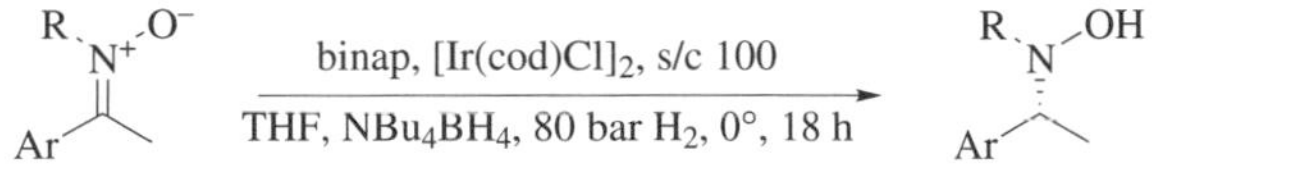
R
O⁻
N⁺
Ar
binap, [Ir(cod)Cl]2, s/c 100
THF, NBu4BH4, 80 bar H2, 0°, 18 h
R
OH
N
Ar

(Eq. 23)

Ar = (subst)Ph, 2-Np
R = Me, Bn

(17–82%) 69–86% ee

High enantioselectivities are obtained for the hydrogenation of a variety of *N*-tosyl imines (Eq. 24) and cyclic analogs (Eq. 25). Ru-binap complexes,[126,127] Pd catalysts with tangphos,[128] segphos,[129,130] and synphos[129,130] are effective hydrogenation catalysts, and several Ru-dpenTs-catalyzed transfer hydrogenations[131–133] have been described. In general, ee values are high and good chemical yields are obtained both for linear and cyclic sulfonylated imines, albeit with low s/c ratios for all catalytic systems.

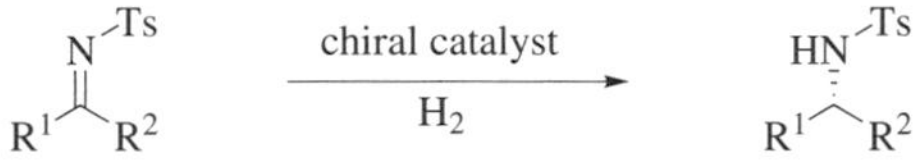
N
Ts
R1
R2
chiral catalyst
H2
HN
Ts
R1
R2

(Eq. 24)

R^1 = alkyl, aryl
R^2 = Me, Et

Chiral catalyst, reaction conditions		Ref.
Pd(tangphos)(CF_3CO_2)$_2$, s/c 100, DCM, 75 bar, 40°, 24 h	>99% conv., 75 to >99% ee	128
Pd(synphos)(CF_3CO_2)$_2$, s/c 50, CF_3CH_2OH, 4 Å MS, 40 bar, rt, 12 h	(84–98%) 88–97% ee	129

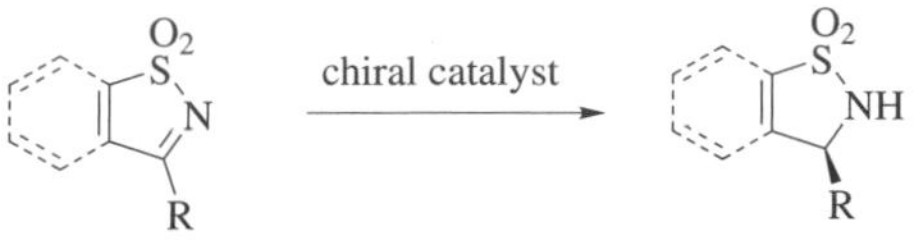

(Eq. 25)

R = alkyl, aryl, Bn, $ROCH_2$

Chiral catalyst, reaction conditions		Ref.
binap, $Ru(cod)Cl_2$, s/c 100, toluene, NEt_3, 4 bar H_2, 22°, 12 h	(84%) 99% ee	126
$Pd(tangphos)(CF_3CO_2)_2$, s/c 100, DCM, 75 bar H_2, 40°, 24 h	>99% conv., 94% ee	128
$Pd(segphos)(CF_3CO_2)_2$, s/c 50, CF_3CH_2OH, 4 Å MS, 40 bar H_2, rt, 12 h	(93–99%) 79–93% ee	129
HCO_2H/NEt_3, $dpenTs_{immob}$, $[Ru(cymene)Cl_2]_2$, s/c 100, neat, 40°, 1.5 h	(99%) 93% ee	131
HCO_2H/NEt_3, $(C_6H_6)Ru((S,S)\text{-dpenTs})Cl$, s/c 200, DCM, rt, 17 h	(—) 91–93% ee	132
HCO_2H/NEt_3, $(R,R)\text{-dpenTs}_{dend}$, $[(cymene)RuCl_2]_2$, s/c 100, DCM, 28°, 10 h	(90%) 96% ee	133
HCO_2Na, $(R,R)\text{-dpenTs}_{sulf}$, $[(cymene)RuCl_2]_2$, s/c 100, H_2O, CTAB, 28°, 10 h	(95%) 94% ee	99

N-Tosylimines of cyclic ketones can be hydrogenated in moderate to high enantioselectivities using a Ru-binap[127] or a Pd-tangphos[128] catalyst (Eq. 26).

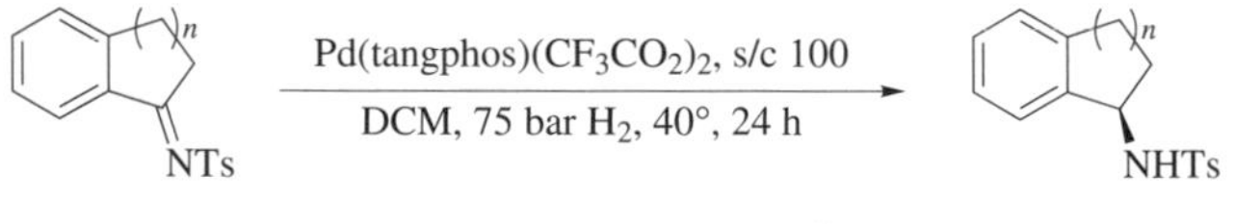

(Eq. 26)

n		Ref.
1	>99% conv., 98% ee	128
2	>99% conv., 94% ee	128

The enantioselectivities that can be achieved for Rh-duphos-catalyzed hydrogenation of *N*-acyl hydrazones, which were quite impressive at the time of the original report,[134,135] confirm the hypothesis that the presence of a second coordinating group in the substrate enhances enantioselectivity (Eq. 27). Whereas ee values are modest for alkyl methyl acyl hydrazones, good to very good enantioselectivities are obtained for the Rh-duphos-catalyzed hydrogenation of acyl hydrazones derived from aryl methyl ketones and α-keto esters. Other Rh-diphosphine complexes give ee values of $\leq 67\%$.[136,137] The resulting *N*-acyl hydrazines can be reduced to the primary amines using SmI_2, but a practical economic method for the cleavage of the N–N bond to obtain the primary amine without racemization is still lacking.

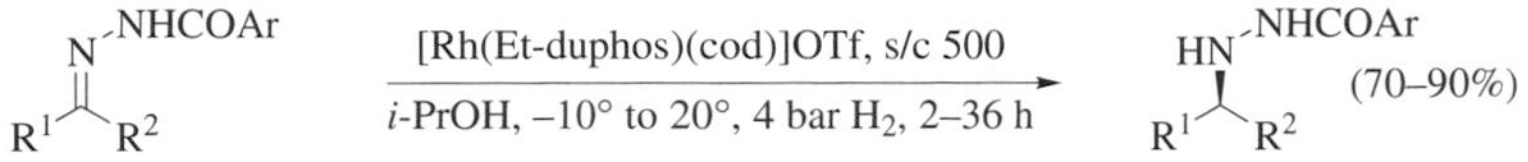
NHCOAr
N
R1
R2
[Rh(Et-duphos)(cod)]OTf, s/c 500
i-PrOH, −10° to 20°, 4 bar H2, 2–36 h
HN
NHCOAr
(70–90%)
R1
R2

R^1	R^2	
alkyl	Me	45–73% ee
aryl	alkyl	88–97% ee
alkyl, aryl	CO_2Et	83–91% ee
Ph	$P(O)(OEt)_2$	90% ee

(Eq. 27)

Phosphinyl imines (Eq. 28) can be hydrogenated in moderate to excellent enantioselectivities with Rh-josiphos,[138] Pd-synphos,[130] as well with cp*Rh (dpenTs)Cl.[90] Whereas s/c ratios for the transfer hydrogenation and the Pd-synphos catalyst are rather low, up to 500 turnovers can be obtained with Rh-josiphos.

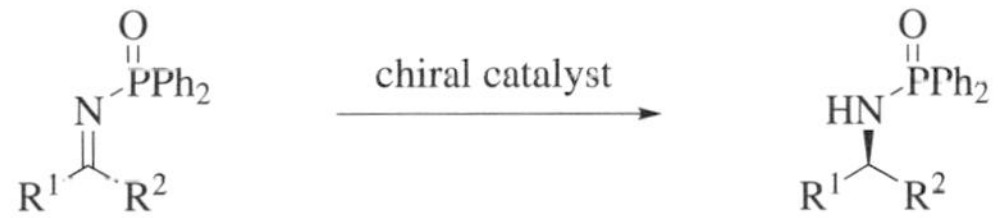
O
PPh2
N
R1
R2
chiral catalyst
O
PPh2
HN
R1
R2

(Eq. 28)

R^1 = (subst)Ph, 2-Np, n-C_6H_{13}
R^2 = Me, Et

Chiral catalyst, reaction conditions		Ref.
HCO_2H/NEt_3, cp*Rh(dpenTs)Cl, s/c 50, MeCN, rt, 2–3 h	100% conv., 86 to >99% ee	90
Pd(segphos)$(CF_3CO_2)_2$, s/c 50, CF_3CH_2OH, 4 Å MS, 70 bar H_2, rt, 8 h	(29–93%) 87–93% ee	130
josiphos, $[Rh(nbd)_2]BF_4$, s/c 100–500, MeOH, 70 bar H_2, 60°, 1–21 h	93–100% conv., 62–99% ee	138

α- and β-Carboxy Imines. α- and β-Functionalized imine derivatives are obvious precursors to α- and β-amino acids. However, effective catalytic systems for this transformation have only recently been developed, with selected examples presented in Eq. 29. α-Amino acid derivatives are accessible via the Ir-dpampp-catalyzed reduction of an oxime[124] (93% ee, very low yields) or the Rh-duphos-catalyzed hydrogenation of acyl hydrazones[134,135] in moderate to good enantioselectivity. In both cases, the primary amino acid can be obtained by reductive cleavage of the N−O or N−N bond, respectively. The hydrogenation of 4-MeO-phenyl imines using either Rh-tangphos,[139] Pd-binap,[140] or a binol-P(O)OH catalyst (**7e**, **21**) in the presence of Hantzsch ester[141,142] also provides the corresponding amino ester in good yields and moderate to high enantioselectivities. The resulting products can be deprotected under mild conditions with cerium ammonium nitrate.[143] Reductive amination using a Rh-deguphos catalyst[144] can be achieved in medium to very high enantioselectivities (Eq. 30). Both yield and ee strongly depend on the nature of R.

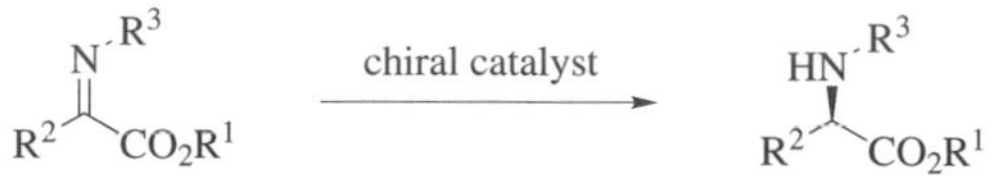

R^1 = Et, *i*-Pr

(Eq. 29)

R^3	R^2	Chiral catalyst, reaction conditions		Ref.
4-$MeOC_6H_4$	alkyl, aryl	Hantzsch ester, **21**, s/c 20, toluene, rt–50°, 19–22 h	(85–99%) 94–99% ee	141
4-$MeOC_6H_4$	alkyl, (het)aryl	Hantzsch ester, **7e,** s/c 10, toluene, 60°, 48 h	(46–95%) 84–98% ee	142
4-$MeOC_6H_4$	alkyl, aryl	[Rh(tangphos)(cod)]BF_4, s/c 100, DCM, 50 bar H_2, 50°, 24 h	85–99% conv., 83–95% ee	139
NHCOPh	Ph, alkyl	[Rh(Et-duphos)(cod)]OTf, s/c 500, *i*-PrOH, 0°, 4 bar H_2, 36 h	(70–90%) 83–91% ee	134, 135
OH	$Ph(CH_2)_2$	$[Ir(dpampp)Cl]_2$, s/c 100, C_6H_6/MeOH, BI_3, 48 bar H_2, rt, 46 h	19–22% conv., 93% ee	124

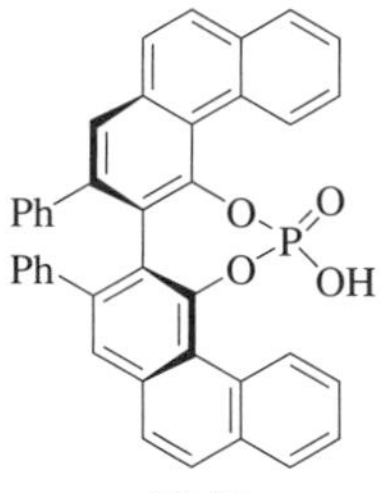

(*S*)-**21**

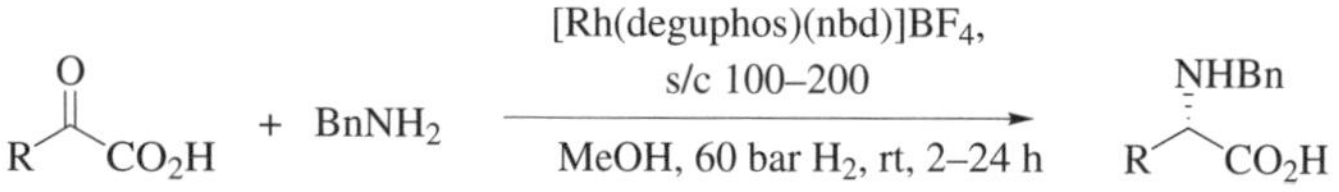

(Eq. 30)

R = alkyl, Bn (80–99%) 81–98% ee

R = Me, $HO_2C(CH)_n$ (19–43%) 60–78% ee

By analogy, β-amino acid derivatives have been prepared in high yields and good to very high enantioselectivities via the Rh-tangphos-catalyzed hydrogenation of β-imino esters (Eq. 31).[145] An interesting new development is the hydrogenation of primary enamines/imines leading to β-amino acid derivatives (Eq. 32), a reaction with considerable synthetic and industrial potential.[146] Whereas the hydrogenation of the analogous *N*-acylated derivatives is a well-known transformation, it was quite unexpected that the unprotected substrates would be amenable to enantioselective hydrogenation. Very good enantioselectivities are achieved for several different primary β-imino esters with Rh-josiphos[146] and Ru-binap (and analogs)[147] in trifluoroethanol, but for both catalyst activity is an issue. While the Rh-josiphos complexes give high conversion at an s/c of 330 after 6–20 hours, the Ru-binap catalysts at an s/c of 100 (in some cases

1000) do not give full conversion even after 15–88 hours. Interestingly, deuteration experiments indicate that it is not the enamine C=C bond that is reduced but the tautomeric primary imine. For the Rh-josiphos-catalyzed hydrogenation of β-imino *N*-aryl amides, the best enantioselectivities are obtained in methanol (Eq. 33).[146] In the presence of $(Boc)_2O$, the hydrogenation with Rh-josiphos leads directly to the N-protected β-amino acid derivatives with improved chemical yields and up to 99% ee (Eq. 34).[148,149]

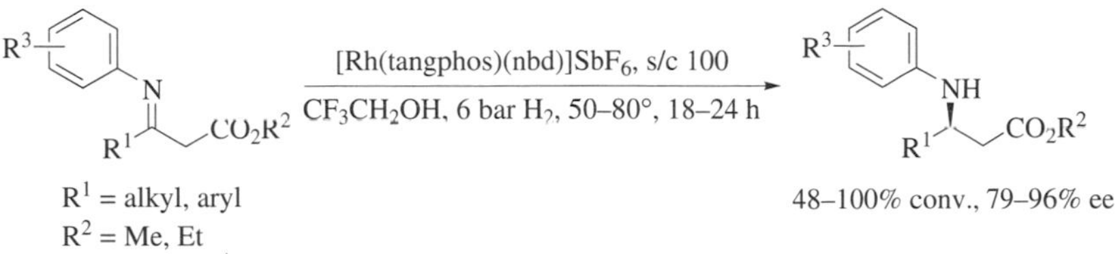
[Rh(tangphos)(nbd)]SbF6, s/c 100
CF3CH2OH, 6 bar H2, 50–80°, 18–24 h
R1 = alkyl, aryl
R2 = Me, Et
48–100% conv., 79–96% ee

(Eq. 31)

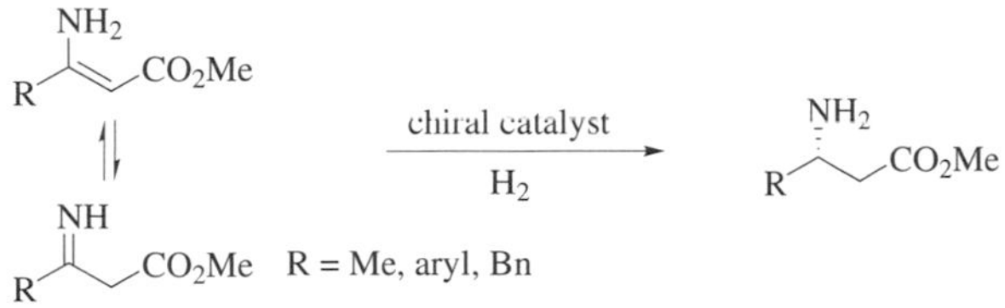
chiral catalyst
H2
R = Me, aryl, Bn

(Eq. 32)

Chiral catalyst, reaction conditions		Ref.
(R,S_{Fc})-josiphos (4-CF_3Ph/*t*-Bu), $[Rh(cod)Cl]_2$, s/c 300, CF_3CH_2OH, 6 bar, 50°, 16–20 h	(88–96%) 93–96% ee	146
$[Ru((S)\text{-segphos})(OAc)_2]$, s/c 100, CF_3CH_2OH, 30 bar, 80°, 15–88 h	(54–85%) 96–97% ee	147

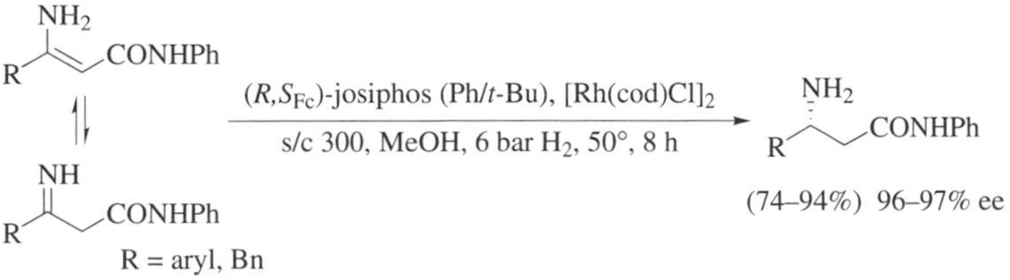
(R,SFc)-josiphos (Ph/t-Bu), [Rh(cod)Cl]2
s/c 300, MeOH, 6 bar H2, 50°, 8 h
R = aryl, Bn
(74–94%) 96–97% ee

(Eq. 33)

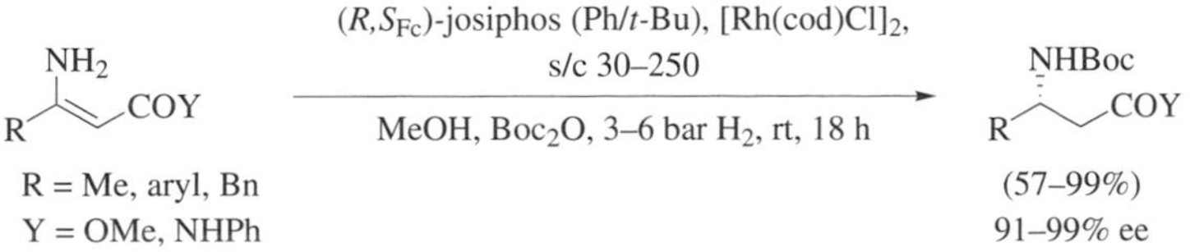
(R,SFc)-josiphos (Ph/t-Bu), [Rh(cod)Cl]2,
s/c 30–250
MeOH, Boc2O, 3–6 bar H2, rt, 18 h
R = Me, aryl, Bn
Y = OMe, NHPh
(57–99%)
91–99% ee

(Eq. 34)

Reductive Amination. The reductive amination of ketones,[20] that is, in situ formation of the imine followed by hydrogenation, is an especially attractive variant for industrial applications, because isolation and purification of the C=N compound is not required. However, suitable catalysts and reaction conditions for this process have only recently been developed. In general, the same metal precursor–ligand combination identified for the isolated imine can be used, but often either the solvent has to be adjusted or additives such as molecular sieves are necessary for good results. In general, the reaction works best with aryl ketones but aliphatic ketones have also been used.

Of special interest is the preparation of primary amines because, with the exception of β-dehydro acid derivatives illustrated above, primary imines cannot usually be isolated. Ruthenium complexes with tol-binap[150] as well as ClMeO-biphep[151] give excellent enantioselectivities and good to high yields for the reductive amination of a variety of aryl methyl ketones with ammonium acetate (Eq. 35).

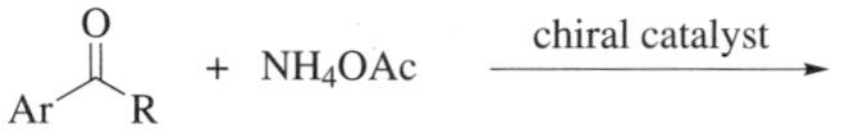
O
Ar
R
+ NH_4OAc
chiral catalyst
NH_2
Ar
R

(Eq. 35)

Ar = (subst)Ph, 1-Np, 2-Np

R	Chiral catalyst, reaction conditions		Ref.
Me, Et	NH_3/HCO_2H, Ru(tol-binap)Cl_2, s/c 100, MeOH, 85°, 21–48 h	(74–93%) 89–95% ee	150
EtO_2CCH_2	(Cymene)Ru(ClMeO-biphep)Cl_2, s/c 100, CF_3CH_2OH, 30 bar H_2, 80°, 16 h	(79–88%) 96–99% ee	151

A number of ketones can be reductively aminated with a variety of aryl amines (Eqs. 36–38) in up to 96% ee, using the phosphoric acid-based transfer hydrogenation catalyst **7a**[104] or an iridium f-binaphane complex.[152] Whereas conversions with the iridium complex are quantitative after 10 hours, the transfer hydrogenation takes longer and yields are around 70–90%. The reaction of α-methoxyacetone with 2-ethyl-6-methyl aniline (Eq. 39) catalyzed by an Ir-josiphos complex[153] must be carried out in a non-polar solvent like cyclohexane to attain high turnover numbers.

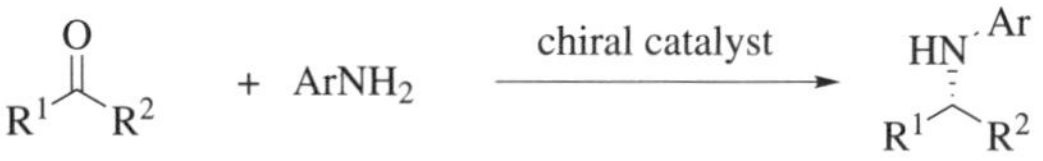
O
R^1
R^2
+ $ArNH_2$
chiral catalyst
HN
Ar
R^1
R^2

(Eq. 36)

Ar = Ph, 4-$MeOC_6H_4$
R^1 = Me, Et

R^2	Chiral catalyst, reaction conditions		Ref.
alkyl, aryl	Hantzsch ester, **7a**, s/c 10, C_6H_6, 5 Å MS, 50°, 24–72 h	(49–82%) 81–96% ee	104
(het)aryl	f-binaphane, $[Ir(cod)Cl]_2$, s/c 100, DCM, I_2, Ti(*i*-PrO)$_4$, 70 bar H_2, rt, 10 h	(>99%) 44–96% ee	152

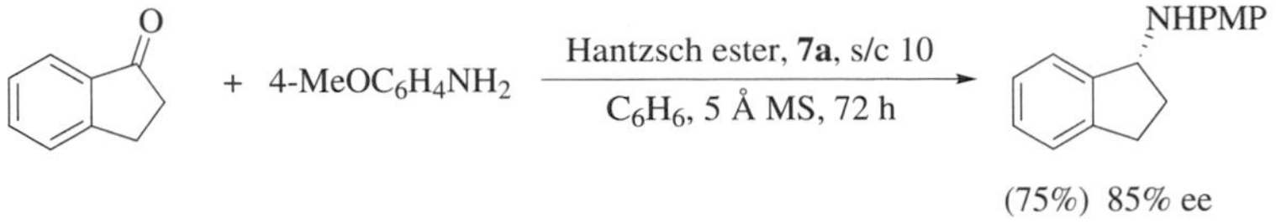
O
+ 4-MeOC6H4NH2
Hantzsch ester, 7a, s/c 10
C6H6, 5 Å MS, 72 h
NHPMP
(75%) 85% ee

(Eq. 37)

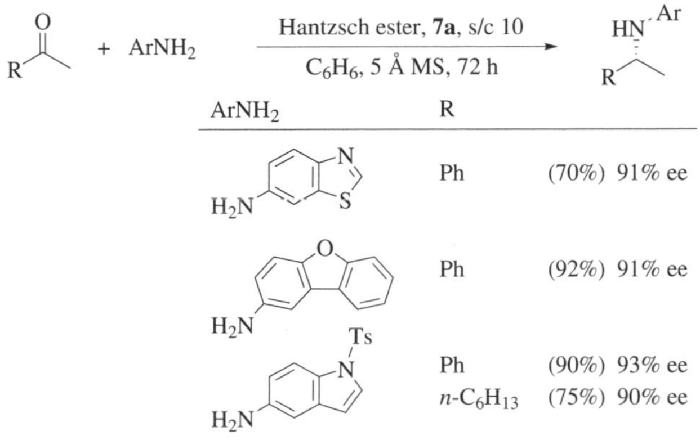
O
R
+ ArNH2
Hantzsch ester, 7a, s/c 10
C6H6, 5 Å MS, 72 h
HN Ar
R
ArNH2
R
H2N
N
S
Ph
(70%) 91% ee
O
H2N
Ph
(92%) 91% ee
Ts
N
H2N
Ph
(90%) 93% ee
n-C6H13
(75%) 90% ee

(Eq. 38)

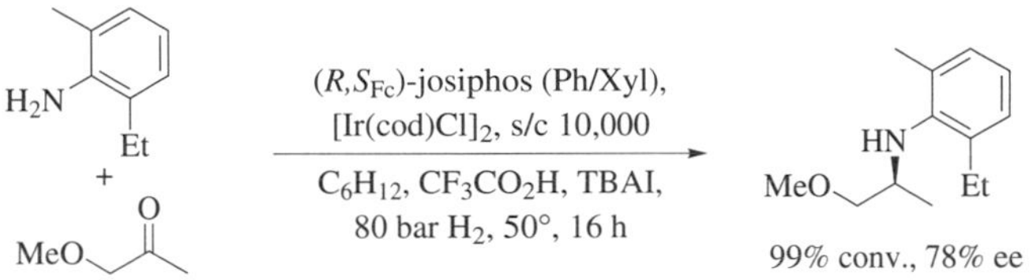
H2N
Et
+
O
MeO
(R,SFc)-josiphos (Ph/Xyl),
[Ir(cod)Cl]2, s/c 10,000
C6H12, CF3CO2H, TBAI,
80 bar H2, 50°, 16 h
HN
MeO
Et
99% conv., 78% ee

(Eq. 39)

Only a small number of enantioselective reductive aminations with aliphatic amines have been described, and with few exceptions, enantioselectivities are lower than for the reaction with ammonium acetate or with aromatic amines. Benzylated α-amino acids can be prepared via Rh-deguphos-catalyzed reductive amination of α-keto acids[144] (Eq. 40) with substrate-dependent yields and enantioselectivities. The Ru-dpenTs-catalyzed transfer hydrogenation of racemic 2-methylcyclohexanone (Eq. 41) occurs with acceptable enantio- and diastereoselectivity, due to a dynamic kinetic asymmetric transformation.[77]

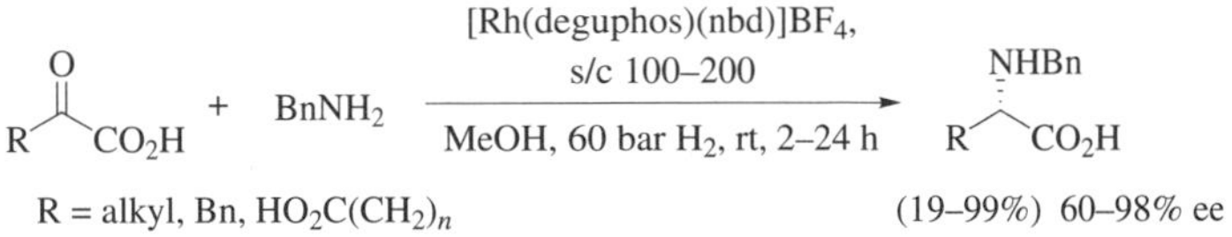
O
R
CO2H
+ BnNH2
[Rh(deguphos)(nbd)]BF4,
s/c 100–200
MeOH, 60 bar H2, rt, 2–24 h
NHBn
R
CO2H
R = alkyl, Bn, HO2C(CH2)n
(19–99%) 60–98% ee

(Eq. 40)

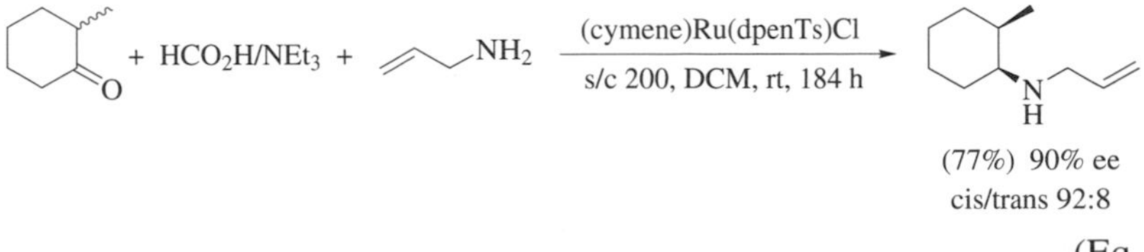
O
+ HCO2H/NEt3 +
NH2
(cymene)Ru(dpenTs)Cl
s/c 200, DCM, rt, 184 h
N
H
(77%) 90% ee
cis/trans 92:8

(Eq. 41)

Cyclohexylamines can be obtained from a reaction cascade involving imine formation, enamine aldol condensation, and transfer hydrogenation (Eq. 42).[154] The reaction is catalyzed by binol-P(O)OH **7c** in the presence of Hantzsch ester and furnishes products in 72–89% yield in very good enantioselectivities and medium to very high diastereomeric ratios. An interesting domino reaction is catalyzed by a Pd complex of duphos or the Trost ligand **22** between an aryl iodide, CO, and cyclohexylamine (Eq. 43). An initial double carbonylation yields an α-keto amide, which reacts with a second equivalent of amine, leading eventually to the corresponding α-amino amide in good to excellent enantioselectivities and modest yields.[155]

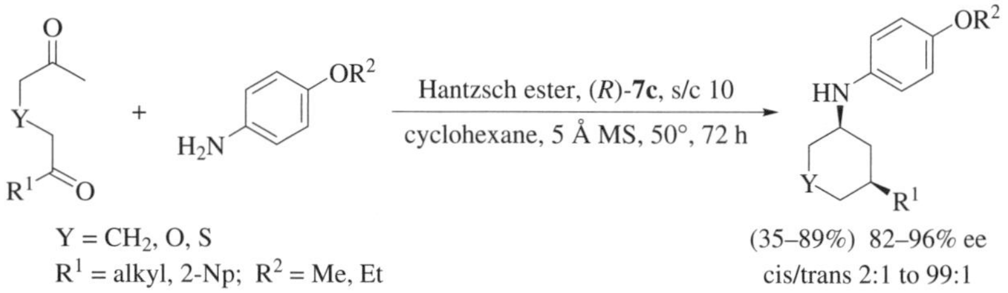
Hantzsch ester, (R)-7c, s/c 10
cyclohexane, 5 Å MS, 50°, 72 h
Y = CH2, O, S
R1 = alkyl, 2-Np; R2 = Me, Et
(35–89%) 82–96% ee
cis/trans 2:1 to 99:1

(Eq. 42)

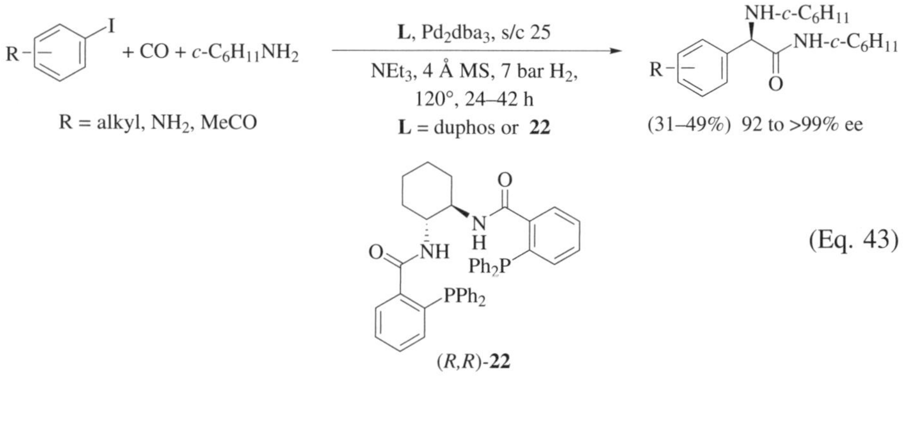
+ CO + c-C6H11NH2
L, Pd2dba3, s/c 25
NEt3, 4 Å MS, 7 bar H2,
120°, 24–42 h
L = duphos or 22
R = alkyl, NH2, MeCO
(31–49%) 92 to >99% ee
(R,R)-22

(Eq. 43)

MECHANISM AND STEREOCHEMISTRY

Only a few detailed studies of the reaction mechanism of the homogeneous hydrogenation of imines have been published. Generalizations about this process are very difficult to make for two reasons. First, different catalyst types are effective and probably act by different mechanisms. Second, the effect of certain additives (especially iodide, and acid or base) is often critical for optimum enantioselectivities and reaction rates, but a promoter in one case can be a deactivator in another. Most catalytic systems described in this review most likely promote the addition of dihydrogen directly to the C=N bond and not to the tautomeric enamine C=C bond,[70,146,135] even though enamines can also be hydrogenated enantioselectively.[156,157]

Rhodium Catalysts

Kinetic studies on the rhodium-catalyzed *N*-aryl imine hydrogenations have led to the conclusion that, by analogy with C=C hydrogenation, the so-called hydride route is preferred.[7,158] As depicted in Scheme 1, it is assumed that the imine is first η^1-coordinated by the nitrogen lone pair to a Rh^{III}-dihydride species. The isomerization to the π-coordinated intermediate thought to be necessary for the reduction might be assisted by a bound alcohol molecule. After two hydrogen transfer steps wherein the first determines the absolute configuration, a Rh^{I}-amine complex results that either directly or in a dissociative manner reacts with dihydrogen to form the Rh^{III}-dihydride species again.

The Rh-duphos-catalyzed hydrogenation of acyl hydrazones has also been studied in some detail.[135] These substrates were intended to provide an additional stabilizing interaction between the substrate and the rhodium center.[134] Such a secondary interaction is considered to be the major reason for the excellent enantioselectivities observed for the hydrogenation of enamides, itaconates, or β-ketoesters.[159] The stronger coordination may also be the reason why many of these hydrogenation reactions are tolerant of functional groups such as halogen, formyl, or cyano. Deuteration studies indicate that the insertion of the C=N moiety into the Rh–H bond occurs irreversibly as proposed for the imine hydrogenation in the catalytic cycle depicted in Scheme 1.

A recent kinetic study[160] of the cp*Rh-dpenTs-catalyzed transfer hydrogenation of an endocyclic imine with formic acid–ammonia indicates that a similar sequence of steps occurs as suggested for the hydrogenation depicted in Scheme 1. The data show that a cp*Rh(dpenTs)-H species is likely to be the

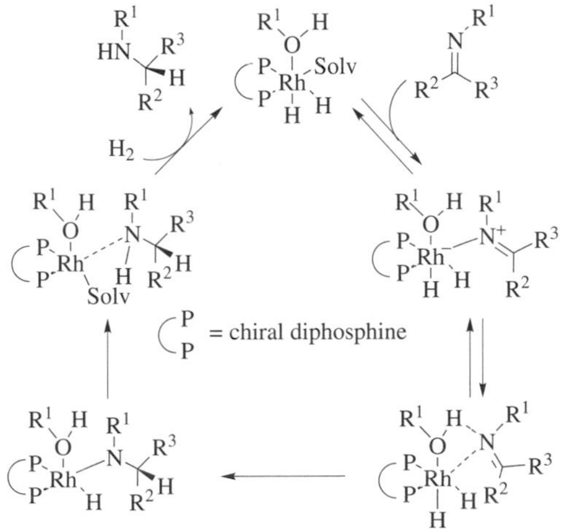

Scheme 1

resting state of the catalytic cycle. This rhodium hydride species is formed by the transfer of a hydride from formic acid and reacts further by coordinating the C=N bond followed by insertion into the Rh–H bond and protonation to give the chiral amine.

Iridium Catalysts

The Ir-diphosphine-catalyzed hydrogenation of *N*-aryl imines has been studied in some detail.[83] Ir^{III} complexes of the type $[Ir(diphosphine)I_4]^-$, $[Ir(diphosphine)I_2]_2$, and $[Ir(diphosphine)I_3]_2$ have been isolated and characterized. All three types of complexes are catalytically active, suggesting the formation, by splitting the iodo bridge, of the same active monomeric iridium species as for the catalyst formed in situ from $[Ir(cod)Cl]_2$, diphosphine, and iodide. Similar results have been reported for the dimethylaniline imine/Ir-josiphos system[161,162] and a number of reaction intermediates have been isolated. On the basis of these results, the catalytic cycle shown in Scheme 2 can be postulated.

The starting species is an Ir^{III}-H complex that coordinates the imine by the nitrogen lone pair in a η^1-manner (as proposed above for the rhodium-catalyzed reaction). A η^1,η^2-migration leads to two diastereomeric adducts with a π-coordinated imine, which then inserts into the Ir–H bond to give the corresponding iridium amide complexes. The last step involves the hydrogenolysis of the Ir–N bond and the formation of an Ir–H bond, presumably via heterolytic splitting of the dihydrogen bond. In contrast to the Rh-diphosphine-catalyzed hydrogenation of C=C bonds, which most likely occurs via Rh^I and Rh^{III} species, the cycle in Scheme 2 consists exclusively of Ir^{III} species, that is, the halides X and Y remain on the iridium during the cycle. However, this basic catalytic cycle explains neither the mode of enantioselection nor the sometimes dramatic effects of additives, for example, the major rate enhancement in the presence of iodide

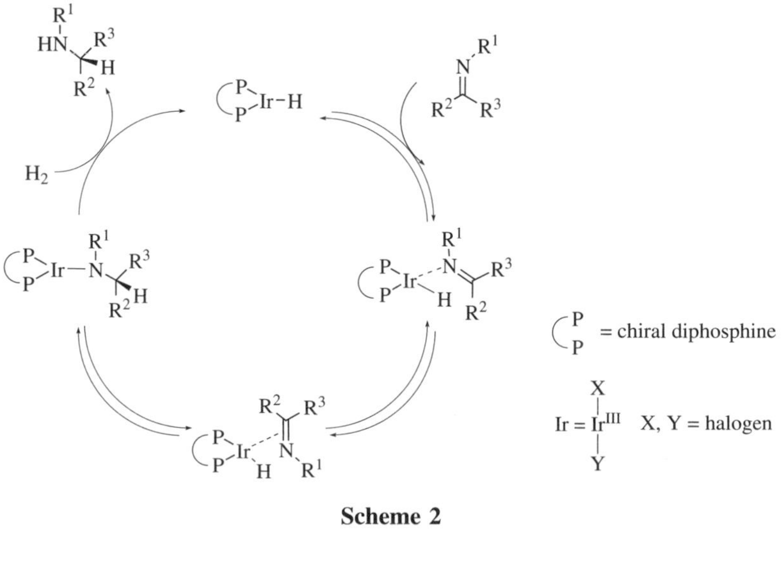

Scheme 2

ion and acids observed for the Ir-josiphos-catalyzed hydrogenation of *N*-aryl imines.[16]

Catalyst deactivation is often a serious problem, especially when hydrogenating relatively basic imines. Such a deactivation is indicated when the initial reduction rate is high but then slows significantly even at low conversion. Several studies[46,163,164] show that the formation of triply hydrogen-bridged dinuclear species is the probable cause of deactivation for many iridium systems (Eq. 44). These catalytically inactive species are formed irreversibly, and their formation is accelerated by base and by the absence of imine substrate. It has been shown that with the josiphos ligand the catalysts are much less prone to deactivation than with many other ligands, and that the presence of iodide ions is often beneficial for the reaction.

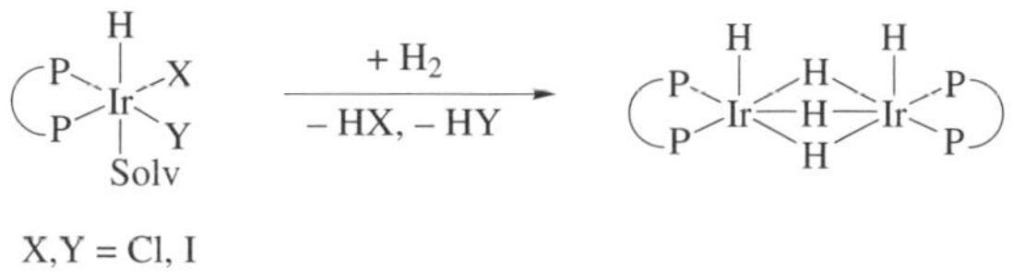

(Eq. 44)

The imine hydrogenation reaction using iridium phosphine oxazoline complexes has not yet been investigated mechanistically, but a plausible stereochemical model was proposed for the iridium-catalyzed hydrogenation of *N*-aryl imines with spirophosphino oxazoline **3**.[41] The prediction is based on the same stereochemical model developed for the (ebthi)Ti catalysts discussed in more detail below. For **3** the hindered quadrants are occupied by the spiroindane backbone and by one of the *P*-phenyl groups, leading to the preferred coordination of the *N*-aryl imine in which the large groups point towards the empty quadrants.

Titanium Catalysts

A mechanism similar to that described for the iridium-catalyzed reactions has been proposed for the titanium-catalyzed reactions (Scheme 3).[34,78] The active catalyst produced by reacting (ebthi)TiX_2 (X_2 = Cl_2 or binol) with *n*-BuLi followed by phenylsilane is assumed to be the monohydride species (ebthi)Ti-H. Kinetic and deuterium-labeling studies are in agreement with the following reaction sequence: (ebthi)Ti-H reacts with the imine via a 1,2-insertion reaction to form two diastereomeric titanium–amide complexes. These intermediate amide complexes then react irreversibly via a σ bond metathesis reaction with dihydrogen, as proposed for the iridium-catalyzed reaction, to regenerate the titanium hydride and form the two amine enantiomers.

For most imine hydrogenations, the resulting absolute configuration of the major enantiomer cannot be predicted and must be determined experimentally. Although this situation is scientifically unsatisfactory, from an experimental standpoint it poses no major problems, because all relevant ligands are available in both enantiomeric forms. An exception is the (ebthi)Ti catalyst, where the absolute

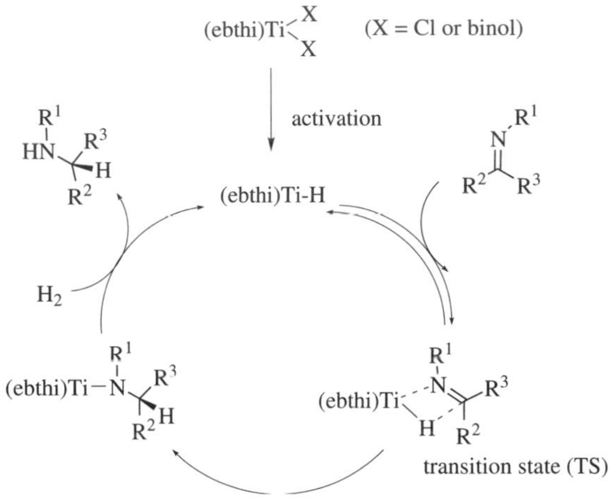

Scheme 3

configuration of the major enantiomer can be predicted by a simple stereochemical model.[34,35,70] As schematically depicted in Fig. 4, two of four quadrants (bold) are occupied by the six-membered ring of the tetrahydroindenyl ligand, whereas the other two quadrants are much less crowded. It is assumed that the imine coordinates horizontally and therefore, for the anti isomer, the preferred coordination can be easily predicted using simple steric arguments. For the syn isomer, the situation is more complicated because there might be a competition between R^3 and R_{large} for an empty quadrant, explaining the lower enantioselectivities usually realized for the hydrogenation of non-cyclic imines, and the pressure dependency of those reactions.

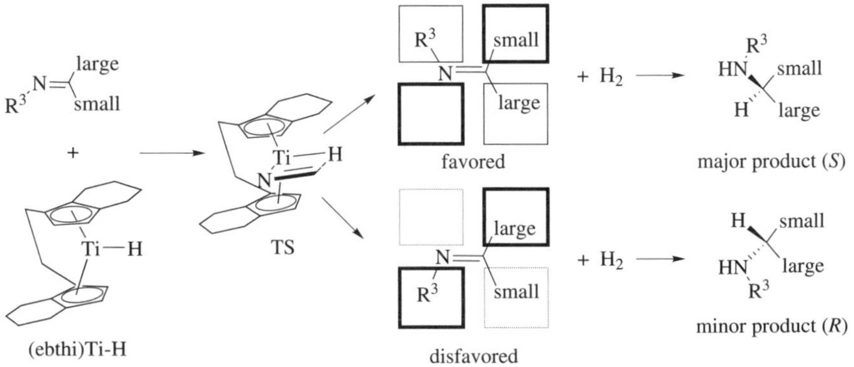

Figure 4. Quadrant model for predicting product stereochemistry based on the transition state with (ebthi)Ti-H as the catalyst.

Ruthenium Catalysts

Although, to date, no mechanistic studies of imine hydrogenation with a chiral ruthenium complex have been reported, two recent reviews discuss possible mechanisms of the reduction of polar C=Y bonds both for reactions with dihydrogen as well as for hydrogen transfer reactions.[165,166] As in the hydrogenation of ketones, two rather different mechanisms likely operate. The first is a classical inner-sphere mechanism as described above for rhodium, iridium, and titanium complexes. This model proposes that all reactants are coordinated to the metal center, and bond-breaking and bond-making occur in the first coordination sphere. The second mechanism is an outer-sphere variant of the first, where the ketone does not coordinate to the ruthenium center but rather interacts with Ru–H and a coordinated N–H group. This mechanism has been convincingly demonstrated to occur in enantioselective ketone reductions catalyzed by Ru-diamine complexes. By analogy, it is quite likely that C=N reductions with Ru-diphosphine complexes occur via a classical inner-sphere mechanism, whereas Ru-diamine complexes react via an outer-sphere mechanism as schematically depicted in Scheme 4.

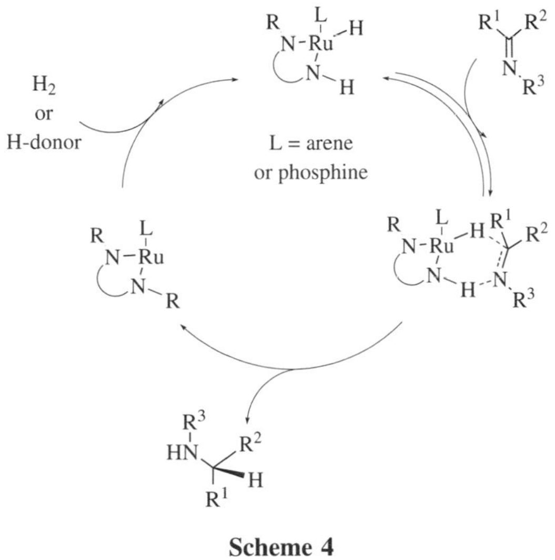

Scheme 4

Miscellaneous Catalysts

No mechanistic information is available for the recently described Pd-diphosphine complexes or the binol-P(O)OH-catalyzed transfer hydrogenation with Hantzsch esters as donors. In the latter method, the Hantzsch ester is probably involved in the stereochemistry-determining step, most likely the transfer of a hydride to the chiral iminium$^+$/binol-P(O)O$^-$ ion pair.[48,55] The very large substituents needed in the ortho positions of the binol probably form a chiral pocket in which the iminium cation is bound, selectively exposing one of the faces for

H-addition and thus leading to high enantioselectivities.[104] For the transfer hydrogenation of heteroarenes, a stepwise addition/isomerization/addition as illustrated in Scheme 5 has been proposed, where the configuration-determining step is the reduction of the C=N bond.[111]

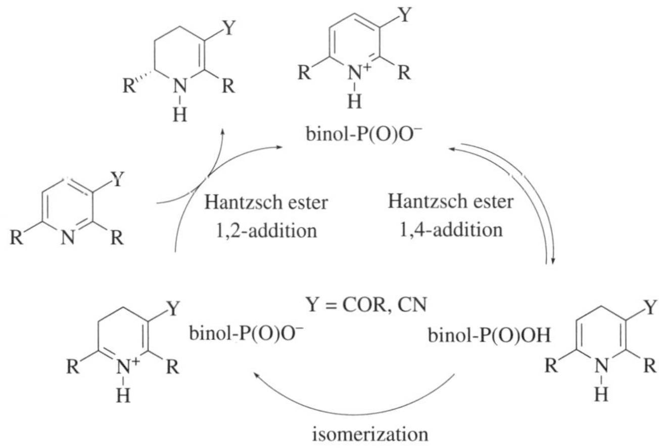

Scheme 5

APPLICATIONS TO SYNTHESIS

As mentioned in the Introduction, most reactions described in this review have been carried out with simple model substrates. In this section, applications of asymmetric C=N reductions to the synthesis of more complex molecules of industrial or biological interest are summarized. The (*S*)-metolachlor process is by far the most important application of the catalytic hydrogenation of C=N functions, and for this reason it is described in detail.

Production Process for (*S*)-Metolachlor (DUAL Magnum)

Metolachlor is the active ingredient of Dual, one of the most important grass herbicides for use in maize and a number of other crops.[167] The active compound is an *N*-chloroacetylated, *N*-alkoxyalkylated *o*-disubstituted aniline. The commercial product was introduced in the market in 1976 as a racemic mixture of two diastereomers (Fig. 5), then in 1982 it was found that about 95% of the herbicidal activity of metolachlor resides in the two (1′*S*)-diastereomers. In 1997, after years of intensive research, Dual Magnum was introduced into the market. The product contains approximately 90% of the (1′*S*)-diastereomers and has the same biological effect of the racemate at about 65% of its use rate. Today, with a production volume of >10,000 tons per year, Dual Magnum represents, by far, the most significant application of enantioselective catalysis in terms of output.

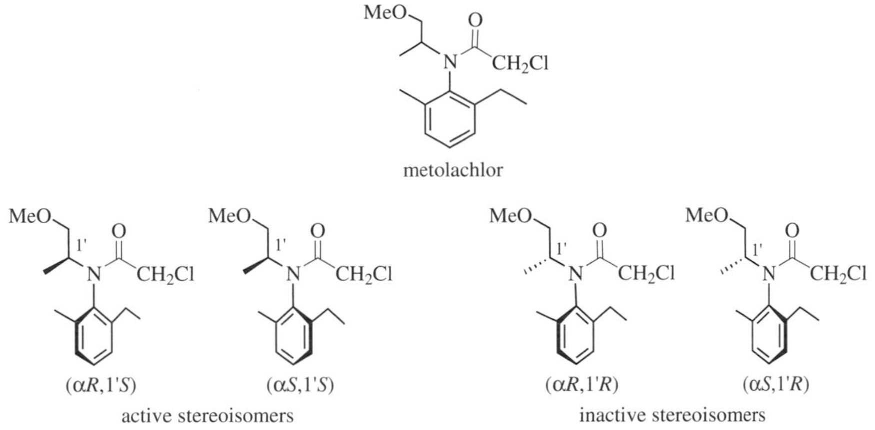
MeO
O
N
CH_2Cl
metolachlor
1'
(αR,1'S)
(αS,1'S)
active stereoisomers
(αR,1'R)
(αS,1'R)
inactive stereoisomers

Figure 5. Structures of metolachlor and its individual stereoisomers.

A key step in the new process is the enantioselective hydrogenation of the distilled 2-methyl-6-ethyl aniline (MEA) imine substrate (Eq. 45). The optimized process operates at 80 bar hydrogen and 50° with a catalyst generated in situ from $[Ir(cod)Cl]_2$ and (R,S_{Fc})-josiphos (Ph/Xyl) at an s/c ratio of 2,000,000. Complete conversion is reached within 3–4 hours, initial TOF exceeds 1,800,000 h^{-1}, and enantioselectivity is approximately 80% ee.[16] Key success factors for the process are the novel, very active Ir-josiphos catalyst, the use of iodide and acid as additives, and the high purity of MEA imine. Alternative processes such as direct reductive amination[153] as well as the application of immobilized josiphos[168] in order to avoid the distillation of the *N*-alkylated aniline were investigated as well. Whereas both catalyst systems reach respectable turnover numbers of 10,000–100,000, these variants are not competitive.

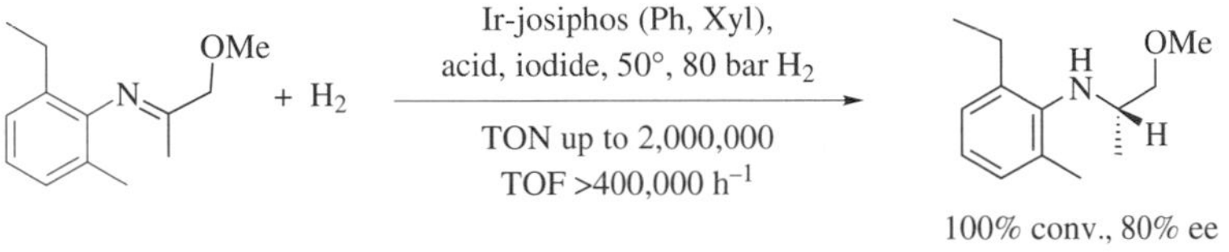
OMe
N
+ H_2
Ir-josiphos (Ph, Xyl),
acid, iodide, 50°, 80 bar H_2
TON up to 2,000,000
TOF >400,000 h^{-1}
H
N
OMe
H
100% conv., 80% ee

(Eq. 45)

Production Process for Sitagliptin

Merck has developed a process for the manufacture of Sitagliptin, a DPP-IV inhibitor for type 2 diabetes (Eq. 46). The key imine reduction reaction is carried out with a Rh-josiphos catalyst with up to 98% ee albeit with low to medium TON and TOF.[146,169] Success is dependent on the choice of the ligand, the solvent, and the presence of trace amounts of ammonium chloride. Interestingly, deuteration experiments indicate that it is not the enamine C=C bond that is reduced but the tautomeric imine. The reaction is now carried out on a multiton scale.[170]

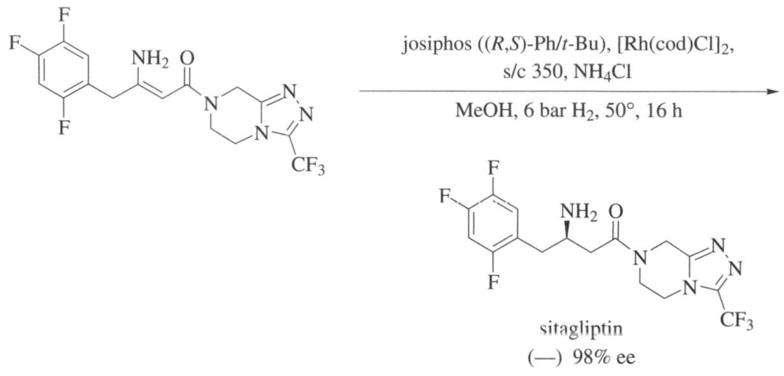

(Eq. 46)

Pilot Process for Dextromethorphane

A pilot process for the preparation of an intermediate in the synthesis of dextromethorphane, a traditional antitussive agent, was developed using an Ir-josiphos catalyst in a two-phase system (toluene/water) (Eq. 47).[171] Key success factors are ligand fine-tuning, the use of the phosphoric acid salt of the imine, the reaction medium, and the addition of base and iodide. Chemoselectivity with respect to C=C hydrogenation is high but the turnover number is somewhat low for an economical technical application.

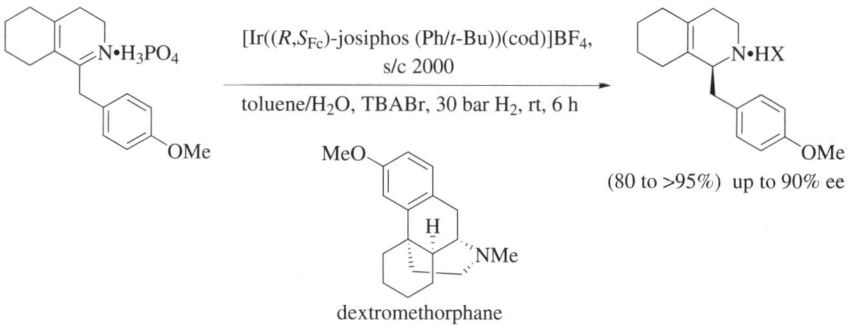

(Eq. 47)

Industrial Feasibility Studies

Noyori's Ru-PP-NN catalyst system was successfully applied in a feasibility study for the hydrogenation of a sulfonyl amidine, an intermediate for S 18986, an AMPA receptor modulator (Eq. 48).[172] Considering that the substrate is an amidine, the catalytic activity is surprisingly high, but at 87% ee, the enantioselectivity of the reaction is modest. A factorial experimental design was used to optimize reaction conditions showing the importance of the nature and amount of base.

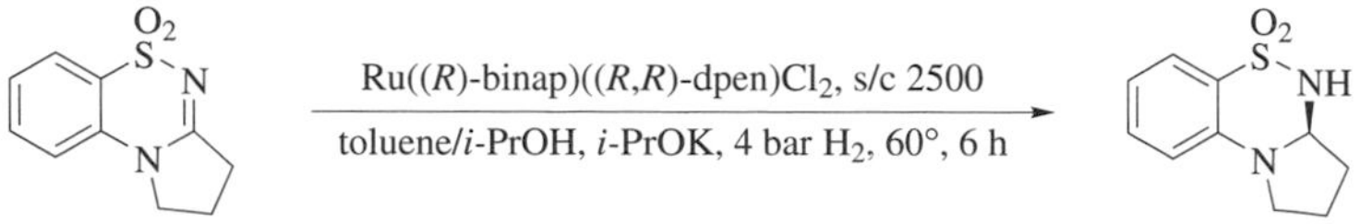
Ru((R)-binap)((R,R)-dpen)Cl2, s/c 2500
toluene/i-PrOH, i-PrOK, 4 bar H2, 60°, 6 h

S 18986
(97%) 87% ee

(Eq. 48)

The commercial viability of the CATHy catalysts based on a cp*Rh complex has been demonstrated for the transfer hydrogenation of phosphinyl imines (Eq. 49).[173] The reaction with the 2-naphthyl derivative was scaled up to the multikilo level. Bubbling nitrogen through the reaction solution increases reaction rates significantly, due to the faster removal of the CO_2 byproduct.

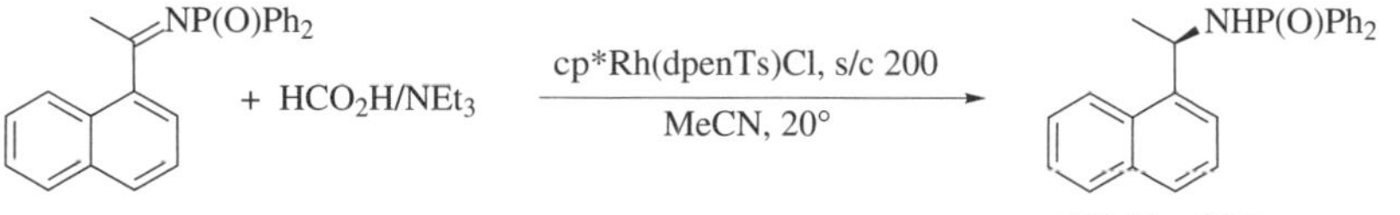
NP(O)Ph2
+ HCO2H/NEt3
cp*Rh(dpenTs)Cl, s/c 200
MeCN, 20°
NHP(O)Ph2

(100%) >99% ee

(Eq. 49)

Up to 90% ee was achieved in the hydrogenation of an intermediate in the process to the antibiotic levofloxacin using Ir-diphosphine complexes (Eq. 50).[174] The best results were obtained with bppm and mod-diop in the presence of bismuth iodide at low temperature.

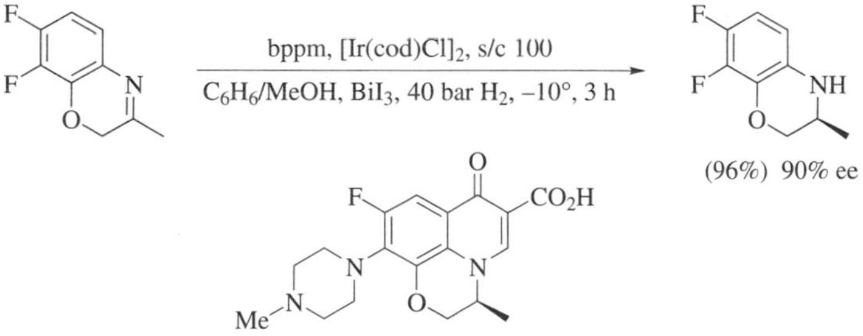
bppm, [Ir(cod)Cl]2, s/c 100
C6H6/MeOH, BiI3, 40 bar H2, –10°, 3 h
(96%) 90% ee
CO2H
Me

(Eq. 50)

levofloxacin

The hydrogenation of folic acid, formally a diastereoselective reaction as depicted in Eq. 51, has been claimed to proceed with up to 90% ee with an Ir-bppm complex adsorbed on silica gel, a claim that later had to be retracted.[175] A major problem is the insolubility of folic acid in most organic solvents. Functionalized catalysts offer the opportunity to perform this reaction in water and it was shown that a rhodium complex of the functionalized josiphos ($josiphos_{funct}$) can achieve diastereoselectivities of up to 49%.[176] An alternative is the hydrogenation of the corresponding bis(methyl) ester, which can be carried out in methanol.

However, the best catalyst, $[Rh(cod)_2]BF_4$/(*R*)-binap, achieved a diastereoselectivity of <44%.[177] Even though for both processes s/c ratios up to 1000 were possible, selectivity and activity are not sufficient for commercial applications.

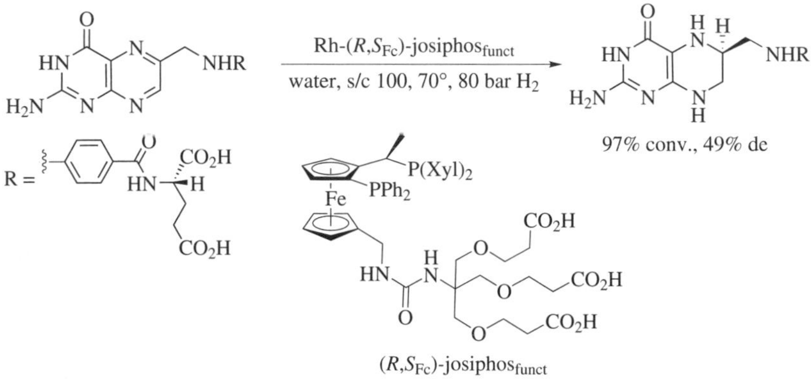

(Eq. 51)

Synthesis of Tetrahydroisoquinoline Alkaloids

Several tetrahydroisoquinoline alkaloids such as salsoline, laudanosine, and cryptostyline have been synthesized starting from the corresponding endocyclic dihydroisoquinolines. With very few exceptions,[34,87,102,103] Noyori's transfer hydrogenation catalyst system was applied.[91,94,96,99–101,106,107,178,179] A few selected examples are described to illustrate the scope and limitations of the technology. The first results were reported for the synthesis of the closely related cryptostyline II, norlaudanosine, and tetrahydrohomopapaverine alkaloids (Eq. 52) using Ir-diphosphine catalysts.[103] Good results (up to 88% ee) are obtained with the bcpm ligand in the presence of phthalimides for norlaudanosine and tetrahydrohomopapaverine, whereas poor enantioselectivity and yield are reported for cryptostyline II.

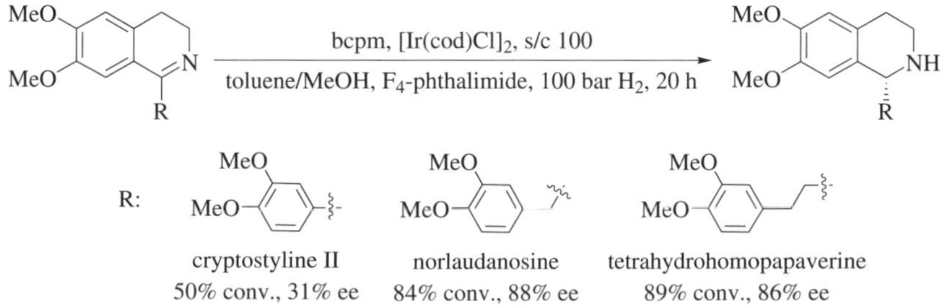

(Eq. 52)

Noyori's transfer hydrogenation technology is also effective in the synthesis of tetrahydroisoquinoline alkaloids.[22,96,106,107,180] The synthesis of a homoprotoberine alkaloid was achieved in good yield and 99% ee (Eq. 53).[96] The key step for the synthesis of cryspine A[106] (Eq. 54), harmicine,[107] and desbromoarborescidine[107] (Eq. 55) is the transfer hydrogenation of tri- and tetracyclic endocyclic iminium species, which occurs in all three syntheses in satisfactory yields and modest to good enantioselectivities, albeit at high catalyst loadings.

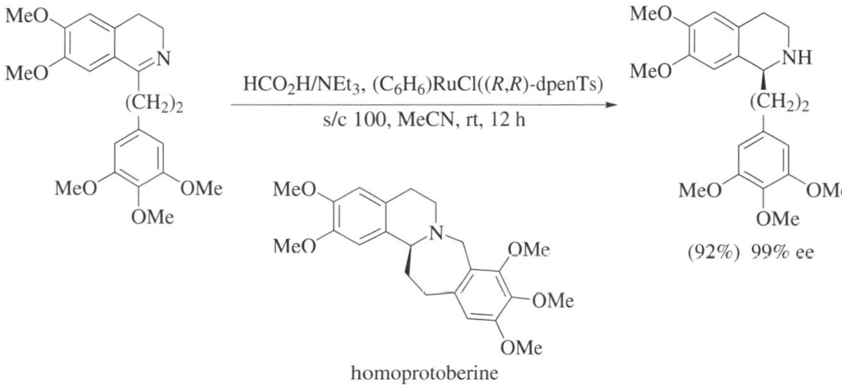

(Eq. 53)

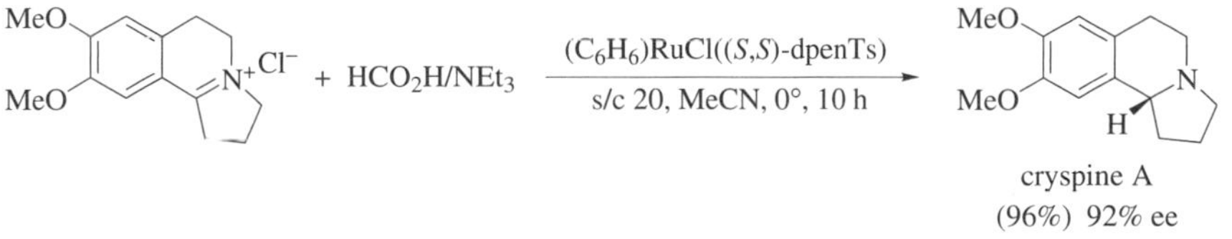

(Eq. 54)

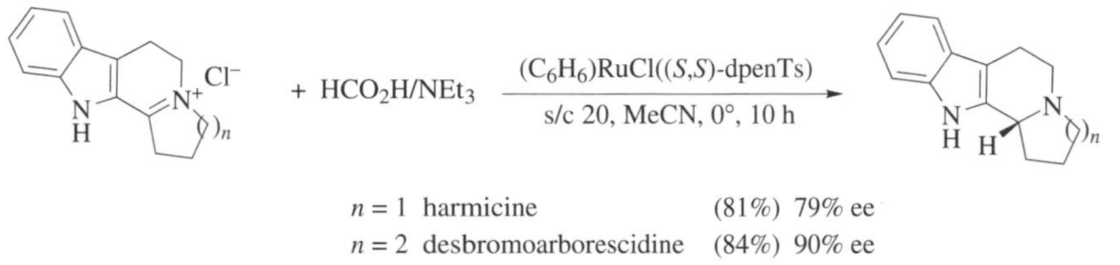

(Eq. 55)

The (*S*)-cryptostyline moiety of the short-acting neuromuscular blocker GW 0430 has been prepared via transfer hydrogenation of an appropriate dihydroisoquinoline derivative (Eq. 56).[95] Classical Noyori conditions using a Ru-dpen(1-Nps) catalyst affords the tetrahydro derivative in 83% ee, which was enriched by crystallization to 99% ee.

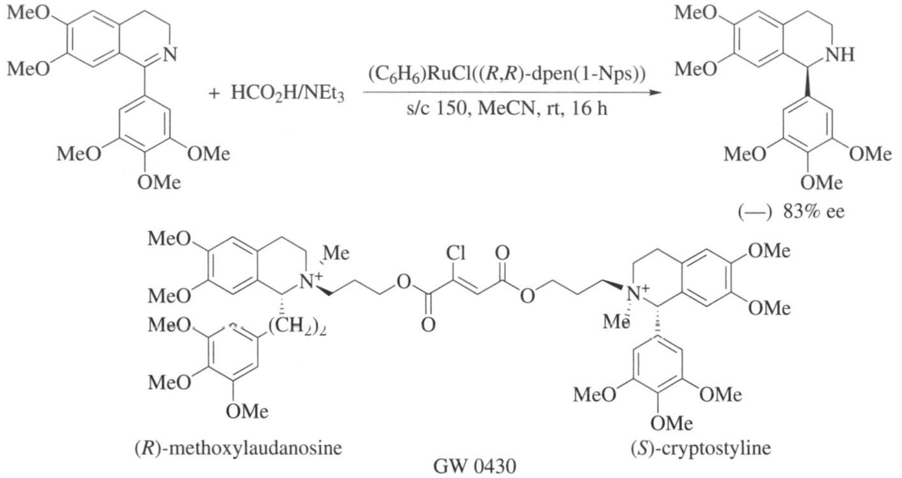

(Eq. 56)

All possible stereoisomers of emetine, an Ipecacuanha alkaloid, have been prepared via two consecutive asymmetric transfer hydrogenations under standard conditions (see Eq. 57, where the synthesis of the natural stereoisomer, (–)-emetine, is depicted).[98] The predominant absolute configuration of the stereogenic centers is controlled by the choice of the appropriate dpenTs ligands. For the first reduction, 2.5% of the catalyst was needed to give the desired tetrahydroquinolines in 93% yield and >95% ee. The second, diastereoselective reduction required 10% catalyst and, depending on the relative absolute configuration of the catalyst and the substrate, yields of 71–82% and diastereoselectivities of 81 to >96% de were obtained.

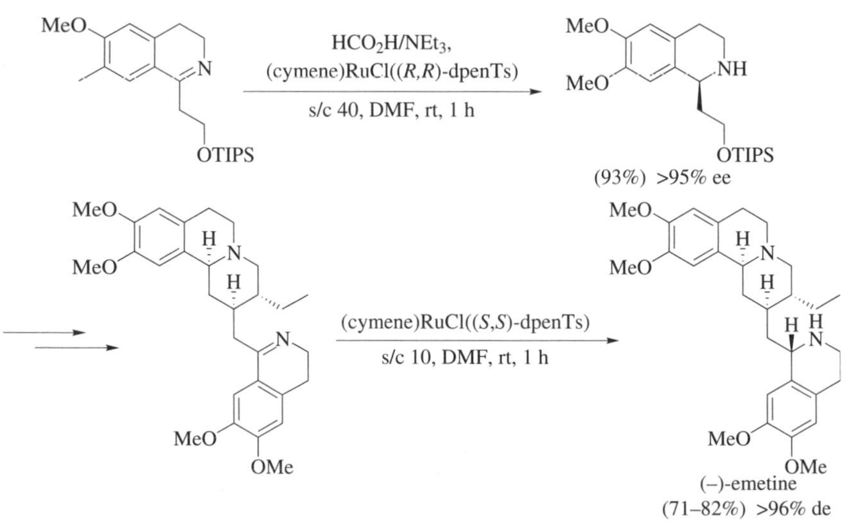

(Eq. 57)

ALTERNATIVE REDUCTION SYSTEMS

Chiral Hydrides

Hydride reductions of C=N groups are well known in organic chemistry and several chiral reducing agents derived from BH_3, $LiAlH_4$, or $NaBH_4$ and relatively cheap amino alcohols or diols have been developed for the reduction of imines and oxime derivatives.[25,26,181–183] Enantioselectivities are medium to high. Most of the effective chiral auxiliaries can be prepared in one or two steps from rather inexpensive starting materials such as binol, amino acids, tartaric acid, or sugars, and can potentially be recycled. A major drawback of most hydride reduction methods is the fact that stoichiometric or higher amounts of chiral reagents are needed, and that disposal of the hydrolyzed borate and aluminate byproducts leads to increased costs for the reduction step. Chiral hydrides are currently useful on a laboratory scale but their potential for commercial applications is medium to low. Hydroboration of the C=N function catalyzed by chiral oxazaborolidines has also been reported.[184,185]

Hydrosilylation

Because the silane has to be used in stoichiometric amounts, reactions involving hydrosilylation of C=N functions have cost and disposal issues similar to those noted for hydride reductions, except that fewer effective reduction systems have been developed.[27,28,186] Despite some recent progress with highly selective Ti-[187] and Cu-based[188] catalysts using cheap polymethylhydrosiloxane as the reducing agent, and of organocatalysts able to activate trichlorosilane,[189,190] hydrosilylation will probably have major application only in small-scale laboratory syntheses.

Biocatalysis

Chiral amines can also be produced using aminotransferases either by kinetic resolution of the racemic amine or by asymmetric synthesis from the corresponding prochiral ketone.[191] A variety of chiral amines can be obtained with good to excellent enantioselectivities. Several transformations have been developed and can be carried out on a 100-kg scale.[29] At the moment, application to synthetic problems, especially to more elaborate targets, is challenging because optimization of the enzyme and reaction conditions is time-consuming.

EXPERIMENTAL CONDITIONS

Note: *Hydrogen forms explosive mixtures with air (explosion limit in air: 4.75 vol %). The apparatus must be tested at elevated pressure for leaks before the reaction (H_2-tightness). For large-scale applications or with higher pressures, a detection system for H_2 leaks during the run is recommended.*

Choice of Metal, Anion, Ligands, and Solvents

Unfortunately, there are no general guidelines for any of the substrate classes described in this review. The specific combination of these process variables must be optimized for each transformation. Nevertheless, existing results allow the identification of catalytic systems with the best chance of success and these are listed in Table A.

As a rule, relatively high pressures are needed to achieve acceptable reaction times, but, in general, pressure does not significantly affect enantioselectivities. For transfer hydrogenations, the azeotropic mixture formic acid–ammonia (5:2) is generally used in organic solvents and sodium formate in water, whereas Hantzsch esters are required for organocatalytic reactions.

Table A. Successful catalyst systems for selected substrate classes.

Substrate Class	Metal (anion)	Chiral Ligand	Solvent	Pressure
N-aryl	Ir (H^+/ iodide)	josiphos	toluene	10–50 bar
	Ir (BARF)	phox (PN ligands)	DCM	10–50 bar
	Ir (Cl / I_2)	PN=SO	toluene	20 bar
	Rh (BF_4)	tangphos	DCM	50 bar
N-alkyl	Ti	ebthi	THF	140 bar
	Ru (Cl)	dpenTs	DCM	H-donor
	Rh	bdpp	MeOH or biphasic	50–70 bar
Endocyclic	Ti	ebthi	THF	5–30 bar
	Ru (Cl)	dpenTs	H_2O, DCM	H-donor
	cp*Rh (Cl)	dpenTs	MeOH or biphasic	H-donor
	Ir (halide)	diphosphine	variable	30–100 bar
	—	binol-P(O)OH	benzene, $CHCl_3$	H-donor
Heteroarene	Ir (iodide)	diphosphine	THF, toluene	30–50 bar
	binol-P(O)OH	binol-P(O)OH	benzene	H-donor
C=N–Ts	Ru (Cl)	binap	toluene	4 bar
	Pd (CF_3CO_2)	segphos, tangphos	CF_3CH_2OH	40–70 bar
C=N–NHAc	Rh (OTf)	duphos	*i*-PrOH	4 bar
C=N–P(O)Ph$_2$	Rh (BF_4)	josiphos	MeOH	70 bar
	Pd (CF_3CO_2)	synphos	CF_3CH_2OH	40–70 bar
	cp*Rh (Cl)	dpenTs	MeCN	H-donor
α-CO$_2$R	Rh (BF_4)	deguphos	MeOH	60 bar [a]
	—	binol-P(O)OH	toluene	H-donor
β-CO$_2$R	Ru (OAc)	segphos	CF_3CH_2OH	30 bar [a]
	Rh (Cl)	josiphos	CF_3CH_2OH, MeOH	6 bar
	Rh (SbF_6)	tangphos	CF_3CH_2OH	6 bar

[a] reductive amination

Temperatures are usually between room temperature and 50° with a varying effect on enantioselectivity. Catalyst loadings vary greatly, depending on all components of the reaction system. Of special importance is the purity of the starting material. Impurities such as traces of amines (from the preparation of the C=N function), acid, or anions can have a strong negative effect on catalytic activity, and sometimes also on enantioselectivity.

Many of the most active ligands are commercially available (indicated by the pound (#) symbol in Chart 1 of the Tabular Survey) in small quantities from Aldrich and Strem, and in larger quantities from ChiralQuest, Dow, Johnson Matthey, Solvias, or Takasago. Most diphosphines should be handled with care, especially ligands with alkylphosphino groups, which are very air-sensitive. If a hydrogenation reaction is carried out using an in situ approach, either Schlenk techniques or a glove box are recommended. An alternative is to use the preformed complexes, which are also available with selected ligands and are usually less air-sensitive.

Preparative reactions at normal pressure can be carried out using two-necked round-bottom flasks with a magnetic stirrer. The dihydrogen can be provided either from a dihydrogen-filled balloon or a gas burette that allows measuring the dihydrogen consumption. Pressures up to 4 bar and measurement of dihydrogen uptake can be handled with the well-known and reliable Parr Shaker, supplied by Labeq.[192] However, temperature control with this apparatus is poor, and prices are on the order of \$3,000.

For higher pressures, the construction of special hydrogenation stations with the necessary safety installations (rupture disc, expansion vessel, reinforced cubicle, etc.) is recommended. Depending on the size and construction material of the autoclave, the safety installations and the accuracy of the measurement of dihydrogen consumption, the price for such a system is between \$20,000 and \$100,000. Suppliers are Autoclave Engineers,[193] Büchi,[194] and others. We would also strongly recommend consulting colleagues who have practical experience with the setup and the operations of a hydrogenation laboratory.

EXPERIMENTAL PROCEDURES

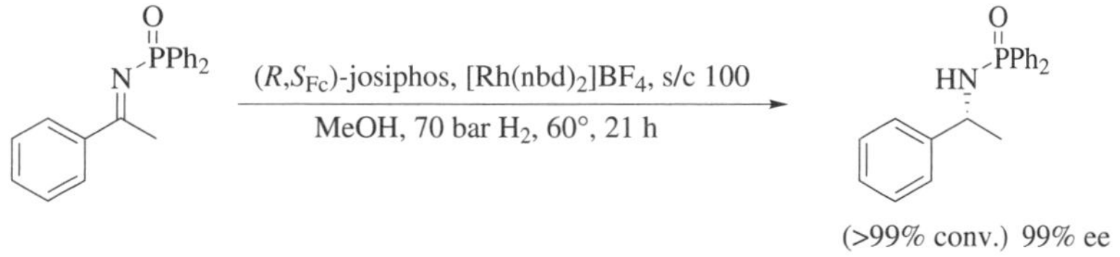

***N*-(1-Phenylethyl)diphenylphosphinamide [Enantioselective Hydrogenation of *N*-Alkylidendiphenylphosphinamides Using Rh-Diphosphine Catalysts].[138]** *N*-(1-phenylethyliden)diphenylphosphinamide (0.5 g, 1.55 mmol) was dissolved in 7 mL of MeOH under argon. A catalyst solution was prepared by dissolving $[Rh(nbd)_2]BF_4$ (5.8 mg, 0.0155 mmol) and (*R*)-(1)-{[((*S*)-2-dicyclohexylphosphino)ferrocenyl]ethyl}dicyclohexylphosphine ((*R*, S_{Fc})-josiphos

(Cy/Cy)) (10.6 mg, 0.0173 mmol) in 8 mL of MeOH under argon. This solution was stirred for 15 minutes at room temperature. The substrate and the catalyst solutions were transferred via steel tubing into a 50-mL stainless steel autoclave. The inert gas was then replaced by H_2 (three cycles of vacuum/H_2) to a pressure of 70 bar, and the reaction temperature set to 60°. After 21 hours, the heating was discontinued, the pressure was released and, once the reaction had reached room temperature, the autoclave was opened. The conversion was determined by GLC [DB-17, 30 m; temperature program: 60°/1 minute to 220°/15 minutes, $\Delta T = 10°$/minute] as compared to a standard. The enantiomeric purity of the *N*-(1-phenylethyl)diphenylphosphinamide was determined by GLC after derivatization with perfluorobutyric acid anhydride [Lipodex-D, 50 m; temperature = 160°, isotherm; carrier He (170 kPa)]. The conversion was determined to be ≥99% with 99% ee (*R*).

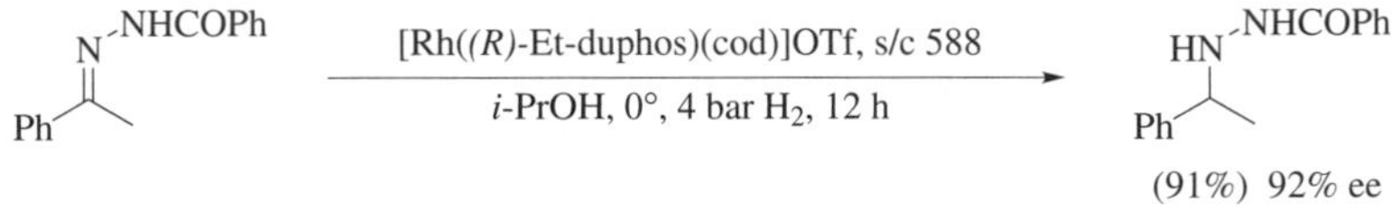

(*S*)-(−)-1-Phenyl-1(2-benzoylhydrazino)ethane [Asymmetric Hydrogenation of *N*-Acyl Hydrazones Using [Rh(Et-Duphos)(cod)]OTf Complexes].[135] In a N_2-filled dry box, a 100-mL Fisher Porter glass pressure vessel [available from Andrews Glass Co., 3740 NW Boulevard, Vineland, NJ 08360; www.andrewsglass.com] was charged with a stirring bar, and acetophenone *N*-benzoylhydrazone (200 mg, 0.80 mmol) was added, followed by degassed 2-propanol (10 mL), and [Rh((*R*)-Et-duphos)(cod)]CF_3SO_3 (1 mg, 0.0014 mmol). The lines were purged of air (six cycles of vacuum/H_2), then the reaction mixture was purged twice more using the same technique. The vessel was pressurized to 4 bar of H_2. The reaction was stirred at 0° until no further H_2 uptake was observed (12 hours). The reaction was evaporated to dryness and the residue subjected to chromatography on a short silica column (6 × 0.5 cm) using 50% EtOAc in hexane as eluent. The appropriate fractions were evaporated to yield the title compound as a colorless solid (182 mg, 91% yield). Chiral analysis (HPLC, Daicel column Chiralcel OJ, 10% *i*-PrOH in hexane, 40°, 0.5 mL/minute) indicated a product of 92% ee: mp 75–76.5°; $[\alpha]_D^{20} = -163.6$ (c 2.72, $CHCl_3$); ^1H NMR (300 MHz, $CDCl_3$) δ 7.7 (m, 2H), 7.6–7.2 (m, 8H), 4.37 (q, $J = 6.7$ Hz, 1H), 1.50 (d, $J = 6.7$ Hz, 3H). Anal. Calcd for $C_{15}H_{16}N_2O$: C, 74.97; H, 6.71; N, 11.66. Found: C, 74.81; H, 6.83; N, 11.61.

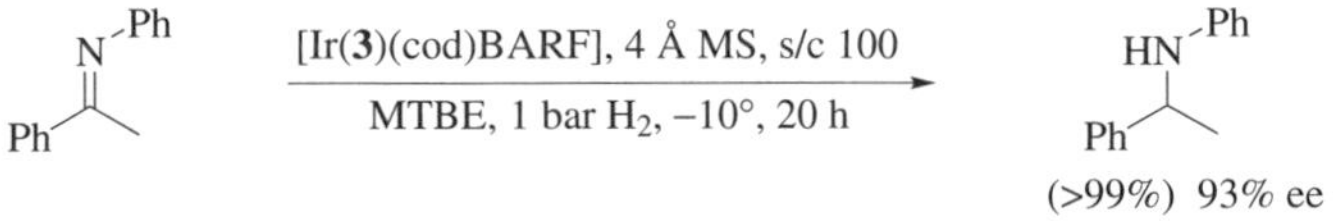

(*R*)-*N*-Phenyl-1-Phenylethylamine [Asymmetric Hydrogenation of *N*-Aryl Imines Using Ir-Phosphino Oxazoline Catalysts].[41] A Schlenk tube was charged with *N*-Phenyl(4-methylphenyl)ethylidene)amine (39 mg, 0.20 mmol), 4 Å molecular sieves (80 mg), and catalyst [Ir(**3**)(cod)]BARF (3.8 mg, 2.0 μmol). Then *tert*-butyl methyl ether (1 mL) was added, and the solution was stirred at room temperature for 10 minutes. The mixture was cooled to −10°, degassed using three freeze/thaw cycles, then placed under a balloon of H_2 (atmospheric pressure), stirring at −10° for 20 hours, after which time the conversion was complete (GC, see below). The solution was evaporated and the residue applied to a silica gel column, eluting with EtOAc/petroleum ether (1:12 v/v). The title compound was obtained in >99% yield as a colorless oil: $[\alpha]_D^{18}$ − 37 (c 0.91, CH_2Cl_2); ^{1}H NMR (300 MHz, $CDCl_3$) δ 7.39–7.06 (m, 7H), 6.64 (t, J = 7.2 Hz, 1H), 6.50 (d, J = 7.5 Hz, 2H), 4.48 (dd, J = 13.5 and 6.6 Hz, 1H), 4.03 (bs, 1H), 1.51 (d, J = 6.9 Hz, 1H). Conversion was determined by GC using an HP-5 column (T = 100–220° at 5°/minute); retention times = 14.77 minutes (product) and 15.22 minutes (starting material). The ee was determined as 93% by HPLC [Chiralcel OD-H column, hexane/*i*-PrOH (98:2), 1.0 mL/minute, λ 254 nm]; retention times = 14.91 minutes (*S*), and 19.10 minutes (*R*).

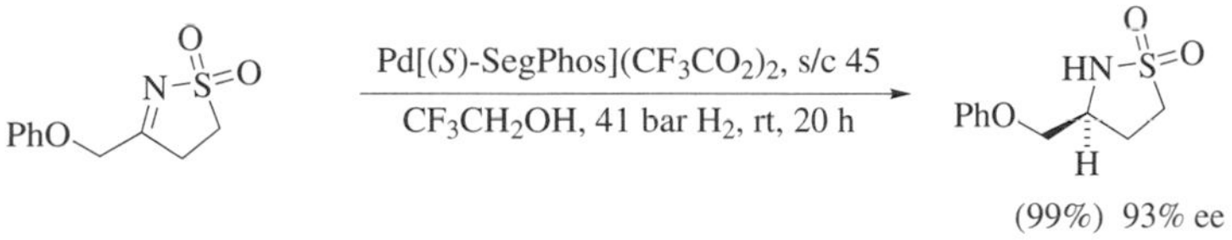

3-Phenoxymethyl-1,2-thiazolidine-1,1-dioxide [Asymmetric Hydrogenation of *N*-Sulfonyl Imines Using a Pd(diphosphine)(CF_3CO_2) Catalyst].[129] (*S*)-Segphos (60.4 mg, 0.099 mmol) and Pd(CF_3CO_2)$_2$ (29.9 mg, 0.09 mmol) were placed in a dried Schlenk tube under a N_2 atmosphere, and degassed anhydrous acetone (8 mL) was added. The mixture was stirred at room temperature for 2 hours. The solvent was removed under vacuum to give the catalyst. This catalyst was transferred into a glove box filled with N_2 and dissolved in dry trifluoroethanol (16 mL). The catalyst solution was added to 3-phenoxymethyl-1,2-thiazoline-1,1-dioxide (1.014 g, 4.50 mmol) and then the mixture was transferred to an autoclave. The autoclave was pressurized to 41 bar with H_2, and the reaction was stirred at room temperature for 20 hours. After the release of H_2, the autoclave was opened and the reaction mixture evaporated. The crude product was purified by chromatography on silica gel using petroleum ether/EtOAc (1:1) as eluent, to yield the title compound (1.013 g, 99% yield, 93% ee), $[\alpha]_D^{30}$ = +14.1 (c 1.12, $CHCl_3$). In order to improve the optical purity, this product was recrystallized from EtOH/water (3:2) to yield a white solid (738 mg, 72%, >99% ee). Enantiomeric excess was determined by HPLC [Chiralcel OD-H column, *i*-PrOH:hexane (20:80), 0.8 mL/minute, λ 254 nm]. ^{1}H NMR (400 MHz, $CDCl_3$) δ 7.28–7.33 (m, 2H), 7.00 (t, J = 7.4 Hz, 1H), 6.79 (d, J = 7.9 Hz, 2H), 4.74 (br s, 1H), 3.99–4.07 (m, 3H), 3.15–3.26 (m, 2H), 2.35–2.38 (m, 1H).

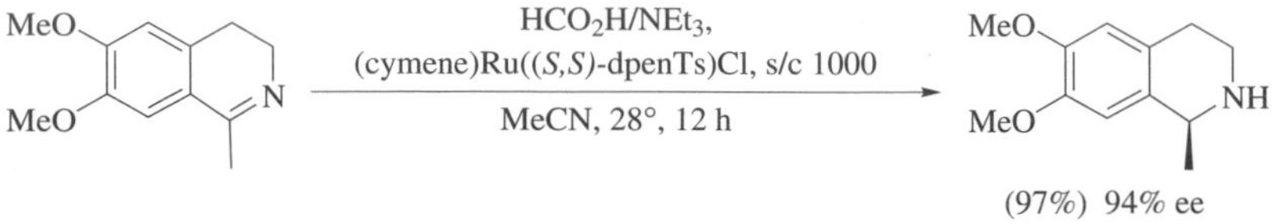

(*R*)-6,7-Dimethoxy-1-methyl-1,2,3,4-tetrahydroisoquinoline [Transfer Hydrogenation Using a Ruthenium Catalyst].[91] *Preparation of the Catalyst.* A mixture of [(*p*-cymene)$_2$RuCl$_2$] (1.53 g, 2.5 mmol), (1*S*,2*S*)-*N*-*p*-toluenesulfonyl-1,2-diphenylethylenediamine (1.83 g, 5.0 mmol) and triethylamine (1.4 mL, 10 mmol) in 2-propanol was heated at 80° for 1 hour. The orange solution was concentrated and the solid Ru complex collected by filtration. The crude material was washed with a small amount of water and dried under reduced pressure to afford [(cymene)Ru(dpenTs)Cl] (2.99 g, 94%).

Transfer hydrogenation. To a solution of 6,7-dimethoxy-1-methyl-3,4-dihydroisoquinoline (4.10 g, 20 mmol) and the preformed ruthenium catalyst (12.7 mg, 0.02 mmol) in MeCN (40 mL), a formic acid/triethylamine (5:2) azeotropic mixture (10 mL) was added. The mixture was stirred at 28° for 12 hours, made basic by addition of aqueous Na_2CO_3, and then extracted with EtOAc. The organic layer was washed with brine, dried over $MgSO_4$, and concentrated under reduced pressure. The crude product was purified by flash chromatography on silica gel using EtOAc/MeOH/NEt_3 (92:5:3) as eluent to afford (*R*)-6,7-dimethoxy-1-methyl-1,2,3,4-tetrahydroisoquinoline (4.02 g, 97%) in 94% ee as determined by HPLC [Daicel Chiralcel OD column (4.6 mm x 25 cm), hexane/isopropanol/diethylamine (90:10:0.1), 0.5 mL/minute]; retention times = 30.2 minutes (*R*), 39.6 minutes (*S*). The identification was confirmed by optical rotation, the title compound having a rotation equal but opposite in sign compared to that reported for the *S*-enantiomer.[195]

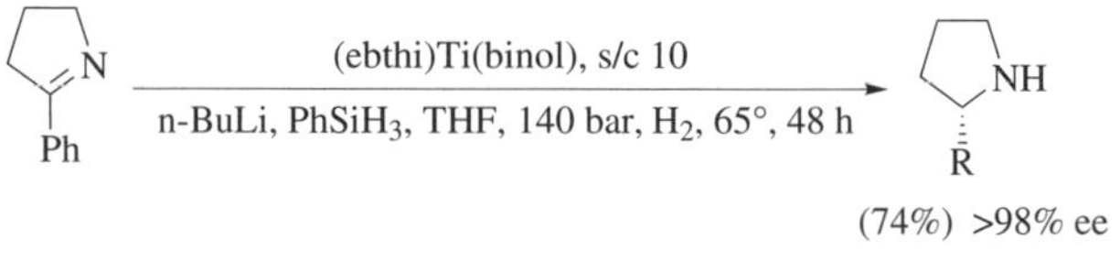

(*R*)-(+)-2-Phenylpyrrolidine [Hydrogenation of Endocyclic Imines with (Ebthi)Ti(binol)].[34] To a dry Schlenk flask under argon was added (*R*, *R*, *R*)-(ebthi)Ti(binol) (50 mg, 0.084 mmol) and dry THF (10 mL). A solution of *n*-BuLi (130 μL, 0.168 mmol, 1.29 M in hexane) was added and after 2–3 minutes the solution became green-brown. Phenylsilane (26 μL, 0.21 mmol) was added and the solution turned dark brown. The mixture was moved to a glove box and transferred to a Parr model 4565 autoclave containing a magnetic stirring bar. 2-Phenylpyrroline (122 mg, 0.84 mmol) was added, and the solution was pressurized to 140 bar with H_2, and stirred for 48 hours at 65°. The solvent was evaporated and the product was dissolved in Et_2O, then extracted with 1 M HCl solution. The aqueous phase was made basic and re-extracted with

Et_2O. The Et_2O layer was dried and evaporated, leaving the title compound as a pure liquid (92 mg, 74% yield): $[\alpha]_D^{22}$ +35 (c = 3.42, MeOH); ^1H NMR (300 MHz, $CDCl_3$) δ 7.38–7.31 (m, 4H), 7.29–7.20 (m, 1H), 4.11 (t, J = 7.5 Hz), 3.25–3.17 (m, 1H), 3.05–2.97 (m, 1H), 2.21–2.13 (m, 1H), 2.01–1.70 (m, 2H), 2.01 (br s, 1H), 1.73–1.61 (m, 1H). GC analysis (Cyclodex B column from J&W Scientific) of the α-methoxy-α-(trifluoromethyl)phenylacetamide derivative indicated >98% ee.

TABULAR SURVEY

Chart 1 presents the structures of ligands and catalysts and the bold numbers that are used to refer to them, or their associated acronyms or abbreviations.

Tables 1–7 list hydrogenation and transfer-hydrogenation reactions that have appeared in the literature up to September 2007. When articles describe the detailed optimization of a specific hydrogenation reaction under a variety of conditions, only the optimal conditions are tabulated in this review. In entries where a variety of effective (>90% ee) ligands are available, single contributions where the ee is <70% are not tabulated.

Entries within each table are arranged according to increasing carbon count of the substrate. The carbon count in Table 7, which covers reductive aminations, is that of the amine and the ketone combined.

The reaction conditions are given as follows:

- Reducing agent (if not dihydrogen).
- Ligand, metal precursor for in situ preparation; for preformed metal complexes the following conventions have been used: π-bound ligands are in front of the metal, the chiral ligand and ligands that are removed (if any) follow in this order, and the coordinated anion is at the end. Non-coordinating anions follow the complex set in [brackets].
- Substrate to catalyst ratio (s/c 100 corresponds to 1 mol% catalyst).
- Solvent, additives (if any), hydrogen pressure (1 bar = 14.5 psig), temperature, and reaction time.

In many publications product yields have not been determined, but rather it is stated that "full conversion" was obtained. In these cases, the conversion is given as (100). Unreported percent conversions or yields are indicated by an em-dash in parentheses (—). If available, the absolute configuration of the products is given for the major enantiomer.

The following abbreviations (excluding those appearing in Chart 1) are used in the tables:

BARF	tetrakis[3,5-bis(trifluoromethyl)phenyl] borate
cod	1,5-cyclooctadiene
cp*	pentamethylcyclopentadienyl
C_{10}mim	1-decyl-3-methylimidazolium
CTAB	cetyltrimethylammonium bromide

cymene	*p*-cymene
DCM	dichloromethane
DMA	2,6-dimethylaniline
DMPEG	poly(ethylene glycol) dimethylether
emim	ethyl methylimidazolium
MEA	2-methyl-6-ethylaniline
MS	molecular sieves
nbd	norbornadiene
Np	naphthyl
Nps	naphthylsulfonyl
PMP	*p*-methoxyphenyl
Py	pyridinyl
s/c	substrate to catalyst ratio
$scCO_2$	supercritical CO_2
TBABr	tetrabutylammonium bromide
TBAI	tetrabutylammonium iodide
TBDMS	*tert*-butyldimethylsilyl
TBME	*tert*-butyl methyl ether
Tf	trifluoromethanesulfonyl
TFA	trifluoroacetic acid
THF	tetrahydrofuran
TIPS	tri-*iso*-propylsilyl
TMP	2,4,6-trimethylphenyl
Tol	*p*-tolyl, 4-methylphenyl
Ts	*p*-toluenesulfonyl
Xyl	3,5-dimethylphenyl

CHART 1. DESIGNATIONS FOR LIGANDS AND CATALYSTS

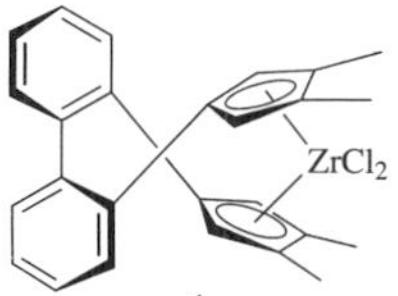

1

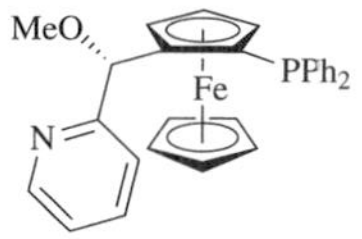

2

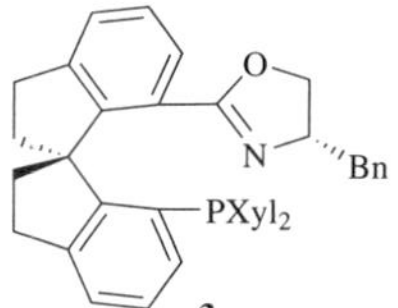

3

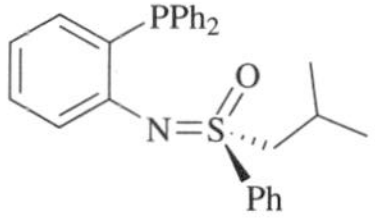

(*R*)-**4**

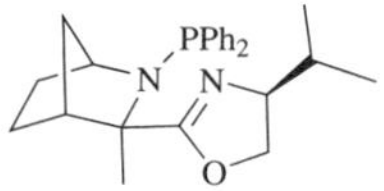

5

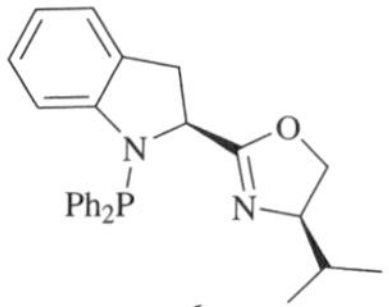

6

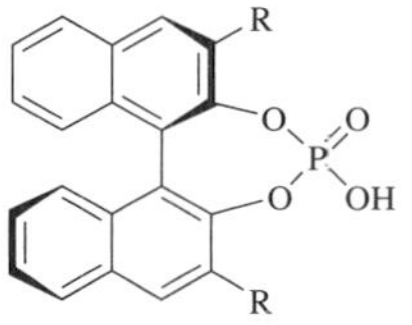

(*R*)-**7**

	R
7a	$(Ph)_3Si$
7b	3,5-$(CF_3)_2C_6H_3$
7c	2,4,6-$(i\text{-}Pr)_3C_6H_2$
7d	9-phenanthryl
7e	9-anthryl

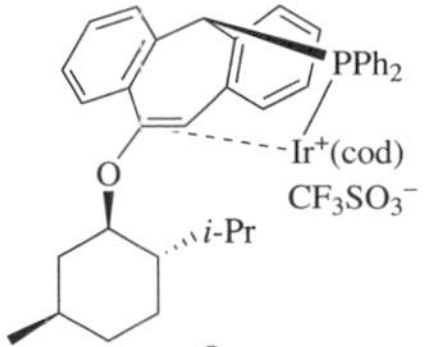

8

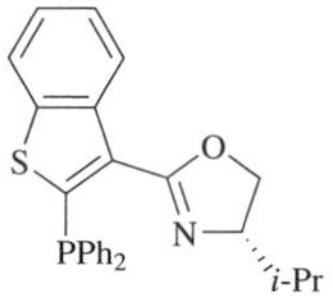

9

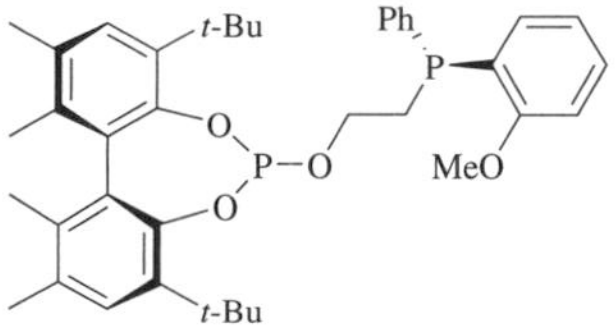

10

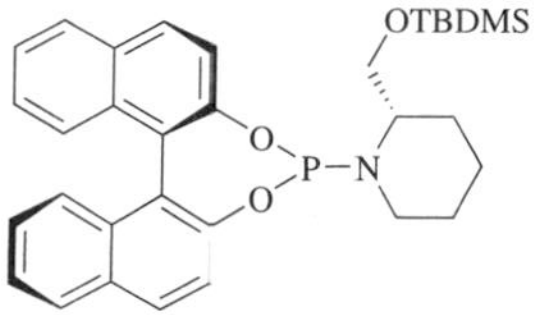

11

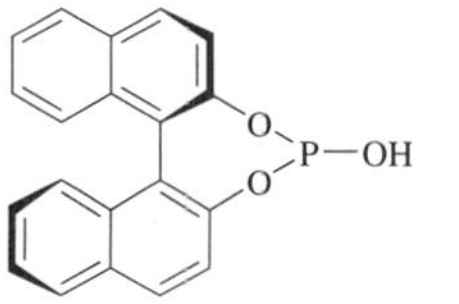

12

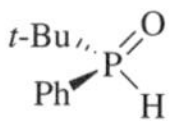

(*R*)-**13**

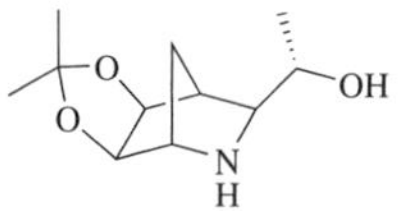

14

CHART 1. DESIGNATIONS FOR LIGANDS AND CATALYSTS (*Continued*)

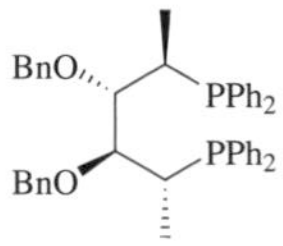
BnO
PPh2
PPh2
BnO

15

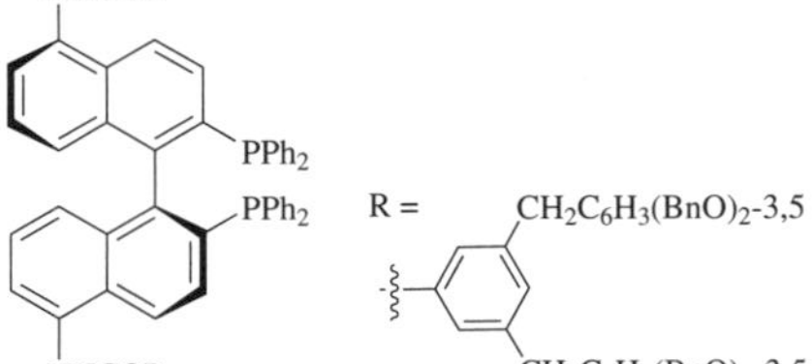
NHCOR
PPh2
PPh2
NHCOR
R =
CH2C6H3(BnO)2-3,5
CH2C6H3(BnO)2-3,5

(*S*)-**16**

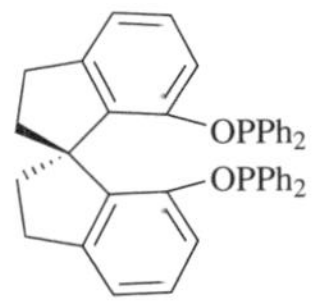
OPPh2
OPPh2

17

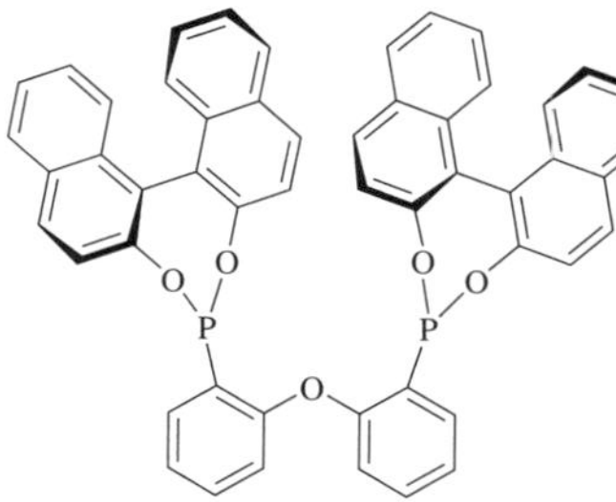
O
O
P
O
O
P
O

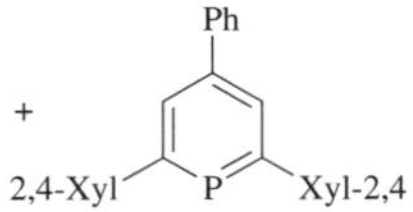
+
Ph
2,4-Xyl
P
Xyl-2,4

18

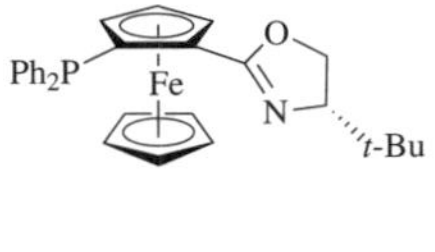
Ph2P
Fe
O
N
t-Bu

(S,S_{Fc})-**19** (#)

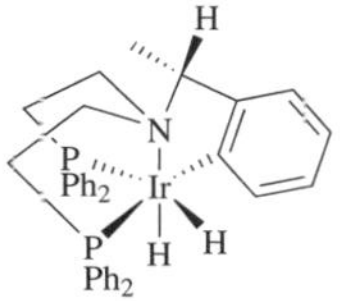
H
N
P
Ph2
Ir
P
Ph2
H
H

20

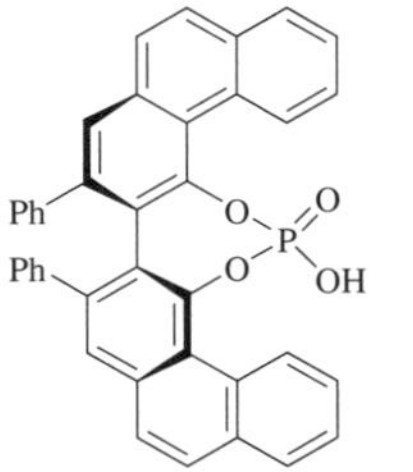
Ph
Ph
O
P
O
O
OH

(*S*)-**21**

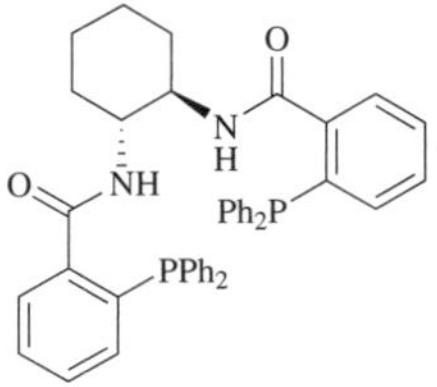
O
N
H
O
NH
Ph2P
PPh2

(*R*,*R*)-**22** (#)

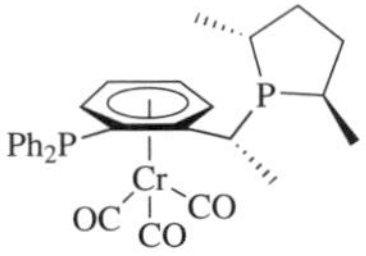
P
Ph2P
Cr
OC
CO
CO

23

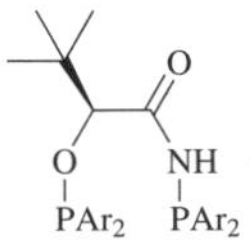
O
O
NH
PAr2
PAr2

24

Ar = 3,5-$(t\text{-Bu})_2C_6H_3$

(#) commercially available

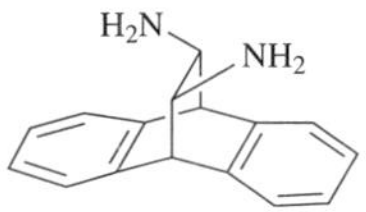
H2N
NH2

anden (#)

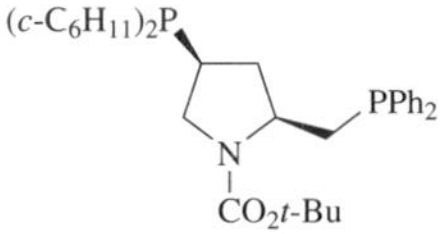
(c-C6H11)2P
PPh2
N
CO2t-Bu

bcpm

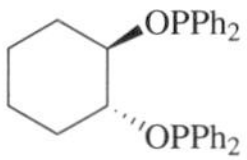
OPPh2
OPPh2

bdpch

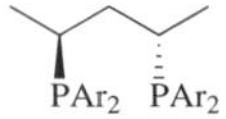
PAr2
PAr2

	Ar
bdpp	Ph
$bdpp_{sulf}$	3-$NaO_3SC_6H_4$

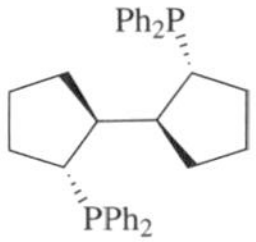
Ph2P
PPh2

bicp

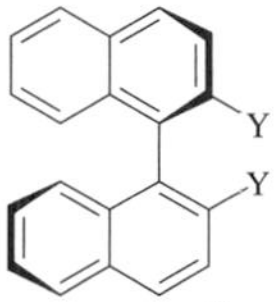
Y
Y

	Y
(*R*)-binap (#)	PPh_2
(*R*)-binol (#)	OH

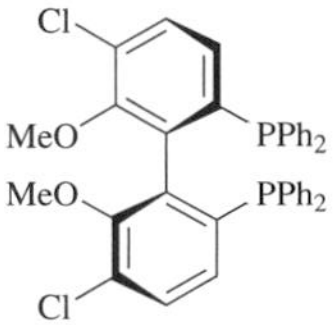
Cl
MeO
PPh2
MeO
PPh2
Cl

(*R*)-ClMeO-biphep (#)

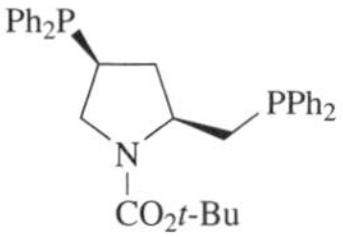
Ph2P
PPh2
N
CO2t-Bu

bppm

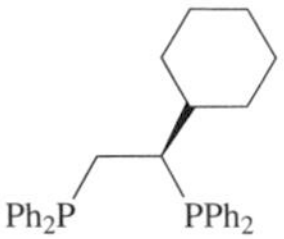
Ph2P
PPh2

cycphos

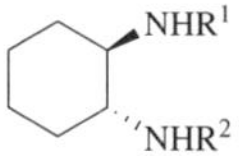
NHR1
NHR2

	R^1	R^2
(*R*,*R*)-dach (#)	H	H
(*R*,*R*)-dachTs	H	SO_2(4-Tol)
(*R*,*R*)-dppach	PPh_2	PPh_2

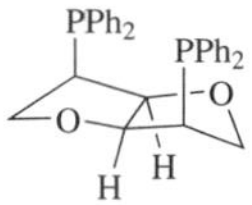
PPh2
PPh2
O
O
H
H

ddppm

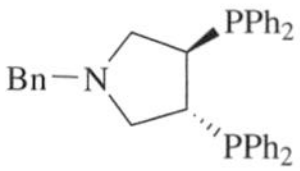
Bn
N
PPh2
PPh2

(*R*,*R*)-deguphos

CHART 1. DESIGNATIONS FOR LIGANDS AND CATALYSTS (*Continued*)

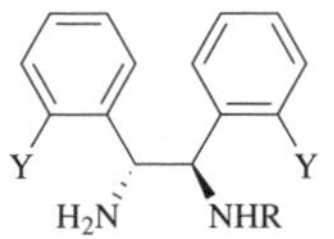

Y
Y
H_2N
NHR

(*S*,*S*)-dpen

	Y	R
dpen (#)	H	H
dpen(1-Nps)	H	SO_2(1-Np)
dpenSO_2TMP	H	SO_2TMP
dpenTs (#)	H	SO_2(4-Tol)
dpenTs_{amin}	NH_2	SO_2(4-Tol)
dpenTs_{sulf}	SO_3Na	SO_2(4-Tol)
dpenTs_{dend}	H	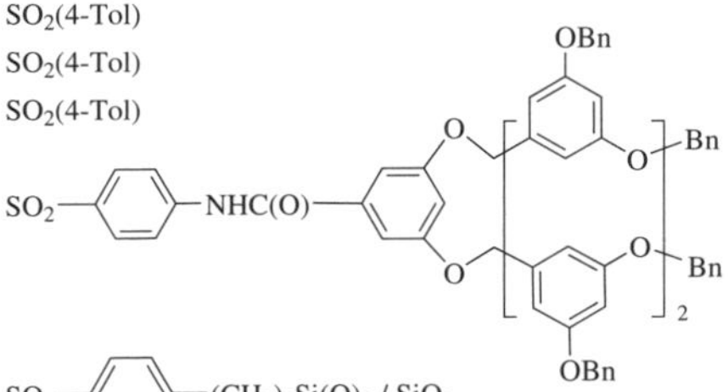 SO_2 NHC(O) O O OBn O Bn O Bn 2 OBn
dpenTs_{immob}	H	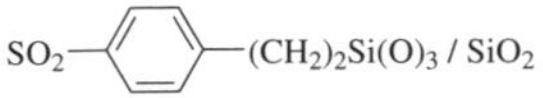 SO_2 $(CH_2)_2Si(O)_3$ / SiO_2
dpenTs_{SMF}	H	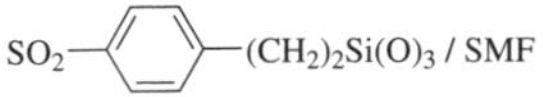 SO_2 $(CH_2)_2Si(O)_3$ / SMF SMF = siliceous mesocellular foam

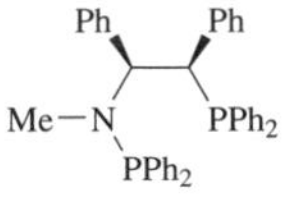

Ph
Ph
Me−N
PPh_2
PPh_2

dpampp

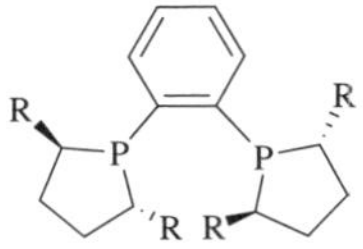

R
P
P
R
R
R

(*R*,*R*)-duphos (#)
R = Me, Et

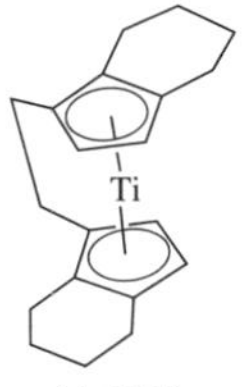

Ti

(ebthi)Ti

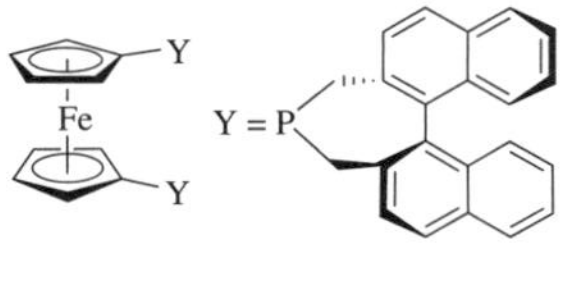

Y
Fe
Y
Y = P

(*R*)-f-binaphane (#)

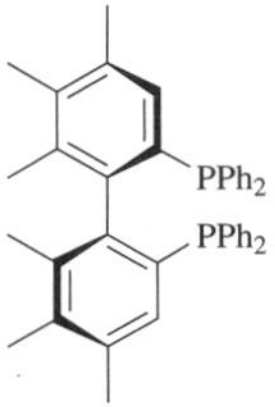

PPh_2
PPh_2

(*R*)-hexaphemp (#)

(#) commercially available

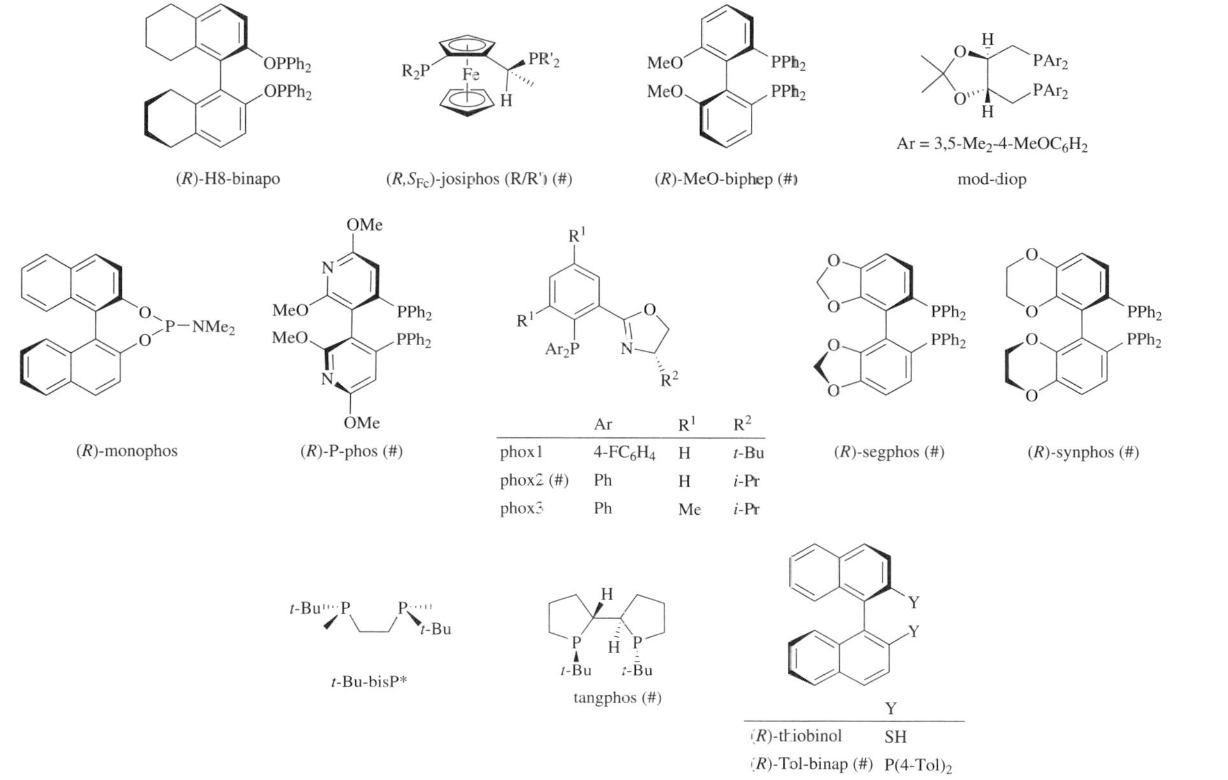
OPPh2
OPPh2
(R)-H8-binapo
R2P
Fe
PR'2
H
(R,SFc)-josiphos (R/R') (#)
MeO
MeO
PPh2
PPh2
(R)-MeO-biphep (#)
O
H
PAr2
PAr2
Ar = 3,5-Me2-4-MeOC6H2
mod-diop
O
P—NMe2
(R)-monophos
OMe
N
MeO
MeO
PPh2
PPh2
(R)-P-phos (#)
R1
Ar2P
N
R2
PPh2
(R)-segphos (#)
PPh2
(R)-synphos (#)
t-Bu
P
t-Bu-bisP*
H
P
t-Bu
tangphos (#)
Y

	Ar	R^1	R^2
phox1	4-FC_6H_4	H	t-Bu
phox2 (#)	Ph	H	i-Pr
phox3	Ph	Me	i-Pr

	Y
(R)-thiobinol	SH
(R)-Tol-binap (#)	$P(4\text{-}Tol)_2$

(#) commercially available

TABLE 1. *N*-ARYL IMINES

Imine	Conditions	Product(s) and Yield(s) (%), % ee	Refs.
C_{10} 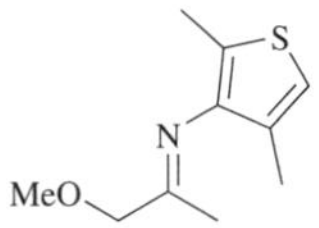	(R,S_{Fc})-Josiphos (Ph/Ph), $[Ir(cod)Cl]_2$, s/c 200, toluene, AcOH, TBAI, 30 bar, rt	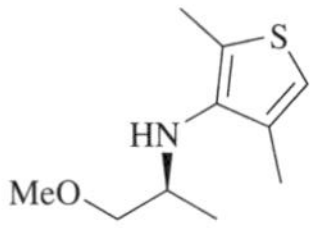 (100)[a], 78	37
C_{12} 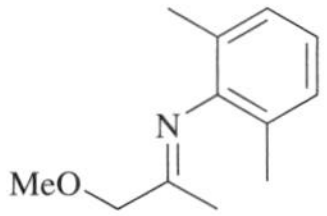	Ir((S,S)-bdpp)$(CF_3CO_2)_3$, s/c 500, THF/DCM, 40 bar, 0°, 145 h	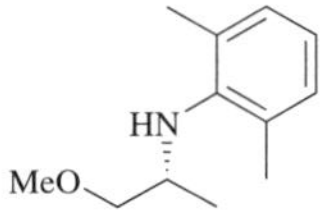 (96)[a], 90	38
C_{12-19} 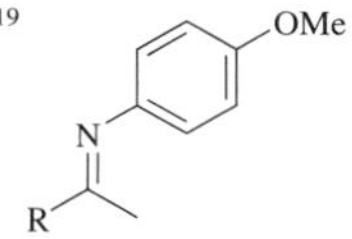	Hantzsch ester, (S)-**7c,** s/c 100, toluene, 35°	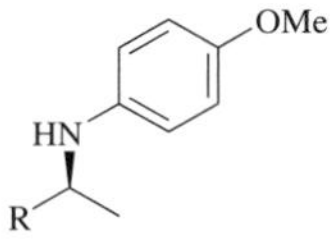 R / Time / % Conv., % ee *i*-Pr / 60 h / (80), 90 2-Np / 42 h / (85), 84	48
C_{13} 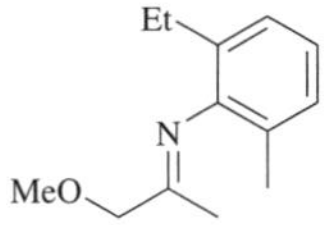	(R,S_{Fc})-Josiphos (R^1/R^2), $[Ir(cod)Cl]_2$, neat, 50 bar, 50°, 3-4 h	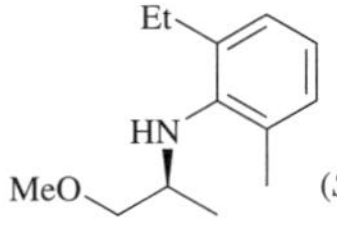 (*S*)-**I** (100)[a]	
		R^1/R^2 / s/c / Additives / % ee	
		Ph/Xyl / 2,000,000 / HI / 80	16
		Ph/4-(*n*-Pr)$_2$NXyl / 100,000 / AcOH, TBAI / 87	37
	23, $[Ir(cod)Cl]_2$, s/c 100, toluene, 80 bar, rt, 16 h	(*S*)-**I** (96)[a], 82	196

C_{14} 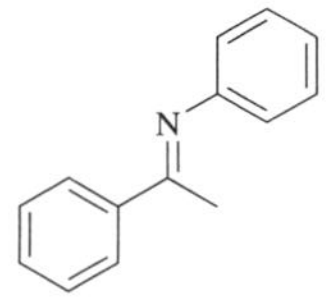	[Ir(**6**)(cod)]BARF, s/c 50, DCM, 20 bar, rt, 12 h	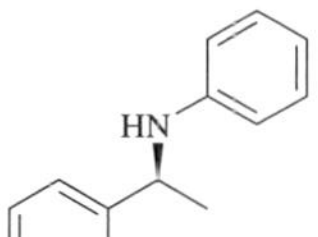 (*S*)-**I** (100)[a], 90	45
	[Ir((*S*)-phox3)(cod)]BARF, s/c 1000, DCM, 100 bar, 45°, 5 h	(*R*)-**I** (99)[a], 89	50
	[Ir((*S*)-phox2)(cod)]BARF, s/c 700, $scCO_2$, 30 bar, 40°, 20 h	(*R*)-**I** (100)[a], 81	49
	8, s/c 100, DCM, 50 bar, 50°, 2 h	(*R*)-**I** (100)[a], 86	51
	Ru((*R,R*)-Et-duphos)((*R,R*)-dach)Cl_2, s/c 100, *t*-BuOH, *t*-BuOK, 15 bar, 65°, 20 h	(*S*)-**I** (92)[a], 92	47
	[Ir(**9**)(cod)]BARF, s/c 1000, DCM, 50 bar, rt, 4 h	(*R*)-**I** (>99)[a], 85	52
	[Ir((*S*)-phox2)(cod)]BARF, s/c 500, [emim]BARF/sCO_2, 30 bar, 40°, 22 h	(*R*)-**I** (>99)[a], 78	65
	11, $[Ir(cod)_2]BF_4$, s/c 100, DCM, 80 bar, rt, 17 h	(*S*)-**I** (—), 73	54
C_{14-15} 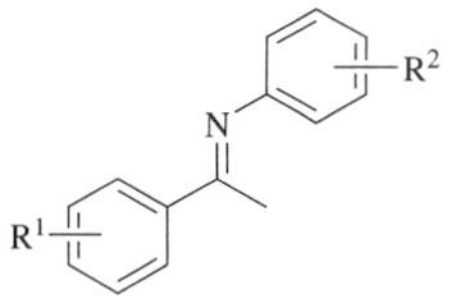	[Ir(ddppm)(cod)]PF_6, s/c 100, DCM, 1 bar, rt, 24 h	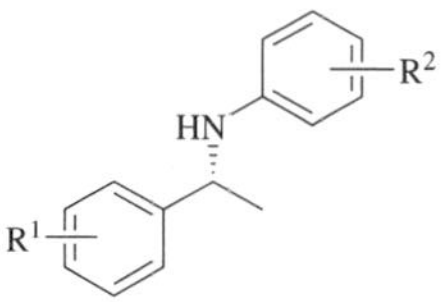	46

R^1	R^2	% Conv., % ee
H	H	(99), 84
4-F	H	(99), 80
4-MeO	H	(100), 81
H	4-MeO	(100), 94

TABLE 1. *N*-ARYL IMINES (*Continued*)

Imine	Conditions	Product(s) and Yield(s) (%), % ee	Refs.
C_{14-15} 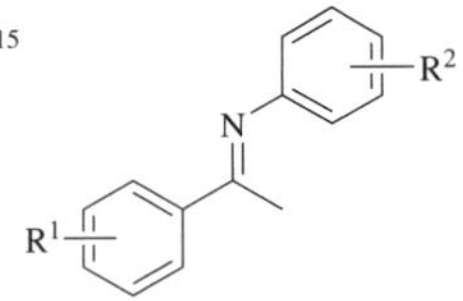	**10**, $[Ir(cod)Cl]_2$, s/c 100, DCM, 30 bar, rt, 24 h	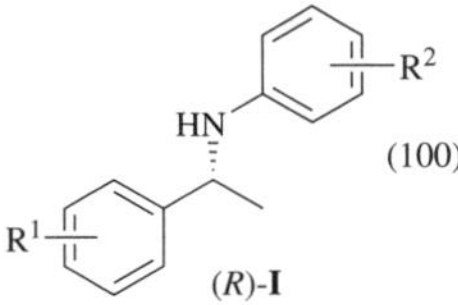 (*R*)-**I** (100)[a]	53
C_{14-16}	[Ir((*S*,*S*)-*t*-Bu-bisP*)(cod)]BARF, s/c 200, DCM, 1 bar, rt	(*R*)-**I**	43
	[Ir(**5**)(cod)]BARF, s/c 200, DCM, 20 bar, rt	(*R*)-**I**	44

C_{14-15}, ref. 53:

R^1	R^2	% ee
H	H	84
4-Me	H	72
4-MeO	H	85
4-F	H	79
4-Cl	H	82
H	4-MeO	81

C_{14-16}, ref. 43:

R^1	R^2	Time	
H	H	1.5 h	(91), 86
4-F	H	1.5 h	(92), 84
H	4-F	12 h	(99), 84
H	4-Cl	12 h	(99), 83
4-MeO	H	2 h	(98), 69
H	4-MeO	2 h	(93), 86
H	4-CF_3	12 h	(95), 99
H	3,5-$(CF_3)_2$	12 h	(97), 90
4-MeO	4-MeO	2 h	(98), 83

C_{14-16}, ref. 44:

R^1	R^2	Time	% Conv., %ee
H	H	2 h	(98), 90
4-F	H	2 h	(99), 89
2-Me	H	12 h	(53), 83
4-MeO	H	3 h	(99), 86
H	2-Me	3 h	(99), 80
H	4-MeO	1.5 h	(99), 89
4-Cl	4-MeO	1.5 h	(99), 89
4-MeO	4-MeO	2 h	(99), 86

C_{14-20}

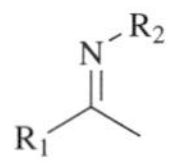

[Ir(**3**)(cod)]BARF, s/c 100, TBME, 4 Å MS, 1 bar, 10°, 20 h

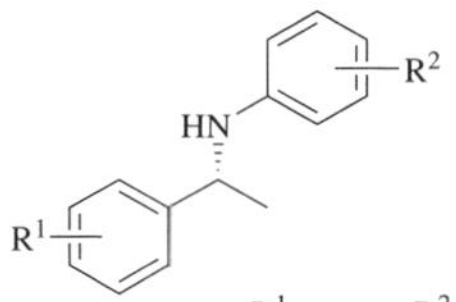

(>99.5)[a]

41

R^1	R^2	% ee
H	H	93
3-Cl	H	93
4-Cl	H	90
3-Br	H	92
4-Br	H	91
4-Me	H	94
4-MeO	H	94
3,4-Me_2	H	94
H	4-Cl	97
H	3-Br	94
H	4-Br	96
H	3-Me	91
H	4-Me	93

f-Binaphane, $[Ir(cod)Cl]_2$, s/c 100, DCM, 70 bar, 14–44 h

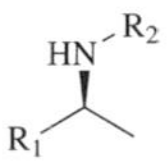

39

R^1	R^2	Additive	Temp	% Conv., % ee
Ph	Ph	I_2	–5°	(100), 94
Ph	4-$MeOC_6H_4$	I_2	–5°	(100), 95
Ph	2-$MeOC_6H_4$	—	rt	(100),81
Ph	2,6-$Me_2C_6H_3$	—	rt	(77), >99
4-$MeOC_6H_4$	2,6-$Me_2C_6H_3$	—	rt	(77), 98
4-$CF_3C_6H_4$	2,6-$Me_2C_6H_3$	—	rt	(80), 99
Ph	2-MeO-6-MeC_6H_3	—	rt	(72), 98
1-Np	2-MeO-6-MeC_6H_3	—	rt	(75), 96

TABLE 1. *N*-ARYL IMINES (*Continued*)

Imine	Conditions	Product(s) and Yield(s) (%), % ee	Refs.
C_{15-16} 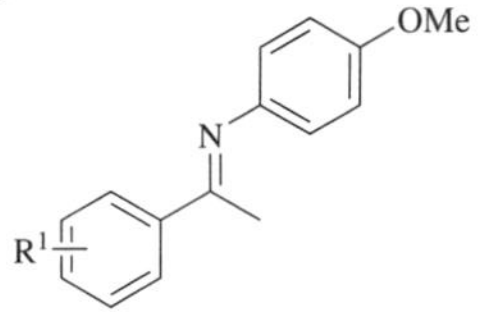	**4**, $[Ir(cod)Cl]_2$, s/c 100, toluene, I_2, 20 bar, rt	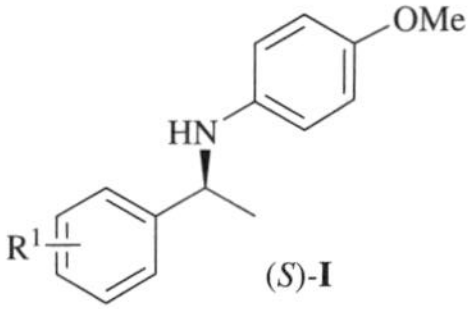 (*S*)-**I** (see table a)	42
C_{15-17}	Hantzsch ester, (*S*)-**7c**, s/c 100, toluene, 35°	(*S*)-**I** (see table b)	48
C_{15-18} 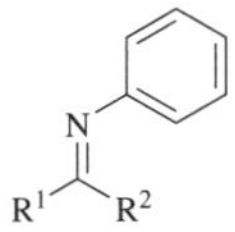	[Ir(**5**)(cod)]BARF, s/c 200, DCM, 20 bar, rt	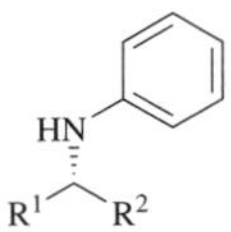 (see table c)	197

(a)

R^1	Time	% Conv., % ee
H	4 h	(99), 96
4-Cl	4 h	(99), 95
2-Me	4 h	(99), 94
3-Me	4 h	(99), 93
4-Me	4 h	(99), 96
2-MeO	6 h	(99), 90
3-MeO	4 h	(99), 96
4-MeO	4 h	(99), 94

(b)

R^1	Time	% Conv., % ee
H	45 h	(96), 88
2-F	45 h	(95), 85
4-NO_2	42 h	(96), 80
2-Me	71 h	(91), 93
4-Me	42 h	(98), 88
2-MeO	45 h	(92), 80
4-CN	42 h	(87), 80
2,4-Me_2	71 h	(88), 92
3,4-$(MeO)_2$	45 h	(84), 89

(c)

R^1	R^2	Time	% Conv., % ee
Ph	Et	6 h	(99), 78
2-Np	Me	1 h	(99), 91

Carbon	Substrate	Conditions	Product(s) and Yield(s) (%), % ee	Refs.
C_{15-19}	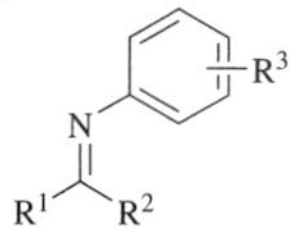	(*S*)-**4**, $[Ir(cod)Cl]_2$, s/c 100, toluene, I_2, 20 bar, rt	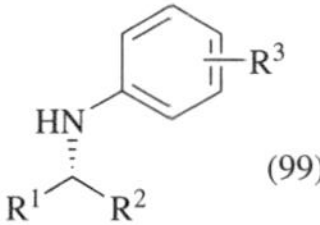 (99)[a]	42

R^1	R^2	R^3	Time	% ee
Ph	Me	2-MeO	12 h	79
Ph	Et	4-MeO	4 h	92
1-Np	Me	4-MeO	6 h	98
2-Np	Me	4-MeO	4 h	61

Carbon	Substrate	Conditions	Product(s) and Yield(s) (%), % ee	Refs.
C_{15-21}	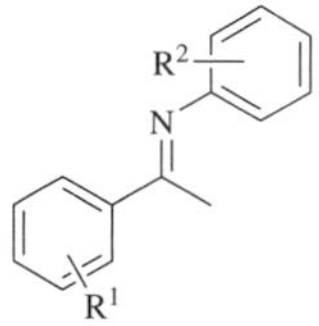	Hantzsch ester, **7b**, s/c 5, C_6H_6, 60°, 72 h	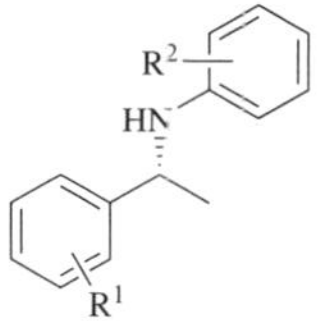	55

R^1	R^2	
$4-CF_3$	H	(58), 70
H	4-MeO	(76), 74
2-F	4-MeO	(82), 84
3-Br	4-MeO	(62), 72
2-Me	4-MeO	(74), 78
2-MeO	4-MeO	(76), 72
$2-CF_3$	4-MeO	(46), 82
$4-CF_3$	4-MeO	(71), 72
$3,5-Me_2$	4-MeO	(91), 78
4-Ph	4-MeO	(71), 74

Carbon	Substrate	Conditions	Product(s) and Yield(s) (%), % ee	Refs.
C_{16}	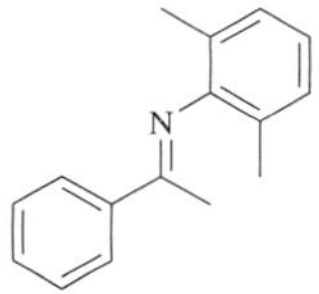	(R,S_{Fc})-Josiphos (Ph/4-$CF_3C_6H_4$), $[Ir(cod)Cl]_2$, s/c 200, toluene, AcOH, TBAI, 30 bar, rt	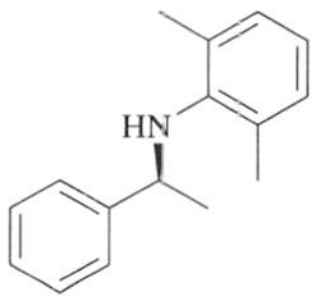 (100)[a], 96	37
C_{17}	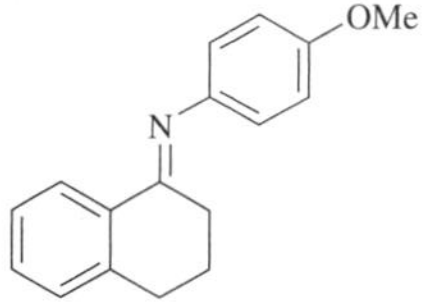	**4**, $[Ir(cod)Cl]_2$, s/c 100, toluene, I_2, 20 bar, rt, 4 h	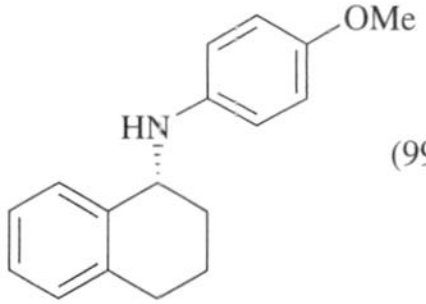 (99)[a], 91	42

TABLE 1. *N*-ARYL IMINES (*Continued*)

Imine	Conditions	Product(s) and Yield(s) (%), % ee	Refs.
C_{17-24} 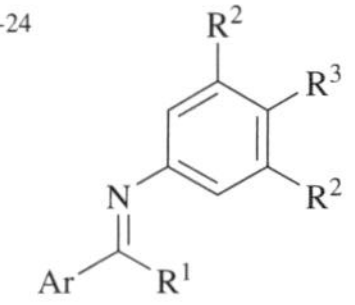	[Ir(**2**)(cod)]BARF, s/c 100, toluene/MeOH, 10 bar, rt, 2–6 h	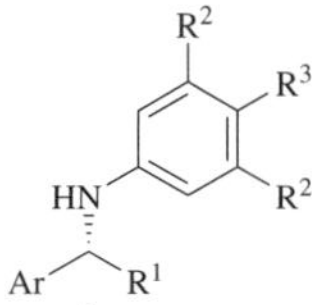 (>99.5)[a]	40
C_{19} 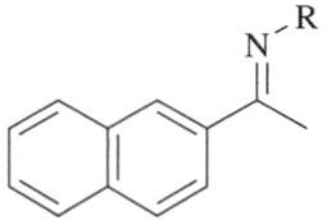	[Ir(**5**)(cod)]BARF, s/c 200, DCM, 20 bar, rt, 2 h	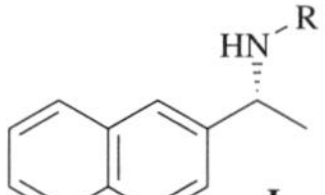 **I** R = H (99)[a], 91[b]	44
	Hantzsch ester, (*R*)-**7b**, s/c 5, C_6H_6, 60°, 72 h	**I** (82), 70; R = PMP	55

Ar	R^1	R^2	R^3	% ee
Ph	Me	H	H	84
Ph	Me	H	MeO	85
Ph	Me	Me	H	94
Ph	Me	Me	MeO	94
3-FC_6H_4	Me	Me	MeO	93
4-ClC_6H_4	Me	Me	MeO	92
2-MeC_6H_4	Me	Me	MeO	94
3-MeC_6H_4	Me	Me	MeO	93
4-$CF_3C_6H_4$	Me	Me	MeO	89
Ph	Et	Me	MeO	94
Ph	*n*-C_5H_{11}	Me	MeO	95
4-$MeO_2CC_6H_4$	Me	Me	MeO	94
4-PhC_6H_4	Me	Me	MeO	92
2-Np	Me	Me	MeO	93
Ph	$Bz(CH_2)_3$	Me	MeO	99

[a] This value is the percent conversion.

[b] The stereochemistry of the product was not reported in the original reference.

TABLE 2. *N*-ALKYL IMINES

Imine	Conditions	Product(s) and Yield(s) (%), % ee	Refs.
C_{9-19}			
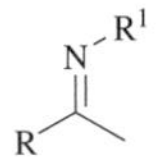	(Ebthi)Ti(binol), THF, 65°, 8–48 h	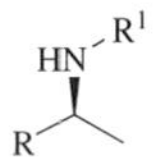	34
C_{11}			
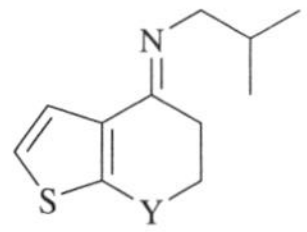	HCO_2H/NEt_3, (cymene)Ru((*S,S*)-dpenSO_2(1-Nps))Cl, s/c 100, MeCN, 28°	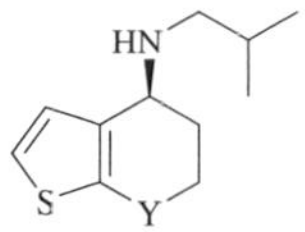	91
C_{12}			
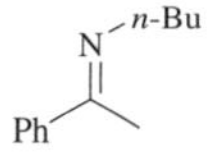	Ru((*R,R*)-dppach)((*R,R*)-dach)HCl, s/c 1500, neat, *i*-PrOK, 3 bar, 20°, 60 h	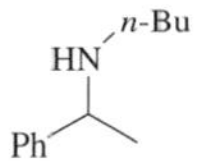 (91)[a], 92[b]	63

Products for ref. 34 (from the (Ebthi)Ti(binol) row):

R	R^1	s/c	Pressure	
c-C_6H_{11}	Me	20	35 bar	(85), 92
c-C_6H_{11}	*n*-Pr	20	140 bar	(70), 79
i-Pr	Bn	10	140 bar	(66), 76
c-C_3H_5	Bn	20	140 bar	(91), 61
n-Bu	Bn	10	140 bar	(68), 58
2-furyl	Bn	20	140 bar	(70), 53
$Me_2C=CH(CH_2)_2$	Bn	10	140 bar	(64), 62
c-C_6H_{11}	Bn	20	140 bar	(93), 76
Ph	Bn	50	140 bar	(93), 85
c-C_6H_{11}	4-MeOBn	20	140 bar	(92), 78
4-MeOC_6H_4	Bn	20	140 bar	(86), 86
2-Np	Bn	20	140 bar	(82), 70

Products for ref. 91 (from the HCO_2H/NEt_3 row):

Y	Time	
S	2 h	(82), 85
SO_2	5 h	(84), 88

TABLE 2. *N*-ALKYL IMINES (*Continued*)

Imine	Conditions	Product(s) and Yield(s) (%), % ee	Refs.
C_{12-16} 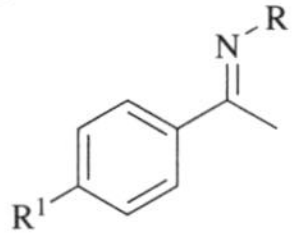	[Ir((*S*)-phox3)(cod)]BARF, s/c 25, DCM, 100 bar, rt, 16 h	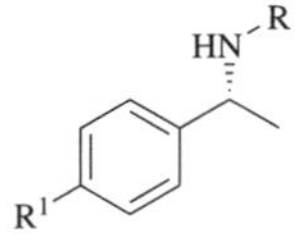 (100)[a] R / R^1 / % ee *n*-Bu / H / 75 Bn / H / 76 Bn / Me / 79	50
C_{15} 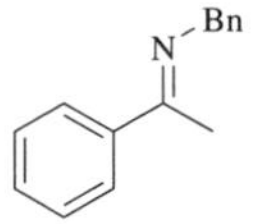	HCO_2Na, $[(C_6Me_6)Ru((R,R)\text{-dachTs})H_2O]BF_4$, s/c 100, H_2O, pH 9, 60°, 2 h	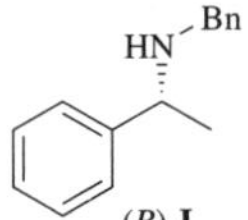 (*R*)-**I** (100)[a], 91	64
	[Ir(**6**)(cod)]BARF, s/c 50, DCM, 20 bar, rt, 12 h	(*S*)-**I** (100)[a], 82	45
	HCO_2H/NEt_3, (cymene)Ru((*S,S*)-dpenSO_2TMP)Cl, s/c 200, DCM, 28°, 36 h	(*S*)-**I** (72), 77	91
	1, s/c 1000, toluene, 150 bar, 80°, 12 h	(*S*)-**I** (95), 76	80
	[Ir(phox2)(cod)]BARF, s/c 500, [emim]BARF/scCO_2, 30 bar, 40°, 22 h	(*R*)-**1** (>99)[a], 78	65
	$(AuCl)_2$((*R,R*)-Me-duphos), s/c 1000, EtOH, 4 bar, 20°, ~1 h	(*S*)-**I** (100)[a], 75	71
	[Rh((*R,R*)-bdpch)(cod)]BF_4, s/c 500, MeOH, 50 bar, rt	(*S*)-**I** (>99)[a], 70–72	69
	(*S*)-Tol-binap, $[Ir(cod)Cl]_2$, s/c 100, C_6H_6, $BnNH_2$, 60 bar, 20°, 18 h	(*R*)-**I** (100)[a], 70	67

$C_{15\text{-}16}$

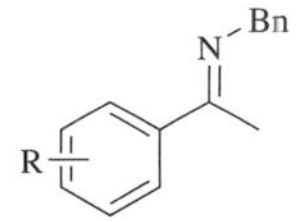

[Rh(bdpp)(nbd)]ClO_4, s/c 100, C_6H_6/reverse micelles, 70 bar, 4–8°, 21–73 h

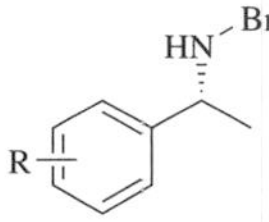

(*R*)-**I**

R	
H	(96), 89
4-MeO	(95), 92

61

Cycphos, [Rh(cod)Cl]$_2$, s/c 100, C_6H_6/MeOH, KI, 70 bar

(*S*)-**I**

R	Temp	Time	
H	20°	90 h	(90), 79
2-MeO	20°	120 h	(>99), 71
4-MeO	–20°	144 h	(>99), 91

62

Bdpp$_{sulf}$, [Rh(cod)Cl]$_2$, s/c 100, H_2O/AcOEt, 70 bar, rt, 16 h

(*R*)-**I**

R	
H	(94), 88–96
2-MeO	(94), 91–92
3-MeO	(93), 86–89
4-MeO	(96), 86–95

60

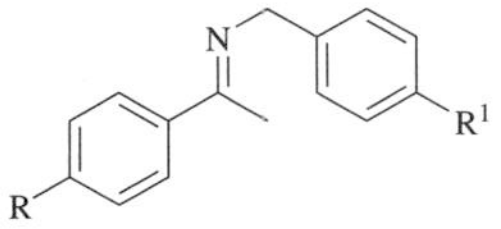

13, [Ir(cod)Cl]$_2$, s/c 100, toluene, 25 bar

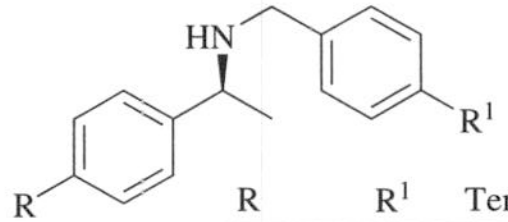

R	R[1]	Temp	Time	% Conv., % ee
H	H	0°	120 h	(75), 82
H	Cl	rt	24 h	(75), 77
MeO	H	0°	120 h	(80), 83
H	MeO	rt	24 h	(85), 76

68

$C_{15\text{-}17}$

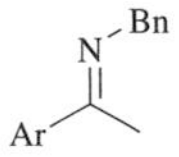

12, [Ir(cod)Cl]$_2$, PPh_3, s/c 100, DCM, 50 bar, rt, 48 h

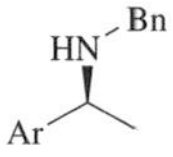

Ar	% Conv., % ee
Ph	(100), 88
4-ClC_6H_4	(99), 90
4-$MeOC_6H_4$	(99), 92
2-Np	(100), 92

66

TABLE 2. *N*-ALKYL IMINES (*Continued*)

Imine	Conditions	Product(s) and Yield(s) (%), % ee	Refs.
C_{16-19}			
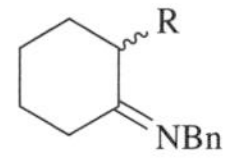	HCO_2H/NEt_3, cp*Ir((*S*,*S*)-dpenTs)Cl, s/c 500, DCM, rt	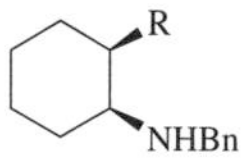 R / Time / % cis: allyl, 24 h, 93, (75), 63; $(CH_2)_2CN$, 144 h, 96, (60), 72; Ph, 24 h, >99, (55), 50	77
C_{17-20}			
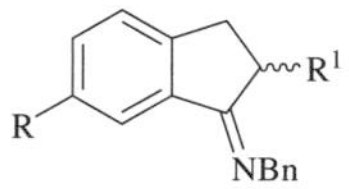	HCO_2H/NEt_3, (cymene)Ru((*R*,*R*)-dpenTs)Cl, s/c 200, DCM, rt, 144 h	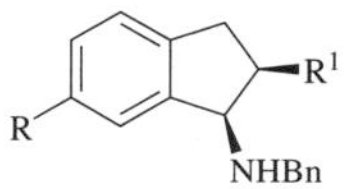 R / R^1: H, Me, (70), 96; Me, Me, (82), 97; Me, allyl, (67), 92	77
C_{17-18}			
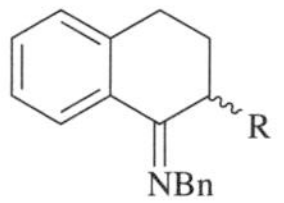	HCO_2H/NEt_3, (cymene)Ru((*S*,*S*)-dpen(1-Nps)Cl, s/c 100, DCM, 28°, 6 h	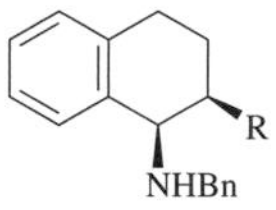 (*R*,*R*)-**I** R = H (90), 89	91
	HCO_2H/NEt_3, (cymene)Ru((*R*,*R*)-dpenTs)Cl, s/c 200, DCM, rt, 120 h	(*S*,*S*)-**I** R = Me (45), 50	77
C_{19}			
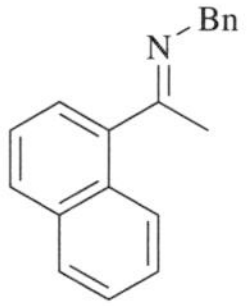	HCO_2H/NEt_3, (cymene)Ru((*S*,*S*)-dpenTs)Cl, s/c 200, DCM, 20°, 6 h	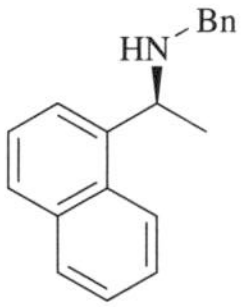 $(80)^a$, 88^b	90

[a] This value is the percent conversion.

[b] The stereochemistry of the product was not reported in the original reference.

TABLE 3. ENDOCYCLIC IMINES

Imine	Conditions	Product(s) and Yield(s) (%), % ee	Refs.
C_{8-9}			
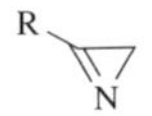	**14**, [(cymene)RuCl]$_2$, s/c 100, *i*-PrOH, *i*-PrOK, 0.1–1 h	 R / Temp: Ph, 0°, (56), 78; 3-Tol, rt, (75), 65; 4-Tol, rt, (83), 72	82
C_9			
	(2*S*,4*S*)-bppm, [Ir(cod)Cl]$_2$, s/c 100, C_6H_6/MeOH, BiI_3, 40 bar, –10°, 3 h	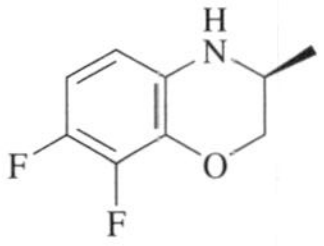 (96), 90	174
C_{9-10}			
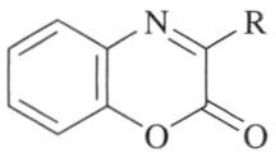	Hantzsch ester, (*R*)-**7a**, s/c 10, C_6H_6, 5 Å MS, 40°	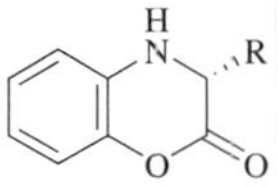 R / Time: Me, 7 h, (82), 97; Et, 50 h, (27), 79	104
C_{10}			
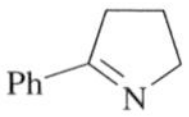	(Ebthi)Ti(binol), s/c 100, THF, 5 bar, 65°, 8–48 h	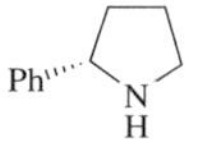 (*S*)-**I** (84), 99	34
	1, s/c 1000, toluene, 150 bar, 80°, 12 h	(*S*)-**I** (96), 98	80
	Ir((*S*)-binap)HBr(OBz), s/c 100, toluene, 60 bar, 20°, 18 h	(*S*)-**I** (38), 89	81

TABLE 3. ENDOCYCLIC IMINES (*Continued*)

Imine	Conditions	Product(s) and Yield(s) (%), % ee	Refs.
C_{10} 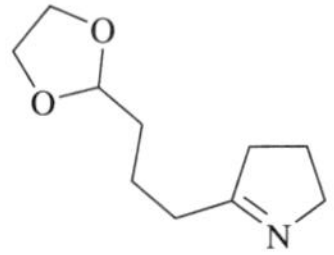	(Ebthi)Ti(binol), s/c 20, THF, 5 bar, 65°, 8–48 h	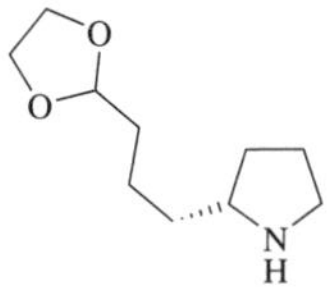 (82), 99	34
C_{10} 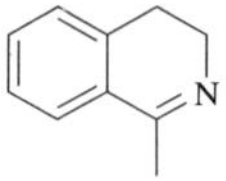	Ru((*R*,*R*)-Et-duphos)((*R*,*R*)-dach)Cl_2, s/c 100, *i*-PrOH, *t*-BuOK, 15 bar, 50–65°, 20 h	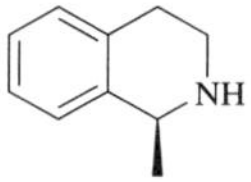 (80), 79	47
$C_{10\text{-}14}$ 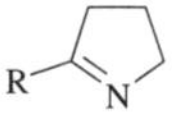	(Ebthi)Ti(binol), s/c 20, THF, 8–48 h	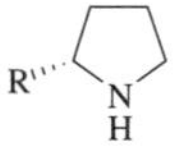 **I** + 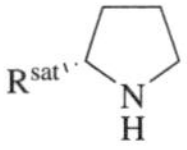 **II** R^{sat} = saturated R group	34

R	Pressure	Temp	I	II		
CH_2=CH(CH_2)$_4$	5 bar	45°	(0), —	(72), 99		
(*Z*)-EtCH=CH(CH_2)$_5$	5 bar	45°	(31–42), 99	(~15), 99	(*E*)-EtCH=CH(CH_2)$_5$	(~16), 99
(*E*)-TMSCH=CH(CH_2)$_4$	5 bar	50°	(65–68), 99	(5–8), 99		
Me_2C=CH(CH_2)$_2$	5 bar	50°	(79), 99	—		
n-C_6H_{13}	138 bar	65°	(81), 98	—		
HO(CH_2)$_7$	5 bar	65°	(84), 99	—		
TBDMSO(CH_2)$_4$	5 bar	65°	(82), 99	—		

C_{11}				
	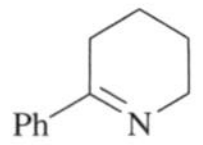	(Ebthi)Ti(binol), s/c 20, THF, 35 bar, 65°, 8–48 h	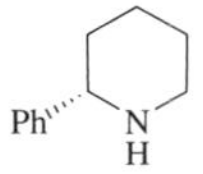 (S)-**I** (78), 98	34
		(S)-Tol-binap, $[Ir(cod)Cl]_2$, s/c 100, C_6H_6, $BnNH_2$, 60 bar, 20°, 18 h	(R)-**I** (100)[a], 90	67
		$[Ir((S)\text{-binap})HI_2]_2$, s/c 1000, toluene, 60 bar, 20°, 3 h	(S)-**I** (99), 91	81
C_{11-16}				
	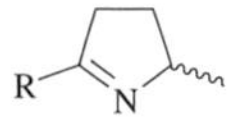	(Ebthi)Ti(thiobinol), s/c 20, THF, 5 bar, 65°, 8–48 h	 **I** + **II**	79

R	**I**	**II**
Ph	(34), 99	(37), 99
$TIPSOCH_2$	(41), 98	(43), 98
$n\text{-}C_{11}H_{23}$	(41), >95	(41), >95

TABLE 3. ENDOCYCLIC IMINES (*Continued*)

Imine	Conditions	Product(s) and Yield(s) (%), % ee	Refs.
C_{11} 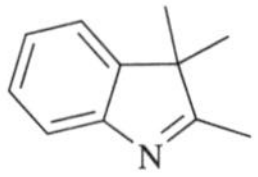	Bicp, $[Ir(cod)Cl]_2$, s/c 100, DCM, phthalimide, 70 bar, 0°, 100 h	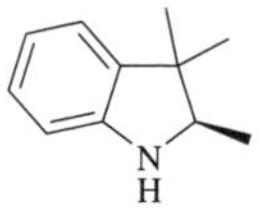 **I** (100)[a], 95	84
	(R,S_{Fc})-Josiphos (Xyl/Xyl), $[Ir(cod)Cl]_2$, s/c 250, toluene, TFA/TBAI, 40 bar, 15°, 47 h	**I** (100)[a], 95	37
	Bcpm, $[Ir(cod)Cl]_2$, s/c 100, C_6H_6/MeOH, BiI_3, 100 bar, –30°, 90 h	**I** (92)[a], 91	87
	Ru((*S*)-MeO-biphep)((*S,S*)-anden)Cl_2, s/c 100, *i*-PrOH, *t*-BuOK, 15 bar, 50–65°, 18 h	**I** (—), 88	47
	(R,S_{Fc})-Josiphos (Ph/Xyl), $[Ir(cod)Cl]_2$, s/c 250, $(C_{10}mim)BF_4$, TFA/TBAI, 40 bar, 50°, 15 h	**I** (100)[a], 86	85
	15, $[Ir(cod)Cl]_2$, s/c 100, DCM, I_2, 75 bar, 0°, 24 h	**I** (97)[a], 85	88
	$[Ir(bdpp)HI_2]_2$, s/c 500, THF/DCM, 40 bar, 30°, 43 h	**I** (100)[a], 80	83
C_{12} 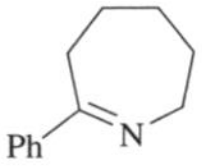	(Ebthi)Ti(binol), s/c 20, THF, 35 bar, 45°, 8–48 h	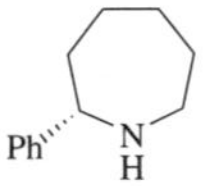 (*S*)-**I** (71), 98	34
	Ir((*S*)-binap)HBr(OBz), s/c 100, toluene, 60 bar, 20°, 18 h	(*R*)-**I** (99),69	81

C_{12-30} 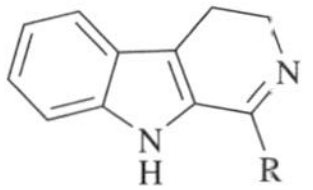	HCO_2Na, (*R,R*)-dpenTs$_{surf}$, $[(cymene)RuCl_2]_2$, H_2O, CTAB, 28°	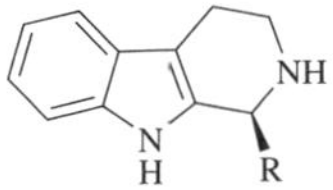 (*S*)-**I**	
		R / s/c / Time: Me, 500, 38 h, (99), 99; Et, 100, 20 h, (94), 99; *i*-Pr, 100, 30 h, (92), 99; *n*-C_6H_{13}, 100, 25 h, (96), 98; Ph, 100, 4 h, (83), 99	99
	HCO_2H/NEt_3, (cymene)Ru((*S,S*)-dpenTs)Cl, s/c 230, MeCN, rt, 12 h	(*R*)-**I** R: Me (84), >98; *n*-Pr (79), >98; *n*-C_8H_{17} (85), >98; $Me(CH_2)_{16}$ (79), >98; (*Z*)-$Me(CH_2)_7CH{=}CHC_8H_{16}$ (84), >98; (*Z*)-$Me(CH_2)_3(CH_2CH{=}CH)_4(CH_2)_3$ (70), >98	92
	HCO_2H/NEt_3, (cymene)Ru((*S,S*)-dpenTs)Cl, s/c 200, DMF, 28°, 5 h	(*R*)-**I** R: Me (86), 97; Ph (83), 96	91
	HCO_2Na, (*S,S*)-dpenTs$_{amin}$, $[cp^*RhCl_2]_2$, s/c 100, H_2O, 28°, 10 h	(*R*)-**I** R = Me (94), 93	100
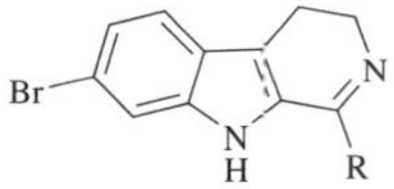	HCO_2H/NEt_3, (cymene)Ru((*S,S*)-dpenTs)Cl, s/c 25, DMF, rt, 12 h	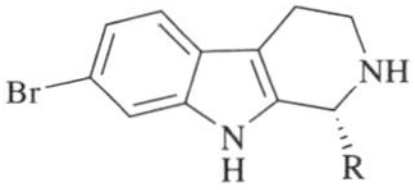 R: $(CH_2)_3CO_2H$ (89), 96; $(CH_2)_3CH{=}CH_2$ (96), 93	97

TABLE 3. ENDOCYCLIC IMINES (*Continued*)

Imine	Conditions	Product(s) and Yield(s) (%), % ee	Refs.
C_{12} 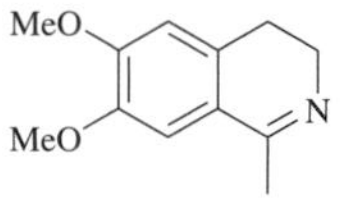	HCO_2H/NEt_3, (cymene)Ru((*S,S*)-dpenTs)Cl, s/c 1000, MeCN, 28°, 12 h	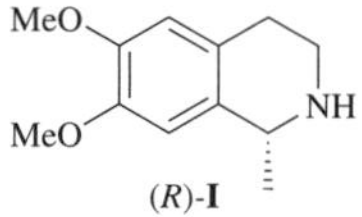 (*R*)-**I** (97), 94	91
	(Ebthi)Ti(binol), s/c 20, THF, 135 bar, 65°, 8–48 h	(*S*)-**I** (82), 98	34
	HCO_2Na, (*S,S*)-dpenTs_{amin}, $[Cp^*RhCl_2]_2$, s/c 100, H_2O, 28°, 8 h	**I** (95), 93[b]	100
	HCO_2H/NEt_3, (cymene)Ru((*S,S*)-dpenTs_{SMF})Cl, s/c 100, DCM, rt, 12 h	(*R*)-**I** (95–100), 90–91	199
	HCO_2Na, $[(cymene)Ru((R,R)dachTs)H_2O]BF_4$, s/c 100, H_2O, pH 9, 60°, 2–5 h	(*R*)-**I** (100)[a], 88	64
C_{12-14} 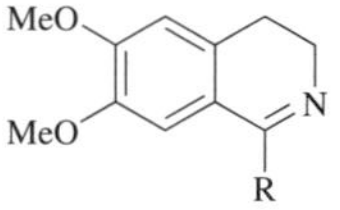	HCO_2Na, (*R,R*)-dpenTs_{sulf}, $[(cymene)RuCl_2]_2$, s/c 100, H_2O, CTAB, 28°	(*S*)-**I** R / Time Me / 10 h / (97), 95 Et / 25 h / (68), 92 *i*-Pr / 15 h / (90), 95	99
C_{12-16} R = Me, Et, *i*-Pr, *n*-C_5H_{11}	HCO_2H/NEt_3, cp*Rh((*S,S*)-dpenTs)Cl, s/c 200, DCM, 20°, 0.15 h	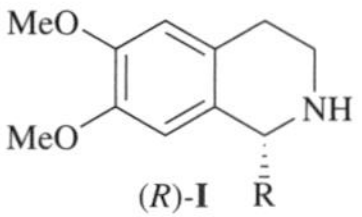 (*R*)-**I** R Me (95), 99 Et (93), 83 *i*-Pr (96), 99 *n*-C_5H_{11} (94), 97	101

C_{14}

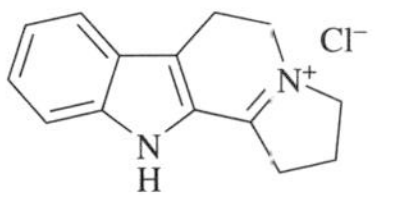

HCO_2H/NEt_3, $(C_6H_6)Ru((S,S)$-dpenTs)Cl, s/c 300, MeCN, 0°, 10 h

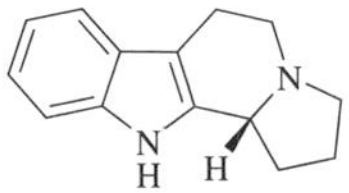

(81), 79

107

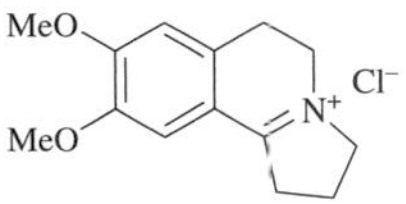

HCO_2H/NEt_3, $(C_6H_6)Ru((S,S)$-dpenTs)Cl, s/c 300, MeCN, 0°, 10 h

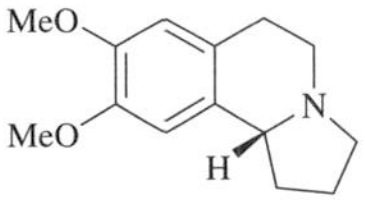

(96), 92

106

C_{14-16}

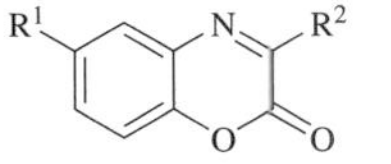

Hantzsch ester, (*R*)-**7d**, s/c 100, $CHCl_3$, rt

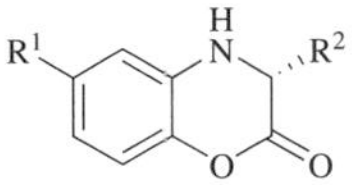

R^1	R^2	
H	Ph	(85), 98
4-Cl	Ph	(55), 96
H	4-BrC_6H_4	(92), >99
H	4-$MeOC_6H_4$	(91), >99
H	3,4-$Me_2C_6H_3$	(90), >99
H	2-thienyl	(81), 90

105

C_{14-20}

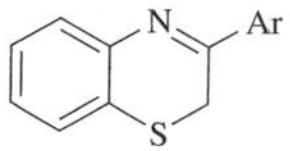

Hantzsch ester, (*R*)-**7d**, s/c 100, $CHCl_3$, rt

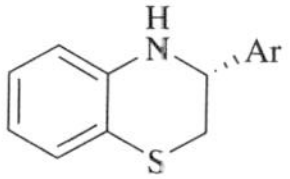

Ar	
3-BrC_6H_4	(51), 94
4-FC_6H_4	(70), >99
4-BrC_6H_4	(87), >99
4-Tol	(50), 96
4-PhC_6H_4	(78), 94
2-Np	(54), 93

105

TABLE 3. ENDOCYCLIC IMINES (*Continued*)

Imine	Conditions	Product(s) and Yield(s) (%), % ee	Refs.
C_{14-20}			
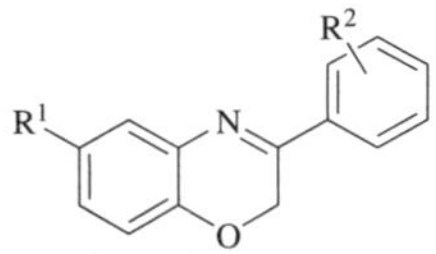	Hantzsch ester, (*R*)-**7d**, $CHCl_3$	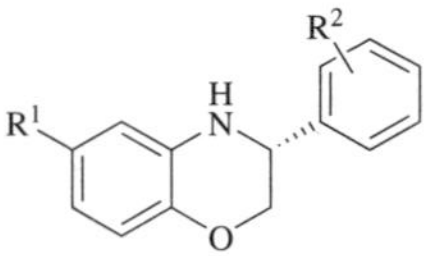	105
		R^1 / R^2 / s/c / Temp:	
		H / H / 10,000 / 60° / (90), 93	
		H / H / 1000 / rt / (95), 98	
		H / 3-Br / 1000 / rt / (93), 98	
		Cl / 4-Br / 1000 / rt / (93), >99	
		H / 4-Me / 1000 / rt / (95), >99	
		H / 4-OMe / 1000 / rt / (92), 98	
		H / 4-Ph / 1000 / rt / (94), 98	
C_{15}			
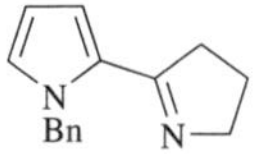	(Ebthi)Ti(binol), s/c 20, THF, 35 bar, 65°, 8–48 h	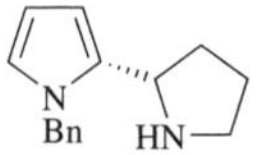 (83), 99	34
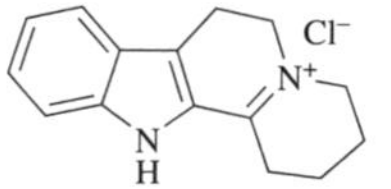	HCO_2H/NEt_3, $(C_6H_6)Ru((S,S)\text{-dpenTs})Cl$, s/c 300, MeCN, 0°, 10 h	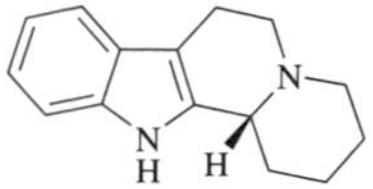 (84), 90	107
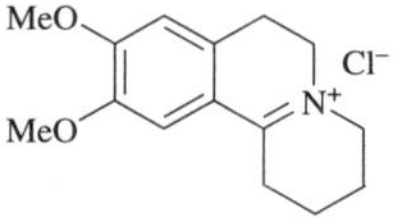	HCO_2H/NEt_3, $(C_6H_6)Ru((S,S)\text{-dpenTs})Cl$, s/c 300, MeCN, 0°, 10 h	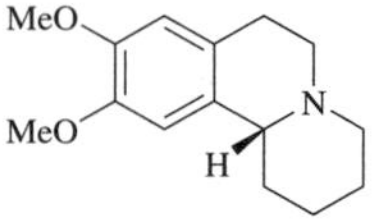 (97), 87	107
C_{17-30}			
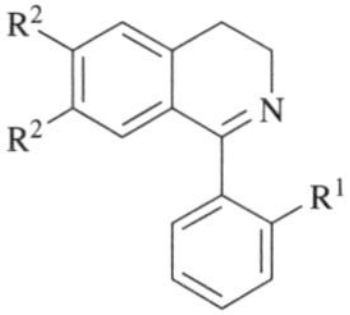	HCO_2H/NEt_3, $(C_6H_6)Ru((S,S)\text{-dpenTs})Cl$, DCM, rt	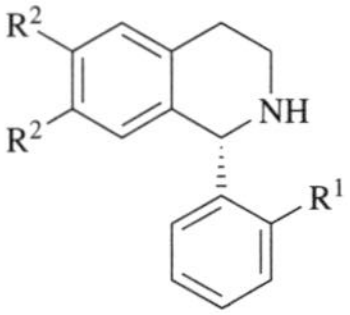	93

R^1	R^2	s/c	Time	% Conv., % ee
H	MeO	200	8 h	(99), 84
Br	H	100	13 h	(41), 94
Br	MeO	150	13 h	(67), 99
NH_2	H	50	16 h	(66), 85
NO_2	MeO	100	13 h	(20), 97
$N(CH_2OMe)Ms$	MeO	14	84 h	(53), 93
NHTs	MeO	14	72 h	(11), 96
$N(CH_2OMe)Ts$	MeO	14	84 h	(58), >99
$N(CH_2OMe)$(1-Nps)	MeO	14	84 h	(53), 97
N(Bn)Ts	MeO	14	72 h	(76), >98

C_{16}

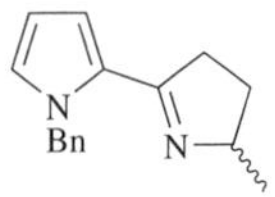

(Ebthi)Ti(thiobinol), s/c 20, THF, 5 bar, 65°, 8–48 h

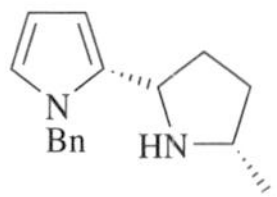

(44), 98 +

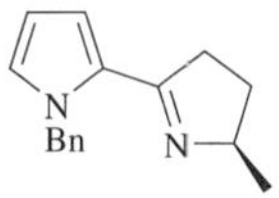

(42), 96

79

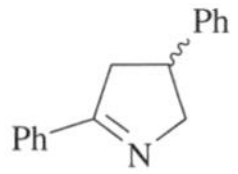

(Ebthi)Ti(thiobinol), s/c 20, THF, 5 bar, 65°, 8–48 h

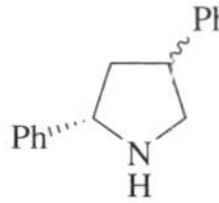

(44), 99 cis/trans 3:1 +

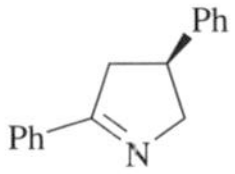

(33), 49

79

C_{17}

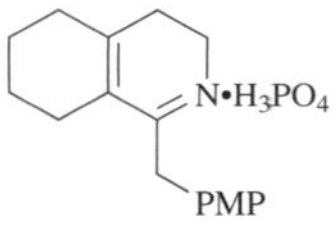

[Ir(cod)((*R*,S_{Fc})-josiphos (4-MeOXyl/*t*-Bu))]BF_4, s/c 1100, toluene/H_2O, NaOH, TBABr, 70 bar, rt, 6 h

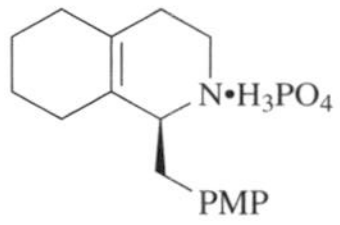

(92), 81

171

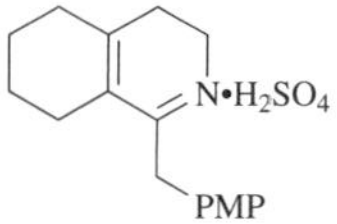

24, $[Ir(cod)Cl]_2$, s/c 100, THF/H_2O, 100 bar, rt, 44 h

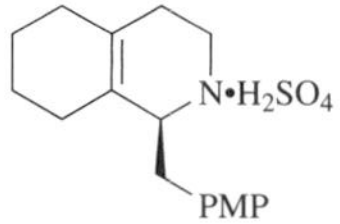

(46), 86

198

TABLE 3. ENDOCYCLIC IMINES (*Continued*)

Imine	Conditions	Product(s) and Yield(s) (%), % ee	Refs.
C_{17-22} 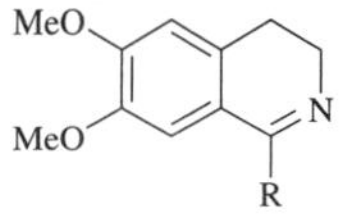	HCO_2H/NEt_3, (arene)Ru((*,*)-dpenSO_2Ar)Cl, s/c 200, DCM or DMF, 28°, 8 h	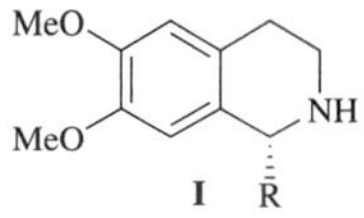 **I**	91

R	arene	*,*	Ar	Time	
Ph	C_6H_6	*S,S*	1-Np	8 h	(99), 84 (*R*)
3,4-$(MeO)_2C_6H_3$	C_6H_6	*R,R*	1-Np	12 h	(>99), 84 (*S*)
(3,4-$(MeO)_2C_6H_3)CH_2$	cymene	*R,R*	TMP	7 h	(90), 95 (*S*)
(3,4-$(MeO)_2C_6H_3)(CH_2)_2$	cymene	*R,R*	TMP	12 h	(99), 92 (*S*)

Imine	Conditions	Product(s) and Yield(s) (%), % ee	Refs.
	HCO_2H/NEt_3, (C_6H_6)Ru((*S,S*)-dpen(1-Nps))Cl, s/c 150, MeCN, rt, 16 h	(*R*)-**I** R = 3,4,5-$(MeO)_3C_6H_2$ (—), 83	95
	HCO_2H/NEt_3, (C_6H_6)Ru((*S,S*)-dpenTs)Cl, s/c 100, MeCN, rt, 12 h	(*R*)-**I** R = (3,4,5-$(MeO)_3C_6H_2)(CH_2)_2$ (92), 99	96
	(*R*)-Binap, $[Ir(cod)Cl]_2$, s/c 200, toluene/MeOH, 100 bar, 2–5°, 72 h	(*S*)-**I** (see below)	102

R	Additive	Time	
$BnOCH_2$	F_4-phthalimide	20 h	(85), 86
$BnO(CH_2)_3$	parabanic acid	72 h	(99), 89

Imine	Conditions	Product(s) and Yield(s) (%), % ee	Refs.
	(*S,S*)-Bcpm, $[Ir(cod)Cl]_2$, s/c 100, toluene/MeOH, 100 bar, 2–5°, 20–40 h	(*S*)-**I** (see below)	103

R	Additive	% Conv., % ee
3,4-$(MeO)_2C_6H_3CH_2$	F_4-phthalimide	(84), 88
3,4-$(MeO)_2C_6H_3(CH_2)_2$	F_4-phthalimide	(89), 86
(*E*)-3,4-$(MeO)_2C_6H_3CH{=}CH$	phthalimide	(79), 86

Imine	Conditions	Product(s) and Yield(s) (%), % ee	Refs.
C_{19} 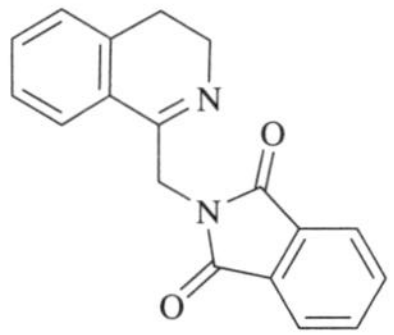	HCO_2H/NEt_3, (C_6H_6)Ru((*R,R*)-dpenTs)Cl, s/c 150, MeCN, rt, 5 h	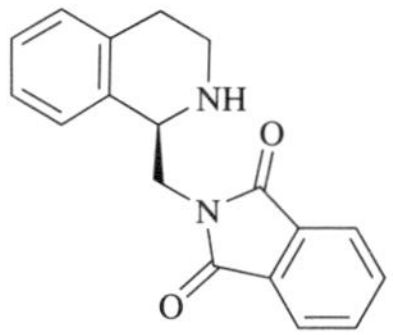 (52), 62	180

C_{18-32}

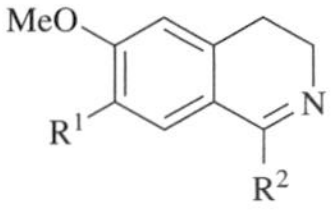

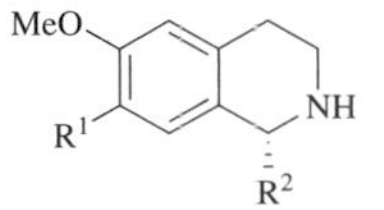

(100)[a]

(*R*)-**I**

	HCO_2H/NEt_3, (cymene)Ru((*S*,*S*)-dpenTs)Cl, s/c 20–40, DMF, 20–30°, 1.5–2 h	(*R*)-**I** $R^1 = H$, $R^2 = 3,5\text{-}(BnO)_2\text{-}4\text{-}MeOC_6H_3$; 86% ee	94
	HCO_2Na, $[(cymene)Ru(dachTs)H_2O]BF_4$, s/c 100, H_2O, pH 9, 60°, 2–5 h	(*R*)-**I** $R^1 = MeO$, $R^2 = 2\text{-}NO_2\text{-}5\text{-}ClC_6H_3$; 68% ee	64

C_{19-24}

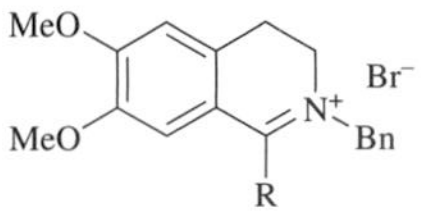

HCO_2Na, (*R*,*R*)-**L**, $[RuCl_2(cymene)]_2$, s/c 100, H_2O, CTAB, 28°

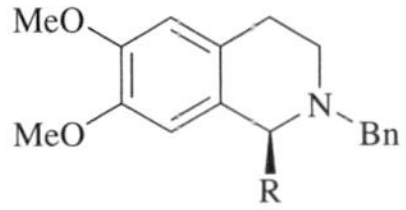

R	**L**	Time	
Me	$dpenTs_{sulf}$	18 h	(86), 90
Me	dpenTs	18 h	(85), 90
Ph	dpenTs	18 h	(98), 98
Ph	$dpenTs_{sulf}$	12 h	(94), 95

99

[a] This value is the percent conversion.

[b] The stereochemistry of the product was not reported in the original reference.

TABLE 4. HETEROAROMATIC SUBSTRATES

Heteroamine	Conditions	Product(s) and Yield(s) (%), % ee	Refs.
C_9			
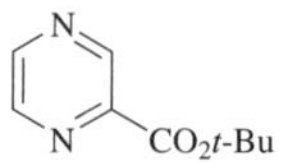	(R,S_{Fc})-Josiphos (Ph/c-C_6H_{11}), $[Rh(nbd)Cl]_2$, s/c 50, MeOH, 50 bar, 70°, 20 h	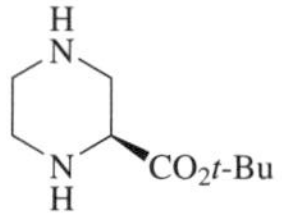 (80)[a], 78	110
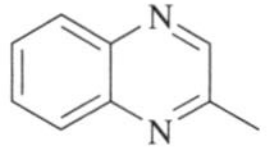	**20**, s/c 100, MeOH, 5 bar, 100°, 24 h	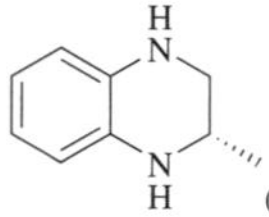 (*R*)-**I** (54)[a], 90	123
	Ru((*S*)-hexaphemp)((*R,R*)-dach)Cl_2, s/c 1000, *t*-BuOH, *t*-BuOK, 30 bar, 50°, 20 h	(*R*)-**I** (100)[a], 69	47
C_{10-15}			
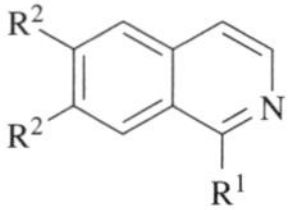	$ClCO_2R^3$, (*S*)-segphos, $[Ir(cod)Cl]_2$, s/c 100, THF, Li_2CO_3, $LiBF_4$, 42 bar, rt, 12–15 h	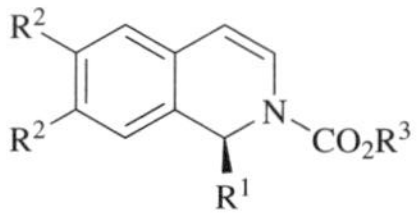	122
C_{10-16}			
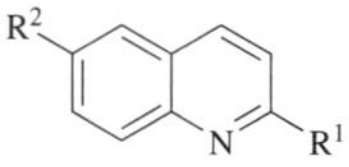	**18**, $[Ir(cod)Cl]_2$, s/c 200, toluene, I_2, 60 bar, rt, 20 h	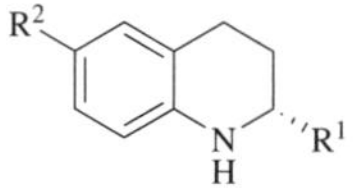	120

Products for ref. 122:

R^1	R^2	R^3	
Me	H	Bn	(87), 83
Et	H	Me	(85), 62
n-Bu	H	Me	(87), 60
Ph	H	Bn	(49), 83
Me	MeO	Bn	(46), 65

Products for ref. 120:

R^1	R^2	% Conv., % ee
Me	H	(>96), 96
Me	F	(>96), 90
Et	H	(>96), 91
Me	Me	(>96), 80
n-Bu	H	(>96), 91
$HOMe_2CCH_2$	H	(>96), 92

C_{10-19}

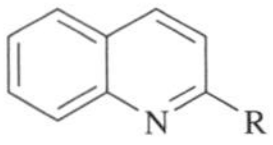

(*S*)-H8-Binapo, [Ir(cod)Cl]$_2$, s/c 100, solvent, I_2, 50 bar, rt, 20 h

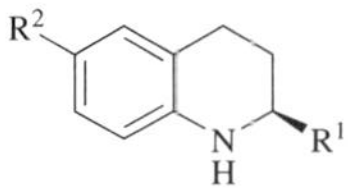

R^1	R^2	Solvent	
Me	H	DMPEG$_{500}$/hexane	(98), 97
Me	F	DMPEG$_{500}$/hexane	(96), 94
Et	H	DMPEG$_{500}$/hexane	(97), 94
Me	Me	DMPEG$_{500}$/hexane	(98), 95
Me	MeO	THF	(90), 94
n-Pr	H	DMPEG$_{500}$/hexane	(99), 95
n-Bu	H	DMPEG$_{500}$/hexane	(99), 94
$HOMe_2CCH_2$	H	DMPEG$_{500}$/hexane	(98), 97
Ph	H	DMPEG$_{500}$/hexane	(98), 87
1-HO(C_6H_{10})CH_2	H	THF	(98), 96

118

Hantzsch ester, (*R*)-**7d**, s/c 50, C_6H_6, 60°

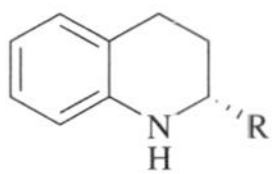

R	Time	
$ClCH_2$	12 h	(91), 88
n-Bu	12 h	(91), 87
2-furyl	12 h	(93), 91
n-C_5H_{11}	12 h	(88), 90
Ph	12 h	(92), 97
2-FC_6H_4	30 h	(93), 98
3-BrC_6H_4	18 h	(92), 98
2-Tol	48 h	(54), 91
4-$MeOC_6H_4$	12 h	(90), 98
4-$CF_3C_6H_4$	30 h	(91), >99
$PhCH_2CH_2$	12 h	(90), 90
2,4-$Me_2C_6H_3$	60 h	(65), 97
2-Np	12 h	(93), >99
3,4-$(MeO)_2C_6H_3(CH_2)_2$	12 h	(95), 90

113

TABLE 4. HETEROAROMATIC SUBSTRATES (*Continued*)

Heteroamine	Conditions	Product(s) and Yield(s) (%), % ee	Refs.
C_{10-23}			
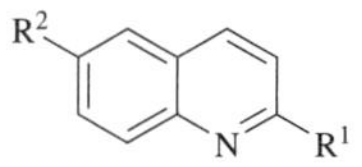	Hantzsch ester, (*S*)-segphos, $[Ir(cod)Cl]_2$, s/c 100, toluene/dioxane, I_2, 40 bar, rt, 42–79 h	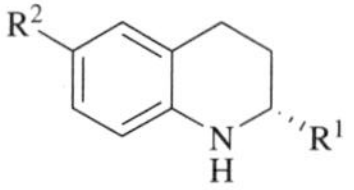 (*S*)-**I**	117
	19, $[Ir(cod)Cl]_2$, toluene, I_2, 40 bar, rt, 12–16 h	(*R*)-**I**	121

(*S*)-**I**:

R^1	R^2	
Me	H	(86), 87
Me	F	(90), 86
Et	H	(92), 87
Me	Me	(82), 86
Me	MeO	(43), 81
n-Bu	H	(98), 81
n-C_5H_{11}	H	(94), 68
$Ph(CH_2)_2$	H	(88), 87
3,4-$(OCH_2O)_2C_6H_3(CH_2)_2$	H	(87), 87
3,4-$(MeO)_2C_6H_3(CH_2)_2$	H	(92), 88
$Ph_2C(OH)CH_2$	H	(76), 78

(*R*)-**I**:

R^1	R^2	s/c	
Me	H	1000	(>95), 86
Me	H	100	(95), 90
Me	F	100	(86), 89
Et	H	100	(95), 91
Me	Me	100	(93), 92
n-C_5H_{11}	H	100	(94), 92
1-$HOC_6H_{10}CH_2$	H	100	(82), 79
$Ph(CH_2)_2$	H	100	(92), 72
3,4-$(MeO)_2C_6H_3(CH_2)_2$	H	100	(82), 87
$HOPh_2CCH_2$	H	100	(89), 80

16, $[Ir(cod)Cl]_2$, s/c 400, THF, I_2, 45 bar, rt, 1.5 h — 116

R^1	R^2	% Conv., % ee
Me	H	(>95), 90
Me	F	(>95), 87
Et	H	(>95), 89
Me	MeO	(87), 89
Me	Me	(77), 87
n-Pr	H	(>95), 89
$HOMe_2CCH_2$	H	(>95), 92
1-$HOC_6H_{10}CH_2$	H	(>95), 93
$Ph(CH_2)_2$	H	(>95), 84
3,4-$(MeO)_2C_6H_3(CH_2)_2$	H	(83), 82
$HOPh_2CCH_2$	H	(77), 76

17, $[Ir(cod)Cl]_2$, s/c 100, THF or $DMPEG_{500}$/hexane, I_2, 50 bar, rt, 18 h — 119

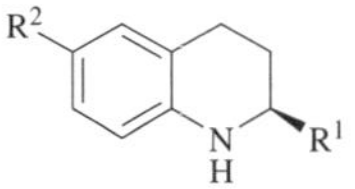

(*R*)-**I**

R^1	R^2	% Conv., % ee
Me	H	(100), 92
Et	H	(100), 87
n-Pr	H	(100), 91
n-Bu	H	(100), 87
n-C_5H_{11}	H	(100), 90
Me	Me	(100), 92
Me	MeO	(66), 92
Me	F	(100), 89
Ph	H	(100), 65
$Ph(CH_2)_2$	H	(100), 83
$HOMe_2CCH_2$	H	(100), 91
1-HO-(*c*-C_6H_{10})CH_2	H	(100), 93

TABLE 4. HETEROAROMATIC SUBSTRATES (*Continued*)

Heteroamine	Conditions	Product(s) and Yield(s) (%), % ee	Refs.
C_{10-23} 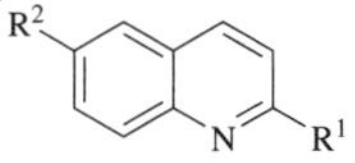	(*R*)-MeO-biphep, $[Ir(cod)Cl]_2$, s/c 100, toluene, I_2, 50 bar, rt, 18 h	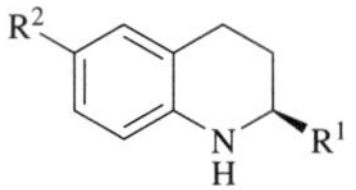 **I**	114
	(*R*)-P-phos, $[Ir(cod)Cl]_2$, s/c 100, THF, I_2, 50 bar, rt, 20 h	**I**	115

Ref. 114:

R^1	R^2	
Me	H	(94), 94
$HOCH_2$	H	(83), 75
Me	F	(88), 96
Et	H	(88), 96
Me	Me	(91), 91
Me	MeO	(89), 84
n-Pr	H	(92), 93
i-Pr	H	(92), 94
$AcOCH_2$	H	(90), 87
n-Bu	H	(86), 92
n-C_5H_{11}	H	(92), 94
$HOMe_2CCH_2$	H	(87), 94
1-HO-(*c*-C_6H_{10})CH_2	H	(89), 92
$Ph(CH_2)_2$	H	(94), 93
3,4-$(MeO)_2C_6H_3(CH_2)_2$	H	(86), 96
$HOPh_2CCH_2$	H	(94), 91

Ref. 115:

R^1	R^2	
Me	H	(97), 91
Me	F	(90), 90
Et	H	(99), 92
n-C_5H_{11}	H	(97), 91
$HOMe_2CCH_2$	H	(99), 91
$Ph(CH_2)_2$	H	(99), 90
$HOPh_2CCH_2$	H	(98), 90

C_{10-25}

$ClCO_2Bn$, (*S*)-segphos, $[Ir(cod)Cl]_2$, s/c 100, THF, Li_2CO_3, 42 bar, rt, 12–15 h

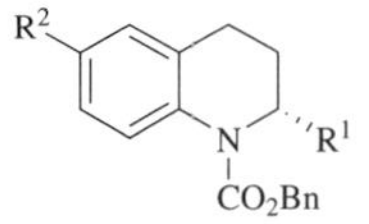

R^1	R^2	
Me	H	(90), 90
Et	H	(85), 90
n-Pr	H	(80), 90
n-Bu	H	(88), 89
n-C_5H_{11}	H	(91), 89
Me	Me	(90), 89
Me	F	(83), 89
Me	MeO	(92), 90
Ph	H	(41), 80
$Ph(CH_2)_2$	H	(86), 90
3,4-$(MeO)_2C_6H_3(CH_2)_2$	H	(80), 90
3-Bn-4-$MeOC_6H_3(CH_2)_2$	H	(88), 88

121

C_{11-17}

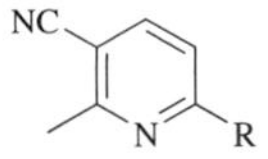

Hantzsch ester, (*R*)-**7e**, s/c 20, C_6F_6, 50°

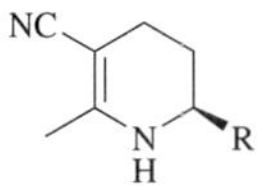

R	
n-Bu	(55), 84
n-C_5H_{11}	(73), 90
$Ph(CH_2)_2$	(47), 86
n-$C_{10}H_{21}$	(68), 89

111

TABLE 4. HETEROAROMATIC SUBSTRATES (*Continued*)

Heteroamine	Conditions	Product(s) and Yield(s) (%), % ee					Refs.
C_{12-18}							
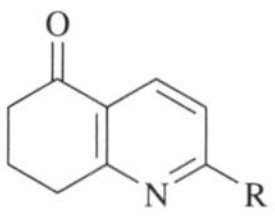	Hantzsch ester, (*R*)-**7e**, s/c 20, C_6H_6, 50°	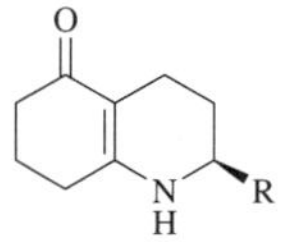	R				111
			n-Pr		(69), 89		
			n-Bu		(72), 91		
			n-C_5H_{11}		(84), 91		
			$Ph(CH_2)_2$		(66), 92		
			n-$C_{10}H_{21}$		(73), 92		
			(*E*,*Z*)-$CH_3(CH_2)_4CH{=}CH(CH_2)_2$		(83), 87		
C_{13-22}							
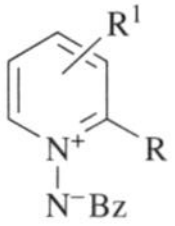	[Ir((*S*)-phox1)(cod)]BARF, s/c 50, toluene, I_2, 27 bar, rt, 6 h	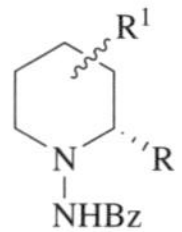	R	R^1			112
			Me	H	(98)[a], 90		
			Et	H	(96)[a], 83		
			n-Pr	H	(98)[a], 84		
			Me	3-Me	(91)[a], 54	>95% cis	
			Me	5-Me	(92)[a], 84–86	57% cis	
			Bn	H	(97)[a], 58		
			$BnOCH_2$	H	(85)[a], 76		
			$BnO(CH_2)_2$	H	(88)[a], 88		
C_{15}							
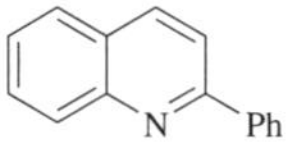	Ir((*S*)-synphos)HI(OAc), s/c 200, THF, 50 bar, 30°, 45 h	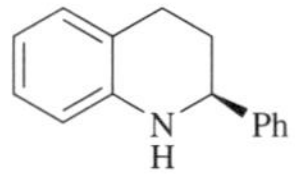	(42), 64				81

C_{18-25}

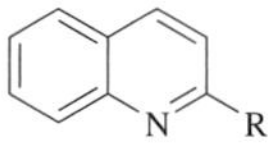

(*)-MeO-biphep, $[Ir(cod)Cl]_2$, s/c 100, toluene, I_2, rt

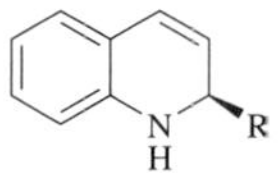

(*R*)-**I**

R	*	Pressure	Time	Enant.		
$(CH_2)_2$- 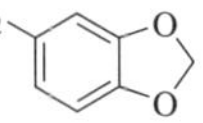	*R*	50 bar	18 h	(*R*)-**I**	(88), 93	114
$(CH_2)_2$- 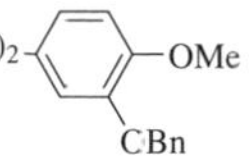	*S*	35 bar	12–15 h	(*S*)-**I**	(94), 96	200

[a] This value is the percent conversion.

TABLE 5. C=N–Y FUNCTIONS

Substrate	Conditions	Product(s) and Yield(s) (%), % ee	Refs.
C_{4-14} 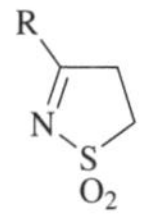	Pd((*S*)-segphos)$(CF_3CO_2)_2$, s/c 50, CF_3CH_2OH, 4 Å MS, 40 bar, rt, 12 h	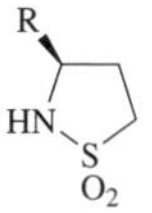 (*S*)-**I** R — %Conv., % ee Me — (91), 88 Ph — (93), 79 n-C_6H_{13} — (99), 90 $PhOCH_2$ — (99), 92 $BnOCH_2$ — (93), 86 4-$CF_3C_6H_4OCH_2$ — (99), 93 2-$TolOCH_2$ — (95), 92 4-$TolOCH_2$ — (93), 91 2-$NpOCH_2$ — (97), 90	129
C_7 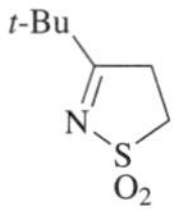	HCO_2H/NEt_3, (*R*,*R*)-dpenTs_{dend}, [(cymene)$RuCl_2]_2$, s/c 100, DCM, 28°, 10 h	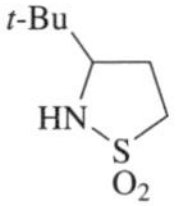 (90), 96[b]	133
C_{8-14} 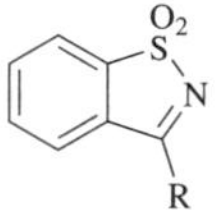	(*R*)-Binap, Ru(cod)Cl_2, s/c 100, toluene, NEt_3, 4 bar, 22°, 12 h	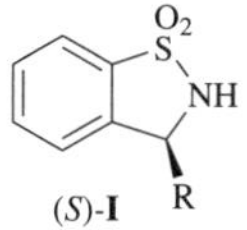 (*S*)-**I** R = Me (84), 99	126
	Pd(tangphos)$(CF_3CO_2)_2$, s/c 100, DCM, 75 bar, 40°, 24 h	(*R*)-**I** R = Me (>99)[a], 94	128
	Pd((*S*)-segphos)$(CF_3CO_2)_2$, s/c 50, CF_3CH_2OH, 4 Å MS, 40 bar, rt, 12 h	(*R*)-**I** R Me (98), 92 n-Bu (98), 90 Bn (93), 88	129
	HCO_2H/NEt_3, dpenTs_{immob}, [(cymene)$RuCl_2]_2$, s/c 100, neat, 40°, 1.5 h	**I** R = n-Bu (>99), 93[b]	131

HCO2H/NEt3 conditions:

HCO_2H/NEt_3, cp*Rh((*,*)-dpenTs)Cl, s/c 200, DCM, 20°, 0.5 h — **I**

R	(*,*)		
Me	*R,R*	(*R*)-**I**	(98), 68
n-Bu	*R,R*	(*R*)-**I**	(98), 67
4-ClC_6H_4	*R,R*	(*S*)-**I**	(96), 81
Bn	*S,S*	(*S*)-**I**	(93), 68

101

HCO_2Na, (*R,R*)-dpenTs$_{sulf}$, [(cymene)$RuCl_2$]$_2$, s/c 100, H_2O, CTAB, 28° — (*R*)-**I**

R	Time	
Me	6 h	(97), 65
n-Bu	10 h	(95), 94

99

HCO_2H/NEt_3, (C_6H_6)Ru((*S,S*)-dpenTs)Cl, s/c 200, DCM, rt, 17 h — (*S*)-**I**

R	
t-Bu	(—), 91
Bn	(—), 93

132

C_{9-13}

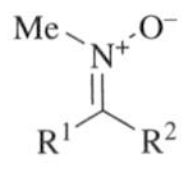

(*S*)-Binap, [Ir(cod)Cl]$_2$, s/c 100, THF, NBu_4BH_4, 80 bar, 0°, 18 h

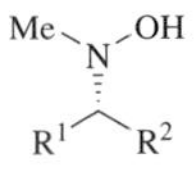

R^1	R^2	
Ph	Bn	(78), 75
Ph	Me	(45), 69
2-ClC_6H_4	Me	(17), 78
3-ClC_6H_4	Me	(68), 81
4-ClC_6H_4	Me	(82), 83
4-BrC_6H_4	Me	(76), 86
2-Np	Me	(64), 80

125

C_{10}

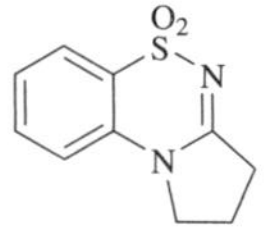

Ru((*R*)-binap)((*R,R*)-dpen)Cl_2, s/c 2500, toluene/*i*-PrOH, *i*-PrOK, 4 bar, 60°, 6 h

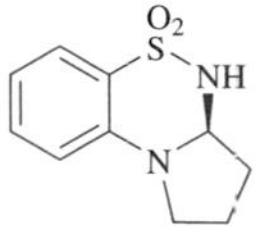

(97), 87

172

C_{11-19}

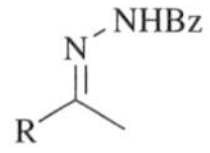

[Rh((*R*)-Et-duphos)(cod)]OTf, s/c 500, *i*-PrOH, 4 bar

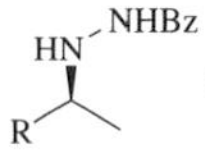

(70–90)

R	Temp	Time	% ee
Et	−10°	36 h	43
i-Pr	−10°	36 h	73
t-Bu	20°	48 h	45
c-C_6H_{11}	−15°	36 h	72
2-Np	0°	12 h	95

134, 135

TABLE 5. C=N–Y FUNCTIONS (*Continued*)

Substrate	Conditions	Product(s) and Yield(s) (%), % ee	Refs.
C_{12}			
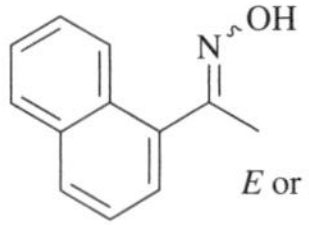 *E* or *Z*	[Rh((*S*)-binap)(nbd)]BF_4, s/c 250, C_6H_6/MeOH, 70 bar, 100°, 5 d	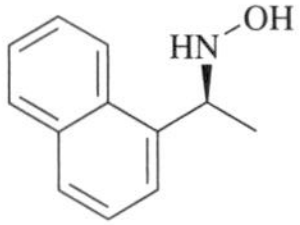 *E* (—), 30; *Z* (—), 66	17
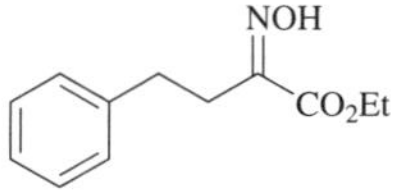	$[Ir(dpampp)Cl]_2$, s/c 100, C_6H_6/MeOH, BI_3 or *n*-Bu_4I, 48 bar, rt, 46 h	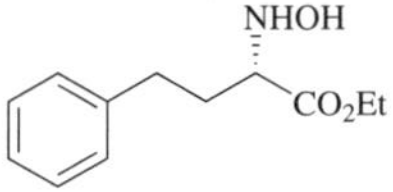 (19–22)[a], 93	124
C_{12-19}			
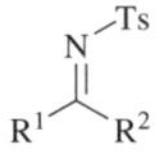	Pd((*S*,*S*)-tangphos)$(CF_3CO_2)_2$, s/c 100, DCM, 75 bar, 40°, 24 h	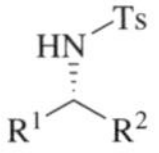 (>99)[a]	128

R^1	R^2	% ee
c-C_3H_5	Me	75
t-Bu	Me	98
Ph	Me	99
4-FC_6H_4	Me	99
3-ClC_6H_4	Me	>99
4-ClC_6H_4	Me	99
4-Tol	Me	96
3-$MeOC_6H_4$	Me	>99
4-$MeOC_6H_4$	Me	99
1-Np	Me	99
2-Np	Me	>99
Ph	Et	93

C_{13-19}

Ru((*R*)-binap)(OAc)$_2$, s/c 20, THF, 75 bar, 40°, 96 h

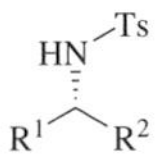

(*R*)-**I**

R^1	R^2	
i-Bu	Me	(48), 48
Ph	Me	(82), 62
Ph	Et	(80), 84
2-Np	Me	(86), 44

127

Pd((*S*)-synphos)(CF_3CO_2)$_2$, s/c 50, CF_3CH_2OH, 4 Å MS, 40 bar, rt, 12 h

(*S*)-**I**

R^1	R^2	
t-Bu	Me	(94), 91
Ph	Me	(84), 96
Ph	Et	(90), 88
4-FC_6H_4	Me	(98), 96
4-$MeOC_6H_4$	Me	(98), 97
3-$MeOC_6H_4$	Me	(86), 93
2-$MeOC_6H_4$	Me	(84), 94
2-Np	Me	(95), 95

129

C_{15-18}

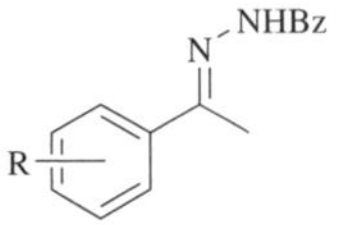

[Rh((*R*)-Et-duphos)(cod)]OTf, s/c 500–1000, *i*-PrOH, 4 bar

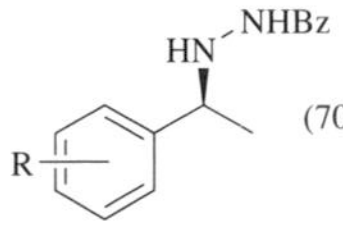

(70–90)

R	Temp	Time	% ee
H	−10°	24 h	95
4-NO_2	0°	12 h	97
4-Br	0°	12 h	96
4-MeO	0°	12 h	88
4-EtO_2C	0°	12 h	96

134, 135

C_{16-18}

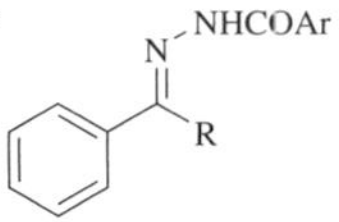

[Rh((*R*)-Et-duphos)(cod)]OTf, s/c 500–1000 *i*-PrOH, 4 bar

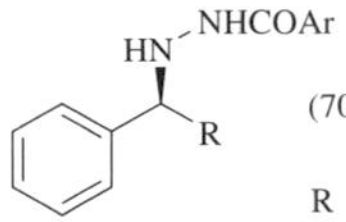

(70–90)

R	Ar	Temp	Time	% ee
Et	Ph	−10°	24 h	85
Me	4-$MeOC_6H_4$	20°	2 h	91
Me	4-$Me_2NC_6H_4$	20°	2 h	92
Me	Ph	0°	12 h	96
Bn	Ph	−10°	24 h	84
CF_3	Ph	20°	2 h	51

134, 135

TABLE 5. C=N–Y FUNCTIONS (*Continued*)

Substrate	Conditions	Product(s) and Yield(s) (%), % ee	Refs.
C_{16-17} 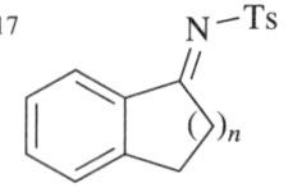	Pd(tangphos)$(CF_3CO_2)_2$, s/c 100, DCM, 75 bar, 40°, 24 h	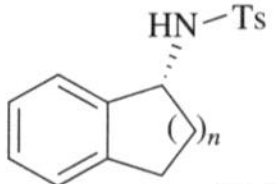 (*R*)-**I** (>99)[a] *n* / % ee 1 / 98 2 / 94	128
	Ru((*R*)-binap)$(OAc)_2$, s/c 20, THF, 75 bar, 40°, 96 h	(*R*)-**I** $n = 2$ (77), 82	127
C_{16-24} 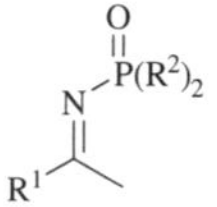	HCO_2H/NEt_3, cp*M(dpenTs)Cl, s/c 50, MeCN, rt, 2–3 h	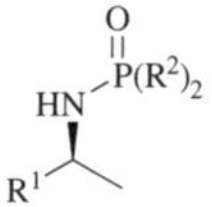 (100)[a] R^1 / R^2 / M / % ee 2-Np / Et / Rh / >90 Ph / Ph / Rh / 86 n-C_6H_{13} / Ph / Ir / 95 2-Np / Ph / Rh / >99	90
C_{18} 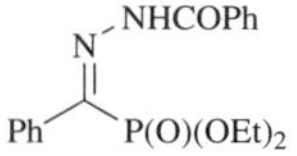	[Rh(Et-duphos)(cod)]OTf, s/c 500, *i*-PrOH, 4 bar, –10°, 48 h	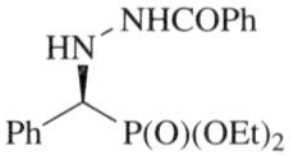 (70–90), 90	132, 135
C_{18-25} 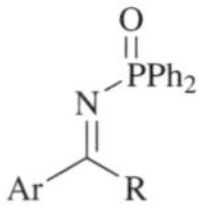	Pd((*S*)-segphos)$(CF_3CO_2)_2$, s/c 50, CF_3CH_2OH, 4 Å MS, 70 bar, rt, 8 h	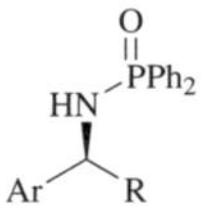 Ar / R 2-furyl / Me / (29), 87 Ph / Et / (93), 87 2-Np / Me / (70), 93	130
C_{20-21} 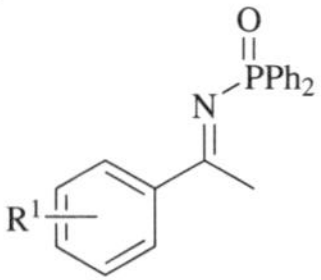	(*R*,S_{Fc})-Josiphos (R^2/R^3), $[Rh(nbd)_2]BF_4$, MeOH, 70 bar, 60°	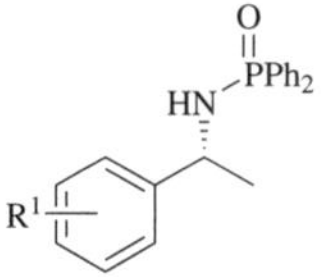 (*R*)-**I**	138

R^1	R^2/R^3	s/c	Time	% Conv., % ee
H	c-C_6H_{11}/c-C_6H_{11}	500	1 h	(100), 99
4-Cl	c-C_6H_{11}/t-Bu	100	18–21 h	(93), 67
4-Me	c-C_6H_{11}/c-C_6H_{11}	100	18–21 h	(100), 97
4-CF_3	c-C_6H_{11}/c-C_6H_{11}	100	18–21 h	(98), 93
4-MeO	c-C_6H_{11}/c-C_6H_{11}	100	18–21 h	(100), 62

Conditions	Product	R	% Conv., % ee	Ref.
Pd((*S*)-segphos)(CF_3CO_2)$_2$, s/c 50, CF_3CH_2OH, 4 Å MS, 70 bar, rt, 8 h	(*R*)-**I**	H	(98), 96	130
		4-F	(87), 94	
		4-Cl	(90), 94	
		4-Me	(93), 97	
		4-MeO	(96), 96	
		3-MeO	(97), 96	
		2-MeO	(80), 99	

[a] This value is the percent conversion.

[b] The stereochemistry of the product was not reported in the original reference.

TABLE 6. α- AND β-CARBOXY IMINES

Imine	Conditions	Product(s) and Yield(s) (%), % ee	Refs.
C_5 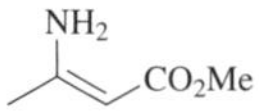	Ru((*S*)-segphos)(AcO)$_2$, s/c 100, CF_3CH_2OH, 30 bar, 80°, 15 h	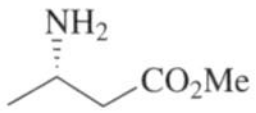 (85), 96	147
$C_{5\text{-}16}$ 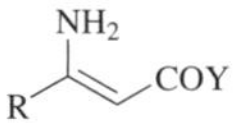	(Boc)$_2$O, (*R*,S_{Fc})-josiphos (Ph/*t*-Bu), [Rh(cod)Cl]$_2$, s/c 30–250, MeOH, 3–6 bar, rt, 18–24 h	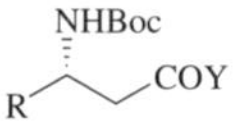 R / Y: Me / OMe (85), 96; *i*-Pr / OMe (75), 95; *t*-Bu / OMe (62), 91; Ph / OMe (57), 97; Bn / OMe (93), 99; Me / NHPh (84), 97; Ph / NHPh (98), 98; Bn / NHPh (99), 97; 2,4,5-$F_3C_6H_2$ / OMe (88), 97	149
$C_{8\text{-}14}$ 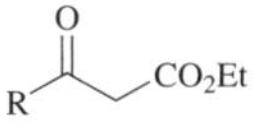 + NH_4OAc	(Cymene)Ru((*R*)-ClMeO-biphep)Cl$_2$, s/c 100, CF_3CH_2OH, 30 bar, 80°, 16 h	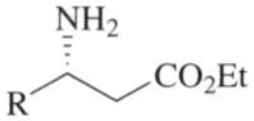 R: Me (80), 96; Ph (88), 98; 3-ClC_6H_4 (81), 98; 4-ClC_6H_4 (79), 99; 4-FC_6H_4 (80), 96; 3-$MeOC_6H_4$ (88), 96; 4-$MeOC_6H_4$ (83), 98	151
$C_{9\text{-}11}$ 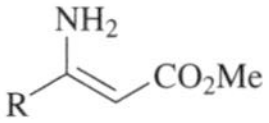	(*R*,S_{Fc})-Josiphos (4-CF_3Ph/*t*-Bu), [Rh(cod)Cl]$_2$, s/c 300, CF_3CH_2OH, 6 bar, 50°	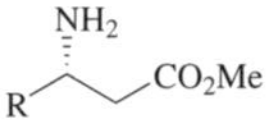 R / Time: 3-Py / 24 h (91), 96; Ph / 6 h (96), 96; 4-FC_6H_4 / 11 h (85), 96; 4-$MeOC_6H_4$ / 11 h (88), 95; Bn / 11 h (94), 93	146

C_{10}

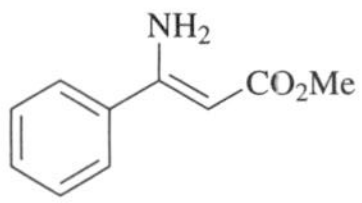

Ru((*S*)-Tol-binap)(AcO)$_2$, s/c 100, CF_3CH_2OH, 30 bar, 50°, 15 h

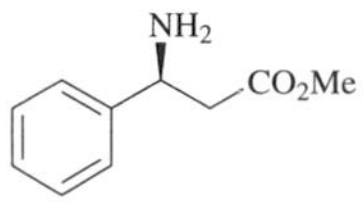

(54), 97

147

C_{10-17}

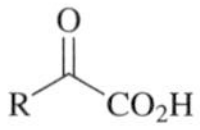

+ $BnNH_2$

[Rh((*R*)-deguphos)(cod)]BF_4, MeOH, 60 bar, rt

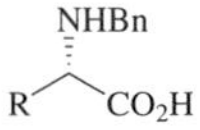

R	s/c	Time	
Me	100	24 h	(43), 78
HO_2CCH_2	100	24 h	(38), 73
$HO_2C(CH_2)_2$	100	24 h	(19), 60
Me_2CHCH_2	200	2 h	(94), 90
Me_3CCH_2	200	24 h	(99), 86
Bn	200	3 h	(99), 98
$Ph(CH_2)_2$	100	24 h	(80), 81

144

C_{11-17}

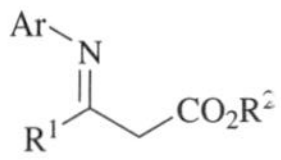

[Rh(tangphos)(nbd)]SbF_5, s/c 100, CF_3CH_2OH, 6 bar

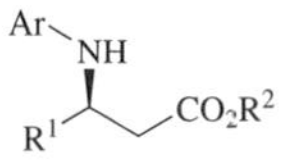

145

R^1	R^2	Ar	Temp	Time	% Conv., % ee
Me	Me	Ph	50°	18 h	(100), 91
Me	Et	Ph	50°	18 h	(100), 95
CF_3	Et	Ph	50°	18 h	(48), 79
Me	Et	4-FC_6H_4	50°	18 h	(100), 96
Me	Et	3-BrC_6H_4	50°	18 h	(83), 96
Et	Et	H	50°	18 h	(100), 95
Me	Et	4-Tol	50°	18 h	(78), 94
Me	Et	3-Tol	50°	18 h	(88), 96
i-C_5H_{11}	Et	Ph	50°	18 h	(100), 90
4-FC_6H_4	Me	Ph	80°	24 h	(100), 95
Ph	Et	Ph	80°	24 h	(100), 92
4-Tol	Me	Ph	80°	24 h	(100), 91
2-$MeOC_5H_4$	Me	Ph	80°	24 h	(100), 90
2-Tol	Me	Ph	80°	24 h	(67), 79

TABLE 6. α- AND β-CARBOXY IMINES (*Continued*)

Imine	Conditions	Product(s) and Yield(s) (%), % ee	Refs.
C_{12} 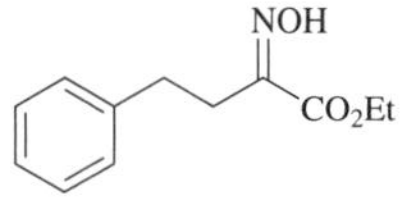	[Ir(dpampp)Cl]$_2$, s/c 100, C_6H_6/MeOH, BI_3 or TBAI, 48 bar, rt, 46 h	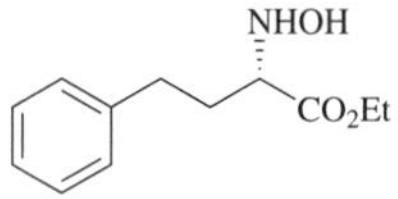 (19–22)[a], 93	124
C_{12-17} 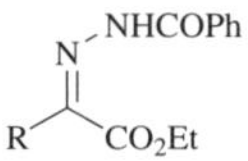	[Rh((*R*)-Et-duphos)(cod)]OTf, s/c 500, *i*-PrOH, 4 bar, 0°, 36 h	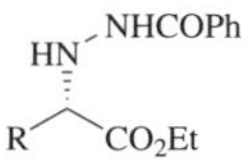 (70–90)	134, 135
C_{12-19} 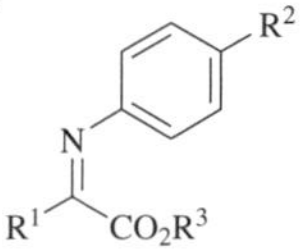	Hantzsch ester, (*S*)-**21**, s/c 20, toluene, 19–22 h	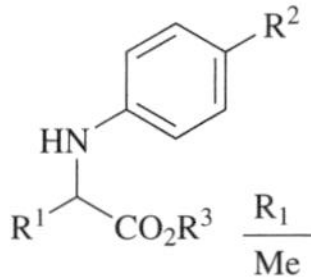	141

R	% ee
Me	89
Et	91
n-Pr	90
n-C_6H_{13}	83
Ph	91

R_1	R^2	R^3	Temp	
Me	OMe	Et	rt	(88), 99 (*S*)
n-C_6H_{13}	OMe	Et	rt	(90), 96[b]
Ph	H	Et	50°	(94), 95[b]
Ph	OMe	Me	50°	(99), 98 (*R*)
Ph	OMe	Et	50°	(93), 96[b]
4-ClC_6H_4	OMe	Et	50°	(95), 98[b]
4-BrC_6H_4	OMe	Et	50°	(93), 98[b]
3,5-$F_2C_6H_3$	OMe	Et	50°	(95), 98[b]
4-$CF_3C_6H_4$	OMe	Et	50°	(98), 96[b]
4-Tol	OMe	Et	50°	(98), 96[b]
4-MeOC_6H_4	OMe	Et	50°	(96), 94[b]
Ph$(CH_2)_2$	OMe	Et	rt	(85), 98[b]

$C_{12\text{-}23}$

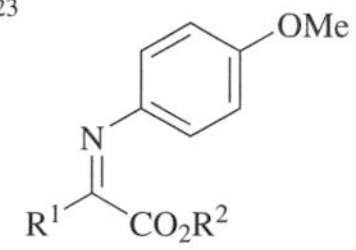

(*R*)-Binap, $Pd(CF_3CO_2)_2$, s/c 25, CF_3CH_2OH, 100 bar, rt, 24 h

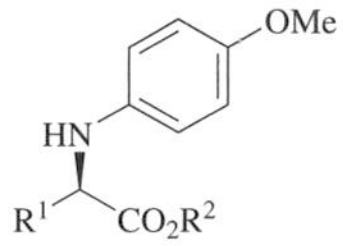

R^1	R^2	
CF_3	Et	(>99), 88
CF_3	*t*-Bu	(92), 85
$CClF_2$	*t*-Bu	(69), 81
CF_3	Bn	(95), 84
n-C_7F_{15}	Bn	(98), 61

140

$C_{14\text{-}15}$

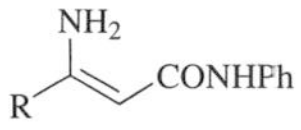

(*R*,S_{Fc})-Josiphos (Ph/*t*-Bu), $[Rh(cod)Cl]_2$, s/c 300, MeOH, 6 bar, 50°, 8 h

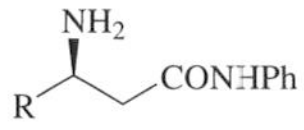

R	
Ph	(75), 96
4-FC_6H_4	(74), 96
4-MeO_6H_4	(82), 96
Bn	(94), 97

146

C_{16}

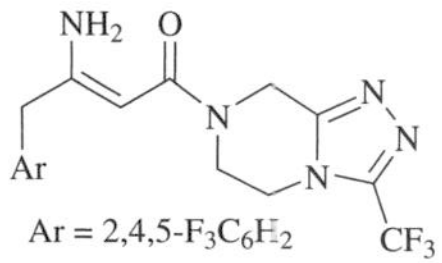

Ar = 2,4,5-$F_3C_6H_2$

(*R*,S_{Fc})-Josiphos (Ph/*t*-Bu), $[Rh(cod)Cl]_2$, s/c 350, CF_3CH_2OH, 6 bar, 50°, 7 h

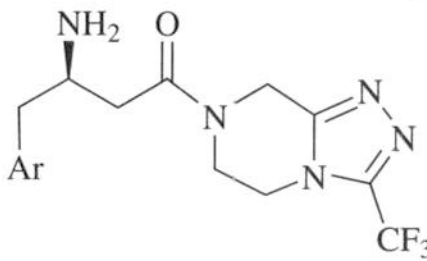

(95)[a], 94

169

$C_{16\text{-}20}$

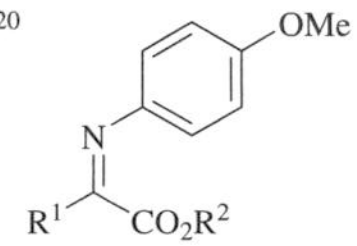

$[Rh(tangphos)(cod)]BF_4$, s/c 100, DCM, 50 bar, 50°, 24 h

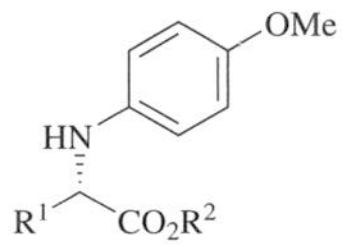

R^1	R^2	% Conv., % ee
c-C_6H_{11}	Me	(85), 94
Ph	Me	(99), 95
2-FC_6H_4	Me	(99), 91
3-FC_6H_4	Me	(95), 94
4-FC_6H_4	Me	(95), 93
4-ClC_6H_4	Me	(99), 92
4-BrC_6H_4	Me	(95), 92
2-$MeOC_6H_4$	Me	(95), 95
3-$MeOC_6H_4$	Me	(99), 93
4-$MeOC_6H_4$	Me	(95), 93
Ph	Et	(>95), 84
4-Tol	Me	(99), 93
3-$O_2NC_6H_4$	Me	(99), 93
2-Np	Me	(99), 90
1-Np	Me	(95), 91

139

TABLE 6. α- AND β-CARBOXY IMINES (*Continued*)

Imine	Conditions	Product(s) and Yield(s) (%), % ee				Refs.
C_{16-22}			R	Y		
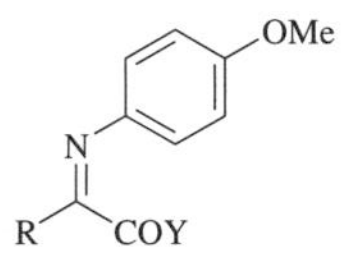	Hantzsch ester, (*S*)-**7e**, s/c 100, toluene, 60°, 48 h	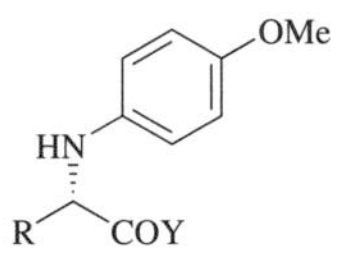	thienyl	*i*-PrO	(78), 84	142
			Ph	EtO	(88), 92	
			Ph	*i*-PrO	(87), 97	
			c-C_6H_{11}	*i*-PrO	(46), 88	
			4-BrC_6H_4	*i*-PrO	(92), 97	
			4-ClC_6H_4	*i*-PrO	(95), 98	
			4-FC_6H_4	*i*-PrO	(82), 97	
			Ph	*t*-BuO	(78), 98	
			Ph	*t*-BuNH	(85), 96	
			3-MeC_6H_4	*i*-PrO	(89), 98	
			4-MeC_6H_4	*i*-PrO	(90), 98	
			4-$MeOC_6H_4$	*i*-PrO	(94), 97	
			2-Np	*i*-PrO	(93), 98	
			Ph	BnO	(86), 95	

[a] This value is the percent conversion.

[b] The stereochemistry of the product was not reported in the original reference.

TABLE 7. REDUCTIVE AMINATION

Ketone	Amine	Conditions	Product(s) and Yield(s) (%), % ee	Refs.
C_{8-14}				
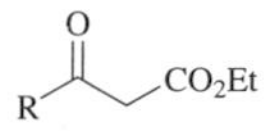	NH_4OAc	(Cymene)Ru((*R*)-ClMeO-biphep)Cl_2, s/c 100, CF_3CH_2OH, 30 bar, 80°, 16 h	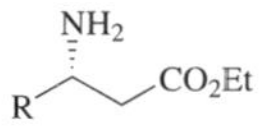	151

R	
Me	(80), 96
Ph	(88), 98
3-ClC_6H_4	(81), 98
4-ClC_6H_4	(79), 99
4-FC_6H_4	(80), 96
3-$MeOC_6H_4$	(88), 96
4-$MeOC_6H_4$	(83), 98

Ketone	Amine	Conditions	Product(s) and Yield(s) (%), % ee	Refs.
C_{10-11}				
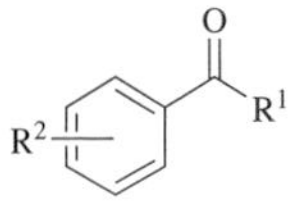	NH_3	1. HCO_2NH_4, Ru((*R*)-Tol-binap)Cl_2, s/c 100, MeOH, 85° 2. HCl, EtOH, reflux	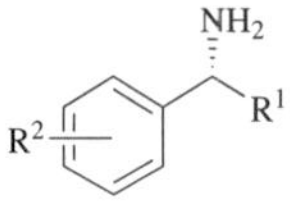	150

R^1	R^2	Time	
Me	H	20 h	(92), 95
Me	4-Cl	24 h	(93), 92
Me	4-Br	48 h	(56), 91
Me	4-NO_2	48 h	(92), 95
Me	3-Me	24 h	(74), 89
Me	4-Me	21 h	(93), 93
Me	4-MeO	25 h	(83), 95
Et	H	21 h	(89), 95

Ketone	Amine	Conditions	Product(s) and Yield(s) (%), % ee	Refs.
C_{10}				
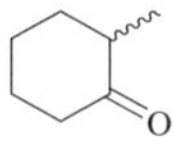		HCO_2H/NEt_3, (cymene)Ru((*S*,*S*)-dpenTs)Cl, s/c 200, DCM, rt, 184 h	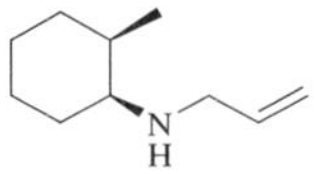 (77), 90–92% cis	77
C_{10-17}				
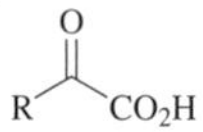	$BnNH_2$	[Rh((*R*)-deguphos)(nbd)]BF_4, MeOH, 60 bar, rt	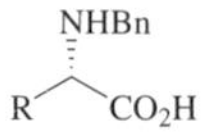	144

R	s/c	Time	
Me	100	24 h	(43), 78
HO_2CCH_2	100	24 h	(38), 73
$HO_2C(CH_2)_2$	100	24 h	(19), 60
Me_2CHCH_2	200	2 h	(99), 90
Me_3CCH_2	200	24 h	(94), 86
Bn	200	3 h	(99), 98
$Ph(CH_2)_2$	100	24 h	(80), 81

TABLE 7. REDUCTIVE AMINATION (*Continued*)

Ketone	Amine	Conditions	Product(s) and Yield(s) (%), % ee	Refs.
C_{11-19}				
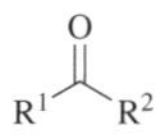	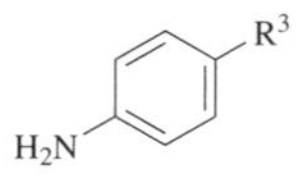	Hantzsch ester, (*R*)-**7a**, s/c 10, C_6H_6, 5 Å MS	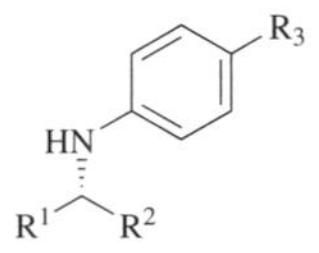	104

R^1	R^2	R^3	Temp	Time	
Et	Me	MeO	40°	72 h	(71), 83
CH_2=CH(CH_2)$_2$	Me	MeO	40°	96 h	(60), 90
Ph	Me	H	50°	24 h	(73), 93
Ph	CH_2F	MeO	5°	7 h	(70), 88
n-C_6H_{13}	Me	MeO	40°	96 h	(72), 91
c-C_6H_{11}	Me	MeO	50°	96 h	(49), 86
Ph	Me	CF_3	40°	24 h	(55), 95
Ph	Me	H	50°	24–72 h	(73), 93
Ph	Me	CF_3	50°	24–72 h	(55), 95
Ph	Me	MeO	50°	24–72 h	(87), 94
2-FC_6H_4	Me	MeO	50°	24–72 h	(60), 83
3-FC_6H_4	Me	MeO	50°	24–72 h	(81), 95
4-FC_6H_4	Me	MeO	50°	24–72 h	(75), 94
4-ClC_6H_4	Me	MeO	50°	24–72 h	(75), 95
4-$O_2NC_6H_4$	Me	MeO	50°	24–72 h	(71), 95
4-MeC_6H_4	Me	MeO	50°	24–72 h	(79), 91
4-$MeOC_6H_4$	Me	MeO	50°	24–72 h	(77), 90
Ph(CH_2)$_2$	Me	MeO	40°	72 h	(75), 94
4-$EtCOC_6H_4$	Me	MeO	50°	24–72 h	(85), 96
$BzOCH_2$	Me	MeO	40°	96 h	(72), 81
2-Np	Me	MeO	50°	72 h	(73), 96

C_{12-13}

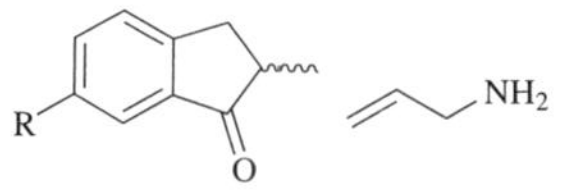

HCO_2H/NEt_3, (cymene)Ru((*R,R*)-dpenTs)Cl, s/c 200, DCM, rt, 144 h

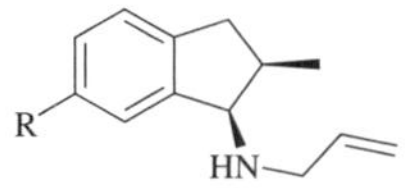

R	
H	(55), 90
Me	(60), >98

77

C_{13}

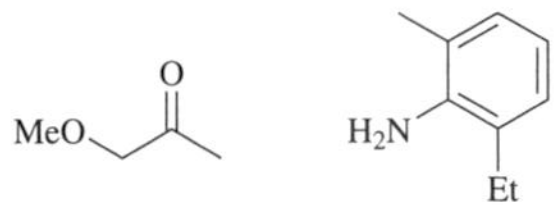

(*R*,S_{Fc})-Josiphos (Ph/Xyl), $[Ir(cod)Cl]_2$, s/c 10,000, C_6H_{12}, CF_3CO_2H, TBAI, 80 bar, 50°, 16 h

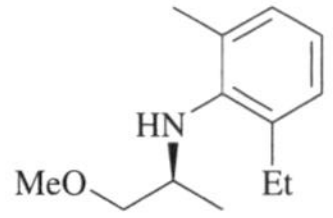

(99)[a], 78

153

C_{13-15}

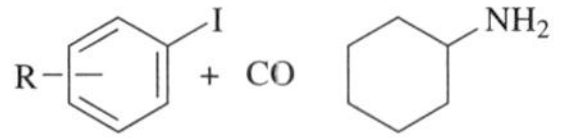

L, Pd_2dba_3, s/c 25, NEt_3, 4 Å MS, 7 bar H_2, 55 bar CO, 120°, 24–42 h

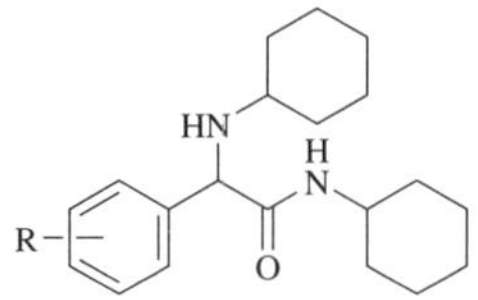

R	L	
H	Me-duphos	(31), >99[b]
H	**22**	(49), 98[b]
4-NH_2	**22**	(45), 90[b]
3-Me	Me-duphos	(43), 93[b]
4-Me	Me-duphos	(45), 92[b]
4-MeCO	**22**	(44), >99[b]
4-Et	Me-duphos	(46), 94[b]

155

C_{13-18}

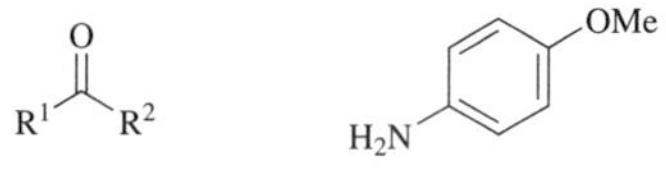

(*S,S*)-f-Binaphane, $[Ir(cod)Cl]_2$, s/c 100, DCM, I_2, Ti(*i*-PrO)$_4$, 70 bar, rt, 10 h

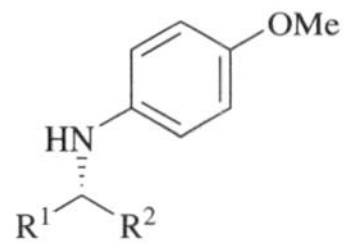

R^1	R^2	
2-furyl	Me	(>99), 92
Ph	Et	(>99), 85
Ph	*n*-Bu	(>99), 79

152

TABLE 7. REDUCTIVE AMINATION (*Continued*)

Ketone	Amine	Conditions	Product(s) and Yield(s) (%), % ee	Refs.
C_{14-24}				
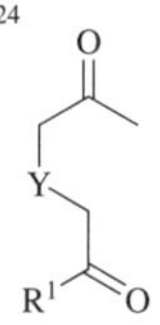	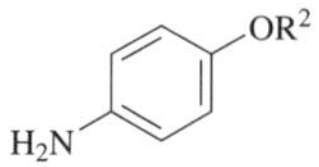	Hantzsch ester, (*R*)-**7c**, s/c 10, cyclohexane, 5 Å MS, 50°, 72 h	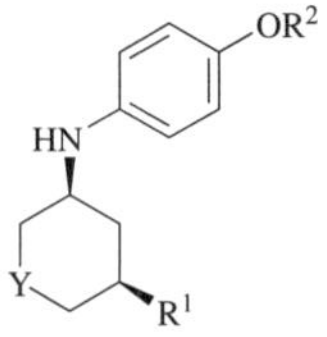	154
C_{15}				
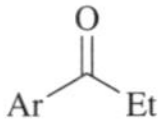	NH_3	1. HCO_2NH_4, Ru((*R*)-Tol-binap)Cl_2, s/c 100, MeOH, 85° 2. HCl, EtOH, reflux	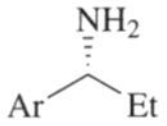	150
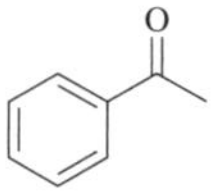	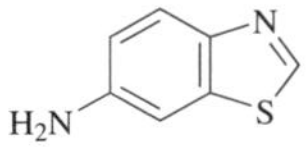	Hantzsch ester, (*R*)-**7a**, s/c 10, C_6H_6, 5 Å MS, 50°, 72 h	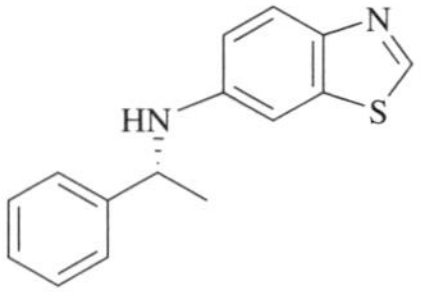 (70), 91	104

Product table for the C_{14-24} entry (ref. 154):

R^1	R^2	Y	cis/trans	
Me	Et	O	99	(72), 92
Me	Et	S	2	(35), 90
Me	Et	CH_2	6	(88), 84
i-Pr	Me	CH_2	3	(76), 92
n-Bu	Et	CH_2	10	(75), 90
i-Bu	Et	CH_2	12	(79), 96
Bn	Et	CH_2	6	(77), 86
$Ph(CH_2)_2$	Et	CH_2	24	(82), 96
$(c\text{-}C_5H_9)CH_2$	Et	CH_2	24	(72), 96
$(c\text{-}C_6H_{11})CH_2$	Me	CH_2	19	(89), 96
$(c\text{-}C_6H_{11})(CH_2)_2$	Et	CH_2	4	(78), 92
2-Np	Et	CH_2	2	(73), 82

Product table for the C_{15} entry (ref. 150):

Ar	
1-Np	(69), 86
2-Np	(91), 95

C_{15-16}	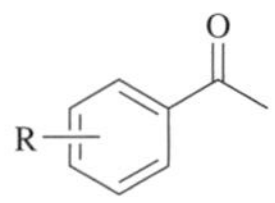	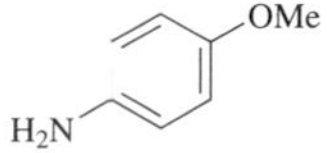	(*S,S*)-f-Binaphane, $[Ir(cod)Cl]_2$, s/c 100, DCM, I_2, $Ti(i\text{-}PrO)_4$, 70 bar, rt, 10 h	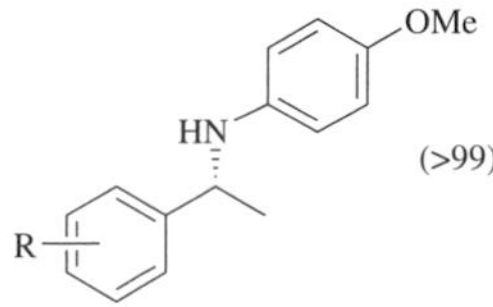	(>99)	152
C_{16}	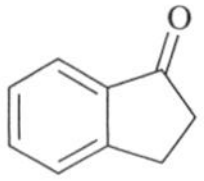	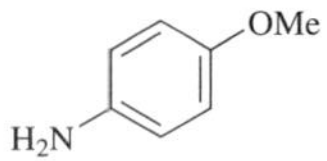	Hantzsch ester, (*R*)-**7a**, s/c 10, C_6H_6, 5 Å MS, 50°, 72 h	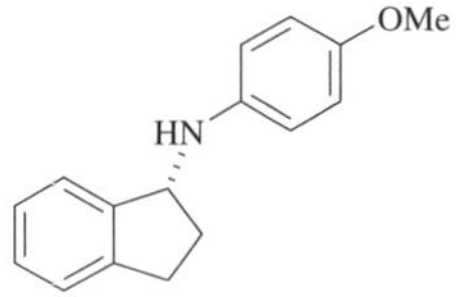	(75), 85	104
C_{20}	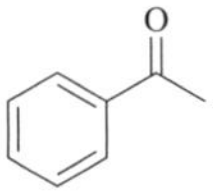	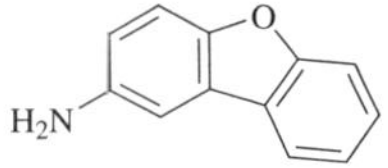	Hantzsch ester, (*R*)-**7a**, s/c 10, C_6H_6, 5 Å MS, 50°, 72 h	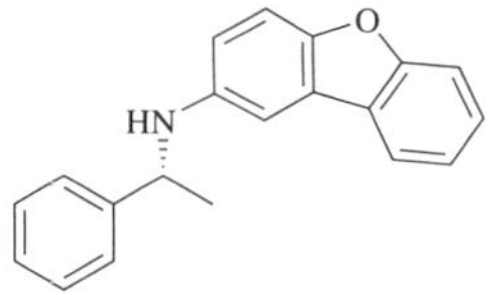	(92), 91	104
C_{23}	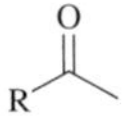	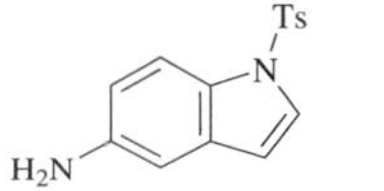	Hantzsch ester, (*R*)-**7a**, s/c 10, C_6H_6, 5 Å MS, 50°, 72 h	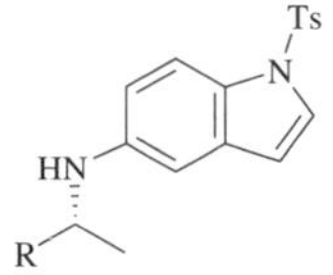	R: Ph (90), 93; n-C_6H_{13} (75), 90	104

R	% ee
H	94
4-F	93
4-Cl	92
4-Br	94
2-Me	44
3-Me	89
4-Me	96
4-MeO	95

R	
Ph	(90), 93
n-C_6H_{13}	(75), 90

[a] This value is the percent conversion.

[b] The stereochemistry of the product was not reported in the original reference.

REFERENCES

[1] Nakamura, Y. *Bull. Chem. Soc. Jpn.* **1941**, *16*, 367.
[2] For an overview on chiral heterogeneous catalysts see Blaser, H. U.; Müller. M. *Stud. Surf. Sci. Catal.* **1991**, *59*, 73.
[3] Botteghi, C.; Bianchi, M.; Benedetti, E.; Matteoli, U. *Chimia* **1975**, *29*, 256.
[4] Kagan, H. B.; Langlois, N.; Dang, T. P. *J. Organomet. Chem.* **1975**, *90*, 353.
[5] Levi, A.; Modena, G.; Scorrano, G. *Chem. Commun.* **1975**, 6.
[6] Vastag, S.; Bakos, J.; Torös, S.; Takach, N. E.; King, R. B.; Heil, B.; Marko, L. *J. Mol. Catal.* **1984**, *22*, 283.
[7] James, B. R. *Catalysis Today* **1997**, *37*, 209.
[8] Ohkuma, T.; Kitamura, M.; Noyori R. In *Catalytic Asymmetric Synthesis*, 2nd ed.; Ojima, I., Ed., Wiley-VCH: Weinheim, 2000; p 1.
[9] Blaser, H. U.; Spindler F. In *Comprehensive Asymmetric Catalysis*; Jacobsen, E. N.; Pfaltz, A.; Yamamoto H., Eds., Springer: Berlin, 1999; p 247.
[10] Spindler, F.; Blaser H. U. In *Transition Metals for Organic Synthesis*, 2nd ed.; Bolm, C.; Beller, M., Eds.; Wiley-VCH: Weinheim, 2004; Vol. 2, p 113.
[11] Brunel, J. M. *Recent Res. Devel. Org. Chem.* **2003**, *7*, 155.
[12] Tang, W.; Zhang, X. *Chem. Rev.* **2003**, *103*, 3029.
[13] Gladiali, S.; Alberico, E. In *Transition Metals for Organic Synthesis*, 2nd ed.; Bolm, C.; Beller, M., Eds.; Wiley-VCH: Weinheim, 2004; Vol. 2, p 145.
[14] Gladiali, S.; Alberico, E. *Chem. Soc. Rev.* **2006**, *35*, 226.
[15] Yurovskaya, M. A.; Krachav, A. V. *Tetrahedron: Asymmetry* **1998**, *9*, 3331.
[16] Blaser, H. U.; Buser, H. P.; Coers, K.; Hanreich, R.; Jalett, H. P.; Jelsch, E.; Pugin, B.; Schneider, H. D.; Spindler, F.; Wegmann, A. *Chimia* **1999**, *53*, 275.
[17] Chan, A. S. C.; Chen, C.-C.; Lin, C.-W.; Lin, Y-C.; Cheng, M.-C.; Peng, S.-M. *Chem. Commun.* **1995**, 1767.
[18] Tararov, V. I.; Kadyrov, R.; Riermeier, T. H.; Fischer, C.; Börner, A. *Adv. Synth. Catal.* **2004**, *346*, 561.
[19] Tararov, V. I.; Börner, A. *Synlett* **2005**, 203.
[20] For a general review on reductive amination, see: Baxter, E.W.; Reitz, A.B. *Org. React.* **2002**, *59*, 1.
[21] Blaser, H.U.; Spindler, F. In *Handbook of Homogeneous Hydrogenation*; de Vries J.G.; Elsevier C. J. Eds.; Wiley-VCH: Weinheim, 2007; p 1193.
[22] Roszkowski, P.; Czarnocki, Z. *Mini-Reviews in Org. Chem.* **2007**, *4*, 190.
[23] Glorius, F. *Org. Biomol. Chem.* **2005**, *3*, 4171.
[24] Zhou, Y.-G. *Acc. Chem. Res.* **2007**, *40*, 1357.
[25] Deloux, L.; Srebnik, M. *Chem. Rev.* **1993**, *93*, 763.
[26] Kobayashi, S.; Ishitani, H. *Chem. Rev.* **1999**, *99*, 1069.
[27] Riant, O.; Mostëfeï, N.; Courmarcel, J. *Synthesis* **2004**, 2943.
[28] Nishiyama, H. In *Transition Metals for Organic Synthesis*, 2nd ed.; Bolm, C.; Beller, M., Eds.; Wiley-VCH: Weinheim, 2004; Vol. 2, p 182.
[29] Bommarius, A.S. In *Enzyme Catalysis in Organic Synthesis*, 2nd ed.; Drauz, K.; Waldmann H., Eds.; Wiley-VCH, Weinheim, 2002; p 1047.
[30] *Handbook of Homogeneous Hydrogenation*; de Vries J. G.; Elsevier C. J. Eds.; Wiley-VCH: Weinheim, 2007.
[31] *Catalytic Asymmetric Synthesis*, 2nd ed.; Ojima, I., Ed.; Wiley-VCH: Weinheim, 2000.
[32] Drexler, H-J.; Baumann, W.; Spannenberg, A.; Fischer, C.; Heller, D. *J. Organomet. Chem.* **2001**, *621*, 89 and references cited therein.
[33] Spindler, F.; Pugin, B.; Blaser, H. U. *Angew. Chem., Int. Ed. Engl.* **1990**, *29*, 558.
[34] Willoughby, C. A.; Buchwald, S. L. *J. Am. Chem. Soc.* **1994**, *116*, 8952.
[35] Willoughby, C. A.; Buchwald, S. L. *J. Am. Chem. Soc.* **1992**, *114*, 7562.
[36] Blaser, H. U. *Adv. Synth. Catal.*, **2002**, *344*, 17.
[37] Blaser, H. U.; Buser, H. P.; Häusel, R.; Jalett, H. P.; Spindler, F. *J. Organomet. Chem.* **2001**, *621*, 34.

[38] Sablong, R.; Osborn, J. A. *Tetrahedron: Asymmetry* **1996**, *7*, 3059.
[39] Xiao, D.; Zhang, X. *Angew. Chem., Int. Ed.* **2001**, *40*, 3425.
[40] Cheemala, M. N.; Knochel, P. *Org. Lett.* **2007**, *9*, 3089.
[41] Zhu, S.-F.; Xie, J.-B.; Zhang, Y.-Z.; Li, S.; Zhou, Q.-L. *J. Am. Chem. Soc.* **2006**, *128*, 12886.
[42] Moessner, C.; Bolm, C. *Angew. Chem., Int. Ed.* **2005**, *44*, 7564.
[43] Imamoto, T.; Iwadate, N.; Yoshida, K. *Org. Lett.* **2006**, *8*, 2289.
[44] Trifonova, A.; Diesen, J. S.; Chapman, C. J.; Andersson, P. G. *Org. Lett.* **2004**, *6*, 3825.
[45] Blanc, C.; Agbossou-Niedercorn, F.; Nowogrocki, G. *Tetrahedron: Asymmetry* **2004**, *15*, 2159.
[46] Dervisi, A.; Carcedo, C.; Ooi, L-l. *Adv. Synth. Catal.* **2006**, *348*, 175.
[47] Cobley, C. J.; Henschke, J. P. *Adv. Synth. Catal.* **2003**, *345*, 195.
[48] Hoffmann, S.; Seayad, A. M.; List, B. *Angew. Chem., Int. Ed.* **2005**, *44*, 7424.
[49] Kainz, S.; Brinkmann, A.; Leitner, W.; Pfaltz, A. *J. Am. Chem. Soc.* **1999**, *121*, 6421.
[50] Schnider, P.; Koch, G.; Prétôt, R.; Wang, G.; Bohnen, F. M.; Krüger, C.; Pfaltz, A. *Chem. Eur. J.* **1997**, *3*, 887.
[51] Maire, P.; Deblon, S.; Breher, F.; Geier, J.; Boehler, C.; Rüegger, H.; Schoenberg, H.; Gruetzmacher, H. *Chem. Eur. J.* **2004**, *10*, 4198.
[52] Cozzi, P. G.; Menges, F.; Kaiser, S. *Synlett* **2003**, 833.
[53] Vargas, S.; Rubio, M.; Suarez, A.; del Rio, D.; Alvarez, E.; Pizzano, A. *Organometallics* **2006**, *25*, 961.
[54] Murai, T.; Inaji, S.; Morishita, K.; Shibahara, F.; Tokunaga, M.; Obora, Y.; Tsuji, Y. *Chem. Lett.* **2006**, *35*, 1424.
[55] Rueping, M.; Sugiono, E.; Azap, C.; Theissmann, T.; Bolte, M. *Org. Lett.* **2005**, *7*, 3781.
[56] Cahill, J. P.; Lightfoot, A. P.: Goddard, R.; Rust, J.; Guiry, P. J. *Tetrahedron: Asymmetry* **1998**, *9*, 4307.
[57] Guiu, E.; Munoz, B.; Castillon, S.; Claver, C. *Adv. Synth. Catal.* **2003,** *345*, 169.
[58] Guiu, E.; Claver, C.; Benet-Buchholz, J.; Castillon, S. *Tetrahedron: Asymmetry* **2004**, *15*, 3365.
[59] Bakos, J.; Orosz, A.; Heil, B.; Laghmari, M.; Lhoste, P.; Sinou, D. *Chem. Commun.* **1991**, 1684.
[60] Lensink, C.; Rijnberg, E.; de Vries, J. G. *J. Mol. Catal. A: Chem.* **1997**, *116*, 199.
[61] Buriak, J. M.; Osborn, J. A. *Organometallics* **1996**, *15*, 3161.
[62] Kang, G.-J.; Cullen, W. R.; Fryzuk, M. D.; James B. R.; Kutney, J. P. *Chem. Commun.* **1988**, 1466.
[63] Abdur-Rashid, K.; Lough, A. J.; Morris R. H. *Organometallics* **2001**, *20*, 1047.
[64] Canivet, J.; Süss-Fink, G. *Green Chemistry* **2007**, *9*, 391.
[65] Solinas, M.; Pfaltz, A.; Cozzi, P. G.; Leitner, W. *J. Am. Chem. Soc.* **2004**, *126*, 16142.
[66] Reetz, M. T.; Bondarev, O. *Angew. Chem., Int. Ed.* **2007**, *46*, 4523.
[67] Tani, K.; Onouchi, J.; Yamagata, T.; Kataoka, Y. *Chem. Lett.* **1995**, 955.
[68] Jiang, X.-B.; Minnaard, A. J.; Hessen, B.; Feringa, B. L.; Duchateau, A. L. L.; Andrien, J. G. O.; Boogers, J. A. F.; de Vries, J. G. *Org. Lett.* **2003**, *5*, 1503.
[69] Tararov, V. I.; Kadyrov, R.; Riermeier, T. H.; Holz, J.; Boerner, A. *Tetrahedron: Asymmetry* **1999**, *10*, 4009.
[70] Willoughby, C. A.; Buchwald, S. L. *J. Am. Chem. Soc.* **1994**, *116*, 11703.
[71] Gonzalez-Arellano, C.; Corma, A.; Iglesias, M.; Sanchez, F. *Chem. Commun.* **2005**, 3451.
[72] Ezhova, M. B.; Patrick, B. O.; James, B. R.; Waller, F. J.; Ford, M. E. *J. Mol. Catal. A: Chem.* **2004**, *224*, 71.
[73] Margalef-Catala, R.; Claver, C.; Salagre, P.; Fernandez, E. *Tetrahedron: Asymmetry* **2000**, *11*, 1469.
[74] Okuda, J.; Verch, S.; Spaniol, T. P.; Stuermer, R. *Chem. Ber.* **1996**, *129*, 1429.
[75] Rethore, C.; Riobe, F.; Fourmigue, M.; Avarvari, N.; Suisse, I.; Agbossou-Niedercorn, F. *Tetrahedron: Asymmetry* **2007**, *18*, 1877.
[76] Fogg, D. E.; James, B. R.; Kilner, M. *Inorg. Chim. Acta* **1994**, *222*, 85.
[77] Ros, A.; Magriz, A.; Dietrich, H.; Ford, M.; Fernandez, R.; Lassaletta, J. M. *Adv. Synth. Catal.* **2005**, *347*, 1917.
[78] Willoughby, C. A.; Buchwald, S. L. *J. Org. Chem.* **1993**, *58*, 7627.
[79] Viso, A.; Lee, N. E.; Buchwald, S. L. *J. Am. Chem. Soc.* **1994**, *116*, 9373.
[80] Ringwald, M.; Stürmer, R.; Brintzinger, H. H. *J. Am. Chem. Soc.* **1999**, *121*, 1524.

[81] Yamagata, T.; Tadaoka, H.; Nagata, M.; Hirao, T.; Kataoka, Y.; Ratovelomana, V.; Genêt, J. P.; Mashima, K. *Organometallics* **2006**, *25*, 2505.
[82] Roth, P.; Andersson, P. G.; Somfai, P. *Chem. Commun.* **2002**, 1752.
[83] Chan, N. G. Y.; Osborn, J. A. *J. Am. Chem. Soc.* **1990**, *112*, 9400.
[84] Zhu, G.; Zhang, X. *Tetrahedron: Asymmetry* **1998**, *9*, 2415.
[85] Giernoth, R.; Krumm, M. S. *Adv. Synth. Catal.* **2004**, *346*, 989.
[86] Reetz, M. T.; Beuttenmüller, E. W.; Goddard, R.; Pasto, M. *Tetrahedron Lett.* **1999**, *40*, 4977.
[87] Morimoto, T.; Nakajima, N.; Achiwa, K. *Synlett* **1995**, 748.
[88] Liu, D.; Li, W.; Zhang, X. *Tetrahedron: Asymmetry* **2004**, *15*, 2181.
[89] Faller, J. W.; Milheiro, S. C.; Parr, J. *J. Organomet. Chem.* **2006**, *691*, 4945.
[90] Campbell, L. A. *Proceedings of the ChiraSource '99 Symposium;* The Catalyst Group: Spring House, USA, 1999.
[91] Uematsu, N.; Fujii, A.; Hashiguchi, S.; Ikariya, T.; Noyori, R. *J. Am. Chem. Soc.* **1996**, *118*, 4916.
[92] Roszowski, P.; Wojtasiewicz, K.; Leniewsky, A.; Maurin, J. K.; Lis, T.; Czarnocki, Z. *J. Mol. Catal. A: Chem.* **2005**, *232*, 143.
[93] Vedeijs, E.; Trapencieris, P.; Suna, E. *J. Org. Chem.* **1999**, *64*, 6724.
[94] Meuzelaar, G. J.; van Vliet, M. C. A.; Leendert, M., Sheldon, R. A. *Eur. J. Org. Chem.* **1999**, 2315.
[95] Samano, V.; Ray, J. A.; Thompson, J. B.; Mook, R. A.; Jung, D. K.; Koble, C. S.; Martin, M. T.; Bigham, E. C.; Regitz, C. S.; Feldman, P. L.; Boros, E. E. *Org. Lett.* **1999**, *1*, 1993.
[96] Szawkalo, J.; Czarnocki, Z. *Monatsh. Chem.* **2005**, *136*, 1619.
[97] Santos, L. S.; Pilli, R. A.; Rawal, V. H. *J. Org. Chem.* **2004**, *69*, 1283.
[98] Tietze, L. F.; Zhou, Y.; Topken, E. *Eur. J. Org. Chem.* **2000**, 2247.
[99] Wu, J.; Wang, F.; Ma, Y.; Cui, X.; Cun, L.; Zhu, J.; Deng, J.; Yu, B. *Chem. Commun.* **2006**, 1766.
[100] Li, L.; Wu, J.; Wang, F.; Liao, J.; Zhang, H.; Lian, C.; Zhu, J.; Deng, J. *Green Chem.* **2007**, *9*, 23.
[101] Mao, J.; Baker, D. C. *Org. Lett.* **1999**, *1*, 841.
[102] Morimoto, T.; Suzuki, N.; Achiwa, K. *Tetrahedron: Asymmetry* **1998**, *9*, 183.
[103] Morimoto, T.; Suzuki, N.; Achiwa, K. *Heterocycles* **1996**, *43*, 2557.
[104] Storer, R. I.; Carrera, D. E.; MacMillan, D. W. C. *J. Am. Chem. Soc.* **2006**, *128*, 84.
[105] Rueping, M.; Antonchick, A. P.; Theissmann, T. *Angew. Chem., Int. Ed.* **2006**, *45*, 6751.
[106] Szawkalo, J.; Zawdzka, A.; Wojtasiewicz, K.; Leniewski, A.; Drabowicz, J.; Czarnocki, Z. *Tetrahedron: Aysmmetry* **2005**, *16*, 3619.
[107] Szawkalo, J.; Czarnocki, S. J.; Zawadzka, A.; Wojtasiewicz, K.; Leniewski, A.; Maurin, J. K.; Czarnocki, Z.; Drabowicz, J. *Tetrahedron: Aysmmetry* **2007**, *18*, 406.
[108] Williams, G. D.; Wade, C. E.; Wills, M. *Chem. Commun.* **2005**, 4735.
[109] Blaser, H. U.; Malan, C.; Pugin, B.; Spindler, F.; Steiner, H.; Studer, M. *Adv. Synth. Catal.* **2003**, *345*, 103.
[110] Fuchs, R. European Patent 0803502 (1997). *Chem. Abstr.* **2004**, *141*, 206915.
[111] Rueping, M.; Antonchick, A. P. *Angew. Chem., Int. Ed.* **2007**, *46*, 4562.
[112] Legault, C. Y.; Charette, A. B. *J. Am. Chem. Soc.* **2005**, *127*, 8966.
[113] Rueping, M.; Antonchick, A. P.; Theissmann, T. *Angew. Chem., Int. Ed.* **2006**, *45*, 3683.
[114] Wang, W.-B.; Lu, S.-M.; Yang, P.-Y.; Han, X.-W.; Zhou, Y.-G. *J. Am. Chem. Soc.* **2003**, *125*, 10536.
[115] Xu, L.; Lam, K. H.; Ji, J.; Wu, J.; Fan, Q.-H.; Lo, W.-H.; Chan, A. S. C. *Chem. Commun.* **2005**, 1390.
[116] Wang, Z.-J.; Deng, G.-J.; Li, Y.; He, Y.-M.; Tang, W.-J.; Fan, Q.-H. *Org. Lett.* **2007**, *9*, 1243.
[117] Wang, D-W.; Zeng, W.; Zhou, Y-G. *Tetrahedron: Asymmetry* **2007**, *18*, 1103.
[118] Lam, K. H.; Xu, L.; Feng, L.; Fan, Q.-H.; Lam, F. L.; Lo, W.-G.; Chan, A. S. C. *Adv. Synth. Catal.* **2005**, *347*, 1755.
[119] Tang, W.-J.; Zhu, S.-F.; Xu, L.-J.; Zhou, Q.-L.; Fan, Q.-H.; Fan, Q.-H.; Zhou, H.-F.; Lam, K.; Chan, A. S. C. *Chem. Commun.* **2007**, 613.
[120] Reetz, M. T.; Li, X. *Chem. Commun.* **2006**, 2159.
[121] Lu, S.-M.; Han, X.-W.; Zhou, Y.-G. *Adv. Synth. Catal.* **2004**, *346*, 905.

[122] Lu, S.-M.; Wang, Y.-Q.; Han, X.-W.; Zhou, Y.-G. *Angew. Chem., Int. Ed.* **2006**, *45*, 2260.
[123] Bianchini, C.; Barbaro, P.; Scapacci, G.; Farnetti, E.; Graziani, M. *Organometallics* **1998**, *17*, 3308.
[124] Xie, Y.; Mi, A.; Jiang, Y.; Liu, H. *Synth. Commun.* **2001**, *31*, 2767.
[125] Murahashi, S.-I.; Tsuji, T.; Ito, S. *Chem. Commun.* **2000**, 409.
[126] Oppolzer, W.; Wills, M.; Starkemann, C.; Bernardinelli, G. *Tetrahedron Lett.* **1990**, *31*, 4117.
[127] Charette, A.; Giroux, A. *Tetrahedron Lett.* **1996**, *37*, 6669.
[128] Yang, Q.; Shang, G.; Gao, W.; Deng, J.; Zhang, X. *Angew. Chem., Int. Ed.* **2006**, *45*, 3832.
[129] Wang, Y.-Q.; Lu, S.-M.; Zhou, Y.-G. *J. Org. Chem.* **2007**, *72*, 3729.
[130] Wang, Y.-Q.; Zhou, Y.-G. *Synlett* **2006**, 1189.
[131] Liu, P.-N.; Gu, P.-M.; Deng, J.-G.; Tu, Y.-Q.; Ma, Y.-P. *Eur. J. Org. Chem.* **2005**, 3221.
[132] Ahn, K. H.; Ham, C.; Kim, S.-K.; Cho, C.-W. *J. Org. Chem.* **1997**, *62*, 7047.
[133] Chen, Y.-C.; Wu, T.-F.; Jiang, L.; Deng, J.-G.; Liu, H.; Zhu, J.; Jiang, Y-Z. *J. Org. Chem.* **2005**, *70*, 1006.
[134] Burk, M. J.; Feaster, J. E. *J. Am. Chem. Soc.* **1992**, *114*, 6266.
[135] Burk, M. J.; Martinez, J. P.; Feaster, J. E.; Cosford, N. *Tetrahedron* **1994**, *50*, 4399.
[136] Ireland, T.; Tappe, K.; Grossheimer, G.; Knochel, P. *Chem. Eur. J.* **2002**, *8*, 843.
[137] Yamazaki, A.; Achiwa, I.; Horikawa, K.; Tsurubo, M.; Achiwa, K. *Synlett.* **1997**, 455.
[138] Spindler, F.; Blaser, H. U. *Adv. Synth. Catal.* **2001**, *343*, 68.
[139] Shang, G.; Yang, Q.; Zhang, X. *Angew. Chem., Int. Ed.* **2006**, *45*, 6360.
[140] Abe, H.; Amii, H.; Uneyama, K. *Org. Lett.* **2001**, *3*, 313.
[141] Li, G.; Liang, Y.; Antilla, J. C. *J. Am. Chem. Soc.* **2007**, *129*, 5830.
[142] Kang, Q.; Zhao, Z.-A.; You, S.-L. *Adv. Synth. Catal.* **2007**, *349*, 1656.
[143] Hasegawa, M.; Taniyama, D.; Tomioka, K. *Tetrahedron* **2000**, *56*, 10153.
[144] Kadyrov, R.; Riermeier, T. H.; Dingerdissen, U.; Tararov, V. I.; Börner, A. *J. Org. Chem.* **2003**, *68*, 4067.
[145] Dai, Q.; Yang, W.; Zhang, X. *Org. Lett.* **2005**, *7*, 5343.
[146] Hsiao, Y.; Rivera, N. R.; Rosner, T.; Krska, S. W.; Njolito, E.; Wang, F.; Sun, Y.; Armstrong, J. D.; Grabowski, E. J. J.; Tillyer, R. D.; Spindler, F.; Malan, C. *J. Am. Chem. Soc.* **2004**, *126*, 9918.
[147] Matsumura, K.; Zhang, X.; Saito, T. European Patent 1386901 (2004); *Chem. Abstr.* **2004**, *141*, 206915.
[148] Kubryk, M.; Hansen, K. B. *Tetrahedron: Asymmetry* **2006**, *17*, 205.
[149] Hansen, K. B.; Rosner, T.; Kubryk, M.; Dormer, P. G.; Armstrong, J. D. *Org. Lett.* **2005**, *7*, 4935.
[150] Kadyrov, R.; Riermeier, T. H. *Angew. Chem., Int. Ed.* **2003**, *42*, 5472.
[151] Bunlaksananusorn, T.; Rampf, F. *Synlett* **2005**, 2682.
[152] Chi, Y.; Zhou, Y.-Z.; Zhang, X. *J. Org. Chem.* **2003**, *68*, 4120.
[153] Blaser, H. U.; Buser, H. P.; Jalett, H. P.; Pugin, B.; Spindler, F. *Synlett* **1999**, 867.
[154] Zhou, J.; List, B. *J. Am. Chem. Soc.* **2007**, *129*, 7498.
[155] Nanayakkara, P.; Alper, H. *Chem. Commun.* **2003**, 2384.
[156] Lee, N.; Buchwald, S. L. *J. Am. Chem. Soc.* **1994**, *116*, 5985.
[157] Tararov, V. I.; Kadyrov, R.; Riermeier, T. H.; Holz, J.; Boerner, A. *Tetrahedron Lett.* **2000**, *41*, 2351.
[158] Becalski, A. G.; Cullen, W. R.; Fryzuk, M. D.; James, B. R.; Kang, G.-J.; Rettig, S. J. *Inorg. Chem.* **1991**, *30*, 5002.
[159] *Handbook of Homogeneous Hydrogenation*; de Vries, J. G., Elsevier C. J., Eds.; Wiley-VCH: Weinheim, 2007.
[160] Blackmond, D. G.; Ropic, M.; Stefinovic, M. *Org. Process Res. Dev.* **2006**, *10*, 457.
[161] Dorta, M.; Broggini, D.; Stoop, R.; Rüegger, H.; Spindler, F.; Togni, A. *Chem. Eur. J.* **2004**, *10*, 267.
[162] Dorta, M.; Broggini, D.; Kissner, R.; Togni, A. *Chem. Eur. J.* **2004**, *10*, 4546.
[163] Smidt, S. P.; Pfaltz, A.; Martinez-Viviente, E.; Pregosin, P. S.; Albinati, A. *Organometallics* **2003**, *22*, 1000.
[164] Blaser, H. U.; Pugin, B.; Spindler, F.; Togni, A. *C. R. Chim.* **2002**, *5*, 379.
[165] Clapham, S. E.; Hadzovic; A.; Morris, R. H. *Coord. Chem. Rev.* **2004**, *248*, 2201.

[166] Samec, J. S. M.; Bäckvall, J.-E.; Andersson, P. G.; Brandt, P. *Chem. Soc. Rev.* **2006**, *35*, 237.
[167] Hofer, R. *Chimia* **2005**, *59*, 10.
[168] Pugin, B.; Landert, H.; Spindler, F.; Blaser, H. U. *Adv. Synth. Catal.* **2002,** *344*, 974.
[169] Clausen, A. M.; Dziadul, B.; Cappuccio, K. L.; Kaba, M.; Starbuck, C.; Hsiao, Y.; Dowling, T. M. *Org. Process Res. Dev.* **2006**, *10*, 723.
[170] Thayer, A. M. *Chem. Eng. News* **2007**, *85 (32)*, 11.
[171] Imwinkelried, R. *Chimia* **1997**, *51*, 300.
[172] Cobley, C. J.; Foucher, E.; Lecouve, J.-P.; Lennon, I. C.; Ramsden, J. A.; Thominot, G. *Tetrahedron: Asymmetry* **2003**, *14*, 3431.
[173] Blacker, J.; Martin J. In *Large-Scale Asymmetric Catalysis*; Blaser, H. U.; Schmidt, E., Eds.; Wiley-VCH: Weinheim, 2003; p 201.
[174] Satoh, K.; Inenaga, M.; Kanai, K. *Tetrahedron: Asymmetry* **1998**, *9*, 2657.
[175] Brunner, H.; Rosenboem, S. *Monatsh. Chem.* **2000**, *131*, 1371.
[176] Pugin, B.; Groehn, V.; Moser, R.; Blaser, H. U. *Tetrahedron: Asymmetry* **2006**, *17*, 544.
[177] Groehn, V.; Moser, R.; Pugin, B. *Adv. Synth. Catal.* **2005,** *347*, 1855.
[178] Tietze, L. F.; Rackelmann, N.; Sekar, G. *Angew. Chem., Int. Ed.* **2003**, *42*, 4254.
[179] Mujahidin, D.; Doye, S. *Eur. J. Org. Chem* **2005**, 2689.
[180] Roszowski, P.; Maurin, J. K.; Czarnocki, Z. *Tetrahedron: Asymmetry,* **2007**, *17*, 1415.
[181] Wallbaum, S.; Martens, J. *Tetrahedron: Asymmetry* **1992**, *3*, 1475.
[182] Yamada, T.; Nagata, T.; Sugi, K. D.; Yoruzu, K.; Ikeno, T.; Ohtsuka, Y.; Miyazaki, D.; Mukaiyama, T. *Chem. Eur. J.* **1993**, *9*, 4485.
[183] Graves, C. R.; Scheidt, K. A.; Nguyen, S. T. *Org. Lett.* **2006**, *8*, 1229.
[184] Glushkov, V. A.; Tolstikov, A. G. *Russ. Chem. Rev.* **2004**, *73*, 581.
[185] Gosselin, F.; O'Shea, P. D.; Roy, S.; Reamer, R. A.; Chen, C.; Volante, R. P. *Org. Lett.* **2005**, *7*, 355.
[186] Carpentier, J. F.; Bette, V. *Curr. Org. Chem.* **2002**, *6*, 913.
[187] Hansen, M. C.; Buchwald, S. L. *Org. Lett.* **2000**, *2*, 713 and references cited therein.
[188] Lipshutz, B. H.; Shimizu, H. *Angew. Chem., Int. Ed.* **2004**, *43*, 2228.
[189] Wang, Z.; Ye, X.; Wie, S.; Wu, P.; Zhang, A.; Sun, J. *Org. Lett.* **2006**, *8*, 999.
[190] Malkov, A. V.; Stoncius, S.; MacDougall, K. N.; Mariani, A.; McGeoch, G. D.; Kocovsky, P. *Tetrahedron* **2006**, *62*, 264.
[191] Breuer, M.; Ditrich, K.; Habicher, T.; Hauer, B.; Kesseler, M.; Stuermer, R.; Zelinski, T. *Angew. Chem., Int. Ed.* **2004**, *43*, 788.
[192] Labeq Laboratory Equipment AG, CH-8006 Zürich, Switzerland.
[193] Autoclave Engineers Europe, F-Nogent sur Olle, Cedex, France.
[194] Büchi AG, CH-8610 Uster, Switzerland.
[195] Battersby, A. R.; Edwards, T. P. *J. Chem. Soc.* **1960**, 1214.
[196] Braun, W.; Salzer, A.; Spindler, F.; Alberico, E. *Appl. Catal., A* **2004**, *274*, 191.
[197] Trifonova, A.; Diesen, J. S.; Andersson, P. G. *Chem. Eur. J.* **2006**, *12*, 2318.
[198] Broger, E. A.; Burkart, W.; Henning, M.; Scalone, M.; Schmid, R. *Tetrahedron: Asymmetry,* **1998**, *9*, 4043.
[199] Huang, X.; Ying, J. Y. *Chem. Commun.* **2007**, 1825.
[200] Yang, P.-Y.; Zhou, Y-G. *Tetrahedron: Asymmetry* **2004**, *15*, 1145.

CHAPTER 2

OXOAMMONIUM- AND NITROXIDE-CATALYZED OXIDATIONS OF ALCOHOLS

JAMES M. BOBBITT and CHRISTIAN BRÜCKNER

University of Connecticut, Storrs, Connecticut 06269, U.S.A.

NABYL MERBOUH

Simon Fraser University, Burnaby, B.C. V5A 1S6, Canada

CONTENTS

james.bobbitt@uconn.edu
Organic Reactions, Vol. 74, Edited by Scott E. Denmark et al.

ACKNOWLEDGEMENTS

We thank Dr. V. A. Golubev and Dr. V. D. Sen' of the Institute of Problems of Chemical Physics, Russian Academy of Sciences, Chernogolovka, Russia, Dr. Jerzy Zakrzewski of the Institute of Industrial Organic Chemistry, Warsaw, Poland, Mrs. Jane Ann Bobbitt, and Dr. Priya Pradhan, and Ms. Ashley L. Bartelson of the University of Connecticut for reading and correcting the manuscript text. Dr. Vijayata Sharma of this Department and Dr. Prydhan checked the references for both the text and the many tables. Professor William F. Bailey and Dr. Arie Bessemer, Amerongen, The Netherlands, assisted with the mechanisms sections and helped us to stay informed about the current literature. Dr. Abay Vase of this Department provided us with Figure 2. Finally, we thank our editors, Dr. Michael Martinelli and Dr. Danielle Soenen for their patience and help in preparing the manuscript.

With respect, this chapter is dedicated to Dr. V. A. Golubev and his coworkers, who discovered and developed the chemistry of oxoammonium salts.

Research in this area at the University of Connecticut has been supported by grants from the Petroleum Research Fund of the American Chemical Society, the University of Connecticut Research Foundation, and Jane Ann and James M. Bobbitt.

INTRODUCTION

The discovery in 1959 of stable nitroxide-based free radicals such as the prototypical 2,2,6,6-tetramethylpiperidine-1-oxyl, universally known as TEMPO (**1**),[1,2] led to their use as electron spin resonance (ESR) spin labels in chemistry, biomedicine, and materials sciences.[3] These nitroxides are prepared by the oxidation of secondary amines that contain no hydrogen atoms on the α-carbons (Eq. 1). If the amines carry α-hydrogens, the oxidation products are nitrones, not free-radical nitroxides.

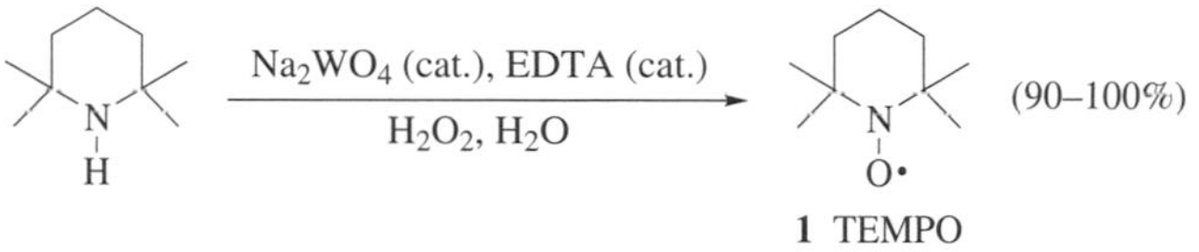

(Eq. 1)

EDTA = ethylenediaminetetraacetic acid

The unique redox properties of nitroxides (Scheme 1) enable their use as oxidants in organic synthesis. A one-electron oxidation converts nitroxide **1** (TEMPO) into the oxoammonium cation **2** ([TEMP=O]$^+$), a strong and specific oxidant. In an alternative preparation, strong acids cause the disproportionation of nitroxide to the oxoammonium cation **2** and the corresponding hydroxylamine (TEMPOH), in its protonated form **3** ([TEMP(OH)H]$^+$).[4] Under basic conditions, the reverse reaction, the comproportionation of one equivalent of TEMPOH with one equivalent of [TEMP=O]$^+$ to give two equivalents of TEMPO takes place. The mechanisms of disproportionation and comproportionation have been discussed.[5–7] A two-electron oxidation of TEMPOH also leads to the formation of [TEMP=O]$^+$.[6] Combined with a range of anions (X$^-$), [TEMP=O]$^+$X$^-$ salts are often stable and can be isolated. Almost all of the nitroxides described in this chapter are structurally derived from TEMPO.

The varied preparations and uses of oxoammonium salts as stoichiometric oxidants for alcohols are one subject of this review. The particular utility of oxoammonium salts in organic synthesis as oxidants for the conversion of primary or secondary alcohols into aldehydes, carboxylic acids, lactones, or ketones is presented. These reactions lead to the reduction of the oxoammonium salt to the corresponding salt of the hydroxylamine (Eq. 2).[8] If an oxidant system capable

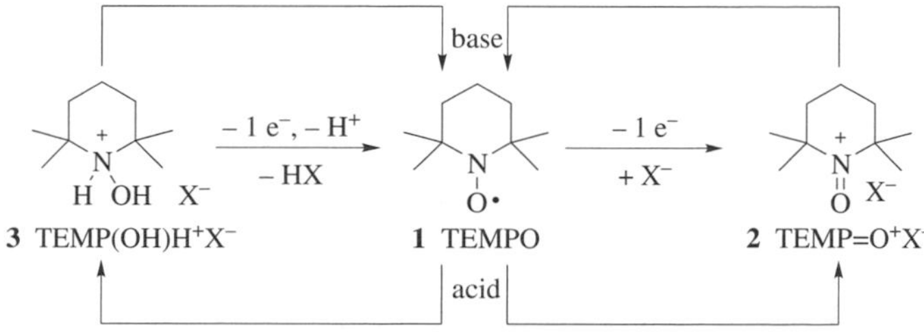

Scheme 1

of oxidizing hydroxylamine **3** back to the oxoammonium salt **2** is present, the nitroxide turns over, giving rise to the second, and major, theme of this review, namely nitroxide-catalyzed oxidations.

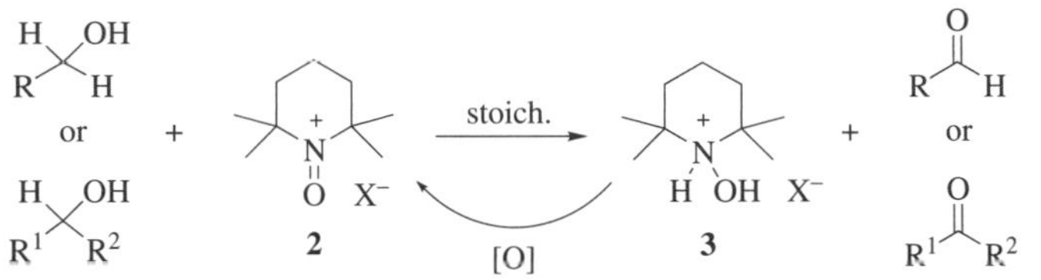

(Eq. 2)

Oxoammonium salts are known under several other names: iminoxyl salts,[8] oxopiperidinium salts,[9] 1-oxopiperidinium salts,[4] nitrosonium salts,[10] and oxo-iminium salts.[11] Although oxoammonium salts are well-known compounds and can be used for oxidations, a more general term describing the reactions in this chapter might be "oxoammonium ion oxidations". This encompasses oxidations both stoichiometric and catalytic in nitroxide.

Oxoammonium ions as oxidants for alcohols have a number of advantages. The method is heavy-metal-free, and some of the reactions can be performed in water or aqueous mixtures. Depending on the particular reaction conditions, primary alcohols can be specifically converted into aldehydes, and neither over-oxidation to carboxylic acids nor carbon–carbon bond cleavages are observed. Furthermore, neither double bond isomerizations nor racemizations of stereogenic centers take place, even when these functional groups are adjacent to the carbon being oxidized. Yields are usually high and frequently the aldehyde products have been used in a subsequent synthetic step without isolation (see Tables).

A few side reactions are associated with oxoammonium ion oxidations that can be generalized. The most serious one is the fast reaction of the oxoammonium ion with free amines, although amides and pyridines are unreactive. The oxoammonium–amine reaction has been little studied and is not well understood. Oxoammonium salts react slowly with activated double bonds, which can cause the formation of significant amounts of side products in reactions where the alcohol oxidation is slow. A few other special side reactions can take place, and these are listed in the section "Miscellaneous Reactions." It should be noted that fewer side reactions seem to take place with the nitroxide-catalyzed oxidations than with stoichiometric reactions.

Oxidation reactions using oxoammonium salts and nitroxides have been carried out in several ways (summarized below), all of which are discussed in detail in the "Scope and Limitations" section:

1. Oxidations using stoichiometric quantities of preformed oxoammonium salts, carried out either under neutral, acidic (Eq. 3) or basic (Eq. 4) conditions.

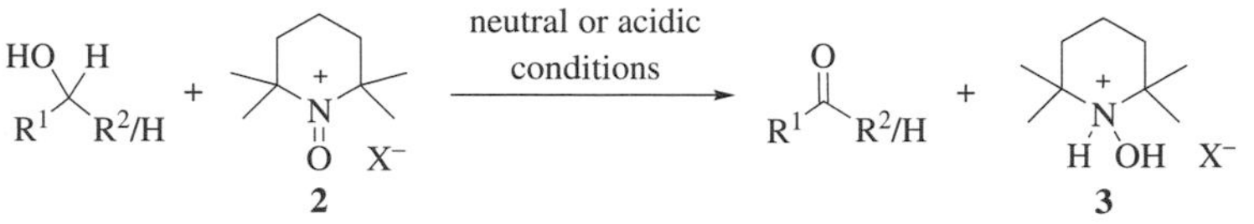

(Eq. 3)

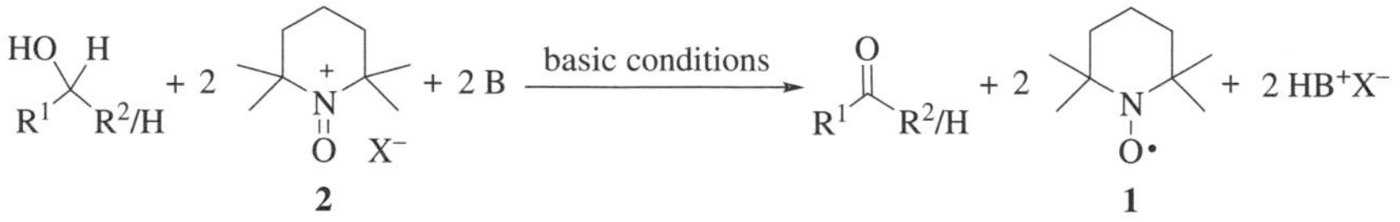

(Eq. 4)

Of all the different protocols, the stoichiometric reactions are the simplest and easiest to use, primarily because of the high selectivity for alcohol groups and the ease of product isolation.

2. Reactions in which stoichiometric quantities of oxoammonium salts are generated in situ by disproportionation of a nitroxide in the presence of a strong acid, such as *p*-toluenesulfonic acid (TsOH) (Eq. 5; compare also to Scheme 1).

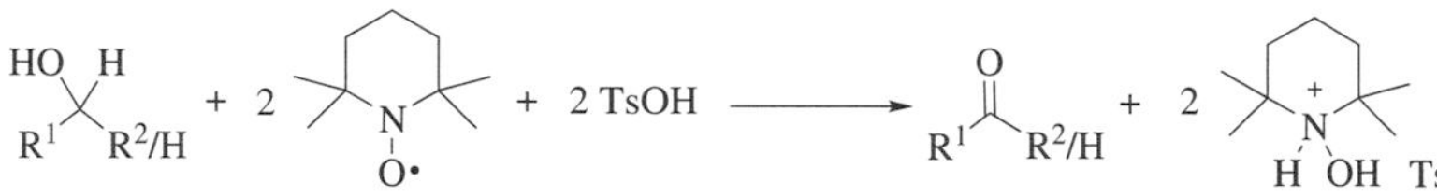

(Eq. 5)

This in situ method has been used less often, but it has some interesting characteristics that merit more attention. Readily available nitroxides may be used, although two equivalents of nitroxide are required to generate one equivalent of oxidant. The hydroxylamine product can be easily recovered and re-oxidized to a nitroxide.

3. Nitroxide-catalyzed oxidations using a secondary oxidant such as aqueous sodium hypochlorite (bleach) that lead, depending on the substrate and the stoichiometric ratio of the secondary oxidant, to aldehydes, ketones, or carboxylic acids (Eq. 6).

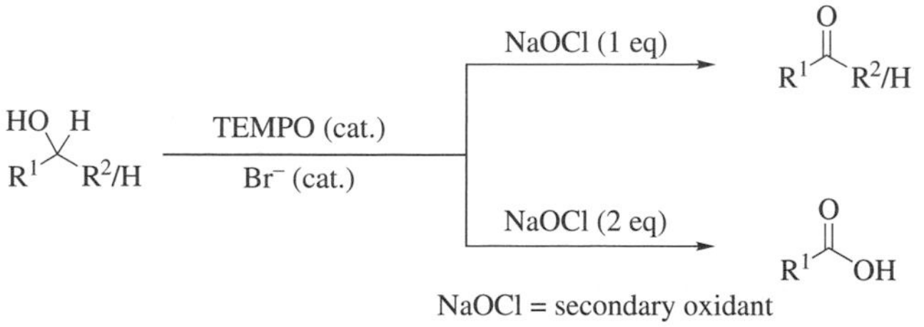

(Eq. 6)

It is important to realize that the active oxidant in the catalyzed oxidation of an alcohol is an oxoammonium ion that is formed in situ. These specific reactions sometimes use bromide ion as co-catalyst. The secondary oxidant, sodium hypochlorite, can be replaced by a variety of other oxidants that will be considered in detail below. Two equivalents of sodium hypochlorite can lead to the oxidation of a primary alcohol to a carboxylic acid whereby the oxidation of the intermediate aldehyde to the acid group is probably caused by the secondary oxidant (NaOCl), not the oxoammonium salt.

Nitroxide-catalyzed oxidations utilizing inexpensive secondary oxidants, most of which are commercially available, are overall cost-effective reactions.

4. Efficient nitroxide-catalyzed oxidations of primary alcohols to carboxylic acids are carried out using sodium chlorite ($NaClO_2$) as the secondary oxidant in the presence of catalytic amounts of sodium hypochlorite (Eq. 7). These reactions also benefit from bromide as a co-catalyst for fast reactions.

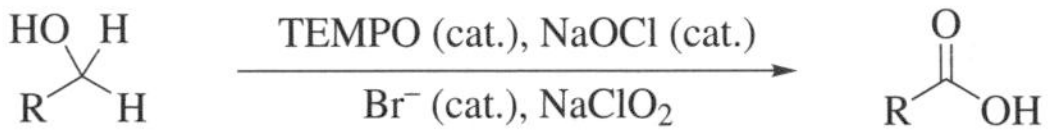
HO H
R H
TEMPO (cat.), NaOCl (cat.)
Br⁻ (cat.), NaClO2
O
R OH

(Eq. 7)

5. Oxidations of primary alcohols or hemiacetals can lead to lactones if a suitable hydroxyl group is present for ring formation (Eq. 8). This reaction occurs under all of the oxidation types listed above and represents a situation where a primary hydroxyl group is transformed into a carboxylic acid derivative. The intermediacy of a hemiacetal (containing the hydroxyl group required for oxidation) appears to be a mechanistic prerequisite in this reaction.

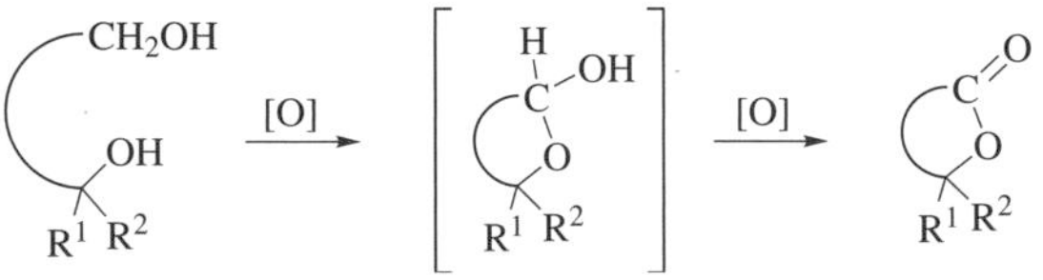
CH2OH
OH
R1 R2
[O]
H
C
OH
O
R1 R2
[O]
C
O
O
R1 R2

(Eq. 8)

6. Other transformations effected by oxoammonium salts include double bond addition, carbon–hydrogen oxidation, aromatization reactions, phenol coupling, and sulfur oxidations. These reactions are discussed below in the "Miscellaneous Reactions" section.

Examples of four major reaction types utilizing oxoammonium salt oxidants and nitroxide-catalyzed oxidations have been published in *Organic Syntheses*. In each of these contributions, concise but thorough discussions of the following methods are given: (1) the preparation of a specific oxoammonium salt ($[TEMP{=}O]^+[BF_4]^-$) and its use for the oxidation of geraniol;[12] (2) nitroxide-catalyzed oxidations using sodium hypochlorite;[13] (3) nitroxide-catalyzed oxidations using a catalytic amount of sodium hypochlorite with sodium chlorite as the secondary oxidant;[14] and (4) nitroxide-catalyzed oxidations using bis(acetoxy)iodobenzene ($PhI(OAc)_2$) as the secondary oxidant.[15]

Comprehensive reviews have appeared on the chemistry of oxoammonium salts[10,16–20] and on oxoammonium ion reactions.[10,21–26] The syntheses of hindered amine precursors leading to nitroxides and oxoammonium salts (with specific, but not tested, procedures from the literature) have been well-summarized.[20,27] Reviews have appeared on the stability of nitroxides;[28] the synthesis and applications of optically active nitroxides;[29] the use of TEMPO in synthesis in general;[30] and particularly the application of nitroxide-catalyzed oxidations in β-lactam chemistry,[31] carbohydrate chemistry,[32–35] and in the synthesis of *N*-protected α-amino aldehydes.[36] Reviews of nitroxide-promoted glycerol oxidations,[37,38] and selective oxidation of secondary alcohols are also available.[39] The catalysis of redox processes by nitroxides in aqueous solutions has also been reviewed.[40] Furthermore, oxoammonium oxidations are included in reviews that summarize the following: use of hypervalent iodine reagents for the oxidation of alcohols and their application to complex molecule synthesis,[41] polymer-supported organic reagents (including polymer-supported nitroxides),[42–44] new developments in catalytic oxidations for fine chemicals synthesis,[45,46] polymer reagents in combinatorial chemistry,[47] and catalyzed air oxidations.[48,49] Lastly, a review on the industrial uses of oxidations, including nitroxide-catalyzed reactions, is available.[50]

MECHANISMS, STEREOCHEMISTRY, AND SITE SELECTIVITY

Mechanistic Considerations

The oxidation of an alcohol to an aldehyde or ketone is a deceptively simple transformation. A number of studies have been published concerning the mechanism of oxoammonium oxidations and nitroxide-catalyzed oxidations of alcohols, although the details of this formal two-proton and two-electron process (Eq. 9) remain unclear.

H
O
R1
R2
H
OH
N
H
O
R1
R2
+
+

(Eq. 9)

Some of the uncertainties of the mechanism arise from the fact that the oxoammonium cation can be formulated in two forms, **2A** and **2B** (Eq. 10). Resonance structure **2A** has a full electron octet on both oxygen and nitrogen, with the charge on nitrogen, as the less electronegative atom. Hence, **2A** should primarily contribute towards its reactivity. However, the existence of resonance structure **2B** cannot be ruled out and represents a rare example of an electrophilic oxygen.[9,51]

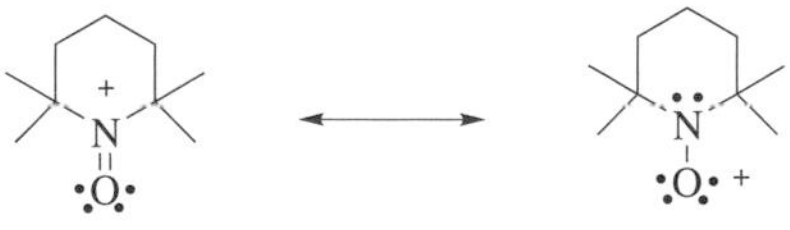
N
O
N
O

(Eq. 10)

2A **2B**

Eq. 11 shows the expected intermediates resulting from the attack of a nucleophile on the nitrogen or oxygen of the nitrosonium ion. In base-catalyzed oxidations, the initial attack is on nitrogen.[52] However, as the space-filling representation of the single crystal X-ray structure of oxoammonium salts reveals,[9] the nitrogen is buried deep in the ion, but the oxygen is quite accessible for reactions

(Fig. 1).[53] In fact, an electrophilic oxygen has been invoked to explain reactions of oxoammonium salts with Grignard reagents[54] and activated double bonds (to be discussed in the "Miscellaneous Reactions" section),[51,55–57] and could well be more important in alcohol oxidations than hitherto realized.

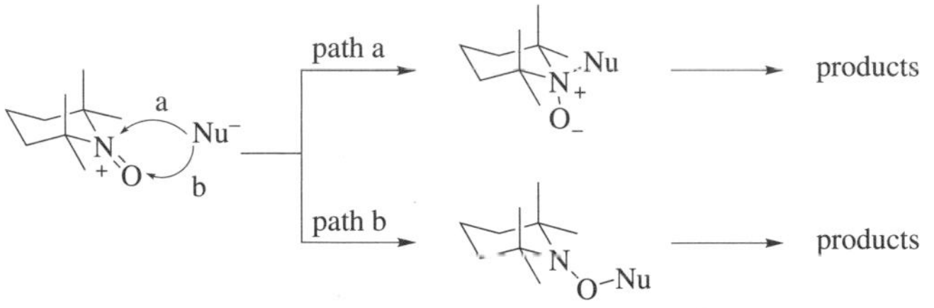

(Eq. 11)

The most comprehensive study of the mechanisms of oxoammonium oxidations of simple alcohols involves oxidations in water and aqueous acetonitrile at various pH values.[59] Two prevailing mechanisms are postulated, one dominating under acidic conditions, and another dominating under basic conditions. They are distinguished in two ways. The reaction under acidic conditions is slower, and secondary alcohols are oxidized faster than primary alcohols. The reactions under basic conditions are faster and show the reverse selectivity; primary alcohols are oxidized faster than secondary alcohols. The pH dependence of the rate constants for the oxidations of methanol, ethanol, and 2-propanol has been determined and is reproduced in Fig. 2.[59]

Three rate regimes for the primary alcohol, ethanol, and the secondary alcohol, 2-propanol, can be distinguished.[59] At pH values below 0, the reaction is greatly retarded. Above pH 0, each alcohol displays a pH-independent regime, with the secondary alcohol oxidizing significantly faster than the primary alcohol. Above pH 4 for 2-propanol and pH 1 for ethanol, the oxidation rate becomes much more pH-dependent. Methanol reacts faster than ethanol above pH 3. Above pH 4, the relative rates for primary and secondary alcohols are reversed. In addition,

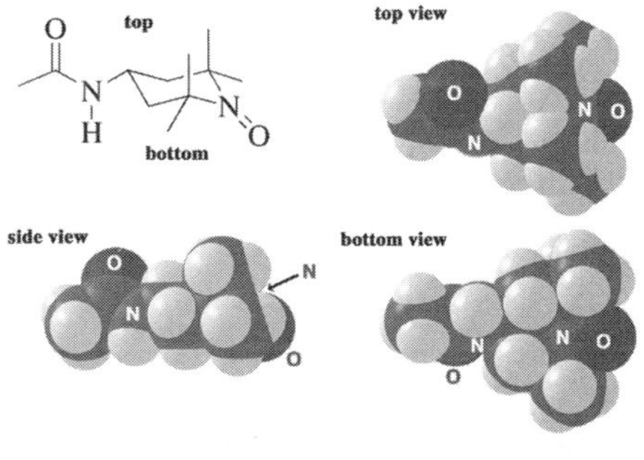

Figure 1. Top and bottom views of a space-filling model of $[AcNH\text{-}TeMP{=}O]^+[BF_4]^-$, as determined by single crystal X-ray diffractometry.[58] The counter-anion is omitted for clarity. The atoms are represented as 100% van der Waals radii.

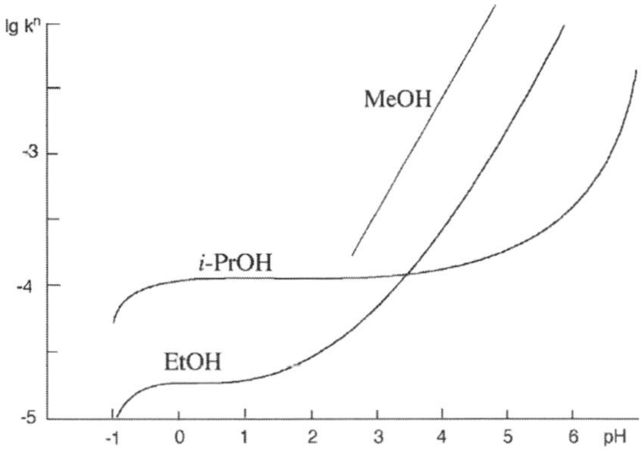

Figure 2. Variation of the log k^n with pH for the oxidation of MeOH, EtOH (primary alcohols), and *i*-PrOH (secondary alcohol) by $[TEMP{=}O]^+Cl^-$ or ClO_4^- at rt. Data reproduced from Golubev et al.[59]

deuterium isotope studies establish that the rate-determining step of the oxidations is the breaking of the carbon–hydrogen bond (Eq. 9), possibly in the form of a hydride abstraction.[59]

The other comprehensive mechanistic study of alcohol oxidations uses an electrochemical system in acetonitrile and 2,6-lutidine to generate the oxoammonium salt.[60] Confirming the earlier results,[59] deuterium isotope effect studies also prove the rate determining step to be the carbon–hydrogen bond cleavage (hydride abstraction). Other mechanistic studies have all been on nitroxide-catalyzed reactions in aqueous or two-phase systems in the presence of base, with pH ranges between 8.5 and 11.5.[61] A coulometric study of alcohol oxidation at pH values ranging from 4 to 5 is in general agreement with the findings of Golubev.[59,62] In the following sections, the various mechanistic proposals are discussed in more detail.

Mechanisms in Acidic or Neutral Media

The original discovery of stoichiometric, oxoammonium-salt alcohol oxidations indicated that the relative reactivity of the substrates is in the order secondary alcohol > primary alcohol > methanol,[8,59] and that the relative reaction rates are in accord with the known strengths of the carbon–hydrogen bonds on the hydroxyl-bearing carbon. A hydride abstraction, possibly assisted by the hydroxyl group, is suggested (Eq. 12). This type of mechanism, which is supported by B3LYP/6-31+G* molecular mechanics calculations, is consistent with the fact that 2-propanol is oxidized faster than methanol.[52]

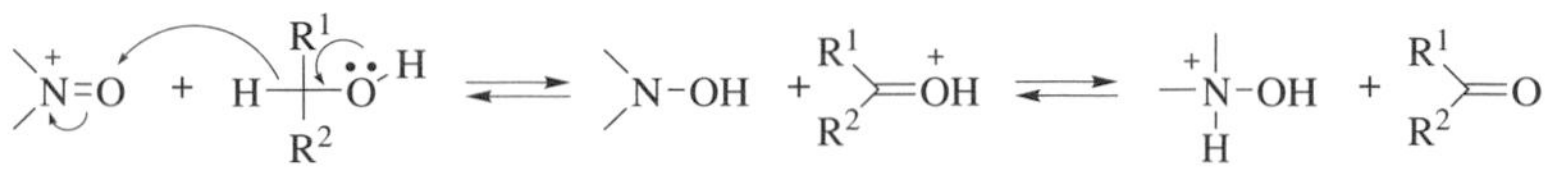

(Eq. 12)

Similar hydride abstraction reactions have been proposed to explain hydrocarbon oxidations and oxidations of alcohols with nitrosonium tetrafluoroborate ($[NO]^+[BF_4]^-$), a close relative of oxoammonium salts.[63–65] Hydride abstractions have also been suggested as the first steps in the nitroxide-catalyzed electrooxidation of double bonds to alkenones[66] and some aromatization oxidations.[67]

A cyclic transition structure can be proposed which contains a hydrogen bond between the hydrogen of the alcohol and the nitrogen lone pair, thus preorganizing the reactants toward the hydride transfer step (Eq. 13).

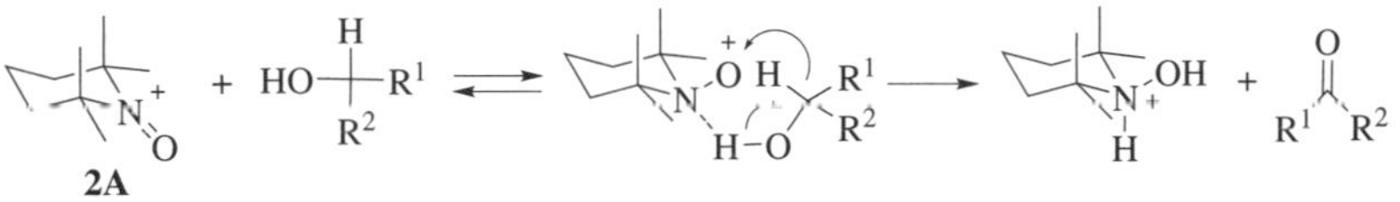

2A

(Eq. 13)

The counter-ions of the oxoammonium salts have an effect on the rate of the oxidation.[18] Chloride ions appear to cause much faster reaction rates than do perchlorate or tetrafluoroborate ions, which, in turn, cause faster reactions than bromide. Explanations of these effects are not clear. The reaction rates using perchlorate or tetrafluoroborate salts are appreciably enhanced in the presence of silica gel, but the mechanism of this rate enhancement is also not known.[68]

A perplexing fact is the almost complete lack of reactivity of alcohols containing a β-oxygen or a β-nitrogen toward oxoammonium salts in acidic media.[68,69] If hydrogen-bonding between the hydroxyl group being oxidized and the oxidant is inhibited by intramolecular hydrogen-bonding with a β-oxygen, as shown in **4**, the reaction would be prevented or significantly retarded. On the other hand, whether such a mechanism would prevail in the strongly acidic or aqueous media is problematic. In oxidations using nitroxides with toluenesulfonic acid as the disproportionation reagent, the β-oxygen problem is present with simple compounds,[56] but not in more complex examples.[70] The β-oxygen problem does not seem to occur in the catalyzed reactions, which are primarily carried out in basic media.

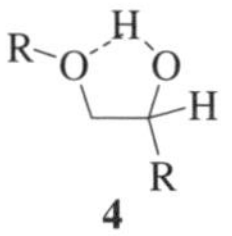

4

Mechanisms in Basic Media

Almost all of the nitroxide-catalyzed oxidations have been carried out in basic media. Attack of a hydroxide ion as the nucleophile on the nitrogen or oxygen of the nitrosonium group (Eq. 11) leads to intermediates **5** and **6**, respectively. There is, in fact, evidence for the formation of both types of intermediates. Reaction of an oxoammonium salt with concentrated aqueous NaOH gives the sodium salt of **5**, although the intermediate hydroxy compound could not be isolated (Eq. 14).[71,72]

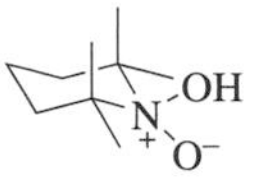

5

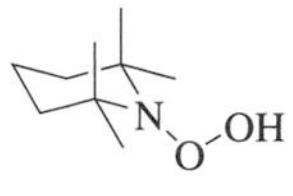

6

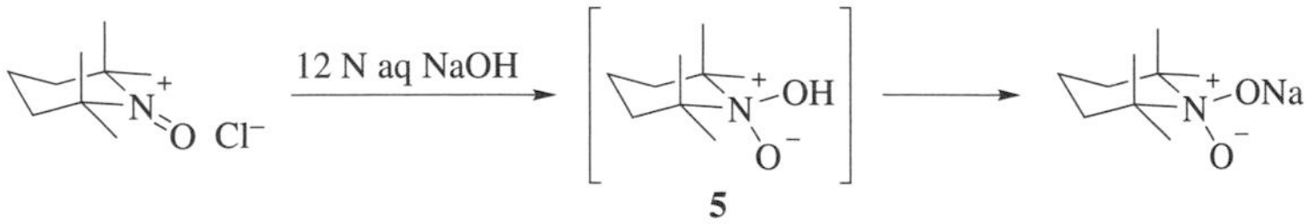

5

(Eq. 14)

With more dilute hydroxide solution, in both aqueous and non-aqueous systems, the reaction produces TEMPO (nitroxide) and hydrogen peroxide (Eq. 15).[73] A mechanism based on an intermediate analogous to that of structure **6** may operate.

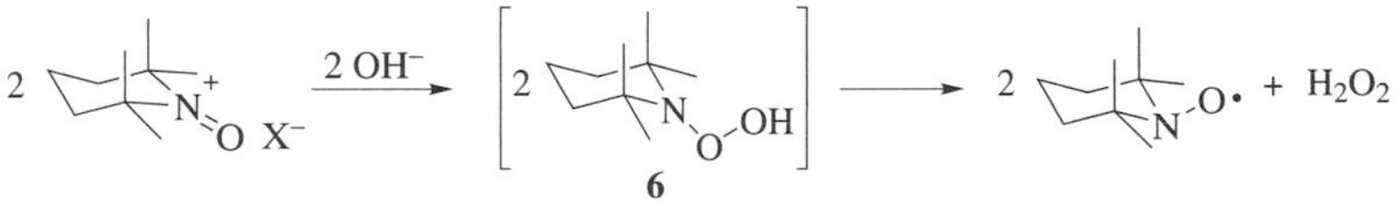

6

(Eq. 15)

Oxidations of alcohols in base involve an alkoxide as the nucleophilic species.[52,59] The alkoxides are formed in equilibrium with the base and may react analogously to the hydroxide ions in Eq. 15. Thus, for electrochemical nitroxide-catalyzed oxidations in base (2,6-lutidine) the mechanisms shown in Scheme 2 have been proposed.[60] Preference was given to the alkoxide attack on the nitrogen (Pathway A), but attack on the oxygen could not be ruled out (Pathway B). It may be that the oxygen attack is reversible and cannot lead to products. Pathway A shows an intramolecular proton transfer but an intermolecular proton transfer mediated by a base is equally likely. Notwithstanding these ambiguities, the mechanism shown in Pathway A (Semmelhack mechanism) is currently favored.

The mechanism in Pathway A (Scheme 2) is supported by the conversion of an *O*-alkylhydroxylamine to a hydroxylamine (with concomitant formation of a ketone) through a similar intermediate (Eq. 16).[74]

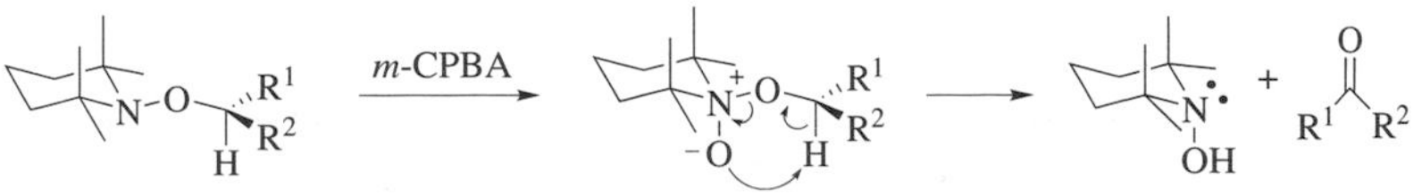

(Eq. 16)

Results from computational studies[52] support the mechanism originally suggested in Pathway A of Scheme 2.[60] It appears that the oxidation of an alcohol

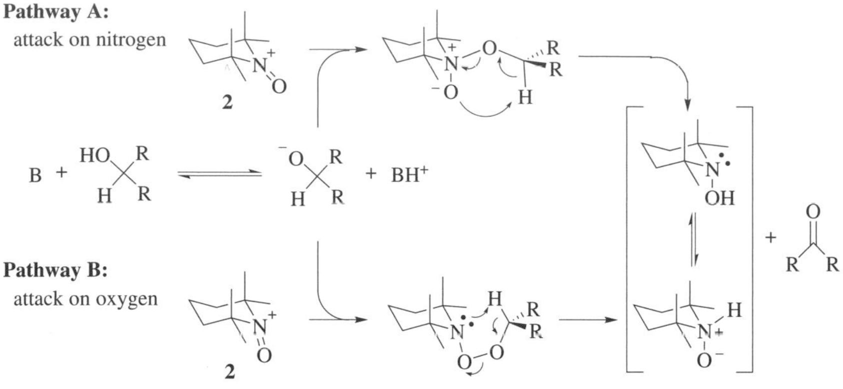
Pathway A:
attack on nitrogen
2
B +
+ BH+
Pathway B:
attack on oxygen
2

Scheme 2

by oxoammonium salt in basic solution involves formation of an intermediate generated by alkoxide attack on the electrophilic nitrogen of the oxoammonium cation. The favorable equilibrium constants for the formation of these complexes decrease as the steric bulk of the alkoxide increases, and the differences in complex stability serve to explain the observation that primary alcohols are oxidized more rapidly than are secondary alcohols.[59] In light of the relative steric hindrance around the nitrogen (Fig. 1), this is quite plausible.[75]

Studies of stoichiometric oxidations in the presence of pyridine bases have been reported, but the mechanisms are not clear.[76,77] The mechanistic rationalization of the reactions is also complicated by the fact that if pyridines are used as bases, profoundly different reaction profiles are observed. For instance, substrates containing β-oxygens react but form dimeric esters as products.[76,77]

Nitroxide-Catalyzed Reactions

Nitroxide-catalyzed oxidations with a secondary oxidant account for the majority of the oxidation reactions in this chapter. These reactions can be classified into two groups depending on whether the secondary oxidant is a two-electron oxidant, such as aqueous sodium hypochlorite (NaOCl), or a one-electron oxidant, such as Cu(II).

Two-Electron Secondary Oxidants. Nitroxide-catalyzed reactions using a two-electron secondary oxidant are all assumed to proceed through an oxoammonium cation. The same mechanistic implications apply as discussed for the stoichiometric oxoammonium oxidations in the preceeding sections. The catalytic cycle can be visualized as shown in Scheme 3.

The secondary oxidant must be a competent oxidant for two reactions. The first is a one-electron oxidation of nitroxide to oxoammonium ion. The mechanism of this first reaction is not clear when a two-electron oxidant, such as hypochlorite, is used. The oxoammonium ion oxidizes the alcohol in a two-electron pathway to the corresponding carbonyl product, and is itself reduced to a hydroxylamine.

In the second oxidation, the hydroxylamine is oxidized back in a two-electron process to the oxoammonium ion to complete the catalytic cycle.

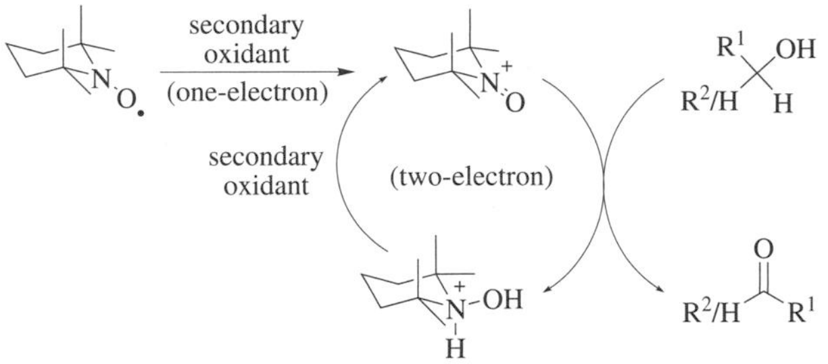

Scheme 3

One-Electron Secondary Oxidants. One-electron secondary oxidants such as Cu(II) in conjunction with oxygen pose some different problems. Mechanistically, the difficulty is to account for the exact role of dioxygen. At least two possibilities can be considered, the oxygen oxidation of Cu(I) to Cu(II)[78] and the oxygen oxidation of hydroxylamine to nitroxide.[79,80]

The first mechanistic possibility involves the oxygen oxidation of the Cu(I)/Cu(II) couple (Eqs. 17–21).[81] This sequence involves four key steps. The first step is the oxidation of a nitroxide by Cu(II) to an oxoammonium salt (Eq. 17). This salt oxidizes the alcohol along any of the mechanisms detailed above, thus generating the carbonyl product and the hydroxylamine (Eq. 18). The hydroxylamine undergoes the well-known comproportionation reaction (Scheme 1) with another equivalent of oxoammonium salt to form two equivalents of nitroxide (Eq. 19), closing the catalytic cycle. Finally, oxygen (air) oxidation of Cu(I) regenerates Cu(II) (Eq. 20), a reaction parallel to a key step in the industrial Wacker process.[78] Thus, overall, oxygen oxidizes the alcohol (Eq. 21).

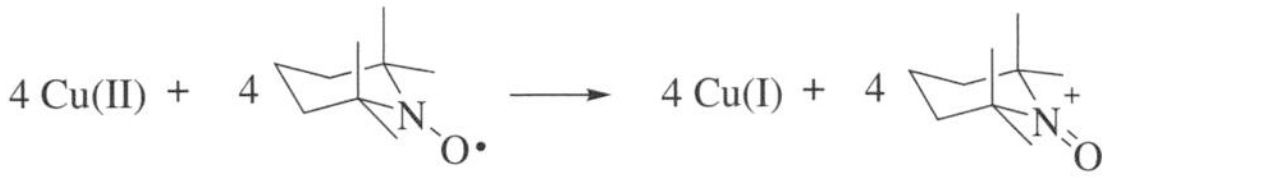 (Eq. 17)

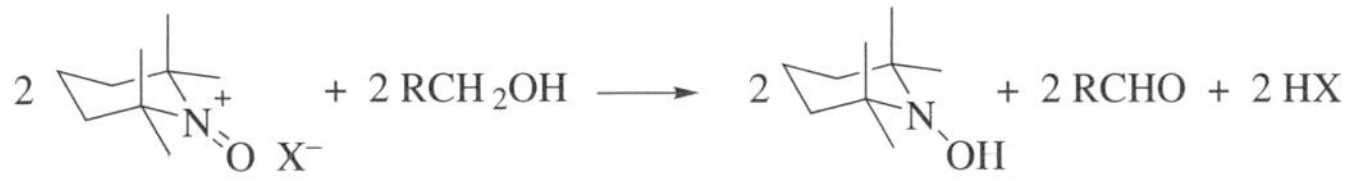 (Eq. 18)

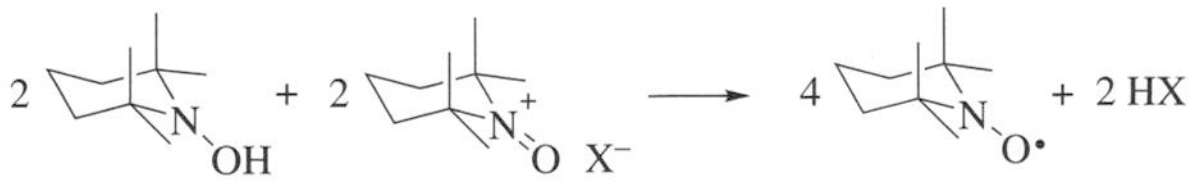 (Eq. 19)

$$4\,Cu(I) + O_2 + 4\,H^+ \longrightarrow 4\,Cu(II) + 2\,H_2O \quad \text{(Eq. 20)}$$

$$2\,RCH_2OH + O_2 \longrightarrow 2\,RCHO + 2\,H_2O \quad \text{(Eq. 21)}$$

A second mechanism has recently been suggested that may rationalize the catalysis by Cu(II) and other metal complexes (Scheme 4).[24,82] In this catalytic cycle, oxidation of the alcohol is brought about by a Cu(II) complex of the nitroxide, rather than an oxoammonium ion.[24,82] Suggested structures for these complexes and the way that they may react are shown in Eq. 22. Next in the cycle, oxygen (air) oxidation causes the oxidation of hydroxylamine to nitroxide, a known reaction.[27]

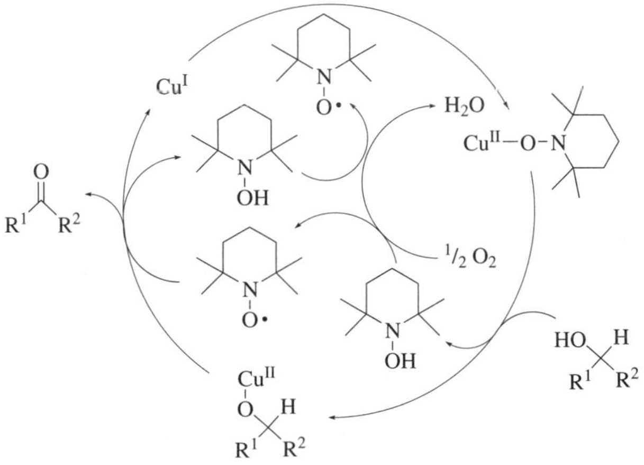

Scheme 4

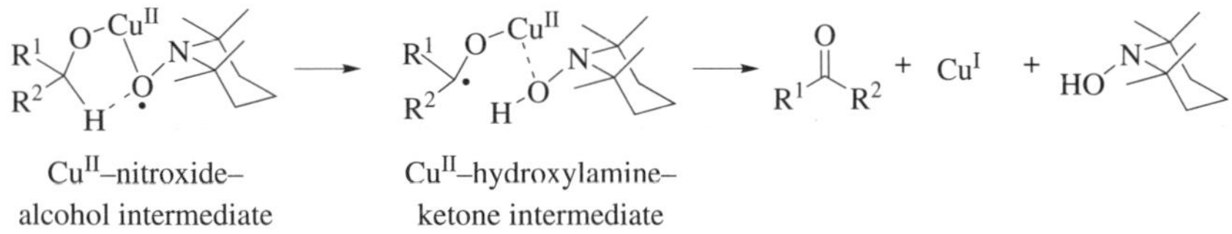

(Eq. 22)

Stereochemistry

In general, oxoammonium ion oxidations do not cause epimerization or double bond isomerization. Alcohol oxidation involves the conversion of an sp^3 carbon to an sp^2 carbon (Eq. 23), thus, if that carbon is a stereocenter, that feature is lost, not created as in most stereoselective reactions. Nevertheless, two possibilities exist in which a stereoselective reaction can take place. Both involve the selective destruction of one stereocenter of an enantiomeric pair.

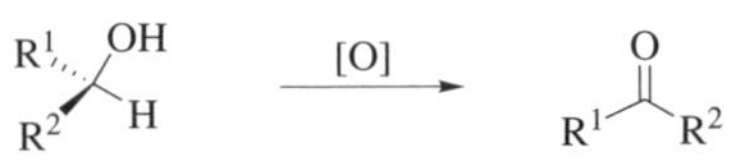

(Eq. 23)

In the first example of a stereoselective oxidation, one enantiomer of a racemic mixture is selectively oxidized in a kinetic resolution (Eq. 24). Kinetic resolutions using chiral nitroxides or chiral auxiliaries have been the major thrust of stereochemical investigations in oxoammonium chemistry, but, in such reactions, the maximum yield of enantiopure material is 50%.

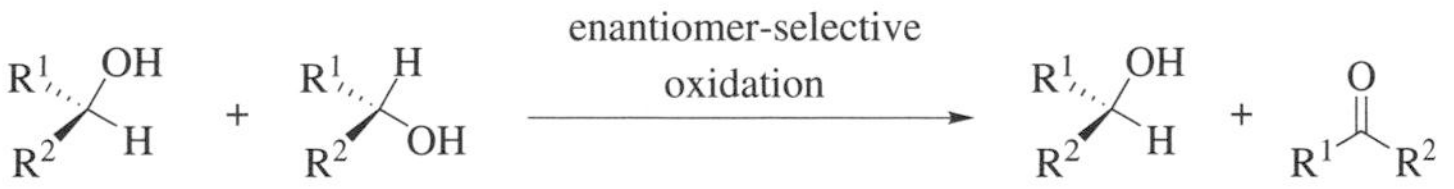

(Eq. 24)

The second example of a stereoselective reaction is the enantiotopic group selective oxidation of one alcohol in a *meso*-diol (Eq. 25). The reaction of one stereocenter in a meso-compound allows, theoretically, a 100% conversion to a chiral product and is therefore of greater interest. A possible extension of this reaction might involve the selective oxidation of one hydroxyl group in an enantiomerically pure polyol. However, this latter reaction has been only briefly explored.[83–85]

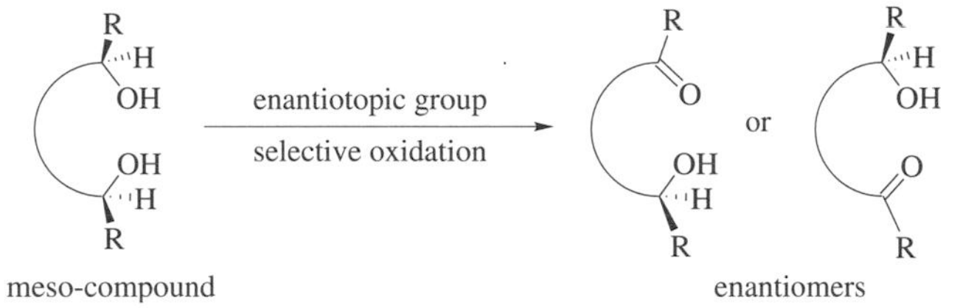

(Eq. 25)

The data on these two reactions are treated in different ways. The success of both types of reactions is expressed in terms of enantiomeric excess (ee), which is defined by Eq. 26, where [R] and [S] are the concentrations of the two enantiomers. The ee is measured by optical activity or calculated using chiral GC or HPLC analysis. For kinetic resolutions, these values refer to the enantiomerically enriched material remaining at any stage of the reaction. For stereoselective reactions, such as the oxidation shown in Eq. 25, the ee is measured at the end of the reaction.

$$\text{ee} = \frac{[R] - [S]}{[R] + [S]} \times 100 \qquad \text{(Eq. 26)}$$

Kinetic Resolutions. Kinetic resolutions represent a special case since the chirality as well as the functionality of one enantiomer is being destroyed.[86,87] In fact, both enantiomers are being oxidized, albeit at different rates, k_1 and k_2. The result of this is that the ee will increase with the extent of the reaction or conversion, represented as C. However, if the rate difference between k_1 and k_2 is small, high ee values are achieved only after the vast majority of both enantiomers have been consumed, thus not presenting a practical solution to the preparation of enantiopure material. This problem of weighing high ee's versus conversion is dealt with using a selectivity factor, or *s* factor. Further, the *s* factor is constant for any given reaction pair and is, in fact, the ratio of k_1 and k_2 (Eq. 27). The *s* factor can be derived from the ee at any given conversion (Eq. 27). The values of C and ee are expressed as fractions in this calculation, rather than as percentages. The rates for the enantioselective reactions are not usually measured. Calculations using the values of ee, C, and *s* can be made using a web-based tool.[88,89] The great value of the *s* factor is that once it is established, it can be used to calculate the best conversion values to give a maximum ee with a maximum yield of enantiomerically enriched material. The *s* value varies from 1 to infinity, but, in theory, a value of 10 will yield an ee of 67% at 50% conversion, an ee of 86% at 60% conversion, and an ee of 98% at 70% conversion. Again, higher conversions imply that one obtains smaller amounts of more enantiomerically enriched product.

$$s = \frac{k_1}{k_2} = \frac{\ln(1-C)(1-ee)}{\ln(1-C)(1+ee)} \qquad \text{(Eq. 27)}$$

In summary, data for kinetic resolutions should be presented either as ee with the corresponding conversion value, or with *s* values alone. It is suggested that the *s* value for any given case be measured at more than one concentration.[89,90] Some authors have given *s* values along with ee and conversion values, and some have not.

Stereoselective oxoammonium salt oxidations can, in principle, be carried out using chiral oxoammonium salts or nitroxide catalysts, or by using an achiral nitroxide in the presence of a chiral auxiliary, or using an electrode surface modified with a chiral nitroxide. Irrespective of the methods used, success in achieving high *s* values has been limited so far.

Only one chiral oxoammonium salt, **7**, has been prepared from the (+)-enantiomer of the corresponding nitroxide,[91] and it showed no enantioselectivity in a number of oxidations.[85,92] Given the large number of known chiral nitroxides,[27,90,93–101] few have been investigated as oxidation catalysts. The (+)-dihydrocarvone-derived nitroxide **8** has been prepared as a mixture of four diastereomers (corresponding to the isomers at the two starred positions), which have been laboriously separated by chromatography.[85] In the partial oxidation of racemic 1-phenylethanol, only one nitroxide isomer, **8**, is sufficiently stable during the oxidation. The nitroxide-catalyzed oxidation of racemic 1-phenylethanol using *m*-chloroperoxybenzoic acid (*m*-CPBA) as a secondary oxidant with nitroxide **8** shows no stereoselectivity.[85] However, the acid-induced in situ disproportionation[56] reaction of **8**, generating the corresponding oxoammonium salt (Eq. 5), shows an approximate 20% ee at 34% conversion of 1-phenylethanol, corresponding to an *s* value of about 3.[85]

Reactions with the chiral binaphthyl derivative **9** have been more successful.[90] Treatment of racemic 1-phenylethanol with sodium hypochlorite/potassium bromide as the secondary oxidant, in the presence of catalytic amounts of nitroxide **9**, results in the *S*-isomer of the phenylethanol being preferentially oxidized with *s* values of 5–7. The values are measured at two conversions, (Eqs. 28, 29) and the *s* value associated with the higher conversion is thought to be more accurate.

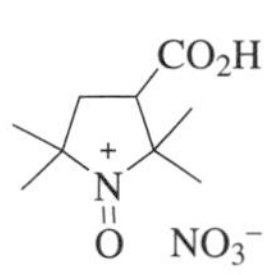

from (+) nitroxide
7

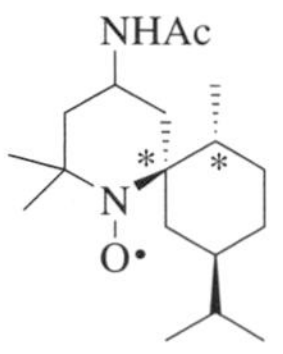

8

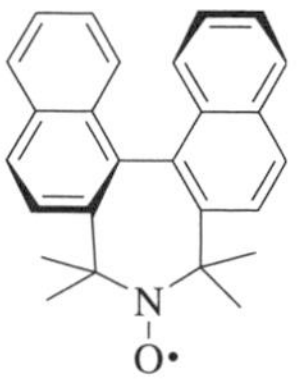

9

The results were rationalized by transition structure calculations.[90] However, compared to reactions with TEMPO, the reactions with nitroxide **9** are slow, and only activated alcohols (benzylic, allylic, or propargylic) are satisfactorily oxidized. For a series of substrates, *s* values vary from 1.5 to 7.1.[90,103,170]

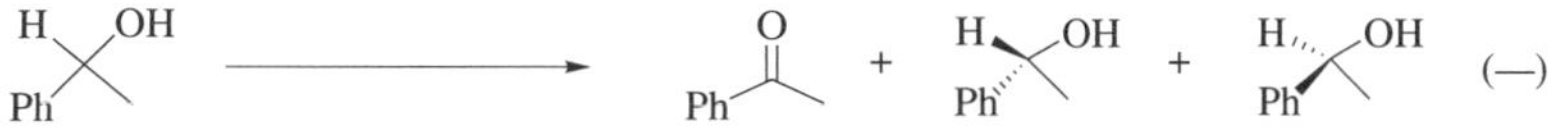

racemate

R-enantiomer *S*-enantiomer

	C	% ee	*s*	Predominant Enantiomer	
9 (cat. 0.5%), NaOCl, KBr (cat.) → CH_2Cl_2, H_2O, 0°, 30 min	69%	81	5.0	*R*	(Eq. 28)
9 (cat. 1.0%), NaOCl, KBr (cat.) → CH_2Cl_2, H_2O, 0°, 30 min	87%	98	7.1	*R*	(Eq. 29)
15 (1.5 eq), NaOCl, KBr (cat.) → CH_2Cl_2, H_2O, 0°, 30 min	83%	67	2.3	*S*	(Eq. 30)
8 (cat.), $NaClO_4$ → 2,6-lutidine, +0.8 V vs. Ag/AgCl, MeCN, H_2O	64%	70	4.6	*R*	(Eq. 31)

From the series of C_2-symmetric nitroxides **10–13**, only nitroxide **10** shows any stereoselectivity when coupled with bis(acetoxy)iodobenzene as secondary oxidant for the oxidation of a series of secondary benzylic alcohols. The *s* values vary between 1.6 and 2.5.[94] A further drawback is that the nitroxides are not very stable and must be prepared just before use by *m*-CPBA oxidation of the appropriate amines.[102]

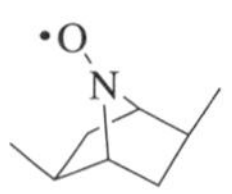

10

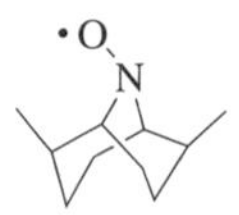

11

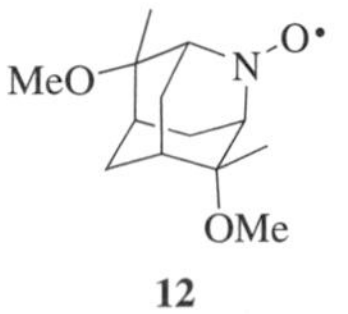

12

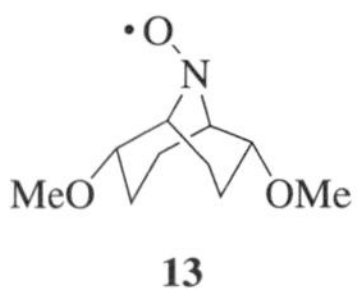

13

A series of peptide derivatives prepared from the 4-amino-4-carboxylic acid derivative of TEMPO **14** (TOAC), has been used in nitroxide-catalyzed oxidations of 1-phenylethanol with NaOCl.[103] Compound **15** gives the best result with reported values for ee, C and *s* (Eq. 30). The *s* values for other similar nitroxides in the same oxidation range from 1.2–2.7.[103] Electrochemical oxidations give similar results.[103] Lower temperatures (−10°) raise the *s* values to around 3. These low *s* values may be explained by the relatively long distance from the chiral environment to the active site of the oxidant, but it is remarkable that any stereoselectivity was observed.

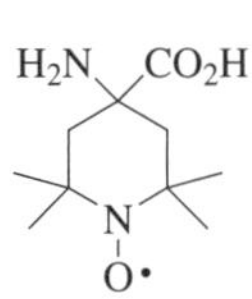

14

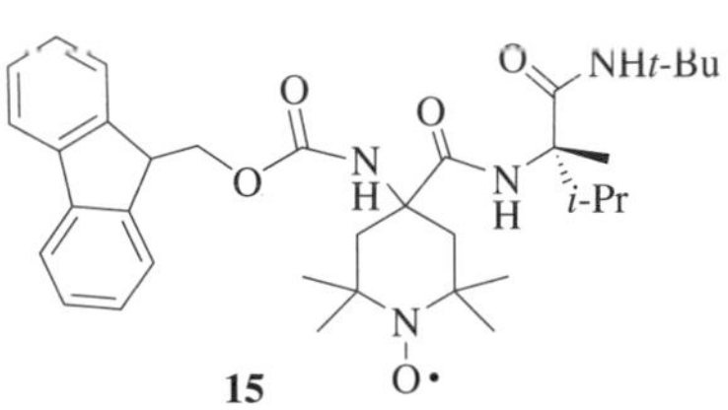

15

A different approach employs a strong, chiral acid, (+)-(1*S*)-camphor-10-sulfonic acid to induce the in situ formation of an oxoammonium salt by disproportionation of a non-chiral nitroxide (see Eq. 5).[56] However, no selectivity is observed.

Many kinetic resolutions have been carried out using electrochemical oxidation as the secondary oxidant. These have been studied using enantiomerically enriched non-nitroxide chiral auxiliaries, enantiomerically enriched nitroxide catalysts in solution, and enantiomerically enriched nitroxides bound to electrode surfaces.

Sparteine (**16**) has been used as a chiral auxiliary in the oxidation of alcohols, aromatic ethers, and phenols using TEMPO bound to a graphite-felt electrode,[104,105] resulting in good ee and *s* values. However, this work has been challenged in that sparteine is oxidized under the experimental conditions.[106]

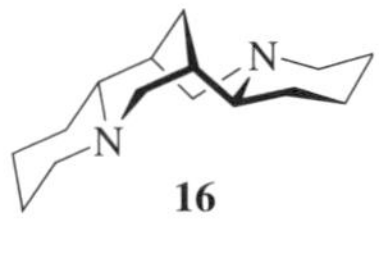

16

Oxidations on a graphite-felt electrode using catalyst **8** in solution in acetonitrile give kinetic resolutions of racemic benzylic alcohols[107] (for example see Eq. 31) and amines[108] with *s* values ranging from 4–6. The amines are oxidized to ketones.

Electrooxidations of secondary benzylic alcohols using **9** as catalyst at low temperatures (−15°) give *s* values ranging from 2 to 20.[109] Nitroxide-catalyzed electrooxidation of 2-(4-chlorophenyl)ethanol using **9** gives an *s* value of 12.[84] Electrooxidations of alcohols[110] and amines[111] using catalyst **17** give satisfactory yields of products, but no stereochemical selectivity is observed. Enantiomerically

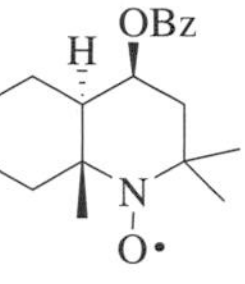
H
OBz
N
O•

17

Bz = benzoyl

enriched nitroxides and chiral electrodes have been shown by voltammetry to oxidize the enantiomers of 1-phenylethanol at different rates, thus showing a possibility for kinetic resolutions.[112,113]

A chiral TEMPO derivative of galactose was embedded into a polymer matrix, and the galactose moiety was removed to yield a polymeric nitroxide, which showed no selectivity toward sugar substrates.[74]

Other Stereoselective Oxidations. Few examples of the selective oxidation of a *meso*-diol of the type shown in Eq. 25 are known. The four diastereomers of nitroxide **8** have been used for the conversion of *cis*-1,2-cyclohexanedimethanol (**18**) into chiral lactones using *m*-CPBA as the secondary oxidant. The specific stereoisomer **8** gives 38% ee (Eq. 32).[85] The other three stereoisomers of **8** give ee's of 6–16% but are unstable under the experimental conditions. A chiral electrode surface prepared from a derivative of **8** was used to oxidize diol **18** to lactone **19** with 82% ee (Eq. 33).[83] Nitroxide **9** has been used in an electrochemical system with diol **18** to give **19** in 89% yield and 74% ee.[84]

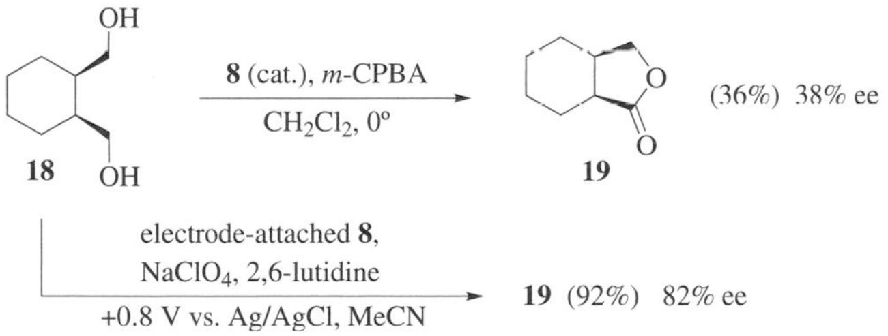
OH
OH
18
8 (cat.), m-CPBA
CH_2Cl_2, 0°
O
O
19
(36%) 38% ee
electrode-attached 8,
$NaClO_4$, 2,6-lutidine
+0.8 V vs. Ag/AgCl, MeCN
19 (92%) 82% ee

(Eq.32)

(Eq.33)

Site Selectivity

The site selectivity is mainly a matter of oxidation rates, i.e. whether primary or secondary alcohols are present and whether they are activated by being in benzylic or allylic positions. Some of these factors are considered above and are detailed below in the "Scope and Limitations" section. However, some generalizations can be made. Activated alcohols such as benzylic or allylic alcohols react faster than unactivated alcohols.[11,68] Under acidic or near acidic conditions (mainly using oxoammonium salts as stoichiometric oxidants), secondary alcohols react a little faster than primary alcohols.[68] Under basic conditions (mainly in the nitroxide-catalyzed reactions), primary alcohols react much faster than secondary alcohols.[23,61] Coulometry can be used to study reaction rates in water.[62]

SCOPE AND LIMITATIONS

Stoichiometric Oxoammonium Oxidations

Oxidations of alcohols and other substrates with stoichiometric quantities of preformed oxoammonium salts have been reported under neutral or slightly acidic conditions (Eq. 3) and under basic conditions (Eq. 4). These reactions have been largely carried out using three TEMPO-derived cations, $[TEMP{=}O]^+$ (**2**),[4] 4-methoxy derivative **20** ($[MeO\text{-}TEMP{=}O]^+$),[11] and 4-acetamido derivative **21** ($[AcNH\text{-}TEMP{=}O]^+$).[68] The anions most used are chloride, bromide, tribromide, perchlorate, and tetrafluoroborate. However, the nitrate salt of **2** can also be prepared and is more stable than the chloride.[114] Chlorite salts have also been prepared and give mixtures of aldehydes and acids from primary alcohols.[115] Other anions are known but their salts have not been explored as oxidants.[10] All of the stoichiometric oxidation rates are enhanced in the presence of silica gel.[68] Nitroxide precursors leading to all three oxidants are commercially available. Specific syntheses are discussed below and in the "Experimental Procedures" section. 4-Acetylamino-2,2,6,6-tetramethyl-1-oxo-piperidinium tetrafluoroborate (**21**, $X = BF_4^-$) is commercially available from TCI America. Because the oxidation reactions in acid and base are distinctly different, they are considered separately.

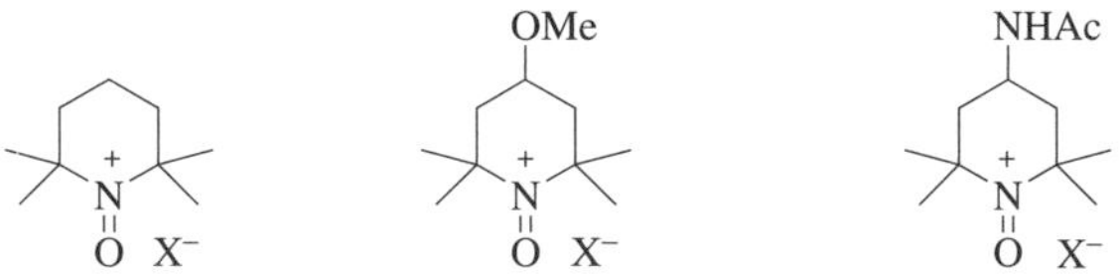

2 $[TEMP{=}O]^+X^-$ **20** $[MeO\text{-}TEMP{=}O]^+X^-$ **21** $[AcNH\text{-}TEMP{=}O]^+X^-$

Common anions (X^-) are Cl^-, Br^-, Br_3^-, ClO_4^-, BF_4^-, NO_3^-, ClO_2^-, and SbF_6^-.

Preparation of Oxoammonium Salts. The oxoammonium chlorides and bromides (and tribromides) are prepared by oxidation of the nitroxides with chlorine or bromine.[11,116] The perchlorates and tetrafluoroborates are made by variations of the acid-disproportionation/oxidation reaction sequence (Scheme 1).[4,5,12,68] The nitrate salt of **20** is prepared using the same method.[114,117] Oxoammonium nitrate salts have also been prepared by reaction of the nitroxides with nitrogen dioxide in nitrogen.[85,92,118] Tetrafluoroborates have been prepared, but not isolated, using nitrosonium tetrafluoroborate.[119] These procedures have been reviewed,[10,20] and two representative examples are provided in the "Experimental Procedures" section. Recently, the hexafluoroantimonate salts were also prepared using fluoroantimonic acid ($HSbF_6$).[120]

The ideal oxoammonium oxidant with respect to product isolation and recovery of the oxidant is attached to a polymeric, swellable but insoluble matrix. To our knowledge, the only such reagent is the polystyrene-based oxoammonium salt

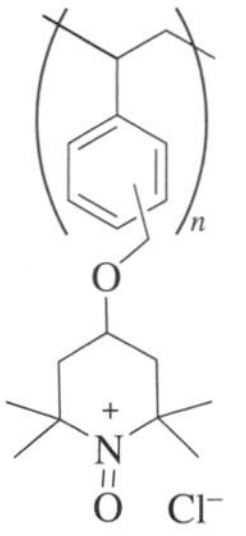

22

22.[121] However, it has a half-life of only about one week. Polymeric nitroxides are discussed in detail below.

The specific anion used in these reactions plays an important, but as yet not understood role. The reaction rates of a given oxoammonium salt/substrate combination are faster if halides are used as compared to the perchlorate or tetrafluoroborate.[18] The size of the anion plays a role in the rearrangement and oxidation of tertiary allylic alcohols.[120] Large ions such as tetrafluoroborate and hexafluoroantimonate function in these reactions whereas small anions such as chloride and tribromide do not. In an unexplained observation,[11,18] bromides or tribromides promote faster reactions with secondary alcohols than with primary alcohols, whereas chlorides show faster reactions with primary alcohols. Although the perchlorate salts are widely used, they are unsafe, and the use of tetrafluoroborates or halides is recommended.[122] Practical considerations such as solubility and the diminished hygroscopicity of the tetrafluoroborates as compared to the halides play a major role in their selection.[123,124] Although salts of many other anions are known,[10] they have not been used, with the exception of *p*-toluenesulfonates in conjunction with the oxoammonium salts that are prepared by acid-catalyzed disproportionation reactions (Eq. 5; see also below).

The bromides of oxoammonium salts present some problems. In at least two cases, monobromides are reported as products of the bromination of nitroxides.[8,11] However, in other cases, only tribromides are isolated, regardless of the amount of bromine used.[125,126] The tribromides are reported to oxidize secondary alcohols exclusively over primary alcohols (in acetonitrile).[125] Finally, the tribromides can be a source of bromine, and thus are known to brominate acetone.[127]

Oxoammonium salts can usually be recycled because the hydroxylamine can be recovered from the oxidation reaction and the oxoammonium salt can be regenerated by oxidation.[20,68]

Stability of Oxoammonium Salts. Although oxoammonium perchlorates are well known and have been used in many oxidations,[68] the perchlorate of compound **21** detonated on vacuum drying.[122,123] However, the tetrafluoroborate salt of **21** possesses the same oxidative properties as the perchlorate, is nonhygroscopic, and is stable indefinitely.[12] Generally, oxoammonium salts have a reputation for being hygroscopic and unstable upon storage for long periods of time.[11,124,128,129] These traits are probably more true for the halide salts of simple TEMPO derivatives. The tetrafluoroborates appear to be more stable and non-hygroscopic.[4,68,124]

Oxidations Using Neutral or Mildly Acidic Reaction Conditions. Although various reaction conditions have been used, a specific reaction using [AcNH-TEMP=O]$^+$[BF_4]$^-$ (**23**) is shown in Scheme 5 that also illustrates its practical advantages.[68] The reactions are usually carried out in methylene chloride in which this specific oxoammonium salt is slightly soluble whereas the reduced byproduct, the hydroxylamine salt **24**, is insoluble. The reactions are carried out with a slight excess of oxidant. Although this reaction takes place with no catalyst, the rates are greatly increased in proportion to the amount of silica gel used.[68] A preformed mixture of silica gel and **23** (1:1 mixture by weight, made by simply grinding them together) is ideal. The reaction has three advantages: It is colorimetric in that the yellow oxoammonium salt **23** is converted into the white hydroxylammonium salt **24**. Product isolation is usually a simple filtration through a thin pad of silica gel and salt **24** can be readily recovered and converted back into the oxoammonium salt **23**. The alcohol oxidation reaction does not require anhydrous solvents or dry salt, but also works equally well under anhydrous conditions. In fact, this reaction is ideal for the preparation of small amounts of volatile aldehydes under anhydrous conditions. Normally, the reaction is high yielding and the resulting aldehydes are quite pure such that the solution of crude aldehyde can be carried on to the next step of a reaction sequence without purification.

Scope. The oxidation shown in Scheme 5 is applicable to many types of alcohols.[68] Reaction rates vary, with allylic and benzylic alcohols being the fastest, followed by propargylic alcohols and aliphatic alcohols. Secondary alcohols are oxidized about twice as fast as comparable primary alcohols although the rate difference is not sufficient to oxidize secondary alcohols selectively. Double bonds in aliphatic alcohols do not isomerize during the reaction (Eq. 34).[12,68,130]

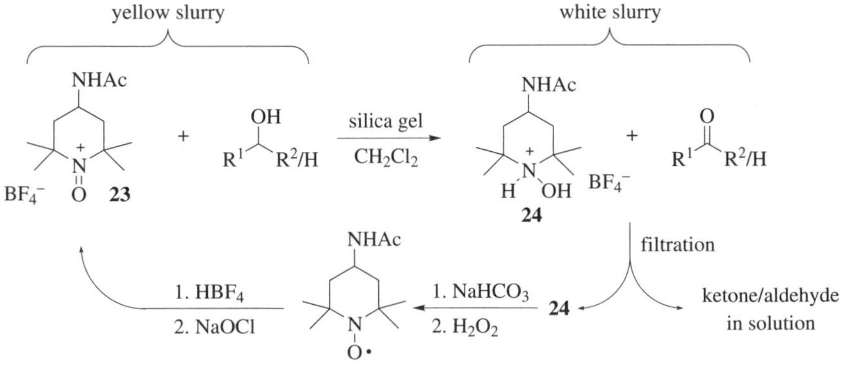

Scheme 5

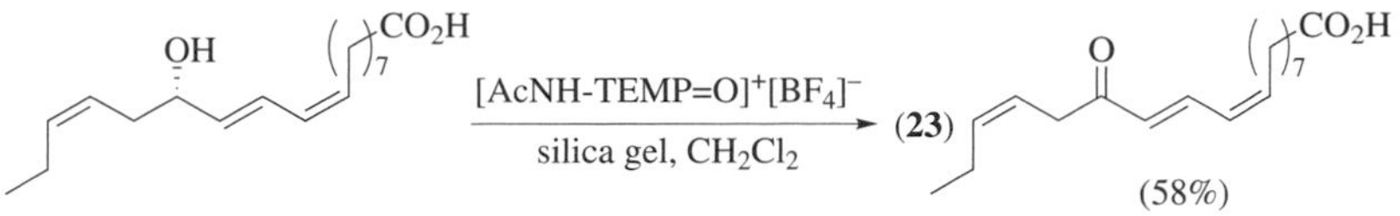

(Eq. 34)

A series of long-chain primary alcohols containing various degrees of unsaturation were oxidized to aldehydes using $[AcNH\text{-}TEMP{=}O]^+[BF_4]^-$.[131] The method was compared directly with pyridinium chlorochromate (PCC) oxidation. In general, yields with the oxoammonium salt were better and allowed a more convenient product isolation. Problems were encountered with polyunsaturated alcohols, which can be partially alleviated by carrying out the reactions in the presence of pyridine (see below).

Oxidations of benzylic alcohols can be carried out in the presence of one free phenol group, but not two.[68] An example is shown in Eq. 35.[132] It is not known whether aliphatic alcohols can be oxidized in the presence of phenols.

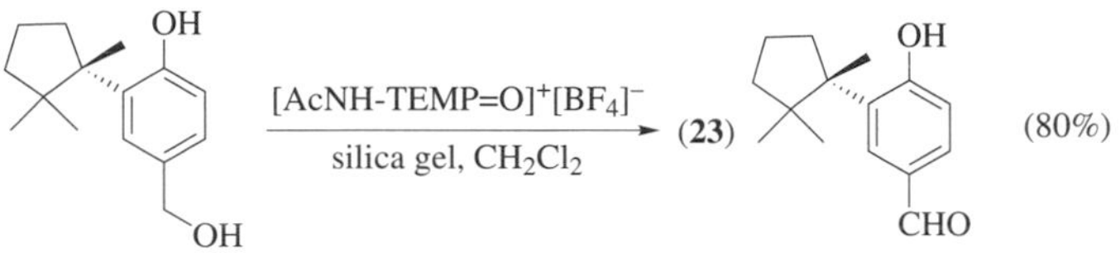

(Eq. 35)

Tertiary allylic alcohols undergo allylic rearrangement and oxidation reaction with oxoammonium tetrafluoroborates and hexafluoroantimonates (Eq. 36).[120] Salts with small anions such as chloride or tribromide do not react, and water plays an important role in some cases. The normal nitroxide-catalyzed reactions also do not take place. A suggested mechanism postulates an allylic alcohol rearrangement that is catalyzed by the oxoammonium ion, followed by oxidation of the rearranged alcohol.[120]

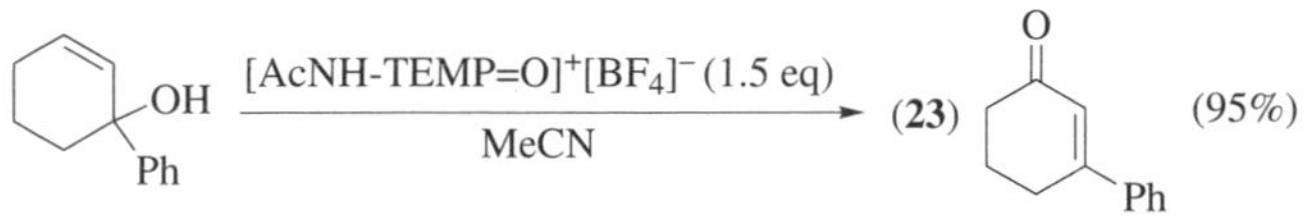

(Eq. 36)

Diols offer some special problems,[133] for example, ethylene glycol is not oxidized in methylene chloride.[56,68] However, this outcome may be a matter of solubility since ethylene glycol is rather insoluble in this solvent, and 2,3-butanediol is reported to be oxidized in methylene chloride to 3-hydroxy-2-butanone (acetoin).[133] 1,3-Diols tend to undergo oxidation followed by a dimerization reaction to form acetals (Eq. 37).[133] If a hydroxyl group is located in a position that allows formation of a stable, cyclic hemiacetal, as is the case for 1,4- and 1,5-diols, then oxidation to the corresponding lactone is observed

(Eq. 38).[133] This type of reaction is common to all of the oxoammonium-salt and nitroxide-catalyzed reactions and is quite useful (see Tables).[134]

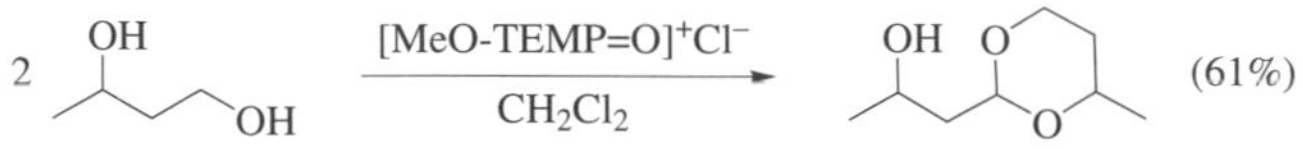

(Eq. 37)

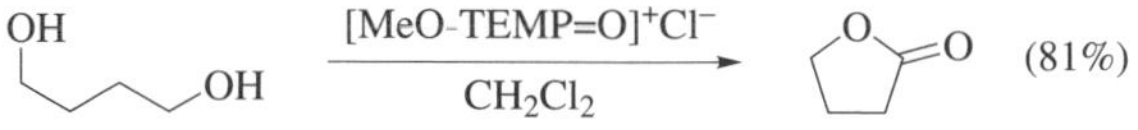

(Eq. 38)

Long-chain terminal diols such as 1,12-dodecanediol and *p*-xylene glycol are oxidized to dialdehydes with [MeO-TEMP=O]$^+$Cl$^-$,[135] as are hydrogenated polybutadiene polymers containing terminal hydroxyl groups,[117] poly(ethylene glycol),[135] and similar polymers.[135] Polyvinyl alcohol and partially saponified polyvinyl acetate polymers are oxidized to materials with varying degrees of ketone functionalization, as shown by ^{1}H NMR spectroscopy.[136]

Limitations. A number of important functional incompatibilities are noted. Oxoammonium salts react rapidly with amines (discussed below) and slowly with benzylic ethers and some double bonds in methylene chloride. *tert*-Butyldimethylsilyl (TBDMS) ethers are slowly oxidized by [AcNH-TEMP=O]$^+$ [BF_4]$^-$,[68] but the *tert*-butyldiphenylsilyloxy (TBDPS) ethers are stable. However, both groups appear to be stable in nitroxide-catalyzed bleach reactions (discussed below).[137] These and similar reactions are discussed in the "Miscellaneous Reactions" section. Problems with double bonds and benzylic ethers are mainly a matter of rates. To illustrate, geraniol (**25**) can be cleanly oxidized to geranial because the allylic alcohol oxidation is fast, but with citronellol (**26**) the less active alcohol oxidation is slow and complicated by appreciable reactions of the trisubstituted double bond (Eq. 39).[57,68] For the same reason, benzylic alcohols containing benzylic ethers can be oxidized satisfactorily, whereas deactivated alcohols containing benzylic ethers partially lose the benzyl group.[138] Phenols are sometimes coupled to furnish biphenyls or are oxidized to quinones under the reaction conditions,[139–141] although all of these reactions have been carried out in acetonitrile, rather than the more conventional methylene chloride.

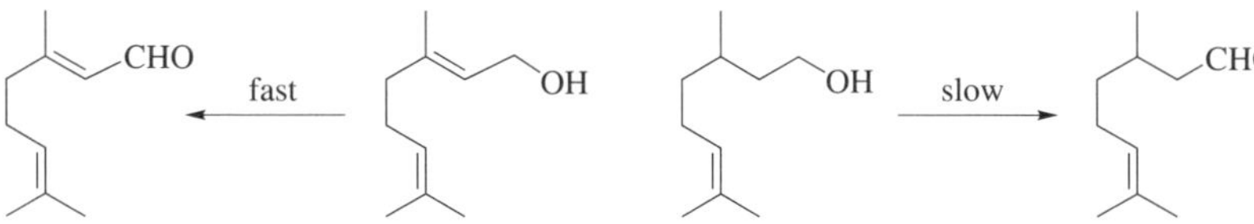

25 **26**

(Eq. 39)

Another significant problem with these oxidations is that alcohols containing a β-oxygen or a β-nitrogen react very slowly, if at all.[11,56,68,142] In fact, any strongly electron-withdrawing group on the β-carbon seems to inhibit oxidation. This limitation can be used as a means of introducing a unique selectivity to this oxidation method. For instance, the two C_2- and C_4-terephthalate side chains carrying a primary alcohol can be differentiated based on the absence or presence of a β-oxygen (Eq. 40).[69] Note, however, that the reaction conditions are not acidic. The use of sodium carbonate as an acid scavenger appears to be unique for these reactions and it is not known what role the sodium carbonate plays, as traditional basic reaction conditions involve pyridine-derived bases.

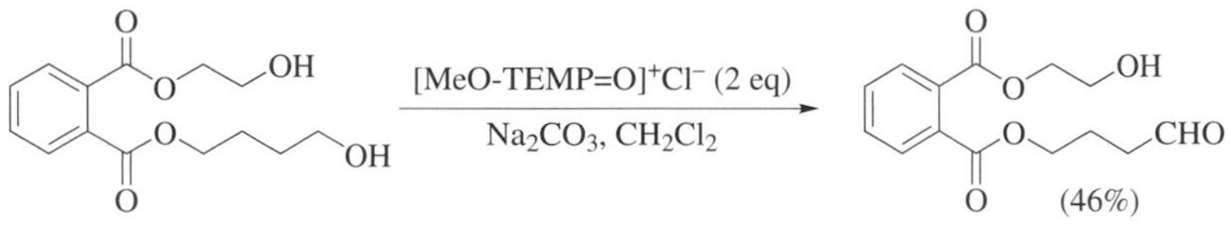

(Eq. 40)

Oxidations Using Basic Reaction Conditions. As is discussed later, bases such as 2,6-lutidine are used in many catalyzed electrochemical reactions, but stoichiometric reactions of oxoammonium salts in conjunction with these bases have been only briefly explored.[76,77] Stoichiometric oxoammonium salt oxidations in pyridine have been carried out indicating that at least one oxoammonium salt, $[AcNH\text{-}TEMP{=}O]^{+}[BF_4]^{-}$, does not react with pyridine itself.[143] Oxidations of unsaturated primary alcohols give better yields in the presence of pyridine than under acidic conditions.[131] In these reactions, two equivalents of salt and two equivalents of pyridine are required, and the products are one equivalent of aldehyde, two equivalents of pyridinium tetrafluoroborate, and two equivalents of nitroxide (see Eq. 4). This situation arises because the hydroxylammonium salt that is formed from the reduction of one equivalent of oxoammonium salt by the alcohol is neutralized by the pyridine. The hydroxylamine thus freed comproportionates with the second equivalent of oxoammonium ion to give two equivalents of nitroxide, as shown in Scheme 1. This side reaction complicates the product isolation somewhat, but conditions have been worked out to accomplish this.[77]

Alcohols containing a β-oxygen (which do not react in simple oxoammonium salt oxidations, see above) do react under these conditions, but surprisingly give rise to dimeric esters in good yields (Eq. 41).[77] Evidently, initial oxidation of the primary alcohol to the corresponding aldehyde is followed by dimerization, which is followed by oxidation to an ester. Intermediate aldehydes are detected in the reaction mixtures only in minute amounts.[77] The mechanism of these reactions is under investigation. These oxidative dimerization reactions in pyridine tolerate the presence of allylic double bonds, acrylic esters, benzylic ethers, cyclopropanes, and dialkylsulfides.[77]

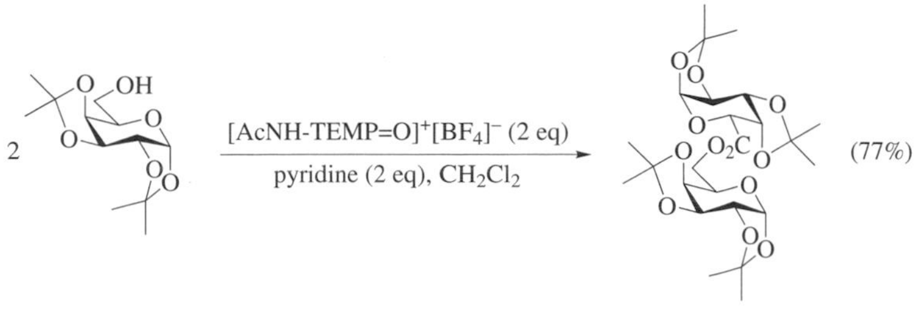

(Eq. 41)

An oxoammonium salt can be used in dry dimethylformamide in the presence of 2,6-lutidine to oxidize C-6 of a methyl glucoside to an aldehyde, isolated as its hydrate (Eq. 42).[144] The use of dimethylformamide rather than methylene chloride as solvent is important and suggests a number of possibilities for the oxidation of polar compounds. This particular reaction is discussed in context in the "Carbohydrates" section.

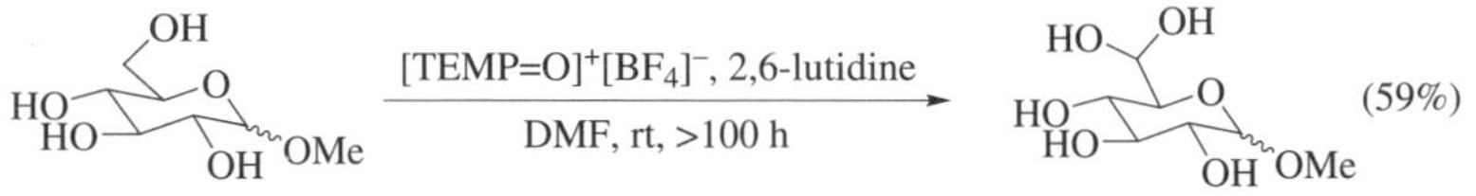
[TEMP=O]+[BF4]−, 2,6-lutidine
DMF, rt, >100 h
(59%)

(Eq. 42)

Stoichiometric Oxidations with Oxoammonium Salts Formed in Situ

The stoichiometric quantities of oxoammonium salt required for this alcohol oxidation method can be generated by an acid-induced disproportionation reaction of nitroxides (Scheme 1 and Eq. 5). The reaction was originally devised using *p*-toluenesulfonic acid[56] and possesses some distinct advantages: (1) suitable nitroxides are commercially available and they can be easily regenerated after the reaction, (2) isolation of the reaction products involves only a filtration step (using the original conditions), followed by the removal of solvent,[56] (3) the yields in this reaction are generally high and, significantly, the reaction works well with 1,2-diols (Eqs. 43 and 44) or α-hydroxyketones with no cleavage of the carbon–carbon bond, and (4) by controlling the stoichiometry of oxidant to diol, either a keto alcohol or a diketone may be obtained.[70,145–147]

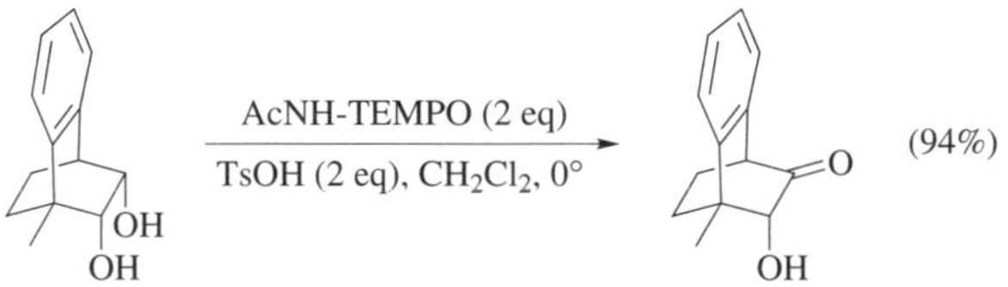
AcNH-TEMPO (2 eq)
TsOH (2 eq), CH2Cl2, 0°
(94%)

(Eq. 43)

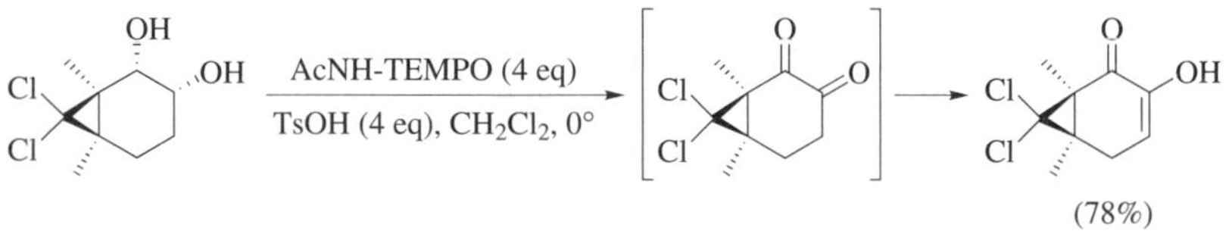
AcNH-TEMPO (4 eq)
TsOH (4 eq), CH2Cl2, 0°
(78%)

(Eq. 44)

The obvious disadvantage and resulting limitation of this method is that the substrates need to be stable under strongly acidic reaction conditions. The scope of these reactions has not been investigated in depth.

Nitroxide-Catalyzed Oxidations

Nitroxide-catalyzed reactions are those in which a secondary oxidant is used in the presence of a nitroxide catalyst to effect alcohol oxidation (Eqs. 6 and 7).[13,102,134,148–151] The nitroxide or more specifically its corresponding oxoammonium salt, provides the specificity of the oxidation reactions and the secondary oxidant sodium hypochlorite (bleach) provides the ultimate oxidation equivalents. Many nitroxides are commercially available, and a wide variety of secondary oxidants (and co-catalysts) are suitable.

The nitroxide-catalyzed reactions are extensively used on many types of alcohols in the presence of a large number of other functional groups. In fact, they account for the bulk of the entries in the Tables. It is of note that some of the side reactions of stoichiometric use of oxoammonium salts have not been reported for the catalyzed reactions. However, two general disadvantages of nitroxide-catalyzed oxidations are noted: (1) the secondary oxidants may lead to unwanted side-reactions (these are noted in the discussion of the specific methods), and (2) the catalyst and byproducts from the secondary oxidant must be separated from the product. To facilitate the purification of products, solid-phase-bound nitroxides and secondary oxidants have been developed, and are discussed below.

Stoichiometric oxidations and nitroxide-catalyzed oxidations do not always give the same results. In fact, some of the side reactions of stoichiometric oxoammonium salts, such as reactions with activated double bonds, phenols, benzylic ethers or with electron-rich heterocyclic systems (see "Miscellaneous Reactions" section) have not been reported for the catalyzed reactions, but in only a few cases have the two methods been directly compared. Nitroxide-catalyzed oxidation conditions do not effect the rearrangement and oxidation of tertiary allylic alcohols (Eq. 36).[120] Electrochemical oxidations, however, do give results similar to stoichiometric oxidations.[106,144,152]

From a practical viewpoint several important aspects to each method must be considered. First, does the reaction take place in the absence of nitroxide? This clarifies whether the specificity of the oxoammonium salt is necessary as opposed to an uncatalyzed oxidation using the secondary oxidant. Second, what are the oxidizing properties of the secondary oxidant that might cause undesired reactions, such as over-oxidation or halogenation? Third, what are the products of the reduction of the secondary oxidant, and can the product be isolated readily from the reaction mixture? These aspects are addressed, where possible, for the reactions described below.

Two critical variables must be considered for these catalyzed reactions: the secondary oxidant and the nitroxide. Secondary oxidants, both monomeric and polymeric, are discussed in the first section. Nitroxides in general, their stabilities, and their monomeric and polymeric forms are discussed in the second section.

Oxoammonium Salt and Nitroxide-Catalyzed Hypochlorous Acid and Chlorine Dioxide Oxidations. In kinetic, spectroscopic studies, oxoammonium perchlorates and nitroxides catalyze the oxidation of methanol and formic acid to carbon dioxide.[40,153]

Nitroxide-Catalyzed Sodium Hypochlorite (Bleach) Oxidations. A typical reaction, described in *Organic Syntheses*, is shown in Eq. 45.[13] Interestingly, no epimerization of the stereogenic center of the aldehyde is observed. In another report, little oxidation of nonanol by the secondary oxidant is observed in the absence of TEMPO.[134]

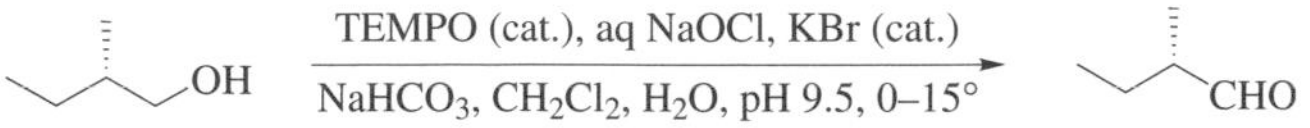
OH
TEMPO (cat.), aq NaOCl, KBr (cat.)
NaHCO3, CH2Cl2, H2O, pH 9.5, 0–15°
CHO

(82–84%)

(Eq. 45)

These reactions are generally carried out in a two-phase, water–methylene chloride system, although it has been suggested that using trifluorotoluene rather than methylene chloride improves the yield in some oxidations.[154] Acetone has been recommended to give "milder" reaction conditions.[155] The sodium hypochlorite oxidant is generally added in the form of commercial bleach of varying concentrations, anywhere from 0.3 M to 2 M.[13] It is recommended that fresh bleach be used.[156,157] The organic phase contains the nitroxide catalyst and the substrate, whereas the aqueous phase contains the bleach, buffers and catalytic amounts of potassium bromide (see below for its role). The two-phase system must be stirred vigorously; the use of a Morton flask is suggested.[158] Control of the pH is crucial, and the aqueous bleach layer is buffered to pH 9.5 with sodium bicarbonate (or other buffers). The reaction becomes prohibitively slow at pH above about 10. Only the lower pH values provide sufficient quantities of the strong oxidant, hypochlorous acid.[159]

Reaction temperatures are usually between 0° and 15°, but lower temperatures (−5° to 0°) are reported to minimize side reactions such as chlorination.[160] The reaction becomes slower at higher temperature, possibly because of the reaction between oxoammonium salt and hydroxide ion in the aqueous phase (Eq. 14).[134] The reaction times are short, and the oxidation is normally complete within minutes. The catalyst as well as the reduction product (sodium chloride) are removed by washing the organic layer with dilute aqueous hydrochloric acid and potassium iodide (to reduce the oxoammonium salt), followed by aqueous sodium thiosulfate (to reduce the iodine formed from the oxidation of the iodide). If the catalyst is TEMPO, it can be recovered from an aqueous solution by codistillation with water under reduced pressure.[161]

The bromide ion is important. It is proposed that sodium hypochlorite oxidizes the bromide to hypobromite and that this actually oxidizes the nitroxide (or hydroxylamine) to the oxoammonium ion. Thus, Scheme 3 must be modified as shown in Scheme 6.[46]

Scope. A wide range of alcohols have been successfully oxidized with this system. The oxidations are, by a factor of ~9 in simple cases, much faster for primary alcohols than for secondary alcohols.[134] This selectivity is substantiated by a study of the oxidation of diols,[151] and is illustrated in Eq. 46 by the oxidation of 1,9-undecanediol to either 9-hydroxyundecanal or 9-oxoundecanal, depending on the amount of bleach, although complete selectivity could not be achieved. About 10% of the other products are present in each reaction mixture. As described in Eq. 8, 5- and 6-membered lactones are formed upon oxidation of the appropriate diols.

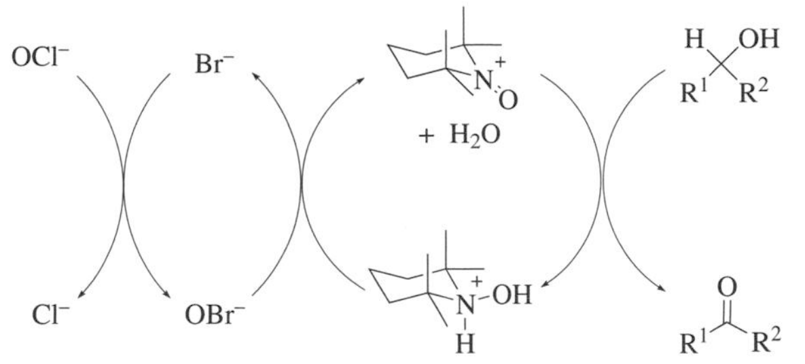
OCl⁻
Br⁻
+ H2O
H OH
R1 R2
Cl⁻
OBr⁻
N-OH
H
O
R1 R2

Scheme 6

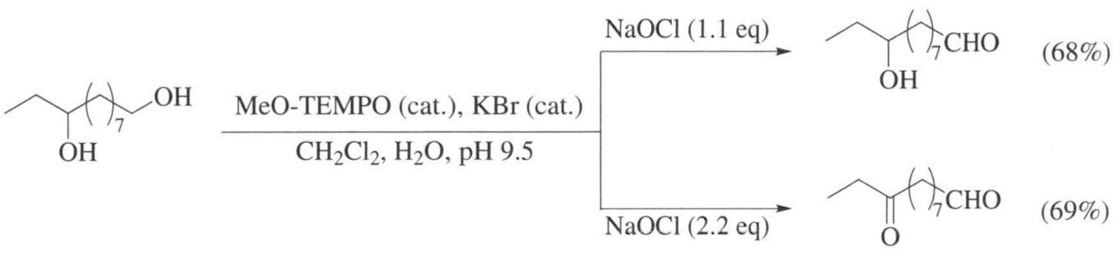
OH
OH
MeO-TEMPO (cat.), KBr (cat.)
CH2Cl2, H2O, pH 9.5
NaOCl (1.1 eq)
CHO
OH
(68%)
NaOCl (2.2 eq)
CHO
O
(69%)

(Eq. 46)

In another example, optically active 4-methyl-5-hexene-1,3-diol is oxidized to the hydroxy aldehyde without racemization and in essentially quantitative yield (Eq. 47).[137] This reaction is of special interest for two reasons. First, there is no reaction with the carbon–carbon double bond (although cinnamyl alcohol leads to poor yields of aldehyde[13,134]). Second, a primary hydroxyl group is selectively oxidized in the presence of a secondary alcohol (Eq. 47).

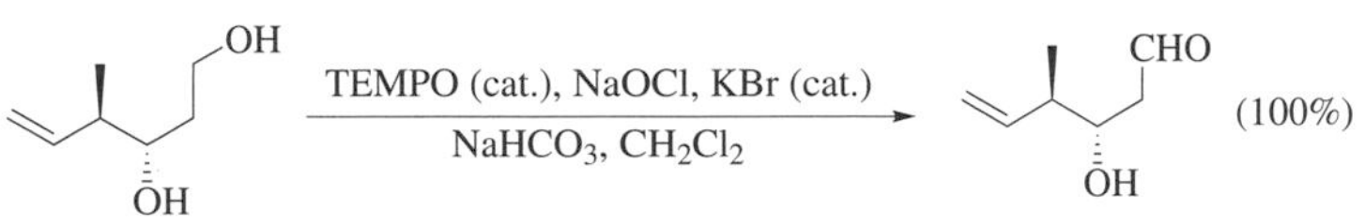
OH
OH
TEMPO (cat.), NaOCl, KBr (cat.)
NaHCO3, CH2Cl2
CHO
OH
(100%)

(Eq. 47)

1,2-Diols represent an important functionality. As with in situ stoichiometric oxidations,[70] internal 1,2-diols are oxidized to diketones without bond cleavage.[162] For simple 1,2-diols, cyclic products have been isolated (Eq. 48).[163] Oxidations of 3-amino-1,2-diols and their *N*-acylated variants provide the corresponding α-hydroxy carboxylic acids (Eq. 49).[164]

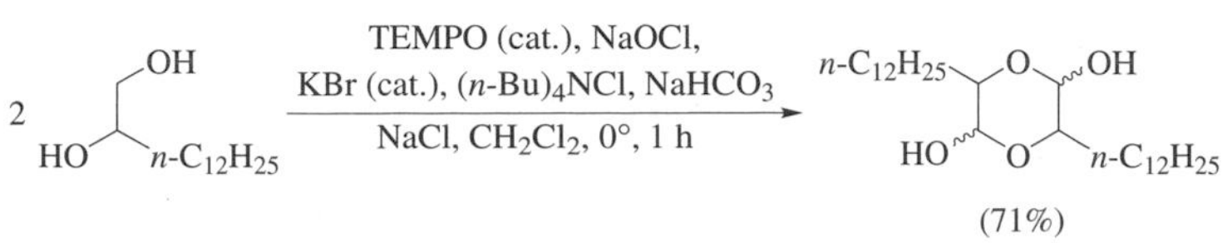
2
OH
HO
n-C12H25
TEMPO (cat.), NaOCl,
KBr (cat.), (n-Bu)4NCl, NaHCO3
NaCl, CH2Cl2, 0°, 1 h
n-C12H25
O
OH
HO
O
n-C12H25
(71%)

(Eq. 48)

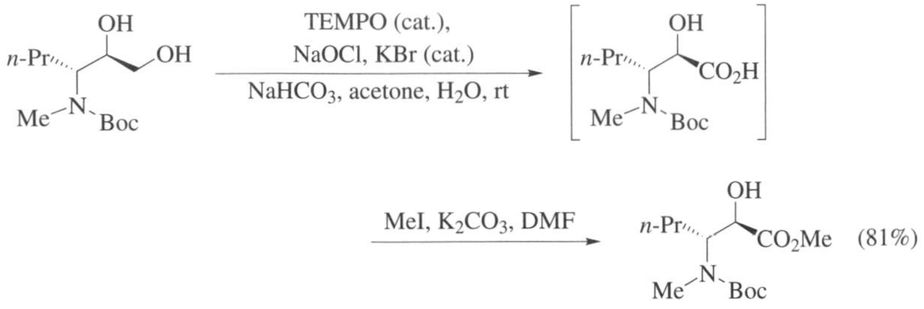

(Eq. 49)

Although alcohols containing free amino groups cannot be selectively oxidized, their *N*-acyl derivatives can be easily converted into aldehydes in excellent yields (Eq. 50).[158] Again, no epimerization at the stereocenter is noted. However, about 5% epimerization occurs in the oxidation reactions of *N*-Fmoc phenylglycinol and *N*-Fmoc (*m*-methoxyphenyl)alaninol with TEMPO, and the Dess–Martin oxidation is preferred.[165] β-Benzyloxy alcohols and β-hydroxy alcohols can also be efficiently oxidized with this system, whereas these substrates are not susceptible to oxidation using the normal stoichiometric oxidation conditions (β-oxygen effect, see above).[68]

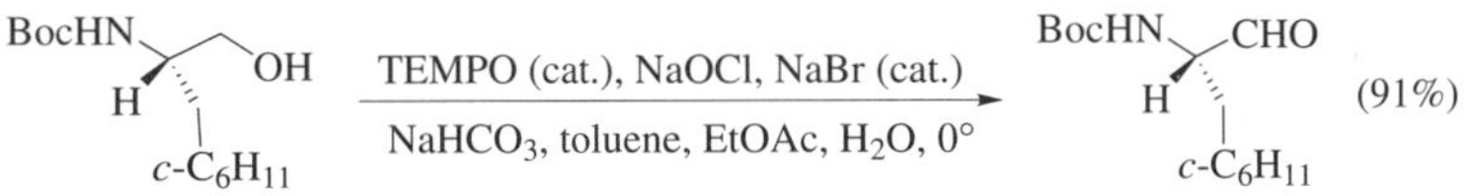

(Eq. 50)

The nitroxide-bleach system can be used in water alone for the oxidation of water-soluble alcohols. Thus, glycerol can be oxidized to the sodium salt of ketomalonic acid in high yields (Eq. 51).[166] The same reaction can also be carried out using a sol-gel-entrapped TEMPO.[167]

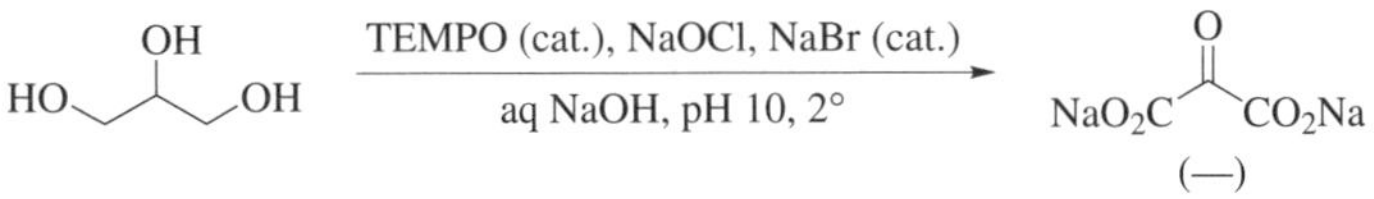

(Eq. 51)

Reactions on a large scale have been described,[156,168,169] and continuous-process oxidations have been devised.[170] Nitroxide-catalyzed bleach reactions have been considered and are likely used commercially on a large scale.[50]

Several variants of the nitroxide-bleach system are used. One is the addition of a phase-transfer agent such as Aliquat 336 (methyl(tri-*n*-octyl)ammonium chloride). The phase-transfer reagent accelerates slow reactions, as, for instance, has been shown for the oxidation of *p*-methoxybenzyl alcohol to the corresponding aldehyde.[134,171] The same phase-transfer agent also allows for the formation of carboxylic acids from alcohols or aldehydes (Eq. 52).[134]

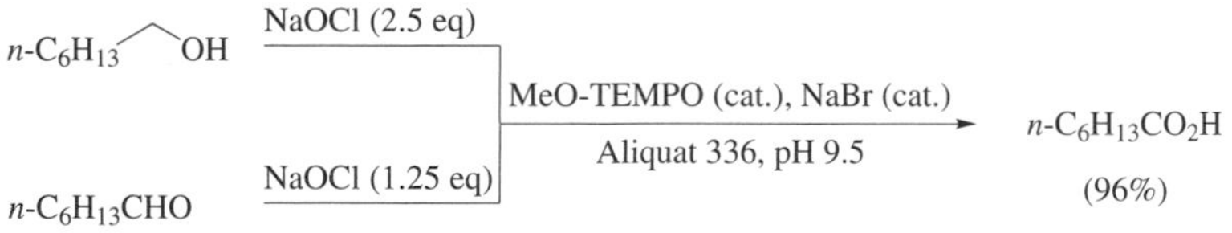
n-C6H13 OH
NaOCl (2.5 eq)
n-C6H13CHO
NaOCl (1.25 eq)
MeO-TEMPO (cat.), NaBr (cat.)
Aliquat 336, pH 9.5
n-C6H13CO2H
(96%)

(Eq. 52)

Another variant is the use of a concentrated (20%) solution of lithium hypochlorite instead of sodium hypochlorite.[151] This oxidant is more suitable for hydrophilic alcohols such as small diols. Lactones are prepared from 1,4-butanediol and 1,5-pentanediol using this oxidant as well.[151]

Calcium hypochlorite ($Ca(OCl)_2$) is believed to be a better secondary oxidant because it is a solid and readily available whereas sodium hypochlorite is not a stable solid.[172–174] *tert*-Butyl hypochlorite has been used to oxidize a phenyl glucoside to its corresponding glucuronide.[175,176]

Limitations. The main restrictions to the use of bleach as a secondary oxidant are that free amines are oxidized (see above) and that double bonds may react slowly,[13] although the latter may be so slow that it does not represent a problem.[137,177] *N*-Chlorination of amides containing an NH group can take place but can be avoided by the use of acetone as the solvent.[155] Benzylic ethers may react under the standard oxidation conditions,[178,179] but in other cases the benzyloxy group appears to be stable.[180,181] The method is also not satisfactory for the preparation of small, anhydrous amounts of low-molecular-weight aldehydes.

A hydroxy-β-lactam is oxidized to an *N*-carboxyanhydride (Eq. 53).[182,183] Presumably, the intermediate keto lactam product undergoes a Baeyer–Villiger-like oxidation. The reaction is carried out at near-neutral pH and is also a rare example in which a carbon–carbon bond is broken during an oxoammonium oxidation.

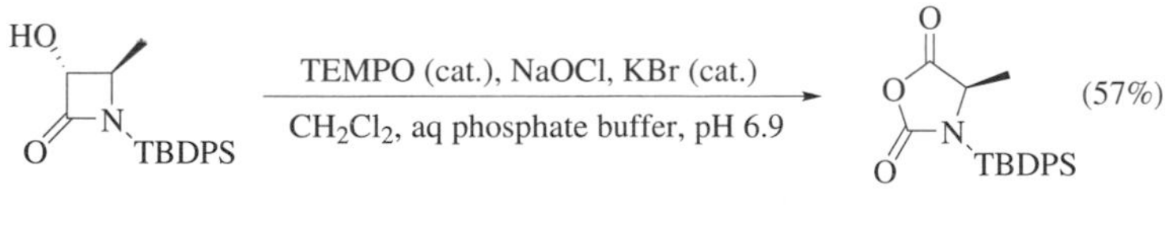
HO
O
N
TBDPS
TEMPO (cat.), NaOCl, KBr (cat.)
CH2Cl2, aq phosphate buffer, pH 6.9
O
O
N
O
TBDPS
(57%)

(Eq. 53)

Some chlorination of electron-rich aromatic rings occurs when the reaction shown in Eq. 54 is attempted with bleach as the secondary oxidant.[184] However, when sodium chlorite is used with only catalytic amounts of sodium hypochlorite, satisfactory results are obtained.

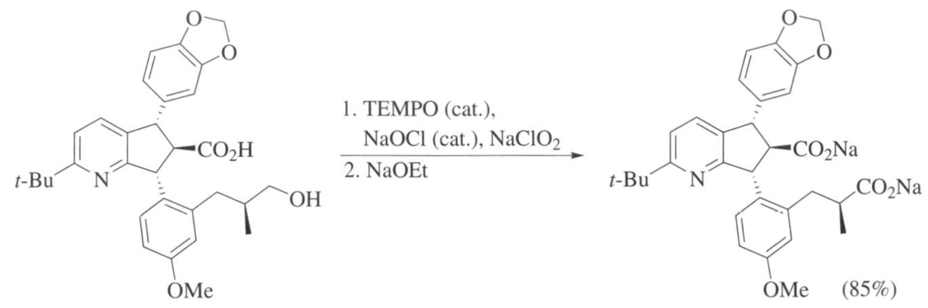

1. TEMPO (cat.),
NaOCl (cat.), NaClO2
2. NaEt
t-Bu
N
CO2H
OH
OMe
CO2Na
CO2Na
OMe
(85%)

(Eq. 54)

One drawback of this oxidation method is that it is carried out in two-phase systems with water. Thus, phase transfer and stirring rate become important. More importantly, the presence of the aqueous phase makes the oxidations of hydrophilic alcohols difficult and can interfere with the isolation and drying of the products. In contrast, the stoichiometric oxidations can be carried out in anhydrous organic media.

Nitroxide- and Hypochlorite-Catalyzed Sodium Chlorite Oxidations. The use of sodium chlorite allows for the preparation of carboxylic acids from primary alcohols while avoiding chlorination side reactions.[184,185] A procedure has been published in *Organic Syntheses*, along with an excellent discussion (Eq. 55).[14]

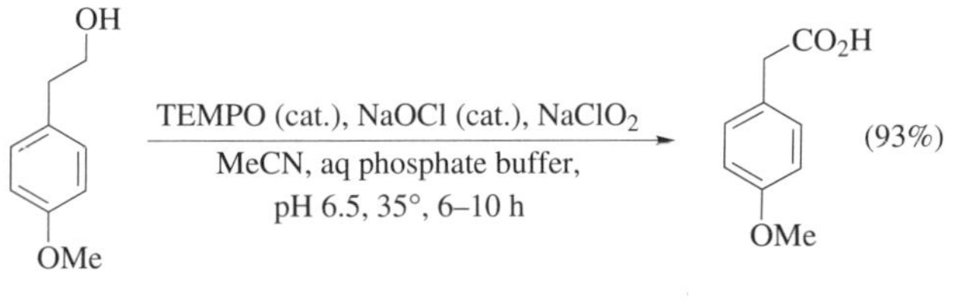

OH
TEMPO (cat.), NaOCl (cat.), NaClO2
MeCN, aq phosphate buffer,
pH 6.5, 35°, 6–10 h
OMe
CO2H
(93%)
OMe

(Eq. 55)

The reactions are carried out in aqueous acetonitrile, thus avoiding the two-phase system of the bleach oxidations, and under near neutral conditions compared to the basic conditions in bleach reactions. The authors note that one should not mix sodium chlorite and the bleach catalyst before reaction since the mixture is not stable. The reactions take 6–10 hours to reach completion at 35°. The oxidation of alcohol to aldehyde is the first step and is similar to that shown in Eq. 45, except that no bromide ion is needed as a co-catalyst. It is believed that in the second step, $NaClO_2$ oxidizes the aldehyde to the acid. The procedure is not applicable to compounds containing carbon–carbon double bonds.[14] Sodium chlorite oxidations to carboxylic acids are also carried out in water and acetonitrile in the absence of sodium hypochlorite.[186]

Scope. Nitroxide/hypochlorite- and chlorite-mediated oxidations have been carried out in two steps, albeit in the same reaction vessel.[187] The primary alcohol is oxidized to the aldehyde with sodium hypochlorite and TEMPO. Then, sodium chlorite is added to the reaction mixture to complete the oxidation. A slightly lower pH of 4–6 enhances the reactivity. The oxidation has been combined with a Sharpless asymmetric dihydroxylation reaction to yield enantiomerically enriched α-hydroxy acids in good yields (Eq. 56).[188,189]

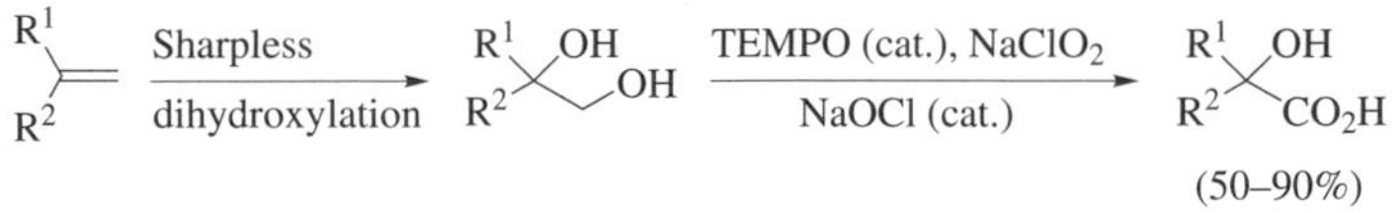
Sharpless
dihydroxylation
TEMPO (cat.), NaClO2
NaOCl (cat.)
(50–90%)

(Eq. 56)

Sodium bromite ($NaBrO_2$) is used in conjunction with a nitroxide to prepare a number of aldehydes and under different conditions, carboxylic acids.[174] Compounds with three consecutive carbonyl groups can be prepared by oxidation of an α,β-dihydroxy ester without carbon–carbon bond cleavage (Eq. 57).[190] Nitroxide-catalyzed electrochemical oxidation methods can also be used for these types of reactions.[190]

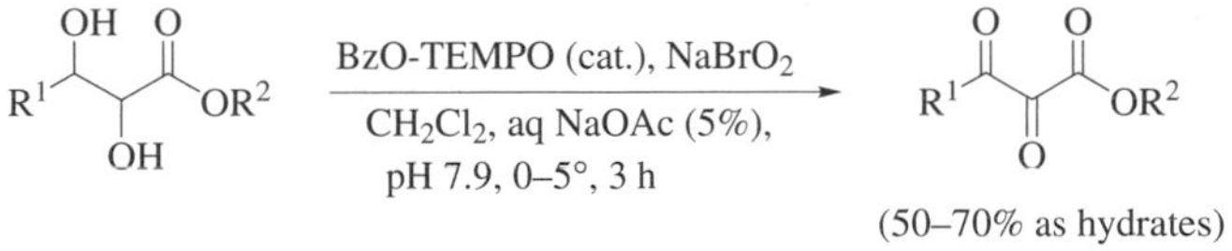
BzO-TEMPO (cat.), NaBrO2
CH2Cl2, aq NaOAc (5%),
pH 7.9, 0–5°, 3 h
(50–70% as hydrates)

(Eq. 57)

Polymers containing the chlorite ion, such as **27**, have been used in these reactions.[44,191] In fact, a polymeric nitroxide (**60**, discussed below), polymeric secondary oxidant **27**, and a polymeric dihydrogen phosphate buffer (**28**) have been used to carry out oxidations in a multiphase system (see also "Experimental Procedures" section).[191]

Amberlite 900
[NMe3]+[ClO2]
Amberlite 900
[NMe3]+[H2PO4]−

27 **28**

Limitations. The limitations of the sodium chlorite–TEMPO oxidations are not clear. While triple bonds are compatible with this reaction, alcohols containing carbon–carbon double bonds and very electron-rich aromatic compounds (such as furfuryl alcohol and alkoxybenzyl alcohols) give poor results.[14]

Nitroxide-Catalyzed Bis(acetoxy)iodobenzene Oxidations. The iodine (III) reagent bis(acetoxy)iodobenzene ($PhI(OAc)_2$) has been introduced as a secondary oxidant for nitroxide-catalyzed oxidations (Eq. 58). In control reactions, no oxidation is noted in the absence of TEMPO.[192]

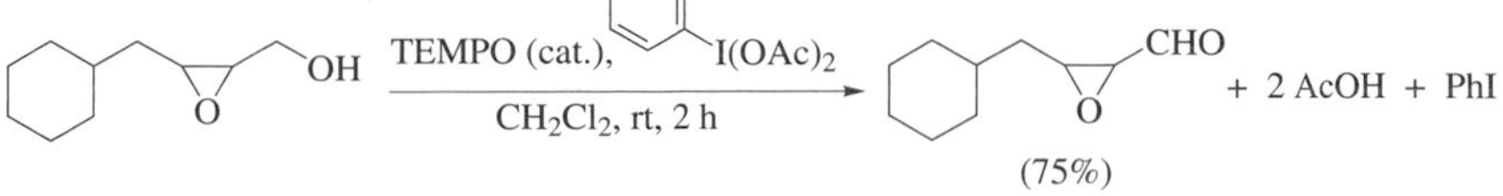
TEMPO (cat.),
I(OAc)2
CH2Cl2, rt, 2 h
+ 2 AcOH + PhI
(75%)

(Eq. 58)

Scope. These reactions have a number of advantages. They can be carried out with substrates that bear carbon–carbon double bonds, epoxides, sulfides, selenides, and electron-rich aromatic rings. Most importantly, secondary alcohols are oxidized slowly. Thus, primary alcohols can be selectively oxidized in the presence of secondary alcohols, even when the latter are allylic or benzylic.[192] The reactions are conducted in methylene chloride, thus allowing anhydrous conditions, but they can also be carried out in a number of other solvents such acetonitrile–water mixtures and they require no buffers.[193] Normally, only the aldehyde is isolated, but in one case, a primary alcohol is oxidized completely to a carboxylic acid in the presence of water.[194] The β-oxygen problem noted with stoichiometric reactions is not observed.[195]

The byproducts of the reaction are iodobenzene, acetic acid, and TEMPO. TEMPO is removed by washing with sodium thiosulfate, which reduces TEMPO to the water-soluble hydroxylammonium sulfate and the acetic acid is removed by washing with aqueous sodium bicarbonate. In small-scale reactions, the iodobenzene may be removed by silica gel chromatography. On a larger scale, at least some iodobenzene separates on vacuum distillation.[196] The reaction rate can be increased with a small amount of acetic acid[192] or water[197] and the reaction has been carried out on a 1.3-mole scale (Eq. 59).[197,198]

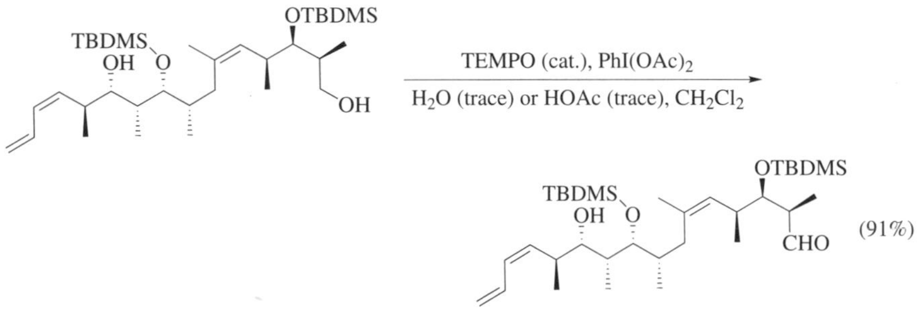

(Eq. 59)

If the diacetoxyiodo group is attached to bi- or terphenyl groups, the iodoaryl products precipitate and can be removed by filtration.[199] The bis-(acetoxy)iodobenzene moiety has been attached to an ionic liquid[200] and used with another ionic liquid containing TEMPO units for oxidation.[201] In the same type of system, an adamantane nucleus with four diacetoxyiodophenyl groups has been used as an oxidant.[41] A polystyrene-bound bis(acetoxy)iodobenzene derivative is known (see also "Experimental Procedures" section).[202] The use of polymeric bis(acetoxy)iodobenzene derivatives solves the problem of removing the iodobenzene byproduct, as product isolation involves filtration to remove the poly(iodostyrene) polymer, which can be recycled. This polymeric system can also be used for the TEMPO-catalyzed oxidation of alcohols to aldehydes with potassium nitrite and air,[203] and for the oxidation of primary alcohols to carboxylic acids, although larger amounts of TEMPO are required.[204]

A procedure for the oxidation of nerol to neral using this oxidation system is published in *Organic Syntheses*.[15] An interesting one-pot reaction in which a primary alcohol is converted into an aldehyde which subsequently is condensed with a Wittig reagent is shown in Eq. 60.[205] Another one-pot method can be used to convert 2,3-epoxy alcohols into isoxazole derivatives.[206]

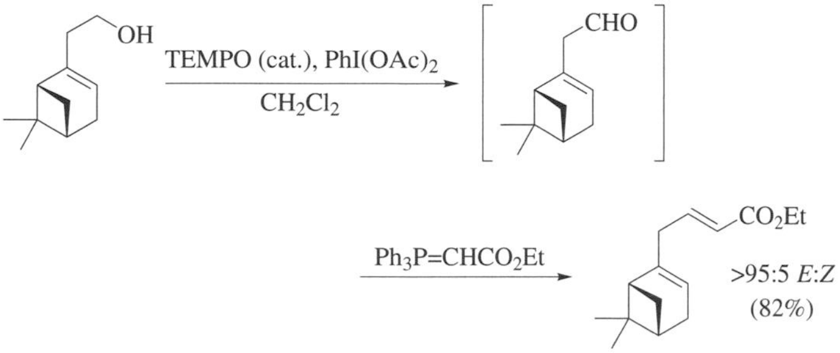
OH
TEMPO (cat.), PhI(OAc)2
CH2Cl2
CHO
Ph3P=CHCO2Et
CO2Et
>95:5 E:Z
(82%)

(Eq. 60)

The mechanism for these reactions differs from that of the normal nitroxide-catalyzed reactions, since $PhI(OAc)_2$ does not oxidize TEMPO to the reacting oxoammonium salt. The reaction pathway shown in Scheme 7 has been suggested.[192] Thus, the initial oxoammonium salt is formed by the acid catalyzed disproportionation of the nitroxide (Scheme 1). This hypothesis is supported by the fact that the reaction is faster with a trace of acetic acid.[192] All of the oxoammonium ion may be formed in this fashion. However, it is also possible that, after the initial formation of some oxoammonium ion by disproportionation, $PhI(OAc)_2$ oxidizes the hydroxylamine to an oxoammonium ion, as shown by the dotted line in Scheme 7. This modification would make the reaction sequence similar to that shown in Scheme 3. Since this reaction is carried out in an essentially acidic medium (acetic acid) one would think that secondary alcohols would be, as discussed earlier, oxidized faster than primary alcohols. This is clearly not the case since primary alcohols can be cleanly oxidized in the presence of secondary alcohols (Eq. 59).[192,197] This observation is general when using $PhI(OAc)_2$ as a secondary oxidant. No explanation of the origin of the deviation from the general reactivity pattern of other nitroxide-catalyzed reactions has been put forward.

An interesting ionic-liquid-derivatized bis(acetoxy)iodobenzene oxidant is described in the "Supported Nitroxides" section.[200] The use of $PhI(OAc)_2$ as a co-catalyst, rather than a reagent, with sodium nitrite allows oxidations to be carried out with oxygen.[203] As is typical, benzylic alcohols give better results than alkyl alcohols with oxygen as the secondary oxidant.

Limitations. Only three potential disadvantages of this method are noted. The first is that 1,2-diols are known to be cleaved by $PhI(OAc)_2$.[197,207] Remarkably, this cleavage has not been observed in nitroxide-catalyzed reactions. The second is that large amounts of iodobenzene must be dealt with. The third is that acetic acid is formed which may cause undesired side reactions.[208] In one case, a β-bromoacetyloxy group is eliminated during the reaction, and a Dess–Martin oxidation is preferred.[209]

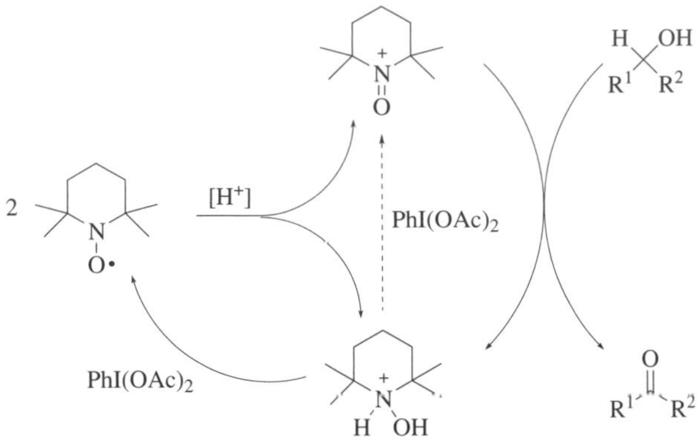

Scheme 7

Nitroxide-Catalyzed Hypervalent Bromine Oxidations. Resins containing diacetoxy Br(I) ate complexes are used as secondary oxidants, somewhat like the iodine(III) derivatives discussed above. They are prepared as shown in Eq. 61,[210] and used as demonstrated with Eq. 62.[43,44,211,212] The reaction can be applied to a number of enantiomerically enriched primary and secondary alcohols with complete preservation of enantiopurity. The polymeric reagents are stable if stored at 0°.

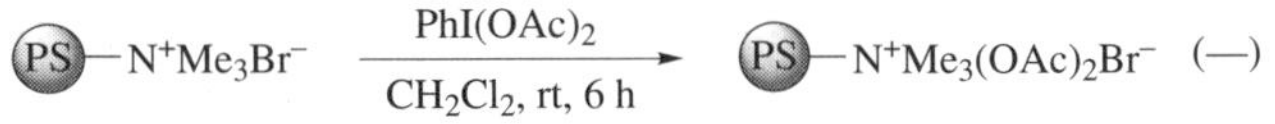

(Eq. 61)

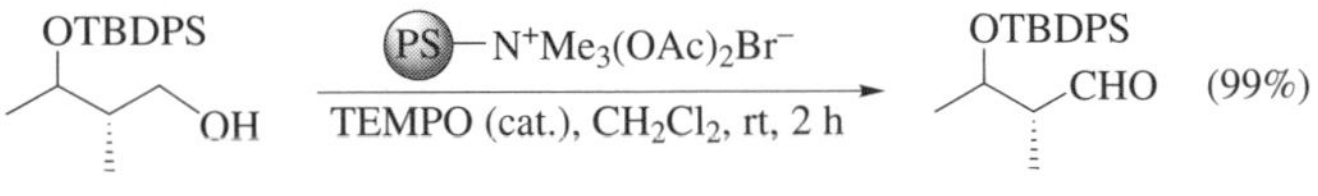

(Eq. 62)

PS = polystyrene

Nitroxide-Catalyzed *m*-Chloroperoxybenzoic Acid Oxidations. Nitroxide-catalyzed reactions using *m*-chloroperoxybenzoic acid (*m*-CPBA) as a secondary oxidant were the first nitroxide-catalyzed reactions to be discovered.[102,148–150] On one hand, the reactions are somewhat simplified in that the *m*-CPBA also oxidizes 2,2,6,6-tetramethylpiperidine to TEMPO, which then serves as the catalyst for the oxidation of alcohols. Thus, reactions can be carried out using the piperidine hydrochloride (Eq. 63),[149] though the reactions can also be carried out using preformed TEMPO.[102] Halide ions, preferably bromide, appear to be necessary for this reaction.[213]

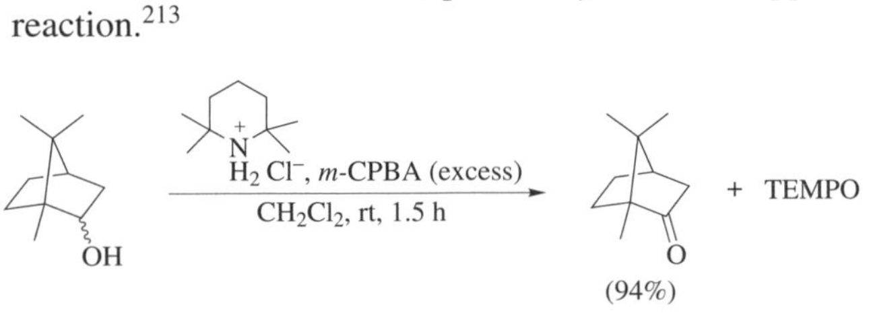

(Eq. 63)

On the other hand, the reaction is complicated by several factors: First, it works well only for the oxidation of secondary alcohols to ketones, since some over-oxidation of primary alcohols to carboxylic acids may be observed.[149] Furthermore, known peracid oxidations such as the Baeyer-Villiger oxidation of ketones to esters or lactones and the epoxidation of alkenes may occur. Of course, these secondary reactions can also be taken advantage of, as in the combined epoxidation-oxidation of 5-norbornene-2-ol (Eq. 64).[149] The reaction is carried out in two steps, albeit in the same flask.

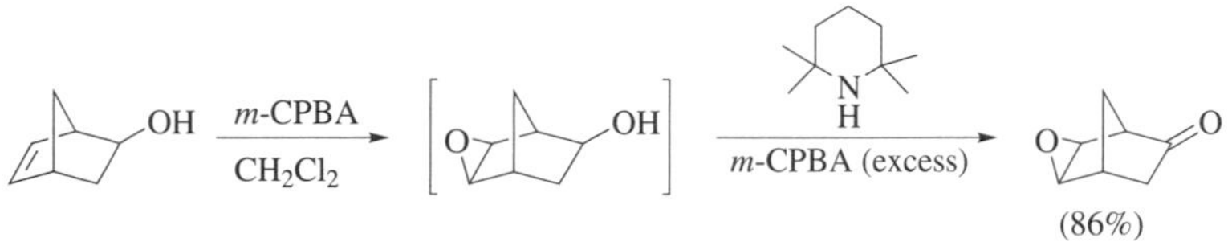
OH
m-CPBA
CH2Cl2
O
OH
N
H
m-CPBA (excess)
O
O
(86%)

(Eq. 64)

The nitroxide-catalyzed *m*-CPBA reaction has not been used extensively, even though it can be carried out in an anhydrous medium, utilizes a commercially available oxidant, and involves simple isolation procedures. It has, however, been used with chiral nitroxide catalysts.[85,94]

Nitroxide-Catalyzed Oxone and Other Peroxyacid Oxidations. Oxone (DuPont) is a commercially available solid oxidant having the approximate composition 2 $KHSO_5 \bullet KHSO_4 \bullet K_2SO_4$. The active ingredient is potassium peroxymonosulfate, $KHSO_5$, which is reduced to sulfate during the reaction.[214] In large-scale reactions, the inorganic salts can be removed by filtration.[215] A typical example of the use of Oxone as a secondary oxidant in a nitroxide-catalyzed secondary alcohol oxidation is shown in Eq. 65.[216]

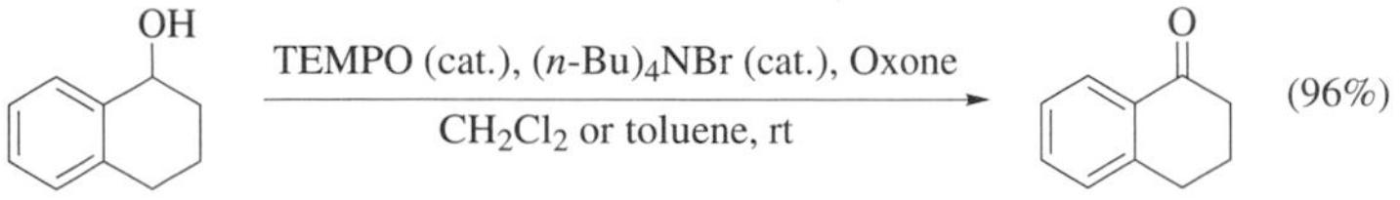
OH
TEMPO (cat.), (n-Bu)4NBr (cat.), Oxone
CH2Cl2 or toluene, rt
O
(96%)

(Eq. 65)

Oxone has also been used in conjunction with TEMPO and a silver carbonate–silica co-catalyst for the oxidation of carbohydrates.[35] The use of Oxone, and other peracid ions such as peracetate, perborate, percarbonate, and peroxydisulfate, have been explored for the oxidation of carbohydrates.[217]

Nitroxide-Catalyzed Trichloroisocyanuric Acid Oxidations. Primary and secondary alcohols can be converted into aldehydes or ketones (Eq. 66),[218] or even to carboxylic acids using trichloroisocyanuric acid (TCC) as a secondary oxidant.[219] Primary alcohols are oxidized selectively over secondary alcohols, and some of the reactions can be carried out in anhydrous solvents.[218] The fate of the TCC in the reactions is not stated.

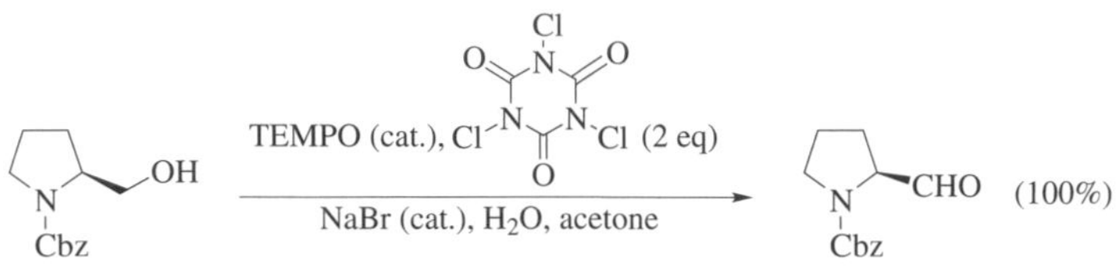

(Eq. 66)

Nitroxide-Catalyzed *N*-Chlorosuccinimide Oxidations. *N*-Chlorosuccinimide (NCS) is an alternate source for the positive chlorine used in bleach and sodium chlorite oxidations.[220] The reaction provides aldehydes with no over-oxidation to carboxylic acids and is highly selective for primary alcohols (Eq. 67). Alcohols containing double bonds give somewhat lower yields, although no reason is cited for this.[220]

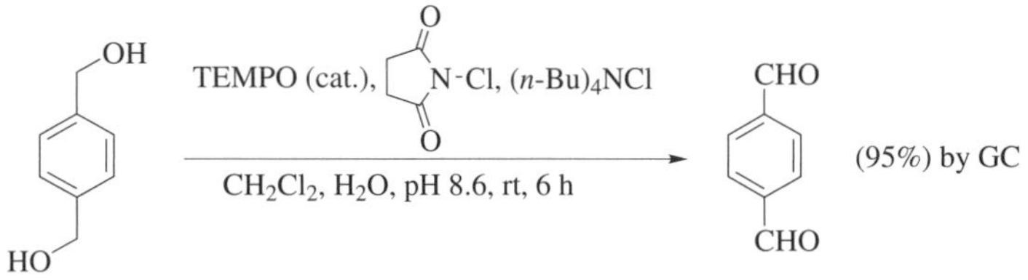

(Eq. 67)

Nitroxide-Catalyzed Electrochemical Oxidations. Nitroxide-catalyzed electrochemical oxidations combine the chemical specificity of oxoammonium ion oxidations with the benefits of electrochemical synthesis, namely the option to vary the oxidation potential and the lack of byproducts resulting from a primary chemical oxidant. However, they also suffer from some difficulties involving: (1) the choice of the specific electrode and its surface modification, (2) the use of a polar solvent that will also dissolve an electrolyte to carry the current, and (3) the possible use of one- or two-compartment cells. The required equipment for an electrochemical oxidation may also be less available.

In the earliest examples of these reactions, a series of alcohols are oxidized in acetonitrile containing TEMPO as a catalyst and lithium perchlorate as the electrolyte in the presence of 2,6-lutidine.[221] The reactions are carried out in a divided cell with a large excess of 1,2-dibromoethane to be reduced at the counter electrode. The base, 2,6-lutidine, is used to capture the protons formed, although it is said to slowly deactivate the oxoammonium ion.[221] The electrodes are platinum for the working electrode and platinum or copper for the counter electrode. A typical reaction is shown in Eq. 68.[221] Primary alcohols are oxidized much more rapidly than secondary alcohols.

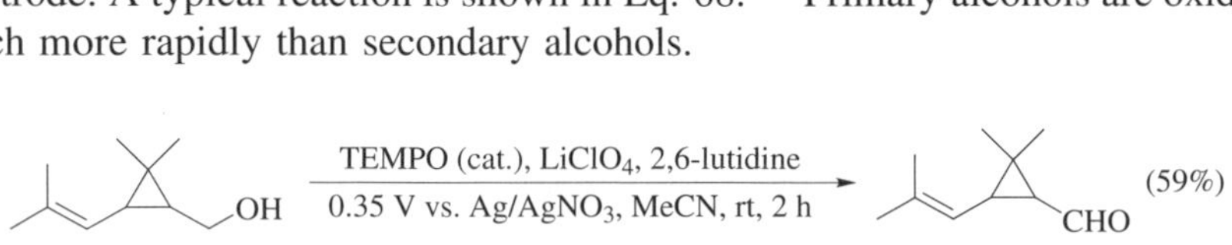

(Eq. 68)

Bromide has been used as a co-catalyst in conjunction with nitroxides for electrochemical oxidations of alcohols (Eq. 69).[222,223] A possible schematic of this reaction is presented in Scheme 8. In this scheme, note that the oxoammonium salts can also react with hydroxylamine to give nitroxide (Eq. 19). In similar work, tetraalkylammonium tribromide, generated electrochemically, has been used with nitroxides to oxidize alcohols.[224]

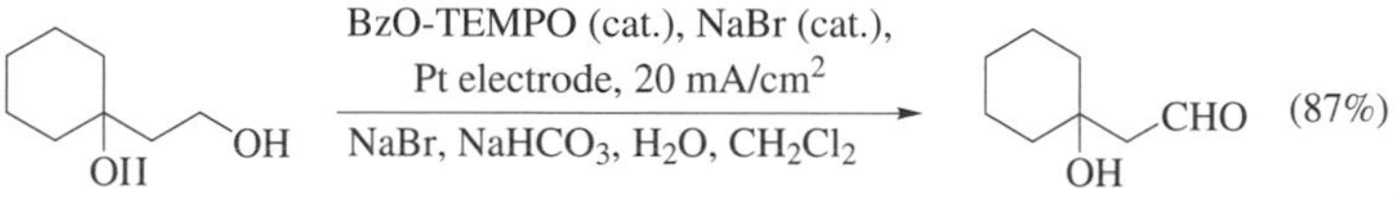

(Eq. 69)

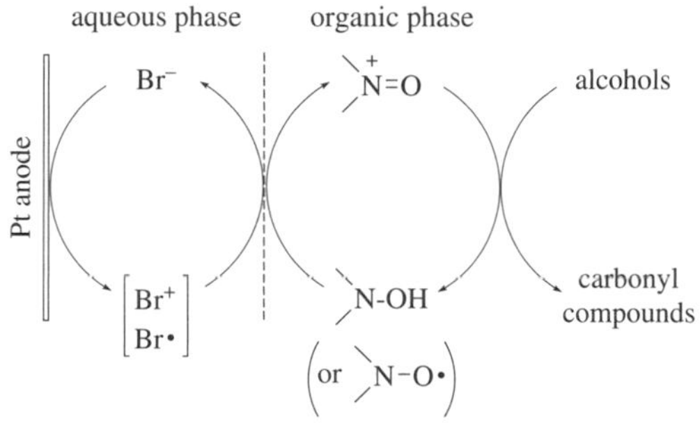

Scheme 8

BzO-TEMPO = 4-benzoyl TEMPO

Relatively few reactions have been carried out with simple substrates. However, oxidations have been carried out with chiral nitroxides and chiral auxiliaries attached to electrode surfaces. Chiral auxiliaries in solution and on surfaces have been discussed in the "Stereochemistry" section and have been summarized elsewhere.[225,226] Here, we discuss simple non-chiral electrode surface modifications.

If an oxidation catalyst can be attached to the electrode surface (or any other insoluble polymer, see below), continuous reactions are possible. In electrochemical reactions, the number of coulombs of electricity passed is used as a measure of the quantity of (alcoholic) substrate oxidized. This calculation has been done for a number of electrooxidations on nitroxide-modified electrodes.[83,141,227] The major problem with coated electrodes is their often limited stability. Stability is sometimes measured as "turn-over numbers" (TON) or the number of times a catalyst molecule can be effectively rejuvenated, as calculated from the amount of catalyst on the electrode and the amount of substrate that is oxidized before the current drops to zero.

Platinum and graphite-felt electrodes are initially modified by reactions such as shown in Eq. 70.[228] The two pyrrole moieties of the specially prepared nitroxide

29 were electropolymerized onto the electrode surface. Several benzylic alcohols are satisfactorily oxidized, but the electrode films are not very stable.

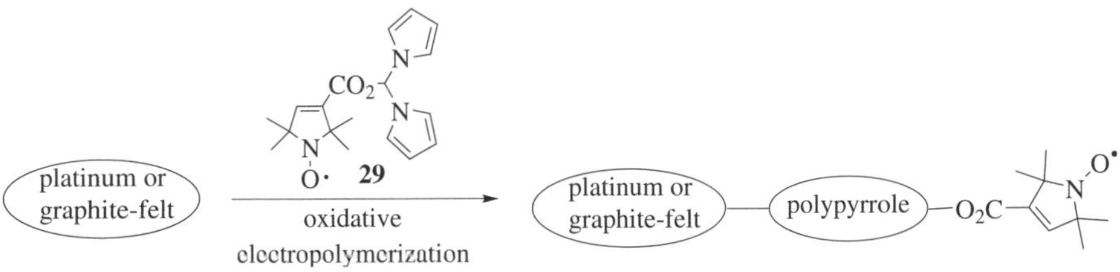

(Eq. 70)

Other similar coatings have been prepared from compounds **30–32**.[112,229] Coatings on platinum have been prepared from nitroxides **30** and **31**, co-polymerized with 2,2′-bithiophene (**33**). The stability of the modified electrodes for electrooxidations is limited for both cases but the layers derived from nitroxide **31** are more stable than those derived from nitroxide **30**.[229] The layers are used more successfully in non-electrochemical experiments using sodium hypochlorite as a secondary oxidant. The latter method is suggested for the comparative study of nitroxide catalysts.[229] Layers prepared from the chiral molecule **32** show faster oxidation rates with (*S*)-1-phenylethanol than with (*R*)-1-phenylethanol, but no preparative reactions have been carried out.[112]

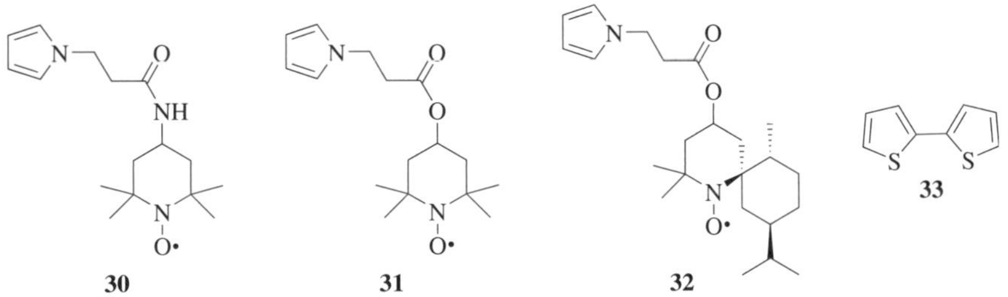

30 **31** **32** **33**

Many of the nitroxide-coated electrodes are prepared on graphite felt using polyacrylic acid, cross-linked with 1,6-hexanediamine. The nitroxide NH_2-TEMPO is attached to the polymer through amide bonds (Eq. 71).[230,231] These electrodes are used for the oxidation of alcohols as well as other functional groups.[108,111,282,141,232]

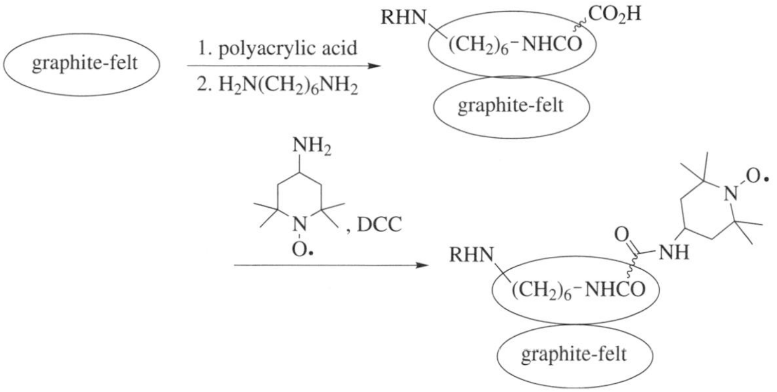

(Eq. 71)

Various nitroxide-electrodes are used for the analytical determination of alcohols and carbohydrates.[112,233–236]

Nitroxide-Catalyzed Oxidations Involving Copper. These reactions can be divided into those in which copper ion is used stoichiometrically as the secondary oxidant and those in which a complexed form of copper is used in conjunction with air or oxygen as the primary catalyst.

Copper(II) as Secondary Oxidant. One representative example uses stoichiometric amounts of copper(II) chloride in acetonitrile as a secondary oxidant, in the presence of calcium hydride as a drying agent (Eq. 72).[81] The key step is the oxidation of the nitroxide to the oxoammonium ion by Cu(II). Alcohol oxidation yields the hydroxylamine which then reacts with the oxoammonium ion to give the nitroxide (Eqs. 17–21). Similar work is reported by others.[117,237,238,239]

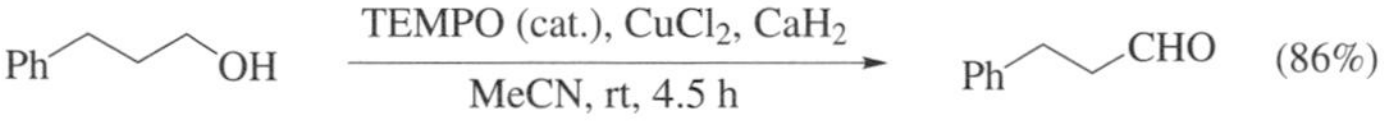

(Eq. 72)

Copper Complexes in Combination with Air as Secondary Oxidants. The most interesting reactions with copper are those in which air or oxygen is used as secondary oxidants.[81] The oxidant Cu(II), either in "free" or complexed form, is particularly interesting as it can be used in a catalytic fashion, with oxygen (air) as the ultimate oxidant. Air oxidations are environmentally friendly and these reactions therefore take on a special importance.[24,46] The oxidations are carried out with copper(II) chloride in dimethylformamide in the presence of TEMPO and oxygen (Eq. 73). Catalytic amounts of copper perchlorate are also used with air and AcNH-TEMPO for the oxidation of alcohols.[240] Copper(II)-catalyzed aerobic oxidations of primary alcohols to aldehydes have also been carried out in ionic liquids.[240] Furthermore, polymeric nitroxides can also be used in the nitroxide-catalyzed oxidation of benzylic and allylic alcohols with copper(I) chloride and air in dimethylformamide.[241] Nitroxide-catalyzed oxidations of aliphatic alcohols are very slow, whether mediated by catalytic or stoichiometric quantities of Cu(II).

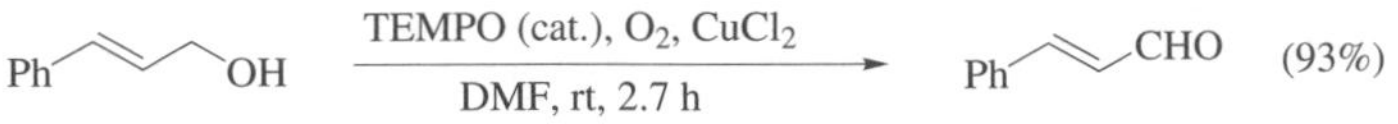
Ph
OH
TEMPO (cat.), O2, CuCl2
DMF, rt, 2.7 h
Ph
CHO
(93%)

(Eq. 73)

More recently, copper(II) complexes have been used; the reactions again involve air as the secondary oxidant. Examples of these copper complexes include **34**,[242] **35**,[243] and **36**.[244] Copper chelates of **37**[245] and **38**, a compound containing an ionic liquid function, have been used.[246] A copper complex of DABCO (**39**) has also been described.[247] A typical reaction is shown in Eq. 74 (see also Scheme 4).[244] As discussed in the "Mechanisms, Stereochemistry, and Site Selectivity" section, these reactions appear to take place by a mechanism different from the normal oxoammonium oxidations in that the species that performs the alcohol oxidation is an oxoammonium–copper complex rather than an uncomplexed oxoammonium salt (Scheme 4 and Eq. 22). The reaction appears to be general for primary alcohols only.

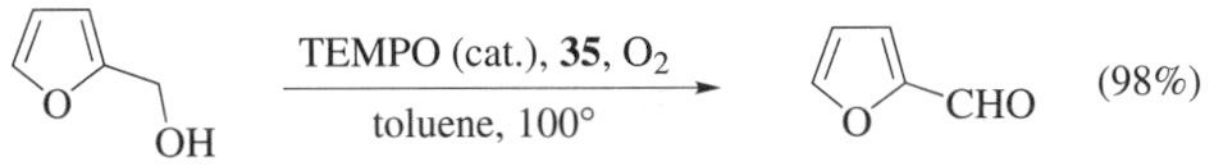
OH
TEMPO (cat.), 35, O2
toluene, 100°
CHO
(98%)

(Eq. 74)

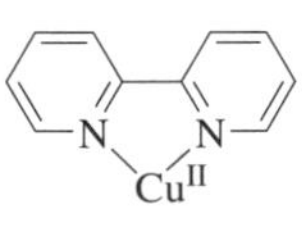

34

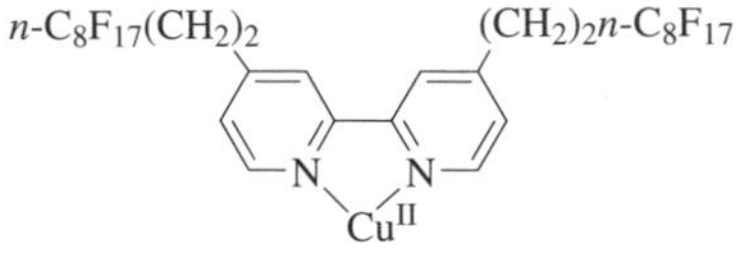
n-C8F17(CH2)2
(CH2)2n-C8F17

35

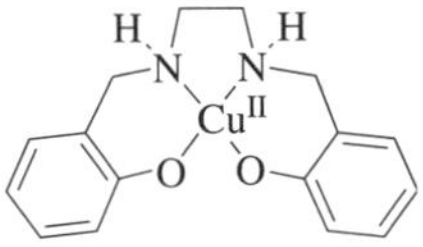

36

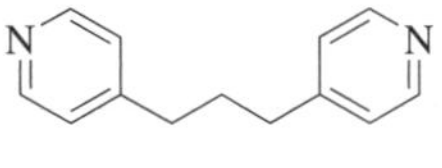

37

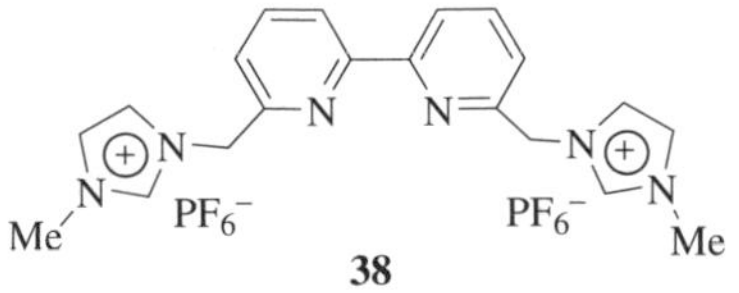
PF6−
PF6−
Me
Me

38

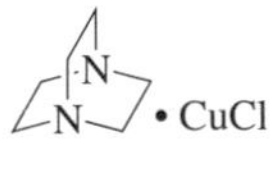
• CuCl

39

Laccase Oxidations. Laccases are a class of enzymes expressed by white rot fungi which cause the breakdown of lignin materials.[248] These enzymes have two interesting properties; they can oxidize nitroxides to oxoammonium ions, and they can be rejuvenated by air.[249] A typical reaction is shown in Eq. 75.[249] Detailed mechanistic studies of this reaction have confirmed that oxoammonium ions are the oxidants but indicate that the laccase oxidizes TEMPOH to TEMPO in a one-electron reaction.[250–253] The reaction has been used for the oxidation of sugar derivatives.[254]

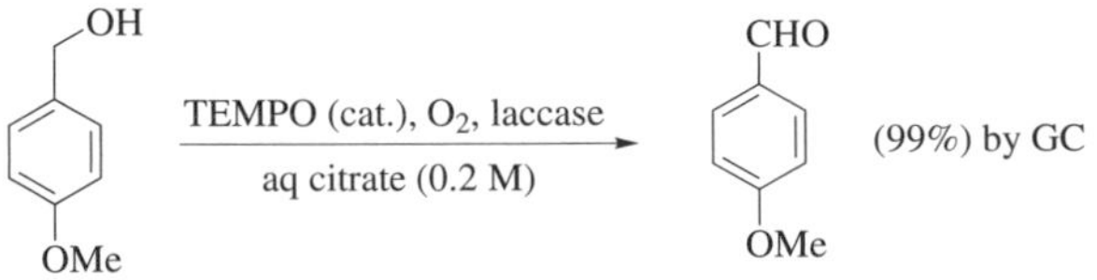
OH
OMe
TEMPO (cat.), O2, laccase
aq citrate (0.2 M)
CHO
OMe
(99%) by GC

(Eq. 75)

Miscellaneous Nitroxide-Catalyzed Oxidations. A number of nitroxide-catalyzed oxidations using secondary oxidants that are not widely used have been recorded. Sorted according to the particular secondary oxidants used, they are listed below with minimal comment.

Elemental Halogens. Elemental iodine, chlorine, and bromine have been used as secondary oxidants in nitroxide-catalyzed oxidations. This transformation is expected since the oxoammonium salts can be prepared from nitroxides with chlorine and bromine.[8,11,116,255] An example of an oxidation using stoichiometric quantities of iodine is shown in Eq. 76.[256] The main reason for using iodine rather than chlorine or bromine is that it prevents the oxidative degradation of electron-rich and heteroaromatic rings in the substrate.

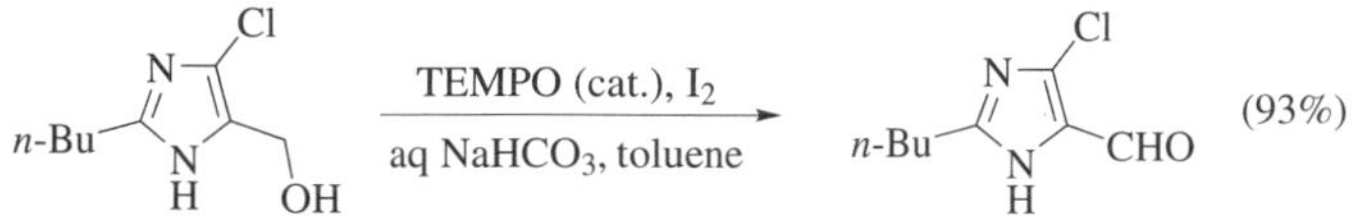

Cl
N
n-Bu
N
H
OH
TEMPO (cat.), I_2
aq $NaHCO_3$, toluene
Cl
N
n-Bu
N
H
CHO
(93%)

(Eq. 76)

The oxidation of various sugars to glycaric acids with chlorine or bromine have been carried out (Eq. 77).[257] In this reaction, the main byproduct is sodium bromide, and the product can be precipitated from water to remove any ionic impurities.

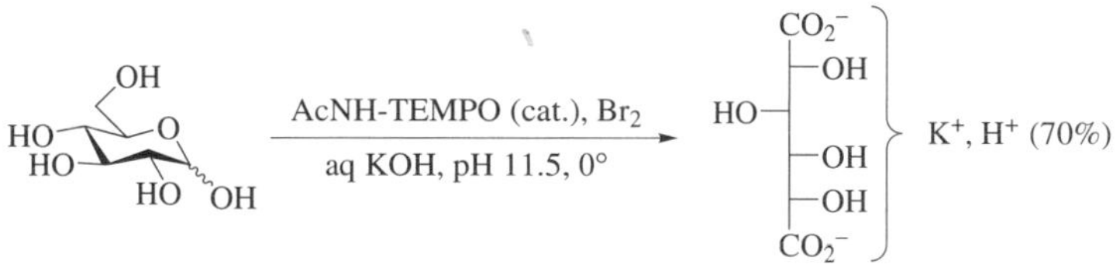

OH
HO
O
HO
HO
OH
AcNH-TEMPO (cat.), Br_2
aq KOH, pH 11.5, 0°
CO_2^-
OH
HO
OH
OH
CO_2^-
K^+, H^+ (70%)

(Eq. 77)

Bromine Compounds. TEMPO with sodium nitrite and bromine[258] or 1,3-dibromo-5,5-dimethylhydantoin has been used to catalyze air oxidation of benzylic alcohols.[259] Similarly, TEMPO with *tert*-butyl nitrite and potassium bromide has been used to catalyze the oxidation of various alcohols.[260]

Periodate. Periodic acid (H_5IO_6)[261] and sodium metaperiodate ($NaIO_4$)[262] are also used as secondary oxidants for nitroxide-catalyzed alcohol oxidation. The reactions are fairly slow (reaction times of 8 to 70 hours), but yields are good (Eq. 78).[262] Select reactions need to be carried out under reflux conditions in methylene chloride, while others require only ambient temperature. A possible problem in this reaction is the well-known periodate-induced cleavage of 1,2-diols.[263] For reactions in organic solvents, sodium periodate can be deposited on silica gel.[264]

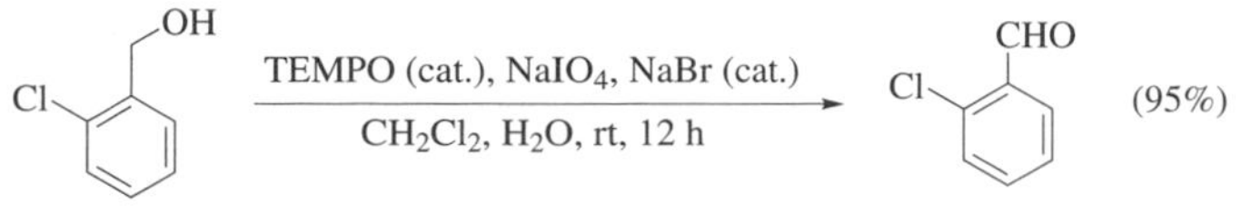

OH
Cl
TEMPO (cat.), $NaIO_4$, NaBr (cat.)
CH_2Cl_2, H_2O, rt, 12 h
CHO
Cl
(95%)

(Eq. 78)

Iron Salts. Oxidations with potassium ferricyanide [$K_3Fe(CN)_6$], have been carried out in conjunction with monomeric and polymeric nitroxide systems.[265] An example of such a reaction catalyzed by (monomeric) BzO-TEMPO in aqueous KOH and acetonitrile is shown in Eq. 79. TEMPO in combination with ferric chloride and sodium nitrite has been used to catalyze air oxidation of various alcohols.[266] Most recently, the use of ferric chloride hexahydrate as secondary oxidant was demonstrated.[267]

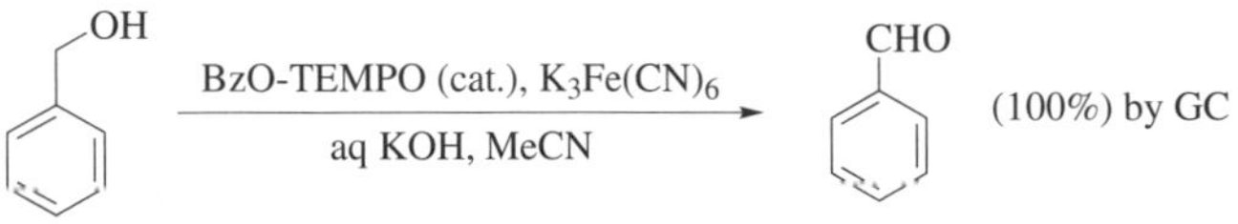
OH
BzO-TEMPO (cat.), K3Fe(CN)6
aq KOH, MeCN
CHO
(100%) by GC

(Eq. 79)

Hydrogen Peroxide in Ionic Liquids. Hydrogen peroxide oxidations catalyzed by AcNH-TEMPO have been carried out in an ionic liquid ([bmim]PF_6).[268] The reaction is confined to benzylic alcohols in the absence of electron-donating groups (Eq. 80). It is suggested that AcNH-TEMPO is more convenient to use than TEMPO.[268]

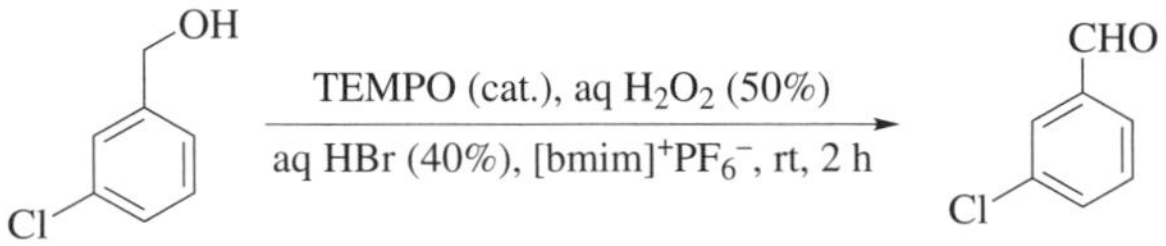
OH
TEMPO (cat.), aq H2O2 (50%)
aq HBr (40%), [bmim]+PF6−, rt, 2 h
Cl
CHO
Cl
(93%)

(Eq. 80)

$[bmim]^+$ = 1-*n*-butyl-3-methylimidazolium

Combination of Ceric Ammonium Nitrate and Oxygen. Ceric ammonium nitrate (CAN) with TEMPO has been used for a number of oxygen-mediated oxidations of primary and secondary alcohols (Eq. 81).[269]

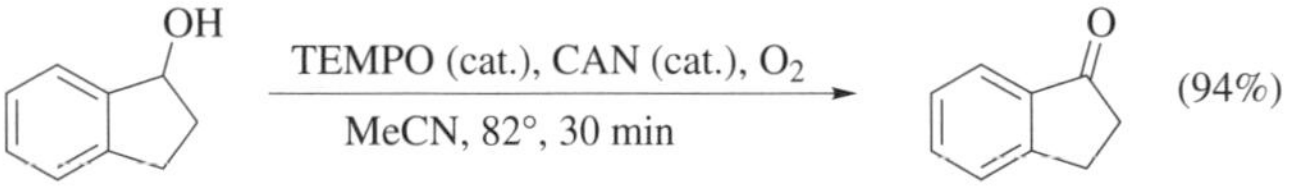
OH
TEMPO (cat.), CAN (cat.), O2
MeCN, 82°, 30 min
O
(94%)

(Eq. 81)

Combination of Cobalt and Manganese Salts and Oxygen. TEMPO in the presence of cobalt and manganese nitrates catalyzes the oxygen or air oxidation of alcohols to aldehydes or ketones in good yield (Eq. 82).[270,271] Similar reactions have been carried out with TEMPO on a polyethylene glycol polymer[272] and on polystyrene.[273]

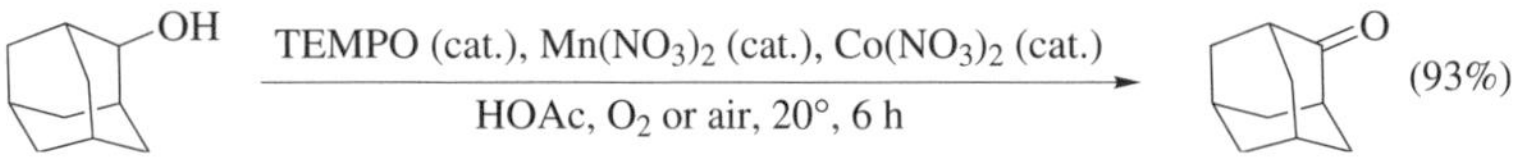
OH
TEMPO (cat.), Mn(NO3)2 (cat.), Co(NO3)2 (cat.)
HOAc, O2 or air, 20°, 6 h
O
(93%)

(Eq. 82)

Combination of Polyoxometalates and Oxygen. TEMPO has been used as a catalyst in conjunction with $H_5PV_2Mo_{10}O_{40}$ and oxygen for alcohol oxidations.[274]

Combination of Methyltrioxorhenium and Hydrogen Peroxide. The catalyst TEMPO has been used to oxidize alcohols in acetic acid using the secondary oxidant hydrogen peroxide in the presence of methyltrioxorhenium ($MeReO_3$) and hydrobromic acid.[275,276]

Nitroxide Catalysts

It is tempting to speculate that of the many known nitroxides,[27,99,277,278] some will have interesting selectivities and properties for various oxidations. However, this search for selective catalysts has not been very rewarding. Essentially all of the reported nitroxide-catalyzed reactions have used the TEMPO moiety, either as such, or modified or attached to polymeric supports, and their reactivity profiles are very similar to each other.

The three most widely used monomeric nitroxides, TEMPO (**1**), MeO-TEMPO, and AcNH-TEMPO, are all stable and commercially available. TEMPO is low melting (mp = 36°), volatile, possesses a strong odor at room temperature, and readily sublimes. Its 4-methoxy and 4-acetamido derivatives are stable, odorless, crystalline substances that are easier to work with. The acetamido derivative AcNH-TEMPO has some desirable solubility properties, it is very soluble in methylene chloride, but much less so in diethyl ether (0.5 g in 100 mL) and water (3.2 g in 100 mL). Hence it has been recommended for carbohydrate oxidations[217] because it can be extracted from ether with water, or it can be extracted from water with methylene chloride.[143]

Reasons to explore other nitroxide systems include finding reagents that are more stereoselective (see "Stereochemistry" section), more site selective, more efficient, or are longer-lasting oxidation catalysts. A series of bicyclic and tricyclic nitroxides, **40, 41, 42**, and **43**, have been synthesized and investigated.[279–281] In direct comparisons with TEMPO, the oxidation yields were appreciably better with **41**, especially with sterically hindered alcohols. It is reasoned that the reactive site is sterically more available for reaction. However, the preparation of these compounds involves several steps. Compounds **44** and **17** are stable, effective catalysts.[233]

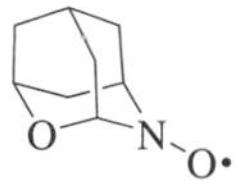

40

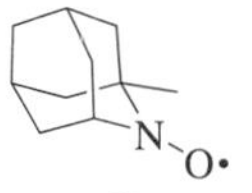

41

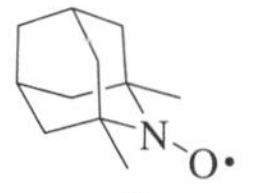

42

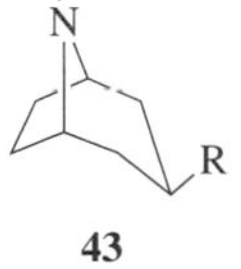

43

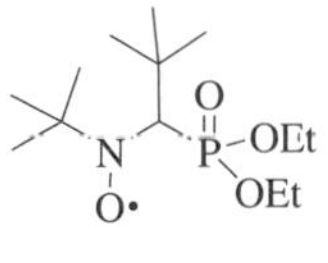

44

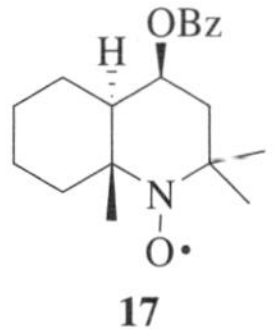

17

Catalyst Stability. The stability problem does not lie with the nitroxides, as many of these are known to be remarkably stable.[27,99,277,278] The problem is whether the oxoammonium ion formed during the catalytic process is stable, or at least stable long enough to carry out the desired oxidation. Several unstable nitroxides are discussed in the Stereochemistry section, namely some isomers of **8**,[85] and **10–13**.[94] Fortunately, such instability is readily detected by cyclic voltammetry.

Cyclic voltammetric (CV) studies have been used extensively to explore the suitability of nitroxide catalysts.[85,94,230,279,282–286] In CV, if the oxidation product, the oxoammonium ion, is stable, the reduction current will equal the oxidation current, and the reaction is said to be reversible. If the current in the reduction mode is less than in the oxidation mode, the reaction is not (fully) reversible because all or a portion of the oxoammonium ion has decomposed within the time frame of the scan. Any nitroxide that does not provide fully reversible oxidation couples would be a poor catalyst. The CV can then be measured in the presence of a substrate. In this case the reverse current would be much less, indicating that the oxoammonium ion was used in an oxidation.

The reasons for oxoammonium instability are unknown in most cases.[10,28] However, a base-promoted Hofmann-like elimination can be observed with certain salts (Eq. 83).[150,287,288] The decomposition is especially prominent in the presence of a 4-keto function in the piperidine framework.

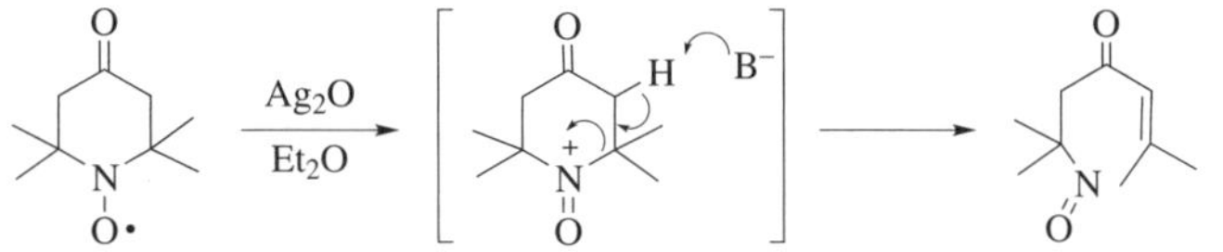

(45%) (Eq. 83)

Only a small fraction of the known nitroxides have been investigated as oxidation catalysts.[10,252,289] The so-called doxyl radical **45** decomposes on oxidation with *m*-CPBA, and the pyrroline and pyrrolidine radicals **46, 47**, and **48** do not react with alcohols in the presence of a peracid[150] or in electrochemical reactions in the presence of bromide.[223] However, similar nitroxides have been converted to oxoammonium salts such as **7** and used for oxidations.[85,290] Nitroxides **49–54** were found to be unsatisfactory catalysts for bleach oxidations.[95,285]

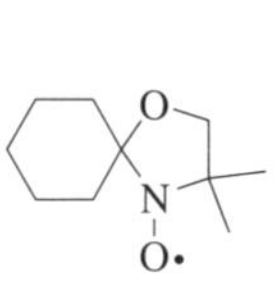

45

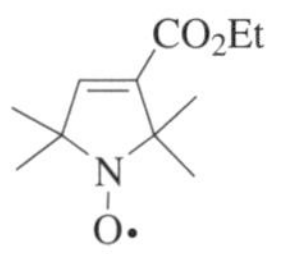

46

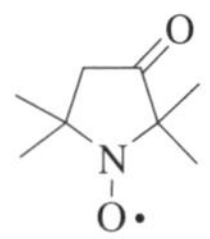

47

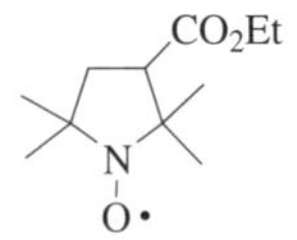

48

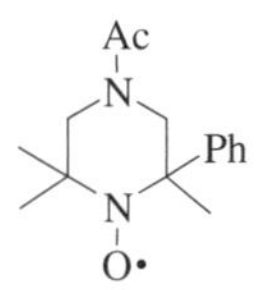

49

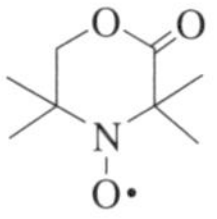

50

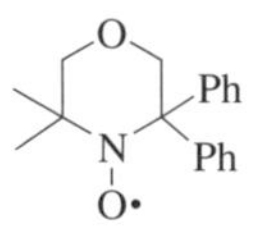

51

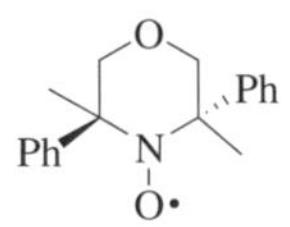

52

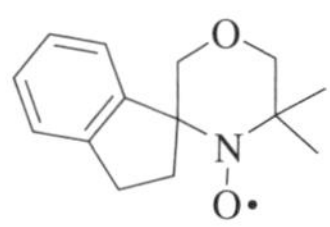

53

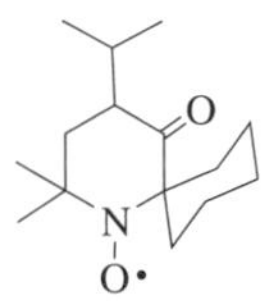

54

A series of fluorinated nitroxides have been prepared and used in oxidations.[291] TEMPO has been linked with a porphyrin and treated with a metal ion to give a metalloporphyrin catalyst used with bleach to oxidize alcohols to carbonyl compounds and sulfides to sulfones.[292] TEMPO has also been linked to a substituted 1,3,5-triazine with fluorinated side chains and used as a catalyst.[293]

Supported Nitroxides. Supported nitroxides can be defined as nitroxides attached to an entity that facilitates the removal of the catalyst from a reaction mixture. It can be as simple as an ionic liquid, but is usually a polymeric material such as an organic polymer, or a silica gel. Finding such systems has been a major thrust in current nitroxide catalysis research. The advantage of these materials is that they can be removed by filtration or phase separation and reused. TEMPO units bound in numerous ways to electrode surfaces have been discussed in the "Nitroxide-Catalyzed Electrochemical Oxidations" section.

Ionic Liquid Nitroxides. Several nitroxide–ionic liquid compounds have been made, such as **55**,[294] **56**,[295] and **57**.[200] Reactions are carried out in an aqueous solution of **56** in ionic liquid **58** ($[bmim]^+PF_6^-$).[295] Both solvent and reagent are water-soluble, allowing the separation of the organic products by extraction with organic solvents.[295] The reactions with **55** are electrochemical oxidations, and those with **56** use bleach as a secondary oxidant. An interesting ionic liquid secondary oxidant, **59**, has been used in conjunction with ionic liquid-modified nitroxide **57**.[200]

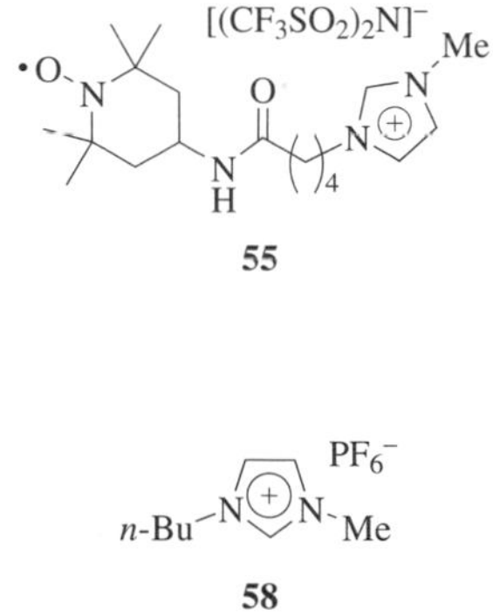

55

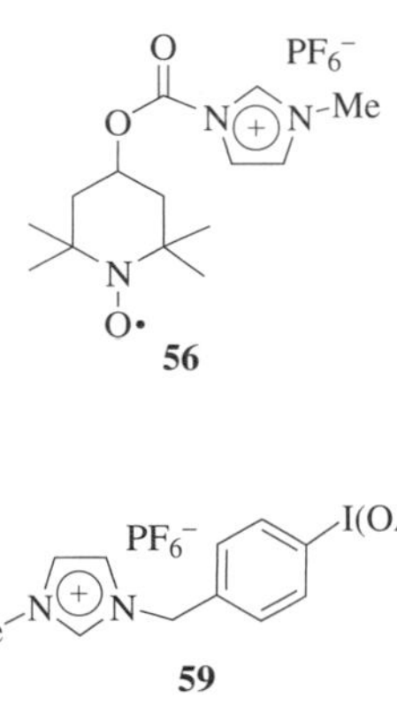

56

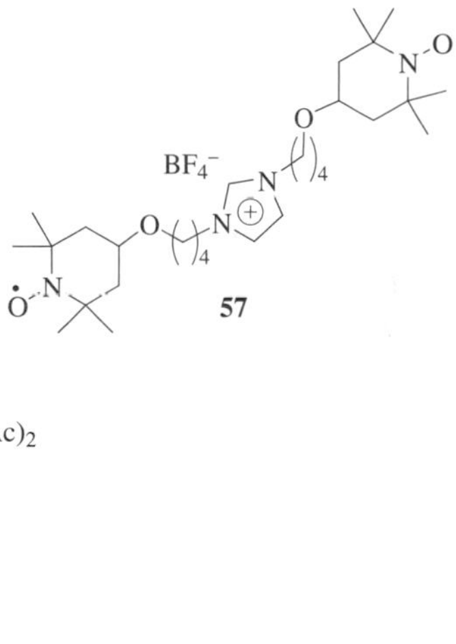

57

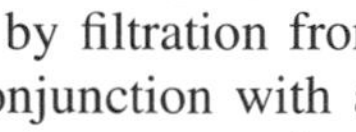

58

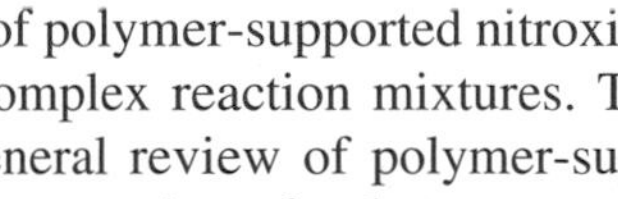

59

Organic Polymer-Supported Nitroxides. A major drive in nitroxide catalysis has been the preparation of polymer-supported nitroxides that can be easily recovered by filtration from complex reaction mixtures. The field has been reviewed in conjunction with a general review of polymer-supported organic catalysts.[42] Important aspects in the preparation of polymer-supported nitroxides are the ease of their preparation, and their reactivity, shelf-life, and stability toward the oxidation conditions.

One organic polymeric material named PS-TEMPO (**60**), originally provided by Novabiochem, is commercially available. The exact structure of the polymer portion of PS-TEMPO is not known. However, directions for the preparation of a similar material are available (Eq. 84).[121] Polymer **60** has been converted into an oxoammonium chloride and used as a stoichiometric reagent.[121] The polymer has been used for the oxidation of a number of alcohols (see the "Experimental Procedures" section).[191]

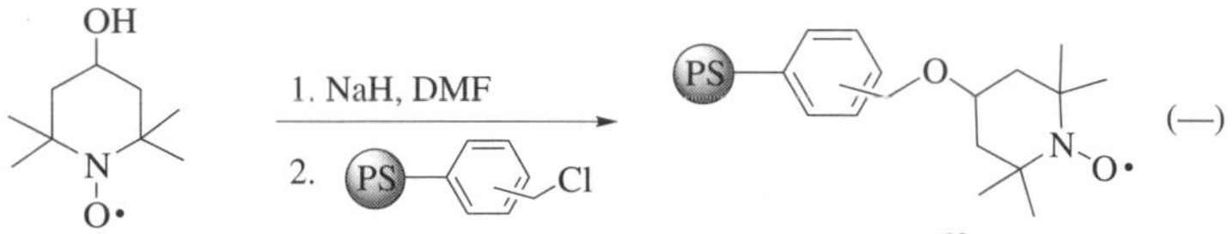

60

(Eq. 84)

Another material, the 4-hydroxy-TEMPO derivative that is linked through an ester moiety to a polymer matrix, is also known as PS-TEMPO (**61**).[273] Similarly linked TEMPO derivatives, such as **62**,[296] **63**,[265] and **64**[297,298] have been prepared by (co)-polymerization of the appropriate vinyl monomers.

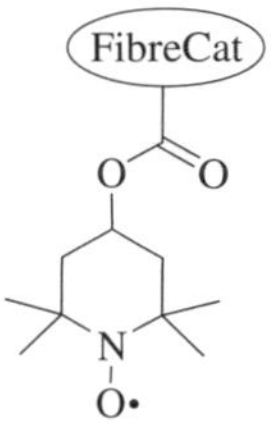

61

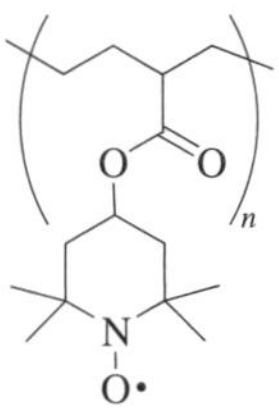

62

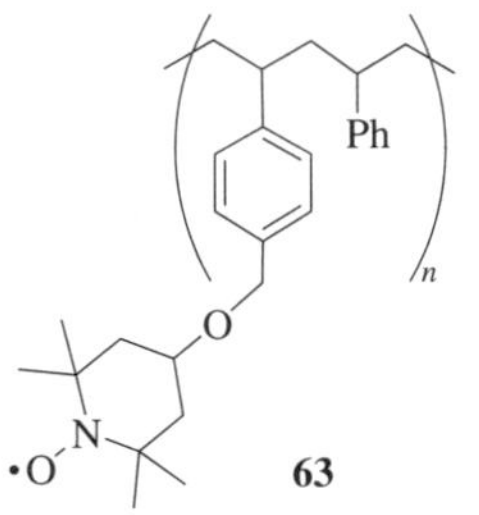

63

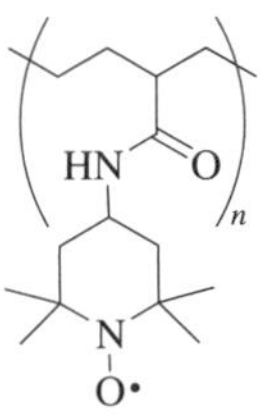

64

Still another polymeric material is based on the commercially available polymer, Chimassorb 944, which contains 4-amino-2,2,6,6-tetrahydropiperidine residues that can be oxidized to a nitroxide polymer, designated PIPO (for Polyamine Immobilized Piperidinyl Oxyl) (Eq. 85).[24,45,46,241,299] PIPO has been used for the catalyzed bleach oxidations of numerous alcohols, as well as methyl α-D-glucoside (to the corresponding uronic acid) and does not require the usual bromide catalyst. It can be used in hexane[46] or *tert*-butyl methyl ether[299] rather than the chlorinated solvents traditionally used. PIPO can also be used in the nitroxide-catalyzed oxidation of benzylic and allylic alcohols with copper(I) chloride and air in dimethylformamide.[241]

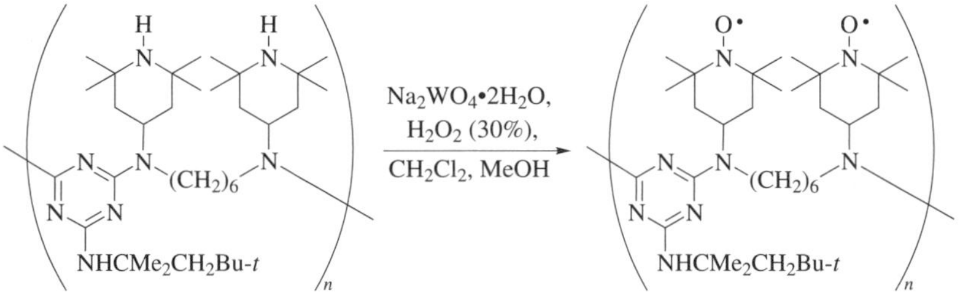

Chimassorb 944 PIPO (—)

(Eq. 85)

A number of other polymer-supported TEMPO systems differ in the nature of the polymer and the attachment of TEMPO. The structures of **65**,[300] **66**,[301] and **67** are shown.[302]

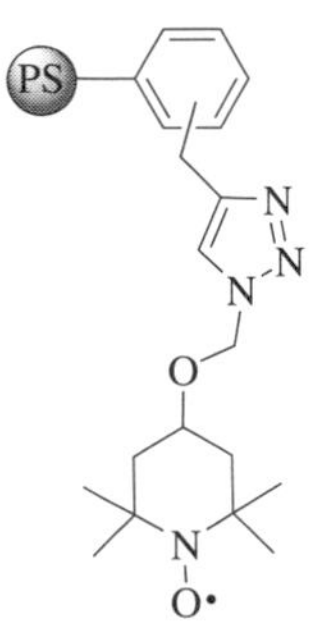

65

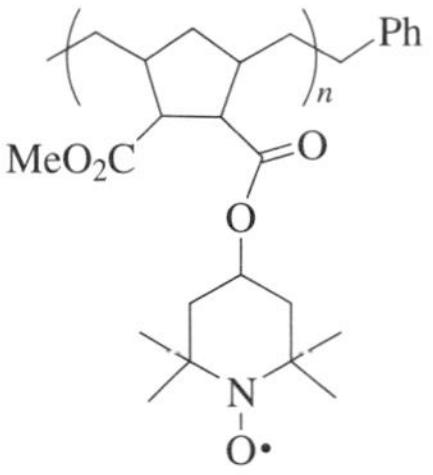

66

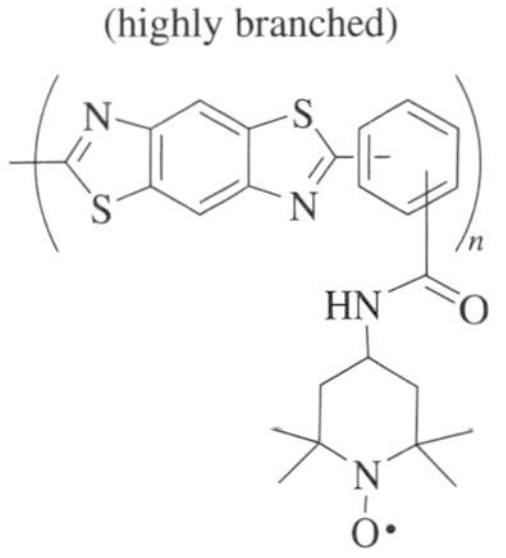

67

Finally, poly(ethylene glycol) polymers have been used to support TEMPO (**68, 69**)[303,304] and PROXYL (**70**)[305] units on the end of the polymer chain. Short polyglycol chains have been explored,[303–305] though systems with longer tethers of molecular weights of <5000 Daltons are also known.[272,306] These reagents are soluble in many polar solvents but can be precipitated with ether.

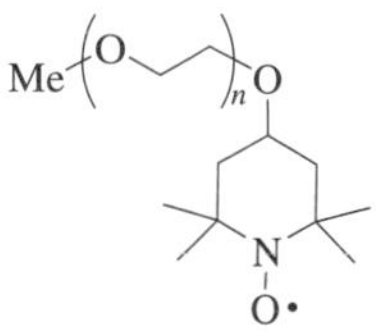

68

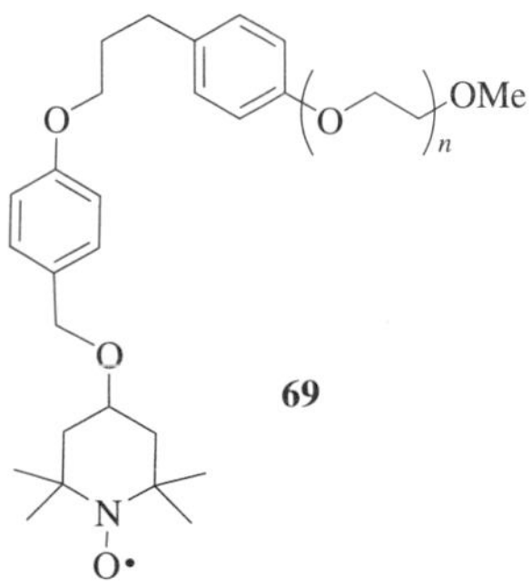

69

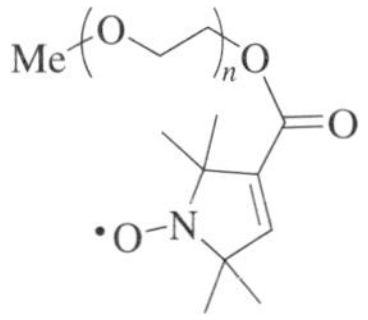

70

Silica-Based Nitroxides. TEMPO residues have been bonded in a multitude of ways to silica gel surfaces or incorporated into ormosil (organically modified silica gels) sponges.

TEMPO and its derivatives can be bonded to a silica surface as shown in Eq. 86.[161,307,308] This TEMPOH derivative has been used in catalytic experiments rather than the nitroxide, thus supporting the catalytic mechanism shown in Scheme 3. A commercial version of the oxidized (radical) form of this material is known as Silica Bond TEMPO or Si-TEMPO, and is manufactured and sold by SiliCycle (Quebec, Canada). It is available (as Si-TEMPO, **71**) from Aldrich and other suppliers.[156] Other silica-bound materials are ester-linked **72**,[273] amine-linked **73** (to MCM-41 or SBA-15, mesoporous silicas)[309,310], and ester-linked

74.[239] Most remarkably, the catalyst bound to SBA-15 can be used for air oxidations in the presence of sodium nitrite; no metal ion is required as co-catalyst.[310]

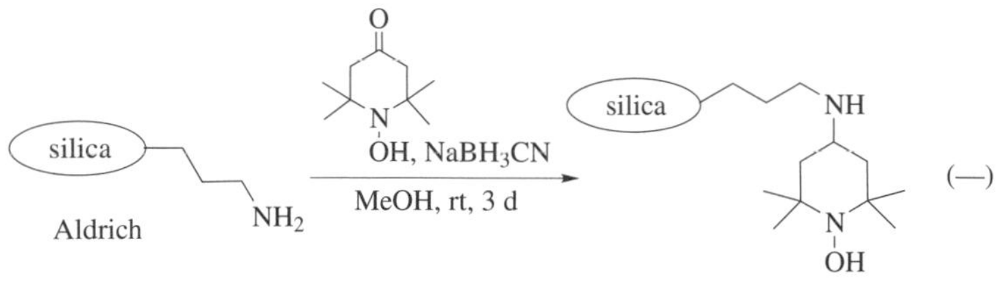

(Eq. 86)

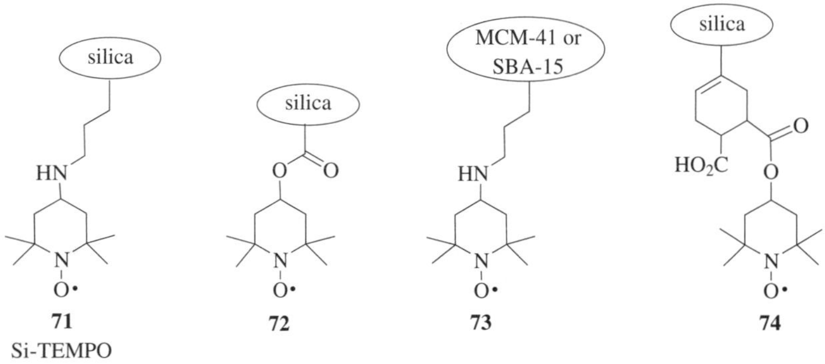

Another type of TEMPO-derived material is one in which the TEMPO moiety is incorporated into a silica matrix, forming an ormosil.[311] These materials are porous silicates formed by the hydrolysis of alkylalkoxysilicates in the presence of trialkoxysilyl derivatives of TEMPO (Eq. 87).[167,312,313] A commercially available form of the material is called SiliaCat TEMPO, from SiliCycle (Quebec, Canada).[156]

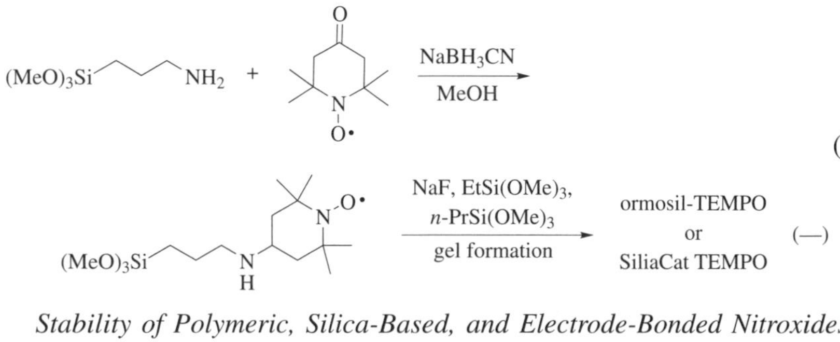

(Eq. 87)

Stability of Polymeric, Silica-Based, and Electrode-Bonded Nitroxides. The ability to recycle the solid-phase nitroxides adds to the advantage of their removability by filtration from the reaction mixtures. However, only a few studies have appeared in which the viability of recovered catalyst is tested in a series of successive oxidations. The silica-TEMPOH derivative shown in Eq. 86 was investigated over ten runs for the bleach oxidation of nonanol to nonanal.[308] Under the standard conditions,[134] the yield of nonanal is reduced from the initial 89% to 80%. In a competition experiment of a mixture of nonanol and, for instance, diphenylmethanol, essentially identical results over ten runs are recorded (90% nonanal and 5% benzophenone). These results are comparable to results of stability experiments obtained with an ormosil-TEMPO catalyst.[313]

Carbohydrate and Nucleoside Oxidations

Nitroxide-catalyzed, and to a much lesser extent, stoichiometric oxoammonium-salt oxidations, have become important tools for the modification of carbohydrates (Eq. 88)[33,61,314,315] and nucleosides (Eq. 89)[193]. In both cases, the primary hydroxyl group can be selectively oxidized to a carboxylic acid without degradation of the structure. During carbohydrate oxidations the secondary alcohols are rarely oxidized, and particularly when strict pH and temperature controls are employed, carbon–carbon bond cleavage is not observed.[316,317] Thus, under suitable conditions, the oxoammonium-mediated oxidations of carbohydrates and nucleosides are quite selective and mild, with little to no over-oxidation.

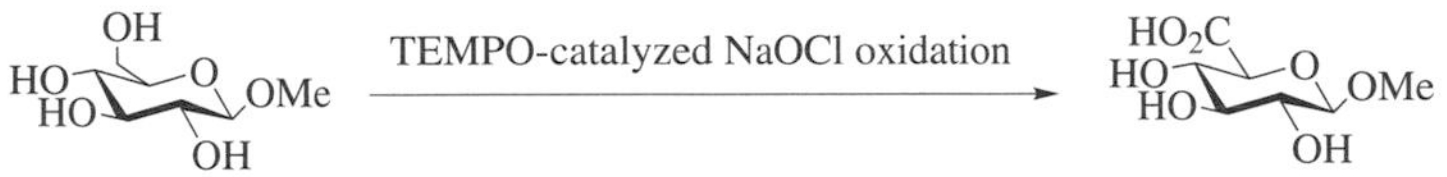

(—)

usually converted to a derivative for isolation

(Eq. 88)

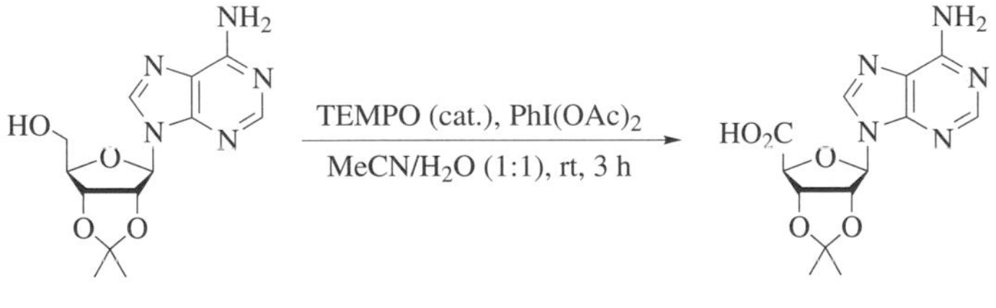

(90%) (Eq. 89)

Carbohydrates. Monomeric and polymeric carbohydrates are the most abundant highly oxygenated carbon compounds in nature. Since they are so important as renewable industrial feedstock, there are many patents on the various aspects of their nitroxide-catalyzed oxidations. We do not consider the patent literature in this chapter, however, it is well covered in several reviews and papers.[23,33,75,318]

Carbohydrates can be oxidized in three possible ways. When entirely unprotected, the first and easiest carbon to oxidize is C-1, the anomeric carbon. This oxidation converts the carbohydrate to the corresponding lactone. In cases where the alcohol on the anomeric carbon is protected as in a glycoside, oxidation on C-6 (or any other primary alcohol group) takes place. The oxidation of this primary alcohol produces in rare cases an isolable aldehyde or, most commonly, a carboxylic acid. If both the anomeric carbon and the primary alcohol group are unprotected, concomitant oxidation at both positions can occur, forming a diacid. Secondary alcohols in carbohydrates do not appear to be oxidized.

Stoichiometric Oxidations. Oxidations of unprotected glycosides using stoichiometric quantities of preformed oxoammonium salts have been observed only in anhydrous dimethylformamide solutions in the presence of 2,6-lutidine (Eq. 42).[144] The C-6 hydrate is isolated. Whether this reaction is made possible by the anhydrous nature of the dimethylformamide or the 2,6-lutidine is not clear.

Oxidations of protected sugars generally have no solubility problems in methylene chloride and proceed smoothly in the presence of pyridine. This reaction in the presences of base is not surprising since 2,6-lutidine has been used extensively in catalyzed electrochemical oxidations (see above).[60] Simple sugars in which C-6 is protected and C-1 is free are easily oxidized to glyconic lactones (Eq. 90, see also the "Experimental Procedures" section).[76] It is of interest that the unprotected secondary hydroxyl group on C-3 is not oxidized. On the other hand, if C-1 is blocked and C-6 is open, oxidation in the presence of pyridine produces dimeric esters (Eq. 41).[77]

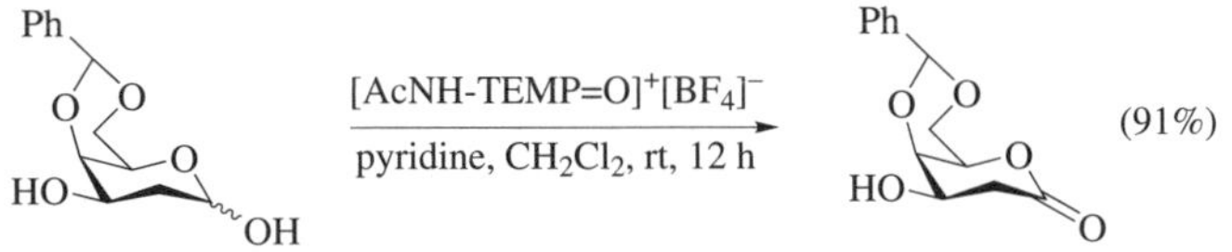
Ph
O
O
O
HO
OH
[AcNH-TEMP=O]+[BF4]−
pyridine, CH2Cl2, rt, 12 h
Ph
O
O
O
HO
O
(91%)

(Eq. 90)

Catalyzed Oxidations. Almost all reported oxidations of carbohydrates have been catalyzed reactions, mostly with the TEMPO–sodium hypochlorite–potassium bromide system. As previously discussed, this oxidation normally employs a two-phase system with water as one phase. Most carbohydrate oxidations are carried out in monophasic aqueous solutions, as both TEMPO and the substrate are water-soluble. Glycoside bond cleavage is rare but has been observed in galactose derivatives[319] and polysaccharides.[317]

The most studied oxidations have been of glycosides in which the C-6 alcohol (or a corresponding primary alcohol group) is unprotected and is oxidized to the corresponding C-6 aldehyde (a dialdose sugar derivative) or, most frequently, to a carboxylic acid (a uronic acid). Oxidations to the aldehyde stage must be carried out under anhydrous conditions in dimethylformamide. In one example, the base is sodium bicarbonate, used with TEMPO and trichloroisocyanuric acid (TCC) as secondary oxidant (Eq. 91).[320]

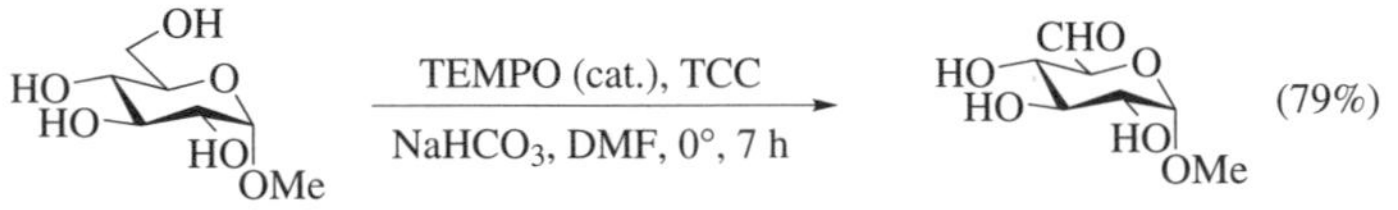
OH
HO
O
HO
HO
OMe
TEMPO (cat.), TCC
NaHCO3, DMF, 0°, 7 h
CHO
HO
O
HO
HO
OMe
(79%)

(Eq. 91)

Intermediate aldehydes have been intercepted in a Wittig reaction (Eq. 92, see also the "Experimental Procedures" section, compare to Eq. 60).[205] A number of similar oxidations of protected sugars with free C-6 alcohols have also been carried out using Swern oxidations. The aldehyde is isolated in those cases.[321]

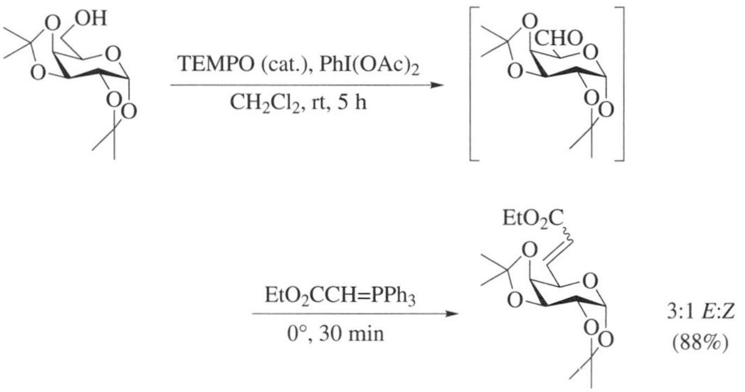

(Eq. 92)

Oxidation of a glycoside to the uronic acid is a four-electron oxidation that takes place with surprising selectivity. Although such a four-electron oxidation must proceed through an aldehyde (Eq. 93), such an intermediate has not been isolated in aqueous conditions, although its hydrated form has been observed in reaction mixtures.[61]

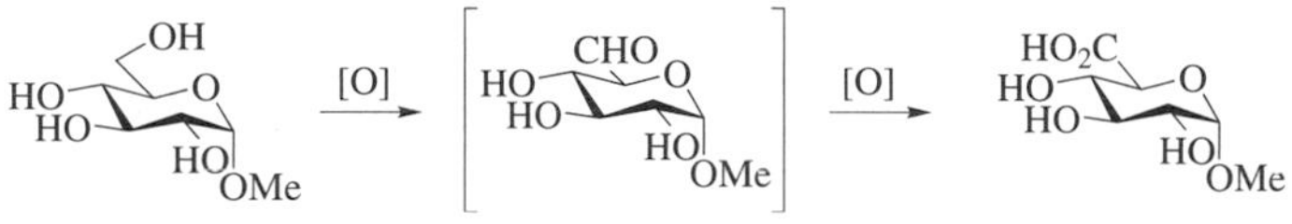

(Eq. 93)

The reason for the surprising selectivity of the oxidation for the primary hydroxyl group is thought to be steric hindrance. This observation is in accord with the arguments presented in the "Mechanisms, Stereochemistry, and Site Selectivity" section.[75] Better selectivity is observed with pyranosides than with furanosides.[75] Cycloalkanone model reactions exhibit similar reactivity. Cyclopentanol reacts four times faster than cyclohexanol under the normal conditions, and 3-methylcyclohexanol reacts faster than 2-methylcyclohexanol;[61] rate differences are attributed to steric factors.

The carboxylic acid groups formed on C-6 are often converted into the corresponding methyl esters with methanol/acid or diazomethane before isolation. Lactones are formed from a series of partially blocked thioglucosides, as shown in Eq. 94 for a galactose derivative.[322] Similar 6,3-lactones are prepared from glucose and mannose derivatives. A 6,1-lactone has been prepared from a glucose derivative, and a 6,2-lactone from a mannose derivative.[322] These lactones have subsequently been converted to methyl esters. No sulfur oxidation in the two-phase methylene chloride/water reactions is seen if the reaction times are carefully controlled. Anhydrous methylene chloride gives lower yields.

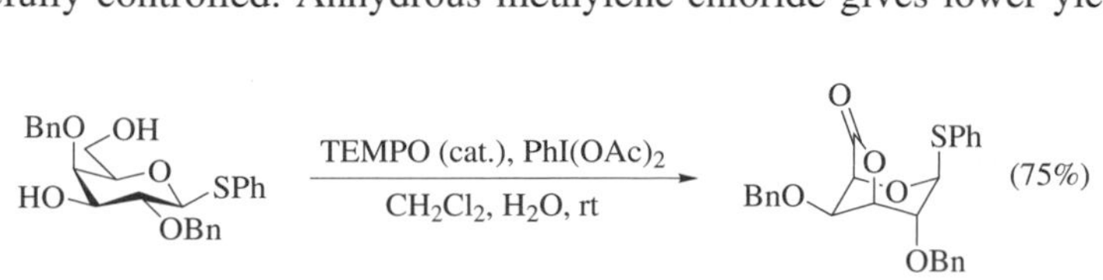

(Eq. 94)

A series of methyl 2-deoxy-2-acetamido-D-glycopyranosyl-1-azide uronates has been prepared (Eq. 95).[173] The intermediate carboxylic acids were esterified and acetylated to yield the final products. As would be expected, the acetamido group is resistant to oxidation.

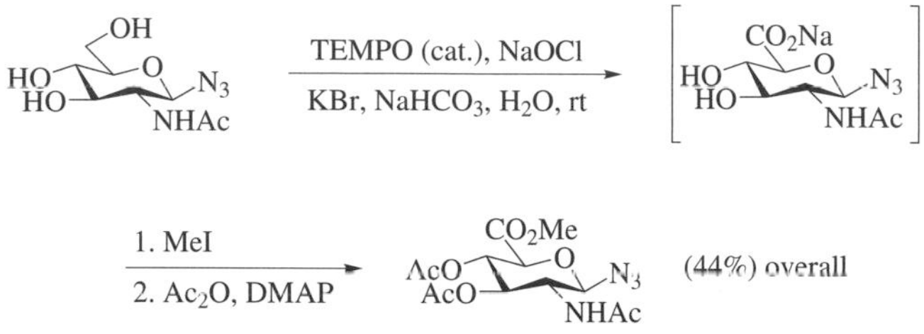

(Eq. 95)

Glucose is oxidized electrochemically in the presence of TEMPO to give mixtures of gluconic and glucaric acid (mostly the latter).[323] Simple alkyl glycosides, some disaccharides, and cyclodextrin have been subjected to nitroxide-catalyzed electrooxidation using TEMPO to give uronic acid derivatives in good yields with a platinum anode in an undivided cell.[283,324,325] A TEMPO-impregnated Nafion membrane has also been used in an electrochemical cell for carbohydrate oxidations.[326] The oxidation of sucrose is shown in Eq. 96.[283] Ultrasonic irradiation in the catalyzed oxidation of methyl α-D-glucopyranoside and sucrose increases the oxidation rates appreciably.[327,328]

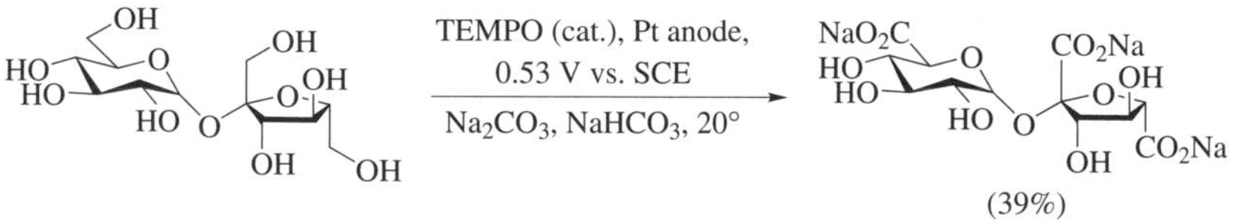

SCE = standard calomel electrode

(Eq. 96)

The nitroxide-catalyzed oxidation of simple aldose sugars to glycaric acids is illustrated in Eq. 77 for D-glucose.[257,316,317,329] In this work, the secondary oxidants used are sodium hypochlorite, chlorine, or bromine. The substrates are D-glucose, D-mannose, and D-galactose, and the yields of sodium salts (from D-glucose and D-mannose) or free acid (from D-galactose) vary between 70 and 80%. Mannaric acid sodium salt is not obtained in a pure form. If the desired product is the glucaric acid salt, potassium hydroxide rather than sodium hydroxide can be used to obtain the water-insoluble monopotassium glucarate.[257,329]

It is suggested that these conversions take place in three steps.[316,330] The first is the oxidation of C-1 to an acid salt as is normal with bromine and chlorine oxidations even without a catalyst. The second reaction, the oxidation of C-6 to an aldehyde is nitroxide-catalyzed and is the slow step. The third step, the further oxidation of the C-6 aldehyde to a carboxylic acid likely does not require any nitroxide catalysis as the secondary oxidants are all known to oxidize aldehydes to acids.

These glycaric acid preparations are carried out between 0° and 5° with careful pH control, held above 11.5. When the pH drops below 11.5, appreciable carbon–carbon cleavage is observed.[316,317] It is interesting that in these cleavage reactions, no five-carbon fragments are observed during a careful kinetic study of the reaction; that is, no C-1/C-2 or C-5/C-6 cleavage products are observed.

A number of polysaccharides have been oxidized using the general sodium hypochlorite conditions.[32,33,75,331,332,333] In general, these oxidations closely parallel the oxidations of monomeric glycosides. In fact, for the 1,6-linked aldopyranosides, methyl glucoside has served as a model for the reactions.[75,317,334] Thus, amylose, for instance, is oxidized under basic conditions on C-6 to the corresponding carboxylic acid salt (Eq. 97, see also the "Experimental Procedures" section). The salt can be neutralized to obtain polyglycuronic acids (uronans).[332]

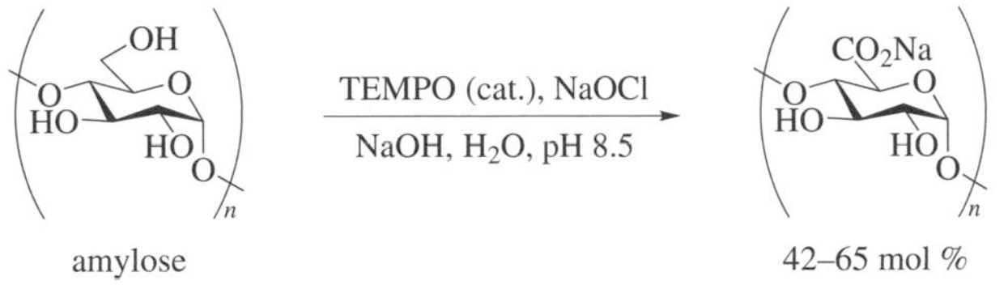
OH
O
O
HO
HO
O
n
amylose
TEMPO (cat.), NaOCl
NaOH, H2O, pH 8.5
CO2Na
O
O
HO
HO
O
n
42–65 mol %

(Eq. 97)

As in the oxidations of simple glycosides, an aldehyde on C-6 is postulated as an intermediate, but is not found in the polysaccharide oxidation products.[33,275] The polypyranosides give better results than the polyfuranosides.[75,319] Control of the pH is essential, although there is not complete agreement on the optimum value. A pH of 9.5 for polysaccharides and 11.5 for monosaccharides, oligosaccharides, and polysaccharides has been recommended.[33,317]

Water-insoluble polysaccharides such as cellulose present an additional problem.[335,336] Native celluloses are oxidized only on the edges of the fibers and crystalites.[337,338] Pretreated celluloses give a more uniform oxidation.[339,340] Cellulose that has been rendered water soluble by partial acetylation can be oxidized to partially acetylated polyglucuronic acids.[341]

β-Cyclodextrin is partially oxidized under the normal bleach conditions.[283,342] If the secondary alcohol groups on cyclodextrin are protected, the material can be almost completely oxidized to the per(5-carboxy-5-dehydroxymethyl) derivative.[343] Some over-oxidation, attributed to the hypochlorite ion, is noted in these reactions.[75]

4-Substituted TEMPO derivatives have been compared with one another for some of the oxidations. These nitroxides include MeOCO-TEMPO, MeO-TEMPO, AcO-TEMPO, AcNH-TEMPO, and $MeOSO_2$-TEMPO. The AcNH-TEMPO functions best at a pH of 8 rather than 9, and its reactivity is higher than that of the other nitroxides.[33,314] Carbohydrates can be oxidized with TEMPO, methyltrioxorhenium, and potassium bromide using hydrogen peroxide as secondary oxidant in acetic acid.[275]

Nucleosides. The ability to oxidize nucleosides constitutes a special case in which oxoammonium salts are suitable despite the fact that amine groups are present. All nitrogens in nucleosides are on or in heterocycles and, therefore, the normal side reactions are not observed. Typical examples of the oxoammonium-catalyzed oxidations of nucleosides to the corresponding acids are shown in Eq. 89 for an adenosine derivative (see also the "Experimental Procedures" section),[193] and in Eq. 98 for a thymine derivative.[344] The secondary alcohol groups in both substrates are protected to prevent any undesired oxidation. The reaction is not limited to solution-phase chemistry as a solid-phase nucleoside carboxylic acid preparation shows (Eq. 99).[345]

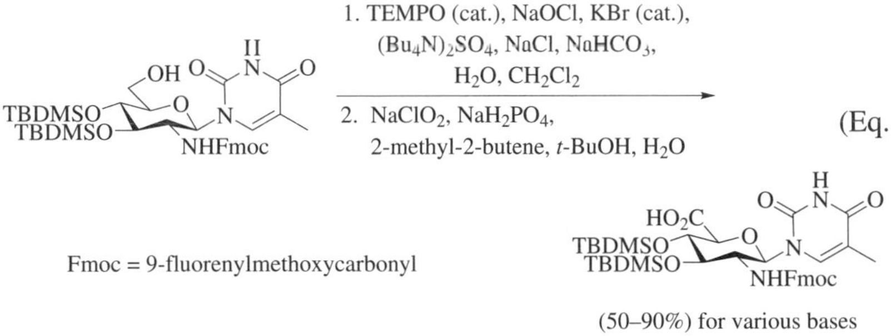

(Eq. 98)

Fmoc = 9-fluorenylmethoxycarbonyl

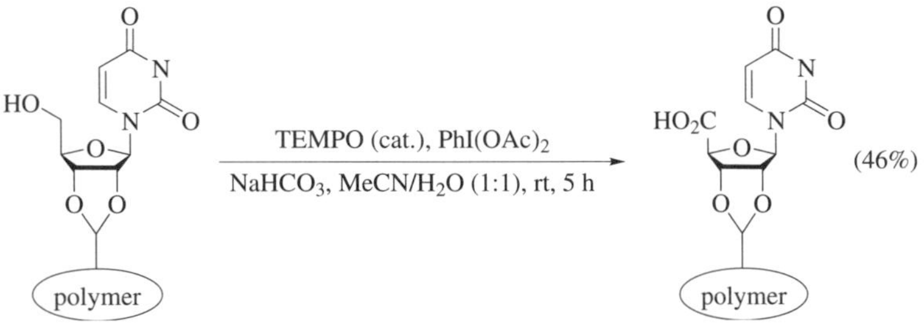

(Eq. 99)

Miscellaneous Reactions

Although this chapter is devoted to the oxidation of alcohols, oxoammonium salts react with various other functional groups. These reactions have been observed mainly during stoichiometric oxidations with preformed oxoammonium salts, often with acetonitrile as solvent. The roles that they might play in nitroxide-catalyzed reactions are not clear. Many unsuccessful, or less successful, nitroxide-catalyzed reactions have been reported, suggestive of the existence of a number of side reactions but the reasons for the reaction failures have rarely been explored. Whether these reactions interfere with a desired oxidation or not is often a matter of relative rates or the use of different solvents. Many of the side reactions are slow compared with an alcohol oxidation. Equations and references for these reactions are given below with minimal comment.

Rearrangement and Oxidation of Tertiary Allylic Alcohols. This reaction is shown in Eq. 36 and may represent a rearrangement of a tertiary allylic alcohol catalyzed by the oxoammonium ion followed by the normal oxidation.[120,120a]

Reactions with Amines. Unprotected and unprotonated amines react rapidly with oxoammonium salts,[8] but the products are not well defined, and the reactions are generally avoided. At least in some cases, the reactions involve an oxidation of amines to yield imines. The imines can either be hydrolyzed to aldehydes or further oxidized to nitriles.[346] The yields are low, however, and much further work is needed to ascertain the scope and limitations of these reactions. The most successful oxidations of amine functional groups are those carried out in electrochemical systems (Eqs. 100 and 101).[108,282,296,346,347] Incidental to a different study, enantiomerically enriched sparteine is oxidized to aphylline, which is identified by GC–MS (Eq. 102).[106]

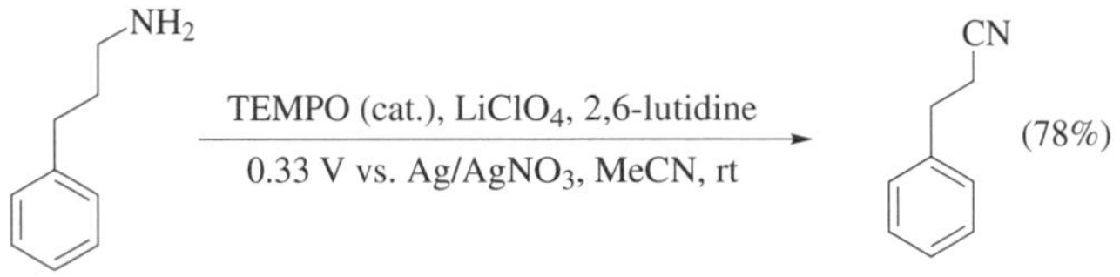
NH2
TEMPO (cat.), LiClO4, 2,6-lutidine
0.33 V vs. Ag/AgNO3, MeCN, rt
CN
(78%)

(Eq. 100)

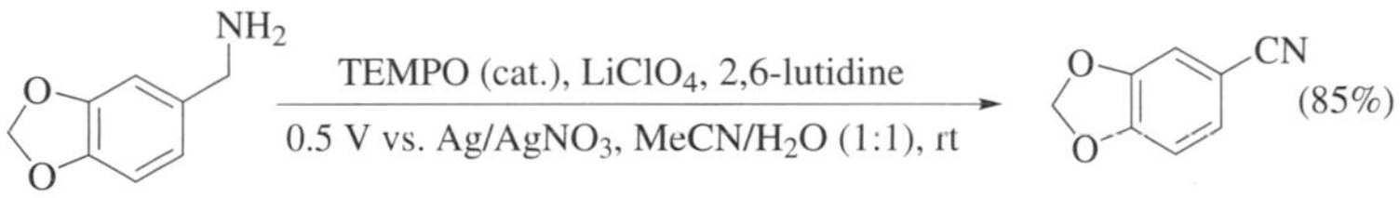
NH2
TEMPO (cat.), LiClO4, 2,6-lutidine
0.5 V vs. Ag/AgNO3, MeCN/H2O (1:1), rt
CN
(85%)

(Eq. 101)

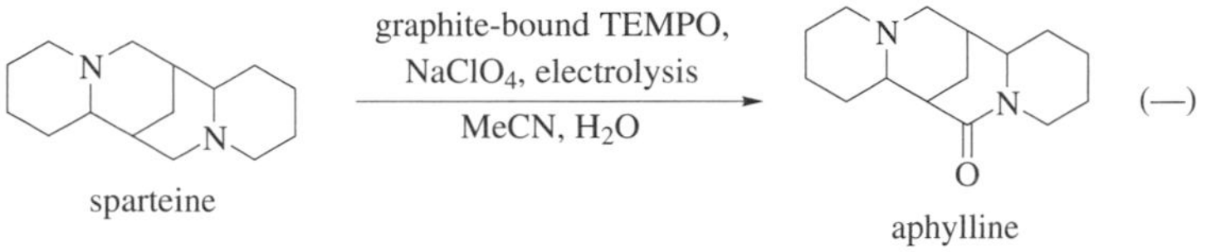
graphite-bound TEMPO,
NaClO4, electrolysis
MeCN, H2O
sparteine
aphylline
(—)

(Eq. 102)

Stoichiometric oxidations of amines with oxoammonium salts are known (Eqs. 103 and 104)[128,139] but are complicated because the amine is not only oxidized but also promotes the comproportionation reaction between oxoammonium ion and hydroxylamine to give nitroxide (Scheme 1).

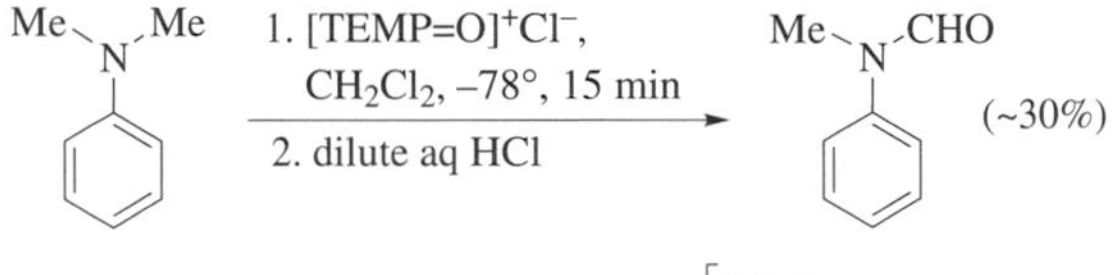
1. [TEMP=O]+Cl−,
CH2Cl2, −78°, 15 min
2. dilute aq HCl
(~30%)

(Eq. 103)

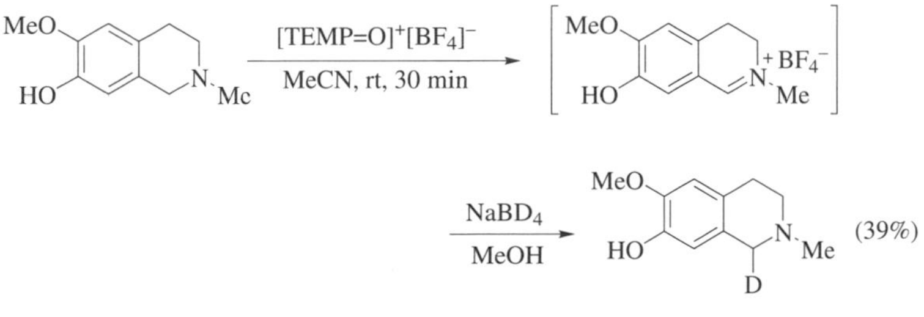
[TEMP=O]+[BF4]−
MeCN, rt, 30 min
NaBD4
MeOH
(39%)

(Eq. 104)

An interesting reaction uses a nitroxide-catalyzed bis(acetoxy)iodobenzene oxidation (Eq. 105).[348] Although the yield is low, the reaction illustrates several possible reactions: lactone formation, resistance of secondary alcohols to oxidation, imine formation, and ring closure. There is also one example of a tertiary amine being oxidized to a nitroxide.[349]

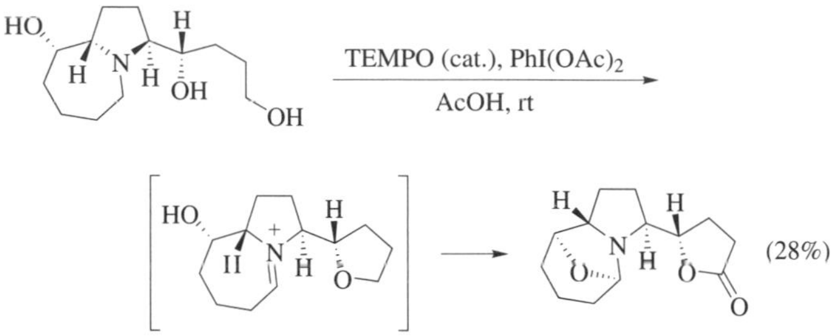

(Eq. 105)

Reactions with Ketones (Enols). Oxoammonium salts produce 1,2-diketones from ketones,[55,121,127,350–352] (Eq. 106)[351] and the intriguing preparation of 1,2,3-triketones from 1,3-diketones (Eq. 107).[350] The underlying reaction is the oxidation of the α-position to a ketone.

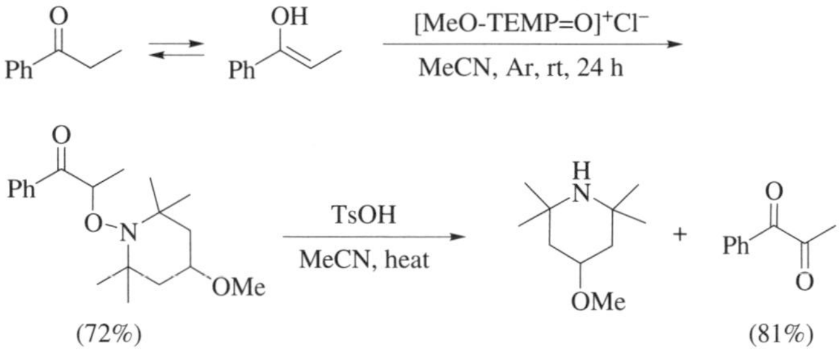

(Eq. 106)

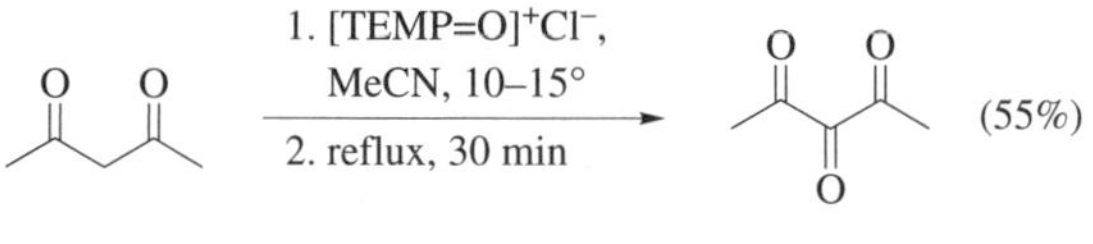

(Eq. 107)

Reactions with Enol Ethers and Enamines. In addition to the reactions of ketones, which can also be considered as reactions of the enol double bond, oxoammonium salts react with enol ethers and enamines. An example of the former is given in Eq. 108.[51] When a nucleophile such as chloride is present, the reaction is a simple double bond addition. The chloride intermediate must be converted into the acetal before isolation. In the presence of a better nucleophile than the counter anion of the oxoammonium salt, such as a thymine derivative, the cationic intermediate can also be trapped (Eq. 109).[114] In essence, these reactions are double bond addition reactions with the oxoammonium salt acting as the electrophile, and chloride, ethoxide, or thymine acting as nucleophiles.

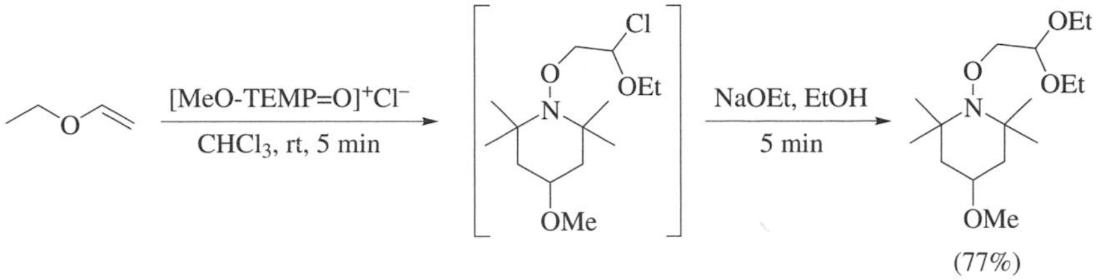

(Eq. 108)

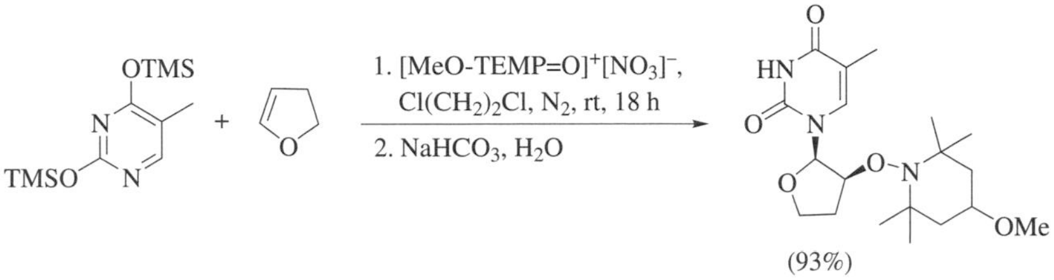

(Eq. 109)

Oxoammonium salts react with enolates[324] and enamines to add the piperidine residue as shown (Eqs. 110 and 111).[152,353] When reactions such as shown in Eq. 110 are carried out electrochemically, the intermediate cation-radical can dimerize.[152] Equation 111 is similar to the reactions with ketones (Eq. 106).

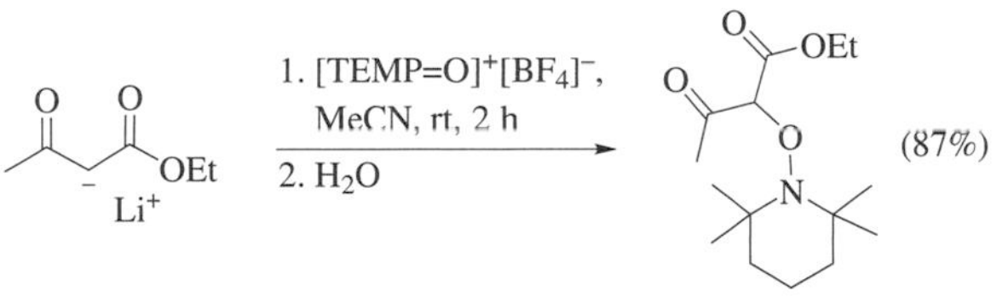

(Eq. 110)

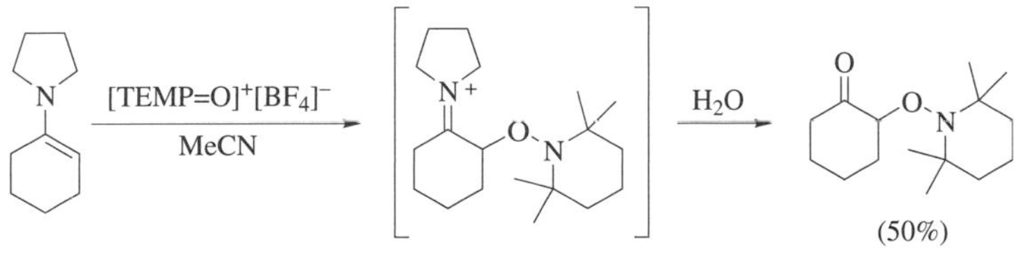

(Eq. 111)

Reactions with Other Activated Double Bonds. Simple alkenes add oxoammonium cations only if they have three or four alkyl substituents (Eq. 112).[57] In this reaction, no subsequent electrophile addition is observed; instead a proton is lost to form an allylic ether.

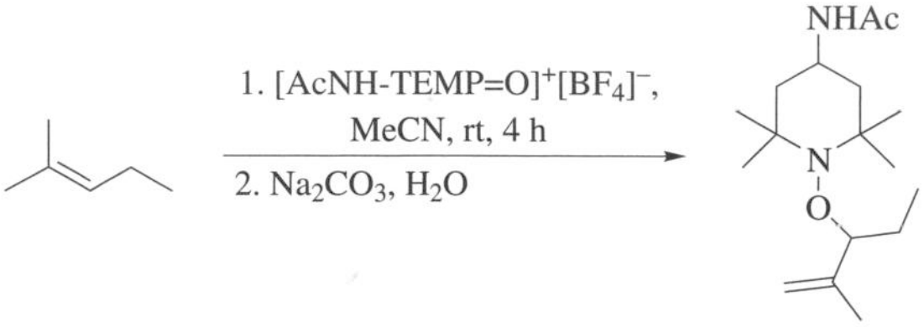

(93%) (Eq. 112)

The unique oxoammonium salt shown in Eq. 113 adds across conjugated double bonds.[354]

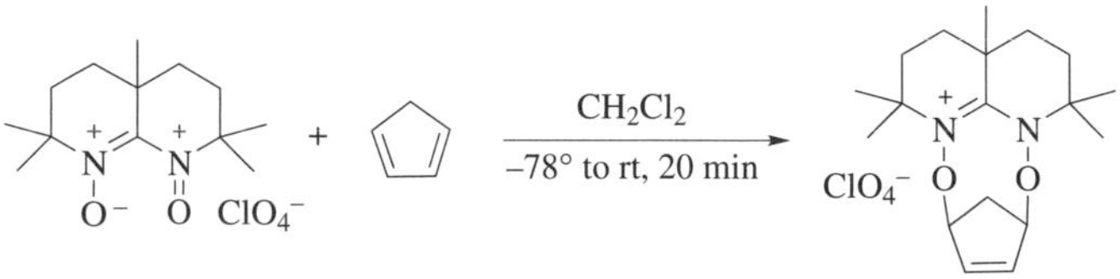

(86%) (Eq. 113)

Activated double bond systems are oxidized, both stoichiometrically and by catalyzed electrooxidation, to provide, depending on the substrates, either enones (Eqs. 114 and 115),[66] or aromatic systems (Eq. 116).[67] Oxidation of tetrahydrocarbazole to the vinylogous amide uses stoichiometric amounts of the oxoammonium salt (Eq. 117).[124] All four of these reactions are carried out in aqueous acetonitrile; water is found to be necessary and, presumably, is the origin of the oxygen in the ketones.[66] Oxoammonium salts can be used to catalyze cationic polymerization of vinyl monomers.[355]

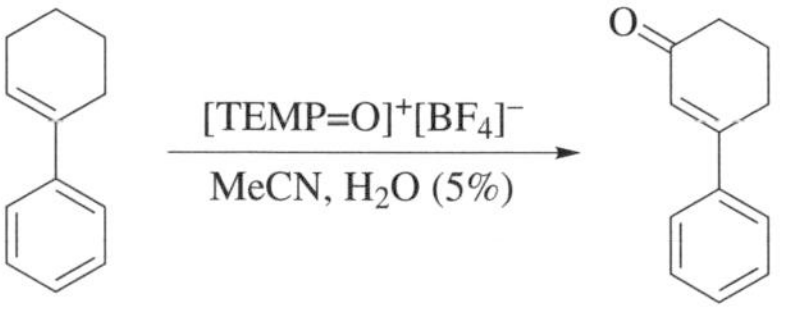

I (95%) (Eq. 114)

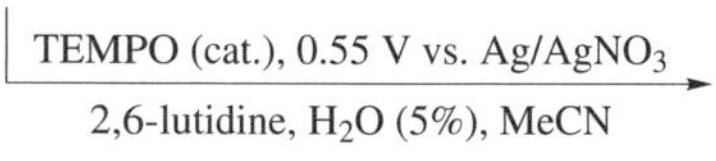

I (96%) (Eq. 115)

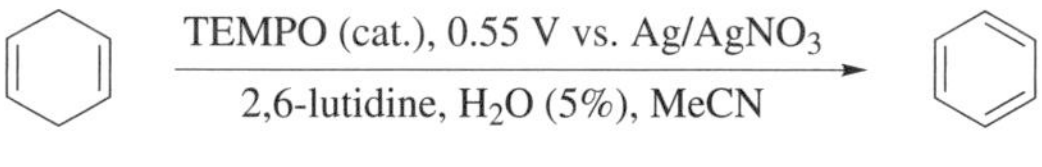

(95%) by GC (Eq. 116)

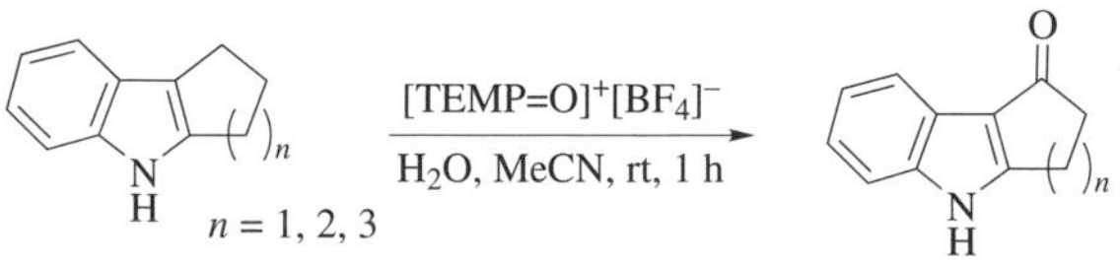

(45–75%) (Eq. 117)

An unexpected oxidation involving a cyclization is shown in Eq. 118.[356] Similar transannular reactions have been reported with sodium chlorite as a secondary oxidant[357] and with TEMPO-catalyzed copper salt reactions.[358] Although the mechanism of these reactions is not entirely clear, they may involve double bond oxidation. Thymine can be oxidized to thymine glycol with $[TEMP{=}O]^+Br^-$.[359] 1-Aryl-1,4-dihydropyridines are oxidized to pyridines with $[TEMP{=}O]^+[BF_4]^-$.[360]

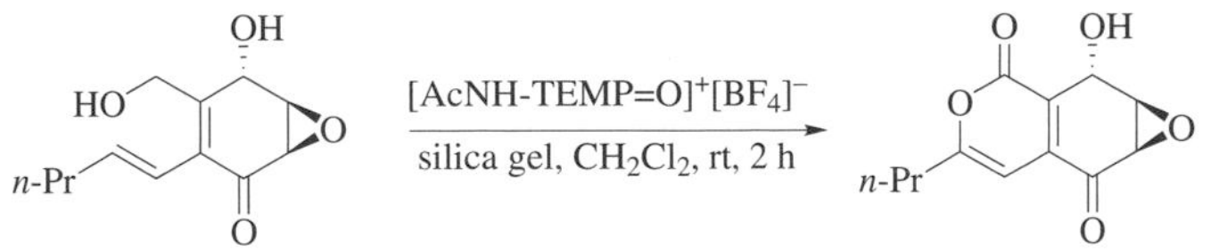

(85%) (Eq. 118)

Reactions with Phenols. As one might anticipate, phenols and phenol ethers can be oxidized to quinones or carbon–carbon coupled to form dimers (Eq. 119).[139–141,352,361,362]

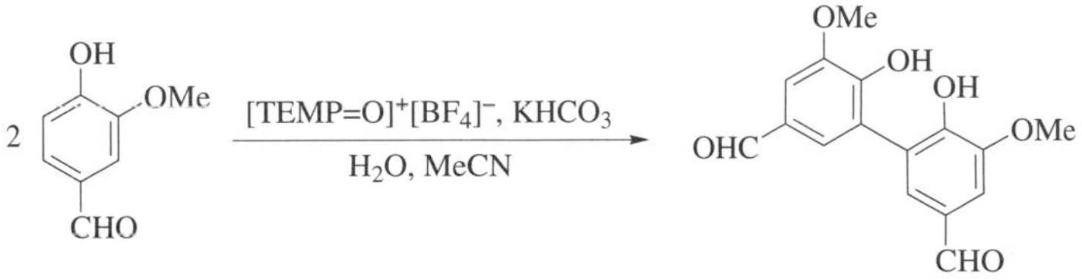

(85%)

(Eq. 119)

Reactions with Sulfur Compounds. Reactions of sulfides with oxoammonium reagents are controversial. It has been reported that sulfides, such as dibenzyl sulfide and benzyl phenyl sulfide, and the sulfone, benzyl phenyl sulfone, do not react with $[TEMP{=}O]^+Cl^-$.[352,363] Alkyl and aryl thioglycosides do not react with TEMPO-catalyzed bis(acetoxy)iodobenzene under carefully controlled conditions[322,364,365] whereas phenyl thioglycosides[292] and phenyl sulfides[292,366,367] give sulfoxides using a TEMPO-catalyzed sodium hypochlorite system. One equivalent of $[TEMP{=}O]^+[BF_4]^-$ reacts with two equivalents of dodecanethiol to give a 79% yield of didodecanyl disulfide,[353] and a series of thiols has been oxidized electrochemically on a TEMPO-modified electrode.[232] A similar reactivity of the thiol groups in peptides under oxoammonium-catalyzed electrochemical reaction conditions has been reported.[368]

The sulfur in 1,3-thiazoles is not oxidized under TEMPO-catalyzed polymer-bound diacetoxybromate conditions.[43] Tetrathiafulvalene is oxidized to the bis-1,3-dithiolium cation radical salt using $[MeO\text{-}TEMP{=}O]^+$ salts with various anions (Eq. 120).[369]

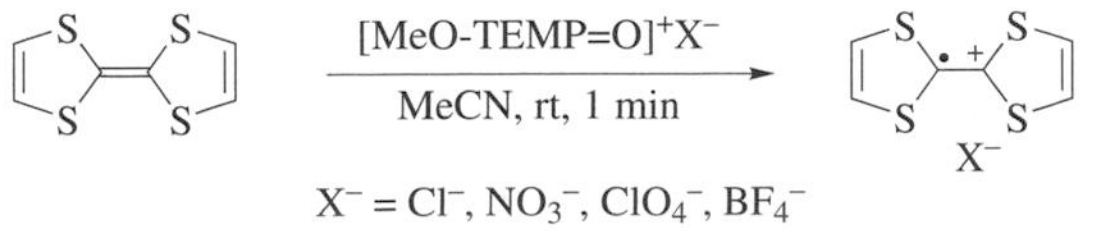

(90–95%) (Eq. 120)

Although sulfoxides have been isolated from catalyzed bleach reactions and should be considered stable to further oxidation,[366,367,370] $[AcNH\text{-}TEMP{=}O]^+$

$[BF_4]^-$ reacts vigorously when dissolved in dimethyl sulfoxide (for NMR analysis, for example), but no products have been isolated.[371]

Reactions with Ethers. Benzylic ethers react with oxoammonium halide salts to give benzaldehydes, although the scope and limitations are not entirely clear.[138] In methylene chloride, the reaction is slow, such that benzylic alcohols containing benzylic ethers can be safely oxidized without ether cleavage.[68] No problems are noted in the nitroxide-catalyzed oxidations of benzylic derivatives, including those in carbohydrates (see Tables).[315] TEMPO-catalyzed laccase oxidations of benzylic ethers give low yields of cleavage products or lactones.[253,372]

Other Reactions. In other minor reactions, oxoammonium salts react with Grignard reagents[54,60] and with a number of polar molecules (in water) to yield undefined products.[373] Most of these reactions have been reviewed.[10,20]

COMPARISON WITH OTHER METHODS OF ALCOHOL OXIDATION

Alcohol oxidations can generally be divided into two types, oxidations involving metal-based reagents and those using purely organic oxidants. The metal-based oxidants (such as chromium(VI) and manganese dioxide) are usually inexpensive and easily prepared, and some are quite selective. However, they almost all involve a (toxic) heavy metal that must be removed from the reaction mixtures and properly disposed of. There are two major non-metallic reagent systems in use, the Dess–Martin oxidation, using a periodinane, and the Moffatt–Swern oxidation, using dimethyl sulfoxide as the oxidant. The oxoammonium-salt and nitroxide-catalyzed reactions represent a third family of purely organic oxidants.

Two excellent comparisons of alcohol oxidation methods have been published in *Organic Reactions*. In the first, chromium-based reagents are compared with Moffatt–Swern oxidations.[321] In the second and more recent review, the chromium(VI)-based oxidations are compared with oxidations by manganese dioxide, Moffatt–Swern oxidations, Dess–Martin oxidations, and catalytic oxidations.[374]

Manganese Dioxide Oxidations

Manganese dioxide oxidations and stoichiometric oxoammonium oxidations are comparable in that they are usually carried out in an organic solvent such as hexane or methylene chloride. Under the proper conditions,[68,375] the reduced oxidant can be removed by a simple filtration to yield an almost pure product. In both reactions, the product solution can often be carried directly to the next step of a reaction sequence.

Manganese dioxide oxidations are carried out by stirring a solution of substrate with an excess of the activated reagent in a non-polar solvent. Activated manganese dioxide can be prepared in many ways and is commercially available, a major advantage over oxoammonium salts, but the exact nature of the manganese dioxide and its activation are not well understood. Commercially available "activated" manganese dioxide (Aldrich) suffices for the reactions. Manganese dioxide reactions can be carried out in many solvents, whereas the oxoammonium salt oxidations are traditionally carried out in methylene chloride, although several other solvents should be suitable (see below). Manganese dioxide oxidations have been extensively reviewed[376,377] and are the basis of the elegant "tandem oxidation processes".[375] These are reactions in which the manganese oxidation is carried out in a reaction mixture containing a reagent which will further react with the oxidized alcohol.

Two major problems are associated with manganese dioxide oxidations. The first is that the oxidations are restricted to allylic, benzylic, and some "semi-activated" alcohols such as cyclopropanemethanol.[378] The second is that the brown managanese dioxide is not soluble and the reactions are dark suspensions so that chromatographic methods must be used to monitor the progress of the reaction. In contrast, stoichiometric oxidations with oxoammonium salts take place with almost all primary or secondary alcohols, although they are slower with saturated aliphatic alcohols.[68] The progress of the reaction can be estimated by observing the conversion of the yellow oxoammonium salt slurry to a white hydroxyammonium salt slurry. Perhaps the most notable difference in the reactivity of the two reagents is that alcohols containing β-oxygens are not oxidized under the standard stoichiometric oxoammonium conditions. Also more side reactions, as described in the "Miscellaneous Reactions" section, may take place with the oxoammonium salts.

Chromium-Based Oxidants

The major advantages of the chromium(VI)-based oxidants are their commercial availability and the large body of knowledge on their use.[374] The major disadvantages are the generation of chromium-containing wastes, their predominant use as stoichiometric reagents, side reactions such as isomerizations and oxygen migrations, and the production of side products such as chromic esters and polymers that must be removed by chromatography. In a direct comparison with pyridinium chlorochromate (PCC) reactions, oxoammonium salts give somewhat better yields.[131] In light of these issues, the oxoammonium and nitroxide-catalyzed oxidations have clear advantages.

Moffatt–Swern Oxidations

Moffatt–Swern oxidations involve the treatment of dimethyl sulfoxide with an activator molecule such as dicyclohexylcarbodiimide (in the Moffatt reaction) or oxalyl chloride (in the Swern reaction).[321] This reagent is allowed to react with the alcohol, and the product is then treated in the same flask with a base such as triethylamine. The reduction product from the oxidant dimethyl sulfoxide is dimethyl sulfide. Several objections to this method have been raised. The reagents and solvents must all be carefully dried in order to get good yields, dimethyl sulfide has an obnoxious smell, and, in most instances, the necessary use of low temperatures. Side reactions are also observed.[321]

The oxoammonium- or nitroxide-catalyzed methods do not generate unpleasant odors and generally do not require anhydrous conditions, although the conditions can be anhydrous, if, for instance, desired for consecutive reactions,[68] or when bis(acetoxy)iodobenzene is used as a secondary oxidant. The in situ, stoichiometric oxidations are preferred over the Swern reaction for diol oxidations.[70] Epimerizations in complex molecules have occurred under Swern conditions that are not observed using nitroxide-catalyzed sodium hypochlorite oxidation conditions.[379] For reasons not understood, some acetylenic alcohols are not satisfactorily oxidized by the Moffatt–Swern method,[321] whereas this functional group is quantitatively oxidized by oxoammonium salts.[68]

Dess–Martin Oxidations

Dess–Martin oxidations are carried out using periodinane **75**, the Dess–Martin reagent.[380,381] The major problems that have arisen with this method are that the reagent itself is not always easy to prepare,[382,383] and there are reports that it may not always be safe.[384] However, it is commercially available.

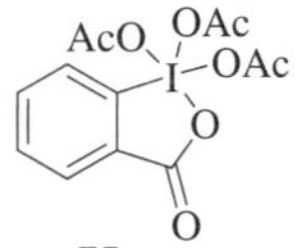

75

It is of interest that the hypervalent iodine reagent, bis(acetoxy)iodobenzene is used as a secondary oxidant for many nitroxide-catalyzed reactions and is thoroughly discussed in a previous section. All of the various hypervalent iodine reagents that have been used for alcohol oxidations have been recently reviewed.[41]

Comparison of the Oxoammonium-Nitroxide Systems with One Another

Stoichiometric oxidations with preformed oxoammonium salts are versatile, convenient, give good yields of pure products, and involve simple isolation procedures. Stoichiometric reactions in bases such as pyridine have not been extensively explored, although there is some promise in the method. Stoichiometric methods using in situ prepared salts have the advantage that the reagents (TEMPO and *p*-toluenesulfonic acid) are readily available. However, the scope and limitations of this reaction have not been extensively studied.

The nitroxide-catalyzed reactions have been extensively explored and account for most of the examples in this chapter. Nitroxides (such as TEMPO) are commercially available (albeit at high costs), but particularly TEMPO derivatives can be easily prepared, even in bulk.[20,27] On the other hand, nitroxides are used in catalytic amounts, and are available in polymer-bound forms. The nitroxide catalysts can be combined with a range of secondary oxidants, some of which are readily available and particularly inexpensive. The major difficulty with all nitroxide-catalyzed reactions is over-oxidation, that is, the secondary oxidant itself reacts with the substrate or its oxidation products to give unwanted products. However, a careful choice of solvents, temperature, and secondary oxidant can inhibit or altogether prevent undesirable over-oxidation.

Aqueous sodium hypochlorite (bleach), has been used most often as the secondary oxidant for the conversion of alcohols to ketones, aldehydes, or carboxylates. When oxidizing substrates that are not water-soluble, the main disadvantage is that the reaction has to be performed in a two-phase system, which may introduce phase-transfer problems. At any rate, the wet products must then be dried. The reaction has also shown some bothersome side reactions such as double bond oxidation and chlorination. The closely related hypochlorite-chlorite method has been most widely used to convert alcohols directly into carboxylates.

The most promising and versatile secondary oxidants for catalyzed reactions are probably bis(acetoxy)iodobenzene or a polymeric derivative thereof.[202] These can be used in aqueous two-phase reactions or in many organic solvents under anhydrous conditions. Little over-oxidation of the aldehydes to the carboxylates is seen. However, the byproducts, acetic acid and iodobenzene, must be dealt with, and 1,2-diols may be cleaved,[192,207] though such cleavage reactions have not been reported in nitroxide-catalyzed chemistry.

EXPERIMENTAL CONDITIONS

Solvents

Neutral or Slightly Acidic Conditions. Stoichiometric oxoammonium oxidation reactions are usually carried out in methylene chloride using a slight stoichiometric excess of oxidant. Acetonitrile (see discussion above), dimethylformamide[144,320] and 1,2-dichloroethane[114] have also been used as solvents for this reaction; water has not been used often.[373] The oxidant (soluble or as a slurry) and the substrate are stirred at ambient temperature until they become white or until a negative starch–iodide paper test is obtained (Scheme 5). The reaction rates are greatly increased by the presence of silica gel.[68] Typical reaction times are 1–12 hours.

Basic Conditions. These reactions are performed in methylene chloride using a two-fold excess of oxidant in the presence of two equivalents of pyridine at ambient temperature.[76,77] The isolation procedure is given in the "Experimental Procedures" section.[76]

Nitroxide-Catalyzed Reactions. Nitroxide-catalyzed oxidations may be carried out in aqueous solutions, aqueous mixtures with acetone or acetonitrile, or in two-phase systems consisting of methylene chloride and a buffered aqueous system. Water is used for its excellent solubility of carbohydrates and other polar substrates.[373] The secondary oxidant used may also necessitate an aqueous solvent. For instance, the most common secondary oxidant is a basic solution of sodium hypochlorite (bleach). Solid sodium hypochlorite is unstable although calcium hypochlorite is stable. Secondary oxidant salts such as sodium chlorite also necessitate an aqueous solvent. However, some secondary oxidants such as bis(acetoxy)iodobenzene can be used in an entirely organic system such as methylene chloride.[192]

Ionic liquids such as 1-butyl-3-methylimidazolium hexafluorophosphate ($[bmim]^+PF_6^-$) have been used for nitroxide-catalyzed oxidations of alcohols with hydrogen peroxide and hydrogen bromide,[268] sodium hypochlorite,[201,246] in electrochemical reactions,[294] and in aerobic oxidations with copper(I) chloride[385].

pH

As noted previously in the Mechanistic Considerations section, the rate and specificity of oxoammonium oxidations, especially nitroxide-catalyzed reactions, are strongly dependent on the pH, which must be carefully controlled. Oxidations using stoichiometric quantities of oxoammonium salts are generally performed under neutral or slightly acidic conditions although there is increasing interest in these reactions in pyridine.[76,77] The in situ stoichiometric reactions are necessarily performed under strongly acidic conditions.

The pH of the various aqueous systems used in nitroxide-catalyzed reactions is extremely important, must be closely controlled, and held in the basic range (typically from 8 to 11). The pH control is usually achieved using aqueous buffers (in purely aqueous systems, in homogenous aqueous solvent systems, as well as in biphasic systems). However, this is not sufficient, especially in large-scale reactions, as base is consumed in the course of the reaction and must be added to maintain the pH at the desired value. Hence, these additions are often conducted with the help of an automated titrator system.[329]

Temperature

Essentially all of the reactions described in this chapter have been carried out at room temperature or below, although generally not lower than $-10°$. Many of the reactions are fast but some may take hours to days. Acceleration of the reaction by raising the temperature is generally not recommended, as this may erode the selectivity of the oxidation reaction. Oxidations at elevated temperatures have not been studied carefully.

EXPERIMENTAL PROCEDURES

Some procedures that follow were extracted from general procedures and were adapted to a specific compound recorded in a table within the research paper. Thus, some of the stated quantities are calculated, rather than taken from actual published experimental procedures. As far as possible, and with the exception for the procedures published previously in *Organic Syntheses*, the procedures herein were submitted to the original authors for editing and approval. We greatly appreciate their cooperation.

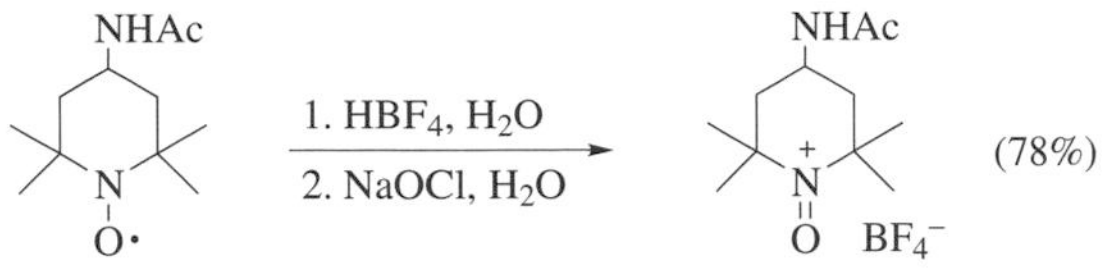

4-Acetamido-2,2,6,6-tetramethylpiperidine-1-oxoammonium Tetrafluoroborate ([AcNH-TEMP=O]$^+$[BF_4]$^-$) [Preparation of an Oxoammonium Salt by Disproportionation/Oxidation of the Corresponding Nitroxide]. This procedure is described in *Organic Syntheses.*[12]

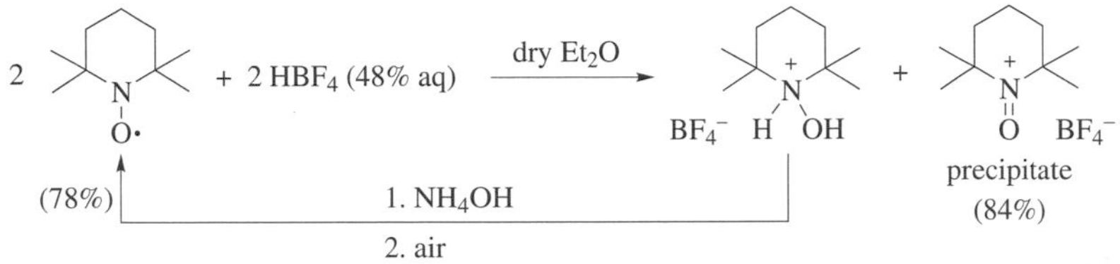

2,2,6,6-Tetramethylpiperidine-1-oxoammonium Tetrafluoroborate ([TEMP= O]$^+$[BF_4]$^-$) [Preparation of an Oxoammonium Salt by Disproportionation and Recovery of Nitroxide].[124] Tetrafluoroboric acid (48% aqueous; 9 mL, 68 mmol) was added dropwise to TEMPO (5.3 g, 34 mmol) in dry Et_2O (20 mL) at 0° with stirring. The solution was allowed to warm to 15° and a yellow precipitate formed. After 10 minutes, the precipitate was collected by filtration and dried under vacuum to give 4.48 g (84%) of 2,2,6,6-tetramethylpiperidine-1-oxoammonium tetrafluoroborate: mp 164–165° (lit.[4]162.5–163.5°). The filtrate was made basic (pH 10) with concentrated aqueous ammonia, and the layers were separated. The aqueous layer was extracted with Et_2O (5 × 15 mL), and the Et_2O layers were dried and evaporated. The residue was allowed to stand in an open flask for three weeks to air-oxidize the hydroxylamine to recovered nitroxide (1.95 g, 78%): mp 35–37° (lit.[386] 36–37°).

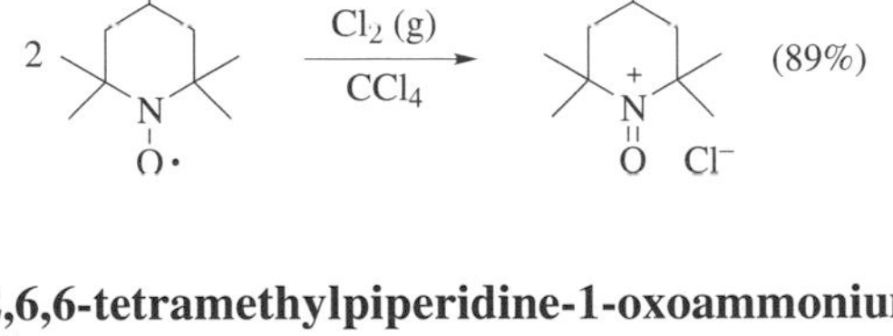

4-Methoxy-2,2,6,6-tetramethylpiperidine-1-oxoammonium Chloride ([4-MeO-TEMP=O]$^+$Cl$^-$) [Preparation of an Oxoammonium Salt by Direct Oxidation of the Corresponding Nitroxide].[11] Anhydrous Cl_2 gas was bubbled into a stirred solution of 4-methoxy-2,2,6,6-tetramethylpiperidiene-1-oxyl

(2.0 g, 10.7 mmol), dissolved in CCl_4 (100 mL). The orange precipitate that was formed was collected by filtration and washed with CCl_4 to give [4-MeO-TEMP=O]$^+$Cl$^-$ (2.1 g, 89%): mp 121–123° dec.; IR (KBr) 2951, 2897, 2827, 1616, 1466, 1446, 1388, 1377, 1219, 1163, 1106 cm^{-1}. Anal. Calcd for $C_{10}H_{20}NO_2Cl$: C, 54.17; H, 9.09; N, 6.32; Cl, 15.99. Found: C, 53.60; H, 9.22; N, 6.29; Cl, 16.88.

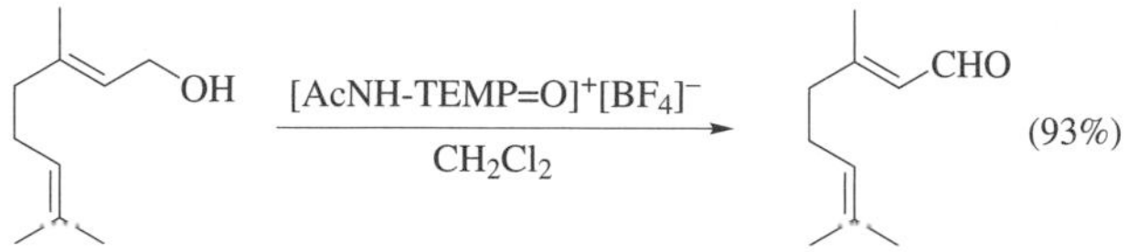

Geranial [Oxoammonium Salt Oxidation of a Primary Alcohol Under Slightly Acidic Conditions]. This procedure is described in *Organic Syntheses.*[12]

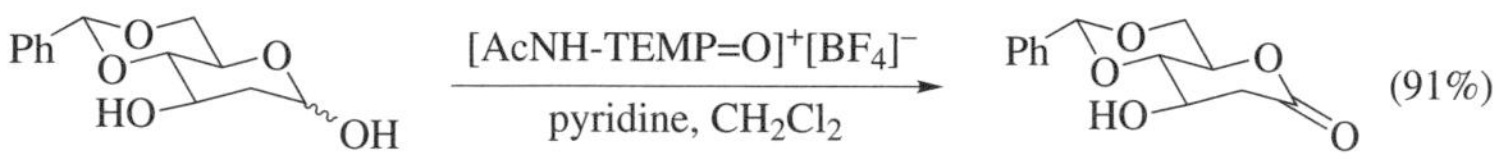

4,6-Benzylidene-2-deoxy-D-gluconolactone [Oxoammonium Salt Oxidation of a Hemiacetal Under Basic Conditions].[76] To a solution of 4,6-benzylidene-2-deoxy-D-glucopyranose (0.50 g, 2 mmol) in CH_2Cl_2 (20 mL) was added pyridine (0.33 g, 4.1 mmol), followed by [AcNH-TEMP=O]$^+$[BF_4]$^-$ (1.23 g, 4.1 mmol). The mixture was stirred overnight, filtered to remove the pyridinium tetrafluoroborate and evaporated to dryness under reduced pressure. The residue was suspended in dry Et_2O (50 mL) and filtered to remove most of the AcNH-TEMPO. The filtrate was concentrated and passed through a short column (5 × 1 cm) of flash-grade silica gel. Ether eluted the product in front of the orange AcNH-TEMPO band. The ether solution was evaporated to dryness, and the residue was recrystallized from H_2O/EtOH to give 0.45 g (91%) of the product: mp 156–157°; ^{1}H NMR (400 MHz, $CDCl_3$) δ 2.72 (dd, 1H), 3.24 (dd, 1H), 3.78 (br t, 1H). 3.87 (t, 1H), 4.14 (m, 1H), 4.29 (m, 3H), 4.49 (dd, 1H), 5.64, (s, 1H), 7.54 (m, 5H); ^{13}C NMR (100 MHz, $CDCl_3$) δ 37.7, 66.5, 68.1, 80.7, 102.1, 126.3, 128.5, 129.6, 136.4, 168.

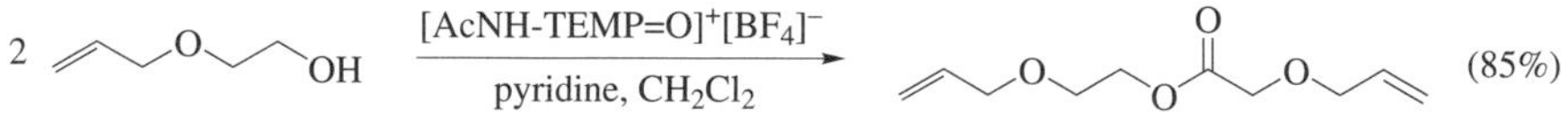

2-(2-Propenyloxy)ethyl-2-(2-propenyloxy)acetate [Oxidative Dimerization of an Alcohol with a β-Oxygen Using an Oxoammonium Salt].[77] Activated 4 Å molecular sieves (4 g, 0.1 g per mL of solvent) and [AcNH-TEMPO]$^+$ [BF_4]$^-$[12] (3.75 g, 12.5 mmol) were stirred with 2-(2-propenyloxy)ethanol (0.51 g,

5 mmol) in CH_2Cl_2 (40 mL). After 30 minutes, pyridine (0.91 g, 11.5 mmol) dissolved in CH_2Cl_2 (5 mL) was added dropwise over 5 minutes. The solution was stirred for 3 hours (turning from bright yellow to orange) and filtered to remove pyridinium tetrafluoroborate. The filtrate was reduced to dryness under vacuum and triturated with dry Et_2O. The solid remaining (AcNH-TEMPO) was washed with Et_2O (4 × 10 mL), and the combined ether solutions were concentrated to a small volume and passed through a column of silica gel. The column was eluted with Et_2O, and all of the fractions eluting before the orange nitroxide band were collected. Removal of the solvent by rotary evaporation provided the title ester (0.43 g, 85%), which required no further purification. The NMR spectrum was identical to that of the known compound.[387]

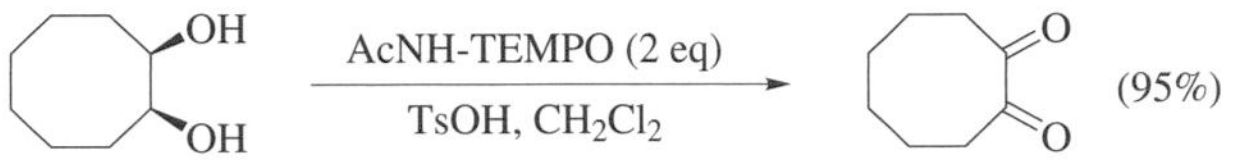

1,2-Cyclooctanedione [In Situ Prepared Oxoammonium Salt Oxidation of a 1,2-Diol to a 1,2-Dione].[70]

A solution of AcNH-TEMPO (1.07 g, 5 mmol) in CH_2Cl_2 (10 mL) was added dropwise to a magnetically stirred suspension of *cis*-cyclooctane-1,2-diol (0.14 g, 1 mmol) and TsOH monohydrate (0.95 g, 5 mmol) in CH_2Cl_2 (5 mL), maintained at 0°. The resulting mixture was stirred for 1 hour at 0°, allowed to warm to room temperature, and stirred for 72 hours for complete reaction (TLC control). EtOH (2 mL) was added to the reaction mixture, which was stirred for further 30 minutes. Water (30 mL) was added, and the phases were separated. The aqueous phase was extracted with CH_2Cl_2 (2 × 10 mL). The combined organic phases were dried ($MgSO_4$), filtered, and concentrated under reduced pressure to give an orange oil containing small amounts of orange AcNH-TEMPO. This oil was dissolved in CH_2Cl_2 and passed through a short pad of flash-grade silica gel. The fractions eluting before the orange band were concentrated under reduced pressure to provide cyclooctane-1,2-dione (0.13 g, 95%) as a light-yellow oil. The NMR spectrum was identical to that of the known compound.[388]

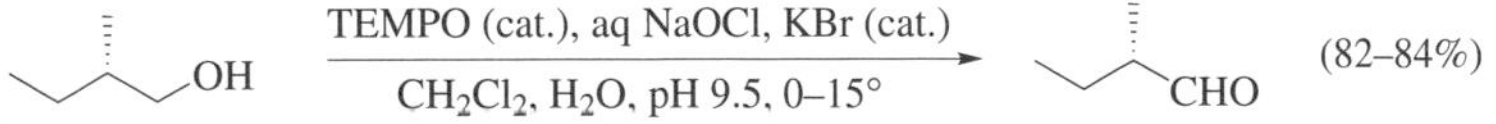

(*S*)-2-Methylbutanal [Nitroxide-Catalyzed Oxidation of a Primary Alcohol to an Aldehyde Using an Aqueous Solution of NaOCl/Br⁻ as Secondary Oxidant]. This procedure is described in *Organic Syntheses.*[13]

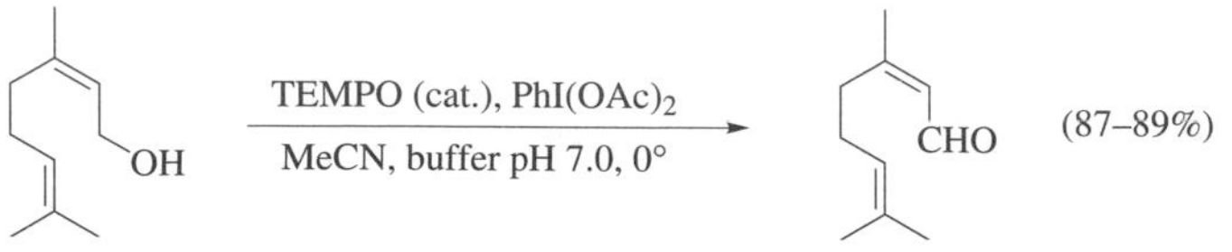
OH
TEMPO (cat.), PhI(OAc)2
MeCN, buffer pH 7.0, 0°
CHO
(87–89%)

Neral [Nitroxide-Catalyzed Oxidation of a Primary Alcohol to an Aldehyde in Buffered Aqueous MeCN Using $PhI(OAc)_2$ as Secondary Oxidant]. This procedure is described in *Organic Syntheses.*[15]

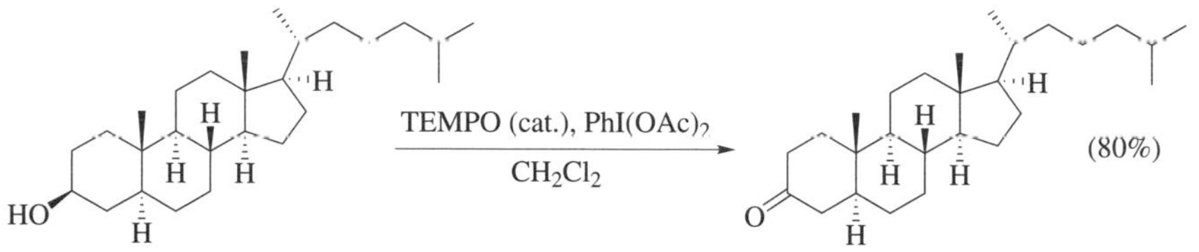
HO
H
TEMPO (cat.), PhI(OAc)2
CH2Cl2
O
(80%)

Cholestan-3-one [TEMPO-Catalyzed Bis(acetoxy)iodobenzene Oxidation of a Secondary Alcohol to a Ketone].[192] Bis(acetoxy)iodobenzene (345 mg, 1.1 mmol) was added to a solution of 0.388 g (1 mmol) of cholestan-3β-ol and TEMPO (0.015 g, 0.1 mmol) in CH_2Cl_2 (1 mL). The reaction mixture was stirred until the alcohol was no longer detectable (TLC control) and diluted with CH_2Cl_2 (5 mL). The mixture was washed with saturated aqueous $Na_2S_2O_3$ (5 mL) and extracted with CH_2Cl_2 (4 × 5 mL). The combined organic extracts were washed with 5% aqueous $NaHCO_3$ (5 mL) and brine (5 mL), dried (Na_2SO_4) and concentrated under vacuum. The product was isolated by flash chromatography to yield cholestan-3-one (0.310 g, 80%). The structures were confirmed by IR, ^1H NMR, and ^{13}C NMR analysis. However, the data were not provided, nor was a secondary reference cited.

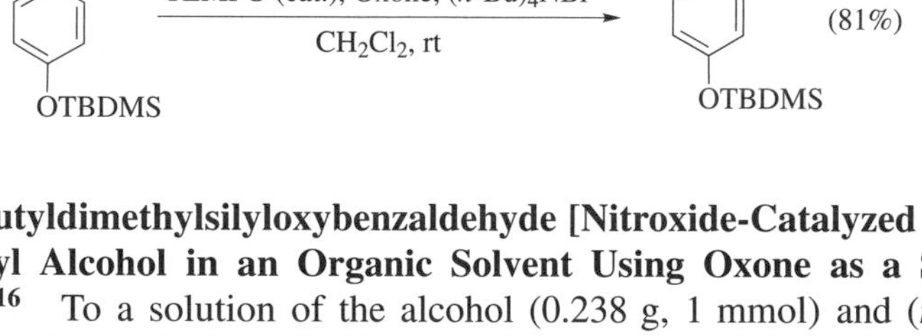
OH
OTBDMS
TEMPO (cat.), Oxone, (n-Bu)4NBr
CH2Cl2, rt
CHO
OTBDMS
(81%)

4-*tert*-Butyldimethylsilyloxybenzaldehyde [Nitroxide-Catalyzed Oxidation of a Benzyl Alcohol in an Organic Solvent Using Oxone as a Secondary Oxidant].[216] To a solution of the alcohol (0.238 g, 1 mmol) and $(n\text{-Bu})_4NBr$ (0.013 g, 0.04 mmol) in CH_2Cl_2 (5 mL) was added TEMPO (0.0015 g, 0.01 mmol) and Oxone (2 $KHSO_5{\bullet}KHSO_4{\bullet}K_2SO_4$, DuPont or Degussa AG; 1.35 g, 2.2 mmol based on a MW of 614 g/mol). The mixture was stirred for 12 hours and monitored by TLC. The solvent was evaporated, and the solid was purified by column chromatography using hexane/EtOAc (10:1) as eluant to give the title aldehyde (0.191 g, 81%). When large amounts were oxidized, the spent oxidant was removed by filtration prior to chromatography. No analytical data or secondary references were provided for the product.

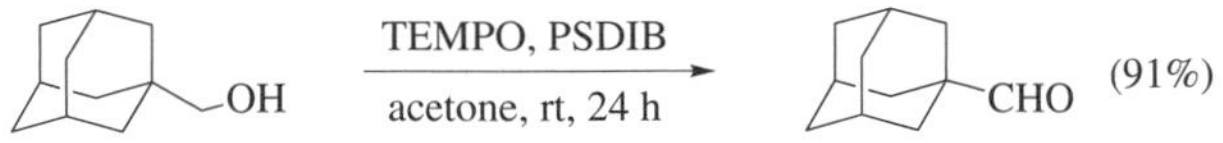

Adamantane-1-carboxaldehyde [Nitroxide-Catalyzed Oxidation of a Primary Alcohol Using a Polymer-Bound Secondary Oxidant].[202] The polymer-bound oxidant PSDIB (see below; 600 mg, 1.2 mmol) was added to a solution of adamantane-1-methanol (166 mg, 1 mmol) and TEMPO (31 mg, 0.2 mmol) in acetone (2 mL), and the mixture was stirred for 24 hours. Et_2O (10 mL) was added, and the mixture was filtered to recover the reduced polymer [poly(4-iodostyrene)], which can be recycled. The filtrate was evaporated to give adamantane-1-carboxaldehyde with a purity of about 95%. The crude material was chromatographed over silica gel to provide the pure aldehyde (149 mg, 91%). No analytical data or secondary references were provided for the product.

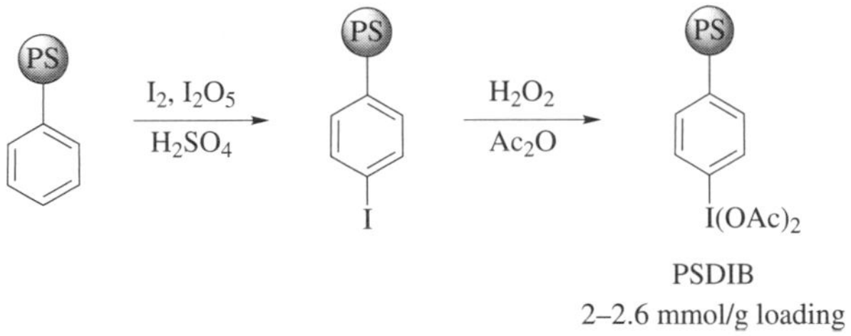

Poly[4-(iododiacetoxy)styrene](PSDIB) [Preparation of a Polymer-Bound Primary Oxidant].[202] A mixture of polystyrene (Aldrich, No. 33165-1; 16 g, 153 mmol), I_2 (18 g, 71 mmol), I_2O_5 (7 g, 21 mmol), CCl_4 (40 mL) and 50% H_2SO_4 (35 mL) in nitrobenzene (200 mL) was heated to 90° for 72 hours. The mixture was diluted with $CHCl_3$ (100 mL), and MeOH (1500 mL) was added. The precipitate was collected by filtration to give poly(4-iodostyrene) with a iodostyrene loading of 4.2 mmol/g. The oxidizing solution was prepared by adding 30% hydrogen peroxide (40 mL) to Ac_2O (145 mL) at 0°. The solution was slowly warmed to room temperature and stirred overnight. To this solution was added poly(4-iodostyrene) (8 g), and the solution was kept at 50° overnight. Ether was added to precipitate the PSDIB (ca. 9 g). The loading rate of the (diacetoxyiodo)phenyl group was found to be 2–2.6 mmol/g by microanalysis: IR (KBr) 1630, 1560, 1480, 1450, 1410, 1260, 1180, 1000 cm^{-1}. Anal. Found: C, 38.86; H, 3.51; I, 37.17.

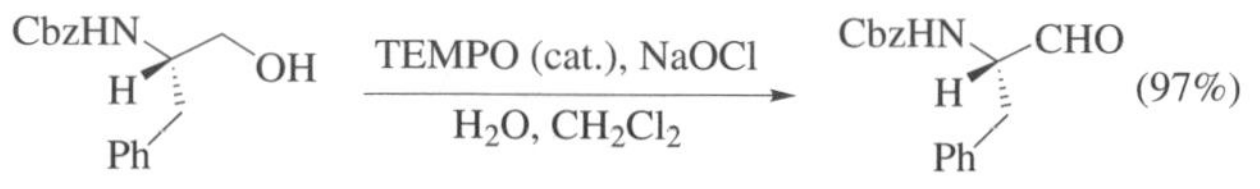

(2*S*)-2-Carbobenzyloxyamino-3-phenylpropanal [Nitroxide-Catalyzed Oxidation of an N-Protected Amino Alcohol].[158] A 1-L 3-necked Morton flask containing *N*-Cbz-L-phenylalaninol (8.56 g, 0.03 mol), TEMPO (0.042 g, 0.3 mmol), and NaBr (3.19 g, 0.031 mol) in a biphasic mixture of toluene (90 mL), EtOAc (90 mL), and water (15 mL) was immersed in a 0° ice-water bath. With rapid mechanical stirring (1200 rpm using an overhead stirrer), an aqueous solution of NaOCl (0.35 M, 94 mL, 0.033 mol) containing $NaHCO_3$ (7.35 g, 0.0875 mol) was added dropwise over a period of one hour, and the mixture was stirred for an additional 10 minutes. The aqueous layer was separated and extracted with toluene (20 mL). The combined organic layers were washed with a solution of KI (0.25 g) dissolved in 10% aqueous $KHSO_3$ (40 mL). The iodine-colored organic layer was then washed successively with 10% aqueous sodium thiosulfate (20 mL) and saturated brine. Drying (Na_2SO_4), filtration, and concentration gave 8.50 g (96.6%) of the title aldehyde as a white solid which was judged to be >98% pure by GC analysis with >99% ee. No analytical data or secondary references were provided for the product.

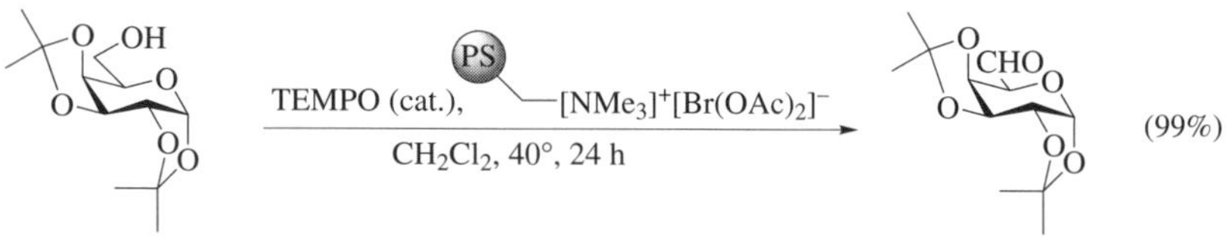

1,2:3,4-Diisopropylidene-6-oxo-α-D-galactopyranose [Nitroxide-Catalyzed Oxidation of a Primary Alcohol Using a Polymer-Bound Bisacetoxybromate(I) Ion].[43] A mixture of 1,2:3,4-diisopropylidene-α-galactopyranose (0.130 g, 0.5 mmol), resin bromoacetate (typically 3 theoretical oxidizing equivalents with reference to loading given by the commercial provider or 2.16 g for 0.69 mmol/g), and TEMPO (2.3 mg, 0.015 mmol) in CH_2Cl_2 (4 mL) was rapidly stirred at 40° for 24 hours. The mixture was filtered, and the filtrate was washed with H_2O (3 × 5 mL). The organic phases were evaporated to give the title product (0.129 mg, 99%): ^1H NMR ($CDCl_3$) δ 1.32 (s, 3H) 1.36 (s, 3H), 1.44 (s, 3H), 1.52 (s, 3H), 4.2 (d, J = 2.2 Hz, 1H), 4.39 (dd, J = 2.2, 4.8 Hz, 1H), 4.65 (2dd, J = 2.2, 8.0 Hz, 2H), 5.67 (d, J = 5.0 Hz, 1H), 9.63 (s, 1H); ^{13}C NMR ($CDCl_3$) δ 24.2 (q), 24.8 (q), 25.8 (q), 25.9 (q), 70.3 (d), 70.4 (d), 71.7 (d), 73.2 (d), 96.2 (d), 109.0 (s), 110.0 (s), 200.3 (d).[389]

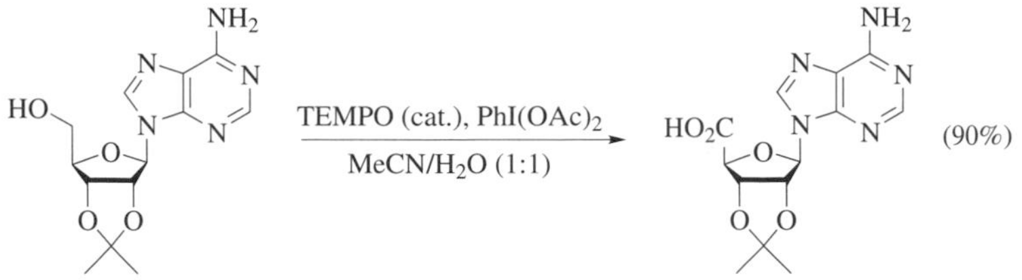

2′,3′-Isopropylideneadenosine-5-carboxylic Acid [Nitroxide-Catalyzed Oxidation of a Nucleoside to the Corresponding Acid Using $PhI(OAc)_2$ as Secondary Oxidant].[193] Bis(acetoxy)iodobenzene (0.709 g, 2.2 mmol), TEMPO (0.032 g, 0.2 mmol), and 2′,3′-isopropylideneadenosine (0.307 g, 1 mmol) were dissolved in $MeCN/H_2O$ (1:1, 2 mL) and stirred for 3 hours at room temperature. The solid precipitate that formed was collected by filtration, triturated sequentially with Et_2O and acetone, and dried under vacuum to yield the title compound (0.275 g, 90%): mp 246–249° dec; ^1H NMR (DMSO-d_6, 50°) δ 1.36 (s, 3H), 1.53 (s, 3H), 4.67 (d, $J = 1.9$ Hz, 1H), 5.48 (d, $J = 6$ Hz, 1H), 5.52 (dd, $J = 1.9$, 6.0 Hz, 1H), 6.32 (s, 1H), 7.15 (s, 2H), 8.10 (s, 1H), 8.23 (s, 1H); ^{13}C NMR (DMSO-d_6, 50°) δ 24.86, 26.39, 83.27, 83.52, 85.14, 89.48, 112.66, 118.70, 140.11, 149.03, 152.15, 155.84, 170.36.

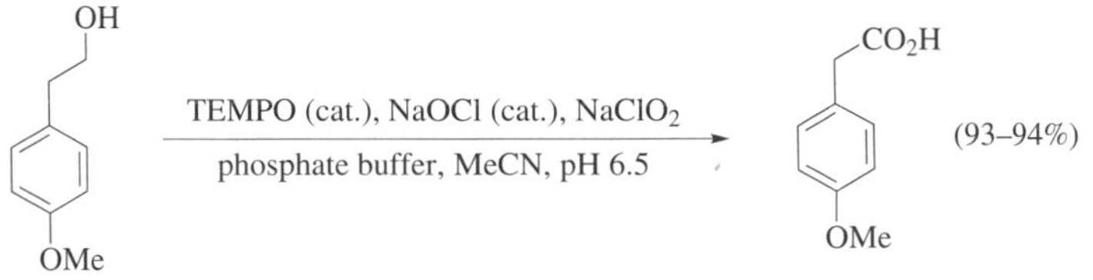

2-(4-Methoxyphenyl)ethanoic Acid [Nitroxide-Catalyzed Oxidation of a Primary Alcohol to the Corresponding Acid Using $NaClO_2/NaOCl$ as Secondary Oxidant]. This procedure is described in *Organic Syntheses*.[14]

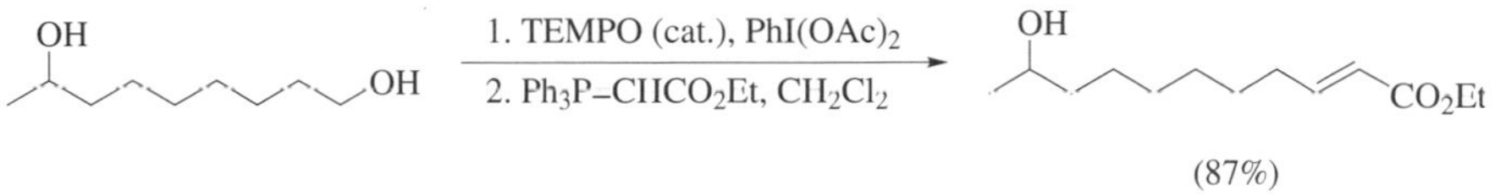

Ethyl 10-Hydroxyundec-2-enoate [Nitroxide-Catalyzed Oxidation Followed by a Wittig Reaction in the Same Flask].[205] To a stirred solution of 1,8-nonadiol (0.15 g, 0.94 mmol) in CH_2Cl_2 (4 mL) were added $PhI(OAc)_2$ (0.346 g, 1.1 mmol) and TEMPO (0.015 g, 0.09 mmol). The yellow solution was stirred for 150 minutes, cooled to 0°, and (carboethoxymethylene)triphenylphosphorane (0.423 g, 1.2 mmol) was added. The reaction mixture was allowed to warm to room temperature and stirred for 1 hour. It was poured into a column of silica gel and eluted with EtOAc/petroleum ether (1:3) to give 0.184 g (87%) of the title ester: ^{1}H NMR ($CDCl_3$, 200 MHz) δ 1.16 (s, J = 6.2 Hz, 3H), 1.23–1.40 (m, 15H), 2.03 (br s, 1H), 2.18 (br q, J = 6.8 Hz, 2H), 3.77 (sextuplet, J = 6.2 Hz, 1H), 4.16 (q, J = 7.1 Hz, 2H), 5.79 (dt, J = 1.5, 15.64 Hz, 1H), 6.94 (tt, J = 6.9, 15.64 Hz, 1H); ^{13}C NMR ($CDCl_3$) δ 14.3, 23.5, 25.6, 27.9, 29.1, 29.4, 32.1, 39.3, 60.1, 68.1, 121.3, 149.4, 166.8.

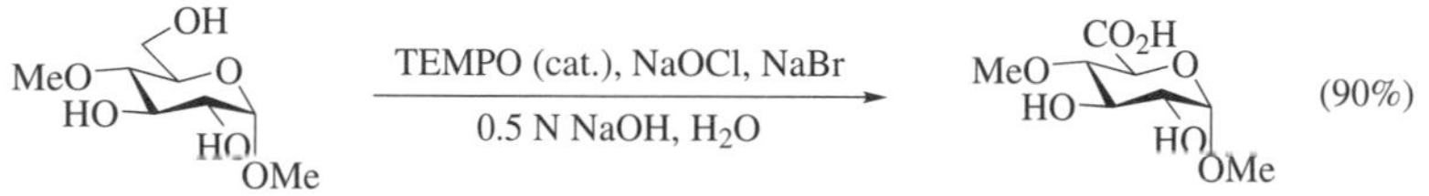

Methyl 4-*O*-Methyl-α-D-glucopyranosideuronic Acid [Nitroxide-Catalyzed Oxidation at the 6-Position of a Non-Reducing Carbohydrate to the Corresponding Uronic Acid Using NaOCl/Br⁻ as Secondary Oxidant].[390] A solution of methyl 4-*O*-methyl-α-D-glucopyranoside (1.00 g, 4.80 mmol), TEMPO (7.5 mg, 0.048 mmol), and NaBr (0.25 g, 2.4 mmol) in deionized water (15 mL) was cooled to 0°, and NaOCl (14.3 mL of a 5% solution, 0.79 g, 10.56 mmol) was added, dropwise, to the solution. The pH was kept at 10–11 by dropwise addition of 0.5 N NaOH. TLC (6:1 $CHCl_3$/MeOH) showed that the starting material had disappeared in 30 minutes. After a reaction time of 60 minutes, MeOH (10 mL) was added to quench the reaction, and the mixture was evaporated to a solid, which was extracted with MeOH. The extract was evaporated to a solid, which was purified by silica gel chromatography (3:1:0.25 $CHCl_3$/MeOH/HOAc) to give 1.2 g (90%) of methyl 4-*O*-methyl-α-D-glucopyranosideuronic acid: ^{1}H NMR (9:1 acetone-d_6/D_2O) δ 3.26 (dd, J = 9.0, 9.9 Hz, 1H), 3.32 (s, 3H), 3.42 (s, 3H), 3.45 (dd, J = 3.8, 9.7 Hz, 1H), 3.67 (t, J = 9.3 Hz, 1H), 3.89 (d, J = 10.1 Hz, 1H), 4.67 (d, J = 3.8 Hz, 1H); ^{13}C NMR (9:1 acetone-d_6/D_2O) δ 55.65, 60.35, 70.60, 72.15, 73.56, 81.83, 100.69, 171.89.

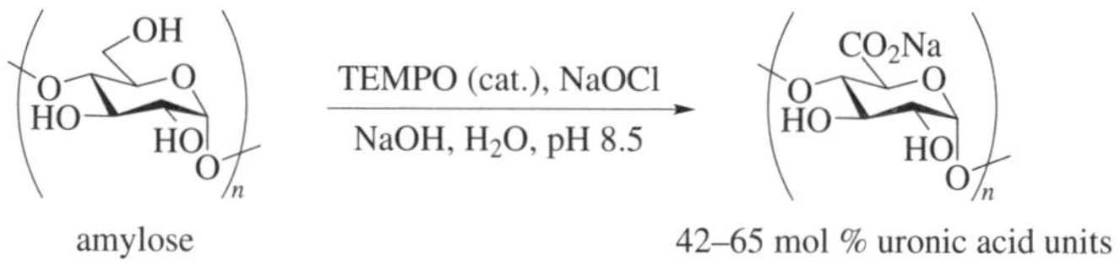

Polycarboxyamylose [Nitroxide-Catalyzed Oxidation at the 6-Position of a Polysaccharide to the Corresponding Polyuronic Acid Using NaOCl as Secondary Oxidant].[333] Potato starch (12.2 g) was gelatinized in deionized water (450 mL) at about 95° and slowly cooled to 20°. The mixture was kept at this temperature during the oxidation. TEMPO (4–8 mg per g of starch) was dissolved in the starch solution, and NaOCl (15%) was added in 2 mL portions. The pH was kept at 8.5 by adding 0.5 M NaOH controlled by a pH meter.

The reaction was continued until no more base was consumed. Any unreacted carbonyl compounds (such as aldehydes) were reduced by $NaBH_4$ (0.2 g). The oxidized materials were precipitated in 2 volumes of EtOH, collected by filtration, rinsed with acetone, and dried in a vacuum oven at 30° for 10–15 hours. The materials were then dissolved in water and passed through an ion-exchange resin (Dowex 50WX8-100) to convert the salts into the acids. The product mixture was freeze-dried and estimated to contain about 42–65 mol % of uronic acid units. ^{13}C spectral analysis of the mixture enabled determination of the relative amount of oxidized starch (C6 to carboxyl) by comparison of C6 of the oxidation product with the unoxidized starting material.

TABULAR SURVEY

Charts 1–4 list the abbreviations for the amines/hydroxylamines, nitroxides, oxoammonium salts, and secondary oxidants used in the Tables, respectively. Tables 1–4 include examples of stoichiometric oxoammonium and nitroxide-catalyzed oxidations of alcohols that appeared in the literature through the Spring of 2008. The alcohols are organized according to increasing carbon count, excluding typical protecting groups. The alkyl groups in simple alkyl esters and alkyl aryl ethers are considered protecting groups in this chapter. The symbol (—) indicates that no yield was reported. The stereochemical purity of starting material and product are included when reported in the literature.

The entries in Tables 1–3 are organized by listing primary alcohols before secondary alcohols before tertiary alcohols within each carbon count. Table 1 contains alcohols which bear no other heteroatom, alkene, or alkyne functional groups, though they may contain a phenyl group. Table 2 lists all molecules in which a (non-heterocyclic) benzylic hydroxyl group is oxidized, irrespective of number and type of other functional groups and heteroatoms present. Table 3 contains alcohols bearing additional functional groups or heteroatoms. Table 4 contains all carbohydrates, and the entries are organized by *N*-glycosides > *C*-glycosides > *O*-glycosides > heavier heteroatom glycosides within each carbon count. For sake of better overview, the tables are divided into Tables 4A to 4G. Tables 4A, 4B, and 4C list glucose-, mannose-, and galactose-derived monosaccharides, respectively. These substrates possess the configuration and oxidation state of the title carbohydrates. Table 4D lists all disaccharides and Table 4E contains remaining saccharides that are not nucleosides or polysaccharides. Table 4F contains all nucleosides and Table 4G lists all C_n carbohydrates.

The following abbreviations are used in the tables:

Ac	acetyl
AcOH	acetic acid
Alloc	allyloxycarbonyl
$[bmim]^+$	1-*n*-butyl-3-methylimidazolium

$[bmpy]^+$	1-*n*-butyl-4-methylpyridinium
Bn	benzyl
Boc	*tert*-butoxycarbonyl
BOM	benzyloxymethyl
BSA	bovine serum albumin
Bz	benzoyl
Cbz	carbobenzyloxy
CSA	camphorsulfonic acid
DABCO	1,4-diazabicyclo[2.2.2]octane
DMAP	4-(dimethylamino)pyridine
DMF	dimcthylformamidc
DMSO	dimethyl sulfoxide
Fmoc	9-fluorenylmethoxycarbonyl
Glu	glucose
m-CPBA	*meta*-chloroperoxybenzoic acid
MEM	β-methoxyethoxymethyl
MOM	methoxymethyl
MTBE	methyl *tert*-butyl ether
NBS	*N*-bromosuccinimide
NMP	*N*-methyl-2-pyrrolidone
Ns	2-nitrobenzenesulfonyl
OSE	triethylsilyloxy
PEG-bipy	methylpolyethylene glycol-derivatized 2,2′-bipyridyl
Phth	phthalimidoyl
Piv	pivaloyl (trimethylacetyl)
PMB	*para*-methoxybenzyl
PMP	*para*-methoxyphenyl
PNP	*para*-nitrophenyl
PS	polystyrene
s	selectivity factor
SCE	standard calomel electrode
TBAB	tetra-*n*-butylammonium bromide
$TBABr_3$	tetra-*n*-butylammonium tribromide
TBAC	tetra-*n*-butylammonium chloride
TBAF	tetra-*n*-butylammonium fluoride
TBAI	tetra-*n*-butylammonium iodide
TBAS	tetra-*n*-butylammonium sulfate
TBDMS	*tert*-butyldimethylsilyl
TBDPS	*tert*-butyldiphenylsilyl
TCA	trichloroacetamide
TCC	trichloroisocyanuric acid
Teoc	2-(trimethylsilyl)ethoxycarbonyl
TES	triethylsilyl
Tf	trifluoromethanesulfonyl
THF	tetrahydrofuran

TIPS	triisopropylsilyl
TMS	trimethylsilyl
TPS	triphenylsilyl
Tr	triphenylmethyl (trityl)
Ts	*para*-toluenesulfonyl
TsOH	*para*-toluenesulfonic acid
Val	valine

CHART 1. AMINES AND HYDROXYLAMINES USED IN TABLES 1–4

Abbreviation	Amine/Hydroxylamine	R^1	R^2	Reference
TEMPH	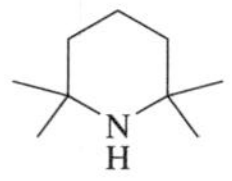			102
TEMPH•HCl	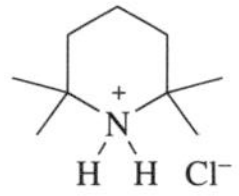			149
TEMPOH	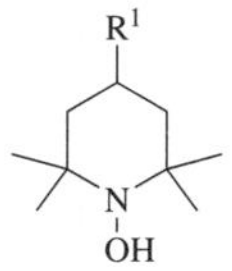	H		4
PCH-TEMPOH	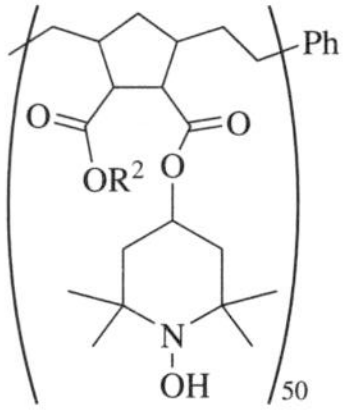		H	301
PCMe-TEMPOH			Me	301
PC-TEMPOH			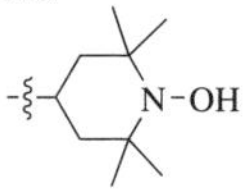	301
SG-NH-TEMPOH	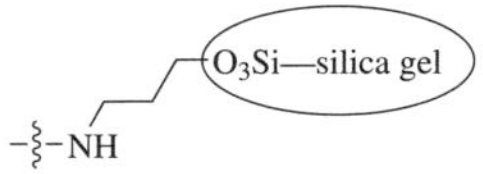			307

CHART 2. NITROXIDES USED IN TABLES 1–4

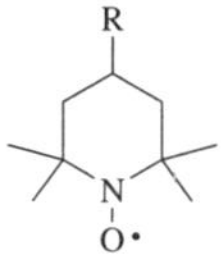

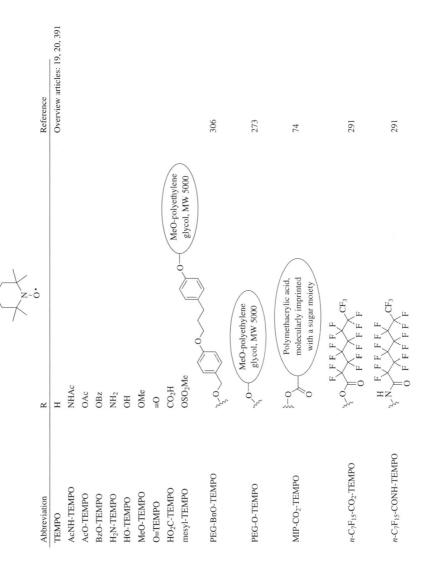

Abbreviation	R	Reference
TEMPO	H	Overview articles: 19, 20, 391
AcNH-TEMPO	NHAc	
AcO-TEMPO	OAc	
BzO-TEMPO	OBz	
H_2N-TEMPO	NH_2	
HO-TEMPO	OH	
MeO-TEMPO	OMe	
O=TEMPO	=O	
HO_2C-TEMPO	CO_2H	
mesyl-TEMPO	OSO_2Me	
PEG-BnO-TEMPO	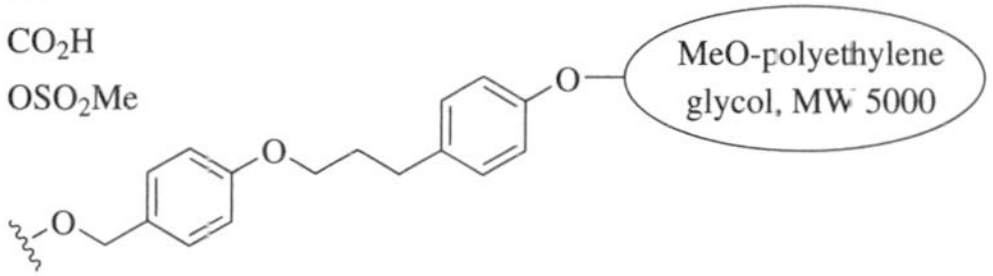MeO-polyethylene glycol, MW 5000	306
PEG-O-TEMPO	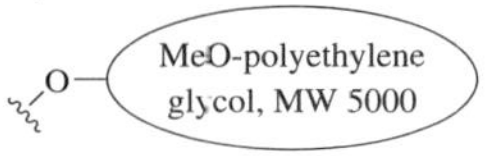MeO-polyethylene glycol, MW 5000	273
MIP-CO_2-TEMPO	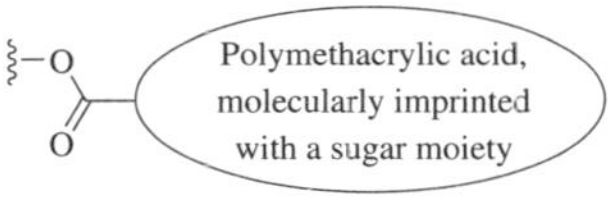Polymethacrylic acid, molecularly imprinted with a sugar moiety	74
n-C_7F_{15}-CO_2-TEMPO	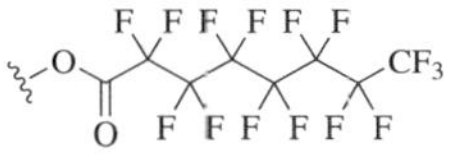	291
n-C_7F_{15}-CONH-TEMPO	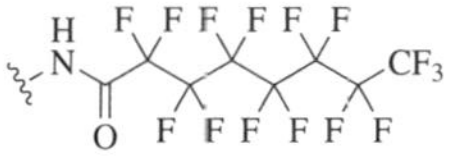	291

CHART 2. NITROXIDES USED IN TABLES 1–4 *(Continued)*

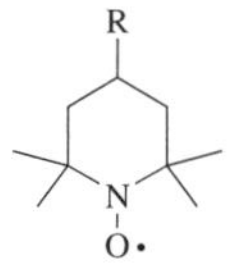

Abbreviation	R	Reference
FT-O-TEMPO	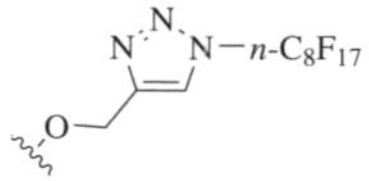	291
ICF-O-TEMPO	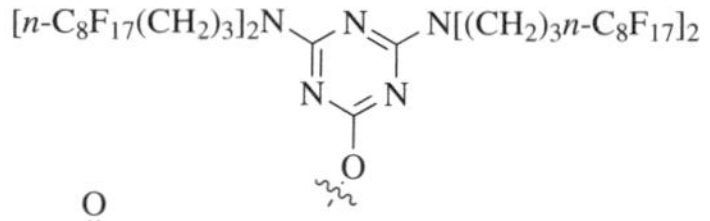	291
$[mim\text{-}CONH\text{-}TEMPO]^+[Tf_2N]^-$	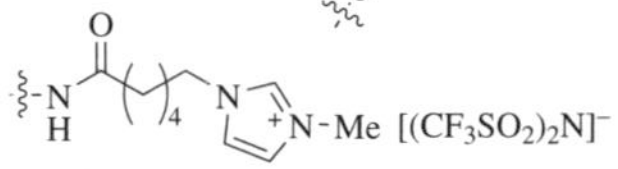	294
$[mim\text{-}O\text{-}TEMPO]^+[BF_4]^-$	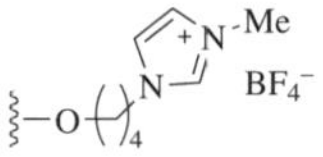	201
$[mim\text{-}CO_2\text{-}TEMPO]^+$	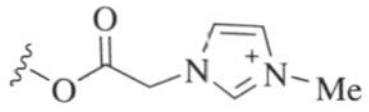 as $[BF_4]^-$ salt as $[PF_6]^-$ salt	201 295
graphite felt-TEMPO	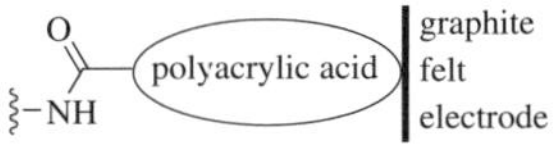	231
Nafion-TEMPO	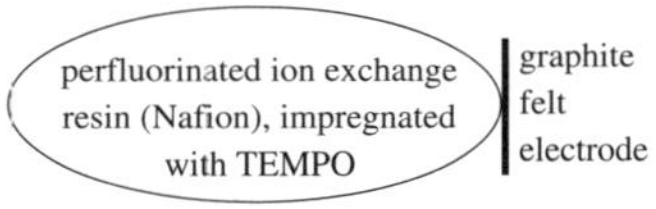	326

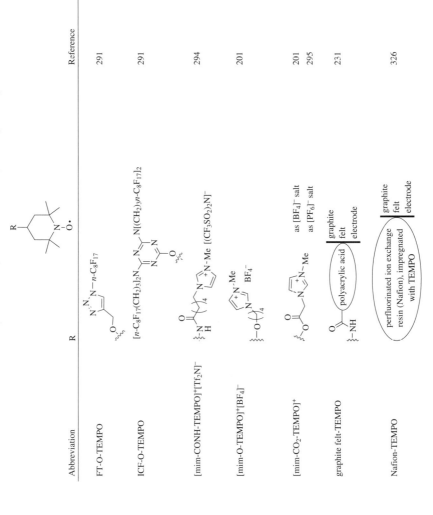

PS-CO_2-TEMPO	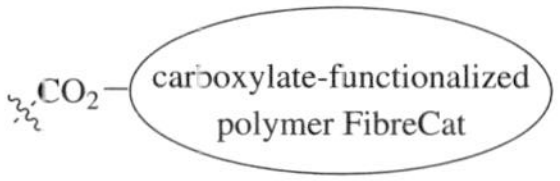 CO_2 carboxylate-functionalized polymer FibreCat	392
PE-co-AA-CONH-TEMPO	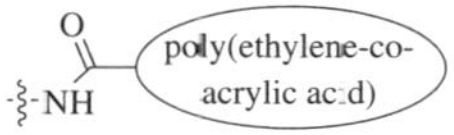 O NH poly(ethylene-co-acrylic acid)	298
PE-CONH-TEMPO	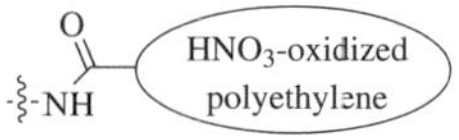 O NH HNO_3-oxidized polyethylene	297
SG-NHCONH-TEMPO	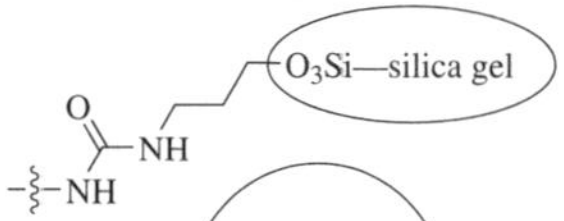 O NH NH O_3Si—silica gel	393
SBA15-NH-TEMPO	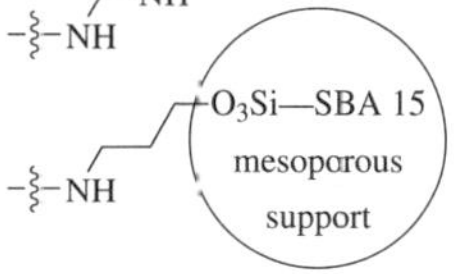 NH O_3Si—SBA 15 mesoporous support	310
MCM41-NH-TEMPO	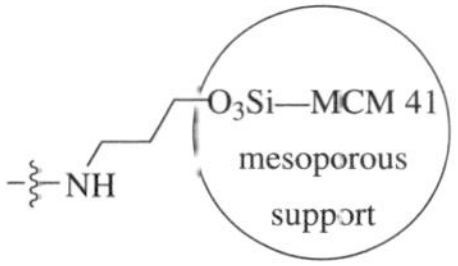 NH O_3Si—MCM 41 mesoporous support	309
ormosil-TEMPO	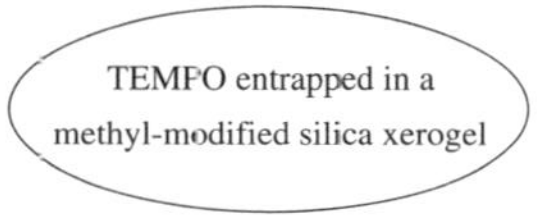 TEMPO entrapped in a methyl-modified silica xerogel	394
SG-NH-TEMPO	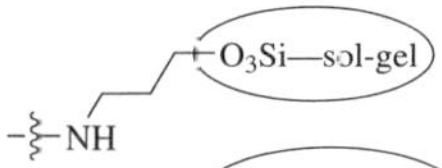 NH O_3Si—sol-gel	313
SG-O-TEMPO	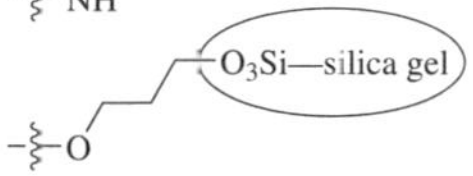 O O_3Si—silica gel	208

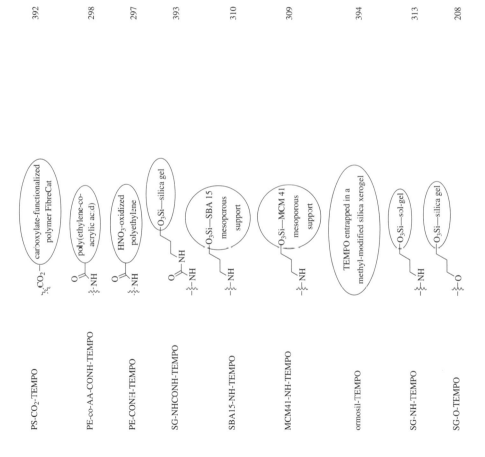

Chart 2. Nitroxides Used in Tables 1–4 *(Continued)*

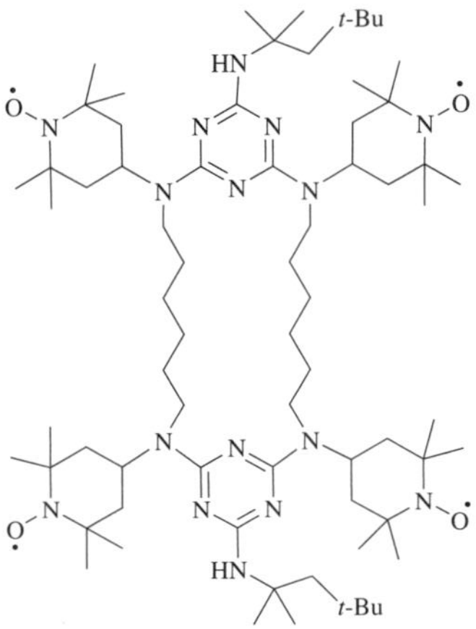

CIBA-TEMPO[395]

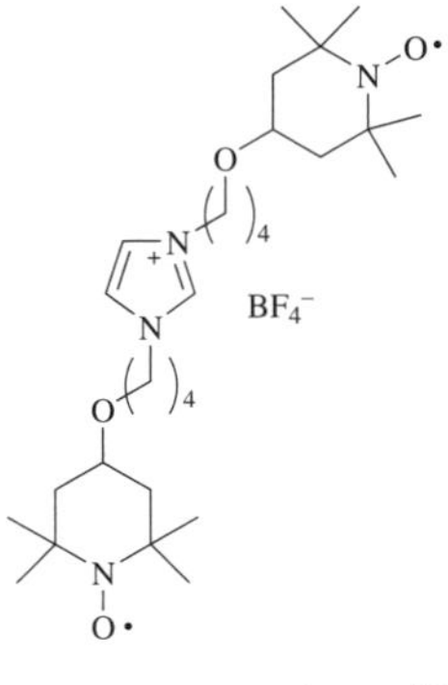

$[Im\text{-}(O\text{-}TEMPO)_2]^+[BF_4]^-$ [201]

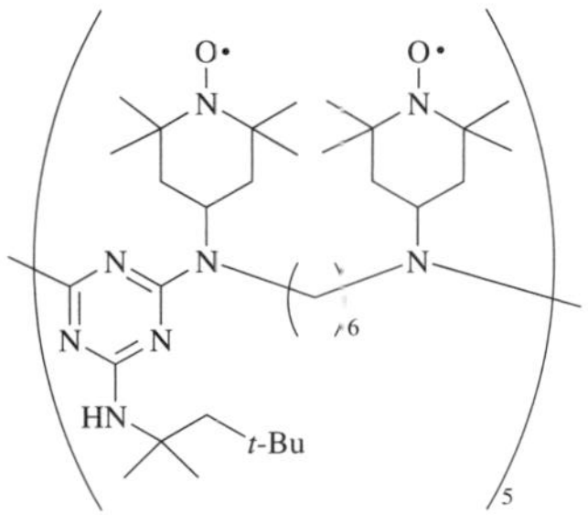

PIPO[5]

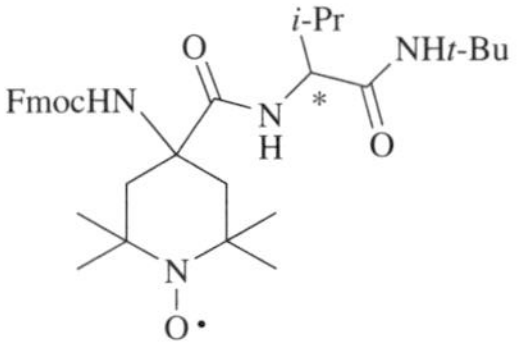

*	
R	(*R*)-Val-TEMPO[103]
S	(*S*)-Val-TEMPO[103]

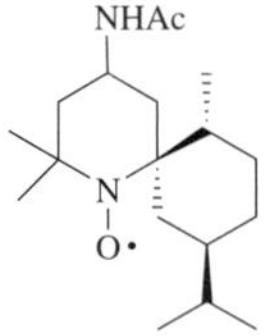

AS-NHAc-TEMPO[85]

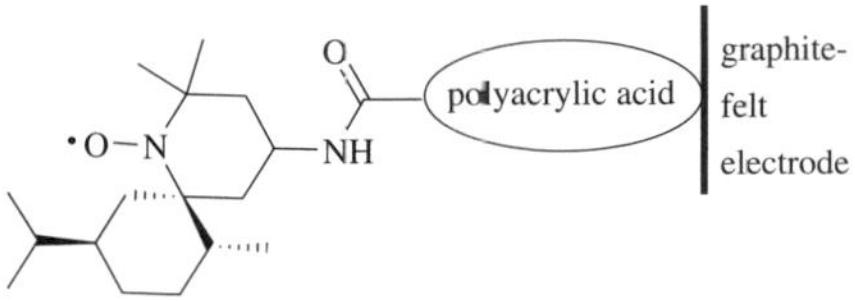

graphite felt-CONH-AS-TEMPO[83]

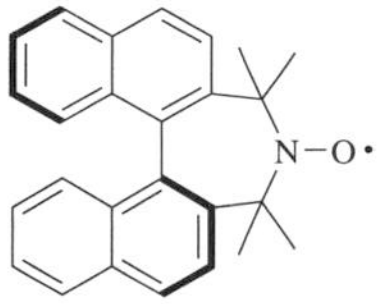

(*S*)-binaphthyl-TEMPO[90]

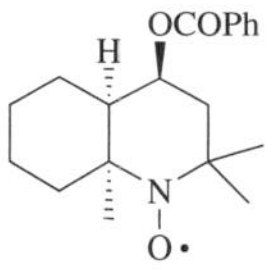

L-DQ-TEMPO[396]

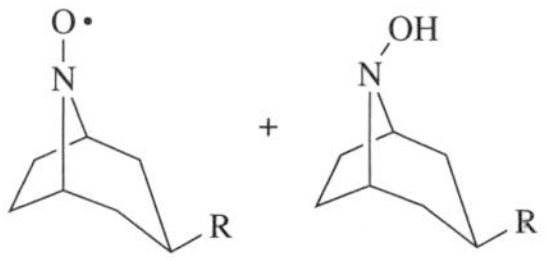

mixtures

R	
H	ABHepO[281]
Cl	ABHep^{Cl}O[281]

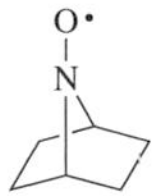

ABHexO[281]

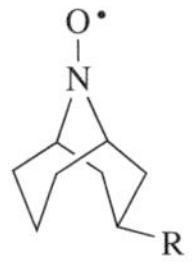

R	
H	ABOctO[281]
Cl	ABOct^{Cl}O[281]

CHART 3. OXOAMMONIUM SALTS USED IN TABLES 1–4

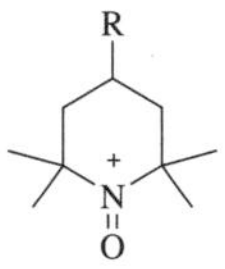

In combination with one of the anions:
Cl^-, ClO_2^-, ClO_4^-, Br^-, Br_3^-, BF_4^-,
PF_6^-, SbF_6^-, NO_3^-.

Abbreviation	R	Reference
$[TEMP{=}O]^+$	H	overview article for R = H to R = O: 20
$[AcNH\text{-}TEMP{=}O]^+$	NHAc	
$[BzO\text{-}TEMP{=}O]^+$	OBz	
$[H_2N\text{-}TEMP{=}O]^+$	NH_2	
$[HO\text{-}TEMP{=}O]^+$	OH	
$[MeO\text{-}TEMP{=}O]^+$	OMe	
$[O{=}TEMP{=}O]^+$	=O	
$[PS\text{-}O\text{-}TEMP{=}O]^+$	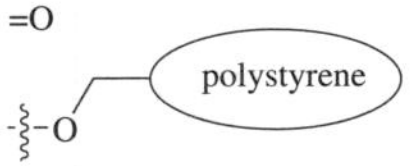	121

CHART 4. SECONDARY OXIDANTS USED IN TABLES 1–4

Cl_2

Br_2

I_2

O_2 or air

H_2O_2

aqueous NaOCl (household bleach)

LiOCl

$Ca(OCl)_2$

t-BuOCl

NaOBr

$NaBrO_2$

$2KHSO_5 \bullet KHSO_4 \bullet K_2SO_4$ (oxone)

$K_3[Fe(CN)]_6$

$CuCl_2$

$Co(NO_3)_2$

$Mn(NO_3)_2$

$(NH_4)_2Ce(NO_3)_6$ (cerium ammonium nitrate, CAN)

H_3IO_5 (periodic acid)

NCS (*N*-chlorosuccinimide)

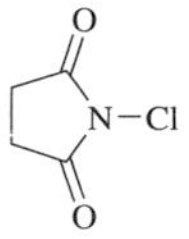

NBS (*N*-bromosuccinimide)

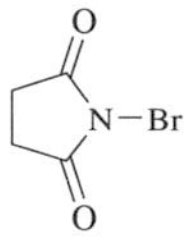

m-CPBA (*m*-chloroperoxybenzoic acid)

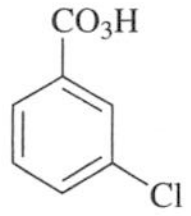

TCC (trichloroisocyanuric acid)

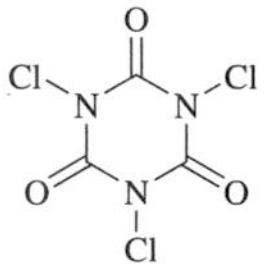

1,3-dibromo-5,5-dimethylhydantoin

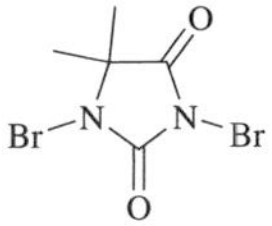

$PhI(OAc)_2$ (bis-acetoxyiodobenzene, diacetoxyiodobenzene)

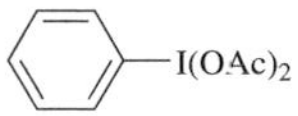

$[mim\text{-}CH_2PhI(OAc)_2]^+[BF_4]^-$

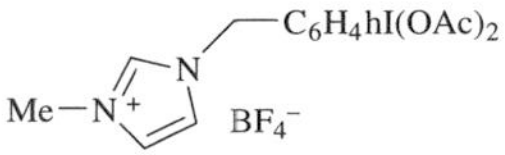

$PS\text{-}I(OAc)_2$

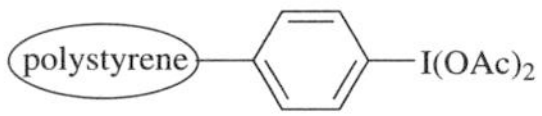

$[PS\text{-}CH_2NMe_3]^+[Br(OAc)_2]^-$

$[PS\text{-}CH_2NMe_3]^+[ClO]^-$

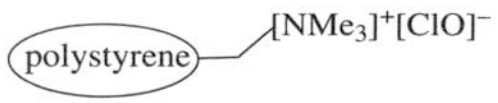

$[PS\text{-}CH_2NMe_3]^+[ClO_2]^-$

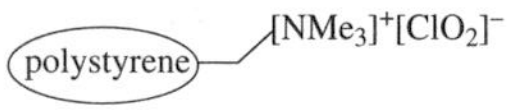

TABLE 1. OXIDATIONS OF MONOFUNCTIONAL ALCOHOLS

	Alcohol	Conditions	Product(s) and Yield(s) (%)	Refs.
C_1	MeOH	HO-TEMPO (cat.), electrolysis, Na_2SO_4, H_2O, acetate buffer, pH 4–5, rt	CH_2O (—) **I**	62
		$[HO\text{-}TEMP{=}O]^+Br^-$, MeOH, reflux	**I** (100), as derivative	8
C_2	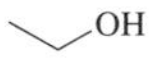	HO-TEMPO (cat.), electrolysis, Na_2SO_4, H_2O, acetate buffer, pH 4–5, rt	MeCHO (—) **I**	62
		$[HO\text{-}TEMP{=}O]^+Br^-$, MeOH, reflux	**I** (97), as derivative	62
C_3	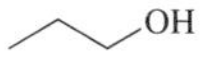	HO-TEMPO (cat.), electrolysis, Na_2SO_4, H_2O, acetate buffer, pH 4–5, rt	 (—)	62
	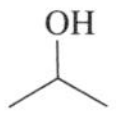	HO-TEMPO (cat.), electrolysis, Na_2SO_4, H_2O, acetate buffer, pH 4–5, rt	 (—)	62
C_4	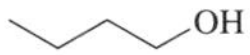	HO-TEMPO (cat.), electrolysis, Na_2SO_4, H_2O, acetate buffer, pH 4–5, rt	 (—) **I**	62
		Hydroxylamine (cat.), KBr (cat.), NaOCl, $NaHCO_3$, CH_2Cl_2, H_2O, pH 9, 0°, 1 h	**I** (—)	301
		Hydroxylamine	% Conversion[a]	
		PC-TEMPOH	71	
		PCH-TEMPOH	75	
		PCMe-TEMPOH	72	
		TEMPO (cat.), NaBr (cat.), NaOCl, H_2O, pH 10, 1.5°	**I** (—)[b]	61
		1. TEMPO (cat.), NaOCl, EtOAc or MeCN, H_2O, pH 8–10 2. $NaClO_2$, H_2O, pH 5	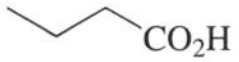 (84)	187

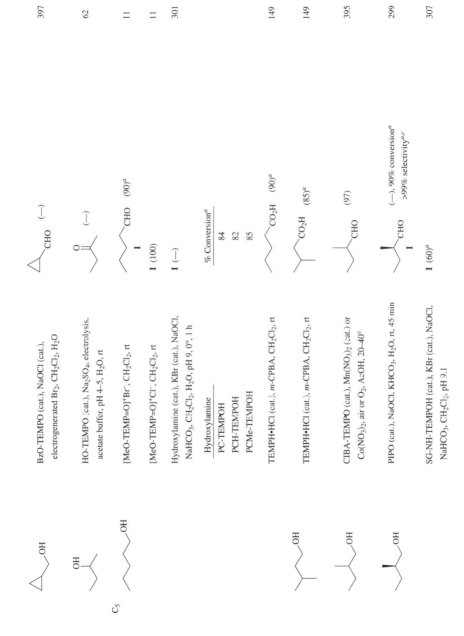

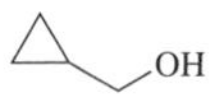	BzO-TEMPO (cat.), NaOCl (cat.), electrogenerated Br_2, CH_2Cl_2, H_2O	 (—)	397
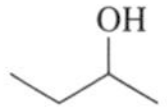	HO-TEMPO (cat.), Na_2SO_4, electrolysis, acetate buffer, pH 4–5, H_2O, rt	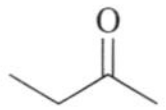 (—)	62
C_5 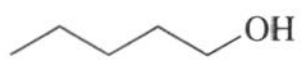	$[MeO\text{-}TEMP{=}O]^+Br^-$, CH_2Cl_2, rt	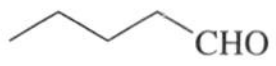 **I** (90)[a]	11
	$[MeO\text{-}TEMP{=}O]^+Cl^-$, CH_2Cl_2, rt	**I** (100)	11
	Hydroxylamine (cat.), KBr (cat.), NaOCl, $NaHCO_3$, CH_2Cl_2, H_2O, pH 9, 0°, 1 h	**I** (—)	301
	Hydroxylamine	% Conversion[a]	
	PC-TEMPOH	84	
	PCH-TEMPOH	82	
	PCMe-TEMPOH	85	
	TEMPH•HCl (cat.), *m*-CPBA, CH_2Cl_2, rt	 (90)[a]	149
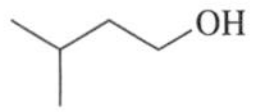	TEMPH•HCl (cat.), *m*-CPBA, CH_2Cl_2, rt	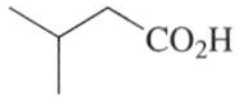 (85)[a]	149
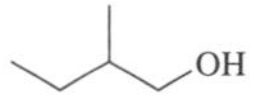	CIBA-TEMPO (cat.), $Mn(NO_3)_2$ (cat.) or $Co(NO_3)_2$, air or O_2, AcOH, 20–40°	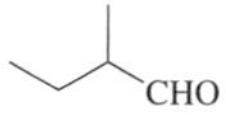 (97)	395
	PIPO (cat.), NaOCl, $KHCO_3$, H_2O, rt, 45 min	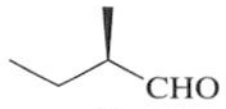 **I** (—), 90% conversion[a] >99% selectivity[a,c]	299
	SG-NH-TEMPOH (cat.), KBr (cat.), NaOCl, $NaHCO_3$, CH_2Cl_2, pH 9.1	**I** (60)[a]	307

TABLE 1. OXIDATIONS OF MONOFUNCTIONAL ALCOHOLS (*Continued*)

Alcohol	Conditions	Product(s) and Yield(s) (%)	Refs.
C_5			
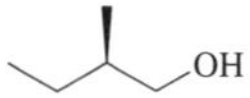	TEMPO (cat.), KBr (cat.), NaOCl, $NaHCO_3$, CH_2Cl_2, H_2O	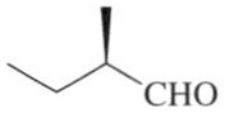 **I** (52)	398, 399
	TEMPO (cat.), KBr (cat.), NaOCl, CH_2Cl_2, H_2O, pH 9.5, 0–15°	**I** (—)	400
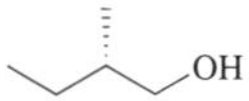	TEMPO (cat.), KBr (cat.), NaOCl, $NaHCO_3$, CH_2Cl_2, H_2O, 0°, 35 min	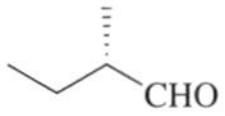 **I** (59)	401
	TEMPO (cat.), KBr (cat.), NaOCl, CH_2Cl_2, H_2O, pH 9.5, 0–15°	**I** (—)	402
	TEMPO (cat.), KBr (cat.), NaOCl, CH_2Cl_2, H_2O, pH 9.5, 0–15°	**I** (82–84)	134
	TEMPO (cat.), KBr (cat.), NaOCl, $NaHCO_3$, CH_2Cl_2, H_2O	**I** (—)	399
	TEMPO (cat.), KBr (cat.), NaOCl, CH_2Cl_2, H_2O, 0–15°	**I** (82)	403
	TEMPO (cat.), KBr (cat.), NaOCl, CH_2Cl_2, H_2O, pH 9.5, 0–15°	**I** (—)	404
	TEMPO (cat.), KBr (cat.), NaOCl, CH_2Cl_2, H_2O	**I** (—)	405
	TEMPO (cat.), $[PS\text{-}CH_2NMe_3]^+[Br(OAc)_2]^-$ (cat.), KNO_2 (cat.), O_2, 80–90°, 15 h	*t*-BuCHO **I** (95)	43

	TEMPO (cat.), PhI(OAc)$_2$ (cat.), KNO$_2$ (cat.), O$_2$, 80–90°, 15 h	**I** (61)[d]	203
	TEMPO (cat.), PS-I(OAc)$_2$ (cat.), KNO$_2$ (cat.), O$_2$, 80–90°, 15 h	**I** (32)[d]	203
	TEMPO (cat.), [PS-CH$_2$NMe$_3$]$^+$[Br(OAc)$_2$]$^-$ (cat.), CH$_2$Cl$_2$, rt, 3.5 h	**I** (95)[a]	211
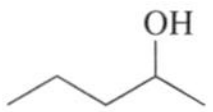	[MeO-TEMP=O]$^+$Cl$^-$, CH$_2$Cl$_2$, rt	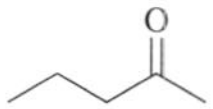 (97)[a]	11
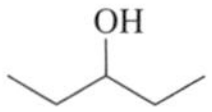	TEMPH (cat.), *m*-CPBA, CH$_2$Cl$_2$, 0° to rt	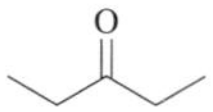 **I** (45)[a]	102
	[MeO-TEMP=O]$^+$Br$^-$, CH$_2$Cl$_2$, rt	**I** (88)	11
	[MeO-TEMP=O]$^+$Cl$^-$, CH$_2$Cl$_2$, rt	**I** (92)[a]	11
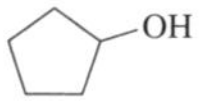	[MeO-TEMP=O]$^+$Cl$^-$, CH$_2$Cl$_2$, rt	 **I** (85[a], 64)	255
	TEMPH•HCl (cat.), *m*-CPBA, CH$_2$Cl$_2$, rt, 1.5 h	**I** (77)	149
	TEMPO, NaOCl, NaBr, H$_2$O, pH 10, 1.5°	**I** (—)[b]	61

TABLE 1. OXIDATIONS OF MONOFUNCTIONAL ALCOHOLS (*Continued*)

	Alcohol	Conditions	Product(s) and Yield(s) (%)	Refs.
C_5	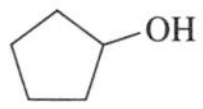	Nitroxide (cat.), NaBr (4 eq), electrolysis, CH_2Cl_2, H_2O, $NaHCO_3$ (satd.), rt		281
		Nitroxide		
		ABHep^{Cl}O	(98)	
		ABHepO	(99)	
		ABOct^{Cl}O	(99)	
		ABOctO	(99)	
		ABHexO	(99)	
		TEMPO	(84)	
C_6		[PS-O-TEMP=O]$^+$Cl$^-$, CH_2Cl_2, rt, 1–2 h	 (95)[a] **I**	121
		Hydroxylamine (cat.), KBr (cat.), NaOCl, $NaHCO_3$, CH_2Cl_2, H_2O, pH 9, 0°, 1 h	**I** (—)	301
		Hydroxylamine	% Conversion[a]	
		PC-TEMPOH	87	
		PCH-TEMPOH	83	
		PCMe-TEMPOH	85	
		[mim-O-TEMPO]$^+$[BF_4]$^-$ (cat.), [mim-CH_2PhI(OAc)$_2$]$^+$[BF_4]$^-$, H_2O, 30°, 4 h	**I** (95)[a]	201
		[Im-(O-TEMPO)$_2$]$^+$[BF_4]$^-$ (cat.), [mim-CH_2PhI(OAc)$_2$]$^+$[BF_4]$^-$, H_2O, 30°, 2 h	**I** (93)	201
	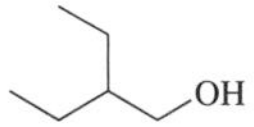	BzO-TEMPO (cat.), electrolysis, NaBr (cat.), $NaHCO_3$, H_2O, CH_2Cl_2, pH 8.6	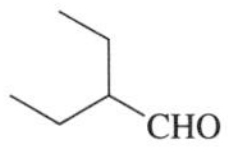 (93)	222

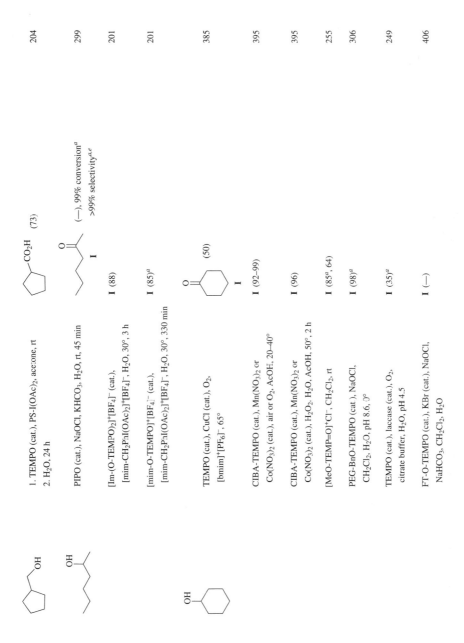

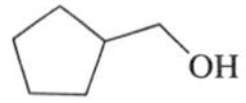	1. TEMPO (cat.), PS-I(OAc)$_2$, acetone, rt 2. H_2O, 24 h	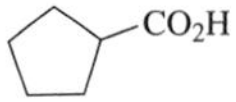 (73)	204
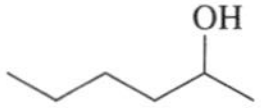	PIPO (cat.), NaOCl, $KHCO_3$, H_2O, rt, 45 min	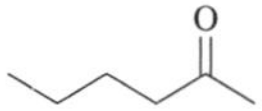 **I** (—), 99% conversion[a] >99% selectivity[a,e]	299
	[Im-(O-TEMPO)$_2$]$^+$[BF$_4$]$^-$ (cat.), [mim-CH$_2$PhI(OAc)$_2$]$^+$[BF$_4$]$^-$, H_2O, 30°, 3 h	**I** (88)	201
	[mim-O-TEMPO]$^+$[BF$_4$]$^-$ (cat.), [mim-CH$_2$PhI(OAc)$_2$]$^+$[BF$_4$]$^-$, H_2O, 30°, 330 min	**I** (85)[a]	201
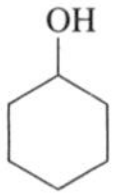	TEMPO (cat.), CuCl (cat.), O_2, [bmim]$^+$[PF$_6$]$^-$, 65°	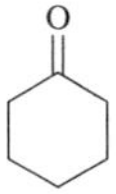 (50) **I**	385
	CIBA-TEMPO (cat.), Mn(NO$_3$)$_2$ or Co(NO$_3$)$_2$ (cat.), air or O_2, AcOH, 20–40°	**I** (92–99)	395
	CIBA-TEMPO (cat.), Mn(NO$_3$)$_2$ or Co(NO$_3$)$_2$ (cat.), H_2O_2, H_2O, AcOH, 50°, 2 h	**I** (96)	395
	[MeO-TEMP=O]$^+$Cl$^-$, CH_2Cl_2, rt	**I** (85[a], 64)	255
	PEG-BnO-TEMPO (cat.), NaOCl, CH_2Cl_2, H_2O, pH 8.6, 0°	**I** (98)[a]	306
	TEMPO (cat.), laccase (cat.), O_2, citrate buffer, H_2O, pH 4.5	**I** (35)[a]	249
	FT-O-TEMPO (cat.), KBr (cat.), NaOCl, $NaHCO_3$, CH_2Cl_2, H_2O	**I** (—)	406

TABLE 1. OXIDATIONS OF MONOFUNCTIONAL ALCOHOLS (*Continued*)

Alcohol	Conditions	Product(s) and Yield(s) (%)	Refs.
C_6 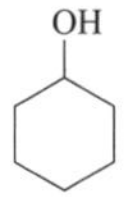	TEMPO (cat.), NaBr (cat.), NaOCl, H_2O, pH 10, 1.5°	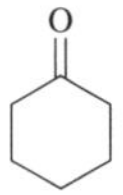 **I** (—)[b]	61
	TEMPO (cat.), I_2, $NaHCO_3$, toluene or CH_2Cl_2, H_2O, 20°, overnight	**I** (85)[f]	256
	TEMPO (cat.), $[PS\text{-}CH_2NMe_3]^+[Br(OAc)_2]^-$, CH_2Cl_2, rt, 2 h	**I** (96)	43
	TEMPO (cat.), $Mn(NO_3)_2$ or $Co(NO_3)_2$ (cat.), O_2 (1 atm), AcOH	**I** (96–98)[a]	407
	TEMPO (cat.), $H_5PV_2Mo_{10}O_{40}$•34 H_2O (cat.), O_2, acetone	**I** (96)[a]	274
	TEMPO (cat.), $NaNO_2$ (cat.), 1,3-dibromo-5,5-dimethylhydantoin, air, H_2O, 80°, 1 h	**I** (—)	259
	TEMPO (cat.), KBr (cat.), NaOCl, CH_2Cl_2, pH 9.5	**I** (98)	134
	TEMPO (cat.), **1** (cat.), CuBr•Me_2S (cat.), O_2, perfluorooctyl bromide, chlorobenzene, 90° 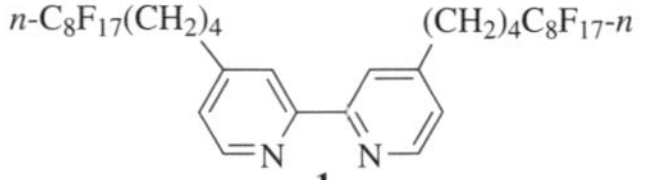 **1**	**I** (74)	243
	TEMPO (cat.), $[PS\text{-}CH_2NMe_3]^+[ClO]^-$, CH_2Cl_2, rt	**I** (28)	44

	TEMPO (cat.), $[PS\text{-}CH_2NMe_3]^+[Br(OAc)_2]^-$, CH_2Cl_2, rt	**I** (96)	211, 44
	TEMPO (cat.), $[PS\text{-}CH_2NMe_3]^+[ClO_2]^-$, CH_2Cl_2, rt	**I** (99)	44
	TEMPO (cat.), Br_2 (cat.), $NaNO_2$ (cat.), air (3.9 atm), CH_2Cl_2, 80°	**I** (89)	258
	$[mim\text{-}CO_2\text{-}TEMPO]^+[PF_6]^-$ (cat.), KBr (cat.), NaOCl, $[bmim]^+[PF_6]^-$, H_2O, 0°	**I** (88)[a]	295
	TEMPO (cat.), **2** (cat.), CuCl (cat.), O_2, $[bmim]^+[PF_6]^-$, 65° 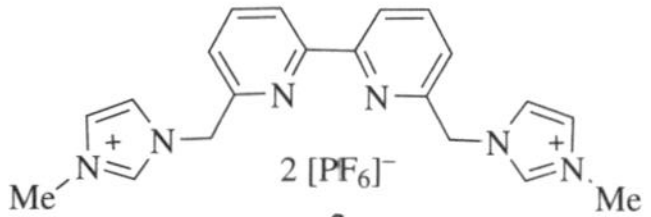 **2**	**I** (83)[a]	246
	TEMPH•HCl (cat.), *m*-CPBA, CH_2Cl_2, H_2O, rt, 1.5 h	**I** (—) + 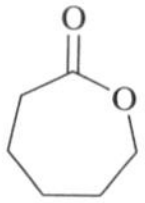 (—)	149
	$[PS\text{-}O\text{-}TEMP{=}O]^+Cl^-$, CH_2Cl_2, rt, 1–2 h	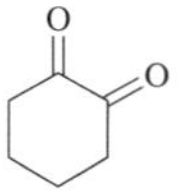 (85)[a]	121
C_7 	TEMPH•HCl (cat.), *m*-CPBA, CH_2Cl_2, H_2O, rt, 1.5 h	 **I** (40)[a]	149
	TEMPO (cat.), KBr (cat.), NaOCl, CH_2Cl_2, H_2O, pH 9.5, 0–15°	**I** (88)	13

TABLE 1. OXIDATIONS OF MONOFUNCTIONAL ALCOHOLS (*Continued*)

Alcohol	Conditions	Product(s) and Yield(s) (%)	Refs.
C_7 	SG-NH-TEMPOH (cat.), KBr (cat.), NaOCl, $NaHCO_3$, CH_2Cl_2, H_2O, pH 9.1	 **I** (90)[a]	307
	TEMPO (cat.), $Mn(NO_3)_2$ or $Co(NO_3)_2$ (cat.), O_2 (1 atm), AcOH	**I** (97)[a]	407
	CIBA-TEMPO (cat.), $Mn(NO_3)_2$ or $Co(NO_3)_2$ (cat.), H_2O_2, H_2O, AcOH, 50°, 2 h	**I** (92)	395
	AcNH-TEMPO (cat.), $Cu(ClO_4)_2$ (cat.), O_2 (1 atm), DMAP, $[bmpy]^+[PF_6]^-$, rt	**I** (54)	240
	TEMPO (cat.), **2** (cat.), CuCl (cat.), O_2, $[bmim]^+[PF_6]^-$, 65°	**I** (70)[a]	246
	Hydroxylamine (cat.), KBr (cat.), NaOCl, $NaHCO_3$, CH_2Cl_2, H_2O, pH 9, 0°, 1 h	**I** (—)	301
	Hydroxylamine	% Conversion[a]	
	PC-TEMPOH	85	
	PCH-TEMPOH	78	
	PCMe-TEMPOH	82	
	TEMPO (cat.), **3** (cat.), O_2, toluene, 100° 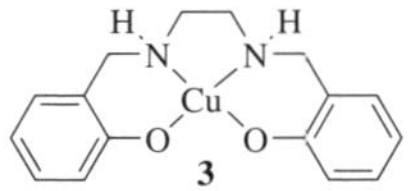 **3**	**I** (90)	244

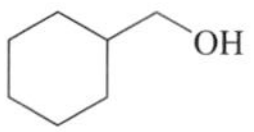	TEMPO (cat.), PhI(OAc)$_2$, CH_2Cl_2, rt	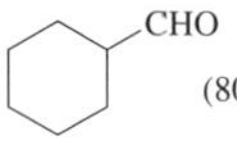 (80) **I**	192
	TEMPO (cat.), laccase (cat.), O_2, H_2O, citrate buffer, pH 4.5	**I** (18)[a]	249
	TEMPO (cat.), **3** (cat.), O_2, toluene, 100°	**I** (75)	244
	Nitroxide (cat.), NaBr (4 eq), electrolysis, CH_2Cl_2, H_2O, $NaHCO_3$ (satd.), rt	**I**	281
	Nitroxide		
	ABHep^{Cl}O	(77)	
	ABHepO	(90)	
	ABOct^{Cl}O	(91)	
	ABOctO	(99)	
	ABHexO	(99)	
	TEMPO	(68)	
	1. TEMPO (cat.), PS-I(OAc)$_2$, acetone, rt 2. H_2O, 24 h	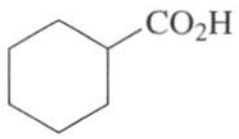 (78)	204
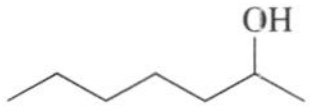	PEG-BnO-TEMPO (cat.), NaOCl, CH_2Cl_2, H_2O, pH 8.6, 0°	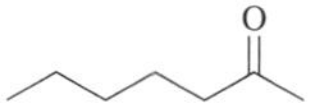 (94)[a] **I**	306
	PIPO (cat.), KBr (cat.), NaOCl, KBr, CH_2Cl_2, H_2O	**I** (—), 99% conversion[a], 99% selectivity[a,e]	299
	[PS-O-TEMP=O]$^+$Cl$^-$, CH_2Cl_2, rt, 1–2 h	**I** (90)[a]	121

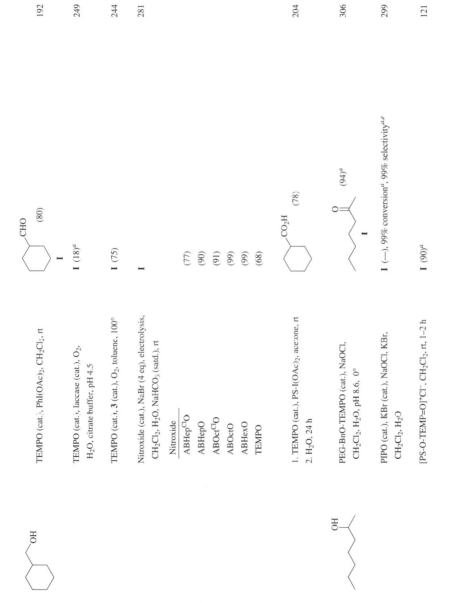

TABLE 1. OXIDATIONS OF MONOFUNCTIONAL ALCOHOLS (*Continued*)

Alcohol	Conditions	Product(s) and Yield(s) (%)	Refs.
C_7 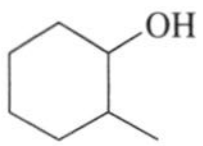	[MeO-TEMP=O]$^+$Cl$^-$, CH_2Cl_2, rt	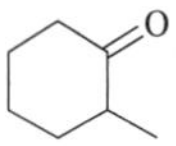 **I** (44)[a]	255
	TEMPO (cat.), NaBr (cat.), NaOCl, H_2O, pH 10, 1.5°	**I** (—)[b]	61
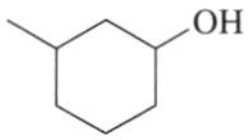	TEMPO (cat.), NaOCl, NaBr, H_2O, pH 10, 1.5°	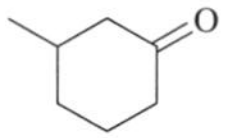 (—)[b]	61
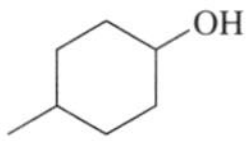	[Im-(O-TEMPO)$_2$]$^+$[BF$_4$]$^-$ (cat.), [mim-CH_2PhI(OAc)$_2$]$^+$[BF$_4$]$^-$, H_2O, 30°, 2 h	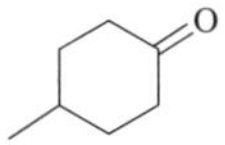 **I** (78)	201
	[mim-O-TEMPO]$^+$[BF$_4$]$^-$ (cat.), [mim-CH_2PhI(OAc)$_2$]$^-$[BF$_4$]$^-$, H_2O, 30°, 5 h	**I** (80)	201
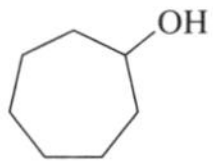	TEMPH•HCl (cat.), *m*-CPBA, CH_2Cl_2, rt, 1.5 h	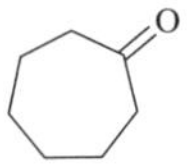 **I** (81)	149
	[MeO-TEMP=O]$^+$Cl$^-$, CH_2Cl_2, rt	**I** (89[a], 79)	255
	TEMPO (cat.), **1** (cat.), CuBr•Me_2S (cat.), O_2, perfluorooctyl bromide, chlorobenzene, 90°	**I** (82)	243
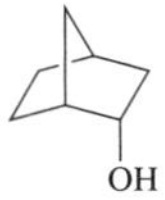	TEMPH•HCl (cat.), *m*-CPBA, CH_2Cl_2, rt, 1.5 h	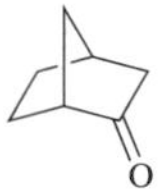 (95)	149

C_8	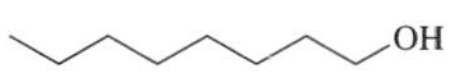	$[mim-CO_2-TEMPO]^+[PF_6]^-$ (cat.), KBr (cat.), NaOCl, $[bmim]^+[PF_6]^-$, H_2O, 0°	 **I** (86)[a]	295
		AcNH-TEMPO (cat.), **4** (cat.), $Cu(ClO_4)_2$ (cat.), O_2, DABCO, DMSO 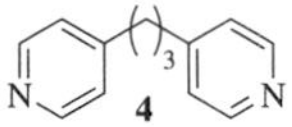 **4**	**I** (33)	245
		TEMPO (cat.), $NaNO_2$ (cat.), 1,3-dibromo-5,5-dimethylhydantoin (cat.), air, H_2O, 80°, 1 h	**I** (86)[g]	259
		PIPO (cat.), NaOCl, $KHCO_3$, CH_2Cl_2, H_2O, 0°	**I** (80–90)[a]	241, 299
		1. TEMPO (cat.), PS-I$(OAc)_2$, acetone, rt 2. H_2O, 24 h	**I** (81)	204
		TEMPO (cat.), KBr (cat.), NaOCl, CH_2Cl_2, H_2O, pH 9.5, 0–15°	**I** (92)	13
		PEG-BnO-TEMPO (cat.), $Co(NO_3)_2 \cdot H_2O$ or $Mn(NO_3)_2 \cdot H_2O$ (cat.), O_2, AcOH, 40°, 4 h	**I** (65–97)[a]	272
		TEMPO (cat.), TCC, NaOAc, H_2O, CH_2Cl_2, 0–5°	**I** (69)	408
		TEMPO (cat.), $H_5PV_2Mo_{10}O_{40} \cdot 34\ H_2O$ (cat.), O_2 (2 atm), acetone	**I** (98.4)[a]	274
		SG-NH-TEMPOH (cat.), KBr (cat.), NaOCl, $NaHCO_3$, CH_2Cl_2, H_2O, pH 9.1	**I** (88)[a]	307
		TEMPO (cat.), $RuCl_2(PPh_3)_3$ (cat.), O_2 (10 atm), chlorobenzene, 100°	**I** (85)[a]	80, 409, 410

TABLE 1. OXIDATIONS OF MONOFUNCTIONAL ALCOHOLS (*Continued*)

Alcohol	Conditions	Product(s) and Yield(s) (%)	Refs.
C_8 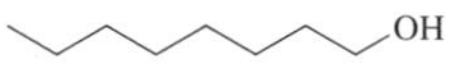	TEMPO (cat.), CuCl, O_2, $[bmim]^+[PF_6]^-$, 65°	 (—) **I**	385
	PS-CO_2-TEMPO (cat.), $Mn(NO_3)_2$ or $Co(NO_3)_2$ (cat.), O_2, AcOH, 40°	**I** (99)[a]	273
	Hydroxylamine (cat.), KBr (cat.), NaOCl, $NaHCO_3$, CH_2Cl_2, H_2O, pH 9, 0°, 1 h	**I** (—)	301
	Hydroxylamine	% Conversion[a]	
	PC-TEMPOH	80	
	PCH-TEMPOH	78	
	PCMe-TEMPOH	77	
	FT-O-TEMPO (cat.), KBr (cat.), NaOCl, $NaHCO_3$, CH_2Cl_2, H_2O	**I** (95)	406
	SBA15-NH-TEMPO (cat.), $NaNO_2$ (cat.), TBAB (cat.), O_2 (1 atm), AcOH, 50–60°, 2.5 h	**I** (100)[a], 76.2% selectivity[c]	310
	SBA15-NH-TEMPO (cat.), $NaNO_2$ (cat.), TBAB (cat.), air (1 atm), AcOH, 50–60°, 3 h	**I** (100)[a], 79% selectivity[c]	310
	TEMPO (cat.), DABCO–CuCl (cat.), O_2, K_2CO_3, toluene, 100°, 72 h	**I** (20)	247
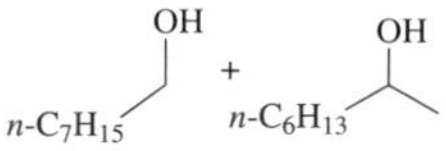	PIPO (cat.), NaOCl, $KHCO_3$, MTBE, H_2O, rt, 45 min	**I** (—), 86% conversion[a] 96% selectivity[a,c] + 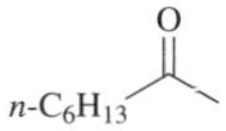 (—), <1% conversion[a]	299

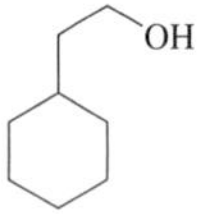	TEMPO (cat.), NaBr (cat.), $NaHCO_3$, NaOCl, CH_2Cl_2, H_2O, 5°	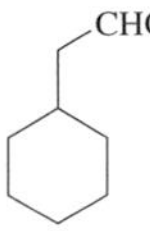 (—) **I**	168
	TEMPO (cat.), laccase (cat.), O_2, H_2O, citrate buffer, pH 4.5	**I** (48)[a]	249
	BzO-TEMPO, O_2, $RuCl_2(PPh_3)_3$, toluene, 70°, 8 h	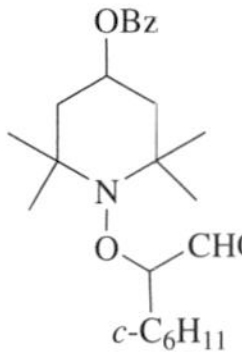 (—)	411
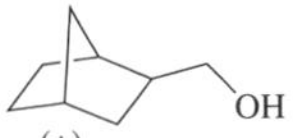 (±)	AcNH-TEMPO, TsOH, CH_2Cl_2, 0°, 2 h	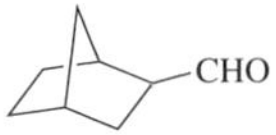 (94) **I**	56
	[PS-O-TEMP=O$^+$]Cl$^-$, CH_2Cl_2, rt, 1–2 h	**I** (72)[a]	121
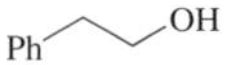	BzO-TEMPO, O_2, $RuCl_2(PPh_3)_3$, toluene, 70°, 8 h	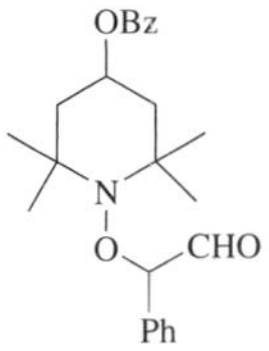 (—)	411
	[AcNH-TEMP=O]$^+$[ClO$_4$]$^-$, silica gel, CH_2Cl_2, rt	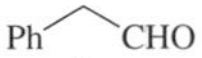 (99) **I**	68
	TEMPO (cat.), $PhI(OAc)_2$, CH_2Cl_2, rt	**I** (—)[h]	205
	Nitroxide (cat.), [mim-$CH_2PhI(OAc)_2$]$^+$[BF$_4$]$^-$, H_2O, 30°	**I** Nitroxide / Time [mim-O-TEMPO]$^+$[BF$_4$]$^-$ 4 h (86) [Im-(O-TEMPO)$_2$]$^+$[BF$_4$]$^-$ 1.5 h (85)	201

TABLE 1. OXIDATIONS OF MONOFUNCTIONAL ALCOHOLS (*Continued*)

	Alcohol	Conditions	Product(s) and Yield(s) (%)	Refs.
C_8	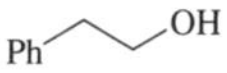	TEMPO (cat.), $[PS\text{-}CH_2NMe_3]^+[ClO_2]^-$, CH_2Cl_2, rt	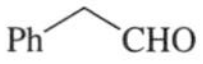 **I** (98)	44
		FT-O-TEMPO (cat.), KBr (cat.), NaOCl, $NaHCO_3$, CH_2Cl_2, H_2O	**I** (97)	406
		PEG-O-TEMPO (cat.), PEG-bipy (cat.), $CuBr_2$ (cat.), O_2, *t*-BuOK, MeCN, H_2O, 80°, 18 h	**I** (21)	392
		TEMPO (cat.), NaOCl (cat.), $NaClO_2$, H_2O, MeCN	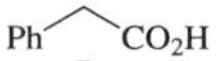 **I** (100)	14, 185
		1. TEMPO (cat.), NaOCl, EtOAc or MeCN, H_2O, pH 8–10 2. $NaClO_2$, H_2O, pH 5	**I** (92)	187
	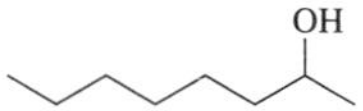	$[AcNH\text{-}TEMP{=}O]^+[ClO_4]^-$, silica gel, CH_2Cl_2, rt	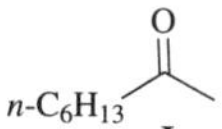 **I** (96)	68
		TEMPO (cat.), KBr (cat.), NaOCl, CH_2Cl_2, H_2O, pH 9.5	**I** (—)	134
		TEMPO (cat.), $H_5PV_2Mo_{10}O_{40}\bullet 34\ H_2O$ (cat.), O_2, acetone	**I** (96)[a]	274
		SG-NH-TEMPOH, KBr (cat.), NaOCl, $NaHCO_3$, CH_2Cl_2, H_2O, pH 9.1	**I** (65)[a]	307
		PIPO (cat.), NaOCl, $KHCO_3$, H_2O, 0°	**I** (99)[a]	299
		Graphite felt-TEMPO (cat.), (–)-sparteine, $NaClO_4$, electrolysis, MeCN	**I** (52.1)[a] + 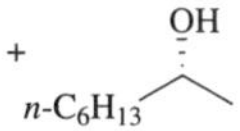 (45)[a], 99% ee	105

	TEMPO (cat.), $RuCl_2(PPh_3)_3$ (cat.), O_2 (10 atm), chlorobenzene, 100°	**I** (85–99)[a]	80, 409, 410
	PIPO (cat.), KBr (cat.), NaOCl, $KHCO_3$, CH_2Cl_2, H_2O, 0° to rt, 20–45 min	**I** (—), 99% conversion[a] 99% selectivity[a,e]	241, 299
	PE-co-AA-CONH-TEMPO, KBr (cat.), electrolysis, $NaHCO_3$, H_2O, 0°	**I** (51)	298
	PS-CO_2-TEMPO (cat.), $Mn(NO_3)_2$ or $Co(NO_3)_2$ (cat.), O_2, AcOH, 40°	**I** (99)[a]	273
	ICF-O-TEMPO (cat.), $PhI(OAc)_2$, CH_2Cl_2	**I** (92)[a]	293
	TEMPO (cat.), Br_2 (cat.), $NaNO_2$ (cat.), air (3.9 atm), CH_2Cl_2, 80°	**I** (98)	258
	TEMPO (cat.), $NaNO_2$ (cat.), 1,3-dibromo-5,5-dimethylhydantoin (cat.), air, H_2O, 80°, 1 h	**I** (94)	259
	TEMPO (cat.), *m*-CPBA, TBAB, CH_2Cl_2	**I** (—), 100% conversion	213
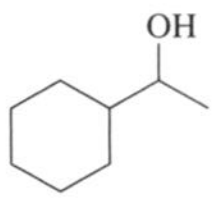	TEMPH•HCl (cat.), *m*-CPBA, CH_2Cl_2, rt, 1.5 h	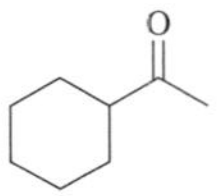 (81)	149
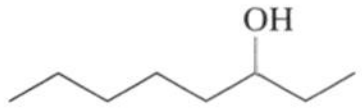	TEMPO (cat.), $H_5PV_2Mo_{10}O_{40}$•34 H_2O (cat.), O_2 (2 atm), acetone	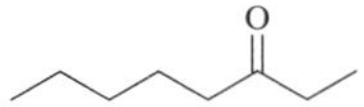 **I** (94)[a]	274
	TEMPO (cat.), Br_2 (cat.), $NaNO_2$ (cat.), air (3.9 atm), CH_2Cl_2, 80°	**I** (88)	258
	PIPO (cat.), NaOCl, $KHCO_3$, CH_2Cl_2, H_2O, rt, 45 min	**I** (—), 70% conversion[a] 99% selectivity[a,e]	299

TABLE 1. OXIDATIONS OF MONOFUNCTIONAL ALCOHOLS (*Continued*)

Alcohol	Conditions	Product(s) and Yield(s) (%)	Refs.
C_8 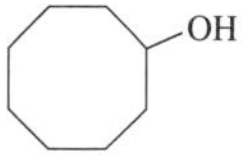	TEMPOH (cat.), *m*-CPBA, CH_2Cl_2, 0° to rt	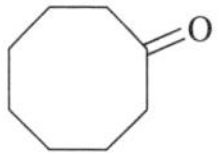 (69) **I**	102
	TEMPO (cat.), $RuCl_2(PPh_3)_3$ (cat.), O_2 (10 atm), chlorobenzene, 100°	**I** (92)[a]	80, 409, 410
	PEG-BnO-TEMPO (cat.), $Co(NO_3)_2$•H_2O (cat.), $Mn(NO_3)_2$•H_2O, O_2, AcOH, 40°, 4 h	**I** (68–99)[a]	272
	TEMPO (cat.), $H_5PV_2Mo_{10}O_{40}$•34 H_2O (cat.), O_2 (2 atm), acetone	**I** (95.5)[a]	274
	TEMPO (cat.), $[PS\text{-}CH_2NMe_3]^+[Br(OAc)_2]^-$, CH_2Cl_2, rt, 3.5 h	**I** (80)	43
	PIPO (cat.), KBr (cat.), NaOCl, CH_2Cl_2, H_2O, rt, 45 min	**I** (—), 100% conversion[a] >99% selectivity[a,e]	299
	PE-co-AA-CONH-TEMPO (cat.), KBr (cat.), electrolysis, $NaHCO_3$, H_2O, 0°	**I** (66–72)	298
	TEMPO (cat.), **1** (cat.), CuBr•Me_2S (cat.), O_2, perfluorooctyl bromide, chlorobenzene, 90°	**I** (85)	243
	TEMPO (cat.), $[PS\text{-}CH_2NMe_3]^+[Br(OAc)_2]^-$, CH_2Cl_2, rt, 24 h	**I** (80)[a]	211
	$[MeO\text{-}TEMP{=}O]^+Cl^-$, CH_2Cl_2, rt	**I** (89[a], 59)	255
	PEG-BnO-TEMPO, NaOCl, CH_2Cl_2, H_2O, pH 8.6, 0°	**I** (98[a], 78)	306
	ICF-O-TEMPO (cat.), $PhI(OAc)_2$, CH_2Cl_2	**I** (91[a], 78)	293

	ICF-O-TEMPO (cat.), KBr (cat.), NaOCl, CH_2Cl_2, 20°, 15 min	**I** (78)	291
	Nitroxide (cat.), KBr (cat.), NaOCl, CH_2Cl_2, 0°, 15 min	**I**	291
	Nitroxide		
	n-C_7F_{15}-CO_2-TEMPO	(80)	
	n-C_7F_{15}-CONH-TEMPO	(77)	
	ICF-O-TEMPO	(81)	
C_9 n-C_8H_{17}CH$_2$OH	TEMPO (cat.), KBr (cat.), NaOCl, CH_2Cl_2, H_2O, pH 9.5, 0–15°	n-C_8H_{17}CHO (88) **I**	13
	SG-NH-TEMPOH (cat.), KBr (cat.), NaOCl, $NaHCO_3$, CH_2Cl_2, pH 9.1	**I** (90[a], 85)	307
	TEMPO (cat.), $PhI(OAc)_2$, CH_2Cl_2, 20°	**I** (62–86)	412
	SG-NH-TEMPO (cat.), NaBr (cat.), NaOCl, H_2O, pH 10	**I** (~60)	313
	$[AcNH\text{-}TEMP{=}O]^+[ClO_4]^-$, silica gel, CH_2Cl_2, rt	 (100)[i]	68
	HO-TEMPO (cat.), CuCl (cat.), O_2, DMF, rt	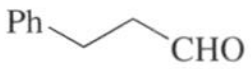 (86) **I**	81
	PS-CO_2-TEMPO (cat.), O_2, $Mn(NO_3)_2$ or $Co(NO_3)_2$ (cat.), AcOH, 40°	**I** (95)[a]	273
	AcNH-TEMPO (cat.), $Cu(ClO_4)_2$ (cat.), O_2 (1 atm), DMAP, $[bmpy]^+[PF_6]^-$, rt	**I** (26–61)	240
	TEMPO (cat.), KBr (cat.), PS-$I(OAc)_2$, acetone or CH_2Cl_2, H_2O, 0°	**I** (93)	202

TABLE 1. OXIDATIONS OF MONOFUNCTIONAL ALCOHOLS (*Continued*)

Alcohol	Conditions	Product(s) and Yield(s) (%)	Refs.
C_9 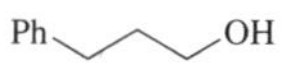	BzO-TEMPO (cat.), silica gel, NaOCl, H_2O, 0°	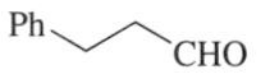 (40) **I**	393
	TEMPO (cat.), $[PS\text{-}CH_2NMe_3]^+[Br(OAc)_2]^-$, CH_2Cl_2, rt, 24 h	**I** (92)	44
	TEMPO (cat.), $[PS\text{-}CH_2NMe_3]^+[OCl]^-$, CH_2Cl_2, rt, 24 h	**I** (89)	44
	SBA15-NH-TEMPO (cat.), $NaNO_2$ (cat.), TBAB (cat.), O_2 (1 atm), AcOH, 50–60°, 2 h	**I** (99)[a], 95% selectivity[c]	310
	SBA15-NH-TEMPO (cat.), $NaNO_2$ (cat.), TBAB (cat.), air (1 atm), AcOH, 50–60°, 3 h	**I** (99)[a], 94.6% selectivity[c]	310
	TEMPO (cat.), NCS, TBAC, $NaHCO_3$, K_2CO_3, H_2O, rt	**I** (—)	413
	$[Im\text{-}(O\text{-}TEMPO)_2]^+[BF_4]^-$, $[mim\text{-}CH_2PhI(OAc)_2]^+[BF_4]^-$, H_2O, 30°	**I** (86)	201
	$[PS\text{-}O\text{-}TEMP{=}O]^+Cl^-$, CH_2Cl_2, rt, 1–2 h	**I** (90)[a]	121
	1. TEMPO (cat.), $PS\text{-}I(OAc)_2$, acetone, rt 2. H_2O, rt, 24 h	**I** (95)	204
	SG-NHCONH-TEMPO (cat.), NaOCl (1.1 equiv), acetone, H_2O, 0°	**I** (45)	393
	PE-co-AA-CONH-TEMPO (cat.), NaBr (cat.), electrolysis, $NaHCO_3$, H_2O, 0°	**I** (21) + 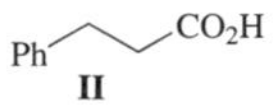 (25) **II**	298
	1. TEMPO (cat.), NaOCl, EtOAc or MeCN, H_2O, pH 8–10 2. $NaClO_2$, H_2O, pH 5	**II** (93)	187

	TEMPO (cat.), $[PS\text{-}CH_2NMe_3]^+[ClO_2]^-$, CH_2Cl_2, rt, 24 h	**II** (90)	44
	SG-NHCONH-TEMPO (cat.), NaOCl (3.5 eq), acetone, H_2O, 0°	**II** (85)	393
	BzO-TEMPO, $RuCl_2(PPh_3)_3$ (cat.), O_2, toluene, 70°, 8 h	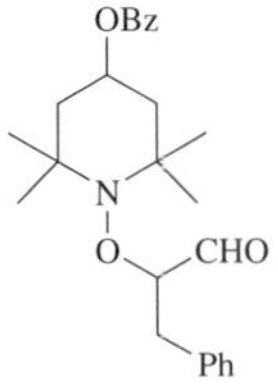 (—)	411
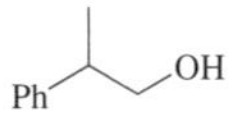	1. TEMPO (cat.), PS-I(OAc)$_2$, acetone, rt 2. H_2O, 24 h	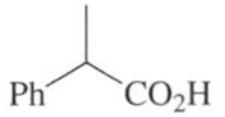 (22)	204
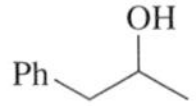	TEMPH•HCl (cat.), *m*-CPBA, CH_2Cl_2, rt	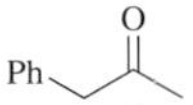 **I** (87)	149
	ABHep^{Cl}O (cat.), NaBr (4 eq), electrolysis, CH_2Cl_2, H_2O, $NaHCO_3$ (satd.), rt	**I** (100)	281
	$[PS\text{-}O\text{-}TEMP{=}O]^+Cl^-$, CH_2Cl_2, rt, 1–2 h	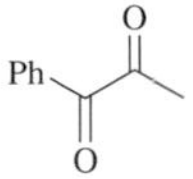 (76)[a]	121
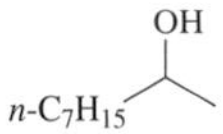	SG-NH-TEMPOH (cat.), KBr (cat.), NaOCl, $NaHCO_3$, CH_2Cl_2, pH 9.1	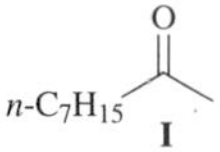 **I** (1)[a]	307
	MeO-TEMPO (cat.), KBr (cat.), NaOCl, CH_2Cl_2, H_2O, pH 9.5	**I** (97)	134
	TEMPO (cat.), O_2 or air (1 atm), $Mn(NO_3)_2$ or $Co(NO_3)_2$ (cat.), AcOH, 20–40°	**I** (100[a], 93)	395, 407

TABLE 1. OXIDATIONS OF MONOFUNCTIONAL ALCOHOLS (*Continued*)

	Alcohol	Conditions	Product(s) and Yield(s) (%)	Refs.
C_9	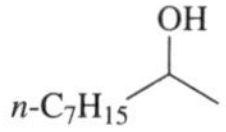	SBA15-NH-TEMPO (cat.), $NaNO_2$ (cat.), TBAB (cat.), O_2 (1 atm), AcOH, 50–60°, 3 h	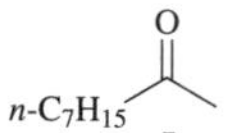 **I** (93)[a], 98% selectivity[e]	310
		SBA15-NH-TEMPO (cat.), $NaNO_2$ (cat.), TBAB (cat.), air (1 atm), AcOH, 50–60°, 4 h	**I** (88)[a], 98% selectivity[e]	310
	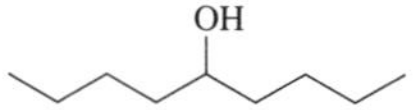	BzO-TEMPO (cat.), silica gel, NaOCl, H_2O, 0°	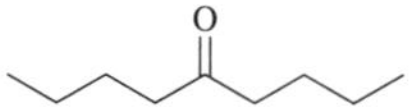 (82)	393
C_{10}	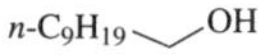	TEMPO (cat.), laccase (cat.), O_2, H_2O, citrate buffer, pH 4.5	n-$C_9H_{19}CHO$ **I** (58)[a]	249
		$[AcNH\text{-}TEMP{=}O]^+[ClO_4]^-$, silica gel, CH_2Cl_2, rt	**I** (99)	68
		TEMPO (cat.), **1** (cat.), CuBr•Me_2S (cat.), O_2, perfluorooctyl bromide, chlorobenzene, 90°	**I** (73)	243, 414
		TEMPO (cat.), **3** (cat.), O_2, toluene, 100°	**I** (84)	244
		MIP-CO_2-TEMPO (cat.), KBr (cat.), NaOCl, $NaHCO_3$, CH_2Cl_2, H_2O, 0°	**I** (78)	74
		1. TEMPO (cat.), NaBr (cat.), NaOCl, TBAB, $NaHCO_3$, CH_2Cl_2, H_2O, 0° to rt, 1 h 2. $NaClO_2$, 2-methyl-2-butene	**I** (90)	415
	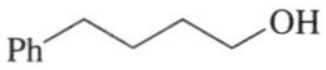	1. TEMPO (cat.), PS-I(OAc)$_2$, acetone, rt 2. H_2O, rt, 24 h	 **I** (92)	204
		TEMPO (cat.), Yb(OTf)$_3$ (cat.), PhI(OAc)$_2$, CH_2Cl_2, 45 min	**I** (83)	416

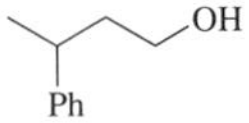	Graphite felt-TEMPO (cat.), (–)-sparteine, electrolysis, $NaClO_4$, MeCN	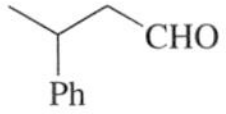 (48.9)[f]	105
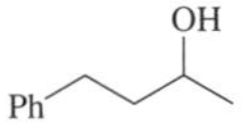	TEMPO (cat.), silica gel, NaOCl, H_2O, 0°	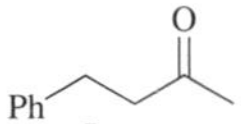 **I** (97)	393
	BzO-TEMPO (cat.), silica gel, NaOCl, H_2O, 0°	**I** (77–93)	393
	PE-CONH-TEMPO (cat.), NaBr (cat.), electrolysis, $NaHCO_3$, H_2O	**I** (83)	297
	PE-co-AA-CONH-TEMPO (cat.), NaBr (cat.), electrolysis, $NaHCO_3$, H_2O, 0°	**I** (94)	298
	BzO-TEMPO (cat.), NaBr (cat.), electrolysis, $NaHCO_3$, dispersed silica gel, H_2O	**I** (86)	84
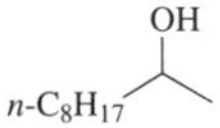	TEMPO (cat.), **1** (cat.), $CuBr \cdot Me_2S$ (cat.), O_2, perfluorooctyl bromide, chlorobenzene, 90°	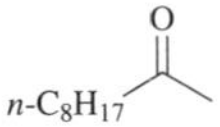 (88)	243
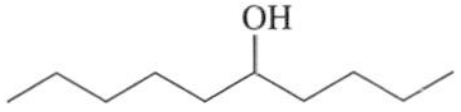	BzO-TEMPO (cat.), $RuCl_2(PPh_3)_3$ (cat.), O_2, toluene, 70°, 8 h	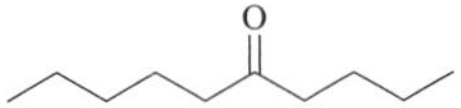 (95)	411
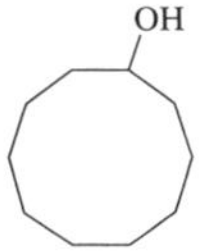	TEMPO (cat.), PS-$I(OAc)_2$, acetone, rt, 30 h	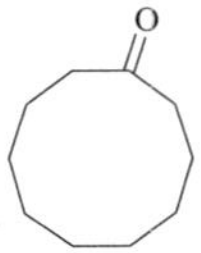 **I** (84)	202
	CIBA-TEMPO (cat.), $Mn(NO_3)_2$ or $Co(NO_3)_2$ (cat.), air or O_2, AcOH, 20–40°	**I** (94)	395
	$[MeO\text{-}TEMP{=}O]^+Cl^-$, CH_2Cl_2, rt	**I** (86^a, 74)	255

TABLE 1. OXIDATIONS OF MONOFUNCTIONAL ALCOHOLS (*Continued*)

Alcohol	Conditions	Product(s) and Yield(s) (%)	Refs.
C_{10} 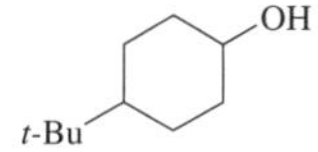	TEMPO (cat.), silica gel, NaOCl, H_2O, 0°	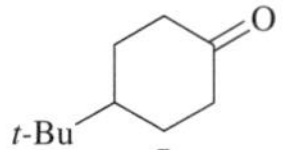 **I** (87)	393
	PE-co-AA-CONH-TEMPO (cat.), NaBr (cat.), electrolysis, $NaHCO_3$, H_2O, 0°	**I** (52–53)	298
	BzO-TEMPO (cat.), electrolysis, TBAB, NaOAc, CH_2Cl_2, H_2O, rt, 1 h	**I** (95)	224
	TEMPO (cat.), $Yb(OTf)_3$ (cat.), $PhI(OAc)_2$, CH_2Cl_2, 1 h	**I** (96)	416
	TEMPH, *m*-CPBA, CH_2Cl_2, 0° to rt, 1.5 h	**I** (70)	102
	TEMPO (cat.), *m*-CPBA, TBAB, CH_2Cl_2	**I** (94)	213
	TEMPO (cat.), $PhI(OAc)_2$, CH_2Cl_2	**I** (90)	192
	BzO-TEMPO (cat.), NaBr (cat.), electrolysis, $NaHCO_3$, CH_2Cl_2, H_2O, pH 8.6	**I** (93)	222
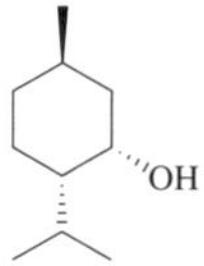	$[MeO\text{-}TEMP{=}O]^+Cl^-$, CH_2Cl_2, rt	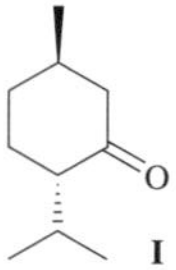 **I** (22)[a]	255
	Nitroxide (cat.), NaBr (4 eq), electrolysis, CH_2Cl_2, H_2O, $NaHCO_3$ (satd.), rt	**I**	281
	Nitroxide		
	ABHep^{Cl}O	(99)	
	ABHepO	(86)	
	ABOct^{Cl}O	(92)	
	ABOctO	(82)	
	ABHexO	(76)	
	TEMPO	(23)	

	Nitroxide (cat.), $NaIO_4$ (1.2 eq), NaBr (cat.), CH_2Cl_2, H_2O, rt, 24 h	**I**	281
	Nitroxide		
	$ABHep^{Cl}O$	(99)	
	ABHepO	(92)	
	$ABOct^{Cl}O$	(99)	
	ABOctO	(99)	
	ABHexO	(99)	
	TEMPO	(22)	
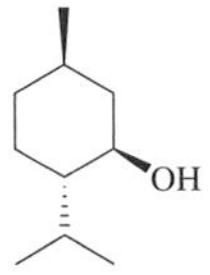	$[MeO\text{-}TEMP{=}O]^+Cl^-$, CH_2Cl_2, rt	**I** (8)[a]	255
	TEMPO (cat.), $[PS\text{-}CH_2NMe_3]^+[ClO]^-$, CH_2Cl_2, rt	 **I** (99)	44
	$[PS\text{-}O\text{-}TEMP{=}O^+]Cl^-$, CH_2Cl_2, rt, 1–2 h	**I** (95)[a]	121
	TEMPO (cat.), $Yb(OTf)_3$ (cat.), $PhI(OAc)_2$, CH_2Cl_2, 2 h	**I** (90)	416
	TEMPO (cat.), silica gel, NaOCl, H_2O, 0°	**I** (76)	393
	TEMPO (cat.), KBr (cat.), $PS\text{-}I(OAc)_2$, acetone, 0°	**I** (92)	202
	AcNH-TEMPO, TsOH, CH_2Cl_2, 0°	**I** (97)	56

TABLE 1. OXIDATIONS OF MONOFUNCTIONAL ALCOHOLS (*Continued*)

Alcohol	Conditions	Product(s) and Yield(s) (%)	Refs.
C_{10} 	TEMPO (cat.), CuCl (cat.), O_2, $[bmim]^+[PF_6]^-$, 65°	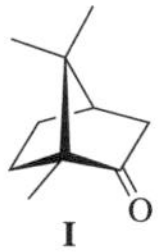 **I** (—)	335
	SBA15-NH-TEMPO (cat.), $NaNO_2$ (cat.), TBAB (cat.), O_2 (1 atm), AcOH, 50–60°, 10 h	**I** (100)[a], 99.6% selectivity[e]	310
	SBA15-NH-TEMPO (cat.), $NaNO_2$ (cat.), TBAB (cat.), air (1 atm), AcOH, 50–60°, 11 h	**I** (99)[a], 99.8% selectivity[e]	310
 (±)	$[AcNH\text{-}TEMP{=}O]^+[ClO_4]^-$, silica gel, CH_2Cl_2, rt	**I** (100)	68
	TEMPO (cat.), KBr (cat.), PS-I(OAc)$_2$, acetone, 0°	**I** (92)	202
	TEMPH•HCl, *m*-CPBA, CH_2Cl_2, rt, 1.5 h	**I** (94)	149
	AcNH-TEMPO, TsOH, CH_2Cl_2, 0°	**I** (97)	56
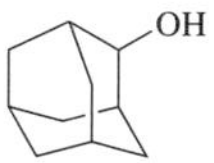	TEMPO (cat.), $RuCl_2(PPh_3)_3$ (cat.), O_2 (10 atm), chlorobenzene, 100°	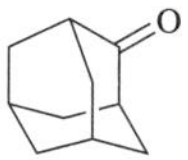 **I** (98)[a]	80, 409, 410
	TEMPO (cat.), $Mn(NO_3)_2$ or $Co(NO_3)_2$ (cat.), O_2 (1 atm), AcOH	**I** (97)[a]	407
	SBA15-NH-TEMPO (cat.), $NaNO_2$ (cat.), TBAB (cat.), O_2 (1 atm), AcOH, 50–60°, 9.5 h	**I** (96)[a], 97% selectivity[e]	310

	Nitroxide (cat.), NaBr (4 eq), electrolysis, CH_2Cl_2, H_2O, $NaHCO_3$ (satd.), rt	**I**	281
	Nitroxide		
	ABHep^{Cl}O	(87)	
	ABHepO	(65)	
	ABOct^{Cl}O	(90)	
	ABOctO	(86)	
	ABHexO	(82)	
	TEMPO	(61)	
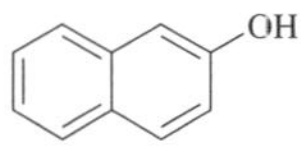	$[TEMP{=}O]^+Cl^-$, CH_2Cl_2, 0°	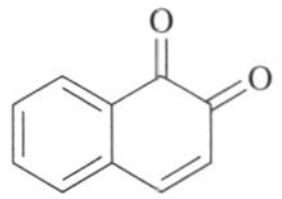 (95)	352
C_{11} 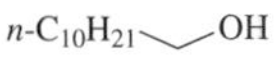	BzO-TEMPO (cat.), electrolysis, $TBABr_3$, NaOAc, CH_2Cl_2, H_2O, rt, 1 h	n-$C_{10}H_{21}CHO$ (95) **I**	224
	PEG-BnO-TEMPO (cat.), NaOCl, CH_2Cl_2, H_2O, pH 8.6, 0°	**I** (93)[a]	293
	PS-CO_2-TEMPO (cat.), O_2, $Mn(NO_3)_2$ or $Co(NO_3)_2$ (cat.), AcOH, 40°	**I** (99)[a]	273
	TEMPO (cat.), KBr (cat.), PS-I$(OAc)_2$, acetone, 0°	**I** (97)	202
	TEMPO (cat.), KBr (cat.), NaOCl, CH_2Cl_2, H_2O, pH 9.5, 0–15°	**I** (93)	13
	PEG-BnO-TEMPO (cat.), $Co(NO_3)_2 \cdot H_2O$ or $Mn(NO_3)_2 \cdot H_2O$ (cat.), O_2, AcOH, 40°, 4 h	**I** (78–99)[a]	272
	ICF-O-TEMPO (cat.), $PhI(OAc)_2$, CH_2Cl_2	**I** (99[a], 80)	293
	BzO-TEMPO (cat.), silica gel, NaOCl, H_2O	n-$C_{10}H_{21}CO_2H$ (79)	393

Alcohol	Conditions	Product(s) and Yield(s) (%)	Refs.
C_{11}			
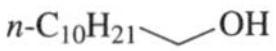	BzO-TEMPO, $RuCl_2(PPh_3)_3$, O_2, toluene, 70°, 8 h	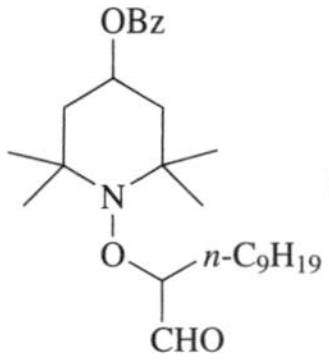 (76)	411
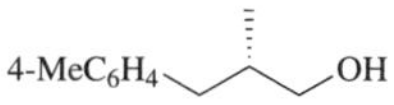	TEMPO (cat.), KBr (cat.), NaOCl, $NaHCO_3$, CH_2Cl_2, H_2O	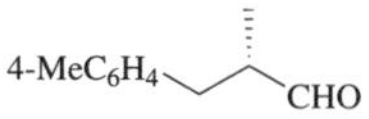 (—)[b]	417
	1. TEMPO (cat.), PS-I(OAc)$_2$, acetone, rt 2. H_2O, rt, 24 h	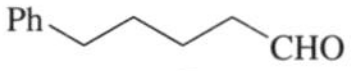 **I** (83)	204
	MIP-CO_2-TEMPO (cat.), KBr (cat.), NaOCl, $NaHCO_3$, H_2O, CH_2Cl_2, 0°	**I** (88)	74
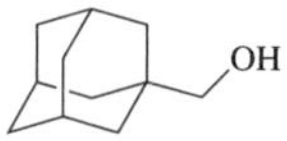	TEMPO (cat.), KBr (cat.), PS-I(OAc)$_2$, acetone, 0°	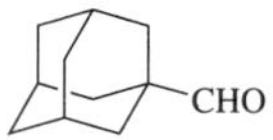 **I** (91)	202
	TEMPO (cat.), $Yb(OTf)_3$ (cat.), $PhI(OAc)_2$, CH_2Cl_2, 3 h	**I** (84)	416
	1. TEMPO (cat.), PS-I(OAc)$_2$, acetone, rt 2. H_2O, rt	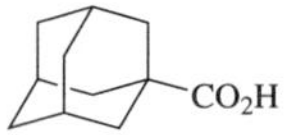 (58)	204
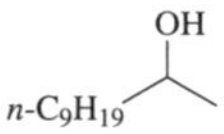	PEG-BnO-TEMPO (cat.), $Co(NO_3)_2$•H_2O or $Mn(NO_3)_2$•H_2O (cat.), O_2, AcOH, 40°, 4 h	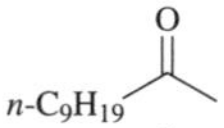 **I** (52)[a]	272
	TEMPO (cat.), **1** (cat.), CuBr•Me_2S (cat.), O_2, perfluorooctyl bromide, chlorobenzene, 90°	**I** (71)	414

		PEG-BnO-TEMPO (cat.), NaOCl, CH_2Cl_2, H_2O, pH 8.6, 0°	**I** (98)[a]	306
		TEMPH, *m*-CPBA, CH_2Cl_2, rt	**I** (50)	102
		ICF-O-TEMPO (cat.), $PhI(OAc)_2$, CH_2Cl_2	**I** (83)[a]	293
		BzO-TEMPO (cat.), O_2, $RuCl_2(PPh_3)_3$, toluene, 70°, 8 h	**I** (—)	411
	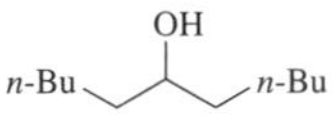	PE-co-AA-CONH-TEMPO (cat.), NaBr (cat.), electrolysis, $NaHCO_3$, H_2O	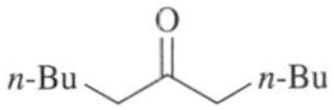 (66–80)	298
C_{12}	*n*-$C_{11}H_{23}$CH$_2$OH	TEMPO (cat.), $[PS\text{-}CH_2NMe_3]^+[Br(OAc)_2]^-$, CH_2Cl_2, rt or toluene, 40°, 3.5 h	*n*-$C_{11}H_{23}$CHO (93)[a] **I**	211
		SG-NH-TEMPOH (cat.), KBr (cat.), NaOCl, $NaHCO_3$, CH_2Cl_2, H_2O, pH 9.1	**I** (81)[a]	307
		TEMPO (cat.), $[PS\text{-}CH_2NMe_3]^+[Br(OAc)_2]^-$, CH_2Cl_2, rt, 24 h	**I** (93)	43
		TEMPO (cat.), KBr (cat.), PS-I$(OAc)_2$, acetone, 0°	**I** (97)	202
		FT-O-TEMPO (cat.), KBr (cat.), NaOCl, $NaHCO_3$, CH_2Cl_2, H_2O	**I** (96)	406
	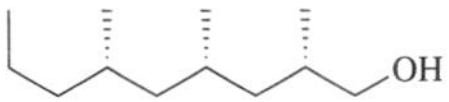	TEMPO (cat.), $PhI(OAc)_2$, CH_2Cl_2, 0.5 h	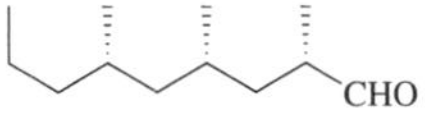 (92)	418
	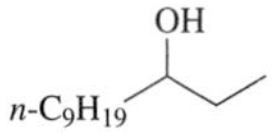	TEMPO (cat.), **1** (cat.), CuBr•Me_2S (cat.), O_2, perfluorooctyl bromide, chlorobenzene, 90°	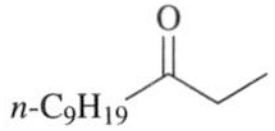 (69)	243, 414

TABLE 1. OXIDATIONS OF MONOFUNCTIONAL ALCOHOLS (*Continued*)

Alcohol	Conditions	Product(s) and Yield(s) (%)	Refs.
C_{12} 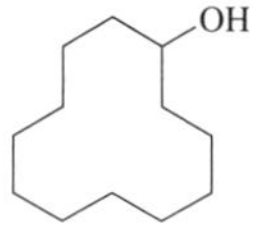	TEMPO (cat.), KBr (cat.), PS-I(OAc)$_2$, acetone, 0°	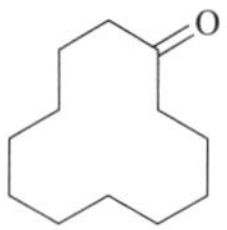 (84)	202
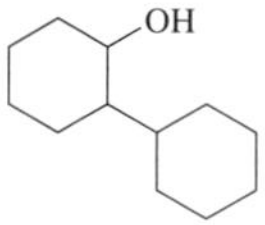	[MeO-TEMP=O]$^+$Cl$^-$, CH_2Cl_2, rt	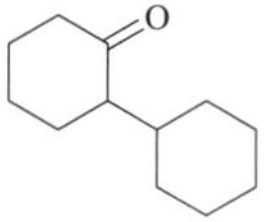 (32)[a]	255
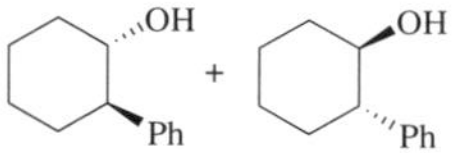	Graphite felt-TEMPO (cat.), (–)-sparteine, electrolysis, $NaClO_4$, MeCN	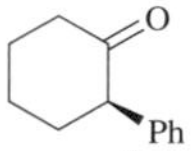 (53.8)[f] + 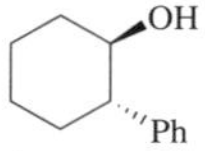 (45.1)[f], 99.4% ee, $s = 49$	105
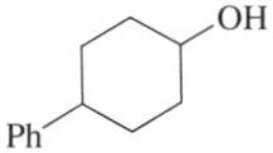	TEMPO (cat.), KBr (cat.), PS-I(OAc)$_2$, acetone, 0°	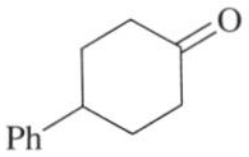 (80)	202
C_{13} n-$C_{12}H_{25}CH_2OH$	1. TEMPO (cat.), PS-I(OAc)$_2$, acetone, rt 2. H_2O, rt, 24 h	n-$C_{12}H_{25}CHO$ (82) **I**	204
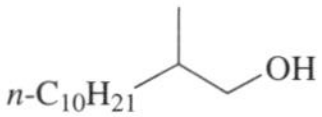	MIP-CO$_2$-TEMPO (cat.), KBr (cat.), NaOCl, $NaHCO_3$, CH_2Cl_2, H_2O, 0°	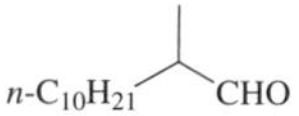 (55)	74
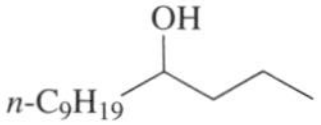	TEMPO (cat.), **1** (cat.), CuBr•Me_2S (cat.), O_2, perfluorooctyl bromide, chlorobenzene, 90°	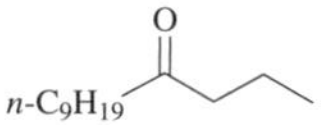 (31)	243, 414
C_{14} 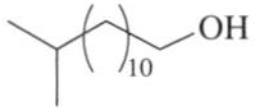	BzO-TEMPO (cat.), KBr (cat.), NaOCl, $NaHCO_3$, CH_2Cl_2, H_2O	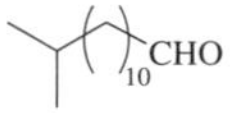 (65–86)	419

C_{16}			
n-$C_{15}H_{31}$⌒OH	AcNH-TEMPO, TsOH, CH_2Cl_2, 0°[c], 2.5 h	n-$C_{15}H_{31}CHO$ (100) **I**	56
	FT-O-TEMPO (cat.), KBr (cat.), NaOCl, $NaHCO_3$, CH_2Cl_2, H_2O	**I** (96)	406
	TEMPO (cat.), $[PS\text{-}CH_2NMe_3]^+[Br(OAc)_2]^-$, CH_2Cl_2, rt, 24 h	**I** (93)	44
	TEMPO (cat.), $[PS\text{-}CH_2NMe_3]^+[ClO]^-$, HCl, CH_2Cl_2, rt, 7 h	**I** (90)	44
	TEMPO (cat.), $[PS\text{-}CH_2NMe_3]^+[ClO_2]^-$, HCl, CH_2Cl_2, rt, 2 h	n-$C_{15}H_{31}CO_2H$ (94)	44
C_{27}			
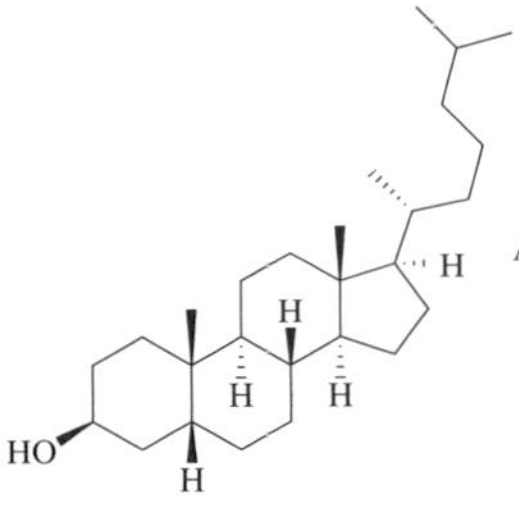	AcNH-TEMPO, TsOH, CH_2Cl_2, 0°	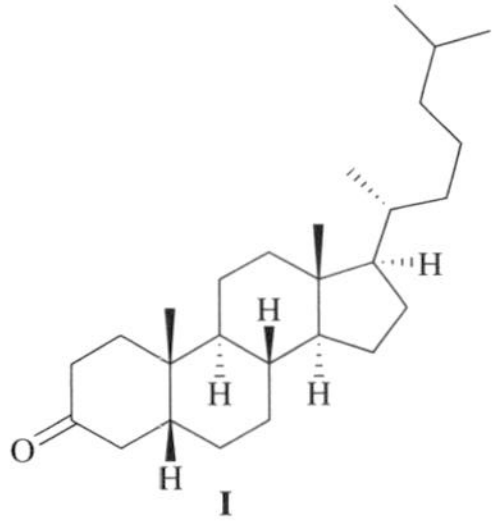 **I** (98)	56
	TEMPH, m-CPBA, CH_2Cl_2, 0° to rt	**I** (52)	102

[a] The reported value was determined by GC analysis.

[b] The reaction rate of this process is provided.

[c] The selectivity refers to formation of the aldehyde versus products with the carboxylic acid oxidation state.

[d] The reported value was determined by NMR analysis.

[e] The selectivity refers to formation of the ketone versus over-oxidation products such as C–C cleavage.

[f] The reported value was determined by HPLC analysis.

[g] Some ester-linked dimer is also produced.

[h] The product was not isolated and was used in a subsequent step without purification.

[i] The product was 92% pure.

TABLE 2. OXIDATIONS OF BENZYLIC ALCOHOLS

Alcohol	Conditions	Product(s) and Yield(s) (%)	Refs.
C_7 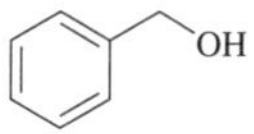	[MeO-TEMP=O]$^+$Br$^-$, CH_2Cl_2, rt	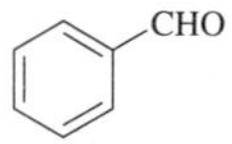 (100)[a] **I**	73
	[MeO-TEMP=O]$^+$Cl$^-$, CH_2Cl_2, rt	**I** (100)[a]	73
	PEG-BnO-TEMPO (cat.), NaOCl, CH_2Cl_2, H_2O, pH 8.6, 0°	**I** (98)[a]	306
	Hydroxylamine (cat.), KBr (cat.), NaOCl, $NaHCO_3$, CH_2Cl_2, H_2O, pH 9, 0°, 1 h	**I** (—)	301
	Hydroxylamine	% Conversion[a]	
	PC-TEMPOH	70	
	PCH-TEMPOH	67	
	PCMe-TEMPOH	69	
	TEMPO (cat.), I_2, $NaHCO_3$, toluene or CH_2Cl_2, 20°, overnight	**I** (96)[b]	256
	TEMPO (cat.), electrolysis, 2,6-lutidine, $NaClO_4$, MeCN, H_2O	**I** (97)[a]	421
	PS-CO_2-TEMPO (cat.), $Mn(NO_3)_2$ or $Co(NO_3)_2$ (cat.), O_2, AcOH, 40°	**I** (99)[a]	273
	MIP-CO_2-TEMPO (cat.), KBr (cat.), NaOCl, $NaHCO_3$, H_2O, CH_2Cl_2, 0°	**I** (79)	74
	ICF-O-TEMPO (cat.), $PhI(OAc)_2$, CH_2Cl_2	**I** (99)[a]	306
	TEMPO (cat.), Br_2 (cat.), $NaNO_2$ (cat.), air (3.9 atm), CH_2Cl_2, 80°	**I** (95)	258

BzO-TEMPO (cat.), NaBr (cat.), electrolysis, $NaHCO_3$, H_2O, CH_2Cl_2, pH 8.6	**I** (97)	222
TEMPO (cat.), **3** (cat.), O_2, toluene, 100° 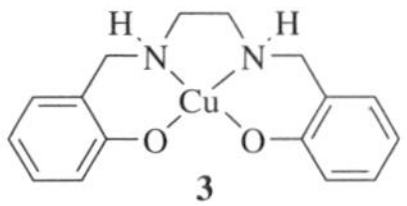 **3**	**I** (99)[a]	244
TEMPO (cat.), $[PS\text{-}CH_2NMe_3]^+[Br(OAc)_2]^-$, CH_2Cl_2, rt or toluene, 40°, 3.5 h	**I** (94)[a]	211
TEMPO (cat.), $(NH_4)_2[Ce(NO_3)_6]$ (cat.), O_2, MeCN, reflux	**I** (71–99)	269
AcNH-TEMPO (cat.), $Cu(ClO_4)_2$ (cat.), DMAP, O_2 (1 atm), $[bmpy]^+[PF_6]^-$, rt	**I** (92)	240
$[PS\text{-}O\text{-}TEMP{=}O]^+Cl^-$, CH_2Cl_2, rt, 1–2 h	**I** (95)[a]	121
TEMPO (cat.), laccase (cat.), O_2, H_2O, citrate buffer, pH 4.5	**I** (92)[a]	249
$[mim\text{-}O\text{-}TEMPO]^+[BF_4]^-$ (cat.), $[mim\text{-}CH_2PhI(OAc)_2]^+[BF_4]^-$, H_2O, 30°, 6 min	**I** (97)[a]	201
TEMPO (cat.), DABCO–CuCl (cat.), O_2, K_2CO_3, toluene, 100°, 24 h	**I** (85)	247
TEMPO (cat.), DABCO–CuCl (cat.), O_2, K_2CO_3, $MeNO_2$, 100°, 8 h	**I** (78)	247
TEMPO (cat.), $PhI(OAc)_2$, CH_2Cl_2, rt	**I** (95)	192
TEMPH•HCl (cat.), *m*-CPBA, CH_2Cl_2, rt	**I** (76)	149
PIPO (cat.), NaOCl, $KHCO_3$, CH_2Cl_2, H_2O, 0°	**I** (100)[a]	241, 299

TABLE 2. OXIDATIONS OF BENZYLIC ALCOHOLS *(Continued)*

Alcohol	Conditions	Product(s) and Yield(s) (%)	Refs.
C_7 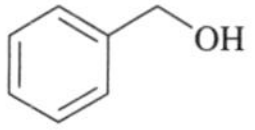	SG-NH-TEMPOH (cat.), KBr (cat.), NaOCl, $NaHCO_3$, CH_2Cl_2, H_2O, pH 9.1	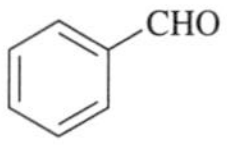 **I** (75)	307
	TEMPO (cat.), $H_5PV_2Mo_{10}O_{40}$•34 H_2O (cat.), O_2 (2 atm), acetone	**I** (99.6)[a]	274
	PEG-BnO-TEMPO (cat.), $Co(NO_3)_2$ or $Mn(NO_3)_2$ (cat.), O_2, AcOH, 40°, 4 h	**I** (99)[a]	272
	TEMPO (cat.), $[PS\text{-}CH_2NMe_3]^+[(Br(OAc)_2]^-$, toluene, rt, 24 h	**I** (94)	43
	MCM41-NH-TEMPO (cat.), *m*-CPBA, HBr, CH_2Cl_2, rt, 1.5 h	**I** (—)	309
	Nitroxide (cat.), NaBr (4 eq), electrolysis, CH_2Cl_2, H_2O, $NaHCO_3$ (satd.), rt	**I**	281
	Nitroxide		
	$ABHep^{Cl}O$	(99)	
	ABHepO	(96)	
	$ABOct^{Cl}O$	(99)	
	ABOctO	(90)	
	ABHexO	(82)	
	TEMPO	(99)	
	TEMPO (cat.), $PhI(OAc)_2$, KNO_2 (cat.), O_2, 80–90°, 24 h	**I** (96)[c]	203
	TEMPO (cat.), PS-$I(OAc)_2$, KNO_2 (cat.), O_2, 80–90°, 24 h	**I** (98)[c]	203
	SBA15-NH-TEMPO (cat.), $NaNO_2$ (cat.), TBAB (cat.), O_2 (1 atm), AcOH, 50–60°, 1.5 h	**I** (100)[a], 99.8% selectivity[d]	310

	SBA15-NH-TEMPO (cat.), $NaNO_2$ (cat.), TBAB (cat.), air (1 atm), AcOH, 50–60°, 2.5 h	**I** (100)[a], 99.1% selectivity[d]	310
	1. TEMPO (cat.), NaBr (cat.), TBAB (cat.), NaOCl, $NaHCO_3$, CH_2Cl_2, H_2O, 0° to rt, 1 h 2. $NaClO_2$, 2-methyl-2-butene	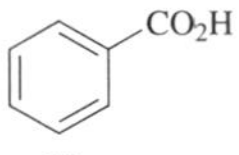 **II** (100)	415
	1. TEMPO (cat.), NaOCl, EtOAc or MeCN, H_2O, pH 8–10 2. $NaClO_2$, H_2O, pH 5	**II** (98)	187
	TEMPO (cat.), NaOCl, $NaClO_2$, MeCN	**II** (98)	185, 420
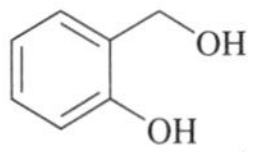	TEMPO (cat.), DABCO–CuCl (cat.), O_2, K_2CO_3, toluene, 100°, 2 h	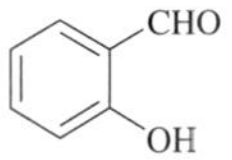 (85)	247
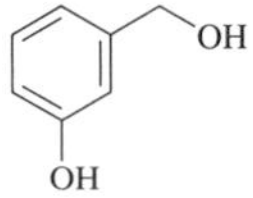	$[AcNH\text{-}TEMP{=}O]^+[ClO_4]^-$, silica gel, CH_2Cl_2, rt	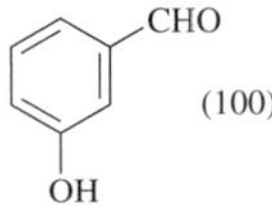 (100)	68
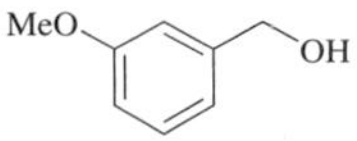	CIBA-TEMPO (cat.), $Mn(NO_3)_2$ or $Co(NO_3)_2$ (cat.), O_2 or air, AcOH, 20–40°	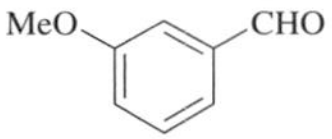 (95)	395
	1. TEMPO (cat.), NaOCl, EtOAc or MeCN, H_2O, pH 8–10, 5°, 1 h 2. $NaClO_2$, H_2O, pH 5, rt, 3 h	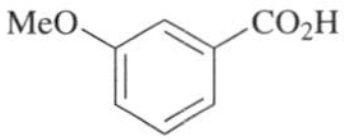 (24)	187
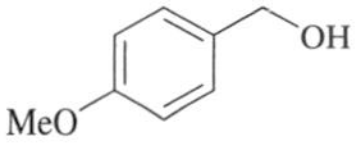	TEMPO (cat.), KNO_2 (cat.), $PhI(OAc)_2$, O_2, 80–90°, 24 h	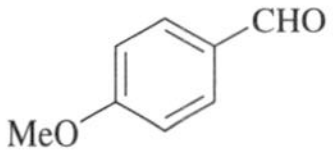 **I** (97)[c]	203
	TEMPO (cat.), KNO_2 (cat.), PS-I$(OAc)_2$, O_2, 80–90°, 24 h	**I** (99)[c]	203

TABLE 2. OXIDATIONS OF BENZYLIC ALCOHOLS *(Continued)*

Alcohol	Conditions	Product(s) and Yield(s) (%)	Refs.
C_7 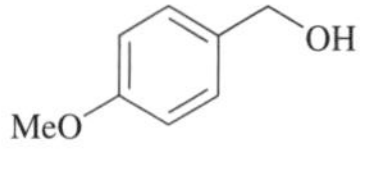	TEMPO (cat.), laccase (cat.), O_2 (or no O_2), H_2O, citrate buffer, pH 5, 24 h	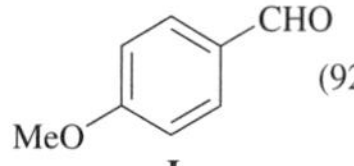 **I** (92–99)	250
	TEMPO (cat.), laccase (cat.), H_2O, citrate buffer, pH 4.5, 24 h	**I** (99)	249
	HO-TEMPO (cat.), CuCl (cat.), O_2, DMF, rt	**I** (96)	81
	TEMPO (cat.), 1,10-phenanthroline (cat.), $CuSO_4$ (cat.), NaOH, O_2, H_2O, 80°, 3 h	**I** (65)[a]	422
	PEG-O-TEMPO (cat.), PEG-bipy (cat.), $CuBr_2$ (cat.), *t*-BuOK, MeCN, H_2O, O_2, 80°, 18 h	**I** (84)	392
	AcNH-TEMPO (cat.), **4** (cat.), $Cu(ClO_4)_2$ (cat.), O_2, DABCO, DMSO 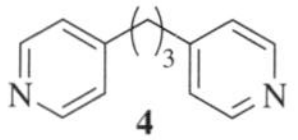 **4**	**I** (92)	245
	TEMPO (cat.), $CuBr_2$ (cat.), **5** (cat.), Et_3N (cat.), air, MeCN 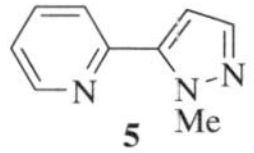 **5**	**I** (98)[a]	423
	TEMPO (cat.), $CuBr_2$ (cat.), **6** (cat.), Et_3N (cat.), air, MeCN 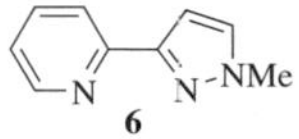 **6**	**I** (30)[a]	423

TEMPO (cat.), $CuBr_2$ (cat.), **7** (cat.), Et_3N (cat.), air, MeCN 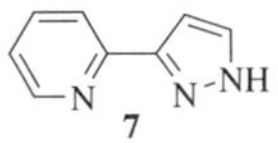 **7**	**I** (33)[a]	423
CIBA-TEMPO (cat.), $Mn(NO_3)_2$ or $Co(NO_3)_2$ (cat.), H_2O_2 or O_2 or air, AcOH, 20–50°, 2 h	**I** (94–99)	395
TEMPO (cat.), I_2, $NaHCO_3$, toluene or CH_2Cl_2, 20°, overnight	**I** (95)[b]	256
TEMPO (cat.), **3** (cat.), O_2, toluene, 100°	**I** (98)	244
BzO-TEMPO (cat.), $TBABr_3$, NaOAc, CH_2Cl_2, H_2O, rt, 1 h	**I** (92)	224
[mim-O-TEMPO]$^+$[BF_4]$^-$ or [mim-CO_2-TEMPO]$^+$[BF_4]$^-$(cat.), [mim-CH_2PhI(OAc)$_2$]$^+$[BF_4]$^-$, H_2O, 30°, 30 min	**I** (98)[a]	201
MeO-TEMPO (cat.), KBr (cat.), NaOCl (1.2 eq), Aliquat 336, CH_2Cl_2, H_2O, pH 9.5	**I** (98)	134
TEMPO (cat.), NaOCl, silica gel, H_2O, 0°	**I** (66)	393
TEMPO (cat.), DABCO–CuCl (cat.), O_2, K_2CO_3, 100°	**I**	247
Solvent / Time		
toluene / 2 h	(89)	
MeCN / 5 h	(59)	
DMSO / 5 h	(64)	
DMF / 2 h	(84)	
1,4-dioxane / 4 h	(51)	
$MeNO_2$ / 5 h	(93)	

TABLE 2. OXIDATIONS OF BENZYLIC ALCOHOLS *(Continued)*

Alcohol	Conditions	Product(s) and Yield(s) (%)	Refs.
C_7 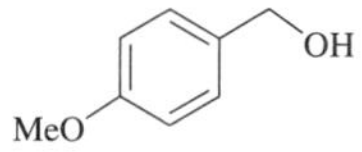	SBA15-NH-TEMPO (cat.), $NaNO_2$ (cat.), TBAB (cat.), O_2 (1 atm), AcOH, 50–60°, 1.5 h	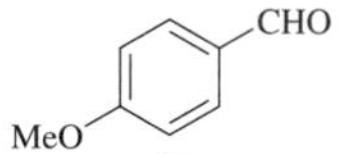 **I** $(100)^a$, 99.5% selectivity[d]	310
	SBA15-NH-TEMPO (cat.), $NaNO_2$ (cat.), TBAB (cat.), air (1 atm), AcOH, 50–60°, 2.5 h	**I** $(100)^a$, 99.2% selectivity[d]	310
	TEMPO (cat.), KBr (cat.), NaOCl (2.4 eq), Aliquat 336, CH_2Cl_2, H_2O, pH 9.5	**I** (95) + 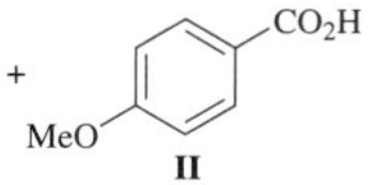 **II** (5)	134
	1. TEMPO (cat.), NaOCl, EtOAc or MeCN, H_2O, pH 8–10, 5°, 1 h 2. $NaClO_2$, H_2O, pH 5, rt, 3 h	**II** (16–17)	187
	TEMPO (cat.), $[PS\text{-}CH_2NMe_3]^+[ClO_2]^-$, CH_2Cl_2, rt, 24 h	**II** (59)	44
	AcNH-TEMPO (cat.), $Cu(ClO_4)_2$ (cat.), O_2 (1 atm), DMAP, $[bmpy]^+[PF_6]^-$, rt	**II** (91)	240
	1. TEMPO (cat.), NaBr (cat.), NaOCl, TBAB, $NaHCO_3$, CH_2Cl_2, H_2O, 0° to rt, 1 h 2. $NaClO_2$, 2-methyl-2-butene	**II** (95)	415
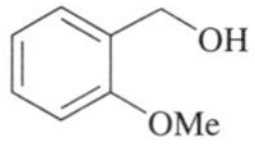	PEG-O-TEMPO (cat.), PEG-bipy (cat.), $CuBr_2$ (cat.), O_2, *t*-BuOK, MeCN, H_2O, 80°, 18 h	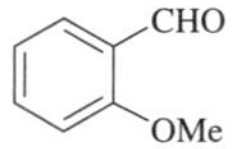 **I** (100)	392
	TEMPO (cat.), $Mn(NO_3)_2$ or $Co(NO_3)_2$ (cat.), O_2, AcOH, 4 h	**I** $(98)^a$	407

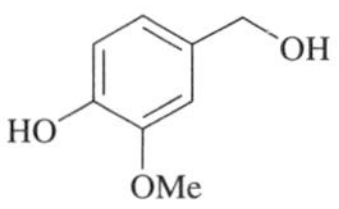	TEMPO (cat.), DABCO–CuCl (cat.), O_2, K_2CO_3, toluene, 100°, 10 h	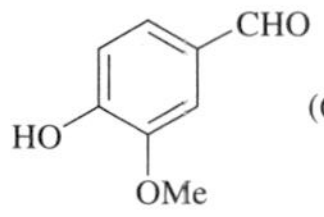 (65)	247
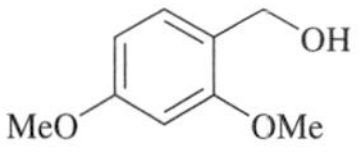	TEMPO (cat.), **1** (cat.), $CuBr \cdot Me_2S$ (cat.), O_2, perfluorooctyl bromide, chlorobenzene, 90° 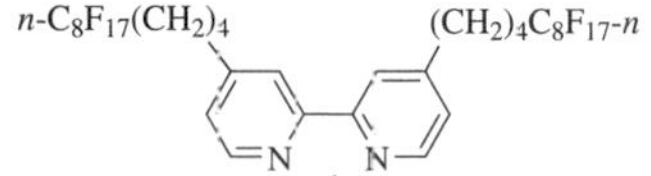 **1**	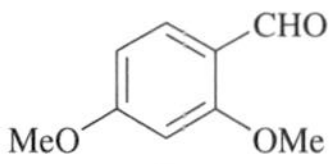 **I** (93)	243, 414
	TEMPO (cat.), 1,10-phenanthroline (cat.), $CuSO_4$ (cat.), O_2, NaOH, H_2O, 80°, 3 h	**I** (99)[a]	422
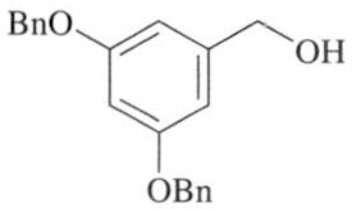	$[AcNH\text{-}TEMP{=}O]^+[ClO_4]^-$, silica gel, CH_2Cl_2, rt	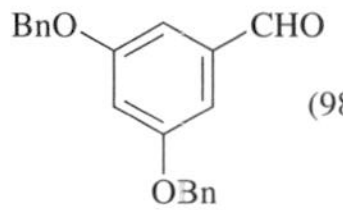 (98)	68
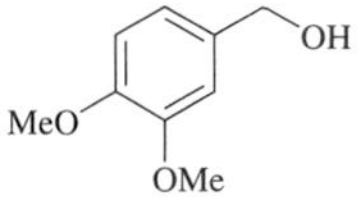	TEMPO (cat.), laccase, O_2 (or no O_2), H_2O, citrate buffer, pH 5, 24 h	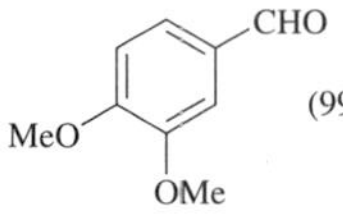 **I** (99)	250
	$[AcNH\text{-}TEMP{=}O]^+[ClO_4]^-$, silica gel, CH_2Cl_2, rt	**I** (98)	68
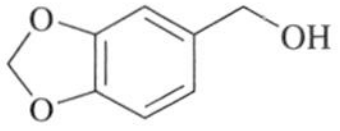	$[AcNH\text{-}TEMP{=}O]^+[ClO_4]^-$, silica gel, CH_2Cl_2, rt	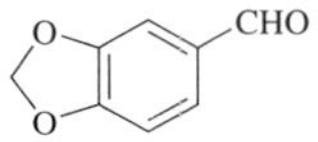 **I** (100)	68
	AcNH-TEMPO, TsOH, CH_2Cl_2, 0°, 1 h	**I** (99)	56
	TEMPO, TsOH, CH_2Cl_2, 0°, 1 h	**I** (90)[a]	56

TABLE 2. OXIDATIONS OF BENZYLIC ALCOHOLS *(Continued)*

	Alcohol	Conditions	Product(s) and Yield(s) (%)	Refs.
C_7	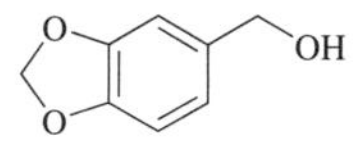	HO-TEMPO (cat.), CuCl (cat.), O_2, DMF, rt	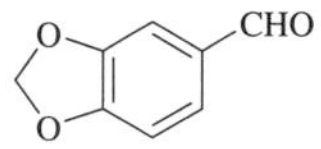 **I** (85)	81
		$[PS\text{-}O\text{-}TEMP{=}O]^+Cl^-$, CH_2Cl_2, rt, 1–2 h	**I** (95)[a]	121
		TEMPO (cat.), $Yb(OTf)_3$ (cat.), $PhI(OAc)_2$, CH_2Cl_2, 15 min	**I** (95)	416
		PEG-O-TEMPO (cat.), PEG-bipy (cat.), $CuBr_2$ (cat.), O_2, *t*-BuOK, MeCN, H_2O, 80°, 18 h	**I** (96)	392
		TEMPH, *m*-CPBA, CH_2Cl_2, 0° to rt	**I** (60)	102
	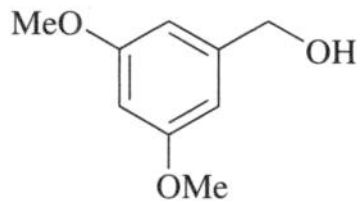	TEMPH•HCl (cat.), *m*-CPBA, CH_2Cl_2, rt	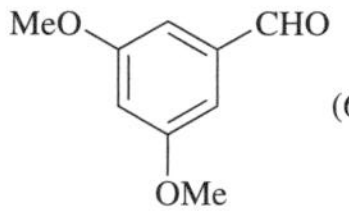 **I** (60)[a]	149
		TEMPO (cat.), 1,10-phenanthroline (cat.), $CuSO_4$ (cat.), O_2, NaOH, H_2O, 80°, 3 h	**I** (55)[a]	422
	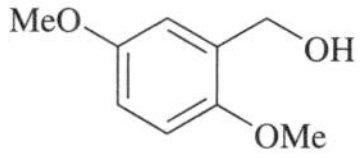	TEMPO (cat.), KNO_2 (cat.), $PhI(OAc)_2$, O_2, 80–90°, 24 h	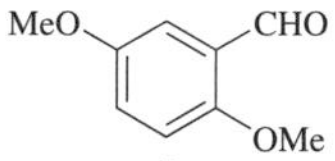 **I** (70)[c]	203
		TEMPO (cat.), KNO_2 (cat.), PS-I$(OAc)_2$, O_2, 80–90°, 24 h	**I** (88)[c]	203
		TEMPO (cat.), $CuSO_4$ (cat.), 1,10-phenanthroline (cat.), O_2, NaOH, H_2O, 80°, 3 h	**I** (96)[a]	422

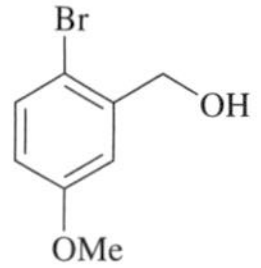	TEMPO (cat.), NaOCl (cat.), $NaClO_2$, MeCN	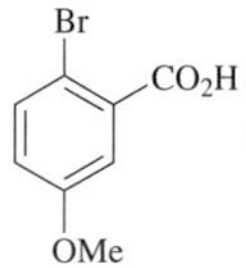 (96)	14, 185
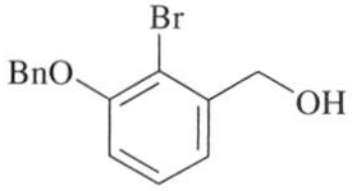	TEMPO (cat.), NaBr (cat.), NaOCl, THF, H_2O, 0°	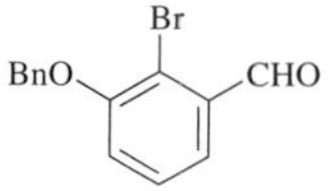 (90)	424
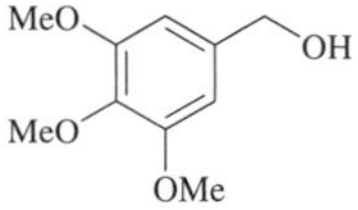	TEMPO (cat.), **3** (cat.), O_2, toluene, 100°	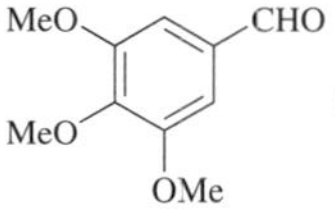 **I** (97)	244
	TEMPO (cat.), 1,10-phenanthroline (cat.), $CuSO_4$ (cat.), O_2, NaOH, H_2O, 80°, 3 h	**I** (77)[a]	422
	TEMPO (cat.), DABCO–CuCl (cat.), O_2, K_2CO_3, toluene, 100°, 2 h	**I** (98)	247
	TEMPO (cat.), DABCO–CuCl (cat.), O_2, K_2CO_3, $MeNO_2$, 100°, 4 h	**I** (94)	247
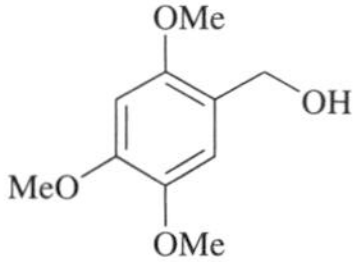	TEMPO (cat.), 1,10-phenanthroline (cat.), $CuSO_4$ (cat.), O_2, NaOH, H_2O, 80°, 3 h	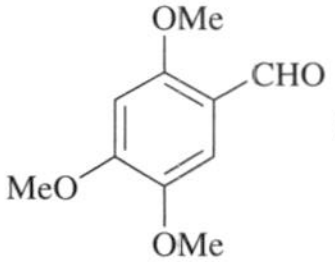 (100)[a]	422
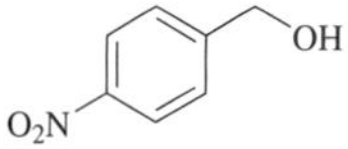	PEG-BnO-TEMPO (cat.), $Co(NO_3)_2$ or $Mn(NO_3)_2$ (cat.), O_2, AcOH, 40°, 4 h	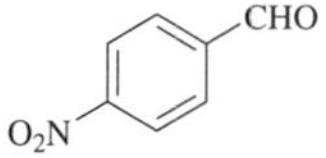 (99)[a]	272

TABLE 2. OXIDATIONS OF BENZYLIC ALCOHOLS *(Continued)*

Alcohol	Conditions	Product(s) and Yield(s) (%)	Refs.
C_7 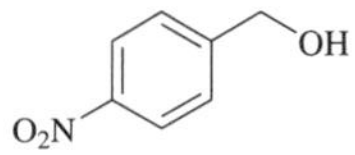	TEMPO (cat.), $H_5PV_2Mo_{10}O_{40}$•34 H_2O (cat.), O_2, acetone, 6 h	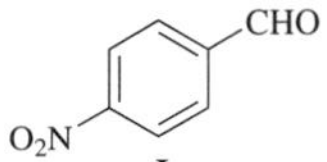 **I** (93.1)[a]	274
	TEMPO (cat.), KNO_2 (cat.), $PhI(OAc)_2$, O_2, 80–90°, 24 h	**I** (59)[c]	203
	TEMPO (cat.), KNO_2 (cat.), PS-$I(OAc)_2$, O_2, 80–90°, 24 h	**I** (55)[c]	203
	TEMPO (cat.), KBr (cat.), NaOCl, CH_2Cl_2, H_2O, pH 9.5, 0–15°	**I** (89)	13
	TEMPO (cat.), $RuCl_2(PPh_3)_3$ (cat.), O_2 (10 atm), chlorobenzene, 100°	**I** (97)[a]	80
	HO-TEMPO (cat.), CuCl (cat.), O_2, DMF, rt	**I** (85)	81
	CIBA-TEMPO (cat.), $Mn(NO_3)_2$ or $Co(NO_3)_2$ (cat.), H_2O_2 or O_2 or air, AcOH, 20–50°, 2 h	**I** (94–98)	395
	TEMPO (cat.), I_2, $NaHCO_3$, toluene or CH_2Cl_2, 20°, overnight	**I** (90)[b]	256
	TEMPO (cat.), **1** (cat.), CuBr•Me_2S (cat.), O_2, perfluorooctyl bromide, chlorobenzene, 90°	**I** (93)	243
	1. TEMPO (cat.), NaOCl, EtOAc or MeCN, H_2O, pH 8–10, 5°, 1 h 2. $NaClO_2$, H_2O, pH 5, rt, 3 h	**I** (100)	187
	TEMPO (cat.), **3** (cat.), O_2, toluene, 100°	**I** (98)	244
	AcNH-TEMPO (cat.), $Cu(ClO_4)_2$ (cat.), DMAP, O_2 (1 atm), $[bmpy]^+[PF_6]^-$, rt	**I** (81)	240

	TEMPO (cat.), $(NH_4)_2[Ce(NO_3)_6]$ (cat.), O_2, MeCN, reflux	**I** (92–97)	269
	BzO-TEMPO (cat.), $TBABr_3$, NaOAc, CH_2Cl_2, H_2O, rt, 1 h	**I** (96)	224
	[mim-CO_2-TEMPO]$^+[PF_6]^-$ (cat.), KBr (cat.), NaOCl, $[bmim]^+[PF_6]^-$, H_2O, 0°	**I** (93)[a]	295
	TEMPO (cat.), **2** (cat.), CuCl (cat.), O_2, $[bmim]^+[PF_6]^-$, 65° 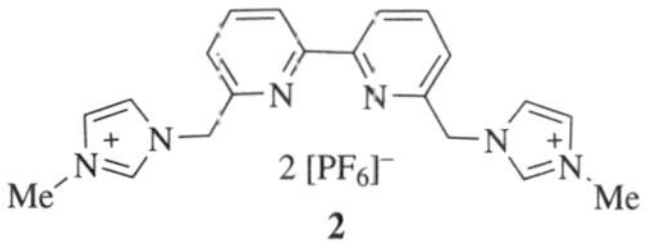 **2**	**I** (98)[a]	246
	PEG-O-TEMPO (cat.), PEG-bipy (cat.), $CuBr_2$ (cat.), O_2, *t*-BuOK, MeCN, H_2O, 80°, 18 h	**I** (94)	392
	SBA15-NH-TEMPO (cat.), $NaNO_2$ (cat.), TBAB (cat.), O_2 (1 atm), AcOH, 50–60°, 1.5 h	**I** (100)[a], 98.8% selectivity[d]	310
	SBA15-NH-TEMPO (cat.), $NaNO_2$ (cat.), TBAB (cat.), air (1 atm), AcOH, 50–60°, 3 h	**I** (100)[a], 99% selectivity[d]	310
	TEMPO (cat.), NaOCl (cat.), $NaClO_2$, MeCN	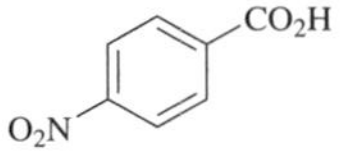 (100)	14, 185
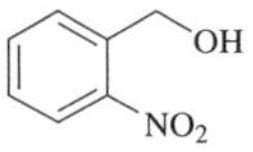	TEMPO (cat.), $Mn(NO_3)_2$ or $Co(NO_3)_2$ (cat.), O_2, AcOH, 6 h	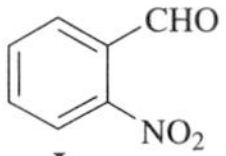 **I** (97)[a]	407
	TEMPO (cat.), KNO_2 (cat.), $PhI(OAc)_2$, O_2, 80–90°, 24 h	**I** (73)[c]	203

TABLE 2. OXIDATIONS OF BENZYLIC ALCOHOLS (*Continued*)

	Alcohol	Conditions	Product(s) and Yield(s) (%)	Refs.
C_7	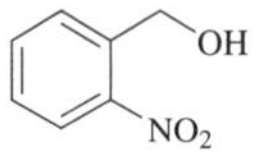	TEMPO (cat.), KNO_2 (cat.), PS-I(OAc)$_2$, O_2, 80–90°, 24 h	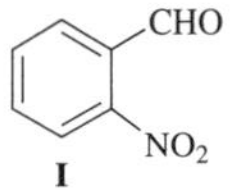 **I** (65)[c]	203
		CIBA-TEMPO (cat.), $Mn(NO_3)_2$ or $Co(NO_3)_2$ (cat.), O_2 or air, AcOH, 20–40°	**I** (92)	395
		[AcNH-TEMP=O]$^+$[ClO$_4$]$^-$, silica gel, CH_2Cl_2, rt	**I** (99)	68
		PEG-O-TEMPO (cat.), PEG-bipy (cat.), $CuBr_2$ (cat.), O_2, *t*-BuOK, MeCN, H_2O, 80°, 18 h	**I** (88)	392
		TEMPO (cat.), DABCO–CuCl (cat.), O_2, K_2CO_3, toluene, 100°, 2 h	**I** (92)	247
	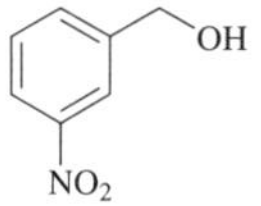	TEMPO (cat.), KBr (cat.), NaOCl, CH_2Cl_2, H_2O, pH 9.5, 0–15°	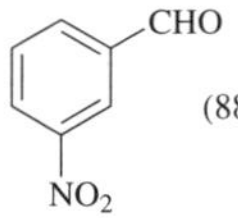 **I** (88)	13
		CIBA-TEMPO (cat.), $Mn(NO_3)_2$ or $Co(NO_3)_2$ (cat.), O_2 or air, AcOH, 20–40°	**I** (94)	395
		[PS-O-TEMP=O]$^+$Cl$^-$, CH_2Cl_2, rt, 1–2 h	**I** (95)[a]	121
		TEMPO (cat.), DABCO–CuCl (cat.), O_2, K_2CO_3, toluene, 100°, 2 h	**I** (91)	247
	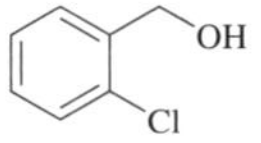	TEMPO (cat.), 1,3-dibromo-5,5-dimethylhydantoin (cat.), $NaNO_2$ (cat.), air, H_2O, 80°, 1 h	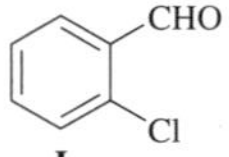 **I** (96)	259
		TEMPO (cat.), Br_2 (cat.), $NaNO_2$ (cat.), air (3.9 atm), CH_2Cl_2, 80°	**I** (96)	258

	[mim-CO_2-TEMPO]$^+$[PF_5]$^-$ (cat.), KBr (cat.), NaOCl, [bmim]$^+$[PF_6]$^-$, H_2O, 0°	**I** (95)[a]	295
	TEMPO (cat.), DABCO–CuCl (cat.), O_2, K_2CO_3, toluene, 100°, 2 h	**I** (96)	247
	1. TEMPO (cat.), NaOCl, EtOAc or MeCN, H_2O, pH 8–10, 5°, 1 h 2. $NaClO_2$, H_2O, pH 5, rt, 3 h	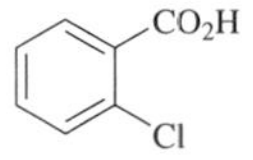 (95)	187
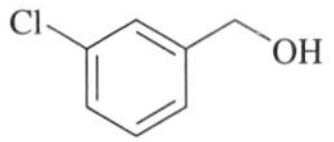	AcNH-TEMPO (cat.), $Cu(ClO_4)_2$ (cat.), O_2 (1 atm), DMAP, [bmpy]$^+$[PF_6]$^-$, rt	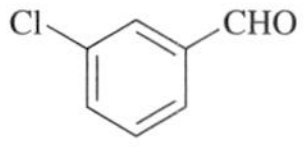 (90) **I**	240
	TEMPO (cat.), Br_2 (cat.), $NaNO_2$ (cat.), air (3.9 atm), CH_2Cl_2, 80°	**I** (96)	258
	TEMPO (cat.), 1,3-dibromo-5,5-dimethylhydantoin (cat.), $NaNO_2$ (cat.), air, H_2O, 80°, 1 h	**I** (96)	259
	TEMPO (cat.), **2** (cat.), CuCl (cat.), O_2, [bmim]$^+$[PF_6]$^-$, 65°	**I** (98)[a]	246
	SBA15-NH-TEMPO (cat.), $NaNO_2$ (cat.), TBAB (cat.), O_2 (1 atm), AcOH, 50–60°, 1.5 h	**I** (99.8)[a], 99.6% selectivity[d]	310
	SBA15-NH-TEMPO (cat.), $NaNO_2$ (cat.), TBAB (cat.), air (1 atm), AcOH, 50–60°, 2.5 h	**I** (99.4)[a], 98.9% selectivity[d]	310
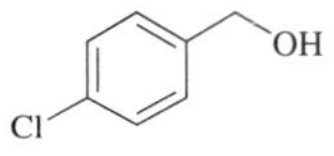	TEMPO (cat.), $H_5PV_2Mo_{10}O_{40}$•34 H_2O (cat.), O_2, acetone, 6 h	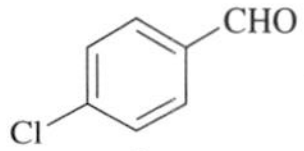 (98.7)[a] **I**	274
	CIBA-TEMPO (cat.), $Mn(NO_3)_2$ or $Co(NO_3)_2$ (cat.), H_2O_2 or O_2 or air, AcOH, 20–40°, 2 h	**I** (93–95)	395

TABLE 2. OXIDATIONS OF BENZYLIC ALCOHOLS *(Continued)*

Alcohol	Conditions	Product(s) and Yield(s) (%)	Refs.
C_7 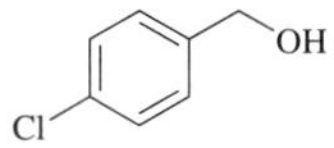	PE-CONH-TEMPO (cat.), NaBr (cat.), electrolysis, $NaHCO_3$, H_2O	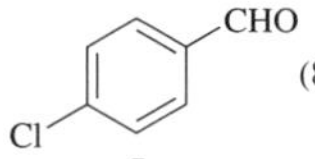 **I** (80)	297
	$[AcNH\text{-}TEMP{=}O]^+[ClO_4]^-$, silica gel, CH_2Cl_2, rt	**I** (96)	68
	$[mim\text{-}O\text{-}TEMPO]^+[BF_4]^-$ (cat.), $[mim\text{-}CH_2PhI(OAc)_2]^+[BF_4]^-$, H_2O, 30°, 15 min	**I** (97)	201
	TEMPO (cat.), DABCO–CuCl (cat.), O_2, K_2CO_3, toluene, 100°, 2 h	**I** (85)	247
	TEMPO (cat.), 1,3-dibromo-5,5-dimethylhydantoin (cat.), $NaNO_2$ (cat.), air, H_2O, 80°, 1 h	**I** (95)	187, 259
	PE-co-AA-CONH-TEMPO (cat.), NaBr (cat.), electrolysis, $NaHCO_3$, H_2O, 0°	**I** (78)	298
	TEMPO (cat.), KBr (cat.), NaOCl, CH_2Cl_2, H_2O, pH 9.5, 0–15°	**I** (90)	13
	TEMPO (cat.), Br_2 (cat.), $NaNO_2$ (cat.), air (3.9 atm), CH_2Cl_2, 80°	**I** (96)	258
	TEMPO (cat.), NaOCl, silica gel, H_2O, 0°	**I** (92)	393
	FT-O-TEMPO (cat.), KBr (cat.), NaOCl, $NaHCO_3$, CH_2Cl_2, H_2O	**I** (99)	406
	1. TEMPO (cat.), NaOCl, EtOAc or MeCN, H_2O, pH 8–10, 5°, 1 h 2. $NaClO_2$, H_2O, pH 5, rt, 3 h	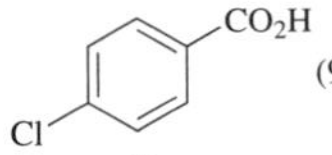 **II** (99)	187

	1. TEMPO (cat.), PS-I(OAc)$_2$, acetone, rt 2. H_2O, 24 h	**II** (74)	204
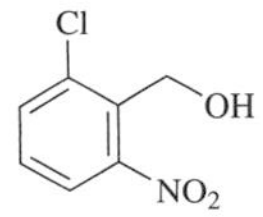	TEMPO (cat.), KBr (cat.), NaOCl, CH_2Cl_2, H_2O, pH 8.6, 0–5°	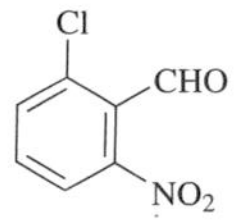 (80)	425
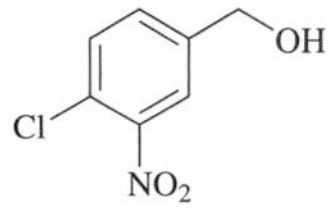	TEMPO (cat.), KNO_2 (cat.), PhI(OAc)$_2$, O_2, 80–90°, 24 h	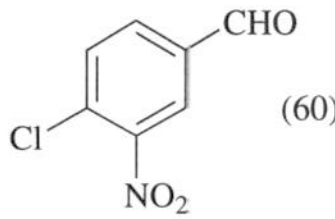 (60)[c]	203
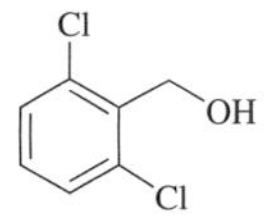	PEG-O-TEMPO (cat.), PEG-bipy (cat.), $CuBr_2$ (cat.), O_2, *t*-BuOK, MeCN, H_2O, 80°, 18 h	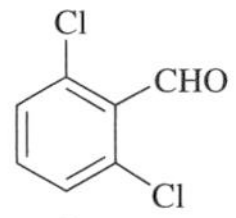 (76) **I**	392
	SBA15-NH-TEMPO (cat.), $NaNO_2$ (cat.), TBAB (cat.), O_2 (1 atm), AcOH, 50–60°, 1.5 h	**I** (100)[a], 99.8% selectivity[d]	310
	SBA15-NH-TEMPO (cat.), $NaNO_2$ (cat.), TBAB (cat.), air (1 atm), AcOH, 50–60°, 3 h	**I** (100)[a], 99.1% selectivity[d]	310
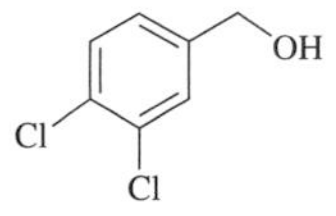	TEMPO (cat.), 1,10-phenanthroline (cat.), $CuSO_4$ (cat.), O_2, NaOH, H_2O, 80°, 3 h	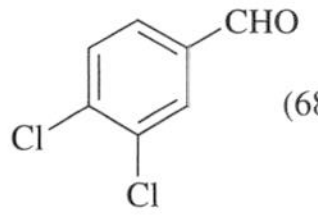 (68)[a]	422
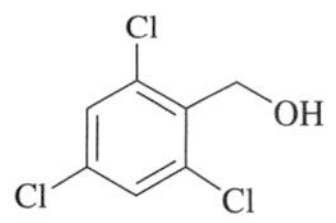	PEG-O-TEMPO (cat.), PEG-bipy (cat.), $CuBr_2$ (cat.), O_2, *t*-BuOK, MeCN, H_2O, 80°, 18 h	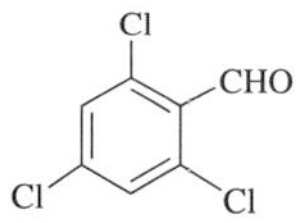 (61)	392

TABLE 2. OXIDATIONS OF BENZYLIC ALCOHOLS (*Continued*)

Alcohol	Conditions	Product(s) and Yield(s) (%)	Refs.
C_7 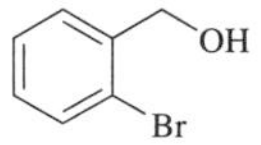	TEMPO (cat.), **1** (cat.), CuBr•Me_2S (cat.), O_2, perfluorooctyl bromide, chlorobenzene, 90°	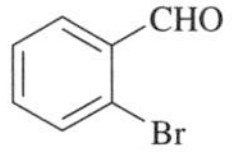 **I** (96)	243, 414
	PEG-O-TEMPO (cat.), PEG-bipy (cat.), $CuBr_2$ (cat.), O_2, *t*-BuOK, MeCN, H_2O, 80°, 18 h	**I** (90)	392
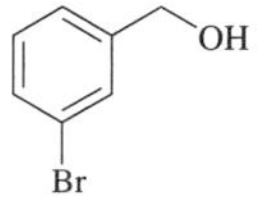	SBA15-NH-TEMPO (cat.), $NaNO_2$ (cat.), TBAB (cat.), O_2 (1 atm), AcOH, 50–60°, 1.5 h	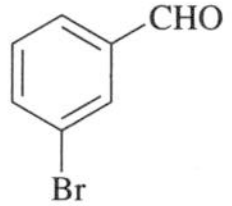 **I** (99.7)[a], 99.4% selectivity[d]	310
	SBA15-NH-TEMPO (cat.), $NaNO_2$ (cat.), TBAB (cat.), air (1 atm), AcOH, 50–60°, 2.5 h	**I** (100)[a], 99.4% selectivity[d]	310
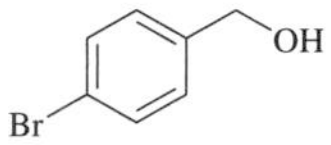	PEG-BnO-TEMPO (cat.), NaOCl, CH_2Cl_2, H_2O, pH 8.6, 0°	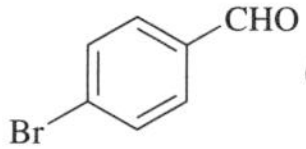 **I** (91)	306
	PEG-BnO-TEMPO (cat.), $Co(NO_3)_2$ or $Mn(NO_3)_2$ (cat.), O_2, AcOH, 40°, 4 h	**I** (72–99)[a]	272
	ICF-O-TEMPO (cat.), $PhI(OAc)_2$, CH_2Cl_2	**I** (97)	293
	FT-O-TEMPO (cat.), KBr (cat.), NaOCl, $NaHCO_3$, CH_2Cl_2, H_2O	**I** (95)	406
	TEMPO (cat.), **3** (cat.), O_2, toluene, 100°	**I** (70)	244
	AcNH-TEMPO (cat.), $Cu(ClO_4)_2$ (cat.), O_2 (1 atm), DMAP, $[bmpy]^+[PF_6]^-$, rt	**I** (84)	240
	PEG-O-TEMPO (cat.), PEG-bipy (cat.), $CuBr_2$ (cat.), O_2, *t*-BuOK, MeCN, H_2O, 80°, 18 h	**I** (94)	392

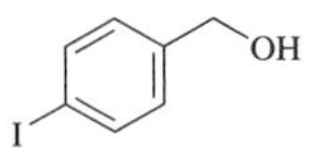	$[AcNH\text{-}TEMP{=}O]^+[ClO_4]^-$, silica gel, CH_2Cl_2, rt	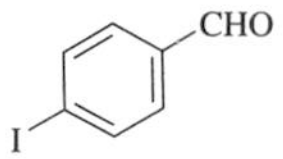 (85)	426
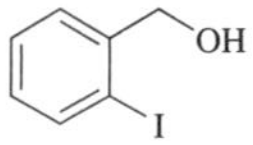	TEMPO (cat.), KNO_2 (cat.), PS-I$(OAc)_2$, O_2, 80–90°, 24 h	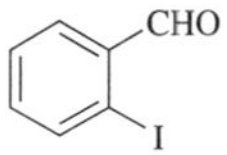 **I** (77)[c]	203
	TEMPO (cat.), KNO_2 (cat.), PhI$(OAc)_2$ or PS-I$(OAc)_2$, O_2, 80–90°, 24 h	**I** (72)[c]	203
$C_{7\text{-}8}$			
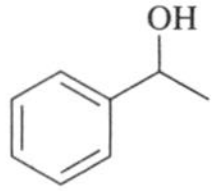 + 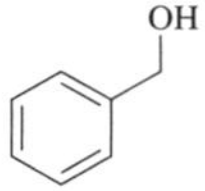	SG-NH-TEMPOH (cat.), KBr (cat.), NaOCl, $NaHCO_3$, CH_2Cl_2, H_2O, pH 9.1	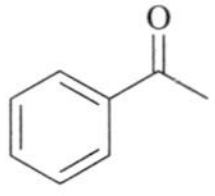 **I** (3)[a] + 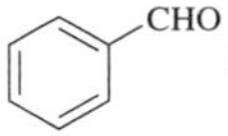 **II** (92)[a]	307
	PIPO (cat.), NaOCl, $KHCO_3$, H_2O, rt, 30 min	**I** (4)[a] + **II** (—), 95% conversion,[a] >99% selectivity[a,d]	299
C_8			
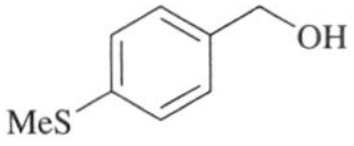	TEMPO (cat.), KNO_2 (cat.), PhI$(OAc)_2$ or PS-I$(OAc)_2$, O_2, 80–90°[c], 24 h	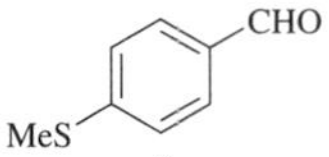 **I** (73–75)[c]	203
	SBA15-NH-TEMPO (cat.), $NaNO_2$ (cat.), TBAB (cat.), O_2 (1 atm), AcOH, 50–60°, 1.5 h	**I** (98.7)[a], 97.1% selectivity[d]	310
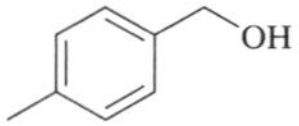	AcNH-TEMPO (cat.), $Cu(ClO_4)_2$ (cat.), O_2 (1 atm), DMAP, $[bmpy]^+[PF_6]^-$, rt	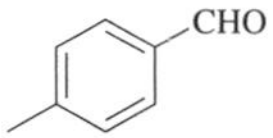 **I** (92)	240
	TEMPO (cat.), $(NH_4)_2[Ce(NO_3)_6]$ (cat.), O_2, MeCN, reflux	**I** (93–99)	269
	TEMPO (cat.), PS-I$(OAc)_2$, acetone, 0°	**I** (93)	202

TABLE 2. OXIDATIONS OF BENZYLIC ALCOHOLS *(Continued)*

Alcohol	Conditions	Product(s) and Yield(s) (%)	Refs.
C_8 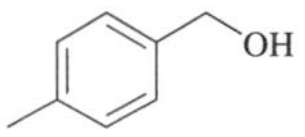	TEMPO (cat.), $H_5PV_2Mo_{10}O_{40}$•34 H_2O (cat.), O_2, acetone, 6 h	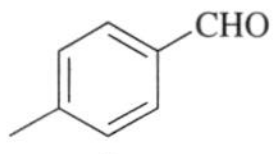 **I** (99.9)[a]	274
	TEMPO (cat.), $Mn(NO_3)_2$ or $Co(NO_3)_2$ (cat.), O_2, AcOH, 6 h	**I** (99)[a]	407
	TEMPO (cat.), Br_2 (cat.), $NaNO_2$ (cat.), air (3.9 atm), CH_2Cl_2, 80°	**I** (94)	258
	TEMPO (cat.), 1,3-dibromo-5,5-dimethyl-hydantoin (cat.), $NaNO_2$ (cat.), air, H_2O, 80°, 1 h	**I** (98)	259
	TEMPO (cat.), KBr (cat.), PS-$I(OAc)_2$, acetone, 0°	**I** (93)	202
	CIBA-TEMPO (cat.), $Mn(NO_3)_2$ or $Co(NO_3)_2$ (cat.), H_2O_2 or O_2 or air, AcOH, 20–40°	**I** (96–98)	395
	$[mim\text{-}CO_2\text{-}TEMPO]^+[PF_6]^-$ (cat.), KBr (cat.), NaOCl, $[bmim]^+[PF_6]^-$, H_2O, 0°	**I** (96)[a]	295
	AcNH-TEMPO (cat.), **4** (cat.), $Cu(ClO_4)_2$ (cat.), O_2, DABCO, DMSO	**I** (92)	245
	FT-O-TEMPO (cat.), KBr (cat.), NaOCl, $NaHCO_3$, CH_2Cl_2, H_2O	**I** (99)	406
	PEG-O-TEMPO (cat.), PEG-bipy (cat.), $CuBr_2$ (cat.), O_2, *t*-BuOK, MeCN, H_2O, 80°, 18 h	**I** (94)	392
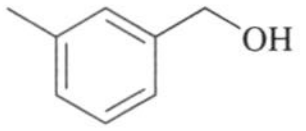	TEMPO (cat.), Br_2 (cat.), $NaNO_2$ (cat.), air (4 atm), CH_2Cl_2, 80°	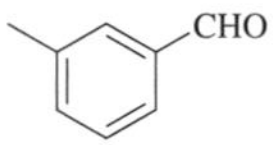 **I** (93)	258
	TEMPO (cat.), $NaNO_2$ (cat.), 1,3-dibromo-5,5-dimethylhydantoin (cat.), air, H_2O, 80°, 1 h	**I** (98)	259

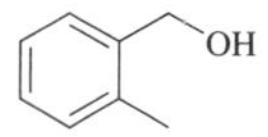	TEMPO (cat.), $NaNO_2$ (cat.), 1,3-dibromo-5,5-dimethylhydantoin (cat.), air, H_2O, 80°, 1 h	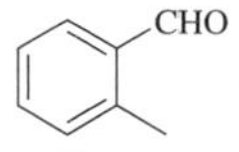 I (95)	259
	AcNH-TEMPO (cat.), $Cu(ClO_4)_2$ (cat.), O_2 (1 atm), DMAP, $[bmpy]^+[PF_6]^-$, rt	I (90)	240
	TEMPO (cat.), Br_2 (cat.), $NaNO_2$ (cat.), air (3.9 atm), CH_2Cl_2, 80°	I (95)	258
	AcNH-TEMPO (cat.), **4** (cat.), $Cu(ClO_4)_2$ (cat.), O_2, DABCO, DMSO	I (90)	245
	PEG-O-TEMPO (cat.), PEG-bipy (cat.), $CuBr_2$ (cat.), O_2, *t*-BuOK, MeCN, H_2O, 80°, 18 h	I (92)	392
	SBA15-NH-TEMPO (cat.), $NaNO_2$ (cat.), TBAB (cat.), O_2 (1 atm), AcOH, 50–60°, 1.5 h	I (99)[a], 99.1% selectivity[d]	310
	SBA15-NH-TEMPO (cat.), $NaNO_2$ (cat.), TBAB (cat.), air (1 atm), AcOH, 50–60°, 3 h	I (94.5)[a], 97.4% selectivity[d]	310
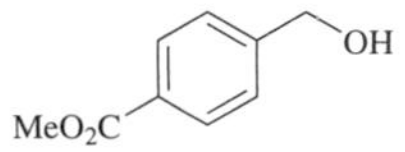	$[PS\text{-}O\text{-}TEMP{=}O]^+Cl^-$, CH_2Cl_2, rt, 1–2 h	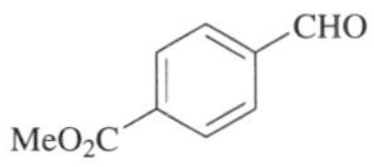 (95)[a]	121
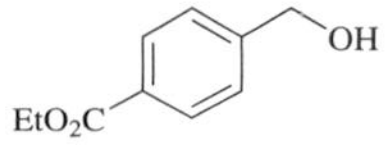	$[mim\text{-}O\text{-}TEMPO]^+[BF_4]^-$ (cat.), $[mim\text{-}CH_2PhI(OAc)_2]^+[BF_4]^-$, H_2O, 30°, 40 min	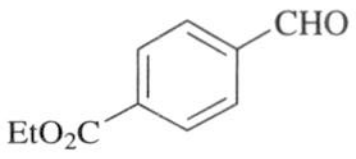 (90)	201
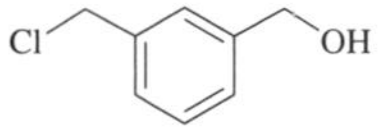	AcNH-TEMPO (cat.), **4** (cat.), $Cu(ClO_4)_2$ (cat.), O_2, DABCO, DMSO	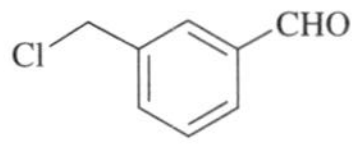 (93)	245

TABLE 2. OXIDATIONS OF BENZYLIC ALCOHOLS *(Continued)*

Alcohol	Conditions	Product(s) and Yield(s) (%)	Refs.
C_8 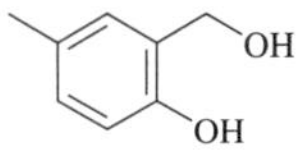	$[AcNH\text{-}TEMP{=}O]^+[ClO_4]^-$, silica gel, CH_2Cl_2, rt	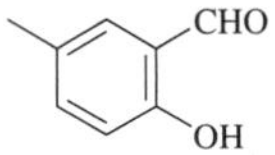 (97)	68
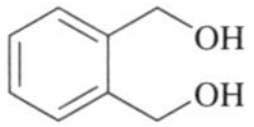	TEMPO (cat.) adsorbed on silica gel, H_2O, NaOCl, 0°	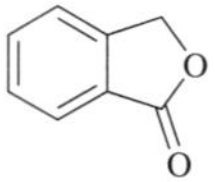 (80) + 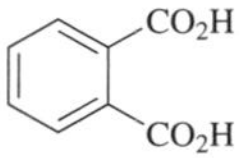 (19)	393
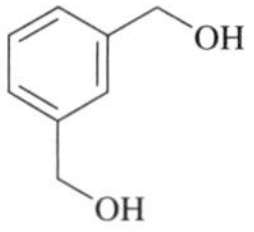	CIBA-TEMPO (cat.), $Mn(NO_3)_2$ or $Co(NO_3)_2$ (cat.), air or O_2, AcOH, 20–40°	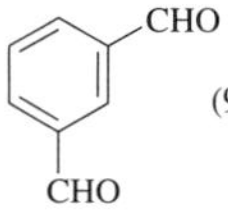 (90)	395
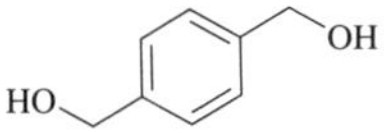	$[MeO\text{-}TEMP{=}O]^+Cl^-$, CH_2Cl_2, dark, rt, 2 h	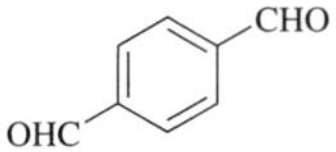 **I** (99)	135
	TEMPO (cat.), DABCO-CuCl (cat.), O_2, K_2CO_3, toluene, 100°, 2 h	**I** (81)	247
	TEMPO (cat.), $PhI(OAc)_2$, CH_2Cl_2, rt	**I** (—)[e]	205
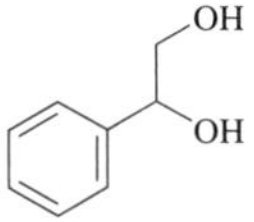	TEMPO (cat.), electrolysis, $NaClO_4$, 2,6-lutidine, MeCN, H_2O	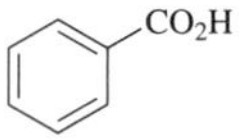 (89)	421
	TEMPO (cat.), $[PS\text{-}CH_2NMe_3]^+[Br(OAc)_2]^-$, CH_2Cl_2, rt, 24 h	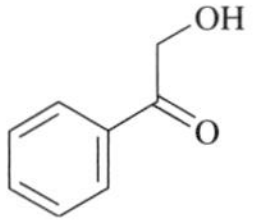 (44)	211

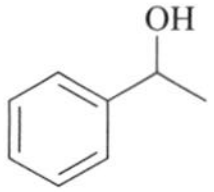	PIPO (cat.), KBr (cat.), NaOCl, $KHCO_3$, CH_2Cl_2, H_2O, rt, 20 min	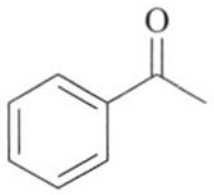 **I** (—), >99% conversion[a] >99% selectivity[a,f]	299
	BzO-TEMPO (cat.), NaBr (cat.), electrolysis, silica gel, $NaHCO_3$, H_2O	**I** (88)	84
	AcNH-TEMPO, TsOH, CH_2Cl_2, 0°, 2 h	**I** (100)	56
	PEG-BnO-TEMPO (cat.), $Co(NO_3)_2$ or $Mn(NO_3)_2$ (cat.), O_2, AcOH, H_2O, 40°, 4 h	**I** (96)[a]	272
	TEMPO (cat.), $H_5PV_2Mo_{10}O_{40}$•34 H_2O (cat.), O_2 (2 atm), acetone	**I** (98.3)[a]	274
	SG-NH-TEMPOH (cat.), KBr (cat.), NaOCl, $NaHCO_3$, CH_2Cl_2, H_2O, pH 9.1	**I** (91)[a]	307
	TEMPO or AcNH-TEMPO, TsOH, CH_2Cl_2, 0°, 1.5 h	**I** (83)	56
	TEMPO (cat.), $[PS\text{-}CH_2NMe_3]^+[Br(OAc)_2]^-$, CH_2Cl_2 or toluene, rt, 24 h	**I** (92–96)	43, 211
	TEMPO (cat.), $(NH_4)_2[Ce(NO_3)_6]$ (cat.), O_2, MeCN, reflux	**I** (72–99)	269
	TEMPO (cat.), Br_2 (cat.), $NaNO_2$ (cat.), air (3.9 atm), CH_2Cl_2, 80°	**I** (98)	258
	TEMPO (cat.), PS-$I(OAc)_2$, acetone, 0°	**I** (93)	202
	TEMPO (cat.), $PhI(OAc)_2$, CH_2Cl_2	**I** (95)	192
	PE-CONH-TEMPO (cat.), NaBr (cat.), electrolysis, $NaHCO_3$, H_2O, 0°	**I** (84–88)	84, 297

TABLE 2. OXIDATIONS OF BENZYLIC ALCOHOLS *(Continued)*

Alcohol	Conditions	Product(s) and Yield(s) (%)	Refs.
C_8 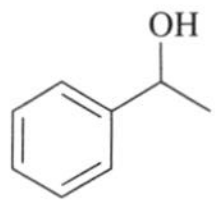	Nitroxide (cat.), NaBr (4 eq), electrolysis, CH_2Cl_2, H_2O, $NaHCO_3$ (satd.), rt	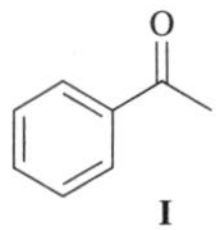 **I**	231
	Nitroxide		
	ABHep^{Cl}O	(99)	
	ABHepO	(99)	
	ABOct^{Cl}O	(99)	
	ABOctO	(99)	
	ABHexO	(99)	
	TEMPO	(72)	
	ICF-O-TEMPO (cat.), $PhI(OAc)_2$, CH_2Cl_2	**I** (88)	293
	TEMPO (cat.), **3** (cat.), O_2, toluene, 100°	**I** (2)	244
	[mim-O-TEMPO]$^+$[BF_4]$^-$ (cat.), [mim-$CH_2PhI(OAc)_2$]$^+$[BF_4]$^-$, H_2O, 30°, 2 h	**I** (90)[a]	201
	TEMPO (cat.), oxidant, CH_2Cl_2, rt, 24 h	**I**	44
	Oxidant		
	[PS-CH_2NMe_3]$^+$[ClO_2]$^-$	(98)	
	[PS-CH_2NMe_3]$^+$[$Br(OAc)_2$]$^-$	(96)	
	[PS-CH_2NMe_3]$^+$[ClO]$^-$	(21)	
	TEMPO (cat.), KNO_2 (cat.), PS-$I(OAc)_2$, O_2, 80–90°, 24 h	**I** (29)[c]	203
	PIPO (cat.), NaOCl, $KHCO_3$, CH_2Cl_2, H_2O, 0°	**I** (100)[a]	241, 299
	PEG-BnO-TEMPO (cat.), NaOCl, CH_2Cl_2, H_2O, pH 8.6, 0°	**I** (93)	306
	TEMPO (cat.), KNO_2 (cat.), $PhI(OAc)_2$, O_2, 80–90°, 24 h	**I** (51–75)[c]	203

TEMPO (cat.), $NaNO_2$ (cat.), 1,3-dibromo-5,5-dimethylhydantoin, air, H_2O, 80°, 1 h	**I** (97)	259
TEMPO (cat.), NaOCl, silica gel, H_2O, 0°	**I** (82)	393
TEMPO (cat.), DABCO–CuCl (cat.), O_2, K_2CO_3, toluene, 100°, 24 h	**I** (69)	247
TEMPO (cat.), KBr (cat.), PS-I$(OAc)_2$, acetone, 0°	**I** (92–93)	202
[mim-CO_2-TEMPO]$^+$[PF_6]$^-$ (cat.), KBr (cat.), NaOCl, [bmim]$^+$[PF_6]$^-$, H_2O, 0°	**I** (96)[a]	295
TEMPO (cat.), **2** (cat.), CuCl (cat.), O_2, [bmim]$^+$[PF_6]$^-$, 65°	**I** (96)[a]	295
PE-co-AA-CONH-TEMPO (cat.), NaBr (cat.), electrolysis, $NaHCO_3$, H_2O, 0°	**I** (92)	298
TEMPO (cat.), 1,10-phenanthroline (cat.), $CuSO_4$ (cat.), O_2, NaOH, H_2O, 80°, 3 h	**I** (82)[a]	422
TEMPO (cat.), *m*-CPBA, TBAB, CH_2Cl_2	**I** (91)	213
TEMPO (cat.), Yb$(OTf)_3$ (cat.), PhI$(OAc)_2$, CH_2Cl_2, 2 h	**I** (98)	416
TEMPO (cat.), **2** (cat.), CuCl (cat.), O_2, [bmim]$^+$[PF_6]$^-$, 65°	**I** (90)	245
CIBA-TEMPO (cat.), Mn$(NO_3)_2$ or Co$(NO_3)_2$ (cat.), air or O_2 (1 atm), AcOH, 20–40°	**I** (96–98)	395, 407
SBA15-NH-TEMPO (cat.), $NaNO_2$ (cat.), TBAB (cat.), O_2 (1 atm), AcOH, 50–60°, 2.5 h	**I** (97.6)[a], 99% selectivity[f]	310

TABLE 2. OXIDATIONS OF BENZYLIC ALCOHOLS *(Continued)*

	Alcohol	Conditions	Product(s) and Yield(s) (%)	Refs.
C_8	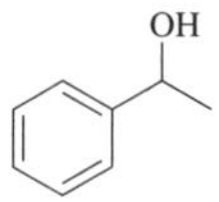	SBA15-NH-TEMPO (cat.), $NaNO_2$ (cat.), TBAB (cat.), air (1 atm), AcOH, 50–60°, 4.5 h	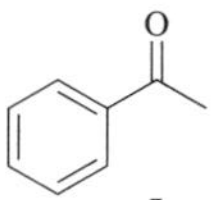 **I** (100)[a], 99% selectivity[f]	310
		L-DQ-TEMPO (cat.), electrolysis, 2,6-lutidine, MeCN	**I** (—), no stereoselectivity	111
		AcNH-TEMPO, CSA, CH_2Cl_2, 0°	**I** (83), no stereoselectivity	56
		PIPO (cat.), KBr (cat.), NaOCl, $KHCO_3$, CH_2Cl_2, H_2O, rt, 20 min	**I** (—), >99% conversion[a], >99% selectivity[a,f]	299
		(*S*)-binaphthyl-TEMPO (cat.), KBr (cat.), electrolysis, silica gel, $NaHCO_3$, H_2O, –15°	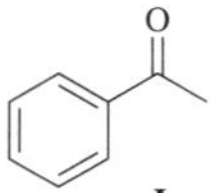 **I** (57) + 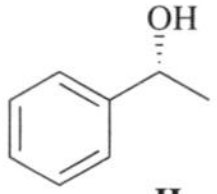 **II** (43), $s = 16$	109
		Graphite felt-TEMPO, (–)-sparteine (cat.), electrolysis, $NaClO_4$, MeCN	**I** (52.9)[b] + **II** (46.2)[b], $s = 48$	105
		(*R*)-Val-TEMPO (cat.), KBr (cat.), NaOCl, CH_2Cl_2, H_2O, 0°	**I** + **II** (—), 83% conversion, $s = 2.3$	103
		(*R*)-Val-TEMPO (cat.), NaBr (cat.), electrolysis, $NaHCO_3$, CH_2Cl_2, H_2O, 0°	**I** + **II** (—), 66% conversion, $s = 1.6$	103
		AS-NHAc-TEMPO (cat.), electrolysis, 2,6-lutidine, $NaOCl_4$, MeCN	**I** + **II** (—), 64% conversion, 70% ee[a], $s = 4.6$	107
		(*S*)-Val-TEMPO (cat.), KBr (cat.), NaOCl, CH_2Cl_2, H_2O, 0°	**I** + 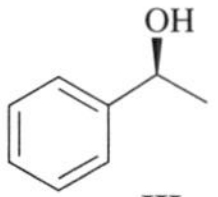 **III** (—), 83% conversion, $s = 2.3$	103

	(*S*)-Val-TEMPO (cat.), NaBr (cat.), electrolysis, $NaHCO_3$, CH_2Cl_2, H_2O, 0°	**I** + **III** (—), 53% conversion, *s* = 1.5	103
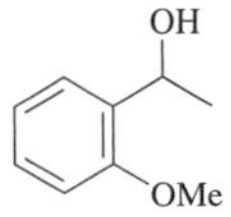	TEMPO (cat.), oxidant, CH_2Cl_2, rt, 24 h	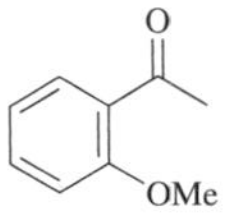	
	Oxidant		
	$[PS\text{-}CH_2NMe_3]^+[Br(OAc)_2]^-$	(96)	43, 44, 211
	$[PS\text{-}CH_2NMe_3]^+[ClO]^-$	(97)	44
	$[PS\text{-}CH_2NMe_3]^+[ClO_2]^-$	(98)	44
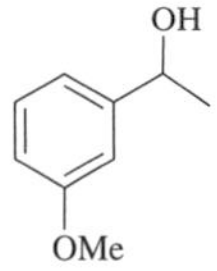	TEMPO (cat.), 1,10-phenanthroline (cat.), $CuSO_4$ (cat.), O_2, NaOH, H_2O, 80°, 3 h	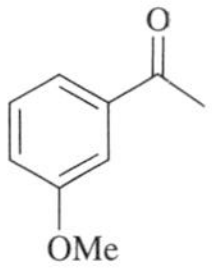 (62)[a]	422
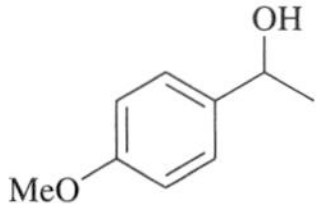	BzO-TEMPO (cat.), NaBr (cat.), electrolysis, silica gel, $NaHCO_3$, H_2O	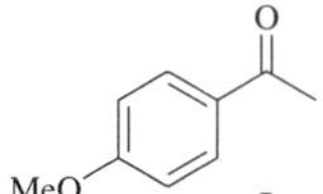 **I** (42)	84
	TEMPO (cat.), NaOCl, silica gel, H_2O, 0°	**I** (30–84)	393
	TEMPO (cat.), laccase (cat.), O_2, H_2O, citrate buffer, pH 4.5	**I** (85)[a]	249
	TEMPO (cat.), DABCO–CuCl (cat.), O_2, K_2CO_3, toluene, 100°, 24 h	**I** (90)	247
	$[mim\text{-}O\text{-}TEMPO]^+[BF_4]^-$ or $[mim\text{-}CO_2\text{-}TEMPO]^+[BF_4]^-$(cat.), $[mim\text{-}CH_2PhI(OAc)_2]^+[BF_4]^-$, H_2O, 30°, 30 min	**I** (98–99)	201

Alcohol	Conditions	Product(s) and Yield(s) (%)	Refs.
C_8 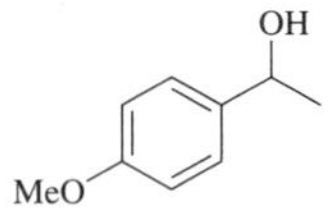	PE-CONH-TEMPO (cat.), NaBr (cat.), electrolysis, $NaHCO_3$, H_2O, 0°	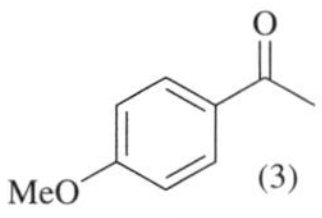 (3) **I** + 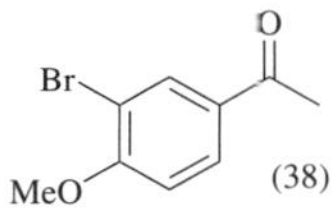 (38) **II** + 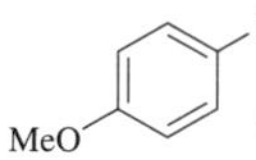 (36) **III**	297
	PE-co-AA-CONH-TEMPO (cat.), NaBr (cat.), electrolysis, $NaHCO_3$, H_2O, 0°	**I** (22) + **II** (19) + **III** (19) + 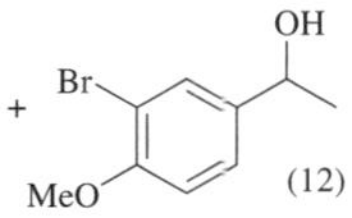 (12)	298
	(*S*)-binaphthyl-TEMPO (cat.), KBr (cat.), electrolysis, silica gel, $NaHCO_3$, H_2O, –15°	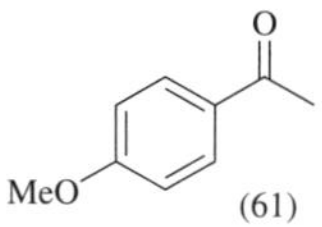 (61) + 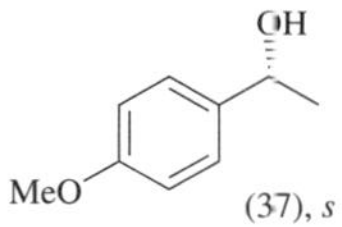 (37), *s* = 5.3	109
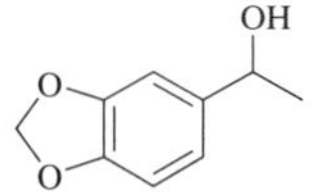	(*S*)-binaphthyl-TEMPO (cat.), KBr (cat.), electrolysis, silica gel, $NaHCO_3$, H_2O, –15°	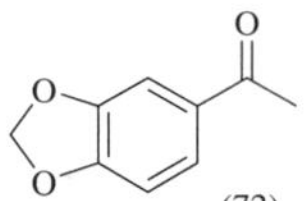 (72) + 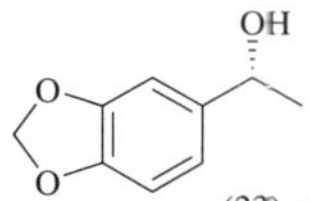 (23), *s* = 2.0	109
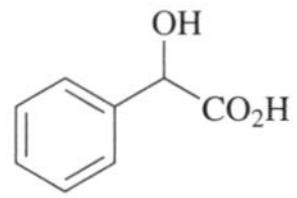	AcNH-TEMPO, TsOH, CH_2Cl_2, 0°	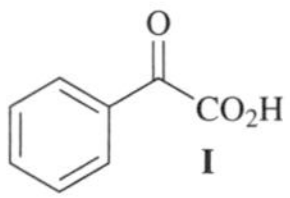 **I** (91)	56
	AcNH-TEMPO, CSA, CH_2Cl_2, 0°	**I** (82)[g], no stereoselectivity	56

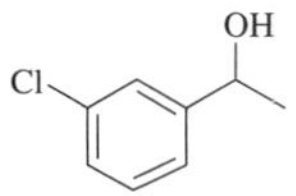	(*S*)-binaphthyl-TEMPO (cat.), KBr (cat.), electrolysis, silica gel, $NaHCO_3$, H_2O, –15°	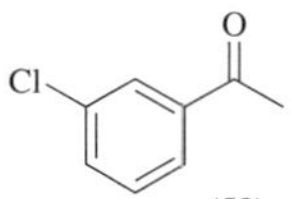 (58) + 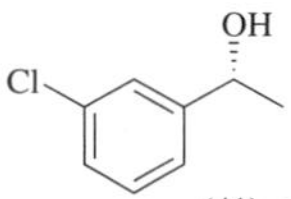 (41), *s* = 13	109
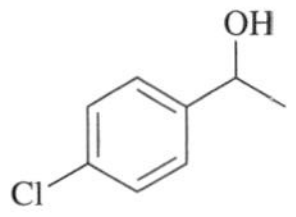	BzO-TEMPO (cat.), NaBr (cat.), electrolysis, silica gel, $NaHCO_3$, H_2O	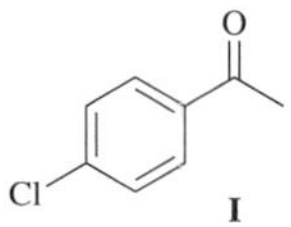 **I** (82)	84
	TEMPO (cat.), $Mn(NO_3)_2$ or $Co(NO_3)_2$ (cat.), O_2 (1 atm), AcOH	**I** (96)[a]	407
	[mim-CONH-TEMPO]$^+$[Tf_2N]$^-$ (cat.) or TEMPO (cat.), electrolysis, [bmim]$^+$[PF_6]$^-$ and/or [bmim]$^+$[Tf_2N]$^-$	**I** (80)	294
	PE-co-AA-CONH-TEMPO (cat.), NaBr (cat.), electrolysis, $NaHCO_3$, H_2O, 0°	**I** (93)	298
	TEMPO (cat.), NaOCl, silica gel, H_2O, 0°	**I** (91)	393
	(*S*)-binaphthyl-TEMPO (cat.), KBr (cat.), electrolysis, silica gel, $NaHCO_3$, H_2O, –15°	**I** (52) + 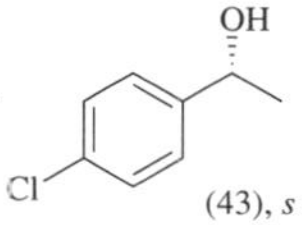 (43), *s* = 12	84
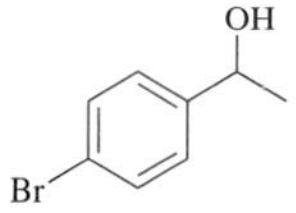	[mim-O-TEMPO]$^+$[BF_4]$^-$ (cat.), [mim-$CH_2PhI(OAc)_2$]$^+$[BF_4]$^-$, H_2O, 30°, 30 min	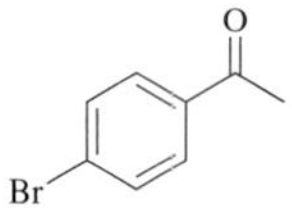 (98)	201

TABLE 2. OXIDATIONS OF BENZYLIC ALCOHOLS *(Continued)*

	Alcohol	Conditions	Product(s) and Yield(s) (%)	Refs.
C_8	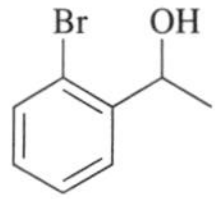	(*S*)-binaphthyl-TEMPO (cat.), KBr (cat.), electrolysis, silica gel, $NaHCO_3$, H_2O, –15°	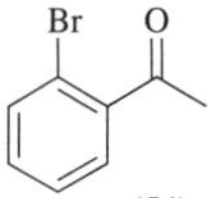 (54) + 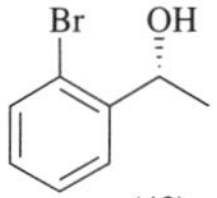 (42), *s* = 13	109
	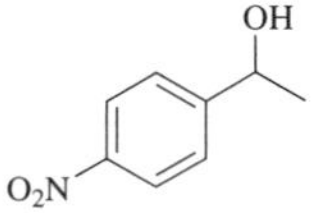	TEMPO (cat.), DABCO–CuCl (cat.), O_2, K_2CO_3, toluene, 100°, 24 h	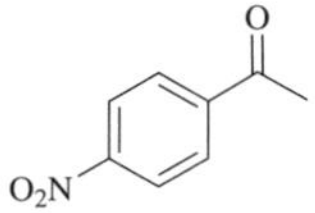 (51)	247
C_9	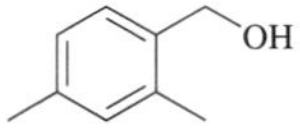	[mim-CO_2-TEMPO]$^+$[PF_6]$^-$ (cat.), KBr (cat.), NaOCl, [bmim]$^+$[PF_6]$^-$, H_2O, 0°	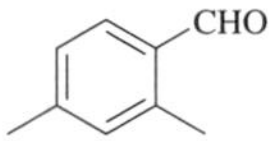 (95)[a]	295
	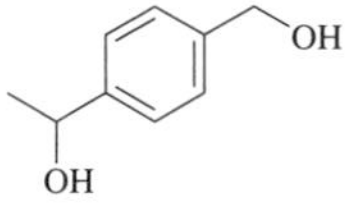	TEMPO (cat.), $RuCl_2(PPh_3)_3$ (cat.), O_2 (10 atm), chlorobenzene, 100°	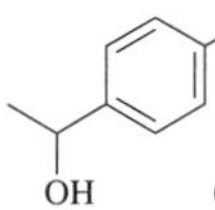 (64) + 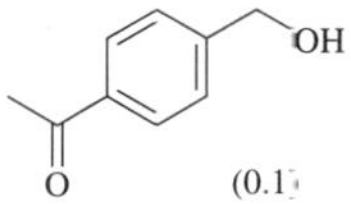 (0.1) + 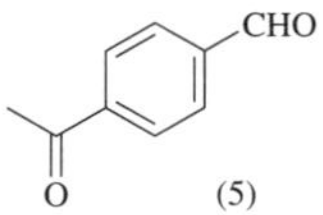 (5)	80, 409, 410
	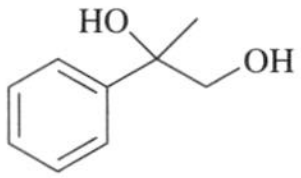	See table.	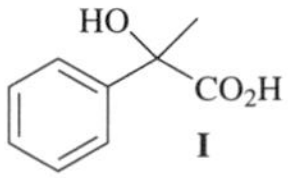 I + 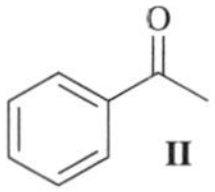 II	394

Conditions	I	II
TEMPO (cat.), NaOCl, $NaHCO_3$, MeCN, H_2O, 0°	(24)	(55)
ormosil-TEMPO (cat.), NaOCl, $NaHCO_3$, MeCN, H_2O, 0°	(60)	(40)

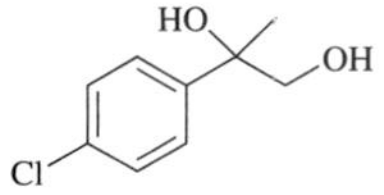	See table.	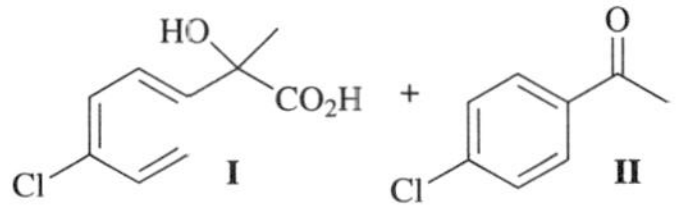 I + II	394
	Conditions	**I** / **II**	
	TEMPO (cat.), NaOCl, $NaHCO_3$, MeCN, H_2O, 0°	(65) / (20)	
	ormosil-TEMPO (cat.), NaOCl, $NaHCO_3$, MeCN, H_2O, 0°	(80) / (5)	
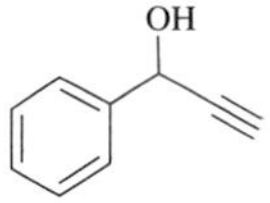	[AcNH-TEMP=O$^+$][ClO_4]$^-$, silica gel, CH_2Cl_2, rt	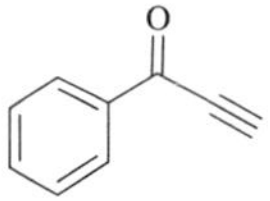 (97)	68
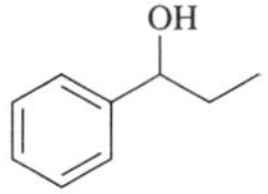	TEMPO (cat.), KBr (cat.), PS-I$(OAc)_2$, CH_2Cl_2, H_2O, 0°	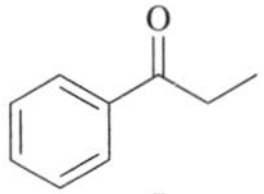 I (94)	202
	TEMPO (cat.), $(NH_4)_2[Ce(NO_3)_6]$ (cat.), O_2, MeCN, reflux	**I** (98)	269
	[PS-O-TEMP=O]$^+$Cl$^-$, CH_2Cl_2, rt, 1–2 h	**I** (95)[a]	121
	TEMPO (cat.), *m*-CPBA, TBAB, CH_2Cl_2	**I** (98)	213
	MIP-CO_2-TEMPO (cat.), KBr (cat.), NaOCl, $NaHCO_3$, H_2O, CH_2Cl_2, 0°	**I** (76)	74
	TEMPO (cat.), PS-I$(OAc)_2$, acetone, 0°	**I** (94)	202
	AS-NHAc-TEMPO (cat.), electrolysis, $NaClO_4$, 2,6-lutidine, MeCN	**I** (—), 53% conversion + 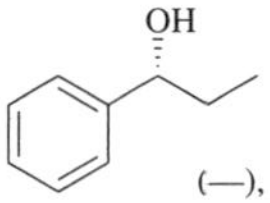 (—), 52% ee, $s = 4.1$[a]	107

TABLE 2. OXIDATIONS OF BENZYLIC ALCOHOLS *(Continued)*

	Alcohol	Conditions	Product(s) and Yield(s) (%)	Refs.
C_9	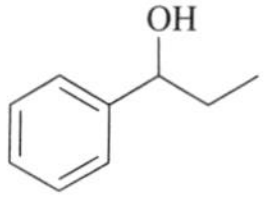	Nitroxide (cat.), NaBr (4 eq), electrolysis, $NaHCO_3$ (satd.), CH_2Cl_2, H_2O, rt	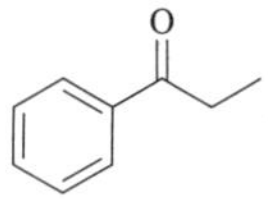	281
		Nitroxide		
		ABHep[Cl]O	(99)	
		ABHepO	(97)	
		ABOct[Cl]O	(86)	
		ABOctO	(90)	
		ABHexO	(99)	
		TEMPO	(77)	
	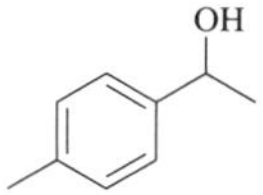	PE-CONH-TEMPO (cat.), NaBr (cat.), electrolysis, $NaHCO_3$, H_2O, 0°	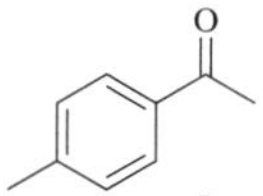 (91) **I**	297
		BzO-TEMPO (cat.), NaBr (cat.), electrolysis, dispersed silica gel, $NaHCO_3$, H_2O	**I** (82)	84
	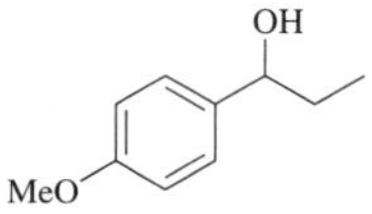	TEMPO (cat.), laccase (cat.), O_2, H_2O, citrate buffer, pH 4.5	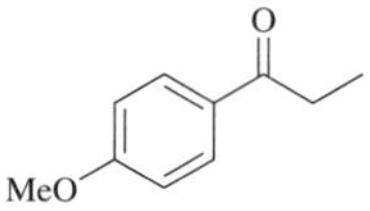 (95)[a]	249
	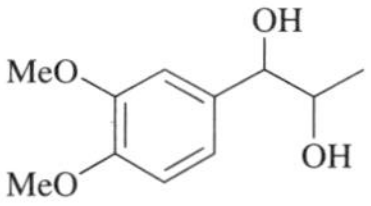	TEMPO (cat.), NaOCl, silica gel, H_2O, 0°	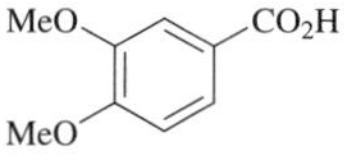 (81)	393

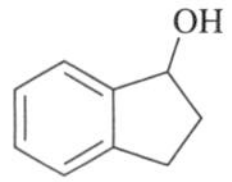	TEMPO (cat.), $(NH_4)_2[Ce(NO_3)_6]$ (cat.), O_2, MeCN, reflux	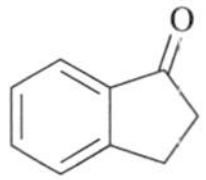 **I** (88–94)	269
	$[PS\text{-}O\text{-}TEMP{=}O]^+Cl^-$, CH_2Cl_2, rt, 1–2 h	**I** (85)[a]	121
	TEMPO (cat.), KBr (cat.), PS-I$(OAc)_2$, CH_2Cl_2, H_2O, 0°	**I** (94)	202
	TEMPO (cat.), $[PS\text{-}CH_2NMe_3]^+[Br(OAc)_2]^-$, CH_2Cl_2, 24 h	**I** (81)	43
	TEMPO (cat.), PS-I$(OAc)_2$, acetone, 0°	**I** (94)	202
	TEMPO (cat.), NaOCl, silica gel, H_2O, 0°	**I** (84)	393
	SG-NHCONH-TEMPO (cat.), NaOCl (2 eq), acetone, H_2O, 0°	**I** (95)	393
	SG-NHCONH-TEMPO (cat.), NaOCl (3.5 eq), acetone, H_2O, 0°	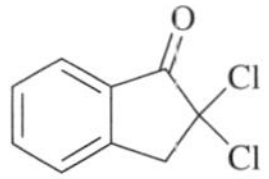 (35)	393
C_{10}			
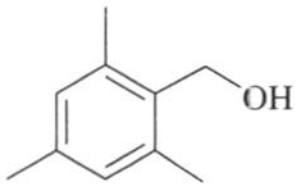	PEG-O-TEMPO (cat.), PEG-bipy (cat.), $CuBr_2$ (cat.), O_2, *t*-BuOK, MeCN, H_2O, 80°, 18 h	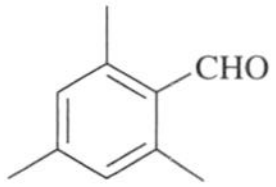 (32)	392
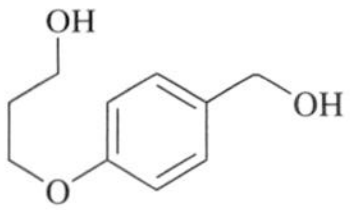	TEMPO (cat.), KNO_2 (cat.), PhI$(OAc)_2$, O_2, 80–90°, 24 h	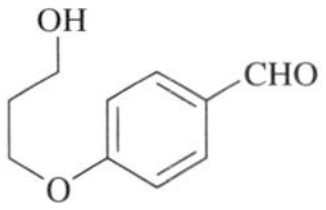 (21)[c]	203

TABLE 2. OXIDATIONS OF BENZYLIC ALCOHOLS *(Continued)*

Alcohol	Conditions	Product(s) and Yield(s) (%)	Refs.
C_{10} 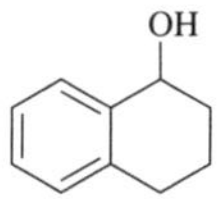	TEMPO, PS-I(OAc)$_2$ (cat.), acetone, rt, 2–30 h	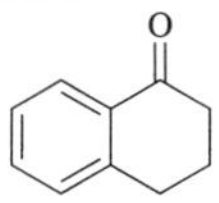 **I** (92)	202
	TEMPO (cat.), [PS-CH$_2$NMe$_3$]$^+$[Br(OAc)$_2$]$^-$, toluene or CH_2Cl_2, rt, 24 h	**I** (94)	43, 211
	TEMPO (cat.), *m*-CPBA, TBAB, CH_2Cl_2	**I** (97)	213
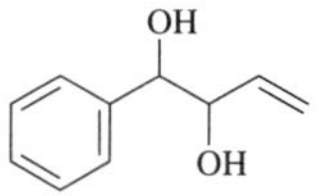	AcNH-TEMPO (5 eq), TsOH (5 eq), CH_2Cl_2, 0°, 4–7 d	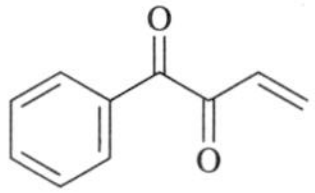 (75)	427
C_{11} 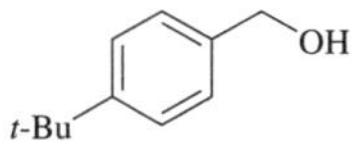	TEMPO (cat.), silica gel, NaOCl, H_2O, 0°	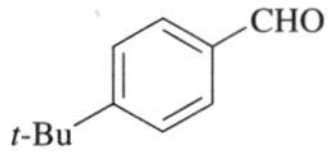 **I** (88)	393
	BzO-TEMPO (cat.), TBABr$_3$, NaOAc, CH_2Cl_2, H_2O, rt, 1 h	**I** (93)	224
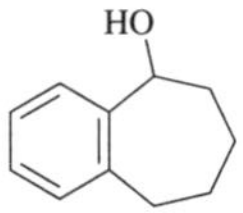	SBA15-NH-TEMPO (cat.), $NaNO_2$ (cat.), TBAB (cat.), O_2 (1 atm), AcOH, 50–60°, 3 h	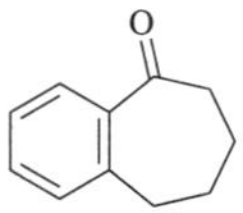 **I** (100)[a], 99.8% selectivity[f]	310
	SBA15-NH-TEMPO (cat.), $NaNO_2$ (cat.), TBAB (cat.), air (1 atm), AcOH, 50–60°, 4.5 h	**I** (100)[a], 99.8% selectivity[f]	310
C_{12} 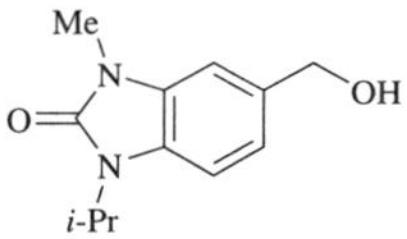	TEMPO (cat.), TBAC, NCS, CH_2Cl_2, H_2O, buffer, pH 8.6, 22°, 16 h	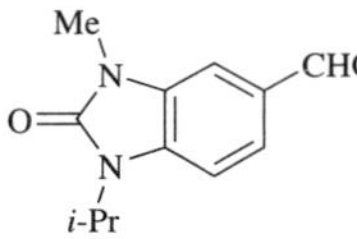 (68)	428

Reactant	Conditions	Product(s) and Yield(s) (%)	Refs.
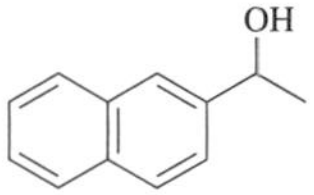	AS-NHAc-TEMPO (cat.), electrolysis, 2,6-lutidine, $NaClO_4$, MeCN	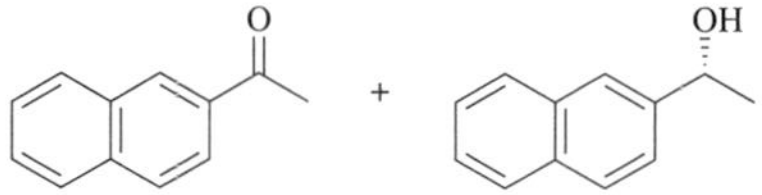 (—), 60% conversion + (—), $s = 4.1$[a]	107
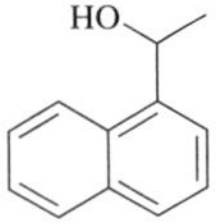	AS-NHAc-TEMPO (cat.), electrolysis, 2,6-lutidine, $NaClO_4$, MeCN	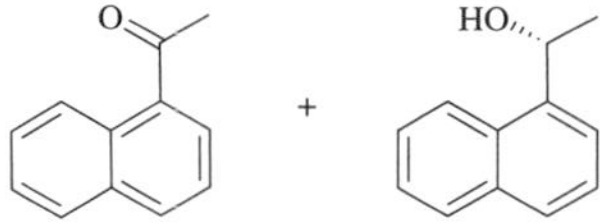 (—), 51% conversion + (—), $s = 4.5$[a]	107
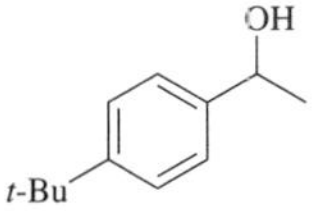	PE-co-AA-CONH-TEMPO (cat.), NaBr (cat.), electrolysis, $NaHCO_3$, H_2O, 0°	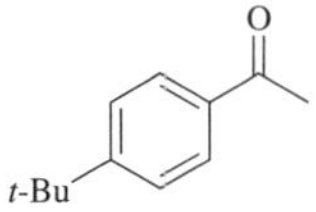 **I** (98)	298
	TEMPO (cat.), NaOCl, silica gel, H_2O, 0°	**I** (82)	393
	(*S*)-binaphthyl-TEMPO (cat.), KBr (cat.), electrolysis, silica gel, $NaHCO_3$, H_2O, –15°	**I** (42) + 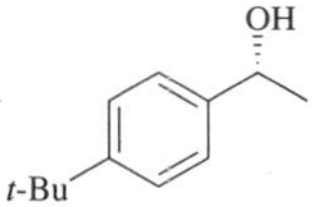 (55), $s = 20$	109
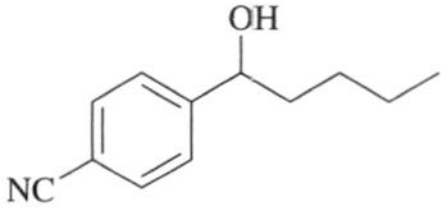	TEMPO (cat.), **1** (cat.), $CuBr{\bullet}Me_2S$ (cat.), O_2, perfluorooctyl bromide, chlorobenzene, 90°	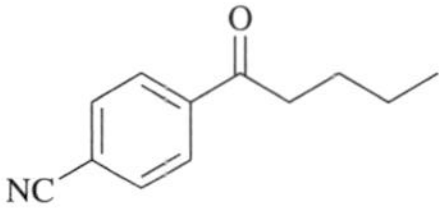 (97)	243
C_{13} 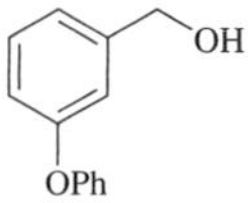	BzO-TEMPO (cat.), Br_2 (cat.), electrolysis, NaOCl, CH_2Cl_2, H_2O	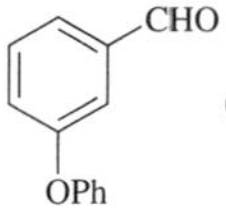 (—)	397

TABLE 2. OXIDATIONS OF BENZYLIC ALCOHOLS *(Continued)*

	Alcohol	Conditions	Product(s) and Yield(s) (%)	Refs.
C_{13}	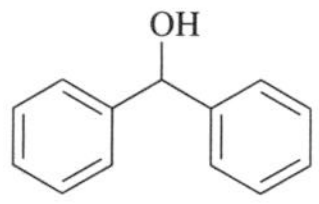	[mim-CO_2-TEMPO]$^+$[PF_6]$^-$ (cat.), KBr (cat.), NaOCl, [bmim]$^+$[PF_6]$^-$, H_2O, 0°	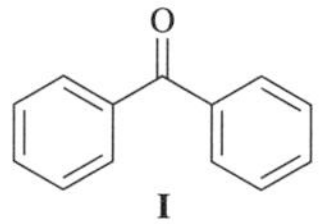 **I** (95)[a]	295
		[MeO-TEMP=O]$^+$Br$^-$, CH_2Cl_2, rt	**I** (97)[a]	11
		TEMPO (cat.), $(NH_4)_2[Ce(NO_3)_6]$ (cat.), O_2, MeCN, reflux	**I** (99)	269
		SBA15-NH-TEMPO (cat.), $NaNO_2$ (cat.), TBAB (cat.), O_2 (1 atm), AcOH, 50–60°, 4.5 h	**I** (100)[a], 99.3% selectivity[f]	310
		SBA15-NH-TEMPO (cat.), $NaNO_2$ (cat.), TBAB (cat.), air (1 atm), AcOH, 50–60°, 6 h	**I** (100)[a], 99.8% selectivity[f]	310
		TEMPO (cat.), DABCO–CuCl (cat.), O_2, K_2CO_3, toluene, 100°, 24 h	**I** (66)	247
	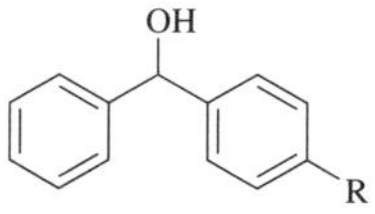	See table.	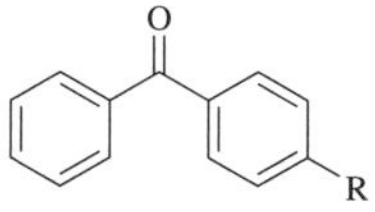	

R	Conditions	Product(s) and Yield(s) (%)	Refs.
OMe	TEMPO (cat.), DABCO–CuCl (cat.), O_2, K_2CO_3, toluene, 100°, 24 h	(78)	247
Cl	TEMPO (cat.), DABCO–CuCl (cat.), O_2, K_2CO_3, toluene, 100°, 24 h	(85)	247
Br	SBA15-NH-TEMPO (cat.), $NaNO_2$ (cat.), TBAB (cat.), air or O_2 (1 atm), AcOH, 50–60°, 4.5–8 h	(100)[a], 99% selectivity[f]	310

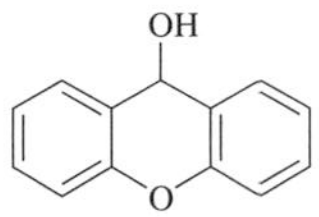	TEMPO (cat.), laccase (cat.), O_2, H_2O, MeCN, citrate buffer, pH 5, rt, 24 h	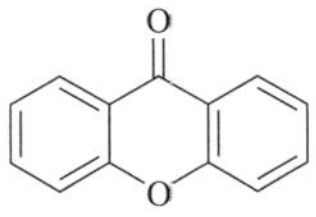 (70)	429
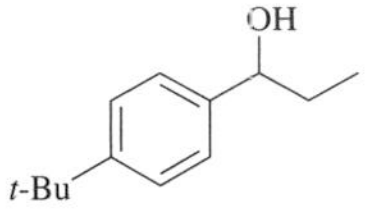	PE-CONH-TEMPO (cat.), NaBr (cat.), electrolysis, $NaHCO_3$, H_2O, 0°	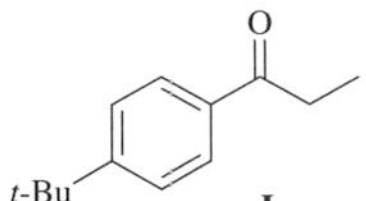 **I** (90)	297
	PE-co-AA-CONH-TEMPO (cat.), NaBr (cat.), electrolysis, $NaHCO_3$, H_2O	**I** (87)	298
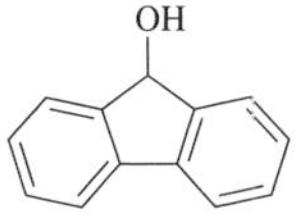	TEMPO (cat.), $(NH_4)_2[Ce(NO_3)_6]$ (cat.), O_2, MeCN, reflux	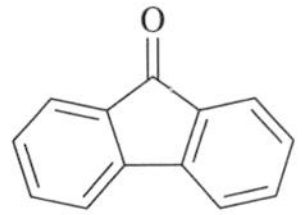 (99)	269
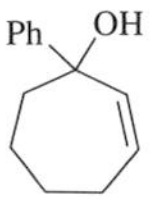	$[TEMP{=}O]^+[BF_4]^-$, MeCN, rt, 0.2 h	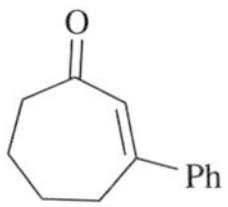 (88)	120
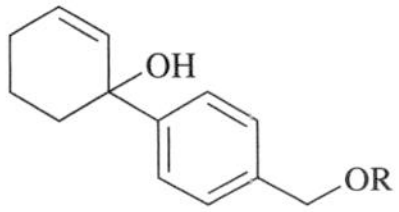 R: TBDMS; TBDPS	$[TEMP{=}O]^+[BF_4]^-$, MeCN, rt, 0.1 h	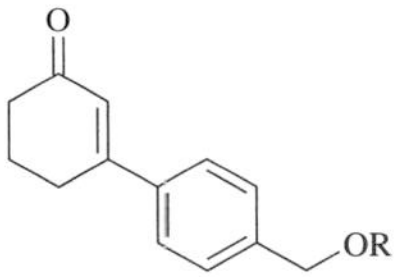 (83); (93)	120

TABLE 2. OXIDATIONS OF BENZYLIC ALCOHOLS *(Continued)*

Alcohol	Conditions	Product(s) and Yield(s) (%)	Refs.
C_{14} 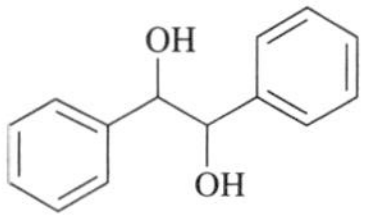	MeO-TEMPO (cat.), KBr (cat.), NaOCl, CH_2Cl_2, H_2O, pH 9.5	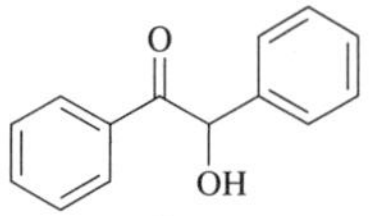 **I** (85)	151
	[mim-O-TEMPO]$^+$[BF_4]$^-$ (cat.), [mim-CH_2PhI(OAc)$_2$]$^+$[BF_4]$^-$, H_2O, 30°, 15 min	**I** (99)	201
	MeO-TEMPO (cat.), KBr (cat.), NaOCl, CH_2Cl_2, H_2O, pH 9.5	**I** (97)	151
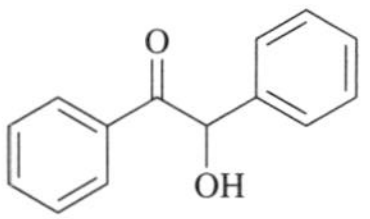	[MeO-TEMP=O]$^+$Br$^-$, CH_2Cl_2, rt	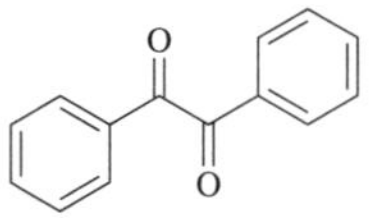 **I** (13)	11
	[MeO-TEMP=O]$^+$Cl$^-$, CH_2Cl_2, rt	**I** (45)[a]	11
	TEMPO (cat.), NaOCl, silica gel, H_2O, 0°	**I** (83)	393
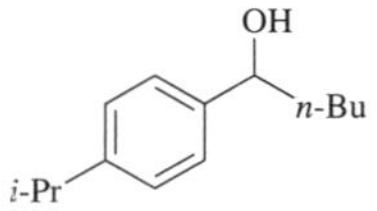	TEMPO (cat.), **1** (cat.), CuBr•Me_2S (cat.), O_2, perfluorooctyl bromide, chlorobenzene, 90°	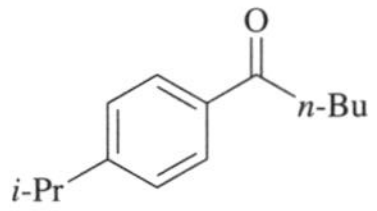 (95)	243
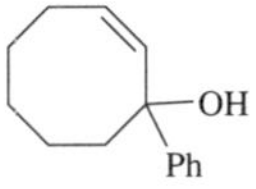	[TEMP=O]$^+$[BF_4]$^-$, MeCN, rt, 0.2 h	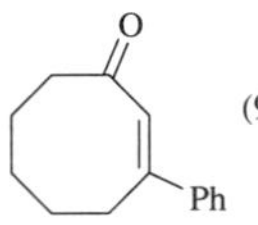 **I** (93)	120
	[TEMP=O]$^+$[SbF_6]$^-$, MeCN, H_2O, rt, 0.2 h	**I** (95)	120

C_{15}	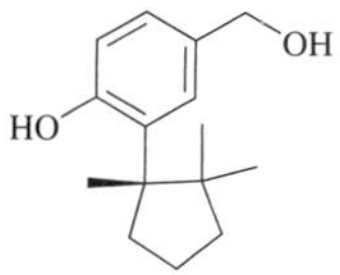	$[AcNH\text{-}TEMP{=}O]^+[ClO_4]^-$, silica gel, CH_2Cl_2, rt	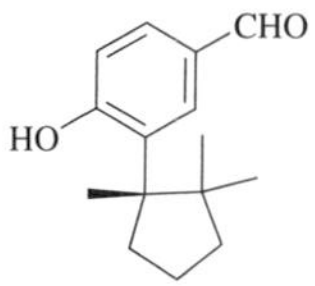 (80)	132
	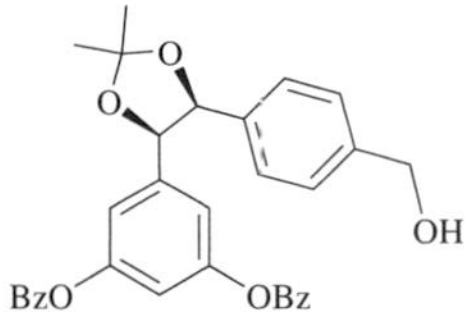	$[AcNH\text{-}TEMP{=}O]^+[ClO_4]^-$, silica gel, CH_2Cl_2, rt	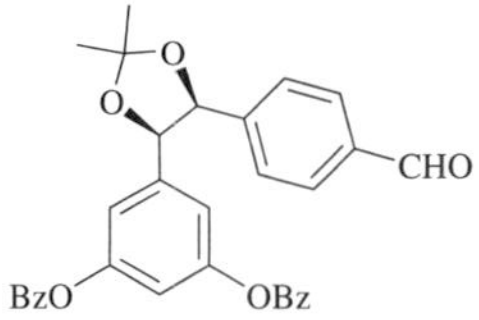 (100)[c]	68
	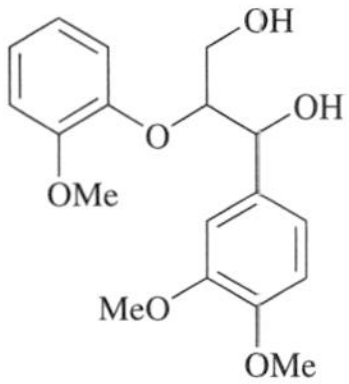	TEMPO (cat.), laccase (cat.), O_2, H_2O, MeCN, citrate buffer, pH 5, rt, 24 h	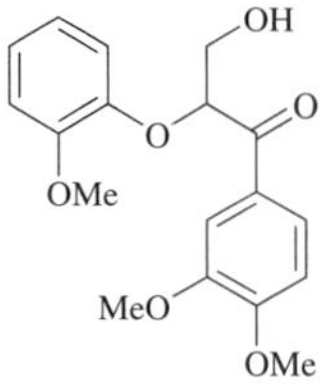 (10)	253
	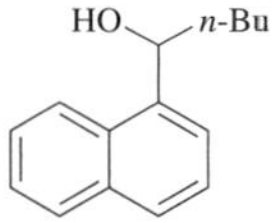	TEMPO (cat.), **1** (cat.), $CuBr{\bullet}Me_2S$ (cat.), O_2, perfluorcoctyl bromide, chlorobenzene, 90°	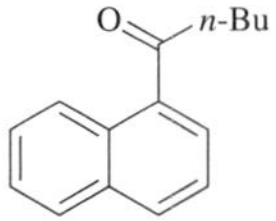 (91)	243, 414
C_{18}	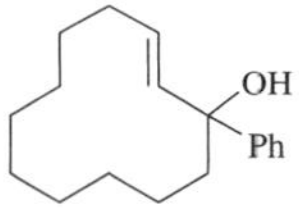	$[TEMP{=}O]^+[BF_4]^-$, MeCN, H_2O, rt, 6 h	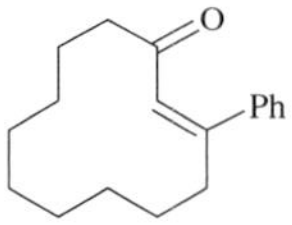 **I** (78), $E{:}Z = 2{:}1$	120
		$[TEMP{=}O]^+[SbF_6]^-$, MeCN, H_2O, rt, 1.5 h	**I** (99), $E{:}Z = 2{:}1$	120

TABLE 2. OXIDATIONS OF BENZYLIC ALCOHOLS *(Continued)*

Alcohol	Conditions	Product(s) and Yield(s) (%)	Refs.
C_{20}			
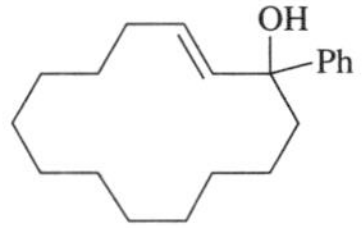	$[TEMP{=}O]^+[BF_4]^-$, MeCN, H_2O, rt, 10 h	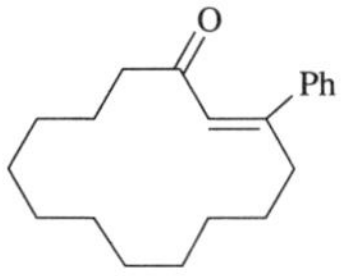 **I** (80), E:Z = 2:1	120
	$[TEMP{=}O]^+[SbF_6]^-$, MeCN, H_2O, rt, 4 h	**I** (98), E:Z = 2:1	120
C_{23}			
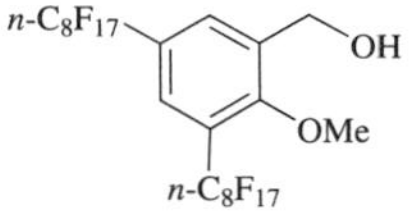	TEMPO (cat.), KBr (cat.), NaOCl, H_2O, trifluorotoluene, 5°	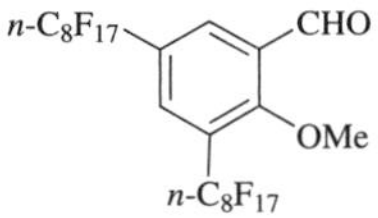 (85)	154
C_{33}			
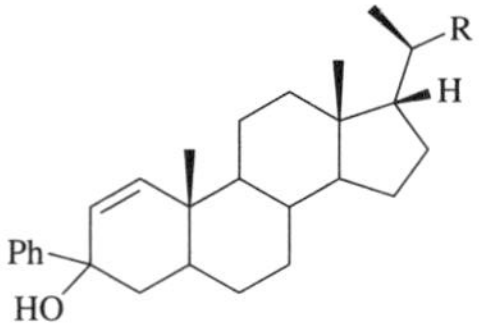 R = $(CH_2)_3$*i*-Pr	$[TEMP{=}O]^+[BF_4]^-$, MeCN, rt, 7 h	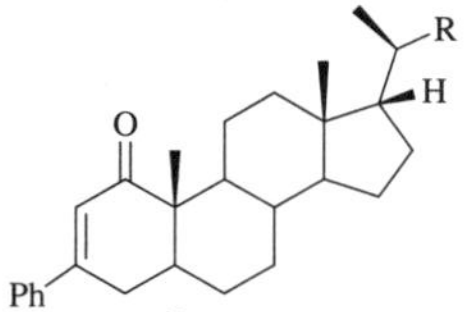 **I** (17) + 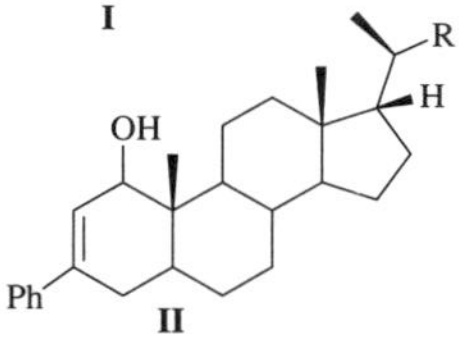 **II** (33)	120
	$[TEMP{=}O]^+[SbF_6]^-$, MeCN, rt, 1 h	**I** (14) + **II** (60)	120

C_{43} 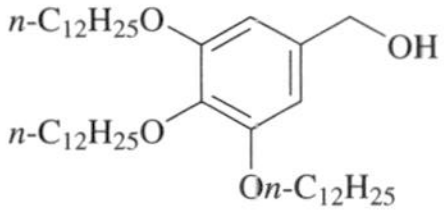	[AcNH-TEMP=O]$^+$[ClO$_4$]$^-$, silica gel, CH_2Cl_2, rt	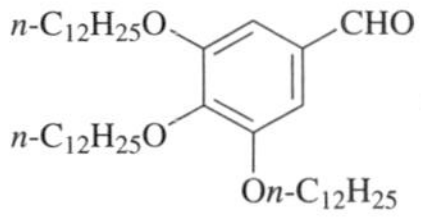	(98)	68

[a] The reported value was determined by GC analysis.

[b] The reported value was determined by HPLC analysis.

[c] The reported value was determined by NMR analysis.

[d] The selectivity refers to formation of the aldehyde versus products with the carboxylic acid oxidation state.

[e] The product was not isolated and was used in a subsequent step without purification.

[f] The selectivity refers to formation of the ketone versus over-oxidation products such as C–C cleavage.

[g] The product was isolated as a derivative.

TABLE 3. OXIDATIONS OF POLYFUNCTIONAL ALCOHOLS

Alcohol	Conditions	Product(s) and Yield(s) (%)	Refs.
C_{2-8}			
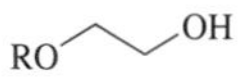	$[AcNH\text{-}TEMP{=}O]^+[BF_4]^-$, pyridine, 4 Å MS, CH_2Cl_2, rt	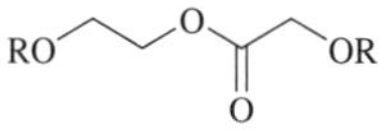	77
R: Ac		(trace)[a]	
R: Ph		(68)	
R: Bn		(76)	
C_3			
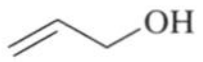	$[mim\text{-}CO_2\text{-}TEMP{=}O]^+[PF_6]^-$ (cat.), KBr (cat.), NaOCl, $[bmim]^+[PF_6]^-$, H_2O, 0°	 **I** (90)[a]	295
	TEMPO (cat.), **2** (cat.), CuCl (cat.), O_2, $[bmim]^+[PF_6]^-$, 65° 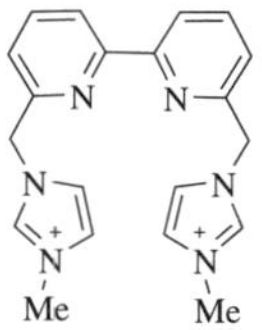 2 $[PF_6]^-$ **2**	**I** (80)[a]	246
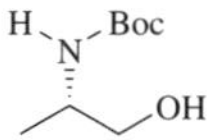	TEMPO (cat.), $[PS\text{-}CH_2NMe_3]^+[Br(OAc)_2]^-$, CH_2Cl_2, rt, 2 h	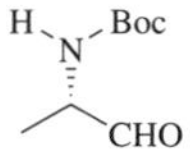 **I** (96)	43
	TEMPO (cat.), NaBr (cat.), NaOCl, $NaHCO_3$, toluene, EtOAc, H_2O, 0°	**I** (51), 99% ee	158

Reactant	Conditions	Product(s) and Yield(s) (%)	Refs.
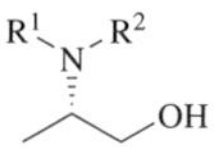	TEMPO (cat.), NaBr (cat.), NaOCl, $NaHCO_3$, toluene, EtOAc, H_2O, 0°	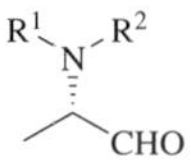	
R^1 R^2			
H Ts		(—)[b]	430
Bn H		(95)	431–433
H Cbz		(51)[b]	158
H Cbz		(—)[b]	431–433
Bn Ts		(—)[b]	430
Bn Cbz		(—)[b]	430
	Conditions, CH_2Cl_2	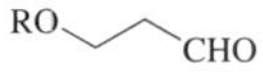	
R	Conditions		
TBDMS	TEMPO (cat.), NaOCl, KBr, H_2O	(90–95)	181
TBDMS	TEMPO (cat.), NaOCl, NaBr, $NaHCO_3$, 0°, 8 min	(98)	434
Bn	TEMPO (cat.), KBr (cat.), NaOCl, H_2O	(90–95)	181
Bz	$[MeO\text{-}TEMP{=}O]^+Cl^-$, rt	(trace)[a]	69, 435
PMB	TEMPO (cat.), $[PS\text{-}CH_2NMe_3]^+[Br(OAc)_2]^-$, rt, 1.5 h	(95)	43
TBDPS	$[AcNH\text{-}TEMP{=}O]^+[ClO_4]^-$, silica gel, rt	(97)	68
	TEMPO (cat.), $PhI(OAc)_2$, CH_2Cl_2, rt	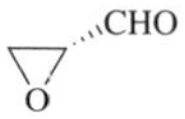 (—)	206
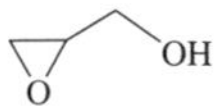	$[AcNH\text{-}TEMP{=}O]^+[BF_4]^-$, pyridine, 4 Å MS, CH_2Cl_2, rt	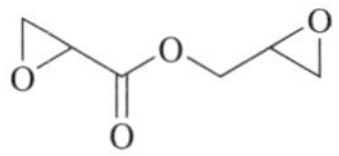 (40)	77

TABLE 3. OXIDATIONS OF POLYFUNCTIONAL ALCOHOLS (*Continued*)

Alcohol	Conditions	Product(s) and Yield(s) (%)	Refs.
C_3 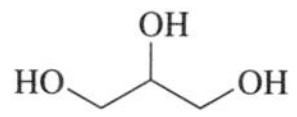	TEMPO or ormosil-TEMPO (cat.), NaBr (cat.), NaOCl, H_2O, pH 10	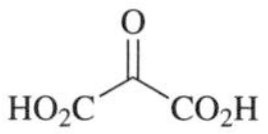 (—)	166
	TEMPO (cat.), electrolysis, $NaHCO_3$, H_2O, short time	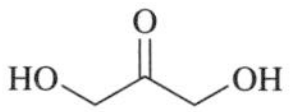 (25)[c]	436
	TEMPO (cat.), electrolysis, $NaHCO_3$, H_2O, long time	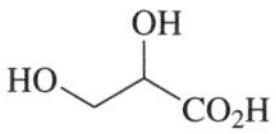 (35)[c]	436
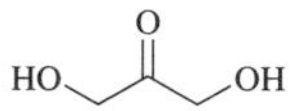	TEMPO (cat.), NaBr (cat.), NaOCl, HCl, H_2O, pH 10, 0°	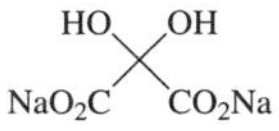 (45)	437
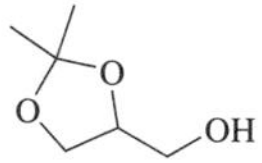	$[AcNH\text{-}TEMP{=}O]^+[BF_4]^-$, pyridine, 4 Å MS, CH_2Cl_2, rt	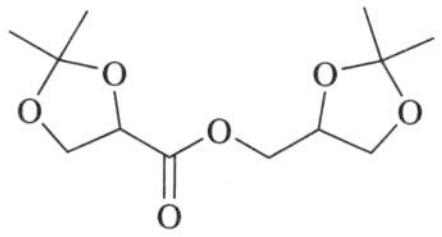 (89)	77
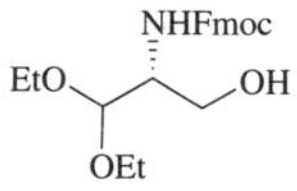	TEMPO (cat.), NaOCl (cat.), $NaClO_2$, $NaHCO_3$, MeCN, H_2O, phosphate buffer	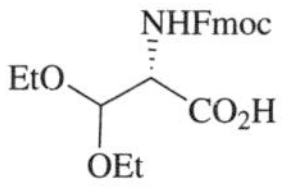 (87)	438
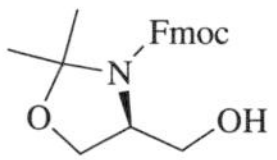	TEMPO (cat.), NaOCl (cat.), $NaClO_2$, $NaHCO_3$, MeCN, H_2O, phosphate buffer	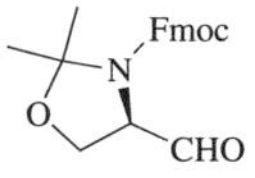 (—)[b]	438

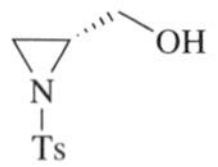	TEMPO (cat.), PhI(OAc)$_2$, CH_2Cl_2, rt, 5 h	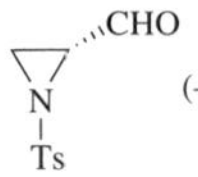 (—)[b]	439
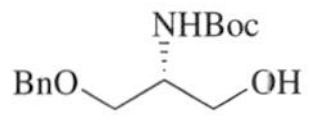	AcNH-TEMPO (cat.), NaBr (cat.), NaOCl, $NaHCO_3$, toluene, EtOAc, H_2O	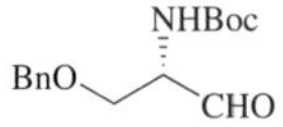 (—)[d]	440, 441
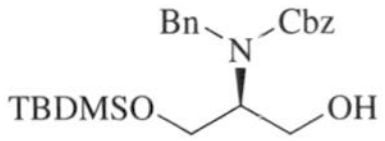	TEMPO (cat.), KBr (cat.), TBAF, NaOCl, CH_2Cl_2, H_2O, 0°	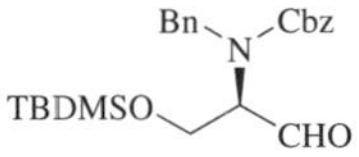 (—)	442
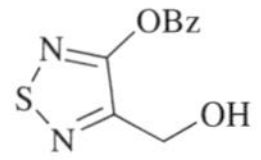	$[AcNH\text{-}TEMP{=}O]^+[BF_4]^-$, CH_2Cl_2, silica gel	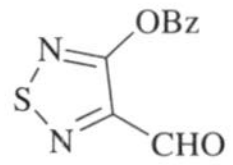 (>90)	443
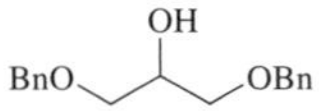	TEMPO (cat.), KBr (cat.), NaOCl, $NaHCO_3$, H_2O, CH_2Cl_2, pH 9.5, −10°	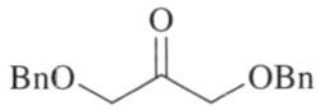 (84)	444
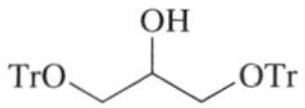	AcNH-TEMPO (cat.), NaBr (cat.), NaOCl, NaBr, $NaHCO_3$, H_2O, CH_2Cl_2, 0°	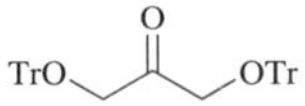 (93)	445

TABLE 3. OXIDATIONS OF POLYFUNCTIONAL ALCOHOLS (*Continued*)

	Alcohol	Conditions	Product(s) and Yield(s) (%)	Refs.
C_3	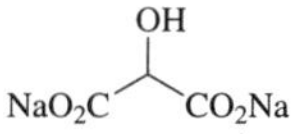	TEMPO (cat.), NaBr (cat.), NaOCl, HCl, H_2O, pH 10, 0°	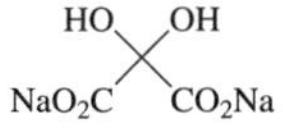 (—)	437
C_4	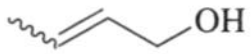	TEMPO (cat.), $PhI(OAc)_2$ (cat.), KNO_2 (cat.), O_2, 80–90°, 15 h	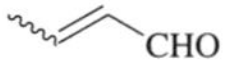 (23)[e]	203
	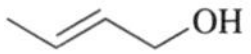	$[MeO\text{-}TEMP{=}O]^+Br^-$, CH_2Cl_2, rt	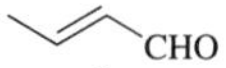 **I** (9)	11
		$[MeO\text{-}TEMP{=}O]^+Cl^-$, CH_2Cl_2, rt	**I** (92)[a]	11
		AcNH-TEMPO (cat.), $Cu(ClO_4)_2$ (cat.), O_2, DMAP, $[bmpy]^+[PF_6]^-$, rt	**I** (89)	240
	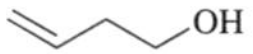	TEMPO (cat.), $PhI(OAc)_2$, pentane, CH_2Cl_2, rt	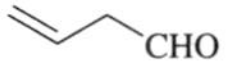 **I** (100)[e]	208
		SG-O-TEMPO (cat.), $PhI(OAc)_2$, pentane, CH_2Cl_2, rt	**I** (98)[e]	208

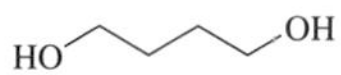	$[PS\text{-}O\text{-}TEMP{=}O]^+Cl^-$, CH_2Cl_2, rt, 1–2 h	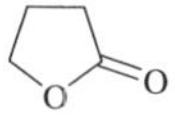 **I** (65)[a]	121
	BzO-TEMPO (cat.), NaBr (cat.), electrolysis, silica gel, $NaHCO_3$, H_2O	**I** (84)	84
	TEMPO or MeO-TEMPO (cat.), LiOCl, $NaHCO_3$, CH_2Cl_2, H_2O, pH 9.5	**I** (69)	151
	$[MeO\text{-}TEMP{=}O]^+Cl^-$, CH_2Cl_2, rt	**I** (100[a], 81)	11
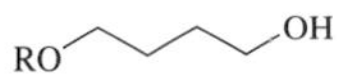	CH_2Cl_2		
R	Conditions		
TBDMS	TEMPO (cat.), NaBr (cat.), NaOCl, $NaHCO_3$, 0°, 8 min	(100)	434
Bn	TEMPO (cat.), NaOCl, H_2O	(77)	446
Bz	$[MeO\text{-}TEMP{=}O]^+Cl^-$, rt	(58)[a]	69, 435
Tr	TEMPO (cat.), $Yb(OTf)_3$ (cat.), $PhI(OAc)_2$, 30 min	(97)	416
	BzO-TEMPO, $RuCl_2(PPh_3)_3$, O_2, toluene, 70°, 8 h	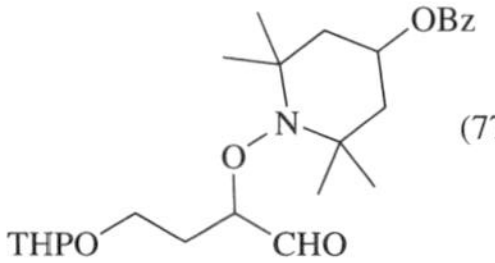 (77)	411
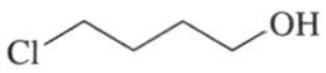	$[AcNH\text{-}TEMP{=}O]^+[BF_4]^-$, CH_2Cl_2, silica gel, rt, 16–24 h	 **I** (17)	131
	$[AcNH\text{-}TEMP{=}O]^+[BF_4]^-$, pyridine, CH_2Cl_2, rt, 16–24 h	**I** (31)	131
	BzO-TEMPO (cat.), NaOCl (cat.), electrogenerated Br_2, CH_2Cl_2, H_2O	**I** (—)	397

TABLE 3. OXIDATIONS OF POLYFUNCTIONAL ALCOHOLS (*Continued*)

	Alcohol	Conditions	Product(s) and Yield(s) (%)	Refs.
C_4	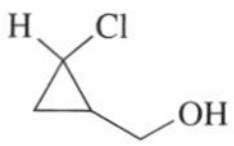	BzO-TEMPO (cat.), NaOCl (cat.), electrogenerated Br_2, CH_2Cl_2, H_2O	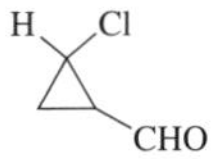 (—)	397
	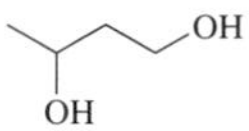	$[MeO\text{-}TEMP{=}O]^+Cl^-$, CH_2Cl_2, rt	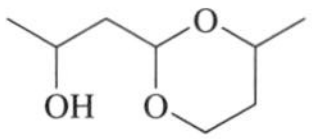 (100^a, 61)	11
	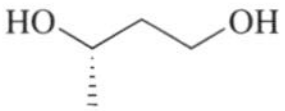	Graphite felt-CONH-AS-TEMPO (cat.), electrolysis, $NaClO_4$, MeCN	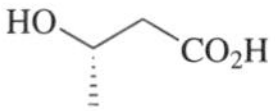 (84.6), 99% ee[a]	83
		TEMPO (cat.), $PhI(OAc)_2$, CH_2Cl_2, rt	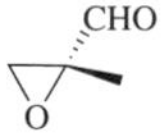 (—)[b]	206
	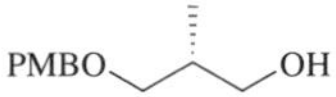	TEMPO (cat.), KBr (cat.), NaOCl, $KHCO_3$, CH_2Cl_2, H_2O	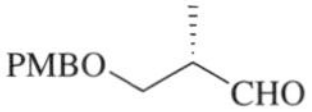 (—)[b]	447
	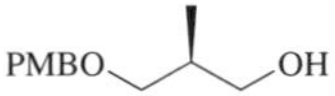	TEMPO (cat.), $[PS\text{-}CH_2NMe_3]^+[Br(OAc)_2]^-$, CH_2Cl_2, rt, 1.5 h	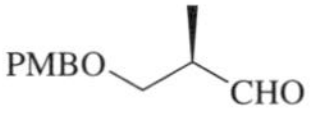 (95)	211
		TEMPO (cat.), oxidant, CH_2Cl_2, rt	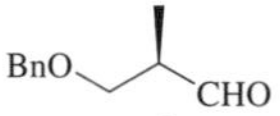 **I** + 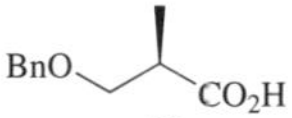 **II**	43, 44, 211

Oxidant	Time	I	II
$[PS\text{-}CH_2NMe_3]^+[Br(OAc)_2]^-$	1.5 h	(98)	(0)
$[PS\text{-}CH_2NMe_3]^+[ClO]^-$	4 h	(71)	(~10)
$[PS\text{-}CH_2NMe_3]^+[ClO_2]^-$	2.5 h	(0)	(99)

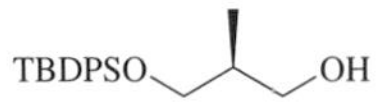	TEMPO (cat.), oxidant, CH_2Cl_2, rt Oxidant / Time $[PS\text{-}CH_2NMe_3]^+[Br(OAc)_2]^-$ / 1.5 h $[PS\text{-}CH_2NMe_3]^+[ClO_2]^-$ / 20 h	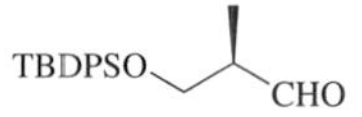 **I** + 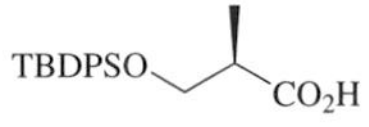 **II** **I** / **II** (99) / (0) (0) / (99)	44, 211
	TEMPO (cat.), PhI(OAc)$_2$, CH_2Cl_2, rt	 (—)[b]	205
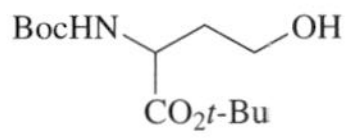	MeO-TEMPO (cat.), KBr (cat.), NaOCl (1.2 eq), CH_2Cl_2, H_2O, pH 9.5	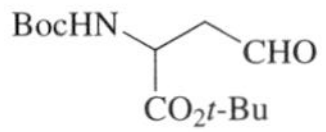 (79)	448
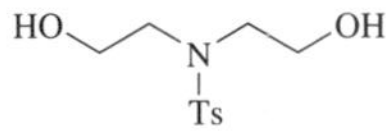	TEMPO or MeO-TEMPO (cat.), KBr (cat.), NaOCl, CH_2Cl_2, H_2O, pH 9.5	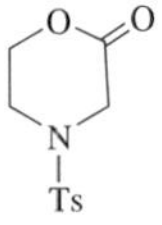 (85)	151
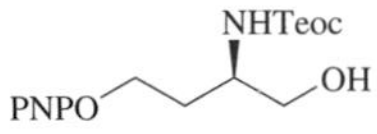	TEMPO (cat.), NaOCl (cat.), $NaClO_2$, MeCN, H_2O, buffer, pH 6.7	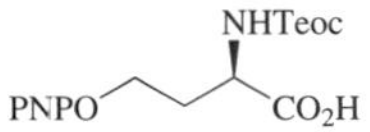 (75)	449
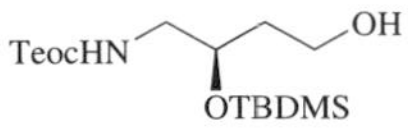	TEMPO (cat.), NaOCl (cat.), $NaClO_2$, MeCN, H_2O, buffer, pH 6.7	 (85)	449
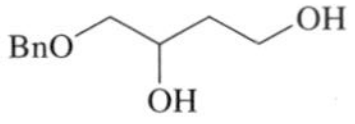	TEMPO (cat.), PhI(OAc)$_2$, CH_2Cl_2, rt	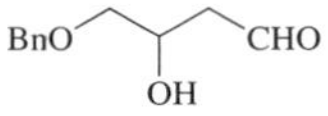 (—)[b]	205
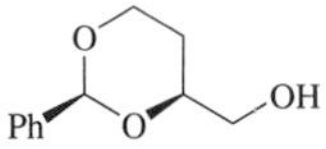	AcNH-TEMPO (cat.), PhI(OAc)$_2$, CH_2Cl_2, 18°, 14 h	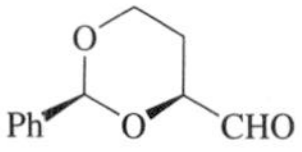 (—)[b]	195

TABLE 3. OXIDATIONS OF POLYFUNCTIONAL ALCOHOLS (*Continued*)

Alcohol	Conditions	Product(s) and Yield(s) (%)	Refs.
C_4			
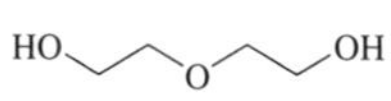	$[AcNH\text{-}TEMP{=}O]^+[BF_4]^-$, pyridine, 4 Å MS, CH_2Cl_2, rt	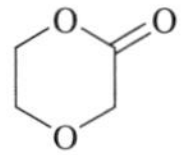 (95)	77
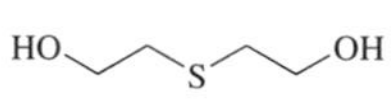	$[AcNH\text{-}TEMP{=}O]^+[BF_4]^-$, pyridine, 4 Å MS, CH_2Cl_2, rt	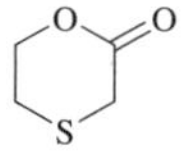 (95)	77
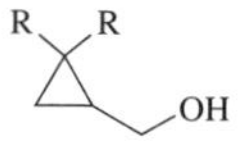	BzO-TEMPO (cat.), NaOCl (cat.), electrogenerated Br_2, CH_2Cl_2, H_2O	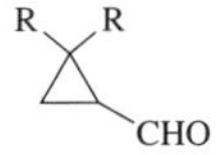 R: Cl (—); Br (—)	397
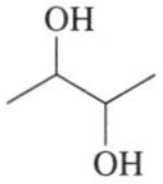	$[MeO\text{-}TEMP{=}O]^+Cl^-$, CH_2Cl_2, rt	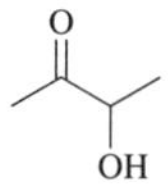 (74)[a]	11
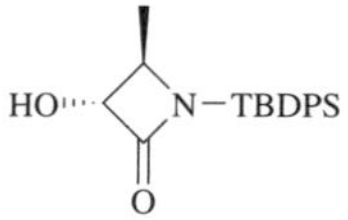	TEMPO (cat.), KBr (cat.), NaOCl, CH_2Cl_2, H_2O, phosphate buffer, pH 6.9	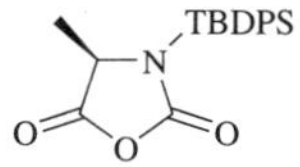 (57)	183
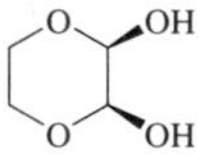	TEMPO (cat.), electrolysis, 2,6-lutidine, $NaClO_4$, MeCN, H_2O	CO_2H–CO_2H (93)	421
C_{4-13}			
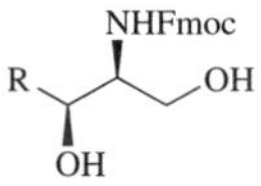	TEMPO (cat.), NaOCl (cat.), $NaClO_2$, MeCN, H_2O, buffer	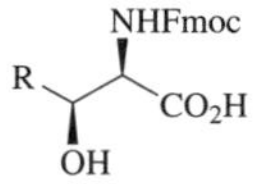 R: Me (95); n-C_6H_{13} (94); n-$C_{10}H_{21}$ (65)	450

C_5 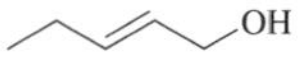	HO-TEMPO (cat.), CuCl (cat.), O_2, DMF, rt	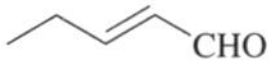 (100)[a]	81
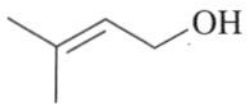	BzO-TEMPO, $RuCl_2(PPh_3)_3$ (cat.), O_2, toluene, 70°, 8 h	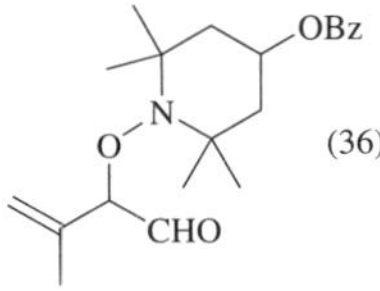 (36)	411
	TEMPO (cat.), $RuCl_2(PPh_3)_3$ (cat.), O_2 (10 atm), chlorobenzene, 100°	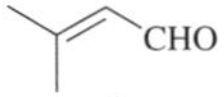 **I** (96)[a]	80, 409, 410
	TEMPO (cat.), Br_2 (cat.), $NaNO_2$ (cat.), air (3.9 atm), CH_2Cl_2, 80°	**I** (—)	258
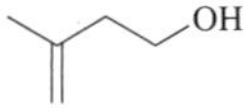	Nitroxide (cat.), $PhI(OAc)_2$, pentane, CH_2Cl_2, rt	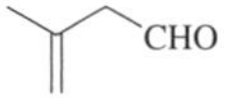 Nitroxide: TEMPO (100)[e]; SG-O-TEMPO (—)	208
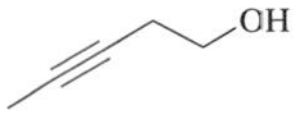	TEMPO (cat.), $PhI(OAc)_2$, pentane, CH_2Cl_2, rt	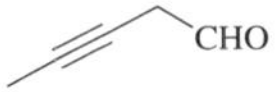 (41)	208
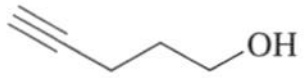	TEMPO (cat.), $PhI(OAc)_2$, CH_2Cl_2, rt	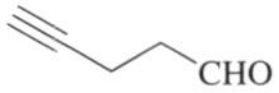 (—)[b]	205
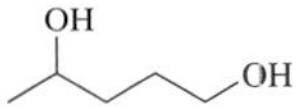	MCM41-NH-TEMPO (cat.), *m*-CPBA, HBr, CH_2Cl_2, rt, 1.5 h	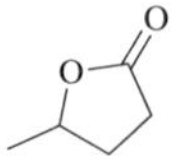 **I** (—), 99% conversion	309
	$[MeO\text{-}TEMP{=}O]^+Cl^-$, CH_2Cl_2, rt	**I** (45[a], 33)	73

TABLE 3. OXIDATIONS OF POLYFUNCTIONAL ALCOHOLS (*Continued*)

Alcohol	Conditions	Product(s) and Yield(s) (%)	Refs.
C_5 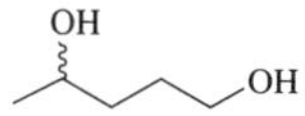	Graphite felt-TEMPO (cat.), (–)-sparteine (cat.), electrolysis, $NaClO_4$, MeCN	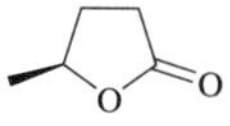 **I** (48.2)[a], 99% ee[a]	451
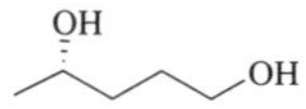	Graphite felt-CONH-AS-TEMPO (cat.), electrolysis, $NaClO_4$, MeCN	**I** (94.8), 99% ee[a]	83
	MeO-TEMPO (cat.), KBr (cat.), NaOCl (1.2 eq), CH_2Cl_2, H_2O, pH 9.5	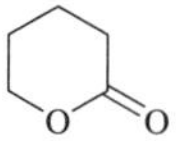 **I** (58)	151
	BzO-TEMPO (cat.), NaBr (cat.), electrolysis, silica gel, $NaHCO_3$, H_2O	**I** (72)	84
	$[MeO\text{-}TEMP{=}O]^+Cl^-$, CH_2Cl_2, rt	**I** (61[a], 40)	73
	TEMPO (cat.), $Yb(OTf)_3$ (cat.), $PhI(OAc)_2$, CH_2Cl_2, 1 h	**I** (86)	416
	$[MeO\text{-}TEMP{=}O]^+Cl^-$, CH_2Cl_2, rt	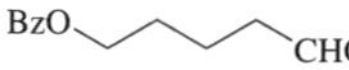 (42)[a]	69, 435
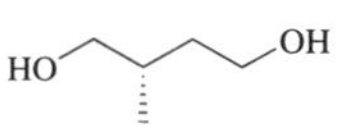	Graphite felt-TEMPO (cat.), (–)-sparteine (cat.), electrolysis, $NaClO_4$, MeCN	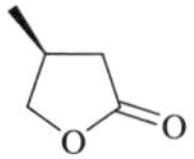 (96.5)[a], 100% ee[a]	451
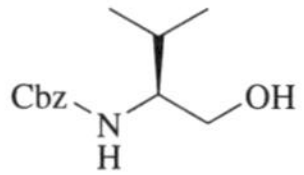	TEMPO (cat.), NaOCl (cat.), $NaClO_2$, MeCN, pH 6.7	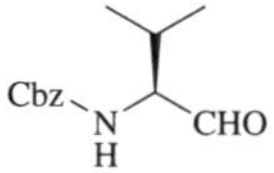 (—)[b]	452
	TEMPO (cat.), NaBr (cat.), NaOCl, $NaHCO_3$, toluene, EtOAc, H_2O, 0°	 (—)[b]	453

	TEMPO (cat.), KBr (cat.), NaOCl, $NaHCO_3$, CH_2Cl_2, 0°	 (85)	454
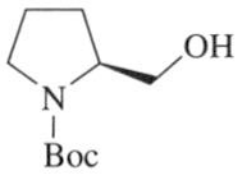	TEMPO (cat.), NaBr (cat.), NaOCl, $NaHCO_3$, toluene, EtOAc, H_2O, 0°	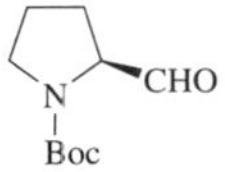 (82), 98% ee	158
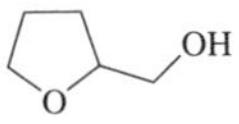	1. TEMPO (cat.), PS-I$(OAc)_2$, acetone, rt 2. H_2O, 24 h	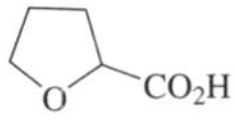 (85)	204
	$[AcNH\text{-}TEMP{=}O]^+[BF_4]^-$, pyridine, 4 Å MS, CH_2Cl_2, rt	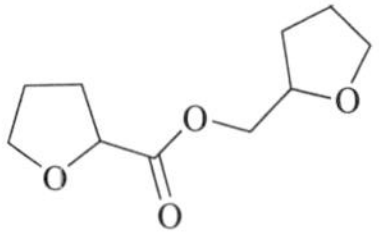 (62)	77
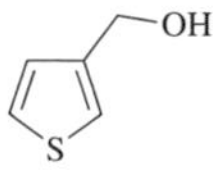	$[AcNH\text{-}TEMP{=}O]^+[ClO_4]^-$, silica gel, CH_2Cl_2	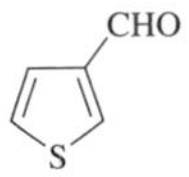 (96)	455
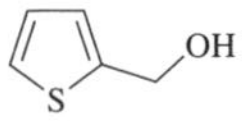	PEG-O-TEMPO (cat.), $CuBr_2$ (cat.), PEG-bipy (cat.), O_2, *t*-BuOK, MeCN, H_2O, 80°, 18 h	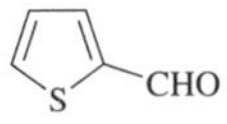 **I** (85)	392
	$[AcNH\text{-}TEMP{=}O]^+[ClO_4]^-$, silica gel, CH_2Cl_2	**I** (92)	455
	AcNH-TEMPO (cat.), **4** (cat.), $Cu(ClO_4)_2$ (cat.), O_2, DABCO, DMSO 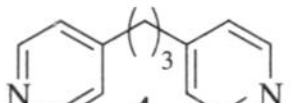 **4**	**I** (92)	245

TABLE 3. OXIDATIONS OF POLYFUNCTIONAL ALCOHOLS (*Continued*)

Alcohol	Conditions	Product(s) and Yield(s) (%)	Refs.
C_5 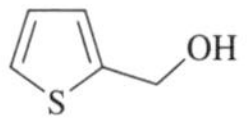	TEMPO (cat.), $NaNO_2$ (cat.), Br_2 (cat.), air, H_2O, 80°, 1 h	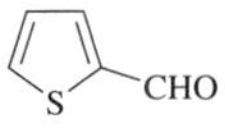 **I** (94)	258
	TEMPO (cat.), **3** (cat.), O_2, toluene, 100° 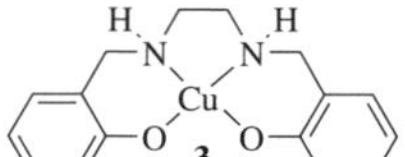 **3**	**I** (94)	244
	TEMPO (cat.), $NaNO_2$ (cat.), 1,3-dibromo-5,5-dimethyl-hydantoin (cat.), air, H_2O, 80°, 1 h	**I** (93)	259
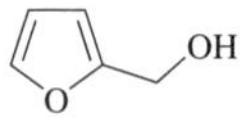	TEMPO (cat.), $(NH_4)_2[Ce(NO_3)_6]$ (cat.), O_2, MeCN, reflux	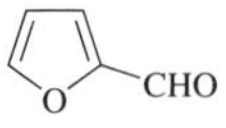 **I** (83)	269
	[mim-O-TEMPO]$^+$[BF_4]$^-$ (cat.), [mim-CH_2PhI(OAc)$_2$]$^+$[BF_4]$^-$, H_2O, 30°, 10 min	**I** (97)[a]	201
	TEMPO (cat.), **3** (cat.), O_2, toluene, 100°	**I** (98)	244
	CIBA-TEMPO (cat.), $Mn(NO_3)_2$ or $Co(NO_3)_2$ (cat.), air or O_2, AcOH, 20–40°	**I** (95–97)	395
	[AcNH-TEMP=O]$^+$[ClO_4]$^-$, silica gel, CH_2Cl_2	**I** (96)	455

	SBA15-NH-TEMPO (cat.), $NaNO_2$ (cat.), TBAB (cat.), O_2 (1 atm), AcOH, 50–60°, 1.5 h	**I** (98)[a], 99% selectivity[f]	310
	SBA15-NH-TEMPO (cat.), $NaNO_2$ (cat.), TBAB (cat.), air (1 atm), AcOH, 50–60°, 3 h	**I** (94.2)[a], 98% selectivity[f]	310
	TEMPO (cat.), DABCO–CuCl (cat.), O_2, K_2CO_3, toluene, 100°, 2 h	**I** (88)	247
	TEMPO (cat.), DABCO–CuCl (cat.), O_2, K_2CO_3, $MeNO_2$, 100°, 10 h	**I** (87)	247
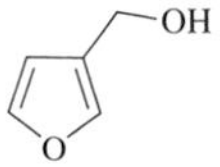	$[AcNH\text{-}TEMP{=}O]^+[ClO_4]^-$, silica gel, CH_2Cl_2	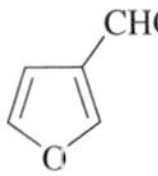 (89)	455
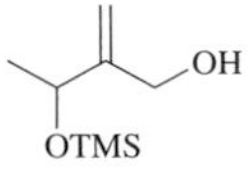	TEMPO (cat.), KBr (cat.), NaOCl, CH_2Cl_2, H_2O, pH 9.5, 0–15°	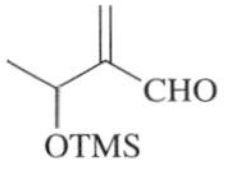 (54)	456
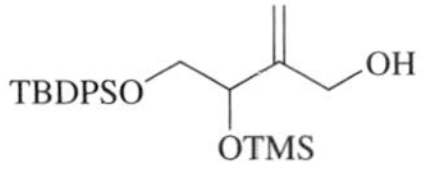	TEMPO (cat.), KBr (cat.), NaOCl, CH_2Cl_2, H_2O, pH 9.5, 0–15°	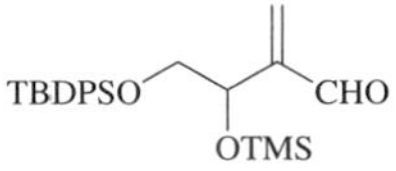 (82)	456
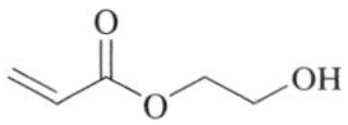	$[AcNH\text{-}TEMP{=}O]^+[BF_4]^-$, pyridine, 4 Å MS, CH_2Cl_2, rt	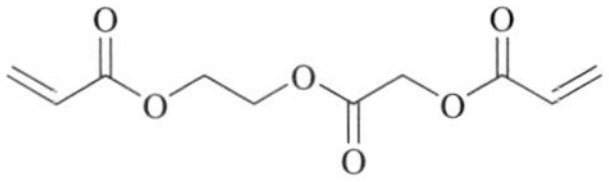 (89)	77
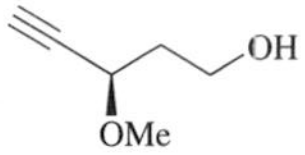	TEMPO (cat.), NaOCl (cat.), $NaClO_2$, MeCN, phosphate buffer, pH 6.7, 45°	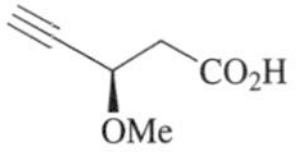 (99)	457

TABLE 3. OXIDATIONS OF POLYFUNCTIONAL ALCOHOLS (*Continued*)

Alcohol	Conditions	Product(s) and Yield(s) (%)	Refs.
C_5			
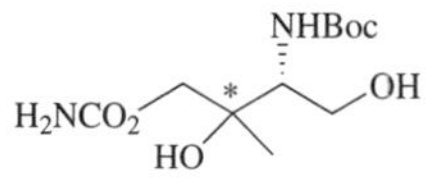	1. $[TEMP{=}O]^+Cl^-$, CH_2Cl_2 2. $NaClO_2$, NaH_2PO_4, H_2O, 2-methyl-2-butene, –20°, 3 h	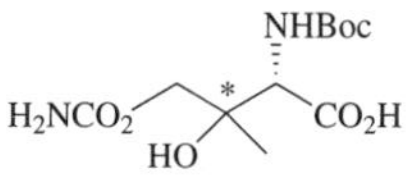 *: *R* (55); *S* (54)	458
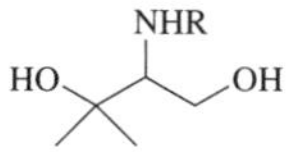	TEMPO (cat.), NaOCl (cat.), $NaClO_2$, MeCN, phosphate buffer, pH 6.7, 35°	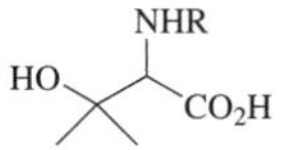 R: Boc (9[illegible]); SO_2Ph (93)	459
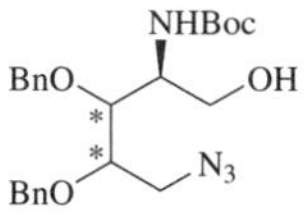	TEMPO (cat.), KBr (cat.), NaOCl, $NaHCO_3$, acetone, H_2O, 0°, 40 min	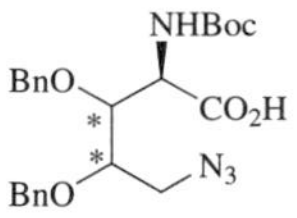 *,*: *R,R* (—)[b]; *S,S* (—)[b]	460
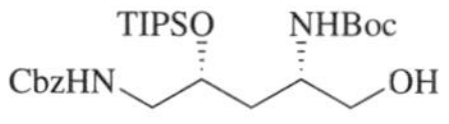	TEMPO (cat.), NaOCl (cat.), $NaClO_2$, MeCN, H_2O, buffer, pH 6.7	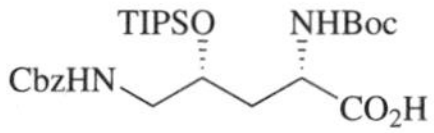 (80)	461
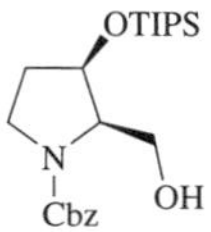	TEMPO (cat.), KBr (cat.), NaOCl, $NaHCO_3$, Et_2O, H_2O	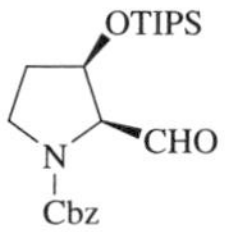 (—)[b]	462
	1. TEMPO (cat.), KBr (cat.), NaOCl, $NaHCO_3$, EtOAc, toluene, H_2O, 0° 2. $RuCl_3$ (cat.), $NaIO_4$, acetone, rt	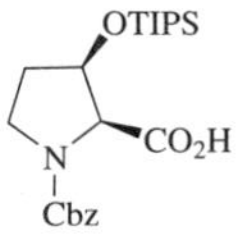 (—)[b]	462

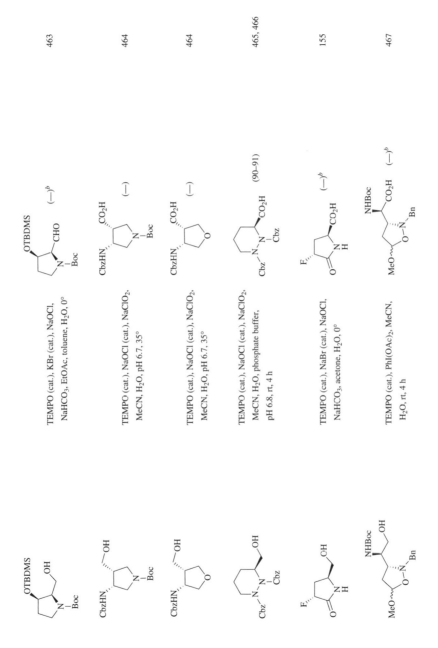

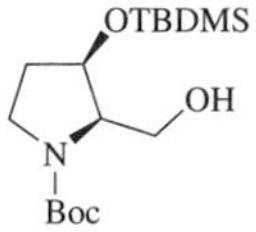	TEMPO (cat.), KBr (cat.), NaOCl, $NaHCO_3$, EtOAc, toluene, H_2O, 0°	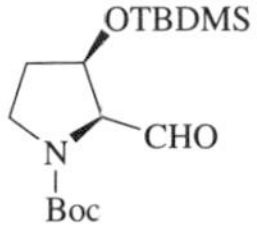	(—)[b]	463
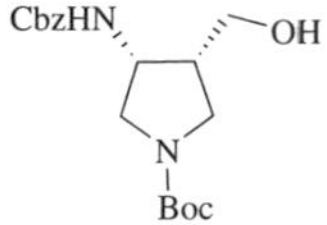	TEMPO (cat.), NaOCl (cat.), $NaClO_2$, MeCN, H_2O, pH 6.7, 35°	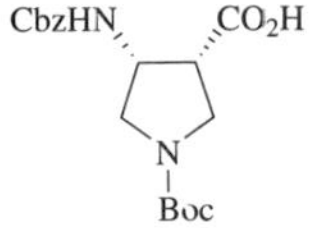	(—)	464
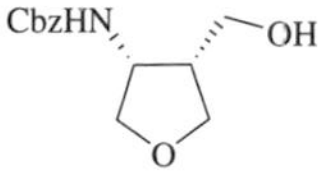	TEMPO (cat.), NaOCl (cat.), $NaClO_2$, MeCN, H_2O, pH 6.7, 35°	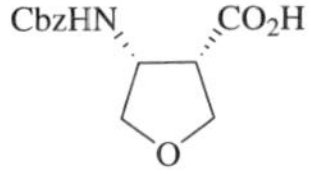	(—)	464
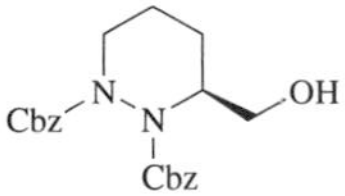	TEMPO (cat.), NaOCl (cat.), $NaClO_2$, MeCN, H_2O, phosphate buffer, pH 6.8, rt, 4 h	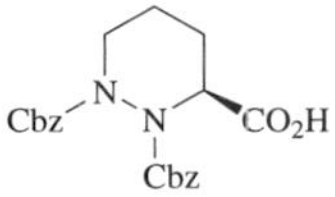	(90–91)	465, 466
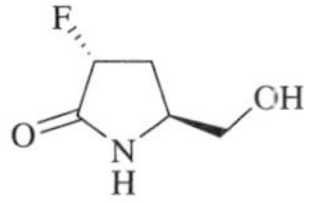	TEMPO (cat.), NaBr (cat.), NaOCl, $NaHCO_3$, acetone, H_2O, 0°	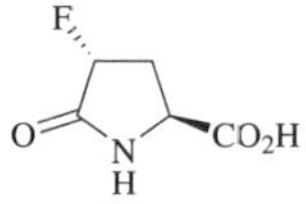	(—)[b]	155
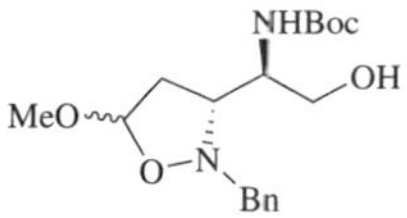	TEMPO (cat.), $PhI(OAc)_2$, MeCN, H_2O, rt, 4 h	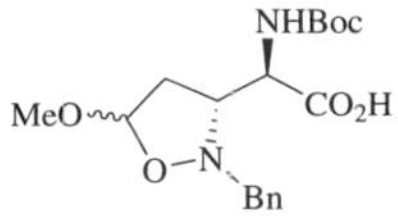	(—)[b]	467

TABLE 3. OXIDATIONS OF POLYFUNCTIONAL ALCOHOLS (*Continued*)

Alcohol	Conditions	Product(s) and Yield(s) (%)	Refs.
C_5 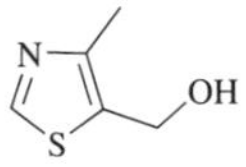	SG-NHCONH-TEMPO (cat.), NaOCl, acetone, H_2O, 0°, 30 min	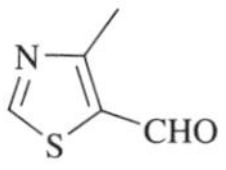 **I** (72)	393
	BzO-TEMPO (cat.), NaOCl, silica gel, acetone, H_2O, 0°, 30 min	**I** (65)	393
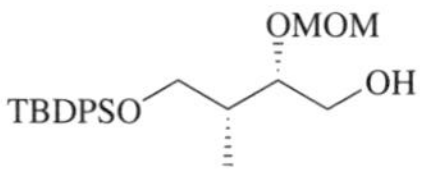	TEMPO (cat.), $[PS\text{-}CH_2NMe_3]^+[Br(OAc)_2]^-$, CH_2Cl_2, rt, 2 h	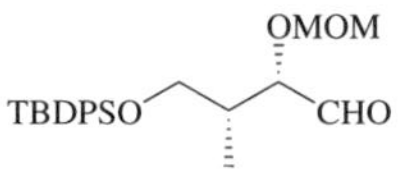 (99)	43, 211
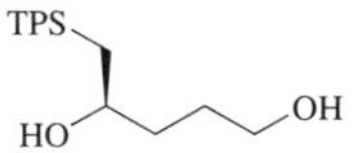	TEMPO (cat.), $PhI(OAc)_2$, CH_2Cl_2, rt, 3 h	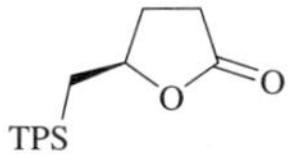 (84), 90% ee	468
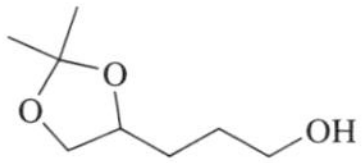	TEMPO (cat.), KBr (cat.), NaOCl, CH_2Cl_2, H_2O, pH 9.5, 0–15°	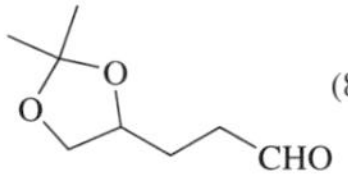 (85)	13
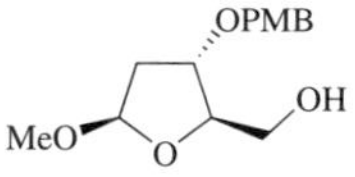	TEMPO (cat.), NaOCl (cat.), $NaClO_2$, MeCN, H_2O, buffer, pH 7.7, 45°	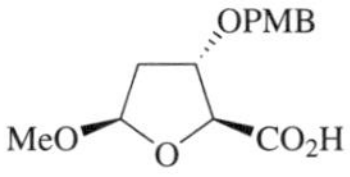 (—)[b]	469
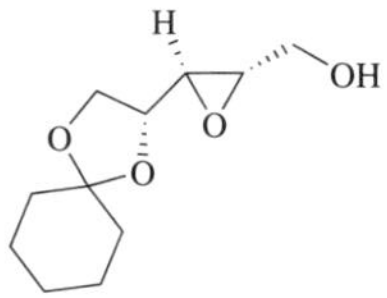	TEMPO (cat.), $PhI(OAc)_2$, CH_2Cl_2, rt	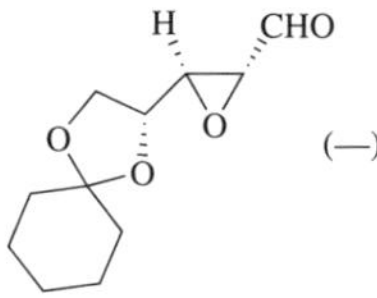 (—)[b]	206

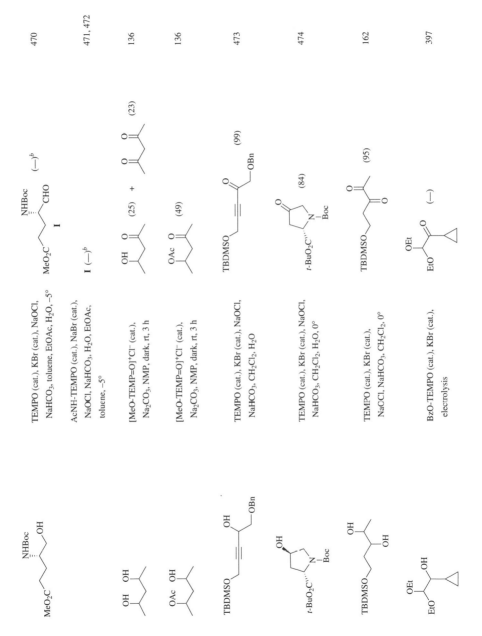

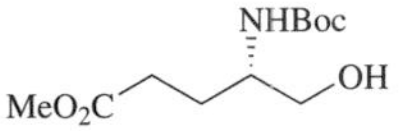	TEMPO (cat.), KBr (cat.), NaOCl, $NaHCO_3$, toluene, EtOAc, H_2O, –5°	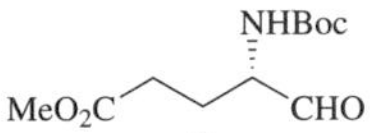 **I** (—)[b]	470
	AcNH-TEMPO (cat.), NaBr (cat.), NaOCl, $NaHCO_3$, H_2O, EtOAc, toluene, –5°	**I** (—)[b]	471, 472
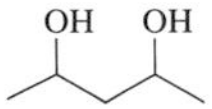	$[MeO\text{-}TEMP{=}O]^+Cl^-$ (cat.), Na_2CO_3, NMP, dark, rt, 3 h	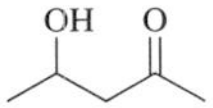 (25) + 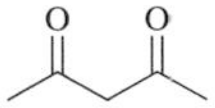 (23)	136
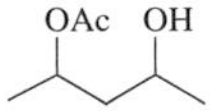	$[MeO\text{-}TEMP{=}O]^+Cl^-$ (cat.), Na_2CO_3, NMP, dark, rt, 3 h	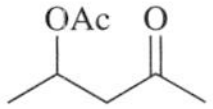 (49)	136
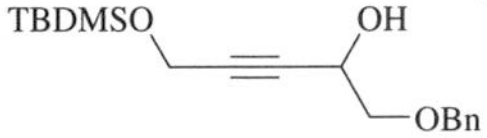	TEMPO (cat.), KBr (cat.), NaOCl, $NaHCO_3$, CH_2Cl_2, H_2O	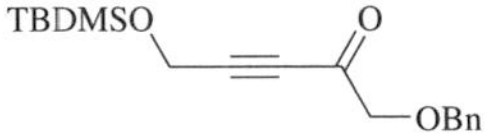 (99)	473
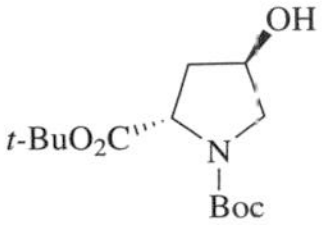	TEMPO (cat.), KBr (cat.), NaOCl, $NaHCO_3$, CH_2Cl_2, H_2O, 0°	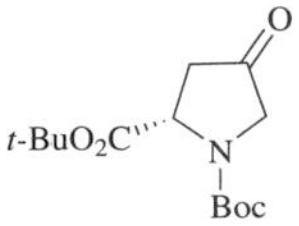 (84)	474
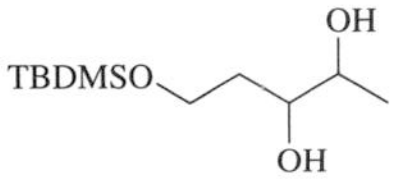	TEMPO (cat.), KBr (cat.), NaCCl, $NaHCO_3$, CH_2Cl_2, 0°	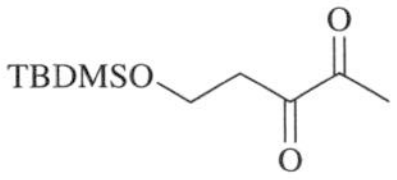 (95)	162
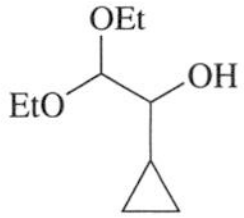	BzO-TEMPO (cat.), KBr (cat.), electrolysis	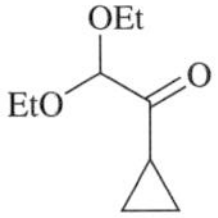 (—)	397

TABLE 3. OXIDATIONS OF POLYFUNCTIONAL ALCOHOLS (*Continued*)

Alcohol			Conditions	Product(s) and Yield(s) (%)			Refs.
C_5 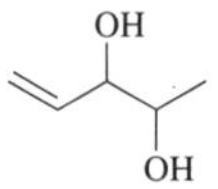			AcNH-TEMPO, TsOH, CH_2Cl_2	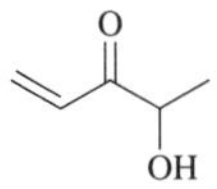 (98)			427
C_{5-12} 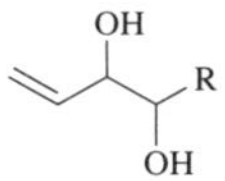			AcNH-TEMPO (5 eq), TsOH (5 eq), CH_2Cl_2, 0°, 4–7 d	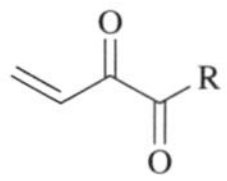	R		427
					Me	(16)	
					i-Pr	(28)	
					n-Bu	(51)	
					n-C_6H_{13}	(50)	
					CH_2Bn	(52)	
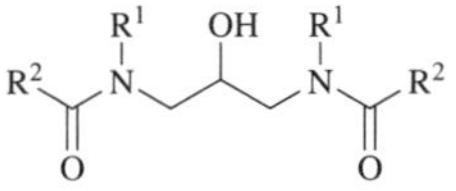			TEMPO (cat.), KBr (cat.), NaOCl, $NaHCO_3$, CH_2Cl_2, H_2O, pH 8	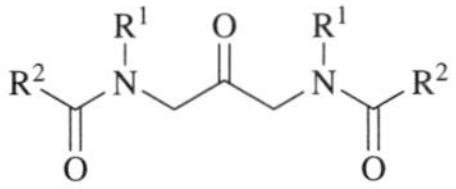			475
	R^1	R^2					
	Bn	OEt		(93)			
	Bn	O*t*-Bu		(93)			
	Bn	OBn		(85)			
	Bn	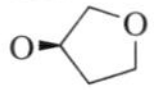		(88)			
	Bn	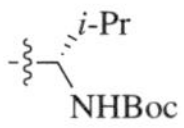		(98)			
	PMB	OCH_2*i*-Pr		(90)			
	H	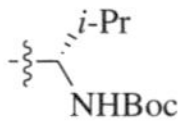		(97)			
	Bn	OCH_2*i*-Pr		(90)			
	Bn	N(Me)Ph		(91)			
	PMB	N(Me)Ph		(93)			

Substrate	Conditions	Product(s) and Yield(s) (%)	Refs.
C_6			
	TEMPO (cat.), laccase (cat.), O_2, H_2O, citrate buffer, pH 4.5	 (19)[a]	249
	TEMPO (cat.), $H_5PV_2Mo_{10}O_{40}$•34 H_2O (cat.), O_2 (2 atm), acetone	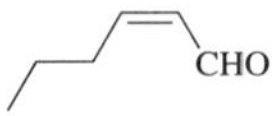 (100)[a]	274
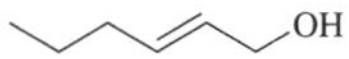	TEMPO (cat.), $H_5PV_2Mo_{10}O_{40}$•34 H_2O (cat.), O_2 (2 atm), acetone	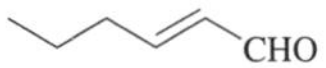 (100)[a]	274
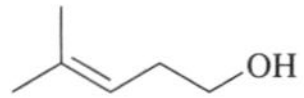	Nitroxide (cat.), $PhI(OAc)_2$, pentane, CH_2Cl_2, rt	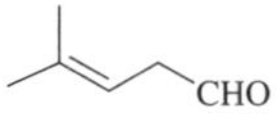 Nitroxide: TEMPO (100)[e]; SG-O-TEMPO (—)	208
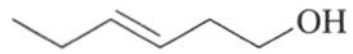	Nitroxide (cat.), $PhI(OAc)_2$, pentane, CH_2Cl_2, rt	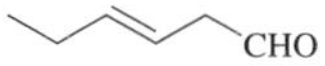 Nitroxide: TEMPO (100)[e]; SG-O-TEMPO (—)	208
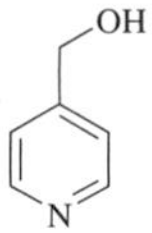	$[AcNH\text{-}TEMP{=}O]^+[ClO_4]^-$, silica gel, CH_2Cl_2, rt	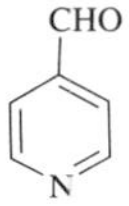 **I** (90)	455
	BzO-TEMPO (cat.), $TBABr_3$, NaOAc, CH_2Cl_2, H_2O, rt, 1 h	**I** (94)	224

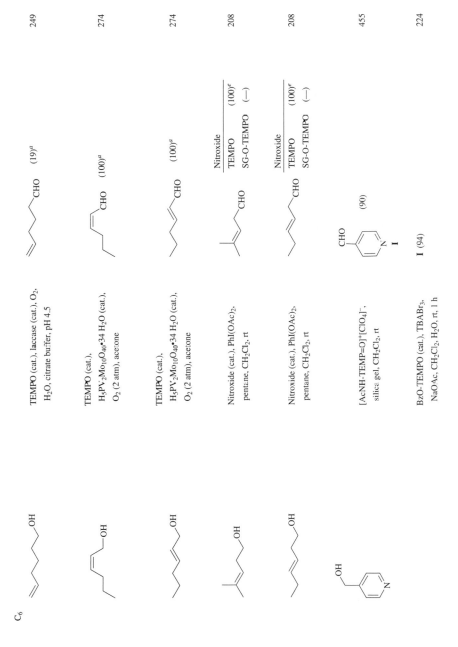

TABLE 3. OXIDATIONS OF POLYFUNCTIONAL ALCOHOLS (*Continued*)

Alcohol	Conditions	Product(s) and Yield(s) (%)	Refs.
C_6 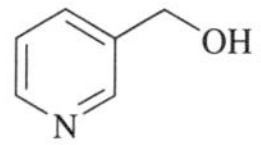	TEMPO (cat.), KBr (cat.), NaOCl, pH 9.5, 0–15°	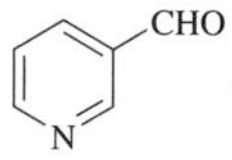 **I** (75)	13
	SBA15-NH-TEMPO (cat.), $NaNO_2$ (cat.), TBAB (cat.), O_2 (1 atm), AcOH, 50–60°, 1.5 h	**I** (92)[a], 99% selectivity[f]	310
	$[AcNH\text{-}TEMP{=}O]^+[ClO_4]^-$, silica gel, CH_2Cl_2, rt	**I** (88)	455
	TEMPO (cat.), $NaNO_2$ (cat.), Br_2 (cat.), air (3.9 atm), CH_2Cl_2, 80°	**I** (98)	258
	TEMPO (cat.), 1,3-dibromo-5,5-dimethylhydantoin (cat.), air, H_2O, 80°, 1 h	**I** (92)	259
	TEMPO (cat.), I_2, $NaHCO_3$, toluene or CH_2Cl_2, 20°, overnight	**I** (86)[c]	256
	PEG-O-TEMPO (cat.), $CuBr_2$ (cat.), PEG-bipy (cat.), O_2, *t*-BuOK, MeCN, H_2O, 80°, 18 h	**I** (80)	392

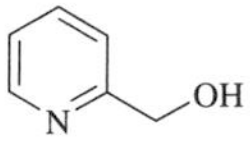	[AcNH-TEMP=O]$^+$[ClO$_4$]$^-$, silica gel, CH_2Cl_2, rt	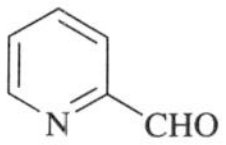 I (85)	455
	CIBA-TEMPO (cat.), $Mn(NO_3)_2$ or $Co(NO_3)_2$ (cat.), air or O_2, AcOH, 20–40°	**I** (98)	395
	TEMPO (cat.), **3** (cat.), O_2, toluene, 100[c]	**I** (92)	244
	AcNH-TEMPO (cat.), **4** (cat.), $Cu(ClO_4)_2$ (cat.), O_2, DABCO, DMSO	**I** (71)	245
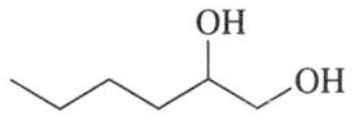	Ormosil-TEMPO (cat.), NaOCl, $NaHCO_3$, MeCN, 0°	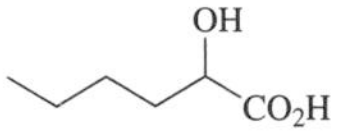 (79)	394
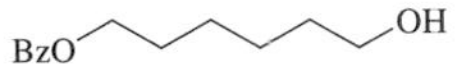	[MeO-TEMP=O]$^+$Cl$^-$, CH_2Cl_2, rt	 (58)[a]	69, 435
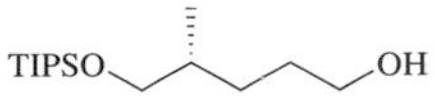	TEMPO (cat.), KBr (cat.), NaOCl, Na_2CO_3, CH_2Cl_2, H_2O, pH 8.6, 0°, 15 min	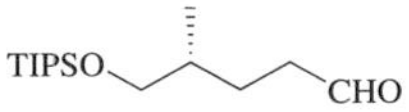 (—)	476
	[AcNH-TEMP=O]$^+$[BF$_4$]$^-$, pyridine, 4 Å MS, CH_2Cl_2, rt	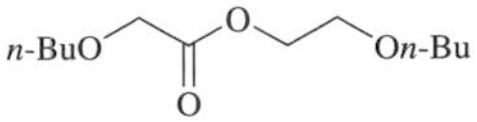 (95)	77
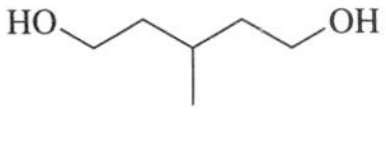	Graphite felt-CONH-AS-TEMPO (cat.), electrolysis, $NaClO_4$, MeCN	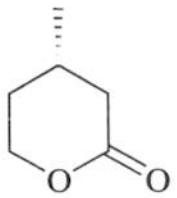 **I** (96), 98% ee[a]	83
	Graphite felt-TEMPO (cat.), (–)-sparteine (cat.), electrolysis, $NaClO_4$, MeCN	**I** (94)[a], 98% ee[a]	451

TABLE 3. OXIDATIONS OF POLYFUNCTIONAL ALCOHOLS (*Continued*)

	Alcohol	Conditions	Product(s) and Yield(s) (%)	Refs.
C_6	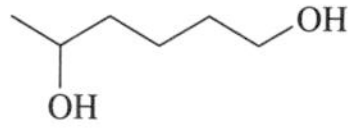	TEMPO (cat.), $RuCl_2(PPh_3)_3$ (cat.), O_2 (10 atm), chlorobenzene, 100°	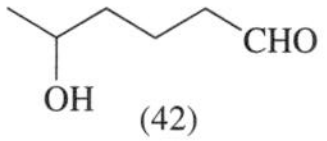 (42) + 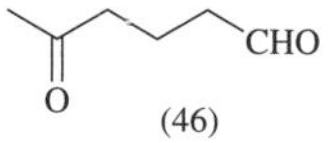 (46)	80, 409, 410
		TEMPO (cat.), $PhI(OAc)_2$, MeCN, H_2O, pH 7 buffer, 0°	 **I** (70)	15
		TEMPO (cat.), $PhI(OAc)_2$, CH_2Cl_2, rt	**I** (70)	192
		TEMPO (cat.), $PhI(OAc)_2$, CH_2Cl_2, rt, 3 h	**I** (60)	477
	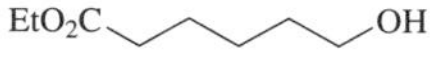	TEMPO (cat.), $CuBr{\bullet}Me_2S$ (cat.), **1** (cat.), O_2, perfluorooctyl bromide, chlorobenzene, 90° 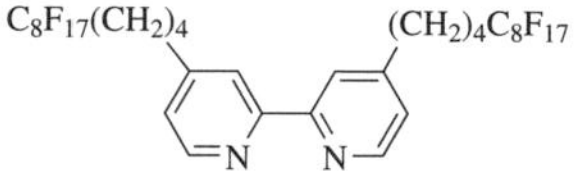 **1**	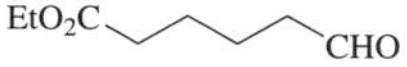 (89)	243
	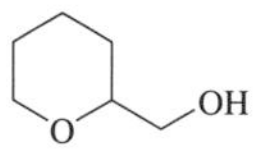	$[AcNH\text{-}TEMP{=}O]^+[BF_4]^-$, pyridine, 4 Å MS, CH_2Cl_2, rt	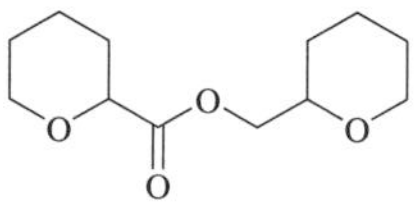 (95)	77
	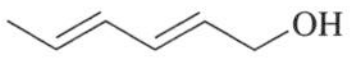	$[AcNH\text{-}TEMP{=}O]^+[BF_4]^-$, CH_2Cl_2, silica gel, rt, 16–24 h	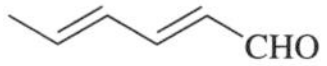 **I** (76)	131
		AcNH-TEMPO (cat.), **4** (cat.), $Cu(ClO_4)_2$ (cat.), O_2, DABCO, DMSO	**I** (85)	245
	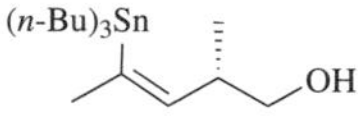	TEMPO (cat.), $PhI(OAc)_2$, CH_2Cl_2, rt, 3 h	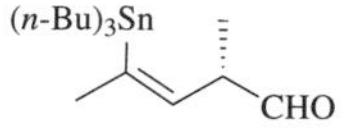 (—)[b]	417

	TEMPO (cat.), $PhI(OAc)_2$, CH_2Cl_2	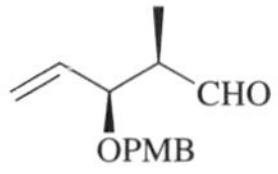 (95)	478
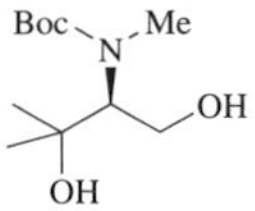	TEMPO (cat.), NaOCl (cat.), $NaClO_2$, MeCN, H_2O, phosphate buffer	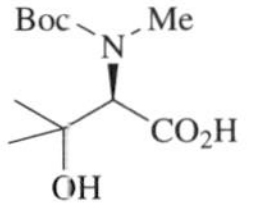 (87)	479
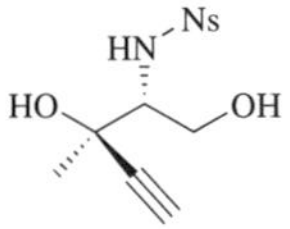	TEMPO (cat.), NaOCl (cat.), $NaClO_2$, NaOH, MeCN, H_2O, phosphate buffer, pH 6.7	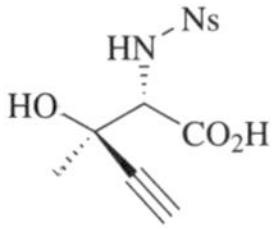 $(—)^b$	480
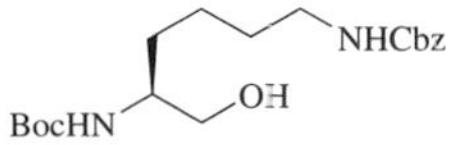	TEMPO (cat.), NaBr (cat.), NaOCl, $NaHCO_3$, toluene, EtOAc, H_2O, 0°	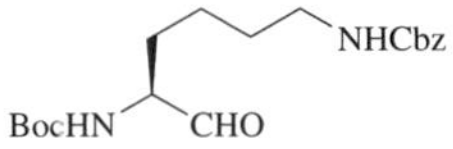 (89), 99% ee	158
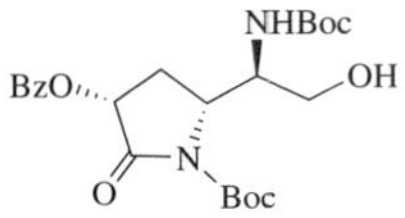	TEMPO (cat.), $PhI(OAc)_2$, MeCN, H_2O, rt	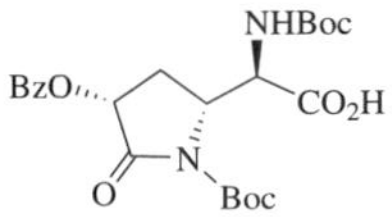 $(—)^b$	481
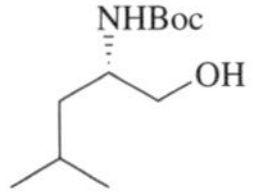	TEMPO (cat.), NaOCl, $NaHCO_3$, CH_2Cl_2, H_2O	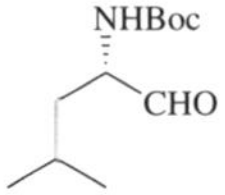 $(—)^b$	482
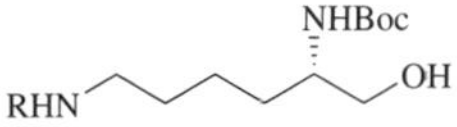	AcNH-TEMPO (cat.), NaBr (cat.), NaOCl, $NaHCO_3$, EtOAc, toluene, H_2O, –5°	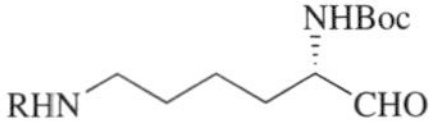 R: Boc $(—)^b$; Cbz $(—)^b$	471, 472

TABLE 3. OXIDATIONS OF POLYFUNCTIONAL ALCOHOLS (*Continued*)

Alcohol	Conditions	Product(s) and Yield(s) (%)	Refs.
C_6			
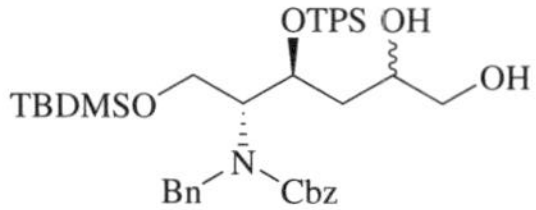	TEMPO (cat.), NaBr (cat.), NaOCl, toluene, EtOAc, H_2O, 0°	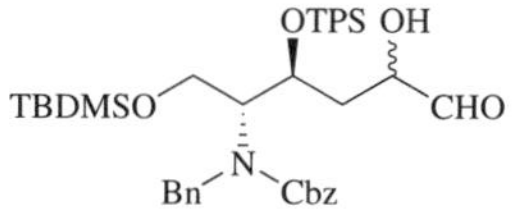 (—)[b]	483
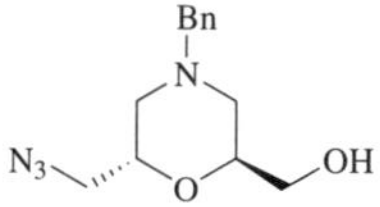	TEMPO (cat.), $PhI(OAc)_2$, CH_2Cl_2, H_2O, 0°, 6 h	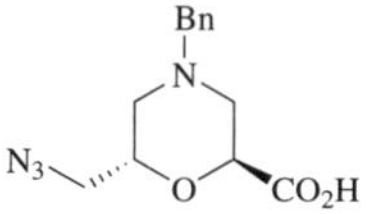 (68)	484
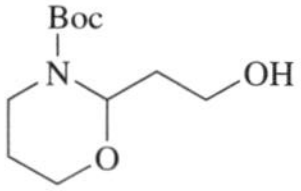	TEMPO (cat.), NaBr (cat.), NaOCl, TBAB, $NaHCO_3$, CH_2Cl_2, 0°, 30 min	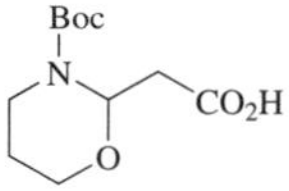 (85)	434
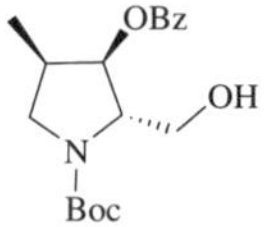	TEMPO (cat.), $PhI(OAc)_2$, MeCN, H_2O, rt, 1 h	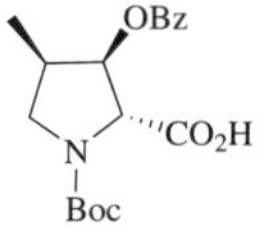 (91)	485
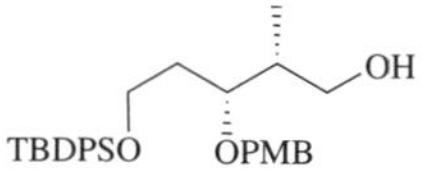	TEMPO (cat.), $PhI(OAc)_2$, CH_2Cl_2	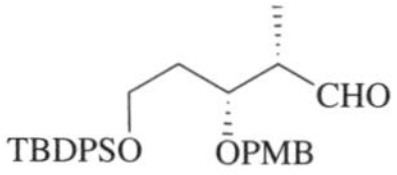 (86)	478
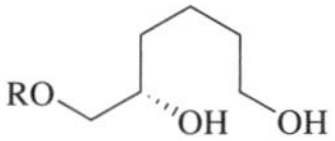	TEMPO (cat.), $PhI(OAc)_2$, CH_2Cl_2, rt	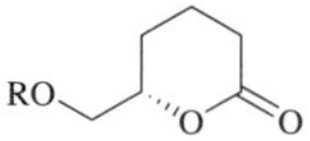 R Bn (96) PMB (77) TBDPS (94)	486
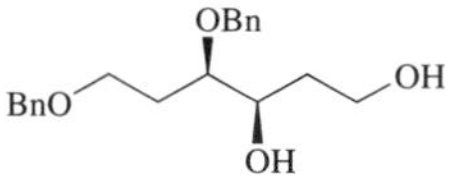	TEMPO (cat.), NaOCl, CH_2Cl_2, H_2O	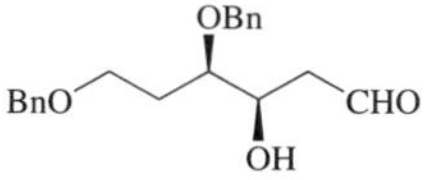 (—)[b]	487

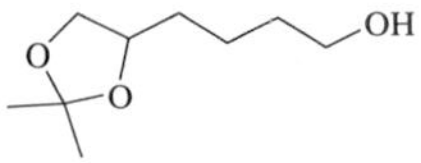	HO-TEMPO (cat.), CuCl (cat.), O_2, DMF, rt	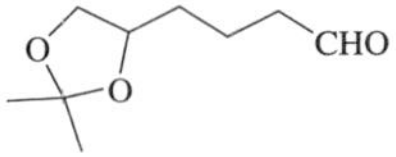 (74)	81
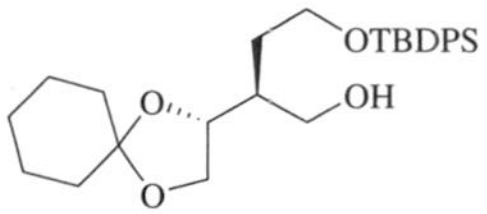	TEMPO (cat.), $PhI(OAc)_2$, CH_2Cl_2, rt, 1.5 h	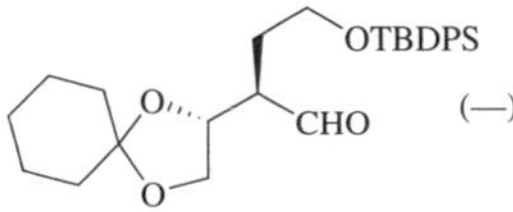 (—)[b]	488
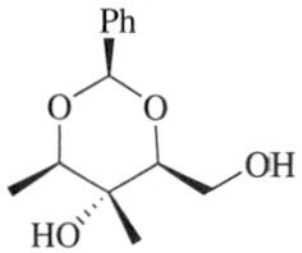	TEMPO (cat.), KBr (cat.), NaOCl, $NaHCO_3$ (satd), CH_2Cl_2, H_2O	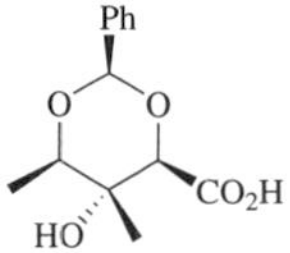 (52)	489
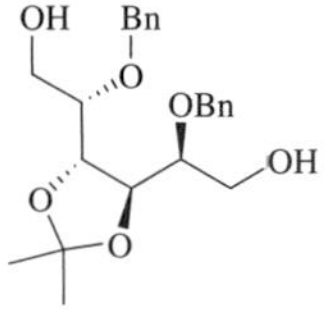	TEMPO (cat.), KBr (cat.), NaOCl, TBAB, $NaHCO_3$, NaCl, CH_2Cl_2, H_2O, 0°	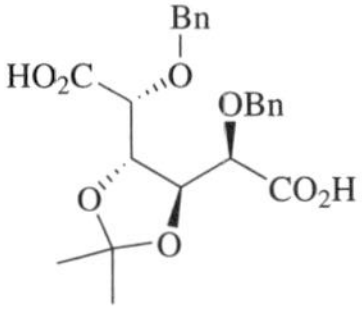 (—)[b]	490
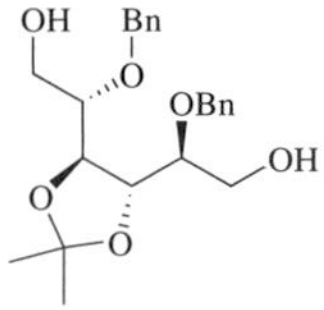	TEMPO (cat.), KBr (cat.), NaOCl, TBAB, $NaHCO_3$, CH_2Cl_2, 0°	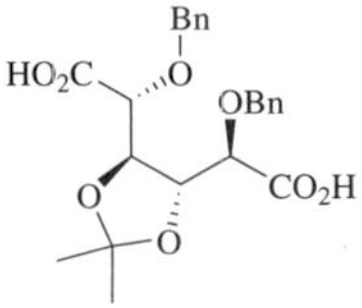 (67)	491
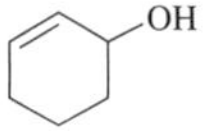	TEMPO (cat.), CuCl (cat.), O_2, $[bmim]^+[PF_6]^-$, 65°	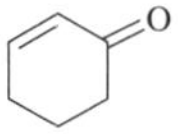 **I** (75)	385
	TEMPO (cat.), $(NH_4)_2[Ce(NO_3)_6]$ (cat.), O_2, MeCN, reflux	**I** (0–86)	269, 385

TABLE 3. OXIDATIONS OF POLYFUNCTIONAL ALCOHOLS (*Continued*)

	Alcohol	Conditions	Product(s) and Yield(s) (%)	Refs.
C_6	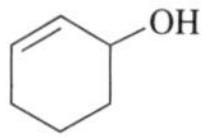	AcNH-TEMPO (cat.), **4** (cat.), $Cu(ClO_4)_2$ (cat.), O_2, DABCO, DMSO	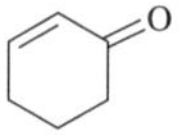 (74) **I**	245
		Graphite felt-TEMPO (cat.), (–)-sparteine (cat.), electrolysis, MeCN, $NaClO_4$	**I** (50)[a] + 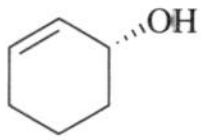 (48)[a], 99.8% ee, S = 116	105
	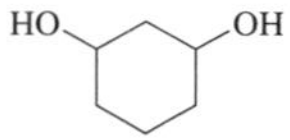	TEMPO (cat.), $PhI(OAc)_2$, CH_2Cl_2, rt	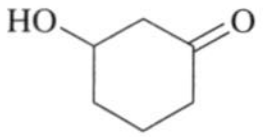 (90)	192
	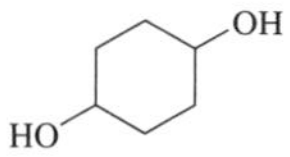	[mim-O-TEMPO]$^+$[BF_4]$^-$ (cat.), [mim-$CH_2PhI(OAc)_2$]$^+$[BF_4]$^-$, H_2O, 30°, 6 h	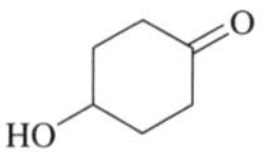 (86)	201
	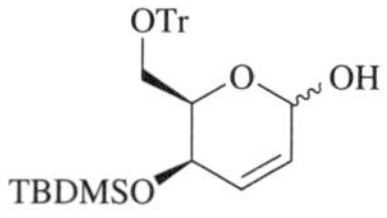	TEMPO (cat.), NCS, TBAC, CH_2Cl_2, H_2O, pH 8.6 buffer, rt, 2 h	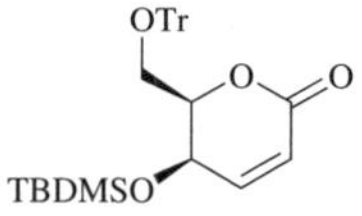 (93)	492
	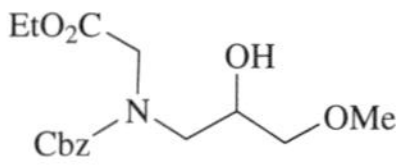	BzO-TEMPO (cat.), KBr (cat.), NaOCl, $NaHCO_3$, CH_2Cl_2, H_2O, 5°	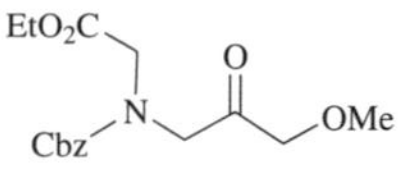 (99)	493
	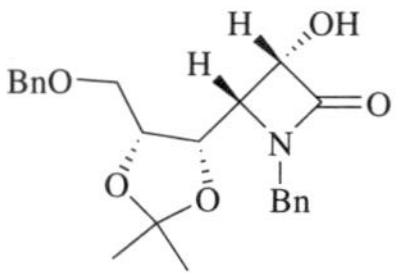	TEMPO (cat.), KBr (cat.), NaOCl, $NaHCO_3$, CH_2Cl_2, phosphate buffer, pH 7.0, 0°, 1 min	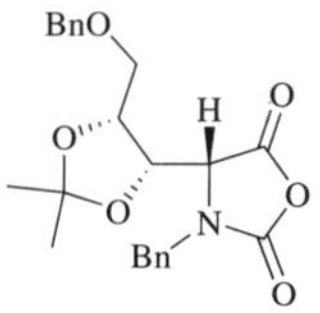 (95)	494, 495

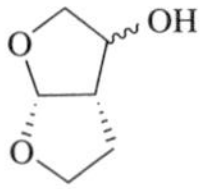 racemic	TEMPO (cat.), KBr (cat.), NaOCl, TBAC, $NaHCO_3$, CH_2Cl_2, H_2O, 0°	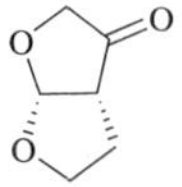 **I** (72–76)	169
	TEMPO (cat.), TCC, CH_2Cl_2, 0° to rt, 15 min	**I** (60–70)	169
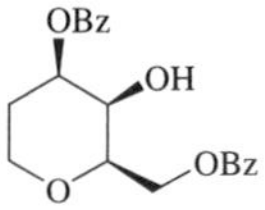	TEMPO (cat.), $PhI(OAc)_2$, CH_2Cl_2, 48 h	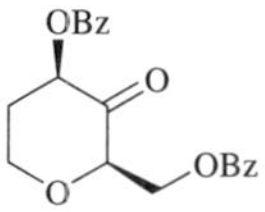 (70)	496
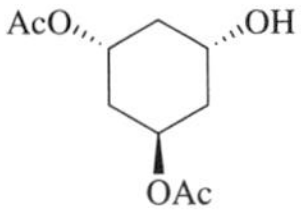	TEMPO (cat.), NaOCl, HCl, CH_2Cl_2, H_2O, pH 6.5–7.5, 0°, 1 h	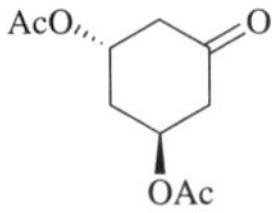 (99)	497
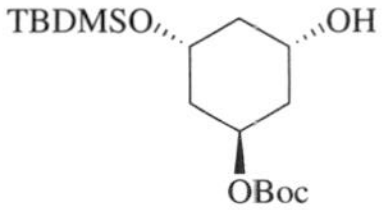	TEMPO (cat.), NaOCl, HCl, CH_2Cl_2, H_2O, pH 6.5–7.5, 0°, 1 h	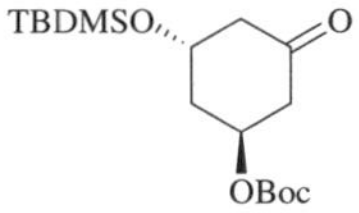 (100)	497
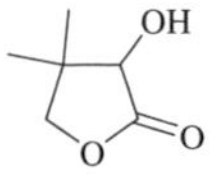	TEMPO (cat.), TCC, NaOAc, CH_2Cl_2, 0–5°	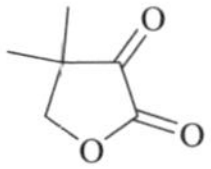 (86)	408
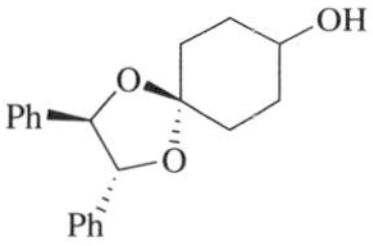	TEMPO (cat.), KBr (cat.), NaOCl, CH_2Cl_2, H_2O, pH 9.5, 0–15°	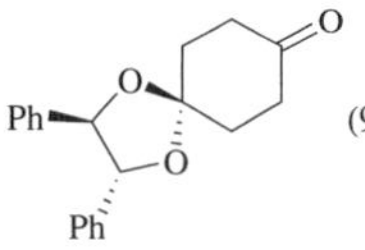 (96)	498

TABLE 3. OXIDATIONS OF POLYFUNCTIONAL ALCOHOLS (*Continued*)

Alcohol	Conditions	Product(s) and Yield(s) (%)	Refs.
C_{6-7} 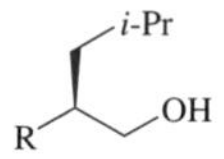	TEMPO (cat.), NaOCl, conditions	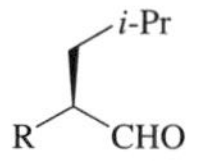	
R	Conditions		
MeO	$NaHCO_3$, CH_2Cl_2, H_2O	(79)	499
BocHN	NaBr (cat.), $NaHCO_3$, toluene, EtOAc, H_2O, 0°	(77), 99% ee	158
BnO	NaBr (cat.), $NaHCO_3$, toluene, EtOAc, H_2O, 0°	(77), 98% ee	158
C_{6-11} 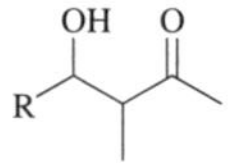	BzO-TEMPO (cat.), KBr (cat.), NaOCl, $NaHCO_3$, CH_2Cl_2, 5°, 2 h	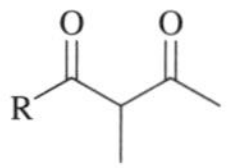 R: Me (60); Et (65); *n*-Pr (83); *n*-Bu (86); *n*-C_5H_{11} (85); *n*-C_6H_{13} (83)	500
C_{6-12} 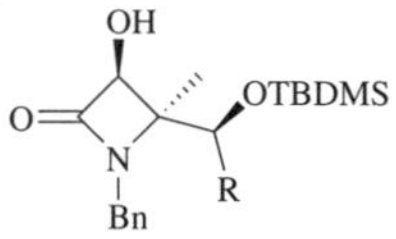	TEMPO (cat.), NaOCl, CH_2Cl_2, H_2O, 0°, 5 min	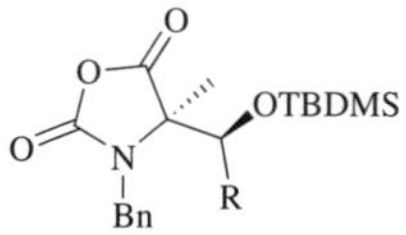 (up to 95) R: Me; *i*-Pr; Bn	501
C_{6-13} 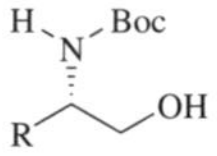	AcNH-TEMPO (cat.), NaBr (cat.), NaOCl, $NaHCO_3$, EtOAc, toluene, H_2O, –5°	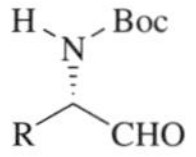 (—)[b] R: *n*-Bu; *i*-$PrCH_2$; Bn; *n*-$C_{10}H_{21}OCH_2$	430

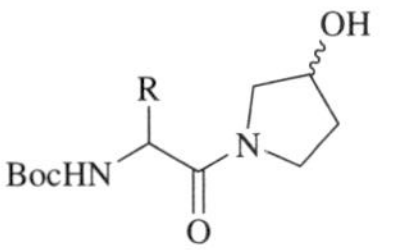	TEMPO (cat.), NaOCl (cat.), $NaHCO_3$, CH_2Cl_2, H_2O, pH 8.2	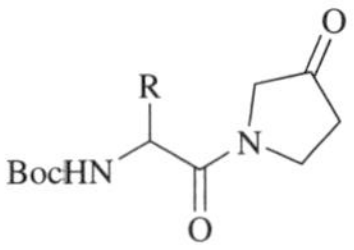 R: H (—)[b]; i-$PrCH_2$ (100); Bn (—)[b]; $(4\text{-}MeOC_6H_4)CH_2$ (60)	502
C_7			
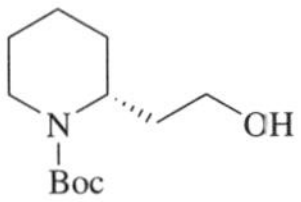	TEMPO (cat.), laccase (cat.), O_2, NaOAc, EtOAc, H_2O, pH 4–5, rt, 2–4 h	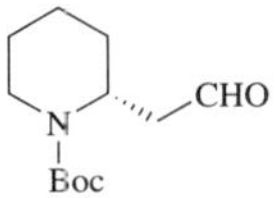 (90)	503
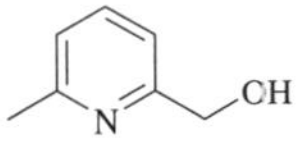	$[AcNH\text{-}TEMP{=}O]^+[ClO_4]^-$, silica gel, CH_2Cl_2, rt	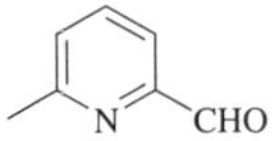 (96)	455
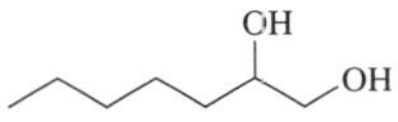	TEMPO (cat.), KBr (cat.), NaOCl, TBAF, CH_2Cl_2, H_2O, 0°	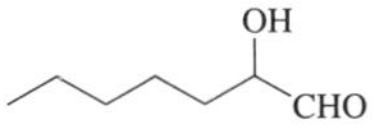 (—)[b]	504
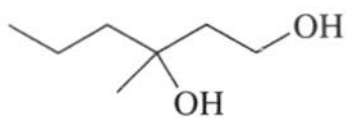	TEMPO (cat.), $Yb(OTf)_3$ (cat.), $PhI(OAc)_2$, CH_2Cl_2, 30 min	 **I** (71)	416
	TEMPO (cat.), $PhI(OAc)_2$, CH_2Cl_2, rt	**I** (—)[b]	205
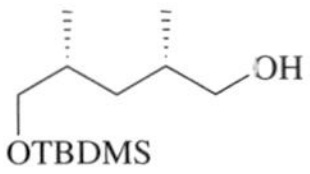	TEMPO (cat.), $PhI(OAc)_2$, CH_2Cl_2, rt	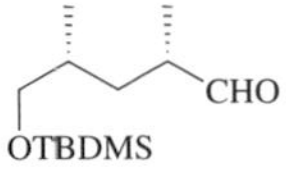 (—)[b]	447
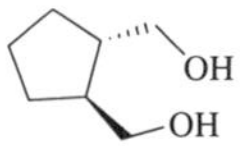	TEMPO (cat.), NaOCl (cat.), $NaClO_2$, MeCN, H_2O	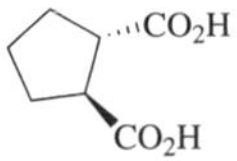 (83)	505

TABLE 3. OXIDATIONS OF POLYFUNCTIONAL ALCOHOLS (*Continued*)

	Alcohol	Conditions	Product(s) and Yield(s) (%)	Refs.
C_7	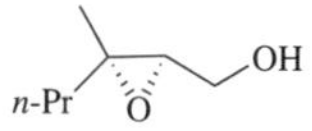	TEMPO (cat.), $PhI(OAc)_2$, CH_2Cl_2, rt	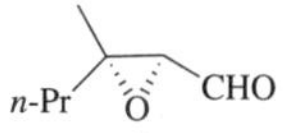 **I** (—)[b]	506
		TEMPO (cat.), $Yb(OTf)_3$ (cat.), $PhI(OAc)_2$, CH_2Cl_2, 45 min	**I** (76)	416
		TEMPO (cat.), $PhI(OAc)_2$, CH_2Cl_2, rt	**I** (—)[b]	205
		$[AcNH\text{-}TEMP{=}O]^+[BF_4]^-$, silica gel, CH_2Cl_2, rt, 16–24 h	 (85)	131
	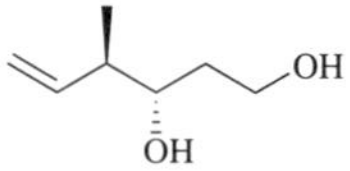	TEMPO (cat.), NaBr (cat.), NaOCl, $NaHCO_3$, CH_2Cl_2, H_2O, pH 9.9–9.2	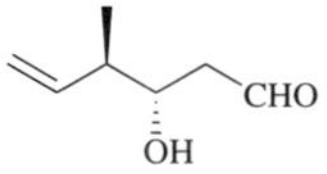 (100)	137
	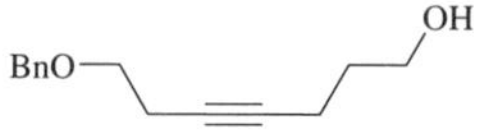	TEMPO (cat.), NaOCl (cat.), $NaClO_2$, MeCN, H_2O, phosphate buffer, pH 6.86, 35°	 (84)	507, 508
	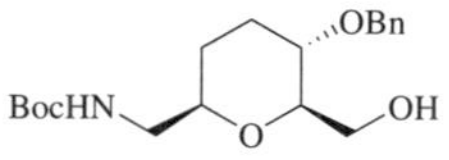	TEMPO (cat.), $PhI(OAc)_2$, CH_2Cl_2, H_2O	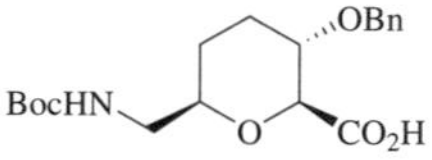 (100)	509
	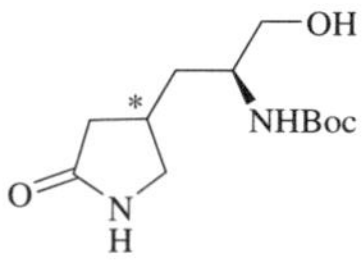	TEMPO (cat.), NaOCl (cat.), $NaClO_2$, MeCN, 35°	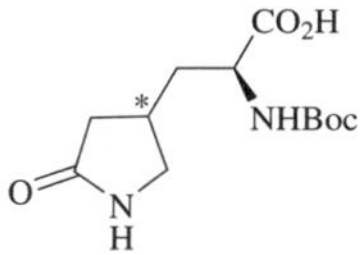 *: *R* (80); *S* (80)	510
	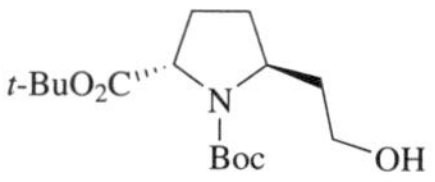	TEMPO (cat.), NaBr (cat.), NaOCl, $NaHCO_3$, toluene, H_2O, 0°	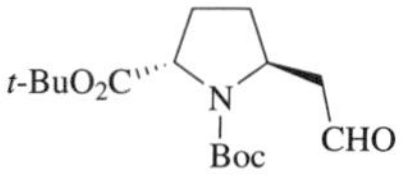 (—)[b]	511

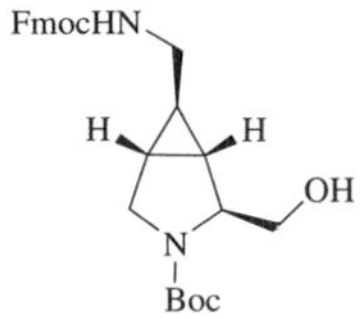	TEMPO (cat.), NaOCl (cat.), $NaClO_2$, NaH_2PO_4, MeCN, H_2O	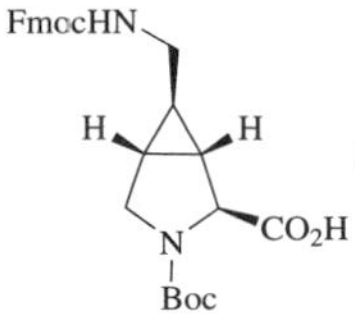 (90)	512
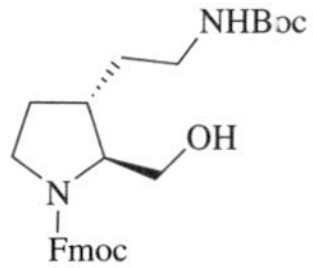	TEMPO (cat.), NaOCl (cat.), $NaClO_2$, NaH_2PO_4, MeCN, H_2O	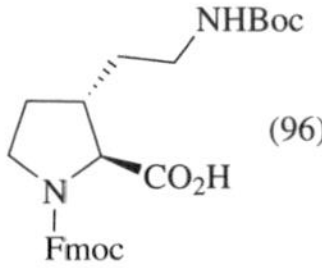 (96)	513
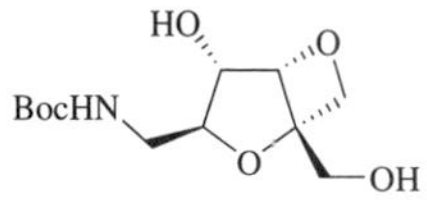	TEMPO (cat.), KBr (cat.), NaOCl, $NaHCO_3$, NaCl, H_2O, 0°, 4 h	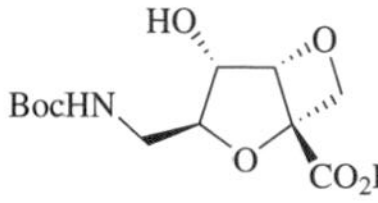 (70)	514
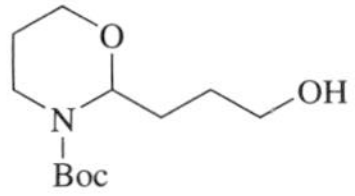	TEMPO (cat.), NaBr (cat.), NaOCl, TBAB, $NaHCO_3$, CH_2Cl_2, 0°, 30 min	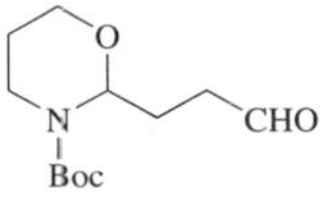 (98)	434
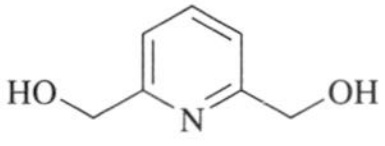	$[AcNH\text{-}TEMP{=}O]^+[ClO_4]^-$, CH_2Cl_2, silica gel, rt	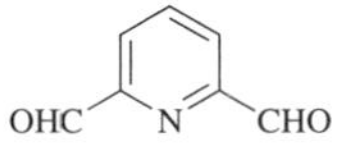 (86)	455
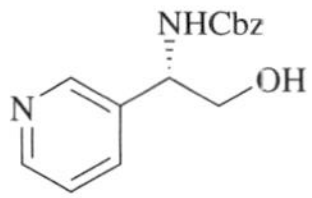	TEMPO (cat.), KBr (cat.), NaOCl, $NaHCO_3$ (5%), H_2O, acetone, 0°, 2 h	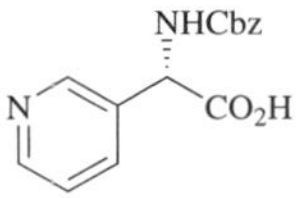 (61)	515
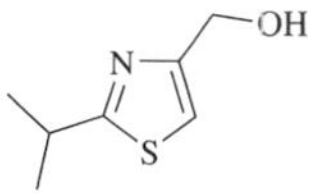	TEMPO (cat.), $[PS\text{-}CH_2NMe_3]^+[Br(OAc)_2]^-$, CH_2Cl_2, 24 h	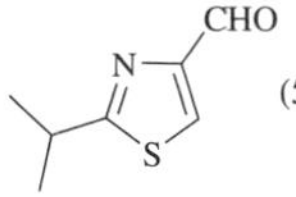 (56)	43

TABLE 3. OXIDATIONS OF POLYFUNCTIONAL ALCOHOLS (*Continued*)

Alcohol	Conditions	Product(s) and Yield(s) (%)	Refs.
C_7			
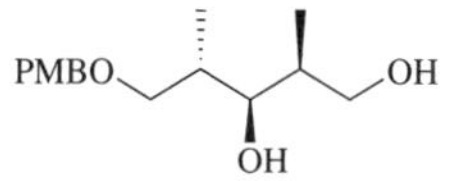	TEMPO (cat.), $PhI(OAc)_2$, CH_2Cl_2	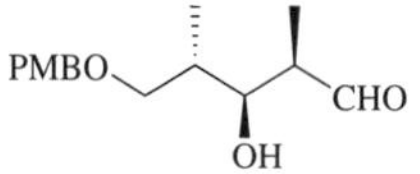 (—)[b]	516–519
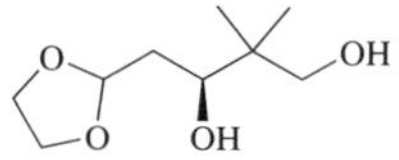	TEMPO (cat.), KBr (cat.), NaOCl, CH_2Cl_2, H_2O, 0°	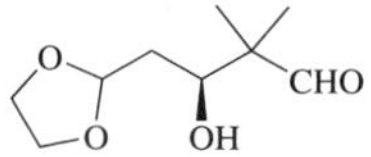 (—)[b]	520
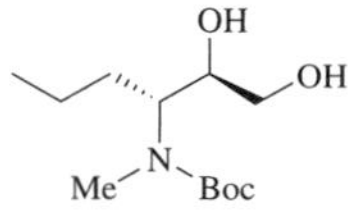	TEMPO (cat.), NaBr (cat.), NaOCl, acetone, H_2O, 0°	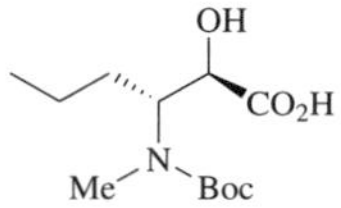 (—)[b]	164
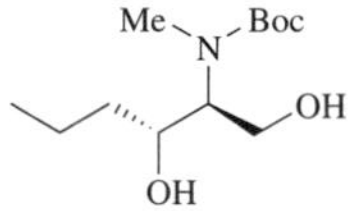	TEMPO or ormosil-TEMPO (cat.), NaBr, NaOCl, H_2O, acetone, 0°	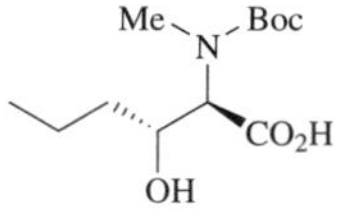 (—)[b]	164
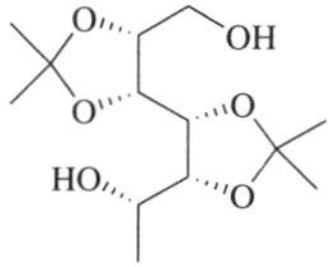	AcNH-TEMPO (cat.), KBr (cat.), NaOCl, TBAI, NaCl, $NaHCO_3$, H_2O, pH ~10, 0°, 2 h	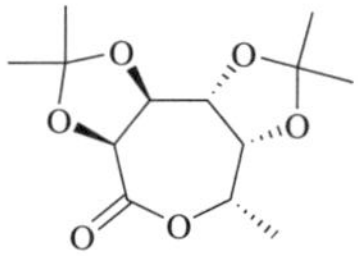 (81)	521
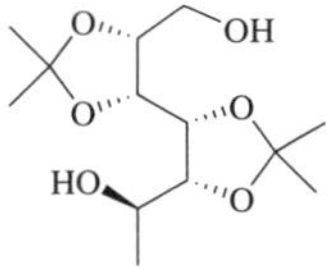	AcNH-TEMPO (cat.), KBr (cat.), NaOCl, TBAI, NaCl, $NaHCO_3$, H_2O, pH ~10, 0°, 2 h	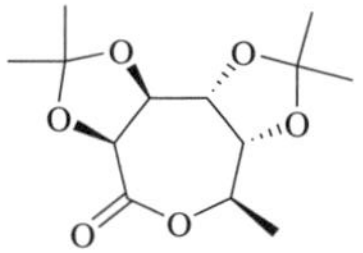 (86)	521

Substrate	Conditions	Product(s) and Yield(s) (%)	Refs.
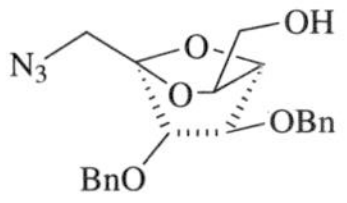	TEMPO (cat.), $PhI(OAc)_2$, CH_2Cl_2, H_2O, 90 min	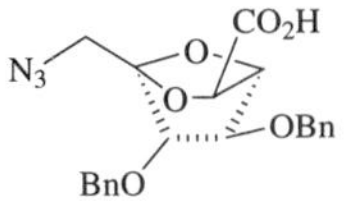 (87)	522
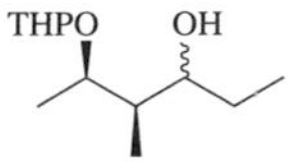	TEMPO (cat.), KBr (cat.), NaOCl, CH_2Cl_2, H_2O, 0°	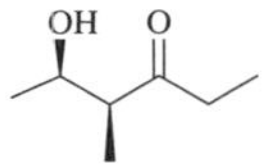 (69)	523
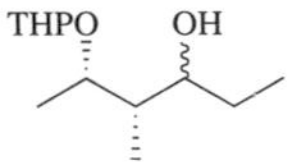	TEMPO (cat.), KBr (cat.), NaOCl, CH_2Cl_2, H_2O, 0°	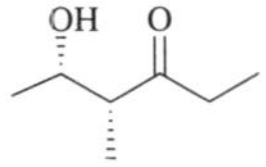 (65)	523
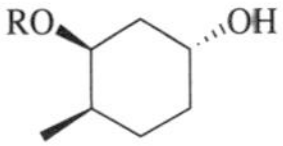	TEMPO (cat.), KBr (cat.), NaOCl, CH_2Cl_2, H_2O, –10°	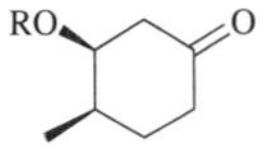 R: Boc (67); TBDMS (87); TBDPS (85)	524
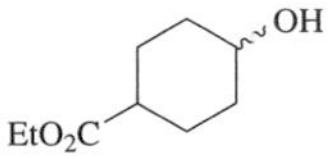	TEMPO (cat.), NaOCl (cat.), KBr (cat.), $NaClO_2$, $NaHCO_3$, CH_2Cl_2, H_2O, 0°	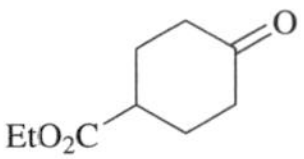 (—)[b]	525
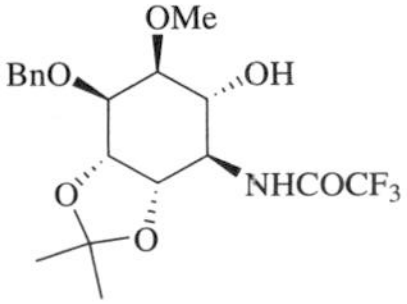	TEMPO (cat.), $NaBrO_2$, $NaHCO_3$, CH_2Cl_2, H_2O, rt	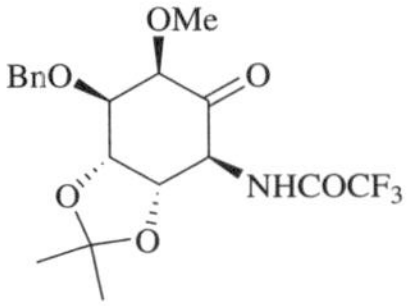 (96)	526
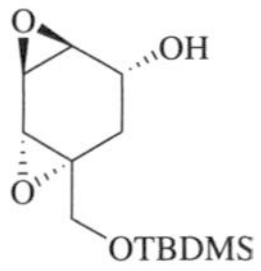	TEMPO (cat.), KBr (cat.), NaOCl, $NaHCO_3$, CH_2Cl_2, H_2O, –10°	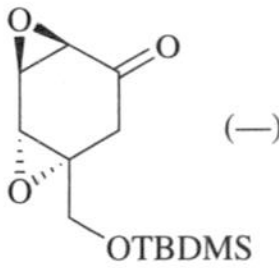 (—)[b]	527

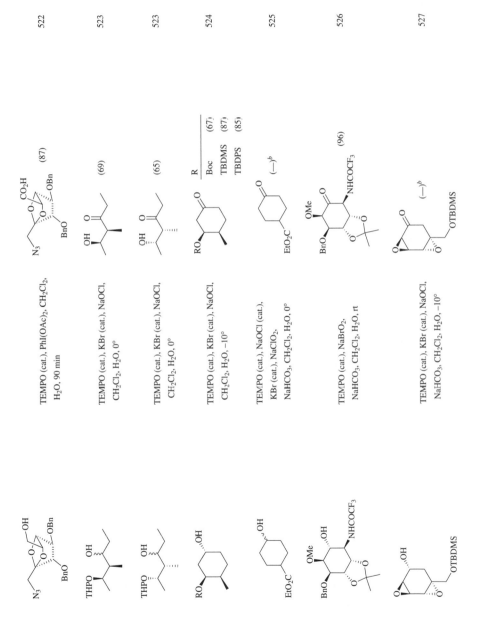

TABLE 3. OXIDATIONS OF POLYFUNCTIONAL ALCOHOLS (*Continued*)

Alcohol	Conditions	Product(s) and Yield(s) (%)	Refs.
C_{7-13} 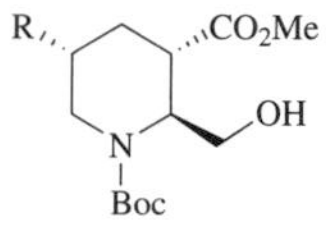	TEMPO (cat.), KBr (cat.), NaOCl, $NaHCO_3$, acetone, H_2O, 4°	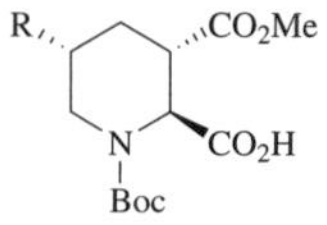 (—)[b] R: H, Me, Ph	528
C_8 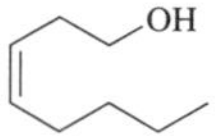	TEMPO (cat.), $PhI(OAc)_2$, CH_2Cl_2, rt	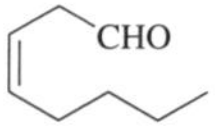 (95)	192
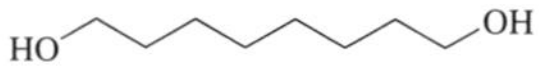	TEMPO (cat.), $PhI(OAc)_2$ (cat.), KNO_2 (cat.), O_2, 80–90°, 15 h	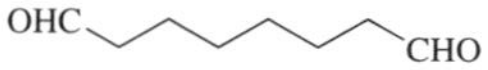 (20)	203
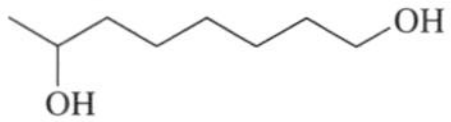	TEMPO (cat.), $PhI(OAc)_2$, CH_2Cl_2, rt	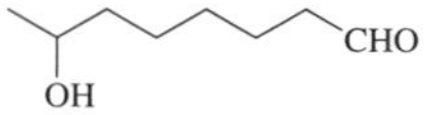 (—)[b]	205
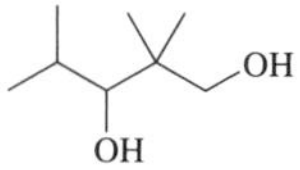	TEMPO (cat.), $PhI(OAc)_2$, CH_2Cl_2, rt	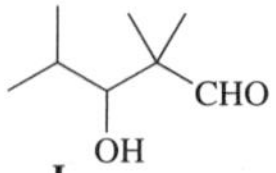 **I** (70)	192
	TEMPO (cat.), $Yb(OTf)_3$ (cat.), $PhI(OAc)_2$, CH_2Cl_2, 7 h	**I** (70)	416
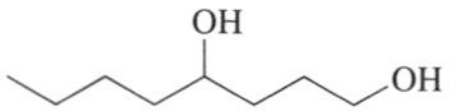	BzO-TEMPO (cat.), NaBr (cat.), electrolysis, silica gel, $NaHCO_3$, H_2O	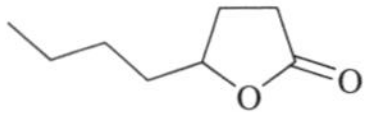 (91)	84
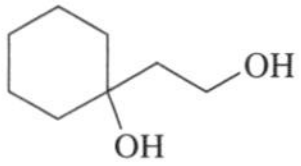	BzO-TEMPO (cat.), NaBr (cat.), electrolysis, $NaHCO_3$, CH_2Cl_2, H_2O, pH 8.6	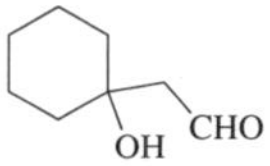 (87)	222
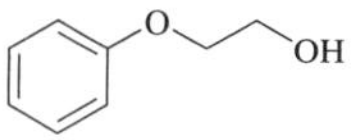	1. TEMPO (cat.), PS-I(OAc)$_2$, acetone, rt 2. H_2O, rt, 24 h	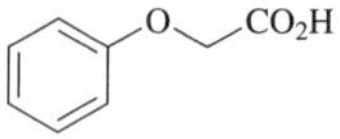 (82)	204

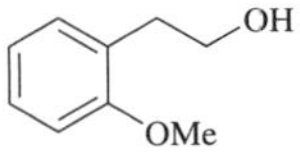	TEMPO (cat.), NaOCl (cat.), $NaClO_2$, MeCN	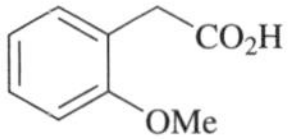 (99)	14, 185
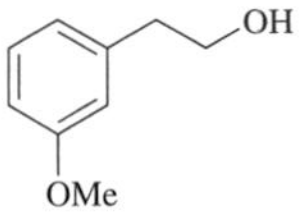	TEMPO (cat.), NaOCl (cat.), $NaClO_2$, MeCN, H_2O	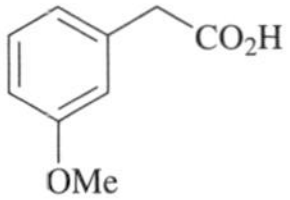 (96)	14, 185
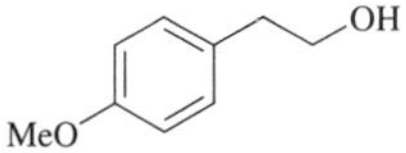	TEMPO (cat.), NaOCl (cat.), $NaClO_2$, MeCN, H_2O	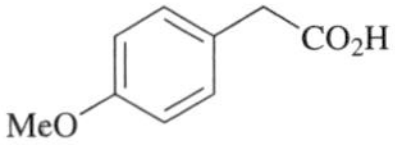 (93)	14, 185
	AcNH-TEMPO, CSA or TsOH, CH_2Cl_2, 0°, 2 h	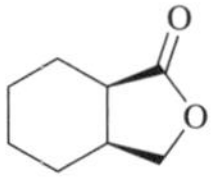 (89) **I** (racemic)	56
	TEMPO (cat.), *m*-CPBA, TBAB, CH_2Cl_2	**I** (89)	213
	BzO-TEMPO (cat.), NaBr (cat.), electrolysis, silica gel, $NaHCO_3$, H_2O	**I** (92)	84
	TEMPO (cat.), $Yb(OTf)_3$ (cat.), $PhI(OAc)_2$, CH_2Cl_2, 1 h	**I** (96)	416
	AS-NHAc-TEMPO (cat.), *m*-CPBA, CH_2Cl_2	**I** (36), 39% ee	85
	(*S*)-Binaphthyl-TEMPO (cat.), NaBr, electrolysis, silica gel, $NaHCO_3$, H_2O, –15°	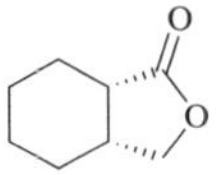 (89), 74% ee **I**	84
	Graphite felt-CONH-AS-TEMPO, electrolysis, $NaClO_4$, MeCN	**I** (92.0), 82% ee[a]	83

TABLE 3. OXIDATIONS OF POLYFUNCTIONAL ALCOHOLS (*Continued*)

Alcohol	Conditions	Product(s) and Yield(s) (%)	Refs.
C_8 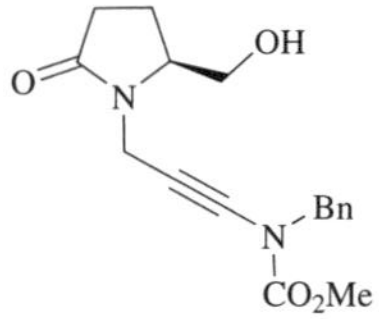	TEMPO (cat.), $PhI(OAc)_2$, CH_2Cl_2, rt	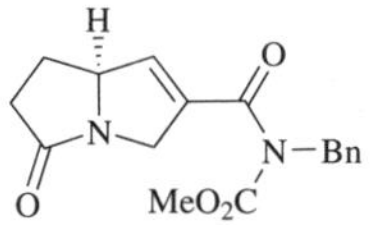 (50)	529
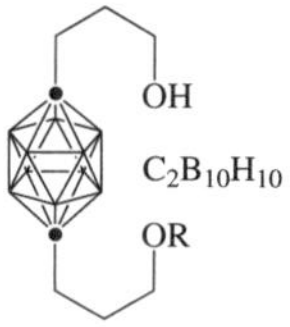	TEMPO (cat.), $PhI(OAc)_2$, CH_2Cl_2, rt	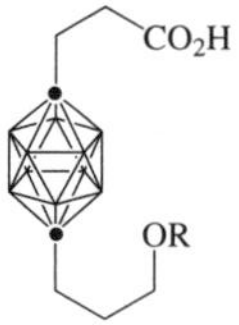 R: TBDMS (85); Bn (91)	530
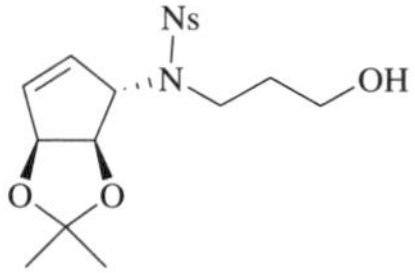	TEMPO (cat.), KBr (cat.), NaOCl, CH_2Cl_2, H_2O, pH 9.5, 0–15°	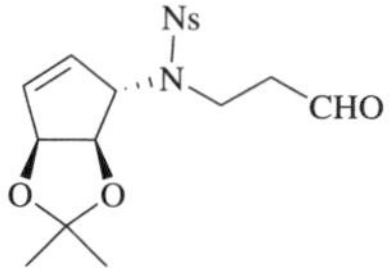 (88)	531
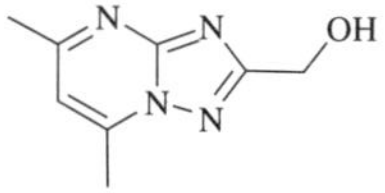	TEMPO (cat.), $PhI(OAc)_2$, CH_2Cl_2	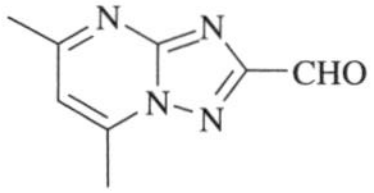 (78)	198
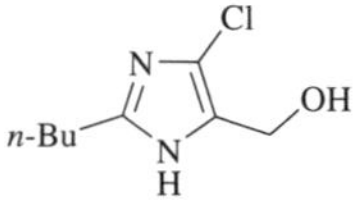	TEMPO (cat.), I_2, $NaHCO_3$, toluene or CH_2Cl_2, 20°, overnight	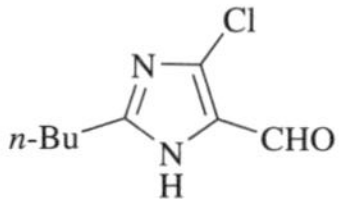 (93)[c]	256

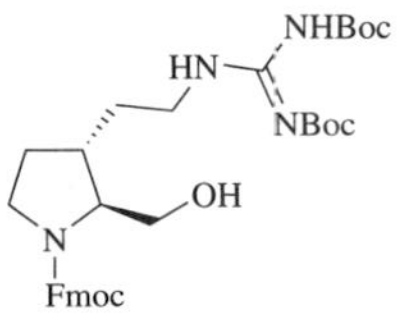	TEMPO (cat.), NaOCl (cat.), $NaClO_2$, NaH_2PO_4, MeCN, H_2O	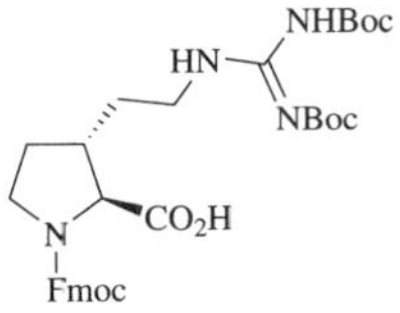	(90)	513
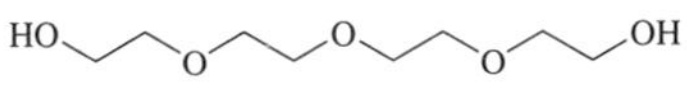	$[PS\text{-}O\text{-}TEMP{=}O]^+Cl^-$, CH_2Cl_2, rt, 1–2 h	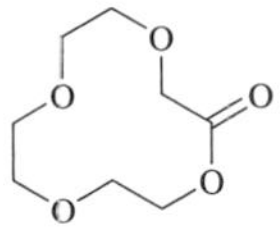	(30)[a]	121
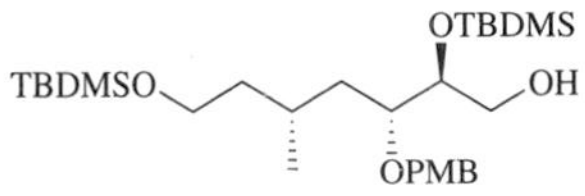	TEMPO (cat.), NaOCl, CH_2Cl_2, H_2O	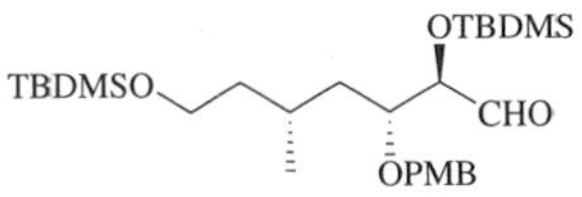	(94)	532
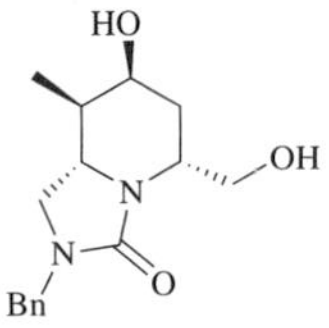	TEMPO (cat.), $PhI(OAc)_2$, $MeSO_3H$, $CDCl_3$	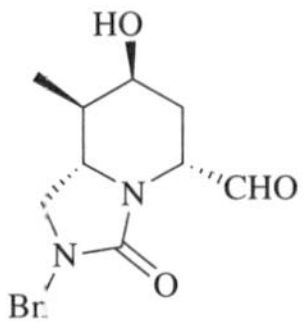	(74)	533
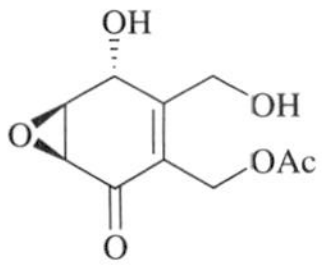	TEMPO (cat.), CuCl (cat.), O_2, DMF, rt	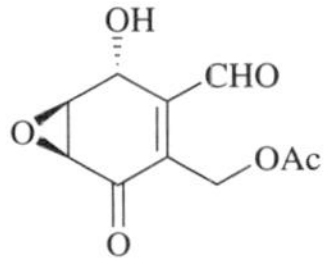	(77)	534
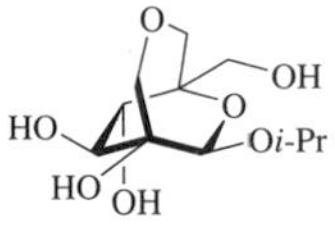	TEMPO (cat.), KBr (cat.), NaOCl, CH_2Cl_2, H_2O, rt	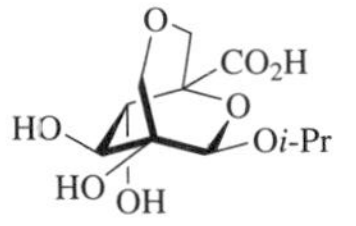	(—)[b]	535

TABLE 3. OXIDATIONS OF POLYFUNCTIONAL ALCOHOLS (*Continued*)

	Alcohol	Conditions	Product(s) and Yield(s) (%)	Refs.
C_8	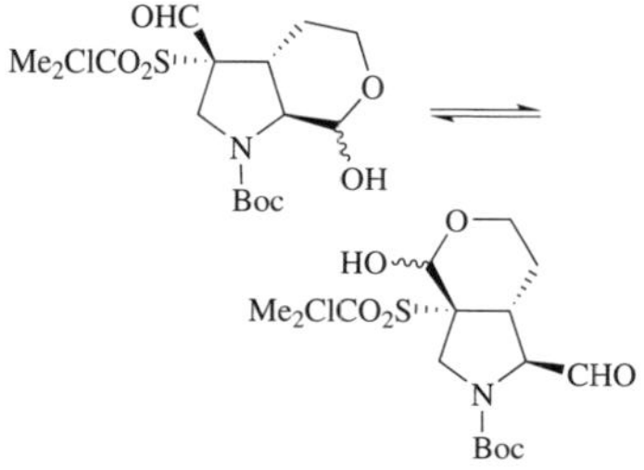	AcNH-TEMPO (cat.), Br_2, KOH, $Na_2S_2O_5$, MeCN, H_2O, pH 11.5 (controlled), 0–95°, 4.5 h	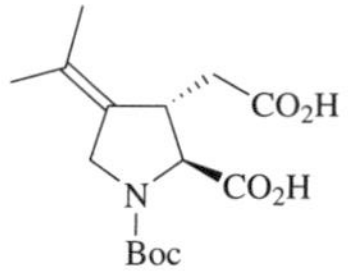 (—)[b]	536, 537
	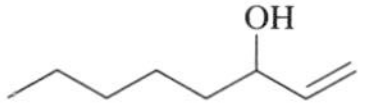	TEMPO (cat.), $PhI(OAc)_2$, CH_2Cl_2, rt	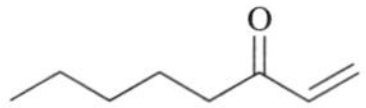 (95)	192
	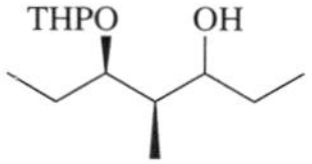	TEMPO (cat.), KBr (cat.), NaOCl, CH_2Cl_2, H_2O, 0°	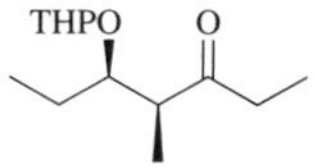 (—)[b]	538
	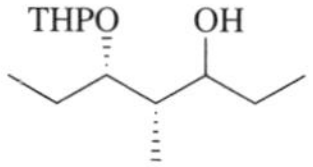	TEMPO (cat.), KBr (cat.), NaOCl, CH_2Cl_2, H_2O, 0°	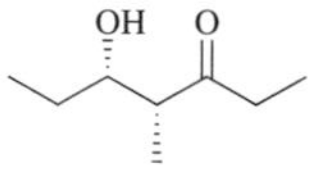 (69)	523
	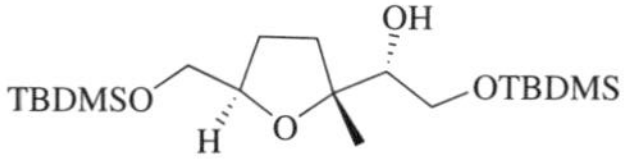	TEMPO (cat.), $[PS\text{-}CH_2NMe_3]^+[ClO_2]^-$ CH_2Cl_2, rt or toluene, 40°, 3.5 h	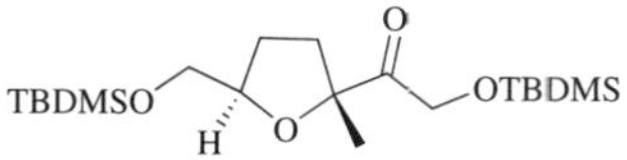 (86)	44
C_{8-9}	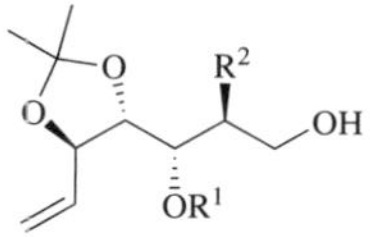	TEMPO (cat.), $PhI(OAc)_2$, MeCN, H_2O, rt, 2 h	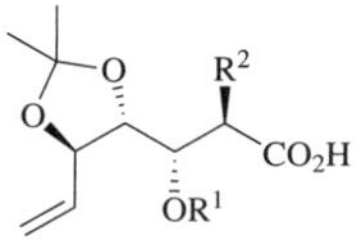 (—)[b] R / R^2 H / OMe TBDMS / OMe TBDMS / Me TBDMS / Et	194

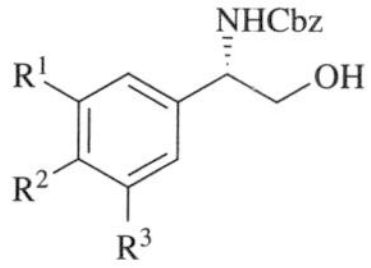	TEMPO (cat.), KBr (cat.), NaOCl, H_2O, $NaHCO_3$ (5%), acetone, 0°, 2 h	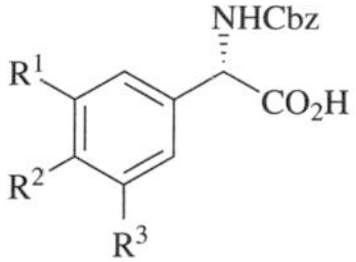 R¹ / R² / R³: H / Br / H (86); O_2N / H / H (82); H / BnO / H (86); MeO / BnO / H (71); MeO / BnO / MeO (75); H / CF_3 / H (77)	515
C_{8-13} 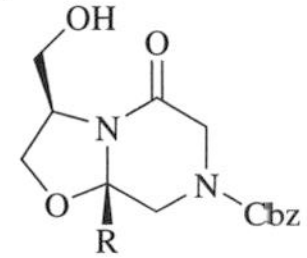	TEMPO (cat.), NaOCl (cat.), $NaClO_2$, MeCN, H_2O, buffer, 60°	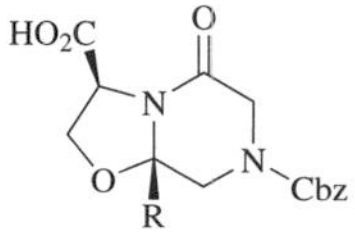 R: Me (94); Ph (92)	539
C_9 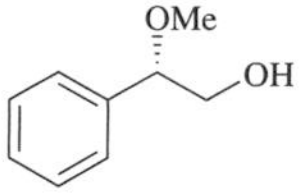	TEMPO (cat.), NaOCl (cat.), $NaClO_2$, MeCN, KH_2PO_4 buffer, pH 6.5, 55°, 4 h	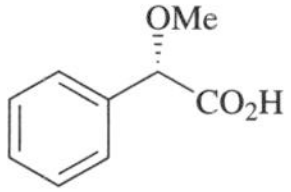 (80)	540
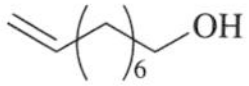	$[AcNH\text{-}TEMP{=}O]^+[BF_4]^-$, CH_2Cl_2, silica gel, rt, 16–24 h	 (74)	540
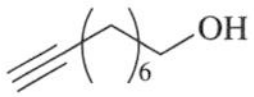	$[AcNH\text{-}TEMP{=}O]^+[BF_4]^-$, CH_2Cl_2, silica gel, rt, 16–24 h	 (67)	540
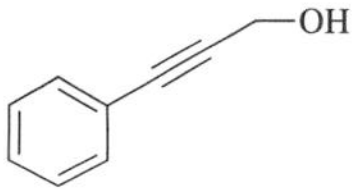	$[AcHN\text{-}TEMP{=}O^+][ClO_4]^-$, silica gel, CH_2Cl_2, rt	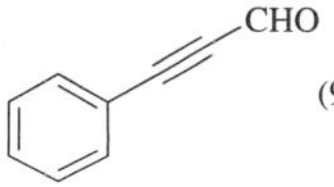 (99)	68
	TEMPO (cat.), NaOCl (cat.), $NaClO_2$, MeCN, H_2O	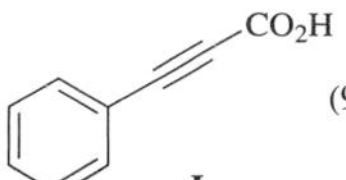 (90) **I**	14, 185

TABLE 3. OXIDATIONS OF POLYFUNCTIONAL ALCOHOLS (*Continued*)

Alcohol	Conditions	Product(s) and Yield(s) (%)	Refs.
C_9 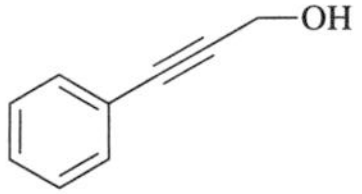	1. TEMPO (cat.), PS-I$(OAc)_2$, acetone, rt 2. H_2O, 24 h	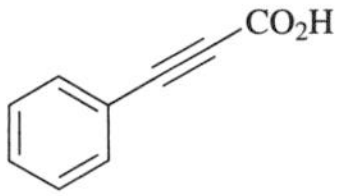 (71)	204
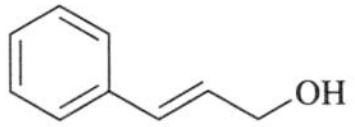	TEMPO (cat.), KBr (cat.), PS-I$(OAc)_2$, acetone, 0°	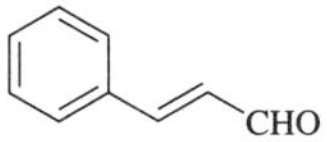 (96) **I**	202
	TEMPO (cat.), $Mn(NO_3)_2$ or $Co(NO_3)_2$ (cat.), O_2 (1 atm), AcOH	**I** (99)[a]	407
	TEMPO (cat.), $CuBr{\bullet}Me_2S$ (cat.), **1** (cat.), O_2, perfluorooctyl bromide, chlorobenzene, 90°	**I** (79)	414
	TEMPO (cat.), $(NH_4)_2[Ce(NO_3)_6]$ (cat.), O_2, MeCN, reflux	**I** (75–99)	269
	AcNH-TEMPO, TsOH, CH_2Cl_2, 0°	**I** (96)	56
	HO-TEMPO (cat.), CuCl (cat.), O_2, DMF, rt	**I** (93)	81
	TEMPO (cat.), laccase (cat.), O_2, H_2O, citrate buffer, pH 4.5	**I** (94)[a]	249
	Nitroxide (cat.), $[mim\text{-}CH_2PhI(OAc)_2]^+[BF_4]^-$, H_2O, 30°	**I**	201
	Nitroxide — Time		
	$[Im\text{-}(O\text{-}TEMPO)_2]^+[BF_4]^-$ — 1 h	(97)	
	$[mim\text{-}CO_2\text{-}TEMPO]^+[BF_4]^-$ — 1 h	(98)	
	$[Im\text{-}(O\text{-}TEMPO)_2]^+[BF_4]^-$ — 20 min	(97)	

TEMPO (cat.), DABCO–CuCl (cat.), O_2, K_2CO_3, toluene, 100°, 3 h	**I** (84)	247
TEMPO (cat.), DABCO–CuCl (cat.), O_2, K_2CO_3, $MeNO_2$, 100°, 10 h	**I** (80)	247
TEMPO (cat.), $PhI(OAc)_2$, CH_2Cl_2, rt	**I** (95)	192
MeO-TEMPO (cat.), KBr (cat.), NaOCl, $NaHCO_3$, Aliquat 336, CH_2Cl_2, H_2O. pH 8.6, 0°	**I** (57)	134
PEG-BnO-TEMPO (cat.), $Co(NO_3)_2$ or $Mn(NO_3)_2$ (cat.), O_2, AcOH, H_2O, 40°, 4 h	**I** (99)[a]	272
TEMPO (cat.), KBr (cat.), PS-$I(OAc)_2$, acetone, 0°	**I** (96)	202
CIBA-TEMPO (cat.), $Mn(NO_3)_2$ or $Co(NO_3)_2$ (cat.), air or O_2, AcOH, 20–40°	**I** (96)	395
[MeO-TEMP=O^+]Cl^-, CH_2Cl_2, rt	**I** (100)[a]	11
PEG-O-TEMPO (cat.), $CuBr_2$ (cat.), PEG-bipy (cat.), O_2, *t*-BuOK, MeCN, H_2O, 80°, 18 h	**I** (100)	392
TEMPO (cat.), **3** (cat.), O_2, toluene, 100°	**I** (98)	244
ICF-O-TEMPO (cat.), $PhI(OAc)_2$, CH_2Cl_2	**I** (99)[a]	293

TABLE 3. OXIDATIONS OF POLYFUNCTIONAL ALCOHOLS (*Continued*)

Alcohol	Conditions	Product(s) and Yield(s) (%)	Refs.
C_9			
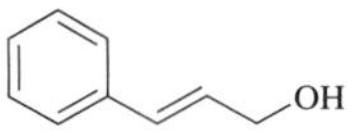	TEMPO (cat.), CuBr•Me_2S (cat.), **1** (cat.), O_2, perfluorooctyl bromide, chlorobenzene, 90°	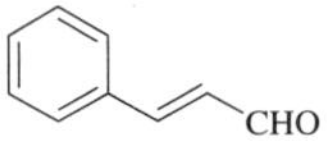 **I** (79)	243
	TEMPO (cat.), I_2, $NaHCO_3$, toluene or CH_2Cl_2, 20°, overnight	**I** (85)[c]	256
	[PS-O-TEMP=O]$^+$Cl$^-$, CH_2Cl_2, rt, 1–2 h	**I** (95)[a]	121
	AcNH-TEMPO (cat.), **4** (cat.), $Cu(ClO_4)_2$ (cat.), O_2, DABCO, DMSO	**I** (92)	245
	PEG-BnO-TEMPO (cat.), $CuBr_2$ (cat.), PEG-bipy (cat.), O_2, *t*-BuOK, MeCN, H_2O, 80°, 18 h	**I** (100)	392
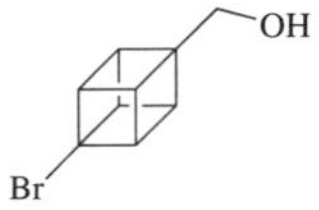	TEMPO (cat.), TCC, $NaHCO_3$, CH_2Cl_2	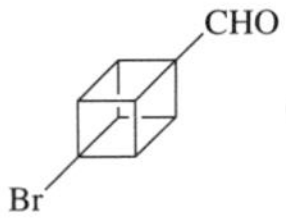 (71)	541
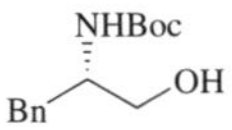	AcNH-TEMPO (cat.), NaBr (cat.), NaOCl, $NaHCO_3$, EtOAc, toluene, H_2O, –5°	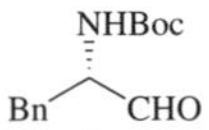 **I** (—)[b]	471, 472
	TEMPO (cat.), [PS-CH_2NMe_3]$^+$[Br(OAc)$_2$]$^-$ CH_2Cl_2, rt, 2 h	**I** (50–99)	43
	TEMPO (cat.), KBr (cat.), NaOCl, $NaHCO_3$, toluene, EtOAc, H_2O, –5°	**I** (—)[b]	470

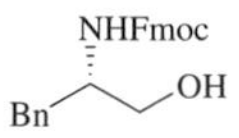	AcNH-TEMPO (cat.), NaBr (cat.), NaOCl, $NaHCO_3$, EtOAc, toluene, H_2O, –5°	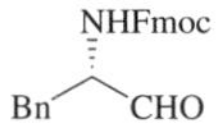 (—)[b]	471, 472
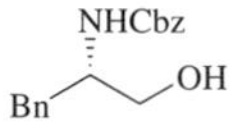	AcNH-TEMPO (cat.), NaBr (cat.), NaOCl, $NaHCO_3$, EtOAc, toluene, H_2O, –5°	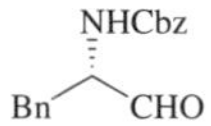 **I** (—)[b]	471, 472
	TEMPO (cat.), $[PS\text{-}CH_2NMe_3]^+[Br(OAc)_2]^-$ CH_2Cl_2, rt, 2 h	**I** (50–99)	43
	1. TEMPO (cat.), NaBr (cat.), NaOCl, $NaHCO_3$, CH_2Cl_2, H_2O, 0° 2. $Na_2S_2O_3$	**I** (90)	542
	TEMPO (cat.), NaOCl (cat.), $NaClO_2$, MeCN, H_2O	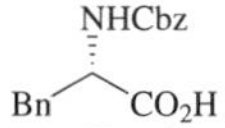 **II** (85)	14, 185
	TEMPO (cat.), NaBr (cat.), NaOCl, $NaHCO_3$, toluene, EtOAc, H_2O, 0°	**II** (96), 99% ee	158
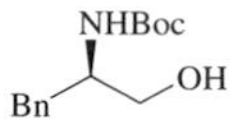	TEMPO (cat.), NaBr (cat.), NaOCl, EtOAc, toluene, H_2O, –10°	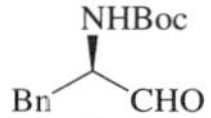 **I** (—)[b]	543
	TEMPO (cat.), NaBr (cat.), NaOCl, $NaHCO_3$, toluene, H_2O, EtOAc, 0°	**I** (—)[b]	452
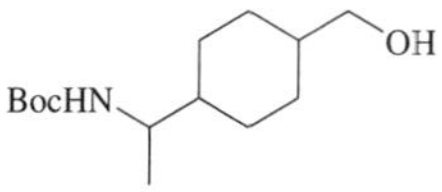	TEMPO (cat.), NaOCl (cat.), $NaClO_2$, MeCN, H_2O, phosphate buffer, 35°	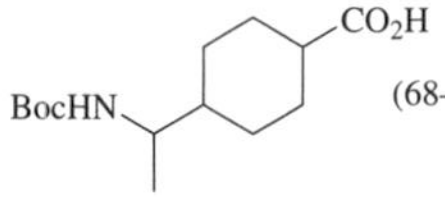 (68–74)	544

TABLE 3. OXIDATIONS OF POLYFUNCTIONAL ALCOHOLS (*Continued*)

Alcohol	Conditions	Product(s) and Yield(s) (%)	Refs.
C_9			
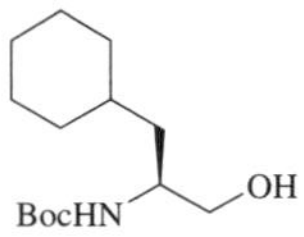	TEMPO (cat.), NaBr (cat.), NaOCl, $NaHCO_3$, toluene, EtOAc, H_2O, 0°	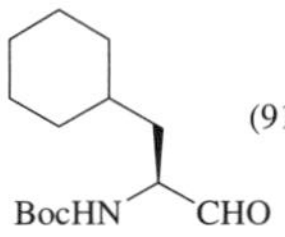 (91), 97% ee	158
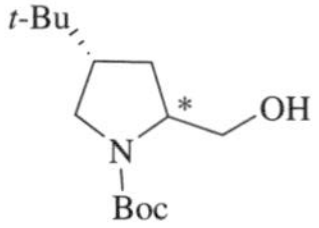	TEMPO (cat.), $PhI(OAc)_2$, MeCN, H_2O	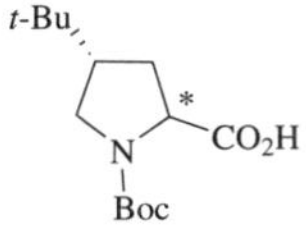 * R (76); S (80)	545
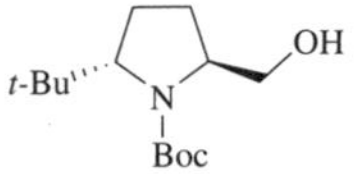	TEMPO (cat.), NaOCl (cat.), $NaClO_2$, MeCN, H_2O, phosphate buffer	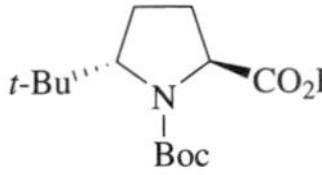 (84)	546
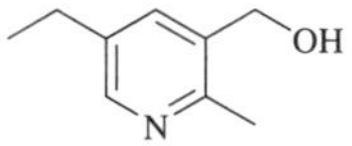	BzO-TEMPO (cat.), $TBABr_3$, NaOAc, CH_2Cl_2, H_2O, rt, 1 h	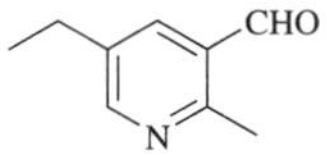 (90)	224
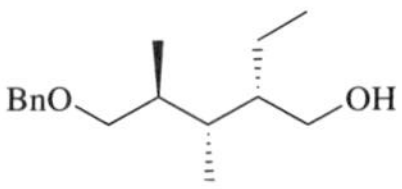	TEMPO (cat.), $PhI(OAc)_2$, CH_2Cl_2, rt, 10 h	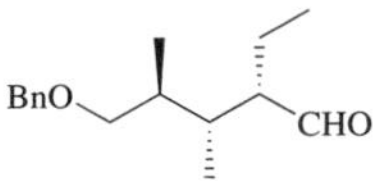 $(—)^b$	547
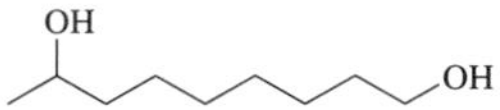	HO-TEMPO (cat.), CuCl (cat.), O_2, DMF, rt	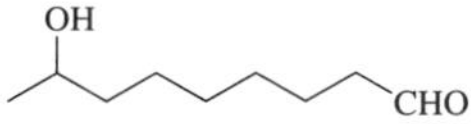 (55)	81
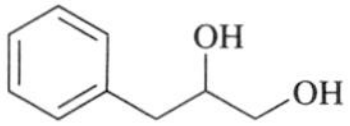	SG-NHCONH-TEMPO (cat.), NaOCl, H_2O, 0°	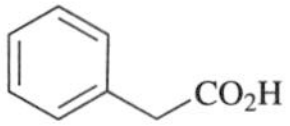 (75)	393
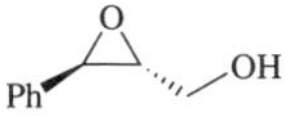	TEMPO (cat.), $PhI(OAc)_2$, MeCN, H_2O, pH 7, 0°	 $(—)^b$	548

	TEMPO (cat.), PhI(OAc)$_2$, CH$_2$Cl$_2$, rt	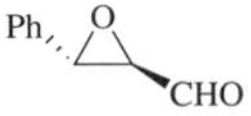 (—)[b]	206
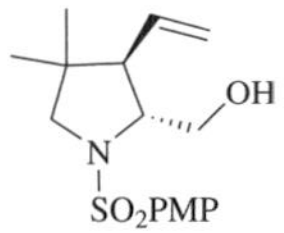	TEMPO (cat.), NaOCl (cat.), NaClO$_2$, MeCN, H$_2$O, phosphate buffer	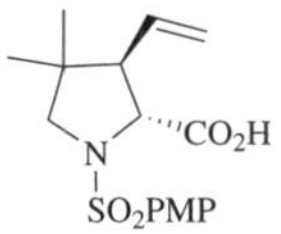 (—)[b]	177
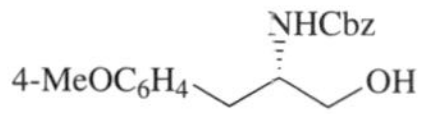	TEMPO (cat.), NaBr (cat.), NaOCl, NaHCO$_3$, toluene, EtOAc, H$_2$O, 0°	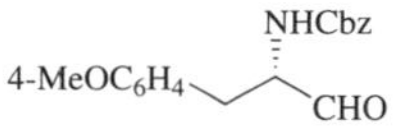 (88)	549
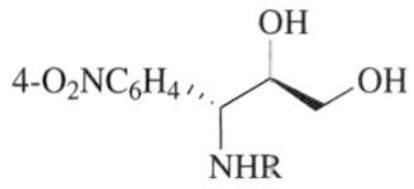	TEMPO, NaBr (cat.), NaOCl, H$_2$O, acetone, 0°	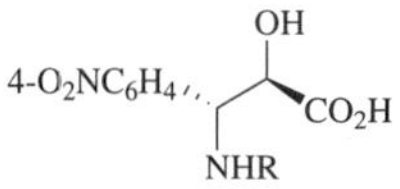 R: C(O)CHCl$_2$ (—)[b]; Boc (—)[b]	164
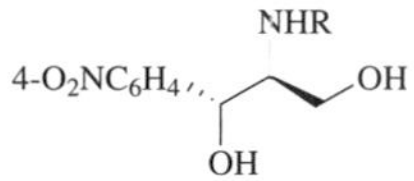	TEMPO or ormosil-TEMPO, NaBr (cat.), NaOCl, H$_2$O, acetone, 0°	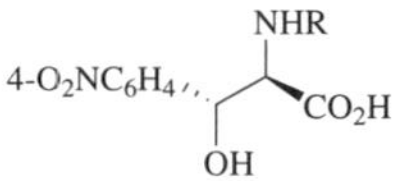 R: C(O)CHCl$_2$ (—)[b]; Boc (—)[b]	164
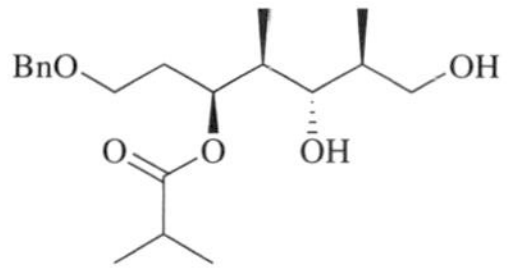	TEMPO (cat.), PhI(OAc)$_2$, CH$_2$Cl$_2$, rt, 2 h	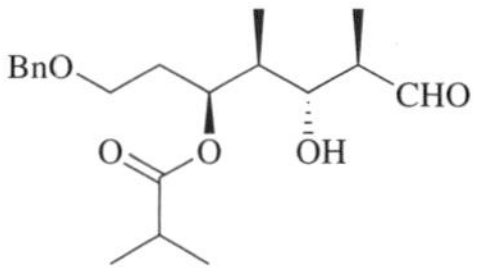 (90)	550
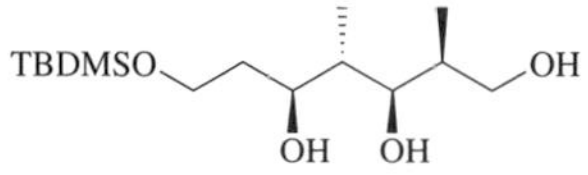	TEMPO (cat.), PhI(OAc)$_2$, CH$_2$Cl$_2$, rt, 2 h	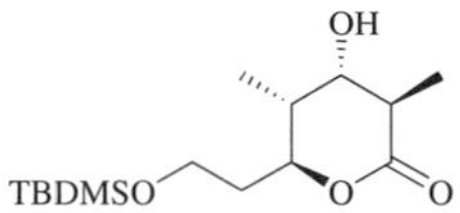 (91)	551

TABLE 3. OXIDATIONS OF POLYFUNCTIONAL ALCOHOLS (*Continued*)

Alcohol	Conditions	Product(s) and Yield(s) (%)	Refs.
C_9			
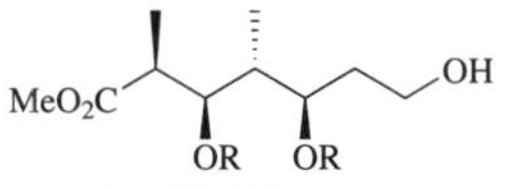 R = TBDMS	TEMPO (cat.), $PhI(OAc)_2$, CH_2Cl_2, rt, 2 h	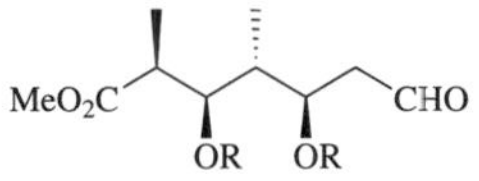 (86)	551
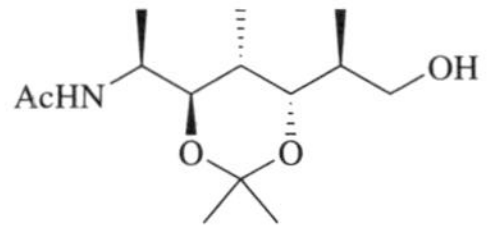	TEMPO (cat.), $PhI(OAc)_2$, CH_2Cl_2, rt, 2 h	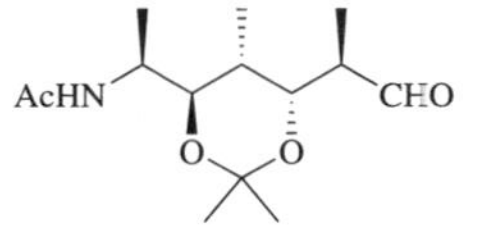 (—)[b]	552
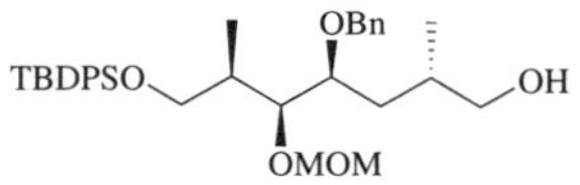	TEMPO (cat.), $[PS\text{-}CH_2NMe_3]^+[Br(OAc)_2]^-$, CH_2Cl_2, 24 h	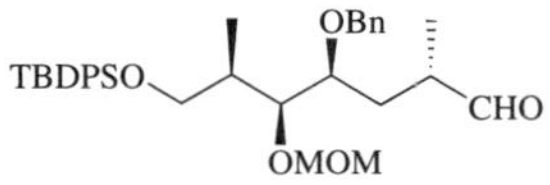 (96)	43
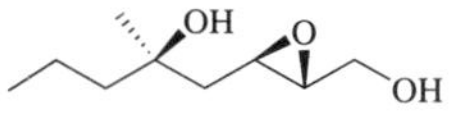	TEMPO (cat.), $PhI(OAc)_2$, CH_2Cl_2, rt, 4 h	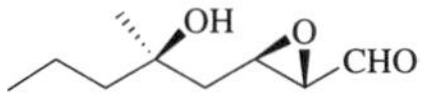 (—)[b]	506
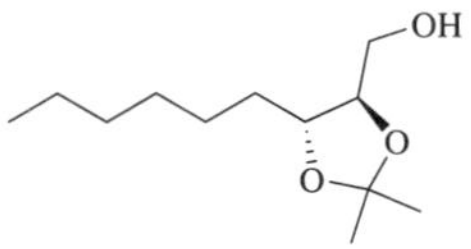	AcNH-TEMPO (cat.), $PhI(OAc)_2$, CH_2Cl_2, 18°, 2 h	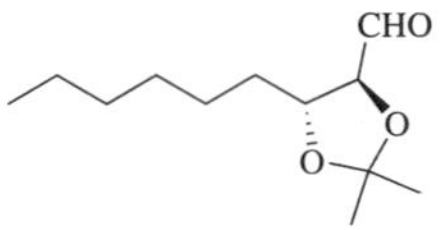 (85)	195
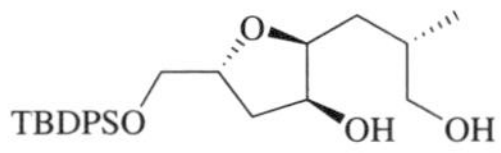	TEMPO (cat.), $PhI(OAc)_2$, CH_2Cl_2, rt	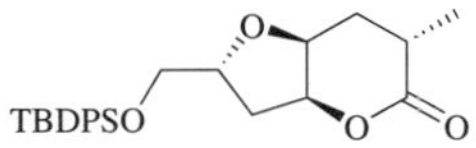 (78)	486
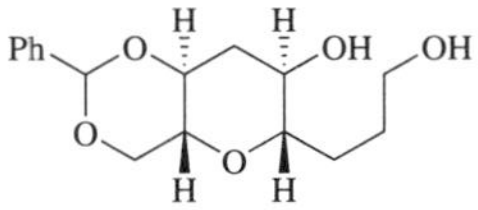	TEMPO (cat.), KBr (cat.), NaOCl, CH_2Cl_2, H_2O, 0°	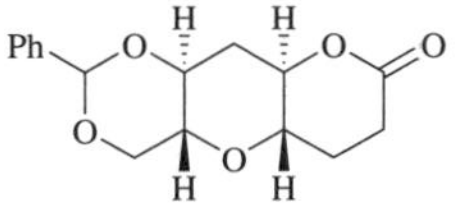 (89)	553

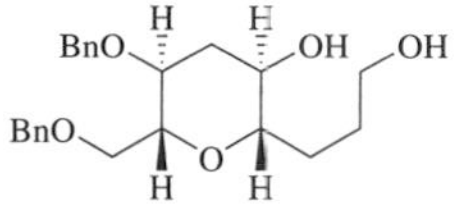	TEMPO (cat.), KBr (cat.), NaOCl, CH_2Cl_2, H_2O, 0°	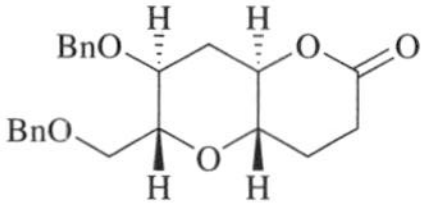 (94)	553
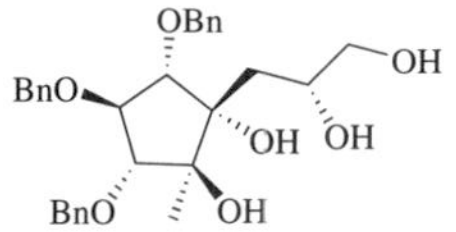	TEMPO (cat.), KBr (cat.), NaOCl, TBAC, $NaHCO_3$ (satd.), H_2O, CH_2Cl_2	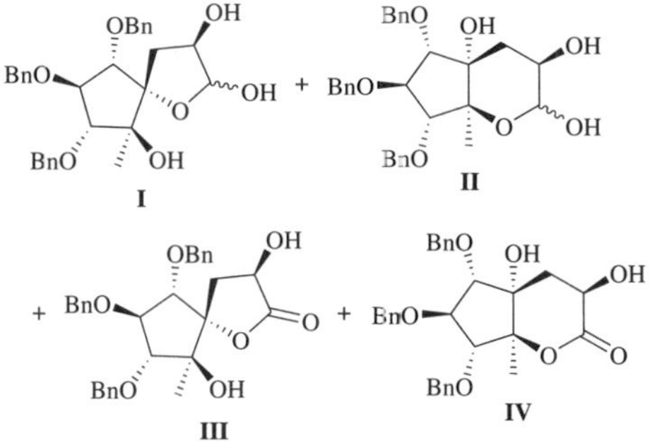 **I** + **II** (27), **III** (trace), **IV** (32)	180
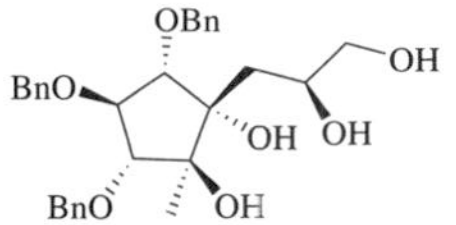	TEMPO (cat.), KBr (cat.), NaOCl, TBAC, $NaHCO_3$ (satd.), H_2O, CH_2Cl_2	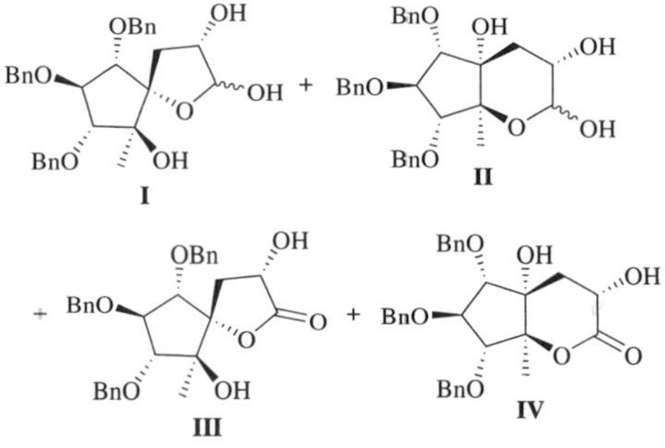 **I** + **II** (31), **III** (trace), **IV** (33)	180

TABLE 3. OXIDATIONS OF POLYFUNCTIONAL ALCOHOLS (*Continued*)

Alcohol	Conditions	Product(s) and Yield(s) (%)	Refs.
C_9			
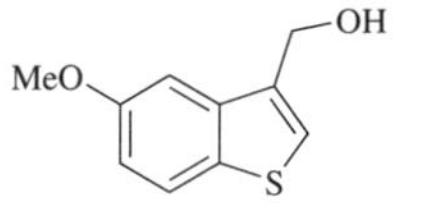	TEMPO (cat.), I_2, $NaHCO_3$, toluene or CH_2Cl_2, 20°, overnight	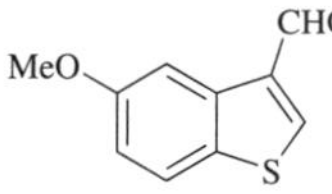 $(92)^c$	256
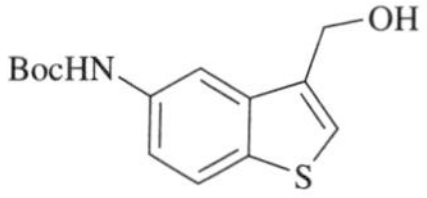	1. TEMPO (cat.), NaOCl, MeCN, 0° 2. $NaClO_2$, H_2O_2	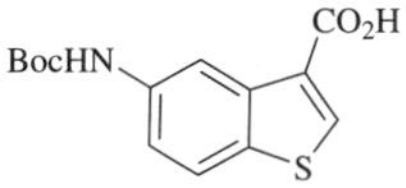 (57)	554
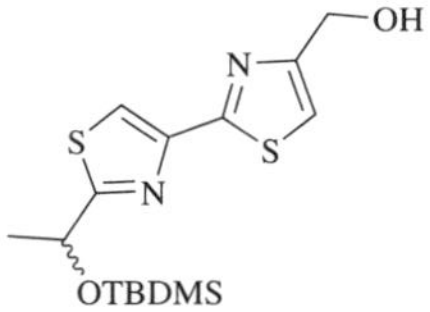	TEMPO (cat.), KBr (cat.), NaOCl, $NaHCO_3$, acetone, H_2O	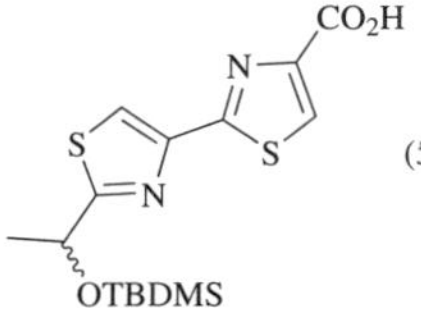 (51)	555
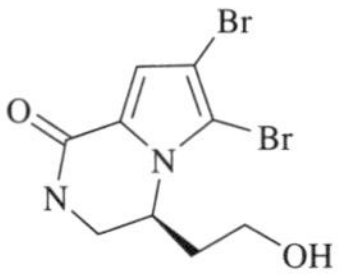	TEMPO (cat.), $PhI(OAc)_2$, CH_2Cl_2	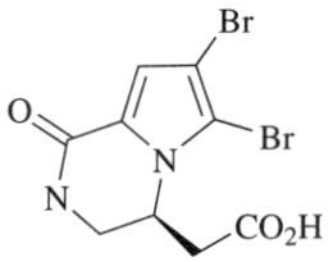 (95)	556
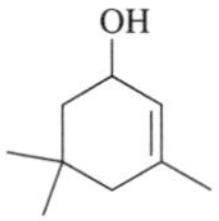	TEMPO (cat.), CuCl (cat.), O_2, $[bmim]^+[PF_6]^-$, 65°	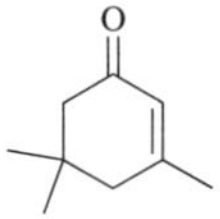 (60)	385
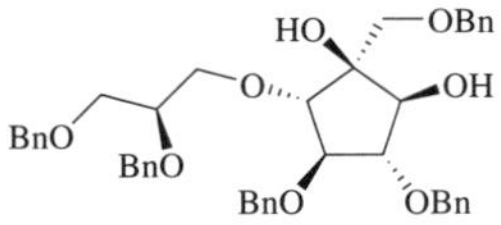	TEMPO (cat.), NaOCl, CH_2Cl_2, H_2O	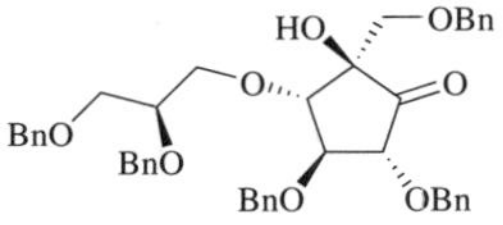 $(—)^b$	557

Substrate	Conditions	Product(s) and Yield(s) (%)	Refs.
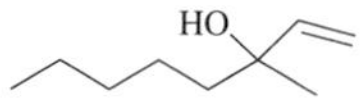	$[TEMPO]^+[BF_4]^-$, MeCN, rt, 6 h	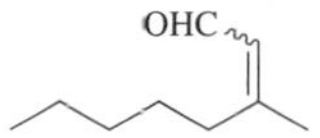 **I** (75), E:Z = 2.4:1	120
	$[TEMPO]^+[SbF_5]^-$, MeCN, rt, 4 h	**I** (80), E:Z = 2.4:1	120
	$[TEMPO]^+[BF_4]^-$, MeCN, H_2O, 50°, 1 h	 **I** (73), E:Z = 3.5:1	120
	$[TEMPO]^+[SbF_6]^-$, MeCN, 50°, 1 h	**I** (85), E:Z = 3.5:1	120
C_{9-10}			
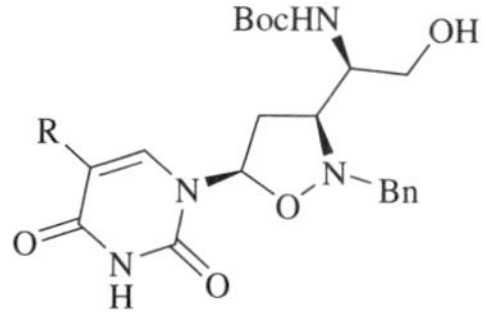	TEMPO (cat.), $PhI(OAc)_2$, MeCN, H_2O, rt, 4 h	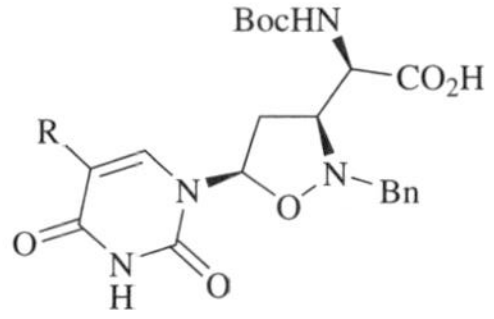 R: H (—)[b]; Me (—)[b]	467
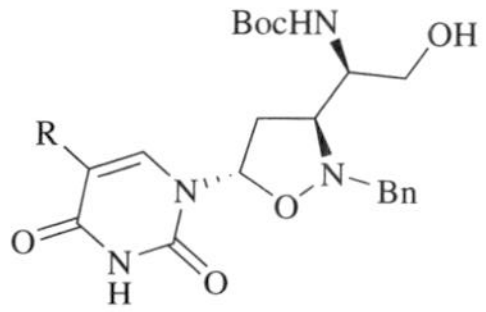	TEMPO (cat.), $PhI(OAc)_2$, MeCN, H_2O, rt, 4 h	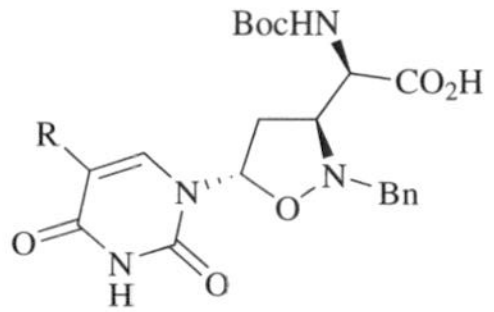 R: H (—)[b]; Me (—)[b]	467
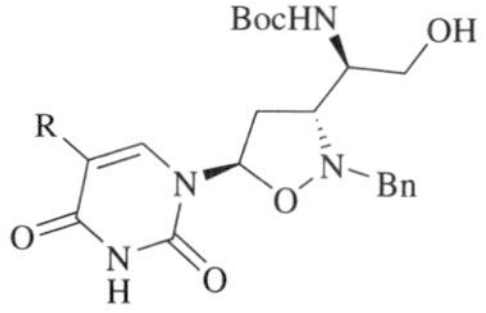	TEMPO (cat.), $PhI(OAc)_2$, MeCN, H_2O, rt, 4 h	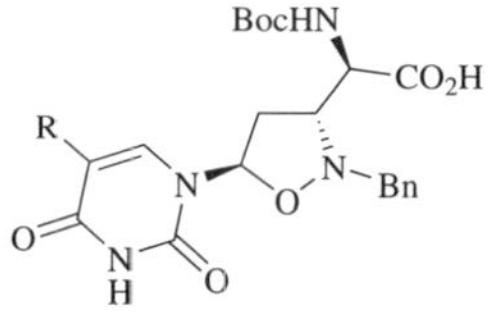 R: H (—)[b]; Me (—)[b]	467

TABLE 3. OXIDATIONS OF POLYFUNCTIONAL ALCOHOLS (*Continued*)

Alcohol	Conditions	Product(s) and Yield(s) (%)	Refs.
C_{9-10} 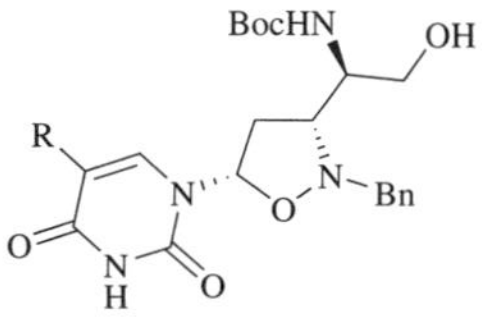	TEMPO (cat.), PhI(OAc)$_2$, MeCN, H_2O, rt, 4 h	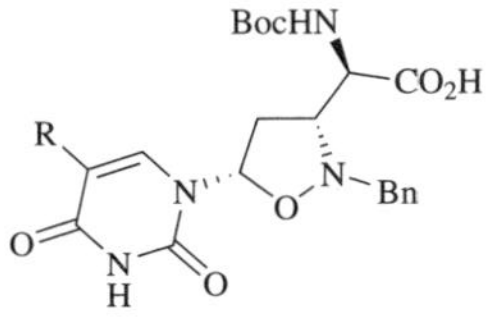 R: H (—)[b]; Me (—)[b]	467
C_{9-12} 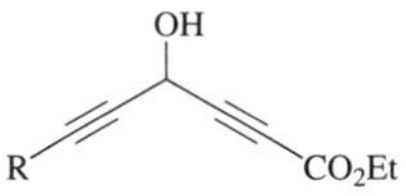	TEMPO (cat.), Oxone, TBAB, toluene, rt, 5 h	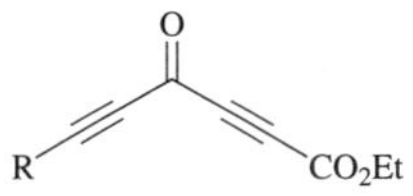 R: *n*-Pr (91); *n*-Bu (93); Ph (95)	558
C_{10} 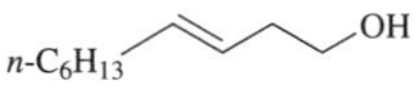	TEMPO (cat.), PhI(OAc)$_2$, pentane, CH_2Cl_2, rt	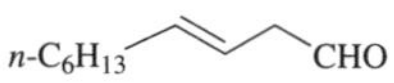 (100)[e]	208
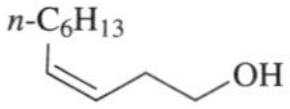	TEMPO (cat.), PhI(OAc)$_2$, pentane, CH_2Cl_2, rt	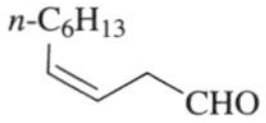 (100)[e]	208
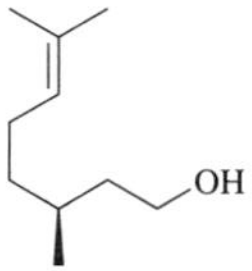	TEMPO (cat.), PhI(OAc)$_2$, CH_2Cl_2, rt	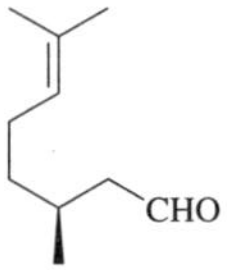 **I** (90)	192

	TEMPO (cat.), $PhI(OAc)_2$, CH_2Cl_2	**I** (70)	559
	TEMPO (cat.), **1** (cat.), $CuBr{\bullet}Me_2S$ (cat.), O_2, perfluorooctyl bromide, chlorobenzene, 90°	**I** (92)	243
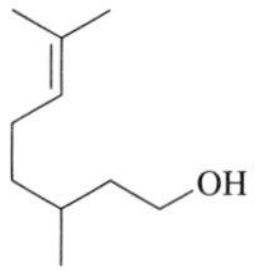	$[PS\text{-}O\text{-}TEMP{=}O]^+Cl^-$, CH_2Cl_2, rt, 1–2 h	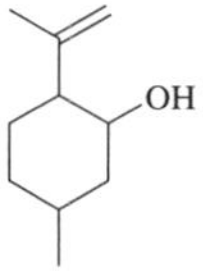 (60)[a]	121
	BzO-TEMPO, $RuCl_2(PPh_3)_3$ (cat.), O_2, toluene, 70°, 8 h	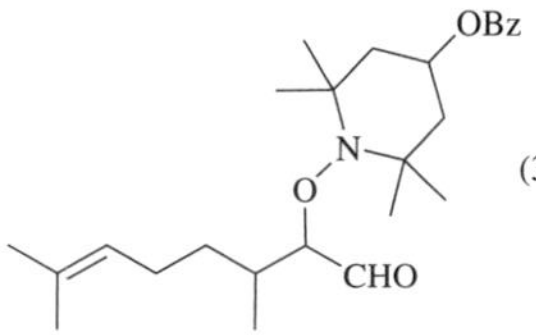 (37)	411
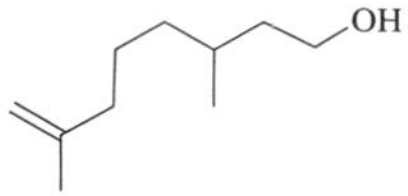	PEG-BnO-TEMPO (cat.), $CuBr_2$ (cat.), PEG-bipy (cat.), O_2, *t*-BuOK, MeCN, H_2O, 80°, 18 h	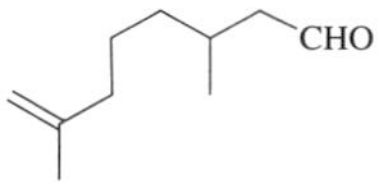 (31)	392
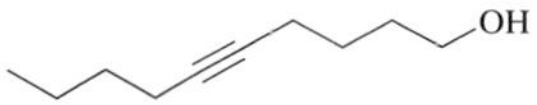	$[AcNH\text{-}TEMP{=}O]^+[BF_4]^-$, CH_2Cl_2, silica gel, rt, 16–24 h	 (71)	131

TABLE 3. OXIDATIONS OF POLYFUNCTIONAL ALCOHOLS (*Continued*)

Alcohol	Conditions	Product(s) and Yield(s) (%)	Refs.
C_{10}			
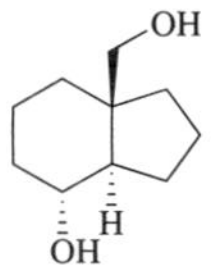	TEMPO (cat.), $PhI(OAc)_2$, CH_2Cl_2, MeCN	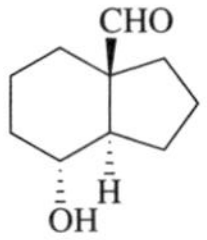 (97)	560
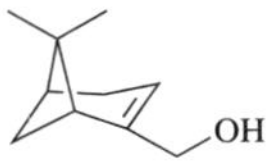	HO-TEMPO (cat.), CuCl (cat.), O_2, DMF, rt	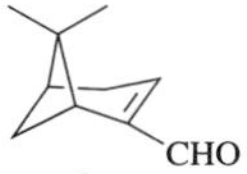 **I** (92)	81
	TEMPO (cat.), **1** (cat.), $CuBr•Me_2S$ (cat.), O_2, perfluorooctyl bromide, chlorobenzene, 90°	**I** (76)	243, 414
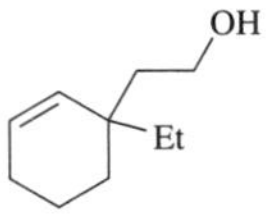	AcNH-TEMPO (cat.), $PhI(OAc)_2$, CH_2Cl_2, 18°, 3 h	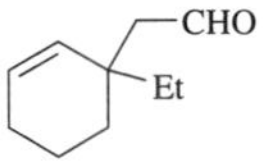 (—)[b]	561
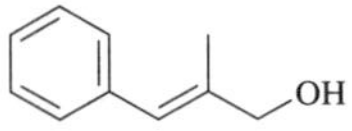	TEMPO (cat.), $(NH_4)_2[Ce(NO_3)_6]$ (cat.), O_2, MeCN, reflux	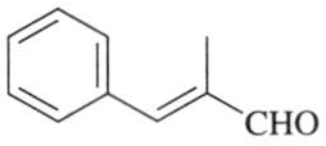 (78–94)	269
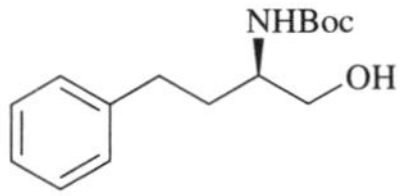	AcNH-TEMPO (cat.), NaBr (cat.), NaOCl, $NaHCO_3$, toluene, EtOAc, H_2O	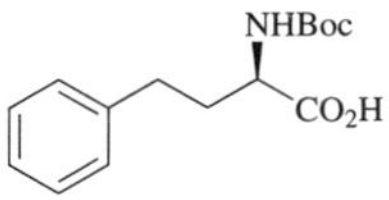 (74)	440

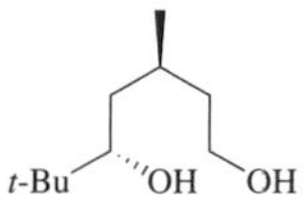	TEMPO (cat.), PhI(OAc)$_2$, CH$_2$Cl$_2$, rt	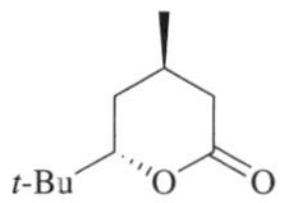 (85)	486
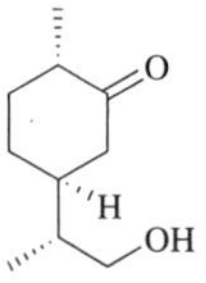	TEMPO (cat.), NaOCl, CH$_2$Cl$_2$, H$_2$O, pH 8	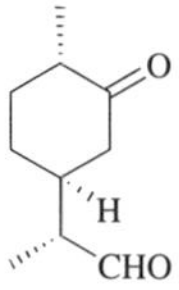 (92)	562
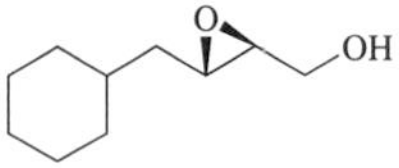	TEMPO (cat.), PhI(OAc)$_2$, CH$_2$Cl$_2$, rt	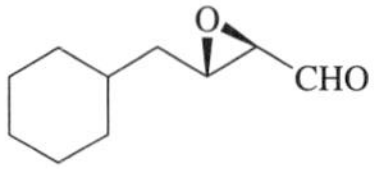 (75)	192
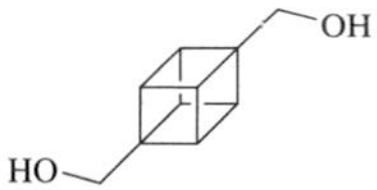	TEMPO (cat.), TCC, NaHCO$_3$, CH$_2$Cl$_2$	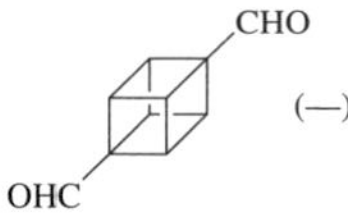 (—)[b]	541
	TEMPO (cat.), PhI(OAc)$_2$, CH$_2$Cl$_2$, 20°	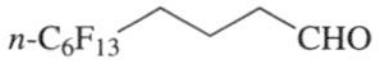 (60)	412
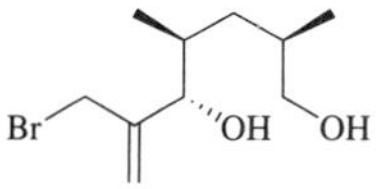	TEMPO (cat.), PhI(OAc)$_2$, CH$_2$Cl$_2$, rt	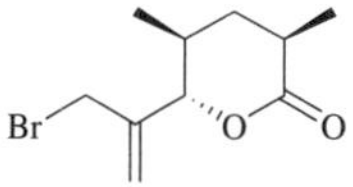 (87)	486

TABLE 3. OXIDATIONS OF POLYFUNCTIONAL ALCOHOLS (*Continued*)

Alcohol	Conditions	Product(s) and Yield(s) (%)	Refs.
C_{10} 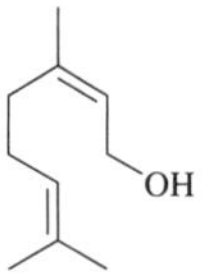	TEMPO (cat.), $PhI(OAc)_2$, CH_2Cl_2, rt	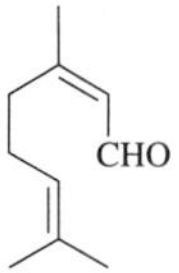 (95) **I**	192
	TEMPO (cat.), $PhI(OAc)_2$, MeCN, H_2O, buffer, pH 7, 0°	**I** (87–89)	15
	PEG-BnO-TEMPO (cat.), $CuBr_2$ (cat.), PEG-bipy (cat.), *t*-BuOK, MeCN, H_2O, O_2, 80°, 18 h	**I** (92)	392
	TEMPO (cat.), $Yb(OTf)_3$ (cat.), $PhI(OAc)_2$, CH_2Cl_2, 15 min	**I** (92)	416
	$[AcNH\text{-}TEMP{=}O]^+[ClO_4]^-$, CH_2Cl_2, rt	**I** (99)	68
	TEMPO (cat.), $PhI(OAc)_2$, CH_2Cl_2, rt	**I** (—)[b]	205
	AcNH-TEMPO, TsOH, CH_2Cl_2, 0°, 2 h	**I** (80)[e] + 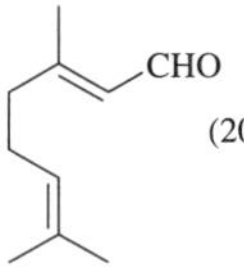 (20)[e] **II**	56
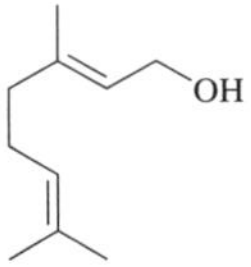	TEMPO (cat.), $PhI(OAc)_2$, CH_2Cl_2, rt	**II** (—)[b]	205

	TEMPO (cat.), KBr (cat.), PS-I(OAc)$_2$, acetone, 0°	**II** (98)	202
	HO-TEMPO (cat.), CuCl (cat.), O_2, DMF, rt	**II** (92)	81
	TEMPO (cat.), PhI(OAc)$_2$, CH_2Cl_2, rt	**II** (95)	192
	TEMPO (cat.), **3** (cat.), O_2, toluene, 100°	**II** (79)	244
	TEMPO (cat.), CuCl (cat.), O_2, [bmim]$^+$[PF$_6$]$^-$, 65°	**II** (96)[a]	385
	TEMPO (cat.), laccase (cat.), O_2, H_2O, citrate buffer, pH 4.5	**II** (85)	249
	[AcNH-TEMP=O]$^+$[ClO$_4$]$^-$, CH_2Cl_2, rt	**II** (99)	68
	AcNH-TEMPO, TsOH, CH_2Cl_2, 0°	**I** (14) + **II** (86)	56
	[PS-C-TEMP=O]$^+$Cl$^-$, CH_2Cl_2, rt, 1–2 h	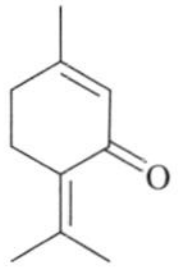 (83)[a]	121
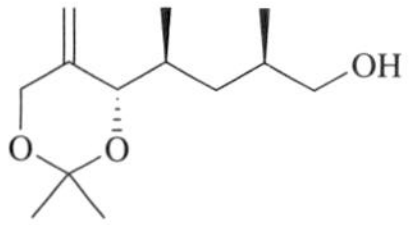	TEMPO (cat.), NaOCl (cat.), $NaClO_2$, MeCN, H_2O, phosphate buffer, pH 6.6, 35°	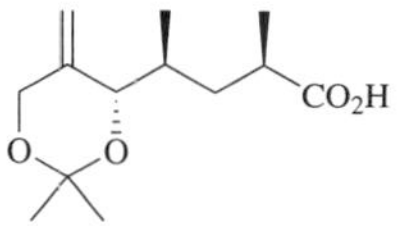 (95)	563

TABLE 3. OXIDATIONS OF POLYFUNCTIONAL ALCOHOLS (*Continued*)

Alcohol	Conditions	Product(s) and Yield(s) (%)	Refs.
C_{10} 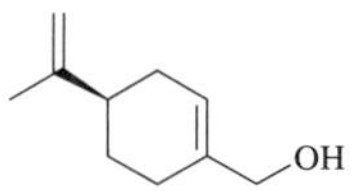	TEMPO (cat.), CuCl (cat.), O_2, $[bmim]^+[PF_6]^-$, 65°	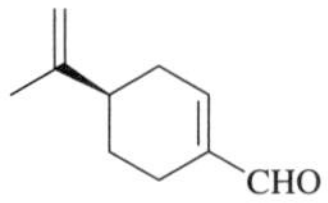 (90)	385
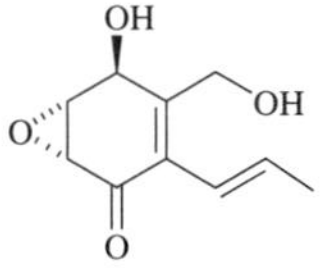	$[AcNH\text{-}TEMP{=}O^+][BF_4]^-$, silica gel, CH_2Cl_2	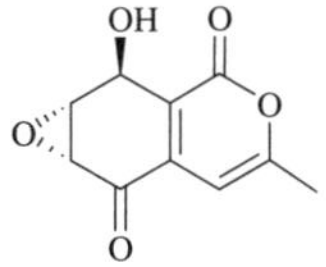 (85)	356
	TEMPO, CuCl, O_2, DMF, rt, 30 min	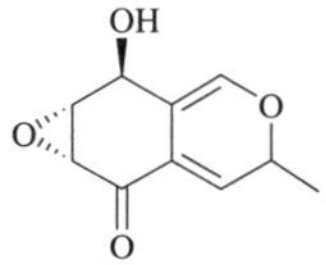 (—)[b]	358
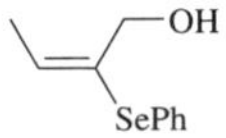	TEMPO, $PhI(OAc)_2$, CH_2Cl_2, rt	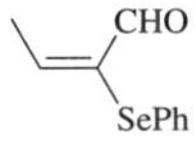 (55)	192
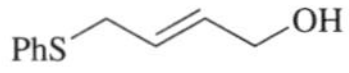	TEMPO (cat.), $PhI(OAc)_2$, MeCN, H_2O, buffer, pH 7, rt, 0°	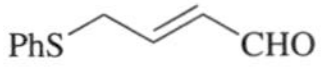 **I** (70)	15
	TEMPO (cat.), $PhI(OAc)_2$, CH_2Cl_2, rt	**I** (70)	192
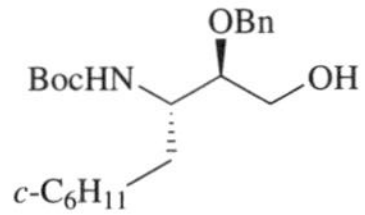	TEMPO (cat.), NaBr (cat.), NaOCl, $NaHCO_3$, toluene, EtOAc, H_2O, 0°	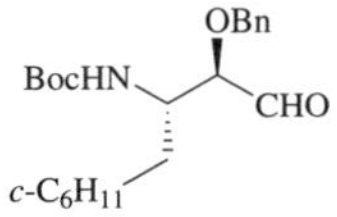 (92), 95% ee	158

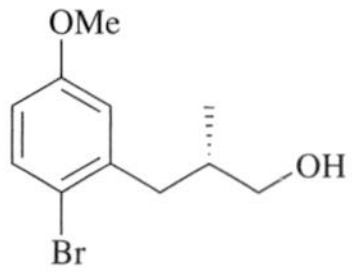	TEMPO (cat.), NaOCl (cat.), $NaClO_2$, MeCN, H_2O	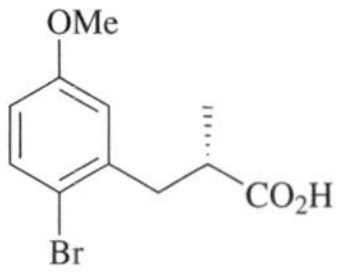 (92)	14, 185
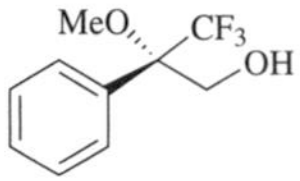	TEMPO (cat.), NaOCl (cat.), $NaClO_2$, MeCN, phosphate buffer, pH 6.5, 55°, 4 h	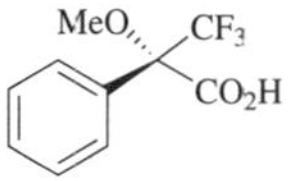 (78)	540
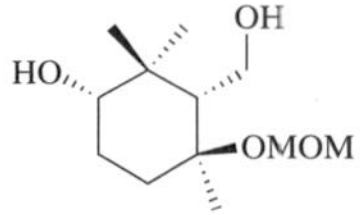	TEMPO (cat.), KBr (cat.), NaOCl, TBAC, $NaHCO_3$, NaCl, CH_2Cl_2, H_2O, 0°	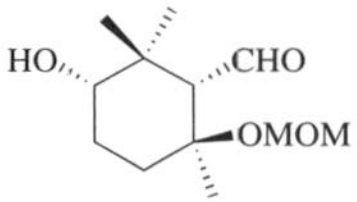 (78)	564
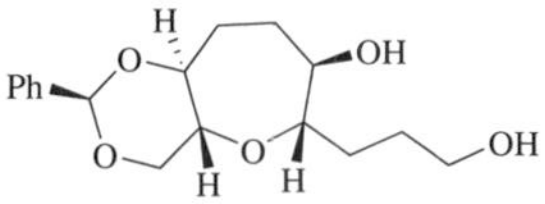	TEMPO (cat.), KBr (cat.), NaOCl, CH_2Cl_2, H_2O, 0°	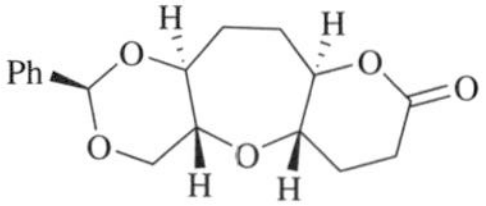 (68)[g]	553
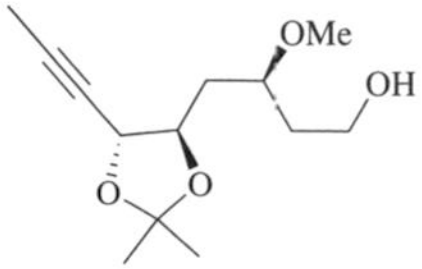	TEMPO (cat.), NaOCl (cat.), $NaClO_2$, MeCN, H_2O, buffer, pH 6.7	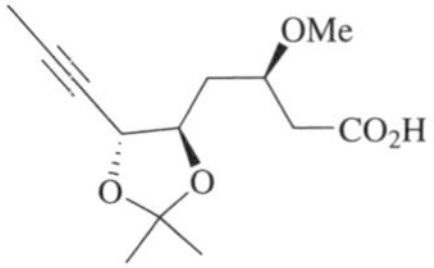 (—)[b]	565, 566
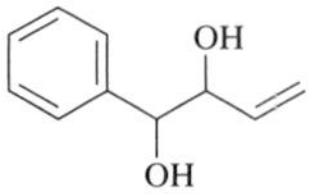	AcNH-TEMPO (1 eq), TsOH (1 eq), CH_2Cl_2	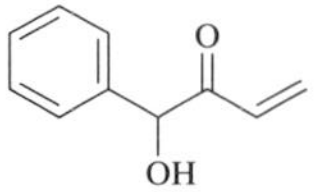 (~100)	427
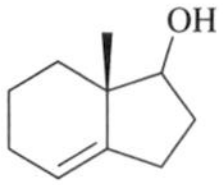	AcNH-TEMPO. TsOH, CH_2Cl_2	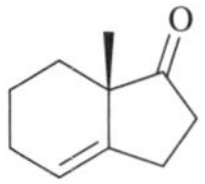 (84)	567

TABLE 3. OXIDATIONS OF POLYFUNCTIONAL ALCOHOLS (*Continued*)

Alcohol	Conditions	Product(s) and Yield(s) (%)	Refs.
C_{10} 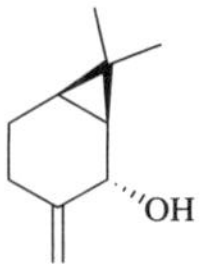	TEMPO (cat.), $PhI(OAc)_2$, CH_2Cl_2, rt, 5 h	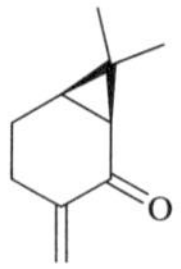 (83)	568
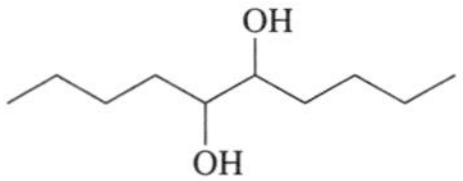	BzO-TEMPO (cat.), NaBr (cat.), electrolysis, $NaHCO_3$, CH_2Cl_2, H_2O, pH 8.6	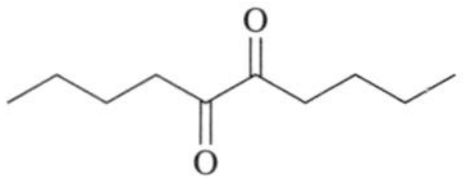 (85	222
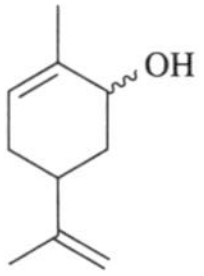	TEMPO (cat.), $(NH_4)_2[Ce(NO_3)_6]$ (cat.), O_2, MeCN, reflux	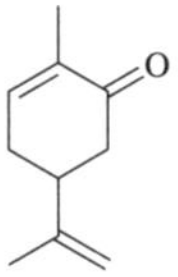 (75)	269
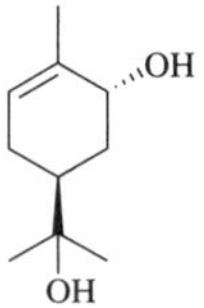	TEMPO (cat.), electrolysis, $NaClO_4$, 2,6-lutidine, MeCN, H_2O	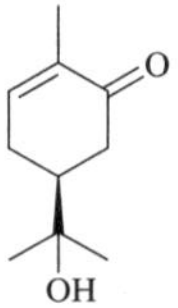 (76)[a]	421
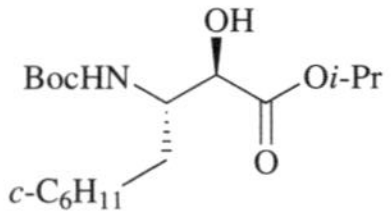	TEMPO (cat.), NaBr (cat.), NaOCl, $NaHCO_3$, toluene, EtOAc, H_2O, 0°	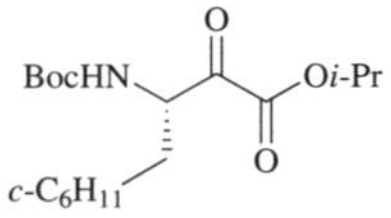 (87)	158
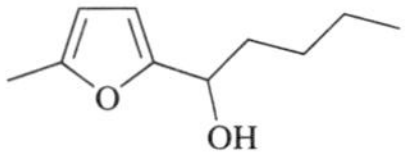	TEMPO (cat.), $PhI(OAc)_2$, MeCN, H_2O, buffer, pH 7, 0°	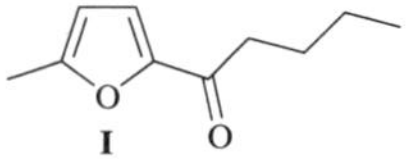 **I** (95)	15
	TEMPO (cat.), $PhI(OAc)_2$, CH_2Cl_2, rt	**I** (95)	192

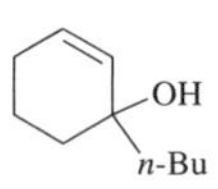	$[TEMP{=}O]^+[BF_4]^-$, MeCN, rt, 0.1 h	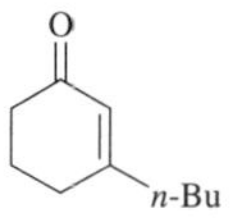 (94)	120
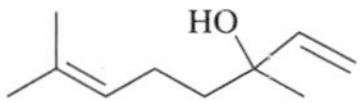	$[TEMP{=}O]^+[BF_4]^-$, MeCN, rt	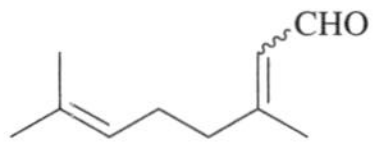 Time: 6 h (trace); 0.3 h (21)	120
C_{10-11}			
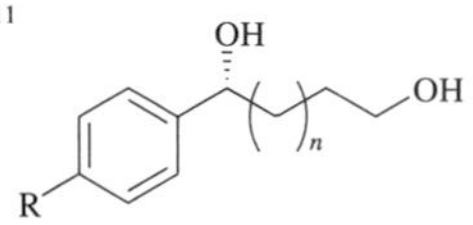	TEMPO (cat.), NCS, TBAI, CH_2Cl_2, 12 h	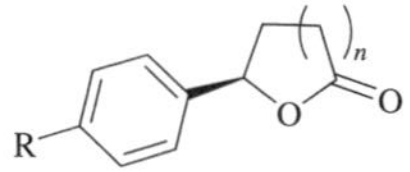 R, *n*, % ee: H, 1, (78), 70; MeO, 1, (83), 99; Cl, 1, (68), 78; H, 2, (72), 90	569
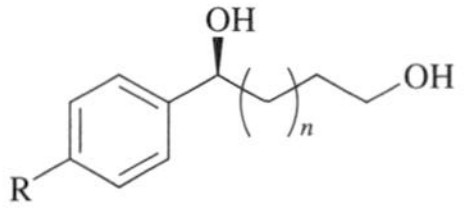	TEMPO (cat.), NCS, TBAI, CH_2Cl_2, 12 h	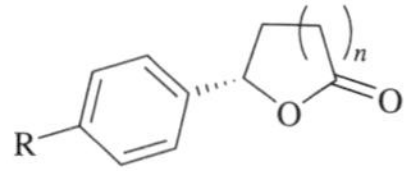 (—)[b] R, *n*: H, 1; MeO, 2; Cl, 2; H, 2	569
C_{11}			
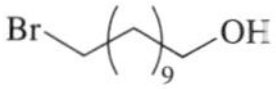	TEMPO (cat.), $CuBr{\bullet}Me_2S$ (cat.), **1** (cat.), O_2, perfluorooctyl bromide, chlorobenzene, 90°	 (81)	243, 414
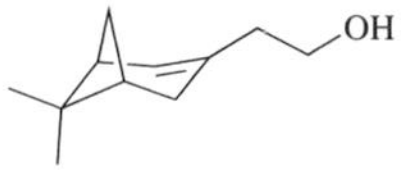	AcNH-TEMPO, TsOH, CH_2Cl_2, 0°, 2 h	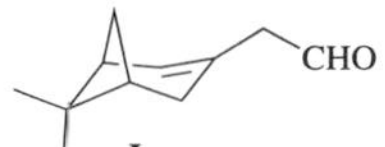 **I** (76)	56
	TEMPO (cat.), $PhI(OAc)_2$, CH_2Cl_2, rt	**I** (—)[b]	205

Alcohol	Conditions	Product(s) and Yield(s) (%)	Refs.
C_{11} 	1. TEMPO (cat.), PS-I(OAc)$_2$ (2 eq), acetone, rt 2. H_2O, 24 h	 **I** (84)	204
	TEMPO (cat.), CuBr•Me_2S (cat.), **1** (cat.), O_2, perfluorooctyl bromide, chlorobenzene, 90°	**I** (78)	243, 414
	BzO-TEMPO, $RuCl_2(PPh_3)_3$ (cat.), O_2, toluene, 78°, 7 h	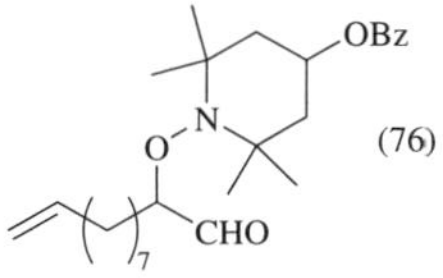 (76)	411
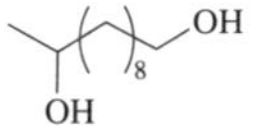	TEMPO (cat.), KBr (cat.), NaOCl (1.1 eq), CH_2Cl_2, H_2O, pH 9.5	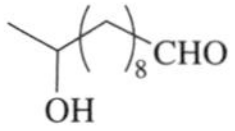 (68)	151
	TEMPO (cat.), KBr (cat.), NaOCl (2.2 eq), CH_2Cl_2, H_2O, pH 9.5	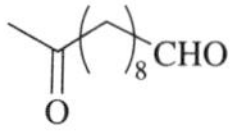 (69)	151
	TEMPO (cat.), KBr (cat.), NaOCl (3.6 eq), CH_2Cl_2, H_2O, pH 9.5	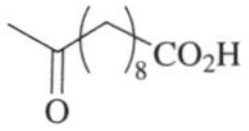 (57)	151
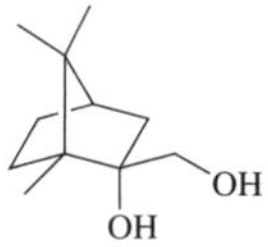	TEMPO (cat.), KBr (cat.), NaOCl, CH_2Cl_2, H_2O	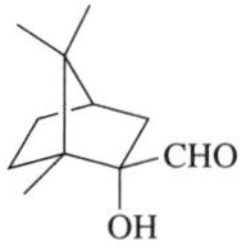 (99)	570
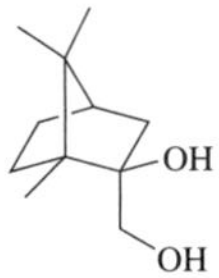	TEMPO (cat.), KBr (cat.), NaOCl, CH_2Cl_2, H_2O	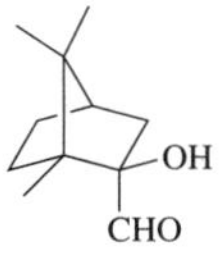 (98)	570

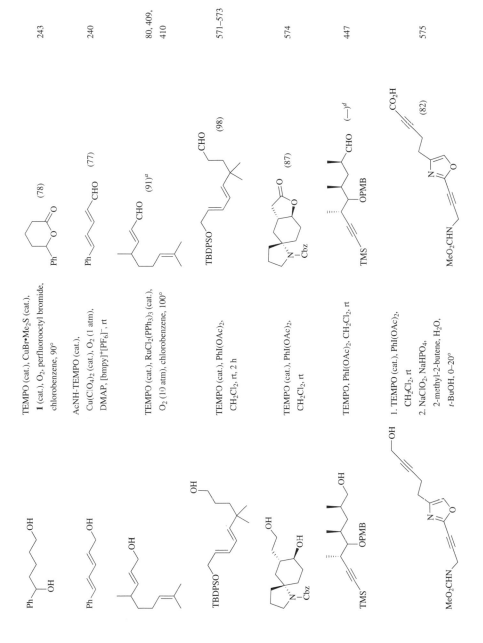

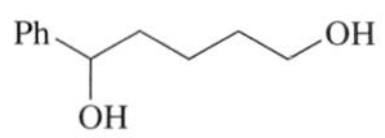	TEMPO (cat.), CuBr•Me_2S (cat.), **1** (cat.), O_2, perfluorooctyl bromide, chlorobenzene, 90°	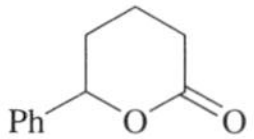 (78)	243
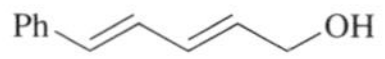	AcNH-TEMPO (cat.), $Cu(ClO_4)_2$ (cat.), O_2 (1 atm), DMAP, $[bmpy]^+[PF_6]^-$, rt	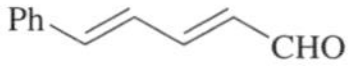 (77)	240
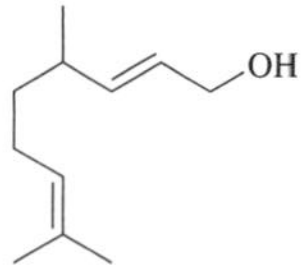	TEMPO (cat.), $RuCl_2(PPh_3)_3$ (cat.), O_2 (10 atm), chlorobenzene, 100°	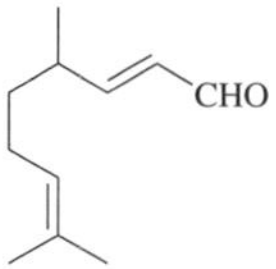 (91)[a]	80, 409, 410
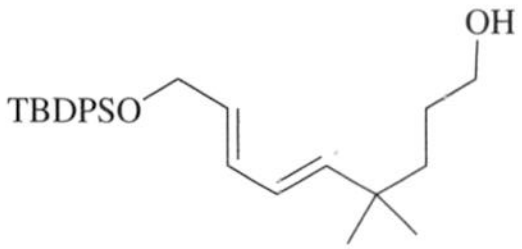	TEMPO (cat.), $PhI(OAc)_2$, CH_2Cl_2, rt, 2 h	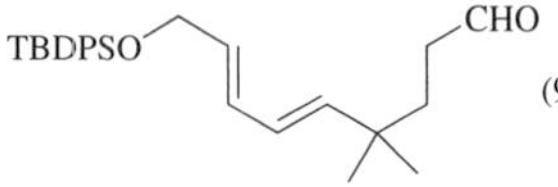 (98)	571–573
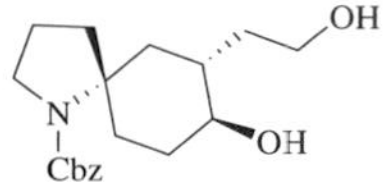	TEMPO (cat.), $PhI(OAc)_2$, CH_2Cl_2, rt	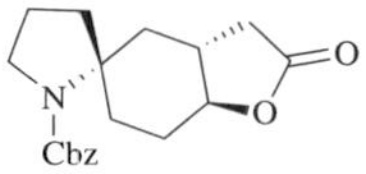 (87)	574
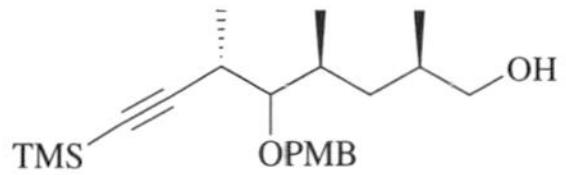	TEMPO, $PhI(OAc)_2$, CH_2Cl_2, rt	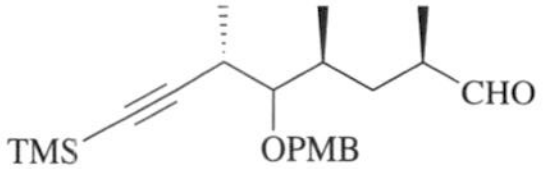 (—)[d]	447
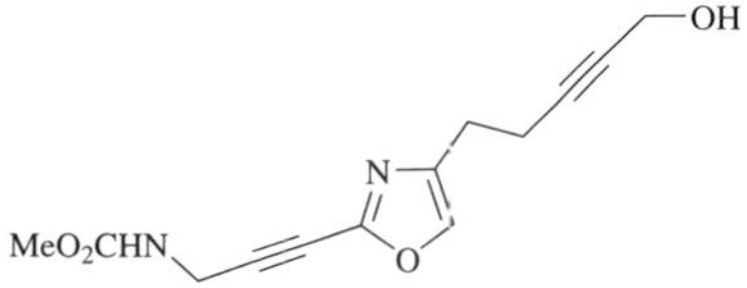	1. TEMPO (cat.), $PhI(OAc)_2$, CH_2Cl_2, rt 2. $NaClO_2$, $NaHPO_4$, 2-methyl-2-butene, H_2O, *t*-BuOH, 0–20°	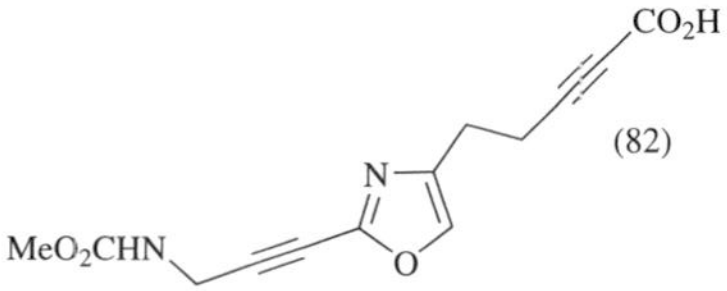 (82)	575

TABLE 3. OXIDATIONS OF POLYFUNCTIONAL ALCOHOLS (*Continued*)

	Alcohol	Conditions	Product(s) and Yield(s) (%)	Refs.
C_{11}	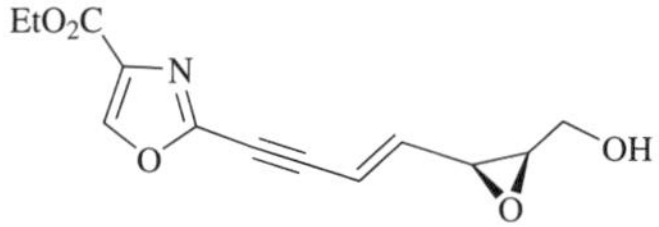	TEMPO (cat.), PhI(OAc)$_2$, CH_2Cl_2, rt	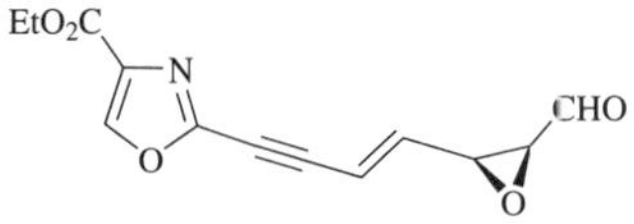 (81)	576
	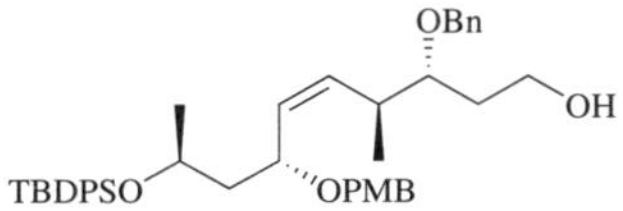	TEMPO (cat.), NaOCl (cat.), $NaClO_2$, MeCN, H_2O, buffer, 35°	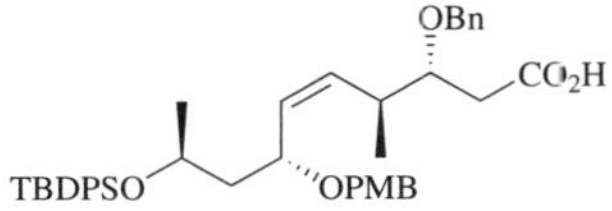 (96)	577
	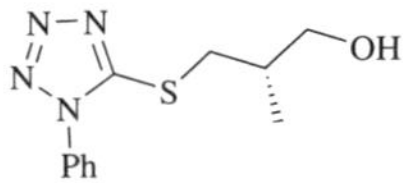	TEMPO (cat.), [PS-CH_2NMe_3]$^+$[Br(OAc)$_2$]$^-$, CH_2Cl_2, 3.5 h	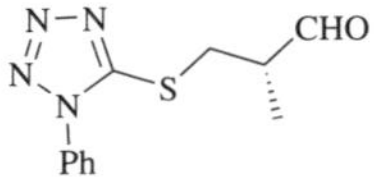 (—)[b]	43
	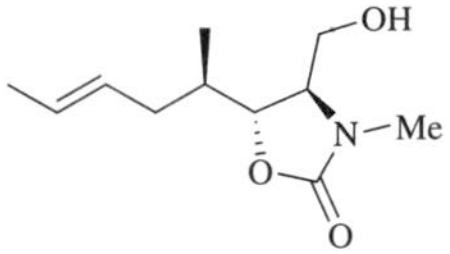	TEMPO (cat.), PhI(OAc)$_2$, MeCN, H_2O, rt, 2 h	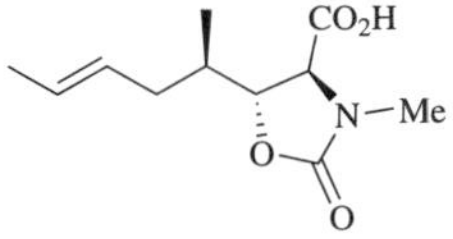 (—)[b]	578
	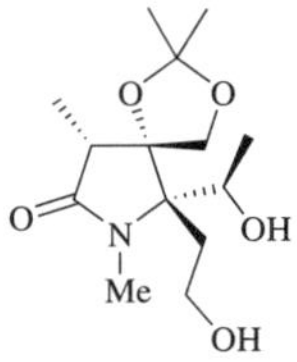	TEMPO (cat.), PhI(OAc)$_2$, CH_2Cl_2	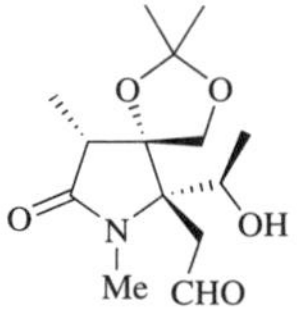 (93)	579
	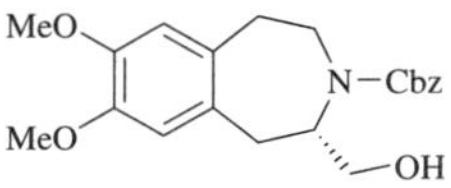	TEMPO (cat.), NaOCl (cat.), $NaClO_2$, MeCN, H_2O, buffer, pH 6.86, 35°	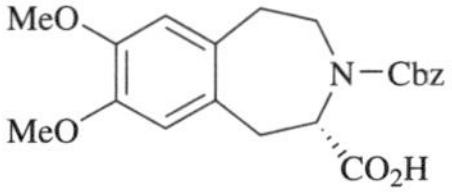 (92)	580

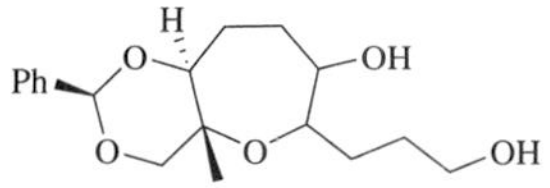	TEMPO (cat.), KBr (cat.), NaOCl, CH_2Cl_2, H_2O, 0°	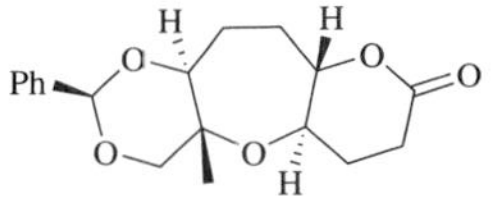 I + 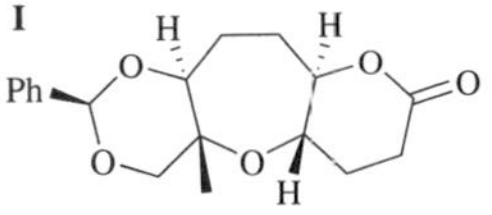 II **I** + **II** (97)^h, **I**:**II** = 57:43	581
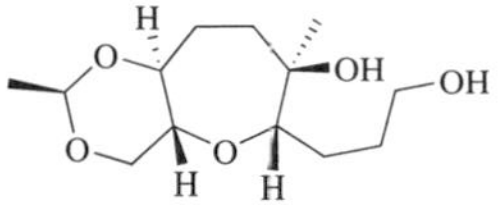	TEMPO (cat.), KBr (cat.), NaOCl, $NaHCO_3$, CH_2Cl_2, H_2O	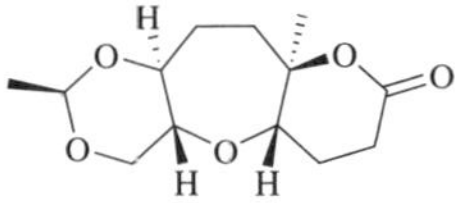 (86)	582
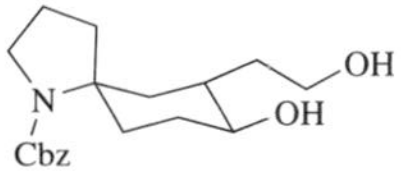	TEMPO (cat.), PhI(OAc)$_2$, CH_2Cl_2, rt, 20 h	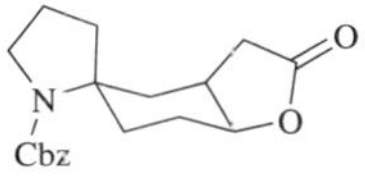 (87)	583
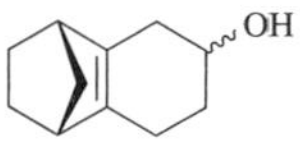	TEMPH•HCl, *m*-CPBA, CH_2Cl_2, rt, 8 h	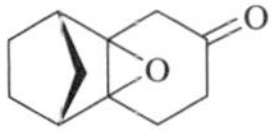 (—)^b	584
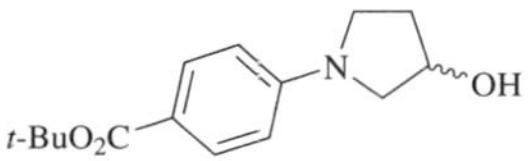	TEMPO (cat.), NaOCl, TBAC, NaCl, $NaHCO_3$, CH_2Cl_2, 0°, 1 h	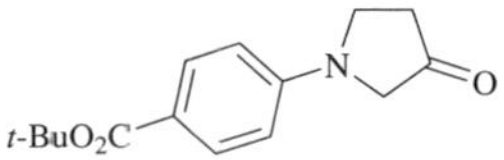 (—)^b	502
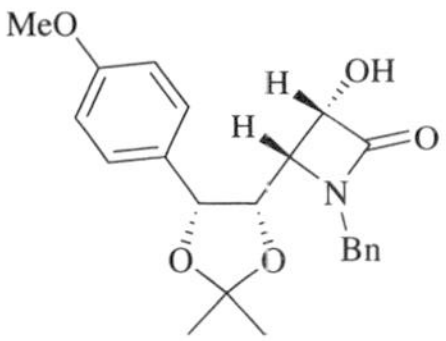	TEMPO (cat.), KBr (cat.), NaOCl, $NaHCO_3$, CH_2Cl_2, H_2O, 3 min	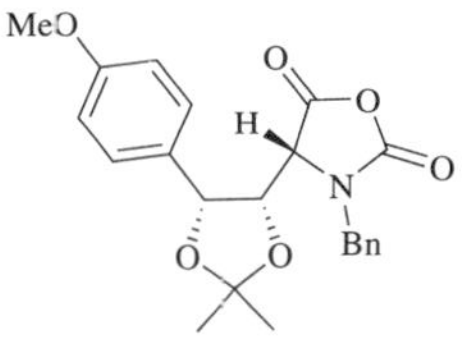 (98)	495, 585

TABLE 3. OXIDATIONS OF POLYFUNCTIONAL ALCOHOLS (*Continued*)

	Alcohol	Conditions	Product(s) and Yield(s) (%)	Refs.
C_{11}	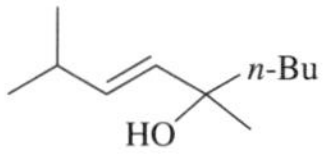	$[TEMP{=}O]^+[BF_4]^-$, MeCN, rt, 6 h	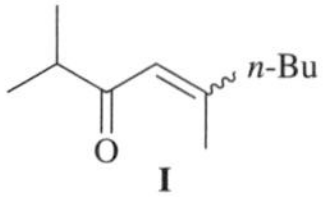 **I** (58), *E*:*Z* = 2 1	120
		$[TEMP{=}O]^+[BF_4]^-$, MeCN, H_2O, rt, 5 h	**I** (78), *E*:*Z* = 2:1	120
		$[TEMP{=}O]^+[SbF_6]^-$, MeCN, H_2O, rt, 4 h	**I** (83), *E*:*Z* = 2:1	120
	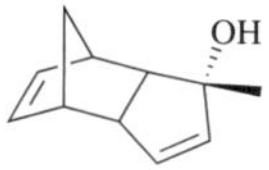	$[TEMP{=}O]^+[BF_4]^-$, MeCN, rt, 0.3 h	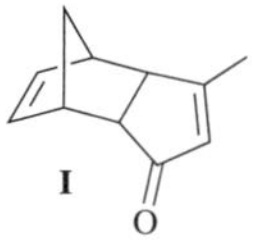 **I** (80)	120
		$[TEMP{=}O]^+[BF_4]^-$, MeCN, H_2O, rt, 0.2 h	**I** (98)	120
C_{12}		$[MeO\text{-}TEMP{=}O]^+Cl^-$, CH_2Cl_2, dark, rt, 2 h	 (99)[i]	135
	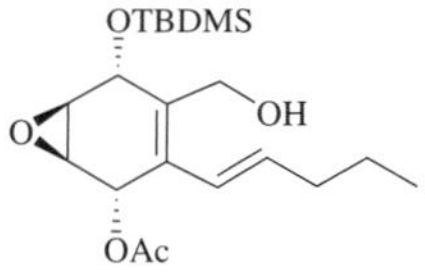	TEMPO (cat.), NaOCl (cat.), $NaClO_2$, NaH_2PO_4, cyclopentene, rt, 15 min	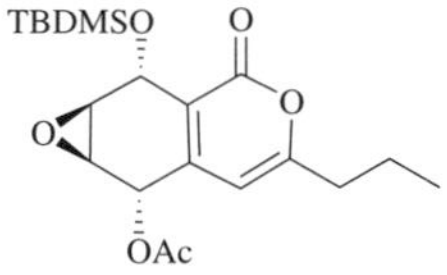 (35)	357
	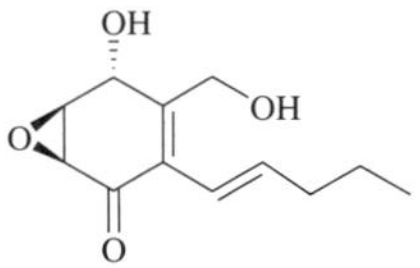	$[AcNH\text{-}TEMP{=}O]^+[BF_4]^-$, silica gel, CH_2Cl_2	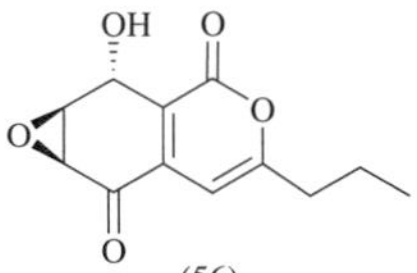 (56) + 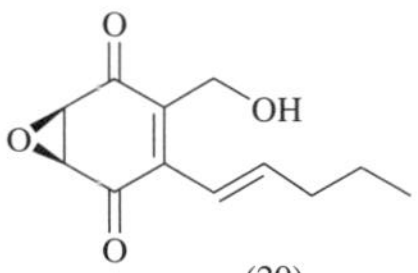 (20)	356

	TEMPO (cat.), CuCl (cat.), O_2, DMF, rt, 30 mi:n	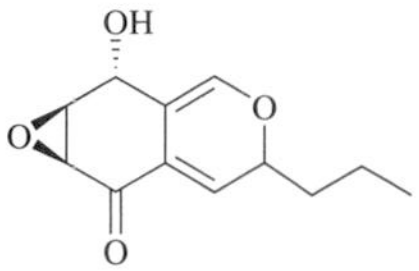 (—)[b]	356
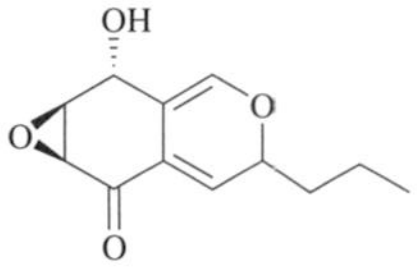	$[AcNH\text{-}TEMP{=}O]^+[BF_4]^-$, TEMPOH, silica gel, CH_2Cl_2, rt, 2 h	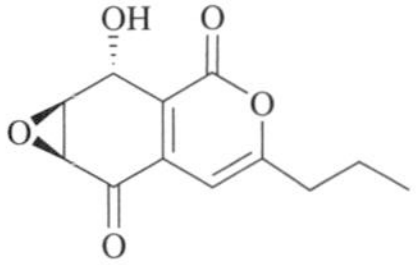 (76)	356
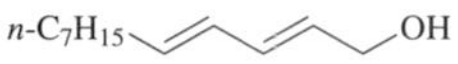	$[AcNH\text{-}TEMP{=}O]^+[BF_4]^-$, CH_2Cl_2, silica gel, rt, 16–24 h	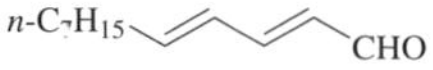 (52)	131
	$[AcNH\text{-}TEMP{=}O]^+[BF_4]^-$, pyridine, CH_2Cl_2, rt, 16–24 h	 (62)	131
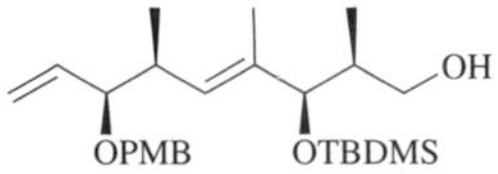	TEMPO (cat.), $PhI(OAc)_2$, CH_2Cl_2	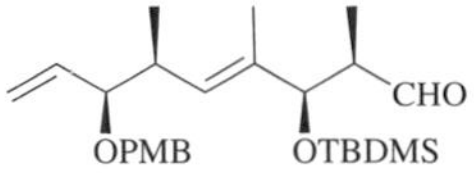 (85)	478
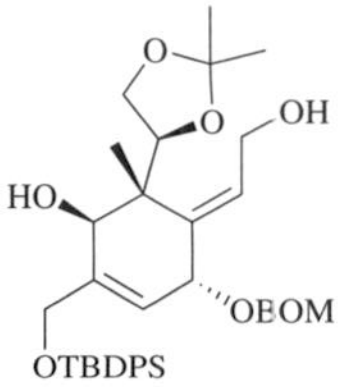	1. TEMPO (cat.;), NCS 2. $NaClO_2$, NaH_2PO_4, 2-methyl-2-butene, H_2O	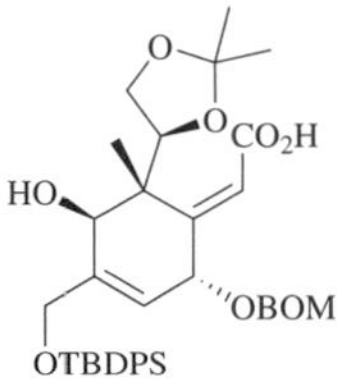 (—)[b]	586
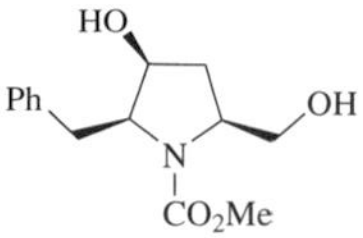	TEMPO (cat.), NaBr (cat.), NaOCl, $NaHCO_3$, EtOAc, toluene, H_2O, –5°	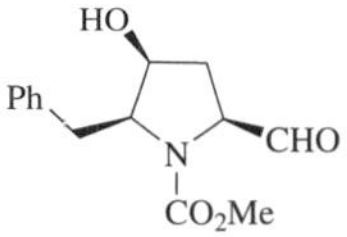 (72)	587

TABLE 3. OXIDATIONS OF POLYFUNCTIONAL ALCOHOLS (*Continued*)

Alcohol	Conditions	Product(s) and Yield(s) (%)	Refs.
C_{12}			
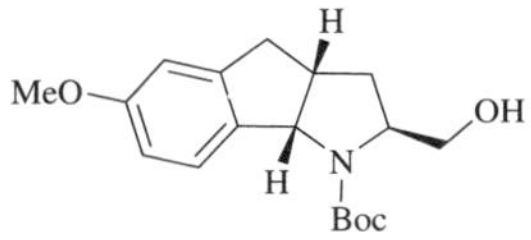	TEMPO (cat.), NaOCl (cat.), $NaClO_2$, MeCN, H_2O, phosphate buffer, 35°	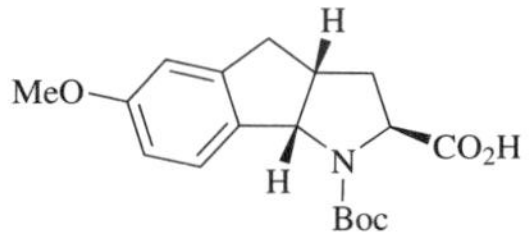 (97)	588
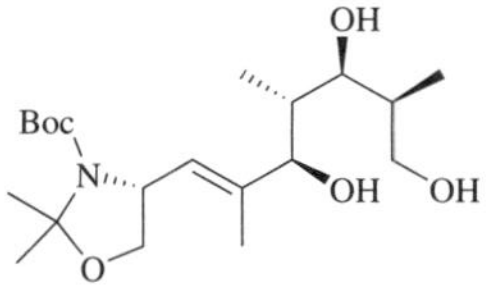	TEMPO (cat.), $PhI(OAc)_2$, CH_2Cl_2, rt	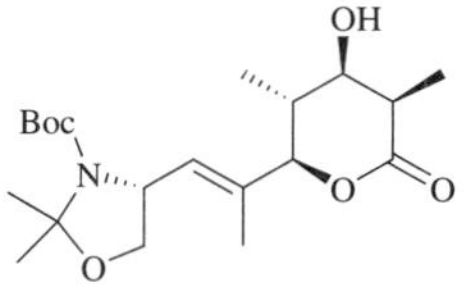 (95)	486
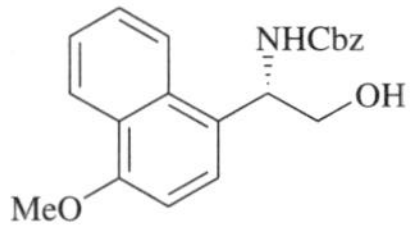	TEMPO (cat.), KBr (cat.), NaOCl, $NaHCO_3$ (5%), H_2O, acetone, 0°, 2 h	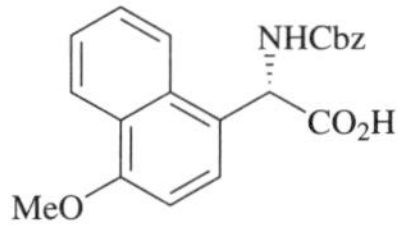 (78)	515
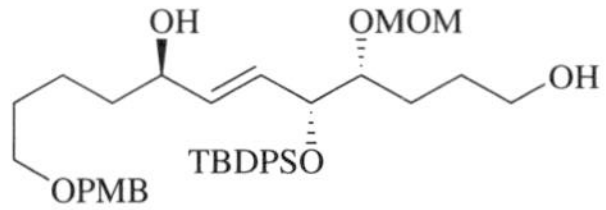	TEMPO (cat.), $PhI(OAc)_2$, CH_2Cl_2, 6 h	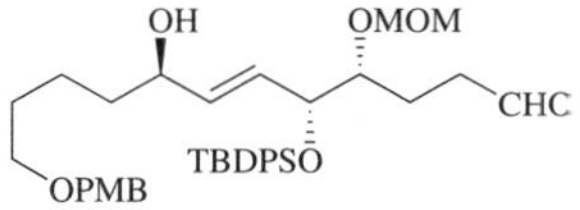 (—)[b]	589
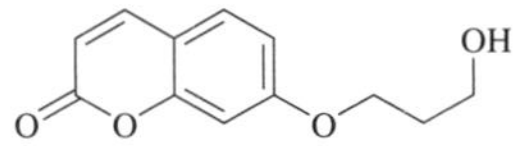	TEMPO (cat.), NaOCl, $NaHCO_3$, CH_2Cl_2, H_2O, BSA buffer, rt	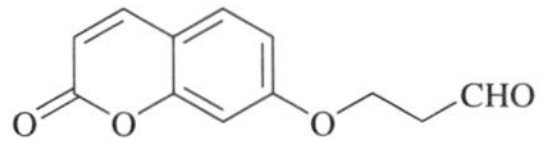 (—)[b]	590

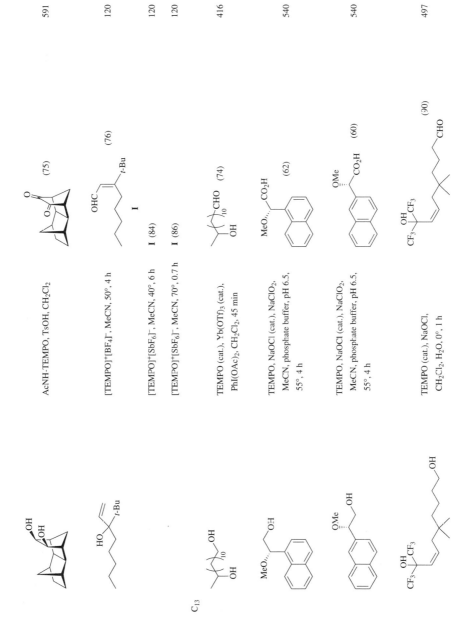

	AcNH-TEMPO, TsOH, CH_2Cl_2	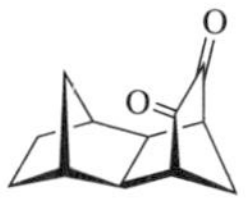 (75)	591
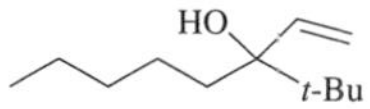	$[TEMPO]^+[BF_4]^-$, MeCN, 50°, 4 h	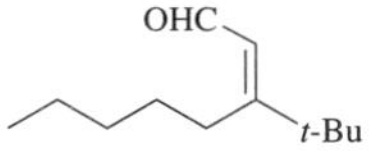 **I** (76)	120
	$[TEMPO]^+[SbF_6]^-$, MeCN, 40°, 6 h	**I** (84)	120
	$[TEMPO]^+[SbF_6]^-$, MeCN, 70°, 0.7 h	**I** (86)	120
C_{13}			
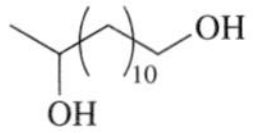	TEMPO (cat.), $Yb(OTf)_3$ (cat.), $PhI(OAc)_2$, CH_2Cl_2, 45 min	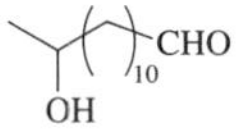 (74)	416
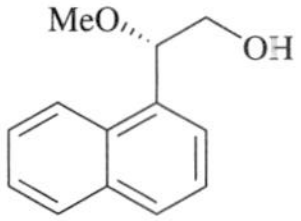	TEMPO, NaOCl (cat.), $NaClO_2$, MeCN, phosphate buffer, pH 6.5, 55°, 4 h	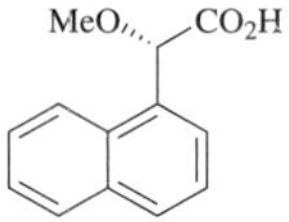 (62)	540
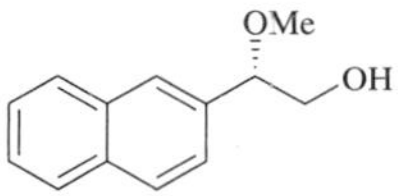	TEMPO, NaOCl (cat.), $NaClO_2$, MeCN, phosphate buffer, pH 6.5, 55°, 4 h	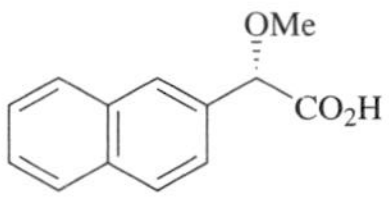 (60)	540
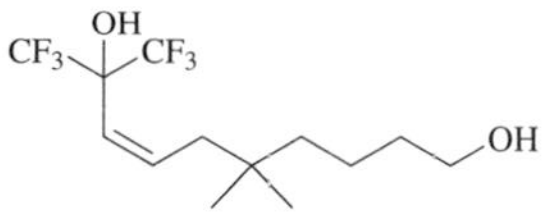	TEMPO (cat.), NaOCl, CH_2Cl_2, H_2O, 0°, 1 h	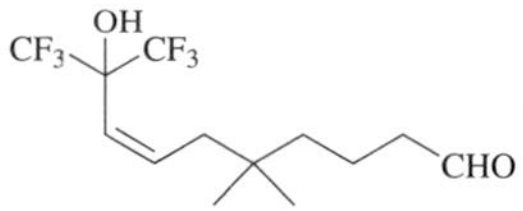 (90)	497

TABLE 3. OXIDATIONS OF POLYFUNCTIONAL ALCOHOLS (*Continued*)

	Alcohol	Conditions	Product(s) and Yield(s) (%)	Refs.
C_{13}	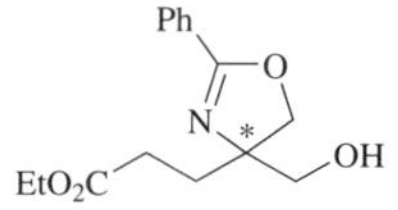	TEMPO (cat.), PhI(OAc)$_2$, CH_2Cl_2, H_2O, rt, 75 min	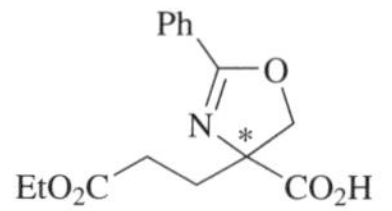 *: *R* (—)[b]; *S* (—)[b]	592
	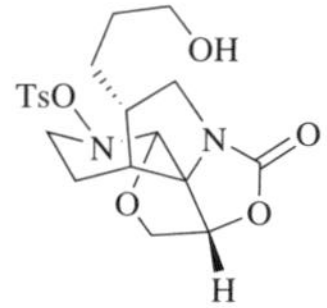	TEMPO (cat.), NaOCl (cat.), $NaClO_2$, MeCN, 37°	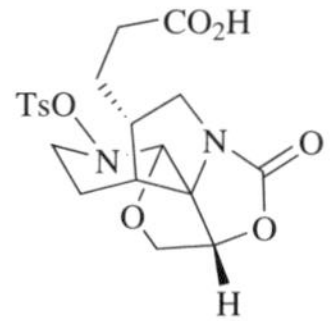 (96)	593
	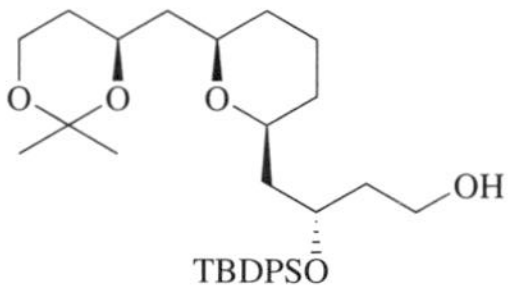	TEMPO (cat.), NaOCl (cat.), $NaClO_2$, MeCN, 50°, 4 h	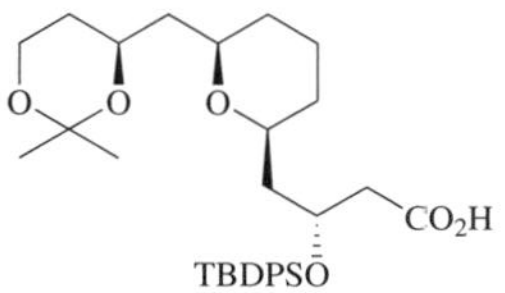 (92)	594
	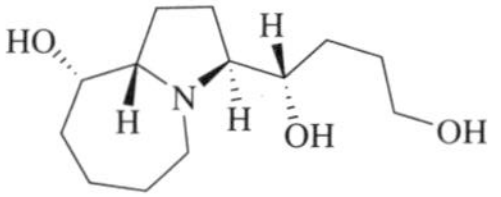	TEMPO (cat.), PhI(OAc)$_2$, AcOH, rt	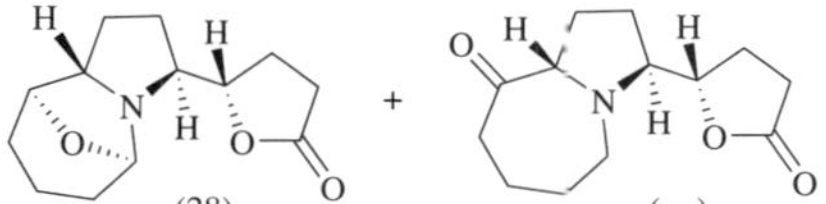 (28) + (—)	348
	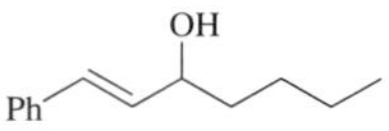	TEMPO (cat.), CuBr•Me_2S (cat.), **1** (cat.), O_2, perfluorooctyl bromide, chlorobenzene, 90°	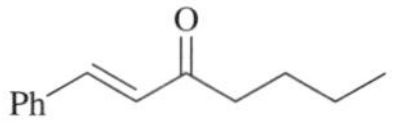 (84)	243, 414
	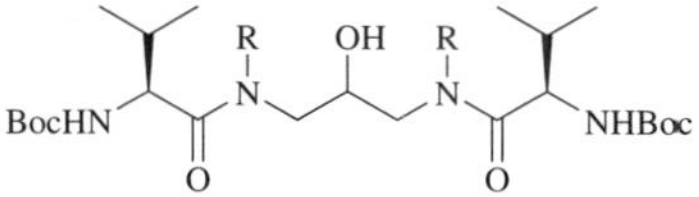	TEMPO (cat.), KBr (cat.), NaOCl, $NaHCO_3$, CH_2Cl_2, H_2O, pH 8	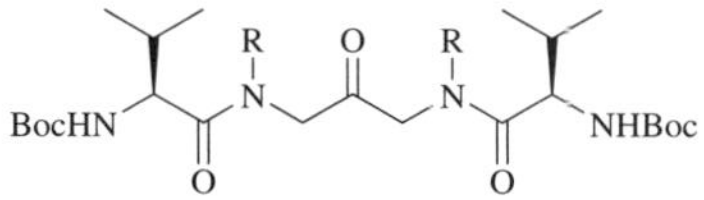 R: H (98); Bn (97)	475

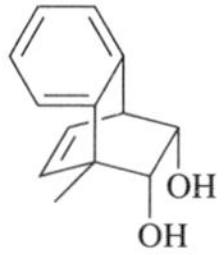	AcNH-TEMPO, TsOH, CH_2Cl_2, 0°	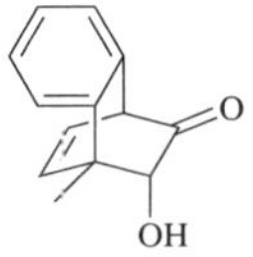 (94)	595
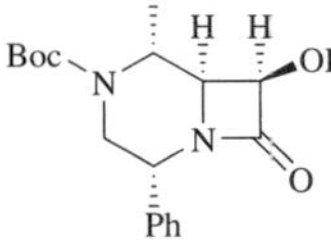	TEMPO (cat.), KBr (cat.), NaOCl, $NaHCO_3$, CH_2Cl_2, H_2O, phosphate buffer, 0°, 1 min	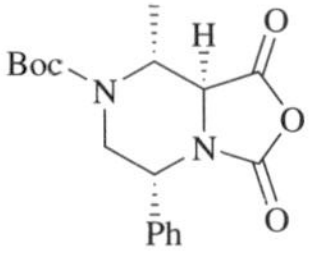 (100)	596
C_{14} 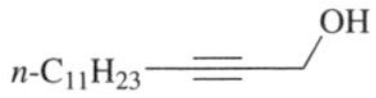	BzO-TEMPO (cat.), NaBr (cat.), electrolysis, $NaHCO_3$, CH_2Cl_2, H_2O, pH 8.6	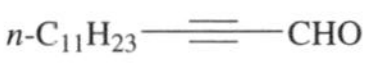 **I** (90)	222
	BzO-TEMPO (cat.), $TBABr_3$, NaOAc, CH_2Cl_2, H_2O, rt, 1 h	**I** (16) + 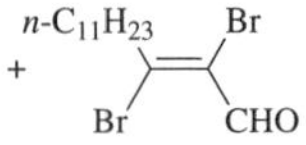 (79)	224
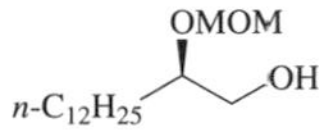	TEMPO (cat.), NaBr (cat.), NaOCl, $NaHCO_3$, toluene, EtOAc, H_2O, 0°	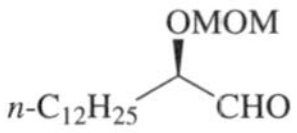 (—)	597
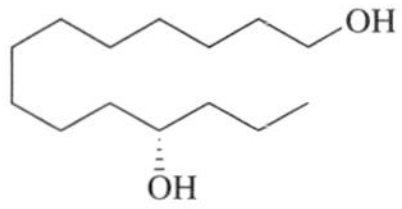	1. TEMPO (cat.), NCS, TBAC, CH_2Cl_2, H_2O, phosphate buffer pH 8.6 2. $NaClO_2$, *t*-BuOH, 2-methyl-2-butene, H_2O 3. CH_2N_2, Et_2O	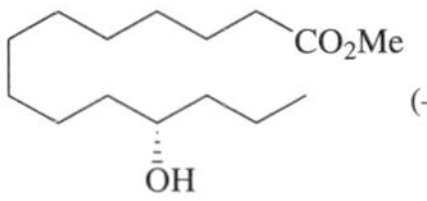 (—)	598
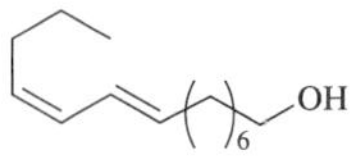	$[AcNH\text{-}TEMP{=}O]^+[BF_4]^-$, silica gel, CH_2Cl_2, rt, 16–24 h	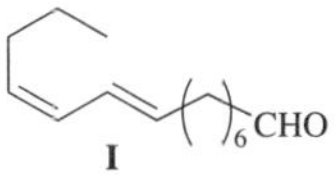 **I** (38)	131
	$[AcNH\text{-}TEMP{=}O]^+[BF_4]^-$, pyridine, CH_2Cl_2, rt, 16–24 h	**I** (75)	131

TABLE 3. OXIDATIONS OF POLYFUNCTIONAL ALCOHOLS (*Continued*)

Alcohol	Conditions	Product(s) and Yield(s) (%)	Refs.
C_{14} 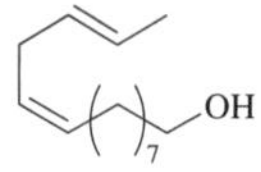	$[AcNH\text{-}TEMP{=}O]^+[BF_4]^-$, pyridine, CH_2Cl_2, rt, 16–24 h	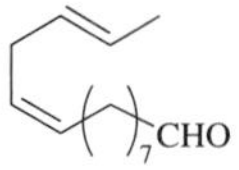 (53)	131
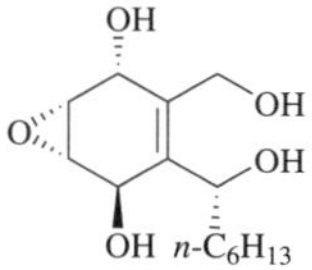	TEMPO (cat.), NaOCl (cat.), $NaClO_2$, NaH_2PO_4, 4 h	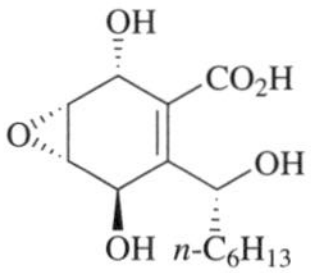 (50)	599
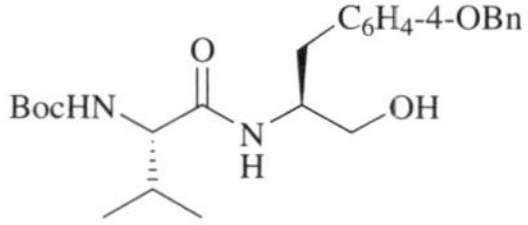	TEMPO (cat.), $PhI(OAc)_2$, CH_2Cl_2, rt, 2 h	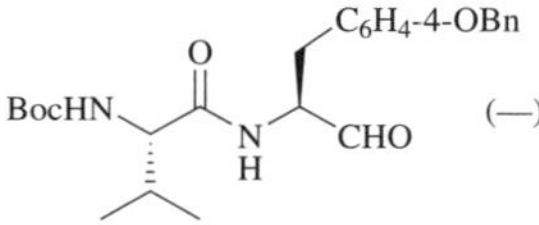 (—)[b]	600
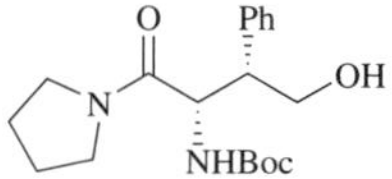	TEMPO (cat.), NaOCl (cat.), $NaClO_2$, MeCN, H_2O, buffer, pH 6.7, 35°, 24 h	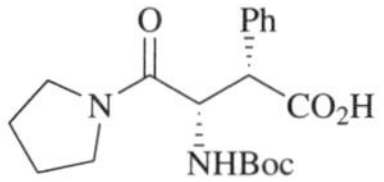 (74)	601
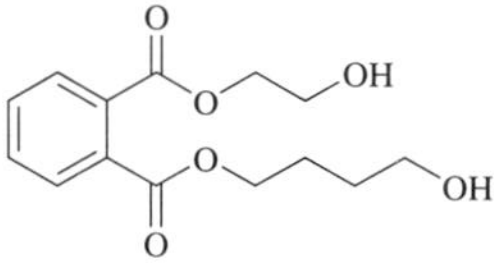	$[MeO\text{-}TEMP{=}O]^+Cl^-$, Na_2CO_3, CH_2Cl_2, 0°	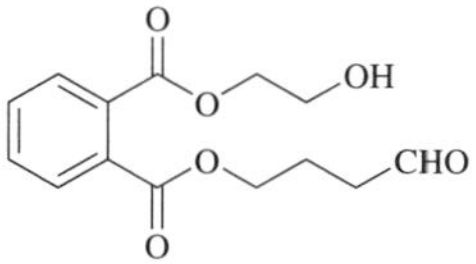 (81[a], 46)	69
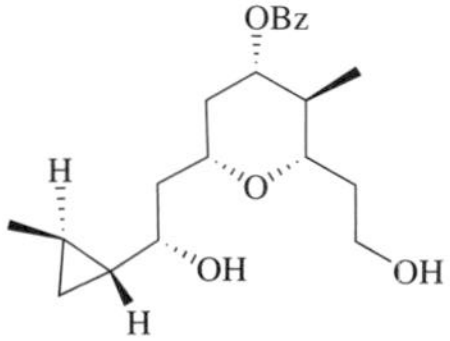	TEMPO (cat.), KBr (cat.), NaOCl, TBAC, NaCl, $NaHCO_3$, CH_2Cl_2, H_2O, 0°	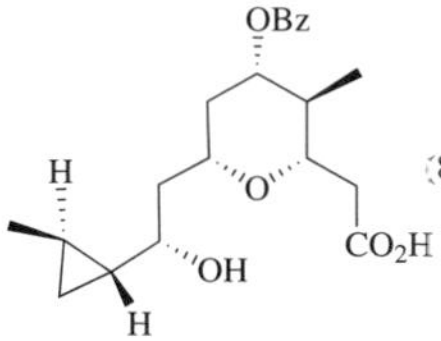 (81)	602

Reactant	Conditions	Product(s) and Yield(s) (%)	Refs.
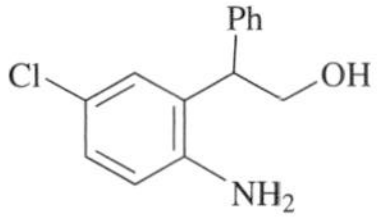	TEMPO (cat.), DABCO–CuCl (cat.), O_2, K_2CO_3, toluene, 100°, 24 h	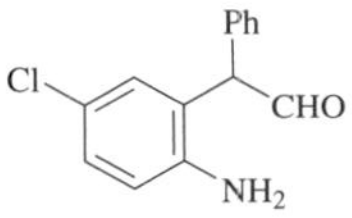 (58)	247
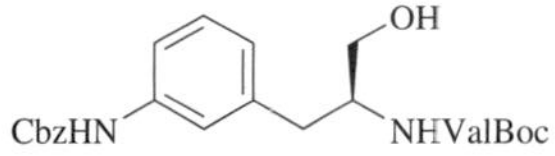	TEMPO (cat.), PhI(OAc)$_2$, CH_2Cl_2, rt, 2 h	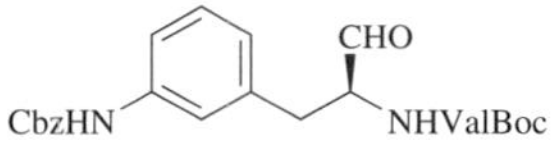 (—)[b]	600
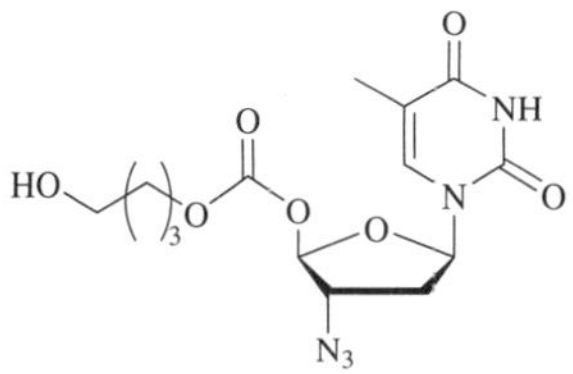	TEMPH (cat.), *m*-CPBA, HCl, CH_2Cl_2	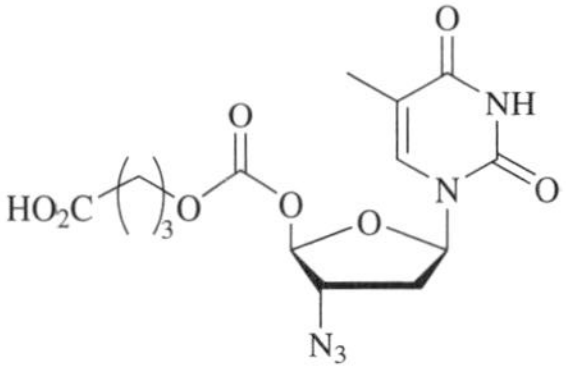 (69)	603
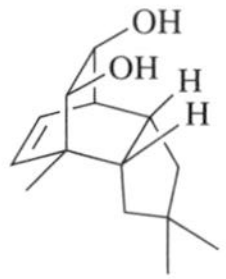	AcNH-TEMPO (2 eq), TsOH (2 eq), CH_2Cl_2, 0–18°, 16 h	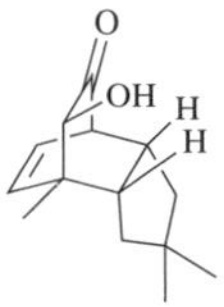 (91)	146, 604
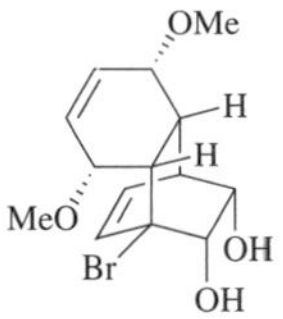	AcNH-TEMPO (2 eq), TsOH (2 eq), CH_2Cl_2, 18°, 16 h	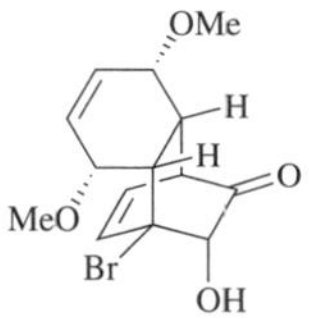 (82)	147
C_{14-17}			
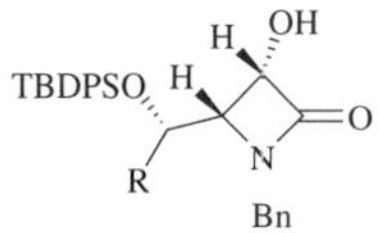	TEMPO (cat.), KBr (cat.), NaOCl, $NaHCO_3$, CH_2Cl_2, H_2O, phosphate buffer, pH 6.9, 1–2 min	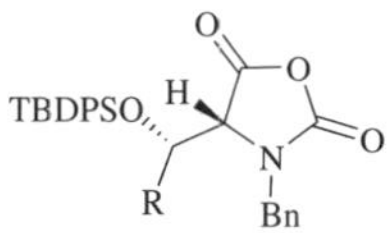 R: *i*-Pr (95); Ph (95)	495

TABLE 3. OXIDATIONS OF POLYFUNCTIONAL ALCOHOLS (*Continued*)

Alcohol	Conditions	Product(s) and Yield(s) (%)	Refs.
C_{15} 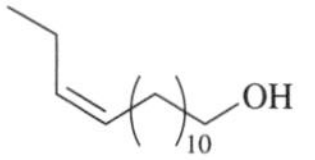	$[AcNH\text{-}TEMP{=}O]^+[BF_4]^-$, silica gel, CH_2Cl_2, rt, 16–24 h	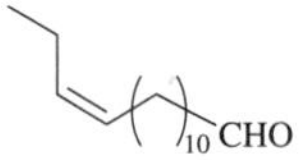 (84)	131
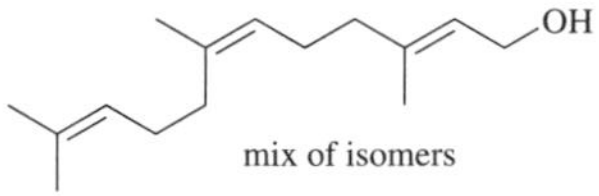 mix of isomers	AcNH-TEMPO (2 eq), TsOH (2 eq), CH_2Cl_2, 0°	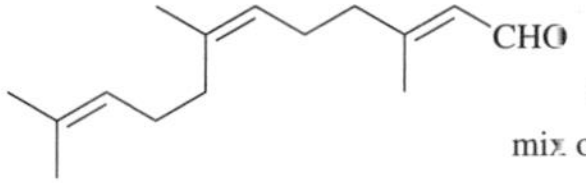 (86) mix of isomers	56
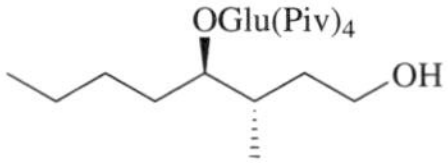	TEMPO (cat.), $PhI(OAc)_2$, MeCN, H_2O, rt, 48 h	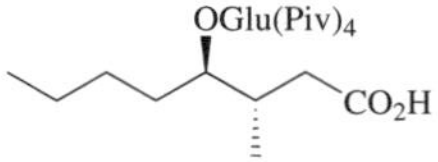 (75)	605
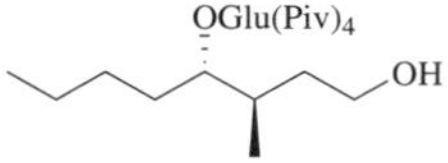	TEMPO (cat.), $PhI(OAc)_2$, MeCN, H_2O, rt, 48 h	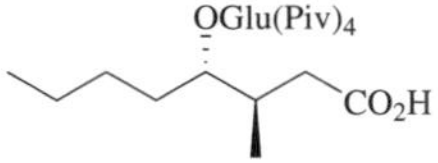 (80)	605
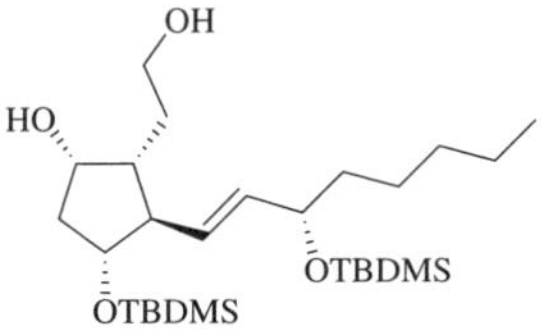	BzO-TEMPO (cat.), gradual addition of $NaBrO_2$, $NaHCO_3$, CH_2Cl_2, H_2O	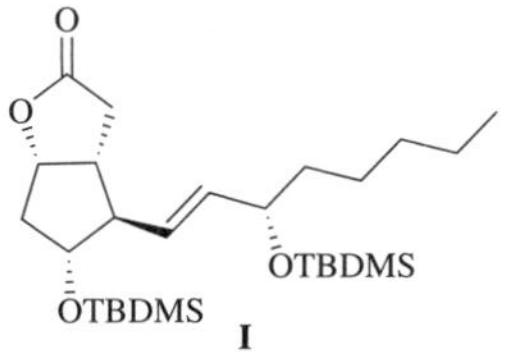 **I** (83)	606
	BzO-TEMPO (cat.), fast addition of $NaBrO_2$, $NaHCO_3$, CH_2Cl_2, H_2O	**I** (71) + 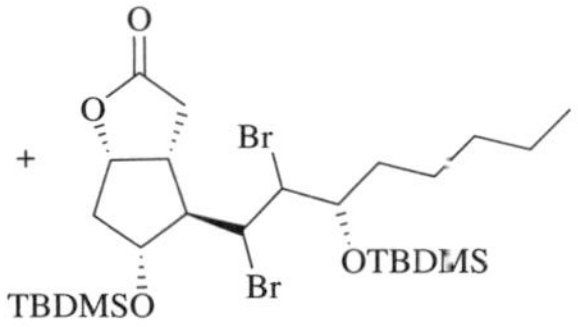 (8)	606

Substrate	Conditions	Product(s) and Yield(s) (%)	Refs.
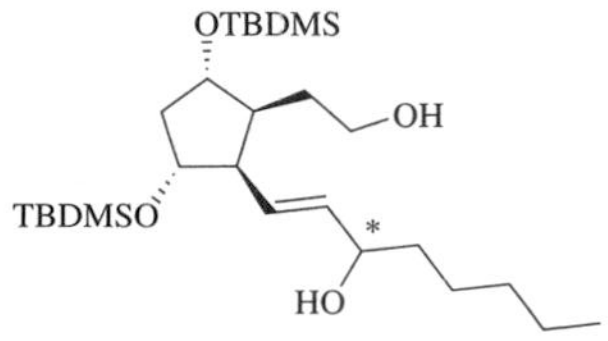	TEMPO (cat.), NCS, TBAC, CH_2Cl_2, H_2O	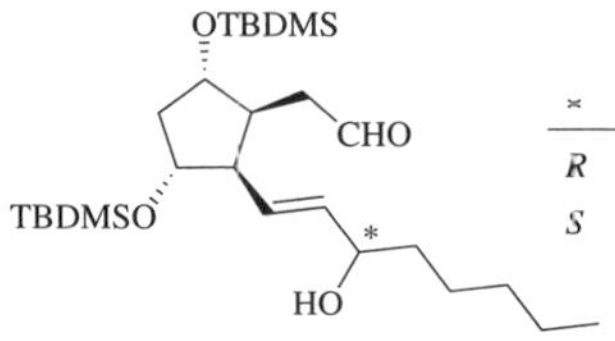 * / *R* (82) / *S* (80)	607
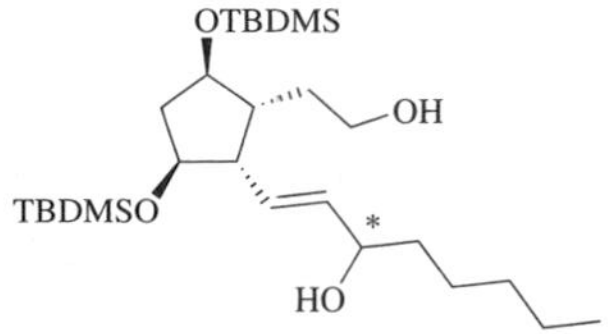	TEMPO (cat.), NCS, TBAC, CH_2Cl_2, H_2O	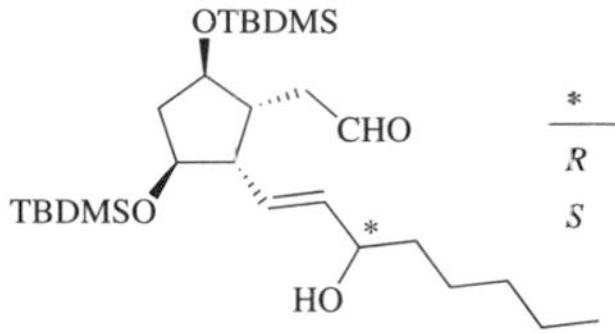 * / *R* (75) / *S* (82)	607
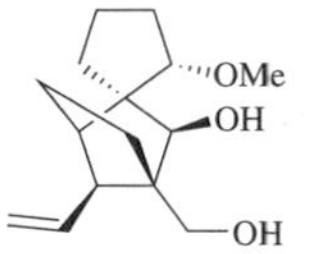	TEMPO (cat.), $PhI(OAc)_2$, CH_2Cl_2, rt, 12 h	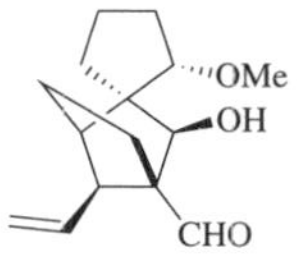 (75)	608
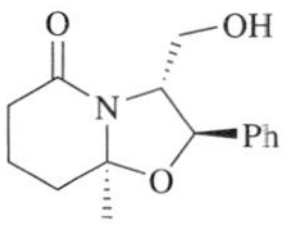	TEMPO (cat.), NaOCl (cat.), $NaClO_2$, MeCN	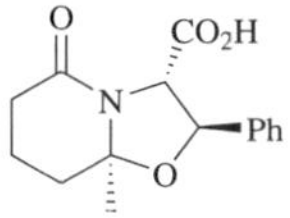 (95)	14, 185
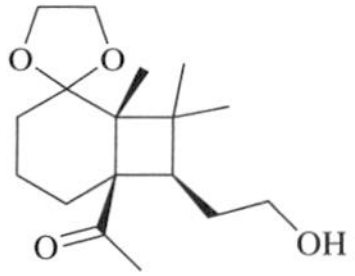	TEMPO (cat.), NCS, TBAC, CH_2Cl_2, H_2O	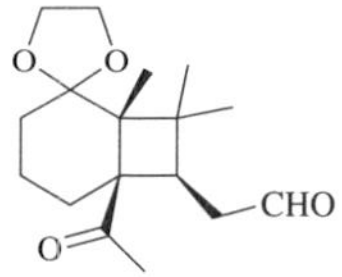 (—)[b]	609

TABLE 3. OXIDATIONS OF POLYFUNCTIONAL ALCOHOLS (*Continued*)

Alcohol	Conditions	Product(s) and Yield(s) (%)	Refs.
C_{15}			
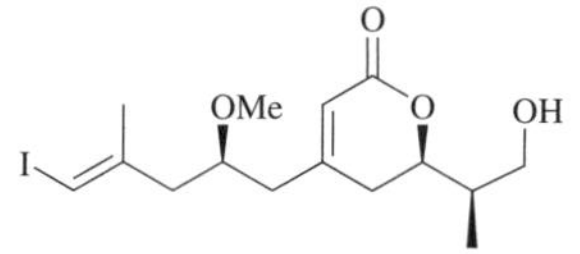	TEMPO (cat.), $PhI(OAc)_2$, CH_2Cl_2	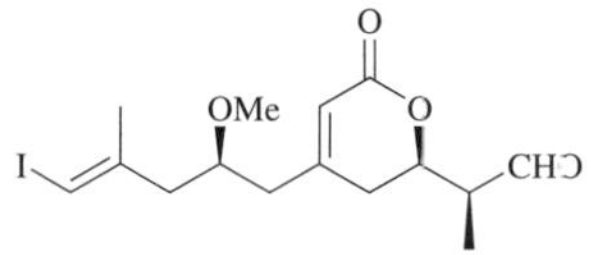 (—)	610
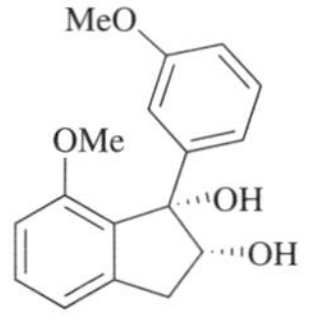	TEMPO (cat.), KBr (cat.), NaOCl, $NaHCO_3$, CH_2Cl_2, H_2O, 0°, 30 min	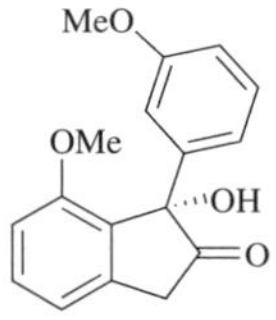 (96)	611
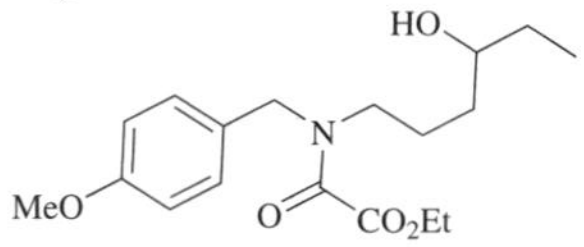	TEMPO (cat.), KBr (cat.), NaOCl, CH_2Cl_2, H_2O, pH 9.5, 0–15°	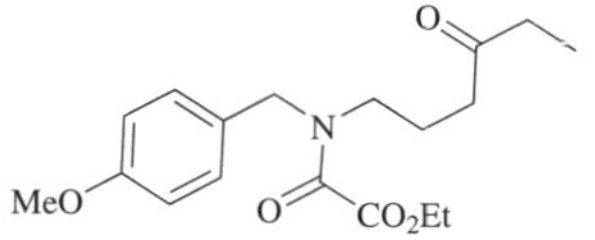 (85)	612
C_{15-19}			
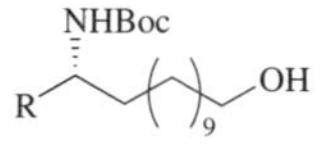	TEMPO (cat.), KBr (cat.), NaOCl (2.5 eq), $NaHCO_3$, toluene, EtOAc, H_2O, –5°	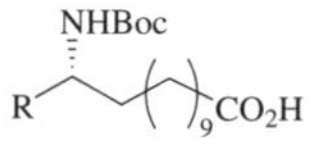 R: $MeO_2C(CH_2)_2$ (77); Bn (91)	470
C_{16}			
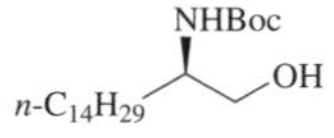	AcNH-TEMPO (cat.), NaBr (cat.), NaOCl, $NaHCO_3$, toluene, EtOAc, H_2O	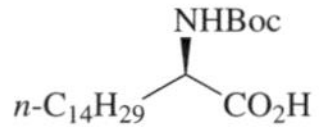 (67)	440
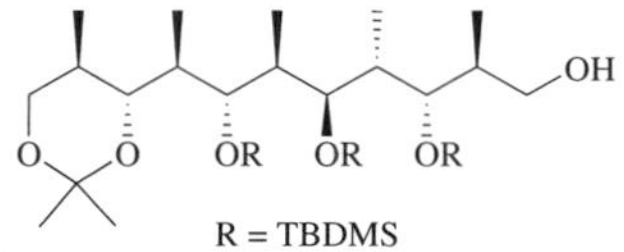 R = TBDMS	TEMPO (cat.), $PhI(OAc)_2$, CH_2Cl_2	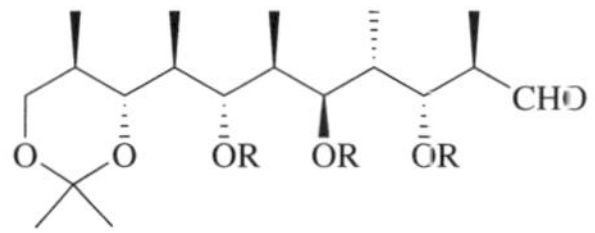 (—)[b]	613

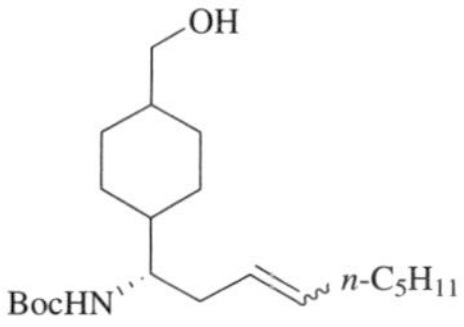	TEMPO (cat.), NaOCl (cat.), $NaClO_2$, phosphate buffer, MeCN, 35°	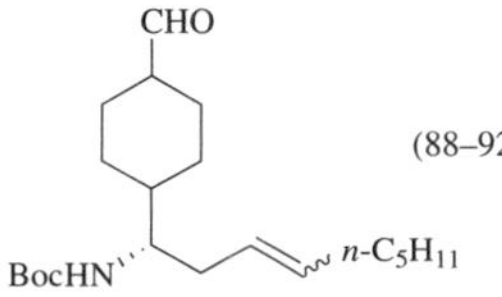 (88–92)	544
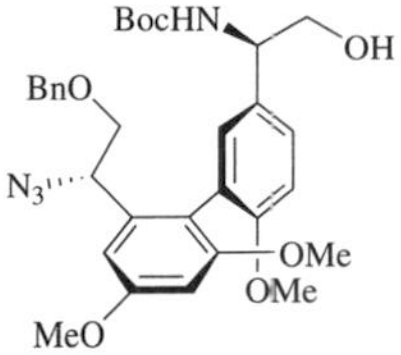	TEMPO (cat.), KBr (cat.), NaOCl, CH_2Cl_2, H_2O, pH 8.6, 0°	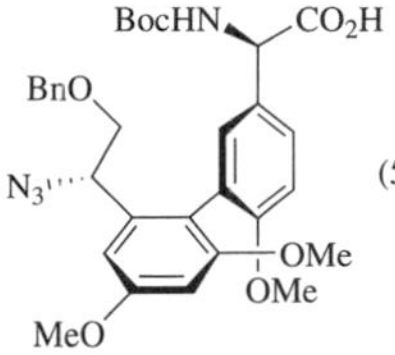 (53)	614
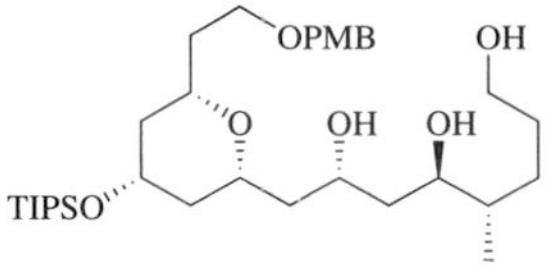	TEMPO (cat.), $PhI(OAc)_2$, CH_2Cl_2, rt	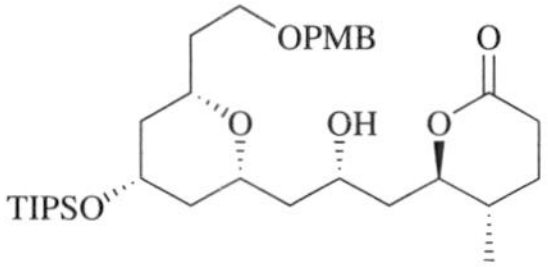 (92)	575
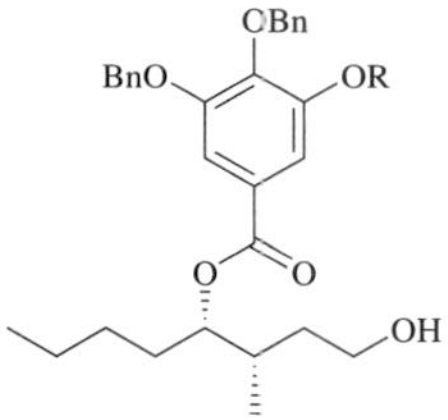	TEMPO (cat.), $PhI(OAc)_2$, MeCN, H_2O	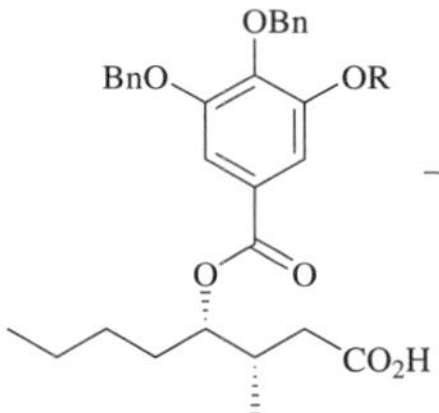 R: Me (92); Bn (84)	615
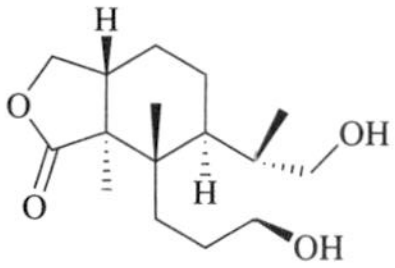	TEMPO (cat.), KBr (cat.), NaOCl, TBAC, NaCl, $NaHCO_3$, CH_2Cl_2, H_2O, 0°	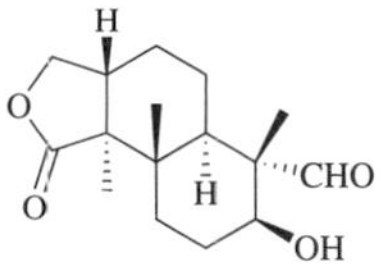 (100)	616

TABLE 3. OXIDATIONS OF POLYFUNCTIONAL ALCOHOLS (*Continued*)

Alcohol	Conditions	Product(s) and Yield(s) (%)	Refs.
C_{16-18} 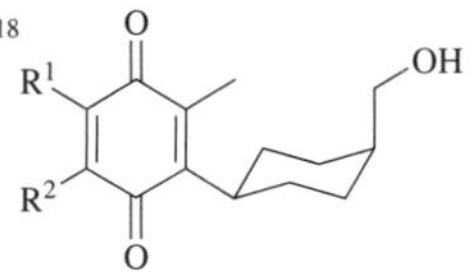	TEMPO (cat.), KBr (cat.), NaOCl, $NaHCO_3$, CH_2Cl_2, H_2O, pH 8.5, 0°	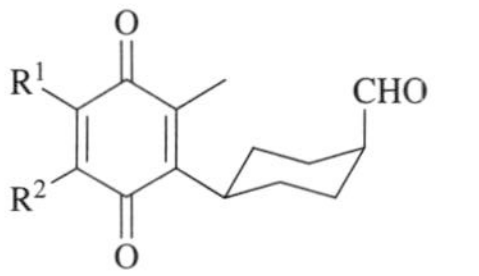	617
R^1 Me, R^2 Me		(98)	
R^1–R^2 —CH=CH—CH=CH—		(85)	
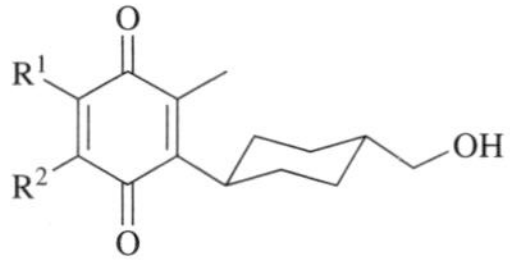	TEMPO (cat.), KBr (cat.), NaOCl, $NaHCO_3$, CH_2Cl_2, H_2O, pH 8.5, 0°	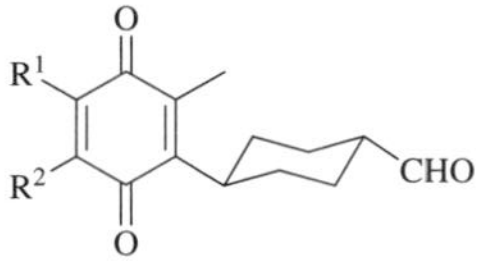	617
R^1 Me, R^2 Me		(86)	
R^1–R^2 —CH=CH—CH=CH—		(85)	
C_{17} 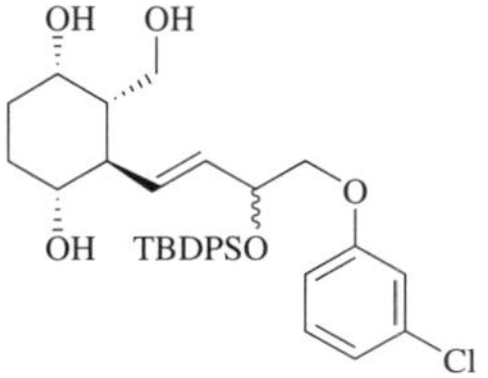	TEMPO (cat.), NCS, TBAC, Na_2CO_3, CH_2Cl_2, H_2O	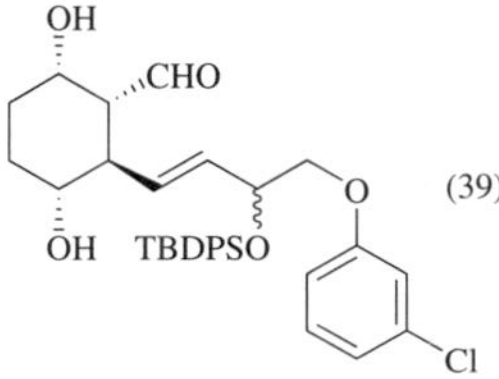 (39)	618
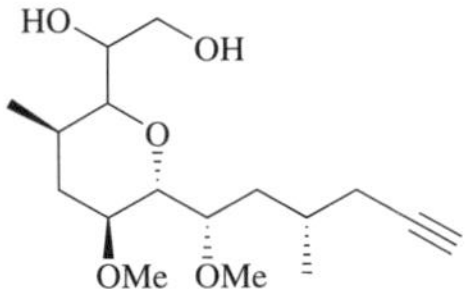	1. MeO-TEMPO (cat.), KBr (cat.), NaOCl, $NaHCO_3$ 2. NaH_2PO_4, $NaClO_2$	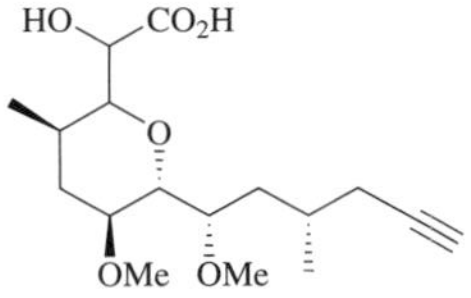 (—)[b]	619

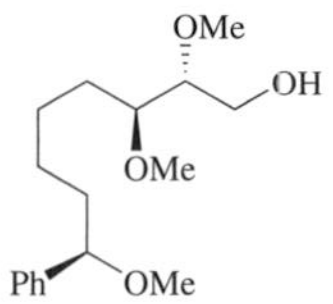	1. TEMPO (cat.), KBr (cat.), NaOCl, $NaHCO_3$, acetone, H_2O 2. $NaClO_2$, NaH_2PO_4, H_2O, 2-methyl-2-butene, 0°, 1.5 h	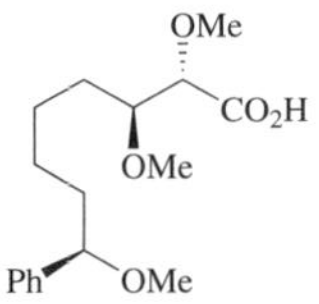	(—)[b]	555
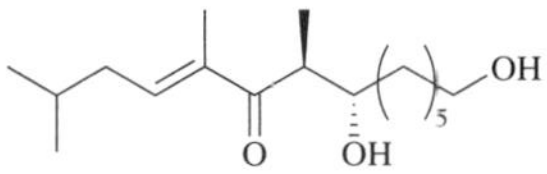	TEMPO (cat.), NaBr (cat.), NaOCl, TBAB, NaCl, $NaHCO_3$, CH_2Cl_2, H_2O, 15 min	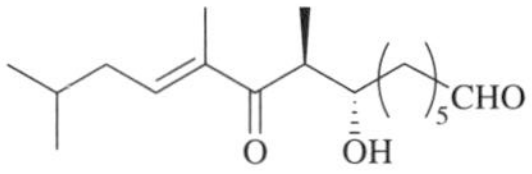	(—)[b]	620
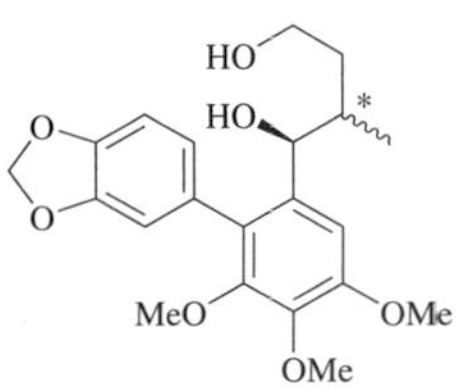	TEMPO (cat.), NCS, TBAI, $NaHCO_3$, K_2CO_3, CH_2Cl_2, rt, 3 h	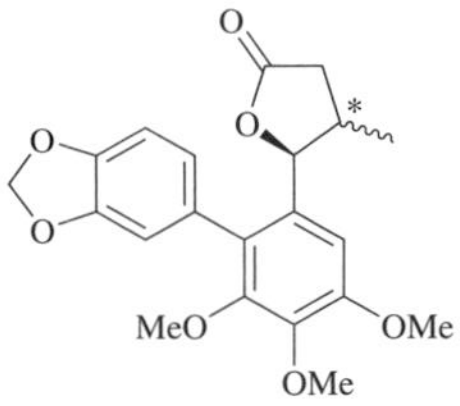	* *R* (81) *S* (85)	621
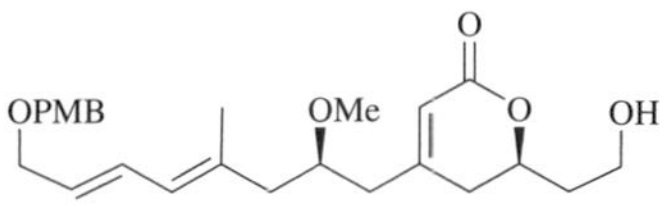	TEMPO (cat.), $PhI(OAc)_2$, CH_2Cl_2, rt	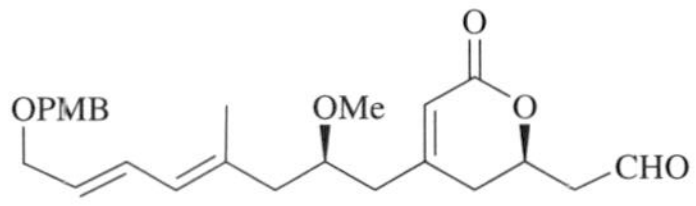	(69)	622
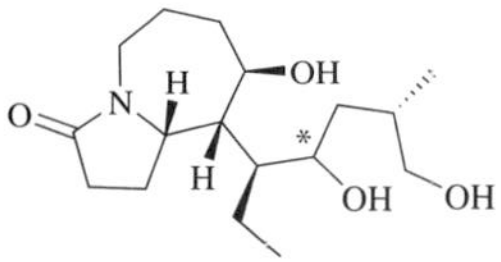	HO-TEMPO (cat.), $Ca(OCl)_2$, acetone, 48 h	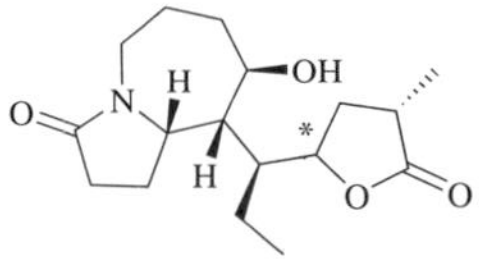	* *R* (52) *S* (27)	623

TABLE 3. OXIDATIONS OF POLYFUNCTIONAL ALCOHOLS (*Continued*)

Alcohol	Conditions	Product(s) and Yield(s) (%)	Refs.
C_{17}			
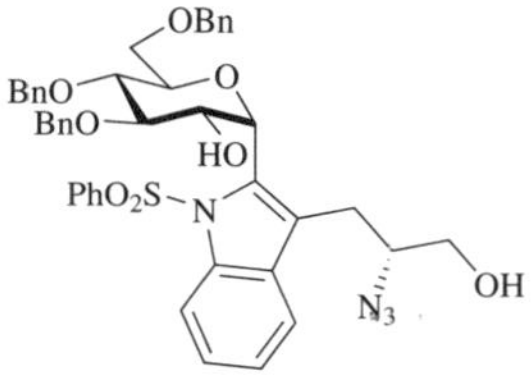	TEMPO (cat.), $PhI(OAc)_2$, MeCN, H_2O, rt, 3 h	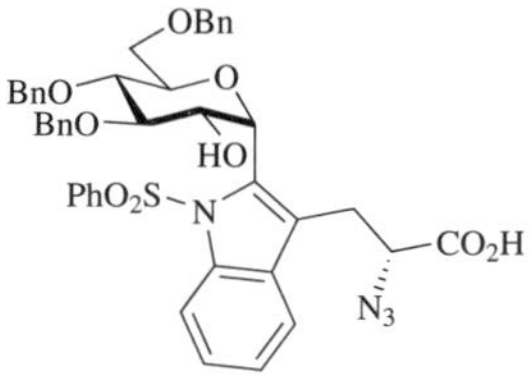 (85)	624
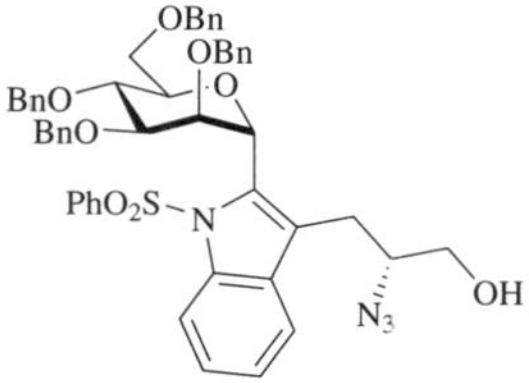	TEMPO (cat.), $PhI(OAc)_2$, MeCN, H_2O, rt, 3 h	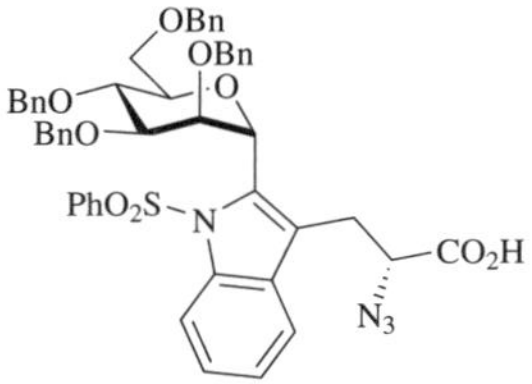 (97)	624, 625
C_{18}			
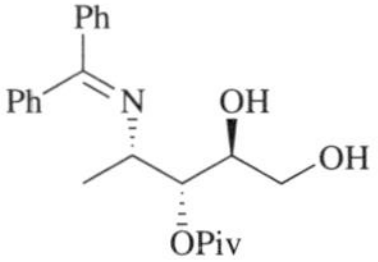	TEMPO (cat.), KBr (cat.), NaOCl, TBAF, $NaHCO_3$, CH_2Cl_2, H_2O, 0°	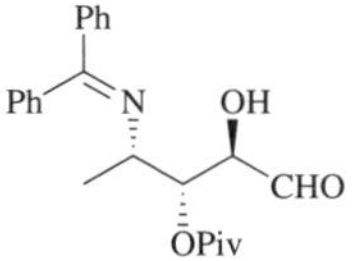 (—)[b]	626
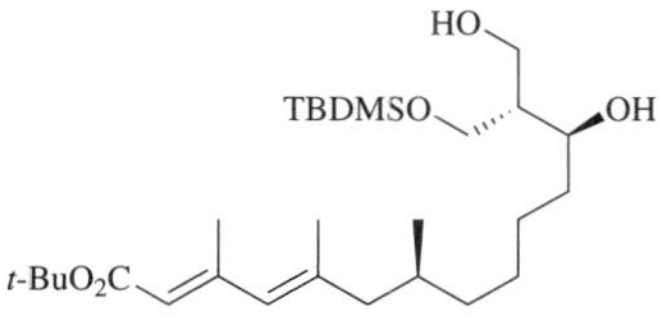	$[TEMP{=}O]^+Cl^-$, CH_2Cl_2	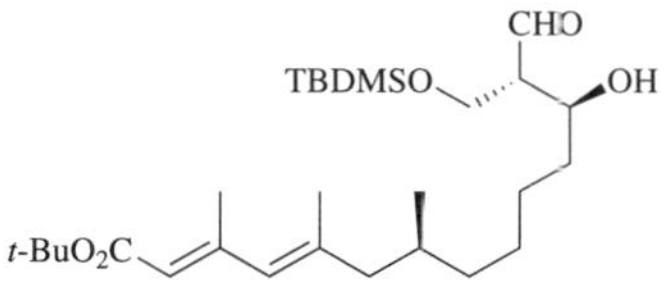 (—)[b]	627

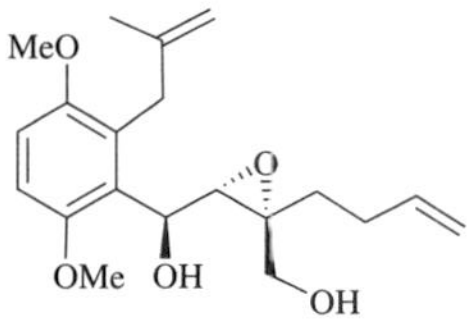

Substrate	Conditions	Product(s) and Yield(s) (%)	Refs.
MeO, OMe, OH, OH (structure)	TEMPO (cat.), $PhI(OAc)_2$, CH_2Cl_2	MeO, OMe, O, O (structure) (80)	628
Alloc–NH, OH, MeO, MeO, N, O, C_6H_4-4-OMe (structure)	TEMPO (cat.), $PhI(OAc)_2$, CH_2Cl_2, 5 h	Alloc–N, OH, H, MeO, MeO, N, O, C_6H_4-4-OMe (structure) (84)	629
OH, OH (structure)	AcNH-TEMPO, TsOH, CH_2Cl_2	O, O (structure) (56)	591
OH, $(CH_2)_7CO_2H$ (structure)	$[AcNH\text{-}TEMP{=}O]^+[BF_4]^-$, silica gel, CH_2Cl_2, rt, 30 min	O, $(CH_2)_7CO_2H$ (structure) (58)	130
n-C_8H_{17}, OH, OH, CO_2H, 7 (structure)	$[AcNH\text{-}TEMP{=}O]^+[ClO_4]^-$, silica gel, CH_2Cl_2	n-C_8H_{17}, O, OH, CO_2H, 7 (structure) (—) + n-C_8H_{17}, OH, O, CO_2H, 7 (structure) (—)	630

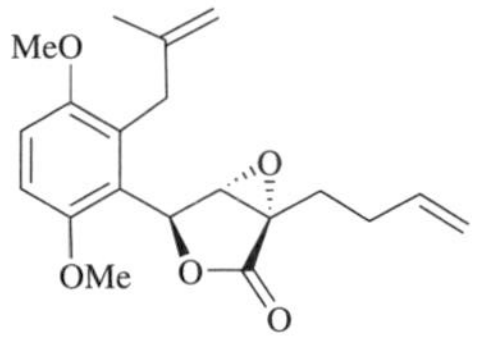

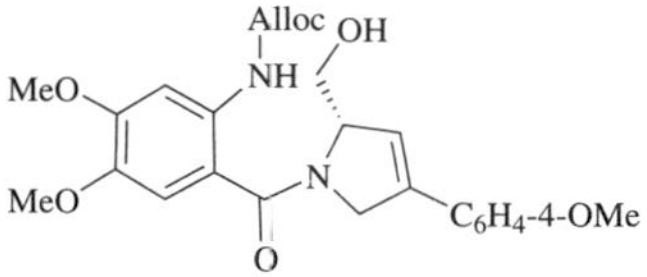

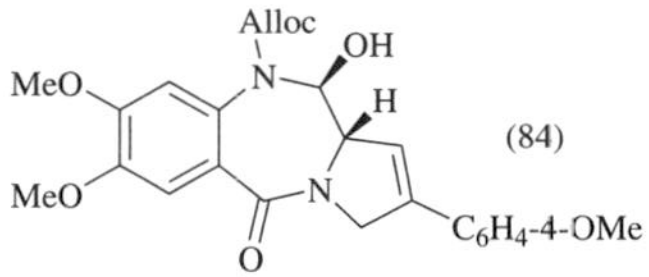

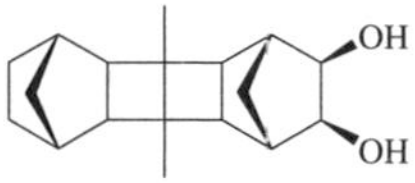

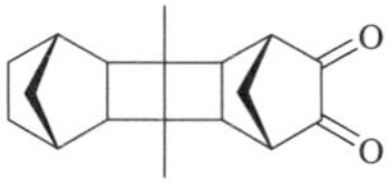

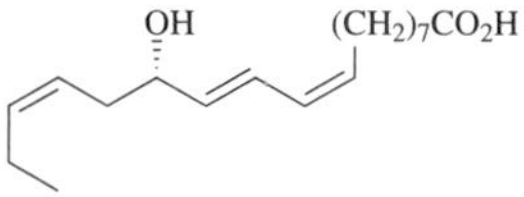

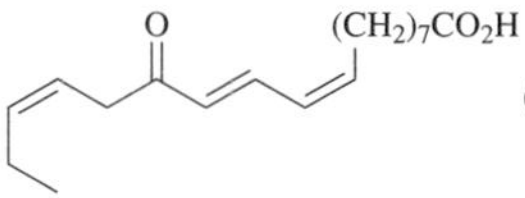

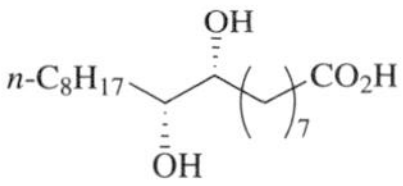

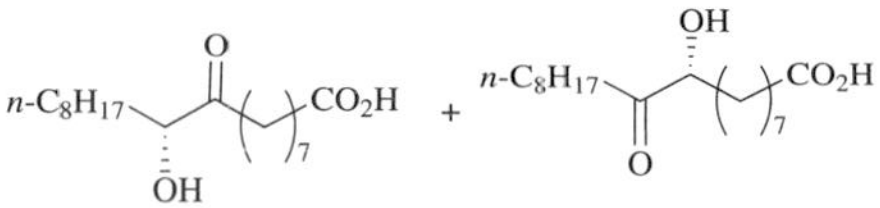

TABLE 3. OXIDATIONS OF POLYFUNCTIONAL ALCOHOLS (*Continued*)

	Alcohol	Conditions	Product(s) and Yield(s) (%)	Refs.
C_{18}	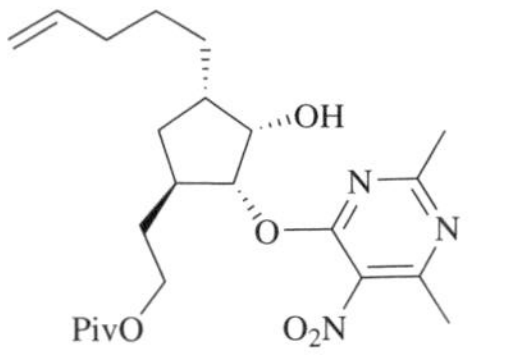	TEMPO (cat.), KBr (cat.), NaOCl, $NaHCO_3$, CH_2Cl_2, –10°, 10 min	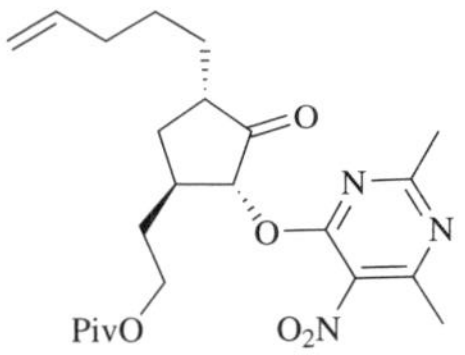 (8€)	379
	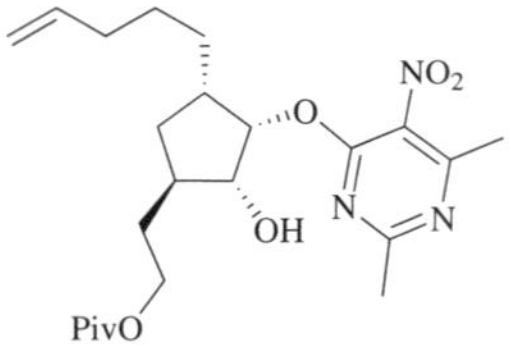	TEMPO (cat.), KBr (cat.), NaOCl, $NaHCO_3$, CH_2Cl_2, –10°, 10 min	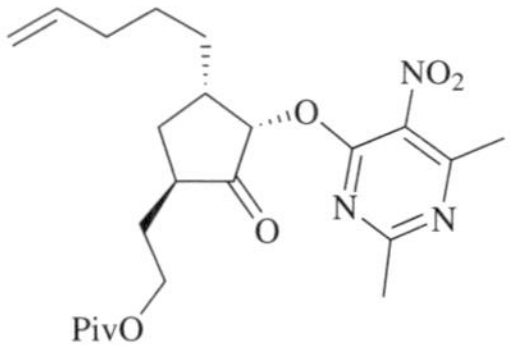 (86)	379
C_{19}	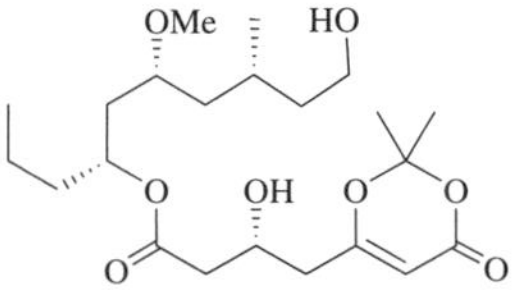	TEMPO (cat.), $PhI(OAc)_2$, MeCN	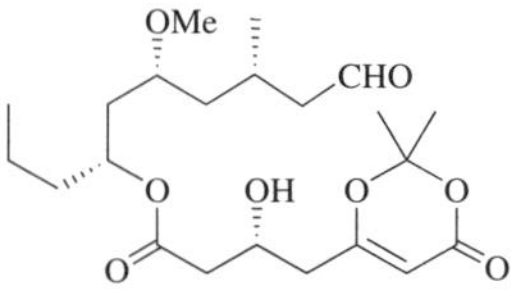 (—)	631
	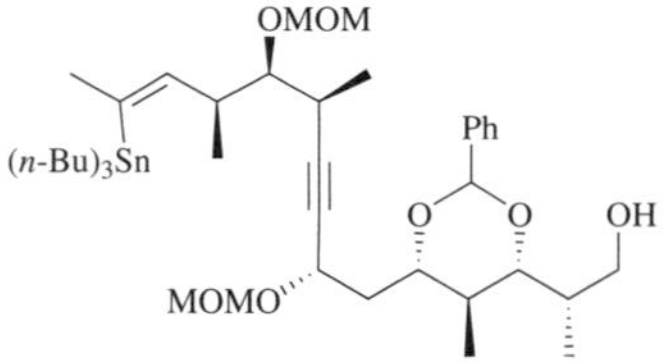	TEMPO (cat.), KBr (cat.), NaOCl, $NaHCO_3$, CH_2Cl_2, H_2O	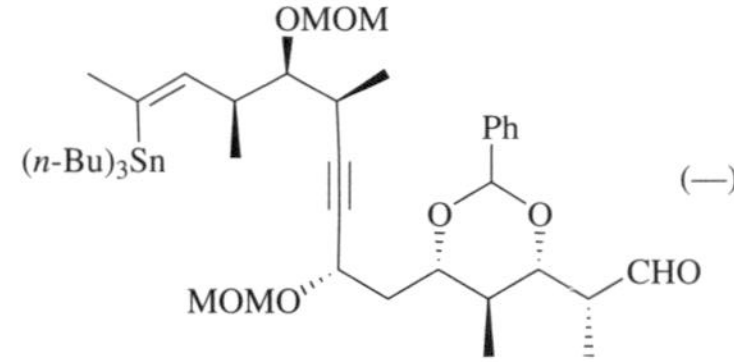 (—)[b]	417
	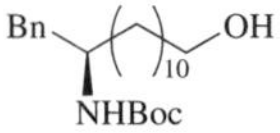	TEMPO (cat.), KBr (cat.), NaOCl, $NaHCO_3$, toluene, EtOAc, H_2O, –5°	 (74)	470

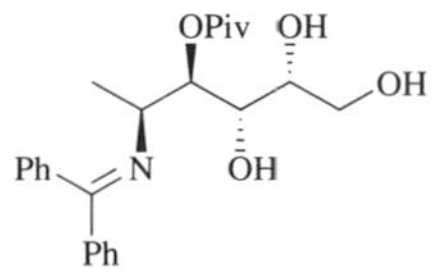

TEMPO (cat.), KBr (cat.), NaOCl, TBAF, CH_2Cl_2, H_2O, 0°

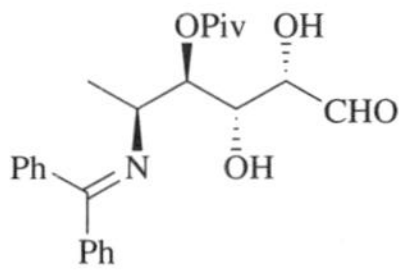

(80)

632

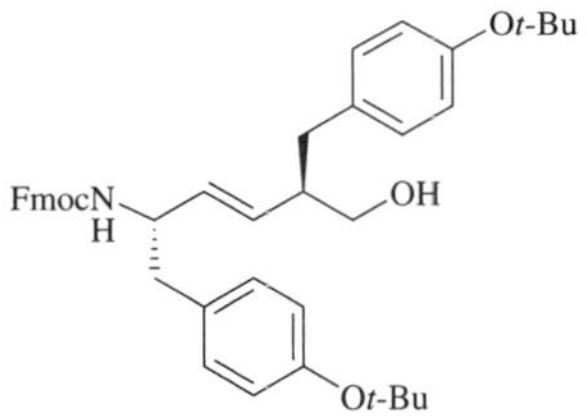

TEMPO (cat.), $PhI(OAc)_2$, CH_2Cl_2, H_2O, 6 h

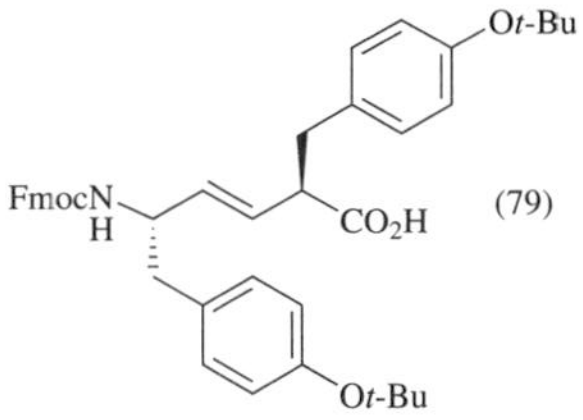

(79)

633

C_{19-20}

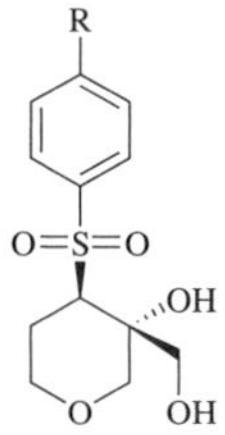

TEMPO (cat.), NaOCl (cat.), $NaClO_2$, MeCN, H_2O, pH 7.0

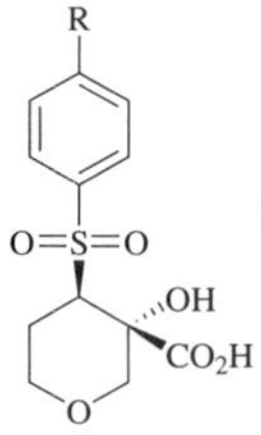

(~80)

R
2-MeC_6H_4
3-MeC_6H_4
4-MeC_6H_4
2-MeC_6H_4O
3-MeC_6H_4O
4-MeC_6H_4O
2-$ClC_6H_4OCH_2O$
3-$ClC_6H_4OCH_2O$
4-$ClC_6H_4OCH_2O$
3-$MeC_6H_4OCH_2O$
4-$MeC_6H_4OCH_2O$

634

TABLE 3. OXIDATIONS OF POLYFUNCTIONAL ALCOHOLS (*Continued*)

Alcohol	Conditions	Product(s) and Yield(s) (%)	Refs.
C_{19-21} 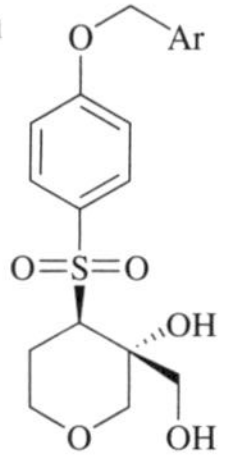	TEMPO (cat.), NaOCl (cat.), $NaClO_2$, MeCN, H_2O, pH 7.0	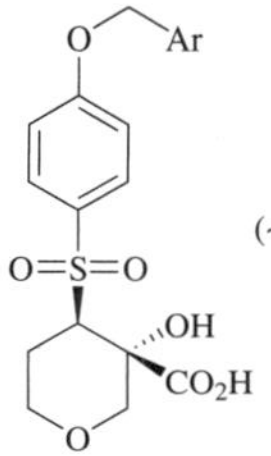 (~80) Ar: 2,3-$Cl_2C_6H_3$; 2,4-$Cl_2C_6H_3$; 2,5-$Cl_2C_6H_3$; 2,6-$Cl_2C_6H_3$; 2-$CF_3C_6H_4$; 2-MeC_6F_4; 2,4-$Me_2C_6H_3$; 2,5-$Me_2C_6H_3$; 2,6-$Me_2C_6H_3$	634
C_{20} 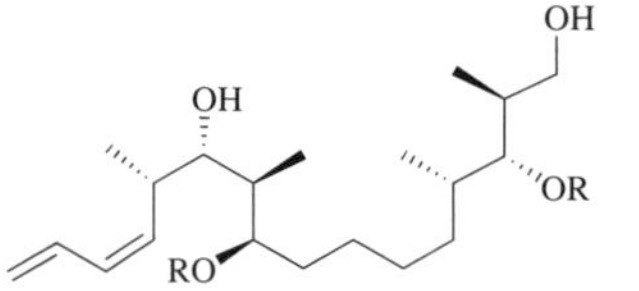 R = TBDMS	TEMPO (cat.), $PhI(OAc)_2$, CH_2Cl_2	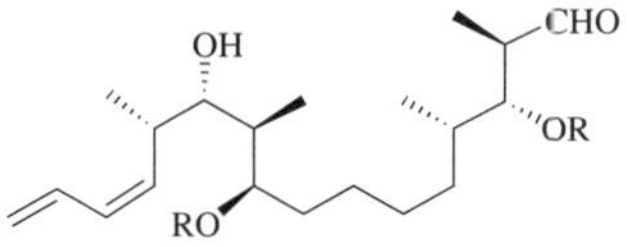 (94)	635
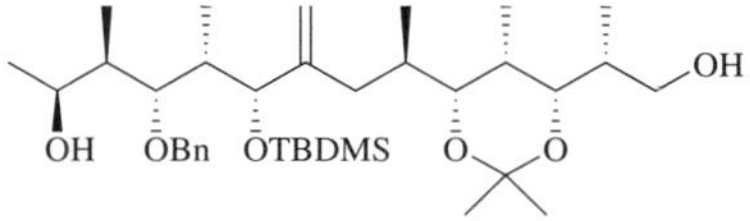	MeO-TEMPO (cat.), KBr (cat.), NaOCl, $NaHCO_3$, H_2O, CH_2Cl_2, 0°	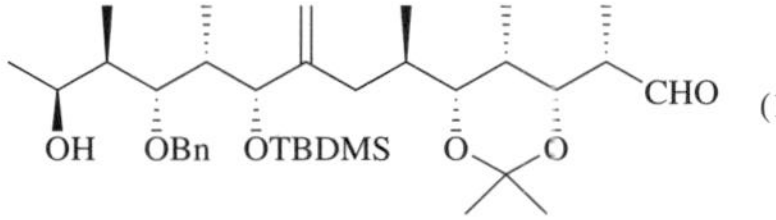 (100)	636
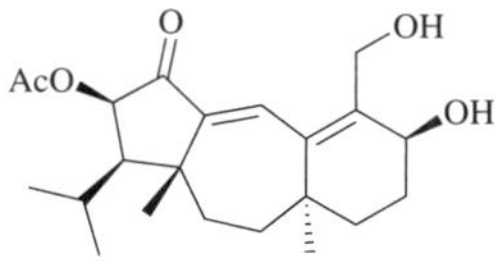	TEMPO (cat.), $PhI(OAc)_2$, CH_2Cl_2	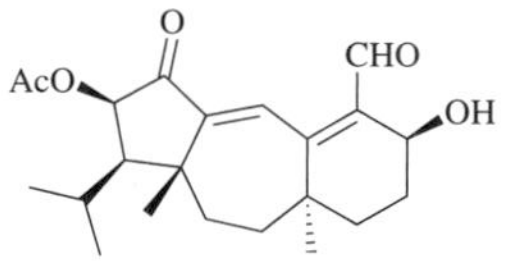 (—)[b]	637, 638

	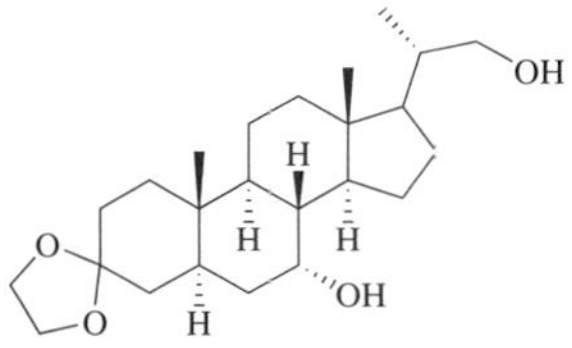	TEMPO (cat.), NaOCl, NaBr, H_2O, CH_2Cl_2	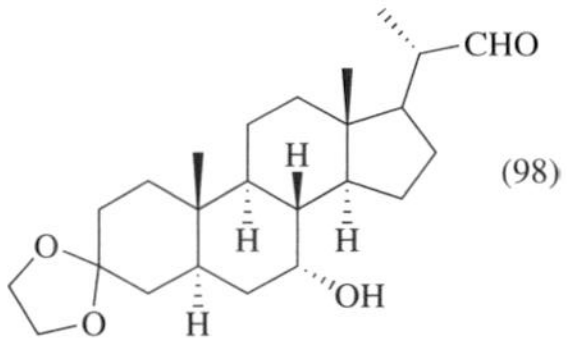 (98)	639
C_{21}	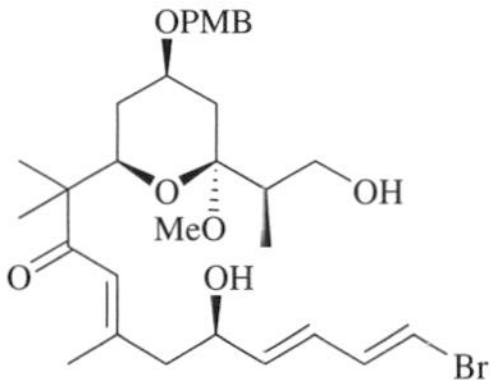	1. TEMPO (cat.), $PhI(OAc)_2$, MeCN, buffer, pH 7, 6 h 2. $NaClO_2$, NaH_2PO_4, *t*-BuOH, H_2O, 2-methyl-2-butene, rt, 30 min	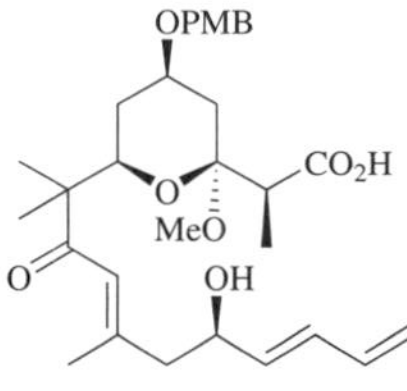 (60)	640
	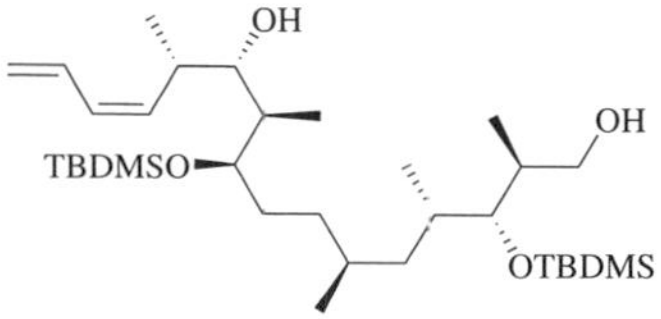	TEMPO (cat.), $PhI(OAc)_2$, CH_2Cl_2, rt, 18 h	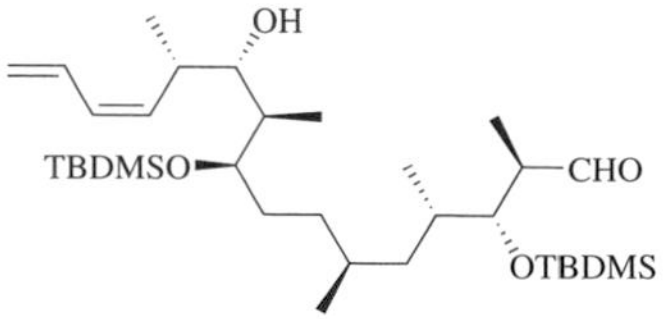 (—)[b]	641
	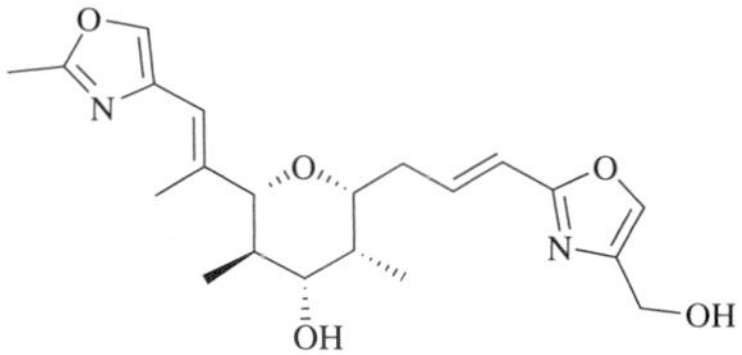	TEMPO (cat.), $PhI(OAc)_2$, CH_2Cl_2, 20°, 30 min	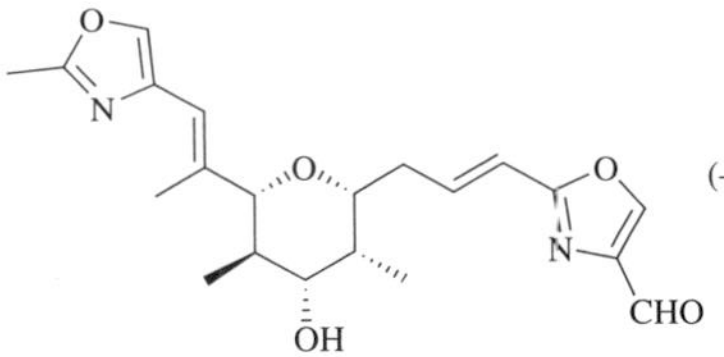 (—)[b]	642
	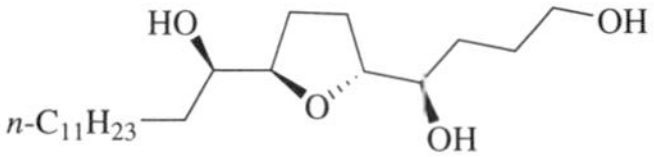	TEMPO (cat.), KBr (cat.), NaOCl, TBAC, NaCl, $NaHCO_3$, H_2O, 0°	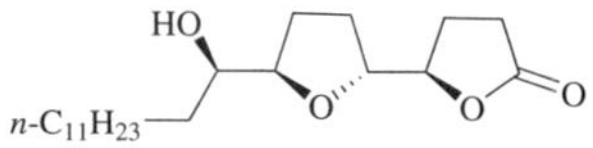 (61)	643

TABLE 3. OXIDATIONS OF POLYFUNCTIONAL ALCOHOLS (*Continued*)

Alcohol	Conditions	Product(s) and Yield(s) (%)	Refs.
C_{21} 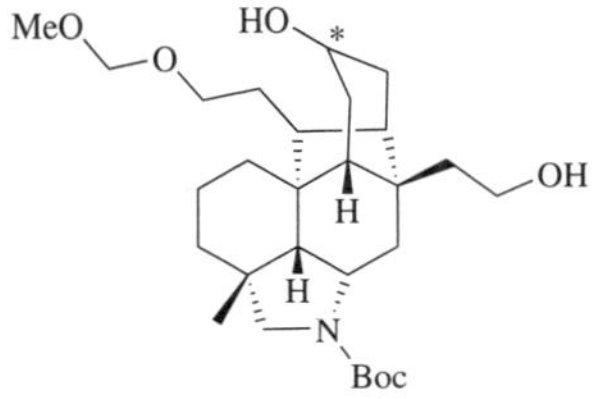	TEMPO (cat.), $PhI(OAc)_2$, CH_2Cl_2	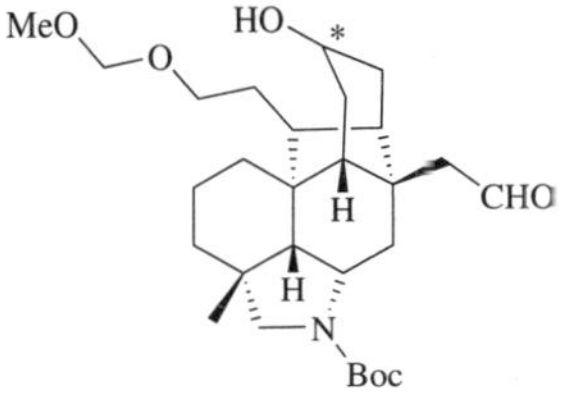 *: *R* (75); *S* (75)	644
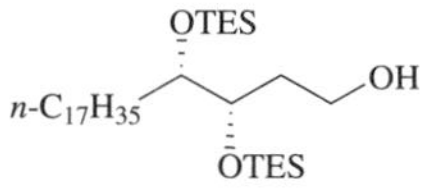	TEMPO (cat.), $PhI(OAc)_2$, CH_2Cl_2, rt	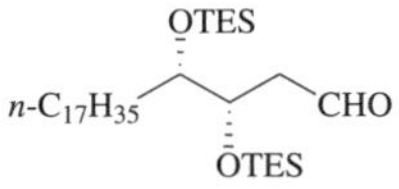 (96)	645
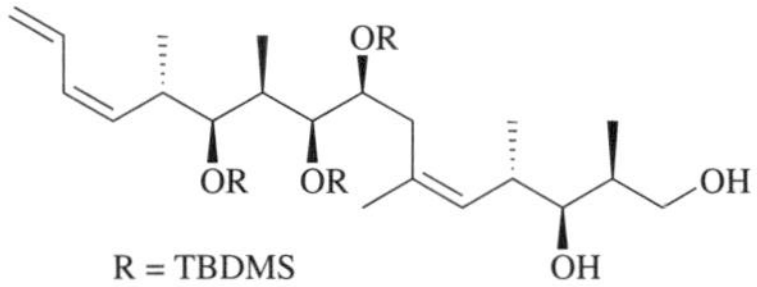 R = TBDMS	TEMPO (cat.), $PhI(OAc)_2$, CH_2Cl_2, 20°, 4 h	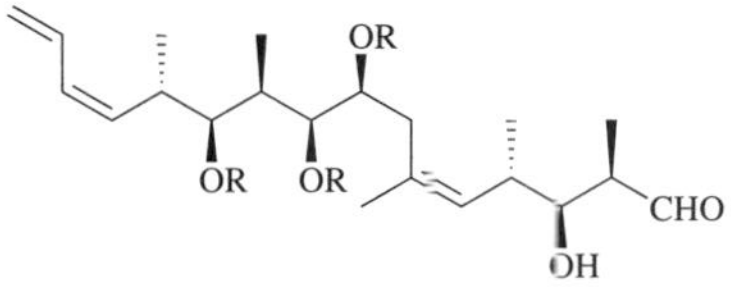 (—)[b]	646
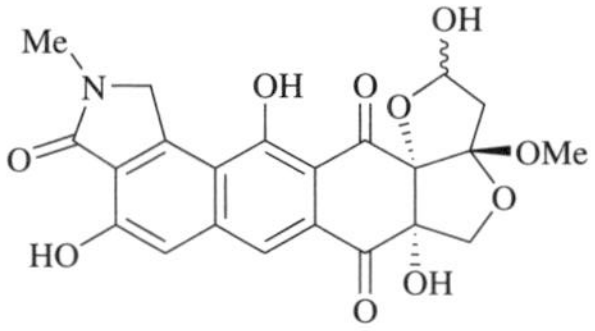	TEMPO (cat.), $PhI(OAc)_2$, CH_2Cl_2	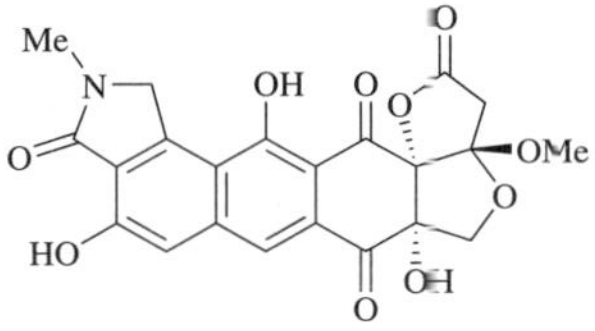 (58)	647

C_{21-22}	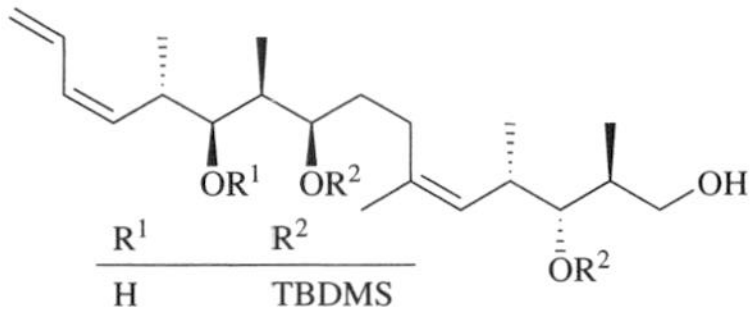	TEMPO (cat.), $PhI(OAc)_2$, CH_2Cl_2, 20°, 4 h	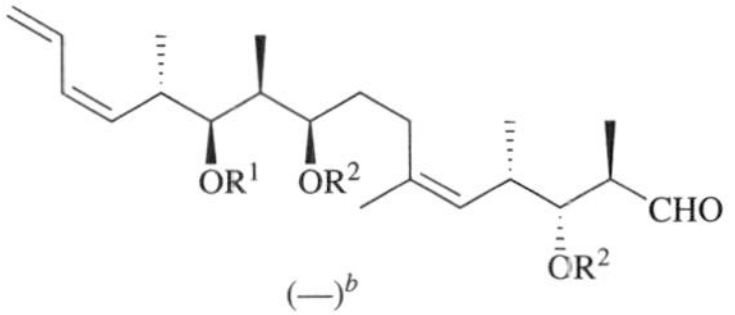	
	R^1 / R^2			
	H / TBDMS		(—)[b]	516, 517, 519
	TBDMS / TBDMS		(—)[b]	516, 517, 519
	$CONH_2$ / TBDMS		(97)	648
C_{22}	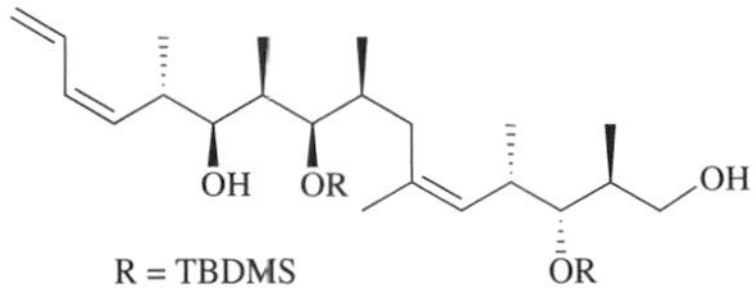 R = TBDMS	TEMPO (cat.), $PhI(OAc)_2$, CH_2Cl_2	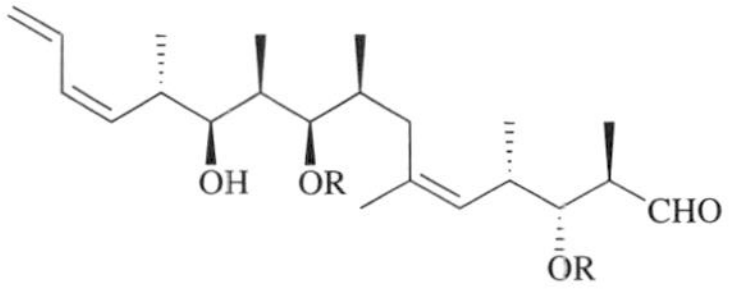 (—)[b]	197
	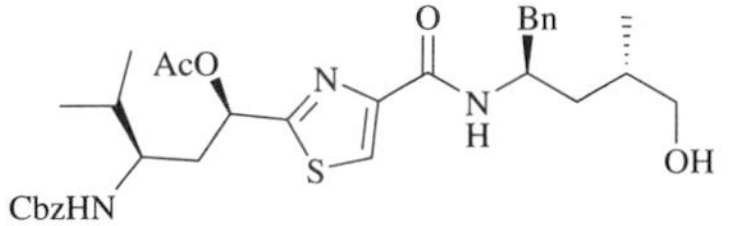	TEMPO (cat.), NaOCl, $NaClO_2$, MeCN, H_2O, pH 6.7	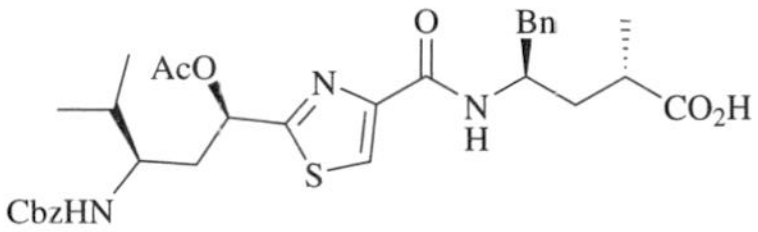 (23)	452
	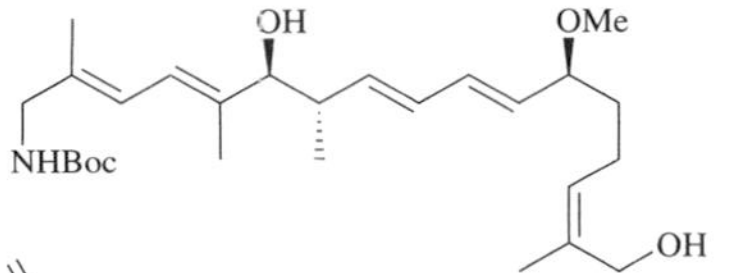	TEMPO (cat.), 2,2'-bipyridine (cat.), $CuBr_2$ (cat.), t-BuOK (cat.), O_2, MeCN, H_2O	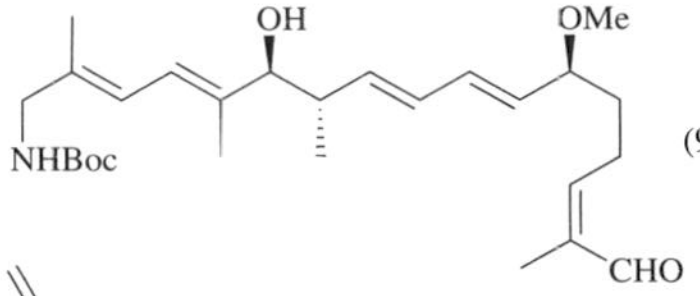 (94)	649
	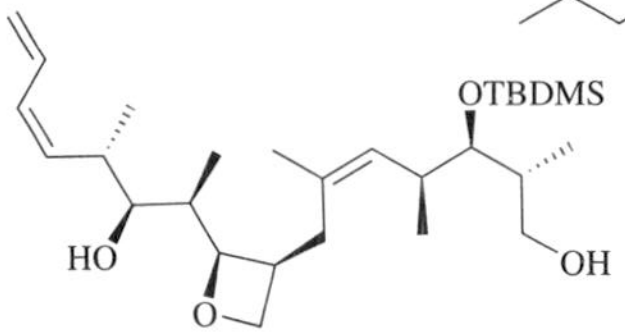	TEMPO (cat.), $PhI(OAc)_2$, CH_2Cl_2, 20°, 4 h	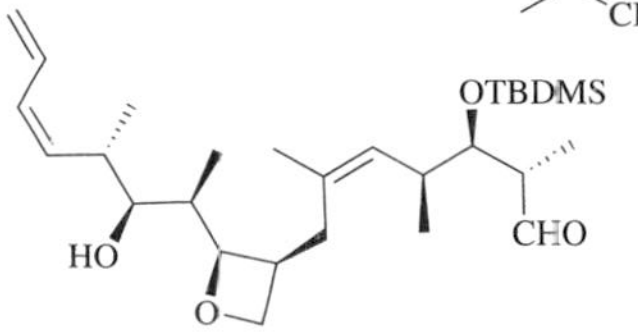 (—)[b]	646

TABLE 3. OXIDATIONS OF POLYFUNCTIONAL ALCOHOLS (*Continued*)

	Alcohol	Conditions	Product(s) and Yield(s) (%)	Refs.
C_{23}	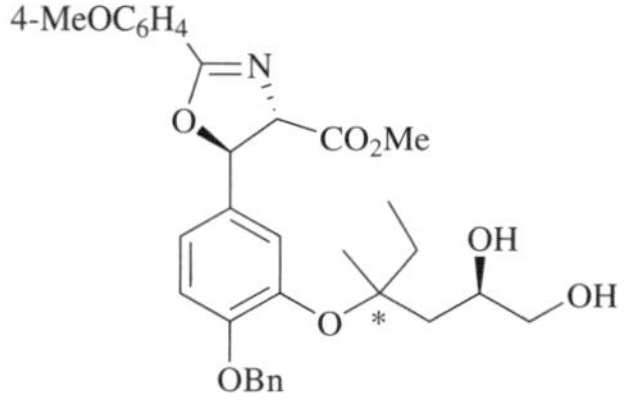	TEMPO (cat.), NaOCl (cat.), $NaClO_2$, KBr, $NaHCO_3$, CH_2Cl_2, H_2O, 0°, 20 min	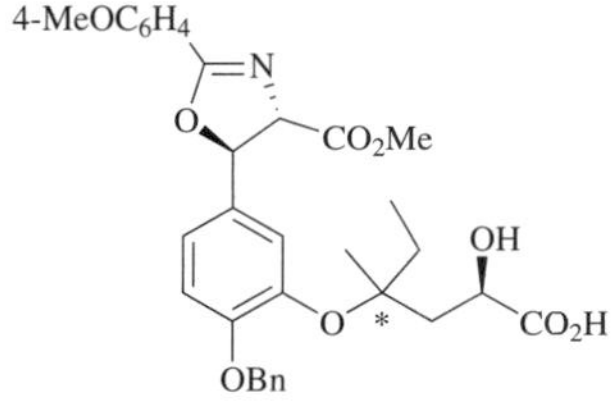 * — *R* (95); *S* (99)	650
C_{24}	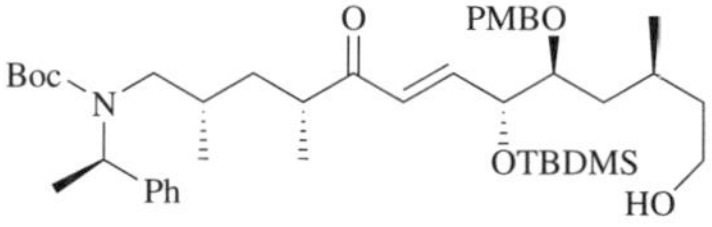	TEMPO (cat.), NaOCl, CH_2Cl_2, H_2O	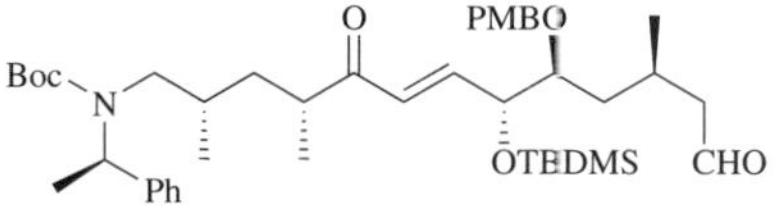 (85)	532
	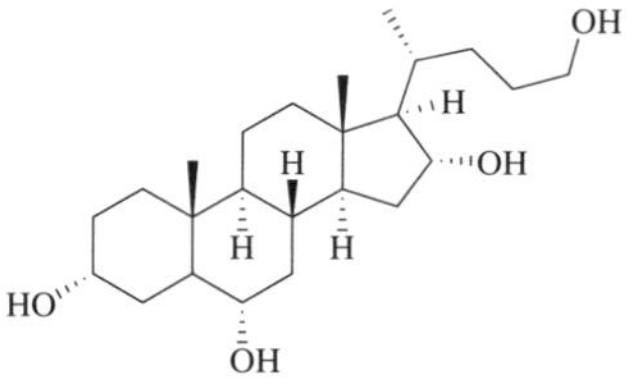	TEMPO (cat.), NCS, TBAC, $NaHCO_3$, CH_2Cl_2, H_2O, buffer, pH 8.6, rt, 4 h	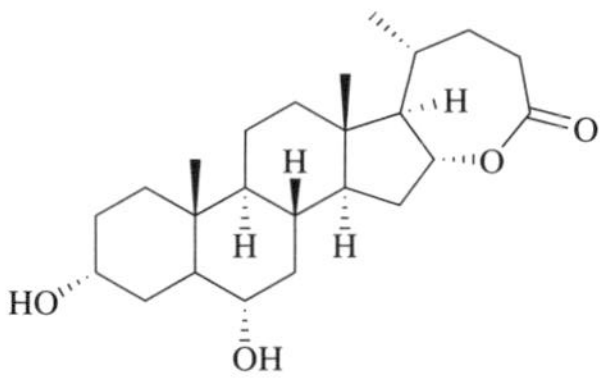 (63)	651
	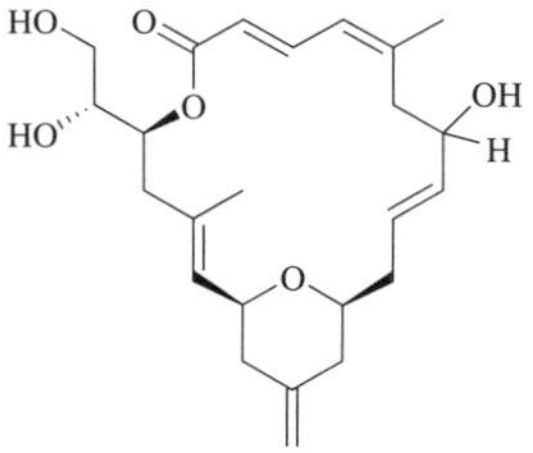	$[AcNH\text{-}TEMP{=}O]^+[BF_4]^-$, silica gel, CH_2Cl_2	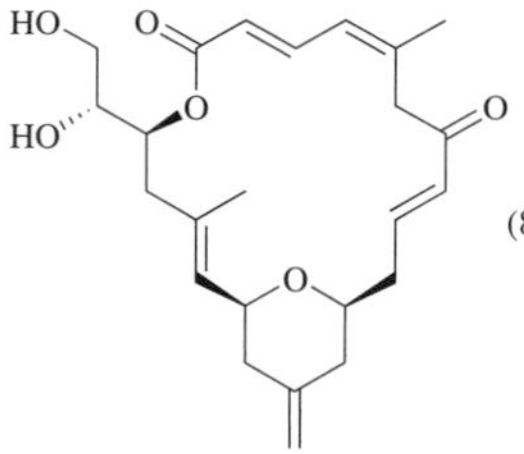 (80)	652

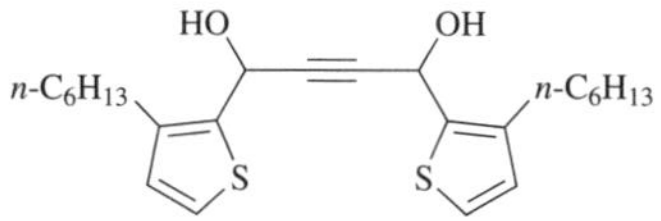	TEMPO (cat.), PhI(OAc)$_2$, CH_2Cl_2, rt, 2 h	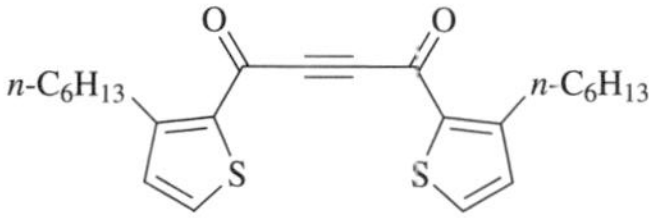	(80)	653
C_{25} 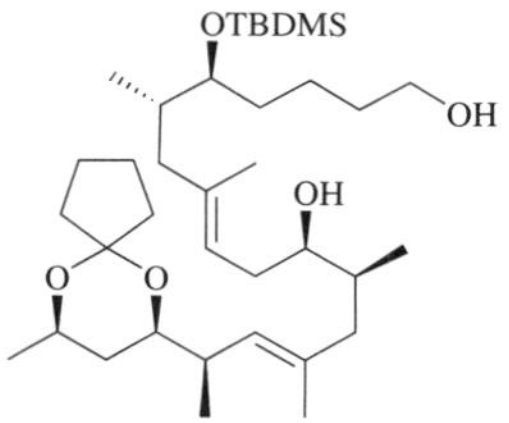	TEMPO (cat.), NCS, TBAC, CH_2Cl_2, H_2O, buffer, pH 8.6	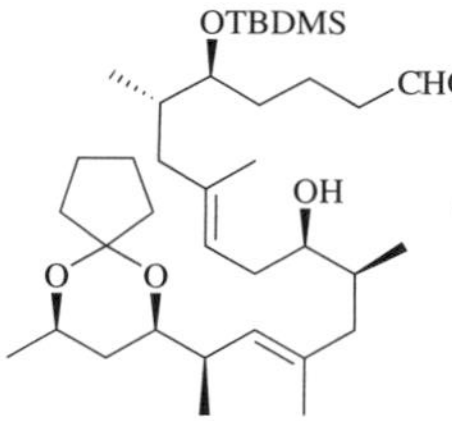	(95)	654
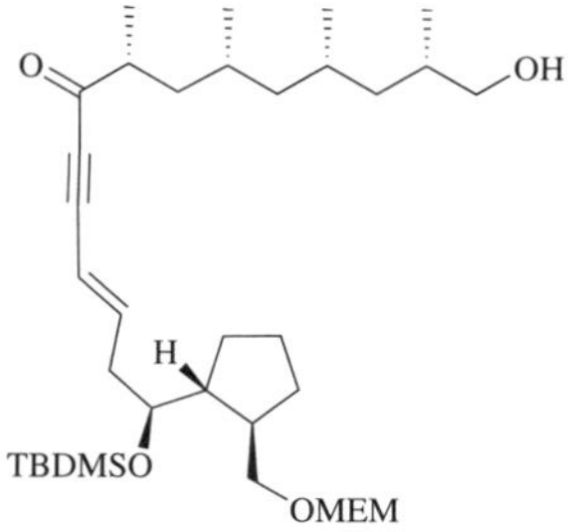	TEMPO (cat.), PhI(OAc)$_2$, CH_2Cl_2, rt, 2 h	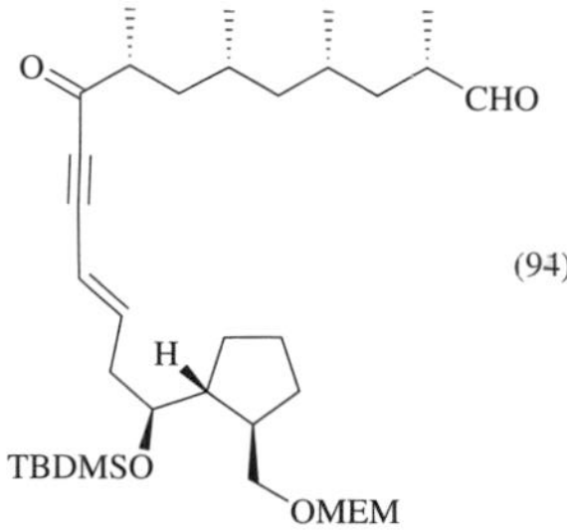	(94)	655
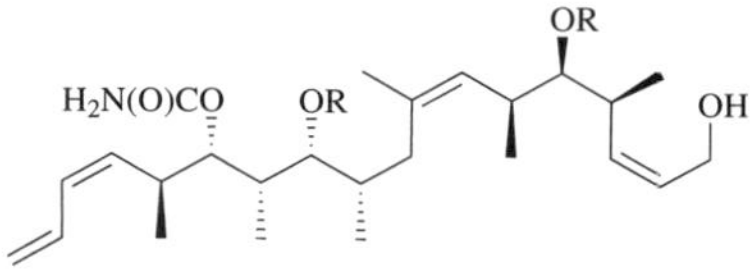 R = TBDMS	TEMPO (cat.), PhI(OAc)$_2$, CH_2Cl_2	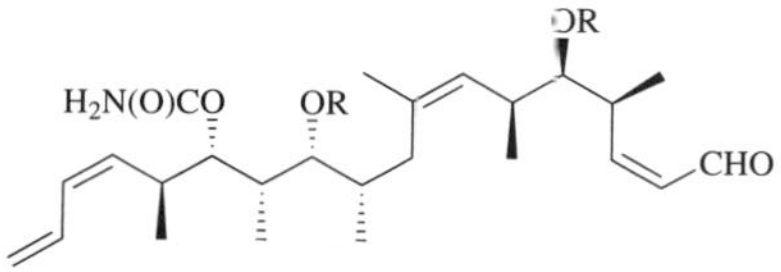	(80)	197

TABLE 3. OXIDATIONS OF POLYFUNCTIONAL ALCOHOLS (*Continued*)

	Alcohol	Conditions	Product(s) and Yield(s) (%)	Refs.
C_{25}	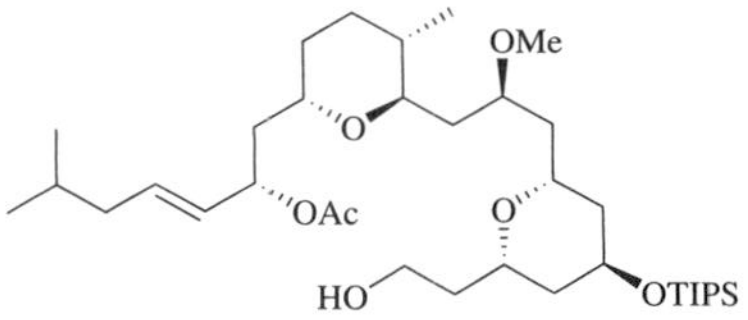	1. TEMPO (cat.), $PhI(OAc)_2$, CH_2Cl_2, rt 2. $NaClO_2$, NaH_2PO_4, H_2O, 2-methyl-2-butene, *t*-BuOH, 0–20°	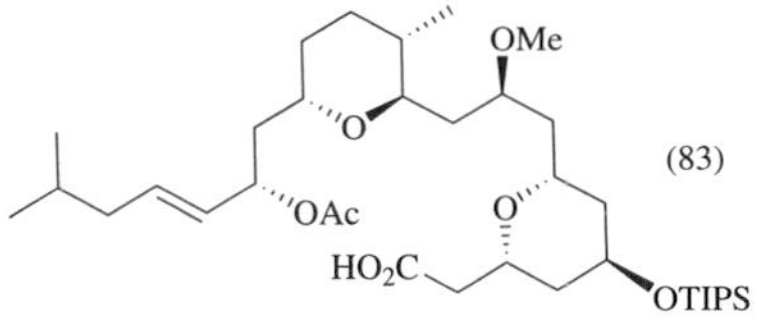 (83)	575
C_{26}	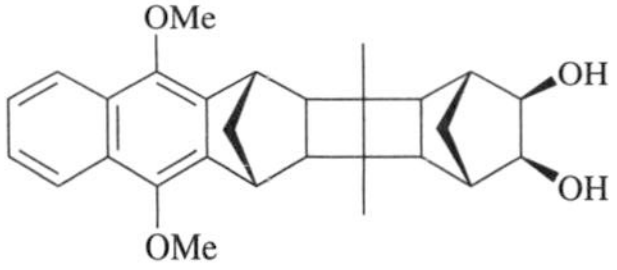	AcO-TEMPO, TsOH, CH_2Cl_2	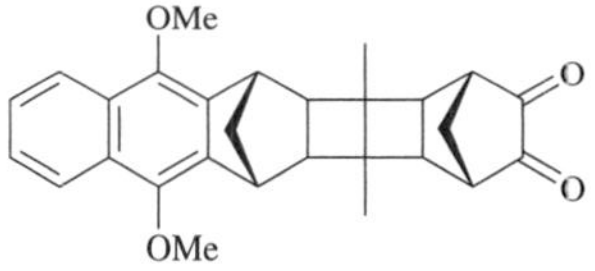 (71)	591
C_{27}	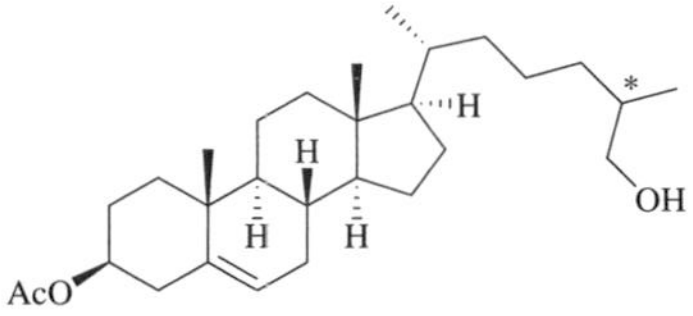	TEMPO (cat.), NaOCl (cat.), $NaClO_2$, THF, H_2O, MeCN, phosphate buffer, pH 6.86	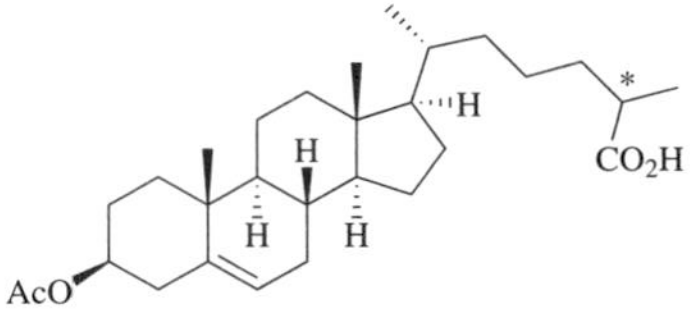 * : *R* (—)[b]; *S* (—)[b]	656
	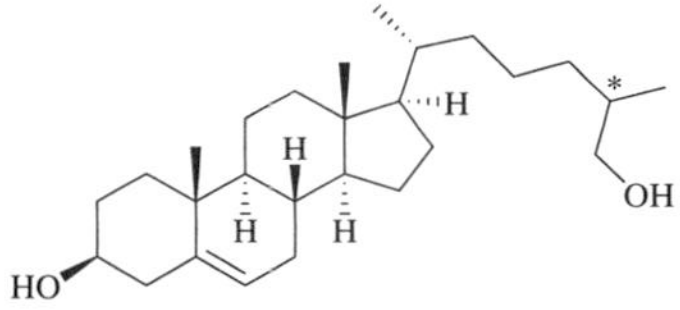	TEMPO (cat.), NaOCl (cat.), $NaClO_2$, THF, H_2O, MeCN, phosphate buffer, pH 6.86	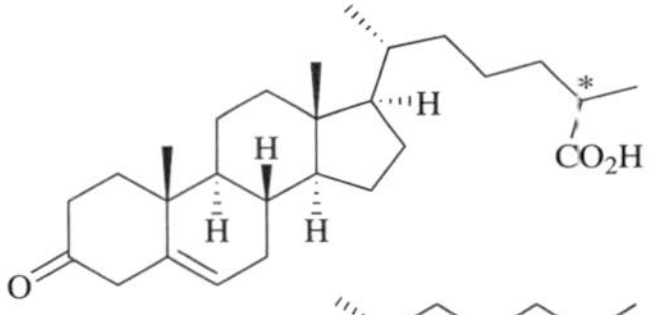 * : *R* (48); *S* (51)	656
	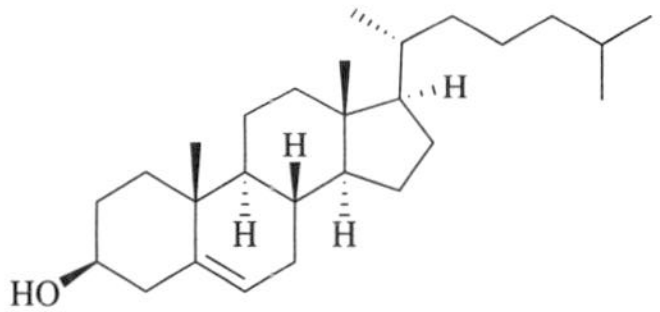	$[PS\text{-}O\text{-}TEMP{=}O]^+Cl^-$, CH_2Cl_2, rt, 1–2 h	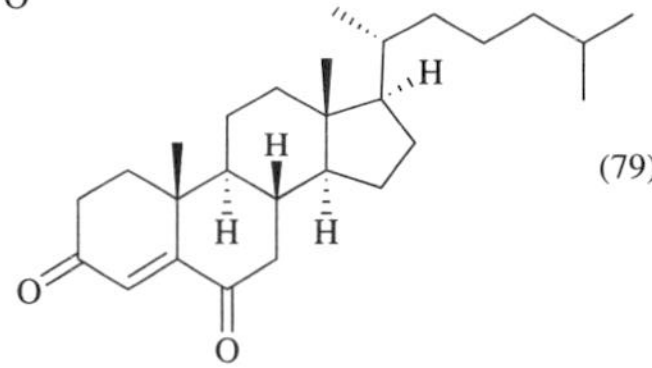 (79)[a]	121

C_{28}				
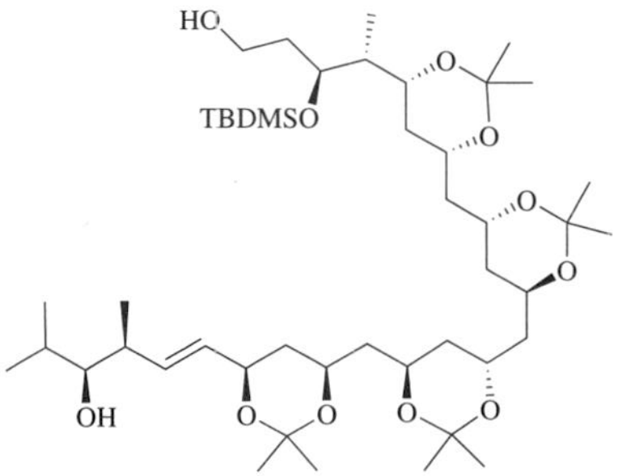	TEMPO (cat.), NCS, TBAC, CH_2Cl_2, H_2O, buffer, pH 8.6, rt	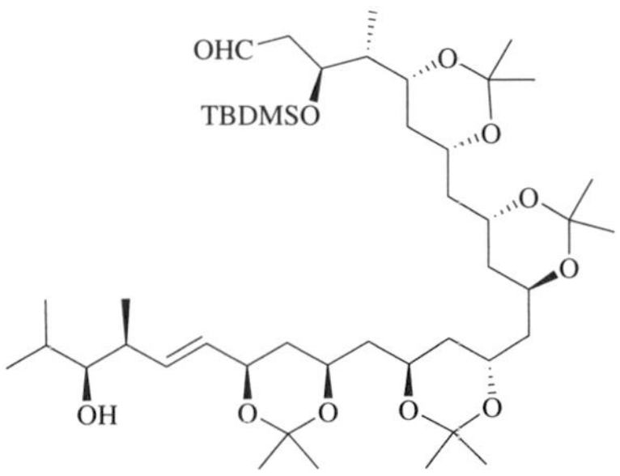	(88)	657, 658
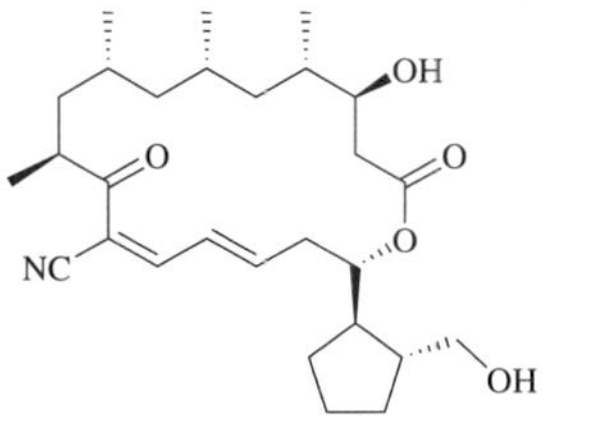	TEMPO (cat.), TCC, CH_2Cl_2, rt, 1 h	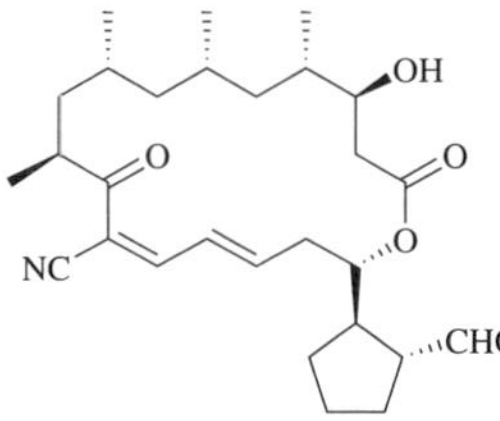	(85)	655
$C_{28\text{–}40}$				
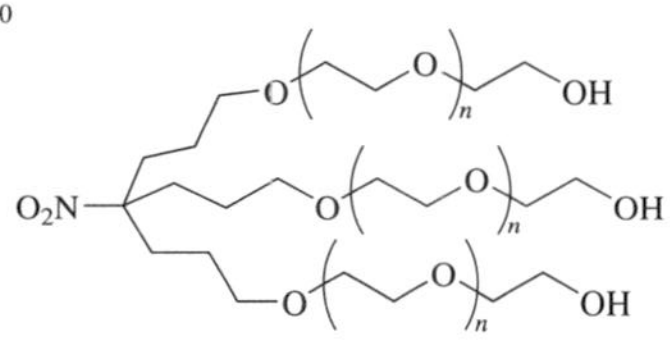 $n = 2, 4$	TEMPO (cat.), KBr (cat.), NaOCl, TBAC, $NaHCO_3$, CH_2Cl_2, H_2O, 0° to rt, 1 h	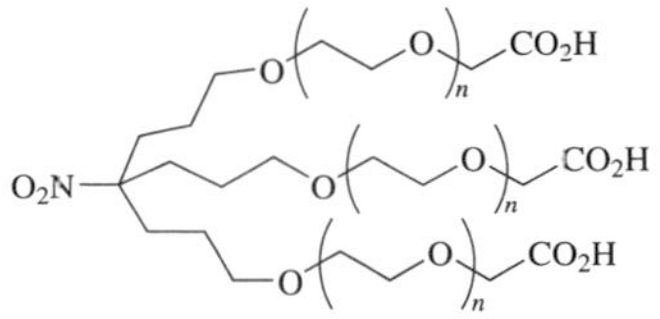	(—)[b]	659

TABLE 3. OXIDATIONS OF POLYFUNCTIONAL ALCOHOLS (*Continued*)

Alcohol	Conditions	Product(s) and Yield(s) (%)	Refs.
C_{30}			
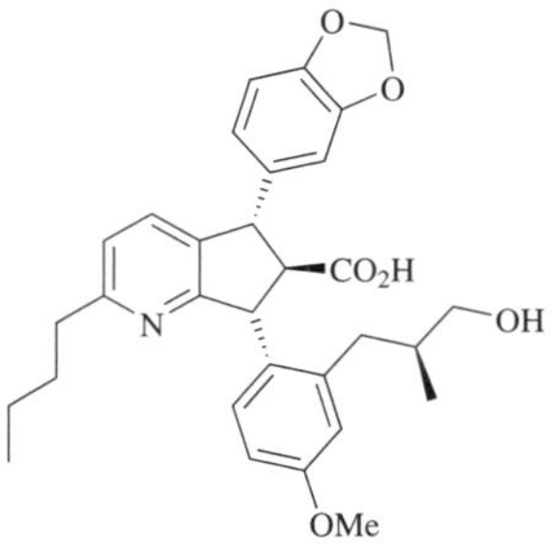	TEMPO (cat.), NaOCl (cat.), $NaClO_2$, MeCN	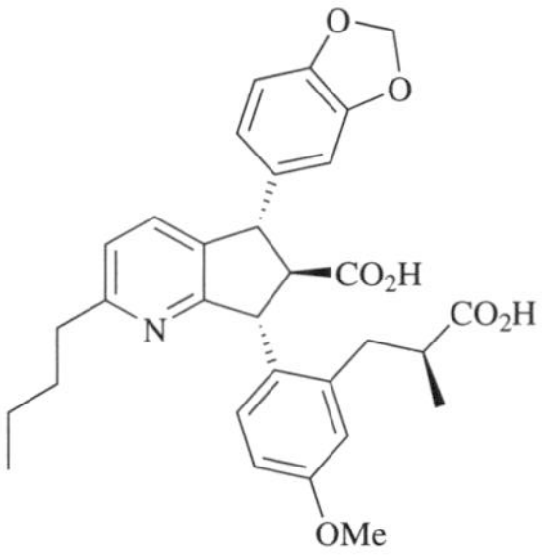 (100)	14
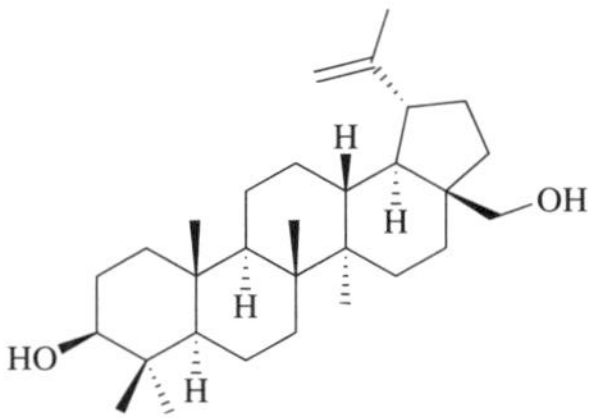	Nitroxide (cat.), NaOCl (cat.), $NaClO_2$, TBAB, CH_2Cl_2, H_2O, phosphate buffer, pH 6.8	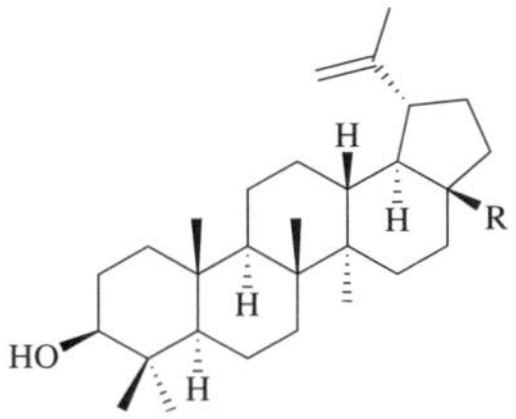	660
	Nitroxide / Temp: TEMPO 35°; AcNH-TEMPO 50°	R: CHO (92); CO_2H (86)	
C_{31}			
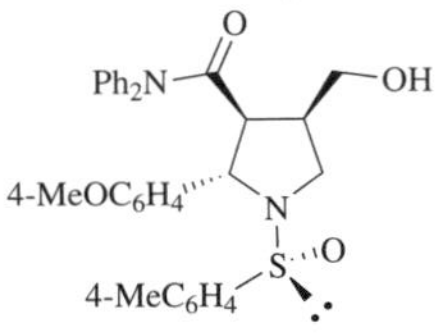	TEMPO (cat.), NaBr (cat.), NaOCl, $NaHCO_3$, EtOAc, toluene, H_2O, 0°	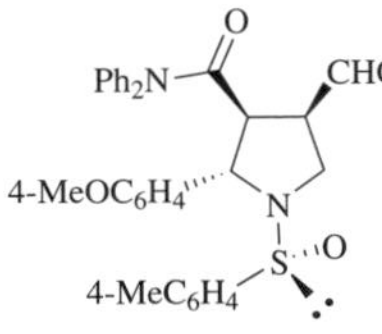 (91)[k]	370

C_{32} 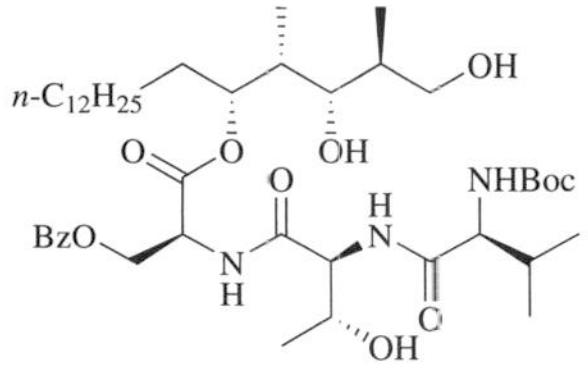	1. TEMPO (cat.), KBr (cat.), NaOCl, $NaHCO_3$, CH_2Cl_2, H_2O 2. $NaClO_2$, $HOSO_2NH_2$, NaH_2PO_4, *t*-BuOH, H_2O	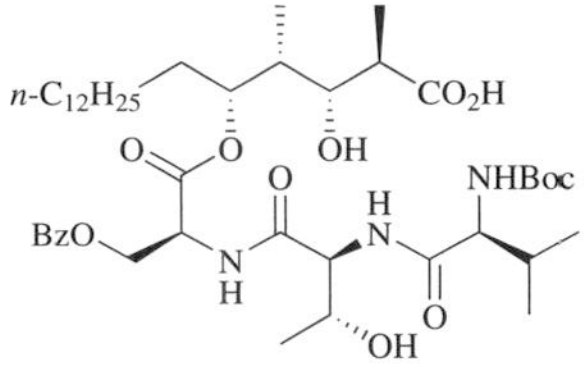	(—)[b]	661
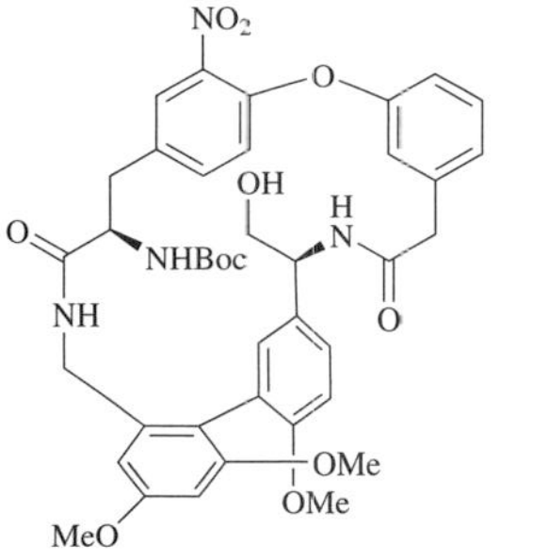	TEMPO (cat.), KBr (cat.), NaOCl, CH_2Cl_2, –5° to 0°	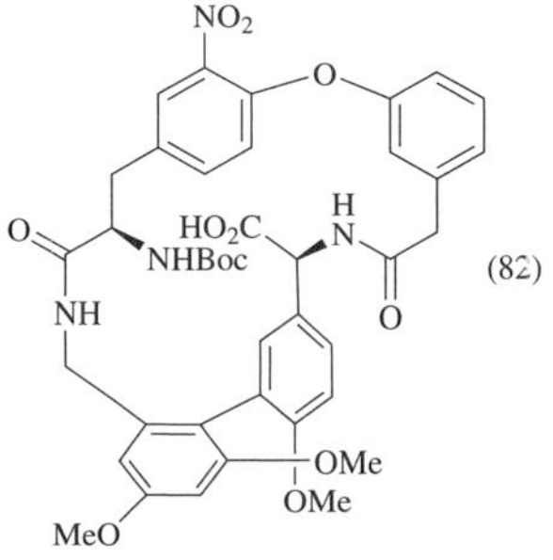	(82)	160
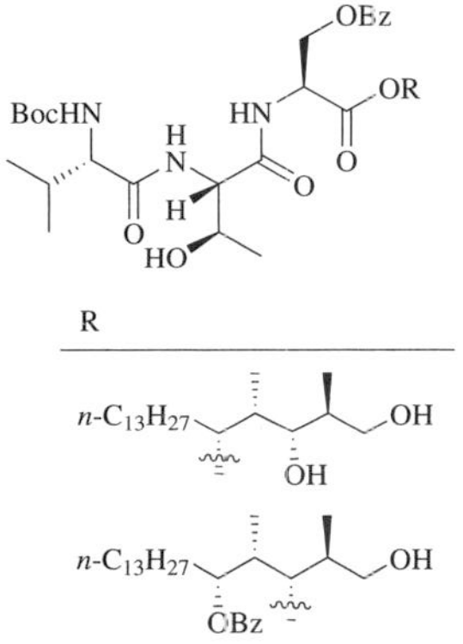	1. TEMPO (cat.), KBr (cat.), NaOCl, $NaHCO_3$, CH_2Cl_2, H_2O, 0° 2. $NaClO_2$, $HOSO_2NH_2$, NaH_2PO_4, *t*-BuOH, H_2O	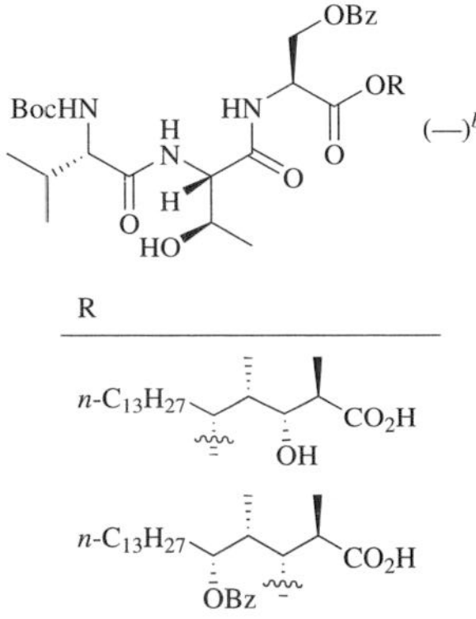	(—)[b]	662

TABLE 3. OXIDATIONS OF POLYFUNCTIONAL ALCOHOLS (*Continued*)

Alcohol	Conditions	Product(s) and Yield(s) (%)	Refs.
C_{45-51} 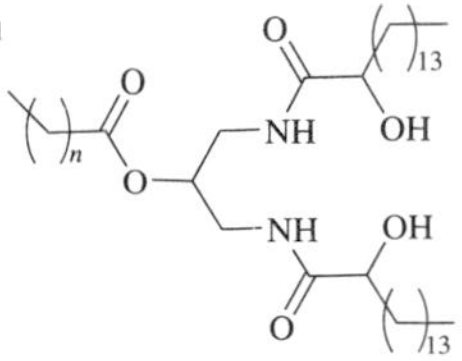	AcNH-TEMPO (cat.), NaBr (cat.), NaOCl, toluene, EtOAc, H_2O	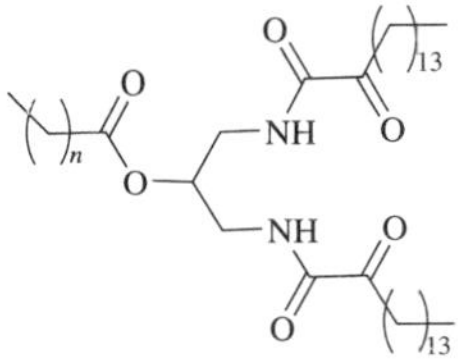 *n*: 8 (60); 14 (51)	663
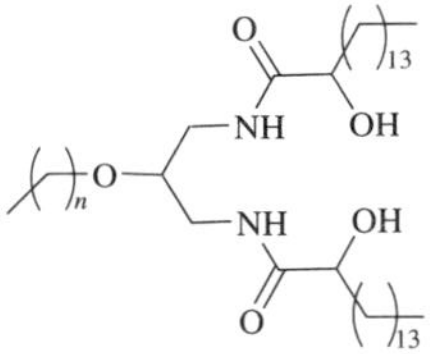	AcNH-TEMPO (cat.), NaBr (cat.), NaOCl, toluene, EtOAc, H_2O	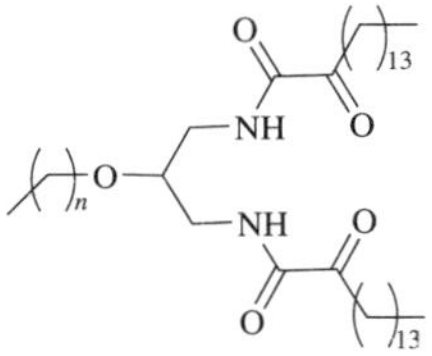 *n*: 9 (41); 15 (55)	663
C_n 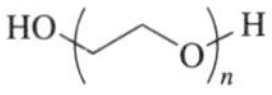	$[MeO\text{-}TEMP{=}O]^+Cl^-$, CH_2Cl_2, dark, rt, 2 h	 **I** (100), 54% conversion	135
	TEMPO (cat.), $PhI(OAc)_2$, CH_2Cl_2, rt	**I** (100)[l]	664
	TEMPO (cat.), $PhI(OAc)_2$, H_2O, MeCN, rt	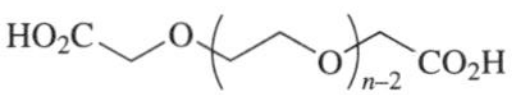 (100)[l]	664
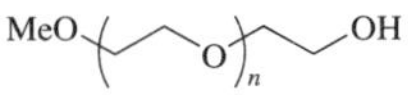	1. TEMPO (cat.), $PhI(OAc)_2$, MeCN, H_2O, rt, 2 h 2. $NaClO_2$, *t*-BuOH, H_2O, rt, 4 h	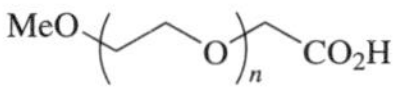 **I** *n*: ~110 (86)	665
	TEMPO (cat.), $PhI(OAc)_2$, H_2O, MeCN, rt	**I** (100)[m]	664
	TEMPO (cat.), $PhI(OAc)_2$, CH_2Cl_2, rt	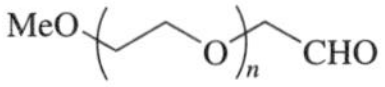 (100)[m]	664

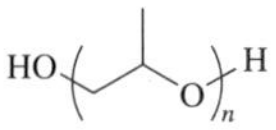	$[MeO\text{-}TEMP{=}O]^+Cl^-$, CH_2Cl_2, dark, rt, 2 h	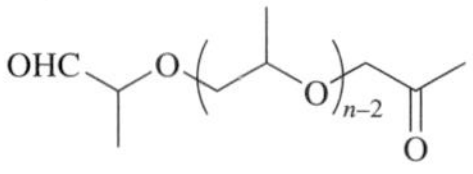 (97), 69% conversion	135
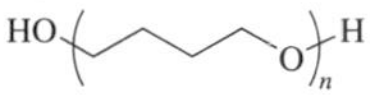	$[MeO\text{-}TEMP{=}O]^+Cl^-$, CH_2Cl_2, dark, rt, 2 h	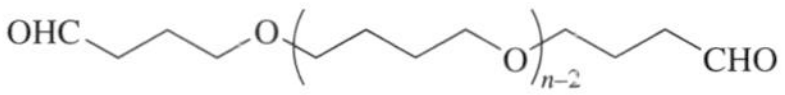 (96), 100% conversion **I**	135
	MeO-TEMPO (cat.), $Cu(NO_3)_2$, CuO, CH_2Cl_2, rt, 14 h	**I** (90), 84% conversion	117
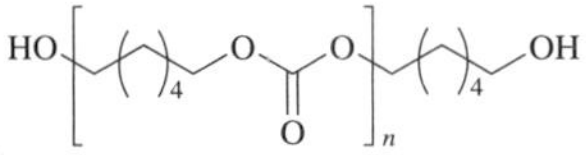	$[MeO\text{-}TEMP{=}O]^+Cl^-$, CH_2Cl_2, dark, rt, 2 h	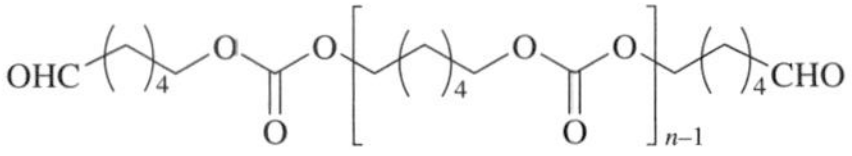 (93), 70% conversion	135
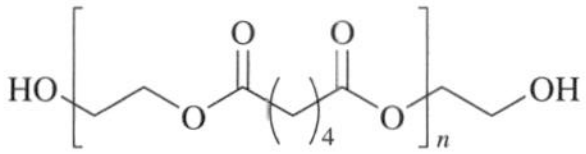	$[MeO\text{-}TEMP{=}O]^+Cl^-$, CH_2Cl_2, dark, rt, 2 h	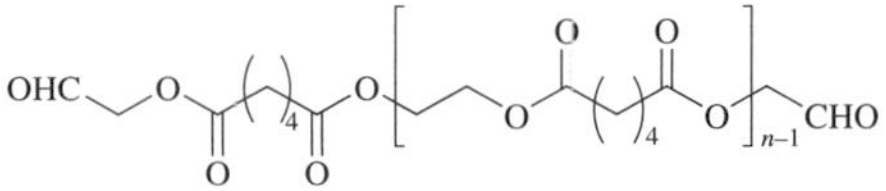 (79), 8% conversion	135
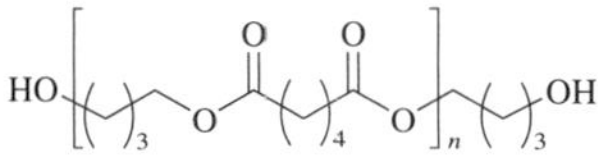	MeO-TEMPO, $Cu(NO_3)_2$, CuO, CH_2Cl_2, rt, 14 h	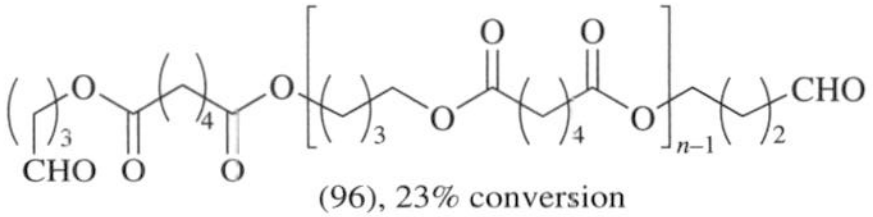 (96), 23% conversion **I**	117
	$[MeO\text{-}TEMP{=}O]^+Cl^-$, CH_2Cl_2, dark, rt, 2 h	**I** (94), 95% conversion	135

TABLE 3. OXIDATIONS OF POLYFUNCTIONAL ALCOHOLS (*Continued*)

Alcohol	Conditions	Product(s) and Yield(s) (%)	Refs.
C_n			
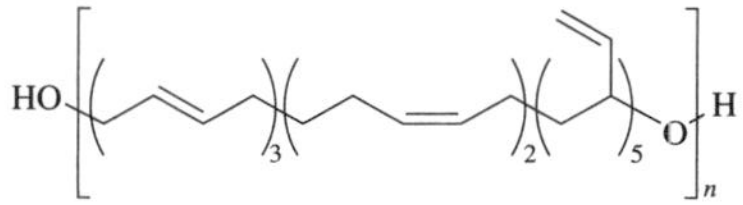	MeO-TEMPO (cat.), $Cu(NO_3)_2$, CuO, CH_2Cl_2, rt, 14 h	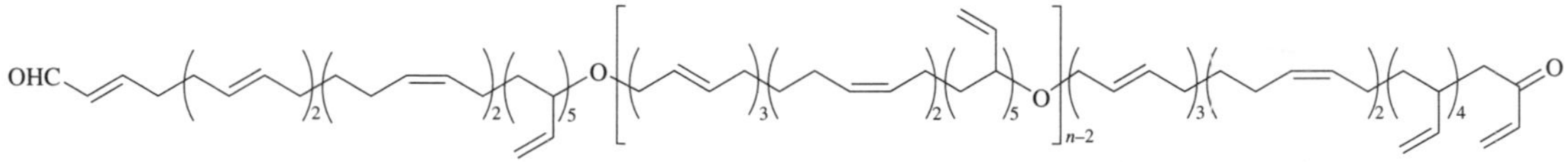 (95), 65% conversion	117
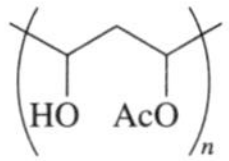 88 mol % degree of saponification	$[MeO\text{-}TEMP{=}O]^+Cl^-$, Na_2CO_3, NMP, dark, rt, 3 h	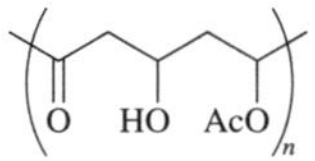	136

Degree of Polymerization	Yield	OH Conv.	Ketone Content
300	(90)	69%	61 mol %
500	(63)	54%	48 mol %
1700	(100)	59%	52 mol %
1400	(94)	67%	59 mol %

[a] The reported value was determined by GC analysis.

[b] The product was not isolated and was used in a subsequent step without purification.

[c] The reported value was determined by HPLC analysis.

[d] The product was not isolated.

[e] The reported value was determined by NMR analysis.

[f] The selectivity refers to formation of the aldehyde versus products with the carboxylic acid oxidation state.

[g] The yield is for several steps.

[h] The yield is for two steps.

[i] The product was isolated as a derivative.

[j] The product was isolated as the methyl ester.

[k] The product formed was a mixture of C-4 epimers in a ratio of 98:2.

[l] The reaction is performed with PEG_{1000}, PEG_{2000}, and PEG_{4600}.

[m] The reaction is performed with MeO-PEG_{1000} and MeO-PEG_{5000}.

TABLE 4A. OXIDATIONS OF GLUCOSE-DERIVED MONOSACCHARIDES

Alcohol	Conditions	Product(s) and Y eld(s) (%)	Refs.
C_6 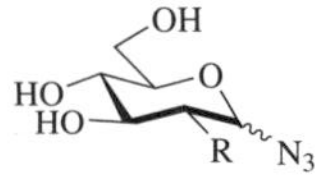	TEMPO (cat.), KBr (cat.), NaOCl or $Ca(OCl)_2$, $NaHCO_3$, H_2O, 15–18°	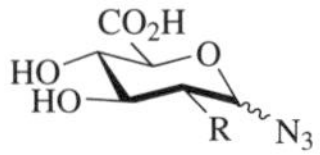 (—)[a] R: OH, NHAc	173
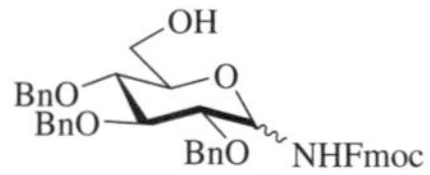	TEMPO (cat.), NaBr (cat.), NaOCl, TBAC, CH_2Cl_2, H_2O	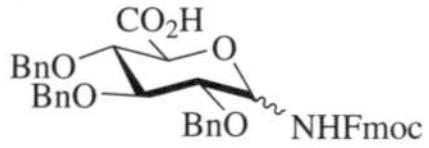 (—)[a]	666
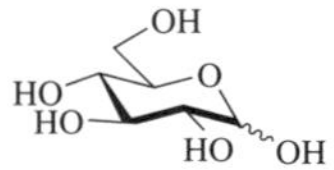	TEMPO at a glassy carbon disk electrode (cat.), electrolysis, 1 M NaOH, H_2O	gluconic acid + glucaric acid + glucuronic acid + formic acid + oxalic acid (—)	323
	AcNH-TEMPO (cat.), NaBr (cat.), NaOCl, NaOH, H_2O, pH 11.5, 0–5°	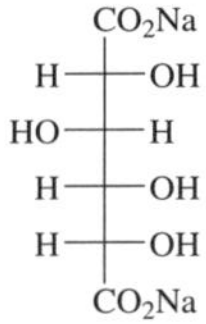 **I** (95^b, 80)	329
	AcNH-TEMPO (cat.), KBr (cat.), Cl_2 or Br_2, H_2O, pH 11.5, NaOH, 0–5°	**I** (79–85)	257
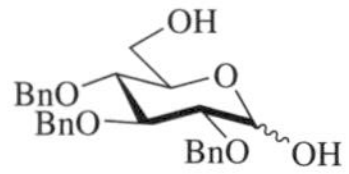	TEMPO (cat.), $PhI(OAc)_2$, CH_2Cl_2, H_2O, rt	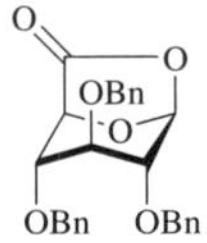 (27)	322

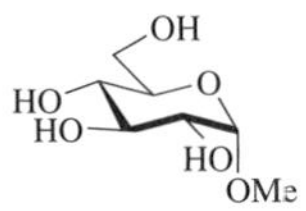	TEMPO, TCC (cat.), $NaHCO_3$, DMF, 0°	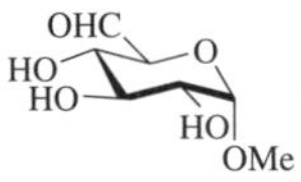 (79)	320
	$[TEMP{=}O]^+[BF_4]^-$, 2,6-lutidine, DMF, 110 h	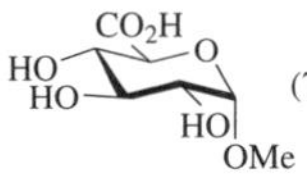 (7) + 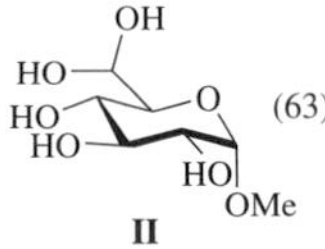 (63) **I** **II**	144
	TEMPO (cat.), electrolysis, $NaHCO_3$, Na_2CO_3, H_2O	**I** (—)[a]	283
	TEMPO (cat.), NaOCl, H_2O, pH 8.5, 20°	**I** (—)	333
	TEMPO or ormosil-TEMPO (cat.), NaBr (cat.), NaOCl, H_2O, pH 10	**I** (—)	167
	PIPO (cat.), NaOCl, H_2O, pH 9.5, rt	**I** (70)[c]	299
	TEMPO (cat.), electrolysis, $NaHCO_3$, Na_2CO_3, H_2O	**I** (—)[a]	325
	TEMPO (cat.), NaBr (cat.), NaOCl, H_2O, pH 10, 1.5°	**I** (—)[d]	61
	Nafion-TEMPO graphite felt electrode (cat.), electrolysis, H_2O, carbonate buffer, pH 10	**I** (63)	326
	MCM41-NH-TEMPO (cat.), NaOCl, NaOH, H_2O, pH 8, 0°	**I** (95)	309
	TEMPO (cat.), NaBr (cat.), NaOCl, H_2O, pH 10.5, sonication, 5°	**I** (63–67)	327, 328
	TEMPO (cat.), NaBr (cat.), NaOCl, H_2O, pH 10, 2°	**I** (100)	75

TABLE 4A. OXIDATIONS OF GLUCOSE-DERIVED MONOSACCHARIDES (*Continued*)

Alcohol	Conditions	Product(s) and Yield(s) (%)	Refs.
C_6 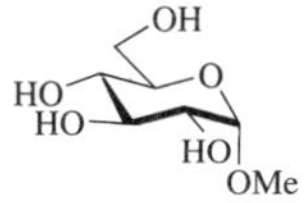	Nitroxide (cat.), NaOCl, H_2O, pH 8.5	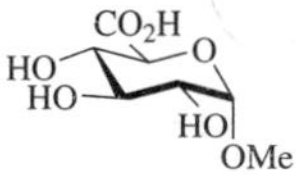 (—) Nitroxide: AcNH-TEMPO, MeO-TEMPO, AcO-TEMPO, mesyl-TEMPO	314
	1. TEMPO (cat.), NaCl (cat.), NaOCl, TBAF, $NaHCO_3$, H_2O 2. CH_2N_2, Et_2O	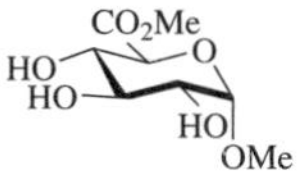 (72)	667
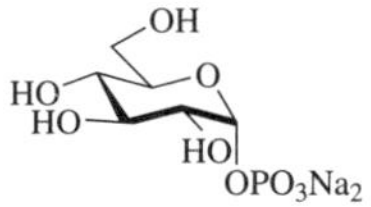	TEMPO (cat.), electrolysis, $NaHCO_3$, Na_2CO_3, H_2O	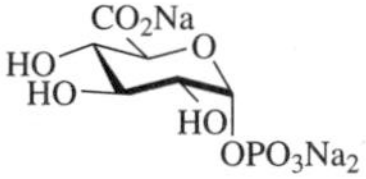 (52)[e]	283
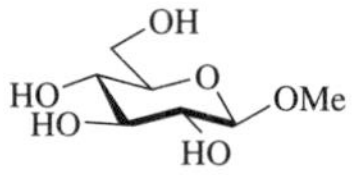	TEMPO (cat.), TCC, $NaHCO_3$, DMF, 0°	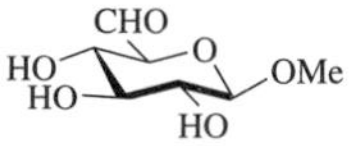 (100)	320
	TEMPO (cat.), electrolysis, $NaHCO_3$, Na_2CO_3, H_2O	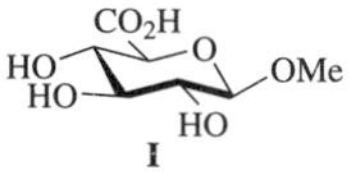 **I** (—)[a]	283
	TEMPO (cat.), NaBr (cat.), NaOCl, H_2O, pH 10, 1.5°	**I** (—)[d]	61
	TEMPO (cat.), NaBr (cat.), NaOCl, H_2O, pH 10, 2°	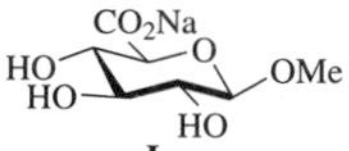 **I** (100)	75
	TEMPO (cat.), *t*-BuOCl, NaOH, H_2O, pH 10–10.5	**I** (34)	175

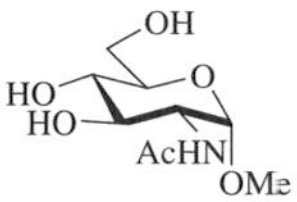	TEMPO (cat.), TCC, $NaHCO_3$, DMF, 0°	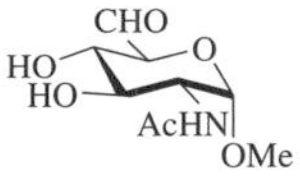 (66)	320
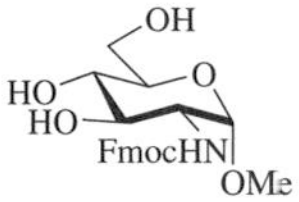	TEMPO (cat.), KBr (cat.), NaOCl, $NaHCO_3$, THF, H_2O, 0°	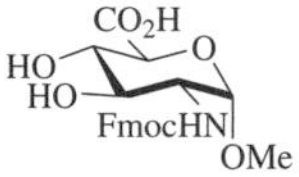 (82)	668
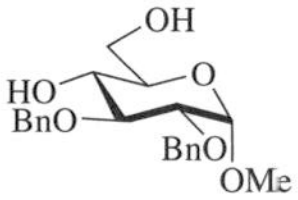	TEMPO (cat.), NaOCl, $NaHCO_3$ (satd.), H_2O, CH_2Cl_2, 0°	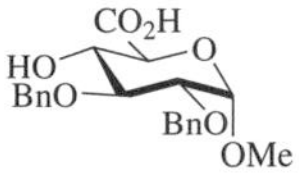 (—)	669
	1. TEMPO (cat.), $PhI(OAc)_2$, CH_2Cl_2, rt 2. $Ph_3P{=}CHCO_2Et$	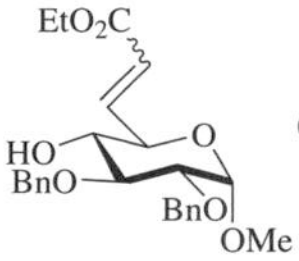 (68), E:Z = 3:1	205
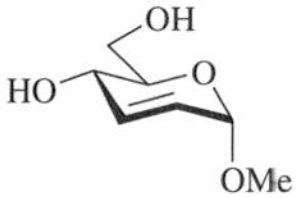	TEMPO (cat.), electrolysis, $NaHCO_3$, Na_2CO_3, H_2O	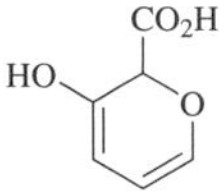 (—)[a]	325
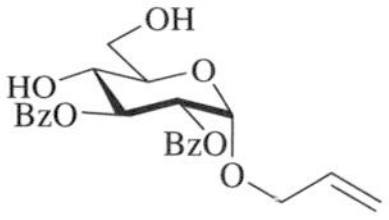	1. TEMPO (cat.), KBr (cat.), $Ca(OCl)_2$, TBAB, $NaHCO_3$ (satd.), CH_2Cl_2, H_2O, 0°, 45 min 2. CH_2N_2, DMF	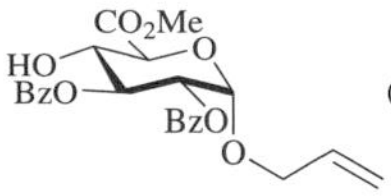 (70)	172

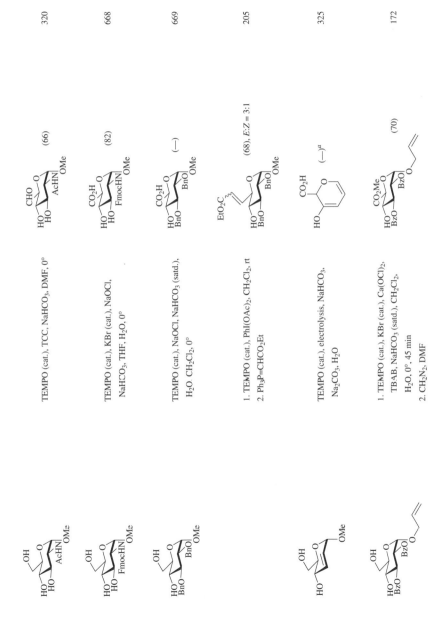

TABLE 4A. OXIDATIONS OF GLUCOSE-DERIVED MONOSACCHARIDES (*Continued*)

Alcohol	Conditions	Product(s) and Yield(s) (%)	Refs.
C_6			
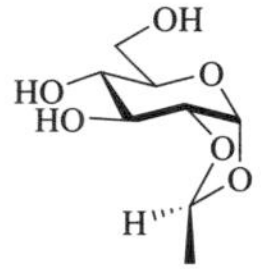	TEMPO (cat.), NaOCl, TBAC, $NaHCO_3$, EtOAc, H_2O	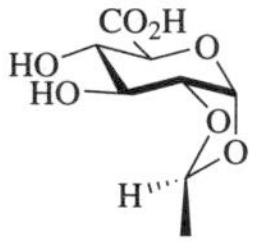 (—)	670
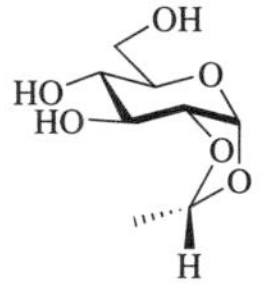	TEMPO (cat.), NaOCl, TBAC, $NaHCO_3$, EtOAc, H_2O	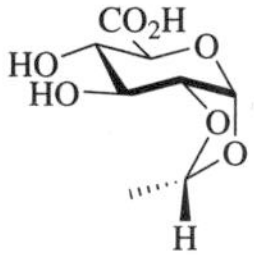 (—)	670
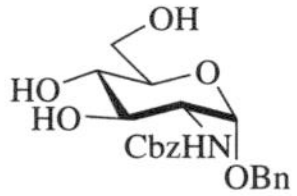	TEMPO (cat.), KBr (cat.), NaOCl, TBAB, CH_2Cl_2, H_2O, 0°, 1.5 h	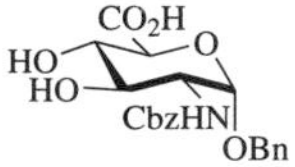 (67)	671
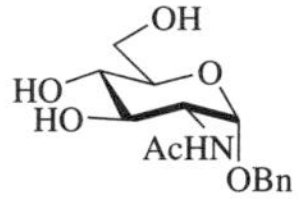	1. TEMPO (cat.), KBr (cat.), $Ca(OCl)_2$, TBAB, $NaHCO_3$ (satd.), CH_2Cl_2, H_2O, 0°, 45 min 2. CH_2N_2, DMF	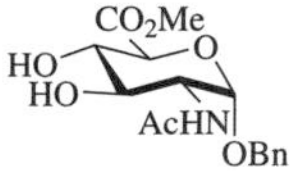 (85)	172
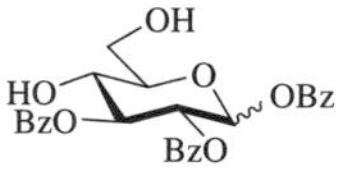	1. TEMPO (cat.), KBr (cat.), $Ca(OCl)_2$, TBAB, $NaHCO_3$ (satd.), CH_2Cl_2, H_2O, 0°, 45 min 2. CH_2N_2, DMF	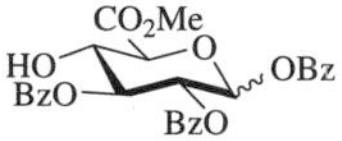 (83)	172
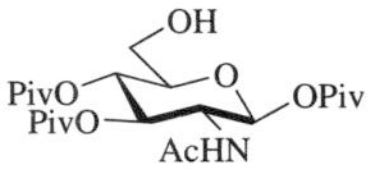	TEMPO (cat.), KBr (cat.), NaOCl, TBAB, $NaHCO_3$ (satd.), NaCl (satd.), H_2O, CH_2Cl_2, rt, 75 min	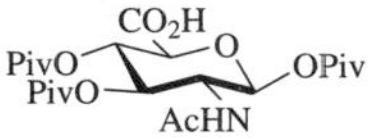 (—)[a]	672

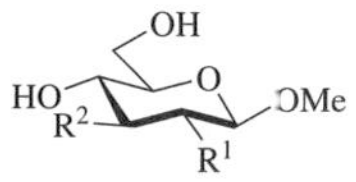

See table.

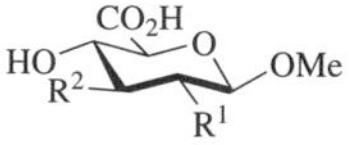

R[1]	R[2]	Conditions	Yield	Refs.
OAc	OAc	TEMPO (cat.), NaBr (cat.), NaOCl, TBAB, $NaHCO_3$, EtOAc, H_2O, 0°	(70)	673
OBn	OH	TEMPO (cat.), KBr (cat.), NaOCl, TBAC, $NaHCO_3$, CH_2Cl_2, H_2O, 0°	(—)[a]	315
OBn	OAc	"	(83)	315
OBn	OBn	"	(—)[a]	315
NPhth	OBn	1. TEMPO (cat.), KBr (cat.), NaOCl, TBAB, $NaHCO_3$, CH_2Cl_2, H_2O, 0° to rt, 1 h 2. $NaClO_2$, 2-methyl-2-butene	(75)	415
OAc	NHFmoc	TEMPO (cat.), KBr (cat.), NaOCl, TBAC, $NaHCO_3$, NaCl, H_2O, 0°	(80)	674
NHCBZ	OMe	"	(87)	674

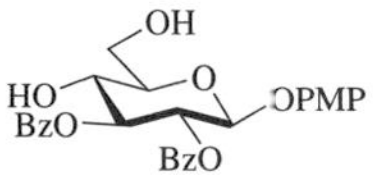

TEMPO (cat.), KBr (cat.), $Ca(ClO)_2$, Aliquat 336, $NaHCO_3$, CH_2Cl_2, H_2O, 0°

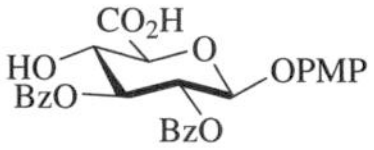

(—)[a] 675

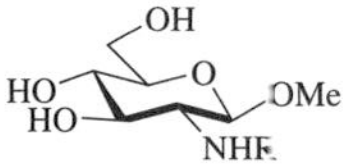

TEMPO (cat.), conditions

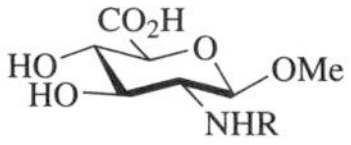

R	Conditions	Yield	Refs.
Ac	electrolysis, $NaHCO_3$, Na_2CO_3, H_2O	(—)[a]	325
Fmoc	KBr (cat.), NaOCl, $NaHCO_3$, THF, H_2O, 0°	(83)	668

TABLE 4A. OXIDATIONS OF GLUCOSE-DERIVED MONOSACCHARIDES (*Continued*)

Alcohol	Conditions	Product(s) and Yield(s) (%)	Refs.
C_6			
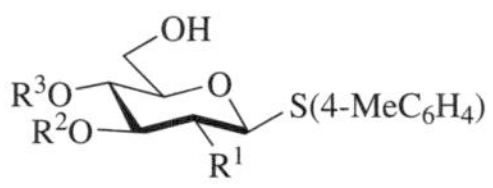	1. TEMPO (cat.), KBr (cat.), NaOCl, TBAB, $NaHCO_3$, CH_2Cl_2, H_2O, 0° to rt, 1 h 2. $NaClO_2$, 2-methyl-2-butene	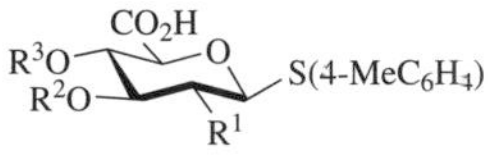	415
R^1 = OBz, R^2 = Bz, R^3 = Bz		(82)	
R^1 = NPhth, R^2 = TBDMS, R^3 = Bn		(86)	
R^1 = NPhth, R^2 = PMB, R^3 = Bn		(85)	
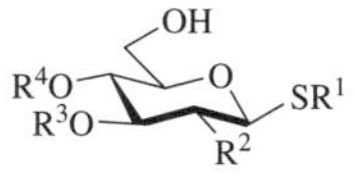	TEMPO (cat.), $PhI(OAc)_2$, CH_2Cl_2, H_2O, rt	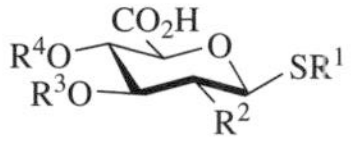	
R^1 = Et, R^2 = N_3, R^3 = Bz, R^4 = H		(88)	364
R^1 = Et, R^2 = OBn, R^3 = Bn, R^4 = H		(92)	364
R^1 = Et, R^2 = OBz, R^3 = Bz, R^4 = H		(76)	364
R^1 = Ph, R^2 = NPhth, R^3 = TBDMS, R^4 = H		(67)	364
R^1 = Ph, R^2 = OBz, R^3 = Bz, R^4 = H		(81)	364
R^1 = Ph, R^2 = OBz, R^3 = Bn, R^4 = H		(83)	676
R^1 = Et, R^2 = OBn, R^3 = Bn, R^4 = Bn		(88)	364
C_{6-10}			
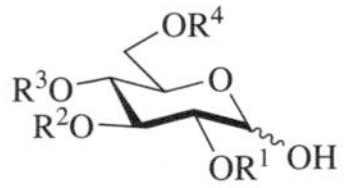	$[AcNH\text{-}TEMP{=}O]^+[BF_4]^-$, pyridine, CH_2Cl_2	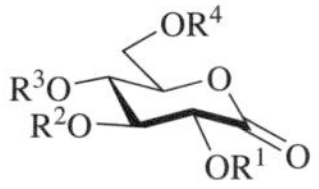	76
R^1 = Me, R^2 = Me, R^3 = Me, R^4 = Me		(95)	
R^1 = Ac, R^2 = Ac, R^3 = Ac, R^4 = Ac		(94)	
R^1 = H, R^2 = Bn, R^3 = Bn, R^4 = Bn		(80)	
R^1 = Bn, R^2 = Bn, R^3 = Bn, R^4 = Bn		(97)	

C_{6-17}	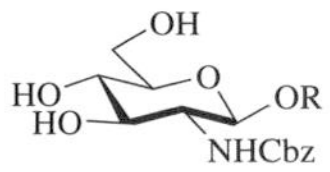	TEMPO (cat.), NaOCl (cat.), $NaClO_2$, $NaHCO_3$, H_2O, MeCN, borax buffer, pH 9	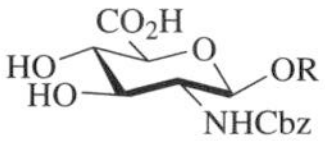	R 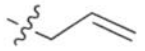 (94) 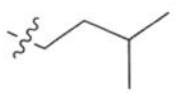 (71–95) n-C_6H_{13} (92–99)  (60) 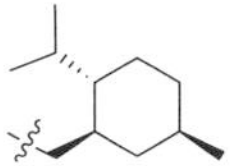 (95) 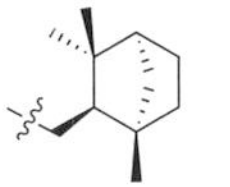 (53)	677
C_7	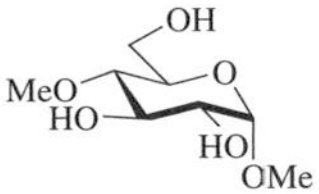	TEMPO (cat.), NaBr (cat.), NaOCl, H_2O, pH 10–11	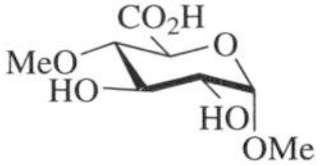	(90)	390
C_8	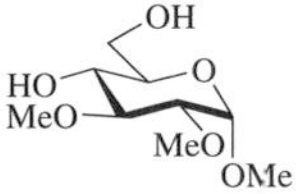	1. TEMPO (cat.), NaCl (cat.), NaOCl, TBAF, $NaHCO_3$ (satd.), H_2O, EtOAc, 0°, 2 h 2. MeI, DMF	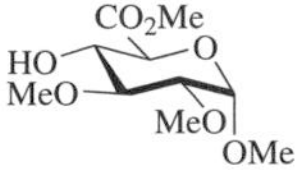	(86)	667
C_9	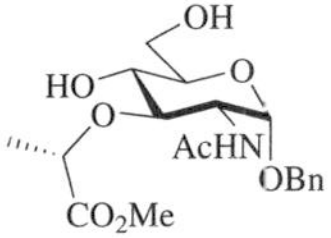	1. TEMPO (cat.), KBr (cat.), $Ca(OCl)_2$, TBAB, $NaHCO_3$ (satd.), CH_2Cl_2, H_2O, 0°, 45 min 2. CH_2N_2, DMF	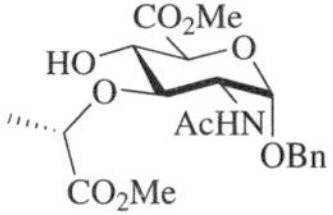	(71)	172
	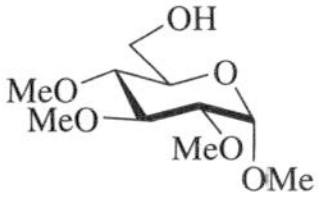	1. TEMPO (cat.), NaCl (cat.), NaOCl, TBAF, $NaHCO_3$ (satd.), H_2O, EtOAc, 0°, 2 h 2. MeI, DMF	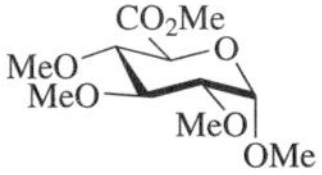	(70)	667

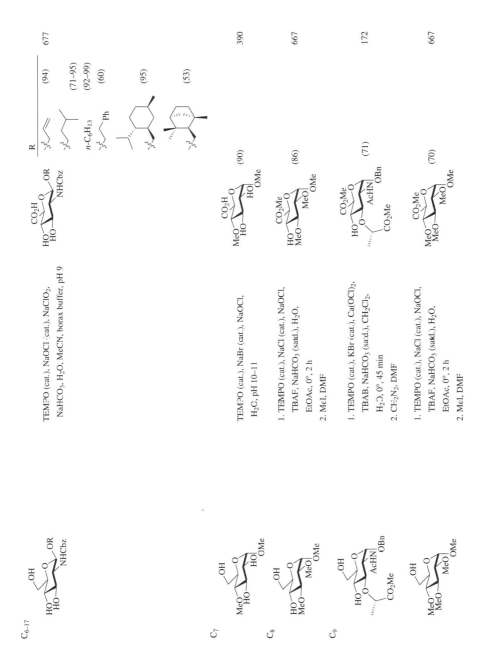

TABLE 4A. OXIDATIONS OF GLUCOSE-DERIVED MONOSACCHARIDES (*Continued*)

Alcohol	Conditions	Product(s) and Yield(s) (%)	Refs.
C_{9-43} 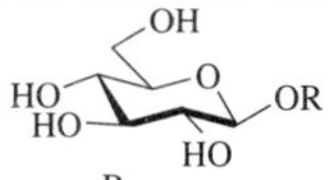	See table.	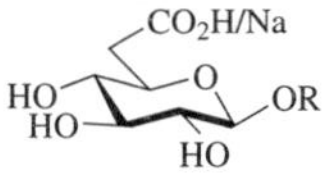	
R	Conditions		
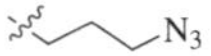	TEMPO (cat.), KBr (cat.), NaOCl, TBAB, NaCl, $NaHCO_3$, CH_2Cl_2, H_2O	(—)	678
Ph	TEMPO (cat.), *t*-BuOCl, NaOH, H_2O, pH 10–10.5	(78–96)	175
2-FC_6H_4	TEMPO (cat.), *t*-BuOCl, NaOH, H_2O, pH 10–10.5	(97)	175
4-$O_2NC_6H_4$	TEMPO (cat.), *t*-BuOCl, NaOH, H_2O, pH 10–10.5	(64)	175
4-MeC_6H_4	TEMPO (cat.), NaBr (cat.), NaOCl, KOH, H_2O, pH 10.5	(74)	679
n-C_7H_{18}	TEMPO (cat.), *t*-BuOCl, NaOH, H_2O, pH 10–10.5	(100)	175
n-C_8H_{17}	TEMPO (cat.), electrolysis, $NaHCO_3$, Na_2CO_3, H_2O	(—)[a]	283
n-C_8H_{17}	TEMPO (cat.), KBr (cat.), NaOCl, TBAC, $NaHCO_3$, CH_2Cl_2, H_2O, 0°	(—)[a]	315
2,6-$Me_2C_6H_3$	TEMPO (cat.), NaBr (cat.), NaOCl, KOH, H_2O, pH 10.5	(74)	679
n-$C_{10}H_{21}$	TEMPO (cat.), electrolysis, 2,6-lutidine, MeCN	(97)	680
n-$C_{10}H_{21}$	TEMPO (cat.), NaOCl (cat.), $NaClO_2$, CH_2Cl_2, H_2O, phosphate buffer, pH 6.7	(90)	680
n-$C_{12}H_{25}$	TEMPO (cat.), KBr (cat.), NaOCl, $NaHCO_3$, H_2O, 20°	(73)	681
n-$C_{12}H_{25}$	TEMPO (cat.), electrolysis, $NaHCO_3$, Na_2CO_3, H_2O	(—)[a]	283

n-$C_{14}H_{29}$	TEMPO (cat.), KBr (cat.), NaOCl, $NaHCO_3$, H_2O, 20°	(76)	681
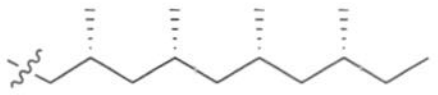	TEMPO (cat.), KBr (cat.), NaOCl, $NaHCO_3$, H_2O, 20°	(73)	681
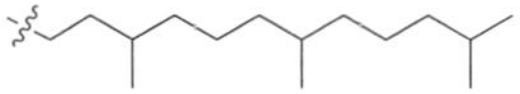	TEMPO (cat.), KBr (cat.), NaOCl, $NaHCO_3$, H_2O, 20°	(54)	681
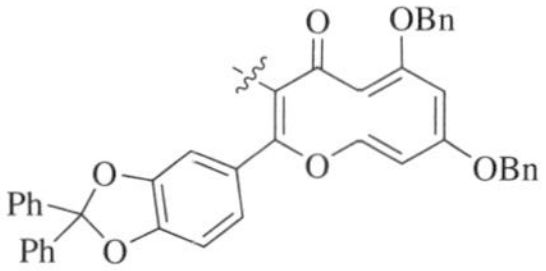	TEMPO (cat.), KBr (cat.), NaOCl, TBAB, $NaHCO_3$, CH_2Cl_2, H_2O	(—)[a]	682
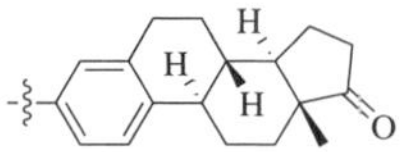	TEMPO (cat.), NaBr (cat.), NaOCl, KOH, H_2O, pH 10.5	(62)	679
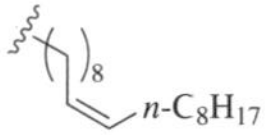	TEMPO (cat.), KBr (cat.), NaOCl, $NaHCO_3$, H_2O, 20°	(80)	681
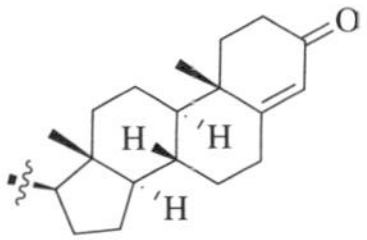	TEMPO (cat.), NaBr (cat.), NaOCl, KOH, H_2O, pH 10.5	(48)	679
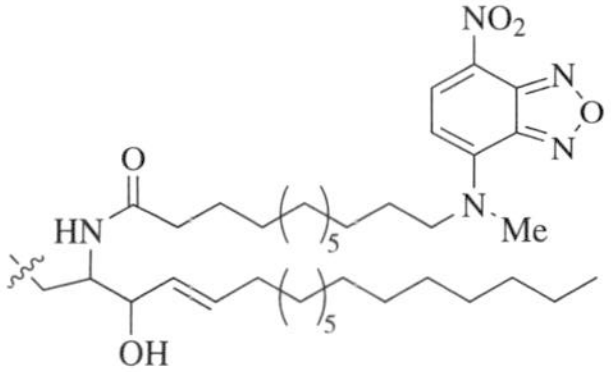	HO_2C-TEMPO (cat.), NaOCl, H_2O, CH_2Cl_2	(—)	683

TABLE 4A. OXIDATIONS OF GLUCOSE-DERIVED MONOSACCHARIDES (*Continued*)

Alcohol	Conditions	Product(s) and Yield(s) (%)	Refs.
C_{10} 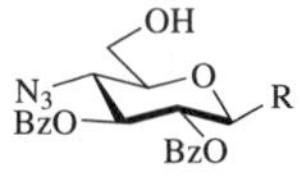	TEMPO (cat.), $PhI(OAc)_2$, MeCN, H_2O	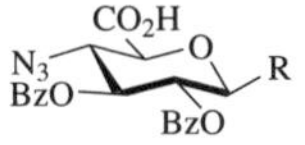 R: 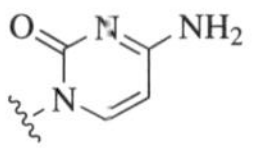 (82); 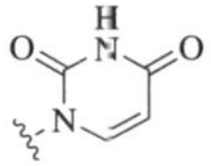 (—)	598
C_{10-15} 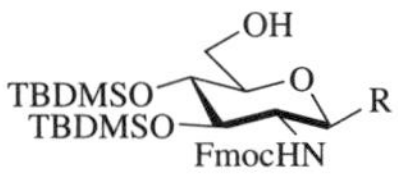	1. TEMPO (cat.), KBr (cat.), TBAS, NaOCl, NaCl, $NaHCO_3$, H_2O, CH_2Cl_2 2. $NaClO_2$, NaH_2PO_4, 2-methyl-2-butene, *t*-BuOH, H_2O	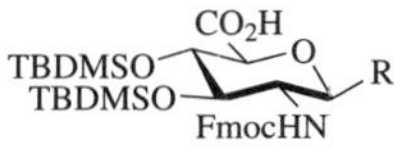 (50–90) R: 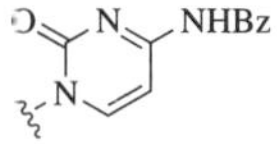 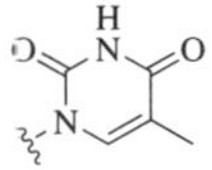 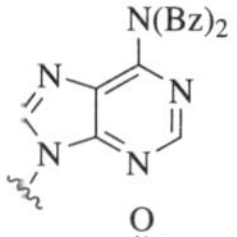 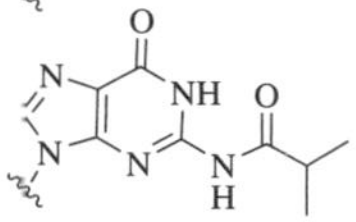	344
C_{12} 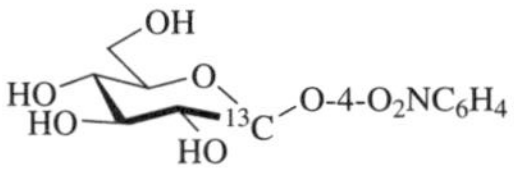	TEMPO (cat.), *t*-BuOCl, H_2O, pH 10–10.5	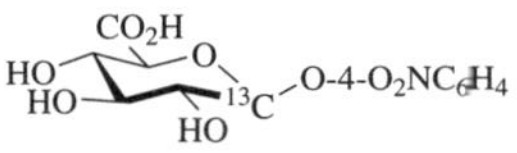 (100)	684
C_{13} 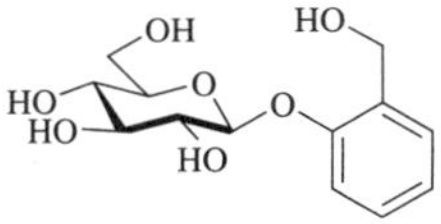	1. TEMPO (cat.), electrolysis, $NaHCO_3$, Na_2CO_3, H_2O 2. Esterification	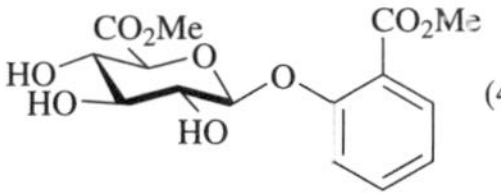 (45)	283

C_{14}				
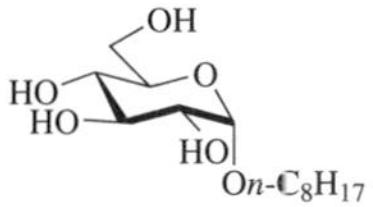	Nafion-TEMPO graphite felt electrode (cat.), electrolysis, H_2O, carbonate buffer, pH 10	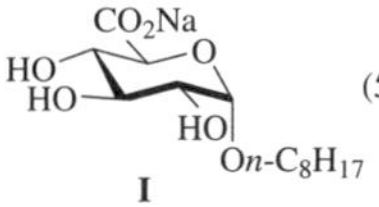 **I**	(58)	326
	TEMPO (cat.), NaOCl, NaBr, H_2O, pH 10, 1.5°	**I** $(—)^d$		61
C_{33}				
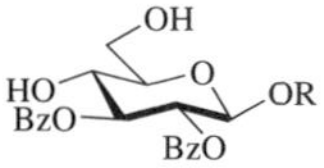	1. TEMPO (cat.), KBr (cat.), $Ca(OCl)_2$, TBAB, $NaHCO_3$ (satd.), CH_2Cl_2, H_2O, 0°, 45 min 2. CH_2N_2, DMF	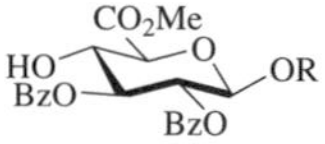		172
R				
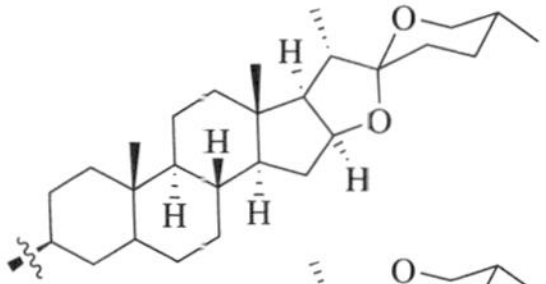			(78)	
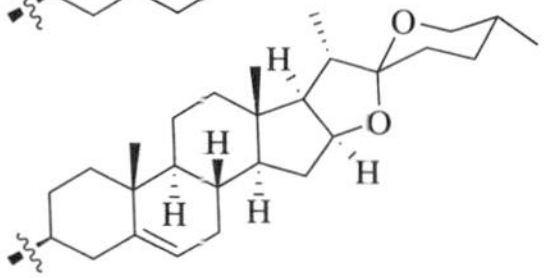			(51)	

TABLE 4A. OXIDATIONS OF GLUCOSE-DERIVED MONOSACCHARIDES (*Continued*)

Alcohol	Conditions	Product(s) and Yield(s) (%)	Refs.
C_{36}			
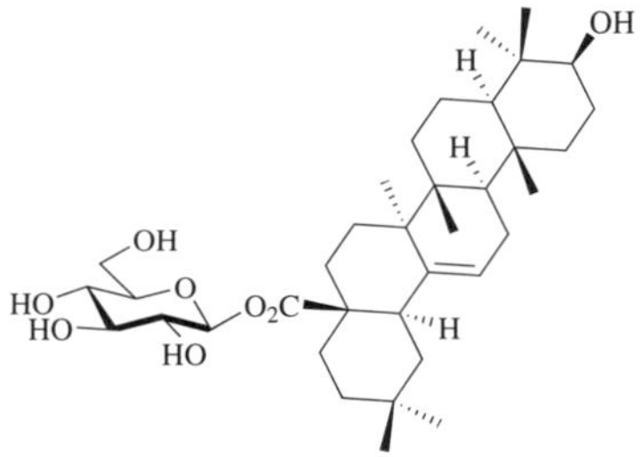	1. TEMPO (cat.), KBr (cat.), $Ca(OCl)_2$, TBAB, $NaHCO_3$ (satd.), CH_2Cl_2, H_2O, 0°, 45 min 2. CH_2N_2, DMF	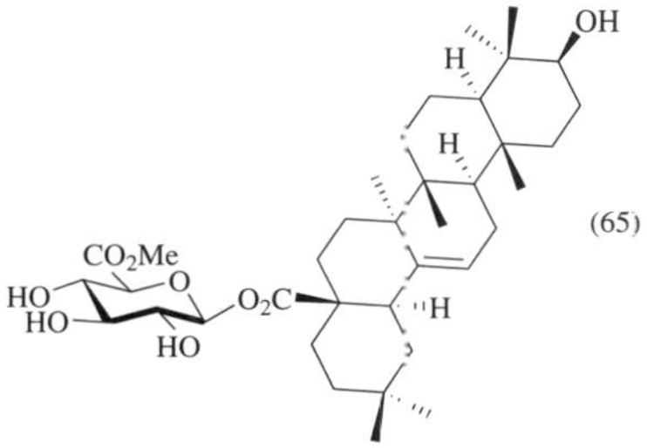 (65)	172

[a] The product was not isolated and was used in a subsequent step without purification.

[b] The reported value was determined by NMR analysis.

[c] The reported value was determined by HPLC analysis.

[d] The reaction rate of this process is provided.

[e] The reported value was determined by GC analysis.

TABLE 4B. OXIDATIONS OF MANNOSE-DERIVED MONOSACCHARIDES

Alcohol	Conditions	Product(s) and Yield(s) (%)	Refs.
C_6 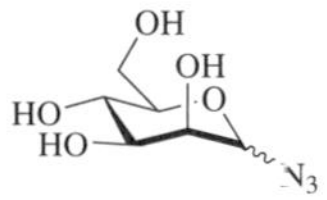	TEMPO (cat.), KBr (cat.), NaOCl or $Ca(OCl)_2$, $NaHCO_3$, H_2O, 15–18°	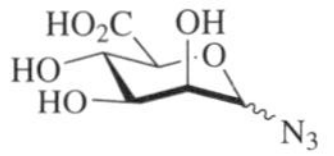 (—)[a]	173
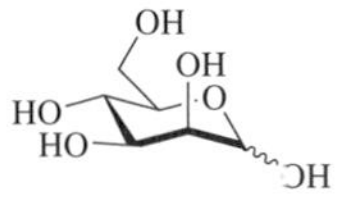	AcNH-TEMPO (cat.), KBr (cat.), Cl_2 or Br_2, NaOH, H_2O, pH 11.5, 0–5°	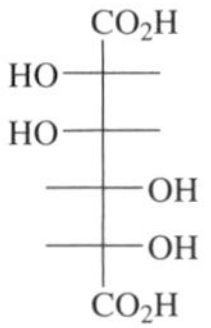 (90[b], 80)	257
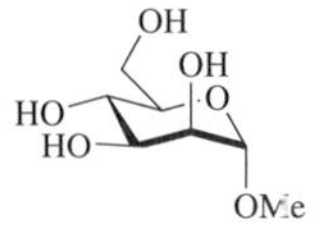	TEMPO (cat.), TCC, $NaHCO_3$, DMF, 0°	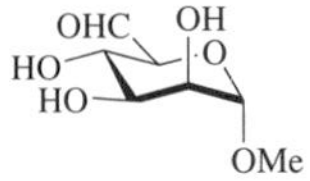 (84)	320
	$[TEMP{=}O]^+[BF_4]^-$, 2,6-lutidine, DMF, 110 h	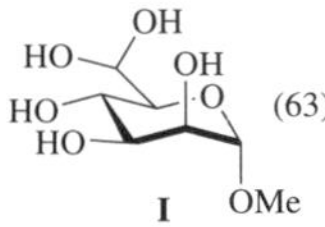 I (63) + 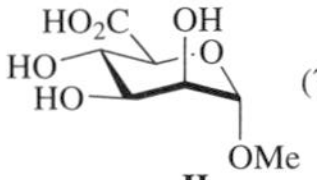 II (7)	144
	TEMPO (cat.), KBr (cat.), NaOCl, TBAC, $NaHCO_3$, CH_2Cl_2, H_2O, 0°	**II** (—)	685
	TEMPO (cat.), NaBr (cat.), NaOCl, NaOH, H_2O, pH 10, 0°[c]	**II** (76)	686
	TEMPO (cat.), electrolysis, $NaHCO_3$, Na_2CO_3, H_2O	**II** (—)[a]	283
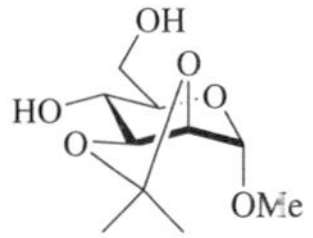	1. TEMPO (cat.), NaCl (cat.), NaOCl, TBAF, $NaHCO_3$, EtOAc, H_2O, 0°, 2 h 2. MeI, DMF	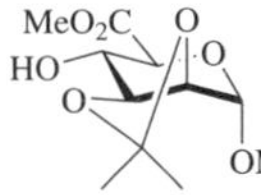 (62)	667

TABLE 4B. OXIDATIONS OF MANNOSE-DERIVED MONOSACCHARIDES (*Continued*)

Alcohol	Conditions	Product(s) and Yield(s) (%)	Refs.
C_6 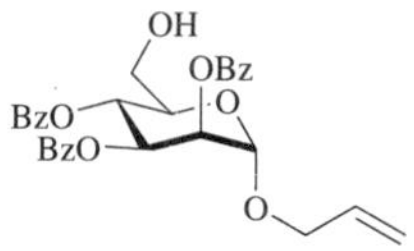	1. TEMPO (cat.), KBr (cat.), NaOCl, TBAB, $NaHCO_3$, CH_2Cl_2, H_2O, 0° to rt, 1 h 2. $NaClO_2$, 2-methyl-2-butene	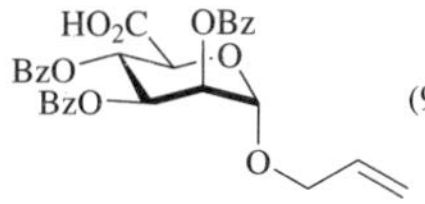 (92)	415
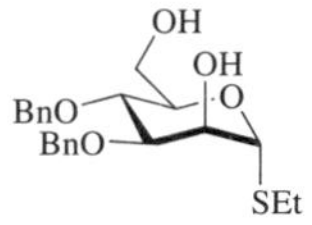	TEMPO (cat.), $PhI(OAc)_2$, CH_2Cl_2, H_2O, rt	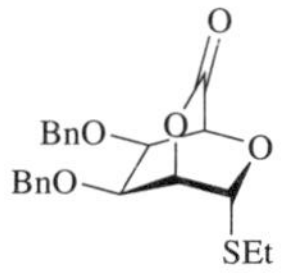 (77)	322
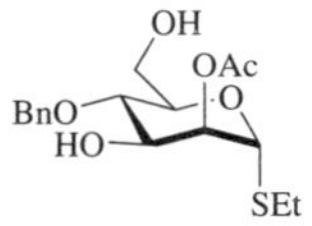	TEMPO (cat.), $PhI(OAc)_2$, CH_2Cl_2, H_2O, rt	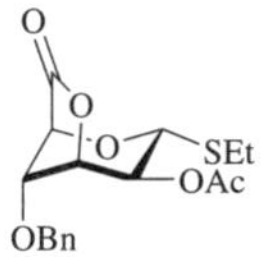 (72)	322
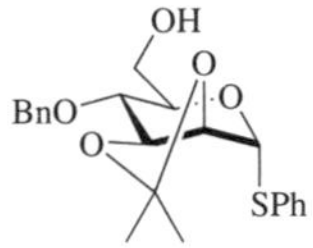	TEMPO (cat.), $PhI(OAc)_2$, CH_2Cl_2, H_2O, rt	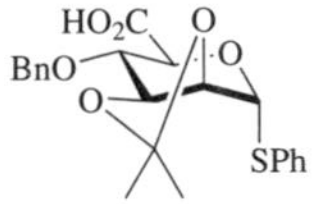 (100)	364

[a] The product was not isolated and was used in a subsequent step without purification.

[b] The reported value was determined by NMR analysis.

Alcohol	Conditions	Product(s) and Yield(s) (%)	Refs.
C_6 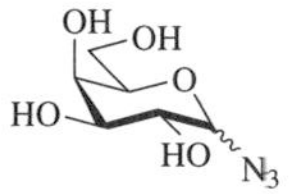	TEMPO (cat.), KBr (cat.), NaOCl or $Ca(OCl)_2$, $NaHCO_3$, H_2O, 15–18°	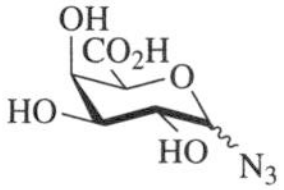 (—)[a]	173
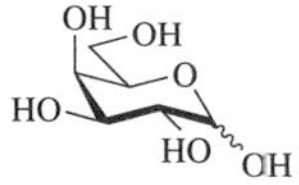	AcNH-TEMPO (cat.), KBr (cat.), Cl_2 or Br_2, NaOH, H_2O, pH 11.4–11.7, 0–5°	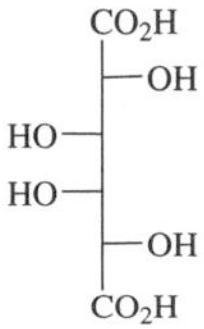 (70–80)	257
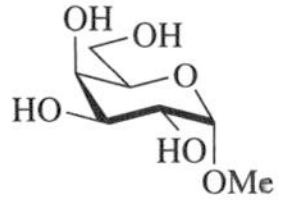	TEMPO (cat.), KBr (cat.), NaOCl, H_2O, pH 10–11, 0°, 2–7 h	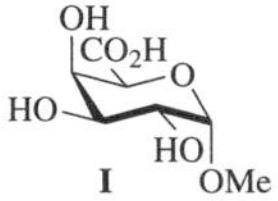 **I** (57)	687
	TEMPO (cat.), electrolysis, $NaHCO_3$, Na_2CO_3, H_2O	**I** (—)[a]	283
	TEMPO (cat.), NaBr (cat.), NaOCl, H_2O, pH 10, 1.5°	**I** (—)[b]	61
	Nafion-TEMPO graphite felt electrode (cat.), electrolysis, H_2O, carbonate buffer, pH 10	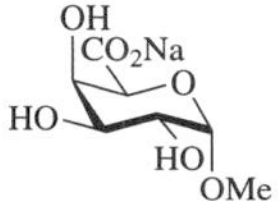 (53)	326
	1. TEMPO (cat.), NaBr (cat.), NaOCl, NaOH, H_2O 2. NaOAc, Ac_2O	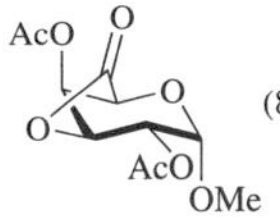 (81)	688

TABLE 4C. OXIDATIONS OF GALACTOSE-DERIVED MONOSACCHARIDES (*Continued*)

	Alcohol	Conditions	Product(s) and Yield(s) (%)	Refs.
C_6	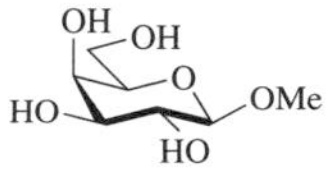	TEMPO (cat.), TCC, $NaHCO_3$, DMF, 0°	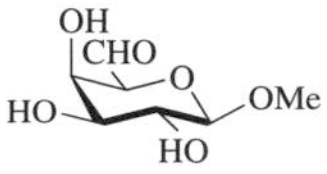 (81)	320
		TEMPO (cat.), KBr (cat.), NaOCl, H_2O, pH 10–11, 0°, 2–7 h	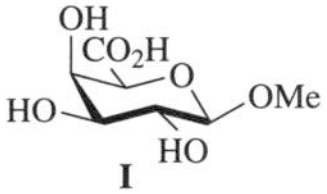 **I** (50)	687
		TEMPO (cat.), electrolysis, $NaHCO_3$, Na_2CO_3, H_2O	**I** (—)[a]	283
	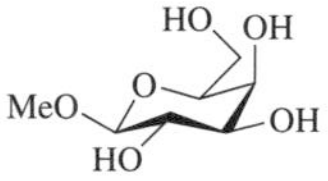	TEMPO (cat.), KBr (cat.), NaOCl, H_2O, pH 10–11, 0°, 2–7 h	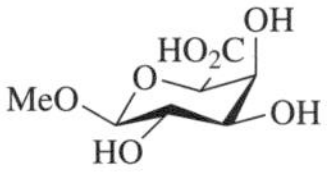 (74)	687
	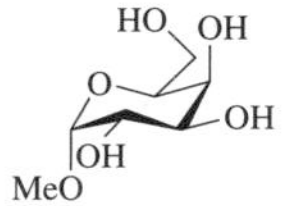	TEMPO (cat.), KBr (cat.), NaOCl, H_2O, pH 10–11, 0°, 2–7 h	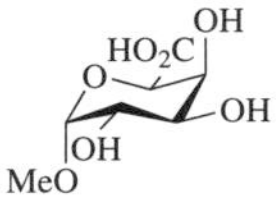 (64)	687
	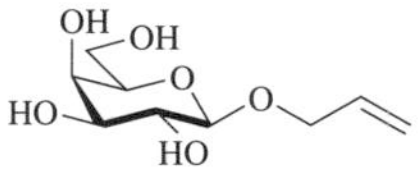	TEMPO (cat.), electrolysis, 2,6-lutidine, MeCN	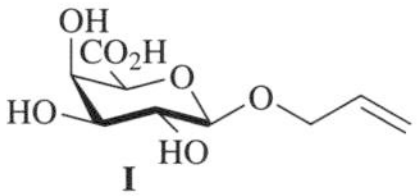 **I** (70)	680
		TEMPO (cat.), NaOCl (cat.), $NaClO_2$, H_2O, CH_2Cl_2, phosphate buffer, pH 6.7	**I** (65)	680
	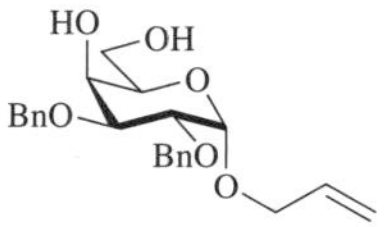	TEMPO (cat.), NaOCl (cat.), $NaClO_2$, H_2O, CH_2Cl_2, phosphate buffer, pH 6.7	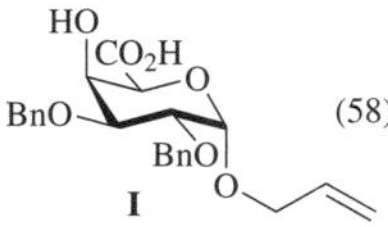 **I** (58)	680
		TEMPO (cat.), electrolysis, 2,6-lutidine, MeCN	**I** (57)	680

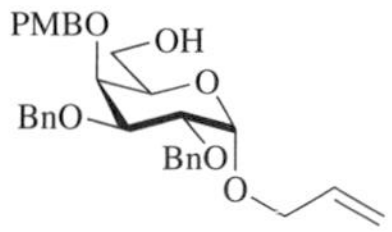	TEMPO (cat.), NaOCl (cat.), $NaClO_2$, H_2O, CH_2Cl_2, phosphate buffer, pH 6.7	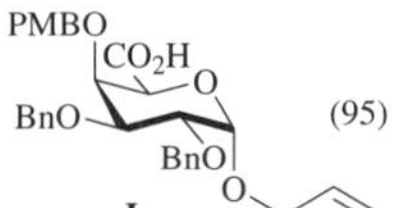 **I** (95)	680
	TEMPO (cat.), electrolysis, 2,6-lutidine, MeCN	**I** (50)	680
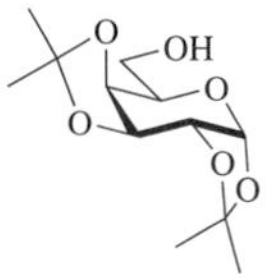	TEMPO (cat.), $[PS\text{-}CH_2NMe_3]^+[Br(OAc)_2]^-$, CH_2Cl_2, rt or toluene, 40°, 3.5 h	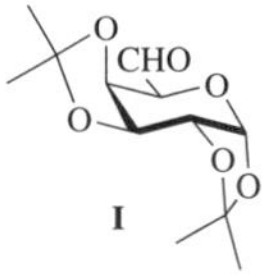 **I** (97)	44, 211
	TEMPO (cat.), $[PS\text{-}CH_2NMe_3]^+[OCl]^-$, CH_2Cl_2, rt or toluene, 40°, 3.5 h	**I** (67)	44
	TEMPO (cat.), electrolysis, 2,6-lutidine, MeCN	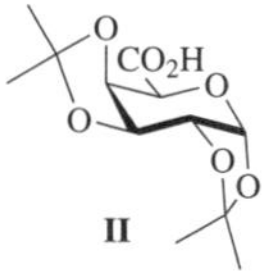 **II** (56)	680
	1. TEMPO (cat.), KBr (cat.), NaOCl, TBAB, $NaHCO_3$, CH_2Cl_2, H_2O, 0° to rt, 1 h 2. $NaClO_2$, 2-methyl-2-butene	**II** (90)	415
	TEMPO (cat.), NaOCl (cat.), $NaClO_2$, CH_2Cl_2, H_2O, phosphate buffer, pH 6.7	**II** (91)	680
	TEMPO (cat.), $[PS\text{-}CH_2NMe_3]^+[Br(OAc)_2]^-$, CH_2Cl_2, rt or toluene, 40°, 3.5 h	**II** (99)	43
	TEMPO (cat.), $[PS\text{-}CH_2NMe_3]^+[ClO_2]^-$, CH_2Cl_2, rt or toluene, 40°, 3.5 h	**II** (96)	44

TABLE 4C. OXIDATIONS OF GALACTOSE-DERIVED MONOSACCHARIDES (*Continued*)

Alcohol	Conditions	Product(s) and Yield(s) (%)	Refs.
C_6 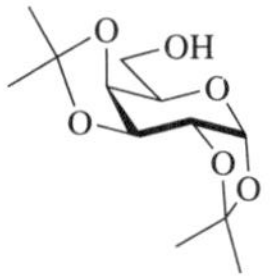	1. TEMPO (cat.), $PhI(OAc)_2$, CH_2Cl_2, rt 2. $Ph_3P{=}CHCO_2Et$	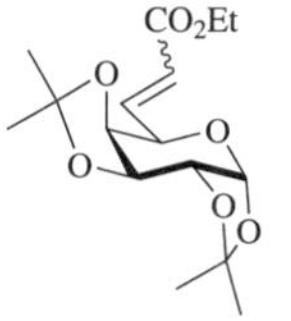 (88), *E*:*Z* = 3:1	205
	$[AcNH\text{-}TEMP{=}O]^+[BF_4]^-$, pyridine, 4 Å MS, CH_2Cl_2, rt	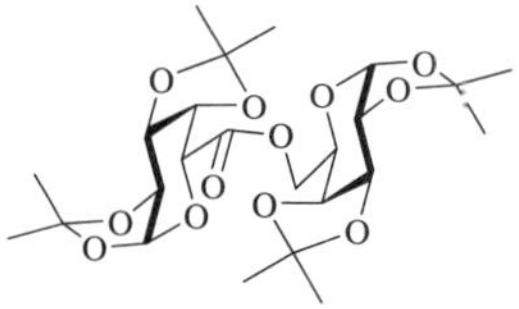 (77)	77
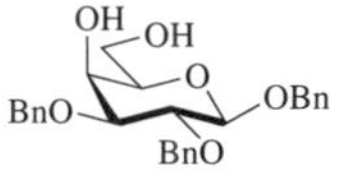	MeO-TEMPO (cat.), KBr (cat.), $Ca(OCl)_2$, Aliquat 336, $NaHCO_3$, CH_2Cl_2, H_2O, pH 7–8	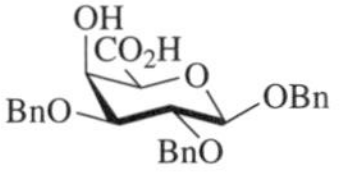 (—)[a]	689
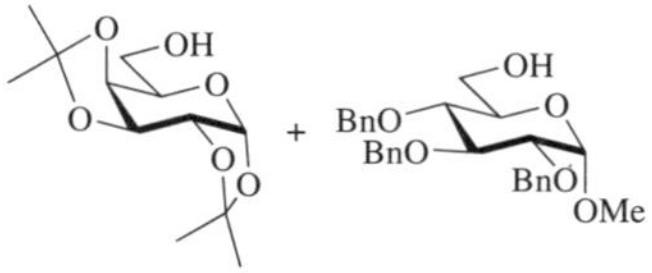	MIP-CO_2-TEMPO (molecularly imprinted for bis(isopropylidene)glucose) (cat.), NaOCl, KBr, $NaHCO_3$, H_2O, CH_2Cl_2, 0°	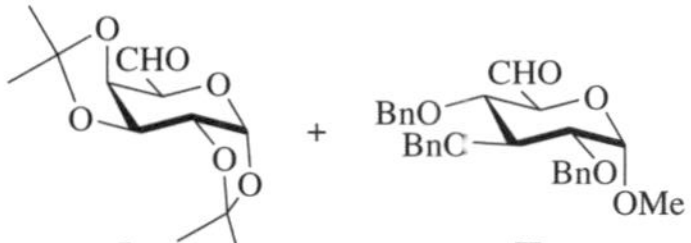 **I** + **II** 8% conversion, **I**:**II** = 4.5:1	74
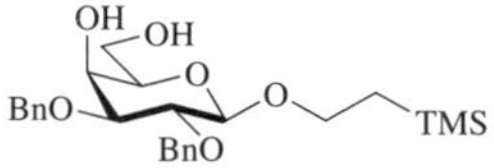	TEMPO (cat.), KBr (cat.), NaOCl, TBAB, NaCl, $NaHCO_3$, H_2O, 0°, 100 min	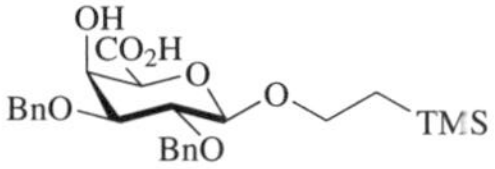 (54)	690
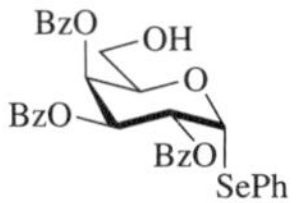	TEMPO (cat.), $PhI(OAc)_2$, CH_2Cl_2, H_2O, rt	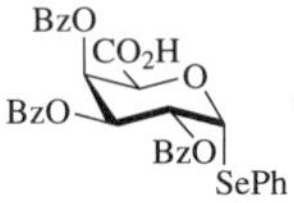 (61)	364

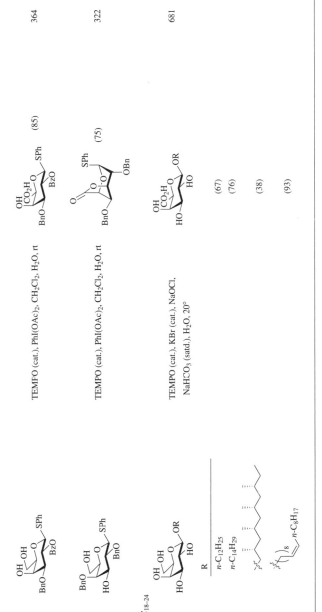

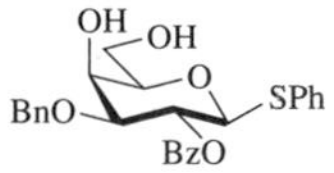	TEMPO (cat.), $PhI(OAc)_2$, CH_2Cl_2, H_2O, rt	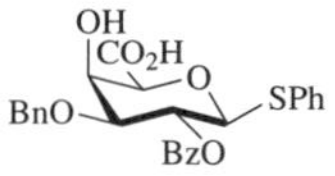 (85)	364
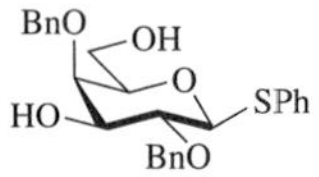	TEMPO (cat.), $PhI(OAc)_2$, CH_2Cl_2, H_2O, rt	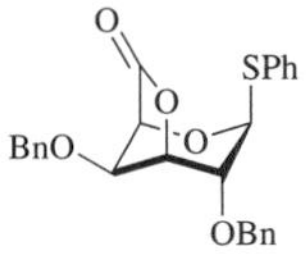 (75)	322
C_{18-24}			
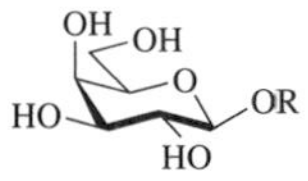	TEMPO (cat.), KBr (cat.), NaOCl, $NaHCO_3$ (satd.), H_2O, 20°	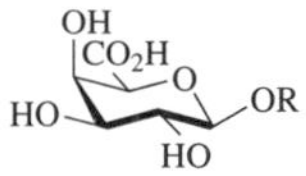	681
R			
n-$C_{12}H_{25}$		(67)	
n-$C_{14}H_{29}$		(76)	
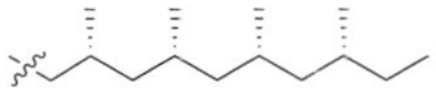		(38)	
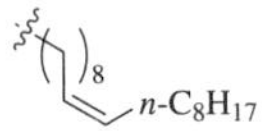		(93)	

[a] The product was not isolated and was used in a subsequent step without purification.

[b] The reaction rate of this process is provided.

TABLE 4D. OXIDATIONS OF DISACCHARIDES

Alcohol	Conditions	Product(s) and Yield(s) (%)	Refs.
C_{11} 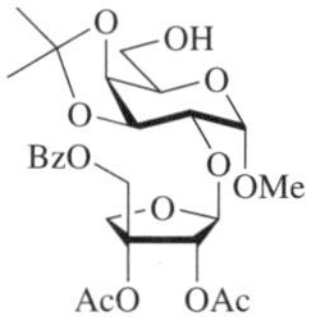	TEMPO (cat.), KBr (cat.), NaOCl, TBAC, $NaHCO_3$, CH_2Cl_2, H_2O	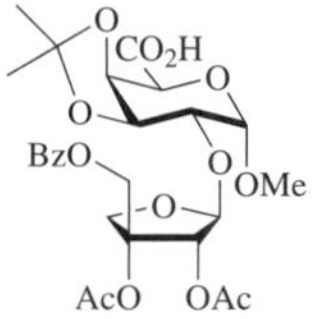 (73)[a]	319
C_{12} 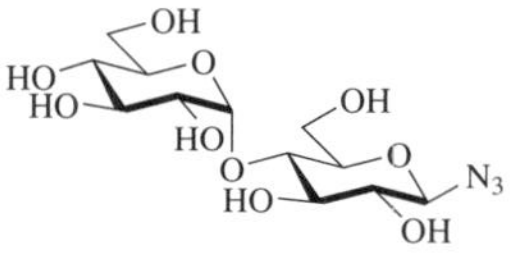	TEMPO (cat.), electrolysis, $NaHCO_3$, Na_2CO_3, H_2O	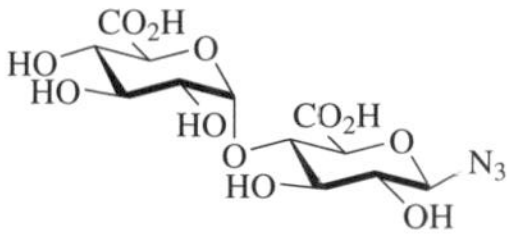 (63)[a]	325
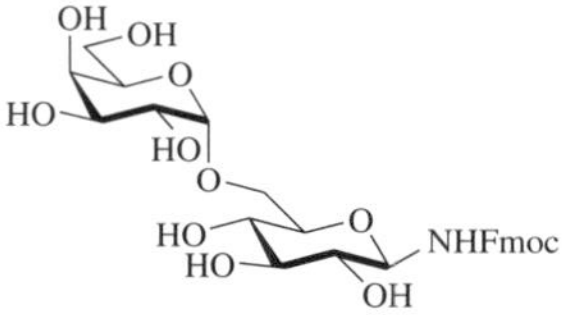	TEMPO (cat.), KBr (cat.), NaOCl, $NaHCO_3$, THF, H_2O	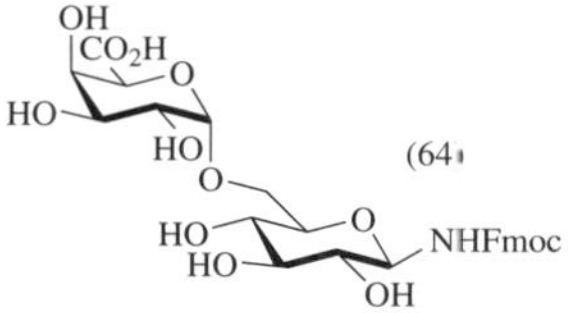 (64)	668
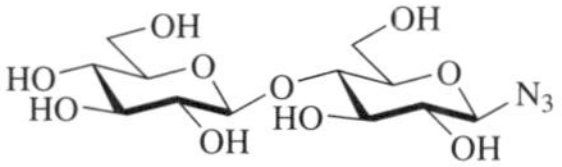	TEMPO (cat.), electrolysis, $NaHCO_3$, Na_2CO_3, H_2O	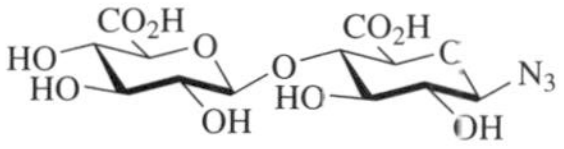 (—)[b]	325
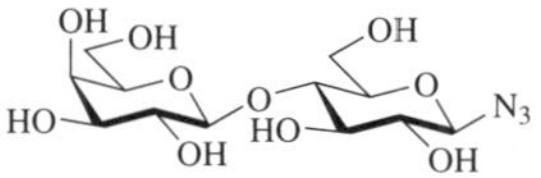	TEMPO (cat.), electrolysis, $NaHCO_3$, Na_2CO_3, H_2O	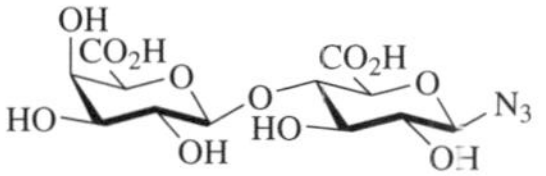 (—)[b]	325
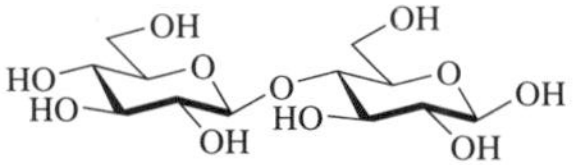	TEMPO (cat.), electrolysis, $NaHCO_3$, Na_2CO_3, H_2O, 2°	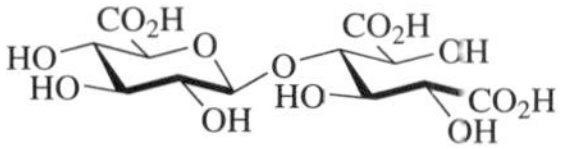 (54)	421

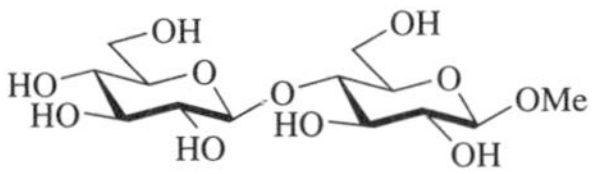	TEMPO (cat.), electrolysis, $NaHCO_3$, Na_2CO_3, H_2O	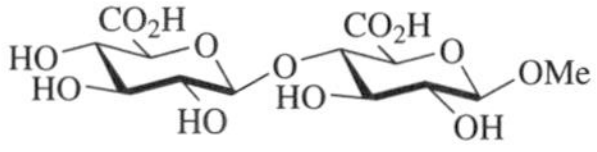	(—)[b]	325
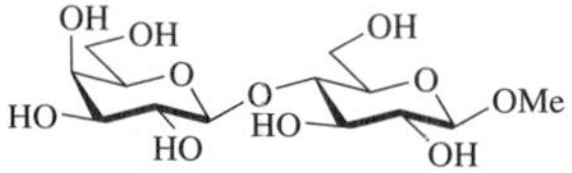	TEMPO (cat.), electrolysis, $NaHCO_3$, Na_2CO_3, H_2O	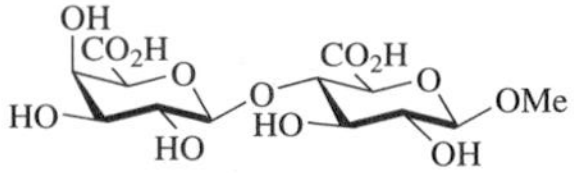	(—)[b]	325
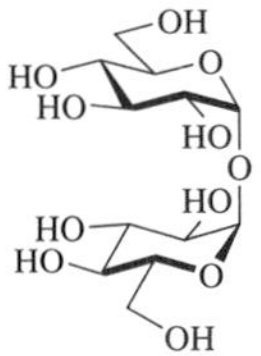	TEMPO (cat.), electrolysis, $NaHCO_3$, Na_2CO_3, H_2O	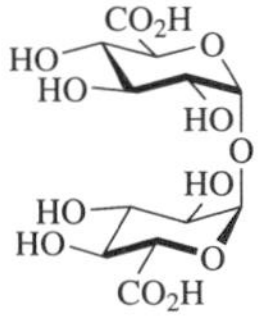 **I**	(61)[a]	283
	TEMPO (cat.), NaBr (cat.), NaOCl, H_2O, pH 10, 2°	**I** (100)		75
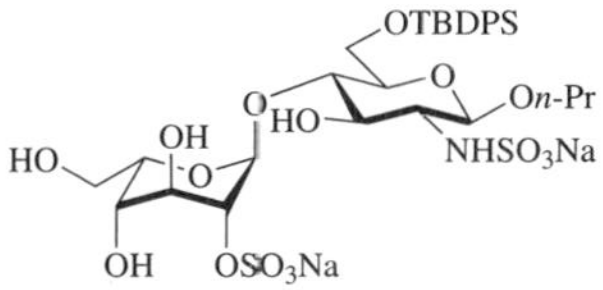	TEMPO (cat.), NaBr (cat.), $NaClO_2$, H_2O	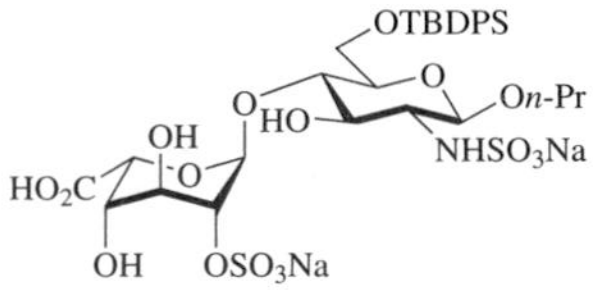	(55)	691
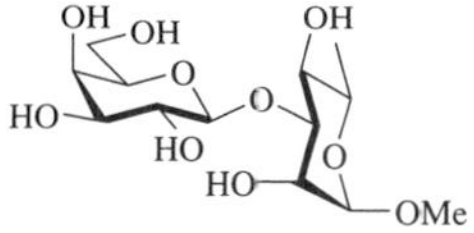	TEMPO (cat.), KBr (cat.), NaOCl, H_2O, 0°	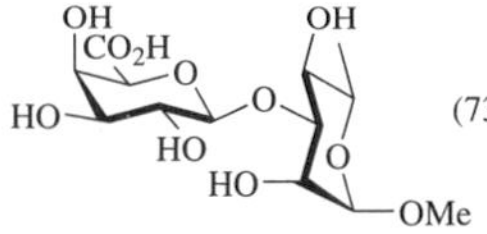	(73)	692

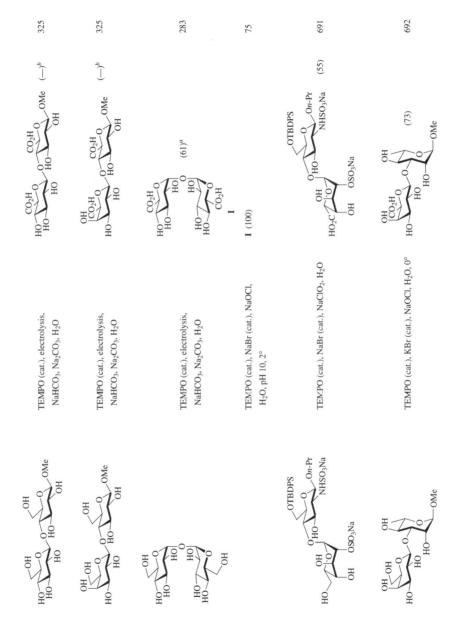

TABLE 4D. OXIDATIONS OF DISACCHARIDES (*Continued*)

Alcohol	Conditions	Product(s) and Yield(s) (%)	Refs.
C_{12} 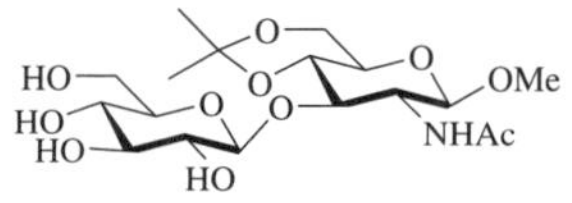	TEMPO (cat.), NaBr (cat.), NaOCl, TBAB, $NaHCO_3$, CH_2Cl_2, H_2O, 0°, 30 min	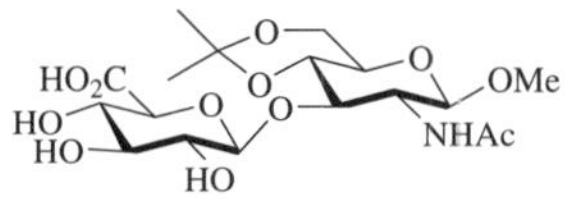 (—)[b]	693
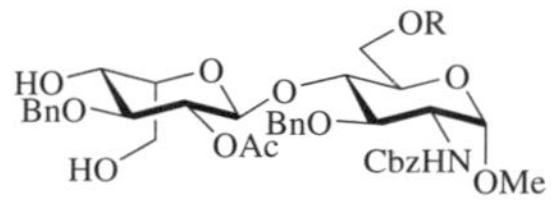	TEMPO (cat.), NaOCl (cat.), $NaClO_2$, KBr, NaH_2PO_4, H_2O, *t*-BuOH	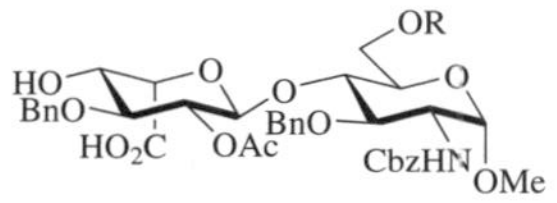 R: Ac (86)[a]; Bn (75)[a]	694
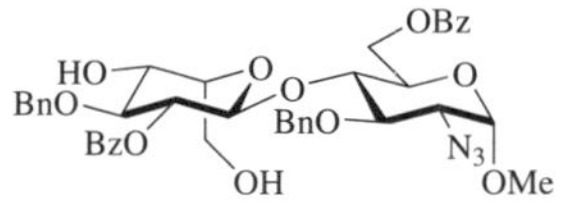	1. TEMPO (cat.), KBr (cat.), $Ca(OCl)_2$, TBAB, $NaHCO_3$, CH_2Cl_2, H_2O, 0°, 45 min 2. CH_2N_2, DMF	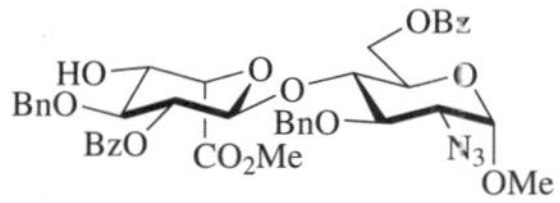 (54)	172
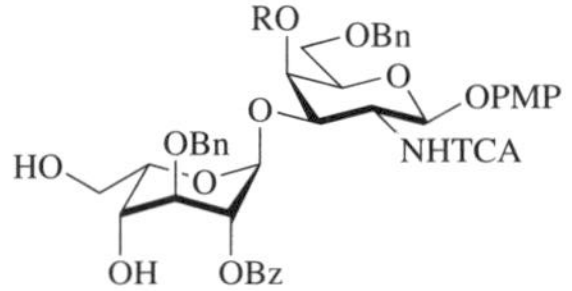	TEMPO (cat.), KBr (cat.), $Ca(OCl)_2$, $NaHCO_3$, H_2O, MeCN, 0°	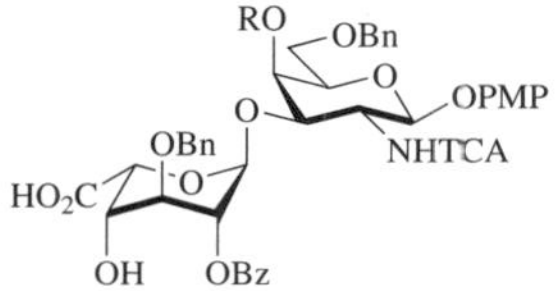 R: Ac (—); Piv (—)	695
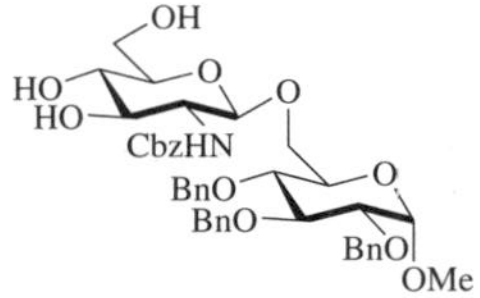	TEMPO (cat.), NaOCl (cat.), $NaClO_2$, MeCN, H_2O, borax buffer, pH 9.0	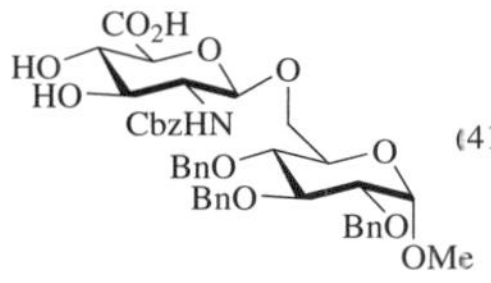 (41)	677
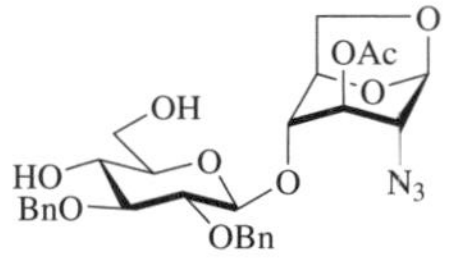	1. TEMPO (cat.), KBr (cat.), $Ca(OCl)_2$, TBAB, $NaHCO_3$, CH_2Cl_2, H_2O, 0°, 45 min 2. CH_2N_2, DMF	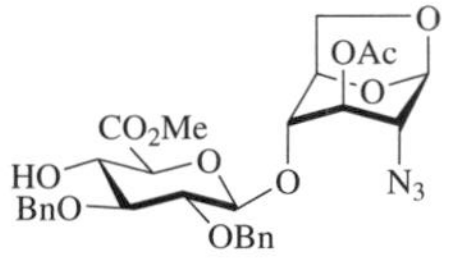 (73)	172

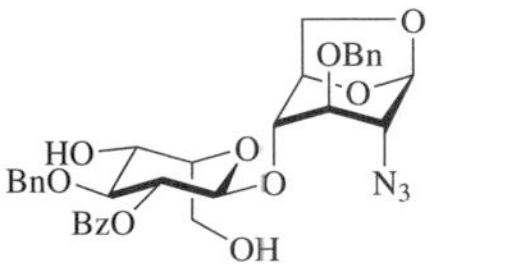	1. TEMPO (cat.), KBr (cat.), $Ca(OCl)_2$, KBr, TBAB, $NaHCO_3$, CH_2Cl_2, H_2O, 0°, 45 min 2. CH_2N_2, DMF	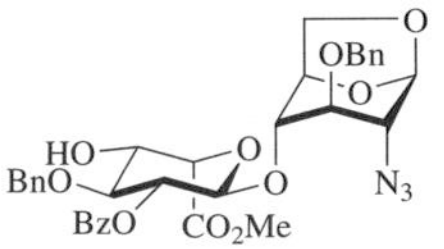 (83)	172
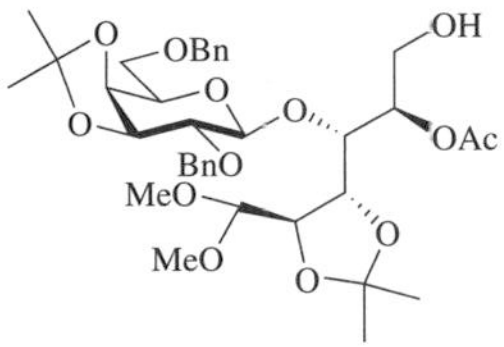	TEMPO (cat.), KBr (cat.), NaOCl, TBAB, $NaHCO_3$, CH_2Cl_2, H_2O, 0°	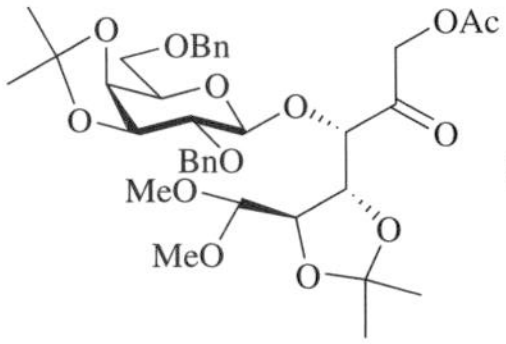 (100)	696
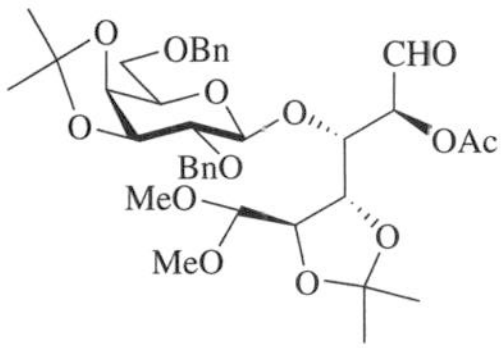	TEMPO (cat.), KBr (cat.), NaOCl, TBAB, $NaHCO_3$, CH_2Cl_2, H_2O, 0°	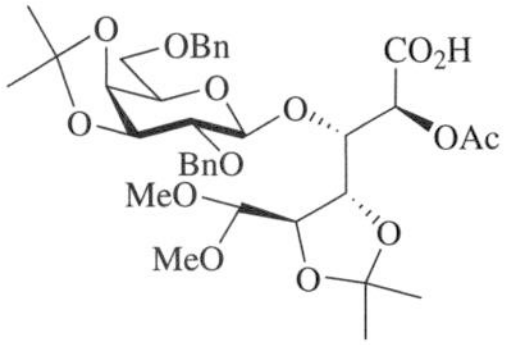 (80)	696
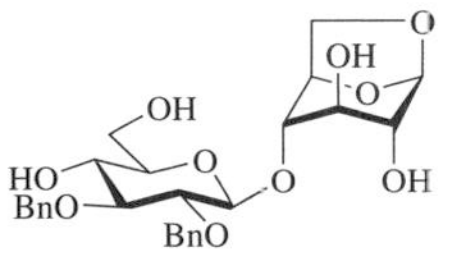	TEMPO (cat.), KBr (cat.), $Ca(OCl)_2$, TBAB, $NaHCO_3$, CH_2Cl_2, H_2O, 0°, 45 min	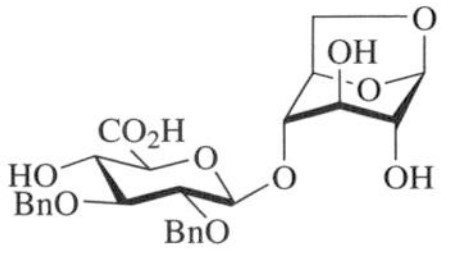 (—)[c]	697
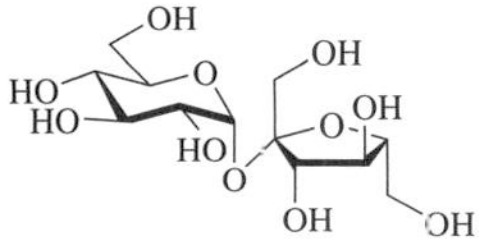	TEMPO (cat.), NaBr (cat.), NaOCl, H_2O, pH 10.5, sonication, 5°	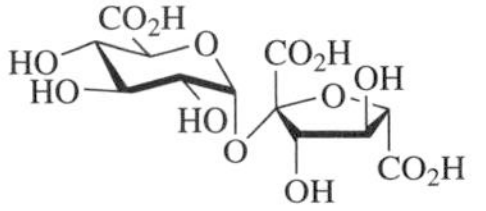 (—)	327, 328
	TEMPO (cat.), electrolysis, $NaHCO_3$, Na_2CO_3, H_2O	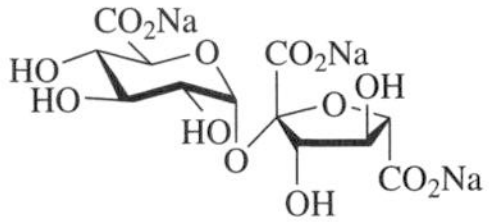 (39)	283

	Alcohol	Conditions	Product(s) and Yield(s) (%)	Refs.
C_{12}	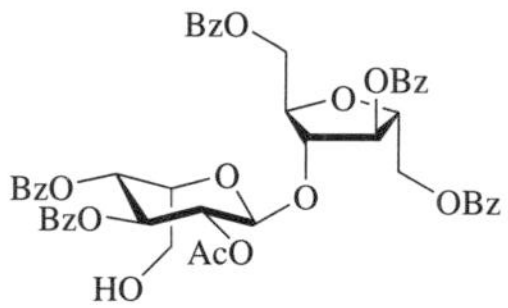	TEMPO (cat.), KBr (cat.), NaOCl, TBAB, $NaHCO_3$, NaCl, H_2O, CH_2Cl_2	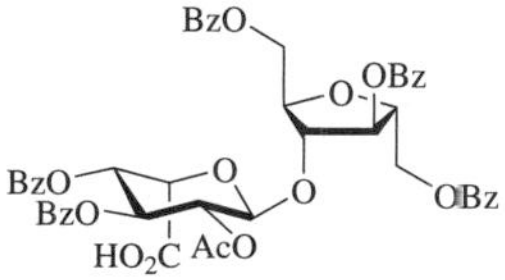 (74)	698
	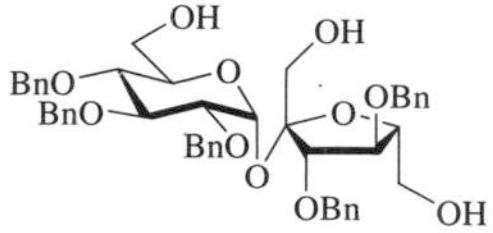	TEMPO (cat.), KBr (cat.), NaOCl, $NaHCO_3$, THF, H_2O, 0°	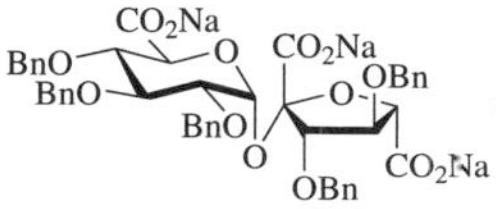 (—)[b]	699
	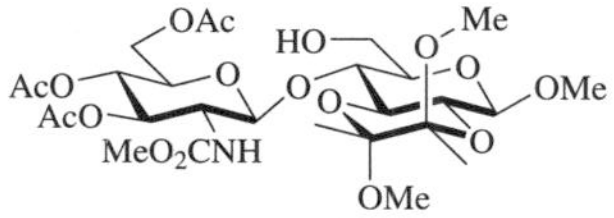	TEMPO (cat.), NaBr (cat.), NaOCl, TBAB, $NaHCO_3$, CH_2Cl_2, H_2O, 0°, 30 min	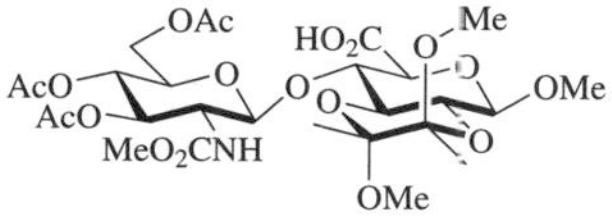 (—)[b]	693
C_{12-53}	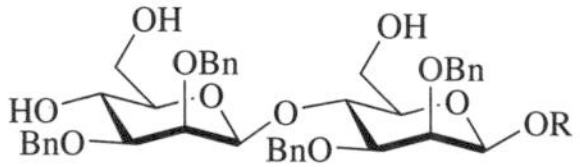	TEMPO (cat.), $PhI(OAc)_2$, CH_2Cl_2, H_2O, rt	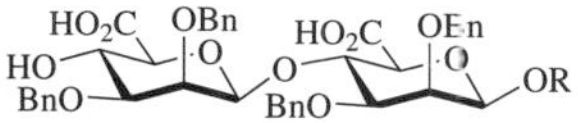	700
	R			
	Me		(—)[b]	
	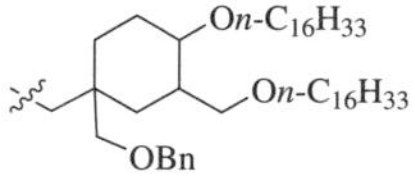		(—)[b]	
C_{14}	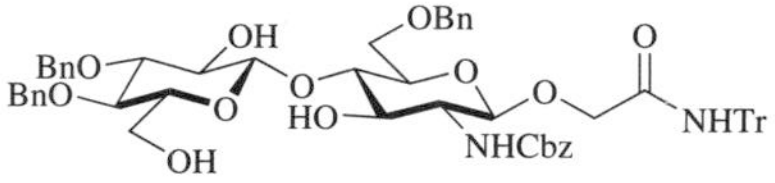	TEMPO (cat.), NaBr (cat.), NaOCl, TBAB, $NaHCO_3$, CH_2Cl_2, H_2O, 0°, 20 min	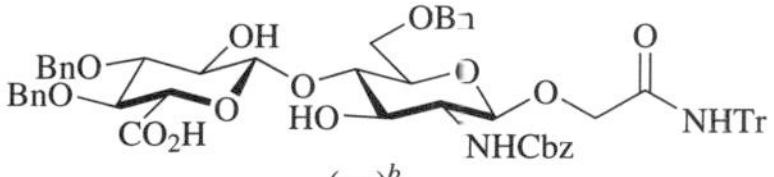 (—)[b]	701

	Substrate	Conditions	Product(s) and Yield(s) (%)	Refs.
	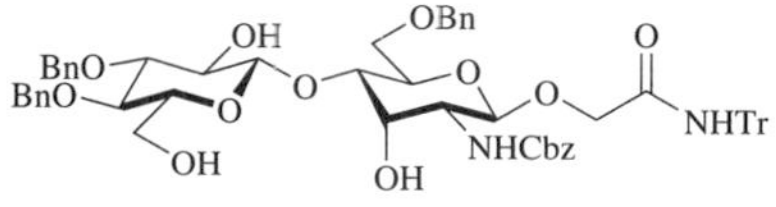	TEMPO (cat.), NaBr (cat.), NaOCl, TBAB, $NaHCO_3$, CH_2Cl_2, H_2O, 0°, 20 min	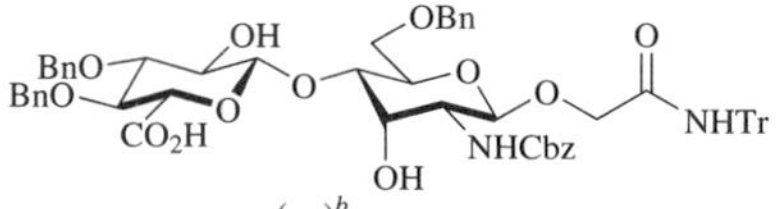 (—)^b	701
	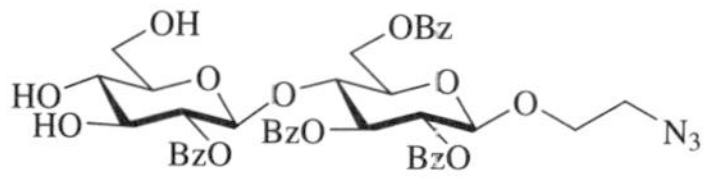	TEMPO (cat.), KBr (cat.), NaOCl, TBAB, NaCl, $NaHCO_3$, CH_2Cl_2, H_2O	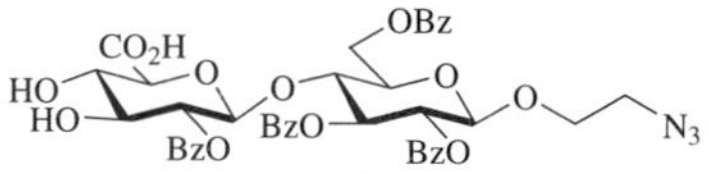 (85)	678
C_{16}	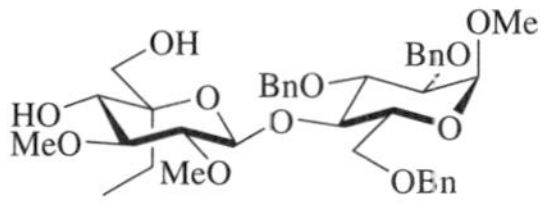	1. TEMPO (cat.), KBr (cat.), NaOCl, TBAC, NaCl, $NaHCO_3$, H_2O, CH_2Cl_2, 0°, 30 min 2. BnBr, TBAC, $NaHCO_3$, rt, 45 min	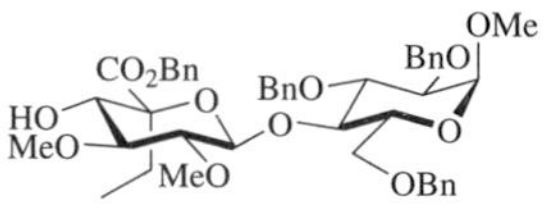 (75)	702
C_{17}	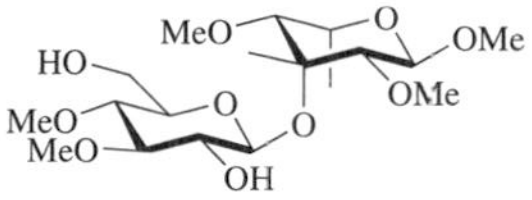	1. TEMPO (cat.), KBr (cat.), NaOCl, $NaHCO_3$, H_2O, 0° to rt, 1 d 2. MeI, DMF	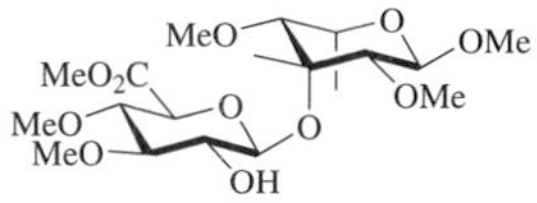 (71)	703
C_{18}	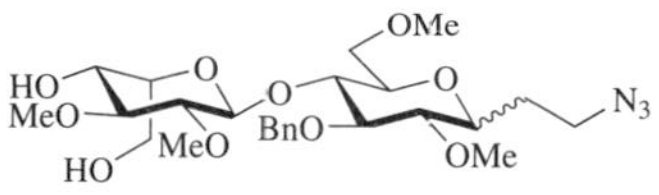	TEMPO (cat.), NaCl (cat.), NaOCl, TBAC, $NaHCO_3$, H_2O, CH_2Cl_2, 0°, 3 h	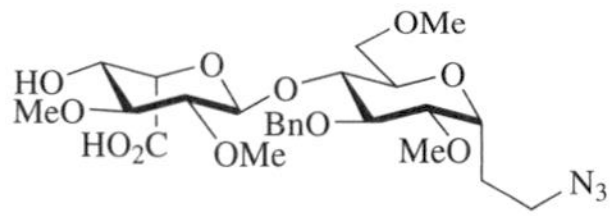 (—)^b	676

TABLE 4D. OXIDATIONS OF DISACCHARIDES (*Continued*)

Alcohol	Conditions	Product(s) and Yield(s) (%)	Refs.
C_{19} 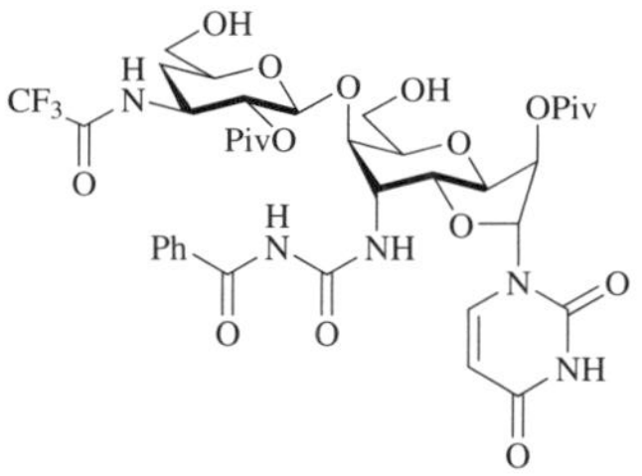	TEMPO (cat.), $PhI(OAc)_2$, $NaClO_2$, $NaHCO_3$, NaH_2PO_4, 2-methyl-2-butene, H_2O, *t*-BuOH, MeCN	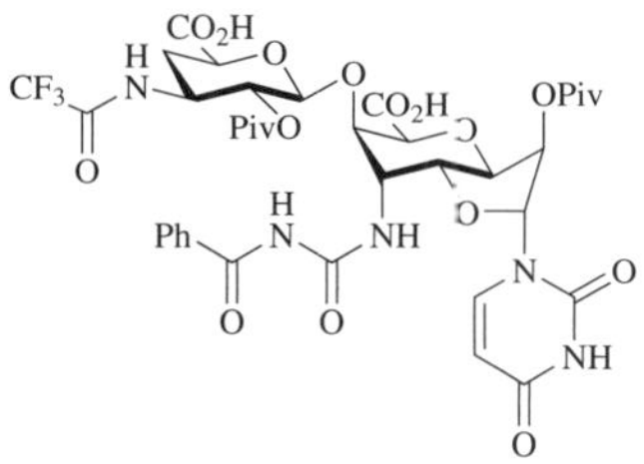 (—)[b]	704

[a] The product was isolated as the methyl ester.

[b] The product was not isolated and was used in a subsequent step without purification.

[c] The product was not isolated.

TABLE 4E. OXIDATIONS OF MISCELLANEOUS MONO- AND OLIGOSACCHARIDES

	Alcohol	Conditions	Product(s) and Yield(s) (%)	Refs.
C_4	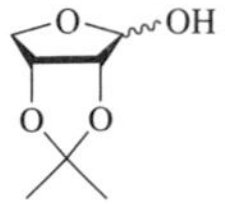	TEMPO (cat.), KBr (cat.), NaOCl, TBAF, $NaHCO_3$, CH_2Cl_2, H_2O, 0°, 45 min	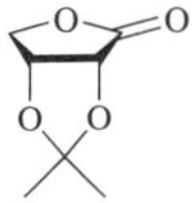 (91)	705
C_5	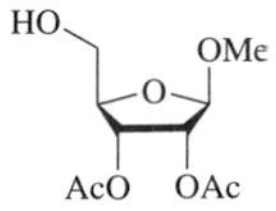	TEMPO (cat.), $PhI(OAc)_2$, MeCN, H_2O, rt	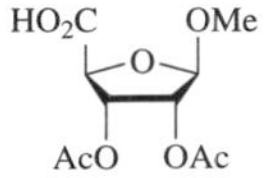 (80)	706
	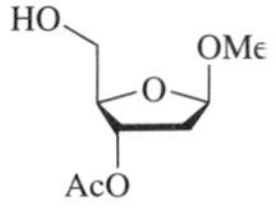	TEMPO (cat.), $PhI(OAc)_2$, MeCN, H_2O, rt	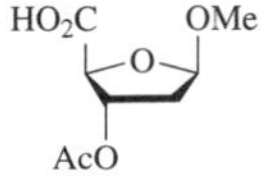 (75)	706
	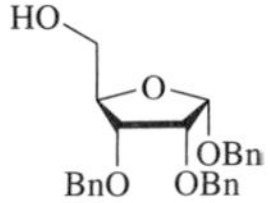	TEMPO (cat.), $PhI(OAc)_2$, MeCN, H_2O	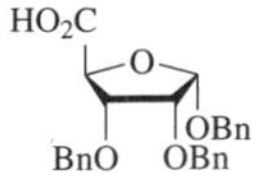 (99)	707
	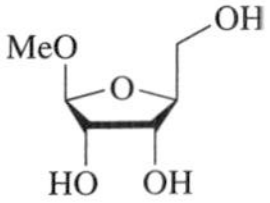	TEMPO (cat.), TCC, $NaHCO_3$, DMF, 0°	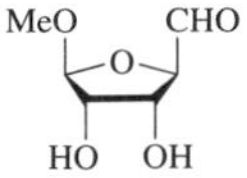 (65)	320
	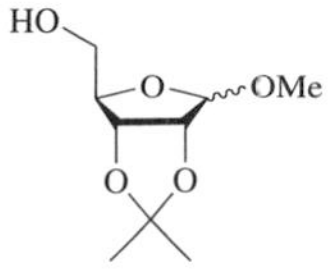	$[AcNH\text{-}TEMP{=}O]^+[BF_4]^-$, pyridine, 4 Å MS, CH_2Cl_2, rt	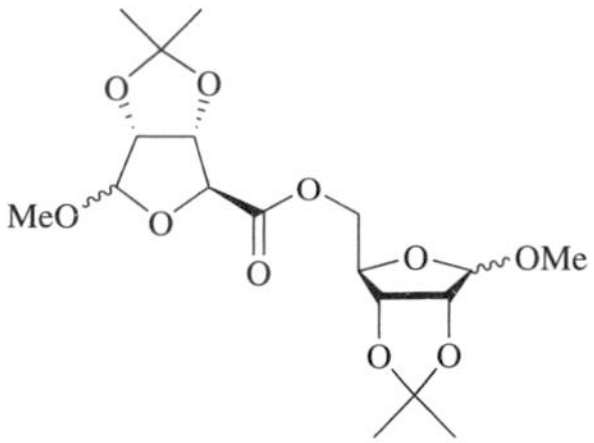 (25)	77

	Alcohol	Conditions	Product(s) and Yield(s) (%)	Refs.
C_5	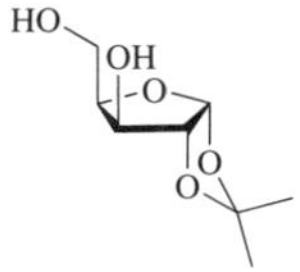	TEMPO (cat.), KBr (cat.), NaOCl, TBAC, $NaHCO_3$, H_2O, CH_2Cl_2, 0°	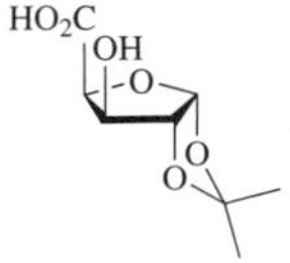 (64)	315
	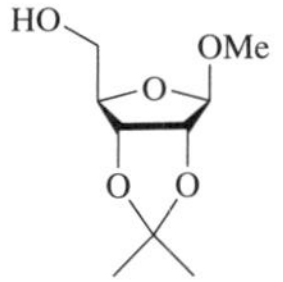	1. TEMPO (cat.), KBr (cat.), NaOCl, TBAB, $NaHCO_3$, CH_2Cl_2, H_2O, 0°, 1 h 2. $NaClO_2$, 2-methyl-2-butene	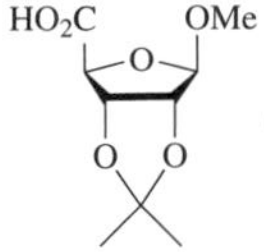 (100)	415
	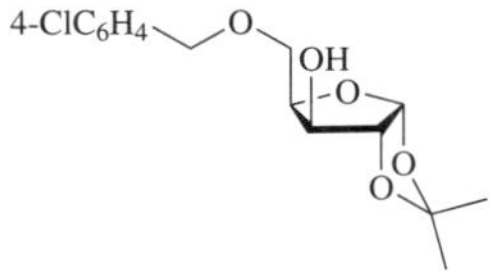	TEMPO (cat.), NaOCl, $NaHCO_3$, H_2O, CH_2Cl_2, 0°, 15 min	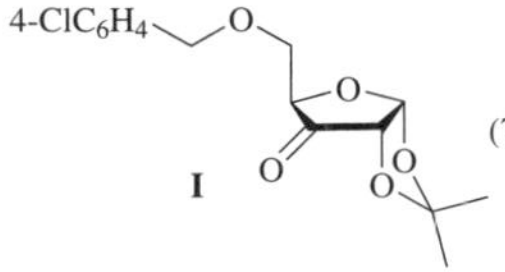 **I** (77)	708
		TEMPO (cat.), $Ca(OCl)_2$, $NaHCO_3$, H_2O, CH_2Cl_2, 0°, 15 min	**I** (43)	703
C_6	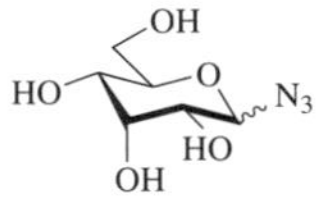	TEMPO (cat.), KBr (cat.), NaOCl or $Ca(OCl)_2$, $NaHCO_3$, H_2O, 15–18°	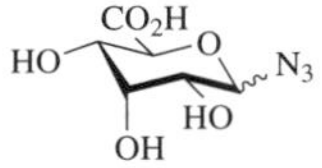 (—)[a]	173
	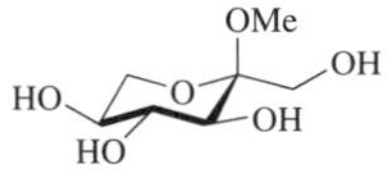	TEMPO (cat.), electrolysis, $NaHCO_3$, Na_2CO_3, H_2O	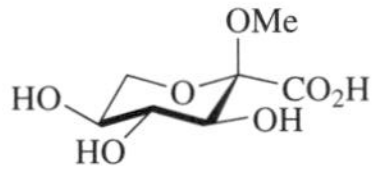 (—)[b]	325
	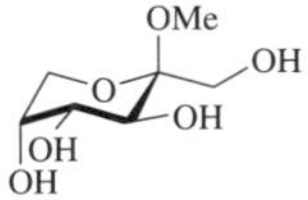	TEMPO (cat.), electrolysis, $NaHCO_3$, Na_2CO_3, H_2O	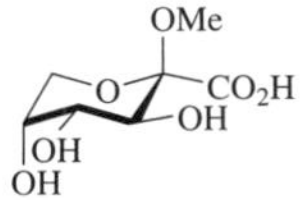 (—)[b]	325

Substrate	Conditions	Product(s) and Yield(s) (%)	Refs.
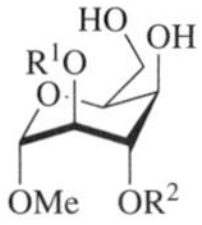	TEMPO (cat.), KBr (cat.), NaOCl, TBAC, $NaHCO_3$, H_2O	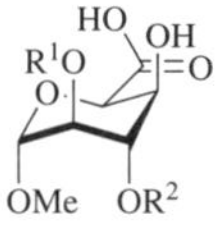 R^1 / R^2: Ac / Bn (45); Bn / Ac (72)	709
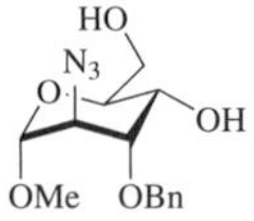	TEMPO (cat.), KBr (cat.), NaOCl, $NaHCO_3$, H_2O, 20°, 2 h	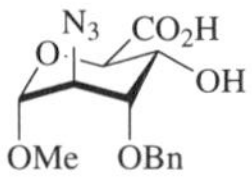 (—)[b]	710
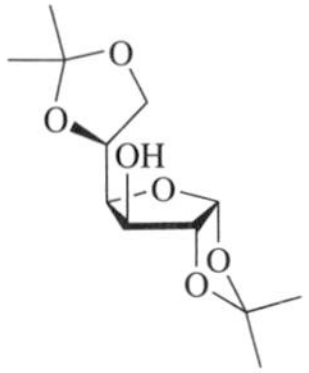	TEMPO (cat.), $[PS\text{-}CH_2NMe_3]^+[ClO_2]^-$, CH_2Cl_2 or toluene, 40°, 3.5 h	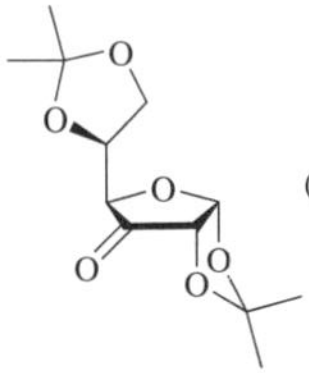 (84)	44
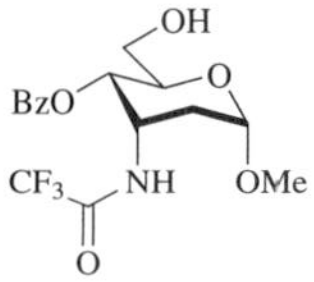	TEMPO (cat.), $[PS\text{-}CH_2NMe_3]^+[ClO_2]^-$, CH_2Cl_2 or toluene, 40°, 3.5 h	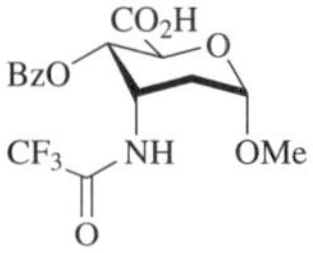 (93)	44
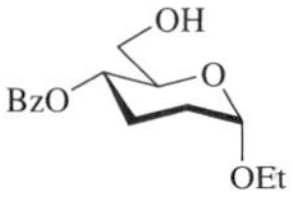	TEMPO (cat.), $[PS\text{-}CH_2NMe_3]^+[ClO_2]^-$, CH_2Cl_2 or toluene, 40°, 3.5 h	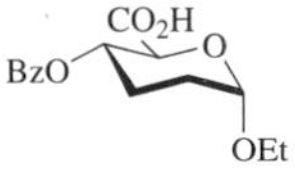 (86)	44
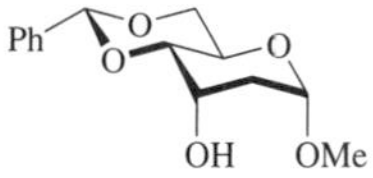	TEMPO (cat.), $[PS\text{-}CH_2NMe_3]^+[ClO_2]^-$, CH_2Cl_2 or toluene, 40°, 3.5 h	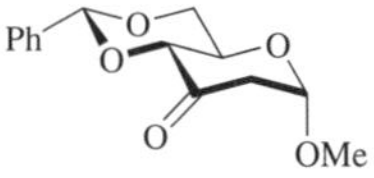 (90)	44

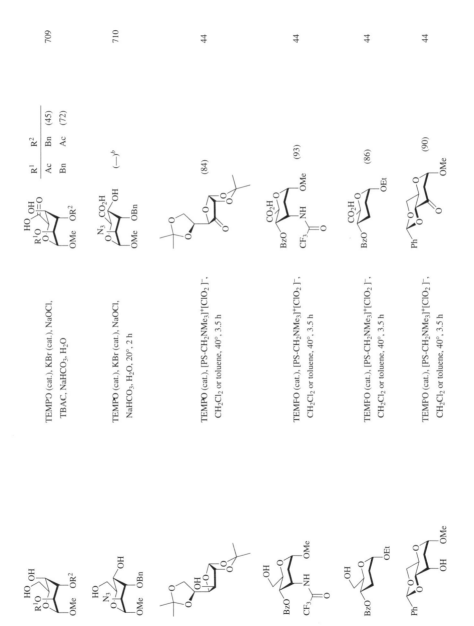

TABLE 4E. OXIDATIONS OF MISCELLANEOUS MONO- AND OLIGOSACCHARIDES (Continued)

Alcohol	Conditions	Product(s) and Yield(s) (%)	Refs.
C_6			
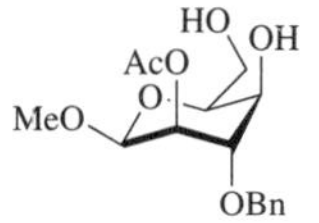	TEMPO (cat.), KBr (cat.), NaOCl, TBAC, $NaHCO_3$, CH_2Cl_2	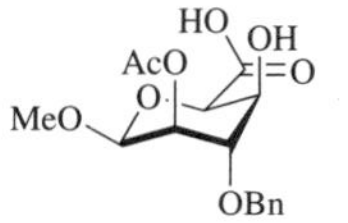 (—)[b]	711
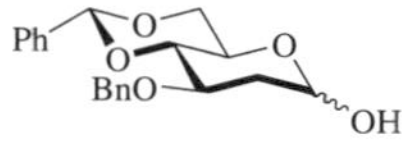	$[AcNH\text{-}TEMP{=}O]^+[BF_4]^-$, pyridine, CH_2Cl_2	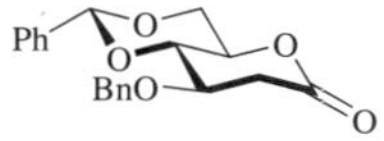 (91)	76
C_7			
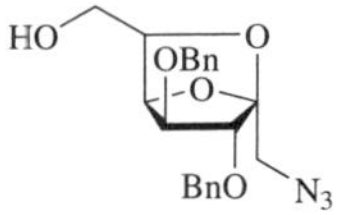	TEMPO (cat.), $PhI(OAc)_2$, wet CH_2Cl_2, rt, 2 h	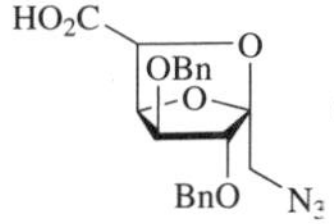 (83)	712
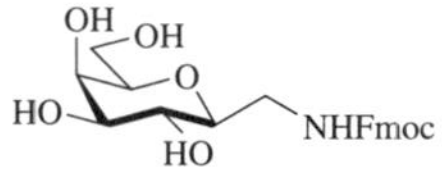	TEMPO (cat.), NaBr (cat.), NaOCl, THF, H_2O	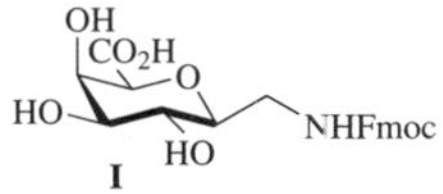 I (65)	713
	TEMPO (cat.), KBr (cat.), NaOCl, $NaHCO_3$, THF, H_2O, 0°	**I** (>74)	663
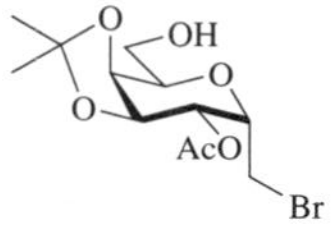	TEMPO (cat.), NaOCl, TBAC, NaCl, $NaHCO_3$, CH_2Cl_2, H_2O, 0° to rt, 3 h	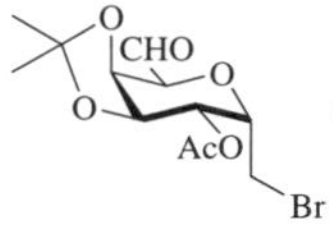 (—)[b]	714
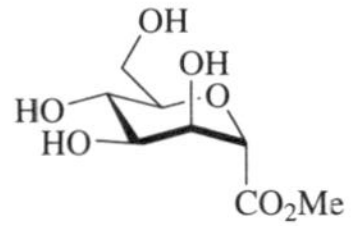	TEMPO (cat.), NaOCl, CH_2Cl_2, H_2O, 0°, 20 h	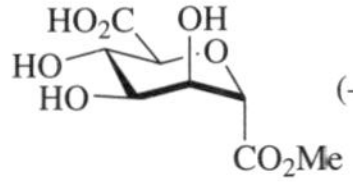 (—)[b]	715, 716
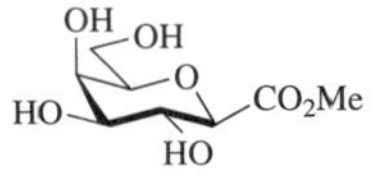	TEMPO (cat.), NaOCl, CH_2Cl_2, H_2O, 0°, 20 h	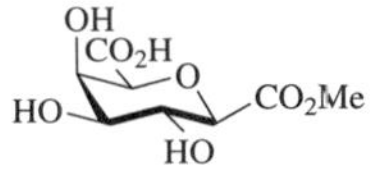 (—)[b]	715, 716

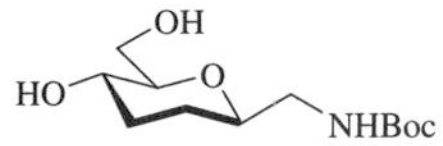	TEMPO (cat.), KBr (cat.), NaOCl, KBr, NaOH, H_2O	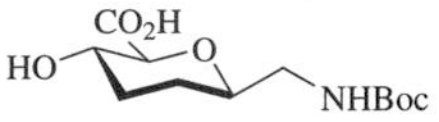 (62)	717
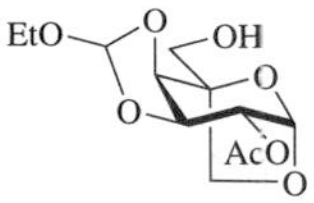	TEMPO (cat.), KBr (cat.), NaOCl, TBAF, $NaHCO_3$, CH_2Cl_2	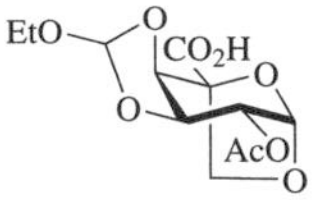 (—)	718

C_{7-15}

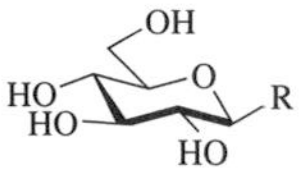

TEMPO (cat.), conditions

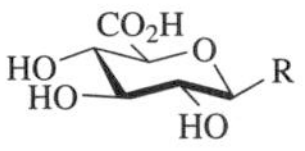

R	Conditions		
CH_2OBn	NaOCl, $NaHCO_3$ (satd.), H_2O, CH_2Cl_2, 0°	(—)	669
$CH_2NHFmoc$	NaOCl, H_2O, THF	(—)	719
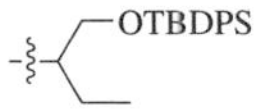	KBr (cat.), NaOCl, TBAC, $NaHCO_3$, NaCl, CH_2Cl_2, H_2O, 30 min	(—)	720
Bn	KBr (cat.), NaOCl, $NaHCO_3$, H_2O, 1.5 h	(—)	721
CH_2-4-$O_2NC_6H_4$	KBr (cat.), NaOCl, $NaHCO_3$, H_2O, 0°, 2 h	(—)[b]	722
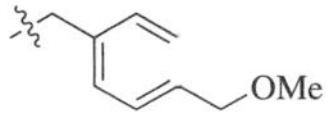	KBr (cat.), NaOCl, $NaHCO_3$, H_2O, 0°, 2 h	(—)[b]	722

C_9

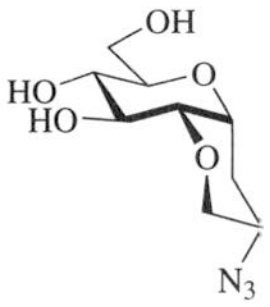	TEMPO (cat.), NaOCl, $NaHCO_3$, H_2O, MeCN, 0°	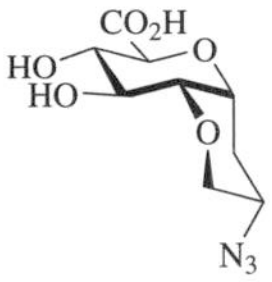 (53)	723

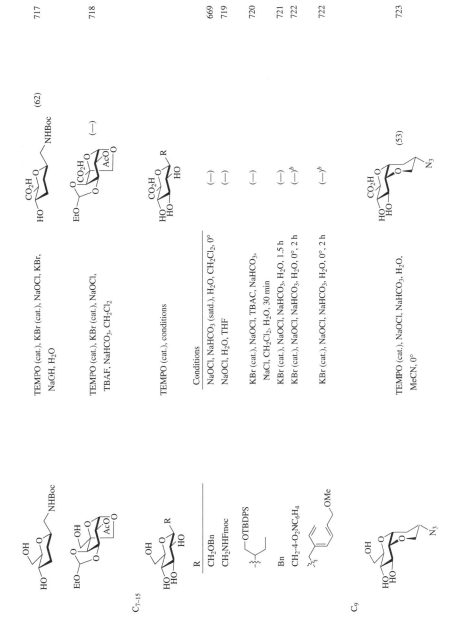

TABLE 4E. OXIDATIONS OF MISCELLANEOUS MONO- AND OLIGOSACCHARIDES (*Continued*)

Alcohol	Conditions	Product(s) and Yield(s) (%)	Refs.
C_9			
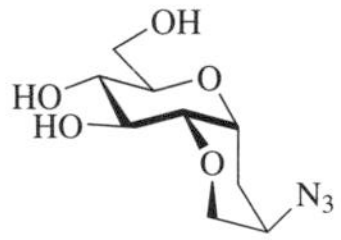	TEMPO (cat.), NaOCl, $NaHCO_3$, H_2O, MeCN, 0°	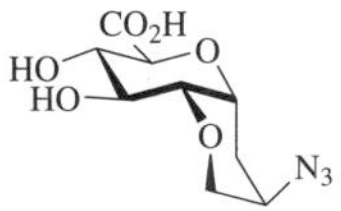 (52)	723
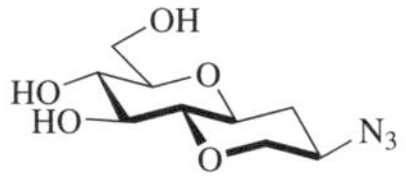	TEMPO (cat.), NaOCl, $NaHCO_3$, H_2O, MeCN, 0°	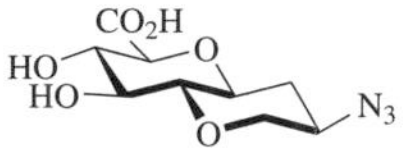 (49)	723
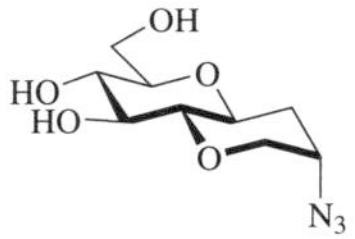	TEMPO (cat.), NaOCl, $NaHCO_3$, H_2O, MeCN, 0°	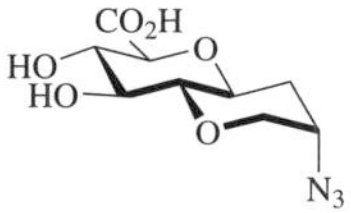 (72)	723
C_{13-14}			
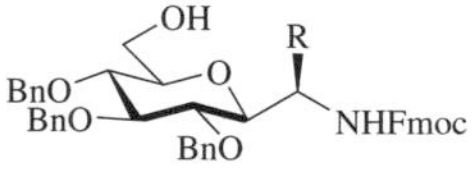	TEMPO (cat.), $PhI(OAc)_2$, CH_2Cl_2, H_2O, rt, 12 h	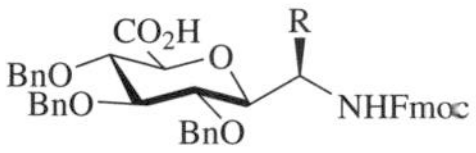 R: Ph (>80); Bn (—)	724
C_{15}			
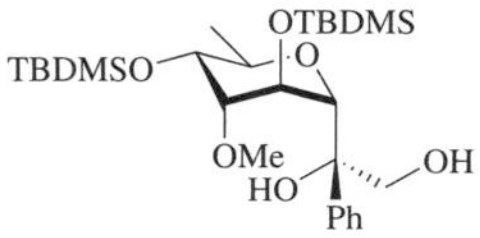	TEMPO (cat.), PS-$I(OAc)_2$, CH_2Cl_2, rt	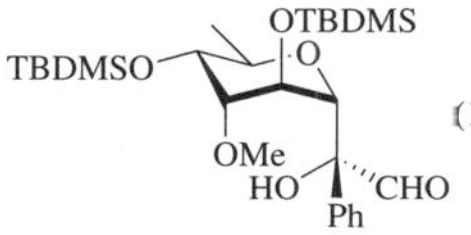 (100)	212
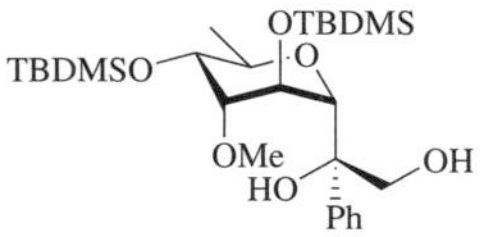	TEMPO (cat.), PS-$I(OAc)_2$, CH_2Cl_2, rt	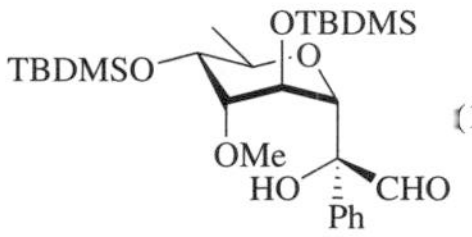 (100)	212

C_{18}	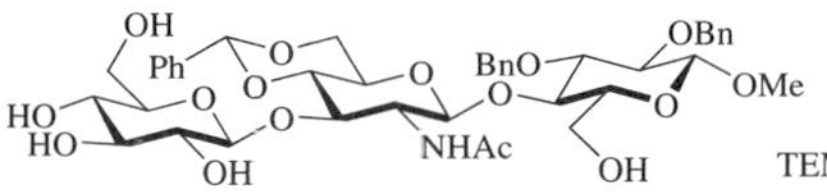	TEMPO (cat.), NaBr (cat.), NaOCl, TBAB, $NaHCO_3$, CH_2Cl_2, H_2O	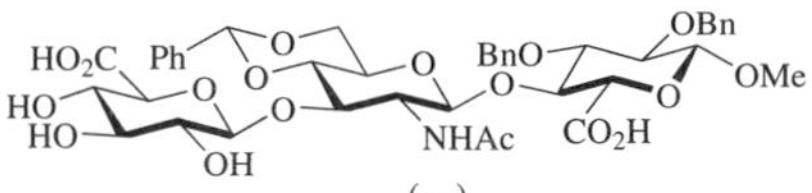 (—)	725
	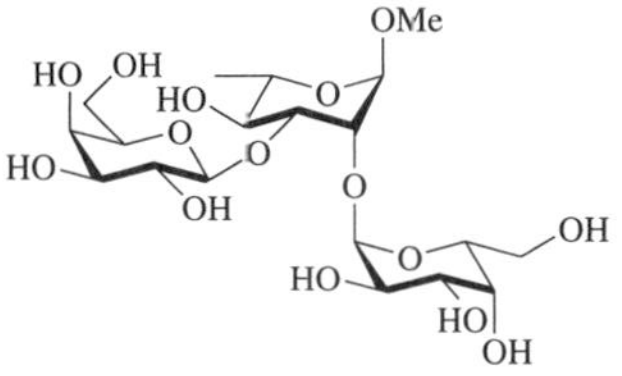	TEMPO (cat.), KBr (cat.), NaOCl, H_2O, 0°	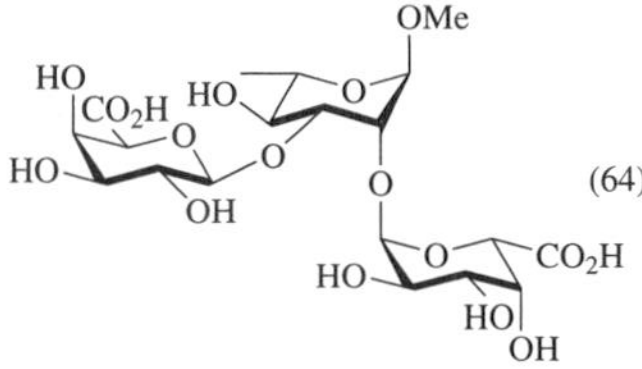 (64)	692
	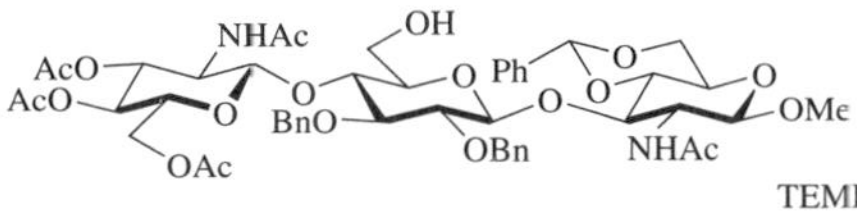	TEMPO (cat.), NaBr (cat.), NaOCl, TBAB, $NaHCO_3$, CH_2Cl_2, H_2O	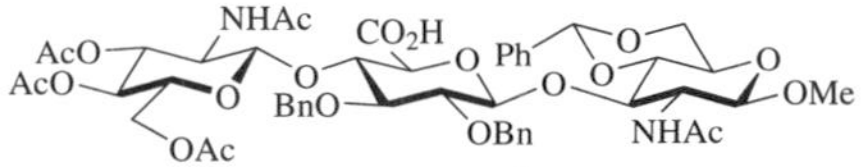 (—)[b]	725
	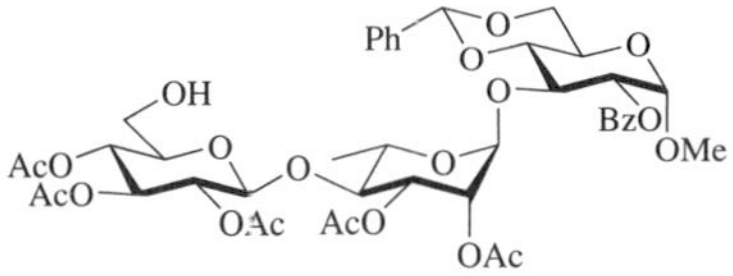	TEMPO (cat.), NaOCl (cat.), $NaClO_2$, CH_2Cl_2, H_2O, rt, 6 h	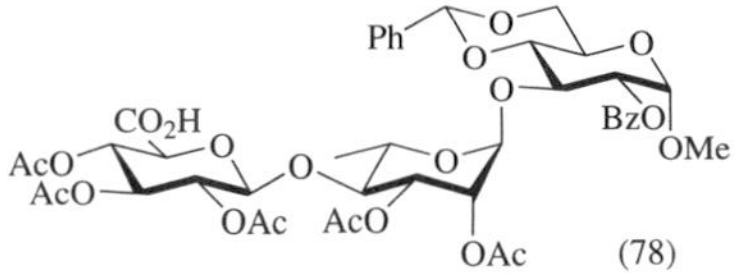 (78)	726
	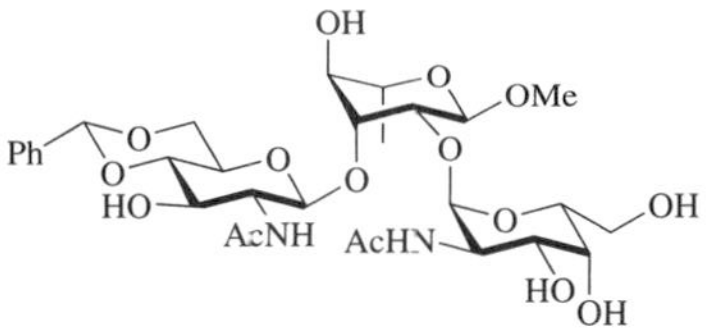	TEMPO (cat.), NaOCl, NaOH, H_2O, rt, 1 h	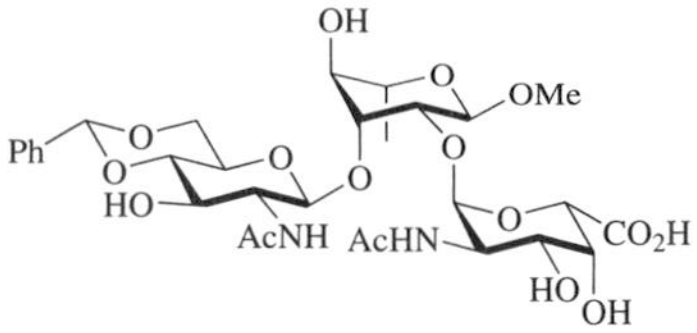 (—)[b]	727

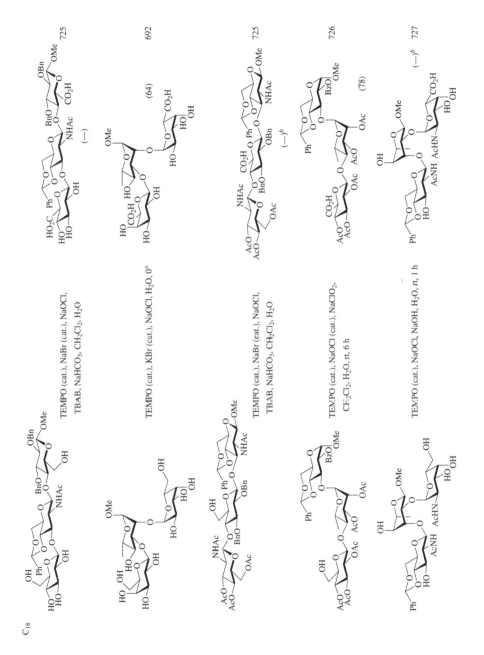

TABLE 4E. OXIDATIONS OF MISCELLANEOUS MONO- AND OLIGOSACCHARIDES (*Continued*)

Alcohol	Conditions	Product(s) and Yield(s) (%)	Refs.
C_{18} 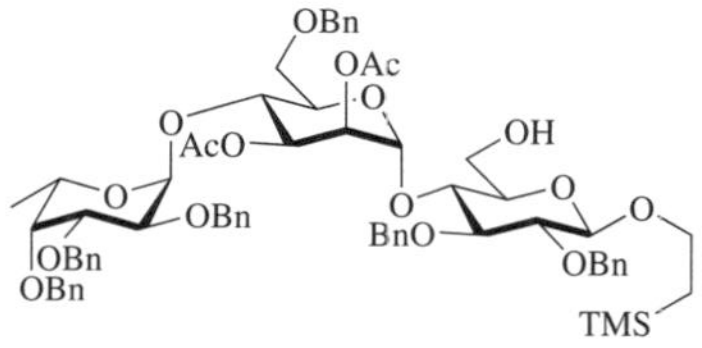	TEMPO (cat.), $PhI(OAc)_2$, MeCN, H_2O	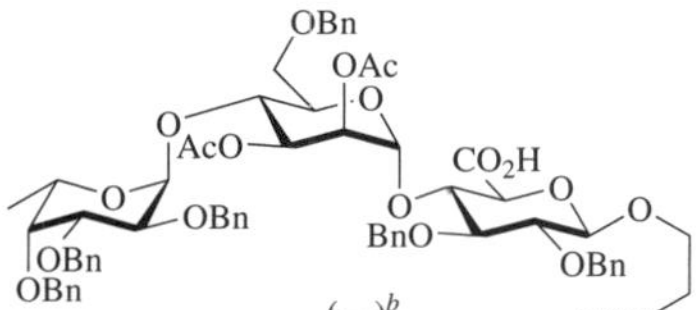 (—)[b]	728
C_{20-21} 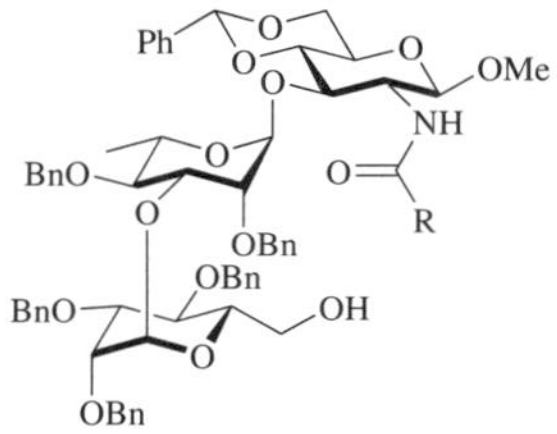	TEMPO (cat.), KBr (cat.), NaOCl, $NaHCO_3$, NaOH, MeCN, H_2O, pH 10.5	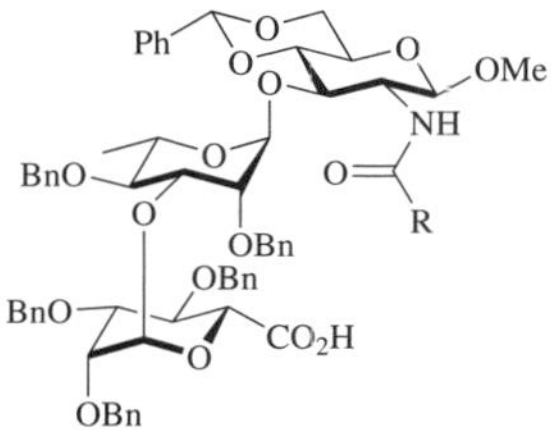	729
R = Me		(—)	
R = $CH_2NHFmoc$		(43)	
R = Et		(—)	
R = $(CH_2)_2NHFmoc$		(63)	
C_{21} 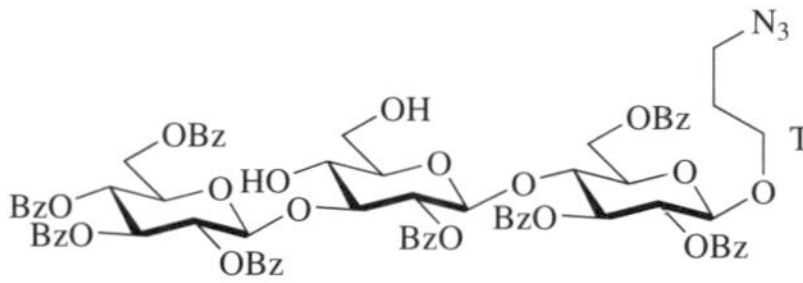	TEMPO (cat.), KBr (cat.), NaOCl, TBAB, NaCl, $NaHCO_3$, CH_2Cl_2, H_2O	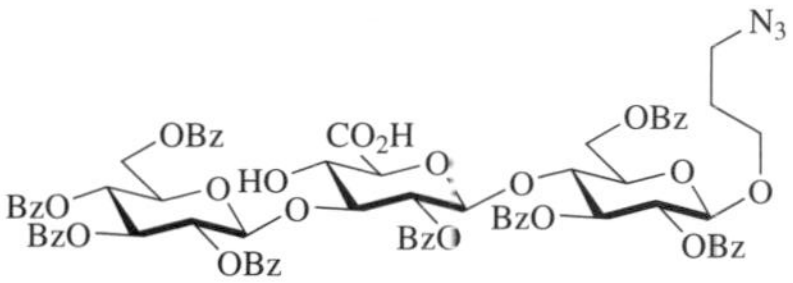 (76)	678

C_{24}

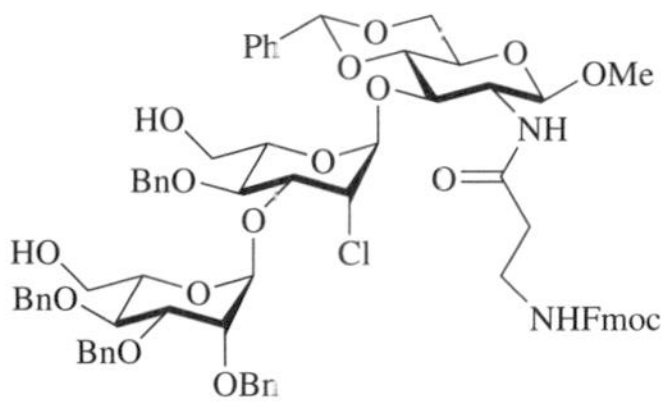	TEMPO (cat.), KBr (cat.), NaOCl, $NaHCO_3$, NaOH, MeCN, H_2O, pH 10.5	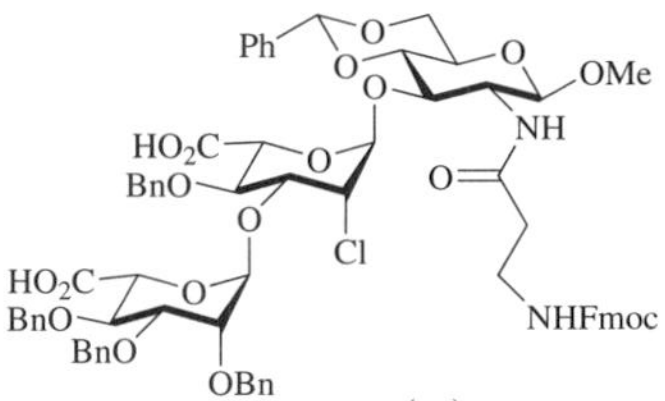 (—)	729
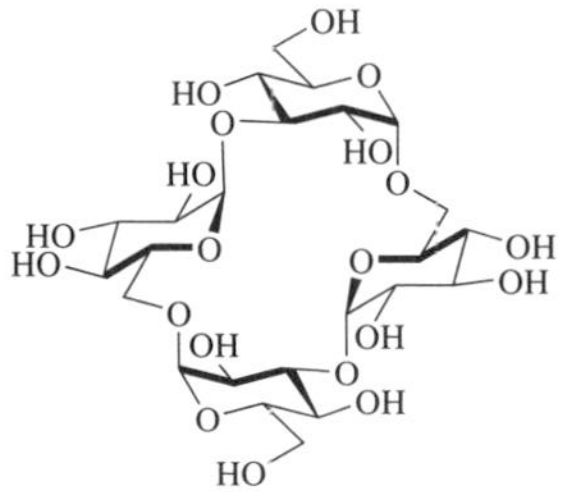	TEMPO (cat.), NaBr (cat.), NaOCl, NaOH, H_2O, pH 10.5–11, 4°	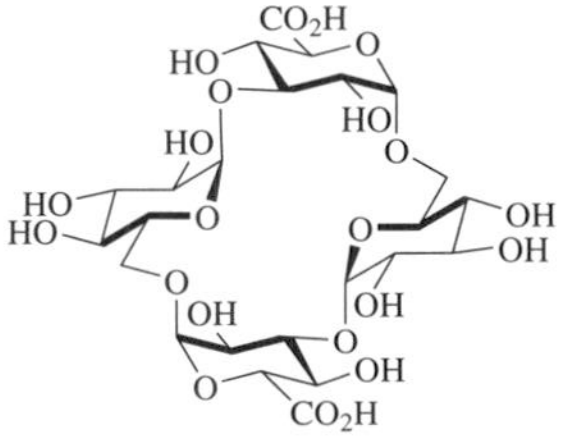 (90–95)	730
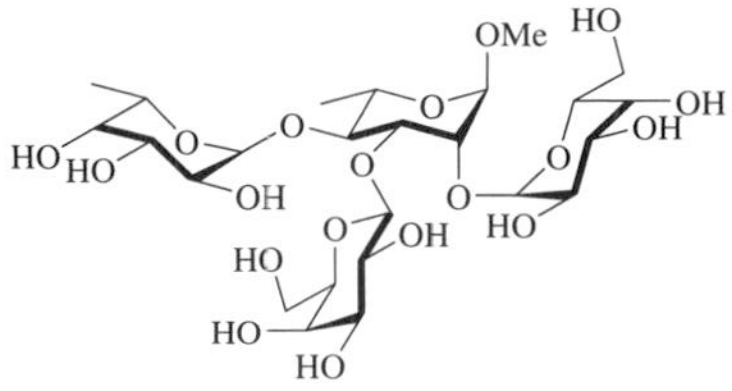	TEMPO (cat.), KBr (cat.), NaOCl, H_2O, 0°	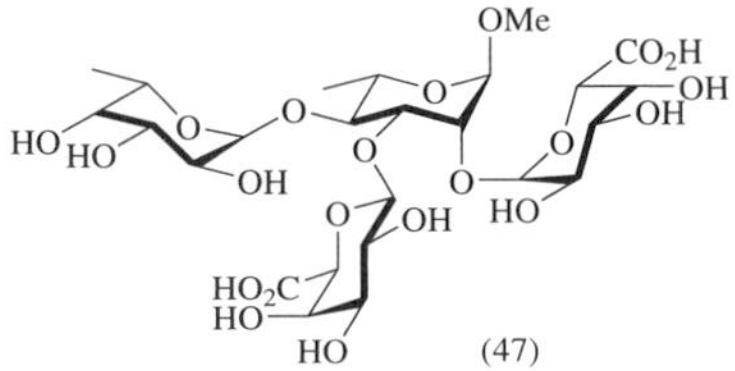 (47)	692

TABLE 4E. OXIDATIONS OF MISCELLANEOUS MONO- AND OLIGOSACCHARIDES (*Continued*)

Alcohol	Conditions	Product(s) and Yield(s) (%)	Refs.
C_{27}			
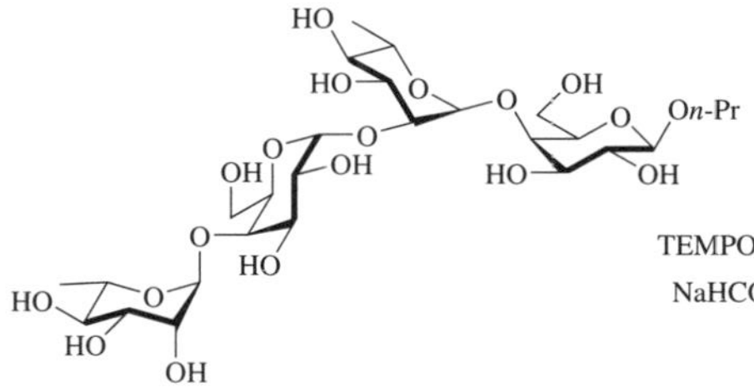	TEMPO (cat.), KBr (cat.), NaOCl, $NaHCO_3$, H_2O, rt	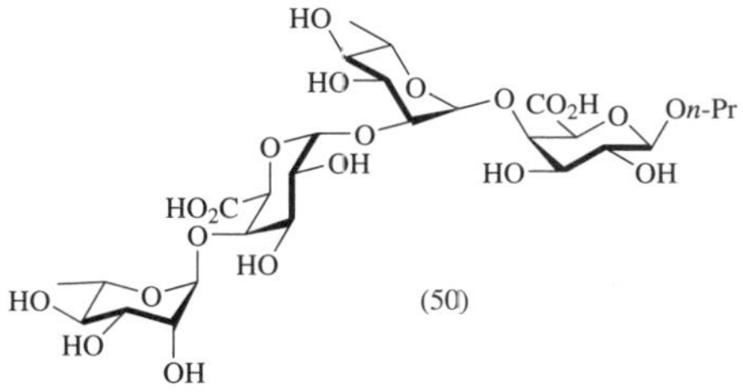 (50)	731
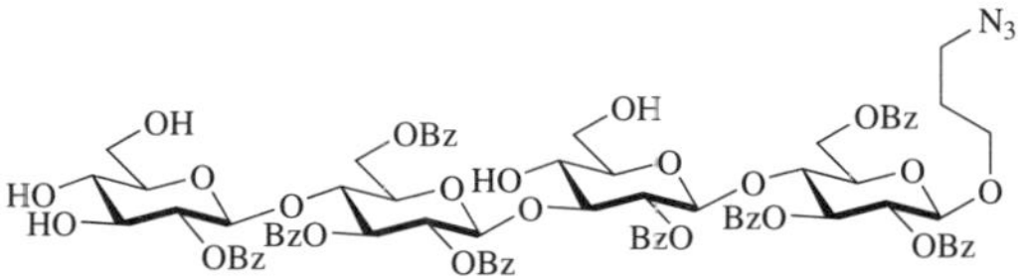	TEMPO (cat.), KBr (cat.), NaOCl, TBAB, NaCl, $NaHCO_3$, CH_2Cl_2, H_2O	(65) 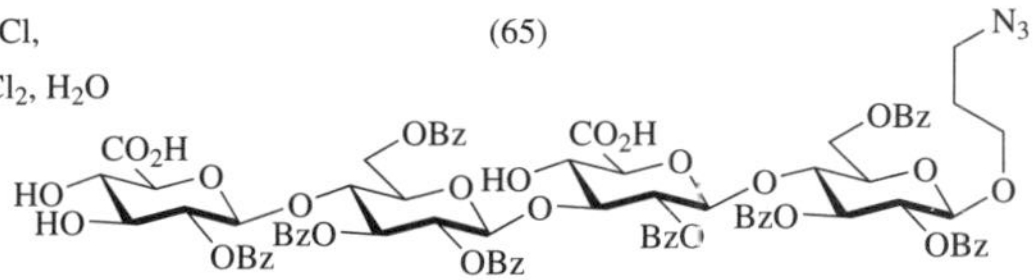	678
C_{30}			
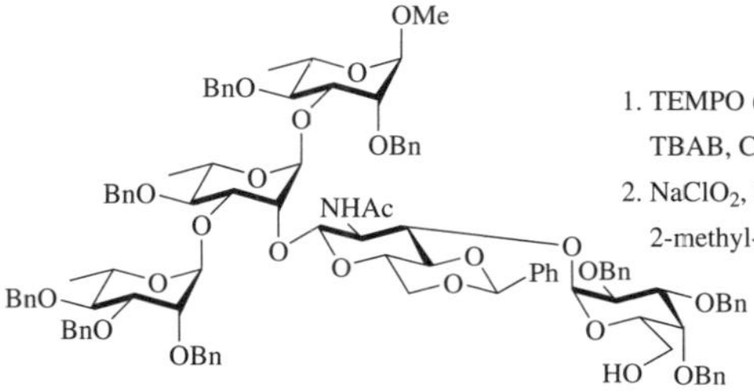	1. TEMPO (cat.), NaBr (cat.), NaOCl, TBAB, CH_2Cl_2, H_2O, 0–5°, 2 h 2. $NaClO_2$, NaH_2PO_4, $NaHCO_3$, 2-methyl-2-butene, CH_2Cl_2, 0°, 5 h	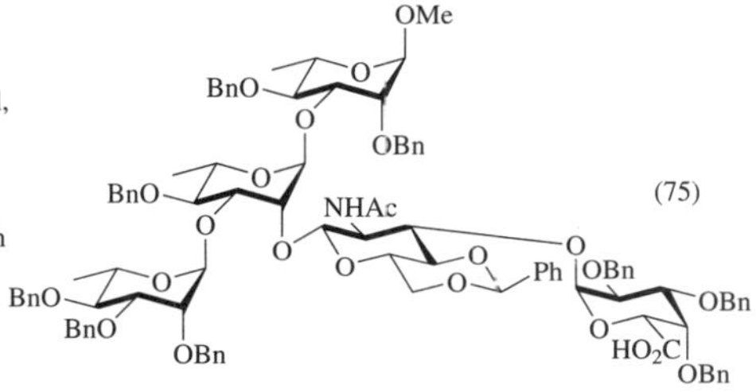 (75)	732

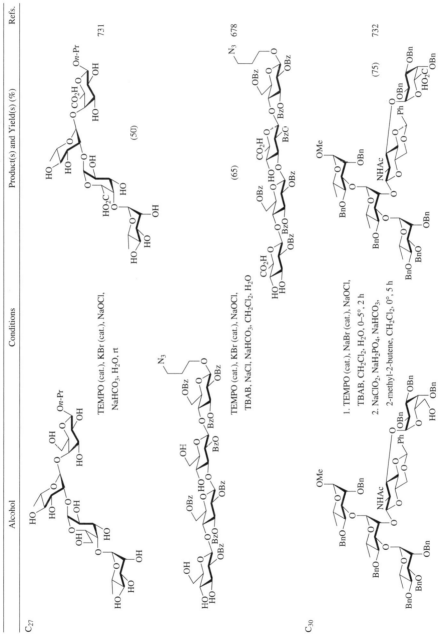

C_{42-91}

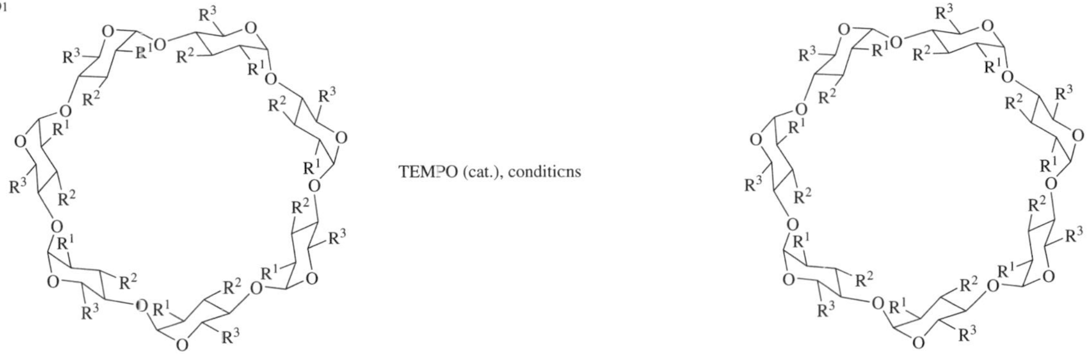

R^1	R^2	R^3	Conditions	R^1	R^2	R^3		
OH	OH	CH_2OH	NaBr (cat.), NaOCl, NaOH, H_2O, pH 10	OH	OH	CH_2OH or CO_2H	(—)	342
OH	OH	CH_2OH	electrolysis, H_2O, $NaHCO_3$ buffer, 20°	OH	OH	CH_2OH or CO_2H	(—)	283
OMe	OMe	CH_2OH	KBr (cat.), NaOCl, $NaHCO_3$, MeCN, H_2O, buffer, pH 10	OMe	OMe	CO_2H	(93)	343
OAc	OAc	CH_2OH	"	OAc	OAc	CO_2H	(68)	733
$O(CH_2)_2OH$	$O(CH_2)_2OH$	OMe	"	OCH_2CO_2H	OCH_2CO_2H	OMe	(94)	343
$O(CH_2)_2OH$	$O(CH_2)_2OH$	O*n*-Bu	"	OCH_2CO_2H	OCH_2CO_2H	O*n*-Bu	(85)	733
$O(CH_2)_2OH$	OMe	$O(CH_2)_2OH$	"	OCH_2CO_2H	OMe	OCH_2CO_2H	(97)	733
$O(CH_2)_2OH$	$O(CH_2)_2OH$	$O(CH_2)_2OH$	"	OCH_2CO_2H	OCH_2CO_2H	OCH_2CO_2H	(94)	733

TABLE 4E. OXIDATIONS OF MISCELLANEOUS MONO- AND OLIGOSACCHARIDES (*Continued*)

Alcohol	Conditions	Product(s) and Yield(s) (%)	Refs.
C_{48-54}			
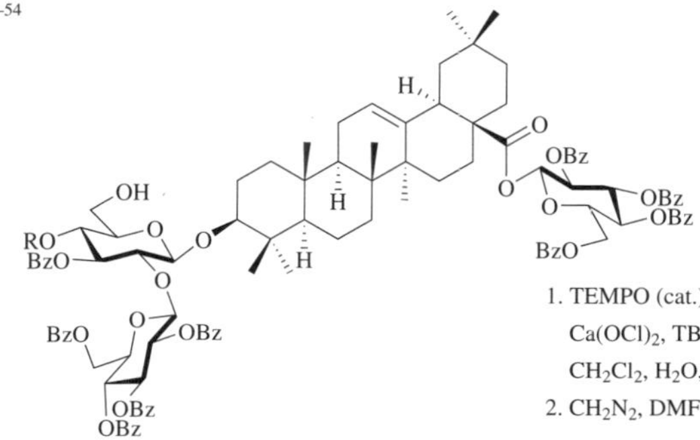	1. TEMPO (cat.), KBr (cat.), $Ca(OCl)_2$, TBAB, $NaHCO_3$, CH_2Cl_2, H_2O, 0°, 45 min 2. CH_2N_2, DMF	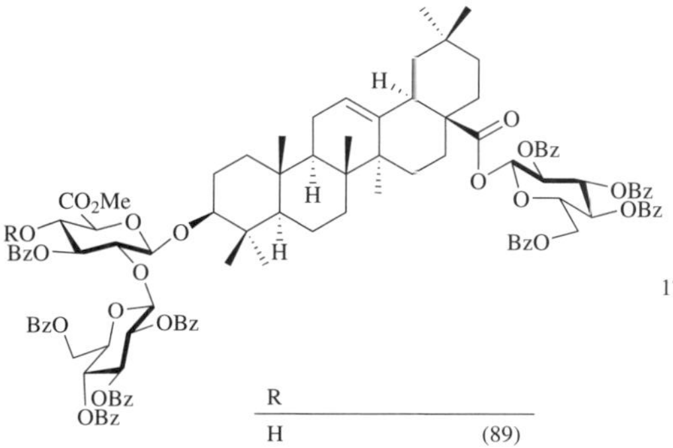 R: H (89); 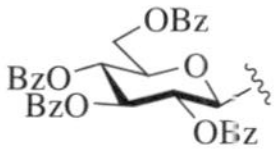 (72)	172
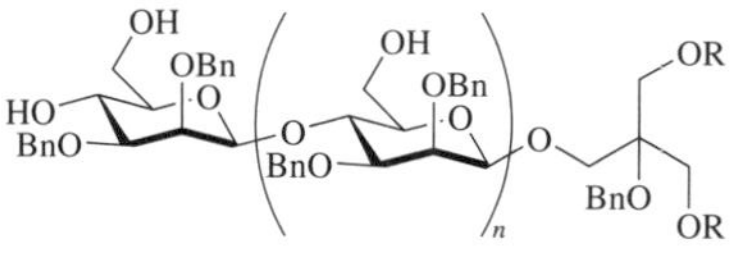 R = n-$C_{16}H_{33}$	TEMPO (cat.), $PhI(OAc)_2$, CH_2Cl_2, H_2O, rt	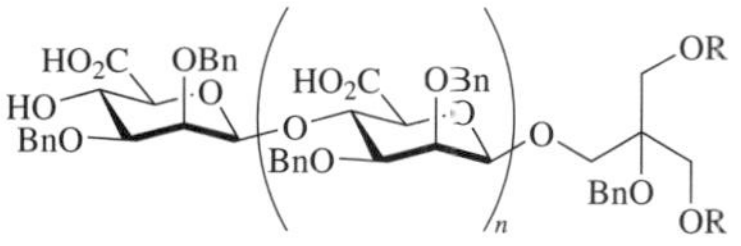 n = 1 (95); n = 2 (70)	734

[a] The product was not isolated.

[b] The product was not isolated and was used in a subsequent step without purification.

TABLE 4F. OXIDATIONS OF NUCLEOSIDES

Alcohol		Conditions	Product(s) and Yield(s) (%)	Refs.
C_9	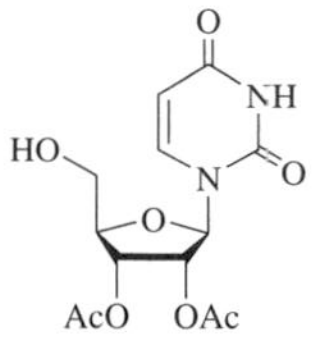	TEMPO (cat.), $PhI(OAc)_2$, MeCN, H_2O, rt	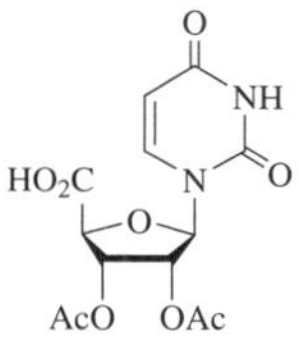 (—)	706
C_{9-18}	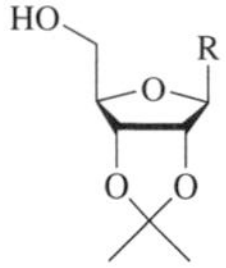	TEMPO (cat.), $PhI(OAc)_2$, MeCN, H_2O, rt, 2–24 h	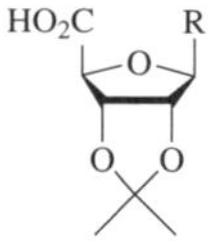	
	R			
	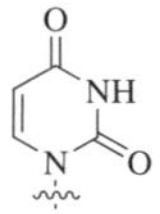		(76)	193
	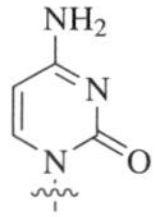		(72)	193
	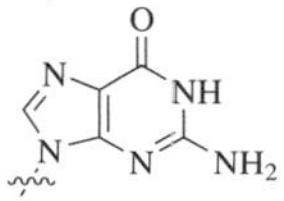		(75)[a]	193
	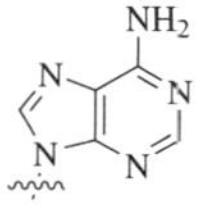		(90)	193

TABLE 4F. OXIDATIONS OF NUCLEOSIDES (*Continued*)

Alcohol	Conditions	Product(s) and Yield(s) (%)	Refs.
C_{9-18} 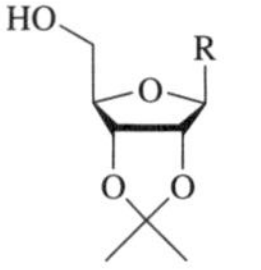	TEMPO (cat.), $PhI(OAc)_2$, MeCN, H_2O, rt, 2–24 h	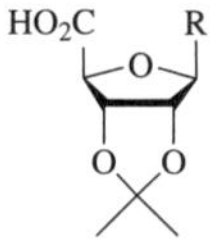	
R			
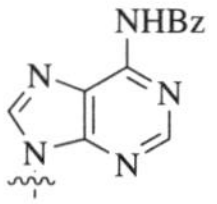		(70)	193
"		(80)	735
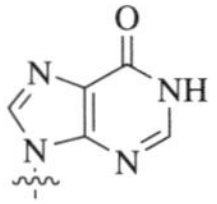		(83)	736
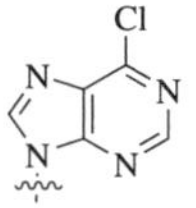		(92)	737
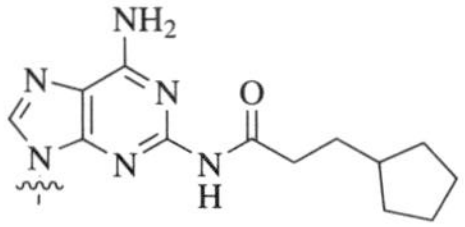		(49)	738
C_{10} 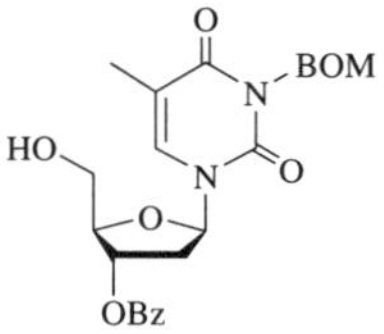	TEMPO (cat.), $PhI(OAc)_2$, MeCN, H_2O, rt, 16 h	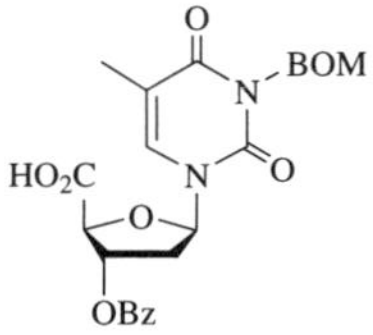 (76)	739

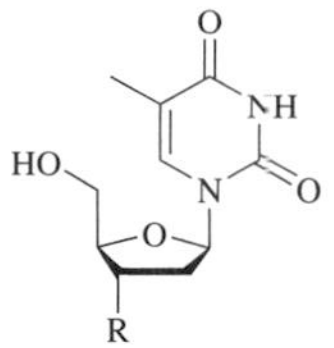

TEMPO (cat.), $PhI(OAc)_2$

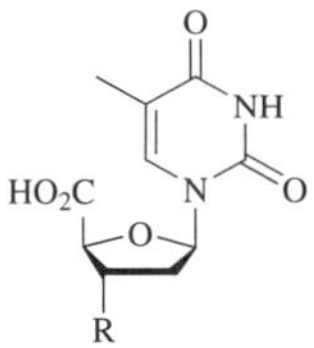

R	Solvent	Temp	Time	Yield (%)	Ref.
OH	CH_2Cl_2	rt	2 h	(100)	740
OTBDMS	MeCN, H_2O	—	—	(67)	741
OAc	MeCN, H_2O	rt	—	(—)	706
NHFmoc	MeCN, H_2O	rt	24 h	(45)	742
N_3	MeCN, H_2O	rt	—	(76)	706
N_3	MeCN, H_2O	rt	—	(—)	743

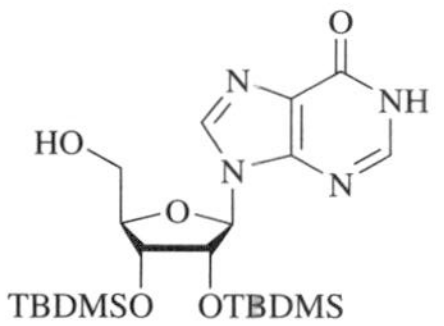

TEMPO (cat.), $PhI(OAc)_2$, CH_2Cl_2, MeCN, H_2O, rt

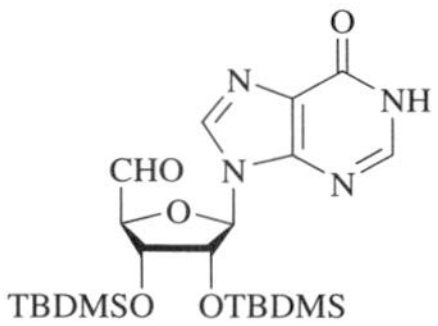

(73) 744

C_{10-11}

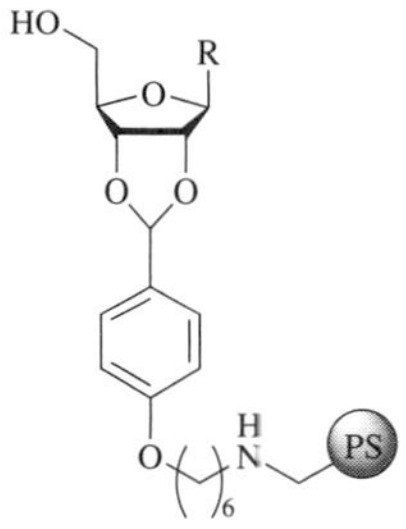

Amberlite XE-305 resin

TEMPO (cat.), $PhI(OAc)_2$, $NaHCO_3$, MeCN, H_2O, rt, 5 h

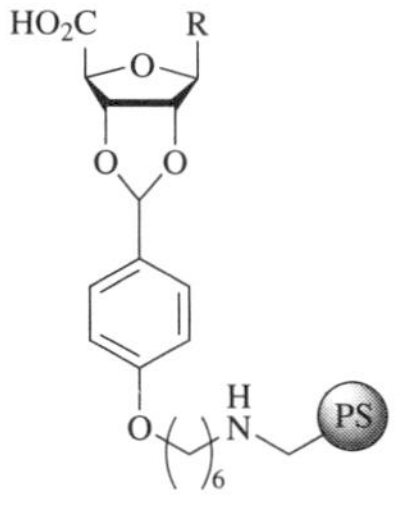

(—)[b]

R

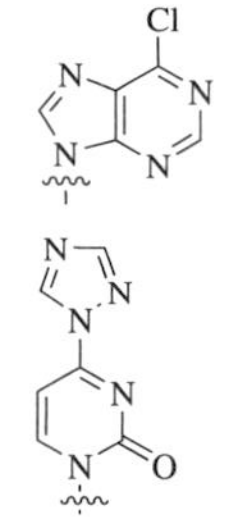

345

TABLE 4F. OXIDATIONS OF NUCLEOSIDES (*Continued*)

Alcohol	Conditions	Product(s) and Yield(s) (%)	Refs.
C_{11-14} 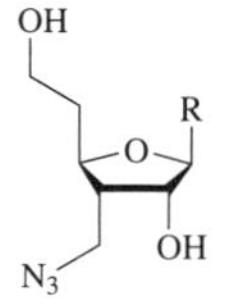	TEMPO (cat.), NaOCl (cat.), $NaClO_2$, TBAC, MeCN, H_2O, phosphate buffer, rt, 5 h	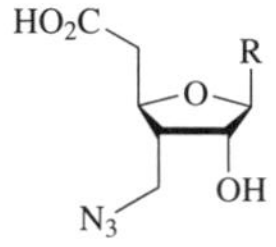 (—)[b] 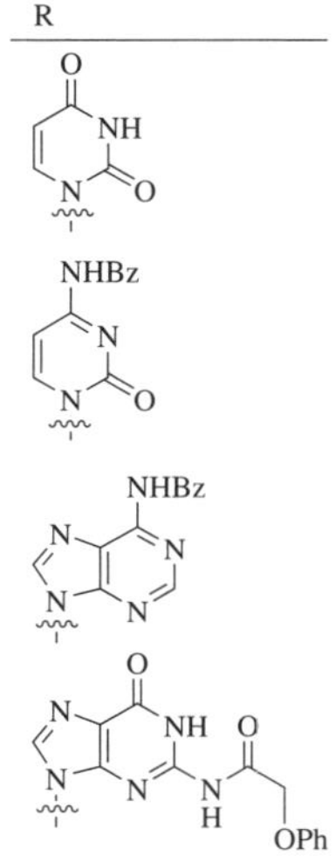	745, 746
C_{11-18} 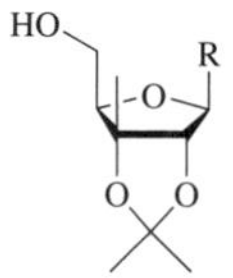 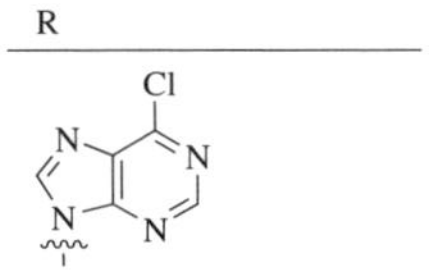	TEMPO (cat.), $PhI(OAc)_2$, MeCN, H_2O, rt, 2 h	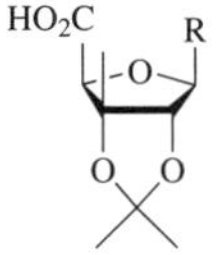 (80)	747

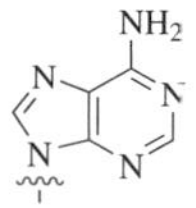		(54)	
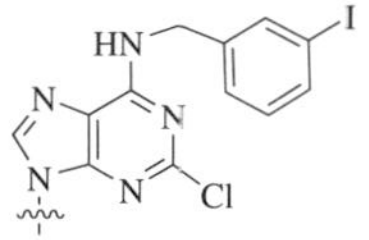		(74)	
C_{12}			
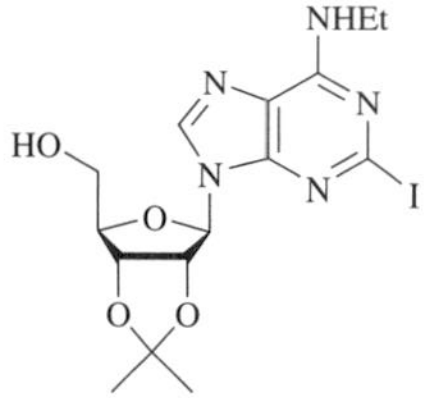	TEMPO (cat.), $PhI(OAc)_2$, MeCN, H_2O, rt	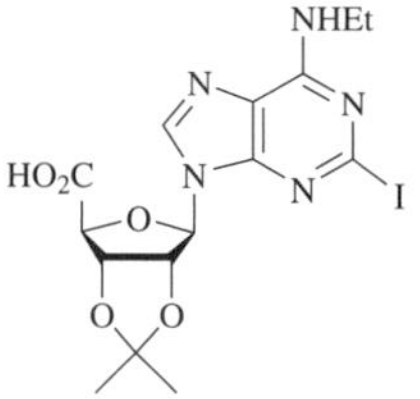 (100)	748
C_{13}			
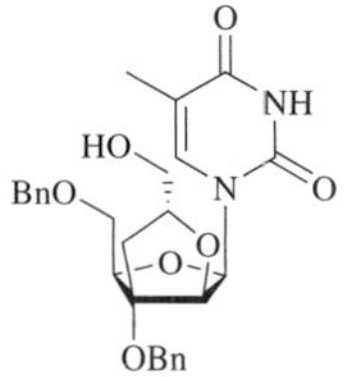	TEMPO (cat.), $PhI(OAc)_2$, H_2O, 5°	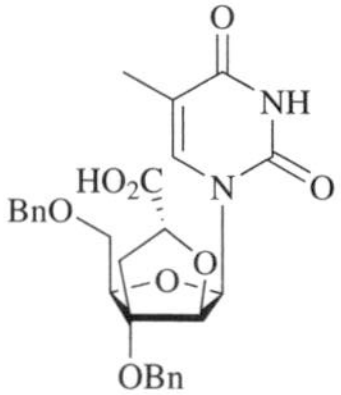 (82)	749
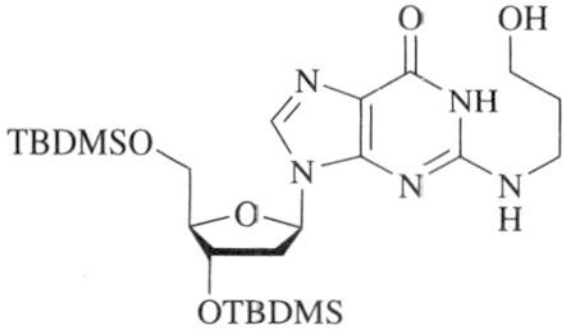	$[AcNH\text{-}TEMPO]^+[BF_4]^-$, silica gel, CH_2Cl_2, rt, 16 h	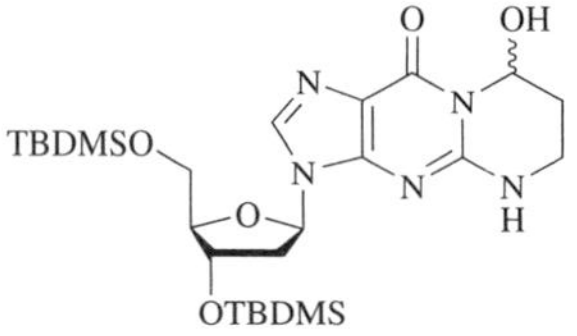 (69)	750

TABLE 4F. OXIDATIONS OF NUCLEOSIDES (*Continued*)

Alcohol	Conditions	Product(s) and Yield(s) (%)	Refs.
C_{18}			
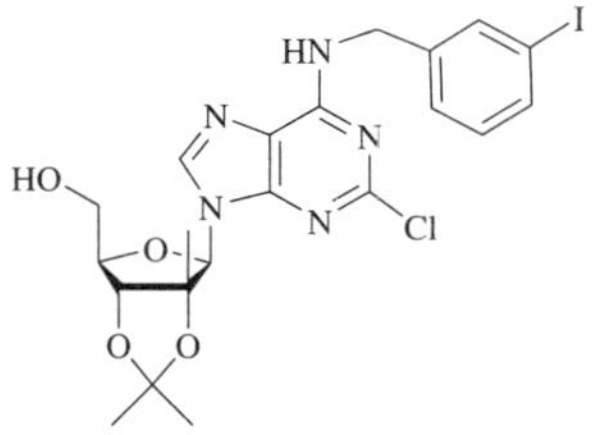	TEMPO (cat.), PhI(OAc)$_2$, MeCN, H_2O, rt, 3 h	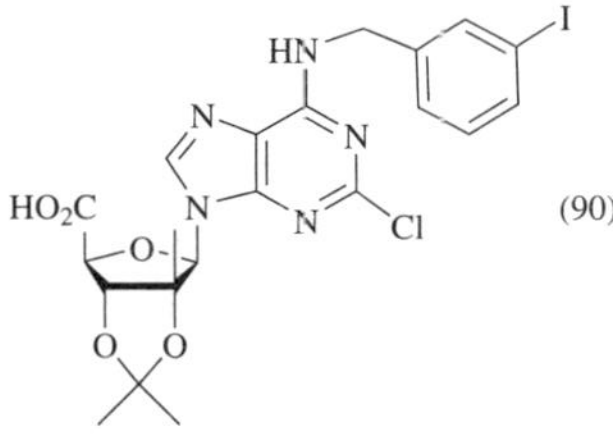 (90)	747
C_{22-26}			
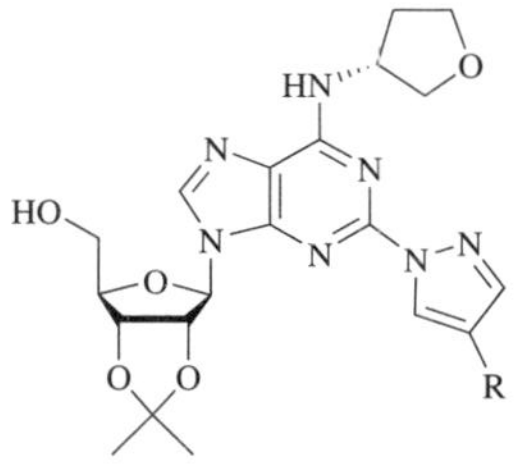	TEMPO (cat.), PhI(OAc)$_2$, MeCN, H_2O	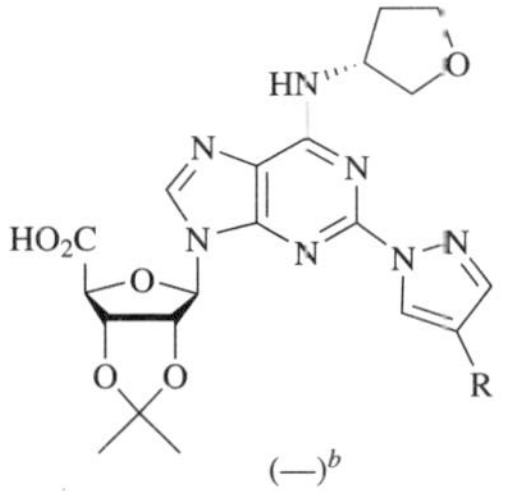 (—)[b] R: 4-pyridyl, 4-MeC$_6$H$_4$, 2-quinolinyl	751

[a] The product was isolated as the sodium salt.

[b] The product was not isolated and was used in a subsequent step without purification.

TABLE 4G. OXIDATIONS OF POLYSACCHARIDES

	Alcohol	Conditions	Product(s) and Yield(s) (%)	Refs.
C_n	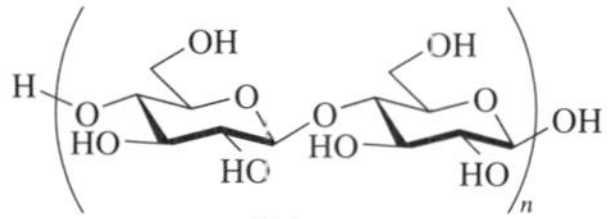 cellulose	See table.	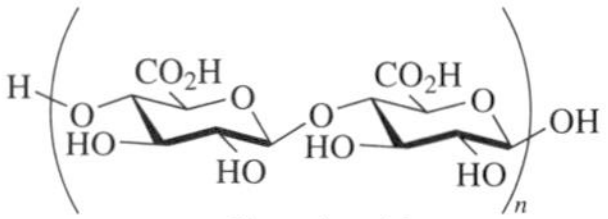 celluronic acid	
	Type	**Conditions**		
	cellulose	TEMPO (cat.), NaBr (cat.), NaOCl, H_2O, pH 10, rt, 2 h	(—)	752
	cellulose	TEMPO (cat.), NaBr (cat.), NaOCl, H_2O, pH 10.8, 0°	(84)	331
	regenerated cellulose	H_2N-TEMPO (cat.), NaOCl, H_2O, pH 10–11	(—)	753
	regenerated cellulose	TEMPO (cat.), NaBr (cat.), NaOCl, $NaHCO_3$, H_2O, pH 10.7, rt, 12 min	(—)	754
	cellulose III	TEMPO (cat.), NaBr (cat.), NaOCl, H_2O, pH 10, 0°, 1 h	(70)	755
	partially hydrolyzed cellulose	TEMPO (cat.), NaBr (cat.), NaOCl, NaOH, H_2O, pH 10.8	(—)	756
	α-cellulose	TEMPO (cat.), NaBr (cat.), NaOCl, H_2O, pH 10.8, 0°	(87)	331
	NH_3(*l*)-treated cellulose	TEMPO (cat.), NaBr (cat.), NaOCl, NaOH, H_2O, pH 10	(—)[a]	757
	cellulose (thermo-mechanical paper pulp fibers)	AcNH-TEMPO (cat.), NaBr (cat.), NaOCl, NaOH, H_2O, pH 10.5	(—)	758
	carboxymethyl cellulose	TEMPO (cat.), NaBr (cat.), NaOCl, H_2O, pH 10.8, 0°	(72)	331
	various water-soluble cellulose acetates	TEMPO (cat.), NaBr (cat.), NaOCl, NaOH, H_2O, pH 10	(—)	341
	cotton linters, ramie, and spruce holocellulose	TEMPO (cat.), KBr (cat.), NaOCl, NaOH, H_2O, pH 10.5	(—)	759

TABLE 4G. OXIDATIONS OF POLYSACCHARIDES (*Continued*)

Alcohol	Conditions	Product(s) and Yield(s) (%)	Refs.
C_n			
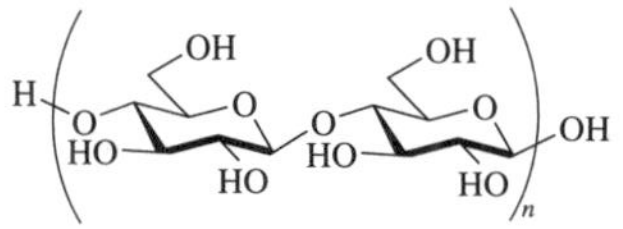 from bleached wood pulp, cotton, tunicin, and bacterial cellulose	TEMPO (cat.), KBr (cat.), NaOCl, NaOH, H_2O, pH 10.5	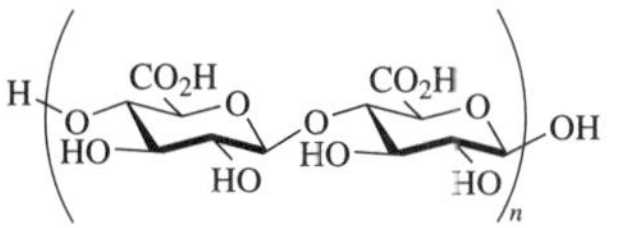 (—) + various polyuronic acid salts	760
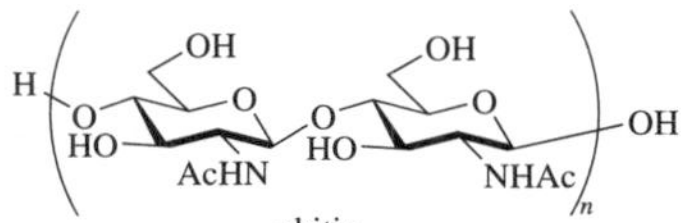 chitin	TEMPO (cat.), NaBr (cat.), NaOCl, H_2O, pH 10.8, 0°	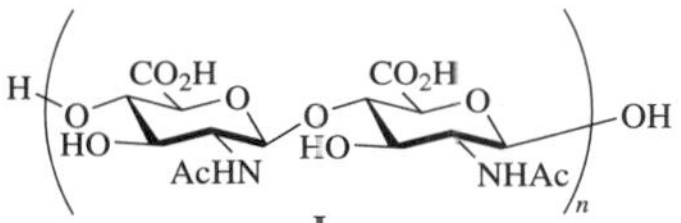 **I** (42)	331
Chitin from *cragnon cragnon, absidia coeulea, absidia orchidis*	TEMPO (cat.), NaBr (cat.), NaOCl, NaOH, H_2O, pH 10.8	**I** (—)	761
Chitin-glucan complex from *aspergillus niger, trichoderma reesei, saprolegnia sp.*	TEMPO, NaBr (cat.), NaOCl, NaOH, H_2O, pH 10.8	**I** (—)	762
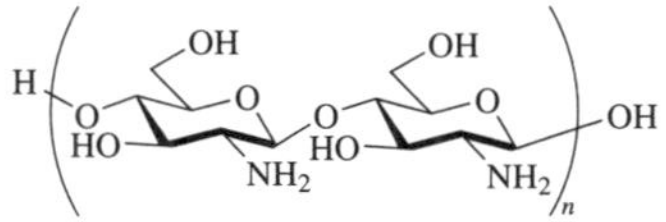 chitosan	TEMPO (cat.), NaBr (cat.), NaOCl, H_2O, pH 10.8, 0°	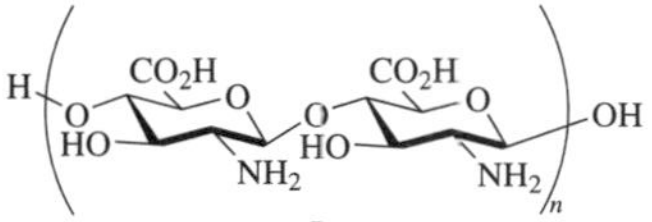 **I** (91)	331
	TEMPO (cat.), NaBr (cat.), NaOCl, NaOH, H_2O, pH controlled	**I** (—)[b]	763
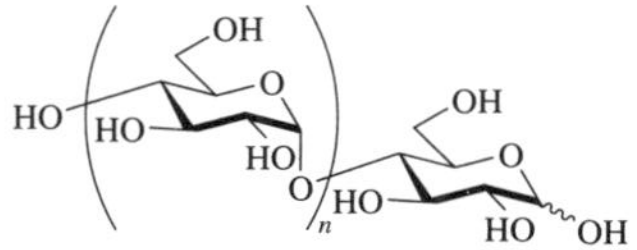 maltodextrin	TEMPO (cat.), NaBr (cat.), NaOCl, NaOH, H_2O, pH varied	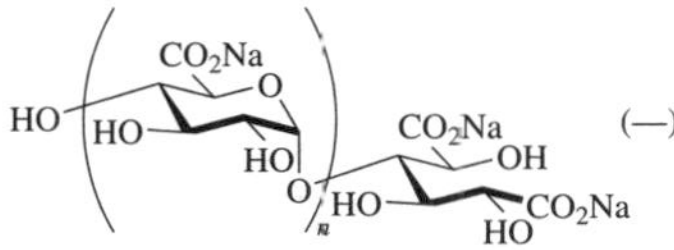 (—)	317

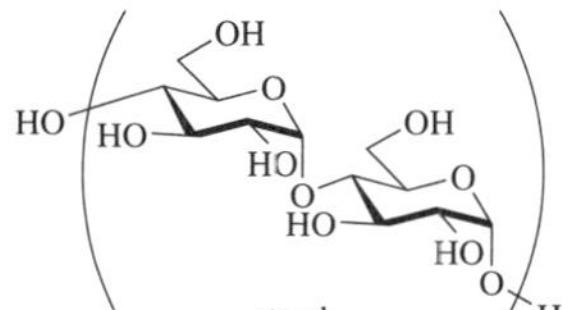
starch

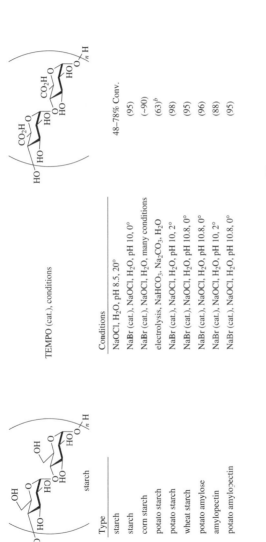

TEMPO (cat.), conditions

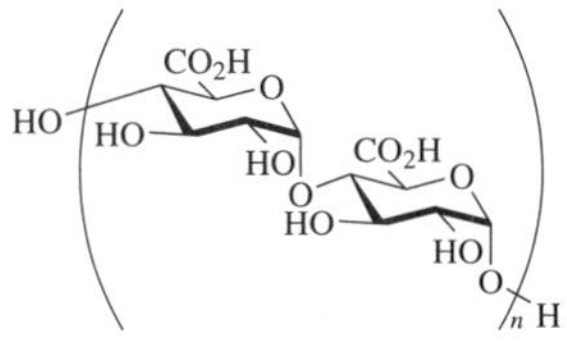

Type	Conditions		
starch	NaOCl, H_2O, pH 8.5, 20°	48–78% Conv.	333
starch	NaBr (cat.), NaOCl, H_2O, pH 10, 0°	(95)	334
corn starch	NaBr (cat.), NaOCl, H_2O, many conditions	(~90)	764
potato starch	electrolysis, $NaHCO_3$, Na_2CO_3, H_2O	(63)[b]	283
potato starch	NaBr (cat.), NaOCl, H_2O, pH 10, 2°	(98)	75
wheat starch	NaBr (cat.), NaOCl, H_2O, pH 10.8, 0°	(95)	331
potato amylose	NaBr (cat.), NaOCl, H_2O, pH 10.8, 0°	(96)	331
amylopectin	NaBr (cat.), NaOCl, H_2O, pH 10, 2°	(88)	75
potato amylopectin	NaBr (cat.), NaOCl, H_2O, pH 10.8, 0°	(95)	331

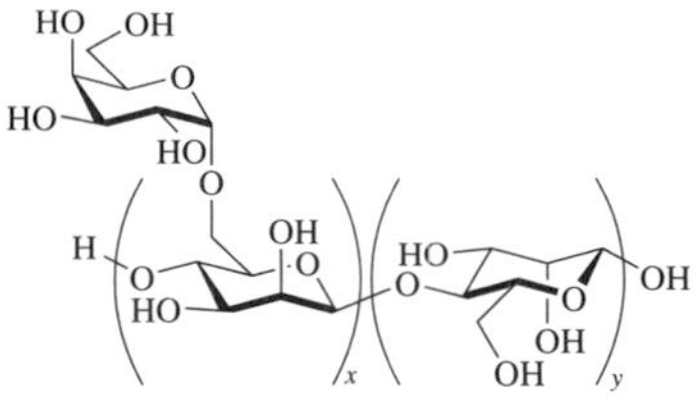
galactomannan

TEMPO (cat.), NaBr (cat.), NaOCl, NaOH, H_2O

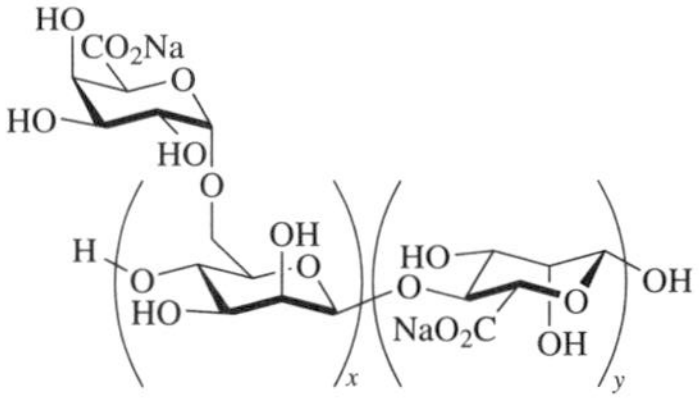
(92)

765

TABLE 4G. OXIDATIONS OF POLYSACCHARIDES (*Continued*)

Alcohol	Conditions	Product(s) and Yield(s) (%)	Refs.
C_n 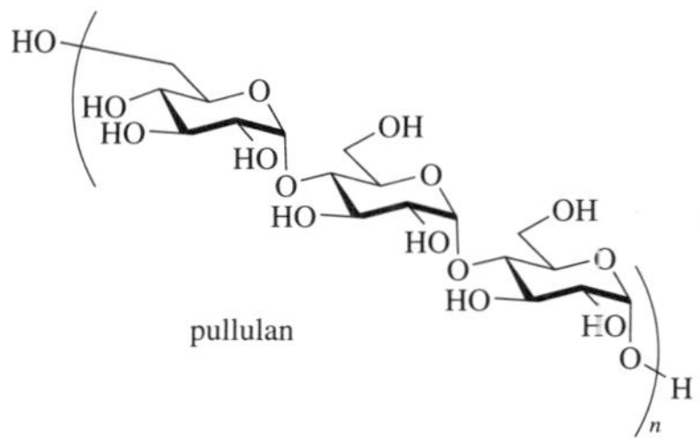 pullulan	TEMPO (cat.), NaBr (cat.), NaOCl, NaOH, H_2O, pH 9.4	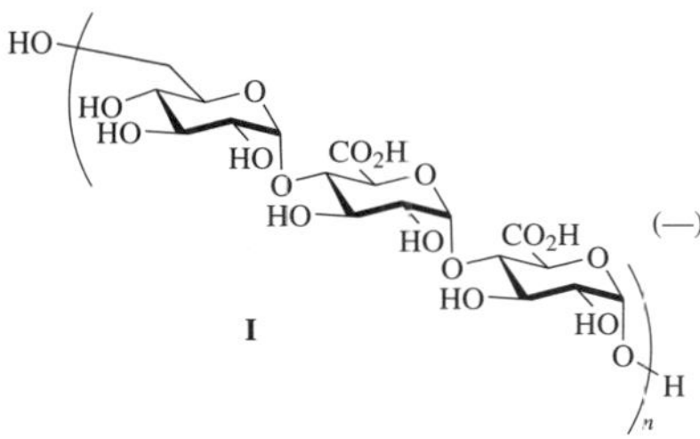 **I** (—)	766
	TEMPO (cat.), NaBr (cat.), NaOCl, H_2O, pH 10, 2°	**I**, 74% conversion	767, 768
	Nitroxide (cat.), NaOCl, H_2O, pH 8.5	**I** (—) Nitroxide: AcNH-TEMPO, MeO-TEMPO, AcO-TEMPO, mesyl-TEMPO	314
	TEMPO (cat.), NaBr (cat.), NaOCl, H_2O, pH 10.8, 0°	**I** (66)	331
	TEMPO (cat.), NaBr (cat.), NaOCl, H_2O, pH 10, 0°	**I** (—)	332
	TEMPO (cat.), NaBr (cat.), NaOCl, H_2O, pH 10, 2°	**I** (95)	75

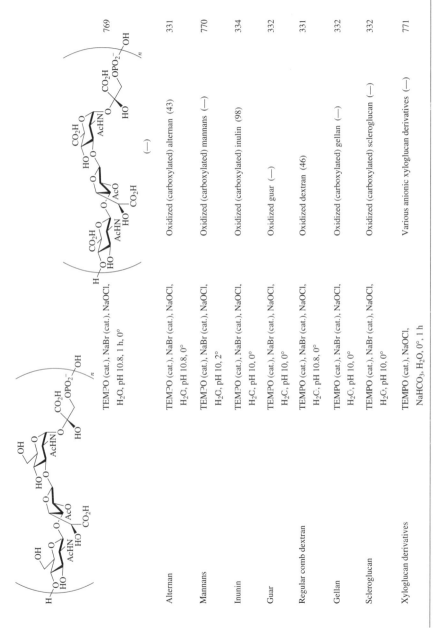

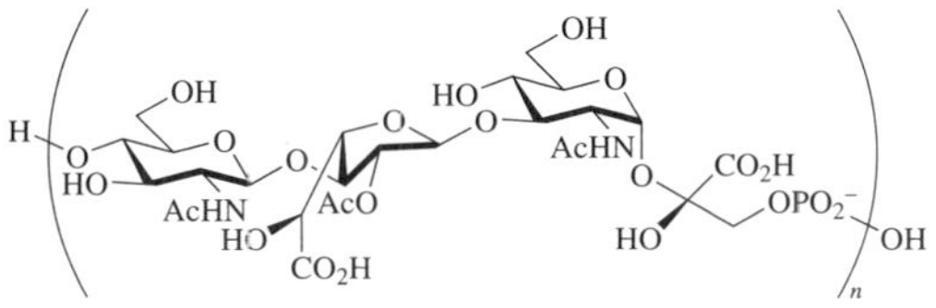	TEMPO (cat.), NaBr (cat.), NaOCl, H_2O, pH 10.8, 1 h, 0°	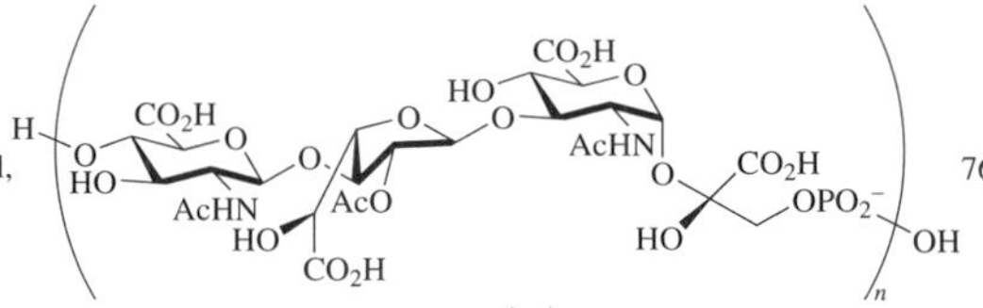 (—)	769
Alternan	TEMPO (cat.), NaBr (cat.), NaOCl, H_2O, pH 10.8, 0°	Oxidized (carboxylated) alternan (43)	331
Mannans	TEMPO (cat.), NaBr (cat.), NaOCl, H_2O, pH 10, 2°	Oxidized (carboxylated) mannans (—)	770
Inunin	TEMPO (cat.), NaBr (cat.), NaOCl, H_2O, pH 10, 0°	Oxidized (carboxylated) inulin (98)	334
Guar	TEMPO (cat.), NaBr (cat.), NaOCl, H_2O, pH 10, 0°	Oxidized guar (—)	332
Regular comb dextran	TEMPO (cat.), NaBr (cat.), NaOCl, H_2O, pH 10.8, 0°	Oxidized dextran (46)	331
Gellan	TEMPO (cat.), NaBr (cat.), NaOCl, H_2O, pH 10, 0°	Oxidized (carboxylated) gellan (—)	332
Scleroglucan	TEMPO (cat.), NaBr (cat.), NaOCl, H_2O, pH 10, 0°	Oxidized (carboxylated) scleroglucan (—)	332
Xyloglucan derivatives	TEMPO (cat.), NaOCl, $NaHCO_3$, H_2O, 0°, 1 h	Various anionic xyloglucan derivatives (—)	771

[a] The product was partially oxidized cellulose.

[b] The product was isolated as the sodium salt.

REFERENCES

[1] Lebedev, O. A.; Kazarnovskii, S. N. *Trudy po Khimii i Khim. Technologii (Gorkii)* **1959**, *8*, 649.
[2] Neiman, M. B.; Rozantzev, É. G.; Mamedova, Y. G. *Nature* **1962**, *196*, 472.
[3] Likhtenshtein, G.; Yamauchi, J.; Nakatsuji, S.; Smirnov, A.; Tamura, R. *Nitroxides: Applications in Chemistry, Biomedicine, and Materials Sciences*; Wiley-VCH: Weinheim, 2008.
[4] Golubev, V. A.; Zhdanov, R. I.; Gida, V. M.; Rozantsev, E. G. *Bull. Acad. Sci. USSR, Div. Chem. Sci.* **1971**, 768.
[5] Golubev, V. A.; Sen, V. D.; Kulyk, I. V.; Aleksandrov, A. L. *Bull. Acad. Sci. USSR, Div. Chem. Sci.* **1975**, *24*, 2119.
[6] Israeli, A.; Patt, M.; Oron, M.; Samuni, A.; Kohen, R.; Goldstein, S. *Free Radical Biol. Med.* **2005**, *38*, 317.
[7] Sen', V. D.; Golubev, V. A. *J. Phys. Org. Chem.* **2008**, *6*, 138.
[8] Golubev, V. A., Rozantsev, E. G.; Neiman, M. B. *Bull. Acad. Sci. USSR, Div. Chem. Sci.* **1965**, 1898.
[9] Atovmyan, L. O.; Golubev, V. A.; Golovina, N. I.; Klitskaya, G. A. *Zh. Strukturnoi Khim.* **1975**, *16*, 92.
[10] Bobbitt, J. M.; Flores, M. C. L. *Heterocycles* **1988**, *27*, 509.
[11] Miyazawa, T.; Endo, T.; Shiihashi, S.; Okawara, M. *J. Org. Chem.* **1985**, *50*, 1332.
[12] Bobbitt, J. M.; Merbouh, N. *Org. Synth.* **2005**, *82*, 80.
[13] Anelli, P. L.; Montanari, F.; Quici, S. *Org. Synth.* **1990**, *69*, 212.
[14] Zhao, M.; Li, J.; Mano, E.; Song, Z.; Tschaen, D. M. *Org. Synth.* **2005**, *81*, 195.
[15] Piancatelli, G.; Leonelli, F. *Org. Synth.* **2005**, *83*, 18.
[16] Miyazawa, T.; Endo, T. *Yuki Gosei Kagaku Kyokaishi* **1986**, *44*, 1134.
[17] Yamaguchi, M.; Miyazawa, T.; Takata, T.; Endo, T. *Pure Appl. Chem.* **1990**, *62*, 217.
[18] Liu, Y.; Guo, H.; Liu, Z. *Huaxue Xuebao* **1991**, *49*, 187.
[19] Merbouh, N. *Synlett* **2003**, 1757.
[20] Merbouh, N.; Bobbitt, J. M.; Brückner, C. *Org. Prep. Proced. Int.* **2004**, *36*, 3.
[21] Barriga, S. *Synlett* **2001**, 563.
[22] Adam, W.; Saha-Möller, C. R.; Ganeshpure, P. A. *Chem. Rev.* **2001**, *101*, 3499.
[23] de Nooy, A. E. J.; Besemer, A. C.; van Bekkum, H. *Synthesis* **1996**, 1153.
[24] Sheldon, R. A.; Arends, I. W. C. E. *Adv. Synth. Catal.* **2004**, *346*, 1051.
[25] De Souza, M. V. N. *Mini-Reviews in Organic Chemistry* **2006**, *3*, 155.
[26] Iwabuchi, Y. *J. Synth. Org. Chem. Jpn.* **2008**, *66*, 1076.
[27] Rozantsev, E. G.; Sholle, V. D. *Synthesis* **1971**, 190.
[28] Nilsen, A.; Braslau, R. *J. Polym. Sci., Part A: Polym. Chem.* **2006**, *44*, 697.
[29] Naik, N.; Braslau, R. *Tetrahedron* **1998**, *54*, 667.
[30] Vogler, T.; Studer, A. *Synthesis* **2008**, 1979.
[31] Palomo, C.; Aizpurua, J. M.; Ganboa, I.; Oiarbide, M. *Synlett* **2001**, 1813.
[32] van Bekkum, H.; Besemer, A. C. *Carbohydrates in Europe* **1995**, 16.
[33] Bragd, P. L.; van Bekkum, H.; Besemer, A. C. *Top. Catal.* **2004**, *27*, 49.
[34] Kochkar, H.; Morawietz, M.; Holderich, W. F. *Stud. Surf. Sci. Catal.* **2000**, 130*A*, 545.
[35] Kochkar, H.; Lassalle, L.; Morawietz, M.; Holderich, W. F. *J. Catal.* **2000**, *194*, 343.
[36] Gryko, D.; Chalko, J.; Jurczak, J. *Chirality* **2003**, *15*, 514.
[37] Pagliaro, M.; Ciriminna, R.; Kimura, H.; Rossi, M.; Della Pina, C. *Angew. Chem., Int. Ed.* **2007**, *46*, 4434.
[38] Zhou, C. H. C.; Beltramini, J. N.; Fan, Y. X.; Lu, G. Q. M. *Chem. Soc. Rev.* **2008**, *37*, 527.
[39] Arterburn, J. B. *Tetrahedron* **2001**, *57*, 9765.
[40] Golubev, V. A.; Kozlov, Y. N.; Petrov, A. N.; Purmal, A. P. *Prog. React. Kinet.* **1991**, *16*, 35.
[41] Tohma, H.; Kita, Y. *Adv. Synth. Catal.* **2004**, *346*, 111.
[42] Benaglia, M.; Puglisi, A.; Cozzi, F. *Chem. Rev.* **2003**, *103*, 3401.
[43] Brünjes, M.; Sourkouni-Argirusi, G.; Kirschning, A. *Adv. Synth. Catal.* **2003**, *345*, 635.
[44] Kloth, K.; Brünjes, M.; Kunst, E.; Jöge, T.; Gallier, F.; Adibekian, A.; Kirschning, A. *Adv. Synth. Catal.* **2005**, *347*, 1423.
[45] Sheldon, R. A.; Arends, I.; Dijksman, A. *Catal. Today* **2000**, *57*, 157.

[46] Sheldon, R. A.; Arends, I. W. C. E.; ten Brink, G. J.; Dijksman, A. *Acc. Chem. Res.* **2002**, *35*, 774.
[47] Rademann, J. Advanced Polymer Reagents Based on Activated Reactants and Reactive Intermediates: Powerful Novel Tools in Diversity-Oriented Synthesis. In *Methods in Enzymology*; Bunin, B. A.; Morales, G., Eds.; Academic Press: San Diego, 2003; Vol. 369, pp 366–390.
[48] Zhan, B. Z.; Thompson, A. *Tetrahedron* **2004**, *60*, 2917.
[49] Lenoir, D. *Angew. Chem., Int. Ed.* **2006**, *45*, 3206.
[50] Caron, S.; Dugger, R. W.; Ruggeri, S. G.; Ragan, J. A.; Ripin, D. H. B. *Chem. Rev.* **2006**, *106*, 2943.
[51] Takata, T.; Tsujino, Y.; Nakanishi, S.; Nakamura, K.; Yoshida, E.; Endo, T. *Chem. Lett.* **1999**, 937.
[52] Bailey, W. F.; Bobbitt, J. M.; Wiberg, K. B. *J. Org. Chem.* **2007**, *72*, 4504.
[53] Hoye, T. R.; Renner, M. K. *J. Org. Chem.* **1996**, *61*, 8489.
[54] Golubev, V. A.; Kobylyanskii, E. V. *Zh. Org. Khim.* **1972**, *8*, 2607.
[55] Golubev, V. A.; Rudyk, T. S.; Sen, V. D.; Aleksandrov, A. L. *Bull. Acad. Sci. USSR, Div. Chem. Sci.* **1976**, 744.
[56] Ma, Z.; Bobbitt, J. M. *J. Org. Chem.* **1991**, *56*, 6110.
[57] Pradhan, P. P.; Bobbitt, J. M.; Bailey, W. F. *Org. Lett.* **2006**, *8*, 5485.
[58] Bobbitt, J. M. University of Connecticut, Storrs, CT. Unpublished work, 2008.
[59] Golubev, V. A.; Borislavskii, V. N.; Aleksandrov, A. L. *Bull. Acad. Sci. USSR, Div. Chem. Sci.* **1977**, *9*, 1874.
[60] Semmelhack, M. F.; Schmid, C. R.; Cortés, D. A. *Tetrahedron Lett.* **1986**, *27*, 1119.
[61] de Nooy, A. E. J.; Besemer, A. C.; van Bekkum, H. *Tetrahedron* **1995**, *51*, 8023.
[62] Chmielewska, B.; Krzyczmonik, P.; Scholl, H. *J. Electroanal. Chem.* **1995**, *395*, 167.
[63] Olah, G. A.; Friedman, N. *J. Am. Chem. Soc.* **1966**, *88*, 5330.
[64] Olah, G. A.; Ho, T. *Synthesis* **1976**, 609.
[65] Olah, G. A.; Ramaiah, P. *J. Org. Chem.* **1993**, *58*, 4639.
[66] Breton, T.; Liaigre, D.; Belgsir, E. M. *Tetrahedron Lett.* **2005**, *46*, 2487.
[67] Breton, T.; Liaigre, D.; Belgsir, E. M. *Electrochem. Commun.* **2005**, *7*, 1445.
[68] Bobbitt, J. M. *J. Org. Chem.* **1998**, *63*, 9367.
[69] Yamaguchi, M.; Takata, T.; Endo, T. *Tetrahedron Lett.* **1988**, *29*, 5671.
[70] Banwell, M. G.; Bridges, V. S.; Dupuche, J. R.; Richards, S. L.; Walter, J. M. *J. Org. Chem.* **1994**, *59*, 6338.
[71] Sen', V. D.; Golubev, V. A.; Kosheleva, T. M. *Bull. Acad. Sci. USSR, Div. Chem. Sci.* **1979**, 1847.
[72] Golubev, V. A.; Sen, V. D.; Rozantsev, E. G. *Bull. Acad. Sci. USSR, Div. Chem. Sci.* **1979**, 1927.
[73] Miyazawa, T.; Endo, T.; Okawara, M. *J. Org. Chem.* **1985**, *50*, 5389.
[74] Anderson, C. D.; Shea, K. J.; Rychnovsky, S. D. *Org. Lett.* **2005**, *7*, 4879.
[75] de Nooy, A. E. J.; Besemer, A. C.; van Bekkum, H. *Carbohydr. Res.* **1995**, *269*, 89.
[76] Merbouh, N.; Bobbitt, J. M.; Brückner, C. *Tetrahedron Lett.* **2001**, *42*, 8793.
[77] Merbouh, N.; Bobbitt, J. M.; Brückner, C. *J. Org. Chem.* **2004**, *69*, 5116.
[78] Luberoff, B. J. *Homogeneous Catalysis*; Adv. Chem. Ser.; American Chemical Society: Washington, 1968; Vol. 70.
[79] Hückel, W.; Liegel, W. *Chem. Ber.* **1938**, *71*, 1442.
[80] Dijksman, A.; Marino-González, A.; Payeras, A. M. I.; Arends, I. W. C. E.; Sheldon, R. A. *J. Am. Chem. Soc.* **2001**, *123*, 6826.
[81] Semmelhack, M. F.; Schmid, C. R.; Cortés, D. A.; Chou, C. S. *J. Am. Chem. Soc.* **1984**, *106*, 3374.
[82] Dijksman, A.; Arends, I. W. C. E.; Sheldon, R. A. *Org. Biomol. Chem.* **2003**, *1*, 3232.
[83] Kashiwagi, Y.; Kurashima, F.; Chiba, S.; Anzai, J.; Osa, T.; Bobbitt, J. M. *Chem. Commun.* **2003**, 114.
[84] Tanaka, H.; Kawakami, Y.; Goto, K.; Kuroboshi, M. *Tetrahedron Lett.* **2001**, *42*, 445.
[85] Ma, Z.; Huang, Q.; Bobbitt, J. M. *J. Org. Chem.* **1993**, *58*, 4837.
[86] Eliel, E. L.; Wilen, S. H. *Stereochemistry of Organic Compounds*; Wiley: New York, 1994; pp 395–402.
[87] Kagan, H. B.; Fiaud, J. C. *Top. Stereochem.* **1988**, *18*, 249.

[88] Kagan, H. B. Kinetic Resolution Calculation Applet. http://www.ch.cam.ac.uk/magnus/KinRes.html; (accessed May 2008).
[89] Goodman, J. M.; Köhler, A.-K.; Alderton, S. C. M. *Tetrahedron Lett.* **1999**, *40*, 8715.
[90] Rychnovsky, S. D.; McLernon, T. L.; Rajapakse, H. *J. Org. Chem.* **1996**, *61*, 1194.
[91] Flohr, K.; Paton, R. M.; Kaiser, E. T. *J. Am. Chem. Soc.* **1975**, *97*, 1209.
[92] Huang, Q. Ph. D. Thesis, University of Connecticut, Storrs, CT, 1992.
[93] Einhorn, J.; Einhorn, C.; Ratajczak, F.; Pierre, J.-L. *Synth. Commun.* **2000**, *30*, 1837.
[94] Graetz, B.; Rychnovsky, S.; Leu, W.-H.; Farmer, P.; Lin, R. *Tetrahedron: Asymmetry* **2005**, *16*, 3584.
[95] Rychnovsky, S. D.; Beauchamp, T.; Vaidyanathan, R.; Kwan, T. *J. Org. Chem.* **1998**, *63*, 6363.
[96] Shibata, T.; Uemae, K.; Yamamoto, Y. *Tetrahedron: Asymmetry* **2000**, *11*, 2339.
[97] Wright, K.; de Castries, A.; Sarciaux, M.; Formaggio, F.; Toniolo, C.; Toffoletti, A.; Wakselman, M.; Mazaleyrat, J. P. *Tetrahedron Lett.* **2005**, *46*, 5573.
[98] Wright, K.; Crisma, M.; Toniolo, C.; Torok, R.; Peter, A.; Wakselman, M.; Mazaleyrat, J. P. *Tetrahedron Lett.* **2003**, *44*, 3381.
[99] Keana, J. F. W. *Chem. Rev.* **1978**, *78*, 37.
[100] Braslau, R.; Kuhn, H.; Burrill II, L. C.; Lanham, K.; Stenland, C. J. *Tetrahedron Lett.* **1996**, *37*, 7933.
[101] Braslau, R.; Chaplinski, V.; Goodson, P. *J. Org. Chem.* **1998**, *63*, 9857.
[102] Ganem, B. *J. Org. Chem.* **1975**, *40*, 1998.
[103] Formaggio, F.; Bonchio, M.; Crisma, M.; Peggion, C.; Mezzato, S.; Polese, A.; Barazza, A.; Antonello, S.; Maran, F.; Broxterman, Q. B.; Kaptein, B.; Kamphuis, J.; Vitale, R. M.; Saviano, M.; Benedetti, E.; Toniolo, C. *Chem.—Eur. J.* **2002**, *8*, 84.
[104] Osa, T.; Kashiwagi, Y.; Yanagisawa, Y.; Bobbitt, J. M. *Chem. Commun.* **1994**, 2535.
[105] Kashiwagi, Y.; Yanagisawa, Y.; Kurashima, F.; Anzai, J.; Osa, T.; Bobbitt, J. M. *Chem. Commun.* **1996**, 2745.
[106] Belgsir, E. M.; Schäfer, H. J. *Chem. Commun.* **1999**, 435.
[107] Kashiwagi, Y.; Kurashima, F.; Kikuchi, C.; Anzai, J.; Osa, T.; Bobbitt, J. M. *Tetrahedron Lett.* **1999**, *40*, 6469.
[108] Kashiwagi, Y.; Kurashima, F.; Kikuchi, C.; Anzai, J.; Osa, T.; Bobbitt, J. M. *Chem. Commun.* **1999**, 1983.
[109] Kuroboshi, M.; Yoshihisa, H.; Cortona, M. N.; Kawakami, Y.; Gao, Z.; Tanaka, H. *Tetrahedron Lett.* **2000**, *41*, 8131.
[110] Kurashima, F.; Kashiwagi, Y.; Kikuchi, C.; Anzai, J.; Osa, T. *Heterocycles* **1999**, *50*, 79.
[111] Kashiwagi, Y.; Kurashima, F.; Kikuchi, C.; Anzai, J.; Osa, T. *Heterocycles* **2000**, *53*, 1583.
[112] Kashiwagi, Y.; Chiba, S.; Anzai, J. *J. Electroanal. Chem.* **2004**, *566*, 257.
[113] Kashiwagi, Y.; Uchiyama, K.; Kurashima, F.; Kikuchi, C.; Anzai, J. *Chem. Pharm. Bull.* **1999**, *47*, 1051.
[114] Church, K. M.; Holloway, L. M.; Matley, R. C.; Brower, R. J, III. *Nucleosides, Nucleotides, Nucleic Acids* **2004**, *23*, 1723.
[115] Ganiev, I. M.; Suvorkina, E. S.; Igoshina, A. V.; Kabalnova, N. N.; Imashev, U. B.; Tolstikov, G. A. *Russian Chem. Bull., Int. Ed.* **2002**, *51*, 982.
[116] Golubev, V. A.; Zhdanov, R. I.; Rozantsev, E. G. *Bull. Acad. Sci. USSR, Div. Chem. Sci.* **1970**, *19*, 186.
[117] Yoshida, E.; Takata, T.; Endo, T. *J. Polym. Sci., Part A: Polym. Chem.* **1992**, *30*, 1193.
[118] Chou, S.; Nelson, J. A.; Spencer, T. A. *J. Org. Chem.* **1974**, *39*, 2356.
[119] Barodkin, G. I.; Elanov, I. R.; Shakirov, M. M.; Shubin, V. G. *Russ. J. Org. Chem.* **2003**, *39*, 1144.
[120] Shibuya, S.; Tomizawa, M.; Iwabuchi, Y. *J. Org. Chem.* **2008**, *73*, 4750.
[120a] Vatéle, J. M. *Synlett* **2008**, 1785.
[121] Weik, S.; Nicholson, G.; Jung, G.; Rademann, J. *Angew. Chem., Int. Ed.* **2001**, *40*, 1436.
[122] Bobbitt, J. M. *Chem. Eng. News* **1999**, *77*, 6.
[123] Bobbitt, J. M. *Molecules* **1999**, *4*, M102.
[124] Bobbitt, J. M.; Guttermuth, M. C. F.; Ma, Z.; Tang, H. *Heterocycles* **1990**, *30*, 1131.
[125] Liu, Y.; Liu, Z.; Guo, H. *Chem. J. Chinese Univ. (Engl. Ed.)* **1988**, *4*, 90.

[126] Zhdanov, R. I.; Golubev, V. A.; Rozantsev, É. G. *Bull. Acad. Sci. USSR, Div. Chem. Sci.* **1970**, 188.
[127] Golubev, V. A.; Zhdanov, R. I.; Protsishin, I. T.; Rozantsev, E. G. *Bull. Acad. Sci. USSR, Div. Chem. Sci.* **1970**, 952.
[128] Hunter, D. H.; Racok, J. S.; Rey, A. W.; Ponce, Y. Z. *J. Org. Chem.* **1988**, *53*, 1278.
[129] Ren, T.; Liu, Y.; Guo, Q. *Bull. Chem. Soc. Jpn.* **1996**, *69*, 2935.
[130] Koch, T.; Hoskovec, M.; Boland, W. *Tetrahedron* **2002**, *58*, 3271.
[131] Zakrzewski, J.; Grodner, J.; Bobbitt, J. M.; Karpińska, M. *Synthesis* **2007**, 2491.
[132] Abad, A.; Agulló, C.; Cunat, A. C.; Perni, R. H. *Tetrahedron: Asymmetry* **2000**, *11*, 1607.
[133] Miyazawa, T.; Endo, T. *J. Am. Chem. Soc.* **1985**, *50*, 3930.
[134] Anelli, P. L.; Biffi, C.; Montanari, F.; Quici, S. *J. Org. Chem.* **1987**, *52*, 2559.
[135] Yoshida, E.; Takata, T.; Endo, T. *Makromol. Chem.* **1993**, *194*, 2507.
[136] Yoshida, E.; Yamaguchi, M.; Takata, T.; Endo, T. *Makromol. Chem.* **1993**, *194*, 1307.
[137] Dhokte, U. P.; Khau, V. V.; Hutchison, D. R.; Martinelli, M. J. *Tetrahedron Lett.* **1998**, *39*, 8771.
[138] Miyazawa, T.; Endo, T. *Tetrahedron Lett.* **1986**, *27*, 3395.
[139] Bobbitt, J. M.; Ma, Z. *Heterocycles* **1992**, *33*, 641.
[140] Liu, Y.; Wang, W.; Guo, Q. *Chin. Chem. Lett.* **1996**, *7*, 790.
[141] Kashiwagi, Y.; Ono, H.; Osa, T. *Chem. Lett.* **1993**, 81.
[142] Deskus, J.; Smith, M. B.; Simandan, T. *Synth. Commun.* **1996**, *26*, 3137.
[143] Bartelson, A.; Bobbitt, J. M. University of Connecticut, Storrs, CT. Unpublished work, 2007.
[144] Breton, T.; Beashiardes, G.; Léger, J.-M.; Kokoh, K. B. *Eur. J. Org. Chem.* **2007**, 1567.
[145] Banwell, M. G.; Collis, M. P.; Mackay, M. F.; Richards, S. L. *J. Chem. Soc., Perkin Trans. 1* **1993**, 1913.
[146] Banwell, M. G.; Edwards, A. J.; Harfoot, G. J.; Jolliffe, K. A. *Tetrahedron* **2004**, *60*, 535.
[147] Banwell, M. G.; Hockless, D. C. R.; Holman, J. W.; Longmore, R. W.; McRae, K. J.; Pham, H. T. T. *Synlett* **1999**, 1491.
[148] Cella, J. A. *J. Org. Chem.* **1977**, *42*, 2077.
[149] Cella, J. A.; Kelley, J. A.; Kenehan, E. F. *J. Org. Chem.* **1975**, *40*, 1860.
[150] Cella, J. A.; Kelley, J. A.; Kenehan, E. F. *Tetrahedron Lett.* **1975**, *33*, 2869.
[151] Anelli, P. L.; Banfi, S.; Montanari, F.; Quici, S. *J. Org. Chem.* **1989**, *54*, 2970.
[152] Schämann, M.; Schäfer, H. J. *Electrochim. Acta* **2005**, *50*, 4956.
[153] Golubev, V. A.; Kozlov, Y. N.; Petrov, A. N.; Purmal, A. P.; Travina, O. A. *Chimicheskaya Fizika* **1985**, *4*, 838.
[154] Pozzi, G.; Cinato, F.; Montanari, F.; Quici, S. *Chem. Commun.* **1998**, 877.
[155] Konas, D. W.; Coward, J. K. *J. Org. Chem.* **2001**, *66*, 8831.
[156] Michaud, A.; Gingras, G.; Morin, M.; Béland, F.; Ciriminna, R.; Avnir, D.; Pagliaro, M. *Org. Process Res. Dev.* **2007**, *11*, 766.
[157] Hajela, S. P.; Johnson, A. R.; Xu, J.; Sunderland, C. J.; Cohen, S. M.; Caulder, D. L.; Raymond, K. N. *Inorg. Chem.* **2001**, *40*, 3208.
[158] Leanna, M. R.; Sowin, T. J.; Morton, H. E. *Tetrahedron Lett.* **1992**, *33*, 5029.
[159] Montanari, F.; Penso, M.; Quici, S.; Vigano, P. *J. Org. Chem.* **1985**, *50*, 4888.
[160] Neuville, L.; Bois-Choussy, M.; Zhu, J. *Tetrahedron Lett.* **2000**, *41*, 1747.
[161] Heeres, A.; van Doren, H. A.; Gotlieb, K. F.; Bleeker, I. P. *Carbohydr. Res.* **1997**, *299*, 221.
[162] Wolf, E.; Kennedy, I. A.; Himmeldirk, K.; Spenser, I. D. *Can. J. Chem.* **1997**, *75*, 942.
[163] Siedlecka, R.; Skarzewski, J.; Mlochowski, J. *Tetrahedron Lett.* **1990**, *31*, 2177.
[164] Testa, M. L.; Ciriminna, R.; Hajji, C.; Garcia, E. Z.; Ciclosi, M.; Arques, J. S.; Pagliaro, M. *Adv. Synth. Catal.* **2004**, *346*, 655.
[165] Myers, A. G.; Zhong, B.; Movassaghi, M.; Kung, D. W.; Lanman, B. A.; Kwon, S. *Tetrahedron Lett.* **2000**, *41*, 1359.
[166] Ciriminna, R.; Pagliaro, M. *Adv. Synth. Catal.* **2003**, *345*, 383.
[167] Ciriminna, R.; Blum, J.; Avnir, D.; Pagliaro, M. *Chem. Commun.* **2000**, 1441.
[168] Krysan, D. J.; Haight, A. R.; Menzia, J. A.; Welch, N. *Tetrahedron* **1994**, *50*, 6163.
[169] Yu, R. H.; Polniaszek, R. P.; Becker, M. W.; Cook, C. M.; Yu, L. H. L. *Org. Process Res. Dev.* **2007**, *11*, 972.

170 Fritz-Langhals, E. *Org. Process Res. Dev.* **2005**, *9*, 577.
171 Markidis, T.; Kokotos, G. *J. Org. Chem.* **2002**, *67*, 1685.
172 Lin, F.; Peng, W.; Xu, W.; Xiuwen, H.; Yu, B. *Carbohydr. Res.* **2004**, *339*, 1219.
173 Györgydeák, Z.; Thiem, J. *Carbohydr. Res.* **1995**, *268*, 85.
174 Inokuchi, T.; Matsumoto, S.; Nishiyama, T.; Torii, S. *J. Org. Chem.* **1990**, *55*, 462.
175 Melvin, F.; McNeill, A.; Henderson, P. J. F.; Herbert, R. B. *Tetrahedron Lett.* **1999**, *40*, 1201.
176 Rye, C. S.; Withers, S. G. *J. Am. Chem. Soc.* **2002**, *124*, 9756.
177 Hanessian, S.; MacKay, D. B.; Moitessier, N. *J. Med. Chem.* **2001**, *44*, 3074.
178 Cho, N. S.; Park, C. H. *J. Korean Chem. Soc.* **1995**, *39*, 657.
179 Cho, N. S.; Park, C. H. *Bull. Korean Chem. Soc.* **1994**, *15*, 924.
180 Adinolfi, M.; Barone, G.; Iadonisi, A.; Mangoni, L.; Manna, R. *Tetrahedron* **1997**, *53*, 11767.
181 Barnett, C. J.; Grubb, L. M. *Tetrahedron Lett.* **2000**, *41*, 9741.
182 Palomo, C.; Aizpurua, J. M.; Cuevas, C.; Urchegui, R.; Linden, A. *J. Org. Chem.* **1996**, *61*, 4400.
183 Durham, T. B.; Miller, M. J. *J. Org. Chem.* **2003**, *68*, 27.
184 Song, Z. J.; Zhao, M.; Desmond, R.; Devine, P.; Tschaen, D. M.; Tillyer, R.; Frey, L.; Heid, R.; Xu, F.; Foster, B.; Li, J.; Reamer, R.; Volante, R.; Grabowski, E. J. J.; Dolling, U. H.; Reider, P. J.; Okada, S.; Kato, Y.; Mano, E. *J. Org. Chem.* **1999**, *64*, 9658.
185 Zhao, M.; Li, J.; Mano, E.; Song, Z.; Tschaen, D. M.; Grabowski, E. J. J.; Reider, P. J. *J. Org. Chem.* **1999**, *64*, 2564.
186 Shibuya, M.; Sato, T.; Tomizawa, M.; Iwabuchi, Y. *Chem. Commun.* **2009**, 1739.
187 Zanka, A. *Chem. Pharm. Bull.* **2003**, *51*, 888.
188 Sharpless, K. B.; Amberg, W.; Bennani, Y. L.; Crispino, G. A.; Hartung, J.; Jeong, K. S.; Kwong, H. L.; Morikawa, K.; Wang, Z. M.; Xu, D.; Zhang, X. L. *J. Org. Chem.* **1992**, *57*, 2768.
189 Aladro, F. J.; Guerra, F. M.; Moreno-Dorado, F. J.; Bustamante, J. M.; Jorge, Z. D.; Massanet, G. M. *Tetrahedron Lett.* **2000**, *41*, 3209.
190 Inokuchi, T.; Liu, P.; Torii, S. *Chem. Lett.* **1994**, 1411.
191 Yasuda, K.; Ley, S. V. *J. Chem. Soc., Perkin Trans. 1* **2002**, 1024.
192 De Mico, A.; Margarita, R.; Parlanti, L.; Vescovi, A.; Piancatelli, G. *J. Org. Chem.* **1997**, *62*, 6974.
193 Epp, J. B.; Widlanski, T. S. *J. Org. Chem.* **1999**, *64*, 293.
194 Sarabia, F.; Sanchez-Ruiz, A. *J. Org. Chem.* **2005**, *70*, 9514.
195 Banwell, M. G.; Loong, D. T. J. *Heterocycles* **2004**, *62*, 713.
196 Mickel, S. J. Novartis Pharma AG, Basel, Switzerland. Personal communication, 2007.
197 Mickel, S. J.; Sedelmeier, G. H.; Niederer, D.; Schuerch, F.; Seger, M.; Schreiner, K.; Daeffler, R.; Osmani, A.; Bixel, D.; Loiseleur, O.; Cercus, J.; Stettler, H.; Schaer, K.; Gamboni, R.; Bach, A.; Chen, G. P.; Chen, W.; Geng, P.; Lee, G. T.; Loeser, E.; McKenna, J.; Kinder, F. R.; Konigsberger, K.; Prasad, K.; Ramsey, T. M.; Reel, N.; Repic, O.; Rogers, L.; Shieh, W. C.; Wang, R. M.; Waykole, L.; Xue, S.; Florence, G.; Paterson, I. *Org. Process Res. Dev.* **2004**, *8*, 113.
198 Camp, D.; Matthews, C. F.; Neville, S. T.; Rouns, M.; Scott, R. W.; Truong, Y. *Org. Process Res. Dev.* **2006**, *10*, 814.
199 Moroda, A.; Togo, H. *Tetrahedron* **2006**, *62*, 12408.
200 Qian, W.; Jin, E.; Bao, W.; Zhang, Y. *Angew. Chem., Int. Ed.* **2005**, *44*, 952.
201 Qian, W.; Jin, E.; Bao, W.; Zhang, Y. *Tetrahedron* **2006**, *62*, 556.
202 Sakuratani, K.; Togo, H. *Synthesis* **2003**, 21.
203 Herrerias, C. I.; Zhang, T. Y.; Li, C. J. *Tetrahedron Lett.* **2006**, *47*, 13.
204 Tashino, Y.; Togo, H. *Synlett* **2004**, 2010.
205 Vatèle, J. M. *Tetrahedron Lett.* **2006**, *47*, 715.
206 Marotta, E.; Micheloni, L. M.; Scardovi, N.; Righi, P. *Org. Lett.* **2001**, *3*, 727.
207 Pausacker, K. H. *J. Chem. Soc.* **1953**, 107.
208 Vugts, D. J.; Veum, L.; al-Mafraji, K.; Lemmens, R.; Schmitz, R. F.; de Kanter, F. J. J.; Groen, M. B.; Hanefeld, U.; Orru, R. V. A. *Eur. J. Org. Chem.* **2006**, 1672.
209 Sawant, K. B.; Ding, F.; Jennings, M. P. *Tetrahedron Lett.* **2006**, *47*, 939.
210 Monenshein, H.; Sourkouni-Arigirusi, G.; Schubothe, K. M.; Kirschning, A. *Org. Lett.* **1999**, *1*, 2101.

[211] Sourkouni-Argirusi, G.; Kirschning, A. *Org. Lett.* **2000**, *2*, 3781.
[212] Pasetto, P.; Franck, R. W. *J. Org. Chem.* **2003**, *68*, 8042.
[213] Rychnovsky, S. D.; Vaidyanathan, R. *J. Org. Chem.* **1999**, *64*, 310.
[214] Kennedy, R. J.; Stock, A. M. *J. Org. Chem.* **1960**, *25*, 1901.
[215] Bolm, C. Institute of Technology, Aachen, Germany. Personal communication, 2006.
[216] Bolm, C.; Magnus, A. S.; Hildebrand, J. P. *Org. Lett.* **2000**, *2*, 1173.
[217] Bragd, P. L.; Besemer, A. C.; van Bekkum, H. *Carbohydr. Polym.* **2002**, *49*, 397.
[218] De Luca, L.; Giacomelli, G.; Porcheddu, A. *Org. Lett.* **2001**, *3*, 3041.
[219] De Luca, L.; Giacomelli, G.; Masala, S.; Porcheddu, A. *J. Org. Chem.* **2003**, *68*, 4999.
[220] Einhorn, J.; Einhorn, C.; Ratajczak, F.; Pierre, J.-L. *J. Org. Chem.* **1996**, *61*, 7452.
[221] Semmelhack, M. F.; Chou, C. S.; Cortés, D. A. *J. Am. Chem. Soc.* **1983**, *105*, 4492.
[222] Inokuchi, T.; Matsumoto, S.; Nisiyama, T.; Torii, S. *Synlett* **1990**, 57.
[223] Inokuchi, T.; Matsumoto, S.; Torii, S. *J. Org. Chem.* **1991**, *56*, 2416.
[224] Inokuchi, T.; Matsumoto, S.; Fukushima, M.; Torii, S. *Bull. Chem. Soc. Jpn.* **1991**, *64*, 796.
[225] Osa, T.; Kashiwagi, Y.; Bobbitt, J. M.; Ma, Z. Electroorganic Synthesis of Catalyst-Coated Electrodes. In *Electroorganic Synthesis*; Baizer, M. M., Little, R. D., Eds.; CRC Press: Boca Raton, 1991.
[226] Bobbitt, J. M.; Ma, Z.; Bolz, D.; Osa, T.; Kashiwagi, Y.; Yanagisawa, Y.; Kurashima, F.; Anzai, J.; Tacorante-Morales, J. E. In *Supported Reagents and Catalysts in Chemistry*; Royal Society of Chemistry: London, 1998; Vol. 216, pp 200–205.
[227] Kashiwagi, Y.; Chiba, S.; Anzai, J. *New J. Chem.* **2003**, *27*, 1545.
[228] Deronzier, A.; Limosin, D.; Moutet, J.-C. *Electrochim. Acta* **1987**, *32*, 1643.
[229] Siu, T.; Yekta, S.; Yudin, A. K. *J. Am. Chem. Soc.* **2000**, *122*, 11787.
[230] Osa, T.; Akiba, U.; Segawa, I.; Bobbitt, J. M. *Chem. Lett.* **1988**, 1423.
[231] Osa, T.; Kashiwagi, Y.; Mukai, K.; Ohsawa, A.; Bobbitt, J. M. *Chem. Lett.* **1990**, 75.
[232] Kashiwagi, Y.; Ohsawa, A.; Osa, T.; Ma, Z.; Bobbitt, J. M. *Chem. Lett.* **1991**, 581.
[233] Kashiwagi, Y.; Nishimura, T.; Anzai, J. *Electrochim. Acta* **2002**, *47*, 1317.
[234] Yamauchi, Y.; Maeda, H.; Ohmori, H. *Chem. Pharm. Bull.* **1996**, *44*, 1021.
[235] Yamauchi, Y.; Maeda, H.; Ohmori, H. *Chem. Pharm. Bull.* **1997**, *45*, 2024.
[236] Maeda, H.; Saka-iri, Y.; Ogasawara, T.; Huang, C. Z.; Yamauchi, Y.; Ohmori, H. *Chem. Pharm. Bull.* **2001**, *49*, 1349.
[237] Miyazawa, T.; Endo, T. *J. Mol. Catal.* **1985**, *32*, 357.
[238] Miyazawa, T.; Endo, T. *J. Mol. Catal.* **1985**, *31*, 217.
[239] Tsubokawa, N.; Kimoto, T.; Endo, T. *J. Mol. Catal. A: Chem.* **1995**, *101*, 45.
[240] Jiang, N.; Ragauskas, A. J. *Org. Lett.* **2005**, *7*, 3689.
[241] Dijksman, A.; Arends, I. W. C. E.; Sheldon, R. A. *Synlett* **2001**, 102.
[242] Geißlmeir, D.; Jary, W. G.; Falk, H. *Monatsh. Chem.* **2005**, *136*, 1591.
[243] Ragagnin, G.; Betzemeier, B.; Quici, S.; Knochel, P. *Tetrahedron* **2002**, *58*, 3985.
[244] Velusamy, S.; Srinivasan, A.; Punniyamurthy, T. *Tetrahedron Lett.* **2006**, *47*, 923.
[245] Jiang, N.; Ragauskas, A. J. *J. Org. Chem.* **2006**, *71*, 7087.
[246] Wu, X. E.; Ma, L.; Ding, M. X.; Gao, L. X. *Chem. Lett.* **2005**, *34*, 312.
[247] Mannam, S.; Alamsetti, S. K.; Sekar, G. *Adv. Synth. Catal.* **2007**, *349*, 2253.
[248] ten Have, R.; Teunissen, P. J. M. *Chem. Rev.* **2001**, *101*, 3397.
[249] Fabbrini, M.; Galli, C.; Gentili, P.; Macchitella, D. *Tetrahedron Lett.* **2001**, *42*, 7551.
[250] Fabbrini, M.; Galli, C.; Gentili, P. *J. Mol. Catal. B: Enzym.* **2002**, *16*, 231.
[251] Kulys, J.; Vidziunaite, R. *J. Mol. Catal. B: Enzym.* **2005**, *37*, 79.
[252] Arends, I. W. C. E.; Li, Y.-X.; Ausan, R.; Sheldon, R. A. *Tetrahedron* **2006**, *62*, 6659.
[253] d'Acunzo, F.; Baiocco, P.; Fabbrini, M.; Galli, C.; Gentili, P. *Eur. J. Org. Chem.* **2002**, 4195.
[254] Marzorati, M.; Danieli, B.; Haltrich, D.; Riva, S. *Green Chem.* **2005**, *7*, 310.
[255] Yamaguchi, M.; Takata, T.; Endo, T. *Bull. Chem. Soc. Jpn.* **1990**, *63*, 947.
[256] Miller, R. A.; Hoerrner, R. S. *Org. Lett.* **2003**, *5*, 285.
[257] Merbouh, N.; Bobbitt, J. M.; Brückner, C. *J. Carbohydr. Chem.* **2002**, *21*, 65.
[258] Liu, R.; Liang, X.; Dong, C.; Hu, X. *J. Am. Chem. Soc.* **2004**, *126*, 4112.
[259] Liu, R.; Dong, C.; Liang, X.; Wang, X.; Hu, X. *J. Org. Chem.* **2005**, *70*, 729.

[260] Xie, Y.; Mo, W.; Xu, D.; Shen, Z.; N., S.; Hu, B.; Hu, X. *J. Org. Chem.* **2007**, *72*, 4288.
[261] Kim, S. S.; Nehru, K. *Synlett* **2002**, 616.
[262] Lei, M.; Hu, R.-J.; Wang, Y.-G. *Tetrahedron* **2006**, *62*, 8928.
[263] Bobbitt, J. M. *Periodate Oxidation of Carbohydrates*; Wolfrom, M. L., Tipson, R. S., Eds.; Advances in Carbohydrate Chemistry; Wiley: New York, 1956; Vol. 11.
[264] Shibuya, M.; Tomizawa, M.; Iwabuchi, Y. *Org. Lett.* **2008**, in press.
[265] Miyazawa, T.; Endo, T. *J. Polym. Sci., Part A: Polym. Chem.* **1985**, *23*, 2487.
[266] Wang, N.; Liu, R.; Chen, J.; Liang, X. *Chem. Commun.* **2005**, 5322.
[267] Kumar, R. S.; Karthikeyan, K.; Perumal, P. T. *Can. J. Chem.* **2008**, *86*, 720.
[268] Jiang, N.; Ragauskas, A. J. *Tetrahedron Lett.* **2005**, *46*, 3323.
[269] Kim, S. S.; Jung, H. C. *Synthesis* **2003**, 2135.
[270] Minisci, F.; Recupero, F.; Pedulli, G. F.; Lucarini, M. *J. Mol. Catal. A: Chem.* **2003**, *204–205*, 63.
[271] Minisci, F.; Recupero, F.; Cecchetto, A.; Gambarotti, C.; Punta, C.; Faletti, R.; Paganelli, R.; Pedulli, G. F. *Eur. J. Org. Chem.* **2004**, 109.
[272] Benaglia, M.; Puglisi, A.; Holczknecht, O.; Quici, S.; Pozzi, G. *Tetrahedron* **2005**, *61*, 12058.
[273] Gilhespy, M.; Lok, M.; Baucherel, X. *Chem. Commun.* **2005**, 1085.
[274] Ben-Daniel, R.; Alsters, P.; Neumann, R. *J. Org. Chem.* **2001**, *66*, 8650.
[275] Herrmann, W. A.; Zoller, J. P.; Fischer, R. W. *J. Organomet. Chem.* **1999**, *579*, 404.
[276] Zauche, T. H.; Espenson, J. H. *Int. J. Chem. Kinet.* **1999**, *31*, 381.
[277] Volodarsky, L. B.; Raznikov, V. A.; Ovcharenko, V. J. *Synthetic Chemistry of Stable Nitroxides*; CRC Press: Boca Raton, 2000.
[278] Rozantsev, E. G.; Sholle, V. D. *Synthesis* **1971**, 401.
[279] Shibuya, M.; Tomizawa, M.; Suzuki, I.; Iwabuchi, Y. *J. Am. Chem. Soc.* **2006**, *128*, 8412.
[280] Dupeyre, R. M.; Rassat, A. *Tetrahedron* **1978**, *34*, 1901.
[281] Demizu, Y.; Shiigi, H.; Oda, T.; Matsumura, Y.; Onomura, O. *Tetrahedron Lett.* **2008**, *49*, 48.
[282] Kashiwagi, Y.; Anzai, J. *Chem. Pharm. Bull.* **2001**, *49*, 324.
[283] Schnatbaum, K.; Schäfer, H. J. *Synthesis* **1999**, 864.
[284] Suga, T.; Pu, Y. J.; Oyaizu, K.; Nishide, H. *Bull. Chem. Soc. Jpn.* **2004**, *77*, 2203.
[285] Rychnovsky, S. D.; Vaidyanathan, R.; Beauchamp, T.; Lin, R.; Farmer, P. J. *J. Org. Chem.* **1999**, *64*, 6745.
[286] Baur, J. E.; Wang, S.; Brandt, M. C. *Anal. Chem.* **1996**, *68*, 3815.
[287] Abakumov, G. A.; Tikhonov, V. D. *Bull. Acad. Sci. USSR, Div. Chem. Sci.* **1969**, 742.
[288] Moad, G.; Rizzardo, E.; Solomon, D. H. *Tetrahedron Lett.* **1981**, *22*, 1165.
[289] Sümmerman, W.; Deffner, U. *Tetrahedron* **1975**, *31*, 593.
[290] Golubev, V. A.; Voronina, G. N.; Rozantsev, E. G. *Bull. Acad. Sci. USSR, Div. Chem. Sci.* **1972**, 146.
[291] Holczknecht, O.; Cavazzini, M.; Quici, S.; Shepperson, I.; Pozzi, G. *Adv. Synth. Catal.* **2005**, *347*, 677.
[292] Huang, J.-Y.; Li, S.-J.; Wang, Y.-G. *Tetrahedron Lett.* **2006**, *47*, 5637.
[293] Pozzi, G.; Cavazzini, M.; Holczknecht, O.; Quici, S.; Shepperson, I. *Tetrahedron Lett.* **2004**, *45*, 4249.
[294] Kuroboshi, M.; Fujisawa, J.; Tanaka, H. *Electrochemistry* **2004**, *72*, 846.
[295] Wu, X. E.; Ma, L.; Ding, M. X.; Gao, L. X. *Synlett* **2005**, 607.
[296] MacCorquodale, F.; Crayston, J. A.; Walton, J. C.; Worsfold, D. J. *Tetrahedron Lett.* **1990**, *31*, 771.
[297] Tanaka, H.; Kubota, J.; Itogawa, S.; Ido, T.; Kuroboshi, M.; Shimamura, K.; Uchida, T. *Synlett* **2003**, 951.
[298] Tanaka, H.; Kubota, J.; Miyahara, S.; Kuroboshi, M. *Bull. Chem. Soc. Jpn.* **2005**, *78*, 1677.
[299] Dijksman, A.; Arends, I. W. C. E.; Sheldon, R. A. *Chem. Commun.* **2000**, 271.
[300] Gheorghe, A.; Matsumo, A.; Reiser, O. *Adv. Synth. Catal.* **2006**, *348*, 1016.
[301] Tanyeli, C.; Gümüs, A. *Tetrahedron Lett.* **2003**, *44*, 1639.
[302] Kubota, J.; Ido, T.; Kuroboshi, M.; Tanaka, H.; Uchida, T.; Shimamura, K. *Tetrahedron* **2006**, *62*, 4769.
[303] Ferreira, P.; Hayes, W.; Phillips, E.; Rippon, D.; Tsang, S. C. *Green Chem.* **2004**, *6*, 310.

[304] Ferreira, P.; Phillips, E.; Rippon, D.; Tsang, S. C.; Hayes, W. *J. Org. Chem.* **2004**, *69*, 6851.
[305] Merz, A.; Bachmann, H. *J. Am. Chem. Soc.* **1995**, *117*, 901.
[306] Pozzi, G.; Cavazzini, M.; Quici, S.; Benaglia, M.; Dell'Anna, G. *Org. Lett.* **2004**, *6*, 441.
[307] Bolm, C.; Fey, T. *Chem. Commun.* **1999**, 1795.
[308] Fey, T.; Fischer, H.; Bachmann, S.; Albert, K.; Bolm, C. *J. Org. Chem.* **2001**, *66*, 8154.
[309] Brunel, D.; Fajula, F.; Nagy, J. B.; Deroide, B.; Verhoef, M. J.; Veum, L.; Peters, J. A.; van Bekkum, H. *Appl. Catal., A* **2001**, *213*, 73.
[310] Karimi, B.; Biglari, A.; Clark, J. H.; Budarin, V. *Angew. Chem., Int. Ed.* **2007**, *46*, 7210.
[311] Avnir, D. *Acc. Chem. Res.* **1995**, *28*, 328.
[312] Gilhespy, M.; Lok, M.; Baucherel, X. *Catal. Today* **2006**, *117*, 114.
[313] Ciriminna, R.; Bolm, C.; Fey, T.; Pagliaro, M. *Adv. Synth. Catal.* **2002**, *344*, 159.
[314] Bragd, P. L.; Besemer, A. C.; van Bekkum, H. *J. Mol. Catal. A: Chem.* **2001**, *170*, 35.
[315] Davis, N. J.; Flitsch, S. L. *Tetrahedron Lett.* **1993**, *34*, 1181.
[316] Ibert, M.; Marsais, F.; Merbouh, N.; Brückner, C. *Carbohydr. Res.* **2002**, *337*, 1059.
[317] Thaburet, J. F.; Merbouh, N.; Ibert, M.; Marsais, F.; Queguiner, G. *Carbohydr. Res.* **2001**, *330*, 21.
[318] *Carbohydrates as Organic Raw Materials*, Proceedings of the Conference on Progress and Prospects in the Use of Carbohydrates as Organic Raw Materials; Lichtenthaler, F. W., Ed; VCH: Weinheim, Germany, 1991.
[319] Buffet, M. A. J.; Rich, J. R.; McGavin, R. S.; Reimer, K. B. *Carbohydr. Res.* **2004**, *339*, 2507.
[320] Angelin, M.; Hermansson, M.; Dong, H.; Ramstrom, O. *Eur. J. Org. Chem.* **2006**, 4323.
[321] Tidwell, T. T. *Org. React.* **1990**, *39*, 297.
[322] van den Bos, L. J.; Litjens, R. E. J. N.; van den Berg, R. J. B. H. N.; Overkleeft, H. S.; van der Marel, G. A. *Org. Lett.* **2005**, *7*, 2007.
[323] Koga, T.; Taniguchi, I. *Electrochemistry* **2004**, *72*, 858.
[324] Schämann, M.; Schäfer, H. J. *Synlett* **2004**, 1601.
[325] Schämann, M.; Schäfer, H. J. *Eur. J. Org. Chem.* **2003**, 351.
[326] Belgsir, E. M.; Schäfer, H. J. *Electrochem. Commun.* **2001**, *3*, 32.
[327] Brochette-Lemoine, S.; Joannard, D.; Descotes, G.; Bouchu, A.; Queneau, Y. *J. Mol. Catal. A: Chem.* **1999**, *150*, 31.
[328] Brochette-Lemoine, S.; Trombotto, S.; Joannard, D.; Descotes, G.; Bouchu, A.; Queneau, Y. *Ultrason. Sonochem.* **2000**, *7*, 157.
[329] Merbouh, N.; Thaburet, J. F.; Ibert, M.; Marsais, F.; Bobbitt, J. M. *Carbohydr. Res.* **2001**, *336*, 75.
[330] Collins, P.; Ferrier, R. *Monosaccharides: Their Chemistry and Their Roles in Natural Products*; Wiley & Sons: Colchester, U. K., 1995; pp 126–138.
[331] Chang, P. S.; Robyt, J. F. *J. Carbohydr. Chem.* **1996**, *15*, 819.
[332] Crescenzi, V.; Delicato, D.; Dentini, M. *J. Carbohydr. Chem.* **1997**, *16*, 697.
[333] Bragd, P. L.; Besemer, A. C.; van Bekkum, H. *Carbohydr. Res.* **2000**, *328*, 355.
[334] de Nooy, A. E. J.; Besemer, A. C.; van Bekkum, H. *Recl. Trav. Chim. Pays-Bas* **1994**, *113*, 165.
[335] Crescenzi, V.; Hartmann, M.; de Nooy, A. E. J.; Rori, V.; Masci, G.; Skjak-Braek, G. *Biomacromolecules* **2000**, *1*, 360.
[336] Lillo, L. E.; Matsuhiro, B. *Carbohydr. Polym.* **2003**, *51*, 317.
[337] Saito, T.; Isogai, A. *Biomacromolecules* **2004**, *5*, 1983.
[338] Saito, T.; Yanagisawa, M.; Isogai, A. *Cellulose* **2005**, *12*, 305.
[339] Isogai, A.; Kato, Y. *Cellulose* **1998**, *5*, 153.
[340] Tahiri, C.; Vignon, M. R. *Cellulose* **2000**, *7*, 177.
[341] Gomez-Bujedo, S.; Fleury, E.; Vignon, M. R. *Biomacromolecules* **2004**, *5*, 565.
[342] Fraschini, C.; Vignon, M. R. *Carbohydr. Res.* **2000**, *328*, 585.
[343] Kraus, T.; Buděšínský, M.; Závada, J. *Eur. J. Org. Chem.* **2000**, 3133.
[344] Goodnow, R. A.; Richou, A. R.; Tam, S. *Tetrahedron Lett.* **1997**, *38*, 3195.
[345] Epple, R.; Kudirka, R.; Greenberg, W. A. *J. Comb. Chem.* **2003**, *5*, 292.
[346] Semmelhack, M. F.; Schmid, C. R. *J. Am. Chem. Soc.* **1983**, *105*, 6732.
[347] Kashiwagi, Y.; Kurashima, F.; Kikuchi, C.; Anzai, J.; Osa, T.; Bobbitt, J. M. *J. Chin. Chem. Soc.* **1998**, *45*, 135.

[348] Pyne, S. G.; Davis, A. S.; Gates, N. J.; Hartley, J. P.; Lindsay, K. B.; Machan, T.; Tang, M. *Synlett* **2004**, 2670.
[349] Schrake, O.; Rahn, V. S.; Frackenpohl, J.; Braje, W. M.; Hoffmann, H. M. R. *Org. Lett.* **1999**, *1*, 1607.
[350] Golubev, V. A.; Miklyush, R. V. *Zh. Org. Khim.* **1972**, *8*, 1376.
[351] Liu, Y.; Ren, T.; Guo, Q. *Chin. J. Chem.* **1996**, *14*, 252.
[352] Hunter, D. H.; Barton, D. H. R.; Motherwell, W. J. *Tetrahedron Lett.* **1984**, *25*, 603.
[353] Ma, Z. Ph.D. Dissertation, University of Connecticut, Storrs, CT, 1991.
[354] Heinzer, F.; Soukup, M.; Eschenmoser, A. *Helv. Chim. Acta* **1978**, *61*, 2851.
[355] Yoshida, E.; Nakamura, K.; Takata, T.; Endo, T. *J. Polym. Sci., Part A: Polym. Chem.* **1993**, *31*, 1505.
[356] Kleinke, A. S.; Li, C.; Rabasso, N.; Porco, J. A., Jr. *Org. Lett.* **2006**, *8*, 2847.
[357] Mehta, G.; Roy, S. *Tetrahedron Lett.* **2005**, *46*, 7927.
[358] Li, C.; Bardhan, S.; Pace, E. A.; Liang, M.-C.; Gilmore, T. D.; Porco, J. A., Jr. *Org. Lett.* **2002**, *4*, 3267.
[359] Kagiya, T.; Komuro, C.; Sakano, K.; Nishimoto, S. *Chem. Lett.* **1983**, 365.
[360] Koop, B.; Straub, A.; Schäfer, H. J. *Tetrahedron: Asymmetry* **2001**, *12*, 341.
[361] Guo, H.-X.; Liu, Y.-C.; Liu, Z.-L.; Li, C.-L. *Res. Chem. Intermed.* **1992**, *17*, 137.
[362] Ding, Y. B.; Yang, L.; Liu, Z. H.; Liu, Y. C. *J. Chem. Res. (S)* **1994**, 328.
[363] Brózda, D.; Gluszyńska, A.; Koociolowicz, A.; Rozwadowska, M. D. *Tetrahedron: Asymmetry* **2005**, *16*, 953.
[364] van den Bos, L. J.; Codee, J. D. C.; van der Toorn, J. C.; Boltje, T. J.; van Boom, J. H.; Overkleeft, H. S.; van der Marel, G. A. *Org. Lett.* **2004**, *6*, 2165.
[365] Codée, J. D. C.; Stubba, B.; Schiattarella, M.; Overkleeft, H. S.; van Boeckel, C. A. A.; van Boom, J. H.; van der Marel, G. A. *J. Am. Chem. Soc.* **2005**, *127*, 3767.
[366] Siedlecka, R.; Skarzewski, J. *Synlett* **1996**, 757.
[367] Siedlecka, R.; Skarzewski, J. *Synthesis* **1994**, 401.
[368] Maeda, H.; Wu, H. Y.; Yamauchi, Y.; Ohmori, H. *J. Org. Chem.* **2005**, *70*, 8338.
[369] Yoshida, E.; Takata, T.; Endo, T.; Ishizone, T.; Hirao, A.; Nakahama, S. *Chem. Lett.* **1994**, 1827.
[370] Scott, M. E.; Lautens, M. *Org. Lett.* **2005**, *7*, 3045.
[371] Pradhan, P.; Bobbitt, J. M. University of Connecticut, Storrs, CT. Unpublished work, 2007.
[372] d'Acunzo, F.; Baiocco, P.; Galli, C. *New J. Chem.* **2003**, *27*, 329.
[373] Dragutan, I.; Mehlhorn, R. J. *Free Radical Res.* **2007**, *41*, 303.
[374] Luzzio, F. A. *Org. React.* **1998**, *53*, 1.
[375] Taylor, R. J. K.; Reid, M.; Foot, J.; Raw, S. A. *Acc. Chem. Res.* **2005**, *38*, 851.
[376] Fatiadi, A. J. *Synthesis* **1976**, 65.
[377] Fatiadi, A. J. *Synthesis* **1976**, 133.
[378] Blackburn, L.; Wei, X.; Taylor, R. J. K. *Chem. Commun.* **1999**, *1999*, 1337.
[379] Busqúe, F.; Hopkins, S. A.; Konopelski, J. P. *J. Org. Chem.* **2002**, *67*, 6097.
[380] Dess, D. B.; Martin, J. C. *J. Org. Chem.* **1983**, *48*, 4155.
[381] Dess, D. B.; Martin, J. C. *J. Am. Chem. Soc.* **1991**, *113*, 7277.
[382] Ireland, R. E.; Liu, L. *J. Org. Chem.* **1993**, *58*, 2899.
[383] Meyer, S. D.; Schreiber, S. L. *J. Org. Chem.* **1994**, *59*, 7549.
[384] Fieser, M. *Reagents for Organic Synthesis*; Wiley-Interscience: New York, 1994; Vol. 17.
[385] Ansari, I. A.; Gree, R. *Org. Lett.* **2002**, *4*, 1507.
[386] Mori, H.; Ohara, M.; Kwan, T. *Chem. Pharm. Bull.* **1980**, *28*, 3178.
[387] Hashimoto, K.; Sumitomo, H.; Kitao, O. J. *Polym. Sci., Polym. Chem. Ed.* **1975**, *13*, 1257.
[388] Amon, C. M.; Banwell, M. G.; Gravatt, G. L. *J. Org. Chem.* **1987**, *52*, 4851.
[389] Lee, H. H.; Hodgson, P. G.; Bernacki, R. J.; Korytnyk, W.; Sharma, M. *Carbohydr. Res.* **1988**, *176*, 59.
[390] Li, K. C.; Helm, R. F. *Carbohydr. Res.* **1995**, *273*, 249.
[391] Bobbitt, J. M.; Merbouh, N. 4-Acetylamino-2,2,6,6-tetramethylpiperidine-1-oxoammonium tetrafluoroborate, 2003. Encyclopedia of Reagents for Organic Synthesis. http://www.mrw.interscience.wiley.com/eros.

[392] Chung, C. W. Y.; Toy, P. H. *J. Comb. Chem.* **2007**, *9*, 115.

[393] Tanaka, H.; Chou, J.; Mine, M.; Kuroboshi, M. *Bull. Chem. Soc. Jpn.* **2004**, *77*, 1745.

[394] Gancitano, P.; Ciriminna, R.; Testa, M. L.; Fidalgo, A.; Ilharco, L. M.; Pagliaro, M. *Org. Biomol. Chem.* **2005**, *3*, 2389.

[395] Minisci, F.; Recupero, F.; Rodino, M.; Sala, M.; Schneider, A. *Org. Process Res. Dev.* **2003**, *7*, 794.

[396] Kashiwagi, Y.; Kurashima, F.; Anzai, J.; Osa, T. *Heterocycles* **1999**, *51*, 1945.

[397] Ogibin, Y. N.; Khusid, A. K.; Nikishin, G. T. *Bull. Acad. Sci. USSR, Div. Chem. Sci.* **1992**, 945.

[398] Tachihara, T.; Ishizaki, S.; Ishikawa, M.; Kitahara, T. *Chemistry & Biodiversity* **2004**, *1*, 2024.

[399] Tachihara, T.; Ishizaki, S.; Kurobayashi, Y.; Tamura, H.; Ikemoto, Y.; Onuma, A.; Yoshikawa, K.; Yanai, T.; Kitahara, T. *Helv. Chim. Acta* **2003**, *86*, 274.

[400] Tabuchi, H.; Hamamoto, T.; Miki, S.; Tejima, T.; Ichihara, A. *J. Org. Chem.* **1994**, *59*, 4749.

[401] Jauch, J.; Czesla, H.; Schurig, V. *Tetrahedron* **1999**, *55*, 9787.

[402] Hammerschmidt, F.; Wuggenig, F. *Tetrahedron: Asymmetry* **1999**, *10*, 1709.

[403] Tyrrell, E.; Skinner, G. A.; Janes, J.; Milsom, G. *Synlett* **2002**, 1073.

[404] Shustov, G. V.; Sun, F.; Sorensen, T. S.; Rauk, A. *J. Org. Chem.* **1998**, *63*, 661.

[405] Castejón, P.; Moyano, A.; Pericas, M. A.; Riera, A. *Chem.—Eur. J.* **1996**, *2*, 1001.

[406] Gheorghe, A.; Cuevas-Yañez, E.; Horn, J.; Bannwarth, W.; Narsaiah, B.; Reiser, O. *Synlett* **2006**, 2767.

[407] Cecchetto, A.; Fontana, F.; Minisci, F.; Recupero, F. *Tetrahedron Lett.* **2001**, *42*, 6651.

[408] Tilstam, U.; Weinmann, H. *Org. Process Res. Dev.* **2002**, *6*, 384.

[409] Dijksman, A.; Arends, I. W. C. E.; Sheldon, R. A. *Chem. Commun.* **1999**, 1591.

[410] Dijksman, A.; Arends, I. W. C. E.; Sheldon, R. A. *Platinum Met. Rev.* **2001**, *45*, 15.

[411] Inokuchi, T.; Nakagawa, K.; Torii, S. *Tetrahedron Lett.* **1995**, *36*, 3223.

[412] Pozzi, G.; Quici, S.; Shepperson, I. *Tetrahedron Lett.* **2002**, *43*, 6141.

[413] Comesse, S.; Piva, O. *Tetrahedron: Asymmetry* **1999**, *10*, 1061.

[414] Betzemeier, B.; Cavazzini, M.; Quici, S.; Knochel, P. *Tetrahedron Lett.* **2000**, *41*, 4343.

[415] Huang, L. J.; Teumelsan, N.; Huang, X. F. *Chem.—Eur. J.* **2006**, *12*, 5246.

[416] Vatèle, J. M. *Synlett* **2006**, 2055.

[417] de Lemos, E.; Poree, F. H.; Commercon, A.; Betzer, J. F.; Pancrazi, A.; Ardisson, J. *Angew. Chem., Int. Ed.* **2007**, *46*, 1917.

[418] Calter, M. A.; Liao, W.; Struss, J. A. *J. Org. Chem.* **2001**, *66*, 7500.

[419] Yuasa, Y.; Tsuruta, H. *Flavour Fragrance J.* **2004**, *19*, 199.

[420] Zhou, Y. L.; De Huo, C.; Miao, S.; Wu, L. M. *Chin. Chem. Lett.* **2004**, *15*, 801.

[421] Liaigre, D.; Breton, T.; Belgsir, E. M. *Electrochem. Commun.* **2005**, *7*, 312.

[422] Figiel, P. J.; Leskela, M.; Repo, T. *Adv. Synth. Catal.* **2007**, *349*, 1173.

[423] Uber, J. S.; Vogels, Y.; van den Helder, D.; Mutikainen, I.; Turpeinen, U.; Fu, W. T.; Roubeau, O.; Gamez, P.; Reedijk, J. *Eur. J. Inorg. Chem.* **2007**, 4197.

[424] Hollinshead, S. P.; Nichols, J. B.; Wilson, J. W. *J. Org. Chem.* **1994**, *59*, 6703.

[425] Banfi, S.; Montanari, F.; Quici, S. *Gazz. Chim. Ital.* **1990**, *120*, 435.

[426] Bieging, A.; Liao, L. X.; McGrath, D. V. *Chirality* **2002**, *14*, 258.

[427] Habel, L. W.; De Keersmaecker, S.; Wahlen, J.; Jacobs, P. A.; De Vos, D. E. *Tetrahedron Lett.* **2004**, *45*, 4057.

[428] McClure, K. F.; Abramov, Y. A.; Laird, E. R.; Barberia, J. T.; Cai, W.; Carty, T. J.; Cortina, S. R.; Danley, D. E.; Dipesa, A. J.; Donahue, K. M.; Dombroski, M. A.; Elliott, N. C.; Gabel, C. A.; Han, S. G.; Hynes, T. R.; LeMotte, P. K.; Mansour, M. N.; Marr, E. S.; Letavic, M. A.; Pandit, J.; Ripin, D. B.; Sweeney, F. J.; Tan, D.; Tao, Y. *J. Med. Chem.* **2005**, *48*, 5728.

[429] d'Acunzo, F.; Galli, C.; Masci, B. *Eur. J. Biochem.* **2002**, *269*, 5330.

[430] Kokotos, G.; Kotsovolou, S.; Six, D. A.; Constantinou-Kokotou, V.; Beltzner, C. C.; Dennis, E. A. *J. Med. Chem.* **2002**, *45*, 2891.

[431] Gryko, D.; Jurczak, J. *Helv. Chim. Acta* **2000**, *83*, 2705.

[432] Gryko, D.; Urbańczyk-Lipkowska, Z.; Jurczak, J. *Tetrahedron: Asymmetry* **1997**, *8*, 4059.

[433] Gryko, D.; Urbańczyk-Lipkowska, Z.; Jurczak, J. *Tetrahedron* **1997**, *53*, 13373.

[434] Groth, T.; Meldal, M. *J. Comb. Chem.* **2001**, *3*, 34.

[435] Yamaguchi, M.; Takata, T.; Endo, T. *J. Org. Chem.* **1990**, *55*, 1490.
[436] Ciriminna, R.; Palmisano, G.; Della Pina, C.; Rossi, M.; Pagliaro, M. *Tetrahedron Lett.* **2006**, *47*, 6993.
[437] Ciriminna, R.; Pagliaro, M. *Tetrahedron Lett.* **2004**, *45*, 6381.
[438] Rush, J.; Bertozzi, C. R. *Org. Lett.* **2006**, *8*, 131.
[439] Righi, P.; Scardovi, N.; Marotta, E.; ten Holte, P.; Zwanenburg, B. *Org. Lett.* **2002**, *4*, 497.
[440] Magrioti, V.; Antonopoulou, G.; Pantoleon, E.; Kokotos, G. *ARKIVOC* **2002**, 55.
[441] Magrioti, V.; Hadjipavlou-Litina, D.; Constantinou-Kokotou, V. *Bioorg. Med. Chem. Lett.* **2003**, *13*, 375.
[442] Krasiński, A.; Jurczak, J. *Tetrahedron Lett.* **2001**, *42*, 2019.
[443] Philipp, D. M.; Muller, R.; Goddard, W. A.; Abboud, K. A.; Mullins, M. J.; Snelgrove, R. V.; Athey, P. S. *Tetrahedron Lett.* **2004**, *45*, 5441.
[444] Csuk, R.; Thiede, G. *Tetrahedron* **1999**, *55*, 739.
[445] Constantinou-Kokotou, V.; Magrioti, V.; Verger, R. *Chem.—Eur. J.* **2004**, *10*, 1133.
[446] Szczepankiewicz, B. G.; Heathcock, C. H. *Tetrahedron* **1997**, *53*, 8853.
[447] Prusov, E.; Rohm, H.; Maier, M. E. *Org. Lett.* **2006**, *8*, 1025.
[448] Tong, J.; Dang, X. J.; Li, H. L.; Yang, M. *Anal. Lett.* **1997**, *30*, 585.
[449] Harding, M.; Bodkin, J. A.; Hutton, C. A.; McLeod, M. D. *Synlett* **2005**, 2829.
[450] Palian, M. M.; Polt, R. *J. Org. Chem.* **2001**, *66*, 7178.
[451] Yanagisawa, Y.; Kashiwagi, Y.; Kurashima, F.; Anzai, J.; Osa, T.; Bobbitt, J. M. *Chem. Lett.* **1996**, 1043.
[452] Wipf, P.; Takada, T.; Rishel, M. J. *Org. Lett.* **2004**, *6*, 4057.
[453] Prashad, M.; Kim, H. Y.; Lu, Y.; Liu, Y.; Har, D.; Repic, O.; Blacklock, T. J.; Giannousis, P. *J. Org. Chem.* **1999**, *64*, 1750.
[454] Jauch, J. *J. Prakt. Chem.* **2000**, *342*, 100.
[455] Kernag, C. A.; Bobbitt, J. M.; McGrath, D. V. *Tetrahedron Lett.* **1999**, *40*, 1635.
[456] Brzezinski, L. J.; Levy, D. D.; Leahy, J. W. *Tetrahedron Lett.* **1994**, *35*, 7601.
[457] Wipf, P.; Graham, T. H. *J. Am. Chem. Soc.* **2004**, *126*, 15346.
[458] Ageno, G.; Banfi, L.; Cascio, G.; Guanti, G.; Manghisi, E.; Riva, R.; Rocca, V. *Tetrahedron* **1995**, *51*, 8121.
[459] Dettwiler, J. E.; Lubell, W. D. *J. Org. Chem.* **2003**, *68*, 177.
[460] Devel, L.; Hamon, L.; Becker, H.; Thellend, A.; Vidal-Cros, A. *Carbohydr. Res.* **2003**, *338*, 1591.
[461] Paintner, F. F.; Allmendinger, L.; Bauschke, G.; Klemann, P. *Org. Lett.* **2005**, *7*, 1423.
[462] Jurczak, J.; Gryko, D.; Kobrzycka, E.; Gruza, H.; Prokopowicz, P. *Tetrahedron* **1998**, *54*, 6051.
[463] Sundram, H.; Golebiowski, A.; Johnson, C. R. *Tetrahedron Lett.* **1994**, *35*, 6975.
[464] Hanselmann, R.; Zhou, J.; Ma, P.; Confalone, P. N. *J. Org. Chem.* **2003**, *68*, 8739.
[465] Makino, K.; Henmi, Y.; Terasawa, M.; Hara, O.; Hamada, Y. *Tetrahedron Lett.* **2005**, *46*, 555.
[466] Henmi, Y.; Makino, K.; Yoshitomi, Y.; Hara, O.; Hamada, Y. *Tetrahedron: Asymmetry* **2004**, *15*, 3477.
[467] Merino, P.; Franco, S.; Merchan, F. L.; Tejero, T. *J. Org. Chem.* **2000**, *65*, 5575.
[468] Tricotet, T.; Brückner, R. *Eur. J. Org. Chem.* **2007**, 1069.
[469] Zammit, S. C.; White, J. M.; Rizzacasa, M. A. *Org. Biomol. Chem.* **2005**, *3*, 2073.
[470] Noula, C.; Loukas, V.; Kokotos, G. *Synthesis* **2002**, 1735.
[471] Loukas, V.; Markidis, T.; Kokotos, G. *Molecules* **2002**, *7*, 767.
[472] Loukas, V.; Noula, C.; Kokotos, G. *J. Pept. Sci.* **2003**, *9*, 312.
[473] Trost, B. M.; Rudd, M. T. *Org. Lett.* **2003**, *5*, 1467.
[474] Chabaud, P.; Pèpe, G.; Courcambeck, J.; Camplo, M. *Tetrahedron* **2005**, *61*, 3725.
[475] Medou, M.; Priem, G.; Rocheblave, L.; Pepe, G.; Meyer, M.; Chermann, J. C.; Kraus, J. L. *Eur. J. Med. Chem.* **1999**, *34*, 625.
[476] Fettes, A.; Carreira, E. M. *J. Org. Chem.* **2003**, *68*, 9274.
[477] Petrova, K. V.; Jalluri, R. S.; Kozekov, I. D.; Rizzo, C. J. *Chem. Res. Toxicol.* **2007**, *20*, 1685.
[478] Nyavanandi, V. K.; Nadipalli, P.; Nanduri, S.; Naidu, A.; Iqbal, J. *Tetrahedron Lett.* **2007**, *48*, 6905.
[479] Dettwiler, J. E.; Bélec, L.; Lubell, W. D. *Can. J. Chem.* **2005**, *83*, 793.

[480] Li, P. X.; Evans, C. D.; Joullie, M. M. *Org. Lett.* **2005**, *7*, 5325.

[481] Merino, P.; Padar, P.; Delso, I.; Thirumalaikumar, M.; Tejero, T.; Kovacs, L. *Tetrahedron Lett.* **2006**, *47*, 5013.

[482] Leanna, M. R.; DeMattei, J. A.; Li, W.; Nichols, P. J.; Rasmussen, M.; Morton, H. E. *Org. Lett.* **2000**, *2*, 3627.

[483] Gryko, D.; Jurczak, J. *Tetrahedron Lett.* **1997**, *38*, 8275.

[484] Grotenbreg, G. M.; Christina, A. E.; Buizert, A. E. M.; van der Marel, G. A.; Overkleeft, H. S.; Overhand, M. *J. Org. Chem.* **2004**, *69*, 8331.

[485] Raghavan, S.; Joseph, S. C. *Tetrahedron Lett.* **2003**, *44*, 6713.

[486] Hansen, T. M.; Florence, G. J.; Lugo-Mas, P.; Chen, J.; Abrams, J. N.; Forsyth, C. J. *Tetrahedron Lett.* **2003**, *44*, 57.

[487] Kadota, I.; Hu, Y.; Packard, G. K.; Rychnovsky, S. D. *Proc. Natl. Acad. Sci. U.S.A.* **2004**, *101*, 11992.

[488] Li, Y.; Hale, K. J. *Org. Lett.* **2007**, *9*, 1267.

[489] Kovács-Kulyassa, A.; Herczegh, P.; Sztaricskai, F. *Tetrahedron* **1997**, *53*, 13883.

[490] Wachtmeister, J.; Muhlman, A.; Classon, B.; Kvarnstrom, I.; Hallberg, A.; Samuelsson, B. *Tetrahedron* **2000**, *56*, 3219.

[491] Alterman, M.; Bjorsne, M.; Muhlman, A.; Classon, B.; Kvarnstrom, I.; Danielson, H.; Markgren, P. O.; Nillroth, U.; Unge, T.; Hallberg, A.; Samuelsson, B. *J. Med. Chem.* **1998**, *41*, 3782.

[492] Domon, D.; Fujiwara, K.; Ohtaniuchi, Y.; Takezawa, A.; Takeda, S.; Kawasaki, H.; Murai, A.; Kawai, H.; Suzuki, T. *Tetrahedron Lett.* **2005**, *46*, 8279.

[493] Saitoh, F.; Mukaihira, T.; Nishida, H.; Satoh, T.; Okano, A.; Yumiya, Y.; Ohkouchi, M.; Johka, R.; Matsusue, T.; Shiromizu, I.; Hosaka, Y.; Matsumoto, M.; Ohnishi, S. *Chem. Pharm. Bull.* **2006**, *54*, 1535.

[494] Palomo, C.; Oiarbide, M.; Esnal, A. *Chem. Commun.* **1997**, 691.

[495] Palomo, C.; Aizpurua, J. M.; Ganboa, I.; Oiarbide, M. *Pure Appl. Chem.* **2000**, *72*, 1763.

[496] Graziani, A.; Passacantilli, P.; Piancatelli, G.; Tani, S. *Tetrahedron Lett.* **2001**, *42*, 3857.

[497] Hilpert, H.; Wirz, B. *Tetrahedron* **2001**, *57*, 681.

[498] Konopelski, J. P.; Deng, H.; Schiemann, K.; Keane, J. M.; Olmstead, M. M. *Synlett* **1998**, 1105.

[499] Horneff, T.; Herdtweck, E.; Randoll, S.; Bach, T. *Bioorg. Med. Chem.* **2006**, *14*, 6223.

[500] Yuasa, Y.; Kato, Y. *Org. Process Res. Dev.* **2002**, *6*, 628.

[501] Palomo, C.; Aizpurua, J. M.; Urchegui, R.; Garcia, J. M. *J. Chem. Soc., Chem. Commun.* **1995**, 2327.

[502] Bouygues, M.; Medou, M.; Quelever, G.; Chermann, J. C.; Camplo, M.; Kraus, J. L. *Bioorg. Med. Chem. Lett.* **1998**, *8*, 277.

[503] Barilli, A.; Belinghieri, F.; Passarella, D.; Lesma, G.; Riva, S.; Silvani, A.; Danieli, B. *Tetrahedron: Asymmetry* **2004**, *15*, 2921.

[504] Zhang, H.; Mootoo, D. R. *J. Org. Chem.* **1995**, *60*, 8134.

[505] Myers, M. C.; Witschi, M. A.; Larionova, N. V.; Franck, J. M.; Haynes, R. D.; Hara, T.; Grajkowski, A.; Appella, D. H. *Org. Lett.* **2003**, *5*, 2695.

[506] Doan, H. D.; Gallon, J.; Piou, A.; Vatele, J. M. *Synlett* **2007**, 983.

[507] Koseki, Y.; Kusano, S.; Ichi, D.; Yoshida, K.; Nagasaka, T. *Tetrahedron* **2000**, *56*, 8855.

[508] Koseki, Y.; Sato, H.; Watanabe, Y.; Nagasaka, T. *Org. Lett.* **2002**, *4*, 885.

[509] El Oualid, F.; Burm, B. E. A.; Leroy, I. M.; Cohen, L. H.; van Boom, J. H.; van den Elst, H.; Overkleeft, H. S.; van der Marel, G. A.; Overhand, M. *J. Med. Chem.* **2004**, *47*, 3920.

[510] Hanessian, S.; Seid, M.; Nilsson, I. *Tetrahedron Lett.* **2002**, *43*, 1991.

[511] Mulzer, J.; Schulzchen, F.; Bats, J. W. *Tetrahedron* **2000**, *56*, 4289.

[512] Mamai, A.; Madalengoitia, J. S. *Org. Lett.* **2001**, *3*, 561.

[513] Mamai, A.; Hughes, N. E.; Wurthmann, A.; Madalengoitia, J. S. *J. Org. Chem.* **2001**, *66*, 6483.

[514] van Well, R. M.; Meijer, M. E. A.; Overkleeft, H. S.; van Boom, J. H.; van der Marel, G. A.; Overhand, M. *Tetrahedron* **2003**, *59*, 2423.

[515] Reddy, K. L.; Sharpless, K. B. *J. Am. Chem. Soc.* **1998**, *120*, 1207.

[516] Paterson, I.; Delgado, O.; Florence, G. J.; Lyothier, I.; O'Brien, M.; Scott, J. P.; Sereinig, N. *J. Org. Chem.* **2005**, *70*, 150.

[517] Paterson, I.; Delgado, O.; Florence, G. J.; Lyothier, I.; Scott, J. P.; Sereinig, N. *Org. Lett.* **2003**, *5*, 35.
[518] Paterson, I.; Florence, G. J.; Gerlach, K.; Scott, J. P. *Angew. Chem., Int. Ed.* **2000**, *39*, 377.
[519] Paterson, I.; Florence, G. J.; Gerlach, K.; Scott, J. P.; Sereinig, N. *J. Am. Chem. Soc.* **2001**, *123*, 9535.
[520] Bode, J. W.; Carreira, E. M. *J. Org. Chem.* **2001**, *66*, 6410.
[521] Banwell, M. G.; Bray, A. M.; Edwards, A. J.; Wong, D. J. *New J. Chem.* **2001**, *25*, 3.
[522] Rommel, M.; Ernst, A.; Koert, U. *Eur. J. Org. Chem.* **2007**, 4408.
[523] Gil, P.; Razkin, J.; Gonzalez, A. *Synthesis* **1998**, 386.
[524] Varie, D. L.; Brennan, J.; Briggs, B.; Cronin, J. S.; Hay, D. A.; Rieck, J. A.; Zmijewski, M. J. *Tetrahedron Lett.* **1998**, *39*, 8405.
[525] Caron, S.; Vazquez, E. *Org. Process Res. Dev.* **2001**, *5*, 587.
[526] Chida, N.; Tanikawa, T.; Tobe, T.; Ogawa, S. *J. Chem. Soc., Chem. Commun.* **1994**, 1247.
[527] Shoji, M.; Imai, H.; Mukaida, M.; Sakai, K.; Kakeya, H.; Osada, H.; Hayashi, Y. *J. Org. Chem.* **2005**, *70*, 79.
[528] Chiou, W. H.; Schoenfelder, A.; Sun, L.; Mann, A.; Ojima, I. *J. Org. Chem.* **2007**, *72*, 9418.
[529] Kurtz, K. C. M.; Hsung, R. P.; Zhang, Y. *Org. Lett.* **2006**, *8*, 231.
[530] Parrott, M. C.; Marchington, E. B.; Valliant, J. F.; Adronov, A. *J. Am. Chem. Soc.* **2005**, *127*, 12081.
[531] Barnett, C. J.; Grubb, L. M. *Tetrahedron* **2000**, *56*, 9221.
[532] Aiguade, J.; Hao, J. L.; Forsyth, C. J. *Tetrahedron Lett.* **2001**, *42*, 817.
[533] Looper, R. E.; Runnegar, M. T. C.; Williams, R. M. *Tetrahedron* **2006**, *62*, 4549.
[534] Mehta, G.; Islam, K. *Org. Lett.* **2004**, *6*, 807.
[535] Herrera, A. J.; Beneitez, M. T.; Amorim, L.; Canada, F. J.; Jimenez-Barbero, J.; Sinay, P.; Bleriot, Y. *Carbohydr. Res.* **2007**, *342*, 1876.
[536] Hodgson, D. M.; Hachisu, S.; Andrews, M. D. *J. Org. Chem.* **2005**, *70*, 8866.
[537] Hodgson, D. M.; Hachisu, S.; Andrews, M. D. *Synlett* **2005**, 1267.
[538] Razkin, J.; Gonzalez, A.; Gil, P. *Tetrahedron: Asymmetry* **1996**, *7*, 3479.
[539] Bencsik, J. R.; Kercher, T.; O'Sullivan, M.; Josey, J. A. *Org. Lett.* **2003**, *5*, 2727.
[540] Moreno-Dorado, F. J.; Guerra, F. M.; Ortega, M. J.; Zubia, E.; Massanet, G. M. *Tetrahedron: Asymmetry* **2003**, *14*, 503.
[541] Shastin, A. V.; Zakharov, V. V.; Bugaeva, G. P.; Eremenko, L. T.; Romanova, L. B.; Lagodzinskaya, G. V.; Aleksandrov, G. G.; Eremenko, L. L. *Russian Chem. Bull., Int. Ed.* **2006**, *55*, 1304.
[542] Pierce, M. E.; Harris, G. D.; Islam, Q.; Radesca, L. A.; Storace, L.; Waltermire, R. E.; Wat, E.; Jadhav, P. K.; Emmett, G. C. *J. Org. Chem.* **1996**, *61*, 444.
[543] Constantinou-Kokotou, V.; Peristeraki, A.; Kokotos, C. G.; Six, D. A.; Dennis, E. A. *J. Pept. Sci.* **2005**, *11*, 431.
[544] Gingras, K.; Avedissian, H.; Thouin, E.; Boulanger, V.; Essagian, C.; McKerracher, L.; Lubell, W. D. *Bioorg. Med. Chem. Lett.* **2004**, *14*, 4931.
[545] Koskinen, A. M. P.; Helaja, J.; Kumpulainen, E. T. T.; Koivisto, J.; Mansikkamaki, H.; Rissanen, K. *J. Org. Chem.* **2005**, *70*, 6447.
[546] Halab, L.; Bélec, L.; Lubell, W. D. *Tetrahedron* **2001**, *57*, 6439.
[547] Keck, D.; Bräse, S. *Org. Biomol. Chem.* **2006**, *4*, 3574.
[548] Scardovi, N.; Casalini, A.; Peri, F.; Righi, P. *Org. Lett.* **2002**, *4*, 965.
[549] Harre, M.; Nickisch, K.; Schulz, C.; Weinmann, H. *Tetrahedron Lett.* **1998**, *39*, 2555.
[550] Paterson, I.; Blakey, S. B.; Cowden, C. J. *Tetrahedron Lett.* **2002**, *43*, 6005.
[551] Paterson, I.; Lyothier, I. *J. Org. Chem.* **2005**, *70*, 5494.
[552] Paterson, I.; Mackay, A. C. *Synlett* **2004**, 1359.
[553] Kadota, I.; Takamura, H.; Sato, K.; Yamamoto, Y. *J. Org. Chem.* **2002**, *67*, 3494.
[554] Mitsumori, S.; Tsuri, T.; Honma, T.; Hiramatsu, Y.; Okada, T.; Hashizume, H.; Kida, S.; Inagaki, M.; Arimura, A.; Yasui, M.; Asanuma, F.; Kishino, J.; Ohtani, M. *J. Med. Chem.* **2003**, *46*, 2446.
[555] Niggemann, J.; Michaelis, K.; Frank, R.; Zander, N.; Hofle, G. *J. Chem. Soc., Perkin Trans. 1* **2002**, 2490.
[556] Trost, B. M.; Dong, G. B. *Org. Lett.* **2007**, *9*, 2357.

[557] Untersteller, E.; Fritz, B.; Bleriot, Y.; Sinay, P. *C. R. Hebd. Acad. Sci. Paris, Serie C* **1999**, *2*, 429.

[558] Adamo, M. F. A.; Adlington, R. M.; Baldwin, J. E.; Pritchard, G. J.; Rathmell, R. E. *Tetrahedron* **2003**, *59*, 2197.

[559] Williams, D. R.; Kammler, D. C.; Donnell, A. F.; Goundry, W. R. F. *Angew. Chem., Int. Ed.* **2005**, *44*, 6715.

[560] Momán, E.; Nicoletti, D.; Mouriño, A. *J. Org. Chem.* **2004**, *69*, 4615.

[561] Banwell, M. G.; Lupton, D. W.; Willis, A. C. *Aust. J. Chem.* **2005**, *58*, 722.

[562] Corey, E. J.; Lazerwith, S. E. *J. Am. Chem. Soc.* **1998**, *120*, 12777.

[563] Hao, J.; Aiguade, J.; Forsyth, C. J. *Tetrahedron Lett.* **2001**, *42*, 821.

[564] Palombo, E.; Audran, G.; Monti, H. *Tetrahedron* **2005**, *61*, 9545.

[565] Smith, III A. B.; Verhoest, P. R.; Minbiole, K. P.; Schelhaas, M. *J. Am. Chem. Soc.* **2001**, *123*, 4834.

[566] Smith, III A. B.; Minbiole, K. P.; Verhoest, P. R.; Schelhaas, M. *J. Am. Chem. Soc.* **2001**, *123*, 10942.

[567] Beckmann, M.; Meyer, T.; Schulz, F.; Winterfeldt, E. *Chem. Ber.* **1994**, *127*, 2505.

[568] Malkov, A. V.; Pernazza, D.; Bell, M.; Bella, M.; Massa, A.; Teply, F.; Meghani, P.; Kocovsky, P. *J. Org. Chem.* **2003**, *68*, 4727.

[569] Kamal, A.; Sandbhor, M.; Shaik, A. A. *Tetrahedron: Asymmetry* **2003**, *14*, 1575.

[570] Yang, J.; Du, Y.; Huang, R.; Wan, Y.; Wen, Y. *Int. J. Biol. Macromol.* **2005**, *36*, 9.

[571] Jauch, J. *Angew. Chem., Int. Ed.* **2000**, *39*, 2764.

[572] Jauch, J. *Eur. J. Org. Chem.* **2001**, 473.

[573] Jauch, J. *Synlett* **2001**, 87.

[574] Simila, S. T. M.; Reichelt, A.; Martin, S. F. *Tetrahedron Lett.* **2006**, *47*, 2933.

[575] Paterson, I.; Tudge, M. *Tetrahedron* **2003**, *59*, 6833.

[576] Hartung, I. V.; Eggert, U.; Haustedt, L. O.; Niess, B.; Schäfer, P. M.; Hoffmann, H. M. R. *Synthesis* **2003**, 1844.

[577] Shiina, I.; Hashizume, M.; Yamai, Y.; Oshiumi, H.; Shimazaki, T.; Takasuna, Y.; Ibuka, R. *Chem.—Eur. J.* **2005**, *11*, 6601.

[578] Raghavan, S.; Rasheed, M. A. *Tetrahedron* **2004**, *60*, 3059.

[579] Donohoe, T. J.; Chiu, J. Y. K.; Thomas, R. E. *Org. Lett.* **2007**, *9*, 421.

[580] Koseki, Y.; Ozawa, H.; Kitahara, K.; Kato, I.; Sato, H.; Fukaya, H.; Nagasaka, T. *Heterocycles* **2004**, *63*, 17.

[581] Kadota, I.; Takamura, H.; Sato, K.; Yamamoto, Y. *Tetrahedron Lett.* **2001**, *42*, 4729.

[582] Clark, J. S.; Grainger, D. M.; Ehkirch, A. A. C.; Blake, A. J.; Wilson, C. *Org. Lett.* **2007**, *9*, 1033.

[583] Simila, S. T. M.; Martin, S. F. *J. Org. Chem.* **2007**, *72*, 5342.

[584] Wiesner, K.; Sanchez, I. H.; Atwal, K. S.; Lee, S. F. *Can. J. Chem.* **1977**, *55*, 1091.

[585] Palomo, C.; Oiarbide, M.; Landa, A. *J. Org. Chem.* **2000**, *65*, 41.

[586] Ohyabu, N.; Nishikawa, T.; Isobe, M. *J. Am. Chem. Soc.* **2003**, *125*, 8798.

[587] Krasiński, A.; Gruza, H.; Jurczak, J. *Heterocycles* **2001**, *54*, 581.

[588] Hanessian, S.; Papeo, G.; Angiolini, M.; Fettis, K.; Beretta, M.; Munro, A. *J. Org. Chem.* **2003**, *68*, 7204.

[589] Jägel, J.; Schmauder, A.; Binanzer, M.; Maier, M. E. *Tetrahedron* **2007**, *63*, 13006.

[590] Gutiérrez, M. C.; Sleegers, A.; Simpson, H. D.; Alphand, V.; Furstoss, R. *Org. Biomol. Chem.* **2003**, *1*, 3500.

[591] Gulyas, P. T.; Langford, S. J.; Lokan, N. R.; Ranasinghe, M. G.; Paddon-Row, M. N. *J. Org. Chem.* **1997**, *62*, 3038.

[592] Miyaoka, H.; Yamanishi, M.; Hoshino, A.; Kinbara, A. *Tetrahedron* **2006**, *62*, 4103.

[593] Becker, M. H.; Chua, P.; Downham, R.; Douglas, C. J.; Garg, N. K.; Hiebert, S.; Jaroch, S.; Matsuoka, R. T.; Middleton, J. A.; Ng, F. W.; Overman, L. E. *J. Am. Chem. Soc.* **2007**, *129*, 11987.

[594] Wender, P. A.; Mayweg, A. V. W.; VanDeusen, C. L. *Org. Lett.* **2003**, *5*, 277.

[595] Banwell, M. G.; Dupuche, J. R.; Gable, R. W. *Aust. J. Chem.* **1996**, *49*, 639.

[596] Palomo, C.; Ganboa, I.; Cuevas, C.; Boschetti, C.; Linden, A. *Tetrahedron Lett.* **1997**, *38*, 4643.

[597] Schaus, S. E.; Branalt, J.; Jacobsen, E. N. *J. Org. Chem.* **1998**, *63*, 4876.

[598] Migawa, M. T.; Risen, L. M.; Griffey, R. H.; Swayze, E. E. *Org. Lett.* **2005**, *7*, 3429.
[599] Mehta, G.; Roy, S. *Chem. Commun.* **2005**, 3210.
[600] Mak, C. C.; Brik, A.; Lerner, D. L.; Elder, J. H.; Morris, G. M.; Olson, A. J.; Wong, C. H. *Bioorg. Med. Chem.* **2003**, *11*, 2025.
[601] Janey, J. M.; Hsiao, Y.; Armstrong, J. D., III *J. Org. Chem.* **2006**, *71*, 390.
[602] Smith, III A. B.; Simov, V. *Org. Lett.* **2006**, *8*, 3315.
[603] Vlieghe, P.; Bihel, F.; Clerc, T.; Pannecouque, C.; Witvrouw, M.; De Clercq, E.; Salles, J. P.; Chermann, J. C.; Kraus, J. L. *J. Med. Chem.* **2001**, *44*, 777.
[604] Banwell, M. G.; Coster, M. J.; Karunaratne, O. P.; Smith, J. A. *J. Chem. Soc., Perkin Trans. 1* **2002**, 1622.
[605] Wilkinson, K. L.; Elsey, G. M.; Prager, R. H.; Tanaka, T.; Sefton, M. A. *Tetrahedron* **2004**, *60*, 6091.
[606] Miyaji, K.; Ohara, Y.; Miyauchi, Y.; Tsuruda, T.; Arai, K. *Tetrahedron Lett.* **1993**, *34*, 5597.
[607] Schrader, T. O.; Snapper, M. L. *J. Am. Chem. Soc.* **2002**, *124*, 10998.
[608] Nicolaou, K. C.; Jennings, M. P.; Dagneau, P. *Chem. Commun.* **2002**, 2480.
[609] Faure, S.; Piva, O. *Tetrahedron Lett.* **2001**, *42*, 255.
[610] Paterson, I.; Ashton, K.; Britton, R.; Cecere, G.; Chouraqui, G.; Florence, G. J.; Stafford, J. *Angew. Chem., Int. Ed.* **2007**, *46*, 6167.
[611] Burns, N. Z.; Baran, P. S. *Angew. Chem., Int. Ed.* **2008**, *47*, 205.
[612] Urban, F. J.; Anderson, B. G.; Orrill, S. L.; Daniels, P. J. *Org. Process Res. Dev.* **2001**, *5*, 575.
[613] Sarabia, F.; Sanchez-Ruiz, A.; Martin-Ortiz, L.; Garcia-Castro, M.; Chammaa, S. *Org. Lett.* **2007**, *9*, 5091.
[614] Kamikawa, K.; Tachibana, A.; Sugimoto, S.; Uemura, M. *Org. Lett.* **2001**, *3*, 2033.
[615] Raunkjaer, M.; Pedersen, D. S.; Elsey, G. M.; Sefton, M. A.; Skouroumounis, G. K. *Tetrahedron Lett.* **2001**, *42*, 8717.
[616] Smith, III A. B.; Davulcu, A. H.; Cho, Y. S.; Ohmoto, K.; Kurti, L.; Ishiyama, H. *J. Org. Chem.* **2007**, *72*, 4596.
[617] Dieks, H.; Senge, M. O.; Kirste, B.; Kurreck, H. *J. Org. Chem.* **1997**, *62*, 8666.
[618] Klimko, P. G.; Davis, T. L.; Griffin, B. W.; Sharif, N. A. *J. Med. Chem.* **2000**, *43*, 3400.
[619] Ireland, R. E.; Gleason, J. L.; Gegnas, L. D.; Highsmith, T. K. *J. Org. Chem.* **1996**, *61*, 6856.
[620] Hulme, A. N.; Howells, G. E. *Tetrahedron Lett.* **1997**, *38*, 8245.
[621] Gurjar, M. K.; Karumudi, B.; Ramana, C. V. *J. Org. Chem.* **2005**, *70*, 9658.
[622] Paterson, I.; Ashton, K.; Britton, R.; Cecere, G.; Chouraqui, G.; Florence, G. J.; Knust, H.; Stafford, J. *Chem. Asian J.* **2008**, *3*, 367.
[623] Kakuta, D.; Hitotsuyanagi, Y.; Matsuura, N.; Fukaya, H.; Takeya, K. *Tetrahedron* **2003**, *59*, 7779.
[624] Manabe, S.; Marui, Y.; Ito, Y. *Chem.—Eur. J.* **2003**, *9*, 1435.
[625] Manabe, S.; Ito, Y. *J. Am. Chem. Soc.* **1999**, *121*, 9754.
[626] Razavi, H.; Polt, R. *Tetrahedron Lett.* **1998**, *39*, 3371.
[627] Wovkulich, P. M.; Shankaran, K.; Kiegiel, J.; Uskokovic, M. R. *J. Org. Chem.* **1993**, *58*, 832.
[628] Larrosa, I.; Da Silva, M. I.; Gomez, P. M.; Hannen, P.; Ko, E.; Lenger, S. R.; Linke, S. R.; White, A. J. P.; Wilton, D.; Barrett, A. G. M. *J. Am. Chem. Soc.* **2006**, *128*, 14042.
[629] Kang, G. D.; Howard, P. W.; Thurston, D. E. *Chem. Commun.* **2003**, 1688.
[630] Schulze, B.; Dabrowska, P.; Boland, W. *ChemBioChem* **2007**, *8*, 208.
[631] Custar, D. W.; Zabawa, T. P.; Scheidt, K. A. *J. Am. Chem. Soc.* **2008**, *130*, 804.
[632] Polt, R.; Sames, D.; Chruma, J. *J. Org. Chem.* **1999**, *64*, 6147.
[633] Bandur, N. G.; Harms, K.; Koert, U. *Synthesis* **2007**, 2720.
[634] Noe, M. C.; Snow, S. L.; Wolf-Gouveia, L. A.; Mitchell, P. G.; Lopresti-Morrow, L.; Reeves, L. M.; Yocum, S. A.; Liras, J. L.; Vaughn, M. *Bioorg. Med. Chem. Lett.* **2004**, *14*, 4727.
[635] Paterson, I.; Gardner, N. M.; Poullennec, K. G.; Wright, A. E. *Bioorg. Med. Chem. Lett.* **2007**, *17*, 2443.
[636] Hu, T.; Takenaka, N.; Panek, J. S. *J. Am. Chem. Soc.* **2002**, *124*, 12806.
[637] Lin, S.; Dudley, G. B.; Tan, D. S.; Danishefsky, S. J. *Angew. Chem., Int. Ed.* **2002**, *41*, 2188.
[638] Mandal, M.; Yun, H.; Dudley, G. B.; Lin, S.; Tan, D. S.; Danishefsky, S. J. *J. Org. Chem.* **2005**, *70*, 10619.

[639] Kinney, W. A.; Zhang, X.; Williams, J. I.; Johnston, S.; Michalak, R. S.; Deshpande, M.; Dostal, L.; Rosazza, J. P. N. *Org. Lett.* **2000**, *2*, 2921.
[640] Paterson, I.; Florence, G. J.; Heimann, A. C.; Mackay, A. C. *Angew. Chem., Int. Ed.* **2005**, *44*, 1130.
[641] Paterson, I.; Britton, R.; Delgado, O.; Meyer, A.; Poullennec, K. G. *Angew. Chem., Int. Ed.* **2004**, *43*, 4629.
[642] Paterson, I.; Luckhurst, C. A. *Tetrahedron Lett.* **2003**, *44*, 3749.
[643] Zhang, H.; Seepersaud, M.; Seepersaud, S.; Mootoo, D. R. *J. Org. Chem.* **1998**, *63*, 2049.
[644] Muratake, H.; Natsume, M.; Nakai, H. *Tetrahedron* **2006**, *62*, 7093.
[645] Narayan, R. S.; Borhan, B. *J. Org. Chem.* **2006**, *71*, 1416.
[646] Paterson, I.; Delgado, O. *Tetrahedron Lett.* **2003**, *44*, 8877.
[647] Siu, T.; Cox, C. D.; Danishefsky, S. J. *Angew. Chem., Int. Ed.* **2003**, *42*, 5629.
[648] Francavilla, C.; Chen, W.; Kinder Jr., F. R. *Org. Lett.* **2003**, *5*, 1233.
[649] Fürstner, A.; Aissa, C.; Chevrier, C.; Teply, F.; Nevado, C.; Tremblay, M. *Angew. Chem., Int. Ed.* **2006**, *45*, 5832.
[650] Sawayama, A. M.; Tanaka, H.; Wandless, T. J. *J. Org. Chem.* **2004**, *69*, 8810.
[651] Mukhopadhyay, S.; Maitra, U. *Org. Lett.* **2004**, *6*, 31.
[652] Hoye, T. R.; Hu, M. *J. Am. Chem. Soc.* **2003**, *125*, 9576.
[653] Barriga, S.; Fuertes, P.; Marcos, C. F.; Rakitin, O. A.; Torroba, T. *J. Org. Chem.* **2002**, *67*, 6439.
[654] Benowitz, A. B.; Fidanze, S.; Small, P. L. C.; Kishi, Y. *J. Am. Chem. Soc.* **2001**, *123*, 5128.
[655] Vong, B. G.; Kim, S. H.; Abraham, S.; Theodorakis, E. A. *Angew. Chem., Int. Ed.* **2004**, *43*, 3947.
[656] Khripach, V. A.; Zhabinskii, V. N.; Konstantinova, O. V.; Khripach, N. B.; Antonchick, A. V.; Antonchick, A. P.; Schneider, B. *Steroids* **2005**, *70*, 551.
[657] Sinz, C. J.; Rychnovsky, S. D. *Angew. Chem., Int. Ed.* **2001**, *40*, 3224.
[658] Sinz, C. J.; Rychnovsky, S. D. *Tetrahedron* **2002**, *58*, 6561.
[659] Patel, A.; Lindhorst, T. K. *Carbohydr. Res.* **2006**, *341*, 1657.
[660] Csuk, R.; Schmuck, K.; Schäfer, R. *Tetrahedron Lett.* **2006**, *47*, 8769.
[661] Kurosawa, K.; Nagase, T.; Chida, N. *Chem. Commun.* **2002**, 1280.
[662] Kurosawa, K.; Matsuura, K.; Nagase, T.; Chida, N. *Bull. Chem. Soc. Jpn.* **2006**, *79*, 921.
[663] Kotsovolou, S.; Chiou, A.; Verger, R.; Kokotos, G. *J. Org. Chem.* **2001**, *66*, 962.
[664] Masson, C.; Scherman, D.; Bessodes, M. *J. Polym. Sci., Part A: Polym. Chem.* **2001**, *39*, 4022.
[665] Oikawa, M.; Ikoma, M.; Sasaki, M. *Tetrahedron Lett.* **2004**, *45*, 2371.
[666] Lohof, E.; Planker, E.; Mang, C.; Burkhart, F.; Dechantsreiter, M. A.; Haubner, R.; Wester, H. J.; Schwaiger, M.; Holzemann, G.; Goodman, S. L.; Kessler, H. *Angew. Chem., Int. Ed.* **2000**, *39*, 2761.
[667] Becher, J.; Seidel, I.; Plass, W.; Klemm, D. *Tetrahedron* **2006**, *62*, 5675.
[668] Ying, L.; Gervay-Hague, J. *Carbohydr. Res.* **2004**, *339*, 367.
[669] Boulineau, F. P.; Wei, A. *J. Org. Chem.* **2004**, *69*, 3391.
[670] Yu, H. N.; Furukawa, J.; Ikeda, T.; Wong, C. H. *Org. Lett.* **2004**, *6*, 723.
[671] Muller, C.; Kitas, E.; Wessel, H. P. *J. Chem. Soc., Chem. Commun.* **1995**, 2425.
[672] Mann, M. C.; Thomson, R. J.; von Itzstein, M. *Bioorg. Med. Chem. Lett.* **2004**, *14*, 5555.
[673] Rye, C. S.; Withers, S. G. *J. Org. Chem.* **2002**, *67*, 4505.
[674] Sofia, M. J.; Hunter, R.; Chan, T. Y.; Vaughan, A.; Dulina, R.; Wang, H.; Gange, D. *J. Org. Chem.* **1998**, *63*, 2802.
[675] Karst, N.; Jacquinet, J.-C. *Eur. J. Org. Chem.* **2002**, 815.
[676] Codée, J. D. C.; van der Marel, G. A.; van Boeckel, C. A. A.; van Boom, J. H. *Eur. J. Org. Chem.* **2002**, 3954.
[677] Kyas, A.; Feigel, M. *Helv. Chim. Acta* **2005**, *88*, 2375.
[678] Lefeber, D. J.; Kamerling, J. P.; Vliegenthart, J. F. G. *Chem.—Eur. J.* **2001**, *7*, 4411.
[679] Desai, R. N.; Blackwell, L. F. *Synlett* **2003**, 1981.
[680] Barbier, M.; Breton, T.; Servat, K.; Grand, E.; Kokoh, B.; Kovensky, J. *J. Carbohydr. Chem.* **2006**, *25*, 253.
[681] Milkereit, G.; Morr, M.; Thiem, J.; Vill, V. *Chem. Phys. Lipids* **2004**, *127*, 47.
[682] Bouktaib, M.; Atmani, A.; Rolando, C. *Tetrahedron Lett.* **2002**, *43*, 6263.

[683] Sillence, D. J.; Raggers, R. J.; Neville, D. C. A.; Harvey, D. J.; van Meer, G. *J. Lipid Res.* **2000**, *41*, 1252.
[684] Patching, S. G.; Brough, A. R.; Herbert, R. B.; Rajakarier, J. A.; Henderson, P. J. F.; Middleton, D. A. *J. Am. Chem. Soc.* **2004**, *126*, 3072.
[685] Patel, A.; Lindhorst, T. K. *J. Org. Chem.* **2001**, *66*, 2674.
[686] Weiss, S. I.; Sieverling, N.; Niclasen, M.; Maucksch, C.; Thunemann, A. F.; Mohwald, H.; Reinhardt, D.; Rosenecker, J.; Rudolph, C. *Biomaterials* **2006**, *27*, 2302.
[687] Baisch, G.; Öhrlein, R. *Carbohydr. Res.* **1998**, *312*, 61.
[688] Adorjan, I.; Jaaskelainen, A. S.; Vuorinen, T. *Carbohydr. Res.* **2006**, *341*, 2439.
[689] Magaud, D.; Grandjean, C.; Doutheau, A.; Anker, D.; Shevchik, V.; Cotte-Pattat, N.; Robert-Baudouy, J. *Carbohydr. Res.* **1998**, *314*, 189.
[690] Hansen, H. C.; Magnusson, G. *Carbohydr. Res.* **1998**, *307*, 233.
[691] Prabhu, A.; Venot, A.; Boons, G. J. *Org. Lett.* **2003**, *5*, 4975.
[692] Chauvin, A. L.; Nepogodiev, S. A.; Field, R. A. *J. Org. Chem.* **2005**, *70*, 960.
[693] Adamski-Werner, S. L.; Yeung, B. K. S.; Miller-Deist, A.; Petillo, P. A. *Carbohydr. Res.* **2004**, *339*, 1255.
[694] Koshida, S.; Suda, Y.; Sobel, M.; Ormsby, J.; Kusumoto, S. *Bioorg. Med. Chem. Lett.* **1999**, *9*, 3127.
[695] Barroca, N.; Jacquinet, J. C. *Carbohydr. Res.* **2002**, *337*, 673.
[696] Attolino, E.; Catelani, G.; D'Andrea, F.; Puccioni, L. *Carbohydr. Res.* **2002**, *337*, 991.
[697] Garegg, P. J.; Oscarson, S.; Tedebark, U. *J. Carbohydr. Chem.* **1998**, *17*, 587.
[698] Kuszmann, J.; Medgyes, G.; Boros, S. *Carbohydr. Res.* **2004**, *339*, 1569.
[699] Gouy, M. H.; Danel, M.; Gayral, M.; Bouchu, A.; Queneau, Y. *Carbohydr. Res.* **2007**, *342*, 2303.
[700] Jiang, Z. H.; Xu, R. S.; Wilson, C.; Brenk, A. *Tetrahedron Lett.* **2007**, *48*, 2915.
[701] Haag, T.; Hughes, R. A.; Ritter, G.; Schmidt, R. R. *Eur. J. Org. Chem.* **2007**, 6016.
[702] Das, S. K.; Mallet, J. M.; Esnault, J.; Driguez, P. A.; Duchaussoy, P.; Sizun, P.; Herault, J. P.; Herbert, J. M.; Petitou, M.; Sinay, P. *Chem.—Eur. J.* **2001**, *7*, 4821.
[703] Fekete, A.; Gyergyoi, K.; Kover, K. E.; Bajza, I.; Liptak, A. *Carbohydr. Res.* **2006**, *341*, 1312.
[704] Knapp, S.; Gore, V. K. *Org. Lett.* **2000**, *2*, 1391.
[705] Munier, P.; Giudicelli, M. B.; Picq, D.; Anker, D. *J. Carbohydr. Chem.* **1996**, *15*, 739.
[706] Cai, L.; Li, Q.; Ren, B.; Yang, Z. J.; Zhang, L. R.; Zhang, L. H. *Tetrahedron* **2007**, *63*, 8135.
[707] Pigro, M. C.; Angiuoni, G.; Piancatelli, G. *Tetrahedron* **2002**, *58*, 5459.
[708] Nomura, M.; Sato, T.; Washinosu, M.; Tanaka, M.; Asao, T.; Shuto, S.; Matsuda, A. *Tetrahedron* **2002**, *58*, 1279.
[709] Davis, N. J.; Flitsch, S. L. *J. Chem. Soc., Perkin Trans. 1* **1994**, 359.
[710] Medgyes, A.; Farkas, E.; Liptak, A.; Pozsgay, V. *Tetrahedron* **1997**, *53*, 4159.
[711] Koshida, S.; Suda, Y.; Fukui, Y.; Ormsby, J.; Sobel, M.; Kusumoto, S. *Tetrahedron Lett.* **1999**, *40*, 5725.
[712] Rommel, M.; Ernst, A.; Harms, K.; Koert, U. *Synlett* **2006**, 1067.
[713] Haubner, R.; Kuhnast, B.; Mang, C.; Weber, W. A.; Kessler, H.; Wester, H. J.; Schwaiger, M. *Bioconjugate Chem.* **2004**, *15*, 61.
[714] Abu Ajaj, K.; Hennig, L.; Findeisen, M.; Giesa, S.; Muller, D.; Welzel, P. *Tetrahedron* **2002**, *58*, 8439.
[715] Lichtenthaler, F. W.; Nakamura, K.; Klotz, J. *Angew. Chem., Int. Ed.* **2003**, *42*, 5838.
[716] Lichtenthaler, F. W.; Klotz, J.; Nakamura, K. *Tetrahedron: Asymmetry* **2003**, *14*, 3973.
[717] Overkleeft, H. S.; Verhelst, S. H. L.; Pieterman, E.; Meeuwenoord, N. J.; Overhand, M.; Cohen, L. H.; van der Marel, G. A.; van Boom, J. H. *Tetrahedron Lett.* **1999**, *40*, 4103.
[718] Witczak, Z. J.; Mielguj, R. *Synlett* **1996**, 108.
[719] Locardi, E.; Stockle, M.; Gruner, S.; Kessler, H. *J. Am. Chem. Soc.* **2001**, *123*, 8189.
[720] Timmers, C. M.; Dekker, M.; Buijsman, R. C.; van der Marel, G. A.; Ethell, B.; Anderson, G.; Burchell, B.; Mulder, G. J.; van Boom, J. H. *Bioorg. Med. Chem. Lett.* **1997**, *7*, 1501.
[721] Wong, M. F.; Weiss, K. L.; Curley, R. W. *J. Carbohydr. Chem.* **1996**, *15*, 763.
[722] Walker, J. R.; Alshafie, G.; Nieves, N.; Ahrens, J.; Clagett-Dame, M.; Abou-Issa, H.; Curley, R. W. *Bioorg. Med. Chem.* **2006**, *14*, 3038.

[723] Risseeuw, M. D. P.; Grotenbreg, G. M.; Witte, M. D.; Tuin, A. W.; Leeuwenburgh, M. A.; Van der Marl, G. A.; Overkleeft, H. S.; Overhand, M. *Eur. J. Org. Chem.* **2006**, 3877.
[724] Raunkjaer, M.; El Oualid, F.; van der Marel, G. A.; Overkleeft, H. S.; Overhand, M. *Org. Lett.* **2004**, *6*, 3167.
[725] Yeung, B. K. S.; Hill, D. C.; Janicka, M.; Petillo, P. A. *Org. Lett.* **2000**, *2*, 1279.
[726] Mandal, S.; Mukhopadhyay, B. *Tetrahedron* **2007**, *63*, 11363.
[727] Soderman, P.; Widmalm, G. *Eur. J. Org. Chem.* **2001**, 3453.
[728] Sarkar, S. K.; Mukhopadhyay, B.; Roy, N. *Indian J. Chem., Sect. B:* **2005**, *44*, 1058.
[729] McGavin, R. S.; Gagne, R. A.; Chervenak, M. C.; Bundle, D. R. *Org. Biomol. Chem.* **2005**, *3*, 2723.
[730] Dunlap, C. A.; Côté, G. L.; Momany, F. A. *Carbohydr. Res.* **2003**, *338*, 2367.
[731] Maruyama, M.; Takeda, T.; Shimizu, N.; Hada, N.; Yamada, H. *Carbohydr. Res.* **2000**, *325*, 83.
[732] Mukherjee, C.; Misra, A. K. *Glycoconjugate J.* **2008**, *25*, 111.
[733] Kraus, T.; Buděšínský, M.; Závada, J. *J. Org. Chem.* **2001**, *66*, 4595.
[734] Xu, R. S.; Jiang, Z. H. *Carbohydr. Res.* **2008**, *343*, 7.
[735] Guillerm, G.; Muzard, M.; Glapski, C. *Bioorg. Med. Chem. Lett.* **2004**, *14*, 5799.
[736] Middleton, R. J.; Briddon, S. J.; Cordeaux, Y.; Yates, A. S.; Dale, C. L.; George, M. W.; Baker, J. G.; Hill, S. J.; Kellam, B. *J. Med. Chem.* **2007**, *50*, 782.
[737] Rodenko, B.; Detz, R. J.; Pinas, V. A.; Lambertucci, C.; Brun, R.; Wanner, M. J.; Koomen, G. J. *Bioorg. Med. Chem.* **2006**, *14*, 1618.
[738] Jagtap, P. G.; Chen, Z.; Szabo, C.; Klotz, K. N. *Bioorg. Med. Chem. Lett.* **2004**, *14*, 1495.
[739] Mackman, R. L.; Zhang, L. J.; Prasad, V.; Boojamra, C. G.; Douglas, J.; Grant, D.; Hui, H.; Kim, C. U.; Laflamme, G.; Parrish, J.; Stoycheva, A. D.; Swaminathan, S.; Wang, K. Y.; Cihlar, T. *Bioorg. Med. Chem.* **2007**, *15*, 5519.
[740] Jung, M. E.; Toyota, A. *J. Org. Chem.* **2001**, *66*, 2624.
[741] Montevecchi, P. C.; Manetto, A.; Navacchia, M. L.; Chatgilialoglu, C. *Tetrahedron* **2004**, *60*, 4303.
[742] Threlfall, R.; Davies, A.; Howarth, N.; Cosstick, R. *Nucleosides, Nucleotides, Nucleic Acids* **2006**, *26*, 611.
[743] Gogoi, K.; Kumar, V. A. *Chem. Commun.* **2008**, 706.
[744] Ashton, T. D.; Scammells, P. J. *Aust. J. Chem.* **2008**, *61*, 49.
[745] Xu, Q.; Katkevica, D.; Rozners, E. *J. Org. Chem.* **2006**, *71*, 5906.
[746] Rozners, E.; Xu, Q. *Org. Lett.* **2003**, *5*, 3999.
[747] Cappellacci, L.; Franchetti, P.; Pasqualini, M.; Petrelli, R.; Vita, P.; Lavecchia, A.; Novellino, E.; Costa, B.; Martini, C.; Klotz, K. N.; Grifantini, M. *J. Med. Chem.* **2005**, *48*, 1550.
[748] Zhu, R.; Frazier, C. R.; Linden, J.; Macdonald, T. L. *Bioorg. Med. Chem. Lett.* **2006**, *16*, 2416.
[749] Raunkjaer, M.; Sorensen, M. D.; Wengel, J. *Org. Biomol. Chem.* **2005**, *3*, 130.
[750] Bae, S.; Lakshman, M. K. *Org. Lett.* **2008**, *10*, 2203.
[751] Elzein, E.; Kalla, R.; Li, X. F.; Perry, T.; Marquart, T.; Micklatcher, M.; Li, Y.; Wu, Y. Z.; Zeng, D. W.; Zablocki, J. *Bioorg. Med. Chem. Lett.* **2007**, *17*, 161.
[752] Alila, S.; Boufi, S.; Belgacem, M. N.; Beneventi, D. *Langmuir* **2005**, *21*, 8106.
[753] Shibata, I.; Yanagisawa, M.; Saito, T.; Isogai, A. *Cellulose* **2006**, *13*, 73.
[754] Konno, N.; Habu, N.; Maeda, I.; Azuma, N.; Isogai, A. *Carbohydr. Polym.* **2006**, *64*, 589.
[755] Habibi, Y.; Vignon, M. R. *Cellulose* **2008**, *15*, 177.
[756] Montanari, S.; Roumani, M.; Heux, L.; Vignon, M. R. *Macromolecules* **2005**, *38*, 1665.
[757] Perez, D. D.; Montanari, S.; Vignon, M. R. *Biomacromolecules* **2003**, *4*, 1417.
[758] Law, K.; Daneault, C.; Guimond, R. *J. Pulp Pap. Sci.* **2007**, *33*, 138.
[759] Saito, T.; Okita, Y.; Nge, T. T.; Sugiyama, J.; Isogai, A. *Carbohydr. Polym.* **2006**, *65*, 435.
[760] Saito, T.; Nishiyama, Y.; Putaux, J. L.; Vignon, M.; Isogai, A. *Biomacromolecules* **2006**, *7*, 1687.
[761] Muzzarelli, R. A. A.; Muzzarelli, C.; Cosani, A.; Terbojevich, M. *Carbohydr. Polym.* **1999**, *39*, 361.
[762] Muzzarelli, R. A. A.; Miliani, M.; Cartolari, M.; Tarsi, R.; Tosi, G.; Muzzarelli, C. *Carbohydr. Polym.* **2000**, *43*, 55.

[763] Yoo, S. H.; Lee, J. S.; Park, S. Y.; Kim, Y. S.; Chang, P. S.; Lee, H. G. *Int. J. Biol. Macromol.* **2005**, *35*, 27.

[764] Suh, D. S.; Chang, P. S.; Kim, K. O. *Cereal Chem.* **2002**, *79*, 576.

[765] Sierakowski, M. R.; Milas, M.; Desbrieres, J.; Rinaudo, M. *Carbohydr. Polym.* **2000**, *42*, 51.

[766] de Nooy, A. E. J.; Besemer, A. C.; van Bekkum, H.; van Dijk, J.; Smit, J. A. M. *Macromolecules* **1996**, *29*, 6541.

[767] Krabi, A.; Stuart, M. A. C. *Macromolecules* **1998**, *31*, 1285.

[768] Paris, E.; Stuart, M. A. C. *Macromolecules* **1999**, *32*, 462.

[769] Rundlof, T.; Widmalm, G. *Anal. Biochem.* **1996**, *243*, 228.

[770] Dŭrana, R.; Lacik, I.; Paulovičová, E.; Bystrický, S. *Carbohydr. Polym.* **2006**, *63*, 72.

[771] Takeda, T.; Miller, J. G.; Fry, S. C. *Planta* **2008**, *227*, 893.

CHAPTER 3

ASYMMETRIC EPOXIDATION OF ELECTRON-DEFICIENT ALKENES

Michael J. Porter

Department of Chemistry, University College London, 20 Gordon Street, London WC1H 0AJ, UK

John Skidmore

Neurosciences Centre of Excellence for Drug Discovery, GlaxoSmithKline, New Frontiers Science Park, Third Avenue, Harlow, Essex CM19 5AW, UK

CONTENTS

---

m.j.porter@ucl.ac.uk

Organic Reactions, Vol. 74, Edited by Scott E. Denmark et al.

INTRODUCTION

Asymmetric epoxidation reactions have the distinction of being among the first enantioselective transformations to be widely used in organic synthesis. The Sharpless asymmetric epoxidation is arguably one of the most important methods for the synthesis of enantiomerically enriched intermediates used en route to a wide range of synthetic targets.[1] To date, most examples of asymmetric epoxidation have employed electrophilic oxidizing agents and are thus applicable primarily to electron-neutral, or electron-rich double bonds.[2,3]

On the other hand, alkenes substituted with electron-withdrawing groups often react inefficiently with electrophilic oxidizing agents; such alkenes are more readily epoxidized using nucleophilic oxidants. The classical nucleophilic epoxidation reaction is the Weitz-Scheffer epoxidation of α,β-unsaturated ketones to the corresponding α,β-epoxy ketones using basic hydrogen peroxide.[4] Similar oxidations can be carried out using hypochlorite ion[5] or basic TBHP (*tert*-butylhydroperoxide).[6]

This chapter considers the more general case of the conversion of an electron-deficient alkene **1** into the corresponding epoxide **2** in an enantioselective fashion (Eq. 1; EWG = electron-withdrawing group).

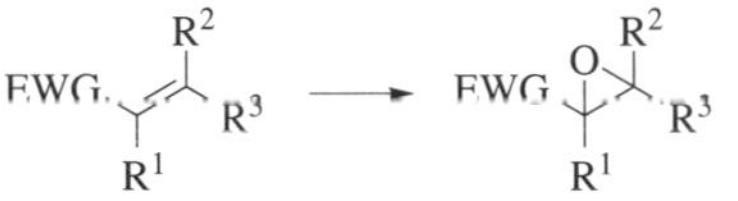

1 **2** (non-racemic) (Eq. 1)

Only methods based on chiral reagents or catalysts are covered by this chapter; thus simple diastereoselective epoxidations, including those where the chiral element is in the form of a chiral auxiliary, are not considered. Also omitted are Sharpless epoxidations of α-(hydroxymethyl)enones,[7–9] enzymatic epoxidations observed in the course of biosynthetic studies,[10] and the enzymatic epoxidation of vinyl phosphonic acids.[11,12]

Several reviews on the asymmetric epoxidation of electron-deficient alkenes have previously been published;[13–15] some more general reviews of asymmetric oxidation reactions also include sections on this topic.[16,17] Where specific methods have been reviewed elsewhere, these reviews are cited in the relevant sections. The present review covers the literature to the end of 2005.

MECHANISM AND STEREOCHEMISTRY

Introduction

Non-Enantioselective Epoxidation. The nucleophilic epoxidation of alkenes is generally a two-step process. Initial attack of the oxidant **3** on the alkene **1** generates a stabilized carbanion **4**. In the second step, the adduct **4** undergoes a displacement at oxygen to generate the epoxide **2** (Eq. 2).

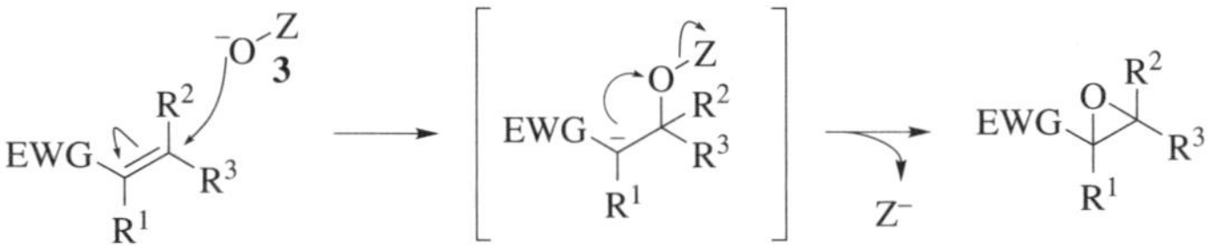

1 **4** **2** (Eq. 2)

Support for this mechanism, in the case where Z = OH, comes from kinetic studies.[18] The epoxidation reaction of mesityl oxide is found to be first-order with respect to both the enone and hydroperoxide ion (Eq. 3). This result indicates that both species are involved in the rate-determining step, but does not indicate whether the reaction proceeds in two steps (as in Eq. 2), or by a concerted mechanism.

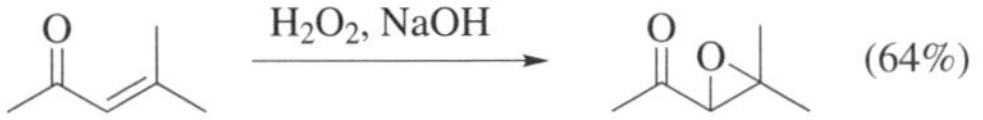

(Eq. 3)

Similar studies on the epoxidation of an unsaturated dinitrile with hypochlorite indicate that the rate-determining step in this case is nucleophilic attack of the hypochlorite ion.[5]

Evidence for a two-step mechanism in nucleophilic epoxidation comes from a study of the stereoselectivity of the reaction. In most cases, the Weitz-Scheffer epoxidation is stereoselective but not stereospecific—both geometric isomers of the starting enone give rise to the same epoxide product (i.e., stereoconvergent). For example, both (*E*)-chalcone (**5**) and (*Z*)-chalcone (**7**) give rise to the same *trans*-epoxide **6** (Eq. 4).[19,20]

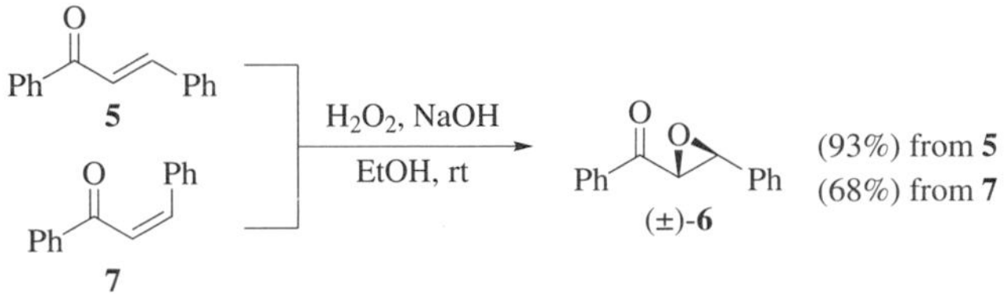

(Eq. 4)

In this example, the observed stereoselectivity could be due to base-catalyzed epimerization of the epoxide products.[21] However, epoxidation of α-substituted enones with basic hydrogen peroxide also occurs in stereoselective fashion. Thus, both geometric isomers of 3-methylpent-3-en-2-one (**8** and **10**) yield the same epoxide **9** (Eq. 5).[21] The lack of an acidic α-proton precludes epimerization of the reaction product.

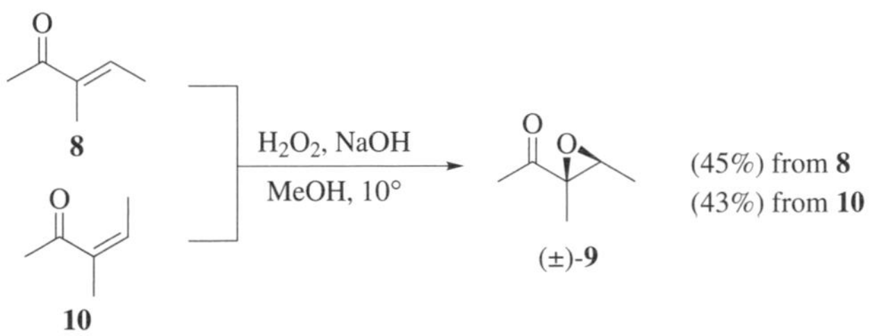

(Eq. 5)

Analogous epoxidations of other classes of electron-deficient alkenes show similar stereoselectivity. Thus the reaction of either (*E*)- or (*Z*)-phenyl styryl sulfone with basic hydrogen peroxide gives the corresponding trans-epoxide;[22] a comparable result is observed for the epoxidation of the *E*- and *Z*-isomers of 2-nitro-1-phenylprop-1-ene.[23]

In contrast to the Weitz-Scheffer reaction, epoxidation with hypochlorite tends to be stereospecific. Treatment of the (*E*)-benzylideneflavanone **11** with hypochlorite gives the epoxides **12** and **13** in which the ketone and phenyl groups are trans on the epoxide ring, whereas the *Z*-isomer **14** yields the epoxide isomers **15** and **16** in which these groups are cis (Eqs. 6 and 7).[24] By contrast, on epoxidation with basic hydrogen peroxide, **11** and **14** both give the trans-substituted epoxides **12** and **13**, in 3:1 and 7:1 ratios respectively.

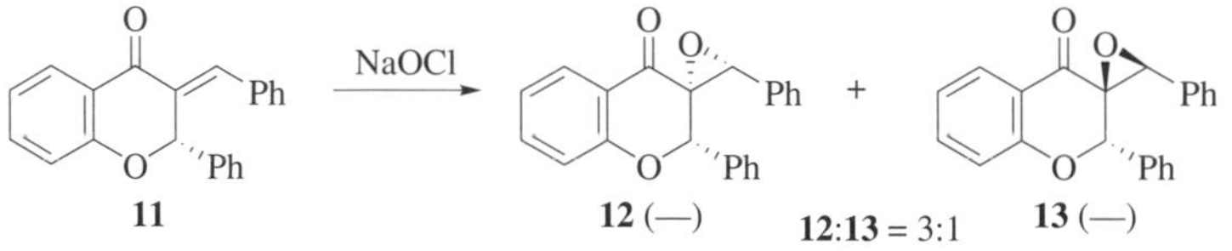

(Eq. 6)

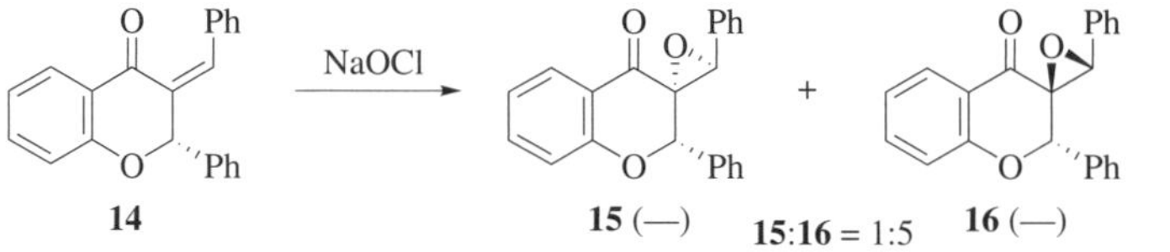

(Eq. 7)

Likewise, epoxidation of (*Z*)-styryl tolyl sulfone with potassium hypochlorite gives the cis-disubstituted epoxide product,[25] whereas epoxidation with basic hydrogen peroxide affords the *trans*-epoxide.[22]

A mechanism accounting for all of these observations is outlined below in Eq. 8.[21] Addition of the nucleophilic oxidant to the alkene isomers **1** and **19** gives different conformations, **17** and **20** respectively, of an anionic intermediate. Cyclization of these intermediates leads to epoxide isomers **18** and **21**. However, intermediates **17** and **20** interconvert by rotation about the central C–C single bond, and the stereospecificity of the reaction is determined by the relative rates of ring-closure and C–C bond rotation. If formation of the three-membered ring is fast compared to the interconversion of **17** and **20**, the reaction is stereospecific, with alkene **1** leading to epoxide **18**, and alkene **19** giving epoxide **21**. Conversely, if the ring-closure is slower than the interconversion of **17** and **20**, the product ratio will be determined simply by the relative rates of the two ring-closure processes; thus both alkenes, **1** and **19**, will give rise to the same ratio of epoxide products.

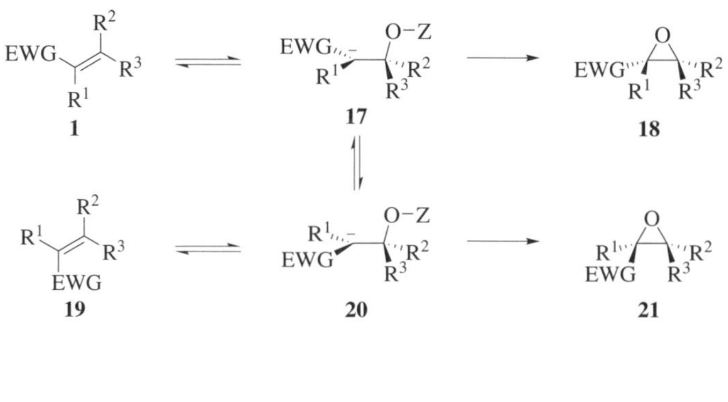

(Eq. 8)

In general, the stereospecificity of the epoxidation reaction is found to increase as (a) the nucleofugality of Z increases (e.g. stereospecificity is higher for Z = Cl than for Z = OH); and (b) the ability of the electron-withdrawing group(s) to stabilize the negative charge decreases.[26] The role of hyperconjugation in determining the stereospecificity has been studied computationally through calculation of the rotational barrier between anions **17** and **20** for a range of substrates.[26]

Enantioselective Epoxidation. Several different strategies for the asymmetric nucleophilic epoxidation of alkenes have been reported. The following methods have all been adopted with some degree of success:

1. Epoxidation using a stoichiometric chiral oxidant.
2. Epoxidation in the presence of a chiral base.
3. Epoxidation in the presence of a metal counterion coordinated by a chiral ligand.
4. Epoxidation in the presence of a chiral phase-transfer catalyst.
5. Epoxidation in the presence of polypeptides.
6. Miscellaneous epoxidation methods.

Within each category there are significant mechanistic similarities and thus the methods are grouped accordingly in the following discussion.

Epoxidation Using a Stoichiometric Chiral Oxidant

Four types of chiral alkyl hydroperoxides have been utilized as nucleophilic alkene oxidants. Enzymatically-resolved (*S*)-1-phenylethyl hydroperoxide (**22**) and its *p*-chloro analogue have been utilized for the epoxidation of acyclic *E*-enones[27,28] and isoflavones;[29] sugar-derived hydroperoxides such as **23** have been used as reagents for the asymmetric epoxidation of quinones;[30] hydroperoxide **24**, based on the TADDOL scaffold,[31] and the (+)-norcamphor-derived reagent **25**[32] have been used as oxidants for a variety of electron-deficient alkenes.

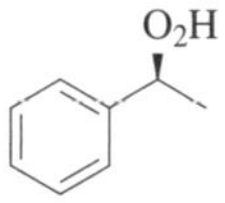

22

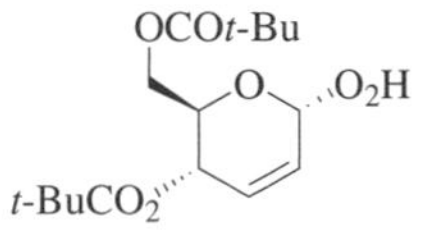

23

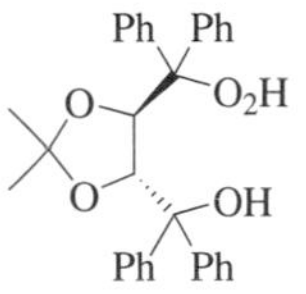

24
TADOOH

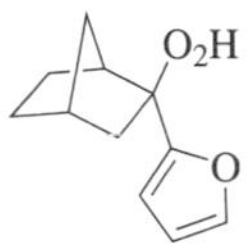

25

Treatment of an acyclic *E*-enone with (*S*)-1-phenylethyl hydroperoxide (**22**) and KOH in acetonitrile leads to preferential formation of the (2*S*,3*R*)-epoxide whereas use of DBU (1,8-diazabicyclo[5.4.0]undec-7-ene) in toluene with the same chiral hydroperoxide tends to give the enantiomeric product.[27,28]

Insight into this stereochemical dichotomy is obtained by varying the size of the β-substituent of the enone. For the KOH-mediated reaction, an increase in size brings about an increase in stereoselectivity; conversely, for the reaction with DBU as base, increasing the size of the β-substituent lowers the facial selectivity of epoxidation, to the extent that in the epoxidation of 4,4-dimethyl-1-phenylpent-2-en-1-one, the (2*S*,3*R*)-product is the major one.

A unified mechanism has been proposed that accounts for these disparate observations.[28] A template effect, in which both the carbonyl oxygen of the enone and the incoming distal oxygen of the peroxide are coordinated to the templating cation (T^+), is deemed to be responsible for the stereochemical induction. Two diastereomeric transition structures, depicted below, can be envisaged. When the templating cation is the small potassium ion, steric interactions with this cation are negligible, and the key interaction is that between the benzylic substituents of the oxidant and the β-substituent (R) of the enone. Due to a steric interaction between this substituent and the benzylic methyl group of the oxidant, transition structure **II** is disfavored, and the (2*S*,3*R*)-epoxide, formed through transition structure **I**, is obtained. As R becomes more sterically demanding, the energy difference between the two transition states increases, and greater stereoselectivity is seen.

By contrast, when the templating cation is a protonated DBU molecule, steric interactions between this moiety and the oxidant become important. For small R groups, the effect of the large template dominates, disfavoring transition structure **I**, and leading to the 2*R*,3*S*-isomer of the product. However, as the size of R increases, transition structure **II** becomes increasingly disfavored, so the stereoselectivity drops and is eventually reversed.[28]

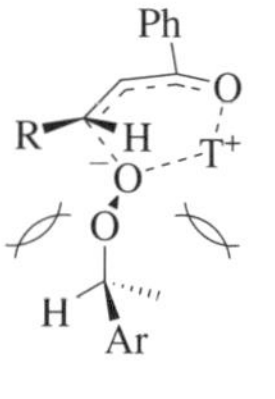

Transition structure **I**
(favored when $T^+ = K^+$)

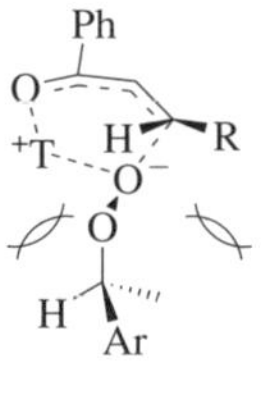

Transition structure **II**
(favored when $T^+ = DBUH^+$)

A series of sugar-derived hydroperoxides has been investigated, under a variety of conditions, as chiral oxidants for the epoxidation of an amido quinone.[30] The pivaloylated species **23** is the most effective oxidant, giving the product in a moderate enantiomeric excess (ee) of 64%. Although some of the oxidants examined generate the enantiomeric epoxide, none is as selective as **23**. Sugar-derived hydroperoxides have also been investigated as chiral oxidants for 2-methyl-1,4-naphthoquinone (**26**).[33,34] Using the most effective oxidants reported in this study, either enantiomer of the corresponding epoxide can be generated in just under 50% ee (Eq. 9).

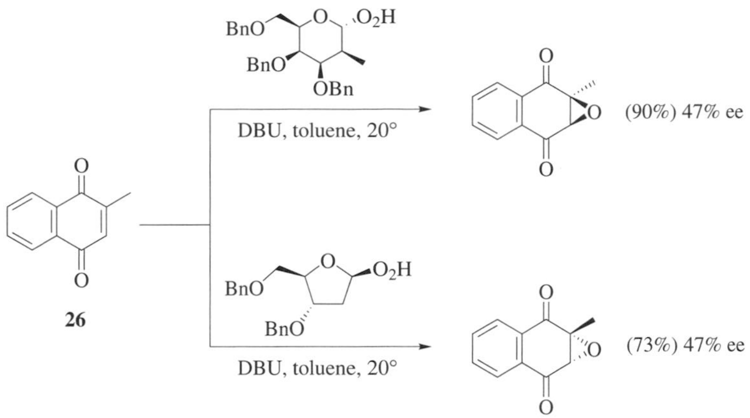

(Eq. 9)

The chiral oxidant TADOOH (**24**) is prepared in a two-step process from the corresponding diol.[31] Two reaction protocols have been developed; the epoxidation can either be performed using *n*-butyllithium, or using an organic base such as DBU or DBN (1,5-diazabicyclo[4.3.0]non-5-ene) in the presence of LiCl. TADOOH (**24**) is capable of effecting a variety of enantioselective oxidation reactions, including the epoxidation of enones. For example, epoxidation of chalcone affords the epoxide (2*S*,3*R*)-**27** in good yield and excellent enantiomeric excess (Eq. 10).[31]

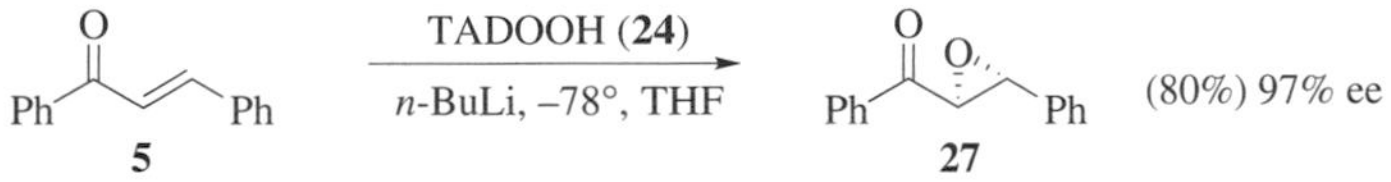

(Eq. 10)

The epoxidation of chalcones by the (+)-norcamphor-derived hydroperoxide **25** has been investigated under a variety of conditions. The highest enantioselectivites are obtained using *n* butyllithium as a base in THF. Moderate improvements in enantioselectivity are obtained when 12-crown-4(1,4,7,10-tetraoxacyclododecane) is added to the chalcone epoxidation.[32]

Epoxidation in the Presence of a Chiral Base

In some cases moderate enantioselectivities and yields are obtained with chiral guanidinium bases; for example chalcone provides epoxy ketone **27** in 53% ee (Eq. 11).[35]

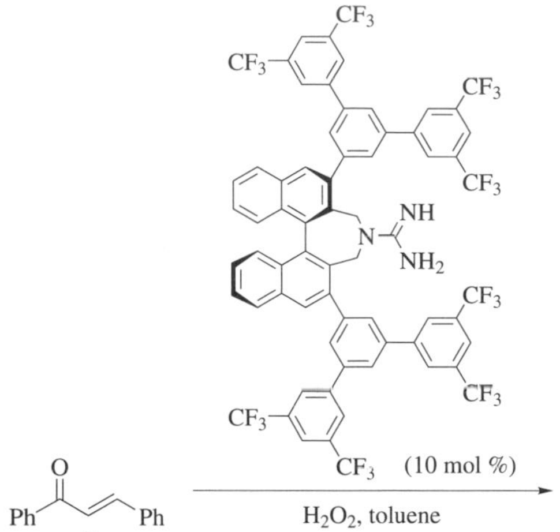

(Eq. 11)

5

27

(48%) 53% ee

A variety of cyclic and acyclic guanidines have also been investigated as chiral bases for the asymmetric epoxidation of the quinone monoketal **28**, giving enantioselectivities up to 60% ee (Eq. 12).[36–39]

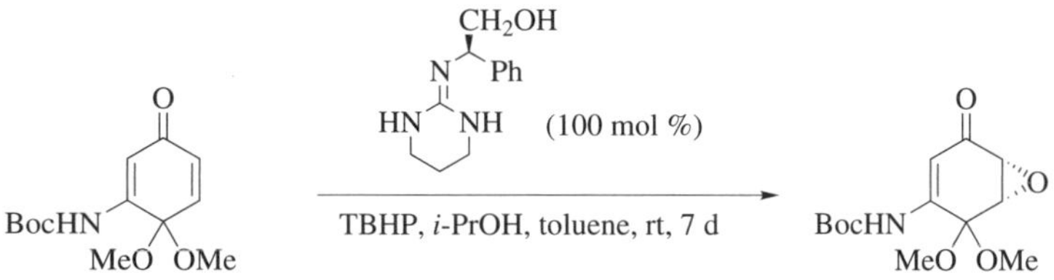

(Eq. 12)

28

(34%), 60% ee

Epoxidation in the Presence of a Metal Counterion Coordinated by a Chiral Ligand

A commonly adopted approach to asymmetric nucleophilic epoxidation is the use of a metal peroxide species modified by a chiral ligand. Four such general approaches have been developed, using, respectively, zinc, a rare-earth metal, a metal of group 1 or 2, or platinum.

Zinc-Mediated Methods. Three classes of chiral alcohol ligands have been developed for the diethylzinc-mediated epoxidation of enones. *N*-Methylpseudoephedrine (**29**), when treated with diethylzinc followed by oxygen gas, forms an oxidant that converts enones into the corresponding epoxides with moderate to good enantioselectivity (Eq. 13).[40,41] The same reagent system has been used for the epoxidation of nitroalkenes.[42] For both classes of substrates, the best results are obtained when more than two equivalents of ligand and diethylzinc are employed.

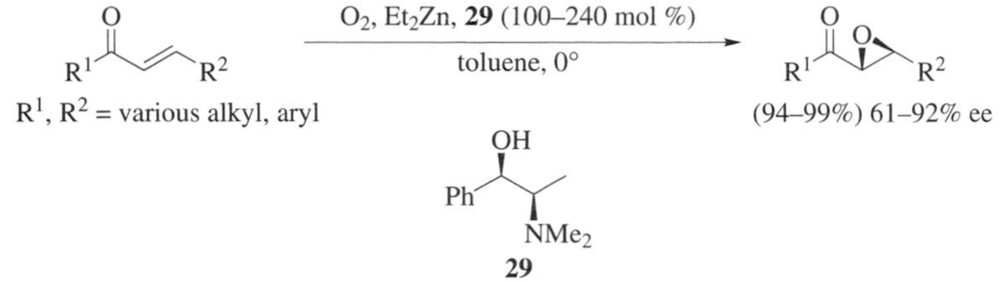
O2, Et2Zn, 29 (100–240 mol %)
toluene, 0°
R1, R2 = various alkyl, aryl
(94–99%) 61–92% ee
OH
Ph
NMe2
29

(Eq. 13)

A related epoxidation system utilizing a binaphthol-containing polymer **30** as the chiral element has been developed. Under conditions similar to those used for ligand **29**, ligand **30** directs the epoxidation of chalcone to afford epoxide (2*S*,3*R*)-**27** in 41% yield and 71% enantiomeric excess (Eq. 14). Again, stoichiometric quantities of ligand and diethylzinc are required.[43]

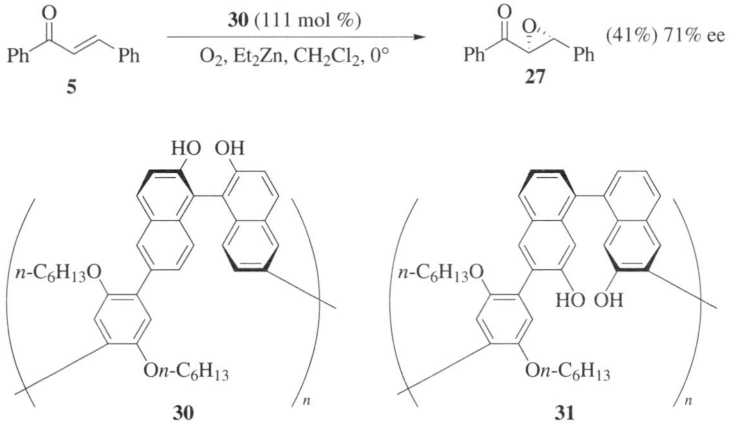
30 (111 mol %)
O2, Et2Zn, CH2Cl2, 0°
(41%) 71% ee
5
27
HO OH
n-C6H13O
On-C6H13
30
31

(Eq. 14)

The mechanism proposed for these two epoxidation systems is depicted in Eq. 15.[41,43] When the chiral alcohol is mixed with diethylzinc, evolution of a gas, presumed to be ethane, is observed, suggesting that the initial step is formation of an ethylzinc alkoxide **32**. On introduction of an oxygen atmosphere, measurement of gas uptake indicates that one equivalent of O_2 is absorbed; thus it is postulated that the ethylperoxyzinc species **33** is formed. Conjugate addition of this zinc peroxide to the enone **34** leads to a β-ethylperoxyzinc enolate **35**, and subsequent formation of epoxide **36** is accompanied by formation of a zinc dialkoxide.

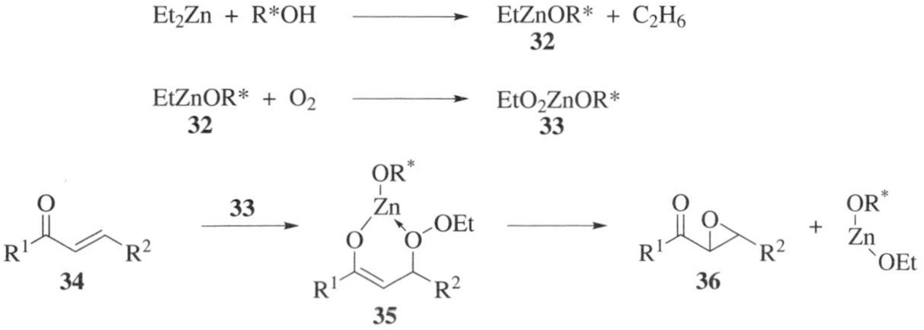
Et2Zn + R*OH → EtZnOR* + C2H6
32
EtZnOR* + O2 → EtO2ZnOR*
32
33
34
35
36
OR*
Zn
OEt

(Eq. 15)

The point at which the configuration is determined is the conjugate addition of ethylperoxyzinc species **33** to the enone; transition structure models accounting for the observed stereochemistries have been put forward for both *N*-methylpseudoephedrine (**29**) (Fig. 1)[41] and poly(binaphthol) ligand **30** (Fig. 2).[43]

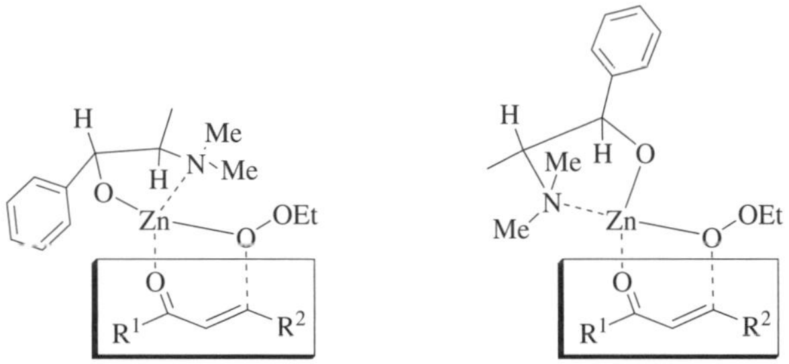

Figure 1. Proposed transition structure models for *N*-methylpseudoephedrine (**29**).

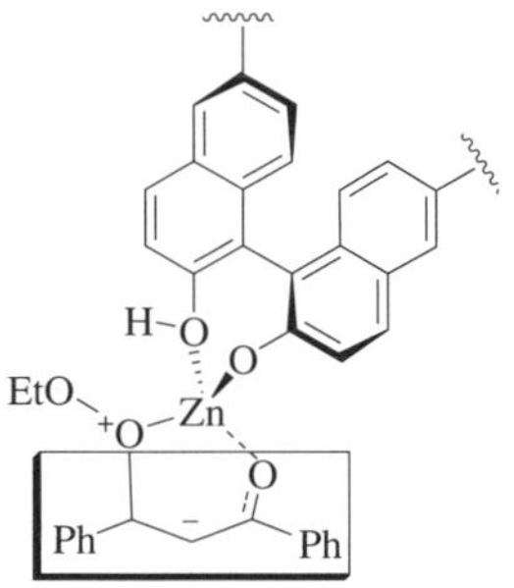

Figure 2. Proposed transition structure models for poly(binaphthol) ligand **30**.

In addition to these stoichiometric systems, various monomeric and polymeric binaphthol ligands have been developed that give reasonable levels of asymmetric induction when used in substoichiometric quantities in combination with TBHP as a stoichiometric oxidant.[43−45] Thus a range of enones may be epoxidized using TBHP as the stoichiometric oxidant, in the presence of polymeric catalyst **31** (20 mol % based on the binaphthol unit) and 36 mol % diethylzinc.[43] Doubling the quantities of ligand and diethylzinc leads to a marginal improvement in yield and enantioselectivity. Similar epoxidations can be carried out using BINOL (1,1′-binaphthalene-2,2′-diol) as the chiral ligand; enantioselectivities of 50–80% ee are obtained for a range of enone substrates.[44]

The mechanism for this catalytic epoxidation is proposed to be similar to that described for the stoichiometric system, as outlined below (Eq. 16).[43] The first step is again formation of a zinc alkoxide **32** through reaction of diethylzinc with the chiral alcohol. Compound **32** then reacts with TBHP to afford the *tert*-butylperoxyzinc species **37**, which effects the epoxidation of **34** to **36** as before. The by-product of the epoxidation is the dialkoxide **38**, which undergoes ligand exchange with further TBHP to regenerate the active oxidant **37**.

$$\underset{\mathbf{32}}{\text{EtZnOR*}} + t\text{-BuO}_2\text{H} \longrightarrow \underset{\mathbf{37}}{t\text{-BuO}_2\text{ZnOR*}} + \text{C}_2\text{H}_6$$

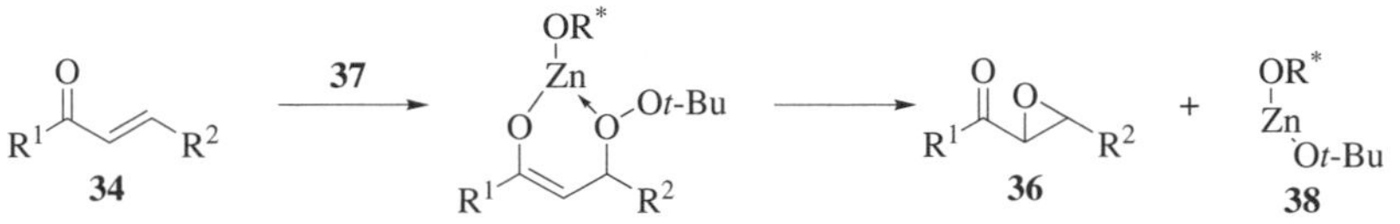

$$\underset{\mathbf{38}}{t\text{-BuOZnOR*}} + t\text{-BuO}_2\text{H} \longrightarrow \underset{\mathbf{37}}{t\text{-BuO}_2\text{ZnOR*}} + t\text{-BuOH}$$

(Eq. 16)

Rare-Earth-Metal-Mediated Methods. Various complexes generated from lanthanide alkoxides and BINOL or a BINOL derivative act as catalysts for the asymmetric epoxidation of α,β-unsaturated ketones with alkyl hydroperoxides.[46–48] Good results are obtained with (*R*)-3-hydroxymethyl-BINOL (**39**), (*S*)-6,6′-diphenyl-BINOL (**40**), and (*S*)-6,6′-dibromo-BINOL (**41**).

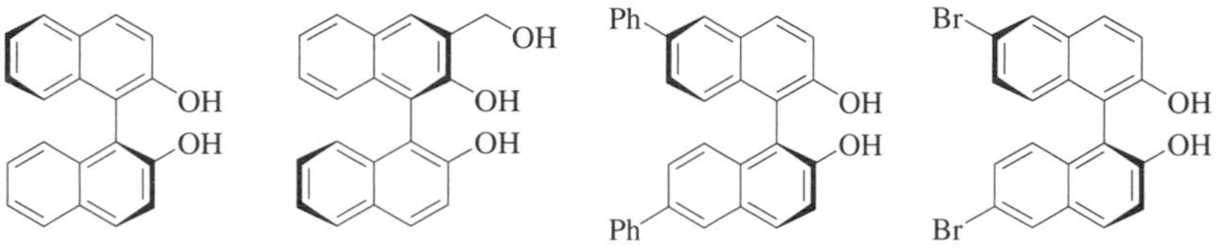

(*S*)-BINOL **39** **40** **41**

For example, the complex prepared from La(O*i*-Pr)$_3$ and **39** (1:1.25) catalyzes the epoxidation of chalcone by CMHP (cumene hydroperoxide) to generate (2*S*,3*R*)-epoxychalcone (**27**) in 93% yield and 91% ee (Eq. 17).[46]

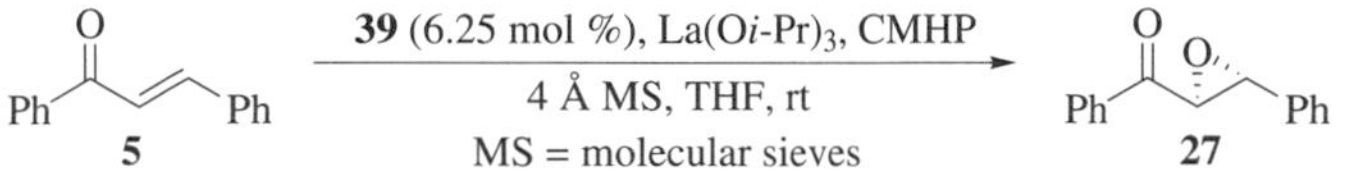

(93%) 91% ee

(Eq. 17)

Higher yields and enantioselectivities may be obtained using a catalyst derived from ligand **40** and Gd(O*i*-Pr)$_3$; complexes prepared from **40** and Yb(O*i*-Pr)$_3$, or **41** and La(O*i*-Pr)$_3$ also give highly enantioselective reactions. The sense of stereoinduction with ligands **40** and **41** is, unsurprisingly, opposite to that observed with **39**.[47,48]

The method is also applicable to *Z*-enones; in this case the combination of ligand **39** with Yb(O*i*-Pr)$_3$ gives the best results. With alkyl ketones (**42**, R^1 = alkyl), levels of stereoselectivity are generally high.[49] It has been suggested that the amphoteric lanthanide ion is able to control the relative orientation of the *Z*-enone and the hydroperoxide, preventing rotation around $C_\alpha-C_\beta$ in intermediate

43 (Eq. 18, path a), which would lead to the *trans*-epoxide. However, with aryl ketones (**42**, R^1 = aryl), the *cis*-epoxide **44** is accompanied by *trans*-epoxide **36** of considerably lower enantiomeric excess. The *trans*-epoxide **36** can be generated either from rotation around C_α–C_β in the β-hydroperoxy enolate intermediate **43** (Eq. 18, path a) or by lanthanide-catalyzed isomerization of the *Z*-enone to the thermodynamically favored *E*-isomer **34** prior to epoxidation (Eq. 18, path b). It is proposed that both pathways a and b operate but favor opposite enantiomers of **45**, thus accounting for the low enantiomeric excess of the *trans*-epoxide **36**.[49]

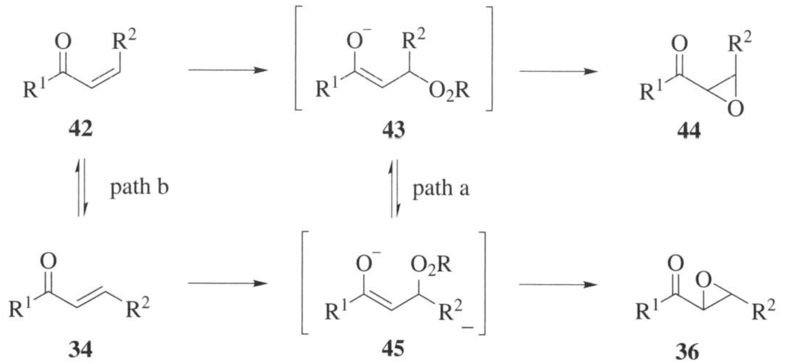

42 **43** **44** **34** **45** **36**

(Eq. 18)

A number of additives that allow highly enantioselective epoxidation of *E*-enones with the commercially available ligand BINOL in place of the more complex and less readily accessible ligands **39, 40**, or **41** have been utilized in the reaction. The addition of a controlled amount of water leads to improved yields and enantioselectivities for epoxidations carried out with the Yb/BINOL/TBHP system; unexpectedly, it is still necessary to add molecular sieves to the reaction.[50] Even more beneficial to the selectivity of these reactions is the addition of TPPO (triphenylphosphine oxide)[51] or tris(4-fluorophenyl)phosphine oxide[52–54] to the Ln/BINOL/TBHP system. It is suggested that these additives serve as ligands that deaggregate the otherwise polymeric catalyst[51] and stabilize it against decomposition.[54] With the latter additive, catalyst loadings as low as 0.5 mol % can be used.[52,53]

Another additive that has been used with some success is TPAO (triphenylarsine oxide).[55] Extensive experimentation has shown the optimum ratio of components for the epoxidation of enones to be a 1:1:1 mixture of La(O*i*-Pr)$_3$/BINOL/TPAO.[55] This catalyst system is effective in the epoxidation of enones and α,β-unsaturated *N*-acylimidazolides.[56–59] A Sm/BINOL/TPAO system effectively catalyzes the asymmetric epoxidation of α,β-unsaturated amides,[60,61] while an octahydro-BINOL ligand **46**, in conjunction with Sm(O*i*-Pr)$_3$ and either TPPO or TPAO, is found to be optimal for the epoxidation of α,β-unsaturated *N*-acylpyrroles.[62,63] A ligand of a slightly different structural type, the biphenyldiol **47**, can be used in conjunction with Y(O*i*-Pr)$_3$ for the epoxidation of α,β-unsaturated esters.[64]

A number of polyvalent BINOL-type ligands have been synthesized that form insoluble adducts upon addition of La(O*i*-Pr)$_3$; these polymeric networks show

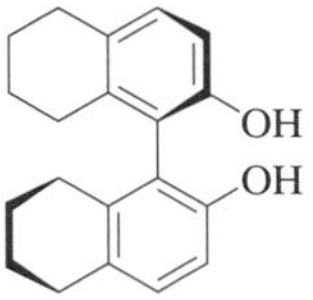

46
(R)-H_8-BINOL

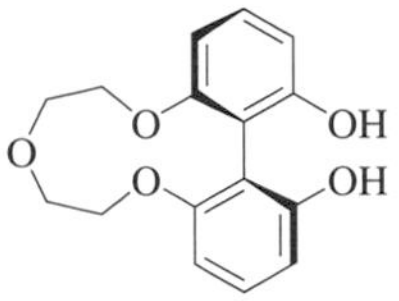

47

high enantioselectivity in the epoxidation of chalcones with CMHP, and offer improved work-up procedures.[65,66]

A detailed understanding of these catalytic systems has not been achieved. However, on the basis of ^{13}C-NMR spectral analysis and asymmetric amplification effects,[67] and of mass spectrometry,[48] it appears that the active Ln–BINOL catalysts are oligomeric. It has been suggested that the catalyst acts as both a Lewis acid and a Brønsted base; thus, the hydroperoxide is deprotonated by a Ln–alkoxide species, while another lanthanide ion acts as a Lewis acid, activating and controlling the orientation of the enone.[46]

The material obtained by combination of La(O*i*-Pr)$_3$, BINOL, and TPAO has been characterized using laser desorption/ionization time-of-flight mass spectrometry,[55] and X-ray analysis of a related complex has also been carried out.[55] Evidence has accrued that supports a structure of composition 1:2:2 La/BINOL/TPAO (**48**). The apparent contradiction between this ratio and the catalytically optimal 1:1:1 ratio is resolved if **48** is not the active catalyst; rather a second molecule of La(O*i*-Pr)$_3$ assists the exchange of a BINOL ligand with TBHP, allowing the formation of the active catalyst **49** (Eq. 19).[55]

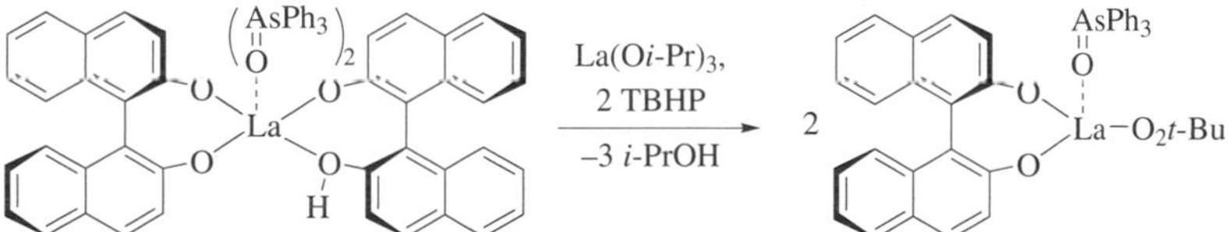

48 **49**

(Eq. 19)

Further insight into the nature of the catalyst in these epoxidations has been obtained from a study of non-linear effects in the TPPO-modified system. Thus, epoxidation of chalcone with the La/BINOL/TPPO/CMHP system, using a 70:30 mixture of (R)-BINOL and (S)-BINOL (i.e. 40% ee), affords (2S,3R)-epoxychalcone with 99% ee.[68] This result has led to the proposal of the dimeric species **50** as the structure of the active catalyst (Eq. 20). On coordination of the substrate, species **51** is formed, in which one of the lanthanum atoms acts as a Lewis acid to activate the chalcone, while the other coordinates the peroxide nucleophile. Following the epoxidation event, the epoxide dissociates and the alcohol by-product is exchanged for a fresh hydroperoxide molecule. No model has yet been proposed to explain the sense of the stereoinduction with Ln–BINOL catalysts.

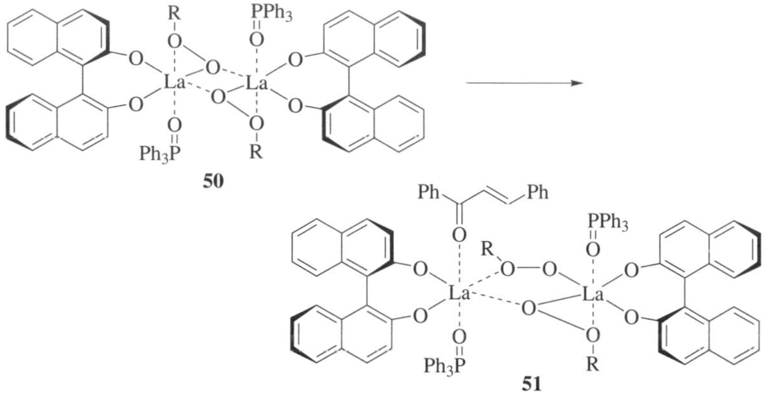

(Eq. 20)

Lithium-, Sodium-, Magnesium-, and Calcium-Mediated Methods. The effect of a number of chiral ligands on the addition of lithium and magnesium alkylperoxides to enones has been investigated. Of the ligands tested, dialkyl tartrates[69,70] and the amino diether ligand **52**[71] result in the highest levels of asymmetric induction.

Epoxidation of chalcone using TBHP, *n*-butyllithium, and *n*-butanol in toluene in the presence of (+)-diethyl tartrate (**53**, 1.1 equivalents) affords (2*R*,3*S*) epoxychalcone in 71–75% yield and 62% ee.[69] Under similar conditions, only 0.2 equivalents of ligand **52** is required to effect an asymmetric epoxidation of enones.[71,72]

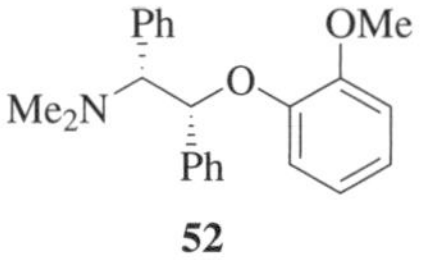

52

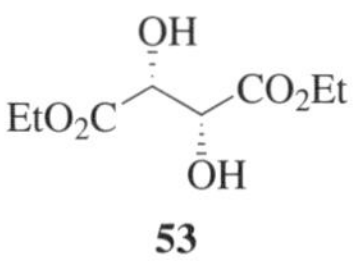

53

A catalytic system utilizing $(n\text{-Bu})_2Mg$ (0.10 equivalent) and diethyl tartrate (**53**, 0.11 equivalent), with TBHP as the oxidant, has also been developed.[69] Interestingly, compared to the stoichiometric method, this system affords the enantiomeric (2*S*,3*R*)-epoxychalcone when the same enantiomer of diethyl tartrate is used as the chiral modifier.

A related magnesium-mediated epoxidation has been reported in which a combination of nanocrystalline magnesium oxide, TBHP, and diethyl tartrate is used to carry out asymmetric epoxidation of chalcone derivatives.[73] A yield of 70% and an ee of 90% is obtained in the epoxidation of chalcone under these conditions.

Three examples have been reported of epoxidation of quinone monoketals using a tartrate-modified peroxide.[74–76] The highest enantiomeric excess is obtained using trityl hydroperoxide, with sodium hexamethyldisilazide as the base. Again, a dependence of product configuration on metal counterion is observed; use of a lithium amide base affords the opposite enantiomer when the same tartrate is used as the ligand.

Epoxidation of a range of chalcone derivatives has been carried out using a catalyst prepared from (*S*)-6,6′-diphenyl-BINOL (**40**), potassium *tert*-butoxide, and calcium chloride (Eq. 21).[77–79] The catalyst has not been fully characterized, but a calcium chelate structure **54** has been proposed.

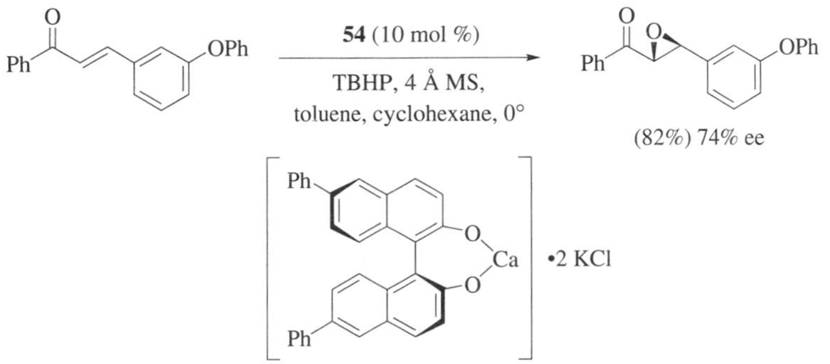

54

(Eq. 21)

Little is known about how these epoxidation systems function. It has been suggested that, in the case of the dibutylmagnesium/diethyl tartrate system, a magnesium bis(alkoxide) is formed from diethyl tartrate; indeed, treatment of diethyl tartrate with dibutylmagnesium gives rise to an isolable amorphous material that is capable of catalyzing the epoxidation reaction.[70] IR spectroscopy of the complex suggests that at least one ester group is coordinated to the magnesium, but further characterization is precluded by the extreme insolubility of the complex. It is posited that the stability of this complex allows effective catalytic asymmetric induction despite the significant reduction in reactivity expected to be associated with chelation.[69]

Platinum-Mediated Methods. Several platinum(II) complexes have been tested as catalysts for the asymmetric epoxidation of enones with hydrogen peroxide. In all cases, conversion of the substrate is extremely slow, and the enantiomeric excess of the epoxide product decreases as the reaction proceeds.[80]

Epoxidation in the Presence of a Chiral Phase-Transfer Catalyst

The possibility of using a chiral quaternary ammonium salt as a phase-transfer catalyst to effect enantioselective Weitz-Scheffer epoxidations was first investigated in the 1970s.[81–85] Subsequently, a few isolated examples of such asymmetric epoxidations have been reported, but enantioselectivities are, at best, moderate.[86–96] Only in the late 1990s were efficient phase-transfer catalysts developed; the most widely studied of these are based on *Cinchona* alkaloids, and have the general structures **55** and **56**, derived from cinchonidine and cinchonine,

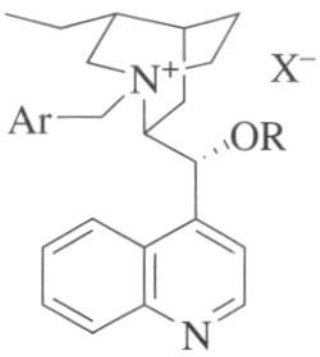

55

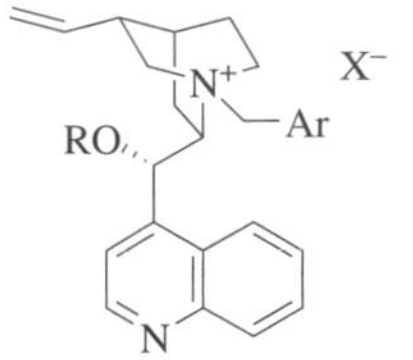

56

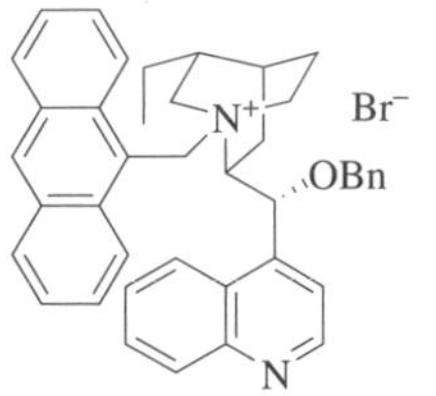

57

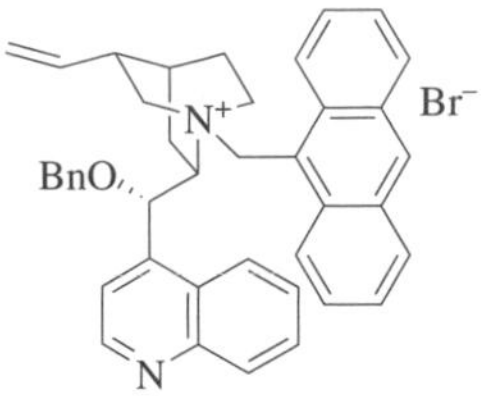

58

respectively. Hydrogen peroxide, alkyl hydroperoxides, and hypochlorites have all been used with some success as oxidants in these reactions.

N-Anthracenylmethyl *O*-benzyl ammonium salts **57** and **58** are among the most efficient and generally applicable catalysts for phase-transfer-catalyzed epoxidation.[97,98] Epoxidation of chalcone with sodium hypochlorite in the presence of catalyst **57** (1 mol %) gives (2*S*,3*R*)-epoxychalcone in 98% yield, with 86% ee.[97] A higher enantioselectivity (93% ee) is obtained if potassium hypochlorite is used in place of the sodium salt, and the reaction temperature is lowered to −40°.[98] Use of the cinchonine-derived catalyst **58** leads to the enantiomeric epoxide product with slightly diminished enantiomeric excess.[99,100] Alternatively, trichloroisocyanuric acid may be used as a convenient source of hypochlorite ion for these epoxidations; the oxidant is released by reaction with potassium hydroxide in either a biphasic toluene/water system[101] or using the solid base in toluene.[102] With chalcone, epoxidation using either set of conditions gives a slightly lower enantiomeric excess than potassium hypochlorite at −40°.

A range of *N*-benzylated cinchoninium salts that have free secondary alcohol groups has also been studied for their potential as epoxidation catalysts.[103–105] The most efficient of these is *N*-(4-iodobenzyl)cinchoninium bromide (**59**), which catalyzes the epoxidation of chalcone to (2*S*,3*R*)-epoxychalcone in 97% yield, with 84% ee.[103,104] Intriguingly, the absolute configuration of the major epoxide in this reaction is opposite that obtained with the closely related catalyst **58**.

Other ammonium salts that have been found to catalyze highly enantioselective epoxidation reactions include the (*p*-trifluoromethyl)benzyl salt **60**, which catalyzes the epoxidation of isoflavones,[29] and the 1-naphthylmethyl salt **61**, which has been used for the epoxidation of naphthoquinones.[103,106]

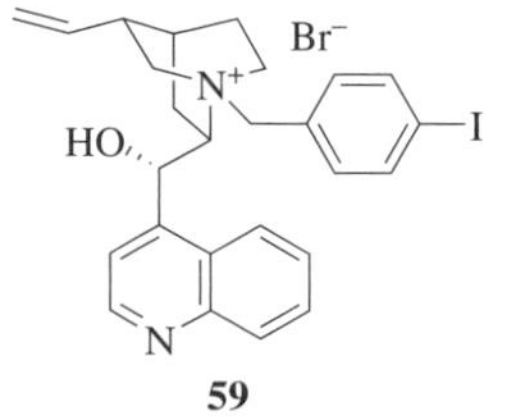

59

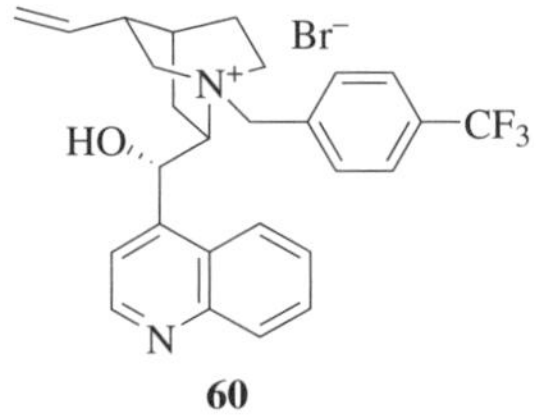

60

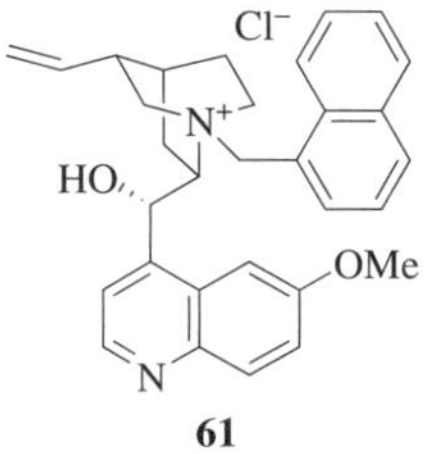

61

The nitrogen atom has also been quaternized with bulky benzyl groups described as "Frechét dendritic wedges." The resulting catalysts give, at best, 56% ee for the epoxidation of chalcone.[107]

The addition of small amounts of a surfactant to epoxidation reactions mediated by the dimeric phase-transfer catalyst **62** enhances the enantioselectivity.[108]

A range of quaternary ammonium phase-transfer catalysts bearing a configurationally rigid binaphthyl unit and a biphenyl group substituted with an alcohol functionality has been reported. These catalysts have been utilized for the epoxidation of a range of enones using sodium hypochlorite in toluene. The most effective catalyst is **63**.[109]

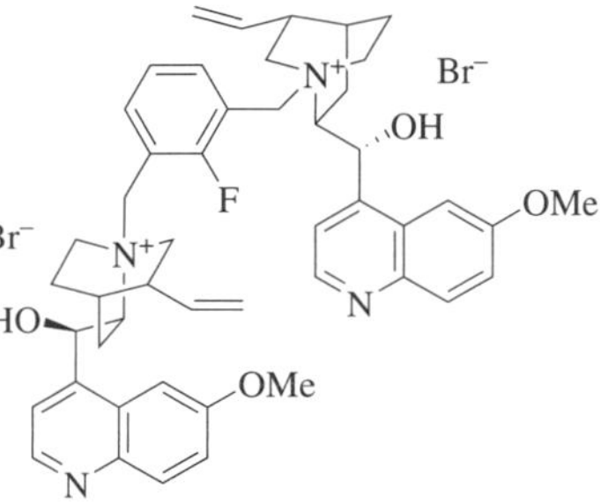

62

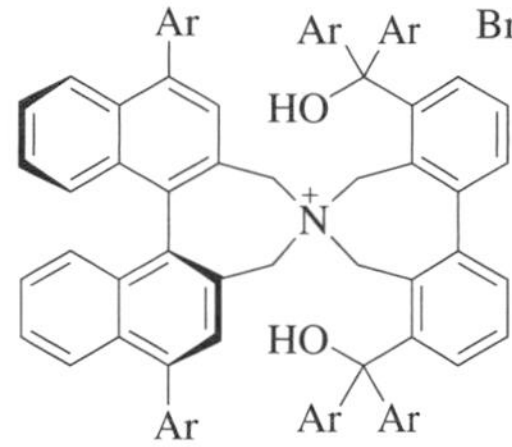

63 (Ar = 3,5-$Ph_2C_6H_3$)

The mechanism of these phase-transfer reactions is presumed to be a nucleophilic addition of hydroperoxide or hypochlorite ion to the enone followed by epoxide formation, and for several of the catalysts, authors have proposed three-dimensional arrangements of the ammonium cation, enone, and oxidant to account for the observed stereoselectivities. For the *O*-benzylated catalyst **57**, the ammonium cation and hypochlorite anion are proposed to form a contact ion-pair, with the substrate carbonyl group held close to the positively charged nitrogen (Fig. 3).[98] With catalysts having a free hydroxy group, transition structures that incorporate hydrogen bonding between the catalyst and substrate have been proposed.[29,110]

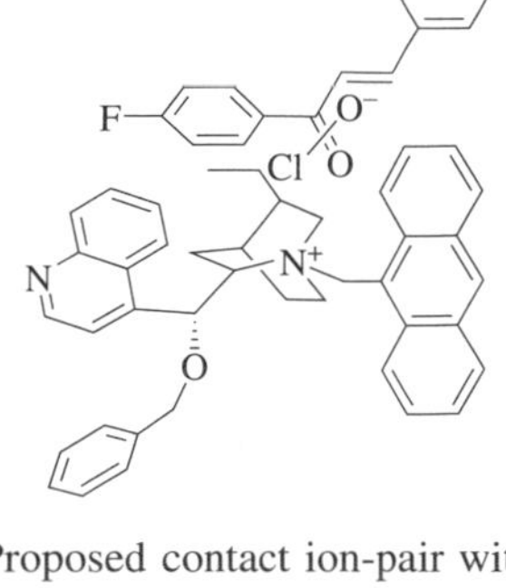

Figure 3. Proposed contact ion-pair with catalyst **57**.

Chiral crown ethers have also been used to catalyze the asymmetric epoxidation of enones.[111–113] The most effective catalysts are sugar-derived lariat ethers. Catalysts have been developed that allow access to either enantiomer of the epoxide; the most effective for the epoxidation of chalcone are **64** (Eq. 22)[112,113] and **66** (Eq. 23),[111] yielding epoxy ketones **65** and **27**, respectively. Enantioselectivities for the epoxidation of related enones using these catalysts tend to be somewhat lower.

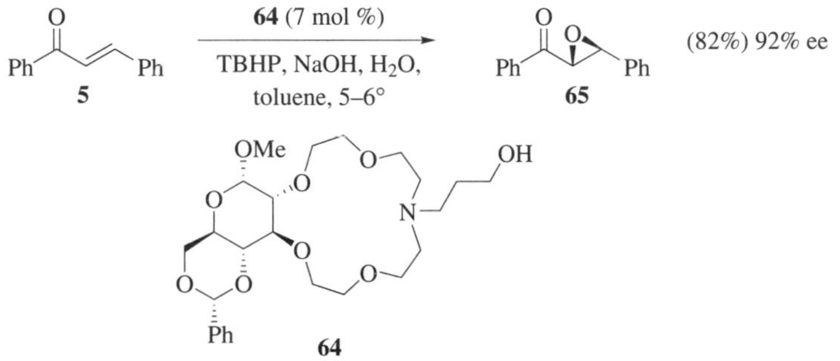

(Eq. 22)

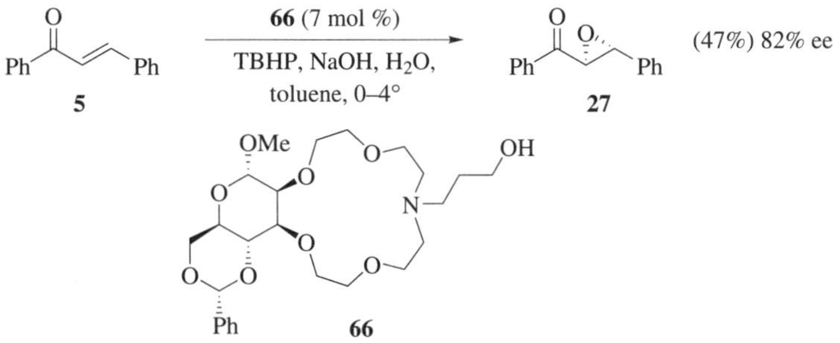

(Eq. 23)

Epoxidation in the Presence of Polypeptides

Epoxidation of electron-deficient alkenes has been performed in the presence of various polypeptides as chiral catalysts. The synthetically most useful catalysts are homopolymers of proteinogenic amino acids. A number of polypeptides, with a defined sequence of amino acid residues, have been used as catalysts with the aim of furthering understanding of the mechanism of the poly(amino acid)-catalyzed epoxidation. In addition, epoxidations have been carried out in the presence of serum albumin proteins.

Catalysis by Homopolymers. Homopolymers of a number of amino acids are found to catalyze the epoxidation of chalcone in good yield and enantiomeric excess.[114,115] The catalysts are typically prepared by polymerization of the *N*-carboxyanhydride of the corresponding amino acid.[114] Polymers of L- or D-leucine are generally found to be the most useful catalysts for synthetic purposes.

The original reaction conditions were triphasic, consisting of an aqueous oxidant, organic solution of substrate, and insoluble catalyst. Recently these conditions have been largely superseded by a number of biphasic reaction protocols. Urea–H_2O_2 can be used as the oxidant with either stoichiometric amounts of

DBU[116] or sub-stoichiometric amounts of BEMP (2-*tert*-butylimino-2-diethylamino-1,3-dimethylperhydro-1,3,2-diazaphosphorine)[117] as a base in THF. Alternatively, sodium percarbonate or a mixture of ammonium bicarbonate and hydrogen peroxide can be used in a DME/water solvent mixture (Eq. 24).[118,119] Improved reaction rates and reduced catalyst loadings can also be achieved by adding a phase-transfer catalyst such as tetrabutylammonium bromide (TBAB) to reactions performed under the original triphasic conditions.[120–122] Under these latter conditions, the range of oxidizing agents has been extended to NaOCl, albeit with a slower rate of reaction and slightly reduced enantioselectivity.[121]

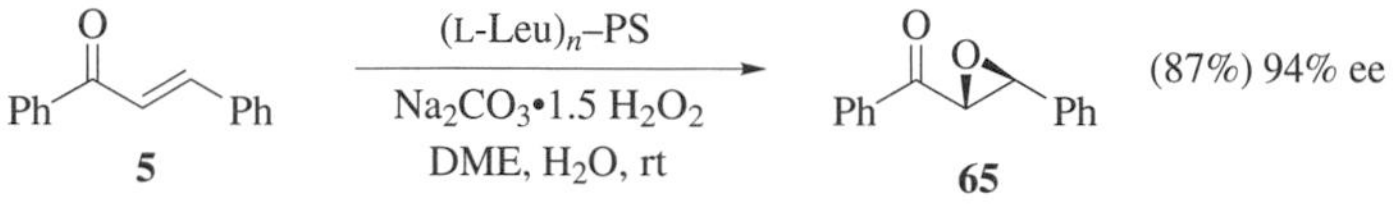

(Eq. 24)

Various modifications of the catalyst have been reported including immobilization on polystyrene, which facilitates rapid filtration,[123,115] adsorption onto silica, which has a similar effect and also allows reduced catalyst loading,[124] and covalent grafting of the catalyst to silica, which also leads to improved filtration properties.[125]

Soluble variants of the catalyst have also been developed, either by adding a polyethylene glycol tail[126,127] or incorporating aminoisobutyric acid residues[128] to solubilize the peptide, thus allowing epoxidations to be carried out under homogeneous conditions. One such catalyst has been used in a membrane reactor for epoxidation of chalcone under continuous-flow conditions.[129]

Although a full understanding of this reaction has not been realized, a number of details pertinent to the mechanism have been reported. The relationship between the secondary structure adopted by a given poly(amino acid) and its effectiveness as a catalyst is not simple. Thus polyalanine (α-helix),[130,131] polyleucine (α-helix),[130,131] polyisoleucine (β-sheet),[130] and polyneopentylglycine (structure not reported)[131] are all reasonably good catalysts; polyvaline (β-sheet),[130,131] polyphenylalanine (β-sheet),[130,131] polyproline (lacking a backbone NH),[115] poly(γ-benzylglutamate) (α-helix),[130] and poly(β-benzylaspartate) (α-helix)[130,131] are poor catalysts, giving relatively low conversions and enantioselectivities.

As the degree of polymerization of the catalysts increases from 5 to 30 there is an increase in the enantioselectivity of the reaction.[132] A similar effect has been seen in catalysts of defined length, prepared by stepwise synthesis.[133,134] However, short oligomers with fewer than 5 leucine residues, when bound to insoluble cross-linked polystyrene[123] or soluble polyethylene glycol,[126] are also highly effective catalysts. In a study of oligomers of L-leucine attached directly to TentaGel S NH_2, a trimer gives the product in greater than 60% ee, and increases in length beyond five amino acids afford no improvement in enantioselectivity; the enantiomeric excess with a pentamer is >95%. On the other hand, the rate of the reaction continues to increase as further residues are added to the catalyst.[134]

The structures of soluble catalysts have been studied using techniques such as infrared spectroscopy[126] or circular dichroism.[127] FT-IR investigations indicate that the catalytically active polyleucine components of soluble PEG-bound catalysts have an α-helical structure.[126] Automated peptide synthesis has allowed the generation of catalysts with modified primary structures; replacement of selected residues in a 20-mer of L-leucine with D-leucine or with glycine has identified the N-terminal region as the site of catalysis.[133,135] However, the presence of a free amino group at the end of the catalyst is not itself a prerequisite for efficient catalysis.[128,135] Catalysts generated by polymerization of *N*-carboxyanhydrides prepared from enantiomerically enriched amino acid monomers show high levels of chiral amplification.[136]

The structure of a possible catalytic complex for this transformation has been suggested. A mechanism in which the enone oxygen forms a hydrogen bond to the terminal amino group, and the penultimate NH group binds to the hydroperoxide anion has been proposed.[134] Other investigators suggest that the three amide NH groups lacking internal hydrogen bonds at the N-terminus of the α-helix are all involved in hydrogen bonding; N2–H and N3–H coordinating to the enone carbonyl and N4–H to the hydroperoxide anion.[136a]

The kinetics of the process have been studied by two groups; although some differences exist between the accounts there is a general agreement as to the order of the key events.[137–139] The polyleucine catalyst sequesters considerable amounts of hydrogen peroxide from solution;[140] the generation of a hydroperoxide-polypeptide complex is believed to be the first step of the mechanism. In the rate-determining step, this complex is then able to react with the enone to generate a hydroperoxy enolate species, which collapses either to regenerate the enone[140] or to give the epoxide product. It has been noted that the proposed enone-polypeptide complexes discussed above are not directly compatible with the observed kinetics.[139]

Catalysis by Serum Albumins. Epoxidation of alkyl-substituted naphthoquinones, using buffered alkaline H_2O_2 or TBHP in the presence of BSA (bovine serum albumin) or HSA (human serum albumin), proceeds with widely varying enantioselectivities.[141–144] The stereochemical outcome of this reaction is difficult to predict. In some cases the different oxidants afford enantiomeric epoxides under otherwise identical conditions, whereas with other substrates they generate products with the same configurations. No rationalization of these variations has been proposed. The effect of varying the substrate structure is similarly difficult to predict; whereas a general trend towards improved asymmetric induction is observed for 2-alkylnaphthoquinones on increasing alkyl chain length, bulky substituents (such as *tert*-butyl) can lead to a racemic product.[144]

Examination of CD spectra shows similarities between complexes giving comparable epoxidation results. It has been proposed that the protein has a number of binding sites for smaller substrates, whereas the exceptionally selective epoxidation of 2-*n*-octylnaphthoquinone (>99% ee) reflects the fact that it binds to a single site on BSA.

Citing evidence from binding studies, it has been proposed that the reaction is subject to product inhibition. This effect can be minimized by the addition of organic cosolvents such as isooctane, which is suggested to prevent the product from binding to the BSA while still allowing the naphthoquinone substrate to coordinate.[143]

Miscellaneous Methods of Epoxidation

Numerous other methods have been employed for the asymmetric epoxidation of electron-deficient alkenes. Of these methods, two—the use of chiral dioxiranes, and the use of manganese-salen catalysts—are applicable to a wide range of alkenes, not simply those that are electron deficient. These will be described first, but the reader is referred elsewhere for detailed discussions of their mechanisms.

Epoxidation Using Chiral Dioxiranes. Several chiral ketones are known to mediate the asymmetric epoxidation of alkenes with Oxone (potassium peroxymonosulfate). Few of these ketones, however, have been employed for the epoxidation of electron-deficient alkenes and several fail to show synthetically useful levels of reactivity and/or enantioselectivity.[145–147]

Ketones **67** and **68**,[148,149] both of which are derived from quinic acid, have been utilized in the epoxidation of enones and enoates, with enantioselectivities in the range of 82–96% ee. More recently, the use of fructose-derived ketone **69** in the epoxidation of a wide range of α,β-unsaturated esters has been reported.[150]

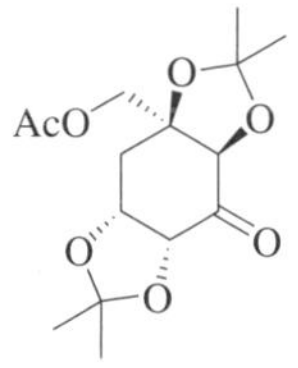

67

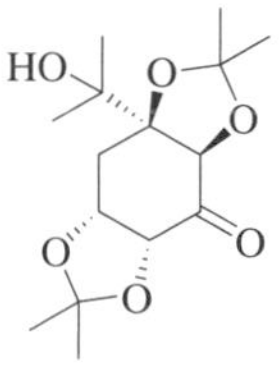

68

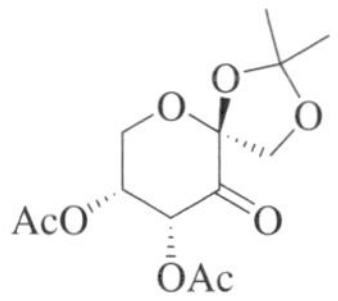

69

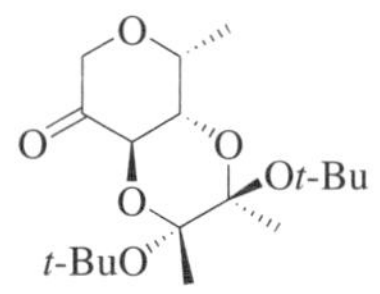

70

Another class of sugar-derived ketones used for cinnamate ester epoxidation contains a butane-2,3-diacetal unit. These catalysts, of which chiral ketone **70** is an example, give somewhat lower enantioselectivities, rising to 86% ee for the epoxidation of *tert*-butyl cinnamate.[151] Several steroidal ketones have been investigated as catalysts for the epoxidation of cinnamic acids.[152,153]

The axially chiral binaphthyl ketone **71** and three related catalysts have been used for the epoxidation of cinnamic esters. Under the optimized reaction conditions, treatment of methyl *p*-methoxycinnamate with 2 mol % of ketone **71** and 1 equivalent of Oxone in aqueous 1,4-dioxane affords the (2*R*,3*S*)-epoxide in 76% ee.[154] The use of this catalyst has been extended to trisubstituted and *Z*-cinnamates, but enantioselectivities are markedly lower.[155] Cinnamamides may also be epoxidized under the same conditions; in this case, levels of asymmetric induction for primary and secondary amides are reasonable (65–71% ee) whereas a number of tertiary amides are epoxidized with low enantioselectivity.[155]

A range of cyclic α-fluoro ketones has been used for the epoxidation of methyl *p*-methoxycinnamate, giving enantiomeric excesses in the range of 2–76% ee;

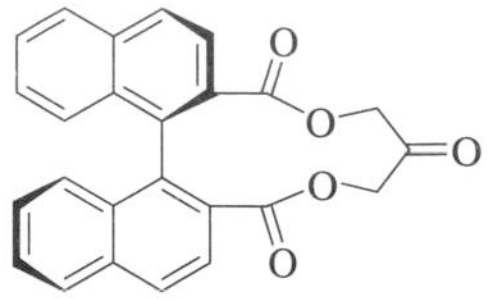

71

however, none of these catalysts has been utilized for the epoxidation of any other electron-deficient alkenes.[156–162]

These epoxidations occur via dioxirane intermediates. Reaction of ketone **72** with the peroxysulfate ion, followed by elimination of a hydrogen sulfate ion, affords the corresponding dioxirane (Eq. 25).

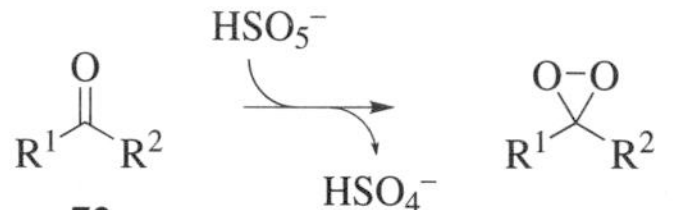

72

(Eq. 25)

Reaction between the dioxirane and the alkene is thought to be concerted and to occur primarily through a spiro transition structure in which the plane of the alkene is perpendicular to the plane of the three-membered dioxirane ring (Eq. 26).[162a] The reaction produces the new epoxide and regenerates the ketone catalyst. Several authors have proposed models based on such a spiro transition structure to explain the asymmetric induction observed with the chiral ketone catalysts discussed above.[146,151,152,161–163]

(Eq. 26)

Epoxidation Using Manganese-Salen Complexes. The manganese(III)-salen complex **73**, its enantiomer, and other related complexes have found widespread application as catalysts for asymmetric epoxidation, particularly of conjugated *Z*-disubstituted alkenes.[163a] A few electron-deficient alkenes have been among those oxidized in this fashion. For example, reaction of (*Z*)-ethyl cinnamate with sodium hypochlorite and catalyst **73** (6.5 mol %) gives a moderate yield of a *cis*-epoxide in excellent enantiomeric excess, and a small amount of the epimeric *trans*-epoxide.[164] The ratio of *cis*- to *trans*-epoxide can be modified in favor of the trans-isomer by the addition of chiral quaternary ammonium salts to the reaction mixture.[165]

The active oxidant in these epoxidations is a manganese(V)-oxo species, which reacts at the α-carbon of the cinnamate to generate radical intermediate **74** (Eq. 27).[164] If this species has a short lifetime—for example, with an electron-rich aromatic group (Ar)—ring closure is fast and *cis*-epoxide **75** is the major product. If the lifetime of the radical is long, rotation around the C_α–C_β

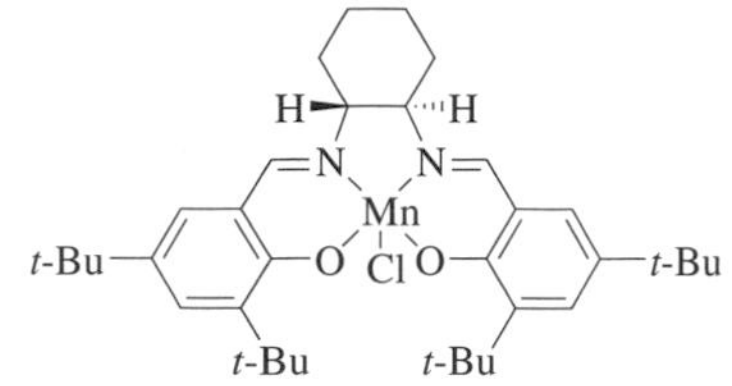

73

bond may occur, resulting in formation of the trans-isomer **76**. In either case, the ring closure is accompanied by regeneration of the manganese(III) catalyst.[164]

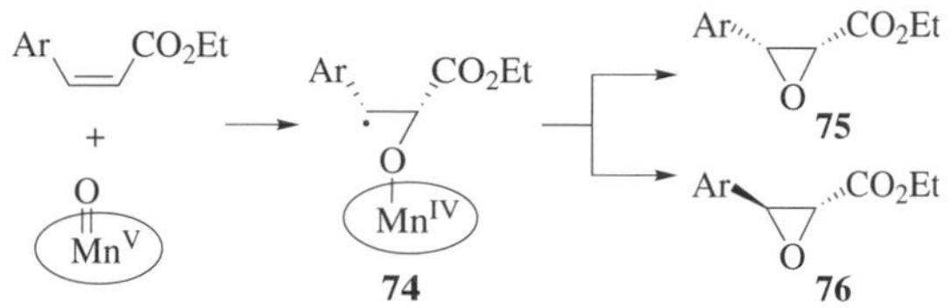

(Eq. 27)

A model has been proposed for the stereoselectivity of the initial C–O bond formation.[166] The enantioselectivity observed in these epoxidations is relatively insensitive to the steric bulk of the aryl group, leading to the suggestion that the alkene adopts a "skewed" rather than a "parallel" approach to the metal-oxo species.[164] No rationale has been presented for the effect of quaternary ammonium salts on the reaction.

Epoxidation Using a Ruthenium Porphyrin Catalyst. A D_4-symmetric ruthenium porphyrin has been employed as a catalyst for the epoxidation of a few electron-deficient alkenes. In all examples, conversion is slow and only low levels of asymmetric induction are observed.[167]

Organocatalysis. Two different catalysts derived from proline are reported to catalyze the epoxidation of electron-deficient alkenes. The silylated proline derivative **77** is an effective catalyst for the epoxidation of a range of α,β-unsaturated aldehydes[168,169] and α,α-diphenylprolinol (**78**) catalyzes the epoxidation of α,β-unsaturated ketones.[170]

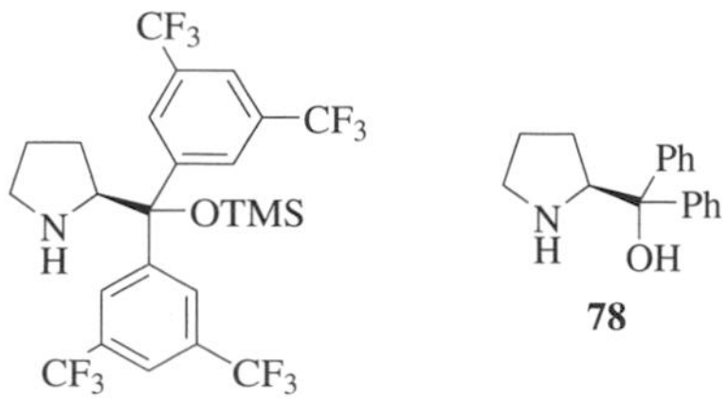

77 **78**

With regard to the enal epoxidation, good enantioselectivities have been reported using a range of peroxides,[168] although 35% aqueous hydrogen peroxide is most commonly employed. Reactions typically use either dichloromethane[168] or ethanol[169] as cosolvent.

The stereospecificity of these reactions is dependent on both the reaction conditions and the structure of the substrate. Increased amounts of the *cis*-epoxide are obtained in the ethanolic solvent system, particularly when alkyl- rather than aryl-substituted enals are the substrates. In the dichloromethane system, the enantiomeric excess of the trans-diastereomer is high for aryl, alkyl, and ester β-substituents.[169]

A mechanism has been proposed for this epoxidation in which the diastereoselective addition of hydrogen peroxide to an α,β-unsaturated iminium ion **79** followed by cyclization of the resultant enamine **80** to generate the epoxide **81** is responsible for the observed enantioselectivity (Eq. 28).

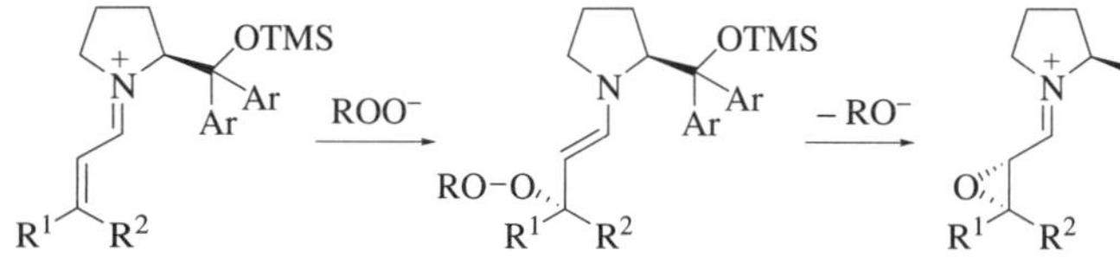

79 **80** **81** (Eq. 28)

The epoxidation of chalcone catalyzed by the prolinol catalyst **78** has been investigated in a range of solvents. With a catalyst loading of 50%, the combination of TBHP with a non-polar solvent such as hexane gives the best yield and enantioselectivity of epoxide **65** (Eq. 29), and the method has been extended to a number of other enones.[170]

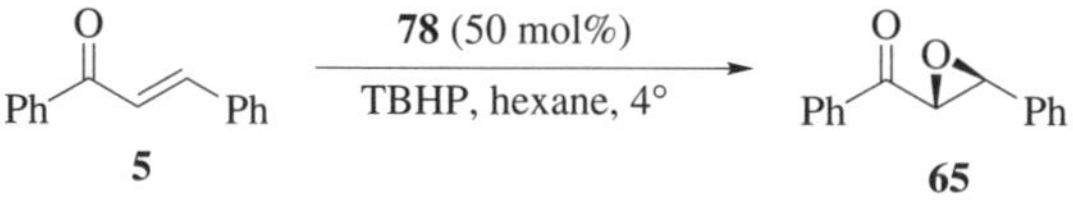

5 **65** (80%) 78% ee (Eq. 29)

Despite the apparent similarities between these methods a very different mechanism has been proposed for the enone epoxidation. Citing the lower reactivity of ketone carbonyls compared to aldehydes, it has been proposed that prolinol **78** acts as a bifunctional catalyst responsible both for deprotonating the TBHP to form a tight ion pair and for activating the enone to nucleophilic attack by hydrogen bonding to the enone carbonyl.[170]

Epoxidation in the Presence of Cyclodextrins. Cyclodextrins have been employed as additives in nucleophilic epoxidations of quinones,[171] enones,[172,173] and enals.[172] In a typical example, epoxidation of 2-methylnaphthoquinone with TBHP in the presence of 5 equivalents of β-cyclodextrin yields the corresponding (2*R*,3*S*)-epoxide in 96% yield and 24% ee.[171] The sense of induction depends on the reaction conditions: use of an aqueous buffer solution rather than sodium carbonate/DMF gives the opposite enantiomer of the product. Similar results are obtained using α-cyclodextrin.

In a related method, cyclodextrin inclusion complexes of chalcone have been epoxidized with sodium hypochlorite. For example, the 1:1 complex of chalcone with β-cyclodextrin is epoxidized to give (2*S*,3*R*)-epoxychalcone in 87% yield and 30% ee. Use of the corresponding γ-cyclodextrin complex yields the enantiomeric epoxide.[173]

Epoxidation with Liposomized *m*-CPBA. An epoxidation protocol has been described in which the oxidant is *m*-CPBA (*m*-chloroperoxybenzoic acid) in liposomized form. This reagent is produced by evaporation of a chloroform solution containing egg phosphatidylcholine and *m*-CPBA, followed by suspension of the resulting residue in water, sonication, and centrifugation.[174] The *m*-CPBA has been shown to be localized in the liposomal bilayer by ^{1}H nuclear Overhauser enhancement experiments.[174]

Oxidation of methyl cinnamate (**82**) with this reagent affords a 75% yield of the epoxide (2*R*,3*S*)-**83**, with 92% ee (Eq. 30).[174]

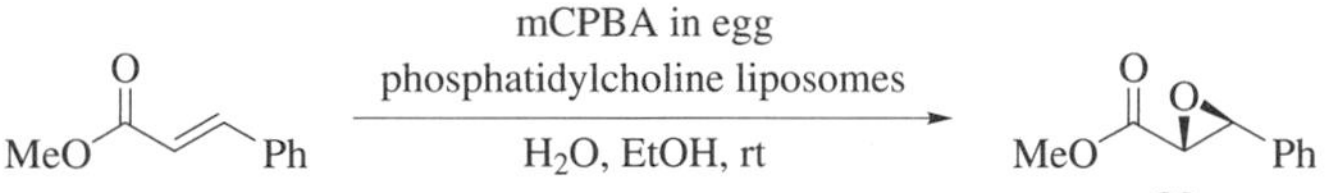

82 **83** (75%) 92% ee (Eq. 30)

SCOPE AND LIMITATIONS

The methods described in the previous section have been applied to a variety of electron-deficient alkenes. For a given substrate class certain approaches are suitable, while others are likely to be less appropriate.

This section is organized by substrate class, and the most effective epoxidation methods for each class are discussed. The classes of substrates are defined in the list below; for classes 1a-1g (various *E*-enones), a less precise but shorter designation is given in square brackets, and is used in the subsequent sections and tables.

1. β-Substituted acyclic enones

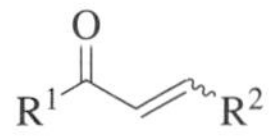

1a. *E*-Enones where R^1, R^2 = alkyl [dialkyl *E*-enones]
1b. *E*-Enones where R^1 = alkyl, R^2 = aryl [alkyl aryl *E*-enones]
1c. *E*-Enones where R^1 = aryl, R^2 = alkyl [aryl alkyl *E*-enones]
1d. *E*-Enones where R^1, R^2 = aryl [diaryl *E*-enones]
1e. *E*-Enones where R^1 or R^2 = heteroaryl [heteroaryl *E*-enones]

1f. E-Enones where R^1 or R^2 = alkenyl or alkynyl [polyunsaturated E-enones]
1g. E-Enones where R^2 is attached via a heteroatom [hetero-substituted E-enones]
1h. Z-Enones
2. Enones bearing a second electron-withdrawing substituent
3. 1,1-Disubstituted and trisubstituted alkenes
4. Endocyclic enones
5. Quinones and quinone monoketals
6. Other α,β-unsaturated carbonyl compounds
   6a. α,β-Unsaturated aldehydes
   6b. α,β-Unsaturated esters
   6c. α,β-Unsaturated acids
   6d. α,β-Unsaturated amides and N-acylated heterocycles
7. Alkenes bearing electron-withdrawing groups other than carbonyl
   7a. α,β-Unsaturated sulfones
   7b. α,β-Unsaturated nitro compounds
   7c. α,β-Unsaturated nitriles
   7d. α,β-Unsaturated phosphonates

Dialkyl *E*-Enones

There have been relatively few asymmetric epoxidations of acyclic E-enones bearing two alkyl substituents. The most widely used protocol uses TBHP in the presence of a sub-stoichiometric quantity of a magnesium tartrate catalyst (Eq. 31). For the epoxidation of a range of dialkyl E-enones these conditions give moderate to excellent enantioselectivities although yields are variable.[70]

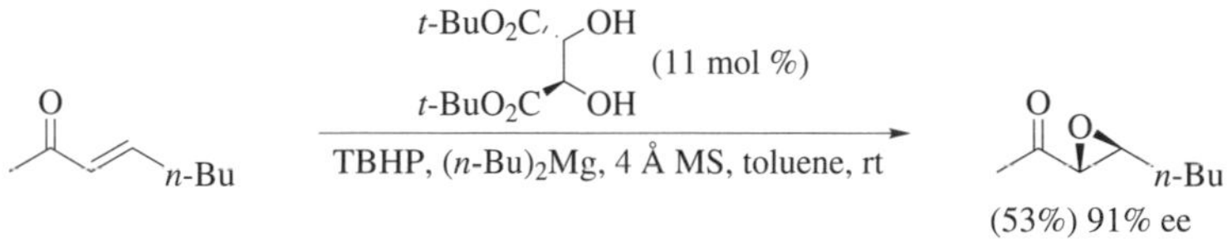

t-BuO2C
OH
(11 mol %)
t-BuO2C
OH
O
n-Bu
TBHP, (n-Bu)2Mg, 4 Å MS, toluene, rt
O
O
n-Bu
(53%) 91% ee

(Eq. 31)

Reports citing the use of alternative epoxidation methods are limited. There is a single example of a substrate epoxidized using stoichiometric $Et_2Zn/O_2/N$-methylpseudoephedrine giving 90% ee,[40,41] whereas other substrates have been epoxidized with a variety of lanthanide-BINOL catalyst systems.

The most effective of these latter epoxidations employs La(Oi-Pr)$_3$/BINOL with an equimolar amount of TPAO as an additive, and gives a 95% ee.[55] Other methods with isolated examples include the use of a chiral phase-transfer catalyst[109] and a diphenylprolinol catalyst.[170]

Alkyl Aryl *E*-Enones

A variety of methods has been applied to the epoxidation of alkyl *E*-enones bearing an aryl group in the β-position. However, of these methods, the lanthanide/BINOL (Ln-BINOL) system and polypeptide catalysis account for the majority of the examples.

Polypeptides have been used to catalyze the epoxidation of a wide range of substrates of this class, under a variety of reaction conditions (Eq. 32).[118] Polyleucine is the most widely employed catalyst.[116,175,176] Yields are generally good and enantioselectivities range from 62–98% ee. In reactions where a variety of different conditions have been applied, the silica-supported variant of the catalyst[124] typically gives the highest enantioselectivity.

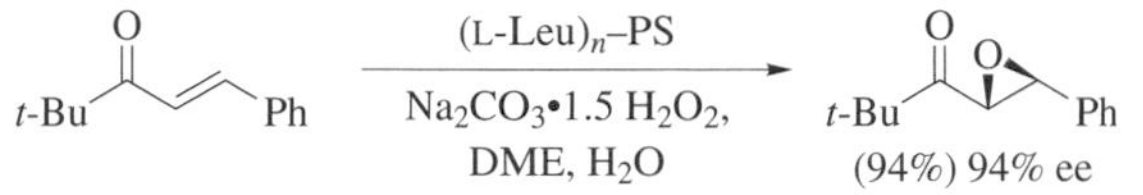
O
t-Bu
Ph
(L-Leu)n–PS
Na2CO3•1.5 H2O2,
DME, H2O
O
O
t-Bu
Ph
(94%) 94% ee

(Eq. 32)

Lanthanide/BINOL catalysis has been applied to several enones of this class. As with polypeptide catalysis, a variety of reaction conditions have been investigated; the best results utilize La(O*i*-Pr)$_3$/BINOL and an additive such as water,[50,51] TPPO,[51,53,67,68] TPAO,[55] or tris(4-fluorophenyl)phosphine oxide (Eq. 33).[54] With substrates where both this method and polyleucine catalysis have been employed, the Ln–BINOL approach affords marginally higher enantioselectivities and yields.

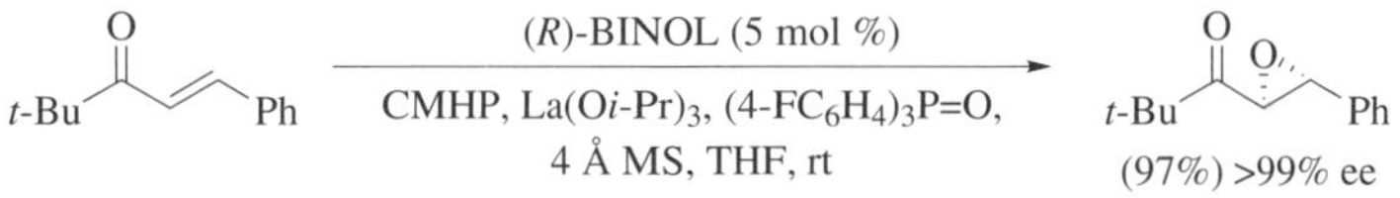
O
t-Bu
Ph
(R)-BINOL (5 mol %)
CMHP, La(Oi-Pr)3, (4-FC6H4)3P=O,
4 Å MS, THF, rt
O
O
t-Bu
Ph
(97%) >99% ee

(Eq. 33)

Substrates of this class have also been epoxidized using phase-transfer catalysis,[99,100,109] a lithium alkylperoxide/chiral ligand system (Eq. 34),[71,72] a chiral hydroperoxide,[31] a ruthenium porphyrin,[167] chiral dioxiranes,[163] a magnesium-tartrate system[70] and an organocatalyst;[170] however, neither scope, enantioselectivity, nor yield is comparable with the methods described above.

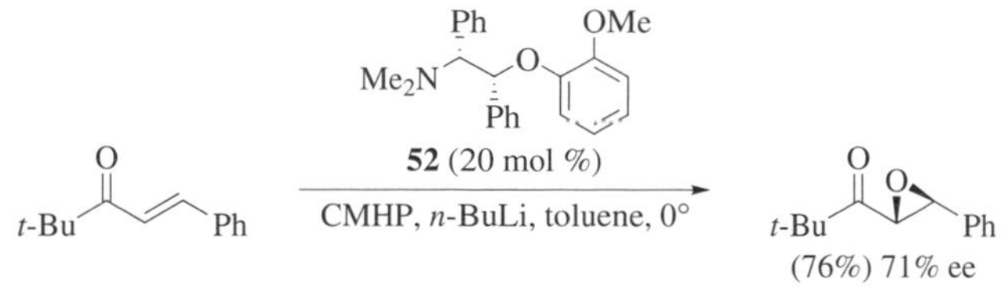
Ph
OMe
Me2N
O
Ph
52 (20 mol %)
O
t-Bu
Ph
CMHP, n-BuLi, toluene, 0°
O
O
t-Bu
Ph
(76%) 71% ee

(Eq. 34)

Aryl Alkyl *E*-Enones

A wide range of methods has been successfully applied to the epoxidation of aromatic *E*-enones bearing an alkyl β-substituent. The most extensively used involve Et$_2$Zn and an oxidant in the presence of a chiral ligand, Ln–BINOL catalysis, chiral phase-transfer catalysis, and polypeptide catalysis. In addition, isolated examples of epoxidation with chiral hydroperoxides[27] and single examples of catalysis by a crown ether[177] and of organocatalysis[170] have been reported.

A variety of phenyl ketones with the general structure **84** has been successfully epoxidized. Where the alkyl group is relatively small and unbranched the use of Et_2Zn and O_2 in the presence of *N*-methylpseudoephedrine (**29**) gives the best enantioselectivities (Eq. 35).[40,41] Zinc-mediated epoxidation using TBHP as the stoichiometric oxidant in the presence of the polymeric ligand **31** is also effective for these substrates, although the enantioselectivities are somewhat lower (Eq. 36).[43]

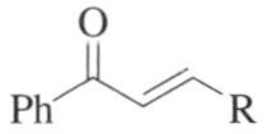

84

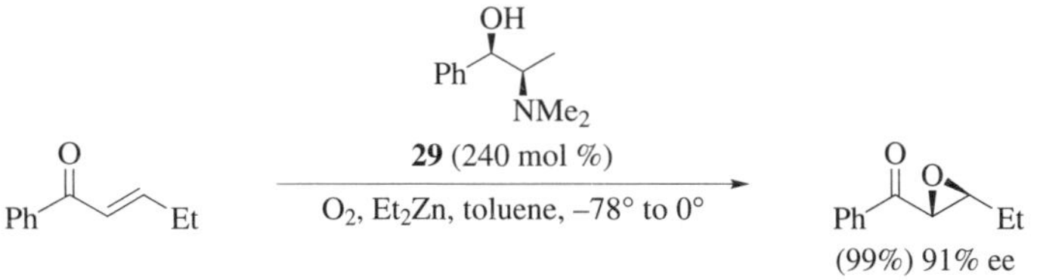

(Eq. 35)

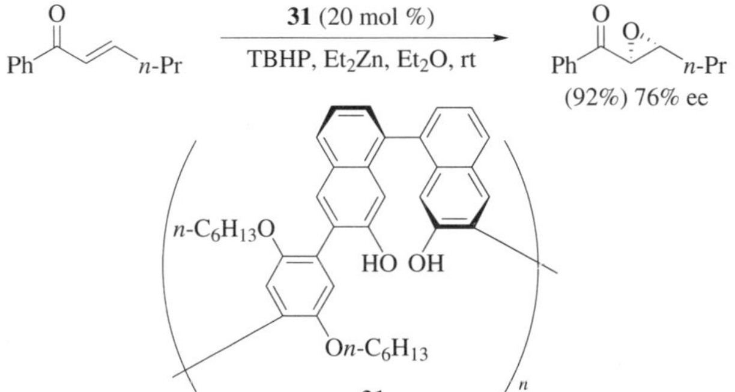

31

(Eq. 36)

With alkenes bearing secondary or tertiary alkyl substituents, catalysis by quaternary ammonium salts,[98] Ln–BINOL (or a related ligand),[67] or polypeptides[117,178,179] proves superior. For simple substrates with secondary or tertiary alkyl groups enantioselectivities in excess of 90% ee have been reported using all three techniques.

A variety of substrates with substituted phenyl groups have been epoxidized with the stoichiometric $Et_2Zn/O_2/N$-methylpseudoephedrine system or under phase-transfer catalysis.[98,99] A particularly effective pair of quaternary ammonium catalysts for these systems is derived from *Cinchona* alkaloids substituted with an anthracenylmethyl group.[98,99] A single example of highly enantioselective phase-transfer-catalyzed epoxidation using the binaphthyl ammonium salt **63** has been reported (Eq. 37). Phase-transfer catalysis and polypeptide catalysis have been successfully applied to 2-naphthyl ketones.[99,178,179]

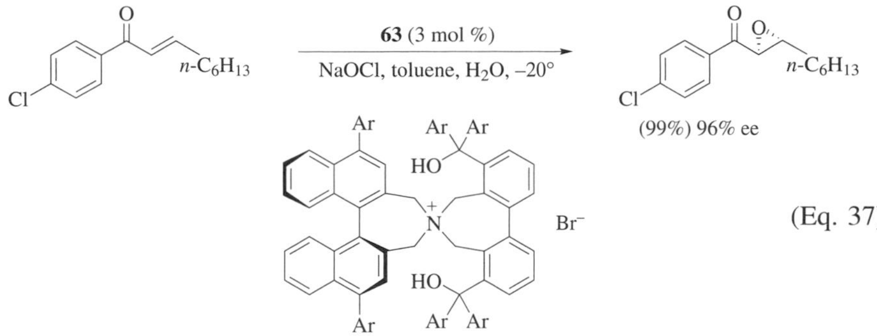

63 Ar = 3,5-$Ph_2C_6H_3$

(Eq. 37)

Polyleucine has been used as a catalyst for a number of enones bearing a γ-stereogenic center.[180,181] For such substrates the epoxidation has an intrinsic diastereoselectivity that a given enantiomer of the catalyst either opposes (mismatched case) or reinforces (matched case). For example, low-temperature epoxidation of enone **85** using sodium percarbonate[118] in the presence of polystyrene-supported poly-D-leucine affords the diastereomeric epoxides **86** and **87** in 97% yield and a ratio of 1:34 (matched). In the mismatched case, use of the poly-L-leucine catalyst gives a 98% yield and a ratio for **86** to **87** of 3.8:1 (Eq. 38).

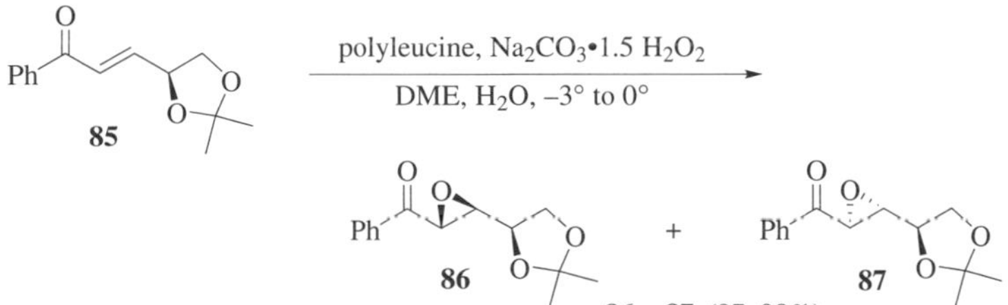

86 + **87** (97–98%)

(Eq. 38)

Diaryl *E*-Enones

E-Enones bearing two aryl groups (chalcones) constitute the most frequently epoxidized class of electron-deficient alkenes. In the subsequent discussion the two aryl groups are referred to as Ar^1 and Ar^2, as shown in structure **88**.

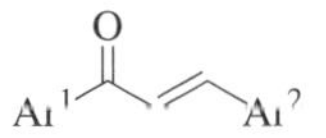

88

Many of the reported methods for the asymmetric epoxidation of chalcones are of limited synthetic use, either due to low enantioselectivity or poor yield. The relatively small number of other reports makes it difficult to assess the generality of some other methods. The remaining methods fall into three groups: catalysis by quaternary ammonium salts, by polypeptides, and by chiral lanthanide complexes.

Several phase-transfer catalysis systems based on chiral quaternary ammonium salts derived from *Cinchona* alkaloids give synthetically useful levels of enantioselectivity across a range of chalcone substrates. Epoxidations catalyzed by the N-anthracenylmethyl ammonium salt **57**, with an alkali metal hypochlorite as the oxidant, have been reported independently by two groups.[97–100] The pseudoenantiomeric catalyst **58** has also been described, and leads to products enantiomeric to those produced by catalyst **57**. The most consistently high enantioselectivities (>91% ee) and yields (70–97%) are obtained by the use of catalyst **57** with potassium hypochlorite (Eq. 39). The method has been successfully applied to substrates in which either aryl ring bears an electron-donating or an electron-withdrawing group.[98] A similar range of substrates has been epoxidized using catalysts **57** and **58** with sodium hypochlorite (Eq. 40)[97,99,100] or trichloroisocyanuric acid/potassium hydroxide[101,102] as the oxidant, or with catalyst **59** and basic hydrogen peroxide (Eq. 41);[103,104] however, slightly lower levels of enantioselectivity are typically observed.

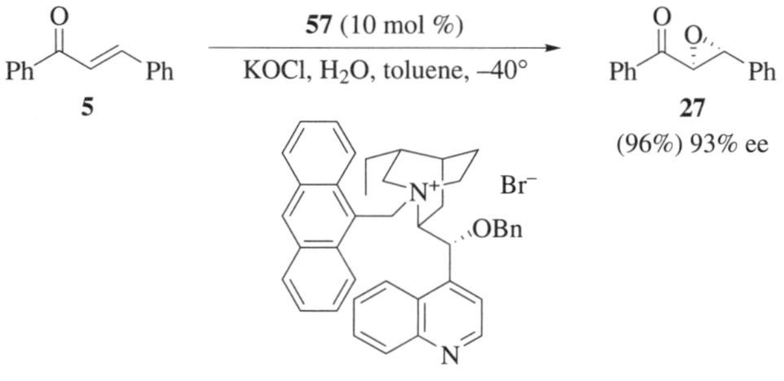

57

(Eq. 39)

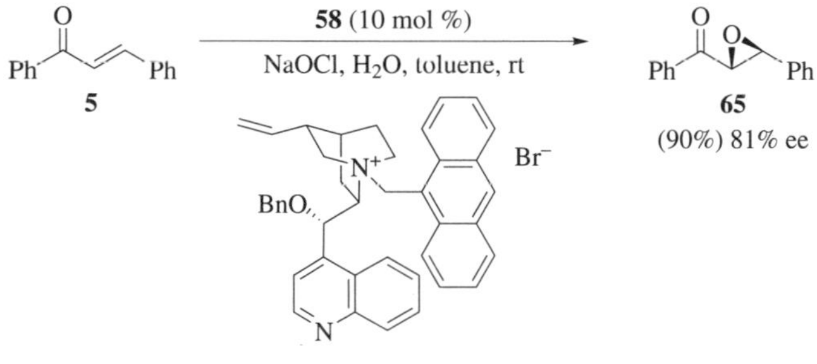

58

(Eq. 40)

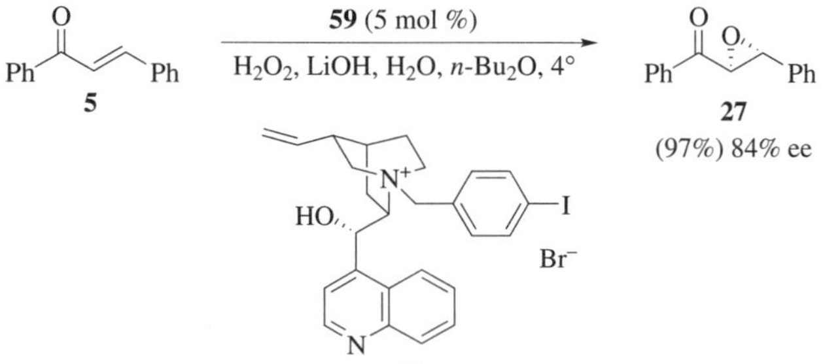

59

(Eq. 41)

Very high enantioselectivities are obtained for the epoxidation of a range of chalcones using dimeric quininium salt **62** in the presence of the surfactant Span 20 (sorbitan monolaurate) (Eq. 42).[108]

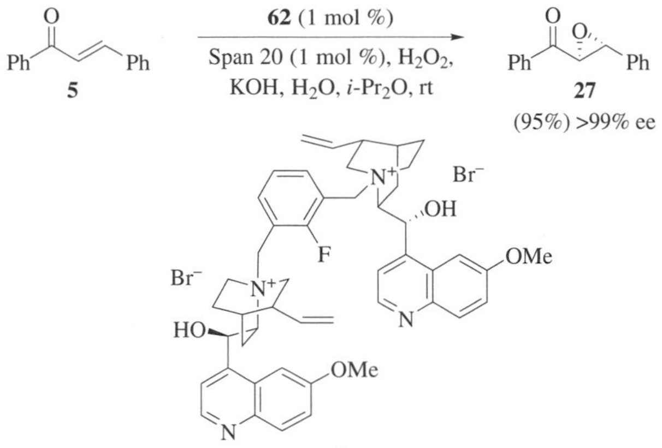

62

(Eq. 42)

A limited number of examples using a different class of quaternary ammonium salt containing a binaphthyl group (**63**, Eq. 37) are on record; enantioselectivities and yields for these epoxidations are generally excellent (Eq. 43).[109]

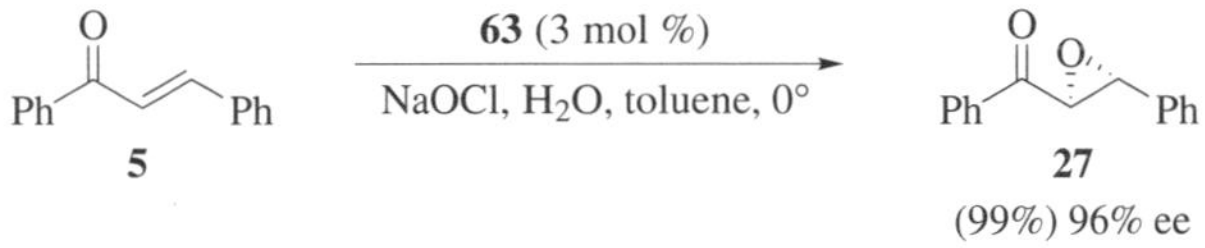

(Eq. 43)

Polypeptide catalysis has been applied to the epoxidation of a large range of substituted chalcones. Substrates with electron-donating and electron-withdrawing groups in a range of positions have been epoxidized with varying levels of enantioselectivity and yield. Both triphasic and biphasic conditions (Eq. 44) have been utilized and a variety of catalysts have been used including polyalanine, polyleucine, and silica-supported polyleucine.

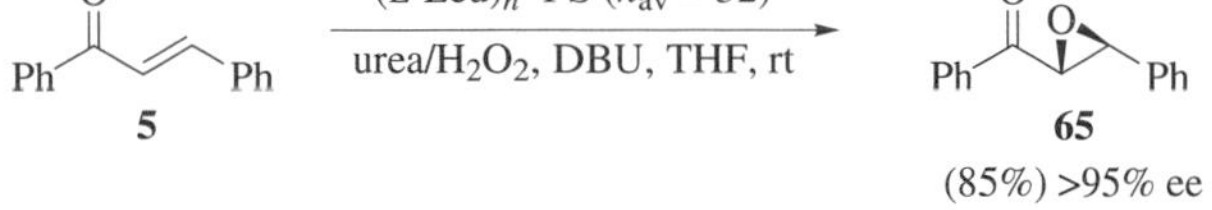

(Eq. 44)

In general, enantioselectivities are high for substrates with electron-withdrawing groups or those with essentially electron-neutral substituents. Ether substituents appear to have a detrimental effect on the enantioselectivity, an effect which seems to be strongest for substituents in the ortho-positions and to be cumulative.[182] This effect apparently does not extend to anilines; chalcones with an *ortho*-amino substituent on Ar^1 prove to be excellent substrates.[183]

Catalysis by Ln–BINOL complexes has been applied to a smaller number of chalcone derivatives. The highest enantioselectivities are obtained using the BINOL/La(O*i*-Pr)$_3$/TPPO system, which gives 99.8% ee in the epoxidation of chalcone.[184] However, because only a small number of chalcone substrates have been epoxidized in this fashion, it is difficult to comment on generality. Replacement of the triphenylphosphine oxide with tris(*p*-fluorophenyl)phosphine oxide allows a catalyst loading of only 0.5 mol % to be used, in which case the enantioselectivity drops to 98% ee (Eq. 45).[53] This approach has been applied to a range of chalcones including examples substituted with both electron-withdrawing and electron-donating groups; yields and enantioselectivities are generally excellent.[54] Other variations on this approach giving synthetically useful enantioselectivities include the use of a diphenyl-substituted BINOL (**40**)/Gd(O*i*-Pr)$_3$[48] catalyst system (Eq. 46) and a ligand for lanthanum consisting of two BINOL units linked by a phenylenediacetylene spacer (**89**).[66] Several other related systems are also found to epoxidize chalcone and substituted chalcones with enantiomeric excesses >90%.[46,50,55,67]

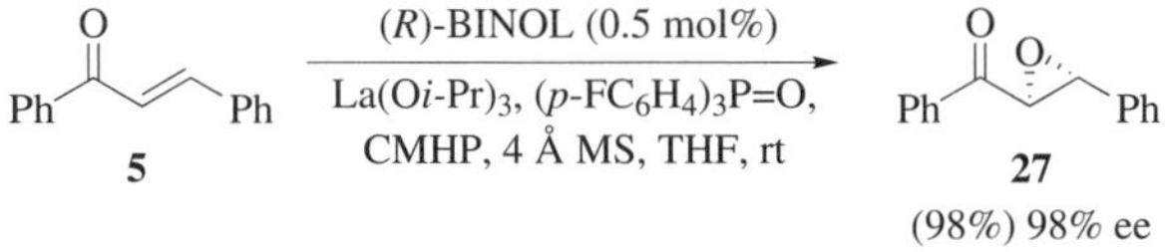

(Eq. 45)

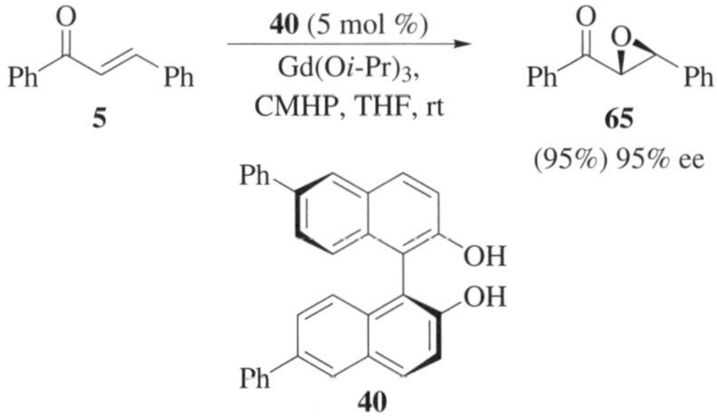

(Eq. 46)

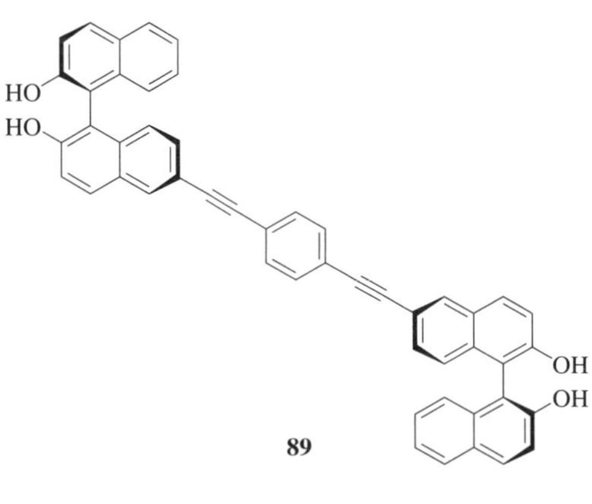

Heteroaryl *E*-Enones

Polypeptide catalysis is the only method that has been systematically applied to enones bearing heteroaryl substituents. Substrates with a thiophene, furan, or pyridine moiety have been successfully epoxidized, mostly under the triphasic reaction conditions and generally in good to excellent enantiomeric excess.[120,132,179,185,186] Isolated examples of heteroaryl enone epoxidation have also been reported using magnesium-tartrate,[73] Ln–BINOL,[55] crown ether,[111] and phase-transfer catalysis.[97,108]

Polyunsaturated *E*-Enones

Enones with alkenyl substituents have been epoxidized under polypeptide catalysis[116,175,178,179,185–190] (and, in one case, Ln–BINOL catalysis).[55] With diene or triene substrates, the double bond adjacent to the ketone is epoxidized preferentially. Selective oxidation of conjugated triene **90**, which contains two α,β-unsaturated ketones, is possible (Eq. 47).[189] Dienone substrates such as **91** can be epoxidized twice in the presence of polypeptides to form C_2-symmetric diepoxides **92** with good levels of enantioselectivity and diastereoselectivity (Eq. 48).[186] An example of an acetylenic substrate has been reported (Eq. 49).[179]

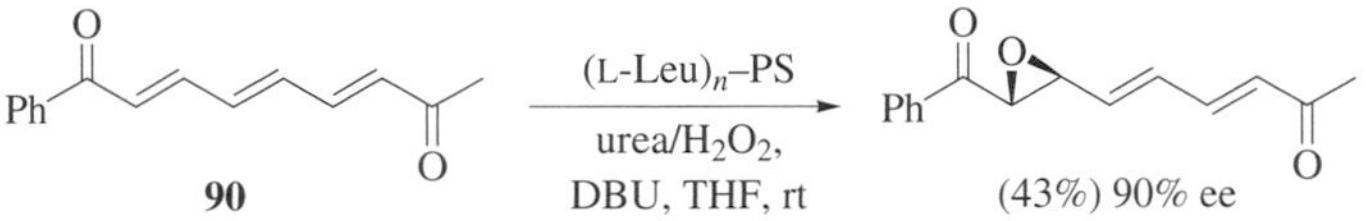

(Eq. 47)

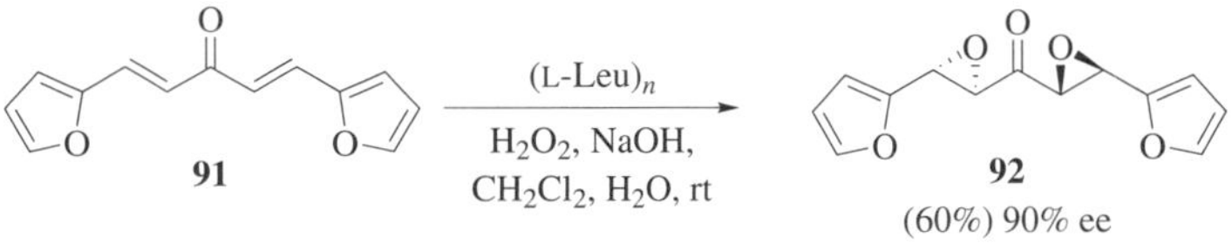

(Eq. 48)

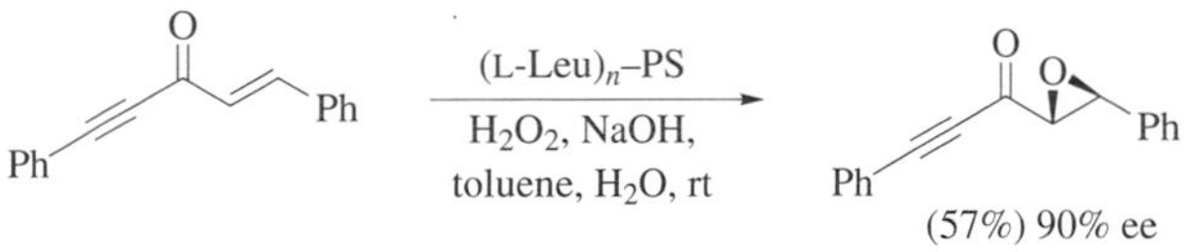

(Eq. 49)

Hetero-Substituted *E*-Enones

Only one enantioselective epoxidation of an enone bearing a heteroatom substituent has been reported; treatment of the vinyl stannane **93** with alkaline hydrogen peroxide and poly-L-leucine in a triphasic system affords the epoxide **94** in good yield and excellent enantiomeric excess (Eq. 50).[191]

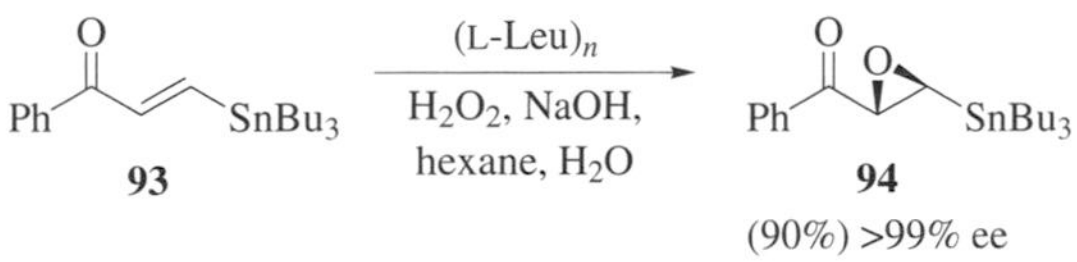

(Eq. 50)

Z-Enones

To date only the Ln–BINOL system has been used for the asymmetric oxidation of acyclic *Z*-enones to the corresponding *cis*-epoxides in synthetically useful yields and enantioselectivities.[49] *Z*-Enones that possess alkyl groups on both the ketone and the alkene have been epoxidized with the Yb(O*i*-Pr)$_3$/**39** system, using TBHP as the oxidant, with enantiomeric excesses ranging from 93–96% (Eq. 51). The system is less effective with aryl ketone substrates, with lower levels of enantioselectivity and diastereoselectivity being observed.

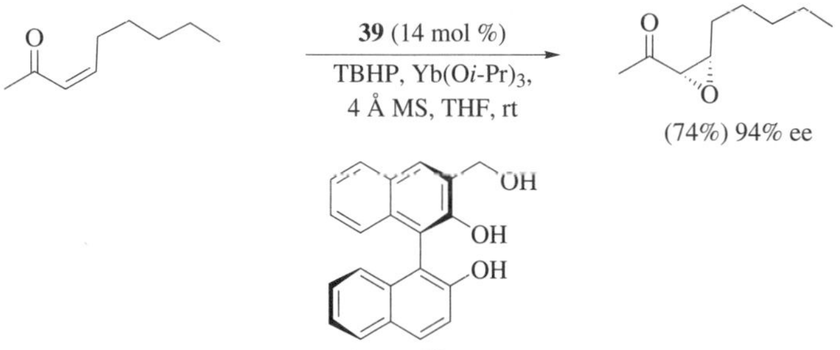

39

(Eq. 51)

The application of other asymmetric epoxidation procedures to *Z*-enones has resulted in the *trans*-epoxide being the major (or sole) product.[180,181]

Enones Bearing a Second Electron-Withdrawing Substituent

Polyleucine has been used as a catalyst for the epoxidation of several substrates of this class; a number of diones and a keto esters have been epoxidized in reasonable yields and with good enantiomeric excess.[178,179] In addition, a keto ester and a dione substrate have been epoxidized using a BINOL/La(O*i*-Pr)$_3$/tri(*p*-fluorophenyl)phosphine oxide system with similar levels of enantioselectivity (Eq. 52).[54]

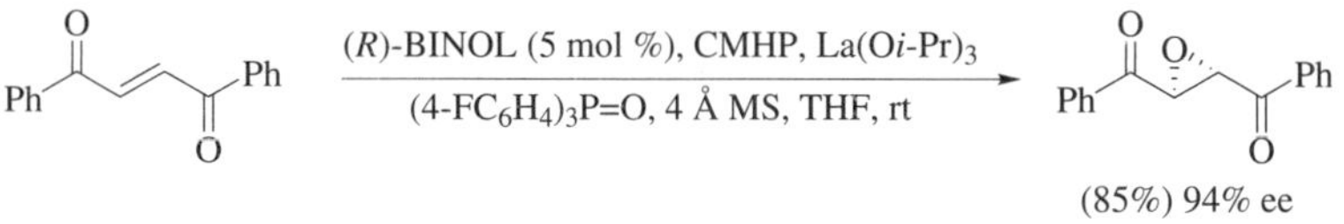

(Eq. 52)

1,1-Disubstituted and Trisubstituted Alkenes

No generally applicable methods exist for the epoxidation of trisubstituted alkenes or for those that are geminally disubstituted, although a few isolated examples have been reported (Eqs. 53–56).[31,49,178,179]

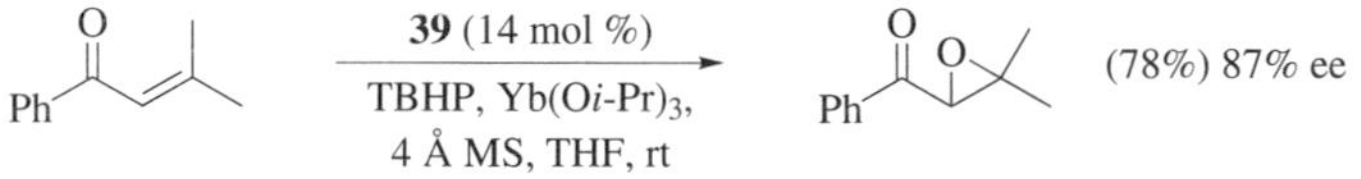

(Eq. 53)

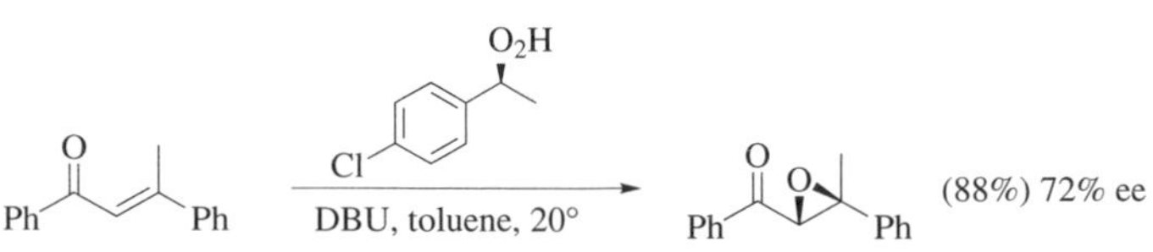

(Eq. 54)

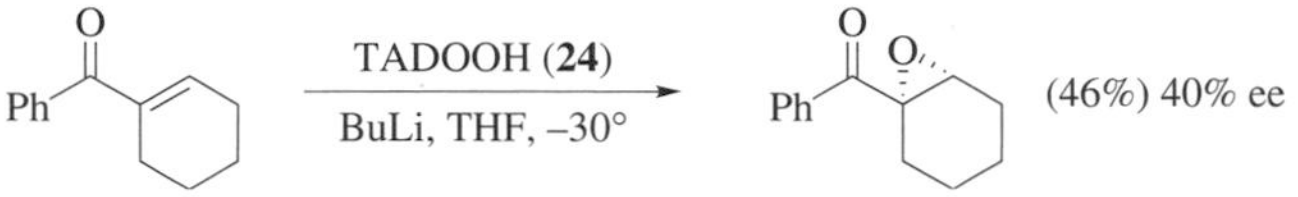
O
Ph
TADOOH (24)
BuLi, THF, –30°
O
O
Ph
(46%) 40% ee

(Eq. 55)

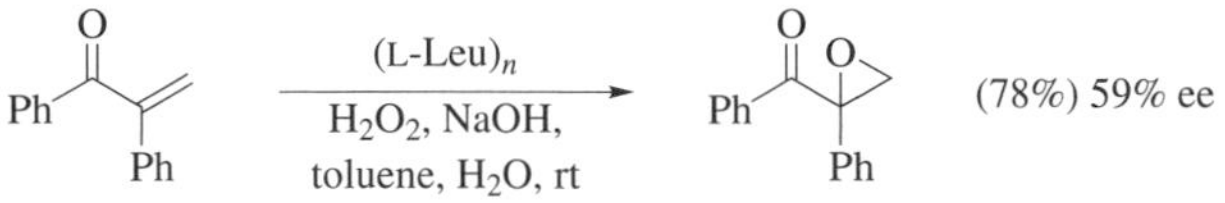
O
Ph
Ph
(L-Leu)$_n$
H_2O_2, NaOH,
toluene, H_2O, rt
O
O
Ph
Ph
(78%) 59% ee

(Eq. 56)

However, alkylidene and arylmethylidene tetralones (**95**) represent one class of trisubstituted enones that have been successfully epoxidized with good enantioselectivities under a number of conditions. Polyleucine catalysis,[192] phase-transfer catalysis,[83,98,105,109] and the $Et_2Zn/O_2/N$-methylpseudoephedrine system[41] all give reasonable to excellent yields and enantiomeric excesses with no single method proving superior. In addition, there is a single example using (*S*)-(1-phenyl)ethyl hydroperoxide as a chiral oxidant, proceeding with good yield and stereoselectivity.[27] Catalysis by polyleucine[192,193] and by chiral quaternary ammonium salts[109] have also been applied to the corresponding indanone derivatives, and the polyleucine method to the 7-membered ring analogues.[192,193]

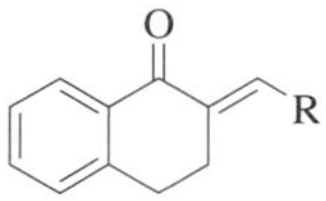
O
R

95

Endocyclic Enones

Epoxidations of simple endocyclic enones tend to be only moderately enantioselective. The majority of such reactions have been carried out under phase-transfer catalysis.[83,84,94,105,194,195] Cyclohex-2-enone is a typical substrate; despite considerable variation of catalyst structure and reagents, enantioselectivities remain moderate.[94] Similar results have been obtained for a number of substituted cyclohexenones.

Cyclohex-2-enone and 3-methylcyclohex-2-enone (**96**) have been epoxidized using the chiral oxidant TADOOH (**24**). While cyclohex-2-enone is epoxidized with poor enantioselectivity, the methyl-substituted epoxide **97** is formed in reasonable yield and enantiomeric excess (Eq. 57).[31]

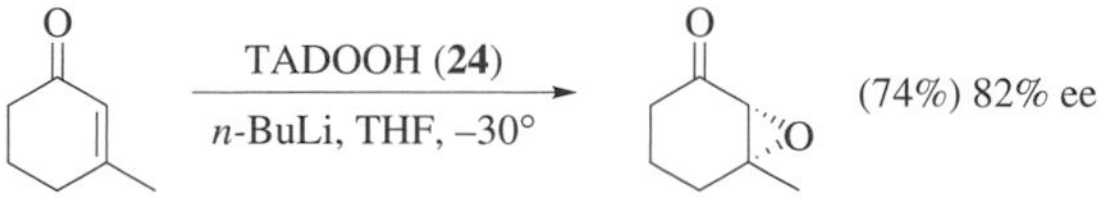
O
TADOOH (24)
n-BuLi, THF, –30°
O
O
(74%) 82% ee

(Eq. 57)

96 **97**

Two methods, namely phase-transfer catalysis and the use of chiral oxidants, have been directly compared for the epoxidation of isoflavones, with the former approach being more successful. Using the most effective catalyst **60**, several

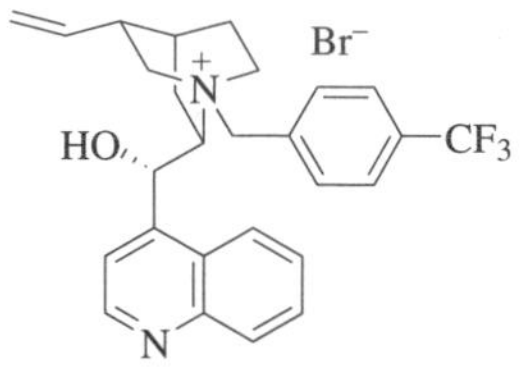

60

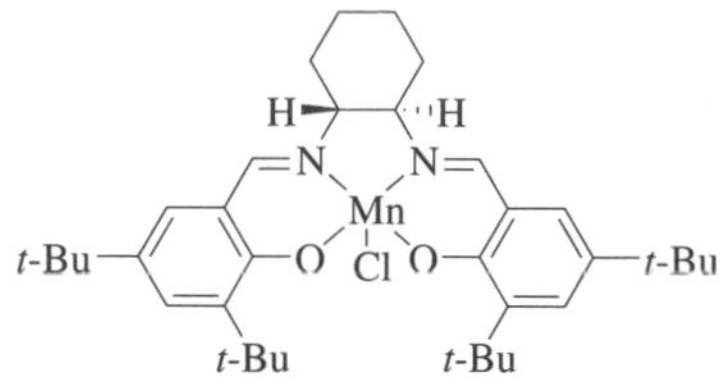

73

different substrates have been epoxidized in good to excellent yields and generally good enantioselectivities.[29]

Isoflavones have also been epoxidized using the manganese-salen catalyst **73**. Enantioselectivities are quite variable and occasionally good, but yields are low.[196]

Quinones and Quinone Monoketals

Almost all of the quinones that have been subjected to asymmetric epoxidation are 2-substituted 1,4-naphthoquinones. Four main approaches have been used: catalysis by serum albumins,[141–144] the use of sugar-derived hydroperoxides as stoichiometric oxidants,[30] phase-transfer catalysis,[81,85,88,103,106] and catalysis by cyclodextrins.[171]

The epoxidations of 2-substituted naphthoquinones in aqueous solution, catalyzed by bovine or human serum albumin, typically proceed with moderate yields (Eq. 58).[141–144] Enantioselectivities range from 0–100%, and are generally increased by the addition of small amounts of an organic solvent such as isooctane.[143] No straightforward correlation exists between enantioselectivity and the size of the 2-substituent, but the best results tend to be obtained with branched or long alkyl chains. It should be noted, however, that 2-*tert*-butylnaphthoquinone is inert under these conditions.[141,142]

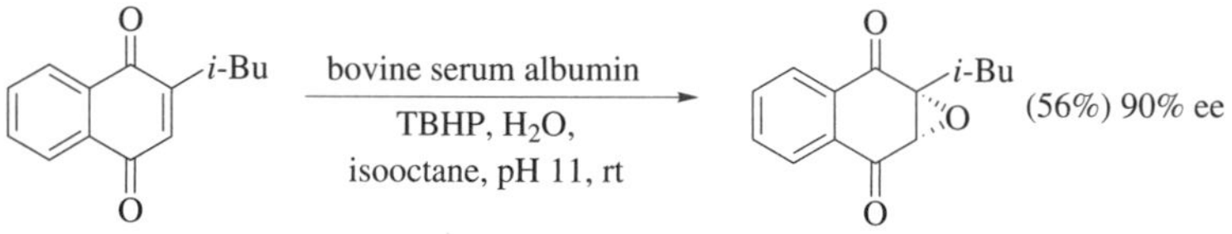

(Eq. 58)

Similar substrates have been subjected to epoxidation with the chiral oxidant **23**. Reasonable yields and enantiomeric excesses ranging from 35–82% are obtained. Again, enantioselectivity improves as the substituent increases in size, with epoxidation of 2-phenylnaphthoquinone (**98**) proving particularly selective (Eq. 59).[30] Various other chiral oxidants have been used for the epoxidation of quinones, however none can match the enantioselectivity or the breadth of examples of the sugar-derived oxidant **23**.[33,34]

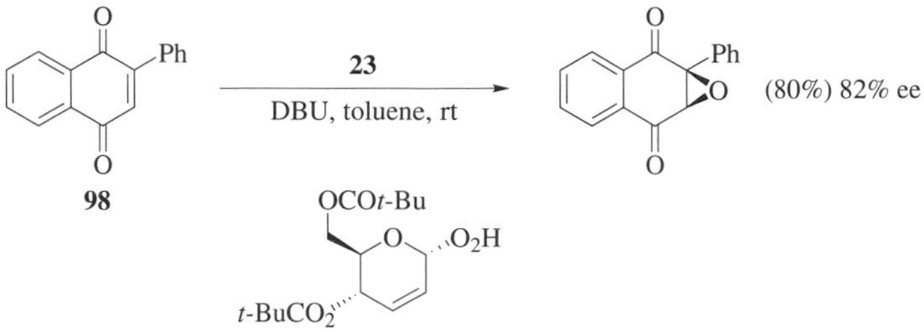

98 **23** (Eq. 59)

A number of different phase-transfer catalysts derived from *Cinchona* alkaloids have been used for the epoxidation of 2-substituted 1,4-naphthoquinones. Although the widest range of substrates has been epoxidized in the presence of benzylquininium chloride (**99**),[81,85,95,197] superior enantiomeric excesses are generally obtained with the 1-naphthyl-substituted quinidinium salt **61**.[103,106] This approach often gives similar enantioselectivities to those observed with the chiral oxidant **23**. With phase-transfer catalysts, the highest enantioselectivities are obtained when the substrate bears an aromatic, secondary alkyl, or tertiary alkyl substituent, but even under optimized conditions, the highest enantiomeric excess obtained is only 76%. Quaternary ammonium salts derived from deaza-analogues of *Cinchona* alkaloids have also been investigated for the epoxidation of 2-isopropylnaphthoquinone. The highest enantiomeric excess, 84%, is obtained with catalyst **100**.[88] The generality of this catalyst for the epoxidation of quinones has not been established.

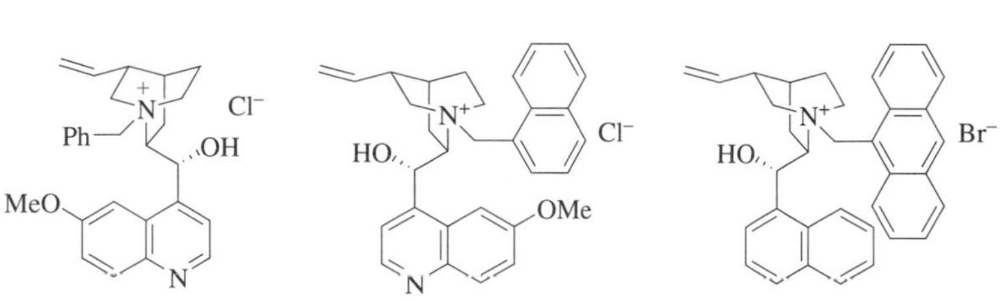

99 **61** **100**

A series of quinone epoxidations has been performed in the presence of 5 molar equivalents of cyclodextrin. Yields and enantiomeric excesses vary considerably with changing substrate structure, however they are all less than 50%.[171–173]

Only one benzoquinone substrate, the amidobenzoquinone **101**, has been subjected to asymmetric epoxidation; treatment with the sugar-derived hydroperoxide **23** generates the epoxide **102**, in which the more electron-deficient alkene has been epoxidized, in 55% yield and 64% ee (Eq. 60).[30]

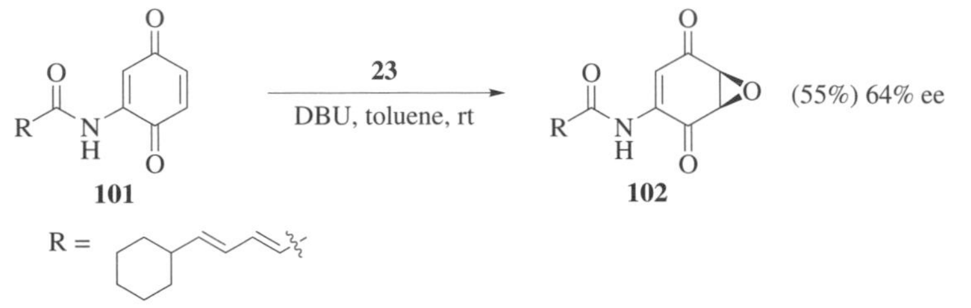

(Eq. 60)

Other sugar-derived hydroperoxide oxidants yield the enantiomeric epoxide **104**; however the enantioselectivity is much lower, the best result being obtained with the dibenzyl derivative **103** (Eq. 61).[30]

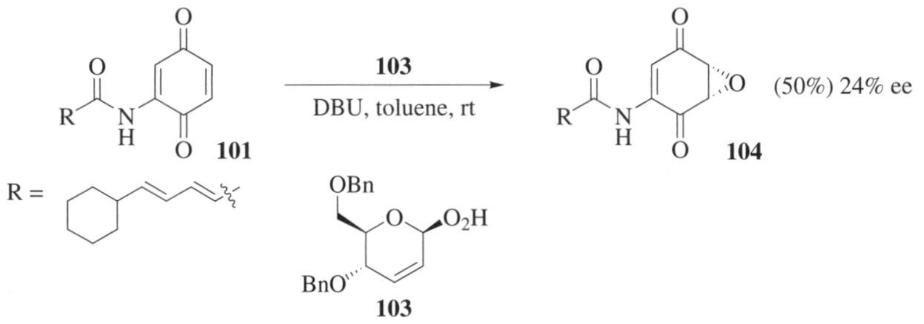

(Eq. 61)

Quinone monoketals have also been epoxidized using a variety of techniques. A substrate of particular interest is the quinone monoketal **28**, a key intermediate in an approach to the manumycin alkaloids.[198] Epoxidation of ketal **28** in the presence of a wide variety of chiral guanidine bases gives only poor to moderate enantiomeric excesses,[36–39] whereas application of phase-transfer catalysis affords epoxide **105** in a low yield but good enantiomeric excess. (Eq. 62).[194,195]

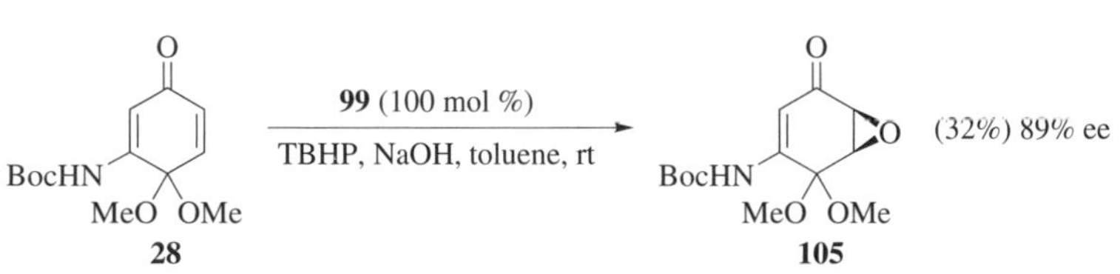

(Eq. 62)

A tartrate-modified sodium alkylperoxide has been used for the epoxidation of four related quinone monoketals giving, in three cases, excellent enantioselectivities (Eq. 63).[74] A benzoquinone monoketal has been epoxidized in good yield and excellent enantiomeric excess using a chiral quaternary ammonium salt as the catalyst.[110]

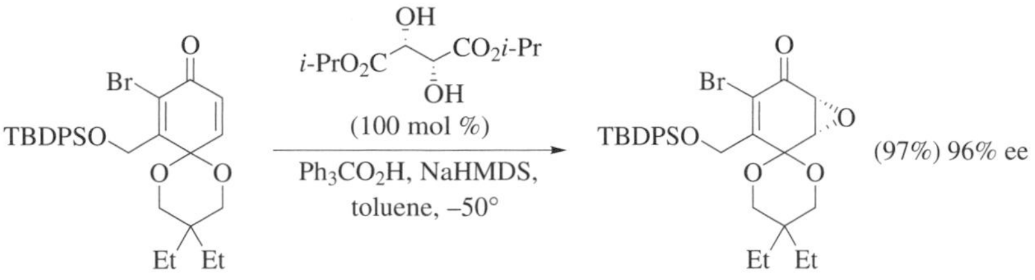

(Eq. 63)

Other α,β-Unsaturated Carbonyl Compounds

α,β-Unsaturated Aldehydes. A single synthetically useful approach to the epoxidation of α,β-unsaturated aldehydes has been reported: a range of substrates has been epoxidized using hydrogen peroxide in the presence of the proline-derived catalyst **77** (Eq. 64).[168,169]

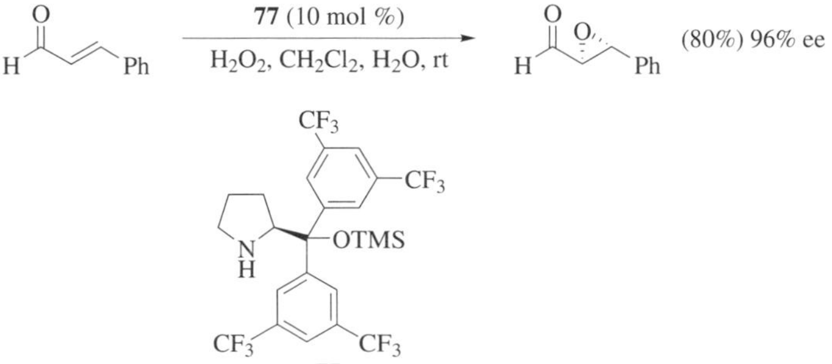

77

(Eq. 64)

A variety of β-substituents is tolerated, with the epoxides generated in reasonable yield and, typically, in excellent enantiomeric excess. A number of substrates bearing a β-aryl group have been epoxidized, including some possessing strongly electron-withdrawing and weakly electron-donating substituents. Moreover, the method has been extended to the epoxidation of a substrate disubstituted on the β-carbon (Eq. 65),[168] and to one where the β-group is an ester.[168,169]

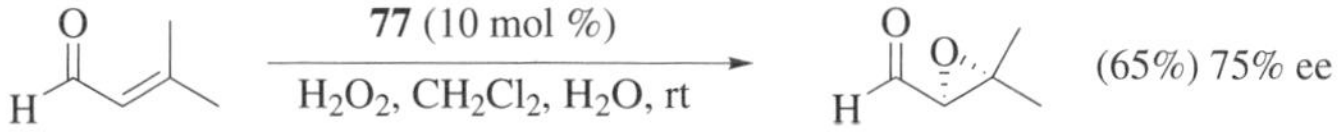

(Eq. 65)

α,β-Unsaturated Esters. Both nucleophilic and electrophilic oxidation methods have been applied to α,β-unsaturated esters. The majority of substrates are disubstituted *E*-alkenes.

The use of chiral metal complexes in the presence of oxidizing agents constitutes probably the most widely applicable method of epoxidizing *E*-enoates. Specifically, chiral biphenyl ligand **47** in the presence of $Y(Oi\text{-}Pr)_3$ and triphenylarsine oxide using TBHP is an epoxidation system that has been successfully applied to cinnamates as well as to substrates where the β-substituent is heterocyclic or alkyl (Eq. 66). These epoxidations invariably proceed in good yield and excellent enantioselectivity.[64]

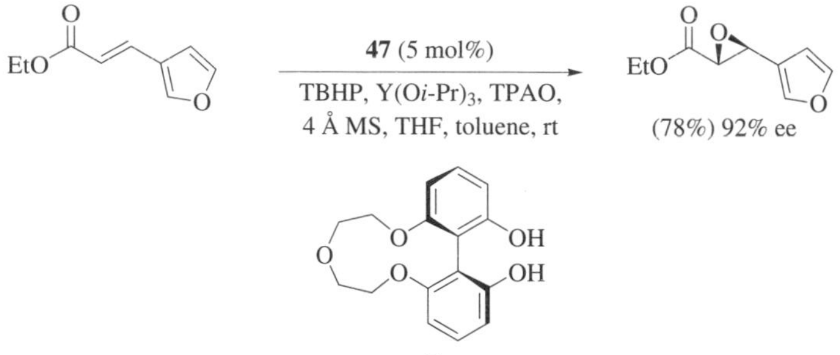

47

(Eq. 66)

Enoates have also been epoxidized using Oxone in the presence of a variety of chiral ketones. The most synthetically useful appear to be ketone **69**, derived from fructose,[150] and the binaphthyl ketone **71**.[154]

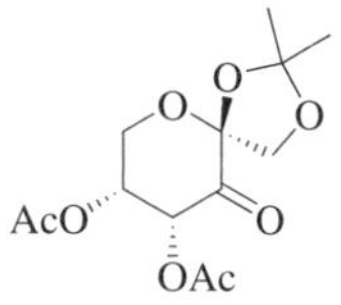

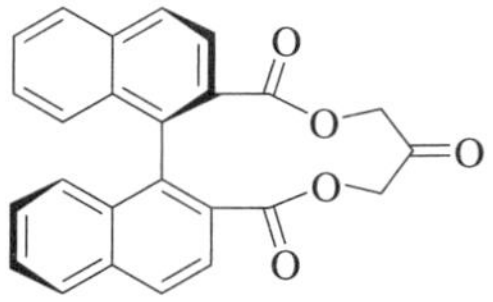

69 **71**

In a typical example, reaction of ethyl cinnamate (**106**) with Oxone in the presence of 30 mol % of ketone **69** leads to epoxide **107** in 73% yield and 96% ee (Eq. 67).[150]

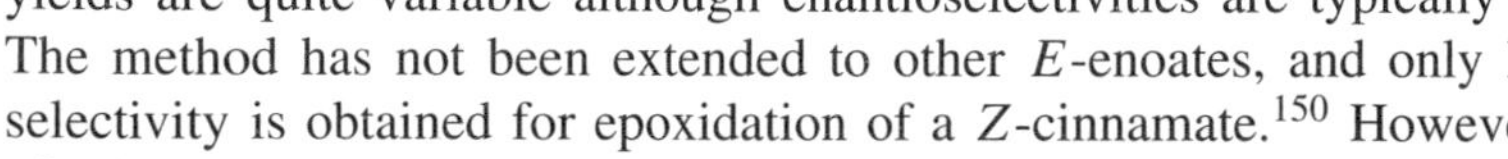

(73%) 96% cc

106 **107**

(Eq. 67)

A wide range of cinnamates has been epoxidized in the presence of ketone **69**; yields are quite variable although enantioselectivities are typically excellent.[150] The method has not been extended to other *E*-enoates, and only low enantioselectivity is obtained for epoxidation of a *Z*-cinnamate.[150] However, a number of trisubstituted epoxides have been successfully prepared in excellent enantiomeric excess using this method including a cyclohexene derivative **108**, β,β-disubstituted examples such as **109**, and α-substituted examples such as **110**.[150]

The binaphthyl ketone **71** has been used to mediate the epoxidation of a large number of cinnamates (Eq. 68). Both yields and enantiomeric excesses are rather variable, but there are a number of synthetically useful examples. Although

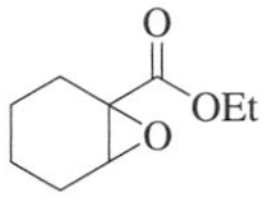
O
OEt
O

108
(77%) 93% ee

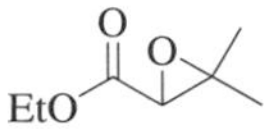
O
O
EtO

109
(64%) 82% ee

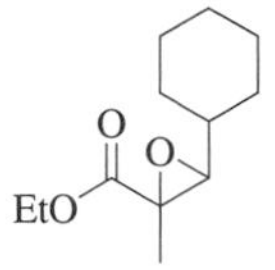
O
O
EtO

110
(77%) 89% ee

this approach has been extended to Z-alkenes, enantioselectivities are low.[154] A variety of other ketones have been used in a similar fashion to epoxidize α,β-unsaturated esters but none of these approaches is as well developed as the methods discussed above.[145,146,149,151,156–158,161–163,199]

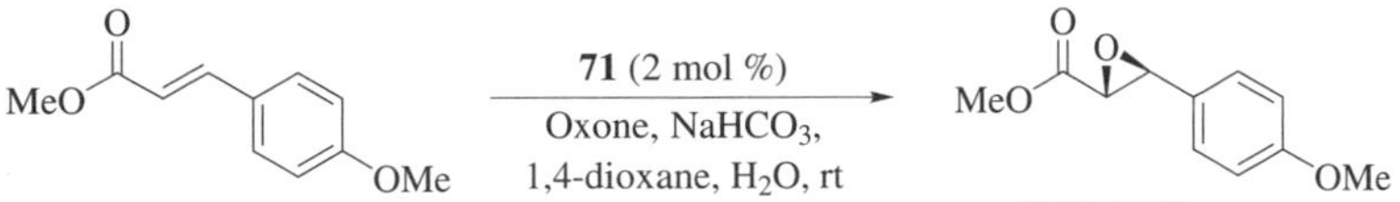
MeO
O
OMe
71 (2 mol %)
Oxone, $NaHCO_3$,
1,4-dioxane, H_2O, rt
MeO
O
O
OMe

(81%) 76% ee

(Eq. 68)

One potential limitation of these essentially electrophilic epoxidations is that if substrates contain more than one double bond, the oxidation tends not to be selective for the α,β-alkene, as evidenced in Eq. 69,[200] where the γ,δ-epoxide **112** predominates over the α,β-epoxidation product **111**.

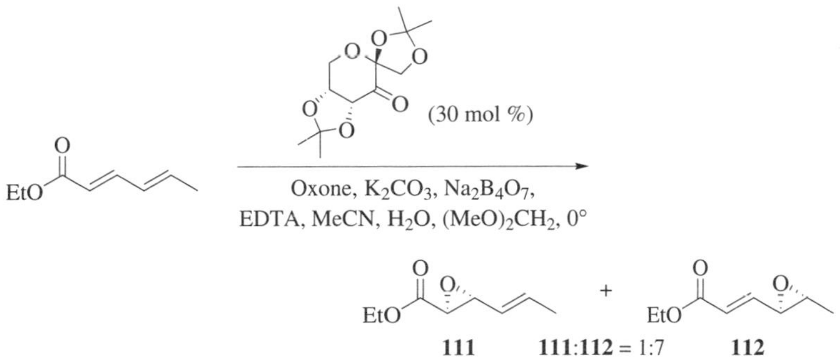
(30 mol %)
EtO
O
Oxone, K_2CO_3, $Na_2B_4O_7$,
EDTA, MeCN, H_2O, $(MeO)_2CH_2$, 0°
EtO
+
EtO

(Eq. 69)

111 **111**:**112** = 1:7 **112**
(7%) 69% ee (41%) 96% ee

A series of Z-β-aryl enoates has been treated with NaOCl in the presence of manganese-salen complex **73** to afford mixtures of *cis*- and *trans*-epoxides in which the cis-isomer predominates (Eq. 70).[293] The highest enantiomeric excesses are obtained when the ester group is bulky, and the aryl group bears an electron-donating substituent. Addition of the chiral ammonium salt **99** to the reaction mixture leads to the isolation of both *trans*- and *cis*-epoxides **113** and **114,** with the *trans*-epoxide as the major product with good enantioselectivity (Eq. 71).[165]

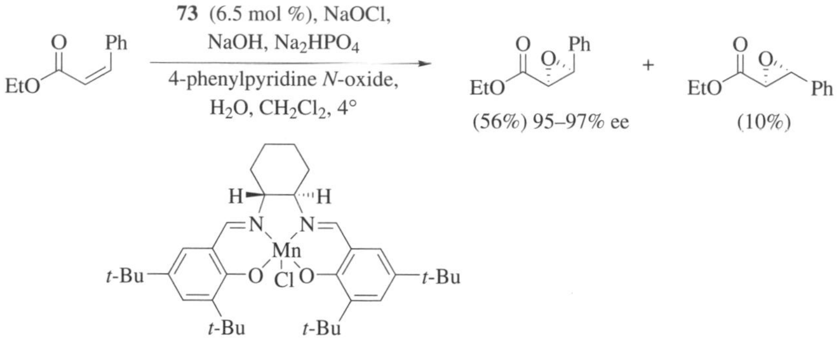
73 (6.5 mol %), NaOCl,
NaOH, Na2HPO4
4-phenylpyridine N-oxide,
H2O, CH2Cl2, 4°
EtO
Ph
(56%) 95–97% ee
(10%)
t-Bu
Mn
Cl

73

(Eq. 70)

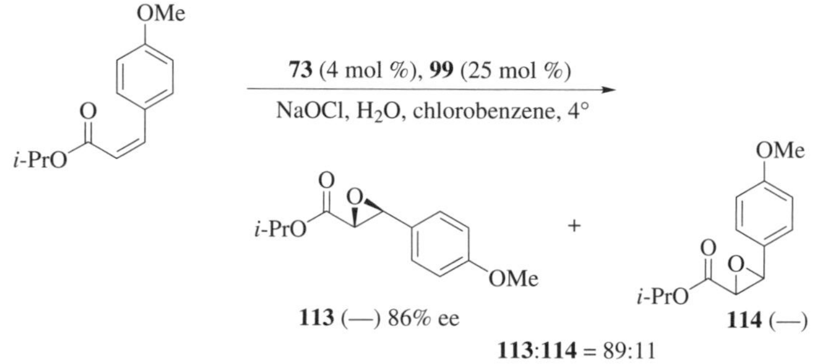
OMe
i-PrO
73 (4 mol %), 99 (25 mol %)
NaOCl, H2O, chlorobenzene, 4°
113 (—) 86% ee
114 (—)
113:114 = 89:11

(Eq. 71)

Other methods that have been successfully applied to the epoxidation of α,β-unsaturated esters but for which the scope has not been established include epoxidations with egg phosphatidylcholine/*m*-CPBA,[174] catalysis by polypeptides,[140,201] and the use of chiral ligands for lithium peroxides.[72]

α,β-Unsaturated Acids. A range of cinnamic acids have been epoxidized in aqueous solution with moderate enantioselectivity by Oxone in the presence of various steroidal ketones.[152,153]

With *p*-methylcinnamic acid (**115**), the highest enantiomeric excess is obtained using one equivalent of ketone **116** (Eq. 72); ketone **117** gives the enantiomeric epoxide in somewhat lower enantiopurity (Eq. 73).[152]

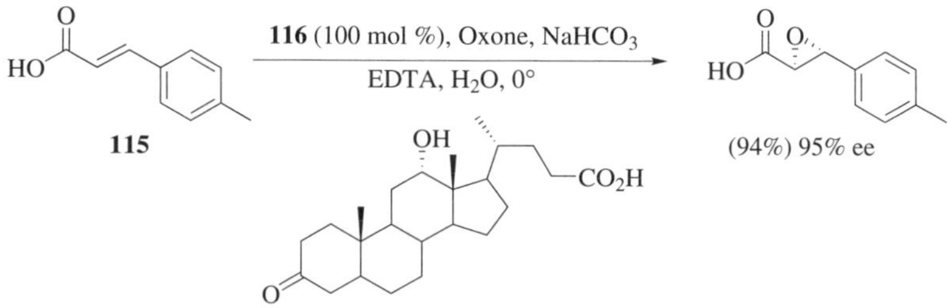
HO
115
116 (100 mol %), Oxone, NaHCO3
EDTA, H2O, 0°
OH
CO2H
(94%) 95% ee

116

(Eq. 72)

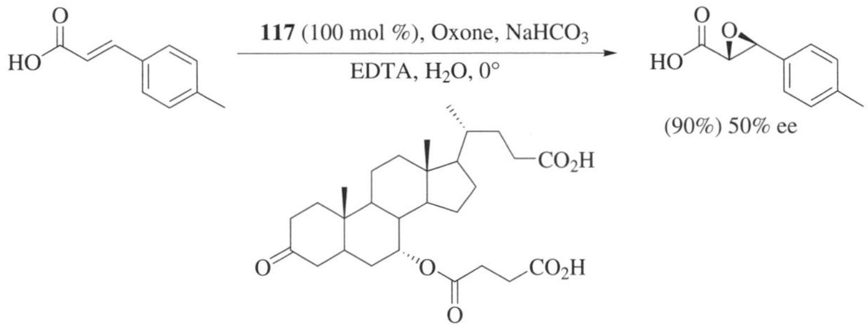

117

(Eq. 73)

α,β-Unsaturated Amides and *N*-Acylated Heterocycles. A number of *E*-α,β-unsaturated amides have been epoxidized using TBHP as an oxidant in the presence of a 1:1:1 Sm(O*i*-Pr)$_3$/BINOL/TPAO complex.[60] Yields and enantioselectivities are excellent. The method is applicable to *N*-alkyl and *N*,*N*-dialkyl amides bearing both aliphatic and aromatic groups on the β-carbon of the double bond. Moreover, the method has been extended to encompass synthetically useful morpholinoamides (Eq. 74).[59,61]

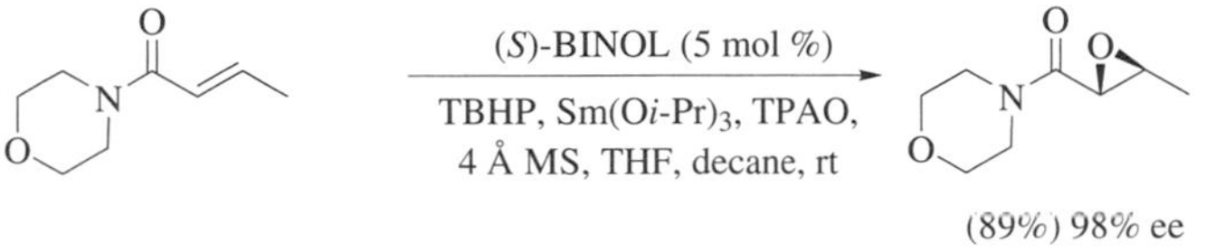

(Eq. 74)

A related system has been applied to the epoxidation of acyl pyrroles. A wide range of substrates, including *E*-alkenes with both saturated and unsaturated β-substituents, has been epoxidized in good yield and excellent enantioselectivities. For this system, the optimal catalyst appears to be the octahydro-BINOL derivative **46** (Eq. 75).[62,63]

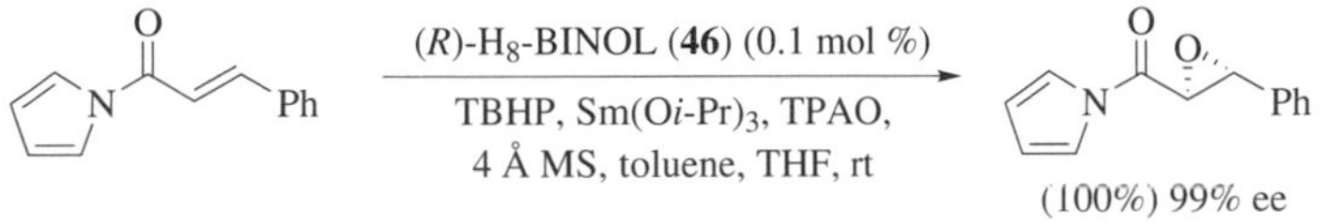

(Eq. 75)

The binaphthyl-substituted chiral ketone **71** has been used with Oxone for the epoxidation of amides; enantioselectivities can be moderate in the case of secondary amides but are poor in the case of tertiary amides.[155]

A method that may prove of general utility for the synthesis of chiral epoxy carbonyl compounds is the asymmetric epoxidation of α,β-unsaturated acyl imidazolides.[56–58] For example, various α,β-unsaturated 4-phenyl-1-acylimidazoles can be epoxidized with TBHP and BINOL/La(O*i*-Pr)$_3$ in the presence of TPAO to give initially the *tert*-butyl peroxyester of the epoxy acid; subsequent addition of methanol affords the corresponding α,β-epoxy ester product with good enantioselectivity (Eq. 76).[56] Substrates containing a second isolated double bond can be epoxidized selectively. Moreover, the intermediate *tert*-butyl peroxyester can be treated with a variety of nucleophilic reagents to generate epoxy amides, aldehydes, and keto esters.

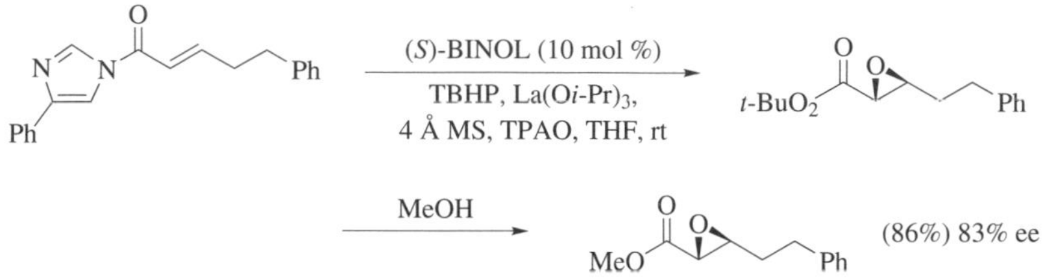

(Eq. 76)

Alkenes Bearing Electron-Withdrawing Groups Other Than Carbonyl

α,β-Unsaturated Sulfones. Polypeptide catalysis is the only method to have been applied to the epoxidation of α,β-unsaturated sulfones. A variety of conditions have been reported, although all feature a toluene/water solvent system and utilize a quaternary ammonium salt additive. In all the examples the sulfone functionality is attached to a phenyl, or substituted phenyl group, and the β-substituent is either a proton or aromatic group. Yields are moderate and enantioselectivities vary from 64–95% (Eq. 77).[120,190,192,201]

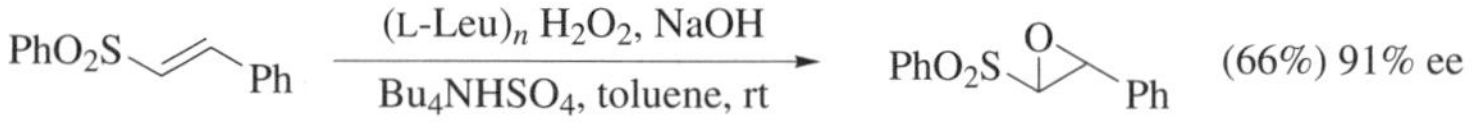

(Eq. 77)

α,β-Unsaturated Nitro Compounds. Epoxidation of a series of *E*-α,β-unsaturated nitro compounds using $Et_2Zn/O_2/N$-methylpseudoephedrine (**29**) proceeds in moderate yields and enantiomeric excesses (36–82%) (Eq. 78).[42] One trisubstituted alkene of this class has been epoxidized in the presence of polyalanine, with very low enantioselectivity.[132]

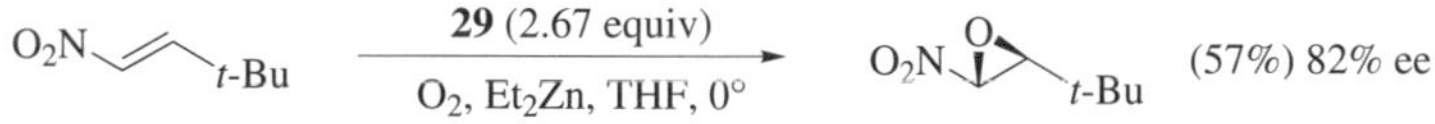

(Eq. 78)

α,β-Unsaturated Nitriles. Although some α,β-unsaturated nitriles have been subjected to asymmetric epoxidation using molecular oxygen and chiral tertiary amines, the levels of enantioselectivity obtained are very low.[202]

α,β-Unsaturated Phosphonates. Vinyl phosphonates are the only class of electron-deficient alkenes to which enzymatic epoxidation has been applied with any success. Epoxidation of (*Z*)-prop-1-enylphosphonic acid using a variety of microorganisms affords the antibiotic fosfomycin.[11,12] Details of this biotransformation are beyond the scope of this review.

COMPARISON WITH OTHER METHODS

Asymmetric epoxidation of electron-deficient alkenes is frequently the preferred method for the synthesis of non-racemic epoxy ketones and related compounds thanks to the directness of the method and the ready accessibility of the alkene precursors. However, three other methods that may be considered are asymmetric variants of the Darzens reaction, chiral-auxiliary-directed epoxidations of electron-deficient alkenes, and oxidation of epoxy alcohols.

Asymmetric Darzens Reactions

The Darzens reaction classically occurs between an α-halo ester and an aldehyde to generate a glycidic ester.[203] This reaction, and variants thereof, have been the subject of extensive study with a view to the development of an efficient asymmetric version. Enantioenriched epoxy esters, amides, acids, and sulfones have been successfully synthesized by this method, along with a limited range of epoxy ketones. A range of approaches has been reported including the use of chiral auxiliaries, asymmetric catalysis, and the use of enantiomerically enriched reagents.

Chiral Auxiliaries. The reaction between the enolate of an α-halo carboxylic acid derivative bearing a chiral auxiliary and an aldehyde affords either an epoxide directly or a halohydrin, depending on the reaction conditions. A range of chiral auxiliaries has been used for such reactions; for example, conversion of oxazolidinone derivative **118** into a boron enolate and reaction with isobutyraldehyde affords the syn-aldol product **119** as a single stereoisomer (Eq. 79).[204] Subsequent ring closure with concomitant removal of the auxiliary affords *cis*-epoxide **120**. The major diastereomer obtained in these reactions depends both on the nature of the aldehyde and the metal enolate generated. Thus aromatic aldehydes tend to give predominantly *anti*-aldol products when lithium, zinc, or tin(IV) enolates are employed, but mainly *syn*-aldols from boron, titanium, or tin(II) enolate species.[205]

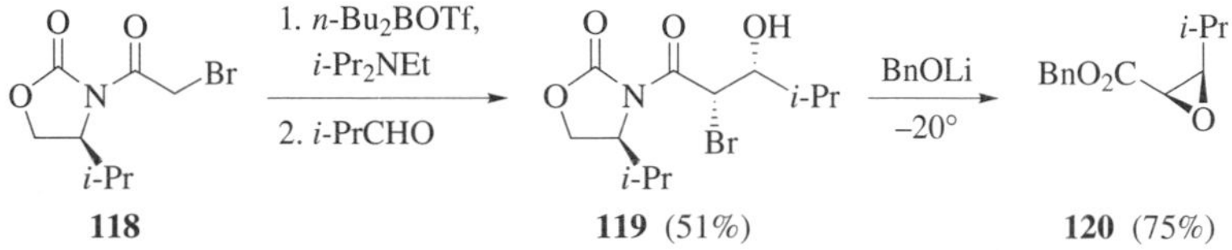

118 **119** (51%) **120** (75%) (Eq. 79)

Some other auxiliaries for asymmetric Darzens reactions are depicted below. These have been used for the preparation of *cis*-epoxy acids, compounds for which no direct asymmetric epoxidation protocol has been developed. *N*-Acyloxazolidinethione **121** has been subjected to a one-pot sequence of bromination followed by aldol reaction with an aliphatic aldehyde to provide *syn*-bromohydrins as single stereoisomers; once again, base treatment effects removal of the auxiliary and ring closure to afford *cis*-epoxy acids.[148]

Reaction of the dianion of hydroxy ketone **122** with aromatic aldehydes leads directly to *trans*-epoxy ketones with ≥94% diastereomeric excess, but lower

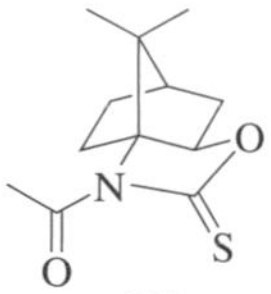
121

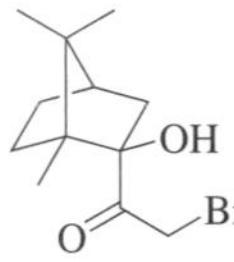
122

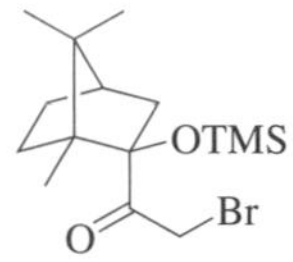
123

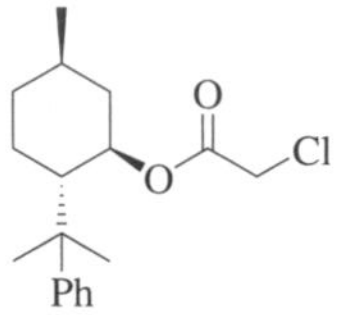
124

(and reversed) stereoselectivity is seen when aliphatic aldehydes are employed. If the silyl ether analogue **123** is used, aliphatic aldehydes react to give *syn*-bromohydrin products in good yields and 84–90% diastereoselectivity excess. Closure to a *cis*-epoxide occurs on treatment either with tetra-*n*-butylammonium fluoride or with hydrochloric acid followed by sodium carbonate, and the auxiliary can be removed oxidatively to give a *cis*-epoxy acid.[206]

8-Phenylmenthyl ester **124** undergoes Darzens reactions with a range of ketones to give trisubstituted epoxy esters with high diastereomeric excess (81–96%),[207–209] whereas other chiral esters have proved to give less stereoselective reactions in the generation of such species.[210,211]

Chiral auxiliaries attached to the electrophilic partner are less prevalent; however, the Darzens reaction of chiral chromium tricarbonyl-complexed aryl aldehydes has been described, with enantiomeric excesses of up to 96% reported following decomplexation.[212,213]

A third method for attachment of a chiral auxiliary is by use of a sulfonium ylide in a reaction mechanistically related to the Darzens reaction. For example, the ylide derived from sulfonium salt **125** reacts with pivalaldehyde to afford glycidic amide **126** with good enantiomeric excess (Eq. 80).[214] Epoxides of similarly high enantiomeric excess can be generated from aromatic aldehydes, but lower levels of asymmetric induction are observed with isobutyraldehyde and a long-chain aliphatic aldehyde.[214]

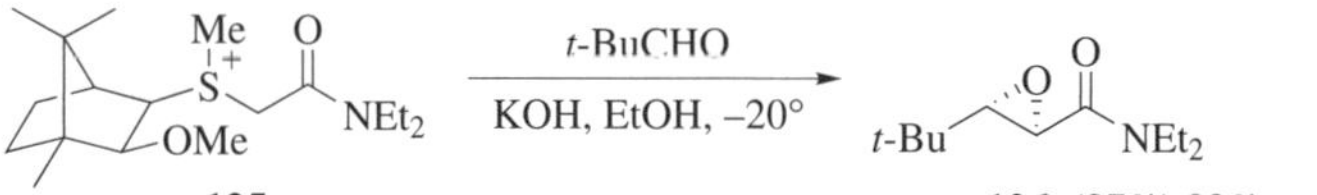
125 **126** (87%) 93% ee (Eq. 80)

Chiral Ligands. The chiral diazaborolidine **128** is a highly effective reagent for the synthesis of *trans*-glycidic esters. Formation of the chiral boron enolate from *tert*-butyl bromoacetate (**127**) followed by reaction with an aldehyde gives an *anti*-bromohydrin in high diastereo- and enantioselectivity. Treatment with base then affords a *trans*-glycidic ester (Eq. 81).[215]

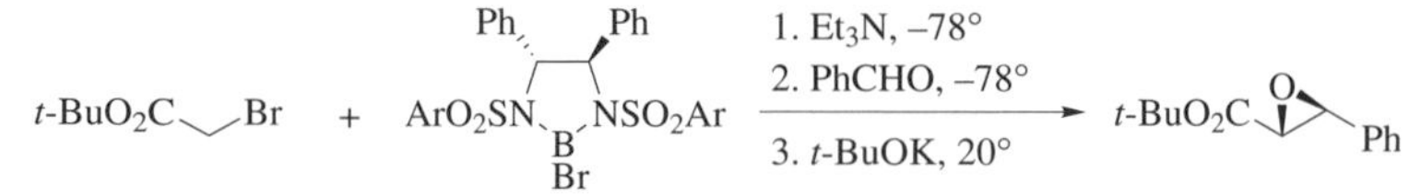
127 **128** (98%) 98% ee (Eq. 81)

Ar = 3,5-$(CF_3)_2C_6H_3$

Asymmetric Catalysis. Asymmetric phase-transfer catalysis has been applied to the Darzens reaction of α-halo ketones; the cinchonine derivative **60** catalyzes the reactions of α-chloroacetophenone[216,217] and β-chloro-α-tetralones[217,218] with aldehydes, giving epoxy ketones with enantiomeric excesses of 42–86%. More significantly, α-chloro sulfone **129** participates in asymmetric Darzens-type reactions catalyzed by quininium salt **130** to give β-aryl epoxy sulfones in 65–74% enantiomeric excess (Eq. 82);[219,220] no asymmetric epoxidation method has been shown to be generally successful for the synthesis of such compounds.

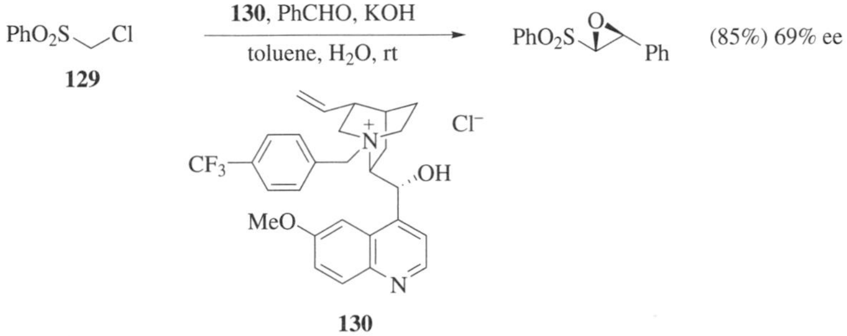

129

130

(Eq. 82)

Moderate to good enantiomeric excesses have also been obtained in Darzens reactions catalyzed by pyrrolidinium salts,[91] chiral aza-crown ethers,[221–224] and BSA.[225]

Chiral Bases. Chiral lithium amides have been used in asymmetric Darzens reactions of *tert*-butyl chloroacetate with up to 84% ee.[226]

Epoxidation of Alkenes Bearing Chiral Auxiliaries

The diastereoselective epoxidation of alkenes bearing chiral auxiliaries has been investigated extensively. However, in most cases, it has not proved possible to effect removal of the auxiliary while preserving the integrity of the epoxide moiety, and thus these reactions cannot be considered general methods for the synthesis of epoxycarbonyl compounds.

Three possible points of attachment exist for a chiral auxiliary on an acyclic electron-deficient alkene: the electron-withdrawing group can contain the controlling chiral element; or, a substituent, either in the α- or β-position, can bear the chiral group. In practice, all of the reported examples are of chiral electron-withdrawing groups and β-substituents. In addition, a number of cyclic enones bearing chiral directing groups have been reported.

Chiral Electron-Withdrawing Groups. The epoxidation of a variety of α,β-unsaturated esters and amides has been examined, but levels of asymmetric induction are generally poor. One exception is the cinnamamide **131**, which when treated with TBHP/*n*-BuLi affords the epoxide (2*R*,3*S*)-**132** with excellent diastereomeric excess (Eq. 83).[227]

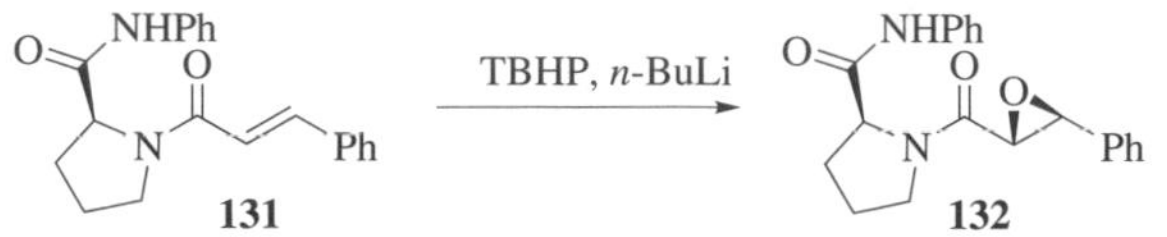

131 → **132** (61%) >99% dr (Eq. 83)

Chiral β-Substituents. Dimethyldioxiranc epoxidation of enones and enoates, substituted at the β-carbon with an acylated glucose-derived auxiliary, proceeds with variable levels of diastereoselectivity in favor of the epoxide (2*R*,3*S*)-**133** (Eq. 84).[228]

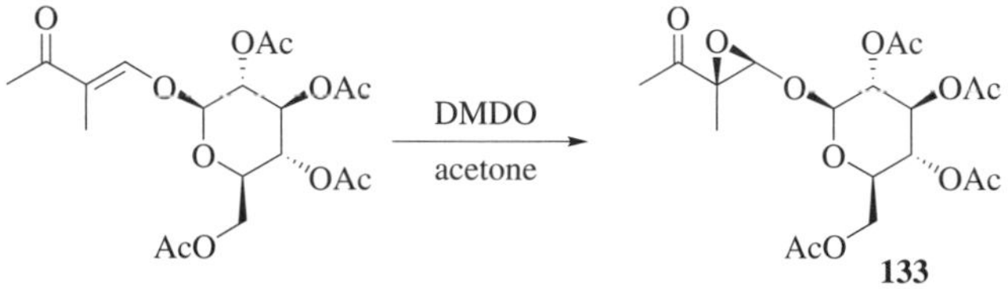

133 (73%) 80% dr

(Eq. 84)

Epoxidation of the norephedrine-derived enal **134** with excess potassium hypochlorite in aqueous THF is accompanied by oxidation of the aldehyde to afford the epoxy carboxylic acid (2*S*,3*S*)-**135** in 90% yield and >90% dr (Eq. 85).[229,230]

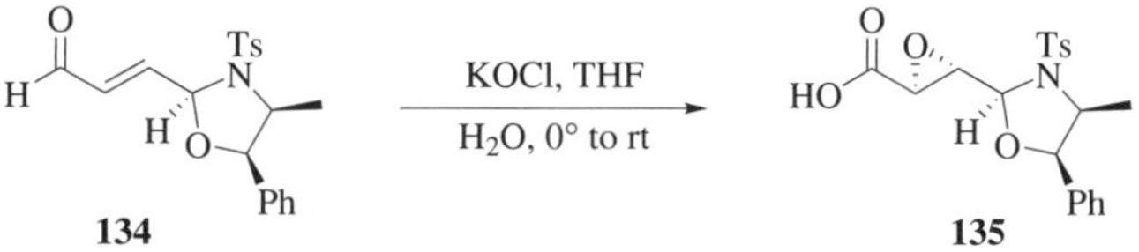

134 → **135** (90%) >90% dr

(Eq. 85)

Cyclic Electron-Deficient Alkenes. Nucleophilic epoxidation of quinone ketals such as **136** affords the corresponding epoxides **137** and **138** with reasonable levels of diastereoselectivity (Eq. 86).[231]

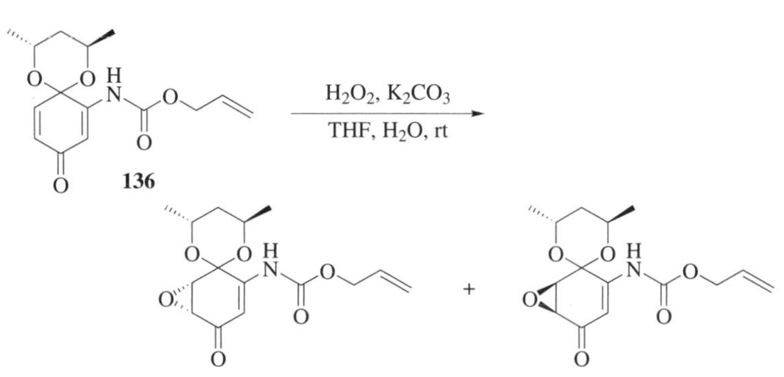

(Eq. 86)

137 (49%) **137**:**138** 4.5:1 **138** (—)

Treatment of lactam **139** with *N*-methylmorpholine *N*-oxide in CH_2Cl_2 affords the epoxide **140** as a single diastereomer (Eq. 87).[232]

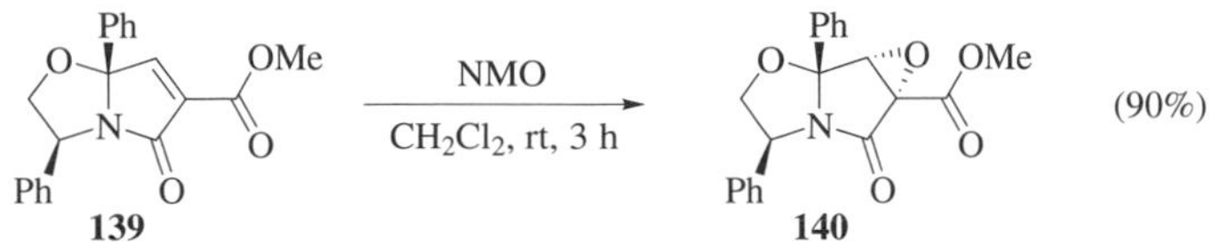

(Eq. 87)

Excellent diastereoselectivities are seen in the epoxidation of Diels-Alder adducts such as **141** using aqueous basic hydrogen peroxide. In this example, epoxidation occurs from the exo-face despite the presence of the ester substituent (Eq. 88).[233]

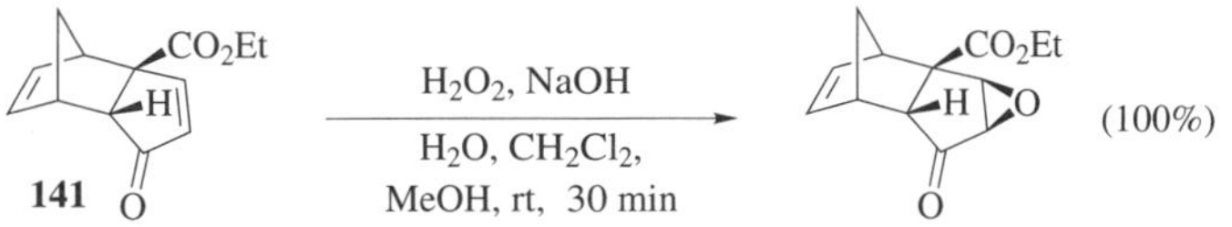

(Eq. 88)

Oxidation of Epoxy Alcohols

The Sharpless asymmetric epoxidation of allylic alcohols is one of the best-established and most reliable methods of asymmetric synthesis, and the resulting alcohols can be oxidized to give α, β-epoxycarbonyl compounds. The epoxides synthesized in this way may be 1,1-, cis-1,2- or trans-1,2-disubstituted, or trisubstituted.[1]

Oxidation of primary epoxy alcohols of this type affords α,β-epoxy aldehydes. A wide range of oxidation methods has been used, including TPAP (tetra-*n*-propylammonium perruthenate)/NMO (*N*-methylmorpholine-*N*-oxide),[234] TEMPO (2,2,6,6-tetramethylpiperidine-*N*-oxyl)/iodobenzene diacetate,[235] Dess-Martin periodinane,[9] manganese dioxide,[236] silver carbonate,[237] various chromium(VI) reagents,[238–240] and, most commonly, the Swern[241] or Parikh-Doering[242] oxidations. As only one efficient method for the asymmetric epoxidation of α,β-unsaturated aldehydes has been reported, this sequence of Sharpless asymmetric epoxidation of allylic alcohols followed by alcohol oxidation is a useful alternative for the preparation of epoxy aldehydes.

More vigorous oxidation of such alcohols may be carried out to generate epoxy acids. By far the most commonly used reagent is catalytic amounts of ruthenium(III) chloride in the presence of sodium periodate,[243] although pyridinium dichromate[244] and platinum/O_2[245] have also been employed for this purpose. Epoxy acids may also be obtained in a two-step process via the epoxy aldehydes referred to previously.

Sharpless epoxidation of secondary allylic alcohols is frequently complicated by the occurrence of kinetic resolution if the substrate is a chiral alcohol present as a racemic mixture; such substrates are nonetheless epoxidized in this manner, and subsequent oxidation to epoxy ketones is carried out either under Swern conditions[246,247] or with methyl(trifluoromethyl)dioxirane.[248]

EXPERIMENTAL CONDITIONS

The following section takes the key methods in turn and comments on the preferred catalysts, solvents, and conditions for the reaction, providing further information regarding the experimental conditions where appropriate.

Zinc-Mediated Epoxidation. The most widely applied approach employs diethylzinc and *N*-methylpseudoephedrine (**29**) in an oxygen atmosphere. In the described procedure, a chiral metal-oxo complex is generated by first mixing *N*-methylpseudoephedrine and a solution of diethylzinc in toluene under an argon atmosphere. After stirring for 80 minutes, oxygen is introduced to the Schlenk flask using a balloon and the oxidant generated is then used to carry out the epoxidation of the enone. Under the described reaction conditions, the chiral ligand can be recovered and reused.[41]

Rare-Earth-Metal-Mediated Epoxidation. A wide range of lanthanide-based methods have been developed for the epoxidation of various classes of α,β-unsaturated carbonyl compounds. All the reported methods take place at room temperature in dry THF as the solvent, and are carried out in the presence of powdered 4 Å molecular sieves; these are generally activated by heating under vacuum immediately prior to the reaction. While substituted BINOL derivatives have been used with some success, the most readily applicable method for the epoxidation of *E*-enones uses the commercially available ligands (*R*)- or (*S*)-BINOL in conjunction with $La(Oi\text{-}Pr)_3$ and CMHP, in the presence of an additive. The most efficient of these, allowing very low loadings of catalyst and ligand, is tris(4-fluorophenyl)phosphine oxide;[54] TPAO also gives excellent results.[55]

For the epoxidation of *Z*-enones, the only conditions that have been shown to give *cis*-epoxides with high enantioselectivity utilize (hydroxymethyl)BINOL **39** (prepared from BINOL in 4 steps)[46] and $Yb(Oi\text{-}Pr)_3$.[49]

The optimal ligand-metal combination for the epoxidation of α,β-unsaturated *N*-acylpyrroles is the reduced BINOL derivative **46** and $Sm(Oi\text{-}Pr)_3$,[62,63] whereas for the epoxidation of α,β-unsaturated esters, the preferred combination is biphenyldiol **47** and $Y(Oi\text{-}Pr)_3$.[64] With both combinations, TPAO is used as an additive. While ligand **46** is commercially available, diol **47** must be prepared from 1,3-dimethoxybenzene via a 5-step sequence and subsequent enantiomeric enrichment.[64,249,250]

Magnesium-Mediated Epoxidation. This type of epoxidation generally utilizes commercially available solutions of TBHP in toluene and dibutylmagnesium in heptane.[69,70] The former reagent is pre-dried over 4 Å molecular sieves, and the reaction is carried out under an inert atmosphere. If the substrate has low solubility in toluene/heptane mixtures, THF may be used as a cosolvent.

For the epoxidation of diaryl enones, diethyl tartrate is a suitable ligand.[69] For alkyl-substituted enones, the best enantiomeric excesses are achieved using di-*tert*-butyl tartrate, although dicycloalkyl tartrates, which are cheaper and more easily prepared, give results that are almost as good.[70] In the epoxidation of such enones, addition of molecular sieves to the reaction mixture has been shown to give the best results.

Epoxidation in the Presence of a Chiral Phase-Transfer Catalyst. The preferred catalysts for this transformation are mostly quaternary ammonium salts derived from *Cinchona* alkaloids. In particular, *N*-anthracenylmethyl cinchoninium salts such as **57** and **58** often prove most effective. These catalysts are not commercially available, but their syntheses have been described starting from cinchonine or dihydrocinchonidine.[99] At the time of writing, a number of intermediates already bearing the *N*-anthracenylmethyl group are commercially available, potentially increasing the ease of the synthesis of these ligands.

There are three distinct experimental procedures for using these catalysts. The first uses aqueous sodium hypochlorite at ambient temperature in the presence of 1 mol % catalyst.[97] The second uses 8 M potassium hypochlorite, which should be prepared shortly before use by reaction of potassium hydroxide with chlorine. This reagent allows the epoxidation to be carried out at −40° in the presence of 10 mol % catalyst and generally gives improved enantioselectivity.[98] Finally, trichloroisocyanuric acid has been proposed as a commercially available, safe, and convenient alternative oxidant for these reactions. Moreover, non-aqueous conditions have been reported for epoxidations using this reagent.[102]

For these reactions the efficiency of agitation can have an influence on the rate of the reaction[97,251] and the concentration of reagents can have a significant impact on the efficiency of the catalysis.[97]

A number of *N* benzylated cinchoninium salts have been used as catalysts; the 4-trifluoromethyl-substituted salt **60** is commercially available, others such as the 4-iodobenzyl catalyst **59** can be prepared by simple alkylation of cinchonine.[104] Likewise, the *N*-1-naphthylmethyl quinidinium catalyst **61** is prepared by alkylation of quinidine.[103] The reported procedure for epoxidation with these catalysts uses readily available aqueous hydrogen peroxide and lithium hydroxide in a biphasic solvent system at 4°.[103]

A related catalyst, bis-quininium salt **62**, has been shown to be efficient in phase-transfer-mediated epoxidations.[108] The catalyst is prepared in a single step by alkylation of quinine with a dibromide. These epoxidations are carried out using aqueous hydrogen peroxide and potassium hydroxide in a biphasic solvent system at ambient temperature. Reaction rates and enantioselectivities are found to increase in the presence of a surfactant, Span 20.

The non-alkaloid-derived binaphthyl-based catalyst **63** gives high enantioselectivity in the epoxidation of a range of enones using sodium hypochlorite as the oxidant. The catalyst is not commercially available, and must be prepared by a multi-step route.[109]

Epoxidation in the Presence of Polypeptides. The catalyst most commonly used for these transformations is polyleucine, which is commercially available. Alternatively, the peptide can be prepared by polymerization of leucine *N*-carboxyanhydride in the presence of an alkylamine initiator[132] or a polymer-supported amine serving the same purpose.[123] Factors relevant to the catalyst preparation, particularly on a large scale, have been discussed, highlighting the importance of preparing the catalyst under the correct conditions in order to reliably obtain good enantioselectivities.[252–254] Catalysts prepared for use under essentially non-aqueous conditions require an activation step in order to be effective.[252] The filtration properties of the catalyst can be enhanced by adsorption[124,254] or grafting[125] onto silica.

As discussed previously, a number of different reaction protocols have been reported. Triphasic reaction conditions, with the addition of a phase-transfer catalyst such as tetrabutylammonium bromide,[120,121] allow lower catalyst loadings than other conditions. The oxidant is most commonly 30% aqueous hydrogen peroxide and the base sodium hydroxide. Under these conditions, catalyst loadings as low as 0.05 mol % can be used as compared to 10 mol % for similar reactions without the added phase-transfer catalyst. It is recommended that these reactions be performed in the dark.

Alternatively, biphasic reaction conditions using commercially available urea hydrogen-peroxide and an organic base such as DBU[176] or BEMP[117] are reported to effect the epoxidation of a wider substrate range than the triphasic conditions and to allow the use of silica-supported catalysts, which are more easily removed from the reaction mixture.[124] While a stoichiometric quantity of DBU is required, the amount of BEMP used can be as low as 3 mol %.

A number of workers have noted that the speed and efficiency of stirring can have an impact on the rate of these reactions and it has been noted that large-scale reactions tend to take longer to reach completion.[252,254]

Organocatalysis. The preferred catalyst for the epoxidation of enones is commercially available α,α-diphenylprolinol (**78**). Reactions are performed in hexane and use commercially available solutions of TBHP in decane as the oxidant.[170] For the epoxidation of α,β-unsaturated aldehydes the recommended catalyst is the silylated proline derivative **77**.[255] The most effective conditions for these reactions use dichloromethane as solvent and 35% aqueous hydrogen peroxide as oxidant.[168]

EXPERIMENTAL PROCEDURES

The examples in the following section have been selected to illustrate a broad range of epoxidation protocols, rather than of substrate types. Readers are directed to the tables that follow for indications of the applicability of each epoxidation method. Methods that are not specific to the epoxidation of electron-deficient alkenes (chiral dioxiranes and Mn–salen methods) are not included in this section.

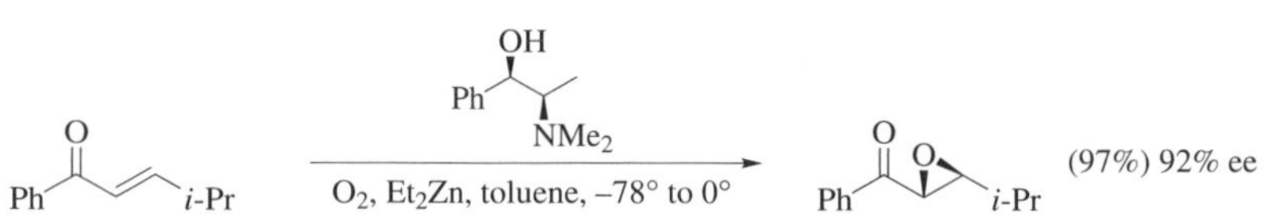

(2*R*,3*S*)-2,3-Epoxy-4-methyl-1-phenylpentan-1-one [Diethylzinc-Mediated Epoxidation].[41] A solution of (*R*,*R*)-*N*-methylpseudoephedrine (430 mg, 2.4 mmol) in toluene (10 mL) was cooled to 0° under argon, and treated with diethylzinc (1.1 M in toluene, 1.0 mL, 1.1 mmol). After stirring the mixture for 80 minutes, the flask was fitted with a balloon containing oxygen. After a further 2.5 hours, the mixture was cooled to −78°, and a solution of (*E*)-4-methyl-1-phenylpent-2-en-1-one (174 mg, 1.0 mmol) in toluene (2 mL) was added. The mixture was stirred for 30 minutes at −78°, then rapidly warmed to 0°, and stirring was continued for 3 hours. Aqueous pH 7 phosphate buffer solution (8 mL) was added, the layers were separated, and the aqueous layer was extracted with CH_2Cl_2. The combined organic layers were dried (Na_2SO_4) and the solvent removed under reduced pressure. The residue was purified by column chromatography (SiO_2; 24:1–9:1 petroleum ether/Et_2O) to afford (2*R*,3*S*)-2,3-epoxy-4-methyl-1-phenylpentan-1-one (184 mg, 97%) in 92% ee (determined by ^{1}H NMR in the presence of Eu(tfc)$_3$) as a yellow oil: $[\alpha]_D^{26}$ + 32.0 (*c* 1.25, CH_2Cl_2); IR 2960, 1690, 1600, 1450, 1230, 700 cm^{-1}; ^{1}H NMR (300 MHz, $CDCl_3$) δ 1.07 (d, J = 6.7 Hz, 3H), 1.10 (d, J = 6.7 Hz, 3H), 1.79 (septet, J = 6.7 Hz, 1H), 2.96 (dd, J = 6.7, 2.0 Hz, 1H), 4.07 (d, J = 2.0 Hz, 1H), 7.37–7.53 (m, 2H), 7.59–7.64 (m, 1H), 8.00–8.04 (m, 2H); ^{13}C NMR (75 MHz, $CDCl_3$) δ 18.3, 18.9, 30.6, 56.5, 65.0, 128.2, 128.4, 133.8, 135.6, 194.7; EI (m/z): [M^+ − O] 174 (2), 147 (50), 105 (100), 77 (41). Anal. Calcd for $C_{12}H_{14}O_2$: C, 75.8, H 7.4. Found: C, 75.8, H 7.5.

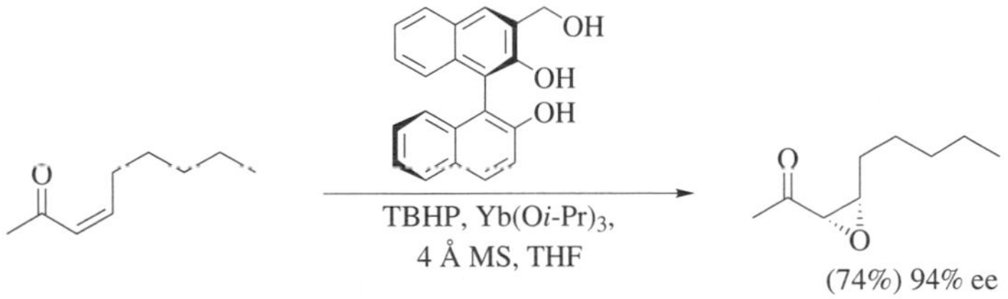
OH
OH
OH
TBHP, Yb(O*i*-Pr)$_3$,
4 Å MS, THF
O
O
(74%) 94% ee

(3*S*,4*S*)-3,4-Epoxynonan-2-one [Ytterbium-Catalyzed Epoxidation of a *cis*-Enone].[49] A solution of Yb(O*i*-Pr)$_3$ (0.1 M in THF, 240 μL, 0.024 mmol) was added to a suspension of 4 Å molecular sieves (48 mg) and (*R*)-3-hydroxy-methyl-BINOL (0.042 M in THF, 800 μL, 0.034 mmol) at room temperature. After the solution was stirred at 40° for 1 hour, TBHP (3–4 M in toluene, 240 μL, 0.72–0.96 mmol) was added and the mixture was stirred at room temperature for a further 5 minutes. To this solution was added (*Z*)-non-3-en-2-one (33.7 mg, 0.24 mmol), and the mixture was stirred at room temperature for 72 hours. The mixture was then treated with saturated aqueous ammonium chloride (1.0 mL). The usual workup and purification gave (3*S*,4*S*)-3,4-epoxynonan-2-one (27.7 mg, 74%) in 94% ee (determined by HPLC; Daicel Chiralpak AS, 100:1 hexane/*i*-PrOH) as a colorless oil: $[\alpha]_D^{25}$ −57.3 (*c* 1.15, $CHCl_3$); ^{1}H NMR (500 MHz, $CDCl_3$) δ 0.88 (t, J = 7.0 Hz, 3H), 1.26–1.58 (m, 8H), 2.23 (s, 3H), 3.20 (dt, J = 4.6, 6.7 Hz, 1H), 3.60 (d, J = 4.6 Hz, 1H); MS (m/z): M^+ 156.

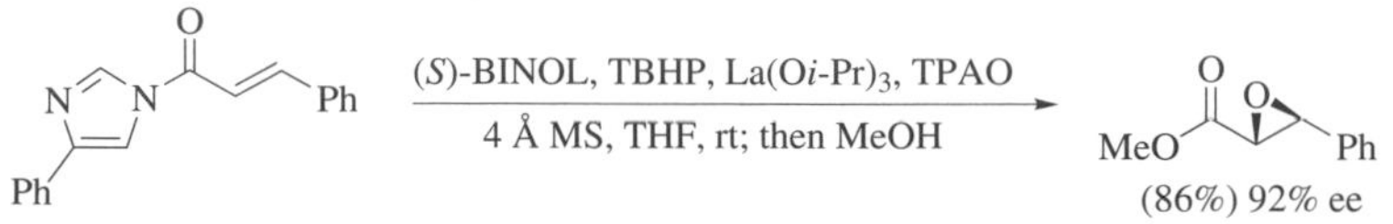

(2*R*,3*S*)-Methyl 2,3-Epoxy-3-phenylpropanoate [Lanthanum-Catalyzed Epoxidation of an Acyl Imidazolide].[56] A solution of La(O*i*-Pr)$_3$ (0.2 M in THF, 125 μL, 0.025 mmol) was added to a suspension of 4 Å molecular sieves (250 mg), (*S*)-BINOL (7.2 mg, 0.025 mmol), and triphenylarsine oxide (8.1 mg, 0.025 mmol) in THF (2.5 mL) at room temperature. After the solution was stirred for 45 minutes, TBHP (5 M in decane, 120 μL, 0.6 mmol) was added and the mixture was stirred at room temperature for an additional 10 minutes. To this solution was added (*E*)-3-phenyl-1-(4-phenylimidazol-1-yl)-prop-2-en-1-one (68.6 mg, 0.25 mmol), and the reaction mixture was stirred at room temperature for 1 hour. Methanol (0.5 mL) was added and the mixture stirred for 3 hours, then treated with 1% aqueous citric acid (2.5 mL) at 0°. The mixture was extracted with EtOAc (2 × 10 mL) and the combined organic layers were washed with 2% aqueous sodium thiosulfate (5 mL) and brine (5 mL), dried (Na_2SO_4), and the solvent was removed under reduced pressure. Purification of the residue by flash chromatography (SiO_2; 50:1 hexane/EtOAc) gave (2*R*,3*S*)-methyl 2,3-epoxy-3-phenylpropanoate (38.5 mg, 86%) in 92% ee (determined by HPLC; Daicel Chiralpak AD, 98:2 hexane/*i*-PrOH) as a colorless oil: $[\alpha]_D^{24}$ −111.8 (*c* 1.17, $CHCl_3$); ^{1}H NMR (400 MHz, $CDCl_3$)[155] δ 3.52 (d, *J* = 1.8 Hz, 1H), 3.83 (s, 3H), 4.10 (d, *J* = 1.8 Hz, 1H), 7.26–7.30 (m, 2H), 7.35–7.39 (m, 3H).

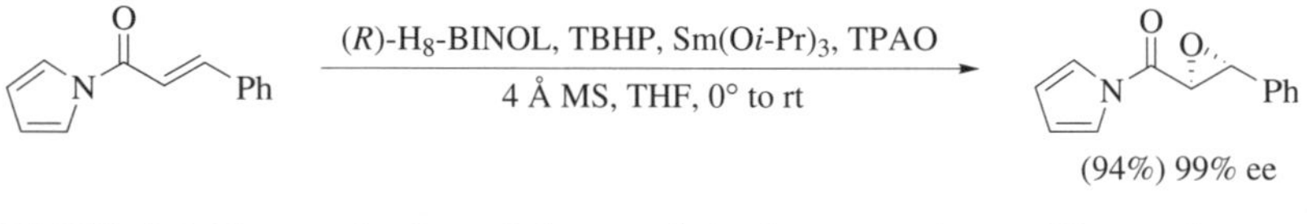

(2*S*,3*R*)-2,3-Epoxy-3-phenyl-1-pyrrol-1-ylpropan-1-one [Samarium-Catalyzed Epoxidation of an *N*-Acyl Pyrrole].[62] To a stirred suspension of 1-cinnamoylpyrrole (1.97 g, 10 mmol), triphenylarsine oxide (0.64 mg, 0.002 mmol), (*R*)-H$_8$-BINOL (0.59 mg, 0.002 mmol) and powdered 4 Å molecular sieves (1.0 g) in THF (2.0 mL) at room temperature was added Sm(O*i*-Pr)$_3$ (0.2 M in THF, 10 μL, 0.002 mmol). The mixture was stirred for 15 minutes at room temperature then cooled to 0°. TBHP (4.5 M in toluene, pre-dried with 4 Å molecular sieves, 3.3 mL, 15 mmol) was added slowly over 60 minutes (45 minutes at 0°, then 15 minutes at room temperature). The mixture was stirred for a further 30 minutes at room temperature and quenched with 2.5% aqueous citric acid. The mixture was filtered through Celite and the filtrate was extracted with EtOAc. The combined organic layers were washed successively with saturated aqueous $NaHCO_3$ and brine, and dried ($MgSO_4$). Evaporation of the solvent and purification of the residue by flash chromatography (SiO_2; 15:1 hexane/EtOAc) gave (2*S*,3*R*)-2,3-epoxy-3-phenyl-1-pyrrol-1-ylpropan-1-one (2.01 g, 94%) in 99% ee (determined by HPLC; Daicel Chiralpak

AD, 98:2 hexane/*i*-PrOH): $[\alpha]_D^{24}$ + 150 (*c* 1.1, $CHCl_3$); IR (KBr) 3149, 1714, 1287, 904 cm^{-1}; ^1H NMR (500 MHz, $CDCl_3$) δ 4.01 (d, *J* = 1.9 Hz, 1H), 4.20 (d, *J* = 1.9 Hz, 1H), 6.30–6.36 (m, 2H), 7.32–7.42 (m, 7H); ^{13}C NMR (125 MHz, $CDCl_3$) 57.3, 58.8, 114.1, 119.0, 125.7, 128.8, 129.2, 134.4, 164.3; ESI-MS (*m*/*z*): $[M + Na]^+$ 236; HRMS (*m*/*z*): $[M + H]^+$ calcd for $C_{13}H_{11}NO_2$, 214.0868; found, 214.0870.

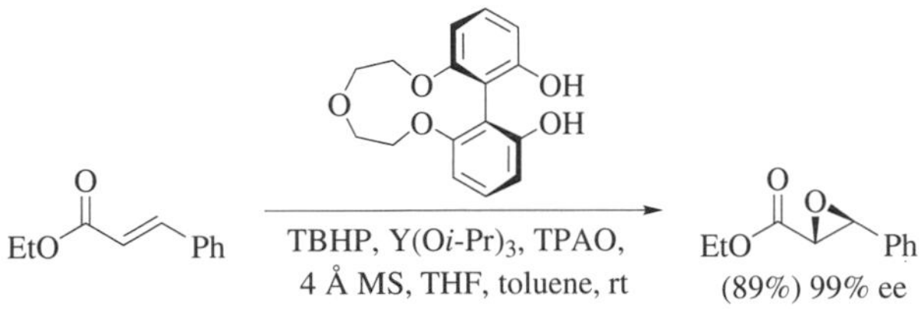

Ethyl (2*R*,3*S*)-2,3-Epoxy-3-phenylpropanoate [Yttrium-Catalyzed Epoxidation of an Ester].[64] To a stirred suspension of 4 Å molecular sieves (250 mg), (*S*)-6,6′-[oxybis(ethylene)dioxy]biphenyl-2,2′-diol (7.2 mg, 0.025 mmol), and triphenylarsine oxide (8.1 mg, 0.025 mmol) in THF (1.125 mL) was added $Y(Oi\text{-}Pr)_3$ (0.2 M in THF, 0.125 mL, 0.025 mmol). After stirring for 45 minutes at room temperature, TBHP (4.0 M in toluene, 0.375 mL, 1.5 mmol) was added and the mixture was stirred for a further 10 minutes. Ethyl cinnamate (220 mg, 1.25 mmol) was added and the mixture was stirred at room temperature for 36 hours, then diluted with EtOAc (10 mL) and quenched with 2% citric acid (2.5 mL). The aqueous layer was extracted with EtOAc (10 mL), then the combined organic layers were washed with brine (5 mL) and dried (Na_2SO_4). Evaporation of the solvent and purification of the residue by flash chromatography (SiO_2; 100:1–50:1 hexane/EtOAc) gave ethyl (2*R*,3*S*)-2,3-epoxy-3-phenylpropanoate (213 mg, 89%) in 99% ee (determined by HPLC; Daicel Chiralpak AD-H, 98:2 hexane/*i*-PrOH) as a clear oil: $[\alpha]_D^{24}$ − 158.8 (*c* 1.1, $CHCl_3$); ^1H NMR (300 MHz, $CDCl_3$)[150] δ 1.34 (t, *J* = 6.9 Hz, 3H), 3.52 (d, *J* = 1.8 Hz, 1H), 4.10 (d, *J* = 1.8 Hz, 1H), 4.25–4.35 (m, 2H), 7.29–7.32 (m, 2H), 7.35–7.39 (m, 3H).

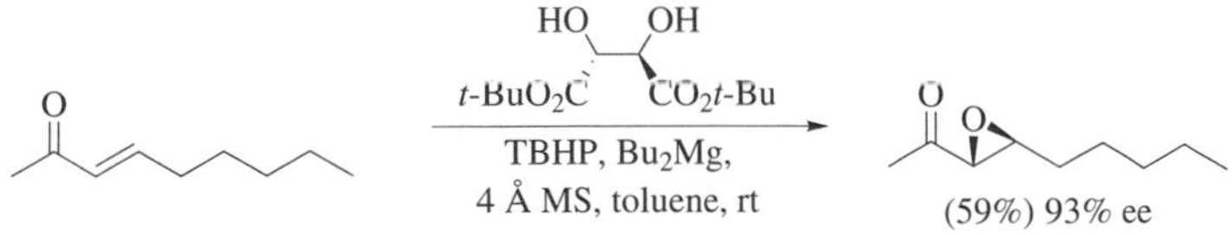

(3*R*,4*S*)-3,4-Epoxynonan-2-one [Dibutylmagnesium-Mediated Epoxidation of an Enone].[70] TBHP (3.7 M in toluene, 1.1 mmol) was dried over activated powdered 4 Å molecular sieves (200 mg) for 2 hours. Dibutylmagnesium (1 M in heptane, 0.1 mmol) was added and the mixture was stirred for 30 minutes, then L-(+)-di-*tert*-butyl tartrate (0.11 mmol) was added. After another 30 minutes, (*E*)-non-3-en-2-one (1 mmol) was added, and the mixture was stirred for 24 hours. Work-up and purification by flash chromatography afforded (3*R*,4*S*)-3,4-epoxynonan-2-one (59%) in 93% ee (determined by GC; heptakis(2,6-di-*O*-methyl-3-*O*-pentyl)-β-cyclodextrin, column temperature 110°) : $[\alpha]_D + 43.9$ (*c* 1.1, $CHCl_3$); ^1H NMR ($CDCl_3$)[46] δ0.89 (m, 3H), 1.20–1.75 (m, 8H), 2.06 (s, 3H), 3.07 (td, $J = 5.0$, 2.0 Hz, 1H), 3.18 (d, $J = 2.0$ Hz, 1H).

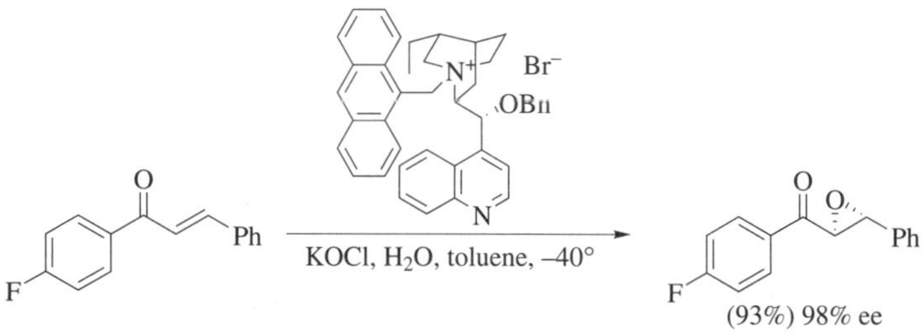

(2*S*,3*R*)-2,3-Epoxy-1-(4-fluorophenyl)-3-phenylpropan-1-one [Epoxidation Mediated by a Chiral Phase-Transfer Catalyst].[98] A mixture of (*E*)-1-(4-fluorophenyl)-3-phenylprop-2-en-1-one (226 mg, 1.0 mmol), *N*-anthracenylmethyl-*O*-benzyldihydrocinchonidinium bromide (66 mg, 0.1 mmol), and toluene (10 mL) was cooled to −40° and treated with an aqueous solution of potassium hypochlorite (8.0 M, 0.63 mL, 5.0 mmol). After stirring at −40° for 12 hours, the solvent was removed under reduced pressure and the solid quaternary ammonium salt was precipitated by addition of 4:1 hexanes/ether. The filtrate was washed with water and brine, concentrated, and purified by chromatography (SiO_2; 6:1 hexanes/EtOAc) to afford (2*S*,3*R*)-2,3-epoxy-1-(4-fluorophenyl)-3-phenylpropan-1-one (225 mg, 93%) in 98% ee (determined by HPLC; Chiralcel OB-H, 95:5 hexanes/*i*-PrOH): mp 86–87°; $[\alpha]^{23}_D + 213.4$ (*c* 2.0, CH_2Cl_2); IR (film) 1682, 1598, 1238, 1229, 1158, 885 cm^{-1}; ^1H NMR (400 MHz, $CDCl_3$) δ 4.07 (d, $J = 1.8$ Hz, 1H), 4.24 (d, $J = 1.8$ Hz, 1H), 7.14–7.18 (m, 2H), 7.41–7.36 (m, 5H), 8.08–8.04 (m, 2H); ^{13}C NMR (100 MHz, $CDCl_3$) δ 59.3, 61.1, 116.2 (d, $J = 21.5$ Hz), 125.8, 128.9, 129.2, 131.2 (d, $J = 9.1$ Hz), 131.9 (d, $J = 3.1$ Hz), 135.4, 166.3 (d, $J = 255.0$ Hz), 191.6; HRMS–EI$^+$ (m/z): $[M + H]^+$ calcd for $C_{15}H_{11}FO_2$, 243.0743; found, 243.0741.

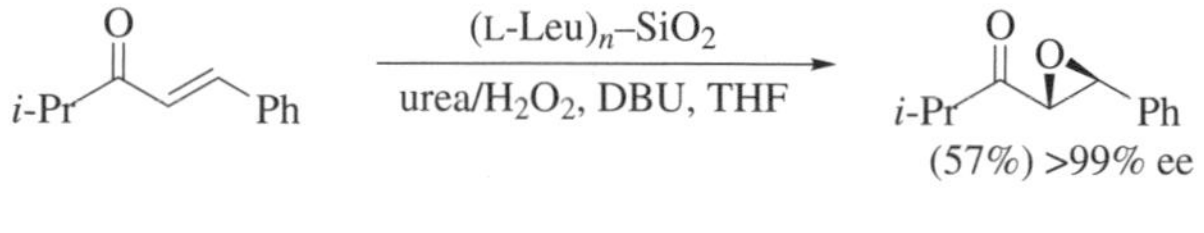

(1*S*,2*R*)-1,2-Epoxy-4-methyl-1-phenylpentan-3-one [Biphasic Polyleucine-Catalyzed Epoxidation].[256] A mixture of (*E*)-4-methyl-1-phenylpent-1-en-3-one (1.0 g, 5.7 mmol), urea/H_2O_2 addition compound (0.64 g, 6.8 mmol), DBU (1.3 mL, 8.5 mmol), and silica-supported poly-L-leucine[124] (10.3 g) in THF (60 mL) was stirred at room temperature for 18 hours. The catalyst was removed by filtration and washed with EtOAc (300 mL). The filtrate was washed with saturated aqueous Na_2SO_3 (2 × 100 mL), water (2 × 100 mL), and brine (100 mL), dried ($MgSO_4$), and concentrated. The residue was purified by chromatography (SiO_2; 6:1 petroleum ether/Et_2O) followed by recrystallization (*n*-hexane) to afford (1*S*,2*R*)-1,2-epoxy-4-methyl-1-phenylpentan-3-one (621 mg, 57%) in >99% ee (determined by HPLC; Chiralpak AD, 95:5 hexanes/EtOH): mp 56–57°; $[\alpha]_D^{22}$ −208 (*c* 1.0, $CHCl_3$); IR (KBr) 1711, 1458, 1417, 1052 cm^{-1}; ^1H NMR (400 MHz, $CDCl_3$) δ 1.16 (d, *J* = 7.0 Hz, 3H), 1.17 (d, *J* = 7.0 Hz, 3H), 2.83 (septet, *J* = 7.0 Hz, 1H), 3.61 (d, *J* = 2.0 Hz, 1H), 3.93 (d, *J* = 2.0 Hz, 1H), 7.29–7.38 (m, 5H); ^{13}C NMR (100 MHz, $CDCl_3$) δ 17.3, 18.1, 37.0, 58.4, 61.9, 125.7, 128.7, 129.0, 135.4, 208.7; EIMS (*m*/*z*): M^+ 190 (4), 174 (57), 148 (88), 147 (77). Anal. Calcd for $C_{12}H_{14}O_2$: C, 75.8; H 7.4. Found: C, 75.8; H 7.4.

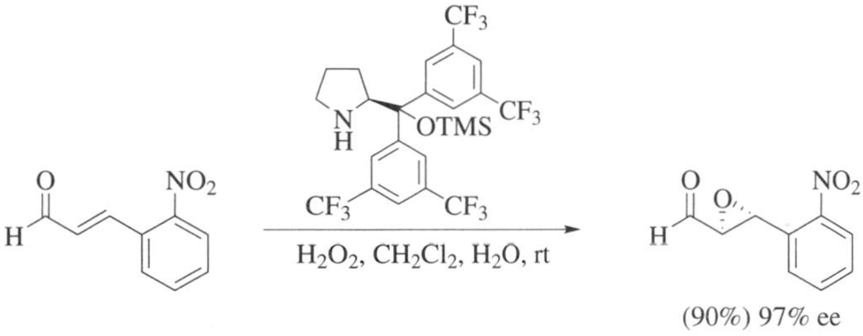

(2*S*,3*R*)-2,3-Epoxy-3-(2-nitrophenyl)propanal [Organocatalytic Epoxidation of an α,β-Unsaturated Aldehyde].[168] To a stirred solution of (*E*)-3-(2-nitrophenyl)propenal (88 mg, 0.5 mmol) in CH_2Cl_2 (1.0 mL) at room temperature were added successively (2*S*)-2-[bis-(3,5-bis-trifluoromethylphenyl) trimethylsilanyloxymethyl]pyrrolidine (0.05 mmol) and H_2O_2 (35% in H_2O, 63 mg, 0.65 mmol). After 4 hours, the reaction mixture was passed through a silica gel column, eluting with Et_2O/pentane, to give (2*S*,3*R*)-2,3-epoxy-3-(2-nitrophenyl)propanal: ^1H NMR (400 MHz, $CDCl_3$) δ 3.32 (dd, *J* = 6.0, 2.1 Hz, 1H), 4.81 (d, *J* = 2.1 Hz, 1H), 7.68 (dd, *J* = 8.2, 1.4 Hz, 1H), 7.71 (dd, *J* = 7.4, 1.5 Hz, 1H), 7.82 (dd, *J* = 8.2, 1.4 Hz, 1H), 8.21 (dd, *J* = 8.2, 1.4 Hz, 1H), 9.30 (d, *J* = 6.0 Hz, 1H); ^{13}C NMR (100 MHz, $CDCl_3$) δ 53.7, 59.9, 123.6, 125.6, 128.1, 130.0, 133.2, 147.2, 194.3. Following $NaBH_4$ reduction to the corresponding alcohol, the enantiomeric excess was determined to be 97% by HPLC (Chiralpak OJ and AS columns coupled in series, 85:15 hexane/*i*-PrOH): $[\alpha]_D^{25}$ −29.7 (*c* 1.3, $CHCl_3$).

TABULAR SURVEY

The following tables are organized by class of substrate, with the same divisions that were used in the "Scope and Limitations" section. Thus Tables 1A-H, 2, and 3 deal with the various substitution patterns of acyclic enones, Tables 4 and 5 cover, respectively, cyclic enones and quinones (including quinone monoketals), tables 6A-D have examples of epoxidation of unsaturated aldehydes, esters, acids, and amides, and Tables 7A-D include all the examples of epoxidation of unsaturated sulfones, nitro compounds, nitriles, and phosphonates. Entries are ordered by the total carbon count, with the exception of Table 5, in which the ketal portion of quinone monoketals is disregarded.

Owing to the nature of the topic under review, several papers detail epoxidation of a single substrate under a wide variety of conditions, in attempts to optimize the yield and/or enantiomeric excess. For such papers, every substrate/catalyst combination is included in these tables, but the details of solvent, temperature, and stoichiometry optimization have been abridged.

The results of each epoxidation are reported in the form (*x*), *y*, where *x* is the percentage yield and *y* the percentage enantiomeric excess. In those instances where a percentage conversion is reported instead of a yield, the figure in parentheses is italicized. A footnote is used to indicate where the enantiomeric excess or the identity of the major enantiomer was not determined in the original paper, but has been deduced from other publications. In all other cases where the identity of the major enantiomer was not determined, the product is shown without stereochemical configuration indicated.

A series of charts precedes the tables. These charts depict catalysts, ligands, and reagents that are indicated by bold numbers in the table entries, and are arranged primarily by structural type. The reader is referred to these charts to locate the structures with which the bold numbers are associated.

The tables cover the literature to the end of 2005.

The following abbreviations are used in the tables:

Aib	2-aminoisobutyrate
BEMP	2-*tert*-butylimino-2-diethylamino-1,3-dimethylperhydro-1,3,2-diazaphosphorine
BINOL	1,1′-bi-2-naphthol
Bn	benzyl
Boc	*tert*-butoxycarbonyl
Bz	benzoyl
CMHP	cumene hydroperoxide
DABCO	1,4-diazabicyclo[2.2.2]octane
DBN	1,5-diazabicyclo[4.3.0]non-5-ene
DBU	1,8-diazabicyclo[5.4.0]undec-7-ene
DCE	1,2-dichloroethane
DMAP	4-(dimethylamino)pyridine
DME	1,2-dimethoxyethane
DMF	*N*,*N*-dimethylformamide
DMSO	dimethylsulfoxide
EDTA	ethylenediaminetetraacetic acid (or acetate)
β^2-Leu	(*R*)-2-(aminomethyl)-4-methylpentanoic acid
β^3-Leu	(*S*)-3-amino-5-methylhexanoic acid
Ln	lanthanide

m-CPBA	3-chloroperoxybenzoic acid
M_r	relative molecular mass
Ms	mesyl
MS	molecular sieves
Na_2EDTA	dihydrogen disodium ethylenediaminetetraacetate
NaHMDS	sodium bis(trimethylsilyl)amide
n_{av}	average chain length of a polydisperse polypeptide
NMU	*N*-nitrosomethylurea
Np	naphthyl
Npg	L-(neopentyl)glycine
Oxone	potassium monopersulfate triple salt
PEG	poly(ethylene glycol)
Piv	pivaloyl
PMB	*p*-methoxybenzyl
PS	polystyrene
Span 20	sorban monolaurate
Succ	succinyl ($COCH_2CH_2CO_2H$)
TBAB	tetra-*n*-butylammonium bromide
TBDPS	*tert*-butyldiphenylsilyl
TBHP	*tert*-butylhydroperoxide
TBME	*tert*-butyl methyl ether
TBS	*tert*-butyldimethylsilyl
TES	triethylsilyl
TFA	trifluoroacetic acid
THF	tetrahydrofuran
TIPS	triisopropylsilyl
TPAO	triphenylarsine oxide
TPPO	triphenylphosphine oxide
Tr	triphenylmethyl, trityl

CHART 1. *CINCHONA* ALKALOID-DERIVED CATALYSTS REFERENCED IN THE TABLES

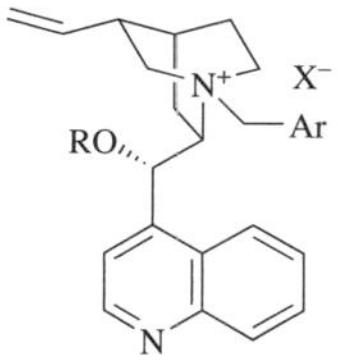

Cinchoninium salts

	Ar	R	X
1a	Ph	H	Cl
1b	Ph	H	Br
1c	4-FC_6H_4	H	Br
1d	4-ClC_6H_4	H	Br
1e	4-BrC_6H_4	H	Br
1f	4-IC_6H_4	H	Br
1g	4-$O_2NC_6H_4$	H	Br
1h	4-$CF_3C_6H_4$	H	Br
1i	4-$CF_3C_6H_4$	Me	Br
1j	4-$MeOC_6H_4$	H	Br
1k	9-anthryl	H	Cl
1l	9-anthryl	Bn	Br
1m	9-acridinyl	H	Br

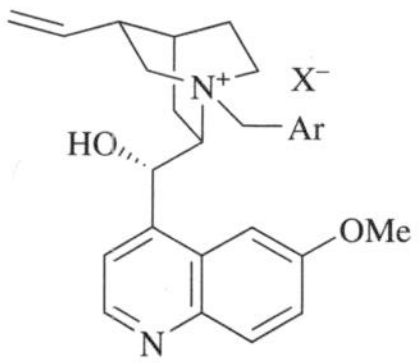

Quinidinium salts

	Ar	X
2a	Ph	Cl
2b	4-$CF_3C_6H_4$	Br
2c	2,6-$Cl_2C_6H_3$	Br
2d	2,4-$Me_2C_6H_3$	Br
2e	1-naphthyl	Cl
2f	2-naphthyl	Br
2g	9-anthryl	Cl

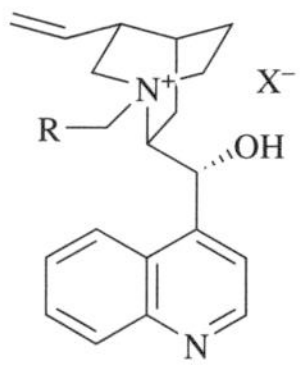

Cinchonidinium salts

	R	X
3a	H	Cl
3b	Ph	Cl
3c	Ph	Br
3d	9-acridinyl	Br
3e	9-anthracenyl	Br

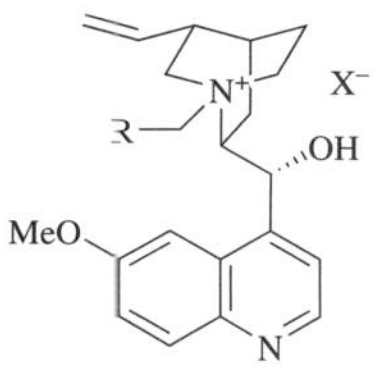

Quininium salts

	R	X
4a	Ph	Cl
4b	Ph	Br
4c	9-fluorenyl	Br
4d	9-acridinyl	Br
4e	PS	Cl
4f	$PS(CH_2)_{11}$	Br

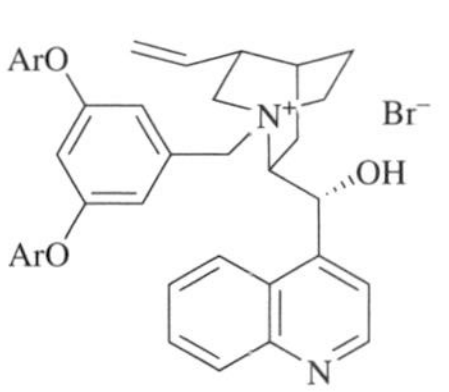

Cinchonidinium salts

	Ar
5a	Bn
5b	3,5-$(BnO)_2C_6H_3CH_2$

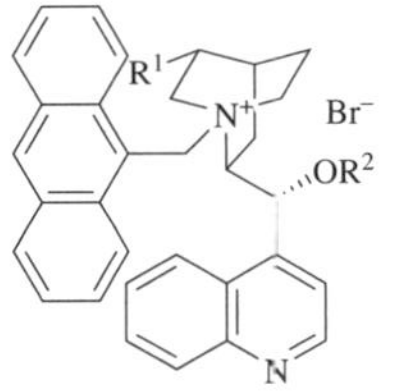

Cinchonidinium salts

	R^1	R^2
6a	$CH=CH_2$	Bn
6b	Et	Bn
6c	Et	$CH_2CH=CH_2$

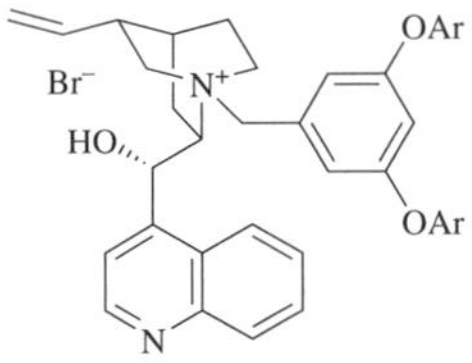

Cinchoninium salts

	Ar
7a	Bn
7b	3,5-$(BnO)_2C_6H_3CH_2$

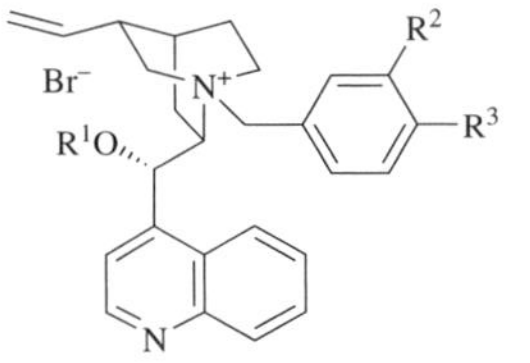

Cinchoninium salts

	R^1	R^2	R^3
8a	H	I	H
8b	$CH_2CH=CH_2$	H	I

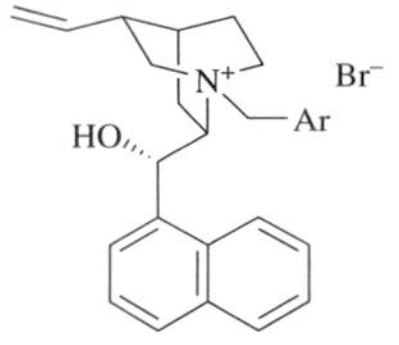

Deazacinchoninium salts

	Ar
9a	Ph
9b	4-$CF_3C_6H_4$

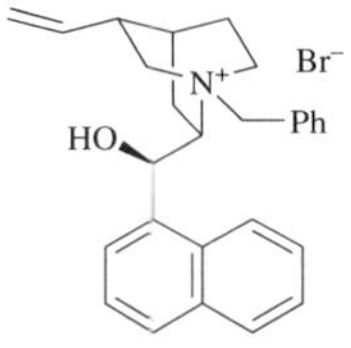

Deazacinchoninium salt

10

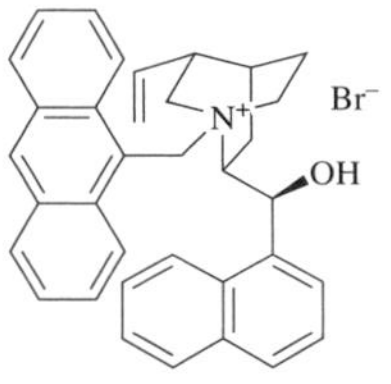

Deazacinchonidinium salt

11

CHART 1. *CINCHONA* ALKALOID-DERIVED CATALYSTS REFERENCED IN THE TABLES (*Continued*)

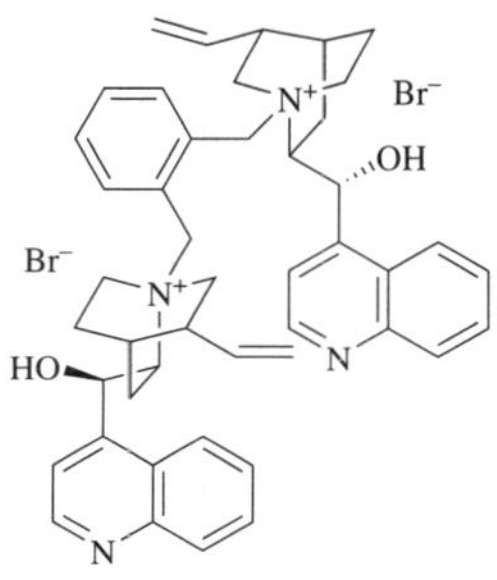

Bis(cinchonidinium) salt
12

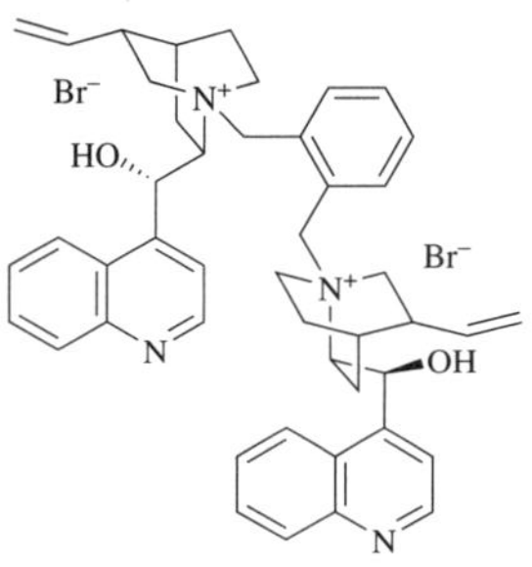

Bis(cinchoninium) salt
13

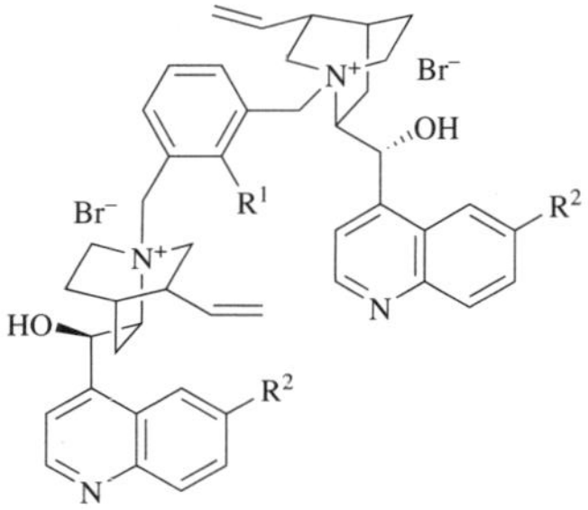

Bis(cinchonidinium) or bis(quininium) salt

	R^1	R^2
14a	H	H
14b	F	H
14c	H	OMe
14d	F	OMe

CHART 2. BINAPHTHYL CATALYSTS / LIGANDS REFERENCED IN THE TABLES

(See also Charts 3 and 5)

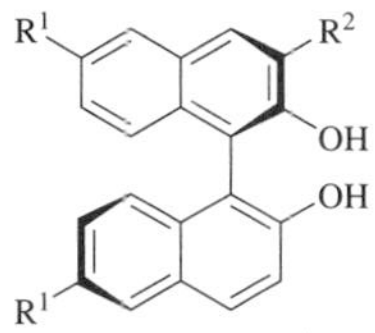

(R)-BINOLs

	R^1	R^2
15a	H	H
15b	naphthoquinon-2-yl	H
15c	Br	H
15d	H	CH_2OH

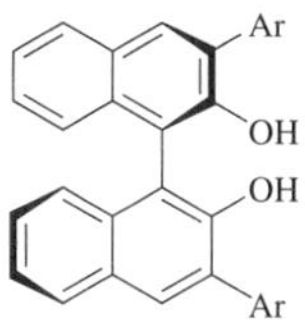

(*R*)-BINOLs

	Ar
16a	2-(n-$C_6H_{13}O$)-5-MeC_6H_3
16b	2,5-(n-$C_6H_{13}O)_2C_6H_3$

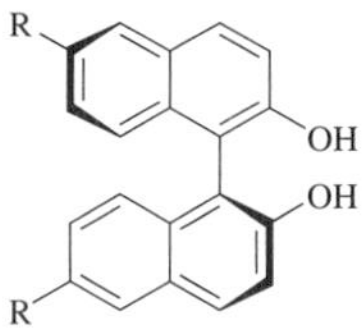

(*S*)-BINOLs

	R
17a	H
17b	Ph
17c	4-$CF_3C_6H_4$
17d	Br

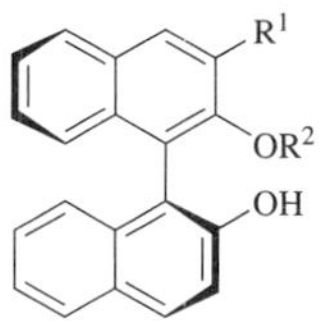

(*S*)-BINOLs

	R^1	R^2
18a	CH_2OH	H
18b	H	CH_2OMe

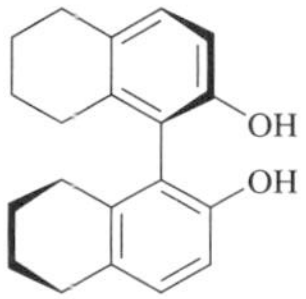

Octahydro-(*R*)-BINOL

19

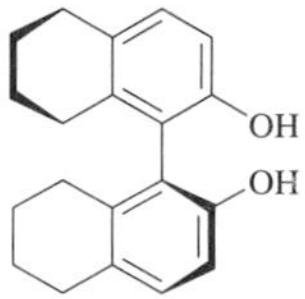

Octahydro-(*S*)-BINOL

20

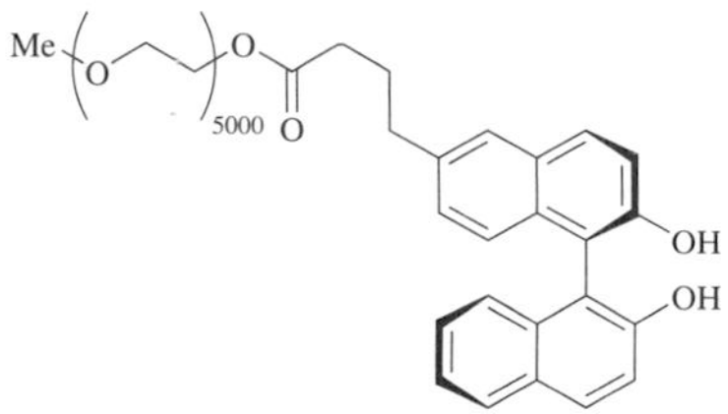

Polymer-bound (*R*)-BINOL

21

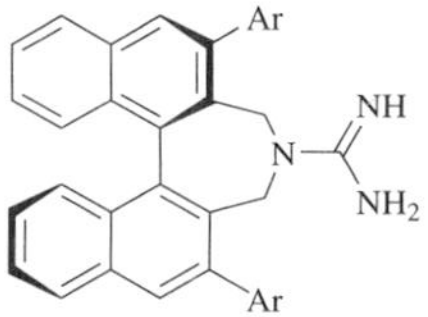

(*R*)-Binaphthyl guanidine

	Ar
22a	3,5-$(CF_3)_2C_6H_3$
22b	3,5-(t-$Bu)_2C_6H_3$
22c	3,5-$Ph_2C_6H_3$
22d	4-$CF_3C_6H_4$
22e	3,5-(n-$C_4F_9)_2C_6H_3$
22f	3,5-$(TMS)_2C_6H_3$
22g	3,5-[3,5-$(CF_3)_2C_6H_3]_2C_6H_3$
22h	3,5-(3,4,5-$F_3C_6H_2)_2C_6H_3$

CHART 2. BINAPHTHYL CATALYSTS / LIGANDS REFERENCED IN THE TABLES (*Continued*)

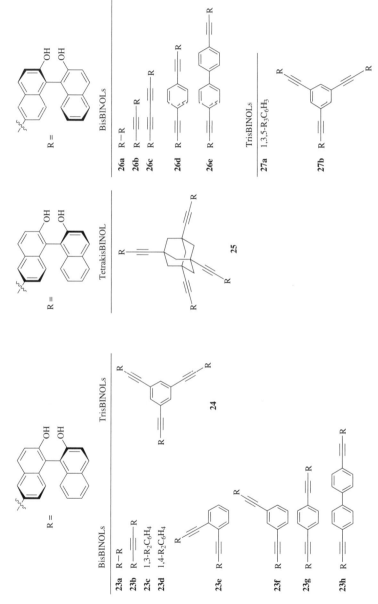

OligoBINOLs

R = 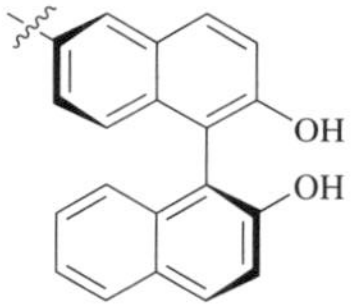

BisBINOLs	
23a	R—R
23b	
23c	1,3-$R_2C_6H_4$
23d	1,4-$R_2C_6H_4$
23e	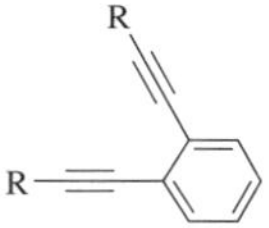
23f	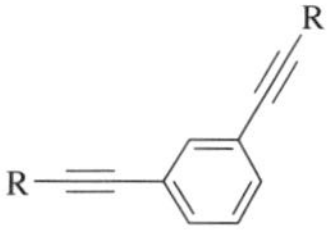
23g	
23h	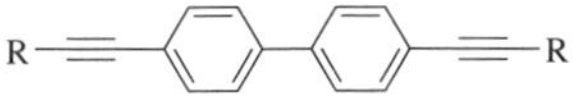

TrisBINOLs

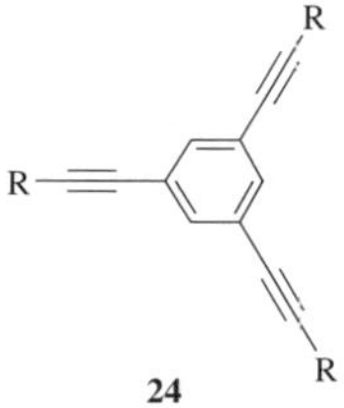

24

R = 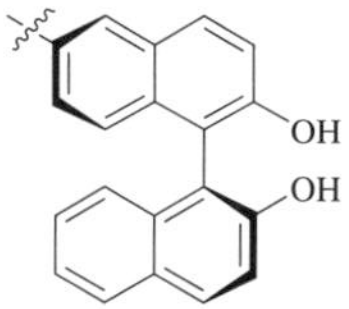

TetrakisBINOL

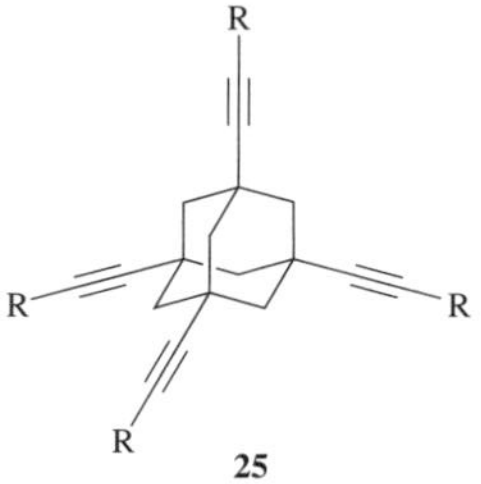

25

R = 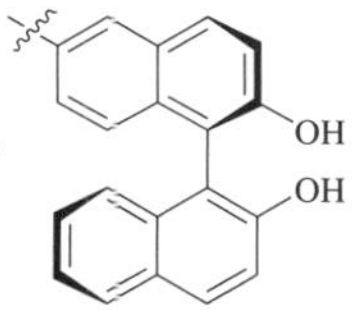

BisBINOLs	
26a	R—R
26b	
26c	
26d	
26e	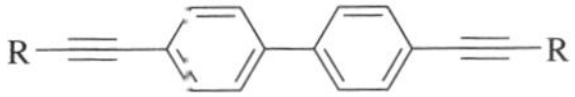

TrisBINOLs	
27a	1,3,5-$R_3C_6H_3$
27b	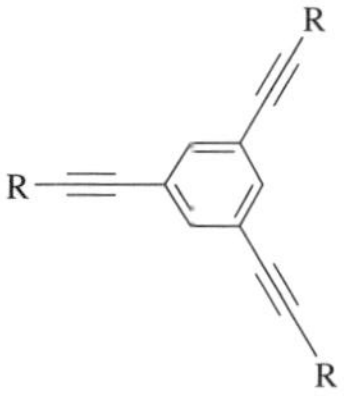

PolyBINOLs (R = n-C_6H_{13})

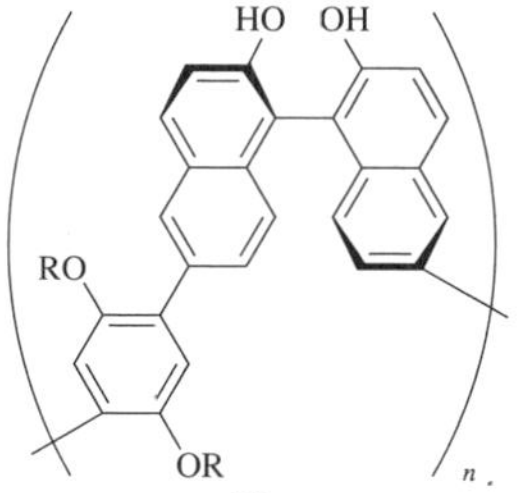

28

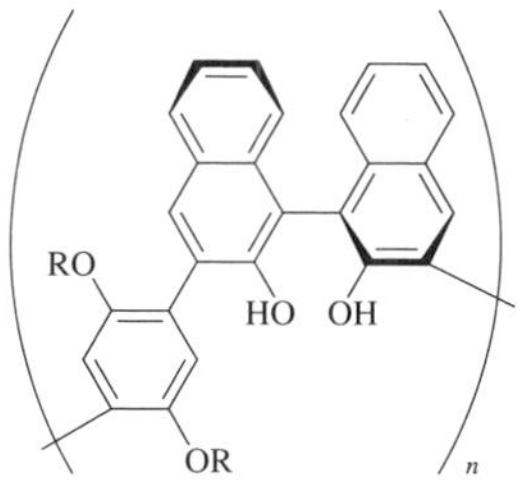

29

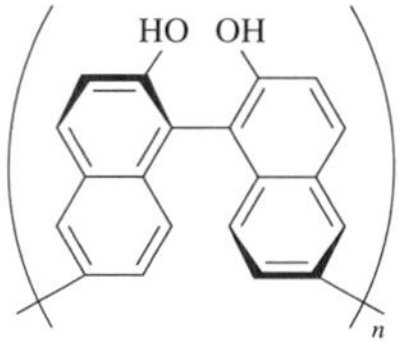

30

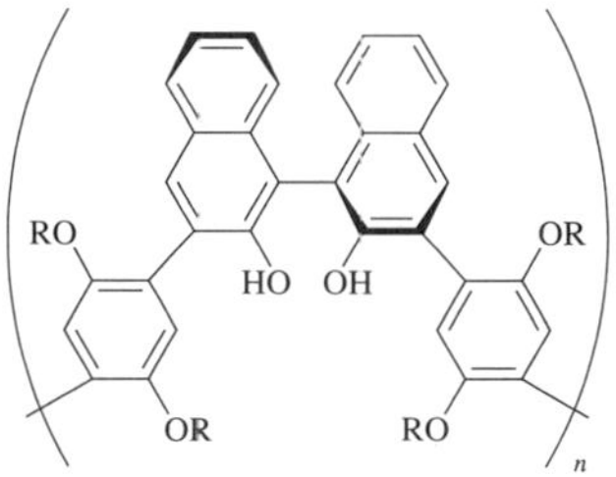

31

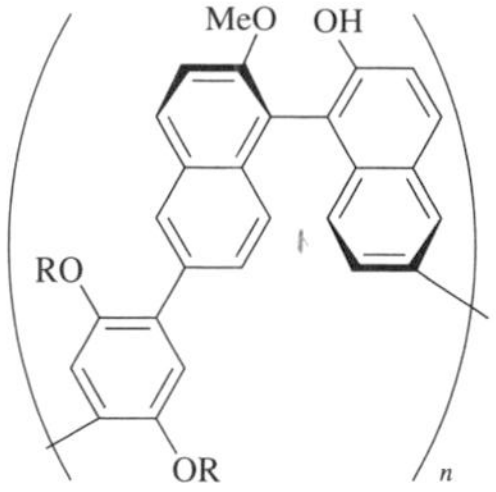

32

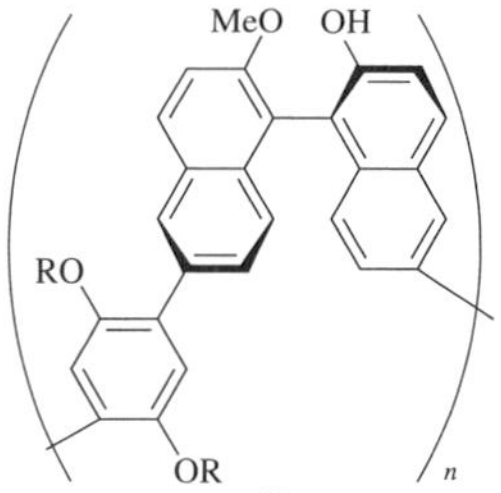

33

La–BINOLS

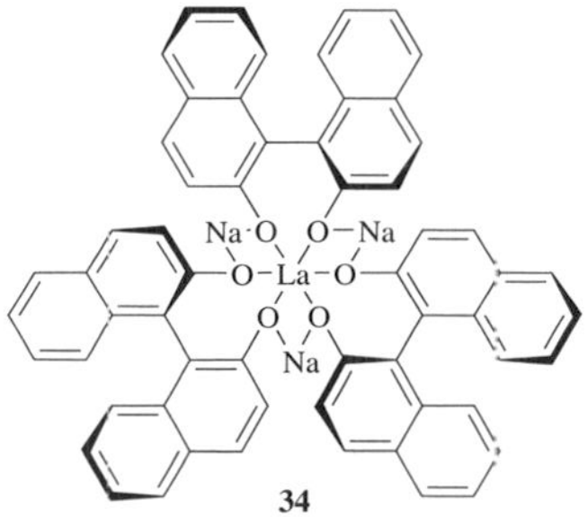

34

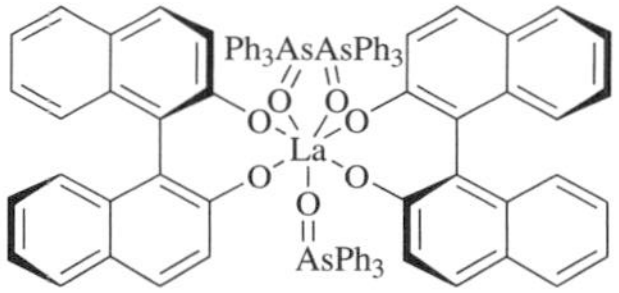

35

CHART 3. CHIRAL KETONE CATALYSTS REFERENCED IN THE TABLES

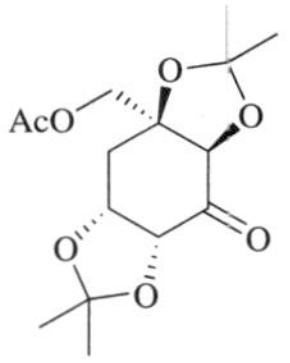

36

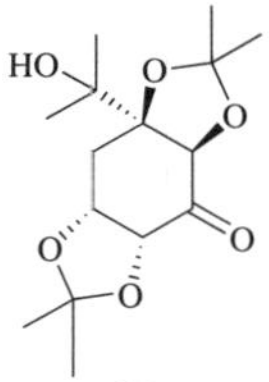

37

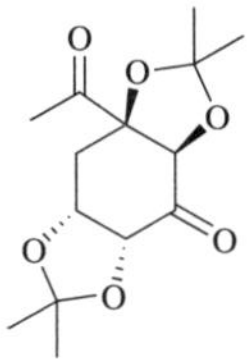

38

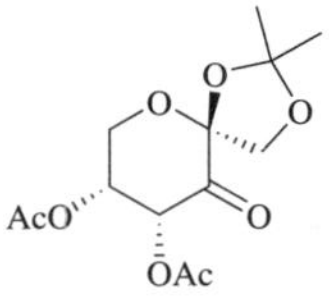

39

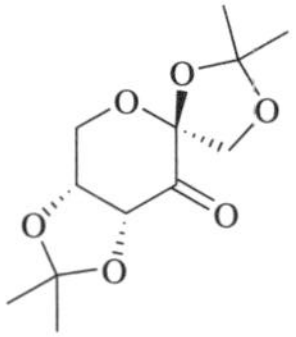

40

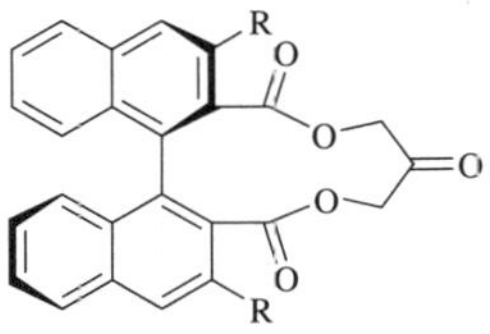

Binaphthyl ketone

	R
41a	H
41b	Cl

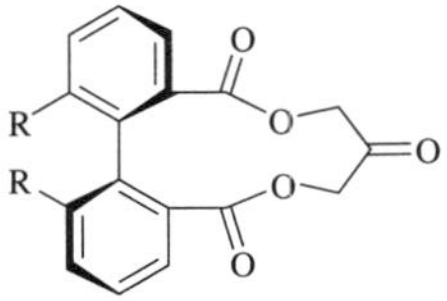

Biphenyl ketone

	R
42a	Cl
42b	NO_2

CHART 4. CROWN AND AZA-CROWN ETHERS REFERENCED IN THE TABLES

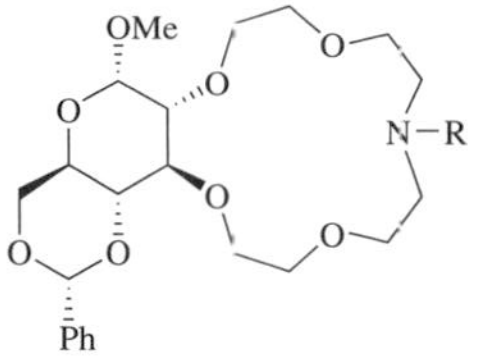

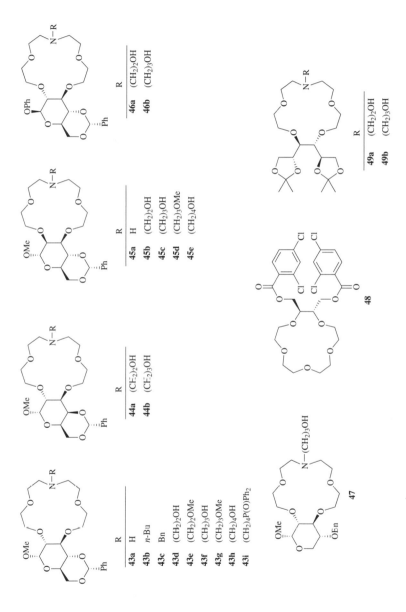

	R
43a	H
43b	*n*-Bu
43c	Bn
43d	$(CH_2)_2OH$
43e	$(CH_2)_2OMe$
43f	$(CH_2)_3OH$
43g	$(CH_2)_3OMe$
43h	$(CH_2)_4OH$
43i	$(CH_2)_4P(O)Ph_2$

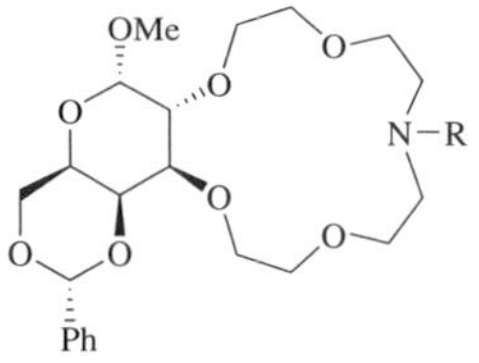

	R
44a	$(CF_2)_2OH$
44b	$(CF_2)_3OH$

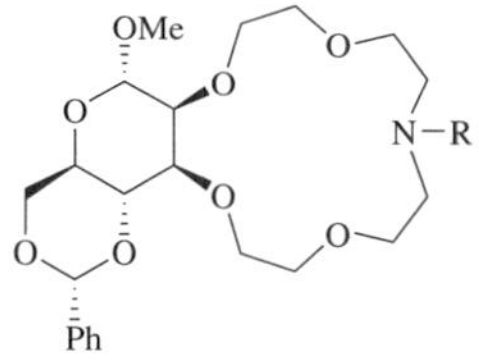

	R
45a	H
45b	$(CH_2)_2OH$
45c	$(CH_2)_3OH$
45d	$(CH_2)_3OMe$
45e	$(CH_2)_4OH$

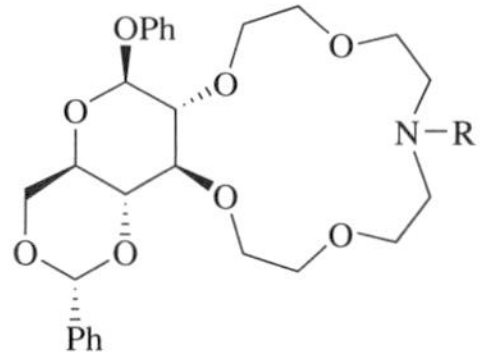

	R
46a	$(CH_2)_2OH$
46b	$(CH_2)_3OH$

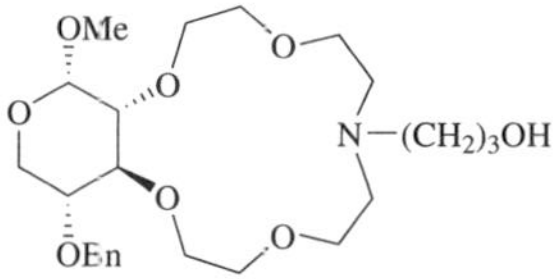

47

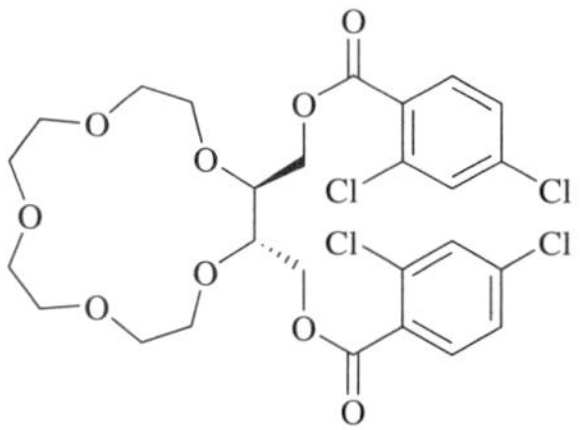

48

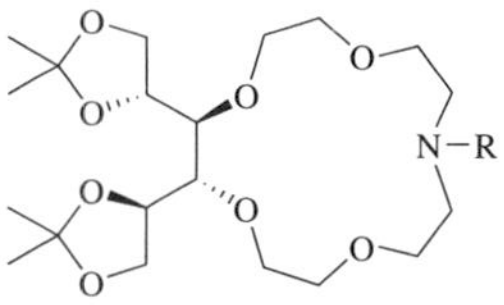

	R
49a	$(CH_2)_2OH$
49b	$(CH_2)_3OH$

CHART 5. QUATERNARY AMMONIUM SALTS REFERENCED IN THE TABLES

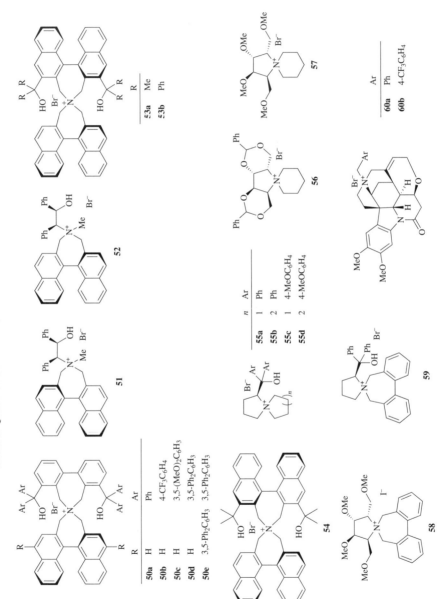

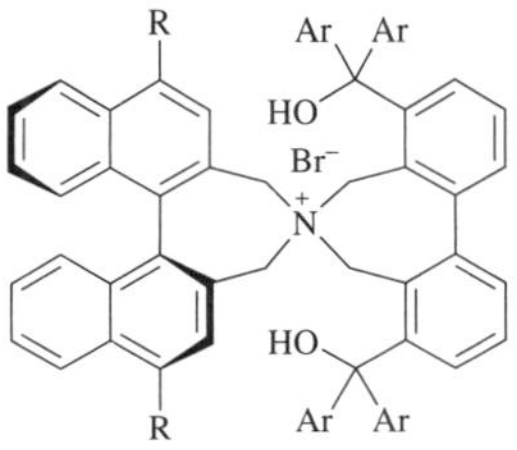

	R	Ar
50a	H	Ph
50b	H	4-$CF_3C_6H_4$
50c	H	3,5-$(MeO)_2C_6H_3$
50d	H	3,5-$Ph_2C_6H_3$
50e	3,5-$Ph_2C_6H_3$	3,5-$Ph_2C_6H_3$

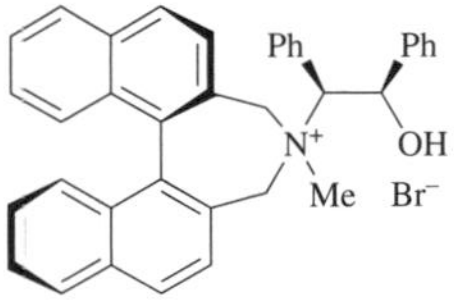

51

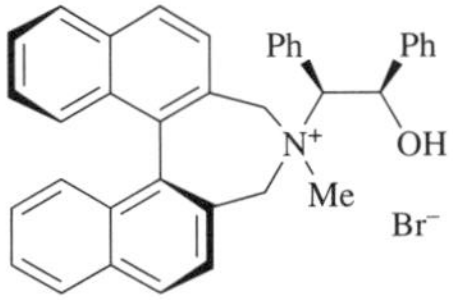

52

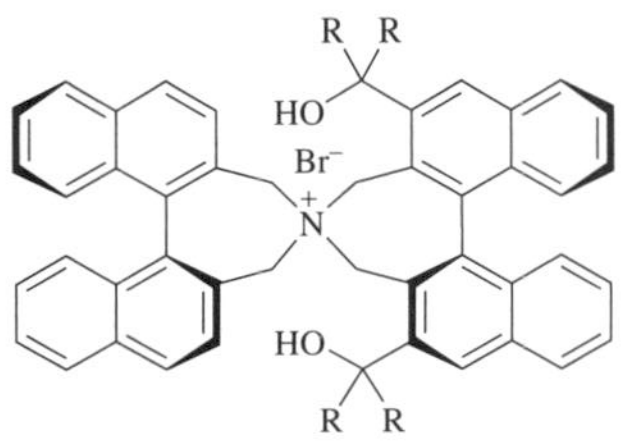

	R
53a	Me
53b	Ph

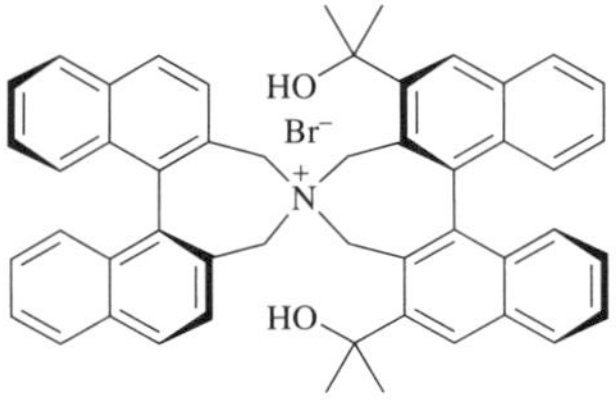

54

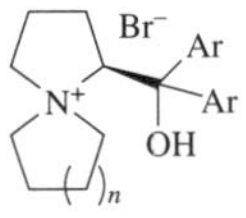

	n	Ar
55a	1	Ph
55b	2	Ph
55c	1	4-$MeOC_6H_4$
55d	2	4-$MeOC_6H_4$

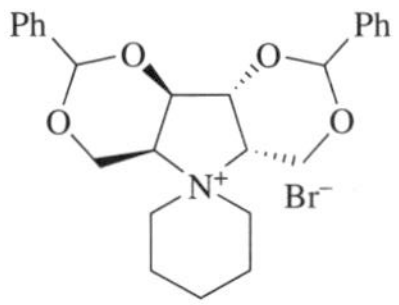

56

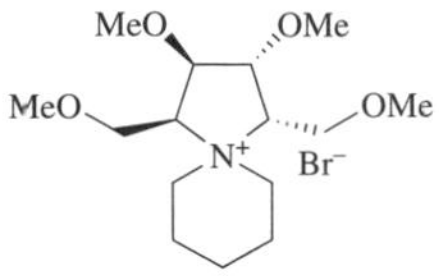

57

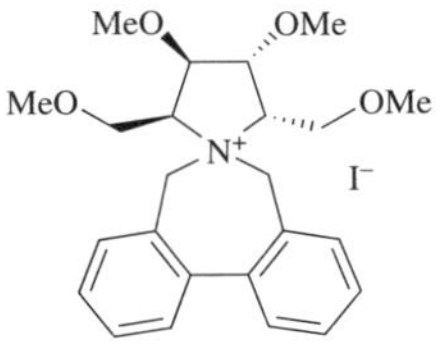

58

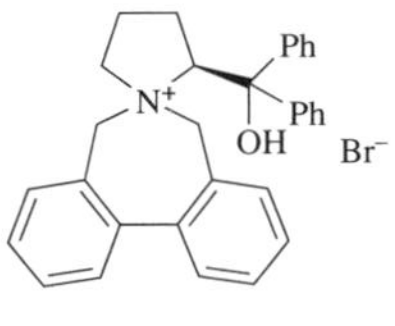

59

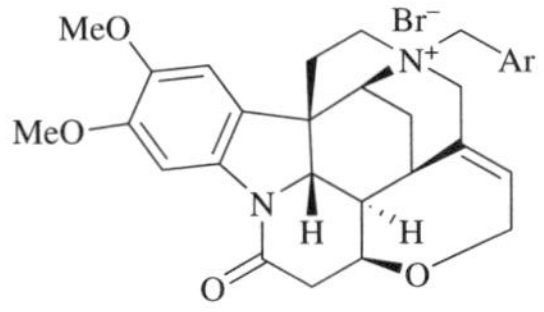

	Ar
60a	Ph
60b	4-$CF_3C_6H_4$

CHART 6. METAL COMPLEXES REFERENCED IN THE TABLES

Salen Complexes

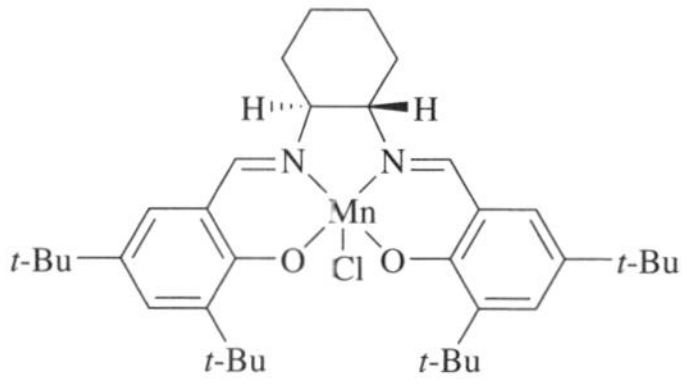

61

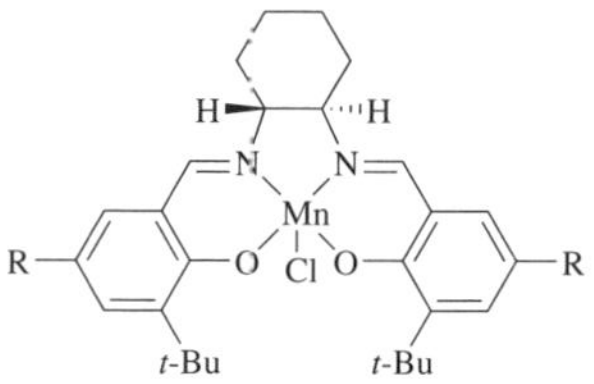

	R
62a	*t*-Bu
62b	OTIPS

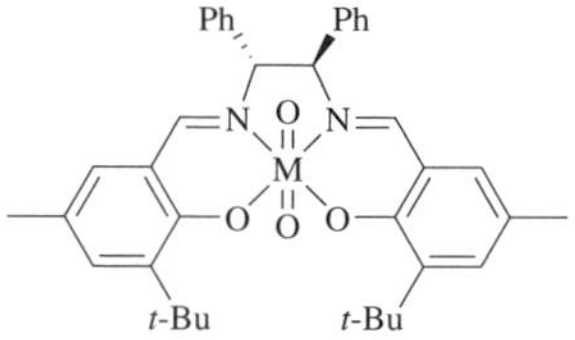

	M
63a	W
63b	Mo

Platinum Complexes

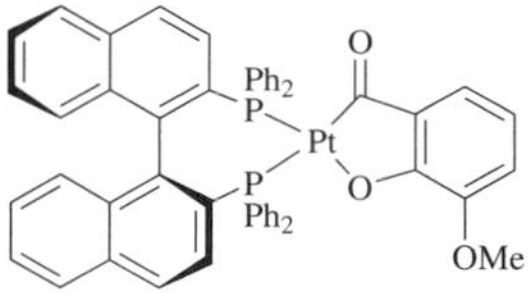

64

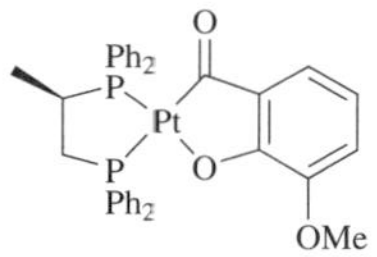

65

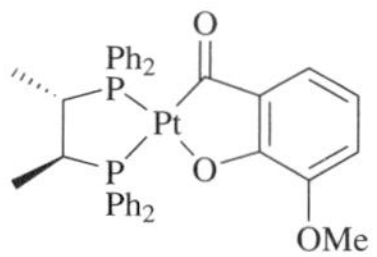

66

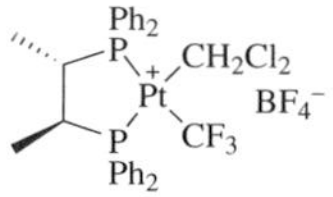

67

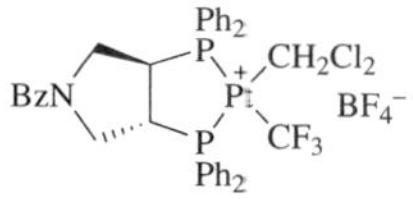

68

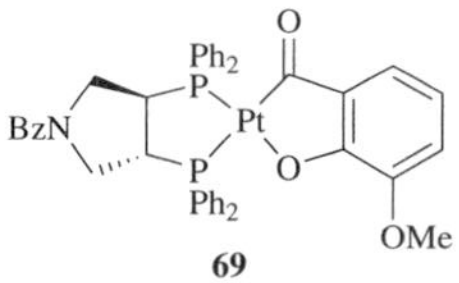

69

CHART 6. METAL COMPLEXES REFERENCED IN THE TABLES (*Continued*)

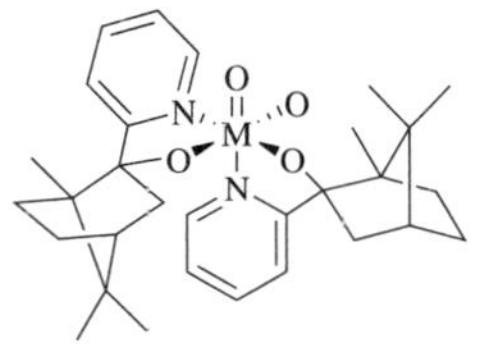

	M
70a	W
70b	Mo

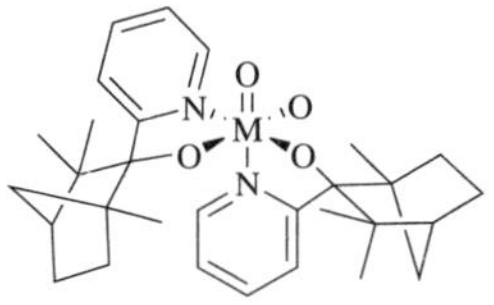

	M
71a	W
71b	Mo

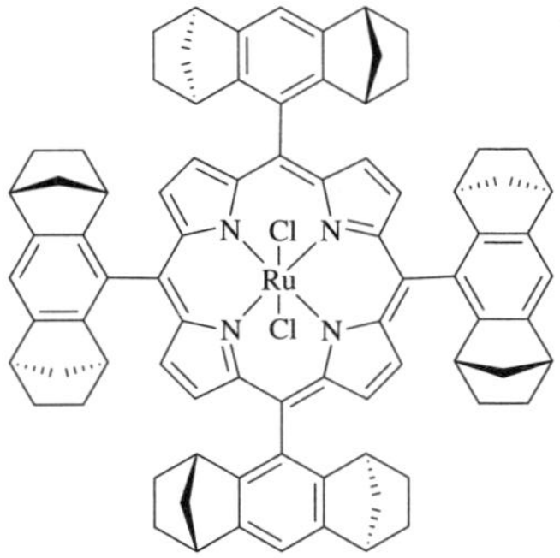

Ru–porphyrin
72

CHART 7. STEROIDAL KETONES REFERENCED IN THE TABLES

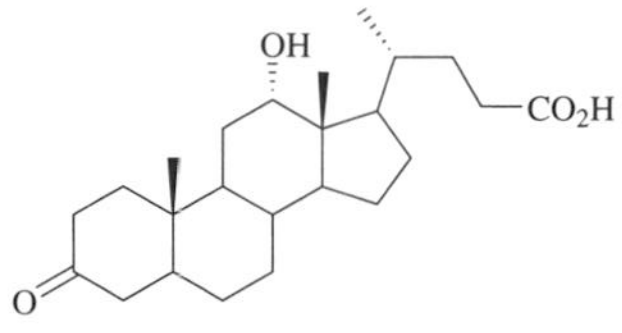

73

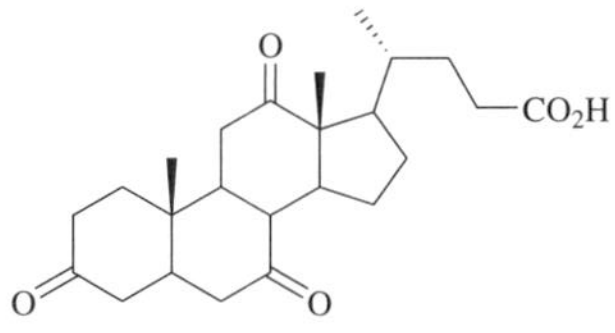

74

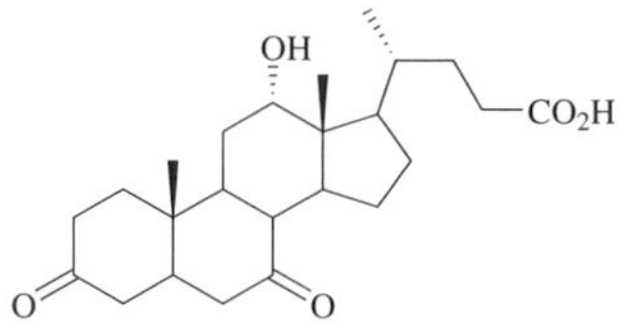

75

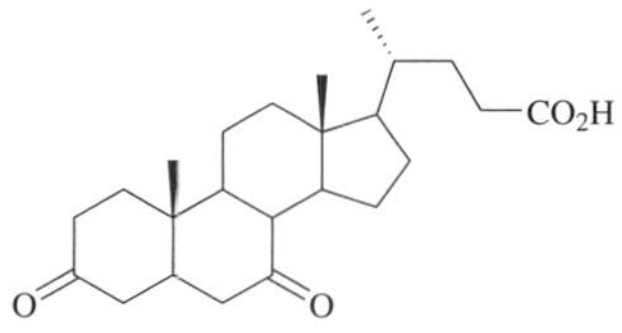

76

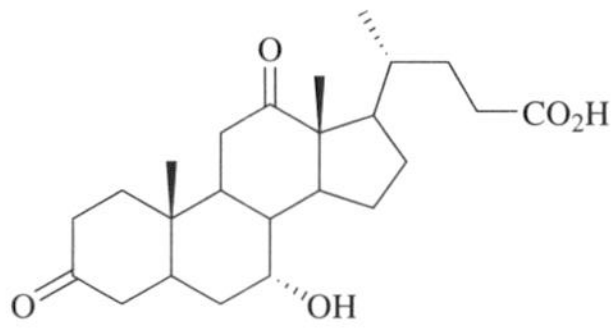

77

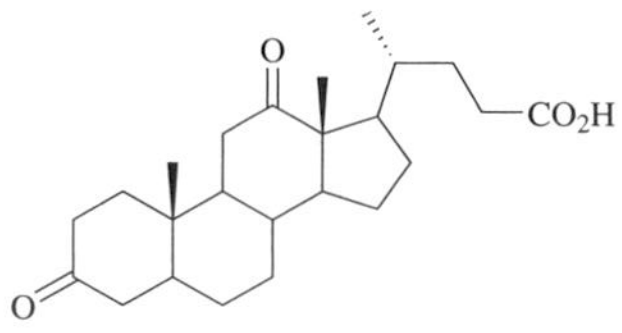

78

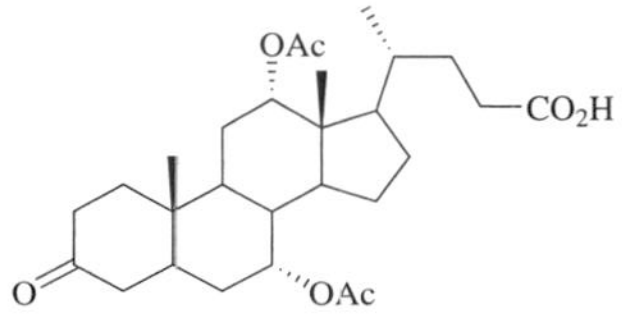

79

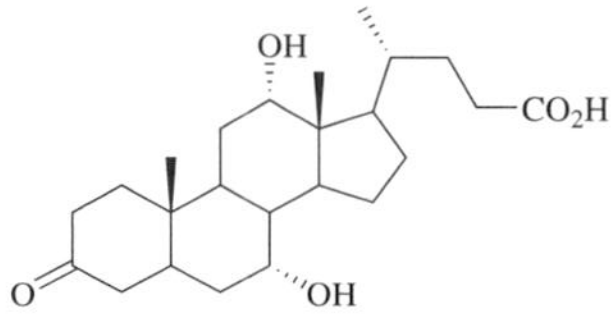

80

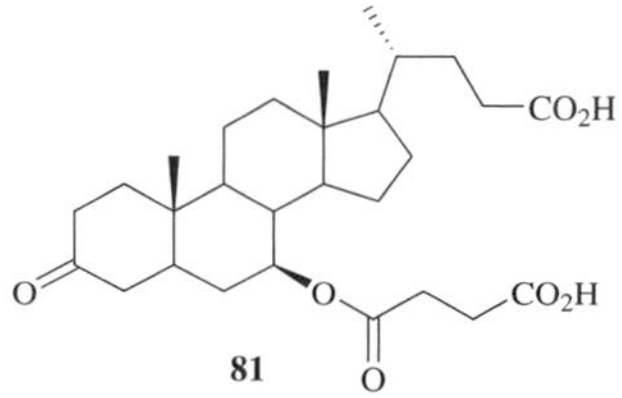

81

CHART 7. STEROIDAL KETONES REFERENCED IN THE TABLES (*Continued*)

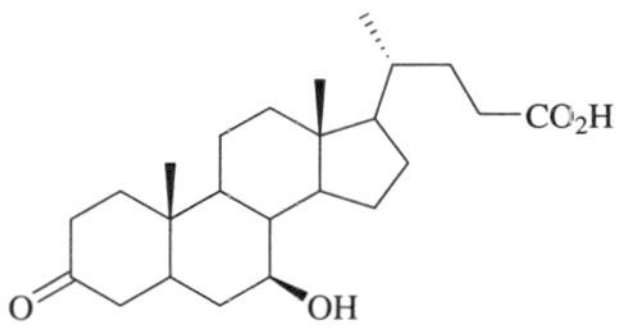

82

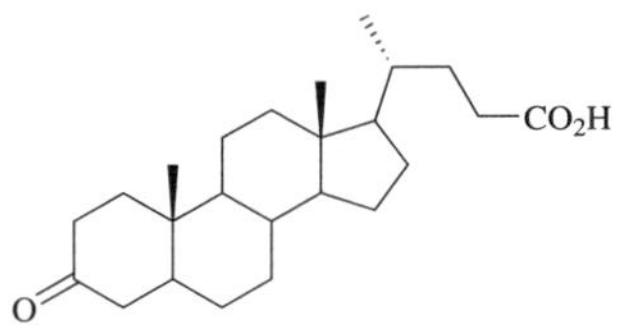

83

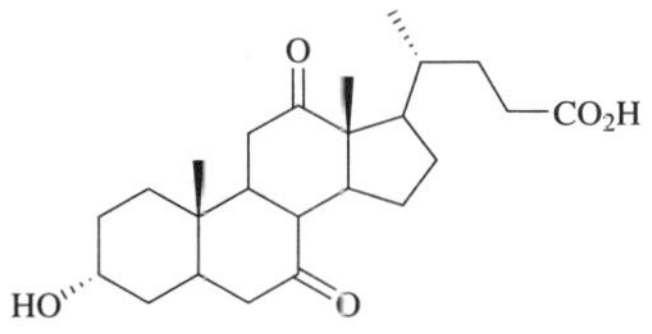

84

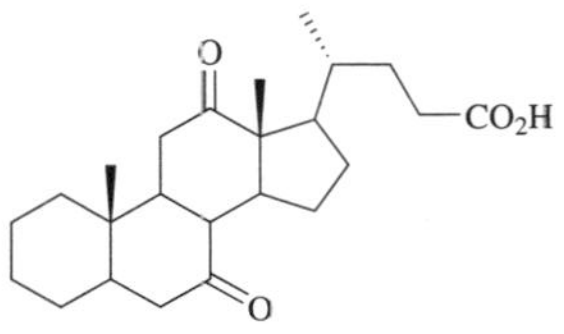

85

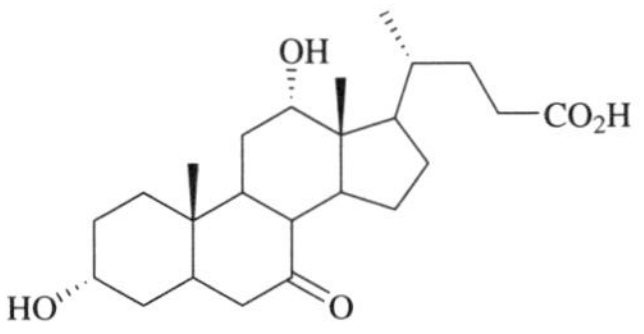

86

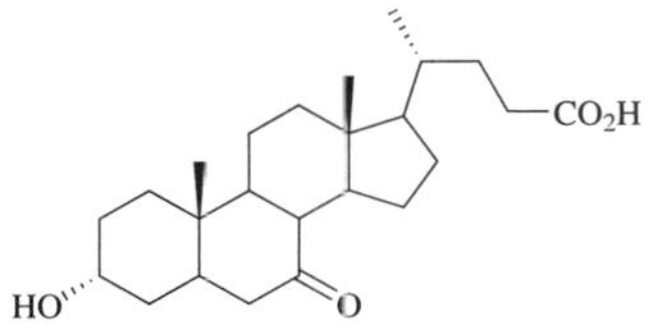

87

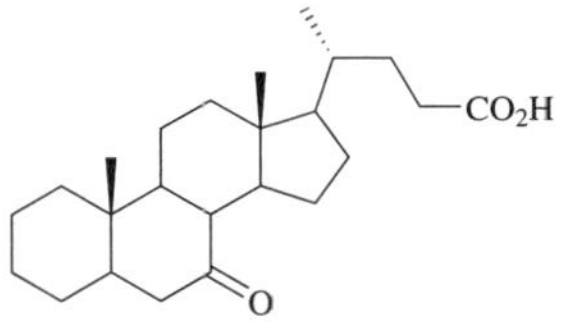

88

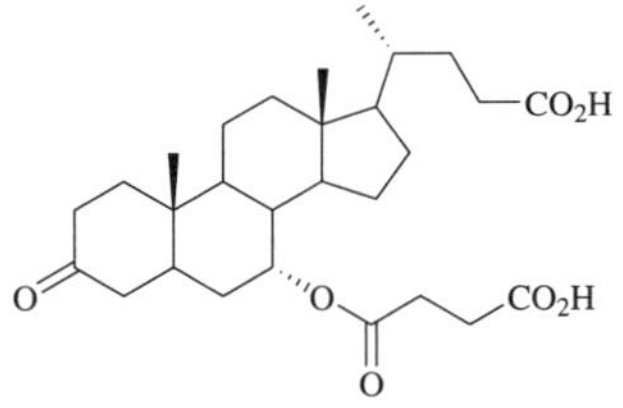

89

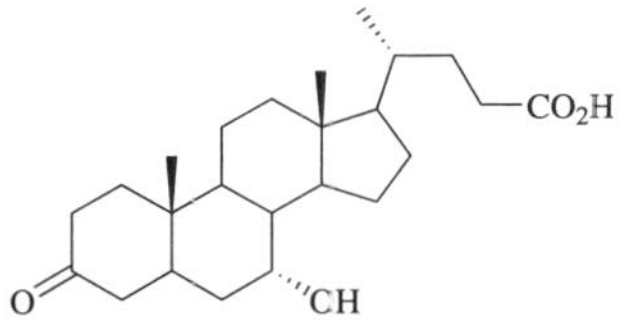

90

Chart 8. Miscellaneous Reagents Referenced in the Tables

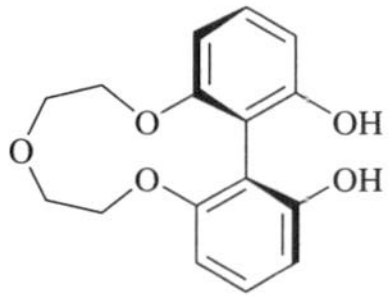

Biphenyl diol

91

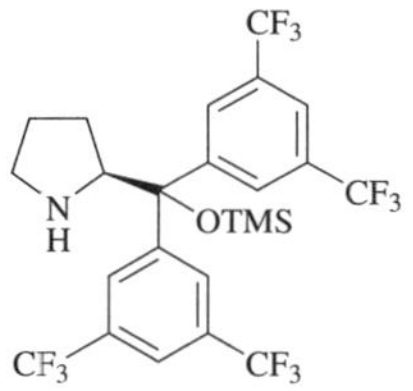

Pyrrolidine

92

TABLE 1A. DIALKYL *E*-ENONES

Substrate	Conditions	Product(s), Yield(s) (%), % ee	Refs.
$C_{6\text{-}9}$ 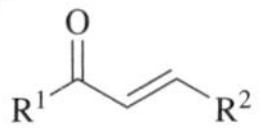	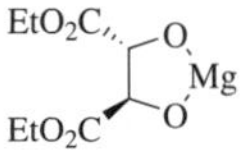 (25 mol %), TBHP, 4 Å MS	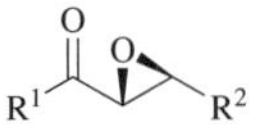	70
R^1 / R^2	Time (h)		
Et / Me	144	(40), 67	
Me / *n*-Bu	120	(54), 75	
Me / *n*-C_5H_{11}	120	(54), 79	
Me / *n*-C_5H_{11}	24[a]	(*82*), 82	
Me / *n*-C_6H_{13}	120	(52), 71	
	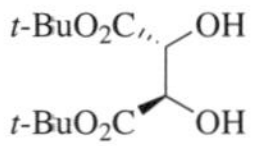 (11 mol %), TBHP, Bu_2Mg, 4 Å MS, toluene, rt, 24 h	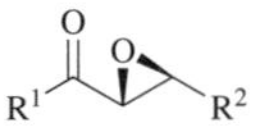	70
R^1 / R^2			
Et / Me		(67), 71	
Me / *n*-Bu		(53), 91	
Me / *n*-C_5H_{11}		(59), 93	
Me / *n*-C_6H_{13}		(63), 92	
C_9 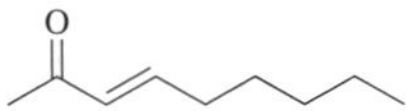	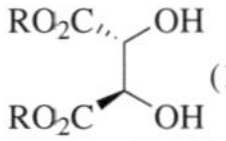 (10 mol %), TBHP, Bu_2Mg, 4 Å MS	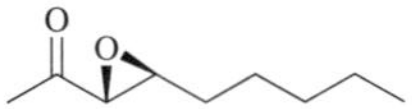	70
	R		
	Me	(*6*), 63	
	Et	(*21–31*), 80–85	
	t-Bu	(*96*), 93–95	
	Bn	(*23*), 85	
	c-C_5H_9	(*80*), 92–93	
	c-C_6H_{11}	(*70*), 92	
	c-$C_{12}H_{23}$	(*88*), 90–93	

C$_{12}$

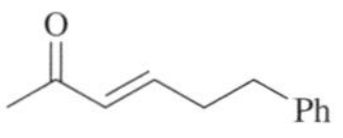

(*R*)-BINOL (x mol %), TBHP, Ln(O*i*-Pr)$_3$, 4 Å MS, THF, rt

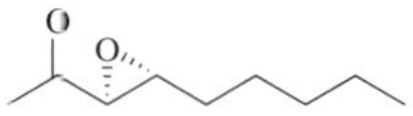

(*R*)-BINOL	x	Ln	Additive	Time (h)		
15a	5	Yb	—	—	(83), 67	46
15d	10	Yb	—	67	(71), 91	46
15a	5	La	TPAO	1.5	(89), 95	55

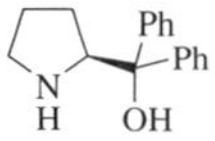 (50 mol %), TBHP, hexane, rt, 185 h

(75), 66 — 170

(*R*)-BINOL (x mol %), oxidant, Ln(O*i*-Pr)$_3$, 4 Å MS, THF, rt

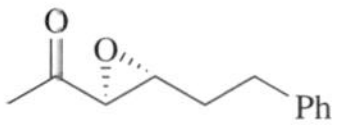

(*R*)-BINOL	x	Oxidant	Ln	Additive	Time (h)		
15a	5	TBHP	Yb	—	—	(87), 53	46
15d	10	TBHP	Yb	—	118	(91), 88	46
15a	5	TBHP	La	TPPO	1	(92), 87	46, 51, 184, 55
15a	5	TBHP	La	TPAO	1.5	(98), 92	55
15a	5	CMHP	La	TPPO	1	(60), 86	67, 55

C$_{13}$

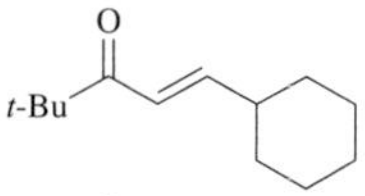

Ammonium salt **50d** (3 mol %), NaOCl, toluene, H_2O, 0°, 48 h

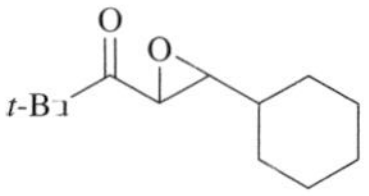

(80), 96 — 109

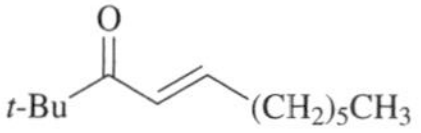

Ammonium salt **50d** (3 mol %), NaOCl, toluene, H_2O, 0°, 112 h

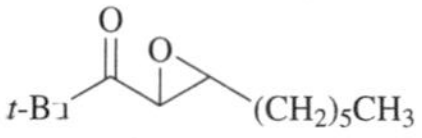

(80), 91 — 109

TABLE 1A. DIALKYL *E*-ENONES (*Continued*)

Substrate	Conditions	Product(s), Yield(s) (%), % ee	Refs.
C_{15} 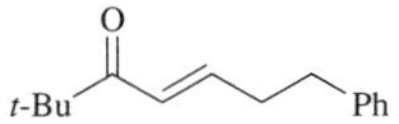	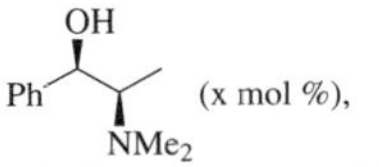 (x mol %), O_2, Et_2Zn, toluene, –78° to 0°	 **I**	
	x = 240, Time (h) = 41	(99), 90	40
	x = 100, Time (h) = 18	(97), 86	41
	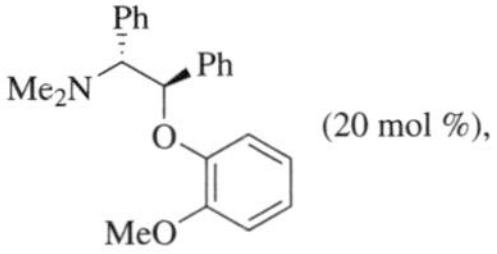 (20 mol %), CMHP, *n*-BuLi, toluene, 0°, 4 h	**I** (55), 37	72
C_{27} 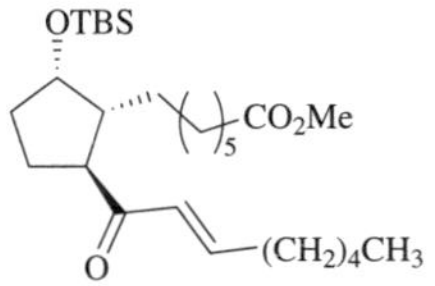	(*S*)-BINOL **18a** (70 mol %), TBHP, La(O*i*-Pr)$_3$, 4 Å MS toluene, 0°, 27 h	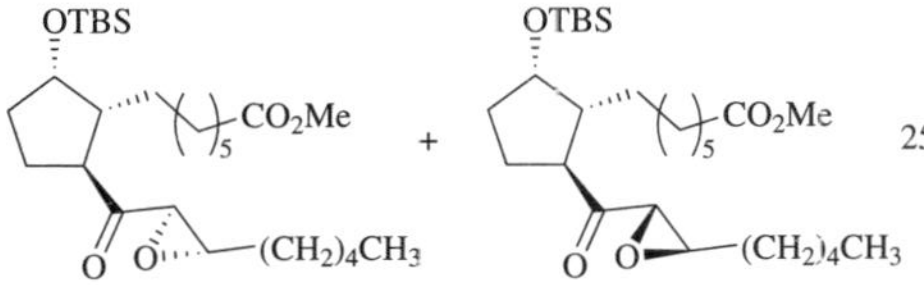 **I** + **II** (89), **I**:**II** = 17:3	257

[a] The catalyst was added in three portions over an eight-hour period.

TABLE 1B. ALKYL ARYL *E*-ENONES

Substrate	Conditions	Product(s), Yield(s) (%), % ee	Refs.
C_{10}			
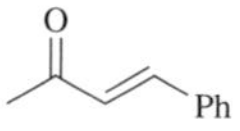	(L-Leu)$_n$–PS (n_{av} = 32), urea/H_2O_2, DBU, THF, rt, 4 h	 **I** (70), 83	116, 176
	(L-Leu)$_n$–PS•SiO_2 (n_{av} = 32), urea/H_2O_2, DBU, TBME, rt, 6 h	**I** (85), >90	175
	(L-Leu)$_n$ (n_{av} = 32; 11 mol %), H_2O_2, NaOH, TBAB, toluene, H_2O, rt, 1 h	**I** (64), 77	120
	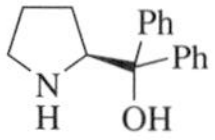 (50 mol %), TBHP, hexane, rt, 186 h	**I** (52), 79	170
	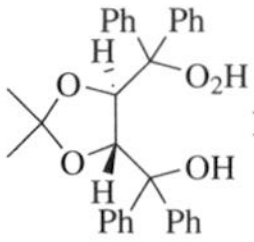 BuLi, THF, –30°	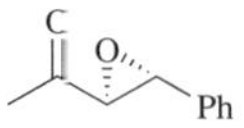 (74), 78	31
	Chiral ketone **37** (10 mol %), Oxone, K_2CO_3, H_2O, DME, AcOH, 0°, 8 h	**I** (75), 82	163

TABLE 1B. ALKYL ARYL *E*-ENONES (*Continued*)

Substrate	Conditions	Product(s), Yield(s) (%), % ee	Refs.
C_{10} 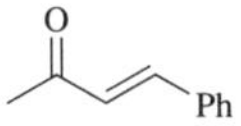	(*R*)-BINOL (x mol %), oxidant, $Ln(Oi\text{-}Pr)_3$, 4 Å MS, THF, rt	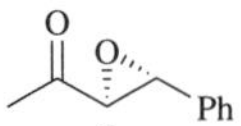 **I**	

(*R*)-BINOL	x	Oxidant	Ln	Additive	Time (h)		Refs.
15a	5	TBHP	Yb	—	—	(94), 88	46
15d	6.25	TBHP	Yb	—	96	(83), 94	46
15a	7.5	TBHP	Yb	H_2O	13	(92), 94	51
15a	5	TBHP	La	TPPO	6	(92), 93	51, 184, 68
15a	5	TBHP	La	TPAO	6	(92), >99	55
15a	5	TBHP	Yb	TPAO	10	(85), 94	55
15a	5	CMHP	La	TPPO	3	(90), >99	67, 68

Conditions	Product(s), Yield(s) (%), % ee	Refs.
(*R*)-BINOL **15a**, TBHP, $Yb(Oi\text{-}Pr)_3$, 4 Å MS, THF	**I** (—)	67

Catalyst % ee[a]	% ee[a]
10	31
43	59
64	76
82	81
100	89

Conditions	Product(s), Yield(s) (%), % ee	Refs.
Ru–porphyrin **72** (0.1 mol %), 2,6-dichloropyridine-*N*-oxide, benzene, rt, 16 h	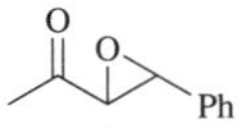 **I** (48), 14	167
PolyBINOL **30** (7.5 mol %), TBHP, $Yb(Oi\text{-}Pr)_3$, TPPO, 4 Å MS, THF, rt, 18 h	**I** (84), 34	258

	PolyBINOL **29** (7.5 mol %), TBHP, Yb(O*i*-Pr)$_3$, TPFO, 4 Å MS, THF, rt, 20 h	**I** (47), 39	258
	PolyBINOL **31** (7.5 mol %), TBHP, Yb(O*i*-Pr)$_3$, TPFO, 4 Å MS, THF, rt, 18 h	**I** (51), 41	258
	PolyBINOL **28** (7.5 mol %), TBHP, Yb(O*i*-Pr)$_3$, TPFO, 4 Å MS, THF, rt, 18 h	**I** (76), 72	258
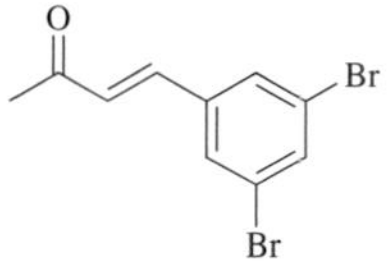	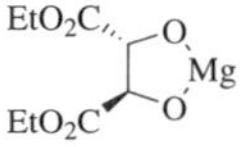 (25 mol %), TBHP, 4 Å MS, toluene, rt, 72 h	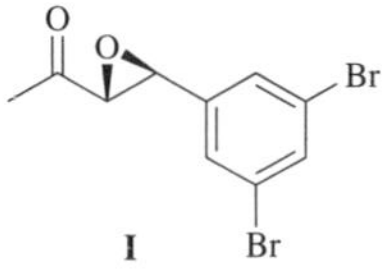 **I** (47), 65	70
	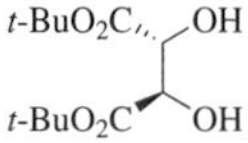 (11 mol %), TBHP, Bu$_2$Mg, 4 Å MS, toluene, rt, 24 h	**I** (62), 81	70
C_{11}			
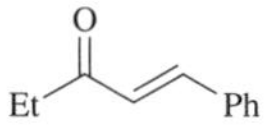	(L-Leu)$_n$–PS (n_{av} = 32), urea/H_2O_2, DBU, THF, rt, 20 h	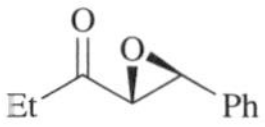 (80), 82	189, 256
C_{12}			
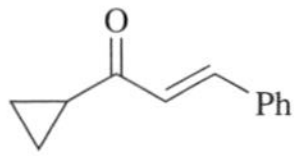	(L-Leu)$_n$, H_2O_2, NaOH, CH_2Cl_2, H_2O, rt, 18 h	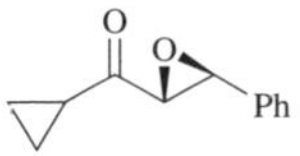 (85), 77	178, 179
	(L-Leu)$_n$ (n_{av} = 33; 11 mol %), H_2O_2, NaOH, TBAB, toluene, H_2O, rt, 5 h	**I** (*40*), 90	120
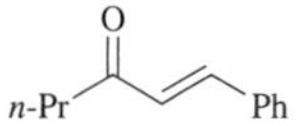	(L-Leu)$_n$–PS (n_{av} = 32), urea/H_2O_2, DBU, THF, rt, 24 h	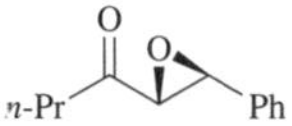 (85), 94	189

Substrate	Conditions	Product(s), Yield(s) (%), % ee	Refs.
C_{12} 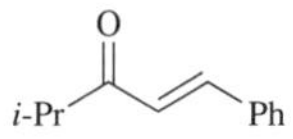	(L-Leu)$_n$, H_2O_2, NaOH, CH_2Cl_2, H_2O, rt, 168 h	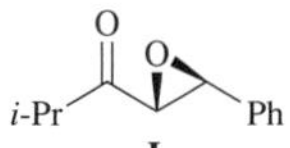 **I** (60), 62	178, 179, 259
	Polypeptide catalyst (n_{av} = 32), urea/H_2O_2, DBU, THF, rt	**I**	
	Catalyst — Time (h)		
	(L-Leu)$_n$–PS — 26	(56), 89	124, 260, 131, 261
	(L-Npg)$_n$–PS — 26	(88), 92	124
	(L-Ala)$_n$–PS — 26	(40), 88	131
	(L-Leu)$_n$–PS•SiO_2 — 18	(57), >99	256, 124, 260, 131
	(L-Npg)$_n$–PS•SiO_2 — 26	(95), 92	131
	(L-Ala)$_n$–PS•SiO_2 — 26	(95), 80	131
	(*S*)-BINOL **17b** (5 mol %), CMHP, Gd(O*i*-Pr)$_3$, THF, rt, 8 h	**I** (82), 82	48
	Chiral ketone **37** (10 mol %), Oxone, K_2CO_3, H_2O, DME, AcOH, 0°, 8 h	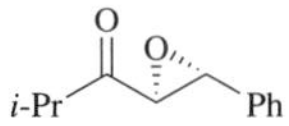 **I** (70), 89	163
	(*R*)-BINOL (x mol %), oxidant, Ln(O*i*-Pr)$_3$, 4 Å MS, THF, rt	**I**	

C$_{13}$

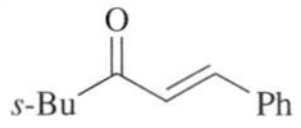

(*R*)-BINOL	x	Oxidant	Ln	Additive	Time (h)		
15a	5	TBHP	Yb	—	—	(84), 83	46, 50
15d	10	TBHP	Yb	—	159	(55), 88	46
15a	5	TBHP	Yb	H_2O	48	(83), 85	50
15a	7.5	TBHP	Yb	H_2O	48	(82), 93	50
15d	7.5	TBHP	Yb	H_2O	48	(55), 88	50
15a	5	TBHP	La	TPPO	12	(67), 96	51, 184, 68
15a	5	TBHP	La	TPAO	8	(72), 95	55
15a	5	CMHP	La	TPPO	4	(72), >99	67, 68
15a	5	CMHP	La	$(4\text{-}FC_6H_4)_3P{=}O$	2	(89), >99	54

(L-Leu)$_n$–PS (n_{av} = 32), urea/H_2O_2, DBU, THF, rt, 24 h

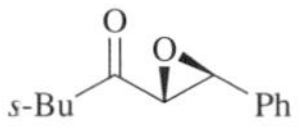

(87), 96 — 189

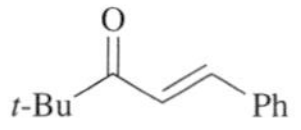

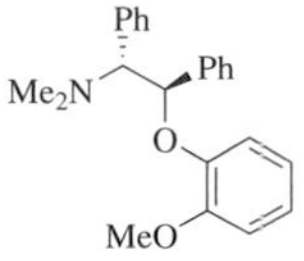

(20 mol %), CMHP, *n*-BuLi, toluene, 0°, 4 h

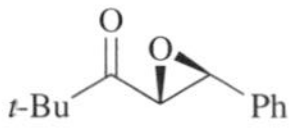

I

(76), 71 — 72

Anthryl cinchoninium salt **1l**, (10 mol %), NaOCl, H_2O, toluene, rt, 48 h — **I** (42), 87 — 100

Polypeptide catalyst, oxidant, base, solvent, rt — **I**

Catalyst	n_{av}	Oxidant	Base	Solvent	Time (h)		
(L-Leu)$_n$	—	H_2O_2	NaOH	CH_2Cl_2/H_2O	18	(92), >98	186, 259, 178, 179
(L-Leu)$_n$–PS	32	urea/H_2O_2	DBU	THF	12	(76), 94	176, 189, 260, 131
(L-Leu)$_n$–PS	32	Na_2CO_3•1.5 H_2O_2	—	DME/H_2O	—	(94), 94	118
(β^3-Leu)$_{20}$–PS	—	urea/H_2O_2	DBU	THF	24	(10), 85	262
(L-Npg)$_n$–PS	32	urea/H_2O_2	DBU	THF	15	(90), 91	131, 260
(L-Ala)$_n$–PS	32	urea/H_2O_2	DBU	THF	15	(31), 86	131
(L-Ala)$_n$–PS•SiO_2	32	urea/H_2O_2	DBU	THF	15	(89), 94	131

TABLE 1B. ALKYL ARYL *E*-ENONES (*Continued*)

Substrate	Conditions	Product(s), Yield(s) (%), % ee	Refs.
C_{13} 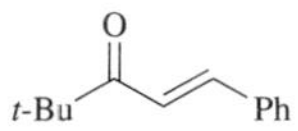	(*R*)-BINOL **15a** (x mol %), oxidant, La(O*i*-Pr)$_3$, 4 Å MS, THF, rt	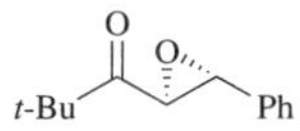 **I**	
	x / Oxidant / Additive / Time		
	5 / CMHP / TPPO / 3 h	(93), 99	67, 68
	5 / CMHP / (4-FC$_6$H$_4$)$_3$P=O / 30 min	(97), >99	54
	5 / TBHP / TPAO / 7 h	(94), 98	55
	10 / TBHP / TPAO / 1 h	(93), 99	55
	Anthryl cinchonidinium salt **6b** (10 mol %), NaOCl, H_2O, toluene, rt, 48 h	**I** (40), 85	100, 99
	Ammonium salt **50d** (3 mol %), NaOCl, toluene, H_2O, 0°, 83 h	**I** (99), 92	109
	Ammonium salt **50e** (3 mol %), NaOCl, toluene, H_2O, 0°, 187 h	**I** (87), 89	109
	(D-Leu)$_n$, H_2O_2, NaOH, CH_2Cl_2, H_2O, rt, 18 h	**I** (90), >86	186, 178, 259
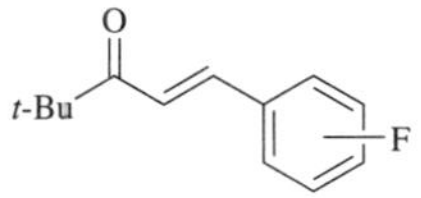	(*R*)-BINOL **15a** (5 mol %), CMHP, La(O*i*-Pr)$_3$, TPPO, 4 Å MS, THF, rt	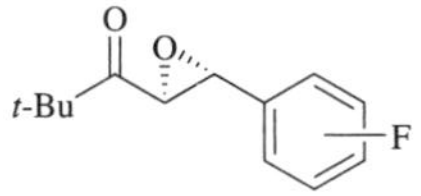	68
	Time		
2-F	20 h	(53), >99	
3-F	12 h	(52), >99	
4-F	12 h	(45), >99	
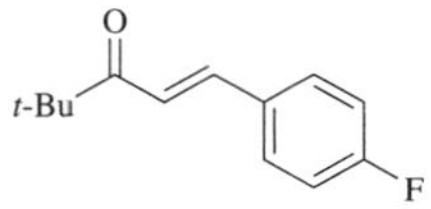	(*R*)-BINOL **15a** (5 mol %), CMHP, La(O*i*-Pr)$_3$, (4-FC$_6$H$_4$)$_3$P=O, 4 Å MS, THF, rt, 20 min	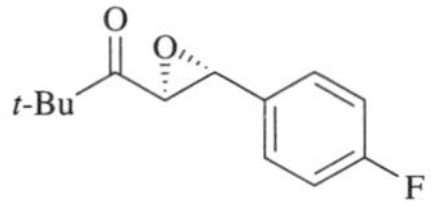 (91), 99	54

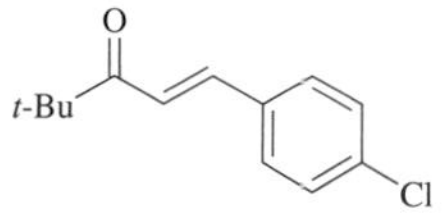	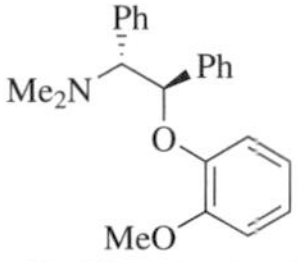 (20 mol %), CMHP, *n*-BuLi, toluene, 0°, 4 h	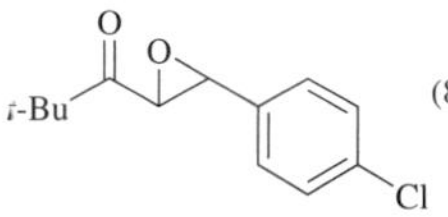 (81), 66	72
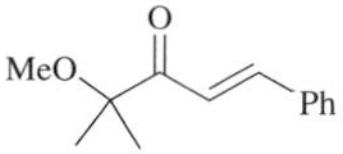	(L-Leu)$_n$, H_2O_2, NaOH, CH_2Cl_2, H_2O, rt, 84 h	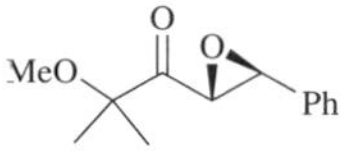 (70), 63	178, 179
C_{14}			
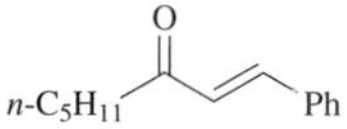	Polypeptide catalyst (n_{av} = 32), urea/H_2O_2, DBU, THF, rt, 24 h	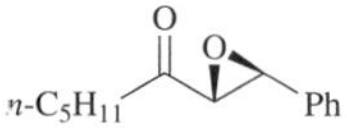	131
	Catalyst		
	(L-Leu)$_n$–PS	(45), 95	
	(L-Npg)$_n$–PS	(97), 90	
	(L-Ala)$_n$–PS	(34), 96	
	(L-Leu)$_n$–PS•SiO_2	(90), 95	
	(L-Npg)$_n$–PS•SiO_2	(100), 97	
	(L-Ala)$_n$–PS•SiO_2	(49), 80	
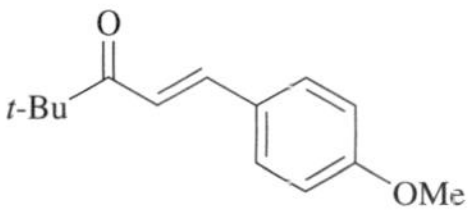	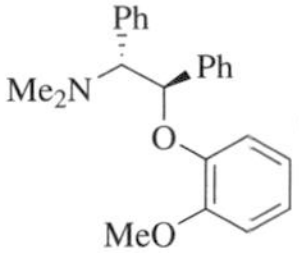 (20 mol %), CMHP, *n*-BuLi, toluene, rt, 36 h	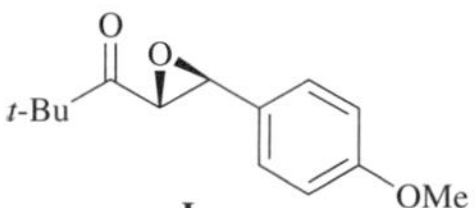 **I** (68), 64	72
	Polypeptide catalyst (n_{av} = 32), urea/H_2O_2, DBU, THF, rt, 28 h	**I**	
	Catalyst		
	(L-Leu)$_n$–PS	(≥90), ≥98	176, 131
	(L-Npg)$_n$–PS	(83), 96	131
	(L-Ala)$_n$–PS	(54), 89	131

TABLE 1B. ALKYL ARYL *E*-ENONES (*Continued*)

Substrate	Conditions	Product(s), Yield(s) (%), % ee	Refs.
C_{15} 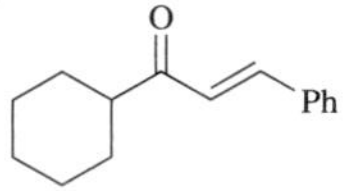	Polypeptide catalyst (n_{av} = 32), urea/H_2O_2, DBU, THF, rt, 12 h	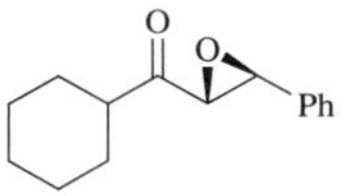	131
	Catalyst: (L-Leu)$_n$–PS	(46), 96	
	(L-Npg)$_n$–PS	(87), 98	
	(L-Ala)$_n$–PS	(32), 75	
C_{16} 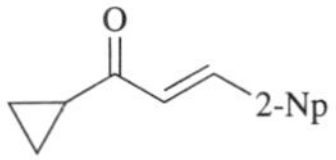	(D-Leu)$_n$, H_2O_2, NaOH, CH_2Cl_2, H_2O, rt, 29 h	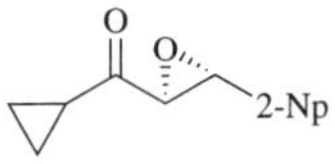 (61), 90	178, 259
C_{17} 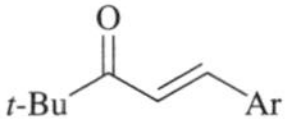	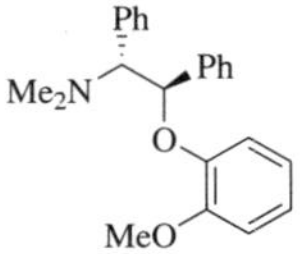 (20 mol %), CMHP, *n*-BuLi, toluene	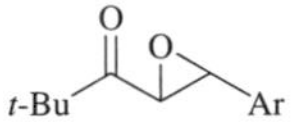	72
Ar: 1-Np	Temp 0°, Time 4 h	(55), 56	
2-Np	rt, 36 h	(74), 58	
C_{28} 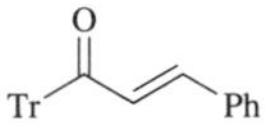	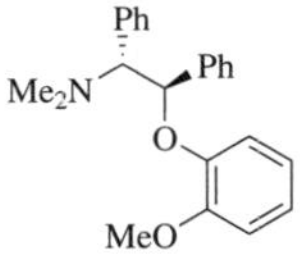 (20 mol %), CMHP, *n*-BuLi, toluene, 0°, 4 h	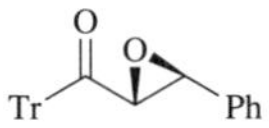 (62), 51	72

[a] These figures are estimated from a graph in the publication.

TABLE 1C. ARYL ALKYL *E*-ENONES

Substrate	Conditions	Product(s), Yield(s) (%), % ee	Refs.
C_9			
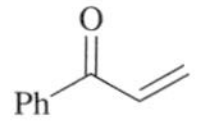	(L-Leu)$_n$–PS (5–10 mol %), urea/H_2O_2, DBU or BEMP, THF, rt, 30 min	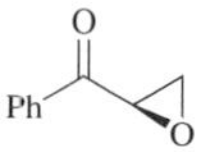 (—), >95	140
C_{10}			
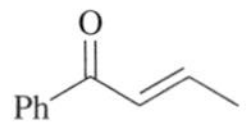	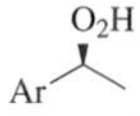 (100 mol %), DBU, toluene, 20°, 24 h	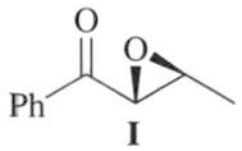 **I**	28
	Ar: Ph	(98), 25	
	Ar: 4-ClC_6H_4	(96), 34	
	Ar: 2-Np	(97), 39	
	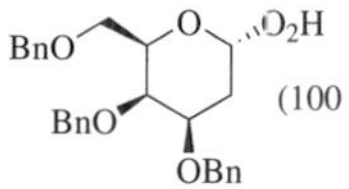 (100 mol %), DBU, toluene, 20°, 24 h	**I** (91), 31	28
	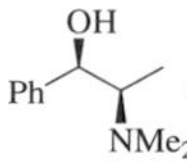 (x mol %), O_2, Et_2Zn, toluene, –78° to 0°	**I** x = 240, Time (h) = 3, (96), 85	40
		x = 100, Time (h) = 17, (94), 78	41
	(*R*)-BINOL **15a** (20 mol %), oxidant, Et_2Zn, Et_2O, 0° to rt, 12 h	**I** Oxidant: TBHP (24), 40	44
		Oxidant: CMHP (60), 54	
	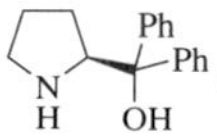 (30 mol %), TBHP, hexane, rt, 114 h	**I** (87), 63	170
	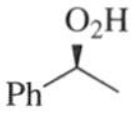 (100 mol %), KOH, MeCN, –40°, 10–20 min	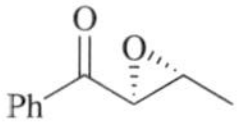 (99), 46	27, 28

TABLE 1C. ARYL ALKYL *E*-ENONES (*Continued*)

Substrate	Conditions	Product(s), Yield(s) (%), % ee	Refs.
C_{10}			
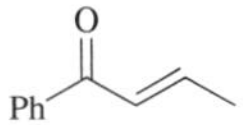	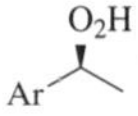 (100 mol %), KOH, MeCN, –40°	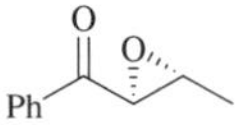 **I**	28
	Ar: 4-ClC_6H_4	(96), 40	
	Ar: 2-Np	(86), 38	
	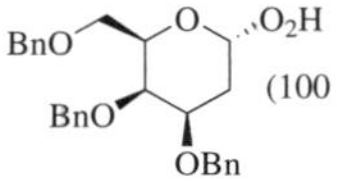 (100 mol %), KOH, MeCN, –40°	**I** (92), 31	28
	(*R*)-BINOL **15d** (14 mol %), TBHP, Yb(O*i*-Pr)$_3$, 4 Å MS, THF, rt	**I** (87), 80	49
C_{11}			
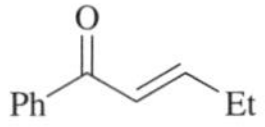	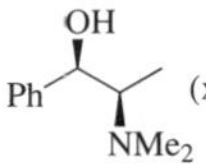 (x mol %), O_2, Et_2Zn, toluene, –78° to 0°	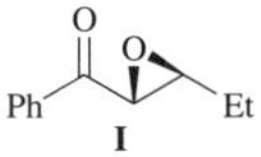 **I**	
		x = 240, Time (h) = 3: (99), 91	40
		x = 100, Time (h) = 17: (94), 87	41
	(*R*)-BINOL **15a** (20 mol %), oxidant, Et_2Zn, Et_2O, 0° to rt, 12 h	**I** Oxidant: TBHP (28), 48; CMHP (79), 50	44
	(L-Leu)$_n$–PS•SiO_2 (n_{av} = 32), urea/H_2O_2, base (x eq), THF, rt	**I**	117
		Base: DBU, x: —, Time (h): 17, (64), 60	
		Base: BEMP, x: 0.03, Time (h): —, (86), 77–79	
		Base: BEMP, x: 0.001, Time (h): —, (—), 74	

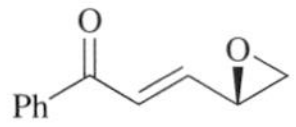	$(\beta^3\text{-Leu})_{20}$–PS, 30% H_2O_2, NaOH, toluene, H_2O, 24 h	**I** (98), 15	262
	$(\beta^3\text{-Leu})_{20}$–PS, urea/$H_2O_2$, DBU, THF	**I** (98), 28	262
	Naphthyl quinidinium salt **2e** (20 mol %), H_2O_2, LiOH, H_2O, $CHCl_3$, –10°, 44 h	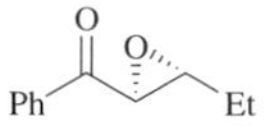 (88), 21	106
	Polyleucine catalyst (n_{av} = 32), urea/H_2O_2, DBU, THF, rt	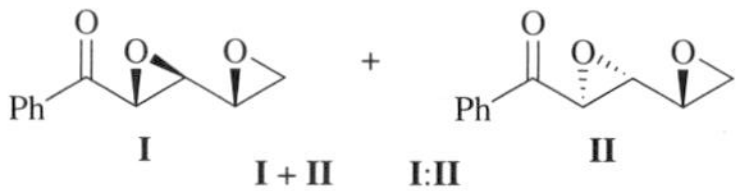 **I** + **II**	181
	Catalyst	**I + II** / **I:II**	
	(D-Leu)$_n$–PS	(95) / 1:10	
	(L-Leu)$_n$–PS	(—) / 3.5:1	
C_{12} 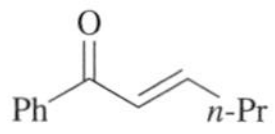	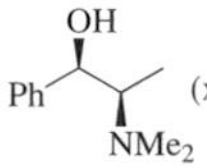 (x mol %), O_2, Et_2Zn, toluene, –78° to 0°	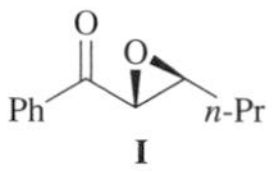 **I**	40
	x = 240, Time (h) = 3	(99), 87	
	x = 100, Time (h) = 17	(95), 80	41
	(*R*)-BINOL **15a** (20 mol %), oxidant, Et_2Zn, Et_2O, 0° to rt, 12 h	**I**	44
	Oxidant: TBHP	(72), 60	
	Oxidant: CMHP	(78), 40	
	(*S*)-BINOL **17b** (5 mol %), CMHP, Yb(O*i*-Pr)$_3$, THF, rt, 8 h	**I** (85), 38	47

TABLE 1C. ARYL ALKYL *E*-ENONES (*Continued*)

Substrate	Conditions	Product(s), Yield(s) (%), % ee	Refs.
C_{12}			
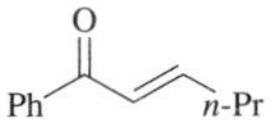	PolyBINOL **29** (x mol %), TBHP, Et_2Zn, rt	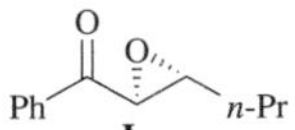 **I**	43
	x / Solvent / Time (h)		
	20 / Et_2O / 5	(92), 76	
	100 / Et_2O / 4.5	(81), 73	
	20 / TBME / 4.5	(89), 71	
	20 / *n*-BuOMe / 3.5	(87), 62	
	20 / PhOMe / 6	(98), 48	
	(*R*)-BINOL, TBHP, Et_2Zn, rt	**I**	43
	(*R*)-BINOL		
	16b	(75), 30	
	16a	(83), 21	
	PolyBINOL **31** (20 mol %), TBHP, Et_2Zn, Et_2O, rt, 5 h	**I** (99), 37	43
	(*R*)-BINOL **15d** (14 mol %), TBHP, Yb(O*i*-Pr)$_3$, 4 Å MS, THF, rt	**I** (81), 71	49
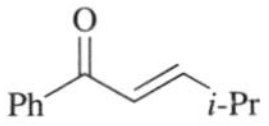	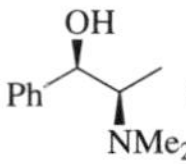 (x mol %), O_2, Et_2Zn, toluene, –78° to 0°	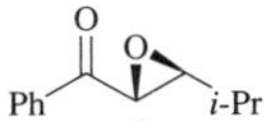 **I**	40
	x / Time (h)		
	240 / 3	(97), 92	
	100 / 17	(97), 86	41

PolyBINOL **29** (x mol %), TBHP, Et_2Zn, Et_2O, rt, 3.5 h	**I**	43
x		
20	(93), 78	
40	(94), 81	
PolyBINOL **31** (20 mol %), TBHP, Et_2Zn (0.36 eq), Et_2O, rt, 5 h	**I** (88), 55	43
(*R*)-BINOL **15a** (20 mol %), oxidant, Et_2Zn, Et_2O, 0° to rt, 12 h	**I**	44
Oxidant		
TBHP	(36), 54	
CMHP	(75), 40	
BisBINOL **23g** (2.5 mol %), CMHP, La(O*i*-Pr)$_3$, TPPO, 4 Å MS, THF, rt, 0.5 h	**I** (91), 84.9	66
PolyBINOL **28** (111 mol %), O_2, Et_2Zn, toluene, 0°	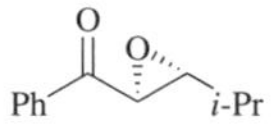 (18), 25	43
(*R*)-BINOL (x mol %), oxidant, La(O*i*-Pr)$_3$, 4 Å MS, THF	**I**	

(*R*)-BINOL	x	Oxidant	Additive	Temp (°)	Time (h)		
15a	5	CMHP	—	rt	12	(93), 86	46
15a	2	CMHP	—	0	5	(92), 96	263
15d	6.25	CMHP	—	rt	7	(95), 94	46, 184
15a	5	CMHP	TPPO	rt	3	(88), 98	67, 68
15a	5	TBHP	TPPO	rt	1	(89), 93	51
15a	5	TBHP	TPAO	rt	1.5	(95), 94	55
15a	5	CMHP	(4-FC_6H_4)$_3$P=O	rt	0.67	(99), 98	54

TABLE 1C. ARYL ALKYL *E*-ENONES (*Continued*)

Substrate	Conditions	Product(s), Yield(s) (%), % ee	Refs.
C_{12} 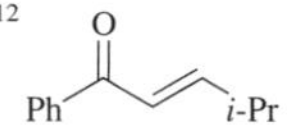	Naphthyl quinidinium salt **2e** (5 mol %), H_2O_2, LiOH, H_2O, $CHCl_3$, –10°, 14 h	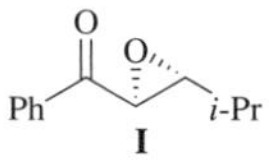 **I** (97), 35	106
	Iodobenzyl cinchoninium salt **1f** (5 mol %), H_2O_2, LiOH, H_2O, $CHCl_3$, 4°, 43 h	**I** (82), 42	104
C_{13} 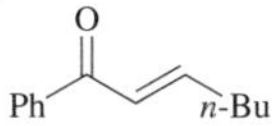	(*R*)-BINOL **15a** (2 mol %), CMHP, La(O*i*-Pr)$_3$, 4 Å MS, THF, 0°, 5 h	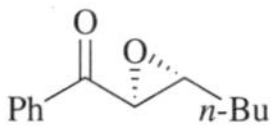 (89), 92	263
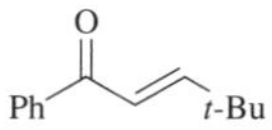	(*R*)-BINOL **15a** (20 mol %), oxidant, Et_2Zn, Et_2O, 0° to rt, 12 h	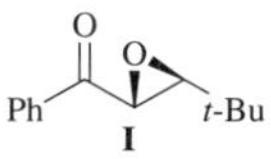 **I** Oxidant: TBHP (<10), 80; CMHP (73), 80	44
	PolyBINOL **29** (20 mol %), TBHP, Et_2Zn, Et_2O, rt, 8 h	**I** (67), 64	43
	Benzyl cinchonidinium salt **3b** (10 mol %), (±)-2-phenylbut-2-yl hydroperoxide, KOH, toluene, 0–20°	**I** (—), 33	105
	(L-Leu)$_n$ (n_{av} = 30), H_2O_2, NaOH, CH_2Cl_2, H_2O, rt, 18 h	**I** (85), 90	186, 179
	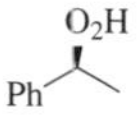 (100 mol %), base	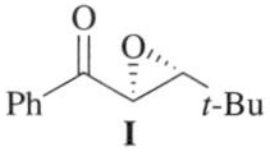 **I**	

Base	Solvent	Temp (°)	Time (h)	Yield (%), % ee	Refs.
KOH	MeCN	–40	4–5	(95), 75	27, 28
KOH/18-crown-6	MeCN	–40	4–5	(91), 11	27, 28
DBU	toluene	20	24	(42), 7	28

Substrate	Conditions	Product(s) and Yield(s) (%), % ee	Refs.
	Trifluoromethylbenzyl cinchoninium salt **1h** (10 mol %), oxidant, KOH, toluene, 0–20°	**I** (—)	105
	Oxidant	% ee **I**	
	(±)-1-phenylethyl hydroperoxide	62	
	(±)-2-phenylbut-2-yl hydroperoxide	36	
	CMHP	28	
	TBHP	27	
	Benzyl cinchoninium salt **1a** (10 mol %), (±)-2-Phenylbut-2-yl hydroperoxide, KOH, toluene, 0–20°	**I** (—), 47	105
	Iodobenzyl cinchoninium salt **1f** (10 mol %), H_2O_2, LiOH, H_2O, $CHCl_3$, 4°, 43 h	**I** (90), 55	104
	(D-Leu)$_n$, H_2O_2, NaOH, CH_2Cl_2, H_2O, rt, 18 h	**I** (86), 76	186
	Crown ether **48** (5 mol %), NaOCl, H_2O, toluene, rt, 140 h	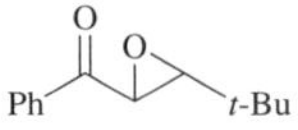 (22), 12	177
C_{14} 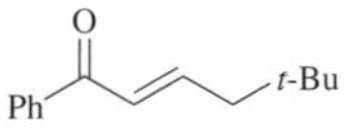	Iodobenzyl cinchoninium salt **1f** (5 mol %), H_2O_2, LiOH, H_2O, n-Bu_2O, 4°, 68 h	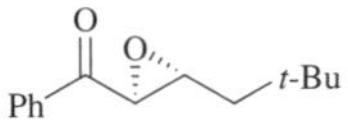 (41), 57	104
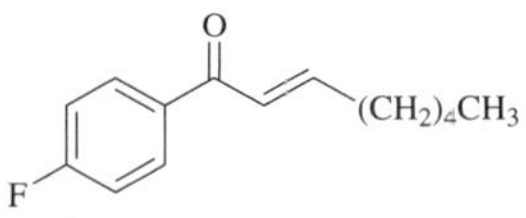	Anthryl cinchonidinium salt **6b** (10 mol %), KOCl, H_2O, toluene, –40°, 12 h	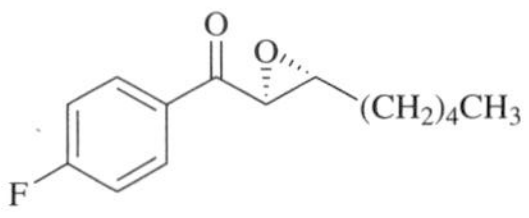 (90), 91	98

TABLE 1C. ARYL ALKYL *E*-ENONES (*Continued*)

Substrate	Conditions	Product(s), Yield(s) (%), % ee	Refs.
C_{14} 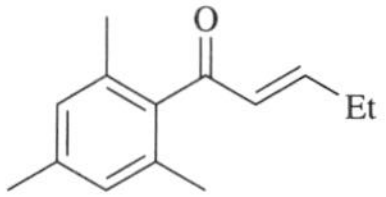	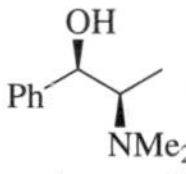 (x mol %), O_2, Et_2Zn, toluene, –78° to 0°	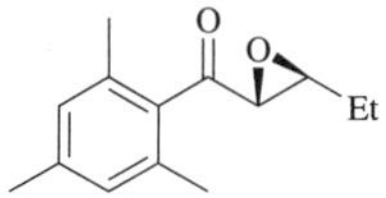 x / Time (h): 240 / 89 / (94), 82; 100 / 72 / (93), 80	40; 41
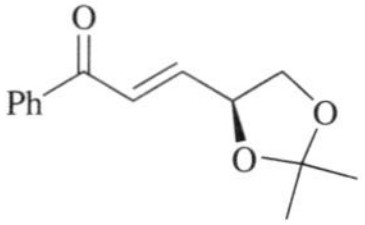	Polyleucine catalyst (n_{av} = 32)	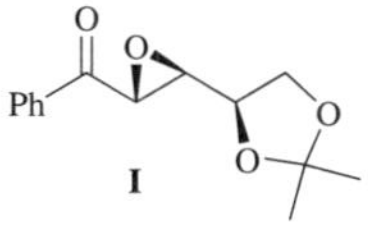 **I** + 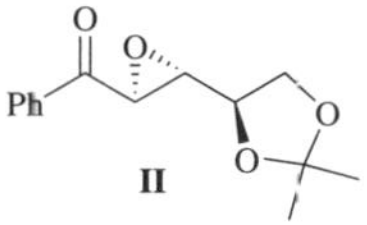 **II**	

Catalyst		Temp (°)	Time (h)	I + II	I:II	Refs.
(L-Leu)$_n$–PS	urea/H_2O_2, DBU, THF	rt	5	(92)	1:1	99, 181
(L-Leu)$_n$–PS	Na_2CO_3•1.5 H_2O_2, DME, H_2O	rt	0.5	(96)	2.4:1	99, 181
(L-Leu)$_n$–PS	Na_2CO_3•1.5 H_2O_2, DME, H_2O	–3 to 0	24	(98)	3.8:1	99, 181
(D-Leu)$_n$–PS	urea/H_2O_2, DBU, THF	rt	5	(95)	1:30	99, 181
(D-Leu)$_n$–PS	Na_2CO_3•1.5 H_2O_2, DME, H_2O	rt	0.5	(97)	1:20	99
(D-Leu)$_n$–PS	Na_2CO_3•1.5 H_2O_2, DME, H_2O	–3 to 0	24	(97)	1:34	99, 181
(L-Leu)$_n$–PS•SiO_2	urea/H_2O_2, DBU, THF	rt	0.5	(96)	2.2:1	181

Substrate	Conditions	Product(s), Yield(s) (%), % ee	Refs.
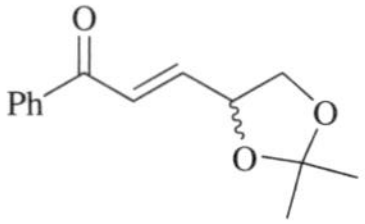	(L-Leu)$_n$–PS (n_{av} = 32), Na_2CO_3•1.5 H_2O_2, DME, H_2O, 0°	**I** + **II** + 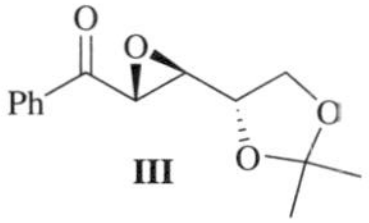 **III** + 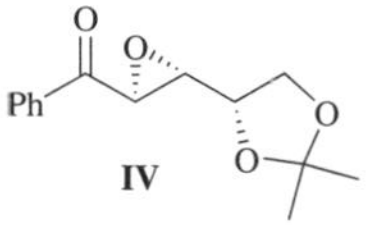 **IV** **I** + **II** + **III** + **IV** (—) **I**:**II** = 3.9:1 **III**:**IV** = 37:1	99

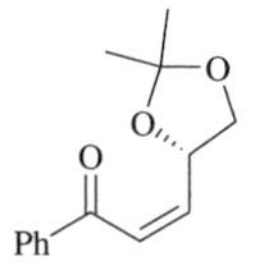

Polyleucine catalyst (n_{av} = 32), urea/H_2O_2, DBU, THF

I + II

99, 181

Catalyst	Temp (°)	Time (min)	I + II	I:II
(L-Leu)$_n$–PS	rt	35 min	(94)	3.6:1
(L-Leu)$_n$–PS	–30	1080 (13 h)	(90)	4.0:1
(D-Leu)$_n$–PS	rt	35 min	(91)	1:8.0

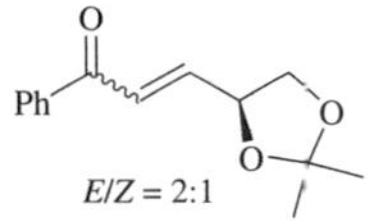

E/*Z* = 2:1

(D-Leu)$_n$–PS (n_{av} = 32), urea/H_2O_2, DBU, THF, rt

I + **II** (—), **I**:**II** = 1:17

99

$C_{15\text{-}19}$

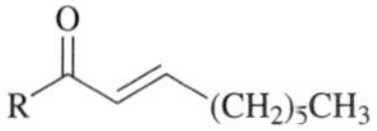

Anthryl cinchonidinium salt **6b** (10 mol %), NaOCl, H_2O, toluene, rt

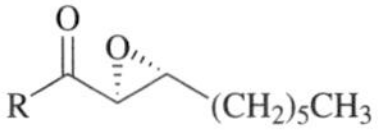

R		
Ph	(92), 77	100, 99
4-BrC_6H_4	(89), 84	99
4-$O_2NC_6H_4$	(79), 90	99
4-MeC_6H_4	(93), 81	99
4-$MeOC_6H_4$	(94), 84	99
3,4-methylenedioxyphenyl	(87), 86	99
2-Np	(83), 78	99

C_{15}

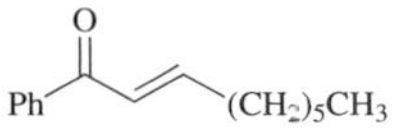

Anthryl cinchoninium salt **1l** (10 mol %), NaOCl, H_2O, toluene, rt, 48 h

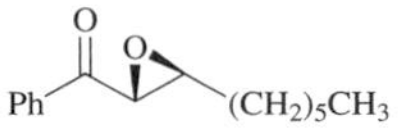

(75), 76

100

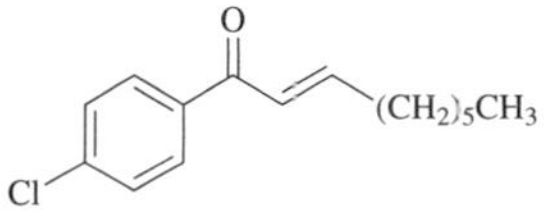

Ammonium salt **50e** (3 mol %), NaOCl, toluene, H_2O, –20°, 41 h

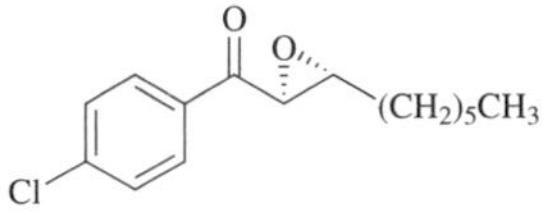

(99), 96

109

Substrate	Conditions	Product(s), Yield(s) (%), % ee	Refs.
C_{15} 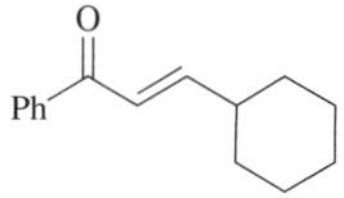	(*R*)-BINOL **15a** (2 mol %), CMHP, La(O*i*-Pr)$_3$, 4 Å MS, THF, 0°, 5 h	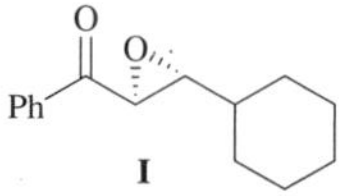 **I** (80), 98	263
	Naphthyl quinidinium salt **2e** (5 mol %), H_2O_2, LiOH, H_2O, $CHCl_3$, –10°, 48 h	**I** (*34*), 26	106
	Benzyl cinchoninium salt **1a** (5 mol %), H_2O_2, LiOH, H_2O, $CHCl_3$, 4°, 37 h	**I** (70), 53	104
	(L-Leu)$_n$–PS (n_{av} = 32), urea/H_2O_2, DBU, THF, rt, 3 h	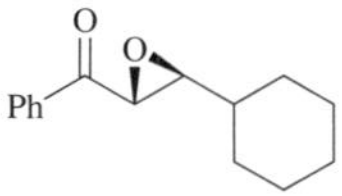 (91), 89	176
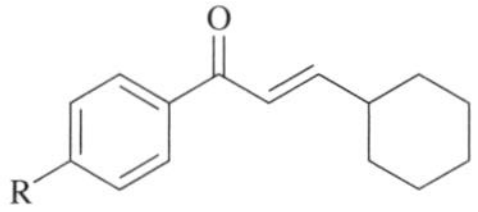	Anthryl cinchonidinium salt **6b** (10 mol %), KOCl, H_2O, toluene, –40°, 12 h	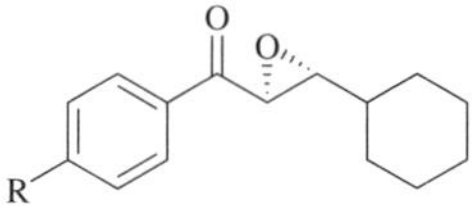 R: H (85), 94; F (87), 95	98
C_{16} 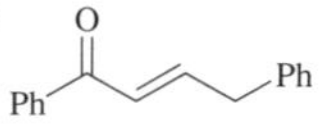	(L-Leu)$_n$–PS•SiO_2 (n_{av} = 32), urea/H_2O_2, BEMP, THF, rt	 (75), >90	117
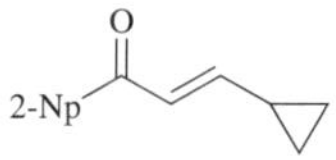	(L-Leu)$_n$, H_2O_2, NaOH, CH_2Cl_2, H_2O, rt, 28 h	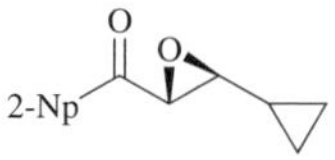 (73), ≥98	178, 179
C_{17} 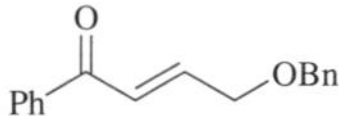	(L-Leu)$_n$–PS•SiO_2 (n_{av} = 32), urea/H_2O_2, BEMP, THF, rt	 (79), 93	117

TABLE 1D. DIARYL *E*-ENONES

Substrate	Conditions	Product(s), Yield(s) (%), % ee	Refs.
C_{15} 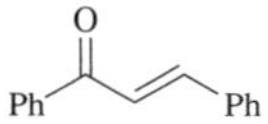	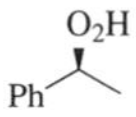 (100 mol %), base, toluene, 20°, 24 h	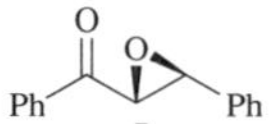 I	28
	Base		
	DBU	(98), 40	
	1,2-dimethyl-1,4,5,6-tetrahydropyrimidine	(94), 38	
	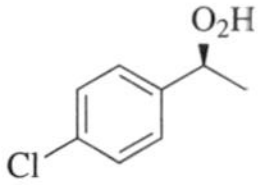 (100 mol %), DBU, toluene, 20°, 24 h	**I** (96), 44	28
	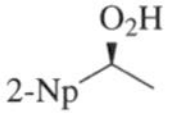 (100 mol %), DBU, toluene, 20°, 24 h	**I** (86), 46	28
	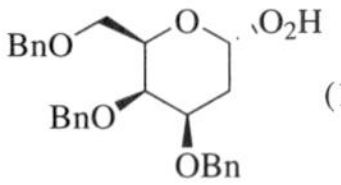 (100 mol %), DBU, toluene, 20°, 24 h	**I** (92), 43	28
	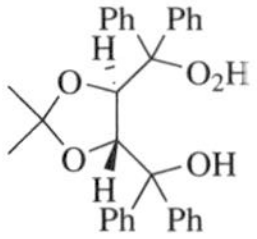, KOH, THF, –78° to rt, 7 h	**I** (96), 34	31

TABLE 1D. DIARYL *E*-ENONES (*Continued*)

Substrate	Conditions	Product(s), Yield(s) (%), % ee	Refs.
C_{15}			
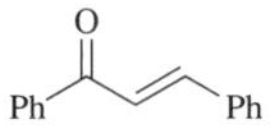	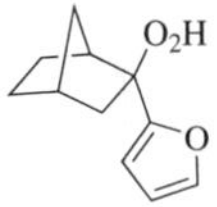 (110 mol %)	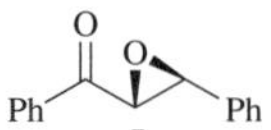 **I**	32

Base	Additive	Solvent	Time (h)	Temp (°)	
n-BuLi	—	THF	3	–20	(66), 43
n-BuLi	—	DME	5	–20	(78), 30
n-BuLi	12-crown-4	THF	3	–20	(86), 49
n-BuLi	12-crown-4	THF	3	rt	(66), 43
DBU	—	THF	47	rt	(38), 43
LiOH	—	THF	28	–20	(54), 46
NaOH	—	THF	2	–20	(100), 37
KOH	—	THF	1.5	–20	(93), 14

Substrate	Conditions	Product(s), Yield(s) (%), % ee	Refs.
	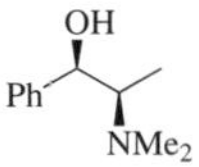 (100 mol %), O_2, Et_2Zn, toluene, –78° to 0°	**I** (92), 61	41, 40
	PolyBINOL **29** (20 mol %), TBHP, Et_2Zn, Et_2O, 0°, 5 h	**I** (95), 74	43
	Binaphthyl guanidine (10 mol %), H_2O_2	**I**	35

Guanidine	Solvent	Time (h)	
22a	CH_2Cl_2	9	(52), 18
22a	toluene	9	(66), 32
22a	Et_2O	9	(48), 9
22b	CH_2Cl_2	9	(76), 7
22c	CH_2Cl_2	9	(43), 4
22c	Et_2O	9	(50), 5
22d	CH_2Cl_2	3	(85), 7
22d	Et_2O	10	(76), 10

Conditions	Product(s) and Yield(s) (%), % ee	Refs.
(*R*)-BINOL **15a** (20 mol %), oxidant, R_2Zn, C° to rt, 12 h	**I**	44

Oxidant	R	Solvent	
TBHP	Et	Et_2O	(80), 47
CMHP	Et	Et_2O	(95), 71
CMHP	Et	CH_2Cl_2	(99), 68
CMHP	Me	Et_2O	(99), 76

Quinone–BINOL **15b** (20 mol %), TBHP, Et_2Zn, 0° to rt, 12 h	**I**	45

Solvent	
CH_2Cl_2	(45), 40
toluene	(27), 38
Et_2O	(99), 32

(*S*)-BINOL (5 mol %), CMHP, Ln(O*i*-Pr)$_3$, THF	**I**

(*S*)-BINOL	Ln	Temp (°)	Time (h)		Refs.
17a	La	rt	8	(90), 82	48, 264
17d	La	0	24	(86), 94	48
17b	La	rt	8	(91), 86	48
17b	La	–20	30	(90), 93	264
17b	Gd	rt	8	(95), 95	48
17b	Yb	rt	8	(91), 95	48, 47
17b	Yb	0	36	(91), 97	47
17c	La	–20	30	(82), 90	264

(*S*)-BINOL **17a** (5 mol %), TBHP, La(O*i*-Pr)$_3$, 4 Å MS, THF, cecane, 0°, 30 min	**I** (87), 98	265

TABLE 1D. DIARYL *E*-ENONES (*Continued*)

Substrate	Conditions	Product(s), Yield(s) (%), % ee	Refs.
C_{15} 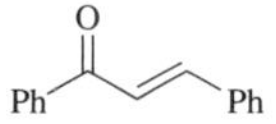	OligoBINOL (x mol %), oxidant, La(O*i*-Pr)$_3$, 4 Å MS, THF, rt	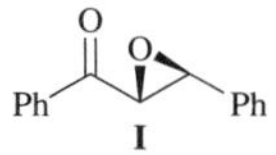 **I**	66

Catalyst	x	Oxidant	Base	Time (h)	
bisBINOL **23a**	2.5	CMHP	TPPO	0.5	(99), 84
bisBINOL **23g**	2.5	CMHP	TPPO	0.5	(99), 98
bisBINOL **23g**	2.5	CMHP	TPAO	0.5	(99), 96
bisBINOL **23g**	2.5	TBHP	TPPO	3	(90), 67
bisBINOL **23g**	0.5	CMHP	TPPO	1	(99), 94
bisBINOL **23b**	2.5	CMHP	TPPO	0.5	(99), 83
bisBINOL **23c**	2.5	CMHP	TPPO	0.5	(99), 93
bisBINOL **23d**	2.5	CMHP	TPPO	0.5	(99), 96
bisBINOL **23e**	2.5	CMHP	TPPO	0.5	(99), 84
bisBINOL **23f**	2.5	CMHP	TPPO	0.5	(99), 95
trisBINOL **24**	1.33	CMHP	TPPO	0.5	(99), 92
tetrakisBINOL **25**	1.25	CMHP	TPPO	0.5	(99), 92

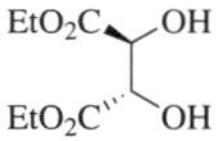 (11 mol %), TBHP, toluene, rt — **I**

Additive	Time (d)		Refs.
Bu_2Mg	1	(53), 89	69
NanoActive MgO Plus	—	(70), 88	73

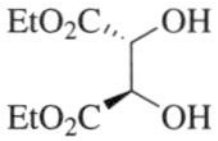 (110 mol %), TBHP, BuLi, *n*-BuOH, toluene, rt, 2 d — **I** (71–75), 62 — 69

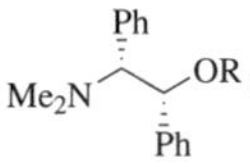

(x mol %), CMHP, BuLi, toluene I

R	x	Temp (°)	Time (h)		
2-$MeOC_6H_4$	150	–40	15	(93), 57	71
2-$MeOC_6H_4$	15	0	4	(99), 40	71, 72
2-$MeOC_6H_4$	180	0	0.5	(75), 47	72
(*S*)-PhCH(OMe)CH_2	20	0	4	(74), 15	72
2-($MeOCH_2$)C_6H_4	20	0	4	(70), 2	72

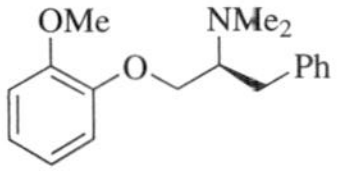

(20 mol %), I (74), 15 72

CMHP, *n*-BuLi, toluene, 0°, 4 h

(*S*)-BINOL (10 mol %), TBHP, KO*t*-Bu, $CaCl_2$, 4 Å MS, toluene, cyclohexane, –10°, 48 h I 79

(*S*)-BINOL	
17a	(88), 55
18a	(70), 10
17d	(60), 40
17b	(85), 70
18b	(89), 5

Polymer-bound (*R*)-BINOL **21** (10 mol %), TBHP, KO*t*-Bu, $CaCl_2$, 4 Å MS, toluene, 0°, 24 h I (92), 45 77

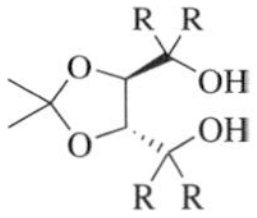

(10 mol %), I 79

TBHP, KO*t*-Bu, $CaCl_2$, 4 Å MS, toluene, cyclohexane, –10°, 48 h

R	
H	(95), 2
Ph	(88), 6

TABLE 1D. DIARYL *E*-ENONES (*Continued*)

Substrate	Conditions	Product(s), Yield(s) (%), % ee	Refs.
C_{15} 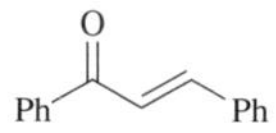	Benzyl quininium salt **4a** (1.7 mol %), H_2O_2, NaOH, H_2O, toluene, rt, 24 h	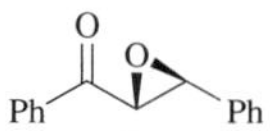 **I** (99), 24[a]	81
	PS-bound quininium salt, H_2O_2, NaOH, H_2O, toluene, rt, 3 d	**I**; Quininium salt: **4e** (48), 0 9; **4f** (30), 3 5	87
	Cinchonidinium salt (5 mol %), H_2O_2, NaOH, *n*-Bu_2O, H_2O, 0°, 48 h	**I**; Cinchonidinium salt: **5a** (45), 56[b]; **5b** (68), 10[b]	107
	Cinchoninium salt (5 mol %), NaOCl, *n*-Bu_2O, H_2O, 0°, 48 h	**I**; Cinchoninium salt: **7a** (62), 40[b]; **7b** (59), 37[b]	107
	Benzyl cinchoninium salt **1a** (5 mol %), NaOCl, *n*-Bu_2O, H_2O, 0°, 48 h	**I** (74), 21[b]	107
	Anthryl cinchoninium salt **1k** (10 mol %), NaOCl, H_2O, rt, 48 h	**I**; Solvent: toluene (65), 39; CH_2Cl_2 (71), 23	100, 99
	Anthryl cinchoninium salt **1l** (10 mol %), oxidant, base, H_2O, rt, 48 h	**I**; Oxidant / Base / Solvent: NaOCl / — / toluene (90), 81; NaOCl / — / CH_2Cl_2 (60), 66; H_2O_2 / KOH / toluene (69), 10; H_2O_2 / KOH / CH_2Cl_2 (<10), 2	100, 99

(*S*)-Binaphthyl ammonium salt **52** (3 mol %), H_2O_2, NaOH, H_2O, toluene, rt, 24 h	**I** (69), 37[b]	96
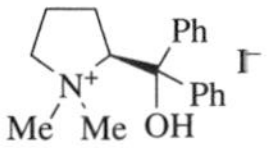 (5 mol %) H_2O_2, NaOH, H_2O, CH_2Cl_2, 0–22°, 22 h	**I** (25), 6.5[b]	92
Ammonium salt **55a** (5 mol %), H_2O_2, NaOH, H_2O, CH_2Cl_2, 0°, 24 h	**I** (70), 5.0[b]	92
Ammonium salt **56** (5 mol %), H_2O_2, NaOH, H_2O, CH_2Cl_2, 0°, 24 h	**I** (69), 1.8[b]	91
Ammonium salt **57** (5 mol %), H_2O_2, NaOH, H_2O, CH_2Cl_2, 0°, 24 h	**I** (70), 6.7[b]	91
Ammonium salt **58** (5 mol %), H_2O_2, NaOH, H_2O, CH_2Cl_2, 0°, 24 h	**I** (82), 3.5[b]	91

TABLE 1D. DIARYL *E*-ENONES (*Continued*)

Substrate	Conditions		Product(s), Yield(s) (%), % ee	Refs.
C_{15} 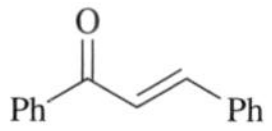	Aza-crown ether (7 mol %), TBHP, NaOH, toluene, H_2O, 5–6°		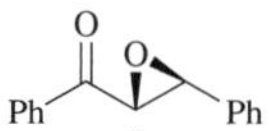 **I**	
	Crown ether	Time (h)		
	43b	10	(59), 11	112, 113
	43c	10	(33), 8	112, 113
	43d	1	(65), 81	112, 113
	43e	3	(58), 20	112, 113
	43f	1	(82), 92	112, 113
	43g	2	(61), 23	112, 113
	43h	1	(65), 41	112, 113
	43i	2	(64), 11	112, 113
	46a	5	(70), 66	113
	46b	6	(51), 76	113
	44a	8	(28), 45	113
	44b	10	(35), 53	113

Polypeptide catalyst, H_2O_2, NaOH, H_2O, rt **I**

Catalyst			Solvent	Time (h)		
(L-Ala)$_n$	$n_{av} = 5$		toluene	48	(9), 11	132
(L-Ala)$_n$	$n_{av} = 7$		toluene	48	(18), 28	132
(L-Ala)$_n$	$n_{av} = 10$		toluene	24	(85), 93	114
(L-Ala)$_n$	$n_{av} = 10$		CCl_4	28	(75), 93	130
(L-Ala)$_n$	$n_{av} = 30$		toluene	48	(57), 93	132
(L-Ala)$_n$	$n_{av} = 30$		CCl_4	24	(96), 96	132, 130, 182
(L-Val)$_n$	$n_{av} = 10$		CCl_4	168	(5.5), 10	130
(L-Val)$_n$	$n_{av} = 30$		CCl_4	144	(4), 33	130
(L-Leu)$_n$	$n_{av} = 8$		toluene	24	(73), 93	266
(L-Leu)$_n$	$n_{av} = 10$		CCl_4	28	(60), 84	130
(L-Leu)$_n$	$n_{av} = 33$		toluene	1.5	(59), 91	120, 253
(L-Leu)$_n$	$n_{av} = 30$		CCl_4	28	(44), 88	130, 267
(L-Ile)$_n$	$n_{av} = 10$		CCl_4	72	(76), 95	130
(L-Phe)$_n$	$n_{av} = 10$		CCl_4	72	(32), 1	130
[L-Glu(OBn)]$_n$	$n_{av} = 10$		toluene	48	(6), 57	114, 130
[L-Glu(OBu-*n*)]$_n$	$n_{av} = 10$		toluene	48	(10), 28	114
[L-Asp(OBn)]$_n$	$n_{av} = 10$		CCl_4	456	(7.5), 3	130
Me_2NCH(*i*-Pr)CO(L-Val)$_n$			toluene	120	(73), 92	115
(L-Ala)$_n$–PS	$n_{av} = 32$		toluene	48	(92), 99	123
(L-Ala)$_m$/(L-Leu)$_n$ random copolymer	$m_{av} = 10$	$n_{av} = 10$	CCl_4	24	(67), 95	130
(L-Ala)$_m$/(L-Val)$_n$ random copolymer	$m_{av} = 10$	$n_{av} = 10$	CCl_4	96	(39), 88	130
(L-Ala)$_m$/(L-Val)$_n$ random copolymer	$m_{av} = 6$	$n_{av} = 14$	toluene	192	(14), 39	130
(L-Ala)$_m$/(L-Val)$_n$ random copolymer	$m_{av} = 2$	$n_{av} = 18$	CCl_4	168	(9), 17	130

TABLE 1D. DIARYL *E*-ENONES (*Continued*)

Substrate	Conditions	Product(s), Yield(s) (%), % ee	Refs.
C_{15} 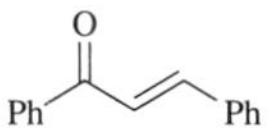	(L-Leu)$_n$, H_2O_2, NaOH, toluene, H_2O, rt, 24 h	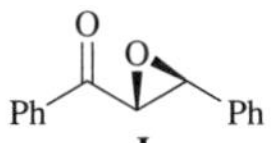 I *n* — 5: (22) 7; 8: (20), 10; 9: (22) 7; 10: (46), 13; 11: (33), 54; 12: (24), 25; 13: (36), 54; 14: (52), 41; 15: (31), 57; 20: (37), 56	268
	TFA•H–(L-Leu)$_{20}$–OBn, H_2O_2, NaOH, toluene, H_2O, rt, 24 h	**I** (92), 91	268
	Boc–(L-Leu)$_{20}$–OBn, H_2O_2, NaOH, toluene, H_2O, rt, 24 h	**I** (11), 41	268
	(L-Leu)$_n$–SiO_2 (n_{av} = 30; 8 mol %), H_2O_2, NaOH, toluene, H_2O, rt, 48 h	**I** (92), 92	125
	(L-Leu)$_n$ (0.1 mol %), H_2O_2, NaOH, Bu_4NHSO_4, toluene, rt, 6 h	**I** (80), 95	190, 201
	(L-Leu)$_n$ (n_{av} = 33; x mol %), H_2O_2, NaOH, TBAB, toluene, H_2O, rt	**I**	
	x — Time (min)		
	11 — 7	(>99), 94	121, 122
	0.2 — 60	(97), 93	120
	0.05 — 60	(89), 90	120
	0.02 — 60	(77), 84	120
	0.01 — 60	(61), 80	120
	(L-Leu)$_n$ (n_{av} = 33; 11 mol %), NaOCl, TBAB, toluene, H_2O, rt, 1.5 h	**I** (32), 90	122, 121

Polypeptide catalyst (n_{av} = 32), oxidant, DBU, THF, rt			**I**	
Catalyst	Oxidant	Time (min)		
(L-Leu)$_n$–PS	urea/H_2O_2	30	(85), >95[c]	131, 116, 189, 252
(L-Leu)$_n$–PS	DABCO•2 H_2O_2	30	(95), 95	116
(L-Leu)$_n$–PS	DABCO•2 H_2O_2	15[d]	(80), 98	269
(L-Leu)$_n$–PS	90% H_2O_2	30	(77), 65	116
(L-Npg)$_n$–PS	urea/H_2O_2	30	(91), >95	131
(L-Ala)$_n$–PS	urea/H_2O_2	30	(60), 80	131
(L-Val)$_n$–PS	urea/H_2O_2	60	(19), 81	131
(L-Phe)$_n$–PS	urea/H_2O_2	60	(36), 21	131
(L-Leu)$_n$–PS•SiO_2	urea/H_2O_2	30	(95), 97	131, 175, 270, 260, 124
(L-Npg)$_n$–PS•SiO_2	urea/H_2O_2	30	(95), 95	131
(L-Ala)$_n$–PS•SiO_2	urea/H_2O_2	30	(95), >95	131
(L-Val)$_n$–PS•SiO_2	urea/H_2O_2	60	(60), 80	131
(L-Phe)$_n$–PS•SiO_2	urea/H_2O_2	60	(34), 30	131
(L-Leu)$_n$–PS (n_{av} = 6), urea/H_2O_2, NaOH, THF, rt, 1 h			**I** (92), 97	129
(L-Leu)$_n$–PS (n_{av} = 6), urea/H_2O_2, NaOH, THF, rt, continuous flow			**I** (80), 91.5[e]	129
(L-Leu)$_n$–SiO_2 (n_{av} = 30; 8 mol %), urea/H_2O_2, DBU, THF, rt, 6 h			**I** (92), 92	125
(L-Leu)$_n$ (n_{av} = 33; 11 mol %), urea/H_2O_2, DBU, TBAB, THF, rt, 30 min			**I** (>99), 78	120
(L-Leu)$_n$–SiO_2 (n_{av} = 33; 11 mol %), urea/H_2O_2, TBAB, THF, rt, 30 min			**I** (>99), 92	120

TABLE 1D. DIARYL *E*-ENONES (*Continued*)

Substrate	Conditions	Product(s), Yield(s) (%), % ee	Refs.
C_{15} 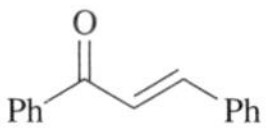	Polypeptide catalyst (n_{av} = 32), Na_2CO_3•1.5 H_2O_2, DME, H_2O, rt	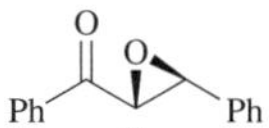 **I**	
	Catalyst / Time (h)		
	(L-Leu)$_n$–PS / —	(87), 94	131, 118
	(L-Npg)$_n$–PS / 0.5	(95), 96	131
	(L-Ala)$_n$–PS / 0.5	(60), 80	131
	(L-Val)$_n$–PS / 1	(19), 81	131
	(L-Phe)$_n$–PS / 1	(36), 21	131
	(L-Leu)$_n$–SiO_2 (n_{av} = 30; 8 mol %), Na_2CO_3•1.5 H_2O_2, DME, H_2O, rt, 2 h	**I** (93), 92	125
	(L-Leu)$_n$–SiO_2 (n_{av} = 30; 8 mol %), H_2O_2, Na_2CO_3, DME, H_2O, rt, 2 h	**I** (93), 93	125
	(L-Leu)$_n$, H_2O_2, NH_4HCO_3, Aliquat 336, DME, H_2O, rt	**I** (92), 94	119
	(L-Leu)$_n$ (n_{av} = 260), $NaBO_4$•4 H_2O, NaOH, toluene, H_2O, sonication, rt, 23 h	**I** (94), 57	271
	Polypeptide catalyst, urea/H_2O_2, base	**I**	

Catalyst	n_{av}	Solvent	Base	Time (h)	Yield (%), % ee	Refs.
(L-Leu)$_n$–PEG	n_{av} = 3.9	THF	DBU	24	(80), 98	126
(L-Leu)$_n$–PEG	n_{av} = 7.5	THF	DBU	24	(80), 97	126
(L-Leu)$_n$–PEG	n_{av} = 11.6	THF	DBU	24	(63), 95	126
(L-Leu)$_n$–PEG	n_{av} = 12.2	THF	DBU	24	(80), 98	126
(L-Leu)$_n$–PEG	n_{av} = 8	THF	NaOH	1	(>99), 94	129
(L-Leu)$_n$–PEG	n_{av} = 8	THF	NaOH	1	(80), 91.5[e]	129
(L-Leu)$_n$–PEG	n_{av} = 5	THF	DBU	3	(30), 81	127
(L-Leu)$_n$–PEG	n_{av} = 10	THF	DBU	3	(95), 97	127
(L-Leu)$_n$–PEG	n_{av} = 10	DME	DBU	3	(99), 96	127
(L-Leu)$_n$–PEG	n_{av} = 10	toluene	DBU	3	(51), 94	127

(L-Ala)$_m$–(L-Leu)$_n$–PEG	$m_{av} = 5$, $n_{av} = 5$	THF	DBU	3	(58), 97	127
(L-Aib)$_m$–(L-Leu)$_n$–PEG	$m_{av} = 5$, $n_{av} = 5$	THF	DBU	3	(20), 78	127
(L-Leu)$_m$–(L-Ala)$_n$–PEG	$m_{av} = 5$, $n_{av} = 5$	THF	DBU	3	(16), 88	127
(L-Ala)$_n$–PEG	$n_{av} = 10$	THF	DBU	3	(10), 28	127
(L-Leu)$_4$–PEG	—	THF	DBU	3	(6), 42	127
(L-Leu)$_5$–PEG	—	THF	DBU	3	(10), 87	127
(L-Leu)$_6$–PEG	—	THF	DBU	3	(28), 95	127
(L-Leu)$_7$–PEG		THF	DBU	3	(31), 96	127

(L-Leu)$_n$–Tentagel (11 mol %), 35% H_2O_2, CH_2Cl_2, H_2O, 20°, 24 h — **I** — 134

n		*n*	
1	(1), 4[f]	11	(73), 97[f]
2	(1), 14[f]	12	(77), 97[f]
3	(4), 66[f]	13	(88), 98[f]
4	(35), 92[f]	14	(96), 98[f]
5	(54), 97[f]	15	(95), 98[f]
6	(30), 96[f]	16	(98), 98[f]
7	(31), 98[f]	17	(99), 98[f]
8	(47), 97[f]	18	(98), 98[f]
9	(48), 98[f]	19	(94), 97[f]
10	(52), 98[f]	20	(100), 98[f]

(Gly)$_m$–(L-Leu)$_n$–Tentagel (11 mol %), 35% H_2O_2, CH_2Cl_2, H_2O, 20°, 24 h — **I** (—) — 134

m	*n*	% ee	*m*	*n*	% ee	*m*	*n*	% ee
1	1	5[f]	1	3	22[f]	1	5	97[f]
2	1	4[f]	2	3	34[f]	2	5	96[f]
3	1	2[f]	3	3	34[f]	3	5	91[f]
4	1	0[f]	4	3	31[f]	4	5	92[f]
5	1	6[f]	5	3	30[f]	5	5	92[f]
1	2	11[f]	1	4	92[f]			
2	2	20[f]	2	4	94[f]			
3	2	15[f]	3	4	92[f]			
4	2	13[f]	4	4	81[f]			
5	2	12[f]	5	4	89[f]			

TABLE 1D. DIARYL *E*-ENONES (*Continued*)

Substrate	Conditions	Product(s), Yield(s) (%), % ee	Refs.
C_{15} 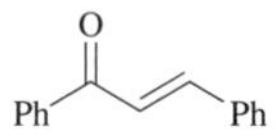	R–(L-Leu)$_m$–(Aib)–(L-Leu)$_n$–OBn, urea/H_2O_2, DBU, rt, 24 h	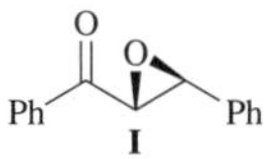 I	128
	R / m / n / Solvent		
	Boc / 4 / 4 / THF	(50), 61	
	TFA•H / 4 / 4 / THF	(61), 68	
	Boc / 4 / 6 / THF	(60), 78	
	TFA•H / 4 / 6 / THF	(54), 73	
	Boc / 6 / 4 / THF	(89), 85	
	Boc / 6 / 6 / THF	(73), 94	
	Boc / 6 / 6 / toluene	(47), 84	
	Boc / 6 / 6 / CH_2Cl_2	(41), 76	
	Boc / 6 / 6 / $CHCl_3$	(72), 15	
	(β^2-Leu)$_{19}$Leu–PS, urea/H_2O_2, DBU, THF, 4 h	**I** (21), 3	262
	(β^3-Leu)$_{19}$Leu–PS, urea/H_2O_2, DBU, THF, 1.5 h	**I** (20), 23	262
	(β^3-Leu)$_{19}$Leu–PS, 30% H_2O_2, NaOH, toluene, H_2O, 24 h	**I** (92), 70	262
	(β^3-Leu)$_{20}$–PS, urea/H_2O_2, DBU, THF, 24 h	**I** (96), 39	262
	(Leu)$_n$ (n_{av} = 32), H_2O_2, NaOH, toluene, H_2O	**I**	136
	L-Leu/D-Leu		
	52.4:47.6	(39), 27	
	54.5:45.5	(51), 45	
	60.0:40.0	(73), 74	
	71.4:28.6	(95), 92	
	100:0	(100), 96	

Polyleucine catalyst, H_2O_2, NaOH, toluene, H_2O, rt — **I** + 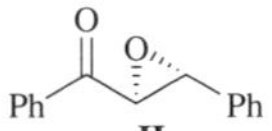 **II** — 135, 272[g]

Catalyst	Time (h)	Major Product	
(L-Leu)$_{10}$–PS	35	**I**	(85), 66
(L-Leu)$_{20}$–PS	9	**I**	(96), 91
(D-Leu)$_5$–(L-Leu)$_{15}$–PS	32	**II**	(85), 45
(D-Leu)$_7$–(L-Leu)$_{13}$–PS	32	**II**	(100), 80
(D-Leu)$_9$–(L-Leu)$_{11}$–PS	32	**II**	(98), 71
(D-Leu)$_1$–(L-Leu)$_9$–PS	35	**I**	(31), 29
(D-Leu)$_3$–(L-Leu)$_7$–PS	35	**I**	(34), 5
(D-Leu)$_5$–(L-Leu)$_5$–PS	35	**II**	(82), 5

Polyleucine catalyst, urea/H_2O_2, DBU, THF, rt — **I** + **II**

Catalyst	Time (h)	Major Product		
(L-Leu)$_6$–PS	1.5	**I**	(33), 8	126
(L-Leu)$_{10}$–FS	1.5	**I**	(62), 54	135
(L-Leu)$_{15}$–FS	1.5	**I**	(89), 87	269
(L-Leu)$_{20}$–FS	1.5	**I**	(66), 89	269, 262
(D-Leu)$_5$–(L-Leu)$_{15}$–PS	1.5	**II**	(68), 52	135
(D-Leu)$_7$–(L-Leu)$_{13}$–PS	1.5	**II**	(81), 83	135
(D-Leu)$_9$–(L-Leu)$_{11}$–PS	1.5	**II**	(87), 81	135
(D-Leu)$_1$–(L-Leu)$_9$–PS	2.5	**I**	(94), 65	135
(D-Leu)$_3$–(L-Leu)$_7$–PS	2.5	**I**	(73), 15	135
(D-Leu)$_5$–(L-Leu)$_5$–PS	2.5	**II**	(96), 2	135

TABLE 1D. DIARYL *E*-ENONES (*Continued*)

Substrate	Conditions	Product(s), Yield(s) (%), % ee	Refs.
C_{15} 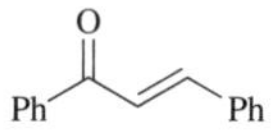	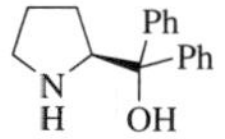 (x mol %), oxidant	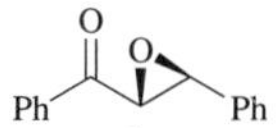 I	170
	x / Oxidant / Solvent / Temp (°) / Time (h)		
	30 / TBHP / toluene / rt / 72	(12), 69	
	30 / TBHP / $CHCl_3$ / rt / 96	(5), 65	
	30 / TBHP / CH_2Cl_2 / rt / 144	(4), 55	
	30 / TBHP / THF / rt / 144	(5), 17	
	50 / TBHP / MeOH / rt / 120	(11), 30	
	30 / TBHP / hexane / rt / 48	(53), 76	
	30 / TBHP / c-C_6H_{12} / rt / 48	(46), 76	
	30 / CMHP / hexane / rt / 94	(51), 55	
	30 / TBHP / hexane / rt / 62	(71), 70	
	50 / TBHP / hexane / 4 / 102	(80), 78	
	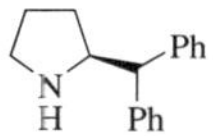 (30 mol %), TBHP, hexane, rt, 140 h	**I** (22), 23	170
	α-Cyclodextrin (100 mol %), H_2O_2, Na_2CO_3, H_2O, 0°, 240 h	**I** (20), 2.5[b]	172
	γ-Cyclodextrin (100 mol %), NaOCl, H_2O	**I** (90), 31	173
	Egg phosphatidylcholine, m-CPBA, H_2O, EtOH, rt, 2 h	**I** (65), 70	174
	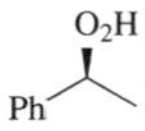 (100 mol %), KOH, MeCN, –40°, 10–20 min	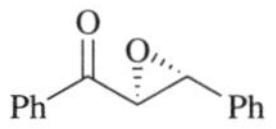 Additive: — (99), 51; 18-crown-6 (94), 6	27, 28

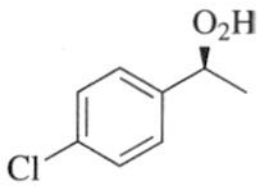
O2H
Cl
(100 mol %), KOH, MeCN, –40° — **I** (95), 48 — 28

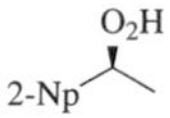
O2H
2-Np
(100 mol %), KOH, MeCN, –40° — **I** (87), 43 — 28

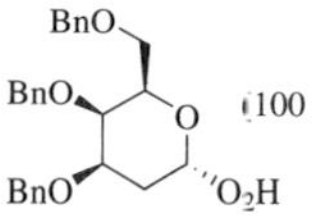
BnO
BnO
BnO
O
O2H
(100 mol %), KOH, MeCN, –40° — **I** (91), 14 — 28

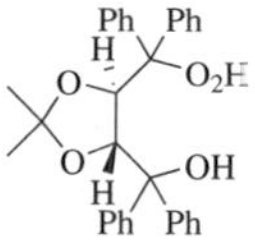
Ph
Ph
H
O
O2H
O
OH
H
Ph
Ph
(150 mol %), base, THF — **I** — 31

Base	Temp (°)	Time (h)	
LiOH	–78 to rt	7	(95), 60
n-BuLi	–78 to rt	7	(85), 90
n-BuLi	0	4	(94), 80
n-BuLi	–30	24	(92), 90
n-BuLi	–78	120	(80), 97
DMAP/LiCl	–78 to rt	7	(65), 54
DBN/LiCl	–78 to rt	7	(96), 76
DBU/LiCl (1.1 eq)	–78 to rt	7	(89), 72
DBU/LiCl (0.11 eq)	–78 to rt	7	(98), 68

TABLE 1D. DIARYL *E*-ENONES (*Continued*)

Substrate	Conditions	Product(s), Yield(s) (%), % ee	Refs.
C_{15} 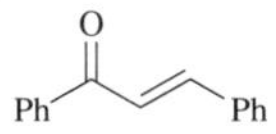	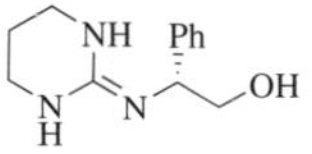 (100 mol %), TBHP, toluene, *i*-PrOH, rt, 7 d	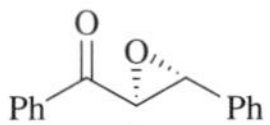 **I** (71), 15	39
	(*R*)-Binaphthyl guanidine catalyst (10 mol %), H_2O_2	**I**	35
	Catalyst / Solvent / Time (h)		
	22e toluene 24	(41), 4	
	22b toluene 9	(55), 11	
	22b Et_2O 9	(93), 4	
	22c toluene 9	(62), 8	
	22f toluene 9	(56), 16	
	22g toluene 9	(48), 53	
	22h toluene 8	(74), 44	
	PolyBINOL **28** (111 mol %), O_2, Et_2Zn, CH_2Cl_2, 0°, 18 h	**I** (41), 71	43
	PolyBINOL **32** (111 mol %), O_2, Et_2Zn, CH_2Cl_2, 0°	**I** (>90), 38	43
	PolyBINOL **28** (5 mol %), TBHP, Et_2Zn, CH_2Cl_2, 0°	**I** (95), 28	43
	(*R*)-BINOL **15a** (x mol %), $Ln(O\textit{i}\text{-}Pr)_3$, oxidant, 4 Å MS, rt	**I**	

x	Ln	Oxidant	Additive	Solvent	Time (min)		
10	La	TBHP	TPAO	THF	3	(95), 97	55
5	La	TBHP	TPAO	THF	15	(99), 96	55
1	La	TBHP	TPAO	THF	180	(97), 89	55
10	La	TBHP	TPAO	DME	30	(94), 96	55
10	La	TBHP	TPAO	benzene	10	(93), 96	55
10	La	TBHP	TPAO	toluene	10	(95), 95	55
10	La	TBHP	TPPO	THF	30	(97), 97	55
5	La	TBHP	TPPO	THF	30	(99), 96	50, 184, 67, 68
5	La	CMHP	TPPO	THF	10	(97), 99.8	184, 53, 68
5	La	CMHP	TPPO	THF	10	(—), >99[h]	67, 68
0.5	La	CMHP	TPPO	THF	90	(77), 98	67, 68
5	Yb	TBHP	H_2O	THF	60	(99), 81	50
0.5	La	CMHP	$(4\text{-}FC_6H_4)_3P{=}O$	THF	60	(98), 98	53, 52, 54

La–BINOL **34** (10 mol %), TBHP, THF, rt, 10 h	**I** (92), 83	46
La–BINOL **35** (10 mol %), TBHP, La(O*i*-Pr)$_3$, 4 Å MS, THF, rt, 1.5 h	**I** (95), 78	55
(*R*)-BINOL **15d** (6.25 mol %), CMHP, La(O*i*-Pr)$_3$, 4 Å MS, THF, rt, 7 h	**I** (93), 91	46

TABLE 1D. DIARYL *E*-ENONES (*Continued*)

Substrate	Conditions	Product(s), Yield(s) (%), % ee	Refs.
C_{15} 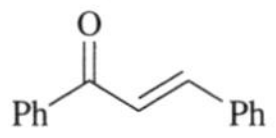	(*R*)-BINOL **15c** (5 mol %), CMHP, La(O*i*-Pr)$_3$, THF, rt, 8 h	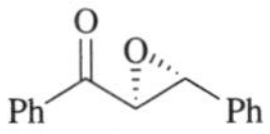 **I** (92), 92	48
	BINOL (x mol %), CMHP, La(O*i*-Pr)$_3$, TPPO, 4 Å MS, THF, rt	**I**	
	BINOL / x / Time (h)		
	polyBINOL **28** / 5 / 1.5	(99), 73[i]	258
	bisBINOL **26a** / 5 / 2	(99), 93	65
	bisBINOL **26b** / 5 / 2	(97), 98	65
	bisBINOL **26c** / 5 / 2	(>99), 97	65
	bisBINOL **26d** / 5 / 2	(>99), 94	65
	bisBINOL **26e** / 5 / 2	(99), 87	65
	trisBINOL **27a** / 3.33 / 2	(>99), 91	65
	trisBINOL **27b** / 3.33 / 2	(92), 98	65
	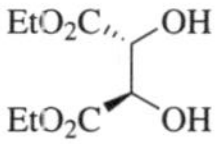 (11 mol %), TBHP, Bu_2Mg, toluene, rt, 1 d	**I** (61), 94	69
	(*S*)-BINOL **17a** (11 mol %), TBHP, NanoActive MgO Plus, toluene, rt	**I** (53), 28	73
	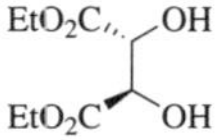 (11 mol %), TBHP, base, toluene	**I**	73
	Base / Temp (°) / Time (h)		
	NanoActive MgO / rt / —	(58), 60	
	NanoActive MgO Plus / 40 / —	(80), 38	
	NanoActive MgO Plus / rt / 36	(70), 90	
	NanoActive MgO Plus / 0 / —	(24), 94	
	NanoActive MgO Plus / –20 / —	(15), 96	

Conditions	Product(s) and Yield(s) (%), % ee		Refs.
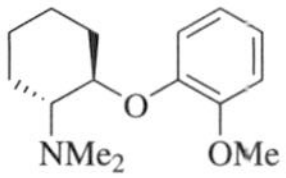 (20 mol %), CMHP, n-BuLi, toluene, 0°, 4 h	**I** (72), 18		72
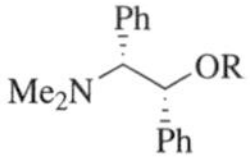 (20 mol %), CMHP, n-BuLi, toluene, 0°, 4 h	**I**		72
	R		
	Me	(97), 7	
	Ph	(69), 7	
	t-Bu(CH_2)$_2$	(88), 22	
	Bn	(85), 42	
	MeO(CH_2)$_2$	(97), 19	
	MeO(CH_2)$_3$	(86), 15	
	(R)-PhCH(OMe)CH_2	(56), 44	
	2-Me$OC_6H_4CH_2$	(95), 23	
Benzyl quininium salt **4a** (1.7 mol %), TBHP, NaOH, toluene, rt, 24 h	**I** (—), 17[a]		81
Benzyl quininium salt **4a** (2.8 mol %), NaOCl, toluene, rt, 72 h	**I** (66), 25[a]		83
Benzyl cinchonidinium salt **3b** (5 mol %), NaOCl, n-Bu_2O, H_2O, 0°, 48 h	**I** (79), 7[b]		107
Cinchonidinium salt (5 mol %), NaOCl, n-Bu_2O, H_2O, 0°, 48 h	**I**		107
Cinchonidinium salt			
5a	(72), 46[b]		
5b	(64), 11[b]		
Anthryl cinchonidinium salt **6b** (10 mol %), KOCl, H_2O, toluene, –40°, 12 h	**I** (96), 93		98
Anthryl cinchonidinium salt **6b** (1 mol %), NaOCl, H_2O, toluene, rt, 24 h	**I** (98), 86		97, 100, 99

TABLE 1D. DIARYL *E*-ENONES (*Continued*)

Substrate	Conditions	Product(s), Yield(s) (%), % ee	Refs.
C_{15} 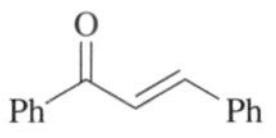	Anthryl cinchonidinium salt (10 mol %), trichloroisocyanuric acid, base	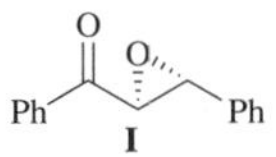 **I**	

Cinchonidinium salt	Base	Solvent	Temp (°)	Time (h)		Refs.
6a	KOH	CH_2Cl_2/H_2O	0	24	(81), 39	101
6a	KOH	CH_2Cl_2/H_2O	–20	48	(49), 67	101
6a	KOH	CH_2Cl_2/H_2O	–30	49	(63), 71	101
6a	KOH	toluene/H_2O	rt	8	(100), 67	101
6a	KOH	toluene/H_2O	0	8	(100), 83	101
6a	KOH	toluene	0	7	(89), 82	102
6a	KOH	toluene/H_2O	–20	16	(100), 80	101
6a	KOH	toluene/H_2O	–30	24	(97), 82	101
6a	NaOH	toluene/H_2O	0	36	(98), 75	101
6a	LiOH	toluene/H_2O	0	48	(94), 76	101
6b	KOH	toluene/H_2O	0	—	(90), 89	101
6b	KOH	toluene	0	7	(90), 87	102
6c	KOH	toluene/H_2O	0	—	(—), 82	101

Substrate	Conditions	Product(s), Yield(s) (%), % ee	Refs.
	Benzyl quinidinium salt **2a** (1.7 mol %), H_2O_2, NaOH, H_2O, CH_2Cl_2, rt, 24 h	**I** (91), 31[b]	92, 81
	Naphthyl quinidinium salt **2e** (5 mol %), H_2O_2, LiOH, H_2O, $CHCl_3$, –10°, 44 h	**I** (85), 17	106
	Benzyl cinchoninium salt **1a** (10 mol %), (±)-1-phenylethyl hydroperoxide, KOH, toluene, 0–20°	**I** (>95), 49	105

Conditions	Product(s) and Yield(s) (%)	Refs.
Trifluoromethylbenzyl cinchoninium salt **1h** (10 mol %), (±)-1-phenylethyl hydroperoxide, KOH, toluene, 0–20°	**I** (—), 32	105
Cinchoninium salt (5 mol %), H_2O_2, NaOH, *n*-Bu_2O, H_2O, 0°, 48 h	**I**	

Cinchoninium salt	Base	Temp (°)	Time (h)		Refs.
7a	NaOH	0	48	(65), 14[b]	107
7b	NaOH	0	48	(75), 8[b]	107
1h	LiOH	4	26	(72), 73	104, 103
1c	LiOH	4	88	(24), 3	104, 103
1d	LiOH	4	41	(68), 65	104, 103
1e	LiOH	4	36	(56), 77	104, 103
1f	LiOH	4	37	(97), 84	104, 103
1g	LiOH	4	37	(61), 72	104, 103
1b	LiOH	4	74	(72), 1	104, 103
1j	LiOH	4	60	(70), 4	103
8a	LiOH	4	38	(94), 7	103
8b	LiOH	4	—	(72), 2	104

Anthryl cinchoninium salt **1k** (10 mol %), H_2O_2, KOH, H_2O, CH_2Cl_2, rt, 4 h	**I** (75), 11	100, 99

TABLE 1D. DIARYL *E*-ENONES (*Continued*)

Substrate	Conditions	Product(s), Yield(s) (%), % ee	Refs.
C_{15}			
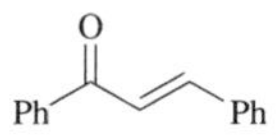	Bis(cinchonidinium) salt (x mol %), H_2O_2, KOH, surfactant, *i*-Pr_2O, H_2O	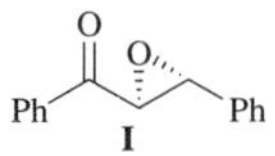 **I**	108

Bis(cinchonidinium) salt	x	Surfactant	Temp (°)	Time (h)	
14a	5	Triton X-100	10	10	(89), 82
14b	5	Triton X-100	10	8	(80), 92
14c	5	Triton X-100	10	8	(90), 92
14d	5	Triton X-100	10	3	(95), 98
14d	1	Triton X-100	rt	3.5	(80), 90
14d	1	Tergitol NP 9	rt	3	(80), 97
14d	1	Brij 78	rt	2.5	(95), 94
14d	1	Tween 20	rt	2	(95), 99
14d	1	Span 20	rt	4	(95), >99

Substrate	Conditions	Product(s), Yield(s) (%), % ee	Refs.
	Ammonium salt (10 mol %), NaOCl, H_2O, toluene, rt, 24 h	**I**	90

Ammonium salt	
60a	(50), 2[b]
60b	(67), 7[b]

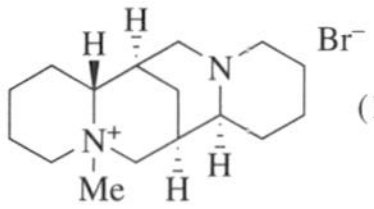 (10 mol %), NaOCl, H_2O, toluene, rt, 24 h	**I** (68), 0.5[b]	90
(*R*)-Binaphthyl ammonium salt **51** (3 mol %), H_2O_2, NaOH, H_2O, toluene, rt, 2 h	**I** (82), 2[b]	96
(*S*)-Binaphthyl ammonium salt (3 mol %), NaOCl, toluene, H_2O, 0°, 24 h	**I**	109
Ammonium salt		
50a	(69), 66	
50b	(50), 61	
50c	(67), 64	
50d	(61), 80	
50e	(99), 96	
Bis(binaphthyl) ammonium salt (3 mol %), NaOCl, toluene, H_2O, 0°, 24 h	**I**	109
Ammonium salt		
53b	(69), 54	
53a	(33), 56	
54	(53), 37	

TABLE 1D. DIARYL *E*-ENONES (*Continued*)

Substrate	Conditions	Product(s), Yield(s) (%), % ee	Refs.
C_{15} 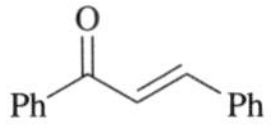	Ammonium salt **59** (5 mol %), H_2O_2, NaOH, H_2O, CH_2Cl_2, 0°, 24 h	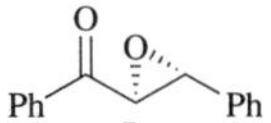 **I**	92
	Ammonium salt		
	59	(90), 7[b]	
	55b	(84), 9[b]	
	55c	(80), 5[b]	
	55d	(72), 5[b]	
	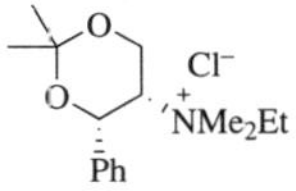 (5 mol %), NaOCl, H_2O, toluene, 140 h	**I** (59), 11[b]	89
	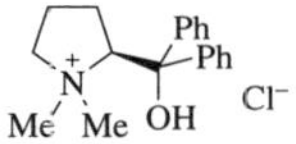 (5 mol %), NaOCl, H_2O, toluene, 140 h	**I** (50), 10[b]	89
	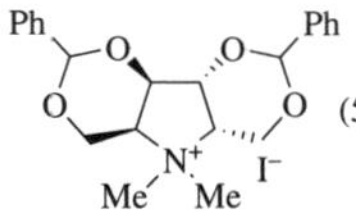 (5 mol %), H_2O_2, NaOH, H_2O, CH_2Cl_2, 0°, 24 h	**I** (14), 1[b]	91

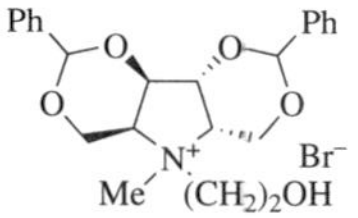 (5 mol %), H_2O_2, NaOH, H_2O, CH_2Cl_2, 0°, 24 h	**I** (20), 1[b]	91
Aza-crown ether **43a** (7 mol %), TBHP, NaOH, toluene, H_2O, 5–6°, 4 h	**I** (47), 28	112, 113
Aza-crown ether (7 mol %), TBHP, NaOH, toluene, H_2O, 0–4°	**I**	
Crown ether / Time (h)		
45a 9	(25), 9	111, 112
45b 3	(50), 72	111, 112
45c 8	(47), 82	111, 112
45d 6	(67), 31	111, 112
45e 3	(61), 51	111
Aza-crown ether (7 mol %), TBHP, NaOH, toluene, H_2O, 5°	**I**	113
Crown ether / Time (h)		
49a 10	(42), 21	
49b 7	(40), 30	
(D-Leu)$_n$–PS (n_{av} = 32), urea/H_2O_2, DBU, THF, 30 min	**I** (85), 94	116
(D-Leu)$_n$–PS•SiO_2 (n_{av} = 32), urea/H_2O_2, DBU, THF, rt, 30 min	**I** (91), —	270
Chiral ketone **36** (10 mol %), Oxone, K_2CO_3, H_2O, DME, AcOH, 0°, 6 h	**I** (80), 94	149, 163

TABLE 1D. DIARYL *E*-ENONES (*Continued*)

Substrate	Conditions	Product(s), Yield(s) (%), % ee	Refs.
C_{15} 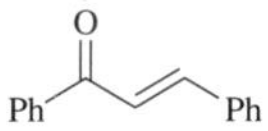	Chiral ketone **37** (10 mol %), Oxone, K_2CO_3, H_2O, DME, AcOH, 0°, 6 h	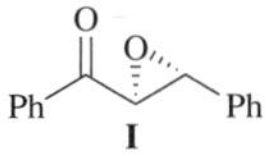 **I** (85), 96	163
	Chiral ketone **38** (10 mol %), Oxone, K_2CO_3, $Na_2B_4O_7$, EDTA, H_2O, MeCN, $(MeO)_2CH_2$, 0°, 6 h	**I** (80), 94	200
	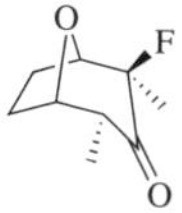 (30 mol %), Oxone, $NaHCO_3$, Na_2EDTA, H_2O, MeCN, 8 h	**I** (24), 67	147
	β-Cyclodextrin (100 mol %), NaOCl, H_2O	**I** (87), 30	173
	Crown ether **48** (5 mol %), NaOCl, H_2O, toluene, rt, 140 h	**I** (31), 7	177
	Bovine serum albumin (5 mol %), H_2O_2, H_2O, pH 11, rt, 6 d	**I** (—), —	141
	Bovine serum albumin (5 mol %), TBHP, H_2O, pH 11, rt, 6 d	**I** (—), —	141
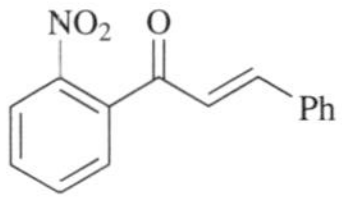	(L-Leu)$_n$–PS (n_{av} = 32), urea/H_2O_2, DBU, THF, rt, 1 h	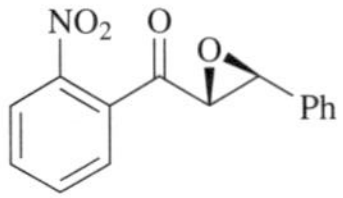 (91), 91	183

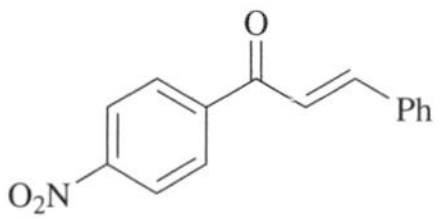	Aza-crown ether **43f** (7 mol %), TBHP, NaOH, toluene, H_2O, rt, 7 h	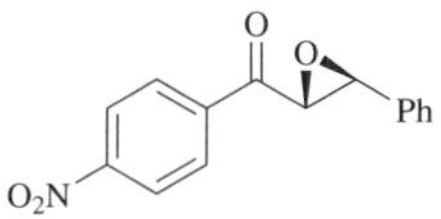 (74), 16	113
	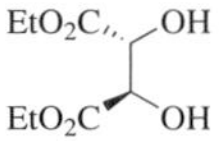 (11 mol %), TBHP, NanoActive MgO Plus, toluene, rt, 36 h	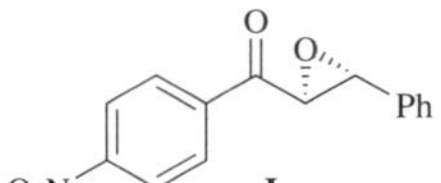 **I** (52), 98	73
	Cinchonidinium salt **5a** (5 mol %), H_2O_2, NaOH, n-Bu_2O, H_2O, 0°, 48 h	**I** (91), 3[j]	107
	Anthryl cinchonidinium salt **6b** (10 mol %), NaOCl, H_2O, toluene, rt	**I** (85), 83	99
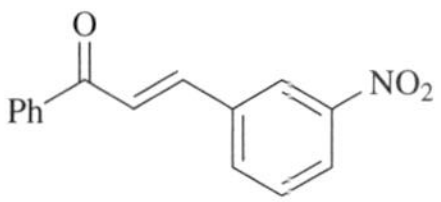	(L-Leu)$_n$, H_2O_2, NH_4HCO_3, Aliquat 336, DME, H_2O, rt	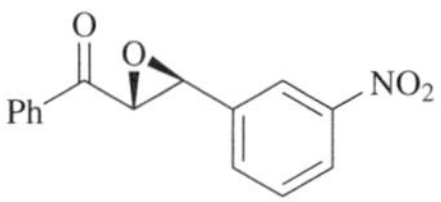 (96), 54	119
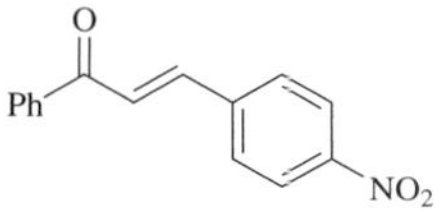	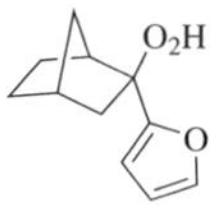 (110 mol %), n-BuLi, THF, –20°, 1 h	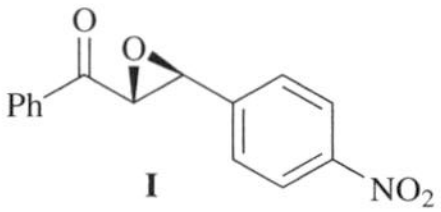 **I** (98), 45	32
	(R)-BINOL **15a** (20 mol %), CMHP, Et_2Zn, Et_2O, 0° to rt, 12 h	**I** (86), 62	44
	BisBINOL **23g** (2.5 mol %), CMHP, La(Oi-Pr)$_3$, TPPO, 4 Å MS, THF, rt, 1 h	**I** (>99), 95.7	66
	Benzyl quininium salt **4a** (1.7 mol %), H_2O_2, NaOH, H_2O, toluene, rt, 24 h	**I** (—), 13[k]	81

TABLE 1D. DIARYL *E*-ENONES (*Continued*)

Substrate	Conditions	Product(s), Yield(s) (%), % ee	Refs.
C_{15} 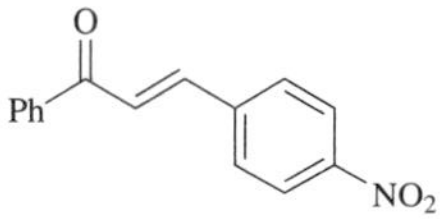	Cinchonidinium salt **5a** (5 mol %), H_2O_2, NaOH, *n*-Bu_2O, H_2O, 0°, 48 h	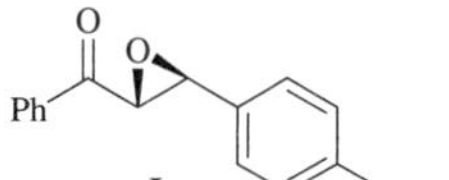 **I** (92), 14[k]	107
	Aza-crown ether **47** (7 mol %), TBHP, NaOH, toluene, H_2O, rt, 2 h	**I** (38), 79	112, 113
	(L-Ala)$_n$ (n_{av} = 10), H_2O_2, NaOH, toluene, H_2O, rt, 48 h	**I** (83), 82[l]	132, 273
	(L-Leu)$_n$–PS (n_{av} = 33), H_2O_2, NaOH, toluene, H_2O, 0° to rt, 2 d	**I** (90), 99	123
	(L-Leu)$_n$–SiO_2 (n_{av} = 30; 6 mol %), Na_2CO_3•1.5 H_2O_2, DME, H_2O, rt, 2 h	**I** (80), 92[m]	125
	(L-Leu)$_n$, H_2O_2, NH_4HCO_3, Aliquat 336, DME, H_2O, rt	**I** (92), 68	118
	(L-Leu)$_n$ (n_{av} = 260), $NaBO_4$•4 H_2O, NaOH, toluene, H_2O, sonication, rt, 7 h	**I** (85), 79	271
	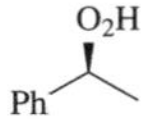 (100 mol %), KOH, MeCN, –40°, 10–20 min	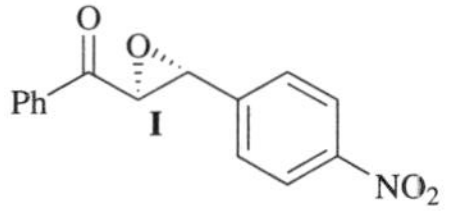 **I** (98), 42	27, 28
	(*R*)-BINOL **15a** (5 mol %), CMHP, La(O*i*-Pr)$_3$, (4-FC_6H_4)$_3$P=O, 4 Å MS, THF, rt, 20 min	**I** (94), >99	54
	Anthryl cinchonidinium salt **6b** (10 mol %), KOCl, H_2O, toluene, –40°, 12 h	**I** (90), 94	98

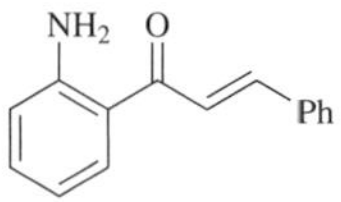

Anthryl cinchonidinium salt **6b** (10 mol %), trichloroisocyanuric acid, KOH, 0°	**I**		
Solvent / Time (h)			
toluene/H_2O / —	(83), 96		101
toluene / 7	(76), 93		102
(*S*)-BINOL **17b** (10 mol %), TBHP, KO*t*-Bu, $CaCl_2$, 4 Å MS, toluene, cyclohexane, 0°, 48 h	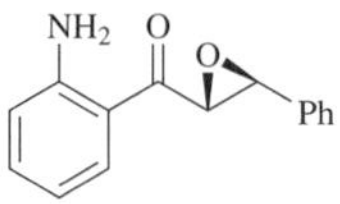 (60), 26[n]		79
(L-Leu)$_n$ (n_{av} = 33; 11 mol %), H_2O_2, NaOH, toluene, H_2O, rt	**I**		120
Additive / Time (min)			
TBAB / 15	(97), 93		
Aliquat 336 / 30	(100), 91		
Polypeptide catalyst (n_{av} = 32), urea/H_2O_2, DBU, THF, rt	**I**		
Catalyst / Time (h)			
(L-Leu)$_n$–PS / 3	(81), >98		183, 260, 131
(L-Npg)$_n$–PS / 17	(97), 97		131
(L-Ala)$_n$–PS / 17	(40), 92		131
(L-Leu)$_n$–PS•SiO_2 / 3	(85), 93		124, 260, 131
(L-Npg)$_n$–PS•SiO_2 / 17	(100), 97		131
(L-Ala)$_n$–PS•SiO_2 / 17	(66), 91		131
(L-Val)$_n$–PS•SiO_2 / 17	(13), 2		131
(L-Phe)$_n$–PS•SiO_2 / 17	(16), 11		131

TABLE 1D. DIARYL *E*-ENONES (*Continued*)

Substrate	Conditions	Product(s), Yield(s) (%), % ee	Refs.
C_{15} 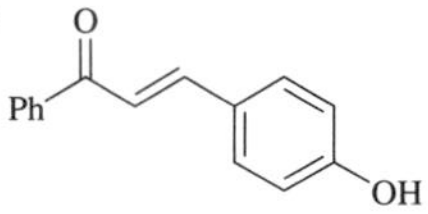	(L-Leu)$_n$ (n_{av} = 260), $NaBO_4$•4 H_2O, NaOH, toluene, H_2O, sonication, rt, 7 h	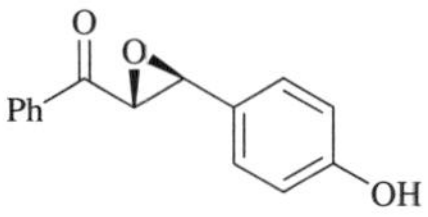 (99), 67	271
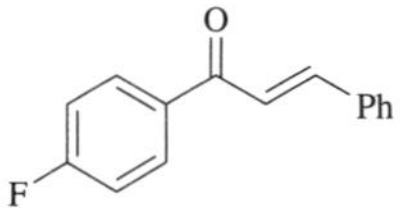	Anthryl cinchonidinium salt **6b** (10 mol %), KOCl, H_2O, toluene, –40°, 12 h	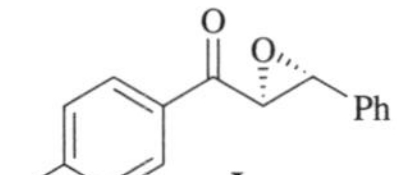 **I** (93), 98	98
	Anthryl cinchonidinium salt **6b** (1 mol %), NaOCl, H_2O, toluene, rt	**I** (75), 87	97
	Anthryl cinchonidinium salt **6b** (10 mol %), trichloroisocyanuric acid, KOH, 0°	**I**	
	Solvent / Time (h)		
	toluene/H_2O / —	(79), 88	101
	toluene / 7	(83), 90	102
	Bis(quininium) salt **14d** (1 mol %), H_2O_2, KOH, Span 20, *i*-Pr_2O, H_2O, rt, 4 h	**I** (94), 98	108
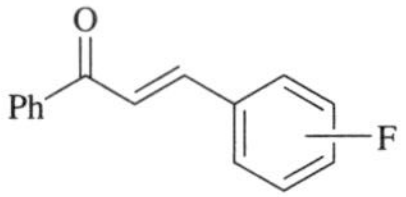	Bis(quininium) salt **14d** (1 mol %), H_2O_2, KOH, Span 20, *i*-Pr_2O, H_2O, rt, 1 h	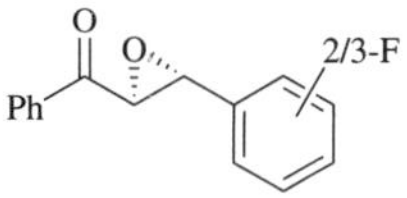 Isomer: 2-F (97), >99; 3-F (96), 98	108
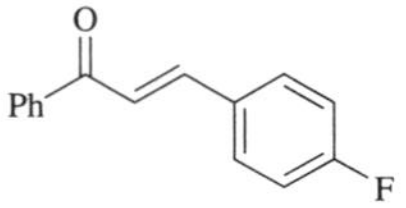	BisBINOL **23g** (2.5 mol %), CMHP, La(O*i*-Pr)$_3$, TPPO, 4 Å MS, THF, rt, 0.5 h	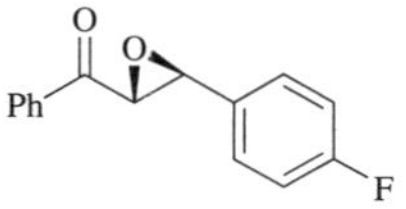 (>99), 96	66

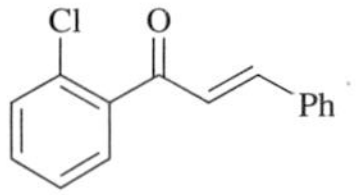	(L-Leu)$_n$ (n_{av} = 8), H_2O_2, NaOH, toluene, H_2O, 0° to rt, 24 h	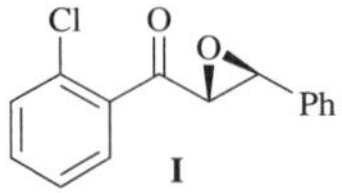 **I** (48), 76	266
	(L-Leu)$_n$–PS (n_{av} = 32), urea/H_2O_2, DBU, THF, rt, 45 min	**I** (90), 89	183
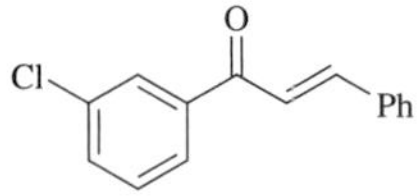	(L-Leu)$_n$ (n_{av} = 8), H_2O_2, NaOH, toluene, H_2O, 0° to rt, 24 h	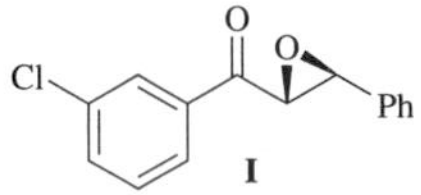 **I** (74), 92	266
	Egg phosphatidylcholine, mCPBA, H_2O, EtOH, rt, 2 h	**I** (67), 62	174
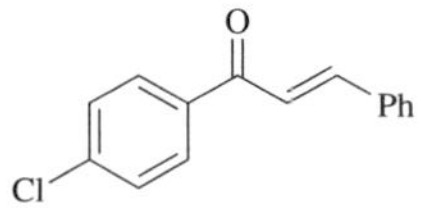	Cinchonidinium salt **5a** (5 mol %), H_2O_2, NaOH, *n*-Bu_2O, H_2O, 0°, 48 h	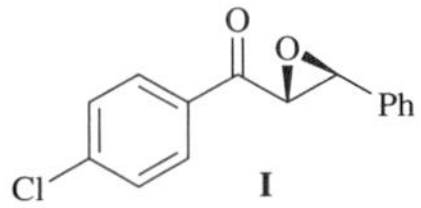 **I** (51), 58[o]	107
	Aza-crown ether **43f** (7 mol %), TBHP, NaOH, toluene, H_2O, rt, 1 h	**I** (57), 80	112, 113
	(L-Leu)$_n$–PS (n_{av} = 33), H_2O_2, NaOH, toluene, H_2O, 0° to rt, 2 d	**I** (98), 99	123
	(L-Leu)$_n$ (M_r 3000–15000), H_2O_2, NaOH, CCl_4, H_2O, rt, 2 d	**I** (79), 88	267
	(L-Leu)$_n$ (n_{av} = 8), H_2O_2, NaOH, toluene, H_2O, 0° to rt, 24 h	**I** (77), 94	266
	(L-Leu)$_n$–SiO_2 (n_{av} = 30; 6 mol %), Na_2CO_3•1.5 H_2O_2, DME, H_2O, rt, 2 h	**I** (88), 93[l]	125

TABLE 1D. DIARYL *E*-ENONES (*Continued*)

Substrate	Conditions	Product(s), Yield(s) (%), % ee	Refs.
C_{15} 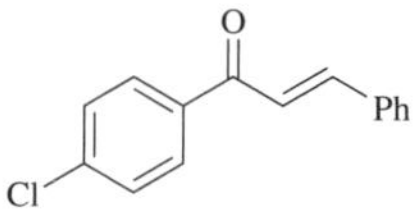	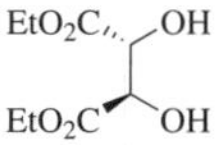 (11 mol %), TBHP, NanoActive MgO Plus, toluene, rt, 36 h	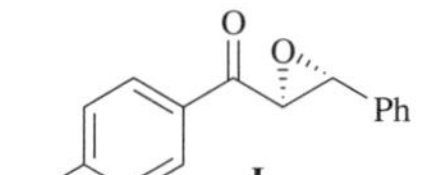 **I** (52), 97	73
	Anthryl cinchonidinium salt **6b** (10 mol %), trichloroisocyanuric acid, KOH, 0°	**I**	
	Solvent: toluene/H_2O; Time (h): —	(85), 93	101
	Solvent: toluene; Time (h): 7	(88), 92	102
	Ammonium salt **50e** (3 mol %), NaOCl, toluene, H_2O, 0°, 24 h	**I** (99), 93	109
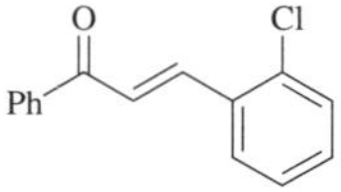	(L-Leu)$_n$ (n_{av} = 8), H_2O_2, NaOH, toluene, H_2O, 0° to rt, 24 h	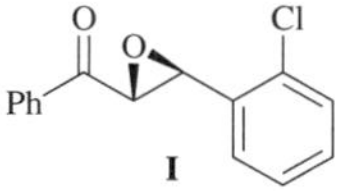 **I** (92), 54	266
	Egg phosphatidylcholine, *m*-CPBA, H_2O, EtOH, rt, 2 h	**I** (63), 68	174
	Iodobenzyl cinchoninium salt **1f** (5 mol %), H_2O_2, LiOH, H_2O, *n*-Bu_2O, 4°, 47 h	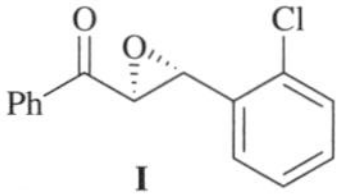 **I** (88), 65	104, 103
	Anthryl cinchonidinium salt **6b** (10 mol %), trichloroisocyanuric acid, KOH, 0°	**I**	
	Solvent: toluene/H_2O; Time (h): —	(89), 64	101
	Solvent: toluene; Time (h): 7	(91), 73	102
	Benzyl quininium salt **4a**, copolymer with acrylamide, phase transfer conditions, 54 h	**I** (28), 5.4[p]	86

	Polystyrene-bound quininium salt, phase transfer conditions, 72 h	**I**	86
	Quininium salt		
	4e	(75), 1.4[p]	
	4f	(43), 3.9[p]	
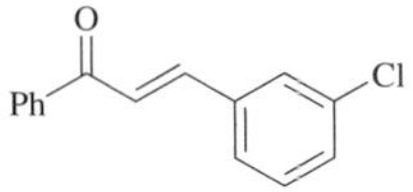	(L-Leu)$_n$ (n_{av} = 8), H_2O_2, NaOH, toluene, H_2O, 0° to rt, 24 h	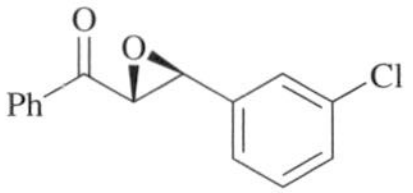 (83), 95	266
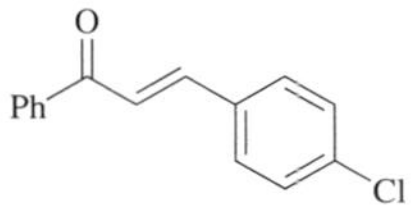	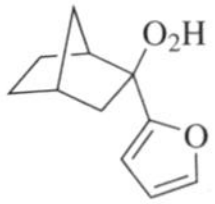 (110 mol %), *n*-BuLi, THF, –20°, 1 h	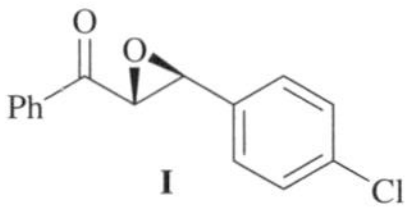 **I** (80), 44	32
	PolyBINOL **29** (20 mol %), TBHP, Et_2Zn, Et_2O, rt, 5.5 h	**I** (81), 79	43
	(*S*)-BINOL (5 mol %), CMHP, Ln(O*i*-Pr)$_3$, THF, rt, 8 h	**I**	
	(*S*)-BINOL — Ln		
	17b — Gd	(85), 94	48
	17d — La	(91), 87	48
	17b — La	(88), 38	264
	17b — Yb	(85), 53	47
	BisBINOL **23g** (2.5 mol %), CMHP, La(O*i*-Pr)$_3$, TPPO, 4 Å MS, THF, rt, 0.5 h	**I** (>99), 96	66
	(*S*)-BINOL **17b** (10 mol %), TBHP, KO*t*-Bu, $CaCl_2$, 4 Å MS, toluene, cyclohexane, –15°, 48 h	**I** (91), 74	79
	Polymer-bound (*R*)-BINOL **21** (10 mol %), TBHP, KO*t*-Bu, $CaCl_2$, 4 Å MS, toluene, 0°, 24 h	**I** (95), 41	77

TABLE 1D. DIARYL *E*-ENONES (*Continued*)

Substrate	Conditions	Product(s), Yield(s) (%), % ee	Refs.
C_{15} 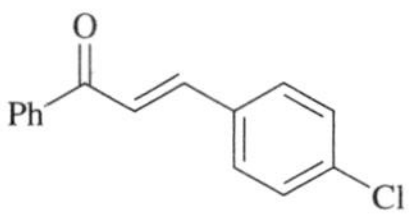	Benzyl quininium salt **4a** (1.7 mol %), H_2O_2, NaOH, H_2O, toluene, rt, 24 h	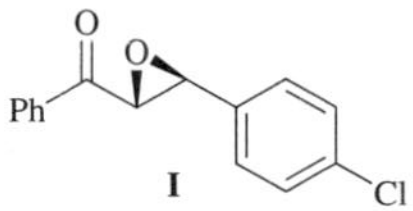 **I** (—), 23[k]	81
	Polystyrene-bound quininium salt, H_2O_2, NaOH, H_2O, toluene, rt, 3 d	**I**	87
	Quininium salt		
	4e	(75), 1.4[k]	
	4f	(43), 3.9[m]	
	Cinchonidinium salt **5a** (5 mol %), H_2O_2, NaOH, *n*-Bu_2O, H_2O, 0°, 48 h	**I** (54), 51[k]	107
	(L-Ala)$_n$ (n_{av} = 10), H_2O_2, NaOH, toluene, H_2O, rt, 48 h	**I** (47), 66	132
	(L-Leu)$_n$–PS (n_{av} = 33), H_2O_2, NaOH, toluene, H_2O, 0° to rt, 2 d	**I** (98), 99	123, 183
	(L-Leu)$_n$ (n_{av} = 8), H_2O_2, NaOH, toluene, H_2O, 0° to rt, 24 h	**I** (87), 94	266
	(L-Leu)$_n$–SiO_2 (n_{av} = 30; 6 mol %), Na_2CO_3•1.5 H_2O_2, DME, H_2O, rt, 2 h	**I** (90), 92[n]	125
	(L-Leu)$_n$, H_2O_2, NH_4HCO_3, Aliquat 336, DME, H_2O, rt	**I** (93), 86	119
	(L-Leu)$_n$ (n_{av} = 260), $NaBO_4$•$4H_2O$, NaOH, toluene, H_2O, sonication, rt, 7 h	**I** (97), 74	271

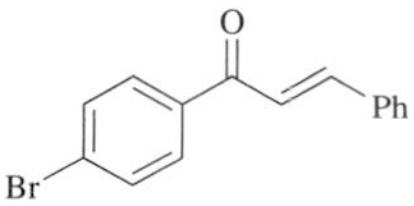	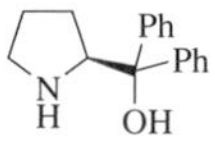 (30 mol %), TBHP, hexane, rt, 105 h	**I** (73), 74	170
	(*R*)-BINOL **15a** (5 mol %), CMHP, La(O*i*-Pr)$_3$, TPPO (0.15 eq), 4 Å MS, THF, rt, 3 h	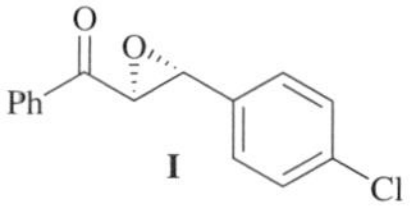 **I** (93), 99.6	274, 68
	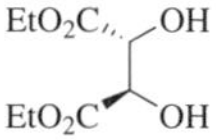 (11 mol %),	**I**	
	TBHP, Bu_2Mg, toluene, THF, rt, 1 d	(54), 81	69
	TBHP, NanoActive MgO Plus, toluene, rt, 36 h	(70), 96	73
	Anthryl cinchonidinium salt **6b** (10 mol %), KOCl, H_2O, toluene, –40°, 12 h	**I** (94), 92	98
	Anthryl cinchonidinium salt **6b** (10 mol %), trichloroisocyanuric acid, KOH, 0°	**I**	
	Solvent / Time (h)		
	toluene/H_2O / —	(97), 84	101
	toluene / 7	(94), 89	102
	Ammonium salt **50e** (3 mol %), NaOCl, toluene, H_2O, 0°, 24 h	**I** (99), 96	109
	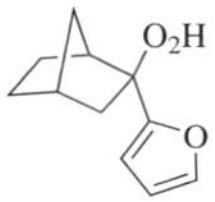 (110 mol %), *n*-BuLi, THF, –20°, 3 h	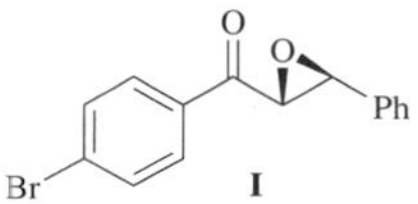 **I** (65), 38	32
	(*S*)-BINOL **17b** (10 mol %), TBHP, KO*t*-Bu, $CaCl_2$, 4 Å MS, toluene, cyclohexane, –15°, 48 h	**I** (82), 32[k]	32

TABLE 1D. DIARYL *E*-ENONES (*Continued*)

Substrate	Conditions	Product(s), Yield(s) (%), % ee	Refs.
C_{15} 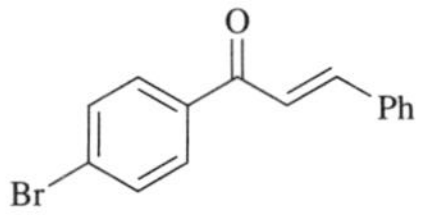	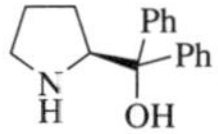 (30 mol %), TBHP, hexane, rt, 98 h	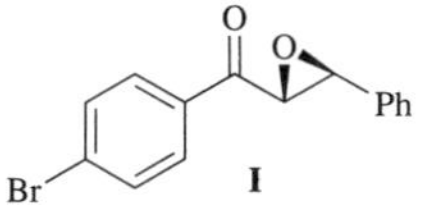 I (72), 74	170
	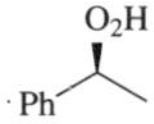 (100 mol %), KOH, MeCN, –40°, 20–30 min	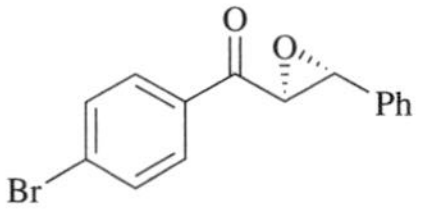 (95), 48	27, 28
	Anthryl cinchonidinium salt **6b** (10 mol %), KOCl, H_2O, toluene, –40°, 12 h	I (92), 93	98
	Anthryl cinchonidinium salt **6b** (1 mol %), NaOCl, H_2O, toluene, rt	I (93), 88	97, 99
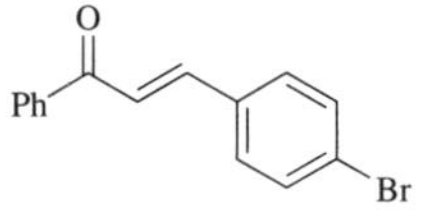	(*R*)-BINOL **15a** (20 mol %), oxidant, Et_2Zn, Et_2O, 12 h	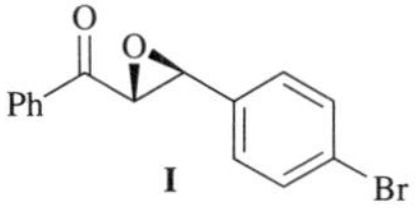 I	44
	Oxidant / Temp (°)		
	TBHP / 0 to rt	(50), 71	
	CMHP / 0 to rt	(95), 72	
	CMHP / 0	(95), 72	
	CMHP / –5	(95), 80	
	CMHP / –10	(79), 72	
	CMHP / –20	(51), 64	
	BisBINOL **23g** (2.5 mol %), CMHP, La(O*i*-Pr)$_3$, TPPO, 4 Å MS, THF, rt, 0.5 h	I (>99), 96	66

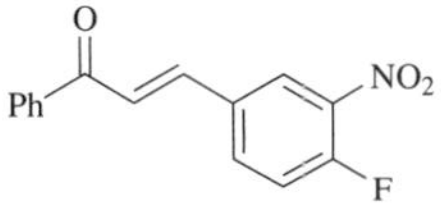	(*R*)-BINOL **15a** (5 mol %), CMHP, La(O*i*-Pr)$_3$, (4-FC$_6$H$_4$)$_3$P=O, 4 Å MS, THF, rt, 20 min	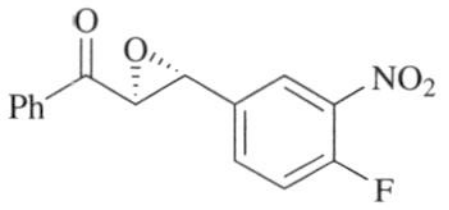 (98), >99	54
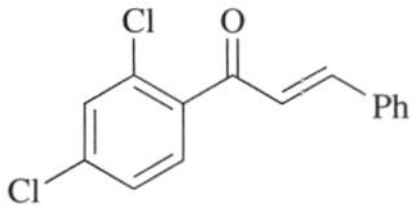	Aza-crown ether **43f** (7 mol %), TBHP, NaOH, toluene, H$_2$O, rt, 1 h	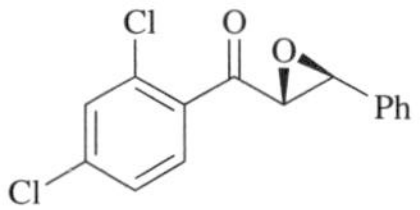 (82), 47	112, 113
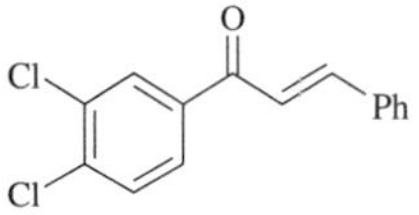	Aza-crown ether **43f** (7 mol %), TBHP, NaOH, toluene, H$_2$O, rt, 0.5 h	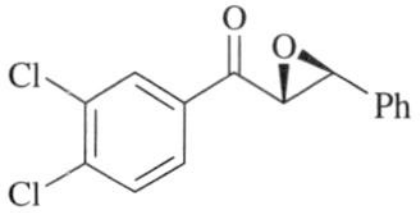 (61), 66	112, 113
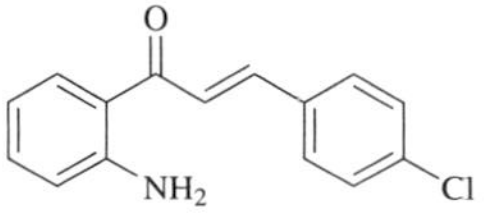	(L-Leu)$_n$–PS (n_{av} = 32), urea/H$_2$O$_2$, DBU, THF, rt, 2 h	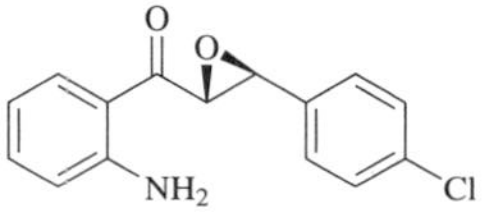 (91), >98	183
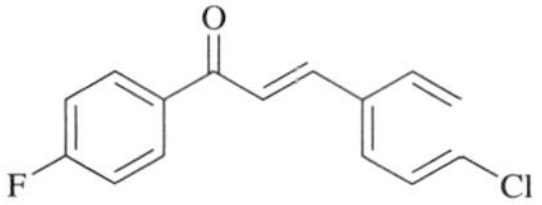	Anthryl cinchonidinium salt **6b** (10 mol %), KOCl, H$_2$O, toluene, –40°, 12 h	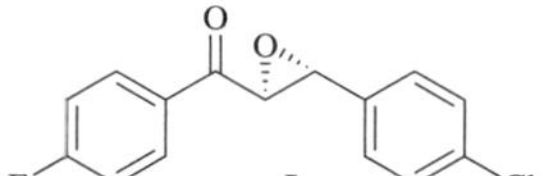 **I** (94), 98.5	98
	Anthryl cinchonidinium salt **6b** (10 mol %), trichloroisocyanuric acid, KOH, 0°	**I**	
	Solvent: toluene/H$_2$O; Time (h): —	(92), 92	101
	Solvent: toluene; Time (h): 6.5	(93), 93	102

TABLE 1D. DIARYL *E*-ENONES (*Continued*)

Substrate	Conditions	Product(s), Yield(s) (%), % ee	Refs.
C_{15} 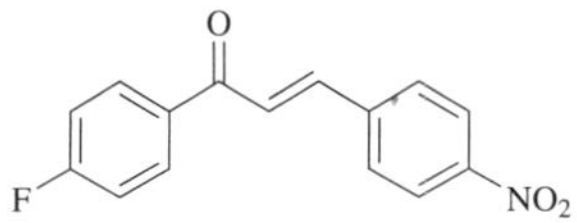	Anthryl cinchonidinium salt **6b** (10 mol %), KOCl, H_2O, toluene, –40°, 12 h	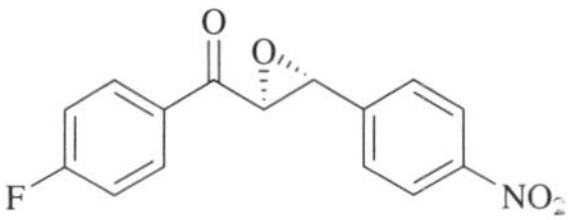 (97), 95	98
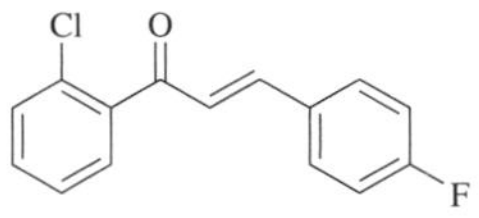	Aza-crown ether **43f** (7 mol %), TBHP, NaOH, toluene, H_2O, rt, 1 h	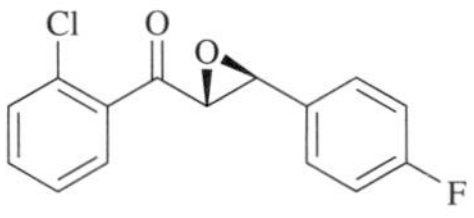 (77), 3	112, 113
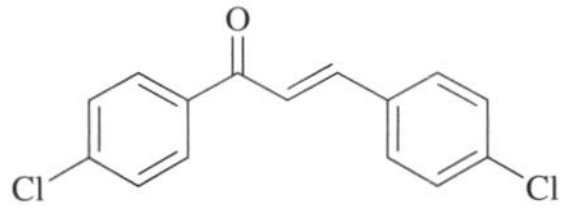	Aza-crown ether **43f** (7 mol %), TBHP, NaOH, toluene, H_2O, rt, 1 h	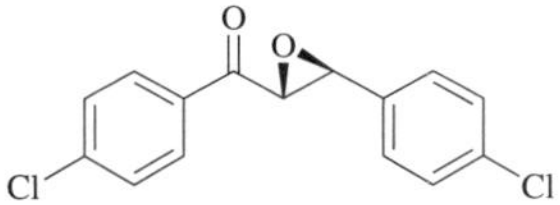 (66), 77	112, 113
	Anthryl cinchonidinium salt **6b** (10 mol %), trichloroisocyanuric acid, KOH, 0°	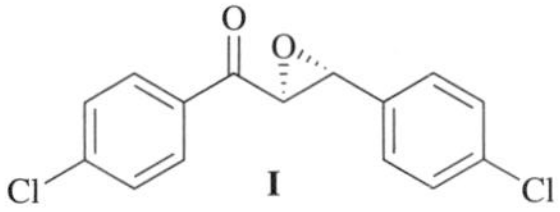 **I**	
	Solvent: toluene/H_2O; Time (h): —	(93), 79	101
	Solvent: toluene; Time (h): 7	(90), 86	102
	Aza-crown ether **45c** (7 mol %), TBHP, NaOH, toluene, H_2O, 0–4°, 4 h	**I** (38), 64	111
C_{16} 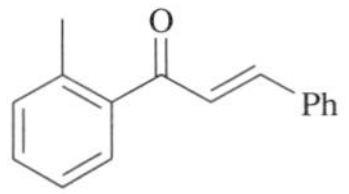	(L-Leu)$_n$ (n_{av} = 8), H_2O_2, NaOH, toluene, H_2O, 0° to rt, 24 h	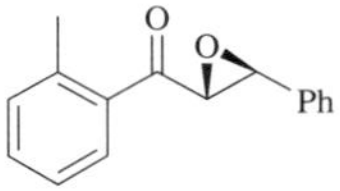 (27), 81	266
	(L-Leu)$_n$–PS (n_{av} = 32), urea/H_2O_2, DBU, THF, rt, 1.5 h	**I** (94), 81	183

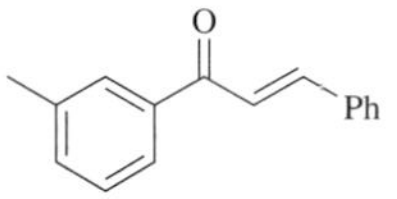	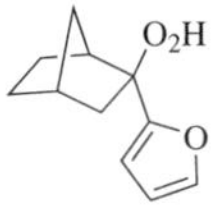 (110 mol %), *n*-BuLi, THF, –20°, 4.5 h	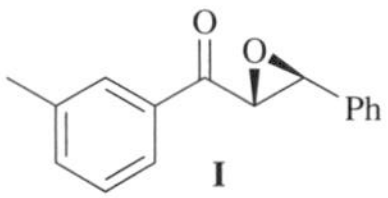 **I** (67), 50	32
	(L-Leu)$_n$ (n_{av} = 8), H_2O_2, NaOH, toluene, H_2O, 0° to rt, 24 h	**I** (47), 94	266
	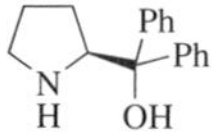 (30 mol %), TBHP, hexane, rt, 135 h	**I** (70), 78	170
	Iodobenzyl cinchoninium salt **1f** (5 mol %), H_2O_2, LiOH, H_2O, *n*-Bu_2O, 4°, 36 h	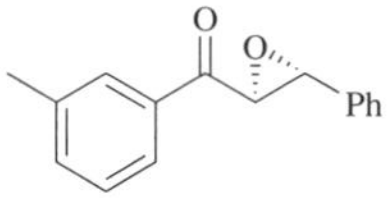 (99), 87	104, 103
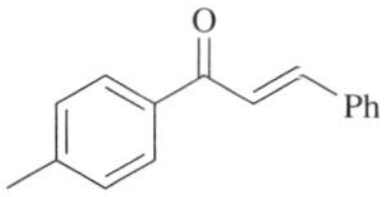	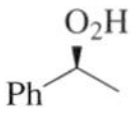 (100 mol %), DBU, toluene, 20°, 24 h	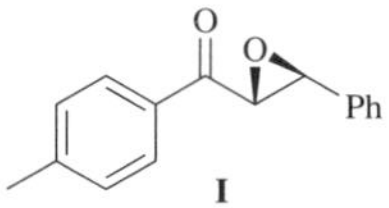 **I** (93), 41	28
	(*S*)-BINOL **17b** (10 mol %), TBHP, KO*t*-Bu, $CaCl_2$, 4 Å MS, toluene, cyclohexane, 5°, 48 h	**I** (68), 62	79
	Polymer-bound (*R*)-BINOL **21** (10 mol %), TBHP, KO*t*-Bu, $CaCl_2$, 4 Å MS, toluene, 0°, 24 h	**I** (95), 40	77
	Aza-crown ether **43f** (7 mol %), TBHP, NaOH, toluene, H_2O, rt, 3 h	**I** (62), 81	112, 113
	(L-Leu)$_n$ (M_r 3000–15000), H_2O_2, NaOH, CCl_4, H_2O, rt, 2 d	**I** (87), 92	267
	(L-Leu)$_n$ (n_{av} = 8), H_2O_2, NaOH, toluene, H_2O, 0° to rt, 24 h	**I** (84), 97	266

TABLE 1D. DIARYL *E*-ENONES (*Continued*)

Substrate	Conditions	Product(s), Yield(s) (%), % ee	Refs.
C_{16} 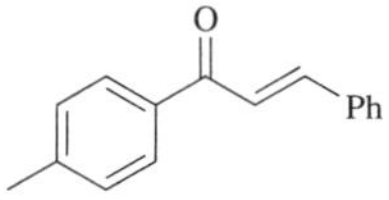	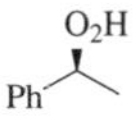 (100 mol %), KOH, MeCN, –40°, 20–30 min	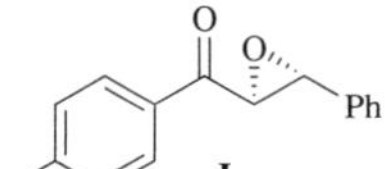 I (98), 54	27, 28
	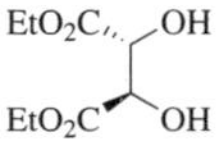 (11 mol %), TBHP, Bu_2Mg, toluene, THF, rt, 1 d	**I** (36), 87	69
	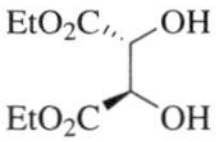 (11 mol %), TBHP, NanoActive MgO Plus, toluene, rt, 36 h	**I** (62), 69	73
	Iodobenzyl cinchoninium salt **1f** (5 mol %), H_2O_2, LiOH, H_2O, *n*-Bu_2O, 4°, 36 h	**I** (95), 89	104, 103
	Bis(quininium) salt **14d** (1 mol %), H_2O_2, KOH, Span 20, *i*-Pr_2O, H_2O, rt, 4 h	**I** (96), 97	108
	Aza-crown ether **43f** (7 mol %), TBHP, NaOH, toluene, H_2O, 0–4°, 7 h	**I** (40), 61	111
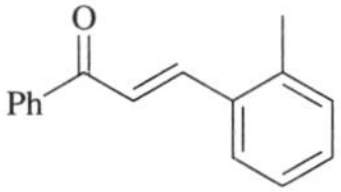	(L-Leu)$_n$ (n_{av} = 8), H_2O_2, NaOH, toluene, H_2O, 0° to rt, 24 h	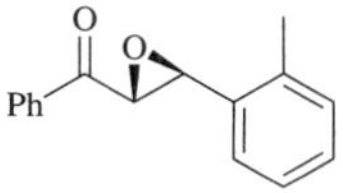 (46), 87	266
	Iodobenzyl cinchoninium salt **1f** (5 mol %), H_2O_2, LiOH, H_2O, *n*-Bu_2O, 4°, 64 h	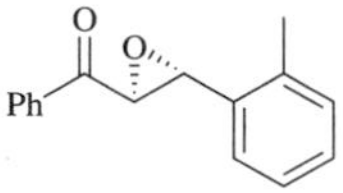 (96), 67	104, 103

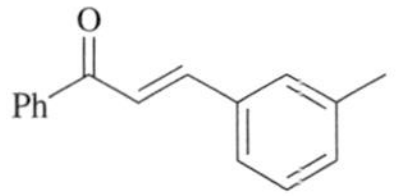	(L-Leu)$_n$ (n_{av} = 8), H_2O_2, NaOH, toluene, H_2O, 0° to rt, 24 h	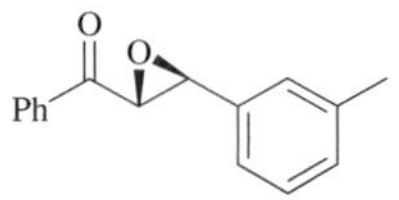	(65), 98	266
	Iodobenzyl cinchoninium salt **1f** (5 mol %), H_2O_2, LiOH, H_2O, n-Bu_2O, 4°, 64 h	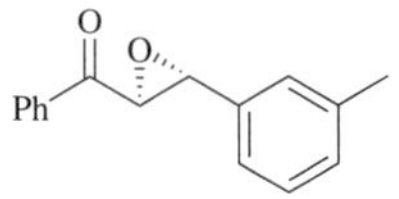	(100), 92	104, 103
	Bis(quininium) salt **14d** (1 mol %), H_2O_2, KOH, Span 20, i-Pr_2O, H_2O, rt, 12 h	**I** (95), 97		108
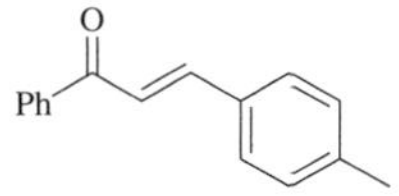	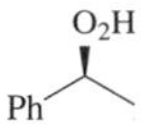 (100 mol %), DBU, toluene, 20°, 24 h	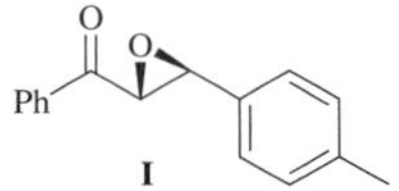 **I**	(97), 37	28
	PolyBINOL **29** (20 mol %), TBHP, Et_2Zn, Et_2O, rt, 8 h	**I** (93), 70		43
	(S)-BINOL (5 mol %), CMHP, Ln(Oi-Pr)$_3$, THF, rt, 8 h	**I**		
	(S)-BINOL / Ln			
	17b / Gd	(83), 93		48
	17d / La	(88), 86		48
	17b / La	(83), 27		264
	17b / Yb	(83), 54		47
	(S)-BINOL **17b** (10 mol %), TBHP, KOt-Bu, $CaCl_2$, 4 Å MS, toluene, cyclohexane, –10°, 48 h	**I** (78), 80		79
	Polymer-bound (R)-BINOL **21** (10 mol %), TBHP, KOt-Bu, $CaCl_2$, 4 Å MS, toluene, 0°, 24 h	**I** (93), 47		79

TABLE 1D. DIARYL *E*-ENONES (*Continued*)

Substrate	Conditions	Product(s), Yield(s) (%), % ee	Refs.
C_{16} 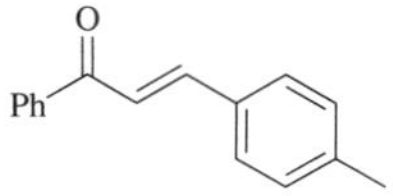	Aza-crown ether **43f** (7 mol %), TBHP, NaOH, toluene, H_2O, rt, 3 h	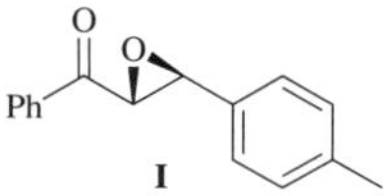 **I** (57), 77	112, 113
	(L-Leu)$_n$ (n_{av} = 8), H_2O_2, NaOH, toluene, H_2O, 0° to rt, 24 h	**I** (66), 93	266
	(L-Leu)$_n$, H_2O_2, NH_4HCO_3, Aliquat 336, DME, H_2O, rt	**I** (85), 92	119
	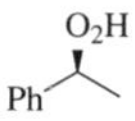 (100 mol %), KOH, MeCN, –40°, 10–20 min	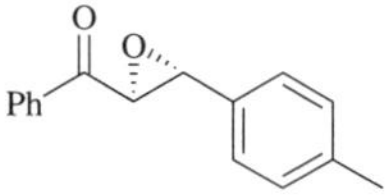 (97), 57	27, 28
	PolyBINOL **28** (111 mol %), O_2, Et_2Zn, CH_2Cl_2, –15°	**I** (34), 54	43
	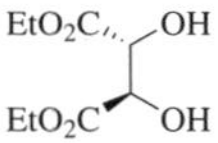 (11 mol %), TBHP, Bu_2Mg, toluene, THF, rt, 3 d	**I** (36), 84	69
	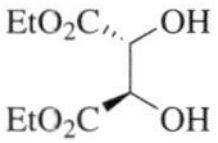 (11 mol %), TBHP, NanoActive MgO Plus, toluene, rt, 36 h	**I** (70), 80	73
	Anthryl cinchonidinium salt **6b** (10 mol %), KOCl, H_2O, toluene, –40°, 12 h	**I** (70), 94	98
	Aza-crown ether **43f** (7 mol %), TBHP, NaOH, toluene, H_2O, 0–4°, 7 h	**I** (37), 55	111

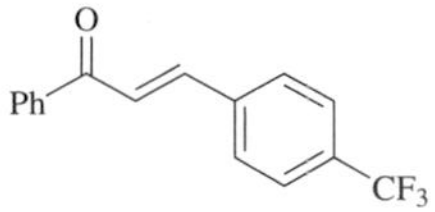	(L-Leu)$_n$–PS (n_{av} = 32), Na_2CO_3•1.5 H_2O_2, DME, H_2O, rt, 18 h	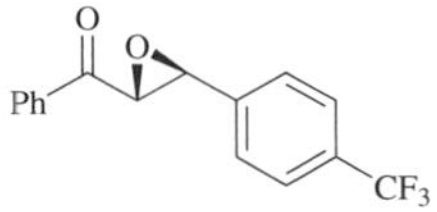 (78), 97	256
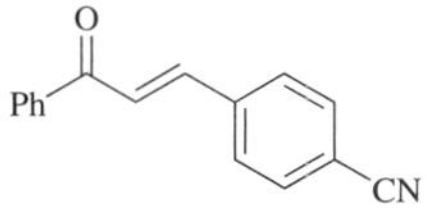	BisBINOL **23g** (2.5 mol %), CMHP, La(O*i*-Pr)$_3$, TPPO, 4 Å MS, THF, rt, 1 h	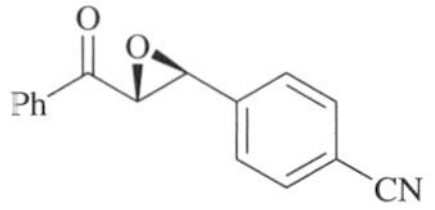 (>99), 94.3	66
	(*R*)-BINOL **15a** (5 mol %), CMHP, La(O*i*-Pr)$_3$, (4-FC_6H_4)$_3$P=O, 4 Å MS, THF, rt, 15 min	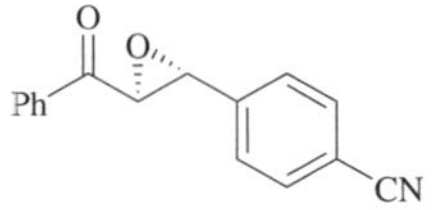 (92), >99	54
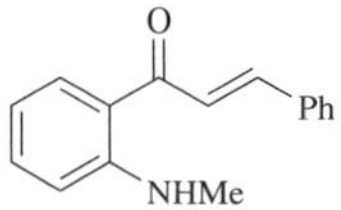	(L-Leu)$_n$–PS (n_{av} = 32), urea/H_2O_2, DBU, THF, rt, 7 h	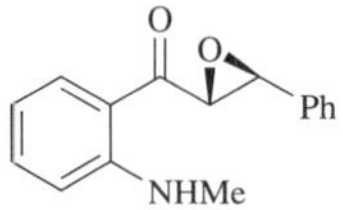 (62), 96	183
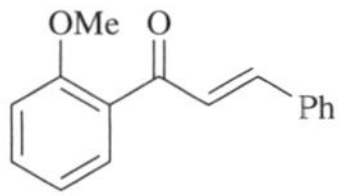	(*S*)-BINOL **17b** (10 mol %), TBHP, KO*t*-Bu, $CaCl_2$, 4 Å MS, toluene, cyclohexane, 0°, 48 h	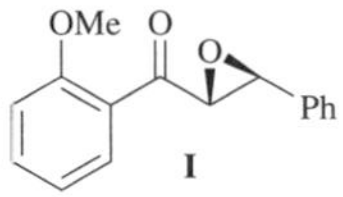 **I** (70), 22	79
	Benzyl quininium salt **4a** (1.7 mol %), H_2O_2, NaOH, H_2O, toluene, rt, 24 h	**I** (—), 25	81
	(L-Ala)$_n$ (n_{av} = 10), H_2O_2, NaOH, toluene, H_2O, rt, 48 h	**I** (54), 50	132
	(L-Leu)$_n$ (n_{av} = 8), H_2O_2, NaOH, toluene, H_2O, 0° to rt, 24 h	**I** (27), 81	266
	(L-Leu)$_n$–SiO_2 (n_{av} = 30; 6 mol %), Na_2CO_3•1.5 H_2O_2, DME, H_2O, rt, 2 h	**I** (70), 80^m	125

TABLE 1D. DIARYL *E*-ENONES (*Continued*)

Substrate	Conditions	Product(s), Yield(s) (%), % ee	Refs.
C_{16} 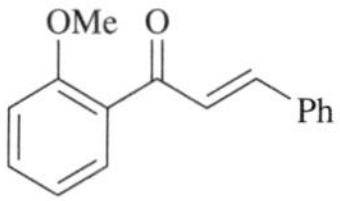	Benzyl quinidinium salt **2a** (1.7 mol %), H_2O_2, NaOH, H_2O, toluene, rt, 24 h	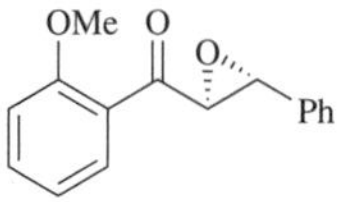 (—), 21	81
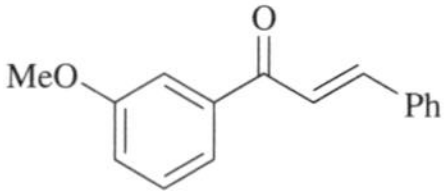	(L-Leu)$_n$ (n_{av} = 8), H_2O_2, NaOH, toluene, H_2O, 0° to rt, 24 h	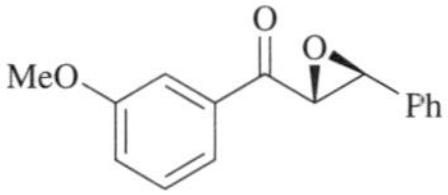 (58), 92	266
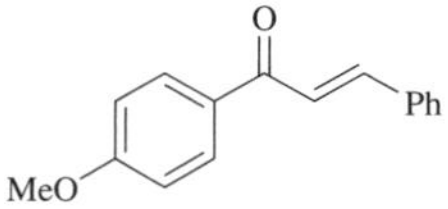	BisBINOL **23g** (2.5 mol %), CMHP, La(O*i*-Pr)$_3$, TPPO, 4 Å MS, THF, rt, 0.5 h	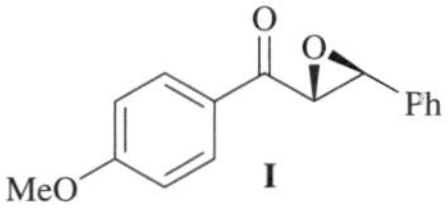 **I** (>99), 95	66
	Cinchonidinium salt **5a** (5 mol %), H_2O_2, NaOH, *n*-Bu_2O, H_2O, 0°, 48 h	**I** (36), 61[o]	107
	Aza-crown ether **43f** (7 mol %), TBHP, NaOH, toluene, H_2O, rt, 3 h	**I** (53), 82	112, 113
	(L-Leu)$_n$–PS (n_{av} = 33), H_2O_2, NaOH, toluene, H_2O, 0° to rt, 2 d	**I** (83), 87	123
	(L-Leu)$_n$ (M_r 3000–15000), H_2O_2, NaOH, CCl_4, H_2O, rt, 2 d	**I** (77), 90	267
	(L-Leu)$_n$ (n_{av} = 8), H_2O_2, NaOH, toluene, H_2O, 0° to rt, 24 h	**I** (72), 97	266
	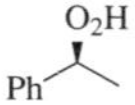 (100 mol %), KOH, MeCN, –40°, 20–30 min	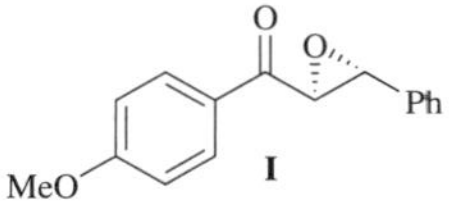 **I** (97), 53	27, 28

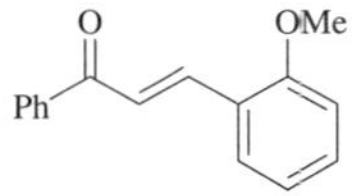	(*R*)-BINOL **15a** (5 mol %), CMHP, La(O*i*-Pr)$_3$, TPPO, 4 Å MS, THF, rt, 1 h	**I** (99), >99	68
	(*R*)-BINOL **15a** (5 mol %), CMHP, La(O*i*-Pr)$_3$, (4-FC$_6$H$_4$)$_3$P=O, 4 Å MS, THF, rt, 15 min	**I** (98), 98	54
	Anthryl cinchonidinium salt **6b** (10 mol %), trichloroisocyanuric acid, KOH, 0°	**I**	
	Solvent — Time (h)		
	toluene/H_2O — —	(82), 90	101
	toluene — 7	(86), 91	102
	Bis(quininium) salt **14d** (1 mol %), H_2O_2, KOH, Span 20, *i*-Pr$_2$O, H_2O, rt, 1.5 h	**I** (95), >99	108
	Aza-crown ether **43f** (7 mol %), TBHP, NaOH, toluene, H_2O, rt, 3 h	**I** (37), 54	111
	(*S*)-BINOL (5 mol %), CMHP, Ln(O*i*-Pr)$_3$, THF, rt, 8 h	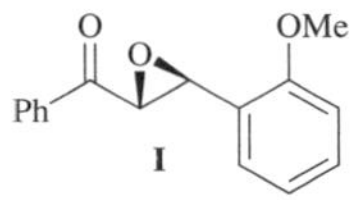 **I**	
	(*S*)-BINOL — Ln		
	17b — Gd	(81), 92	48
	17d — La	(88), 81	48
	17b — La	(83), 28	264
	17b — Yb	(81), 66	47
	(L-Ala)$_n$ (n_{av} = 10), H_2O_2, NaOH, toluene, H_2O, rt, 48 h	**I** (29), 36[q]	132, 273
	(L-Leu)$_n$–PS (n_{av} = 33), H_2O_2, NaOH, toluene, H_2O, 0° to rt, 2 d	**I** (56), 76	123
	(L-Leu)$_n$ (n_{av} = 3), H_2O_2, NaOH, toluene, H_2O, 0° to rt, 24 h	**I** (39), 67	266

TABLE 1D. DIARYL *E*-ENONES (*Continued*)

	Substrate	Conditions	Product(s), Yield(s) (%), % ee	Refs.
C_{16}	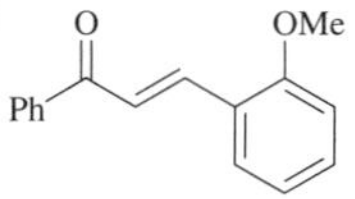	(L-Leu)$_n$–SiO$_2$ (n_{av} = 30; 6 mol %), Na_2CO_3•1.5 H_2O_2, DME, H_2O, rt, 2 h	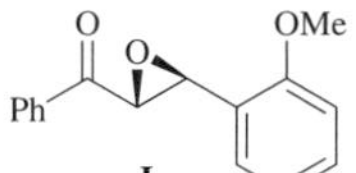 **I** (54), 70[m]	125
		Benzyl quininium salt **4a**, H_2O_2, NaOH, H_2O, rt, 18 h	**I**	82
		Solvent		
		toluene	(92), 48	
		nitrobenzene	(—), 10	
		CH_2Cl_2	(—), 28	
		chlorobenzene	(—), 34	
		o-xylene	(—), 38	
		benzene	(—), 54	
	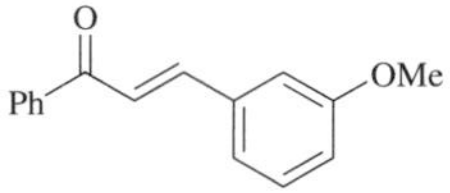	(L-Leu)$_n$ (n_{av} = 8), H_2O_2, NaOH, toluene, H_2O, 0° to rt, 24 h	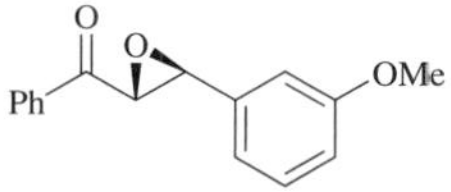 (45), 90	266
	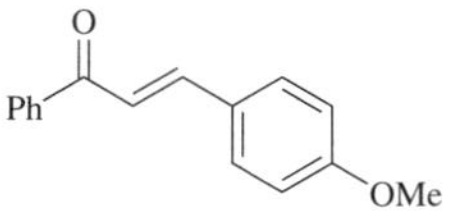	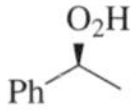 (100 mol %), DBU, toluene, 20°, 24 h	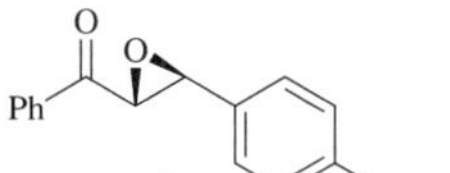 **I** (94), 36	28
		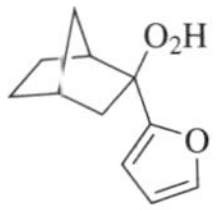 (110 mol %), *n*-BuLi, THF, –20°, 3 h	**I** (30), 42	32

(*S*)-BINOL (5 mol %), CMHP, Ln(O*i*-Pr)$_3$, THF, rt, 8 h	**I**		
(*S*)-BINOL / Ln: **17b** / Gd		(85), 91	48
17d / La		(86), 82	48
17b / La		(85), 31	264
17b / Yb		(85), 54	47
Anthryl cinchoninium salt **11** (10 mol %), NaOCl, H_2O, toluene, rt, 48 h	**I**	(86), 81	100
(L-Ala)$_n$ (n_{av} = 10), H_2O_2, NaOH, toluene, H_2O, rt, 48 h	**I**	(53), —[r]	132
(L-Leu)$_n$–PS (n_{av} = 33), H_2O_2, NaOH, toluene, H_2O, 0° to rt, 2 d	**I**	(89), 90	123
(L-Leu)$_n$ (n_{av} = 8), H_2O_2, NaOH, toluene, H_2O, 0° to rt, 24 h	**I**	(66), 85	266
(L-Leu)*n*–SiO_2 (n_{av} = 30; 6 mol %), Na_2CO_3•1.5 H_2O_2, DME, H_2O, rt, 2 h	**I**	(80), 82[m]	125
(L-Leu)$_n$ (n_{av} = 260), $NaBO_4$•4 H_2O, NaOH, toluene, H_2O, sonication, rt, 7 h	**I**	(88), 89	271
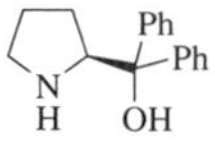 (30 mol %), TBHP, hexane, rt, 190 h	**I**	(46), 80	170
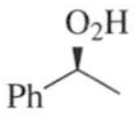 (100 mol %), KOH, MeCN, –40°, 20–30 min	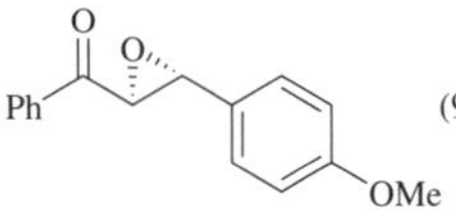	(96), 61	27, 28

TABLE 1D. DIARYL *E*-ENONES (*Continued*)

Substrate	Conditions	Product(s), Yield(s) (%), % ee	Refs.
C_{16}			
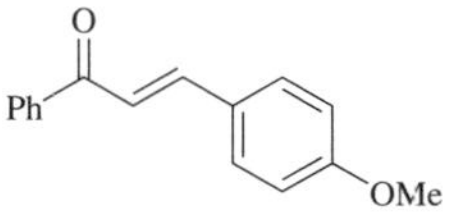	(*R*)-BINOL **15a** (5 mol %), CMHP, La(O*i*-Pr)$_3$, (4-FC$_6$H$_4$)$_3$P=O, 4 Å MS, THF, rt, 15 min	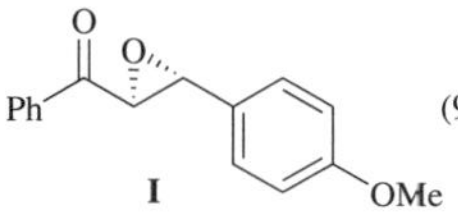 **I** (93), >99	54
	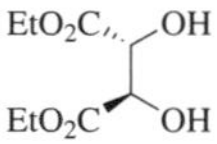 (11 mol %), TBHP, NanoActive MgO Plus, toluene, rt, 36 h	**I** (68), 53	73
	Anthryl cinchonidinium salt **6b** (10 mol %), KOCl, H_2O, toluene, –40°, 12 h	**I** (70), 95	98
	Anthryl cinchonidinium salt **6b** (10 mol %), NaOCl, H_2O, toluene, rt, 48 h	**I** (87), 82	100, 99
	Anthryl cinchonidinium salt **6b** (10 mol %), trichloroisocyanuric acid, KOH, toluene, 0°, 8 h	**I** (80), 91	102
	Ammonium salt **50e** (3 mol %), NaOCl, toluene, H_2O, 0°, 48 h	**I** (83), 96	109
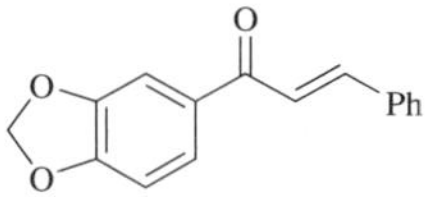	Anthryl cinchoninium salt **1l** (10 mol %), NaOCl, H_2O, toluene, rt, 48 h	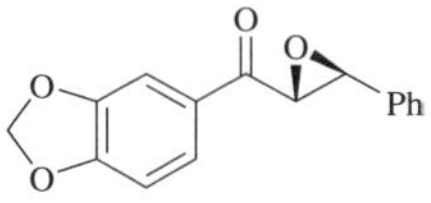 (93), 86	100
	Anthryl cinchonidinium salt **6b** (1 mol %), NaOCl, H_2O, toluene, rt	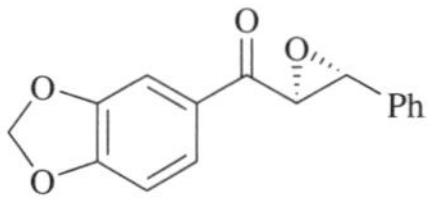 (76), 92	97, 100, 99

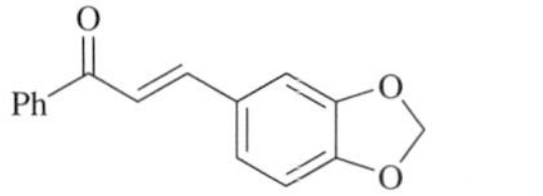	Anthryl cinchoninium salt **1l** (10 mol %), NaOCl, H_2O, toluene, rt, 48 h	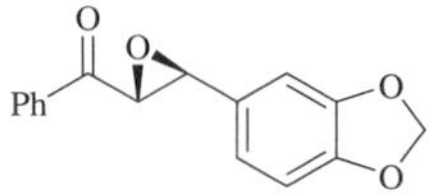	(92), 82	100
	Anthryl cinchonidinium salt **6b** (10 mol %), NaOCl, H_2O, toluene, rt, 48 h	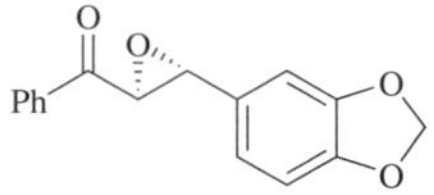	(97), 83	100, 99
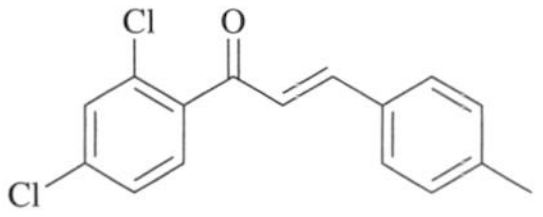	Aza-crown ether **43f** (7 mol %), TBHP, NaOH, toluene, H_2O, rt, 0.5 h	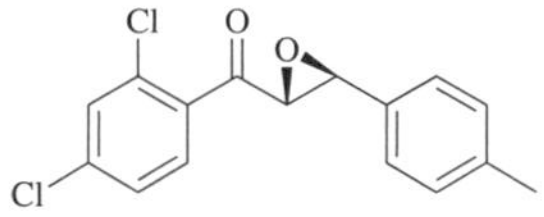	(82), 42	112, 113
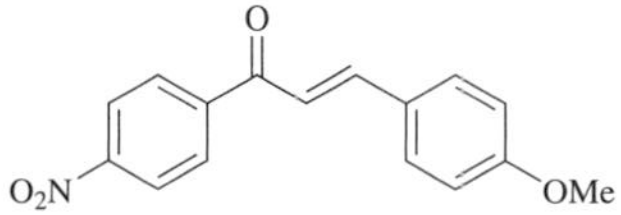	Aza-crown ether **43f** (7 mol %), TBHP, NaOH, toluene, H_2O, rt, 4 h	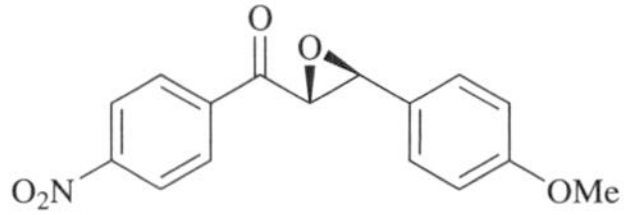	(29), 27	112, 113
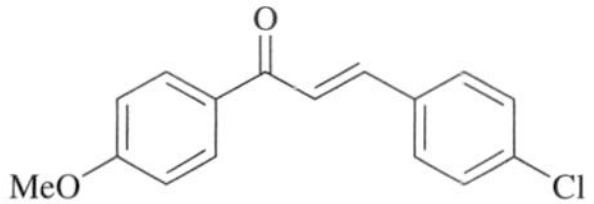	Anthryl cinchonidinium salt **6b** (10 mol %), trichloroisocyanuric acid, KOH, 0°	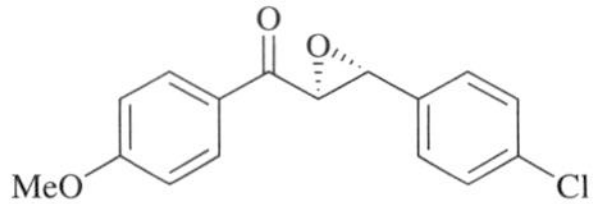		
	Solvent / Time (h)			
	toluene/H_2O / —		(74), 93	101
	toluene / 7		(86), 91	102
C_{17} 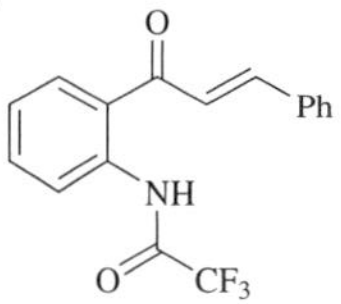	(L-Leu)$_n$–PS (n_{av} = 32), urea/H_2O_2, DBU, THF, rt, 3.5 h	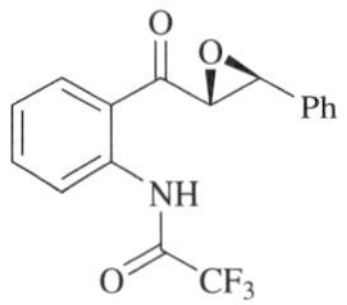	(59), 91	183

TABLE 1D. DIARYL *E*-ENONES (*Continued*)

Substrate	Conditions	Product(s), Yield(s) (%), % ee	Refs.
C_{17} 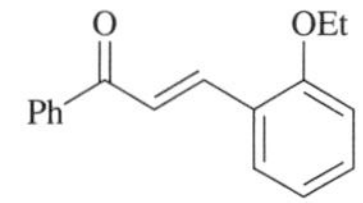	(L-Leu)$_n$–PS (n_{av} = 33), H_2O_2, NaOH, toluene, H_2O, 0° to rt, 2 d	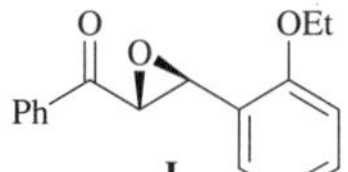 I (56), 83	123
	(L-Leu)*n*–SiO_2 (n_{av} = 30; 6 mol %), Na_2CO_3•1.5 H_2O_2, DME, H_2O, rt, 2 h	**I** (50), 73[m]	125
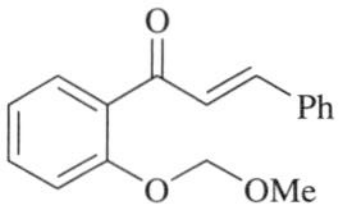	(*S*)-BINOL **17b** (5 mol %), CMHP, Ln(O*i*-Pr)$_3$, THF, rt, 8 h	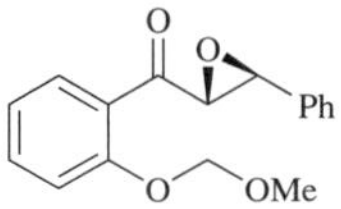 **I**	
	Ln: Yb	(88), 47	47
	Ln: Gd	(85), 73	48
	Benzyl quinidinium salt **2a** (12 mol %), H_2O_2, NaOH, H_2O, toluene, rt, 3 d	**I** (41), 25	93
	(L-Ala)$_n$, H_2O_2, NaOH, CCl_4, H_2O, rt, 48 h	**I** (65), 38	182
	(*R*)-BINOL (x mol %), oxidant, La(O*i*-Pr)$_3$, 4 Å MS, THF, rt	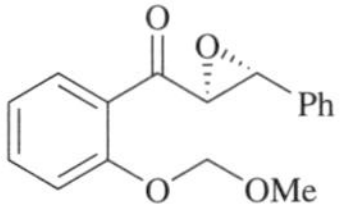 **I**	

(*R*)-BINOL	x	Oxidant	Additive	Time (h)	Yield (%), % ee	Refs.
15a	5	CMHP	—	20	(85), 85	46
15d	6.25	CMHP	—	96	(78), 83	46
15a	5	TBHP	TPAO	4	(91), 95	55

Substrate	Conditions	Product(s), Yield(s) (%), % ee	Refs.
	Benzyl quininium salt **4a** (12 mol %), H_2O_2, NaOH, H_2O, toluene, rt, 22 h	**I** (38), 26	93

		(D-Ala)$_n$, H_2O_2, NaOH, CCl_4, H_2O, rt, 48 h	**I** (57), 53		182
	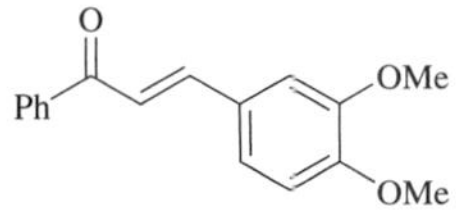	(*R*)-BINOL **15a** (5 mol %), CMHP, La(O*i*-Pr)$_3$, (4-FC_6H_4)$_3$P=O, 4 Å MS, THF, rt, 20 min	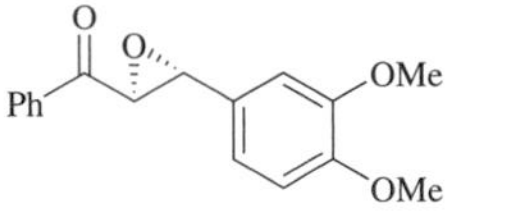	(89), >99	54
	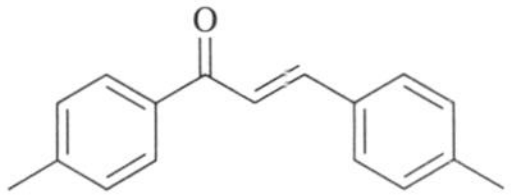	Aza-crown ether **43f** (7 mol %), TBHP, NaOH, toluene, H_2O, rt, 3 h	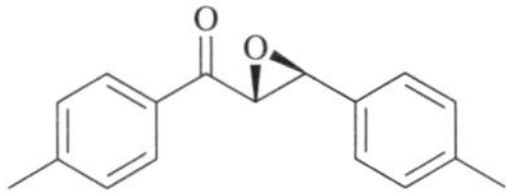	(64), 76	112, 113
		Aza-crown ether **45c** (7 mol %), TBHP, NaOH, toluene, H_2O, rt, 5 h	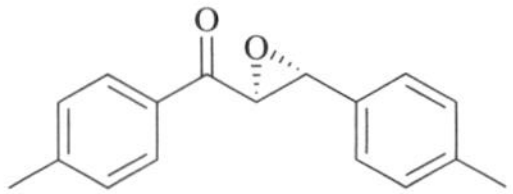	(40), 50	111
C_{18}	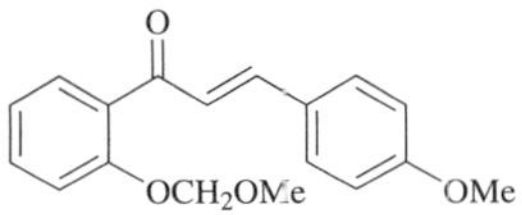	(L-Ala)$_n$, H_2O_2, NaOH, CCl_4, H_2O, rt, 24 h	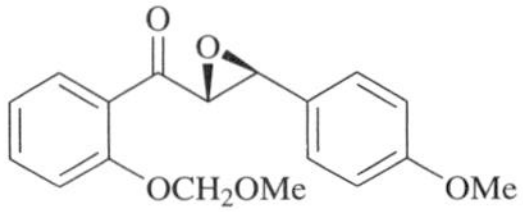 **I**	(99), 84	275, 276, 182
		(L-Leu)$_n$–PS, urea/H_2O_2, DBU, THF, rt, 48 h	**I** (71), 85		277, 278
		(D-Ala)$_n$, H_2O_2, NaOH, CCl_4, H_2O, rt, 24 h	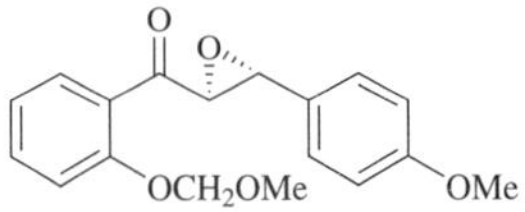	(98), 69	275, 276, 182
		(D-Leu)$_n$–PS, urea/H_2O_2, DBU, THF, rt, 48 h	**I** (69), 81		277, 278

TABLE 1D. DIARYL *E*-ENONES (*Continued*)

Substrate	Conditions	Product(s), Yield(s) (%), % ee	Refs.
C_{19} 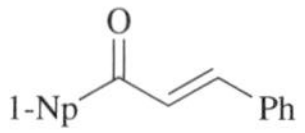	Anthryl cinchoninium salt **1l** (10 mol %), NaOCl, H_2O, toluene, rt, 48 h	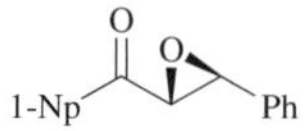 (75), 69	100
	Anthryl cinchonidinium salt **6b** (10 mol %), NaOCl, H_2O, toluene, rt, 48 h	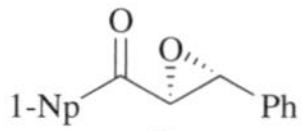 **I** (77), 71	100, 99
	Cinchonidinium salt **5a** (5 mol %), H_2O_2, NaOH, *n*-Bu_2O, H_2O, 0°, 48 h	**I** (72), 58[j]	107
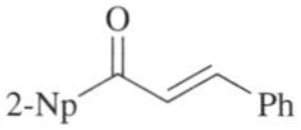	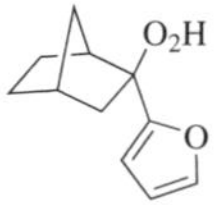 (110 mol %), *n*-BuLi, THF, –20°, 4 h	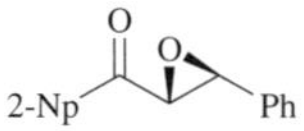 **I** (60), 40	32
	(*S*)-BINOL **17b** (10 mol %), TBHP, KO*t*-Bu, $CaCl_2$, 4 Å MS, toluene, cyclohexane, 0°, 58 h	**I** (72), 73	79
	Benzyl quininium salt **4a** (1.7 mol %), H_2O_2, NaOH, H_2O, toluene, rt, 24 h	**I** (—), —	81
	(L-Leu)$_n$, H_2O_2, NaOH, CH_2Cl_2, H_2O, rt, 30 h	**I** (90), 93	185, 186
	(L-Leu)$_n$, $NaBO_3$•4 H_2O, NaOH, Aliquat 336, CH_2Cl_2, H_2O, rt, 24 h	**I** (98), 88	186
	(L-Leu)$_n$, Na_2CO_3•1.5 H_2O_2, Aliquat 336, CH_2Cl_2, H_2O, rt, 63 h	**I** (87), 84	186

(L-Leu)$_n$, TBHP, NaOH, hexane, H_2O, rt	**I** (71), 47		186
(L-Leu)$_n$–PS (n_{av} = 32), urea/H_2O_2, DBU, THF, rt, 0.5 h	**I** (91), 91		189
(L-Leu)$_n$ (M_r 3000–15000), H_2O_2, NaOH, CCl_4, H_2O, rt, 2 d	**I** (83), 92		267
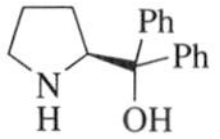 (30 mol %), TBHP, hexane, rt, 133 h	**I** (27), 64		170
PolyBINOL **28** (111 mol %), O_2, Et_2Zn, CH_2Cl_2, 0°	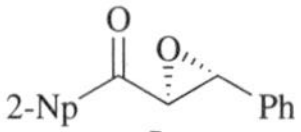O O 2-Np Ph **I**	(91), 47	43
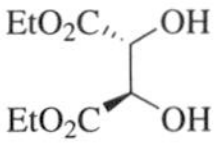EtO_2C OH EtO_2C OH (11 mol %), TBHP, Bu_2Mg, toluene, THF, rt, 1 d	**I** (46), 92		69
Bis(quininium) salt **14d** (1 mol %), H_2O_2, KOH, Span 20, *i*-Pr_2O, H_2O, rt, 6 h	**I** (96), >99		108
(D-Leu)$_n$, H_2O_2, NaOH, CH_2Cl_2, H_2O, rt, 38 h	**I** (67), 93		185, 186
(D-Leu)$_n$, $NaBO_3$•4 H_2O, NaOH, Aliquat 336, CH_2Cl_2, H_2O, rt, 22 h	**I** (96), 90		186

TABLE 1D. DIARYL *E*-ENONES (*Continued*)

Substrate	Conditions	Product(s), Yield(s) (%), % ee	Refs.
C_{19} 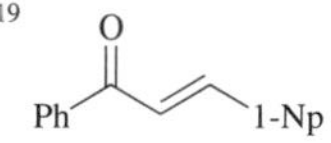	(*S*)-BINOL (5 mol %), CMHP, La(O*i*-Pr)$_3$, THF, rt, 8 h	 **I**	
	(*S*)-BINOL		
	17d	(89), 84	48
	17b	(88), 31	264
	(*S*)-BINOL **17b** (10 mol %), TBHP, KO*t*-Bu, $CaCl_2$, 4 Å MS, toluene, cyclohexane, 0°, 52 h	**I** (76), 25	79
	Anthryl cinchoninium salt **1l** (10 mol %), NaOCl, H_2O, toluene, rt, 48 h	**I** (92), 82	100
	Cinchonidinium salt **5a** (5 mol %), H_2O_2, NaOH, *n*-Bu_2O, H_2O, 0°, 48 h	**I** (46), 58[j]	107
	Anthryl cinchonidinium salt **6b** (10 mol %), NaOCl, H_2O, toluene, rt, 48 h	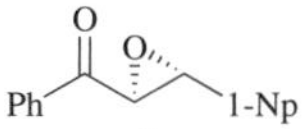 **I** (86), 82	100, 99
	Aza-crown ether **45c** (7 mol %), TBHP, NaOH, toluene, H_2O, 0–4°, 11 h	**I** (49), 42	111
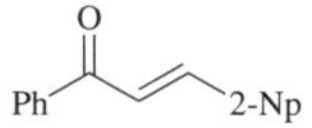	PolyBINOL **28** (111 mol %), O_2, Et_2Zn, CH_2Cl_2, 0°	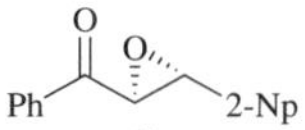 **I** (75), 54	43
	Anthryl cinchonidinium salt **6b** (10 mol %), KOCl, H_2O, toluene, –40°, 12 h	**I** (87), 93	98
	Cinchonidinium salt **5a** (5 mol %), H_2O_2, NaOH, *n*-Bu_2O, H_2O, 0°, 48 h	**I** (85), 54[k]	107

	Ammonium salt **50e** (3 mol %), NaOCl, toluene, H_2O, 0°, 24 h	**I** (99), 97	109
	PolyBINOL **33** (~11 mol %), O_2, Et_2Zn, CH_2Cl_2, 0°	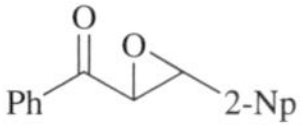 (95), 33	43
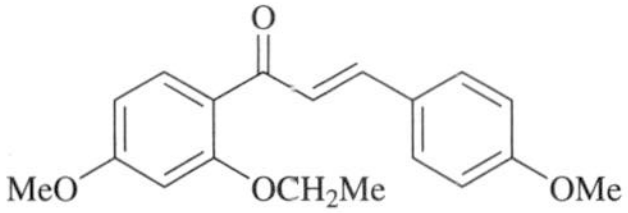	(L-Ala)$_n$, H_2O_2, NaOH, CCl_4, H_2O, rt, 24 h	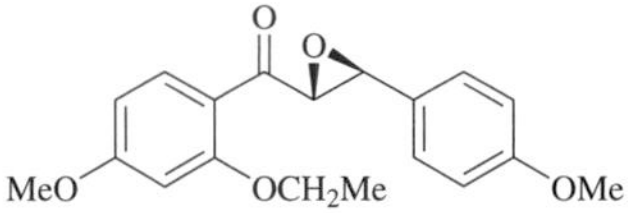 **I** (98), 86	275, 276, 182
	(L-Leu)$_n$–PS, urea/H_2O_2, DBU, THF, rt, 48 h	**I** (80), 95	277, 278
	(D-Ala)$_n$, H_2O_2, NaOH, CCl_4, H_2O, rt, 24 h	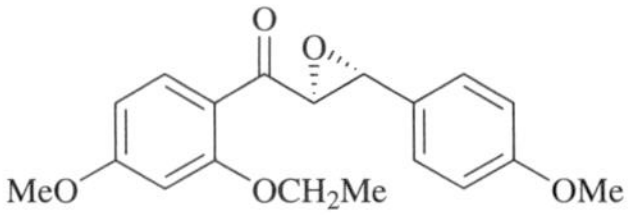 **I** (98), 74	275, 276, 182
	(D-Leu)$_n$–PS, urea/H_2O_2, DBU, THF, rt, 48 h	**I** (76), 90	277, 278
C_{20} 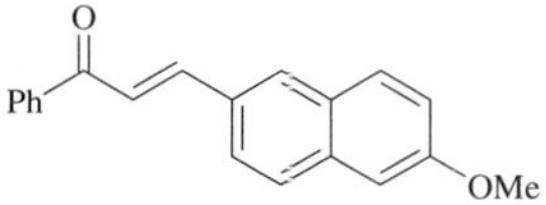	(L-Leu)$_n$–PS•SiO_2 (n_{av} = 32), urea/H_2O_2, DBU, THF, rt, 1.5 h	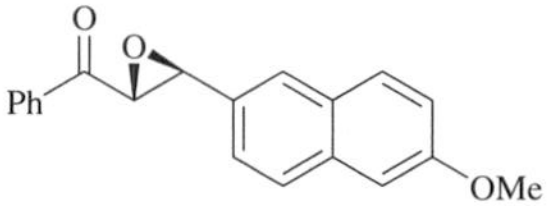 (97), 94	175
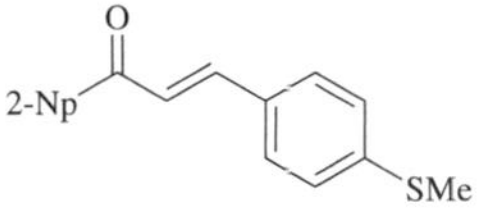	(L-Leu)$_n$, H_2O_2, NaOH, CH_2Cl_2, H_2O, rt	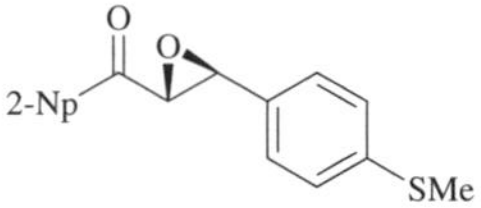 (65), 96 + 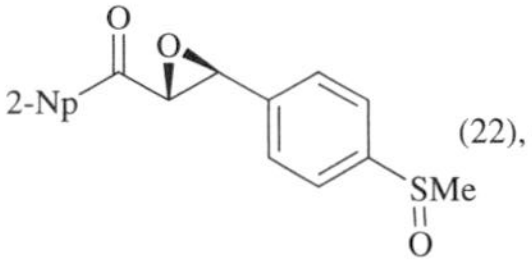 (22), —	185

TABLE 1D. DIARYL *E*-ENONES (*Continued*)

Substrate	Conditions	Product(s), Yield(s) (%), % ee	Refs.
C_{20} 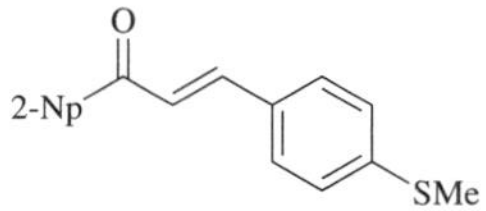O 2-Np SMe	(D-Leu)$_n$, H_2O_2, NaOH, CH_2Cl_2, H_2O, rt, 40 h	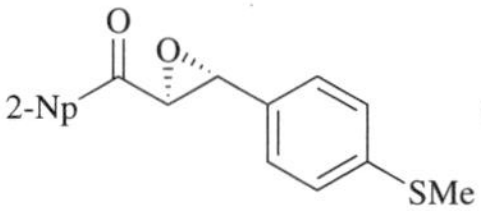O 2-Np SMe (82), 96	185
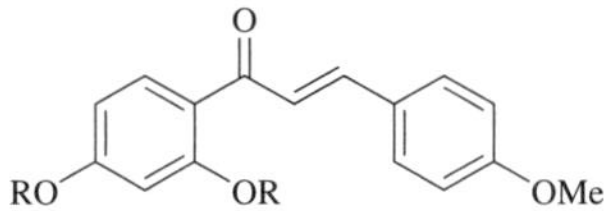O RO OR OMe R = CH_2OMe	(L-Ala)$_n$, H_2O_2, NaOH, CCl_4, H_2O, rt, 72 h	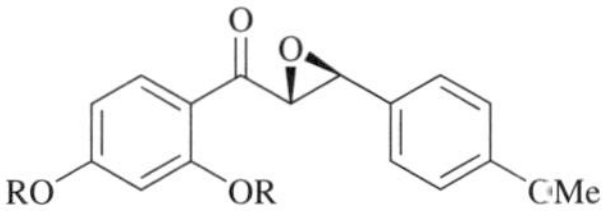O RO OR CMe (—), 70	279
	(D-Ala)$_n$, H_2O_2, NaOH, CCl_4, H_2O, rt, 72 h	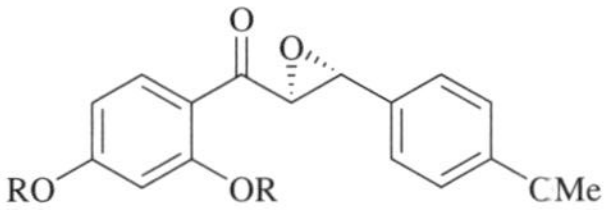O RO OR CMe (—), 36	279
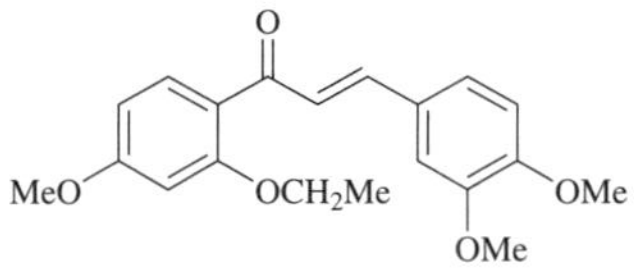O MeO OCH2Me OMe OMe	(L-Ala)$_n$, H_2O_2, NaOH, CCl_4, H_2O, rt	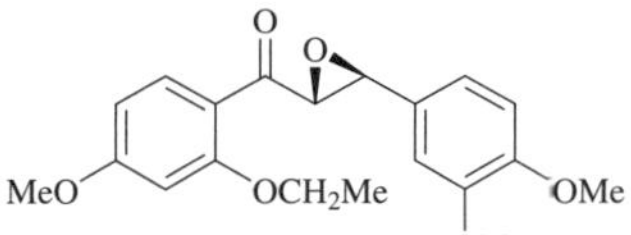O MeO OCH2Me OMe OMe **I** (99), 67	276, 182
	(L-Leu)$_n$–PS, urea/H_2O_2, DBU, THF, rt, 48 h	**I** (64), 8	277, 278
	(D-Ala)$_n$, H_2O_2, NaOH, CCl_4, H_2O, rt	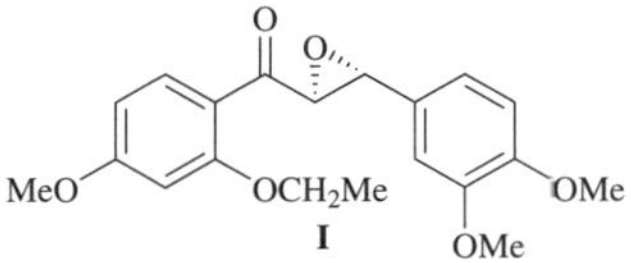O MeO OCH2Me OMe OMe **I** (98), 58	276, 182
	(D-Leu)$_n$–PS, urea/H_2O_2, DBU, THF, rt, 48 h	**I** (61), 87	277, 278

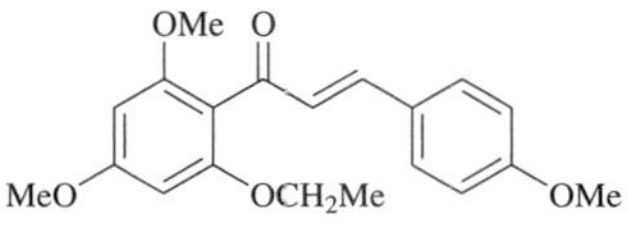	(L-Ala)$_n$, H_2O_2, NaOH, CCl_4, H_2O, rt, 24 h	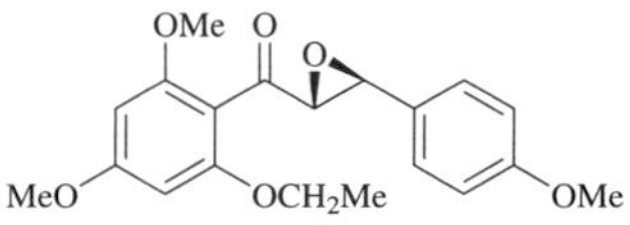 **I** (97), 70	275, 276, 182
	(L-Leu)$_n$–PS, urea/H_2O_2, DBU, THF, rt, 96 h	**I** (36), 60	278
	(D-Ala)$_n$, H_2O_2, NaOH, CCl_4, H_2O, rt, 24 h	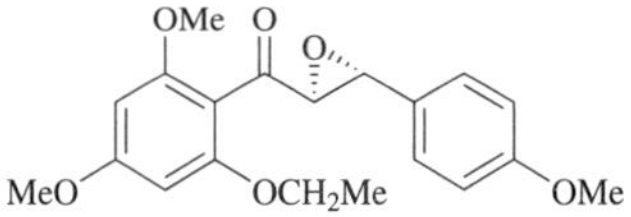 **I** (97), 53	275, 276, 182
	(L-Leu)$_n$–PS, urea/H_2O_2, DBU, THF, rt, 96 h	**I** (33), 61	278
C_{21}			
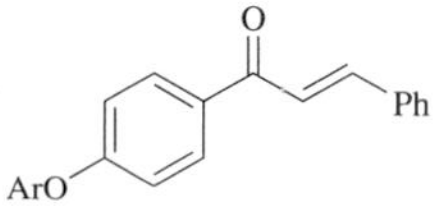	Anthryl cinchonidinium salt **6b** (10 mol %), KOCl, H_2O, toluene, –40°, 12 h	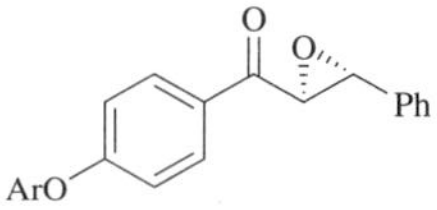 Ar: Ph (89), 93; 2,4-$Br_2C_6H_3$ (90), 98	98
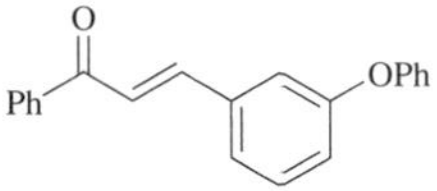	(*S*)-BINOL **17b** (10 mol %), TBHP, KO*t*-Bu, $CaCl_2$, 4 Å MS, toluene, cyclohexane, 0°, 48 h	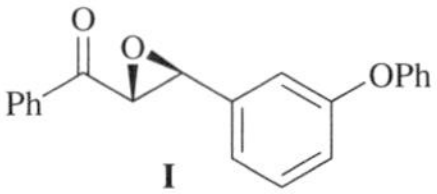 **I** (82), 74	79, 78
	(L-Leu)$_n$–PS•SiO_2 (n_{av} = 32), urea/H_2O_2, DBU, THF, rt, 2 h	**I** (98), 94	270, 175

TABLE 1D. DIARYL *E*-ENONES (*Continued*)

Substrate	Conditions	Product(s), Yield(s) (%), % ee	Refs.
C_{21} 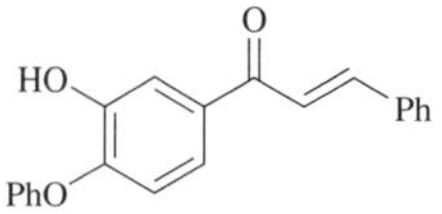	(L-Leu)$_n$–PS, urea/H_2O_2, DBU, THF, rt, 15 h	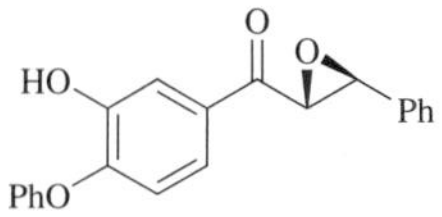 (94), 86	261
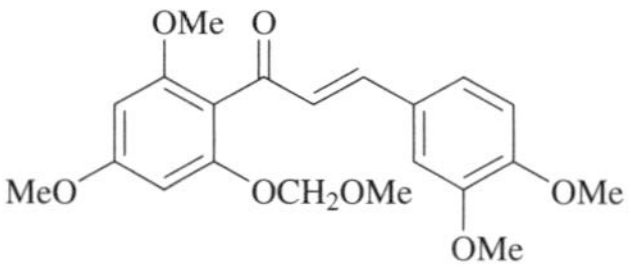	(L-Ala)$_n$, H_2O_2, NaOH, CCl_4, H_2O, rt	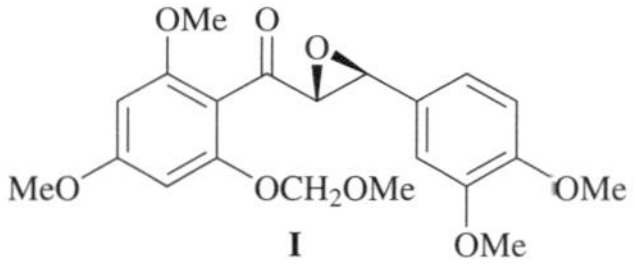 **I** (79), 49	276
	(L-Leu)$_n$–PS, urea/H_2O_2, DBU, THF, rt, 96 h	**I** (21), 53	278
	(D-Ala)$_n$, H_2O_2, NaOH, CCl_4, H_2O, rt	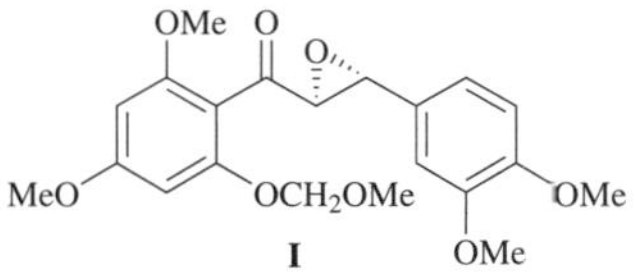 **I** (76), 49	276
	(D-Leu)$_n$–PS, urea/H_2O_2, DBU, THF, rt, 96 h	**I** (19), 50	278
C_{22} 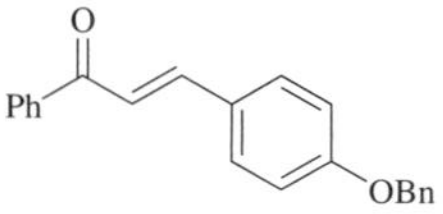	(L-Leu)$_n$ (n_{av} = 40; 0.5 mol %), H_2O_2, NaOH, TBAB, toluene, H_2O, rt, 20 h	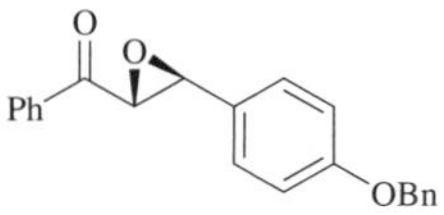 (78), 97.3	254
C_{24} 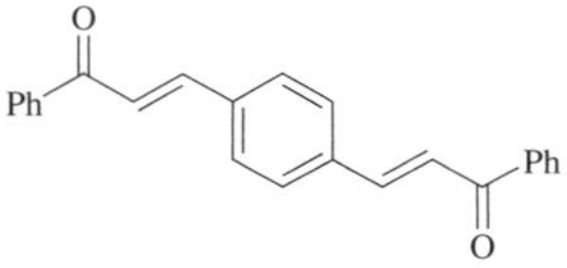	(L-Leu)$_n$, H_2O_2, NaOH, CH_2Cl_2, H_2O, rt, 46 h	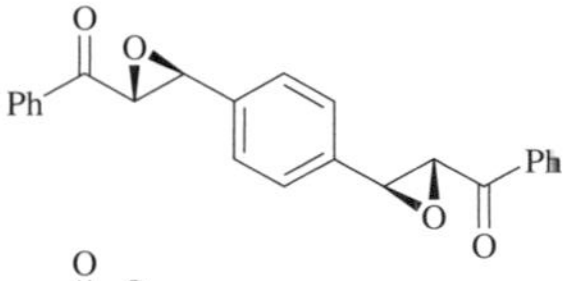 (74), >99[s]	185, 186
	(D-Leu)$_n$, H_2O_2, NaOH, CH_2Cl_2, H_2O, rt, 42 h	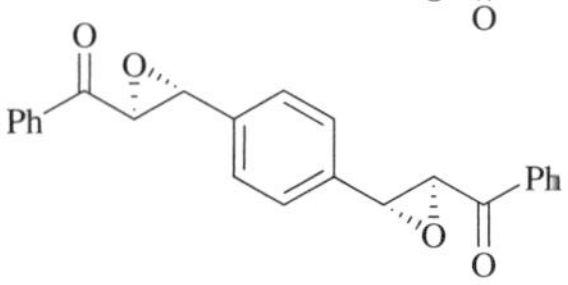 (91), >99[t]	185, 186

C_{27-33}

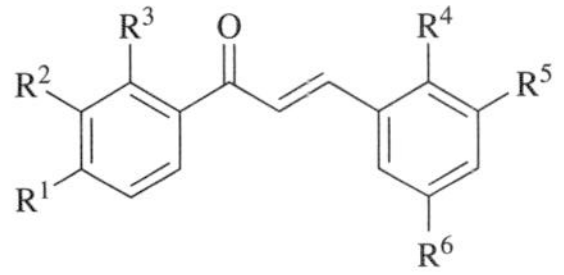

Polypeptide catalyst, oxidant, base, EDTA, solvent, H_2O, rt

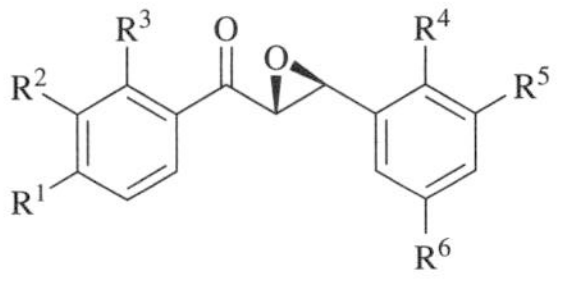

R^1	R^2	R^3	R^4	R^5	R^6	Catalyst	Oxidant	Base	Solvent	Time (h)		
MeO	H	H	n-$C_{11}H_{23}O$	H	NO_2	(L-Ala)$_n$	H_2O_2	KOH	toluene	20	(–), –	280
Cl	H	Cl	4-$CH_3C_6H_4(CH_2)_6$	H	Br	(L-Leu)$_n$	TBHP	KOH	toluene	20	(–), –	280
H	H	H	$Ph(CH_2)_8$	H	H	(L-Ala)$_n$	H_2O_2	NaOH	toluene	44	(–), 80[u]	280
F	H	H	$Ph(CH_2)_8$	H	H	(L-Leu)$_n$	H_2O_2	NaOH	n-hexane	22	(69), >98[u]	280
Br	H	H	$Ph(CH_2)_8$	H	H	(L-Leu)$_n$	H_2O_2	NaOH	n-hexane	48	(70), 99[v]	281, 280
Br	H	H	$Ph(CH_2)_8$	H	H	(L-Leu)$_n$	H_2O_2	NaOH	CCl_4	16	(76), >97[u]	280
Me	H	Me	2-furyl-$(CH_2)_8$	H	H	(L-Ala)$_n$	H_2O_2	NaOH	CCl_4	20	(–), –	280
Me	H	Me	$CF_3(CH_2)_{11}$	H	H	(L-Ala)$_n$	H_2O_2	NaOH	n-hexane	20	(–), –	280
Me	H	H	$Ph(CH_2)_8$	H	H	(L-Leu)$_n$	H_2O_2	NaOH	n-hexane	40	(76), 94[u]	17, 280
MeO	H	H	$Ph(CH_2)_8$	H	H	(L-Leu)$_n$	H_2O_2	NaOH	n-hexane	48	(60), 97	281
H	H	Br	3-$CF_3C_6H_4(CH_2)_8$	H	H	(L-Leu)$_n$	H_2O_2	KOH	n-hexane	20	(–), –	280
Cl	H	Cl	$Ph(CH_2)_8$	H	CF_3	(L-Ala)$_n$	TBHP	KOH	n-hexane	20	(–), –	280
H	CF_3	H	4-$MeOC_6H_4(CH_2)_8$	H	H	(L-Ala)$_n$	H_2O_2	KOH	n-hexane	20	(–), –	280
Et	H	H	$2,4$-$(CH_3)_2C_6H_3(CH_2)_{10}$	NO_2	H	(L-Ala)$_n$	H_2O_2	NaOH	n-hexane	20	(–), –	280

TABLE 1D. DIARYL *E*-ENONES (*Continued*)

Substrate	Conditions	Product(s), Yield(s) (%), % ee	Refs.
C_{28-33} 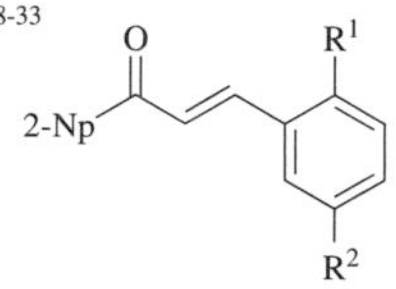	Polypeptide catalyst, H_2O_2, base, EDTA, solvent, H_2O, rt, 20 h	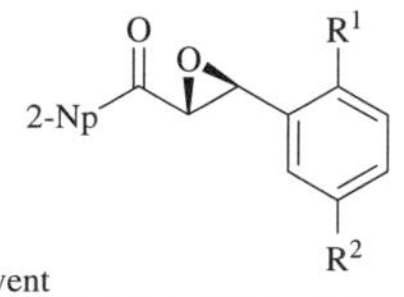	

R^1	R^2	Catalyst	Base	Solvent	Yield, % ee	Refs.
n-$C_9H_{19}O$	H	(L-Ala)$_n$	LiOH	toluene	(–), –	280
n-$C_{11}H_{23}O$	H	(L-Ala)$_n$	LiOH	*n*-hexane	(–), –	280
n-$C_{12}H_{25}$	H	(L-Ala)$_n$	KOH	cyclohexane	(–), –	280
H	$Ph(CH_2)_6$	(L-Ala)$_n$	KOH	*n*-hexane	(–), –	280
$Ph(CH_2)_8$	H	(L-Leu)$_n$	NaOH	*n*-hexane	(82), 96–97	281, 280
c-$C_6H_{11}(CH_2)_8$	H	(L-Ala)$_n$	KOH	*n*-hexane	(–), –	280
4-$ClC_6H_4(CH_2)_7$	H	(L-Leu)$_n$	KOH	petroleum	(–), –	280

Substrate	Conditions	Product(s), Yield(s) (%), % ee	Refs.
C_{29-32} 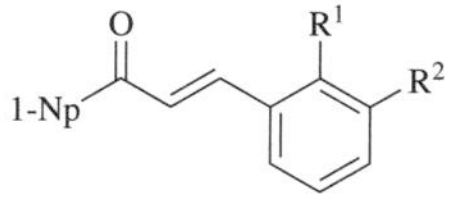	Polypeptide catalyst, H_2O_2, base, EDTA, solvent, H_2O, rt, 20 h	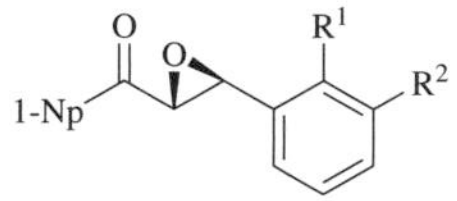	280

R^1	R^2	Catalyst	Base	Solvent	Yield, % ee
$PhO(CH_2)_4$	H	(L-Ala)$_n$	KOH	cyclohexane	(–), –
2-thienyl-$(CH_2)_8$	CF_3	(L-Leu)$_n$	NaOH	*n*-hexane	(–), –

Substrate	Conditions	Product(s), Yield(s) (%), % ee	Refs.
C_{32} 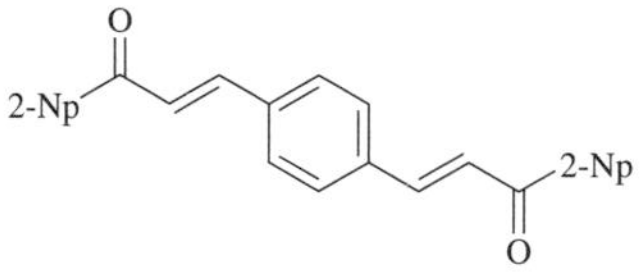	(L-Leu)$_n$, $NaBO_3$•4 H_2O, NaOH, Aliquat 336, CH_2Cl_2, H_2O, rt, 120 h	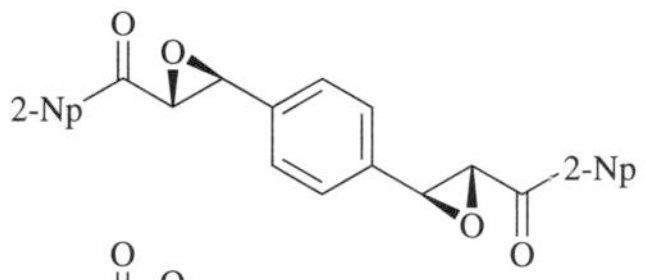 (82), >98[w]	186
	(D-Leu)$_n$, $NaBO_3$•4 H_2O, NaOH, Aliquat 336, CH_2Cl_2, H_2O, rt, 144 h	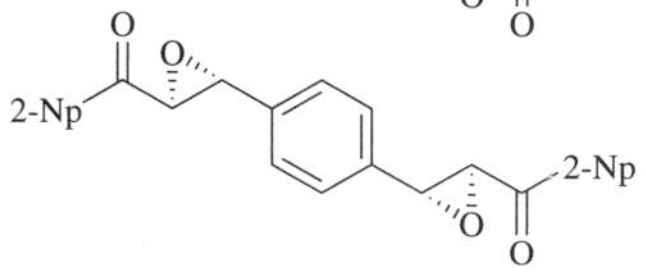 (64), >98[w]	186

C_{33} 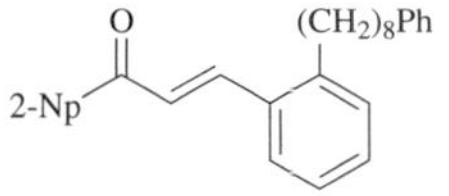	$(D\text{-}Leu)_n$, H_2O_2, NaOH, EDTA, *n*-hexane, H_2O, rt, 198 h	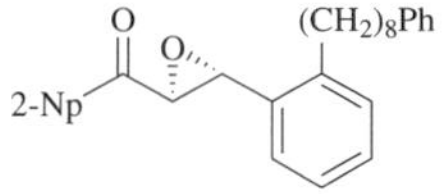	(57), —	280

[a] The absolute configuration and enantiomeric excess were not determined in the original publication, but are assigned by comparison with ref. 282.

[b] The absolute configuration was not determined in the original publication, but is assigned by comparison with ref. 282.

[c] Similar results were obtained when DBU was replaced by DBN in this reaction. Use of diisopropylethylamine gave a slower and less enantioselective reaction.

[d] The reaction was carried out as a flow process, using the catalyst as a fixed bed. The time refers to residence time on the column.

[e] The reaction was carried out in a continuous flow reactor, and the yield and enantiomeric excess quoted are average values over 50 retention times.

[f] These figures are estimated from a graph in the publication.

[g] This paper contains related results from mixed L- and D-leucine polymers, formed through polymerization rather than stepwise synthesis.

[h] This result was obtained using a chiral ligand whose enantiomeric excess was 40%.

[i] The absolute configuration was not determined in the original publication, but is assigned by comparison with ref. 65.

[j] The absolute configuration was not determined in the original publication, but is assigned by comparison with ref. 99.

[k] The absolute configuration and enantiomeric excess were not determined in the original publication, but are assigned by comparison with ref. 98.

[l] This enantiomeric excess is quoted after one recrystallization of the product from ethanol. The enantiomeric excess prior to recrystallization was 76%, and after a second recrystallization was 88%.

[m] The stereoisomer depicted in this publication is the enantiomer of that obtained with all other reported poly-L-leucine epoxidations. The expected stereoisomer is depicted here.

[n] The absolute configuration was not determined in the original publication, but is assigned by comparison with ref. 131.

[o] The absolute configuration was not determined in the original publication, but is assigned by comparison with ref. 123.

[p] The absolute configuration was not determined in the original publication, but is assigned by comparison with ref. 103.

[q] The enantiomeric excess was not determined in the original publication, but is assigned by comparison with ref. 48.

[r] The enantiomeric excess was not determined in the original publication; the reported optical rotation is 144% of that reported in reference 99.

[s] The diastereomeric excess of the product was 88%.

[t] The diastereomeric excess of the product was >88%.

[u] EDTA was not used in this reaction.

[v] This result is quoted after a single recrystallization of the product. The enantiomeric excess prior to recrystallization was 93%.

[w] The diastereomeric excess of the product was >98%.

TABLE 1E. HETEROCYCLIC *E*-ENONES

Substrate	Conditions	Product(s), Yield(s) (%), % ee	Refs.
C_{11} 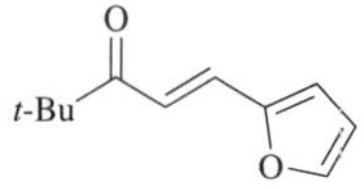	Polypeptide catalyst (n_{av} = 32), urea/H_2O_2, DBU, THF, rt Catalyst — Time (L-Leu)$_n$–PS — 8 h (L-Npg)$_n$–PS — 2 h	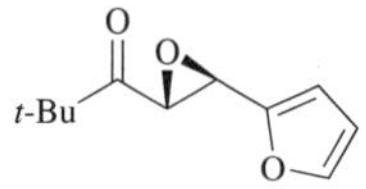 (83), 95 (93), >95	260
C_{12} 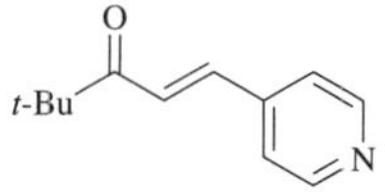	(L-Leu)$_n$ (n_{av} = 30), H_2O_2, NaOH, CH_2Cl_2, H_2O, rt, 30 h	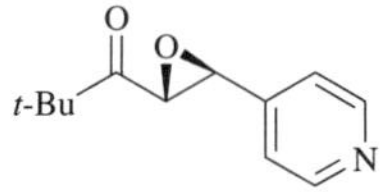 (70), 72	179
C_{13} 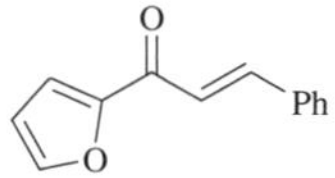	(L-Leu)$_n$, H_2O_2, NaOH, CH_2Cl_2, H_2O, rt, 18 h	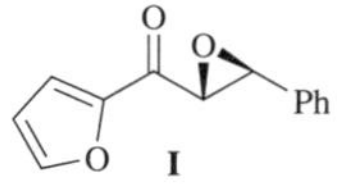 **I** (85), 87	185, 186
	(L-Leu)$_n$, urea/H_2O_2, DBU, THF, rt, 20 min	**I** (>95), >99	283
	(D-Leu)$_n$, H_2O_2, NaOH, CH_2Cl_2, H_2O, rt, 18 h	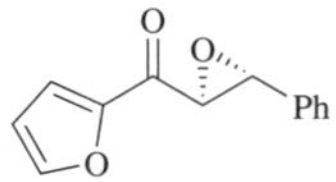 (98), 93	185, 186
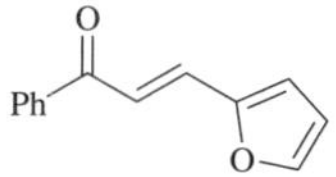	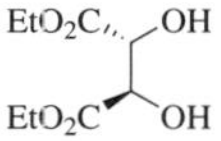 (11 mol %), TBHP, NanoActive MgO Plus, toluene, rt, 36 h	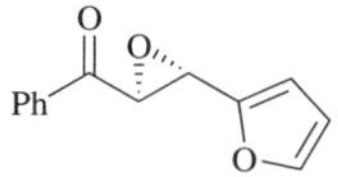 (70), 84	73
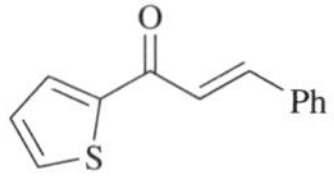	(L-Ala)$_n$ (n_{av} = 10), H_2O_2, NaOH, toluene, H_2O, rt, 48 h	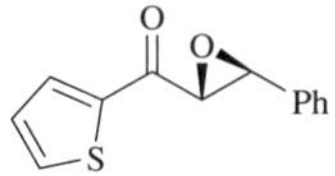 (96), 80	132, 273

		Anthryl cinchonidinium salt **6b** (1 mol %), NaOCl, H_2O, toluene, rt, 2–24 h	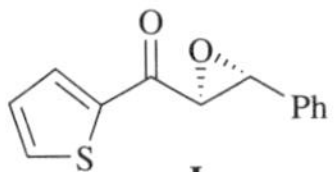 **I**	(97), 84	97, 99
		Bis(quininium) salt **14d** (1 mol %), H_2O_2, KOH, Span 20, *i*-Pr_2O, H_2O, rt, 0.5 h	**I**	(95), 98	108
	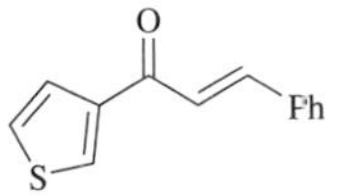	(L-Ala)$_n$ (n_{av} = 10), H_2O_2, NaOH, toluene, H_2O, rt, 48 h	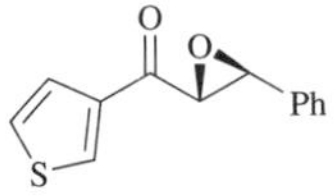	(30), 70	132, 273
	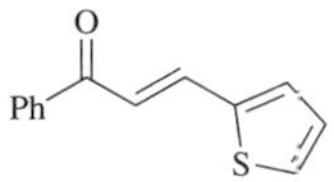	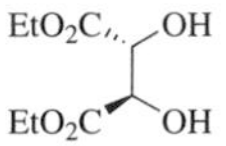 (11 mol %), TBHP, NanoActive MgO Plus, toluene, rt, 36 h	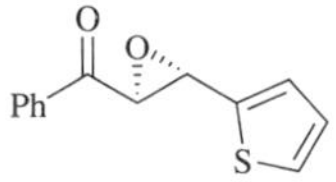	(68), 71	73
C_{14}	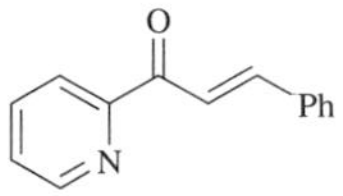	(L-Leu)$_n$, H_2O_2, NaOH, CH_2Cl_2, H_2O, rt, 18 h	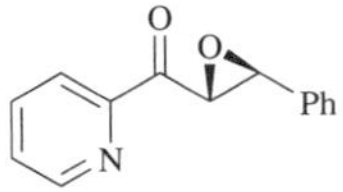	(74), 79	186
		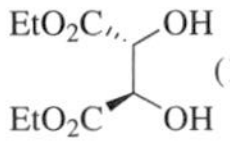 (11 mol %), TBHP, NanoActive MgO Plus, toluene, rt	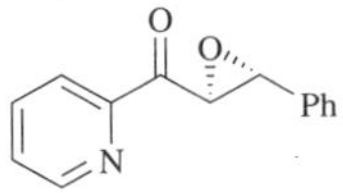	(64), 80	73
	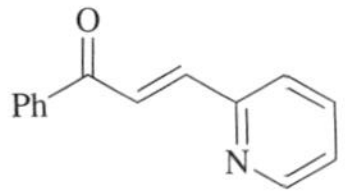	(L-Leu)$_n$, H_2O_2, NaOH, CH_2Cl_2, H_2O, rt, 16 h	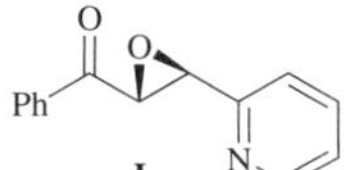 **I**	(84), 72	186
		(L-Leu)$_n$, (n_{av} = 33; 0.5 mol %), H_2O_2, NaOH, TBAB, toluene, H_2O, rt, 30 min	**I**	(>99), 84	120
		Aza-crown ether **45c** (7 mol %), TBHP, NaOH, toluene, H_2O, rt, 5 h	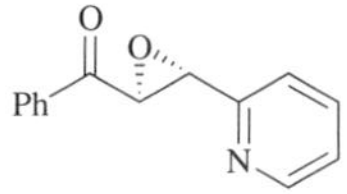	(55), 32	111

TABLE 1E. HETEROCYCLIC *E*-ENONES (*Continued*)

Substrate	Conditions	Product(s), Yield(s) (%), % ee	Refs.
C_{14} 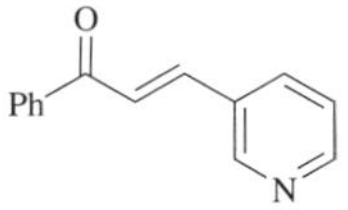	(D-Leu)$_n$, H_2O_2, NaOH, CH_2Cl_2, H_2O, rt, 18 h	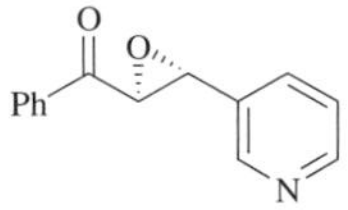 (70), 94	186
C_{15} 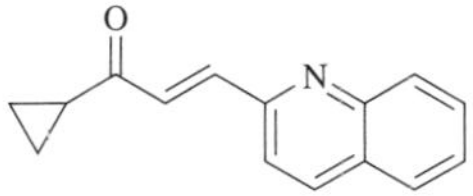	(L-Leu)$_n$, H_2O_2, NaOH, CH_2Cl_2, H_2O, rt, 18 h	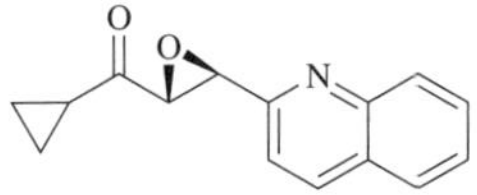 (94), 79	179
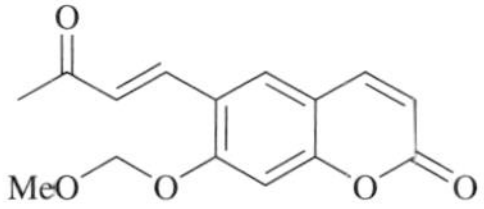	(*R*)-BINOL **15a** (x mol %), TBHP, Ln(O*i*-Pr)$_3$, 4 Å MS, THF, rt	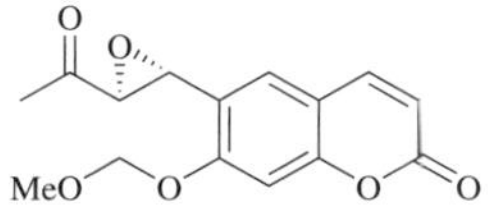	
	x / Ln / Additive / Time (h)		
	25 / Yb / — / 15	(88), 83	284, 55, 285
	25 / La / — / 25	(28), 20	284, 285
	25 / La / TPPO / 2.5	(98), 97	284, 55, 285
	25 / La / TPAO / 2	(94), 96	284, 55, 285
	10 / La / TPAO / 5	(90), 93	284, 55, 285
	7.5 / Yb / H_2O / 48	(65), 85	55
C_{17} 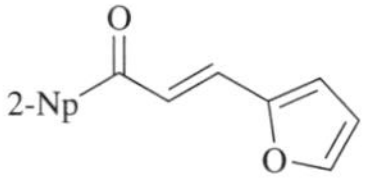	(L-Leu)$_n$, H_2O_2, NaOH, CH_2Cl_2, H_2O, rt, 50 h	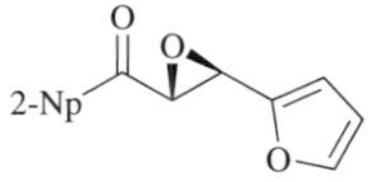 (75), >96	185, 186

C_{13}				
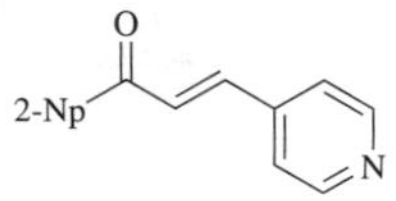	(L-Leu)$_n$, H_2O_2, NaOH, CH_2Cl_2, H_2O, rt, 16 h	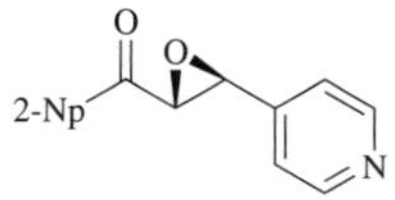	(67), >96	186
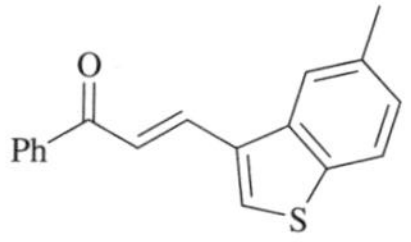	Benzyl quininium salt **4a** (1.7 mol %), H_2O_2, NaOH, toluene, H_2O, rt, 24 h	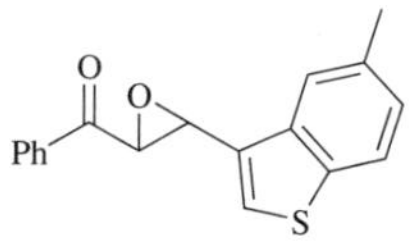	(—), —[a]	81

[a] It cannot be discerned from the original paper whether this enone and epoxide are 2- or 3-substituted benzothiophenes.

TABLE 1F. POLYUNSATURATED *E*-ENONES

Substrate	Conditions	Product(s), Yield(s) (%), % ee	Refs.
C_{11} 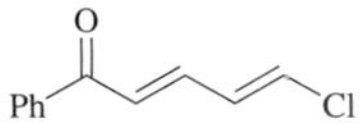	(L-Leu)$_n$–PS (n_{av} = 32), urea/H_2O_2, DBU, THF, rt	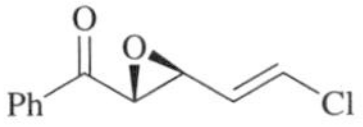 (57), 86	188, 189
C_{13} 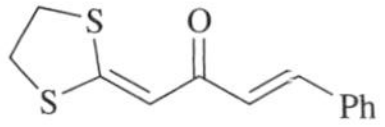	(L-Leu)$_n$, H_2O_2, NaOH, CH_2Cl_2, H_2O, rt, 115 h	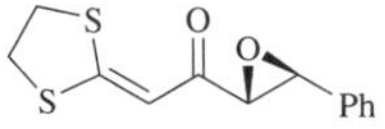 (51), >94	179
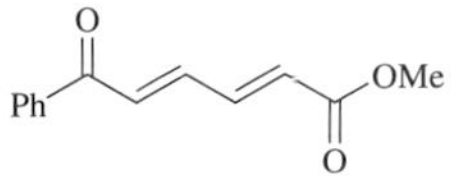	(L-Leu)$_n$–PS (n_{av} = 32), urea/H_2O_2, DBU, THF, rt, 5 h	 **I** (90), 90	188, 189
	(L-Leu)$_n$, (0.1 mol %), H_2O_2, NaOH, Bu_4NHSO_4, toluene, rt, 3 h	**I** (60), >95	190, 201
	(D-Leu)$_n$–PS, urea/H_2O_2, DBU, THF, rt, 5 h	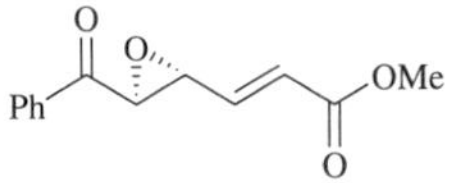 (91), 95	188
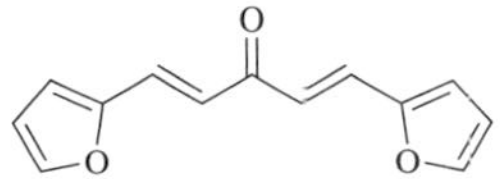	(L-Leu)$_n$, H_2O_2, NaOH, CH_2Cl_2, H_2O, rt, 60 h	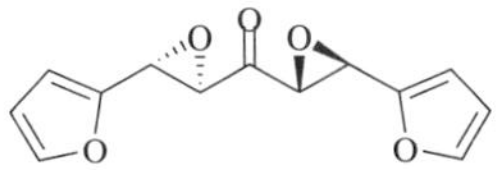 (60), 90	186
C_{14} 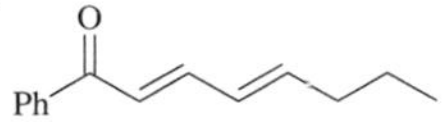	(*R*)-BINOL **15a** (5 mol %), TBHP, La(O*i*-Pr)$_3$, TPAO, 4 Å MS, THF, rt, 3 h	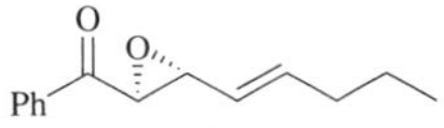 **I** (95), 96	55
	(*R*)-BINOL **15a** (5 mol %), TBHP, Yb(O*i*-Pr)$_3$, 4 Å MS, THF, rt, 2 h	**I** (87), 93	55
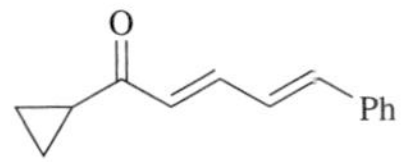	(L-Leu)$_n$, H_2O_2, NaOH, CH_2Cl_2, H_2O, rt, 42 h	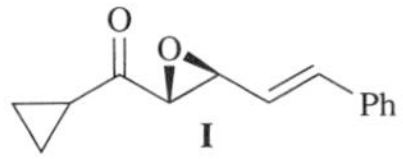 **I** (73), 74	178, 179, 259
	(L-Leu)$_n$, $NaBO_3$•4 H_2O, NaOH, CH_2Cl_2, H_2O, Aliquat 336, rt, 19 h	**I** (52), 98	178, 179

C_{15}				
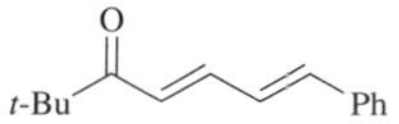	(L-Leu)$_n$, H_2O_2, NaOH, CH_2Cl_2, H_2O, rt, 30 h		(90), >97[a]	186
C_{16}				
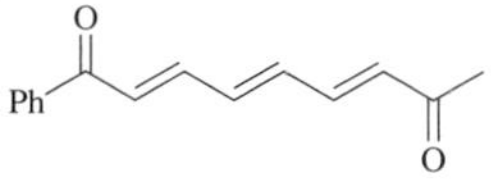	(L-Leu)$_n$–PS (n_{av} = 32), urea/H_2O_2, DBU, THF, rt	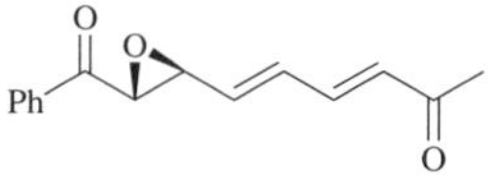	(43), 90	189
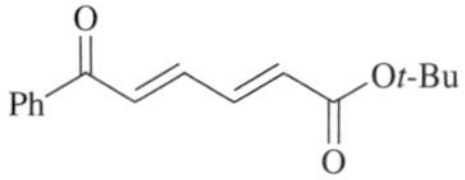	(L-Leu)$_n$–PS (n_{av} = 32), urea/H_2O_2, DBU, THF, rt, 5 h	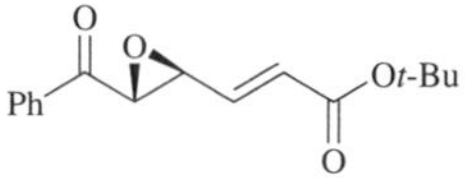	(95), 90	188, 189
C_{17}				
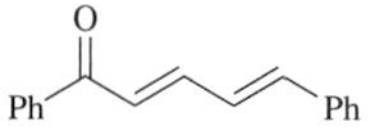	Benzyl quininium salt **4a**, NaOCl, toluene, rt	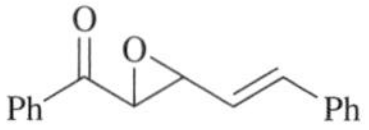	(78), —	83
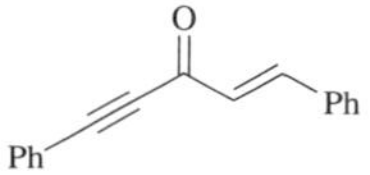	(L-Leu)$_n$–PS (n_{av} = 32), H_2O_2, NaOH, toluene, H_2O, rt, 96 h	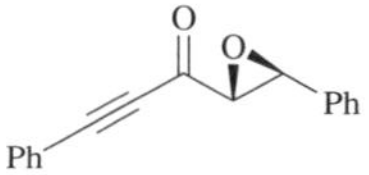	(57), 90	178, 179
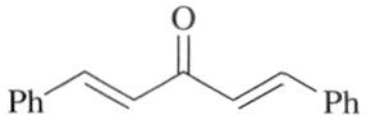	(L-Leu)$_n$, H_2O_2, NaOH, CH_2Cl_2, H_2O, rt	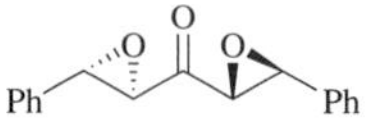 + 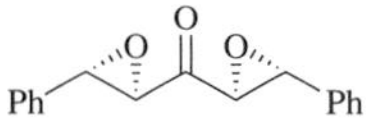 **I** (—), >98 **I**:**II** = 8:1 **II** (—), —		185
	(L-Leu)$_n$–PS•SiO_2 (n_{av} = 32), urea/H_2O_2, DBU, TBME, –10°, 1 h	**I** (90), 99, **II** (—), —; **I**:**II** = 39:1		175
	(L-Leu)$_n$ (0.1 mol %), H_2O_2, NaOH, Bu_4NHSO_4, toluene, rt, 3 h	**I** (51), >95		190, 201
	(D-Leu)$_n$, H_2O_2, NaOH, CH_2Cl_2, H_2O, rt	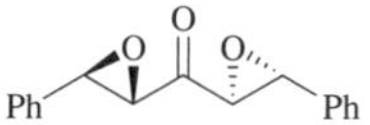 + 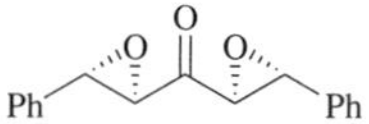 **I** (—), >96 **I**:**II** = 7:2 **II** (—), —		185

TABLE 1F. POLYUNSATURATED *E*-ENONES (*Continued*)

Substrate	Conditions	Product(s), Yield(s) (%), % ee	Refs.
C_{19} 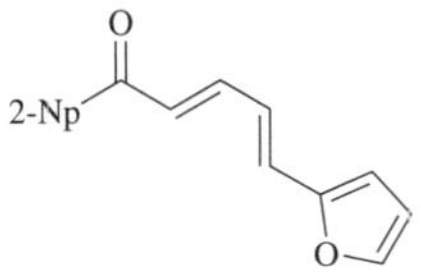	(L-Leu)$_n$–PS (n_{av} = 32), urea/H_2O_2, DBU, THF, rt	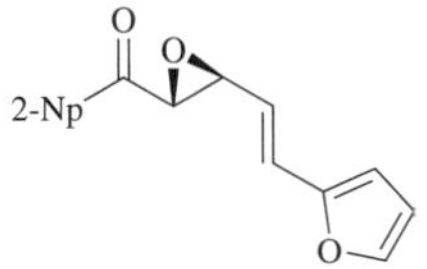 (79), 95	189
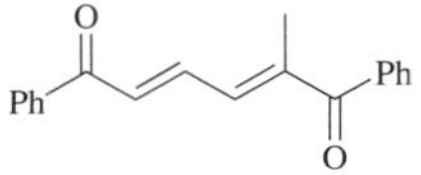	(L-Leu)$_n$–PS (n_{av} = 32), urea/H_2O_2, DBU, THF, rt	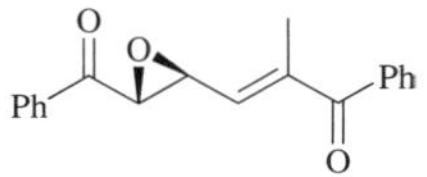 (70), 92	188, 189
C_{21} 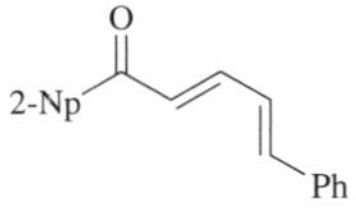	(L-Leu)$_n$, H_2O_2, NaOH, hexane, H_2O, rt, 72 h	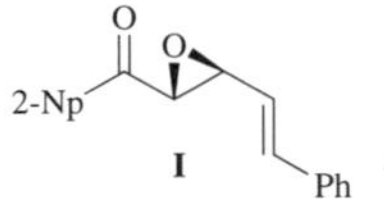 **I** (78), >95	185, 186
	(L-Leu)$_n$, $NaBO_3$•4 H_2O, NaOH, Aliquat 336, CH_2Cl_2, H_2O, H_2O, rt, 120 h	**I** (77), 96	186
	(L-Leu)$_n$–PS (n_{av} = 32), urea/H_2O_2, DBU, THF, rt, 2 h	**I** (85), >98	176, 188, 189, 116
	(D-Leu)$_n$, H_2O_2, NaOH, hexane, H_2O, rt, 74 h	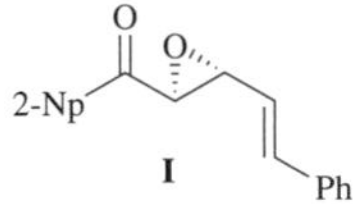 **I** (76), >96	185, 186
	(D-Leu)$_n$–PS (n_{av} = 32), urea/H_2O_2, DBU, THF, rt	**I** (80), 90	188
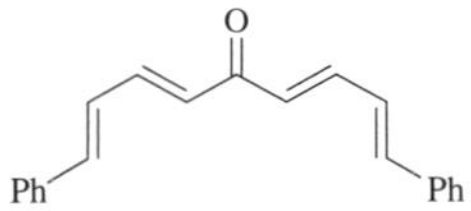	(L-Leu)$_n$, H_2O_2, NaOH, CH_2Cl_2, H_2O, rt	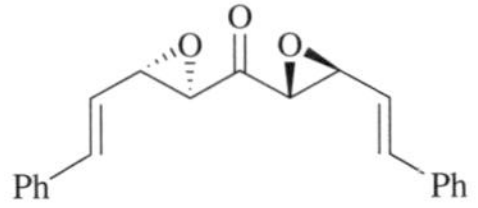 (50), 80	186

[a] This result is quoted after a single recrystallization of the product.

TABLE 1G. HETERO-SUBSTITUTED *E*-ENONES

Substrate	Conditions	Product(s), Yield(s) (%), % ee	Refs.
C_{21}			
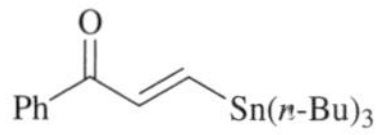	(L-Leu)$_n$, H_2O_2, NaOH, hexane, H_2O	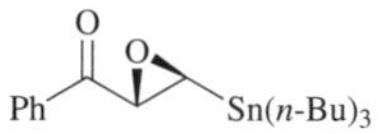 (90), >99	191

TABLE 1H. Z-ENONES

Substrate	Conditions	Product(s), Yield(s) (%), % ee	Refs.
C_{9-14} 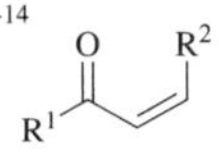	(*R*)-BINOL (x mol %), TBHP, Ln(O*i*-Pr)$_3$, 4 Å MS, THF, rt	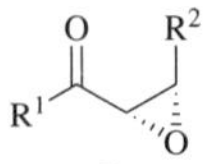 **I** + 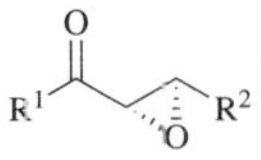 **II**	49

R^1	R^2	(*R*)-BINOL	x	Ln	Time (h)	**I**	**II**
Me	*n*-C_5H_{11}	**15a**	5	La	72	(31), 5	(—), —
Me	*n*-C_5H_{11}	**15d**	14	La	72	(58), 58	(—), —
Me	*n*-C_5H_{11}	**15d**	14	Yb	72	(74), 94	(—), —
Ph	Me	**15d**	14	Yb	81	(60), 82	(32), 10
n-Pr	*n*-C_5H_{11}	**15d**	14	Yb	127	(80), 96	(—), —
Ph	*n*-Pr	**15d**	14	Yb	96	(51), 58	(19), 58
Ph(CH_2)$_2$	*n*-Pr	**15d**	14	Yb	146	(78), 93	(—), —

(*R*)-BINOL **15a** (x mol %), TBHP, Ln(O*i*-Pr)$_3$, 4 Å MS, THF, rt

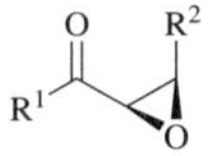

R^1	R^2	x	Ln	Time (h)		Refs.
Me	*n*-C_5H_{11}	5	Yb	72	(60), 4	49
Me	*n*-C_5H_{11}	5	La	20	(61), 59	55
n-Pr	*n*-C_5H_{11}	10	Yb	127	(75), 27	49
Ph(CH_2)$_2$	*n*-Pr	10	Yb	146	(56), 21	49

Substrate	Conditions	Product(s), Yield(s) (%), % ee	Refs.
C_{11} 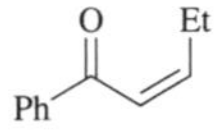	Naphthyl quinidinium salt **2e** (20 mol %), H_2O_2, LiOH, H_2O, $CHCl_3$, –10°, 44 h	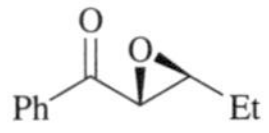 (61), 29	106

TABLE 2. ENONES BEARING A SECOND ELECTRON–WITHDRAWING SUBSTITUENT

Substrate	Conditions	Product(s), Yield(s) (%), % ee	Refs.
C_{12} 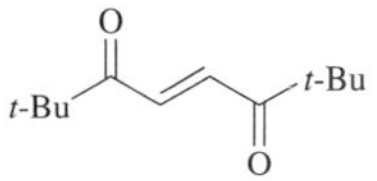	(L-Leu)$_n$–PS (n_{av} = 32), H_2O_2, NaOH, toluene, H_2O, rt	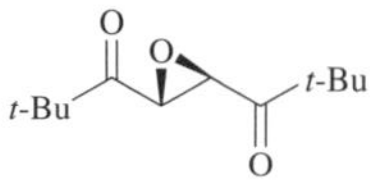 (100), >95	179, 178
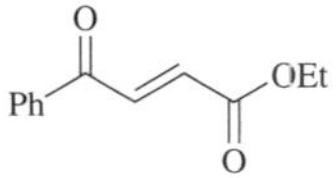	(L-Leu)$_n$ (0.1 mol %), H_2O_2, NaOH, Bu_4NHSO_4, toluene, rt, 3 h	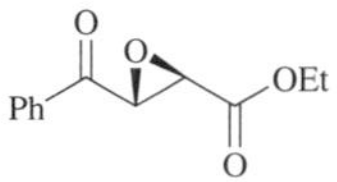 (51), >95	190, 201
	(*R*)-BINOL **15a** (5 mol %), CMHP, La(O*i*-Pr)$_3$, (4-FC_6H_4)$_3$P=C, 4 Å MS, THF, rt, 2 h	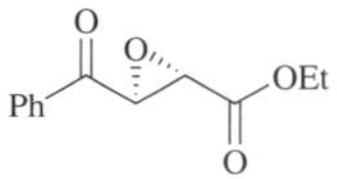 (88), 92	54
C_{14} 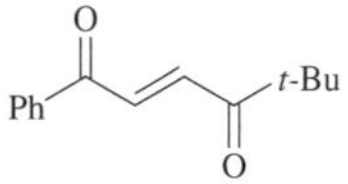	(L-Leu)$_n$ (n_{av} = 30), H_2O_2, NaOH, CH_2Cl_2, H_2O, rt	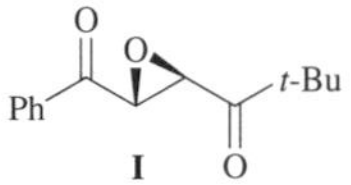 **I** (79), 82	179, 178
	(L-Leu)$_n$–PS (n_{av} = 32), H_2O_2, NaOH, toluene, H_2O, rt	**I** (75), >95	179
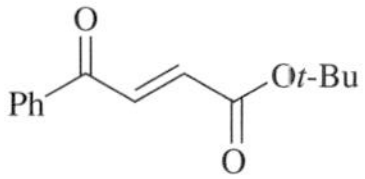	(L-Leu)$_n$–PS (n_{av} = 32), H_2O_2, NaOH, toluene, H_2O, rt	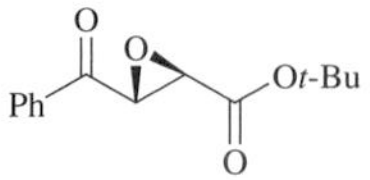 (66), >95	179, 178
C_{16} 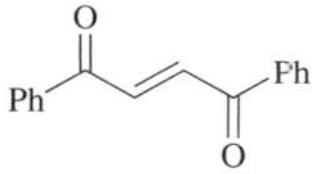	(L-Leu)$_n$ (n_{av} = 30), H_2O_2, NaOH, CH_2Cl_2, H_2O, rt	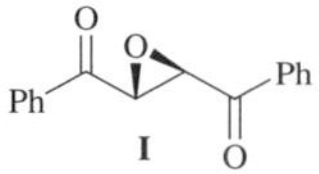 **I** (76), 76	179, 178
	(L-Leu)$_n$ (n_{av} = 33; 11 mol %), H_2O_2, NaOH, TBAB, toluene, H_2O, rt, 8 min	**I** (>99), 92	120
	(L-Leu)$_n$ (n_{av} = 33; 11 mol %), H_2O_2, NaOH, (n-C_8H_{17})$_4$NBr, toluene, H_2O, rt, 15 min	**I** (95), 64	120

TABLE 2. ENONES BEARING A SECOND ELECTRON–WITHDRAWING SUBSTITUENT (*Continued*)

Substrate	Conditions	Product(s), Yield(s) (%), % ee	Refs.
C_{16} 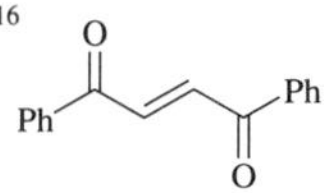	(*R*)-BINOL **15a** (5 mol %), CMHP, La(O*i*-Pr)$_3$, (4-FC$_6$H$_4$)$_3$P=O, 4 Å MS, THF, rt, 3 h	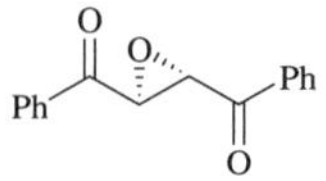 (85), 94	54
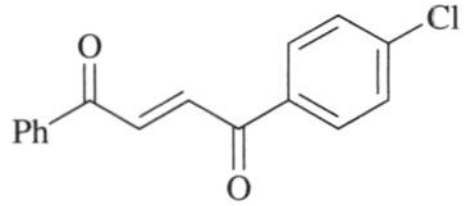	(L-Leu)$_n$, NaBO$_3$•4 H$_2$O, NaOH, CH$_2$Cl$_2$, H$_2$O, rt	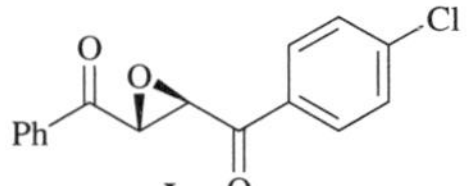 **I** (50), 89	179, 178
	(L-Leu)$_n$–PS (n_{av} = 32), H$_2$O$_2$, NaOH, toluene, H$_2$O, rt	**I** (57), >98	179

TABLE 3. 1,1-DISUBSTITUTED AND TRISUBSTITUTED ALKENES

Substrate	Conditions	Product(s), Yield(s) (%), % ee	Refs.
C_6 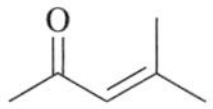	Pt-catalyst **64** (0.5 mol %), H_2O_2, H_2O, 0°, 27 h	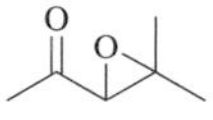 **I** (6), 7	80
	Pt-catalyst **65** (0.5 mol %), H_2O_2, H_2O, 0°, 26 h	**I** (5), 4	80
	Ru–porphyrin **72** (0.1 mol %), 2,6-dichloropyridine-*N*-oxide, benzene, rt, 16 h	**I** (*30*), 28	167
C_{11} 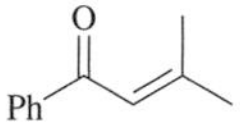	(*R*)-BINOL (x mol %), TBHP, Ln(O*i*-Pr)$_3$, 4 Å MS, THF, rt, 120 h	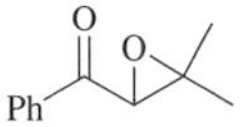	49
	(*R*)-BINOL / x / Ln / Time (h)		
	15d / 14 / Yb / 120	(78), 87	
	15d / 14 / La / 190	(51), 27	
	15a / 5 / Yb / 97	(75), 8	
	15a / 5 / La / 92	(46), 42	
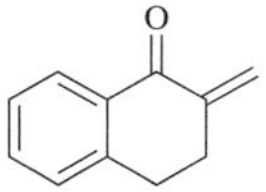	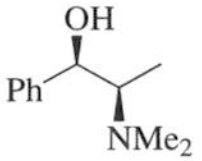 (240 mol %), O_2, Et_2Zn, toluene, –78° to 0°, 4 h	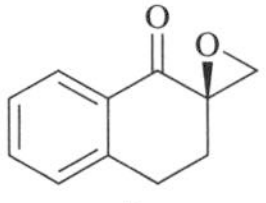 **I** (40), 3	41
	(L-Leu)$_n$–PS (n_{av} = 32), urea/H_2O_2, DBU, *i*-PrOAc, rt, 7 h	**I** (64), 94	192
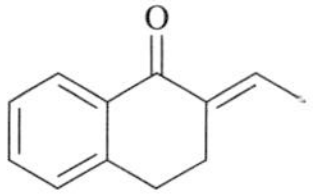	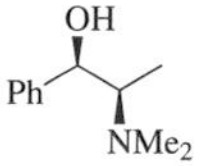 (240 mol %), O_2, Et_2Zn, toluene, –78° to 0°, 16 h	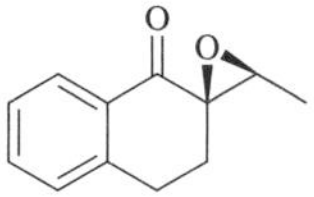 (85), 80	41

TABLE 3. 1,1-DISUBSTITUTED AND TRISUBSTITUTED ALKENES (*Continued*)

Substrate	Conditions	Product(s), Yield(s) (%), % ee	Refs.
C_{12} 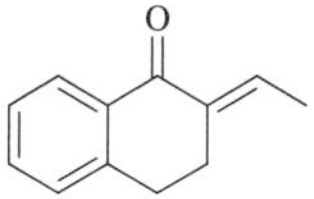	(L-Leu)$_n$–PS (n_{av} = 32), urea/H_2O_2, DBU, *i*-PrOAc, rt, 60 h	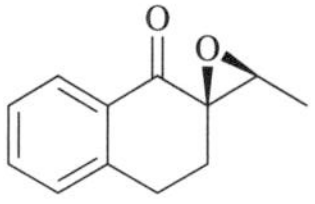 **I** (66), 92	192
	Anthryl cinchonidinium salt **6b** (10 mol %), KOCl, H_2O, toluene, –40°, 12 h	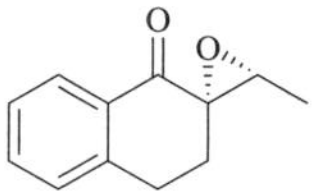 (—), 76	98
C_{13} 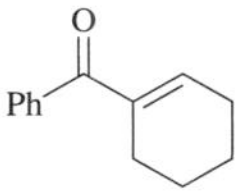	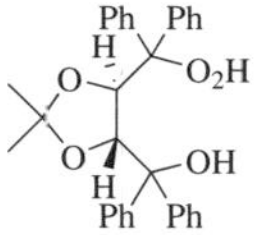 , BuLi, THF, –30°, 24 h	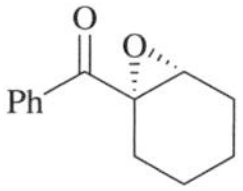 (46), 40[a]	31
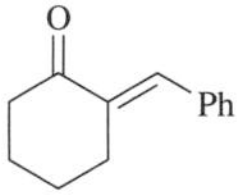	Benzyl cinchonidinium salt **3b**, NaOCl, toluene	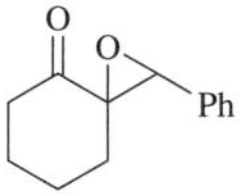 (40), —	83
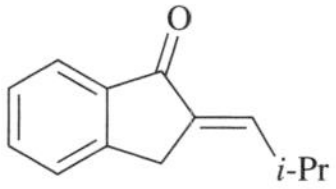	(L-Leu)$_n$–PS•SiO_2, urea/H_2O_2, BEMP, THF, rt	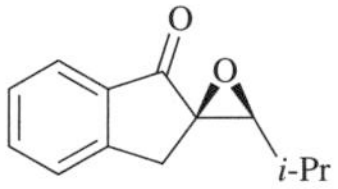 (74), 81	193
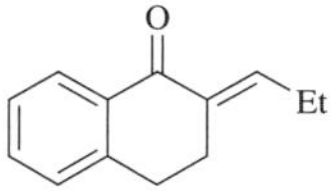	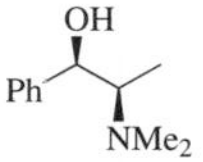 (240 mol %), O_2, Et_2Zn, toluene, –78° to 0°, 16 h	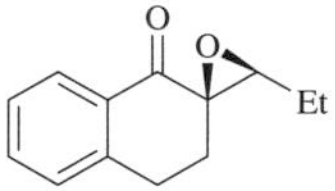 **I** (65), 90	41
	(L-Leu)$_n$–PS•SiO_2, urea/H_2O_2, BEMP, THF, rt	**I** (75), 85	193

C_{14}			
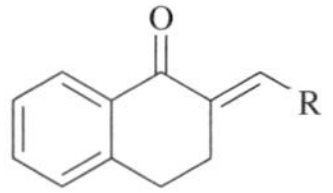	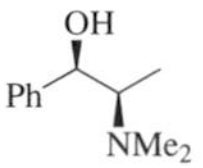 (240 mol %), O_2, Et_2Zn, toluene, –78° to 0°	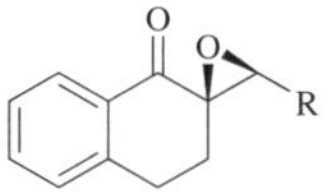	41
C_{15}			
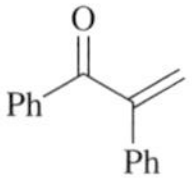	(L-Leu)$_n$–PS (n_{a*} = 30), H_2O_2, NaOH, toluene, H_2O, rt, 17 h	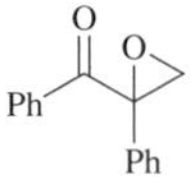 (78), 59	178, 179
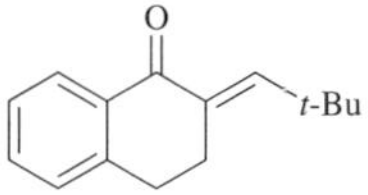	(L-Leu)$_n$–PS (n_{a*} = 32), urea/H_2O_2, DBU, *i*-PrOAc, rt, 192 h	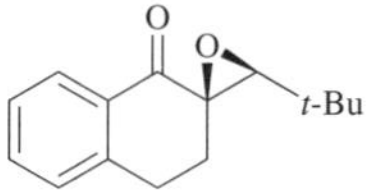 (63), 83	192
	Cinchoninium salt (10 mol %), oxidant, KOH, toluene, 0–20°	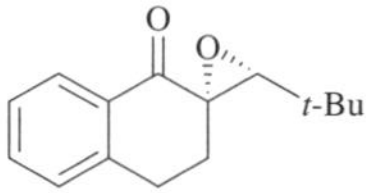 **I**	105
	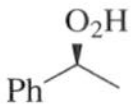 (100 mol %), KOH, MeCN, –40°	**I** (90), 90	206, 28
	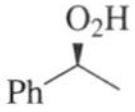 (100 mol %), DBU, toluene, 20°, 24 h	**I** (93), 13	28

R	Time (h)	
n-Pr	17	(99), 83
i-Pr	16	(98), >99

Cinchoninium salt	Oxidant	
1h	(±)-1-phenylethyl hydroperoxide	(—), 84
1h	(±)-2-phenylbut-2-yl hydroperoxide	(—), 92
1h	CMHP	(—), 95
1h	TBHP	(—), 89
1a	(±)-2-phenylbut-2-yl hydroperoxide	(—), 82

Substrate	Conditions	Product(s), Yield(s) (%), % ee	Refs.
C_{16}			
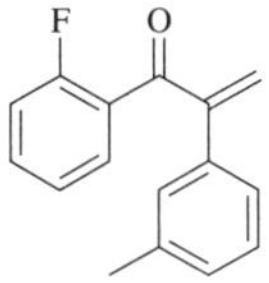	$(D\text{-}Ala)_n$, H_2O_2, NaOH, H_2O, toluene, rt	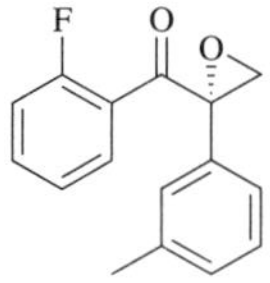 (~50), 20	286
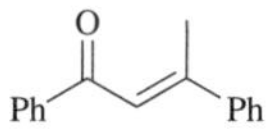	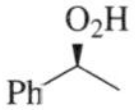 (100 mol %), DBU, toluene, 20°, 24 h	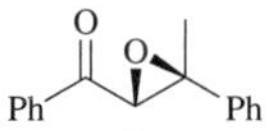 **I** (86), 54	28
	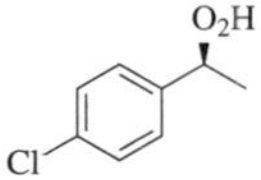 (100 mol %), DBU, toluene, 20°, 24 h	**I** (88), 72	28
	Benzyl quininium salt **4a** (1.7 mol %), H_2O_2, NaOH, H_2O, CH_2Cl_2, rt, 96–120 h	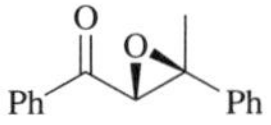 (49), 22 + 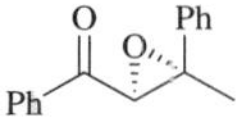 (29), 26	287
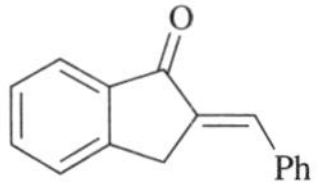	$(L\text{-}Leu)_n$–PS (n_{av} = 32), urea/H_2O_2, DBU, *i*-PrOAc, rt, 48 h	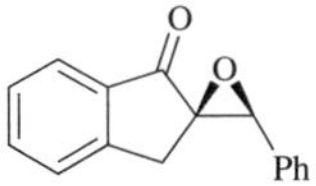 (72), 88	192
	Ammonium salt **50e** (3 mol %), NaOCl, toluene, H_2O, 0°, 70 h	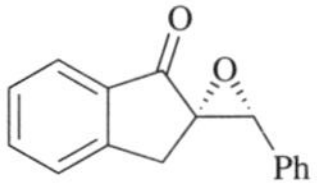 (91), 99	109

C$_{17}$

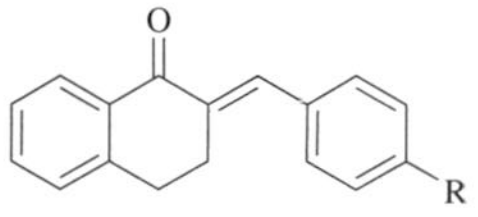

Polypeptide catalyst, oxidant, base, rt

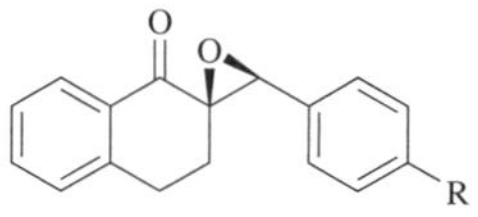

R	Catalyst	n_{av}	Oxidant	Base	Solvent	Time (h)		
H	(L-Leu)$_n$–PS	32	Na_2CO_3•1.5 H_2O_2	—	DME, H_2O	48	(100), 87	118
H	(L-Leu)$_n$–PS	32	urea/H_2O_2	DBU	*i*-PrOAc	90	(76), 84	192
H	(L-Leu)$_n$–PS	32	urea/H_2O_2	DBU	THF	32	(60), 60	260
H	(L-Npg)$_n$–PS	32	urea/H_2O_2	DBU	THF	7	(80), 75	260
H	(L-Leu)$_n$–PS•SiO_2	—	urea/H_2O_2	BEMP	THF	—	(100), 84	193
H	(β^3-Leu)$_{20}$–PS	—	H_2O_2	NaOH	toluene, H_2O	24	(25), 22	262
H	(β^3-Leu)$_{20}$–PS	—	urea/H_2O_2	DBU	THF	24	(35), 9	262
NO_2	(L-Leu)$_n$–PS	32	urea/H_2O_2	DBU	*i*-PrOAc	78	(85), 96	192
Br	(L-Leu)$_n$–PS	32	urea/H_2O_2	DBU	*i*-PrOAc	72	(81), 82	192

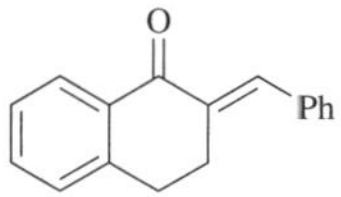

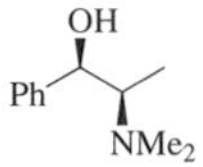

(20 mol %),
O_2, Et_2Zn, toluene, –78° to 0°, 17 h

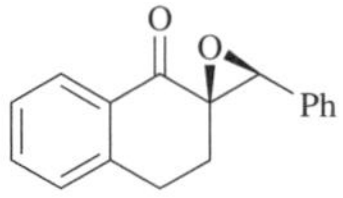

(62), 64 — 41

Anthryl cinchonidinium salt **6b** (10 mol %), KOCl, H_2O, toluene, –40°, 12 h

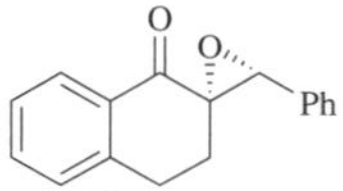

I

(—), 61 — 98

Ammonium salt **50e** (3 mol %), NaOCl, toluene, H_2O, 0°, 70 h

I (98), 96 — 109

TABLE 3. 1,1-DISUBSTITUTED AND TRISUBSTITUTED ALKENES (*Continued*)

Substrate	Conditions	Product(s), Yield(s) (%), % ee	Refs.
C_{18}			
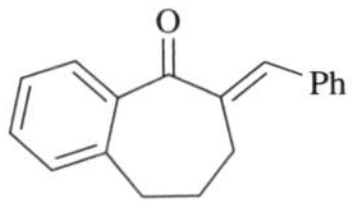	(L-Leu)$_n$–PS (n_{av} = 32), urea/H_2O_2, DBU, *i*-PrOAc, rt, 168 h	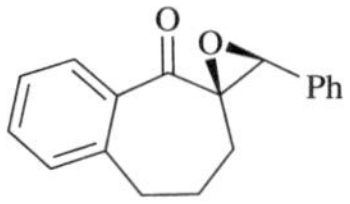 (74), 59	192
C_{20}			
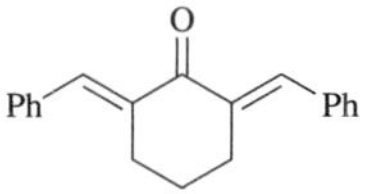	(L-Leu)$_n$, H_2O_2, NaOH, CH_2Cl_2, H_2O, rt, 7 d	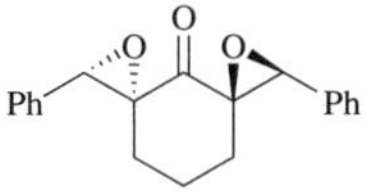 (—),—	178
	(D-Leu)$_n$, H_2O_2, NaOH, CH_2Cl_2, H_2O, rt, 7 d	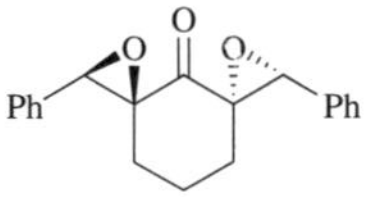 (—),—	178

[a] The authors describe the assignment of absolute stereochemistry as "tentative".

TABLE 4. ENDOCYCLIC ENONES

Substrate	Conditions	Product(s), Yield(s) (%), % ee	Refs.
C_6 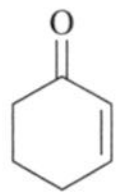	Benzyl quininium salt **4a** (1.1 mol %), TBHP, NaOH, toluene, 0–20°, 4 h	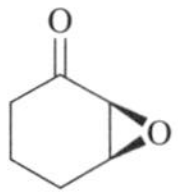 **I** (54), 20	84
	Bis(cinchonidinium) salt **12** (x mol %), oxidant, NaOH, H_2O, toluene, –10° to rt, 6–28 h	**I** [a]	94

x	Oxidant	
0.9	TBHP	(94), 33
1.3	n-$C_6H_{13}O_2H$	(49), 4
0.7	9-hexylfluoren-9-yl hydroperoxide	(78), 61
0.4	9-methylfluoren-9-yl hydroperoxide	(50), 63
0.8	9-phenylfluoren-9-yl hydroperoxide	(50), 38

Substrate	Conditions	Product(s), Yield(s) (%), % ee	Refs.
	Fluorenyl quininium salt **4c** (x mol %), oxidant, base, H_2O, toluene, –10° to rt, 6–28 h	**I** [a]	94

x	Oxidant	Base	
2	TBHP	NaOH	(56), 61
2	n-$C_6H_{13}O_2H$	NaOH	(69), 31
2	9-hexylfluoren-9-yl hydroperoxide	NaOH	(60), 59
6	H_2O_2	NaOH	(49), 0.2
2	TBHP	RbOH	(80), 13
2	TBHP	LiOH	(62), 58
2	TBHP	CsOH	(90), 5
2	TBHP	KOH	(65), 13
2	TBHP	$Ba(OH)_2$	(38), 38

TABLE 4. ENDOCYCLIC ENONES (*Continued*)

Substrate	Conditions	Product(s), Yield(s) (%), % ee	Refs.
C_6 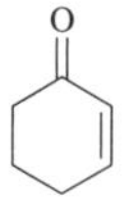	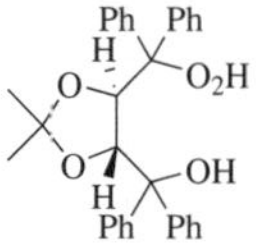 , BuLi, THF, –30°, 24 h	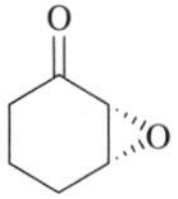 I	31
	Bis(cinchoninium) salt **13** (x mol %),oxidant, NaOH, H_2O, toluene, –10° to rt, 6–28 h	I [b]	94
	x / Oxidant		
	0.7 / TBHP	(86), 13	
	0.7 / n-$C_6H_{13}O_2H$	(82), 17	
	0.4 / 9-hexylfluoren-9-yl hydroperoxide	(75), 35	
	0.4 / 9-methylfluoren-9-yl hydroperoxide	(84), 50	
	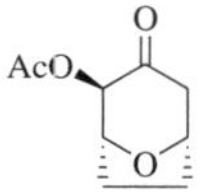 (2 mol %), Oxone, $NaHCO_3$, Na_2EDTA, MeCN, H_2O, rt, 24 h	**I** (51), 2	145
	Pt-catalyst **68** (x mol %), H_2O_2, H_2O	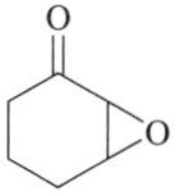	80
	x / Temp (°) / Time (h)		
	0.5 / –10 / 1.4	(0.2), 72	
	0.5 / 0 / 1.1	(0.8), 60	
	0.5 / rt / 0.7	(1.0), 34	
	0.05 / rt / 2.6	(0.8), 72	
	0.05 / rt / 127	(12.5), 1	

C_7 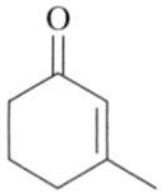	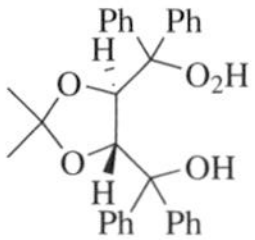 (150 mol %), BuLi, THF, –30°, 24 h	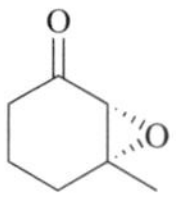 (74), 82	31
C_8 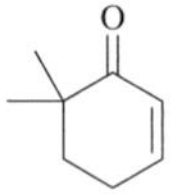	Benzyl quininium salt **4a** (7.4 mol %), TBHP, NaOH, toluene, 20°, 5 h	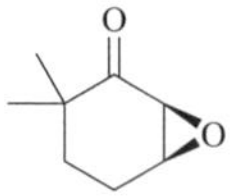 (60), 15	84
	Bis(cinchonidinium) salt **12** (6 mol %), TBHP, NaOH, H_2O, toluene, –10° to rt	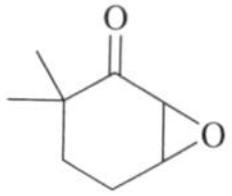 **I** (73), 22	94
	Fluorenyl quininium salt **4c** (2 mol %), 9-hexylfluoren-9-yl hydroperoxide, NaOH, H_2O, toluene, –10° to rt	**I** (100), 14	94
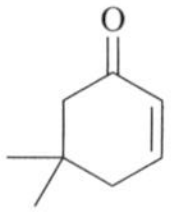	Benzyl quininium salt **4a** (7.4 mol %), TBHP, NaOH, toluene, rt, 5 h	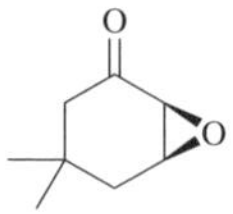 (59), 16	84
	Bis(cinchonidinium) salt **12** (2 mol %), 9-hexylfluoren-9-yl hydroperoxide, NaOH, H_2O, toluene, –10° to rt	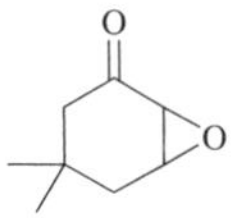 **I** (100), 44	94
	Fluorenyl quininium salt **4c** (7 mol %), TBHP, NaOH, H_2O, toluene, –10° to rt	**I** (60), 45	94

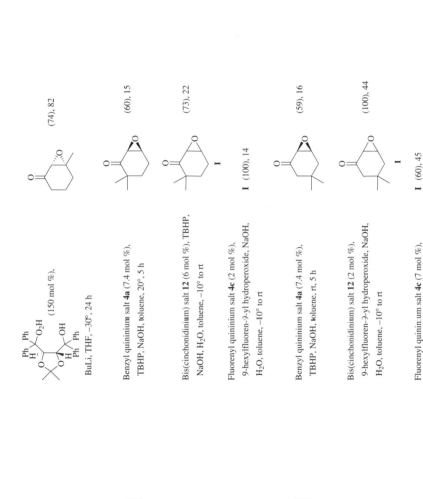

TABLE 4. ENDOCYCLIC ENONES (*Continued*)

Substrate	Conditions	Product(s), Yield(s) (%), % ee	Refs.
C_9 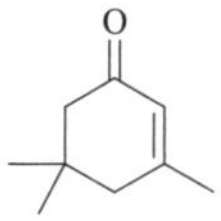	Benzyl quininium salt **4a**, NaOCl, toluene	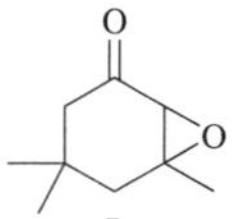 **I** (23), —	83
	Pt-catalyst **69** (0.5 mol %), H_2O_2, H_2O, rt, 1.4 h	**I** (0.5), 47	80
	Pt-catalyst **66** (0.5 mol %), H_2O_2, H_2O, rt, 2.9 h	**I** (0.7), 14	80
	Pt-catalyst **65** (0.5 mol %), H_2O_2, H_2O, 0°, 1.8 h	**I** (0.6), 4	80
C_{10} 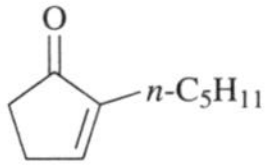	Pt-catalyst **68** (0.5 mol %), H_2O_2, H_2O, 50°, 0.5 h	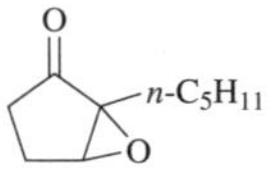 **I** (0.6), 28	80
	Pt-catalyst **67** (0.5 mol %), H_2O_2, H_2O, rt, 0.4 h	**I** (0.6), 38	80
	Pt-catalyst **64** (0.5 mol %), H_2O_2, H_2O, 0°, 1.5 h	**I** (0.9), 41	80
	Pt-catalyst **69** (0.5 mol %), H_2O_2, H_2O, 0°, 0.6 h	**I** (0.4), 42	80
	Pt-catalyst **66** (0.5 mol %), H_2O_2, H_2O, 0°, 1 h	**I** (0.6), 39	80
	Pt-catalyst **65** (0.5 mol %), H_2O_2, H_2O, 0°, 0.2 h	**I** (0.6), 57	80

C_{12} 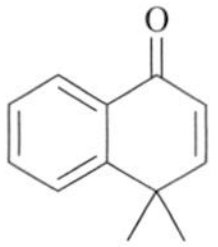	Benzyl cinchonium salt **1a** (10 mol %), (±)-2-phenylbut-2-yl hydroperoxide, KOH, toluene, 0–20°	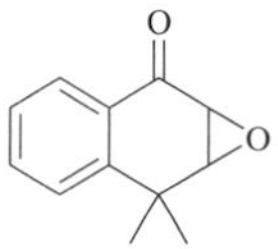	(—), 18	105
	Benzyl cinchonidinium salt **3b** (10 mol %), (±)-2-phenylbut-2-yl hydroperoxide, KOH, toluene, 0–20°	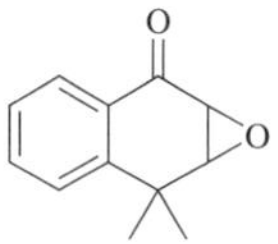	(—), 8[c]	105
C_{15} 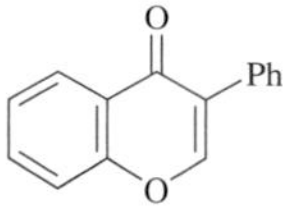	Mn–salen complex **62a** (14 mol %), dimethyldioxirane, acetone, CH_2Cl_2, rt, 10 d	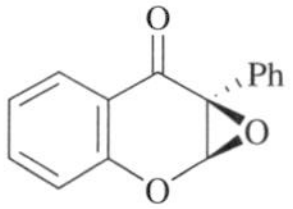	(31), 52	196, 288
	Mn–salen complex **61** (14 mol %), dimethyldioxirane, acetone, CH_2Cl_2, rt, 10 d	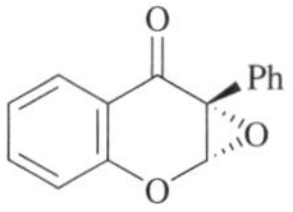 **I**	(34), 56	196, 288
	Mn–salen complex **61** (25 mol %), NaOCl, Na_2HPO_4, 4-phenylpyridine *N*-oxide, H_2O, 24 h	**I** (25), 65		196
C_{16} 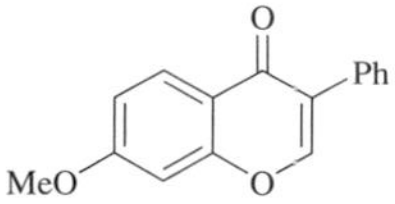	Benzyl cinchonidinium salt **3b** (10 mol %), CMHP, KOH, toluene, H_2O, 0–20°, 20 h	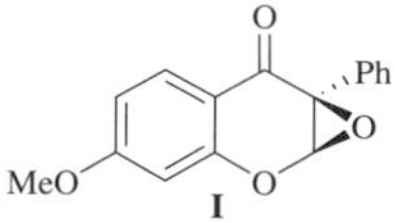 **I**	(94), 90[d]	29
	Mn–salen complex **62a** (14 mol %), dimethyldioxirane, acetone, CH_2Cl_2, rt, 10 d	**I** (27), 37		196, 288
	Mn–salen complex **62a** (25 mol %), NaOCl, Na_2HPO_4, 4-phenylpyridine *N*-oxide, H_2O, 24 h	**I** (30), 71		196

Substrate	Conditions	Product(s), Yield(s) (%), % ee	Refs.
C_{16} 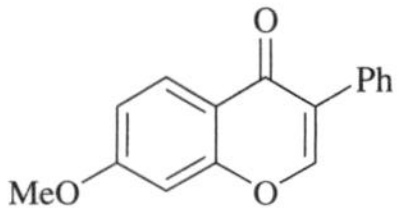	Trifluoromethylbenzyl cinchoninium salt **1h** (10 mol %), CMHP, KOH, toluene, H_2O, 0–20°, 20 h	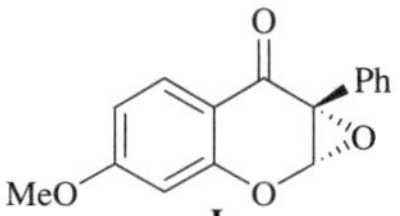 **I** (97), 98[d]	29
	Mn–salen complex **61** (14 mol %), dimethyldioxirane, acetone, CH_2Cl_2, rt, 10 d	**I** (36), 39	196, 288
	Mn–salen complex **61** (25 mol %), NaOCl, Na_2HPO_4, 4-phenylpyridine *N*-oxide, H_2O, 24 h	**I** (26), 77	196
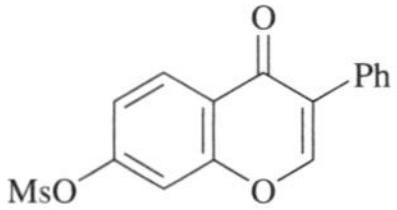	Mn–salen complex **62a** (14 mol %), dimethyldioxirane, acetone, CH_2Cl_2, rt, 10 d	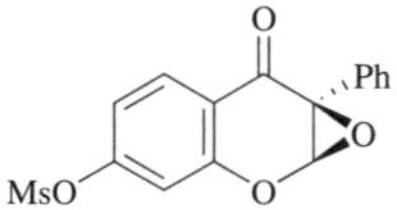 (22), 21	196, 288
	Mn–salen complex **61** (14 mol %), dimethyldioxirane, acetone, CH_2Cl_2, rt, 10 d	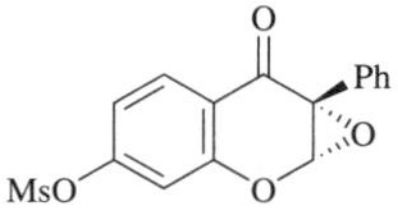 **I** (27), 48	196, 288
	Mn–salen complex **61** (25 mol %), NaOCl, Na_2HPO_4, 4-phenylpyridine *N*-oxide, H_2O, 24 h	**I** (23), 56	196
C_{17} 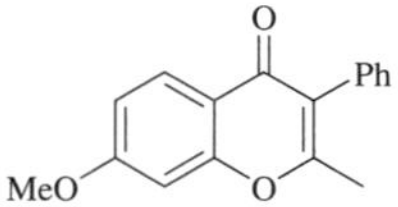	Trifluoromethylbenzyl cinchoninium salt **1h** (10 mol %), CMHP, KOH, toluene, H_2O, 0–20°, 20 h	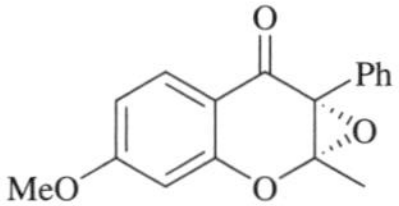 (97), 89[d]	29

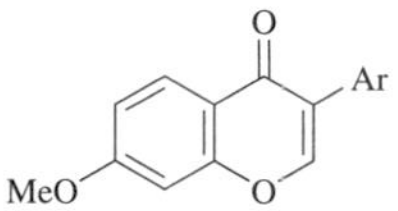	Mn–salen complex **62a** (16 mol %), dimethyldioxirane, acetone, CH_2Cl_2, rt, 10 d	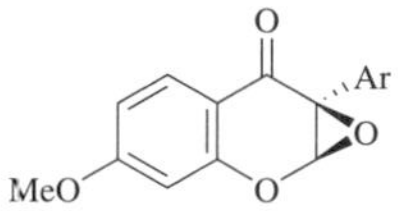	Ar: 2-$MeOC_6H_4$ (31), 82; 4-$MeOC_6H_4$ (39), 52	196, 288
	Mn–salen complex **61** (16 mol %), dimethyldioxirane, acetone, CH_2Cl_2, rt, 10 d	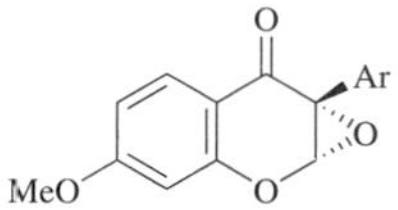 **I**	Ar: 2-$MeOC_6H_4$ (32), 86; 4-$MeOC_6H_4$ (29), 22	196, 288
	Mn–salen complex **61** (25 mol %), NaOCl, Na_2HPO_4, 4-phenylpyridine *N*-oxide, H_2O, 24 h	I	Ar: 2-$MeOC_6H_4$ (30), 90; 4-$MeOC_6H_4$ (30), 76	196
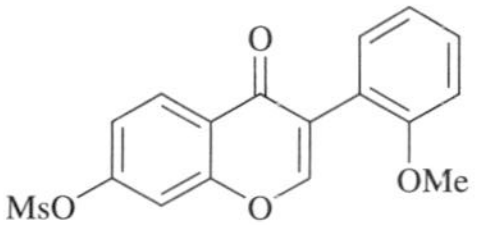	Trifluoromethylbenzyl cinchoninium salt **1h** (10 mol %), CMHP, KOH, toluene, H_2O, 0–20°, 3 d	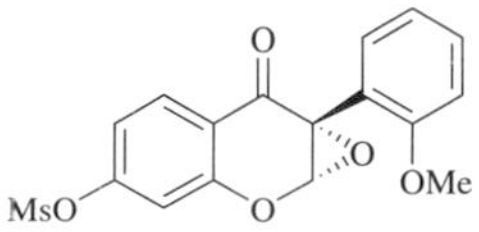 **I**	(84), 80[d]	29
	Mn–salen complex **62a** (16 mol %), dimethyldioxirane, acetone, CH_2Cl_2, rt, 10 d	I (25), 72		196, 288
	Mn–salen complex **61** (16 mol %), dimethyldioxirane, acetone, CH_2Cl_2, rt, 10 d	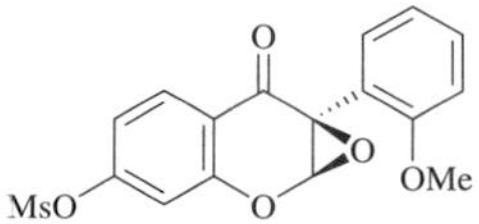 **I**	(23), 90	196, 288
	Mn–salen complex **61** (25 mol %), NaOCl, Na_2HPO_4, 4-phenylpyridine *N*-oxide, H_2O, 24 h	**I** (31), 94		196
C_{18} 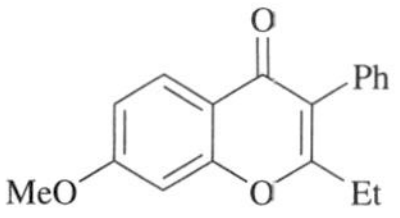	Benzyl cinchonidinium salt **3b** (10 mol %), CMHP, KOH, toluene, H_2O, 0–20°, 20 h	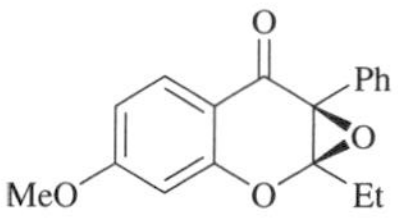	(99), 64[d]	29

TABLE 4. ENDOCYCLIC ENONES (*Continued*)

Substrate	Conditions	Product(s), Yield(s) (%), % ee	Refs.
C_{18} 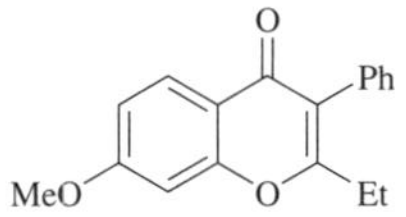	Cinchoninium salt (10 mol %), oxidant, KOH, toluene, H_2O, 0–20°, 20 h	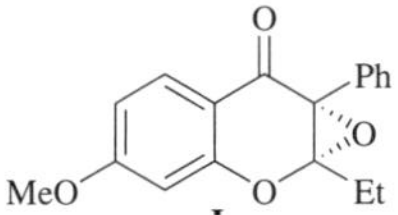 **I**	29
	Cinchoninium salt / Oxidant		
	1a / TBHP	(99), 70[d]	
	1a / CMHP	(99), 83[d]	
	1h / CMHP	(91), 90[d]	
	1i / CMHP	(95), 40[d]	
	1i / 1-phenethyl hydroperoxide	(—), 92[d]	
	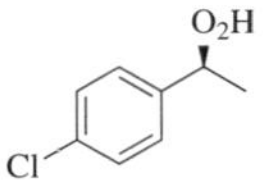 (103 mol %), DBU, toluene, 20°, 26 h	**I** (91), 45	29
C_{19} 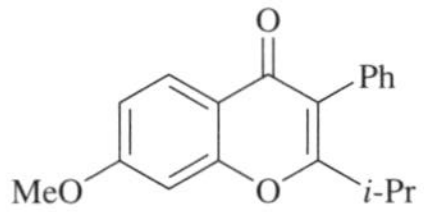	Trifluoromethylbenzyl cinchoninium salt **1h** (10 mol %), CMHP, KOH, toluene, H_2O, 0–20°, 3 d	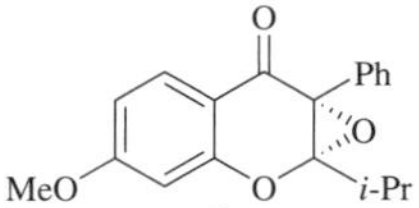 **I** (95), 53[d]	29
	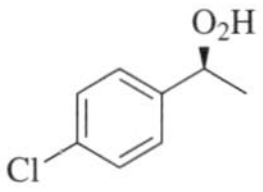 (103 mol %), DBU, toluene, 20°, 26 h	**I** (99), 37	29

[a] The publication refers to the product as the (2*S*,3*R*) isomer. Tentative assignment of the (2*S*,3*S*) structure is based on the stereoselectivity of epoxidation with related catalysts.

[b] The publication refers to the product as the (2*R*,3*S*) isomer. Tentative assignment of the (2*R*,3*R*) structure is based on the stereoselectivity of epoxidation with related catalysts.

[c] The major product is enantiomeric to that in the preceding entry.

[d] The stereochemical descriptors assigned to the products in this publication contradict the structures drawn. The use of the structures drawn in this review is based on comparison of reported optical rotations with those reported in ref. 288.

TABLE 5. QUINONES AND QUINONE MONOKETALS

Substrate	Conditions	Product(s), Yield(s) (%), % ee	Refs.
C_6 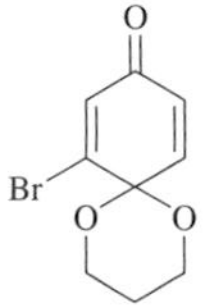	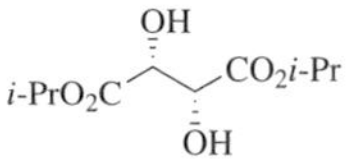 (100 mol %), Ph$_3$CO$_2$H, NaHMDS, 4 Å MS, toluene, –55°, 48 h	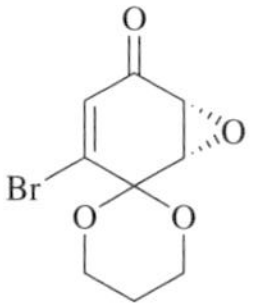 (80), 95	75
C_{10} 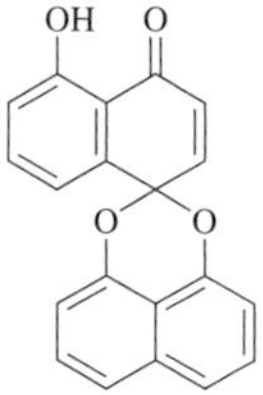	Benzyl cinchoninium salt **1a** (10 mol %), TBHP, NaOH, H_2O, toluene, rt, 10 h	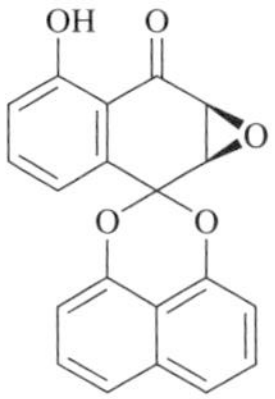 (81), >95	110
C_{11} 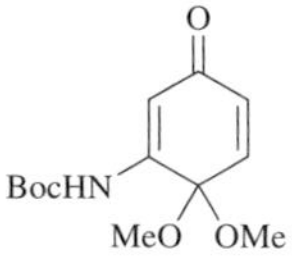	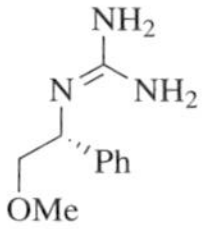 (100 mol %), TBHP, toluene, *i*-PrOH, rt, 7 d	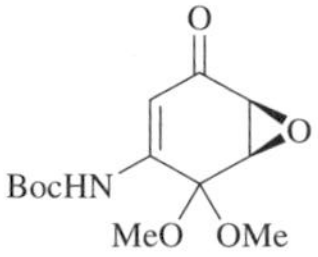 **I** (38), 6	38
	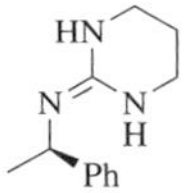 (100 mol %), TBHP, toluene, *i*-PrOH, rt, 7 d	**I** (49), 39	39
	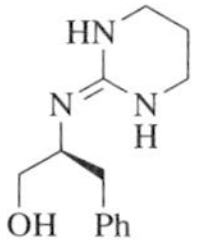 (100 mol %), TBHP, toluene, *i*-PrOH, rt, 7 d	**I** (45), 38	39

TABLE 5. QUINONES AND QUINONE MONOKETALS (*Continued*)

Substrate	Conditions	Product(s), Yield(s) (%), % ee	Refs.
C_{11} 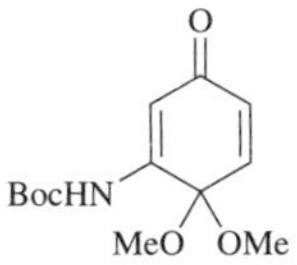	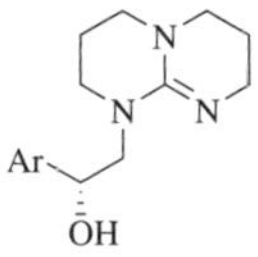 (120 mol %), TBHP, toluene, rt, 8 d	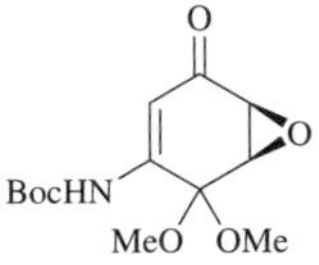 **I**; Ar: Ph (31), 35; 4-ClC_6H_4 (23), 29	36, 37
	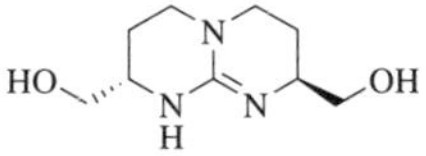 (100 mol %), TBHP, toluene, *i*-PrOH, rt, 2 d	**I** (74), 48	39
	Cinchonidinium or quininium salt (x mol %), TBHP, NaOH, toluene, rt, 7 d	**I**	
	Salt **3b**, x 100	(32), 89	195, 194
	Salt **4a**, x 100	(28), 77	195
	Salt **4a**, x 12	(28), 69	195
	Salt **3a**, x 100	(39), 35	195
	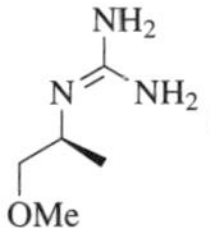 (100 mol %), TBHP, toluene, *i*-PrOH, rt, 7 d	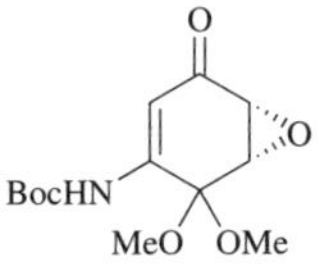 **I** (39), 30	38
	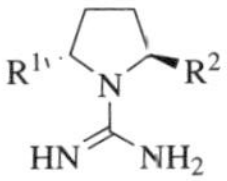 (100 mol %), TBHP, toluene, *i*-PrOH, rt, 7 d	**I**	38
	R^1 = CH_2OH, R^2 = H	(45), 28	
	R^1 = CPh_2OH, R^2 = H	(40), 17	
	R^1 = CH_2OH, R^2 = CH_2OH	(20), 29	
	R^1 = CH_2OMe, R^2 = CH_2OMe	(24), 18	

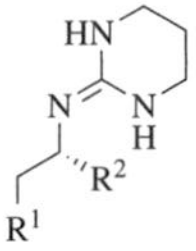
(100 mol %),
TBHP, toluene, *i*-PrOH, rt

I

R^1	R^2	Time (d)	
OH	Ph	7	(34), 60
OMe	Ph	7	(56), 40
H	2-Np	1	(67), 44
H	1-Np	7	(37), 41

39

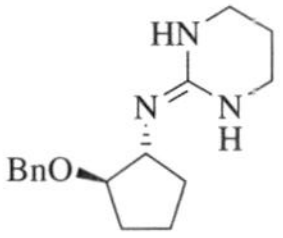
(100 mol %),
TBHP, toluene, *i*-PrOH, rt, 7 d

I (52), 26 — 39

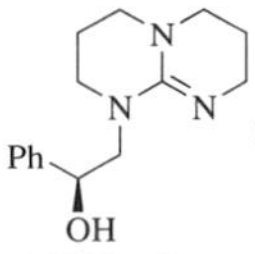
(120 mol %),
TBHP, toluene, rt, 8 d

I (32), 32 — 37, 36

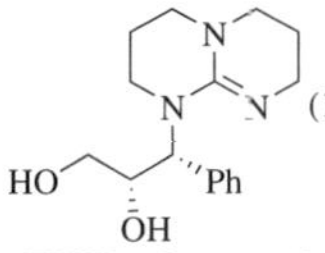
(120 mol %),
TBHP, toluene, rt, 8 d

I (7), 15 — 37

Benzyl cinchoninium salt **1a** (100 mol %),
TBHP, NaOH, toluene, rt, 7 d

I (15), 10 — 195

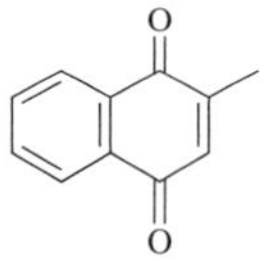

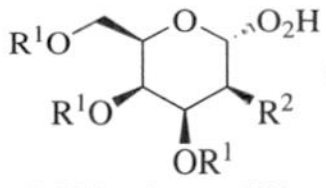
(200 mol %),
DBU, toluene, 20°

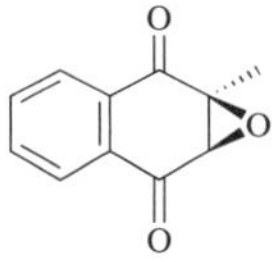

R^1	R^2	Time (h)	
Bn	H	20	(89), 42
Bn	Me	22.5	(90), 47
Piv	H	22	(72), 29

33, 34

TABLE 5. QUINONES AND QUINONE MONOKETALS (*Continued*)

Substrate	Conditions	Product(s), Yield(s) (%), % ee	Refs.
C_{11} 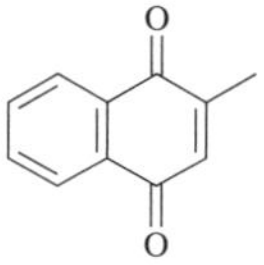	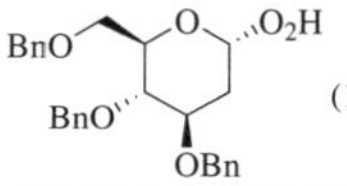 (120 mol %), DBU, toluene, 20°, 26 h	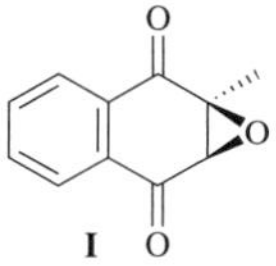 **I** (83), 45	33, 34
	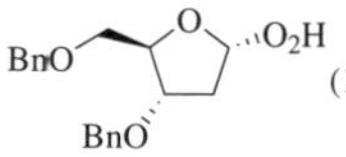 (100 mol %), DBU, toluene, 20°, 25 h	**I** (76), 30	33, 34
	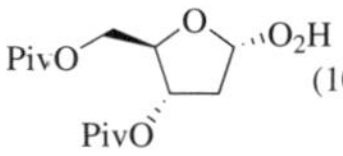 (100 mol %), DBU, toluene, 20°, 30 h	**I** (63), 39	33, 34
	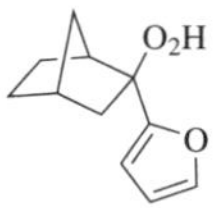 (110 mol %), base, additive, THF	**I**	32

Base	Additive	Time (h)	Temp (°)	
n-BuLi	—	3	0	(70), 13
n-BuLi	—	5	–20	(80), 51
n-BuLi	—	48	–78	(76), 41
NaOH	—	1.5	–20	(100), 29
DBU	LiCl	7	0	(60), 21

Conditions	Product(s) and Yield(s) (%), ee (%)	Refs.
Cinchoninium or quinidinium salt (5 mol %), H_2O_2, LiOH, H_2O, $CHCl_3$	**I**	

Salt	Temp (°)	Time (h)	Product(s) and Yield(s) (%), ee (%)	Refs.
2e	rt	1	(82), 31	106, 103
2e	−10	1	(86), 34	106, 103
2f	rt	1	(84), 27	106
2b	rt	1	(87), 18	106
2c	rt	1	(92), 30	106
2d	rt	1	(93), 26	106, 103
1f	rt	1	(94), 10	103
1h	rt	4	(76), 11	103
2g	rt	1	(91), 28	103

Conditions	Product(s) and Yield(s) (%), ee (%)	Refs.
Bis(cinchonidinium) salt **12** (0.3 mol %), TBHP, NaOH, toluene, H_2O, −10° to rt	**I** (53), 10	94
Fluorenyl quininium salt **4c** (2 mol %), TBHP, NaOH, toluene, H_2O, −10° to rt	**I** (63), 18	94
Bovine serum albumin (5 mol %), H_2O_2, H_2O, pH 11, rt, 3 d	**I** (32), 3	141, 142
Bovine serum albumin (5 mol %), CMHP, H_2O, pH 9, rt, 2 d	**I** (69), 6	142
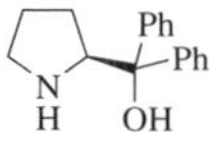 (50 mol %), TBHP, hexane, rt, 182 h	**I** (76), 14	170
β-Cyclodextrin (500 mol %), TBHP, pH 9 borate buffer, H_2O, rt, 4 d	**I** (33), 22	171
Cellulose triacetate, H_2O_2, NaOH, H_2O, 40°, 2 h	**I** (45), 1	289
Cellulose triacetate, H_2O_2, NaOH, TBAB, H_2O, 40°, 4 h	**I** (72), 2.5	289

TABLE 5. QUINONES AND QUINONE MONOKETALS (*Continued*)

Substrate	Conditions	Product(s), Yield(s) (%), % ee	Refs.
C_{11} 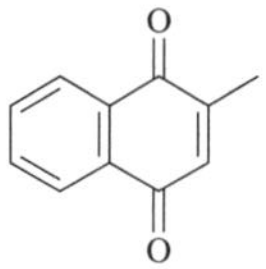	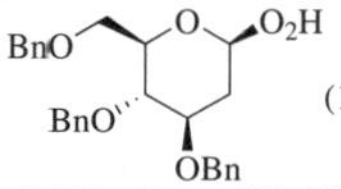 (120 mol %), DBU, toluene, 20°, 26 h	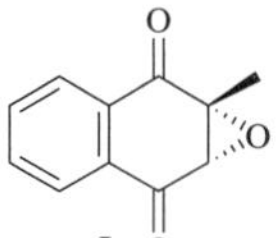 **I** (79), 33	33, 34
	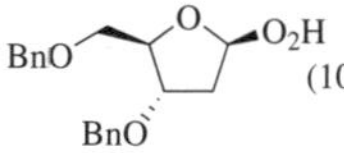 (100 mol %), DBU, toluene, 20°, 25 h	**I** (73), 47	33, 34
	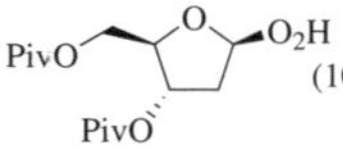 (100 mol %), DBU, toluene, 20°, 30 h	**I** (67), 29	33, 34
	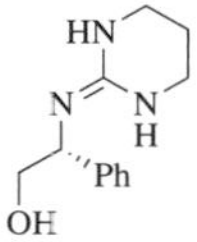 (100 mol %), TBHP, toluene, *i*-PrOH, rt, 7 d	**I** (68), 31	39
	Benzyl quininium salt **4a** (1.7 mol %), TBHP, NaOH, toluene, –5°, 24 h	**I** (—), 6[a]	81
	Bis(cinchonidinium) salt **12** (0.7 mol %), 9-hexylfluoren-9-yl hydroperoxide, NaOH, toluene, H_2O, –10° to rt	**I** (60), 13	94
	β-Cyclodextrin (500 mol %), TBHP, Na_2CO_3, DMF, rt, 2 h	**I** (96), 24	171

Substrate	Conditions	Product(s) and Yield(s) (%), % ee	Refs.
	Bovine serum albumin (5 mol %), TBHP, H_2O, pH 9, rt, 2 d	**I** (27), 36	141, 142, 144
	Bovine serum albumin (5 mol %), TBHP, H_2O, pH 11, rt, 3 d	**I** (34), 20	141, 142, 144

C_{11-23}

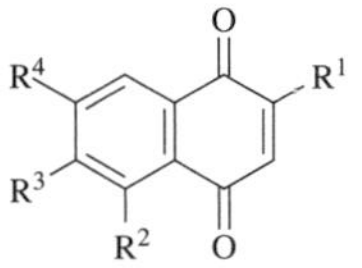

Benzyl quininium salt **4a** (1.9 mol %), H_2O_2, NaOH, H_2O, toluene, 35–40°, 1.5–2 h

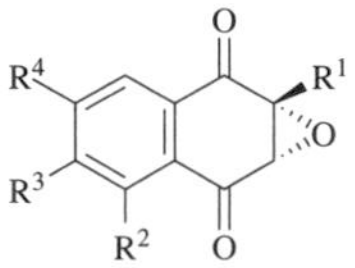

R^1	R^2	R^3	R^4	Yield (%), % ee	Refs.
Me	H	H	H	(60–70), 5	85, 81
Et	H	H	H	(50–60), 10	85
Me	Me	H	H	(40), 18	85
Me	H	Me	H	(70), 5	85
Me	H	H	Me	(65), 5	85
Me	MeO	H	H	(35), 12	85
n-Pr	H	H	H	(76), 14	85
i-Pr	H	H	H	(74), 31	85
n-Bu	H	H	H	(67), 18	85
i-Bu	H	H	H	(60), 16	85
t-Bu	H	H	H	(60), 23	85
n-C_5H_{11}	H	H	H	(64), 19	85
$(CH_3)_2C{=}CHCH_2$	H	H	H	(53), 27	85
n-C_6H_{13}	H	H	H	(60), 24	85
Ph	H	H	H	(92), 45	85
c-C_6H_{11}	H	H	H	(85), 39	85
$PhCH_2$	H	H	H	(66), 23	85
Ph_2CH	H	H	H	(100), 30	85

TABLE 5. QUINONES AND QUINONE MONOKETALS (*Continued*)

Substrate	Conditions	Product(s), Yield(s) (%), % ee	Refs.
C_{11-16} 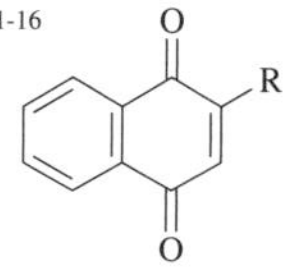	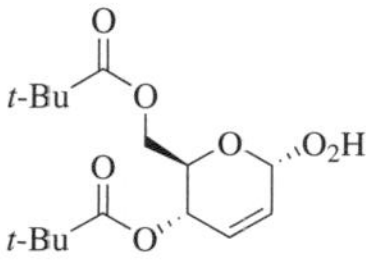 (150–200 mol %), DBU, toluene, rt, 5 h	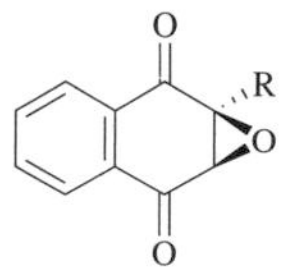 R: Me (71), 45; Et (79), 35; *n*-Pr (30), 69; *i*-Pr (67), 70; Ph (80), 82	30
C_{11} 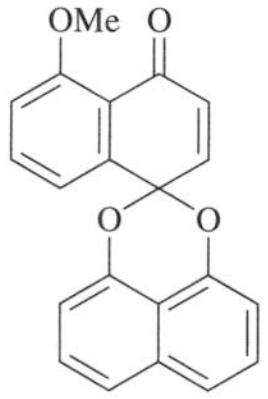	Benzyl cinchoninium salt **1a** (10 mol %), TBHP, NaOH, H_2O, toluene, rt, 9 d	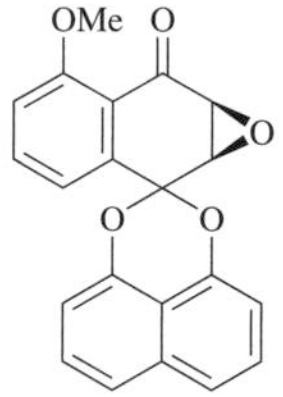 (61), 23	110
C_{12-18} 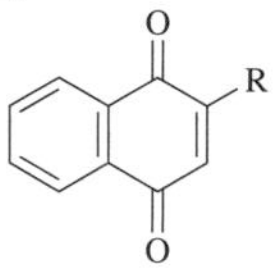	Naphthyl quinidinium salt **2e** (5 mol %), H_2O_2, LiOH, H_2O, $CHCl_3$, –10°	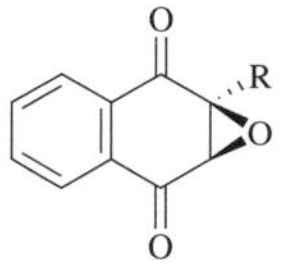	

R	Time (h)	Product(s), Yield(s) (%), % ee	Refs.
Et	16	(99), 41	106, 103
n-Pr	7	(95), 40	106, 103
i-Pr	5	(93), 70	106, 103
n-Bu	21	(87), 44	103
i-Bu	30	(100), 28	106
Ph	23	(47), 76	106
c-C_6H_{11}	23	(60), 64	106, 103
PhC=C	21	(84), 40	106, 103

Reactant	Conditions	Product(s) and Yield(s) (%), ee		Refs.
C_{12} 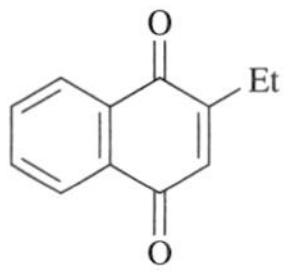	Naphthyl quinidinium salt **2e** (5 mol %), H_2O_2, LiOH, H_2O, $CHCl_3$, –10°, 16 h	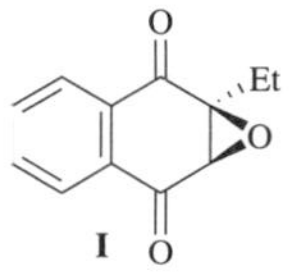 **I**	(99), 41	106, 103
	Bovine serum albumin (5 mol %), H_2O_2, H_2O, pH 11, rt, 2 d	**I** (42), 15		142
	Bovine serum albumin (5 mol %), TBHP, H_2O, pH 9, rt, 2 d	**I** (27), 2		142
	Bovine serum albumin (5 mol %), TBHP, H_2O, pH 11, rt, 2 d	**I** (44), 5		143, 142
	Bovine serum albumin (5 mol %), TBHP, H_2O, isooctane, rt, 1 d	**I** (24), 3.5		143
C_{13} 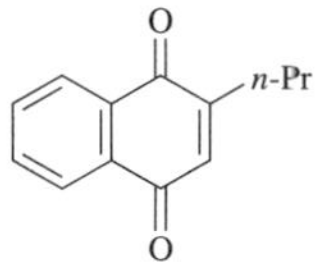	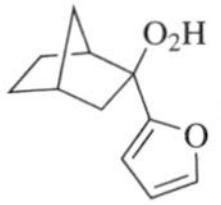 (110 mol %), *n*-BuLi, THF, –20°, 7 h	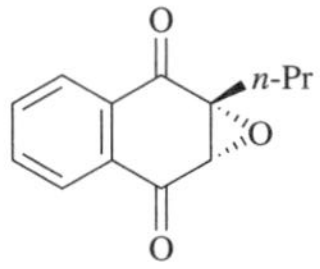	(51), 20	32
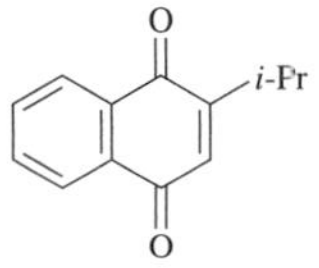	Cinchonine-derived catalyst (5 mol%), H_2O_2, LiOH, H_2O, $CHCl_3$, 0°, 5 h; then rt, 12 h	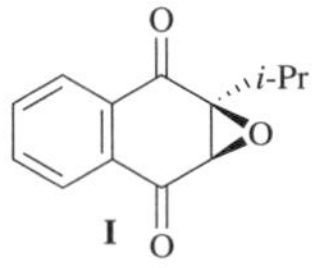 **I**		88
	Catalyst			
	9a	(62), 50[b]		
	9b	(58), 66[b]		
	1b	(80), 42[b]		
	1h	(92), 55[b]		
	1m	(65), 52[b]		

TABLE 5. QUINONES AND QUINONE MONOKETALS (*Continued*)

Substrate	Conditions	Product(s), Yield(s) (%), % ee	Refs.
C_{13} 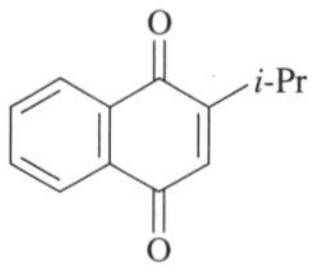	Deazacinchonidinium salt **11** (5 mol %), H_2O_2, LiOH, H_2O, $CHCl_3$, 0°, 5 h; then rt, 12 h	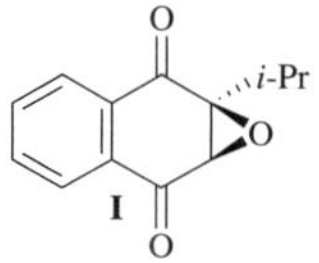 **I** (75), 84[b]	88
	Bovine serum albumin (5 mol %), H_2O_2, H_2O, pH 11, rt, 6 d	**I** (60), 15	141,142
	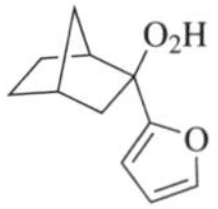 (1 mol %), *n*-BuLi, THF, –20°, 24 h	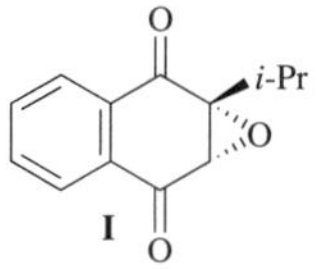 **I** (32), 58	32
	Cinchonidinium or quininium salt (5 mol %), H_2O_2, LiOH, H_2O, $CHCl_3$, 0°, 5 h; then rt, 12 h	**I**	88
	Salt		
	3c	(97), 46[b]	
	3d	(55), 71[b]	
	3e	(97), 74[b]	
	4b	(69), 39[b]	
	4d	(41), 76[b]	
	Deazacinchoninium salt **10** (5 mol %), H_2O_2, LiOH, H_2O, $CHCl_3$, 0°, 5 h; then rt, 12 h	**I** (71), 32[c]	88
	Bovine serum albumin (5 mol %), TBHP, H_2O, pH 11, rt, 6 d	**I** (56), 21	141, 142

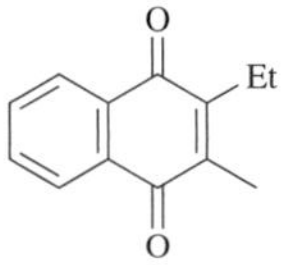	Bovine serum albumin (5 mol %), H_2O_2, H_2O, pH 11, rt, 14 d	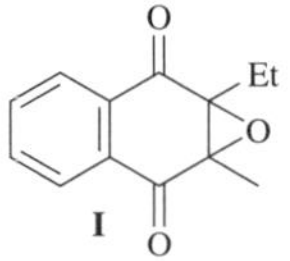 **I** (23), 11	142
	Bovine serum albumin (5 mol %), TBHP, H_2O, pH 11, rt, 12 d	**I** (22), 54	142
C_{14} 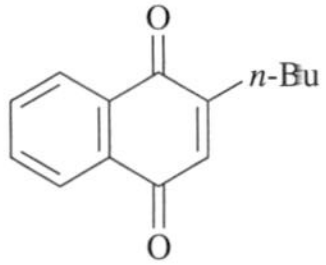	Bovine serum albumin (5 mol %), TBHP, H_2O, pH 11, rt, 2 d	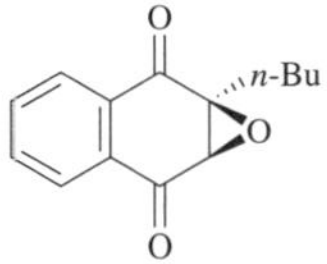 (35), 14	142, 144
	Bovine serum albumin (5 mol %), H_2O_2, H_2O, pH 9, rt, 2 d	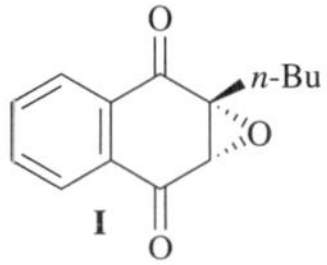 **I** (74), 8	142
	α-Cyclodextrin (500 mol %), TBHP, Na_2CO_3, DMF, rt, 24 h	**I** (79), 39	171
	β-Cyclodextrin (500 mol %), TBHP, Na_2CO_3, DMF, rt, 5 h	**I** (76), 30	171
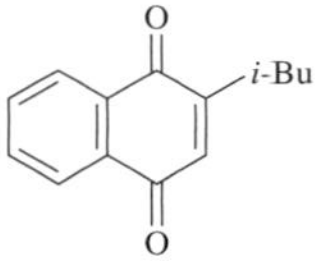	Bovine serum albumin (5 mol %), H_2O_2, H_2O, pH 11, rt, 2 d	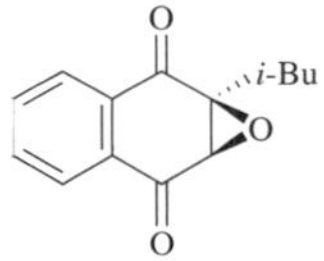 (70), 8	142
	Bovine serum albumin (5 mol %), TBHP, H_2O, pH 9, rt, 2 d	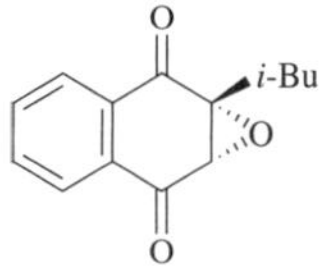 (53), 75	142, 143, 144

TABLE 5. QUINONES AND QUINONE MONOKETALS (*Continued*)

Substrate	Conditions	Product(s), Yield(s) (%), % ee	Refs.
C_{14} 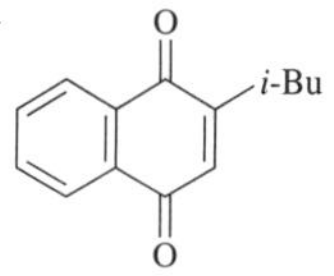	Bovine serum albumin (5 mol %), TBHP, H_2O, cosolvent, pH 11, rt	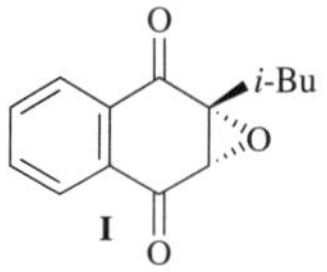 **I**	
	Cosolvent / Time (d)		
	— / 2	(62), 77	144
	isooctane / 2	(56), 90	143
	isooctane / 2	(53), 75	143
	toluene / 2	(30), 35	143
	CCl_4 / 8	(47), 0	143
	EtOH / 2	(76), 66	143
	t-BuOH / 2	(56), 51	143
	DMSO / 2	(81), 68	143
	Bovine serum albumin (5 mol %), CMHP, H_2O, pH 9, rt, 3 d	**I** (57), 36	142
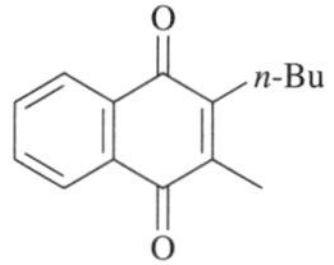	Benzyl quininium salt **4a** (1.9 mol %), H_2O_2, NaOH, H_2O, toluene, 35–40°, 1.5–2 h	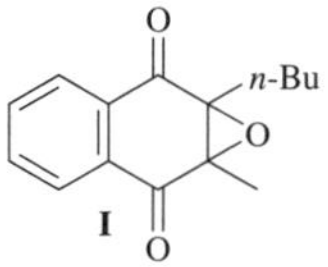 **I** (35), —	85
	Bovine serum albumin (5 mol %), TBHP, H_2O, pH 11, rt, 14 d	**I** (21), 48	142
C_{16} 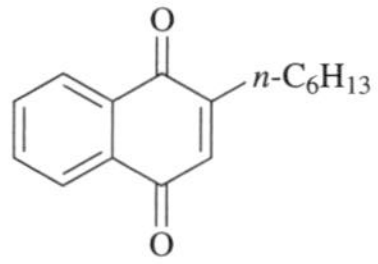	Bovine serum albumin (5 mol %), TBHP, H_2O, pH 11, rt, 2.5 h	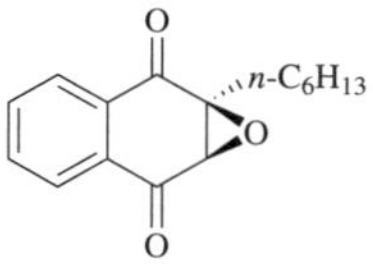 (29), 30	144

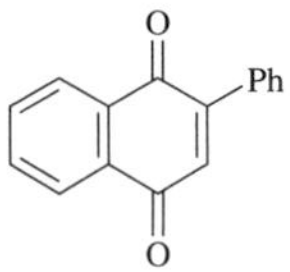	Cellulose triacetate, H_2O_2, NaOH, H_2O, 40°, 2 h	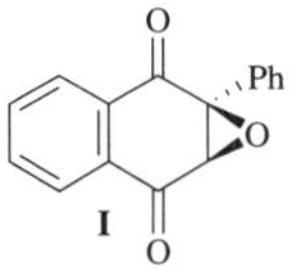 **I** (34), 1.3	289
	Cellulose triacetate, H_2O_2, NaOH, H_2O, TBAB, 40°, 4 h	**I** (49), 4.5	289
	Cellulose tribenzoate, H_2O_2, NaOH, H_2O, 40°, 4 h	**I** (61), 0.4	289
	Cellulose tribenzoate, H_2O_2, NaOH, H_2O, TBAB, 40°, 4 h	**I** (66), 0.2	289
	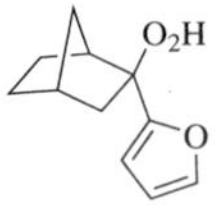 (110 mol %), *n*-BuLi, THF, –20°, 3 h	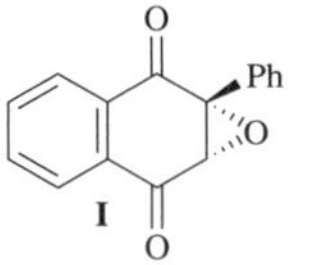 **I** (78), 17	32
	β-Cyclodextrin (500 mol %), TBHP, Na_2CO_3, DMF, rt, 4 h	**I** (88), 5	171
	Bovine serum albumin (5 mol %), TBHP, H_2O, pH 11, rt, 8 d	**I** (46), 50	141[c], 142, 144
	Bovine serum albumin (5 mol %), CMHP, H_2O, pH 9, rt, 7 d	**I** (61), 17	142
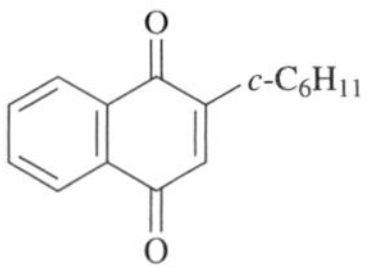	Bovine serum albumin (5 mol %), H_2O_2, H_2O, pH 11, rt, 7 d	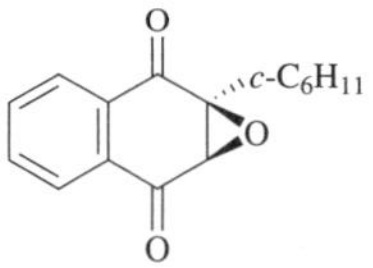 (68), 2	141[c], 142

TABLE 5. QUINONES AND QUINONE MONOKETALS (*Continued*)

Substrate	Conditions	Product(s), Yield(s) (%), % ee	Refs.
C_{16} 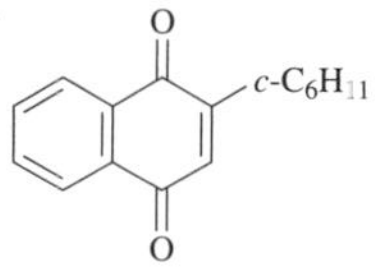	Human serum albumin (5 mol %), TBHP, H_2O, pH 9, rt	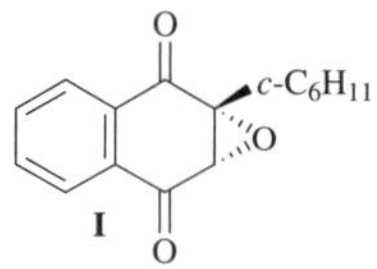 **I** (11), 70	142
	Human serum albumin (5 mol %), TBHP, H_2O, isooctane, pH 11, rt, 5 d	**I** (66), 83	142
	Bovine serum albumin (5 mol %), H_2O_2, H_2O, pH 9, rt, 7 d	**I** (57), 20	142
	Bovine serum albumin (5 mol %), TBHP, H_2O, pH 9, rt, 7 d	**I** (75), 79	142, 144
	Bovine serum albumin (5 mol %), TBHP, H_2O, pH 11, rt, 7 d	**I** (64),70	141[d], 142, 230, 143
C_{17} 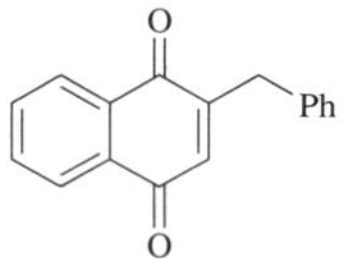	Bovine serum albumin (5 mol %), H_2O_2, H_2O, pH 11, rt, 2 d	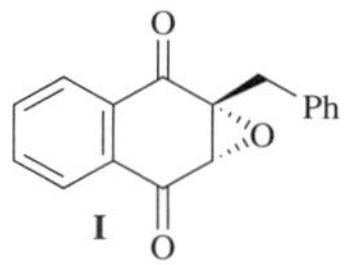 **I** (44), 15	142
	Bovine serum albumin (5 mol %), TBHP, H_2O, pH 11, rt, 10 d	**I** (22), 12	142
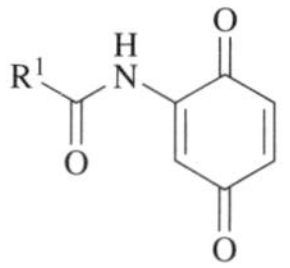 R^1 = 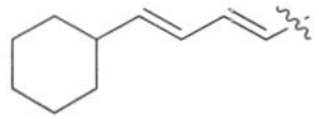	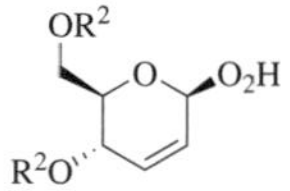 (100 mol %), DBU, toluene, rt	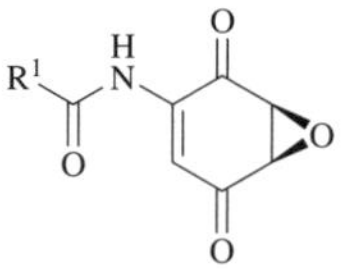 R^2: Bn (50), 24; *t*-BuC(O) (17), 5	30

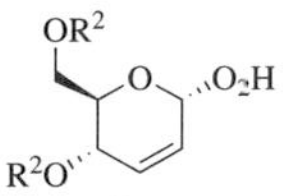
(100 mol %), DBU

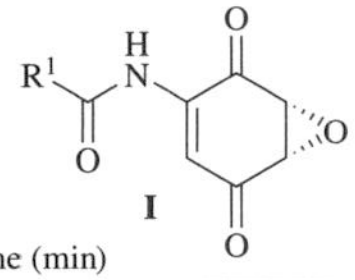
I

30

R^2	Solvent	Temp	Time (min)	
Bn	toluene	rt	—	(38), 36
t-BuC(O)	EtOAc	rt	5	(33), 30
t-BuC(O)	1,2-dichloroethane	rt	<1	(57), 23
t-BuC(O)	CH_2Cl_2	rt	2	(69), 30
t-BuC(O)	MeCN	rt	<1	(35), 38
t-BuC(O)	$CHCl_3$	rt	2	(83), 45
t-BuC(O)	THF	rt	5	(52), 49
t-BuC(O)	*p*-xylene	rt	7	(83), 53
t-BuC(O)	benzene	rt	6	(64), 54
t-BuC(O)	*p*-cymene	rt	45	(47), 61
t-BuC(O)	toluene	rt	5	(55), 64
t-BuC(O)	toluene	–78°	—	(—), 67

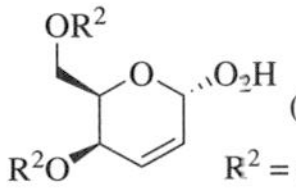
(100 mol %), DBU, toluene, rt **I** (64), 42 30

$R^2 = t$-BuC(O)

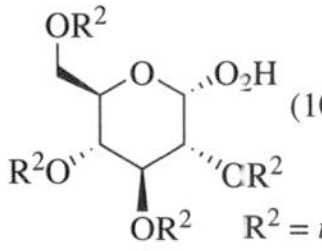
(100 mol %), DBU, toluene, rt **I** (64), 30 30

$R^2 = t$-BuC(O)

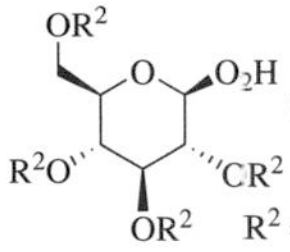
(100 mol %), DBU, toluene, rt **I** (63), 10 30

$R^2 = t$-BuC(O)

TABLE 5. QUINONES AND QUINONE MONOKETALS (*Continued*)

Substrate	Conditions	Product(s), Yield(s) (%), % ee	Refs.
C_{18} 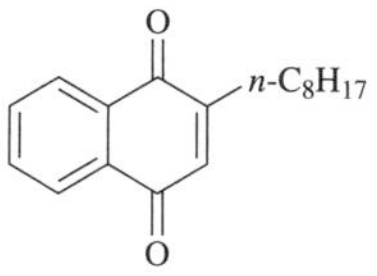	Bovine serum albumin (5 mol %), TBHP, H_2O, pH 11, rt, 7 d	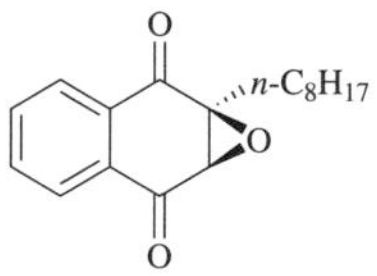 (66), 100	144
	α-Cyclodextrin (500 mol %), TBHP, Na_2CO_3, DMF, rt, 3.5 h	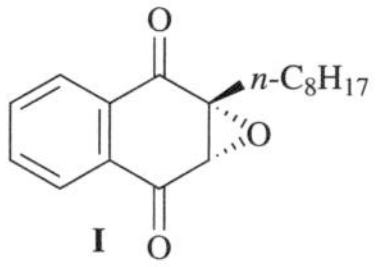 **I** (61), 48	171
	β-Cyclodextrin (500 mol %), TBHP, Na_2CO_3, DMF, rt, 4 h	**I** (90), 41	171
	β-Cyclodextrin (500 mol %), CMHP, Na_2CO_3, DMF, rt, 6 h	**I** (87),19	171
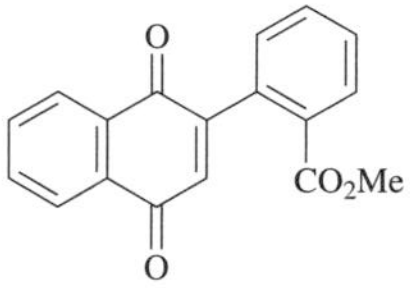	Benzyl quininium salt **4a** (1.7 mol %), TBHP, NaOH, toluene, 0° to rt, 90 min	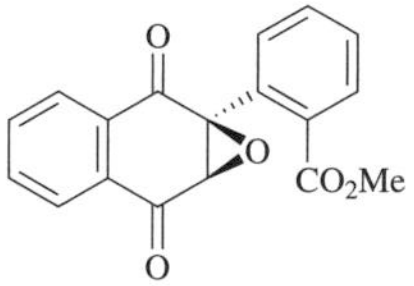 (95), 78	95, 197
	Benzyl quininium salt **4a** (1.7 mol %), H_2O_2, NaOH, H_2O, toluene, rt, 4.5 h	**I** (89), 50	197
C_{23} 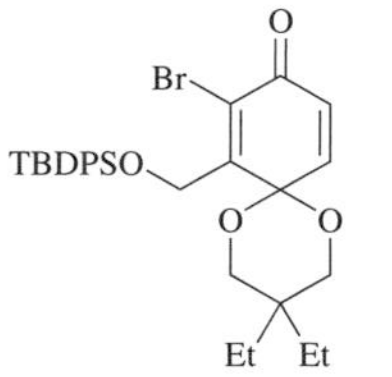	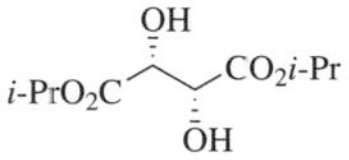 (100 mol %), Ph_3CO_2H, *n*-BuLi, toluene, rt, 24 h	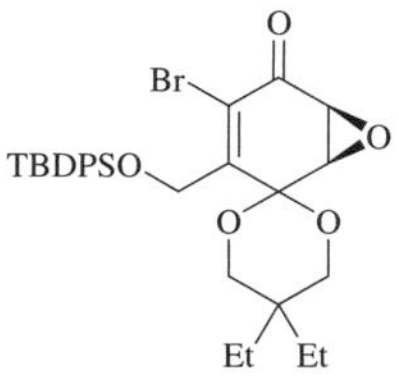 (88), 68	74

		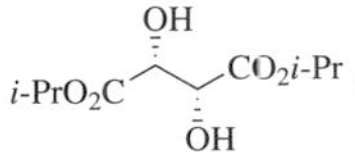 (100 mol %), Ph_3CO_2H, NaHMDS, toluene, –50°, 30 h	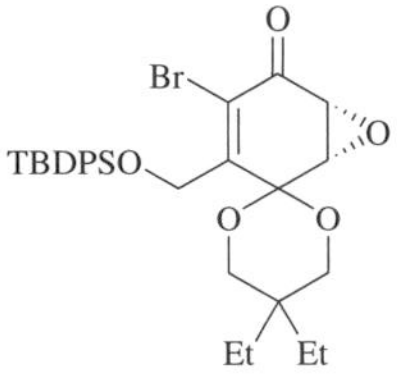	(97), 96	74
C_{25}	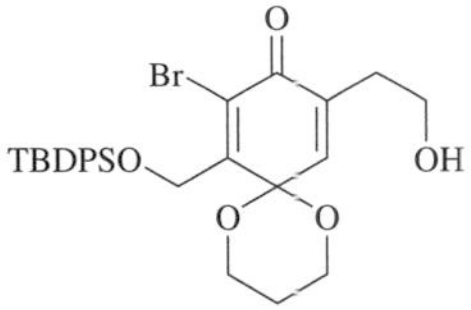	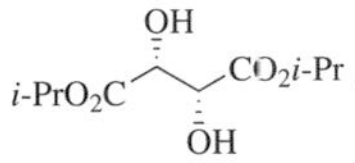 (200 mol %), Ph_3CO_2H, NaHMDS, 4 Å MS, toluene, –40°, 50 h	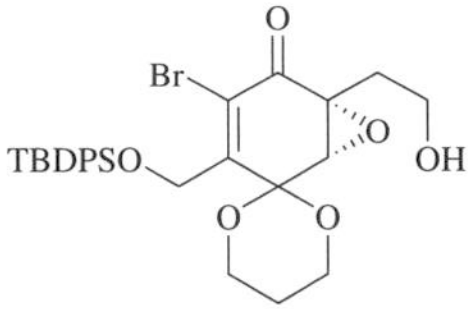	(91), 91	76
C_{28}	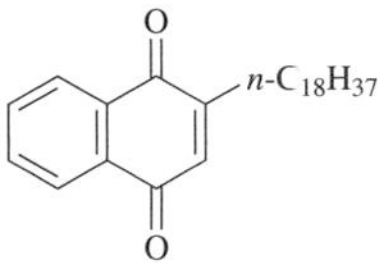	β-Cyclodextrin (500 mol %), TBHP, Na_2CO_3, DMF, rt, 11 d	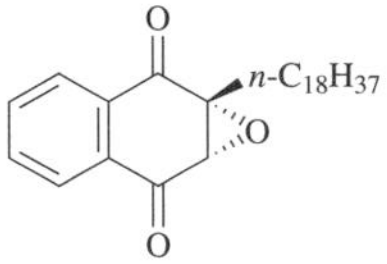	(47), 27	171

[a] The enantiomeric excess was not determined in the original publication, but is calculated by comparison of the reported optical rotation with that in ref. 85.

[b] The absolute configuration was not assigned in the original publication, and only the sign of the optical rotation at an unspecified wavelength is reported. The configuration depicted is assigned by comparison with the results in ref. 85.

[c] This publication assigns the opposite absolute configuration to that depicted here and reported in later publications.

TABLE 6A. α,β-UNSATURATED ALDEHYDES

Substrate	Conditions				Product(s), Yield(s) (%), % ee	Refs.
C_{4-11} 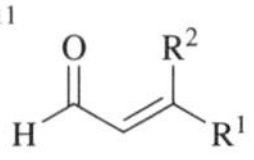	Pyrrolidine **92** (10 mol %), H_2O_2, solvent, H_2O, rt				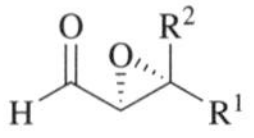	
	R^1	R^2	Solvent	Time (h)		
	Me	H	CH_2Cl_2	4	(>90), 96	168
	Me	Me	CH_2Cl_2	4	(65), 75	168
	i-Pr	H	CH_2Cl_2	4	(75), 96	168
	i-Pr	H	EtOH	16	(48), 91[a,b]	169
	CO_2Et	H	CH_2Cl_2	4	(60), 96	168
	CO_2Et	H	EtOH	16	(53), 88[a]	169
	$4\text{-}FC_6H_4$	H	EtOH	16	(34), 85[a]	169
	$4\text{-}ClC_6H_4$	H	CH_2Cl_2	4	(63), 98	168
	$4\text{-}ClC_6H_4$	H	EtOH	16	(40), 96[a]	169
	$2\text{-}O_2NC_6H_4$	H	CH_2Cl_2	4	(90), 97	168
	$4\text{-}O_2NC_6H_4$	H	EtOH	16	(56), 90[a]	169
	$2\text{-}MeC_6H_4$	H	CH_2Cl_2	4	(65), 96	168
	$BnOCH_2$	H	CH_2Cl_2	4	(84), 94	168
	$BnOCH_2$	H	EtOH	16	(38), 85[a,c]	169
	$Me_2C{=}CH(CH_2)_2$	Me	CH_2Cl_2	4	(73), 85	168
	$Me_2C{=}CH(CH_2)_2$	Me	EtOH	16	(40), 86[a]	169
C_8 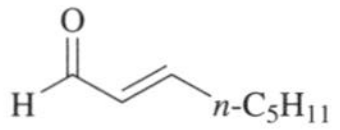	α-Cyclodextrin (20 mol %), H_2O_2, Na_2CO_3, H_2O, 0°, 3 h				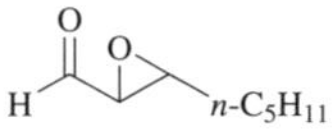 (76), —	172
C_9 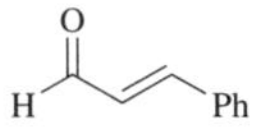	α-Cyclodextrin (100 mol %), H_2O_2, Na_2CO_3, H_2O, 0°, 3 h				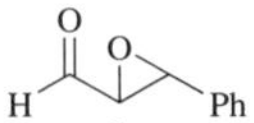 **I** (24), —	172

β-Cyclodextrin (10 mol %), H_2O_2, Na_2CO_3, H_2O, 0°, 3 h	**I** (90), —	172
γ-Cyclodextrin (10 mol %), H_2O_2, Na_2CO_3, H_2O, 0°, 3 h	**I** (80), —	172
Pyrrolidine **92** (10 mol %), oxidant, additive, rt	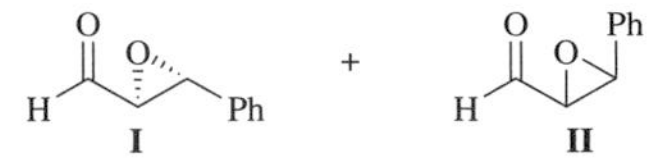	

Oxidant	Additive	Solvent	Time (h)	**I + II**	**I:II**	**I**	**II**	
TBHP	—	CH_2Cl_2	2	(30)	90:10	(—), 93	(—), —	168
CMHP	—	CH_2Cl_2	2	(40)	91:9	(—), 93	(—), —	168
urea/H_2O_2	—	CH_2Cl_2	3	(94)	93:7	(—), 96	(—), —	168
urea/H_2O_2	—	toluene	7	(96)	95:5	(—), 96	(—), —	168
urea/H_2O_2	—	EtOH	7	(92)	87:13	(—), 92	(—), —	168
urea/H_2O_2	—	MeOH	2	(92)	87:13	(—), 92	(—), —	168
urea/H_2O_2	—	THF	2	(94)	93:7	(—), 94	(—), —	168
H_2O_2	—	CH_2Cl_2/H_2O	2	(>95)	93:7	(80), 96	(—), —	168, 169
H_2O_2	—	H_2O	18	(28)	97:3	(—), 90	(—), 10	169
H_2O_2	$KHSO_4$	H_2O	20	(11)	69:31	(—), 93	(—), 18	169
H_2O_2	NaH_2PO_4	H_2O	20	(58)	17:83	(—), 28	(—), 6	169
H_2O_2	$NaHCO_3$	H_2O	20	(40)	93:7	(—), 76	(—), 16	169
H_2O_2	—	$EtOH/H_2O$	9	(94)	86:14	(43), 92	(—), —	169
H_2O_2	—	THF/H_2O	9	(53)	84:16	(—), 92	(—), —	169

[a] The product was isolated as a primary alcohol following reduction with sodium borohydride. The yield and enantiomeric excess quoted are those of the alcohol.

[b] The product was obtained as a 50:50 mixture of diastereomers. It is not clear to which isomer the reported enantiomeric excess refers.

[c] The product was obtained as a 43:57 mixture of diastereomers. It is not clear to which isomer the reported enantiomeric excess refers.

TABLE 6B. α,β-UNSATURATED ESTERS

Substrate	Conditions	Product(s), Yield(s) (%), % ee	Refs.
C_7 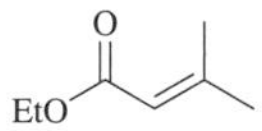	Chiral ketone **39** (30 mol %), Oxone, Bu_4NHSO_4, $NaHCO_3$, Na_2EDTA, H_2O, MeCN, 0° to rt, 24 h	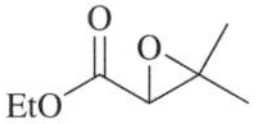 (64), 82	150, 290
C_8 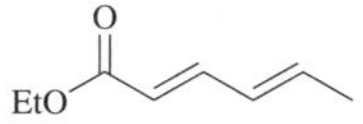	Chiral ketone **40** (30 mol %), Oxone, K_2CO_3, $Na_2B_4O_7$, EDTA, H_2O, MeCN, $(MeO)_2CH_2$, 0°, 4 h	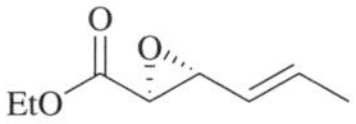 (7), 69 + 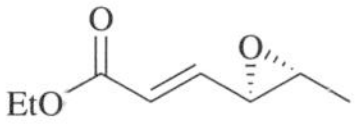 (41), 96	200, 199
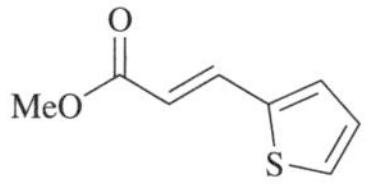	Binaphthyl ketone **41a** (5 mol %), Oxone, $NaHCO_3$, dioxane, H_2O, 10°, 48 h	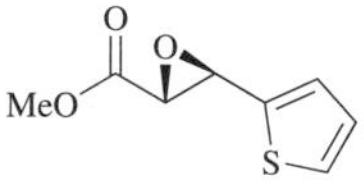 (5), 60	155
C_9 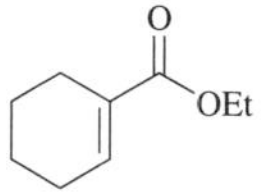	Chiral ketone **39** (25 mol %), Oxone, Bu_4NHSO_4, $NaHCO_3$, Na_2EDTA, H_2O, MeCN, 0° to rt, 24 h	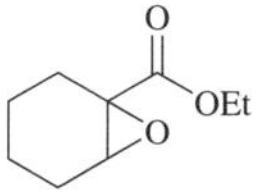 (77), 93	150, 290
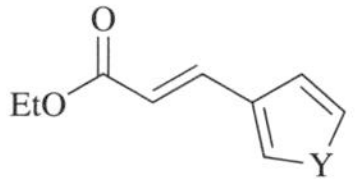	Biphenyl diol **91** (x mol %), TBHP, Y(O*i*-Pr)$_3$, TPAO, 4 Å MS, THF, toluene, rt	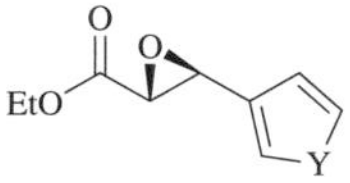 Y / x / Time (h): O / 5 / 27 / (78), 92; S / 3 / 24 / (97), 93	64
C_{10} 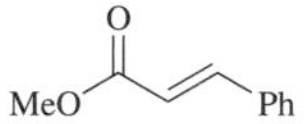	(*S*)-BINOL **17a** (10 mol %),[a] TBHP, Pr(O*i*-Pr)$_3$, TPAO, 4 Å MS, THF, toluene, rt, 72 h	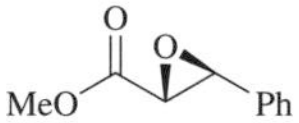 **I** (24), 88	64
	(*S*)-BINOL **17a** (10 mol %),[a] TBHP, Y(O*i*-Pr)$_3$, TPAO, 4 Å MS, THF, toluene, rt, 72 h	**I** (36), 95	64

Conditions	Product(s) and Yield(s) (%), ee (%)	Refs.
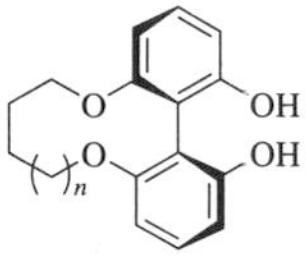 (10 mol %), TBHP, $Y(Oi\text{-}Pr)_3$, TPAO, 4 Å MS, THF, toluene, rt, 144 h	**I** *n* / Time (h) 1 / 144 / (4), 92 2 / 120 / (45), 98 3 / 120 / (49), 99	64
Biphenyl diol **91** (x mol %), TBHP, $Y(Oi\text{-}Pr)_3$, TPAO, 4 Å MS, THF, toluene, rt	**I** x / Time (h) 10 / 48 / (61), 99 5 / 48 / (65), 99 3 / 50 / (79), 99 2 / 65 / (81), 99	64
Binaphthyl ketone **41a** (5 mol %), Oxone, $NaHCO_3$, dioxane, H_2O, 10–27°, 53 h	**I** (75), 74	154, 155
Egg phosphatidylcholine, mCPBA, H_2O, EtOH, rt, 2 h	**I** (75), 92	174
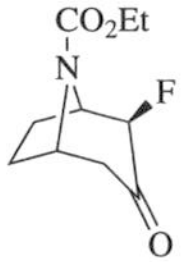 (25 mol %), Oxone, $NaHCO_3$, Na_2EDTA, H_2O, MeCN, 24 h	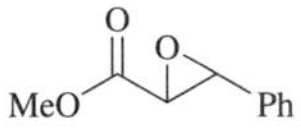 (33), 64	146
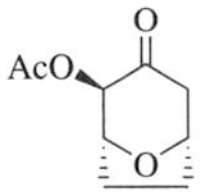 (2 mol %), Oxone, $NaHCO_3$, Na_2EDTA, H_2O, MeCN, rt, 24 h	**I** (*4*), 69	145

Substrate	Conditions	Product(s), Yield(s) (%), % ee	Refs.
C_{10} 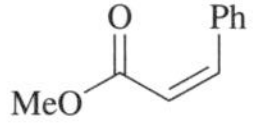	Mn–salen complex **61** (2–10 mol %), NaOCl, NaOH, Na_2HPO_4, H_2O, CH_2Cl_2, 4°, 6 h	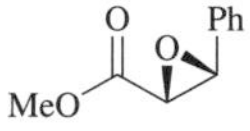 (74), 75	291
	Mn–salen complex **61**[b] (10 mol %), NaOCl, NaOH, Na_2HPO_4, 4-phenylpyridine *N*-oxide, H_2O, CH_2Cl_2, 4°, 3 h	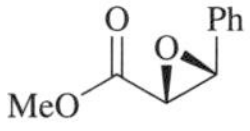 + 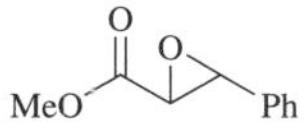 **I** (—), 87–89; **II** (—), 60; **I** + **II** (80); **I**:**II** 4:1	291
	Mn–salen complex **62a** (2 mol %), NaOCl, NaOH, Na_2HPO_4, 4-phenylpyridine *N*-oxide, H_2O, CH_2Cl_2, 4°, 6 h	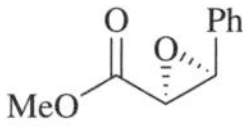 (65), 89	292, 293
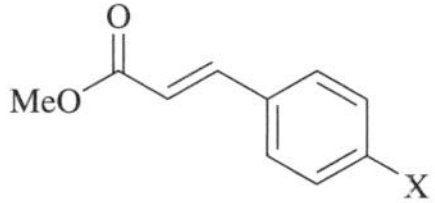	Binaphthyl ketone **41a** (5 mol %), Oxone, $NaHCO_3$, dioxane, H_2O, 10°, 48 h	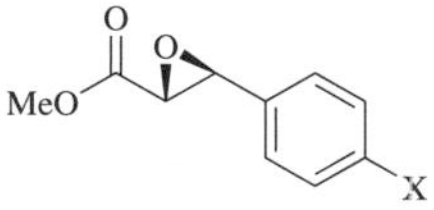 X: F (74), 72; Cl (45), 73	155
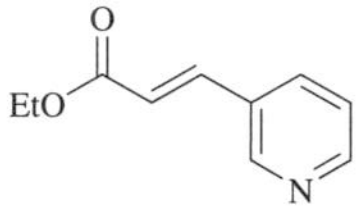	Biphenyl diol **91** (3 mol %), TBHP, Y(O*i*-Pr)$_3$, TPAO, 4 Å MS, THF, toluene, rt, 24 h	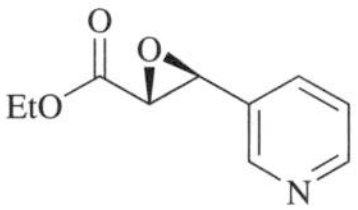 (93), 98	64
C_{10-13} 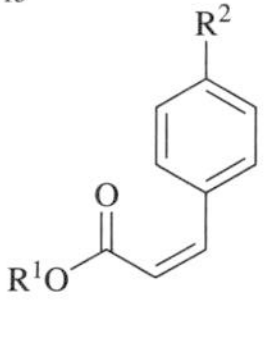	Mn–salen complex **62a** (5 mol %), NaOCl, 4-phenylpyridine *N*-oxide, H_2O, DCE, 4°, 0.5–3 h	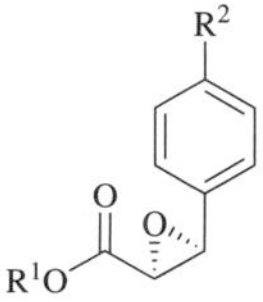 + 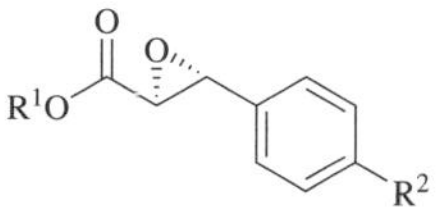 **I** (—) **II** (—)	164

R[1]	R[2]	**I**:**II**	% ee **I**	% ee **II**
Me	H	5.7:1	85	62
Me	F	5.4:1	78	50
Me	Cl	3.0:1	80	53
Me	Br	3.2:1	81	53
Me	NO_2	1:3.7	91	53
Me	OMe	11.7:1	72	66
Me	Me	7.0:1	79	41
Me	CF_3	1:1.25	79	55
Me	CN	1:2.1	84	57
Et	H	—	92	—
i-Pr	H	—	96	—
Et	OMe	—	86	—
i-Pr	OMe	—	96	—

C_{11}

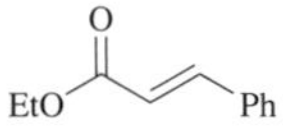

(*S*)-BINOL **17a** (20 mol %), TBHP, La(O*i*-Pr)$_3$,TPAO, 4 Å MS, THF, decane, rt, 4 h

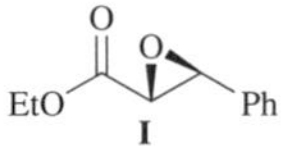

I

(5), 90 — 57, 58

Biphenyl diol **91** (2 mol %), TBHP, Y(O*i*-Pr)$_3$, TPAO, 4 Å MS, THF, toluene, rt, 36 h — **I** (89), 99 — 64

Biphenyl diol **91** (2 mol %), TBHP, Y(O*i*-Pr)$_3$, TPPO, 4 Å MS, THF, toluene, rt, 45 h — **I** (94), 97 — 64

(L-Leu)$_n$ (0.1 mol %), H_2O_2, NaOH, Bu_4NHSO_4, toluene, rt, 3 h — **I** (30), 47 — 201

TABLE 6B. α,β-UNSATURATED ESTERS (*Continued*)

Substrate	Conditions	Product(s), Yield(s) (%), % ee	Refs.
C_{11} 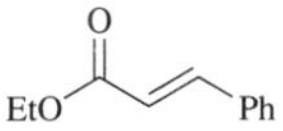	Chiral ketone **37** (10 mol %), Oxone, K_2CO_3, H_2O, DME, AcOH, 0°, 8 h	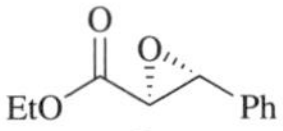 **I** (35), 89	163
	Chiral ketone **36** (10 mol %), Oxone, K_2CO_3, H_2O, DME, AcOH, 0°, 8 h	**I** (34), 86	149, 163
	Chiral ketone **38** (10 mol %), Oxone, K_2CO_3, $Na_2B_4O_7$, EDTA, H_2O, MeCN, $(MeO)_2CH_2$, 0°, 8 h	**I** (34), 86	200
	Chiral ketone **39** (30 mol %), Oxone, Bu_4NHSO_4, $NaHCO_3$, Na_2EDTA, H_2O, MeCN, 0° to rt, 24 h	**I** (73), 96	150, 290
	Ru–porphyrin **72** (0.1 mol %), 2,6-dichloropyridine *N*-oxide, benzene, rt, 16 h	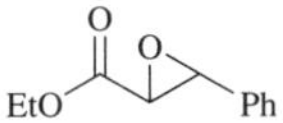 (18), 16	167
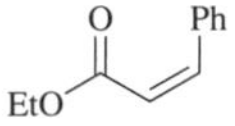	Binaphthyl ketone **41a** (5 mol %), Oxone, $NaHCO_3$, dioxane, H_2O, 10°, 48 h	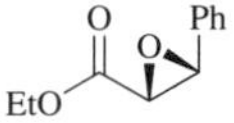 (57), 26	155
	Chiral ketone **39** (30 mol %), Oxone, Bu_4NHSO_4, $NaHCO_3$, Na_2EDTA, H_2O, MeCN, 0° to rt, 24 h	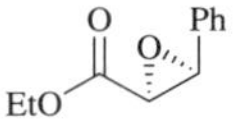 (84), 44	150
	Mn–salen complex **62a** (6.5 mol %), NaOCl, NaOH, Na_2HPO_4, 4-phenylpyridine *N*-oxide, H_2O, CH_2Cl_2, 4°, 12 h	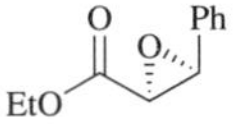 (56), 95–97 + 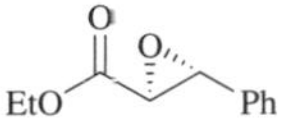 (10), —	293, 291, 164

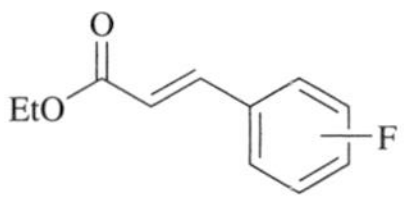	Chiral ketone **39** (30 mol %), Oxone, Bu_4NHSO_4, $NaHCO_3$, Na_2EDTA, H_2O, MeCN, 0° to rt, 24 h	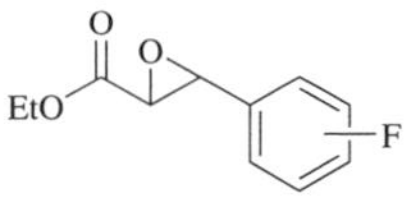	Isomer 2-F (40), 95 3-F (41), 97 4-F (77), 96	150, 290 150 150, 290
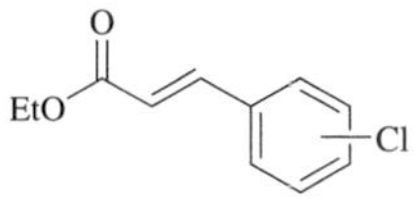	Biphenyl diol **91** (2 mol %), TBHP, $Y(Oi\text{-}Pr)_3$, TPAO, 4 Å MS, THF, toluene, rt, 20 h	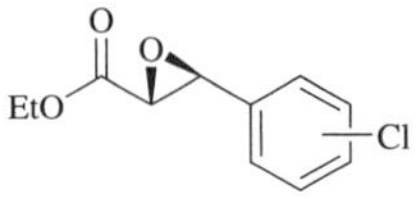	Isomer Time (h) 3-Cl 20 (92), 99 4-Cl 24 (90), 99	64
	Chiral ketone **39** (30 mol %), Oxone, Bu_4NHSO_4, $NaHCO_3$, Na_2EDTA, H_2O, MeCN, 0° to rt, 24 h	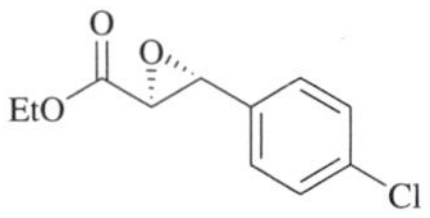	(64), 97[c]	150, 290
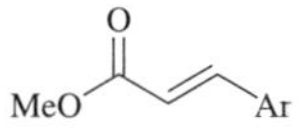	Binaphthyl ketone **41a** (5 mol %), Oxone, $NaHCO_3$, dioxane, H_2O	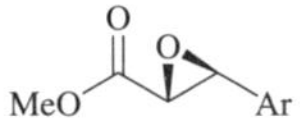		

Ar	Temp (°)	Time (h)		
4-MeC_6H_4	10–27	42	(95), 72	154, 155
4-$CF_3C_6H_4$	10	48	(14), 86	155
2-$MeOC_6H_4$	10	48	(55), 53	155
3-$MeOC_6H_4$	10	48	(11), 47	155

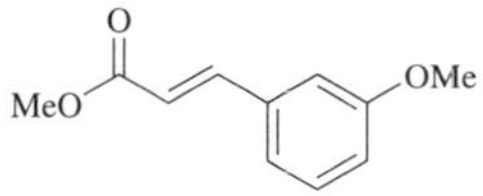	Egg phosphatidylcholine, mCPBA, H_2O, EtOH, rt, 2 h	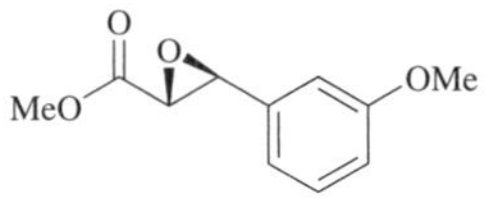	(70), 95	174
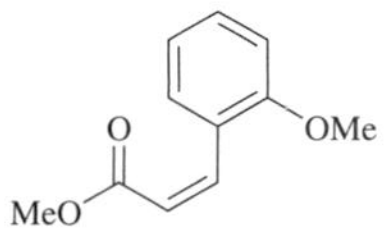	Binaphthyl ketone **41a** (5 mol %), Oxone, $NaHCO_3$, dioxane, H_2O, 10°, 48 h	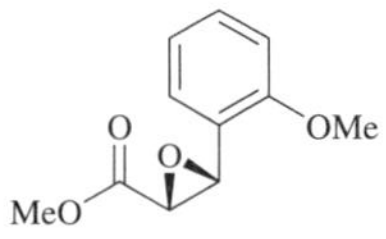	(32), 15	155

Substrate	Conditions	Product(s), Yield(s) (%), % ee	Refs.
C_{11} 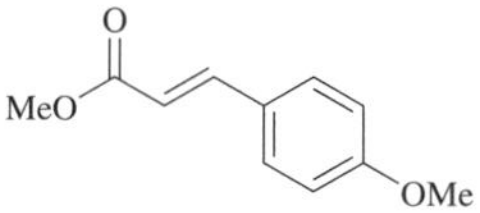	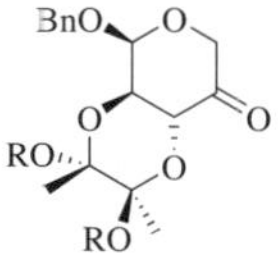 (10 mol %), Oxone, $NaHCO_3$, EDTA, MeCN, H_2O, 12 h	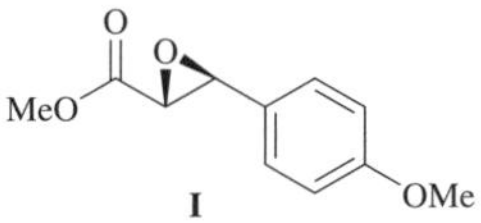 **I**	151
	R		
	i-Bu	(91), 61	
	t-$BuCH_2$	(93), 62	
	Bn	(88), 42	
	c-$C_6H_{11}CH_2$	(91), 59	
	Biphenyl ketone (5 mol %), Oxone, $NaHCO_3$, DME, H_2O, 5–27°, 27 h	**I** Ketone **42a** (52), 67; **42b** (86), 68	154, 155
	Binaphthyl ketone (5 mol %), Oxone, $NaHCO_3$, solvent, H_2O	**I**	

Ketone	Solvent	Temp (°)	Time (h)	Yield (%), % ee	Refs.
41a	DME	27	3.5	(86), 76	155, 154
41b	DME	27	27	(74), 85	155, 154
41a	MeCN	27	24	(33), 72	155, 154
41a	DMF	27	5.5	(45), 70	155
41a	dioxane	27	7	(81), 76	155, 154
41a	1,3-dioxolane	27	6	(23), 75	155
41a	$EtOCH_2CH_2OH$	27	6	(67), 71	155
41a	$MeOCH_2CH_2OH$	27	5.5	(62), 73	155
41a	*i*-PrOH	27	23.5	(19), 73	155
41a	MeOH	27	22.5	(27), 63	155, 154
41a	acetone	27	3.5	(93), 49	155, 154
41a	dioxane	5	48	(92), 80	155, 154
41a	dioxane	40	1.5	(66), 70	154

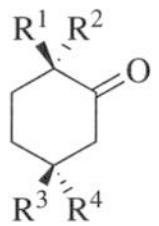

(30 mol %),

Oxone, K_2CO_3, AcOH, H_2O

I

R^1	R^2	R^3	R^4	Solvent	Temp (°)	Time (h)		
Cl	*i*-Pr	H	Me	dioxane	rt	6	(14), 2	162
i-Pr	Cl	H	Me	dioxane	rt	6	(24), 2	162
Cl	Me	H	*i*-Pr	dioxane	rt	6	(20), 5	162
Me	Cl	H	*i*-Pr	dioxane	0	6	(22), 43	162
Me	F	H	*i*-Pr	dioxane	0	6	(97), 37	162, 159, 156
Me	F	H	*i*-Pr	DME	0	8	(71), 46[d]	162
Me	F	H	*i*-Pr	DME	0	8	(82), 46	162
Me	F	H	*i*-Pr	DME	20	8	(99), 40	161
Me	F	H	$CClMe_2$	DME	0	8	(88), 58	161, 159
Me	F	H	$CFMe_2$	DME	0	8	(74), 60	161, 159
Me	F	H	$CFMe_2$	dioxane	rt	6	(74), 60	160
Me	F	Me	Ph	DME	20	8	(90), 66	161
Me	F	Me	Ph	dioxane	rt	6	(90), 66	160, 156
F	Me	Me	Ph	dioxane	0	6	(57), 30	156
F	Me	Bn	Me	dioxane	0	6	(88), 38	156
Me	F	H	2-methyloxiran-2-yl	DME	0	6	(100), 66	159, 158
Me	F	H	2-methyloxiran-2-yl	dioxane	rt	6	(100), 66	158
Me	F	H	$C(OH)Me_2$	dioxane	rt	6	(100), 26	158
Me	F	H	$C(OH)MeCH_2OH$	dioxane	rt	6	(100), 44	158
Me	F	H	(*S*)-2,5,5-trimethyl-1,3-dioxolan-2-yl	dioxane	rt	6	(100), 60	158
Me	F	H	(*R*)-2,5,5-trimethyl-1,3-dioxolan-2-yl	dioxane	rt	6	(100), 60	158
Me	F	H	(*R*,*S*)-2,5,5-trimethyl-1,3-dioxolan-2-yl	DME	0	8	(53), 66	158

TABLE 6B. α,β-UNSATURATED ESTERS (*Continued*)

Substrate	Conditions	Product(s), Yield(s) (%), % ee	Refs.
C_{11} 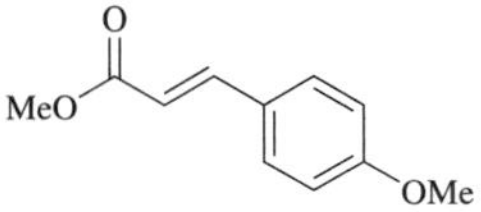	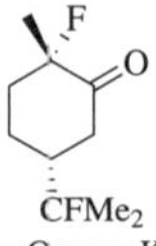 (30 mol %), Oxone, K_2CO_3, $Na_2B_4O_7$, Na_2EDTA, Bu_4NHSO_4, MeCN, H_2O, 0°, 2 h	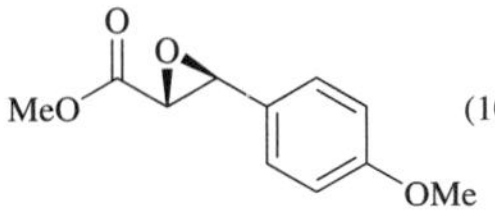 (100), 76	160
	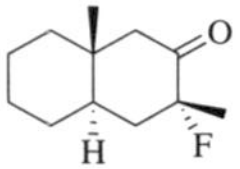 (30 mol %),[e] Oxone, K_2CO_3, AcOH, H_2O, 0°, 6 h	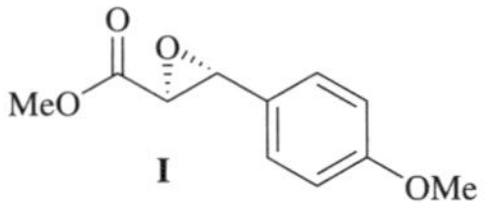 **I**	
	Solvent		
	dioxane	(75), 45[f]	160
	DME	(75), 41[f]	157
	Chiral ketone **40** (30 mol %), Oxone, K_2CO_3, AcOH, dioxane, H_2O, 0°, 6 h	**I** (30), 34	160
	Chiral ketone **40** (30 mol %), Oxone, K_2CO_3, $Na_2B_4O_7$, Na_2EDTA, Bu_4NHSO_4, MeCN, H_2O, 0°, 2 h	**I** (43), 86	160
	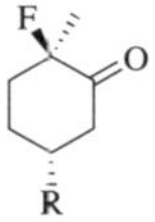 (30 mol %), Oxone, K_2CO_3, AcOH, H_2O	**I**	

R	Solvent	Temp	Time (h)		Refs.
i-Pr	dioxane	rt	8	(43), 6	162, 159
$CFMe_2$	DME	0°	6	(17), 10	159
2-methyloxiran-2-yl	dioxane	rt	8	(20), 8	159

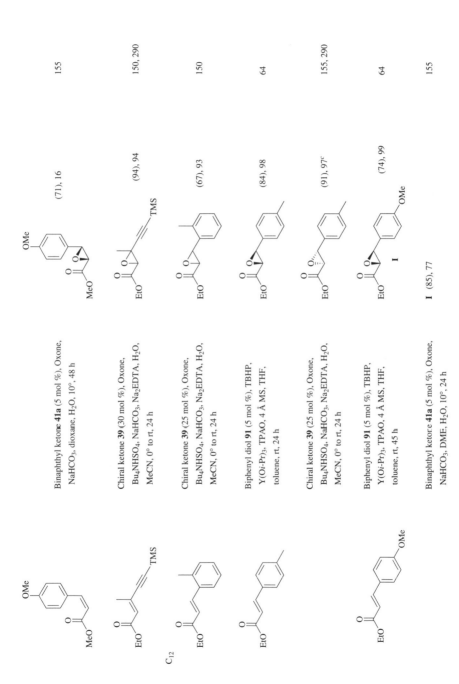

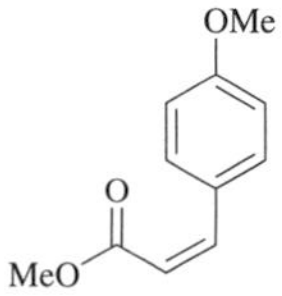	Binaphthyl ketone **41a** (5 mol %), Oxone, $NaHCO_3$, dioxane, H_2O, 10°, 48 h	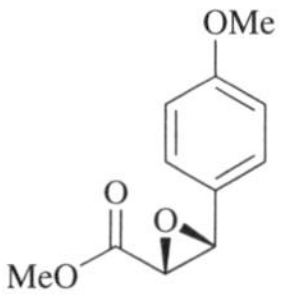 (71), 16	155
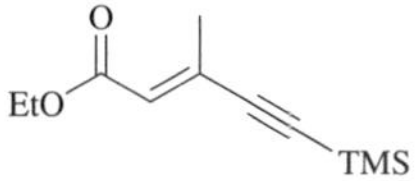	Chiral ketone **39** (30 mol %), Oxone, Bu_4NHSO_4, $NaHCO_3$, Na_2EDTA, H_2O, MeCN, 0° to rt, 24 h	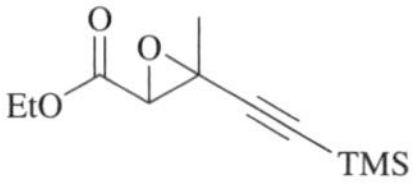 (94), 94	150, 290
C_{12} 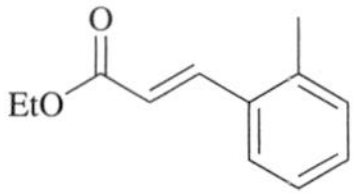	Chiral ketone **39** (25 mol %), Oxone, Bu_4NHSO_4, $NaHCO_3$, Na_2EDTA, H_2O, MeCN, 0° to rt, 24 h	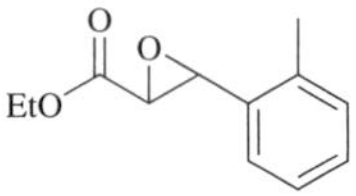 (67), 93	150
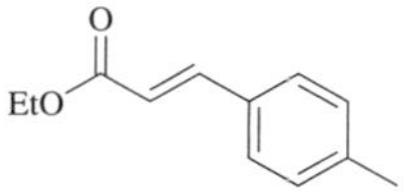	Biphenyl diol **91** (5 mol %), TBHP, Y(O*i*-Pr)$_3$, TPAO, 4 Å MS, THF, toluene, rt, 24 h	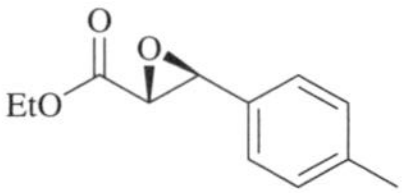 (84), 98	64
	Chiral ketone **39** (25 mol %), Oxone, Bu_4NHSO_4, $NaHCO_3$, Na_2EDTA, H_2O, MeCN, 0° to rt, 24 h	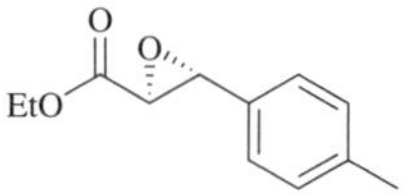 (91), 97[c]	155, 290
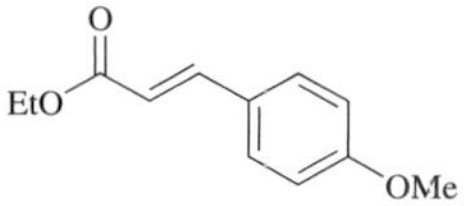	Biphenyl diol **91** (5 mol %), TBHP, Y(O*i*-Pr)$_3$, TPAO, 4 Å MS, THF, toluene, rt, 45 h	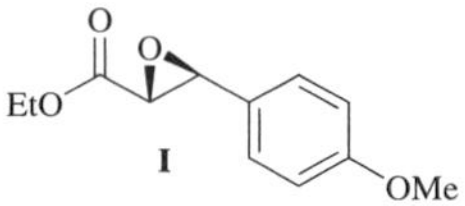 **I** (74), 99	64
	Binaphthyl ketone **41a** (5 mol %), Oxone, $NaHCO_3$, DME, H_2O, 10°, 24 h	**I** (85), 77	155

TABLE 6B. α,β-UNSATURATED ESTERS (*Continued*)

Substrate	Conditions	Product(s), Yield(s) (%), % ee	Refs.
C_{12} 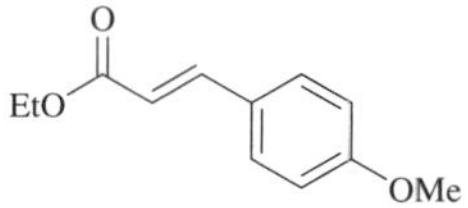	Chiral ketone **39** (20 mol %), Oxone, Bu_4NHSO_4, $NaHCO_3$, Na_2EDTA, H_2O, MeCN, 0°, 12 h	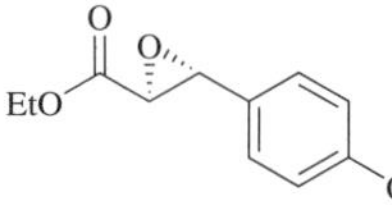 (91), 97	150, 290
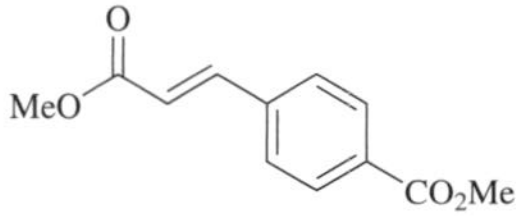	Binaphthyl ketone **41a** (5 mol %), Oxone, $NaHCO_3$, dioxane, H_2O, 10°, 48 h	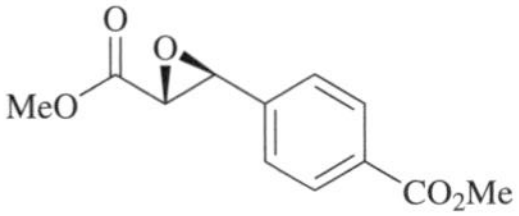 (6), 81	155
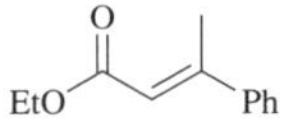	Binaphthyl ketone **41a** (5 mol %), Oxone, $NaHCO_3$, dioxane, H_2O, 10°, 48 h	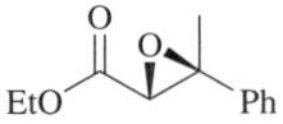 (93), 40	155
	Chiral ketone **39** (25 mol %), Oxone, Bu_4NHSO_4, $NaHCO_3$, Na_2EDTA, H_2O, MeCN, 0° to rt, 24 h	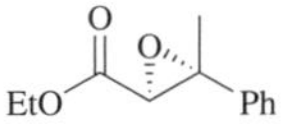 (93), 96	150, 290
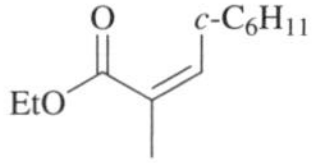	Chiral ketone **39** (30 mol %), Oxone, Bu_4NHSO_4, $NaHCO_3$, Na_2EDTA, H_2O, MeCN, 0° to rt, 24 h	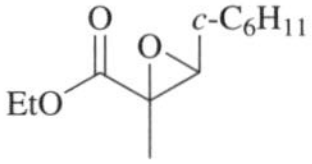 (77), 89	150
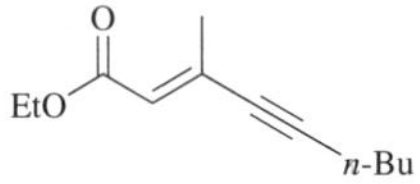	Chiral ketone **39** (25 mol %), Oxone, Bu_4NHSO_4, $NaHCO_3$, Na_2EDTA, H_2O, MeCN, 0° to rt, 24 h	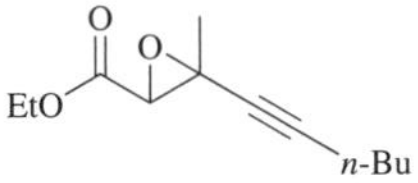 (74), 98	150
C_{13} 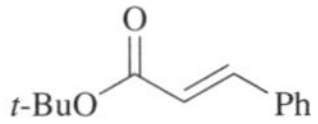	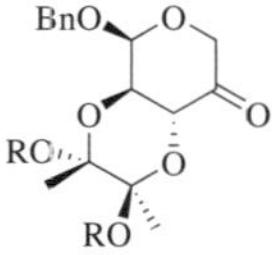 (10 mol %), Oxone, $NaHCO_3$, EDTA, MeCN, H_2O, 12 h	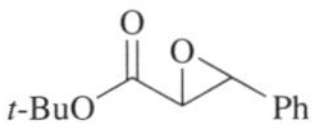 R: *i*-Bu (88), 81; *t*-$BuCH_2$ (90), 86; Bn (92), 66; *c*-$C_6H_{11}CH_2$ (84), 77	151

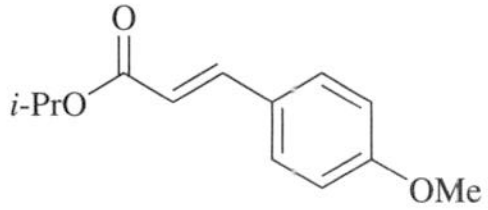	Binaphthyl ketone **41a** (5 mol %), Oxone, $NaHCO_3$, DME, H_2O, 10°, 24 h	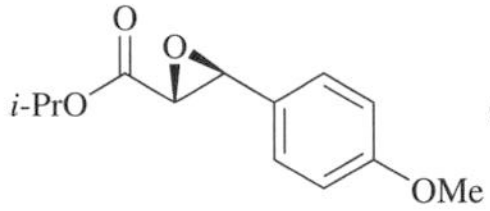 (85), 80	155
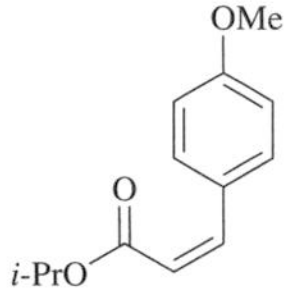	Mn–salen complex **62b** (4 mol %), benzyl quininium salt **4a** (25 mol %), NaOCl, H_2O, chlorobenzene, 4°, 10 h	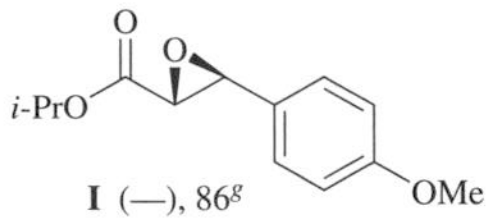 **I** (—), 86[g] + 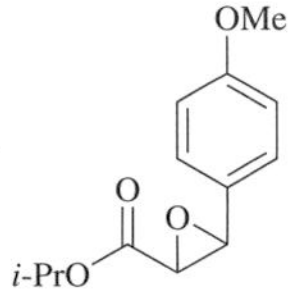 **II** (—), — **I**:**II** = 89:11	165
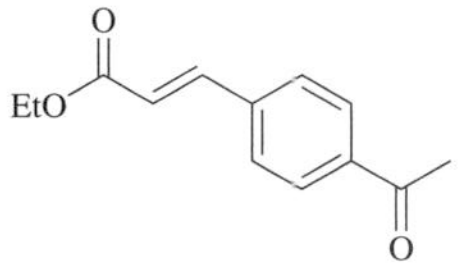	Biphenyl diol **91** (2 mol %), TBHF, Y(O*i*-Pr)$_3$, TPAO, 4 Å MS, THF, toluene, rt, 24 h	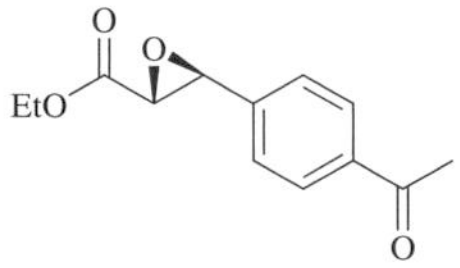 (89), 89	64
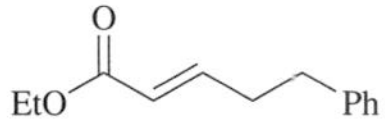	Biphenyl diol **91** (x mol %), TBHF, Y(O*i*-Pr)$_3$, additive, 4 Å MS, THF, toluene, rt x / Additive / Time 10 / TPAO / 47 h 5 / TPPO / 48 h	 (86), 91 (80), 86	64
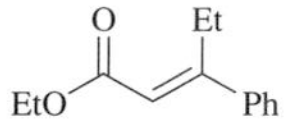	Chiral ketone **39** (25 mol %), Oxone, Bu_4NHSO_4, $NaHCO_3$, Na_2EDTA, H_2O, MeCN, 0° to rt, 24 h	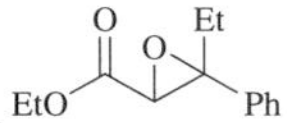 (91), 93	150, 290
C_{14}			
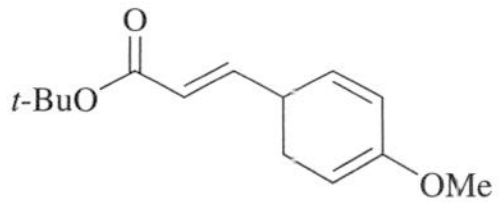	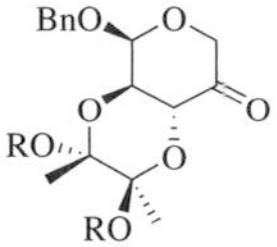 (10 mol %), Oxone, $NaHCO_3$, EDTA, MeCN, H_2O, 12 h	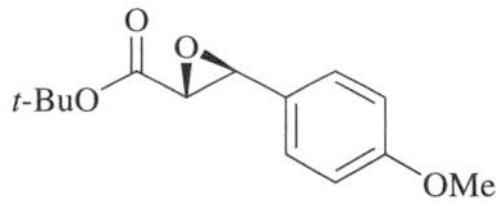 R / *i*-Bu (96), 80 *t*-BuCH$_2$ (92), 81 Bn (89), 63 *c*-C$_6$H$_{11}$CH$_2$ (93), 72	151

TABLE 6B. α,β-UNSATURATED ESTERS (*Continued*)

Substrate	Conditions	Product(s), Yield(s) (%), % ee	Refs.
C_{14}			
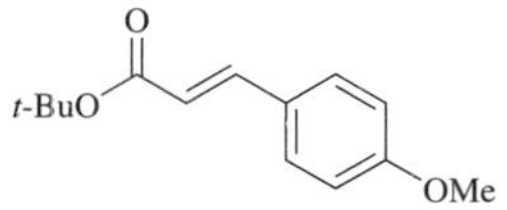	Binaphthyl ketone **41a** (5 mol %), Oxone, $NaHCO_3$, DME, H_2O, 10°, 24 h	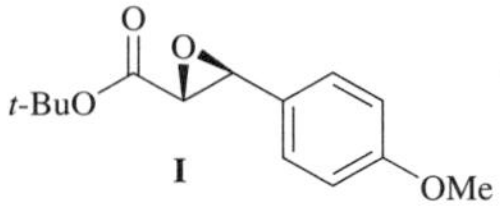 **I** (89), 77	155
	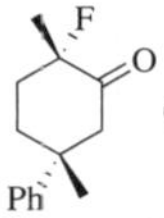 (30 mol %), Oxone, K_2CO_3, AcOH, DME, H_2O, 0°, 8 h	**I** (—), 74	161
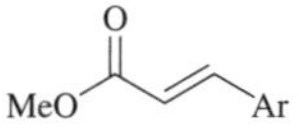	Binaphthyl ketone **41a** (5 mol %), Oxone, $NaHCO_3$, solvent, H_2O, 10°	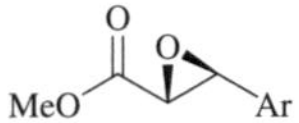 (81), 92	155
Ar	Solvent / Time (h)		
4-(*t*-Bu)C_6H_4	dioxane / 48	(81), 92	
4-(*t*-BuO)C_6H_4	DME / 24	(83), 85	
1-Np	dioxane / 48	(55), 75	
2-Np	dioxane / 48	(47), 63	
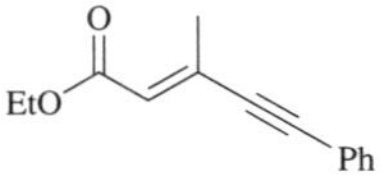	Chiral ketone **39** (25 mol %), Oxone, Bu_4NHSO_4, $NaHCO_3$, Na_2EDTA, H_2O, MeCN, 0° to rt, 24 h	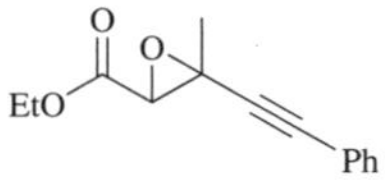 (96), 94	150
C_{15}			
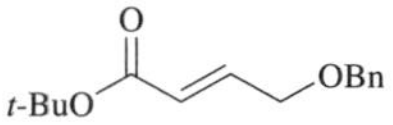	(L-Leu)$_n$–PS (20 mol %), urea/H_2O_2, DBU, THF, rt, 72 h	 (20), >85	140

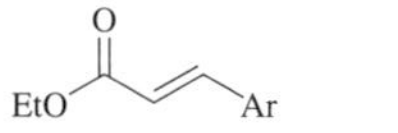	Biphenyl diol **91** (x mol %), TBHP, $Y(Oi\text{-}Pr)_3$, TPAO, 4 Å MS, THF, toluene, rt	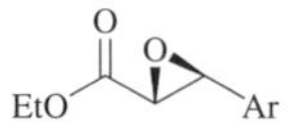	64
Ar: 1-Np	x = 5, Time (h) = 40	(62), 97	
Ar: 2-Np	x = 2, Time (h) = 24	(89), 99	
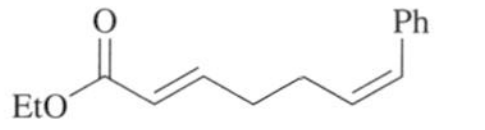	Biphenyl diol **91** (10 mol %), TBHP, $Y(Oi\text{-}Pr)_3$, TPAO, 4 Å MS, THF, toluene, rt, 71 h	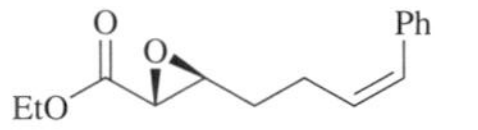 (81), 93	64
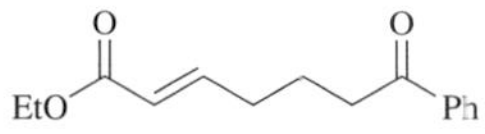	Biphenyl diol **91** (10 mol %), TBHP, $Y(Oi\text{-}Pr)_3$, TPAO, 4 Å MS, THF, toluene, rt, 42 h	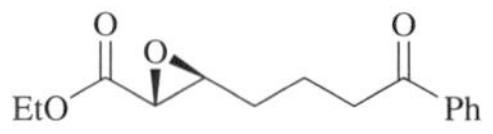 (78), 92	64
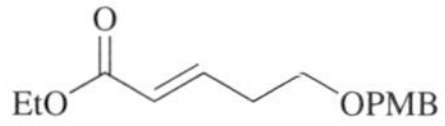	Biphenyl diol **91** (10 mol %), TBHP, $Y(Oi\text{-}Pr)_3$, TPAO, 4 Å MS, THF, toluene, rt, 66 h	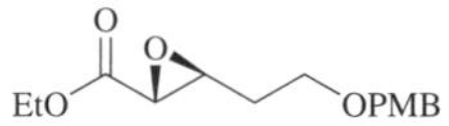 (81), 96	64
C_{16}			
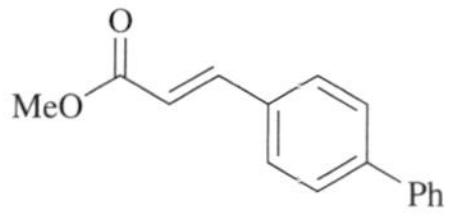	Binaphthyl ketone **41a** (5 mol %), Oxone, $NaHCO_3$, dioxane, H_2O, 10°, 48 h	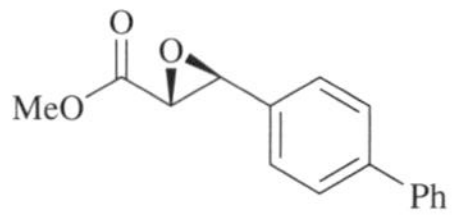 (51), 91	155
C_{17}			
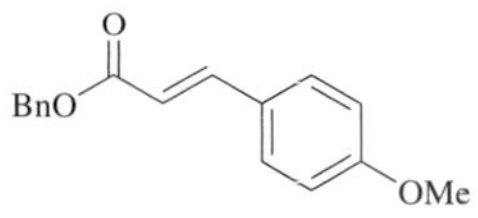	Binaphthyl ketone **41a** (5 mol %), Oxone, $NaHCO_3$, DME, H_2O, 10°, 24 h	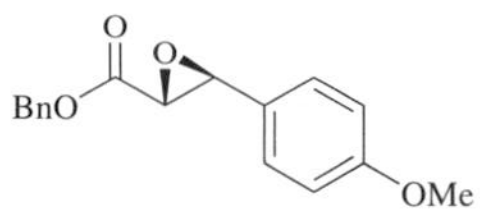 (84), 78	155

TABLE 6B. α,β-UNSATURATED ESTERS (*Continued*)

Substrate	Conditions	Product(s), Yield(s) (%), % ee	Refs.
C_{24} 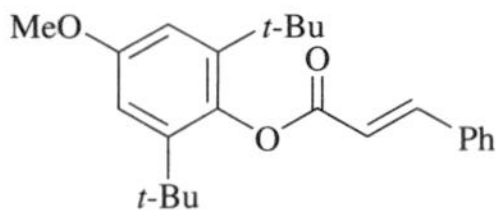	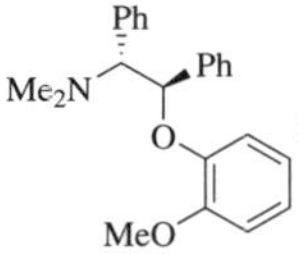 (180 mol %), oxidant, *n*-BuLi, toluene, 0°, 36 h	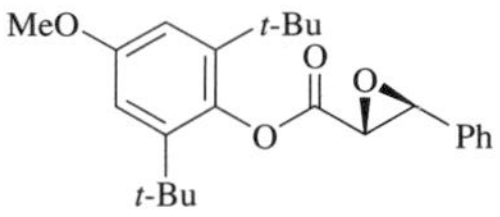	72
	Oxidant		
	TBHP	(62), 72	
	CMHP	(82), 72	
	Ph_3CO_2H	(82), 59	

[a] The configuration of the catalyst used was not specified, but is inferred from other results in the same paper.

[b] The catalyst is incorrectly depicted in this publication with two extra methyl groups at the ring junction.

[c] The absolute configuration was not determined in the original publication, but is assigned by comparison with ref. 64.

[d] This result was obtained using 10 mol % of the catalyst.

[e] This result was obtained using catalyst of 90% ee.

[f] The original publication gives enantiomeric excesses extrapolated to catalyst of 100% ee. The values given here are interpolated back to catalyst of 90% ee.

[g] Product **I** is described in this publication as the (*S*,*S*)-epoxide. The (2*R*,3*S*)-epoxide depicted here is assigned on the basis of the mechanistic rationale presented by the authors of the original publication.

TABLE 6C. α,β-UNSATURATED ACIDS

Substrate	Conditions	Product(s), Yield(s) (%), % ee	Refs.
C_9 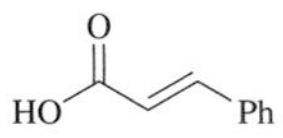	Steroidal ketone **89** (100 mol %), Oxone, $NaHCO_3$, EDTA, H_2O, 0°, 2 h	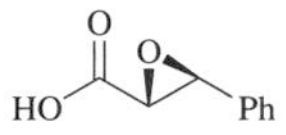 (79), 48	152
	Steroidal ketone **73** (100 mol %), Oxone, $NaHCO_3$, EDTA, H_2O, 0°, 2 h	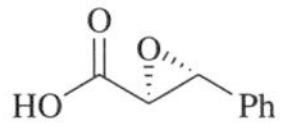 (80), 70	152
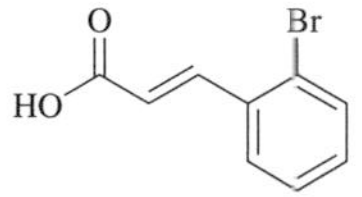	Steroidal ketone **89** (100 mol %), Oxone, $NaHCO_3$, EDTA, H_2O, 0°, 2 h	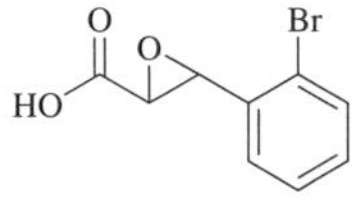 (78), 57	152
	Steroidal ketone **73** (100 mol %), Oxone, $NaHCO_3$, EDTA, H_2O, 0°, 2 h	**I** (75), 82[a]	152
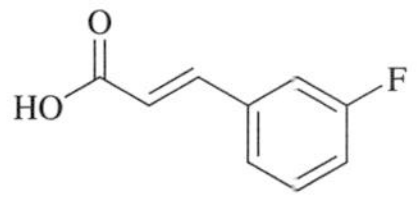	Steroidal ketone **89** (100 mol %), Oxone, $NaHCO_3$, EDTA, H_2O, 0°, 2 h	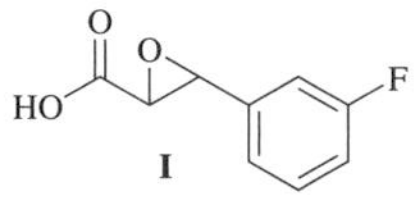 **I** (70), 40	152
	Steroidal ketone **73** (100 mol %), Oxone, $NaHCO_3$, EDTA, H_2O, 0°, 2 h	**I** (63), 46[a]	152

TABLE 6C. α,β-UNSATURATED ACIDS (*Continued*)

Substrate	Conditions	Product(s), Yield(s) (%), % ee	Refs.
C_{9-11} 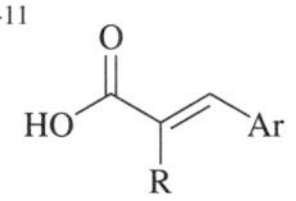	Steroidal ketone **74** (100 mol %), Oxone, $NaHCO_3$, H_2O, 0°	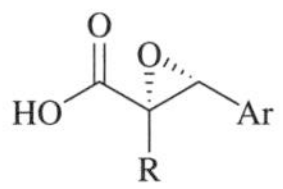	153
Ar / R	Time (min)		
Ph / H	5	(97), 40	
2-BrC_6H_4 / H	5	(98), 55	
3-FC_6H_4 / H	5	(90), 46	
Ph / F	15	(95), 30	
4-MeC_6H_4 / H	5	(60), 75	
4-MeC_6H_4 / H	120	(>99), 68	
4-MeC_6H_4 / H	5	(95), 55[b]	
Ph / Me	5	(98), 12	
2-$HO_2CC_6H_4$ / H	15	(98), 30	
C_{10} 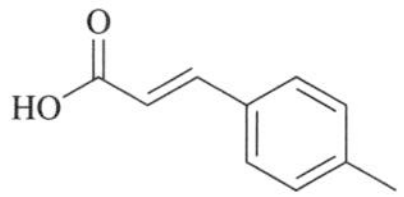	Steroidal ketone (100 mol %), Oxone, $NaHCO_3$, EDTA, H_2O, 0°, 2 h	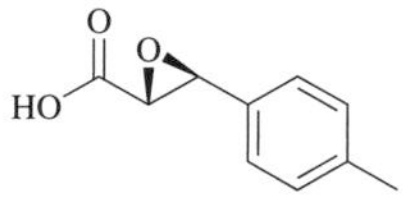	152
	Ketone		
	90	(45), 26[c]	
	89	(90), 50[c]	

Steroidal ketone (100 mol %), Oxone, $NaHCO_3$, EDTA, H_2O, 0°

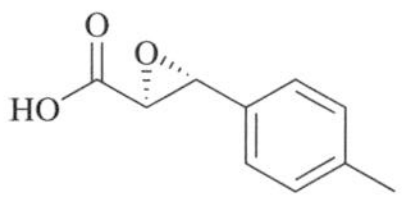

152

Ketone	Time (h)	
74	0.5	(94), 75[c]
75	0.2	(89), 87[c]
76	0.5	(92), 36[c]
77	0.5	(91), 48[c]
78	0.2	(93), 74[c]
79	0.5	(35), 18[c]
80	0.5	(35), 38[c]
81	2	(93), 49[c]
82	2	(91), 16[c]
73	2	(94), 95[c]
83	24	(15), 32[c,d]
84	2	(41), 15[c]
85	24	(38), 8[c]
86	0.5	(6), 41[c]
87	24	(6), 6[c]
88	2	(25), 5[c]

[a] The major product is enantiomeric to that in the preceding entry.

[b] This reaction was carried out at 20°, using 20 mol % catalyst.

[c] The absolute configuration was not determined in the original publication, but is assigned by comparison with ref. 153.

[d] Sodium cholate was added to this reaction.

TABLE 6D. α,β-UNSATURATED AMIDES AND *N*-ACYLATED HETEROCYCLES

Substrate	Conditions	Product(s), Yield(s) (%), % ee	Refs.
C_8			
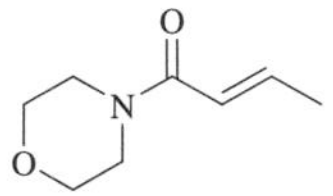	(*S*)-BINOL **17a** (5 mol %), TBHP, $Sm(Oi\text{-}Pr)_3$, TPAO, 4 Å MS, THF, decane, rt	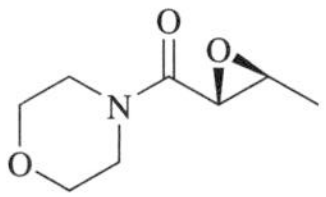 (89), 98	61
C_{10}			
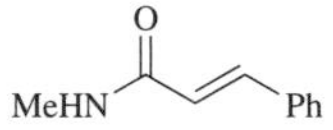	(*S*)-BINOL **17a** (10 mol %), TBHP, $Sm(Oi\text{-}Pr)_3$, TPAO, 4 Å MS, THF, toluene, rt, 18 h	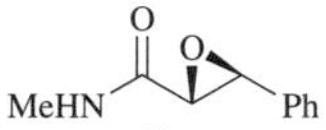 **I** (95), 99	60, 58, 294
	(*S*)-BINOL **17a** (10 mol %), TBHP, $Sm(Oi\text{-}Pr)_3$, TPPO, 4 Å MS, THF, toluene, rt, 9 h	**I** (92), 97	60
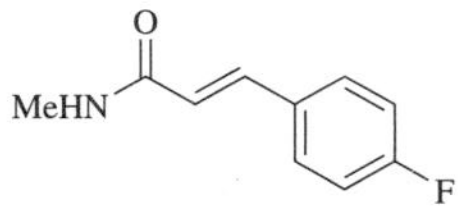	(*S*)-BINOL **17a** (10 mol %), TBHP, $Sm(Oi\text{-}Pr)_3$, TPAO, 4 Å MS, THF, toluene, rt, 20 h	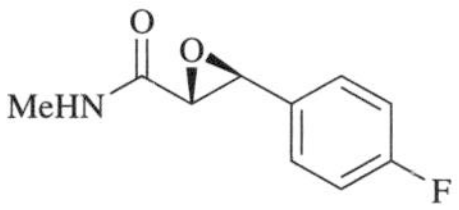 (94), 99	60, 58, 294
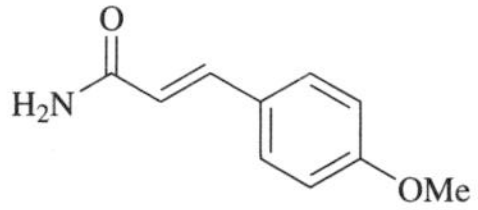	Binaphthyl ketone **41a** (5 mol %), Oxone, $NaHCO_3$, DME, H_2O, 10°, 24 h	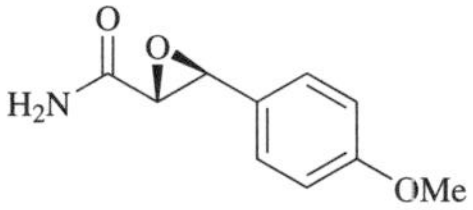 (80), 71	155
C_{11}			
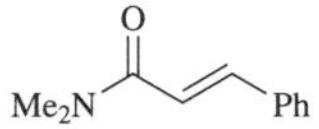	(*S*)-BINOL **17a** (10 mol %), TBHP, $Sm(Oi\text{-}Pr)_3$, TPAO, 4 Å MS, THF, toluene, rt, 9 h	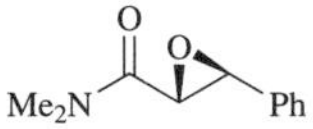 (96), >99	60, 58, 294
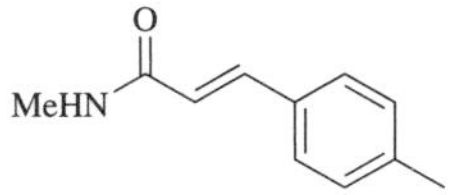	(*S*)-BINOL **17a** (10 mol %), TBHP, $Sm(Oi\text{-}Pr)_3$, TPAO, 4 Å MS, THF, toluene, rt, 21 h	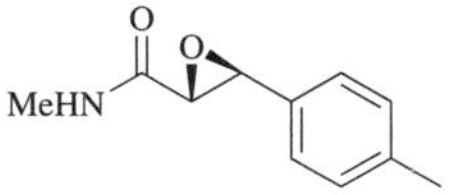 (89), >99	60, 58, 294

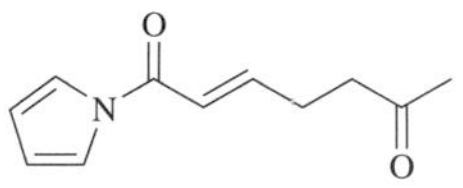	Octahydro-(*R*)-BINOL **19** (5 mol %), CMHP, $Sm(Oi\text{-}Pr)_3$, TPPO, 4 Å MS, THF, toluene, rt, 0.5 h	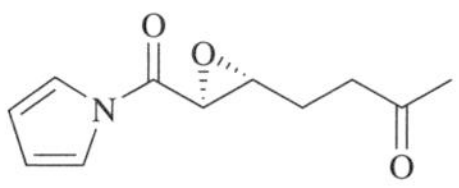 (93), 96[a]	63, 62
C_{12}			
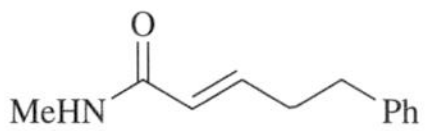	(*S*)-BINOL **17a** (10 mol %), TBHP, $Ln(Oi\text{-}Pr)_3$, 4 Å MS, THF, decane, rt, 8 h	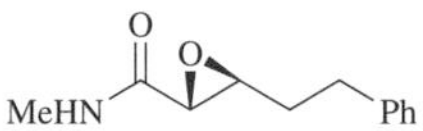	
	Ln / Additive / Time (h)		
	Sm / TPAO / 8	(99), >99	58, 60, 59, 294
	Sm / TPPO / 24	(65), 96	58
	Dy / TPAO / 8	(94), 98	58
	Dy / TPPO / 24	(91), >97	60
	La / TPAO / 48	(93), 97	58
	Pr / TPAO / 8	(96), 99.1	58
	Pr / TPPO / 24	(48), 95	58
	Gd / TPAO / 24	(71), 94	58
	Yb / TPAO / 24	(25), 65	58
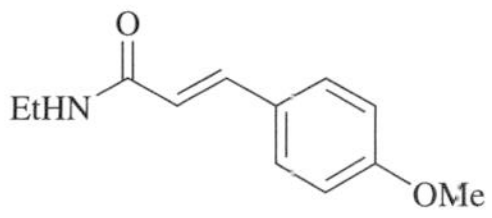	Binaphthyl ketone **41a** (5 mol %), Oxone, $NaHCO_3$, DME, H_2O, 10°, 24 h	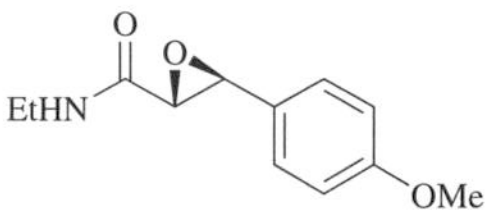 (93), 65	155
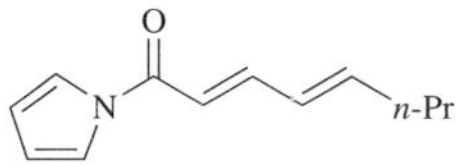	Octahydro-(*R*)-BINOL **19** (10 mol %), TBHP, $Sm(Oi\text{-}Pr)_3$, TPPO, 4 Å MS, THF, toluene, rt, 2 h	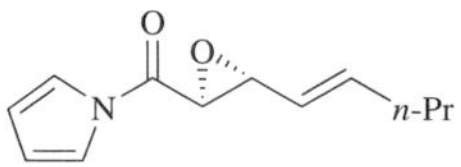 (72), 96[a]	63, 62
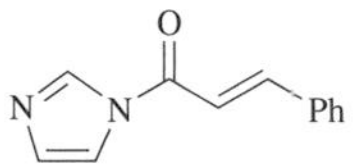	(*S*)-BINOL **17a** (20 mol %), TBHP, $La(Oi\text{-}Pr)_3$, TPAO, 4 Å MS, THF, toluene, rt, 4 h	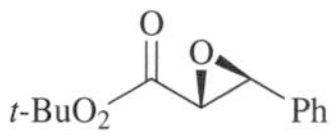 (86), —	56, 57, 58

TABLE 6D. α,β-UNSATURATED AMIDES AND *N*-ACYLATED HETEROCYCLES (*Continued*)

Substrate	Conditions	Product(s), Yield(s) (%), % ee	Refs.
C_{12} 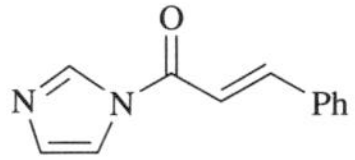	1. (*S*)-BINOL **17a** (20 mol %), TBHP, La(O*i*-Pr)$_3$, TPAO, 4 Å MS, THF, toluene, rt, 4 h 2. MeOH	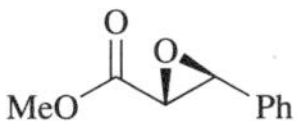 (86), 91	56, 57, 58
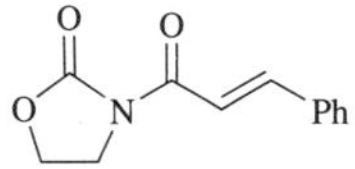	(*S*)-BINOL **17a** (20 mol %), TBHP, La(O*i*-Pr)$_3$, TPAO, 4 Å MS, THF, toluene, rt, 48 h	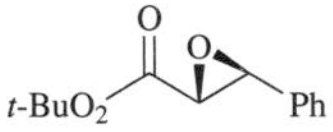 (73), —	57, 58
	1. (*S*)-BINOL **17a** (20 mol %), TBHP, La(O*i*-Pr)$_3$, TPAO, 4 Å MS, THF, toluene, rt, 24 h 2. MeOH	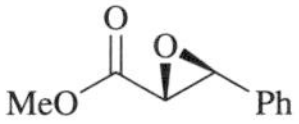 (73), 87	56, 57, 58
C_{13} 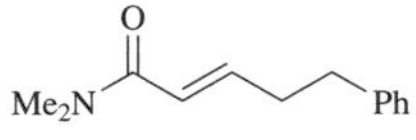	(*S*)-BINOL **17a** (10 mol %), TBHP, Sm(O*i*-Pr)$_3$, TPAO, 4 Å MS, THF, decane, rt, 3 h	 (96), 99	60, 58, 294
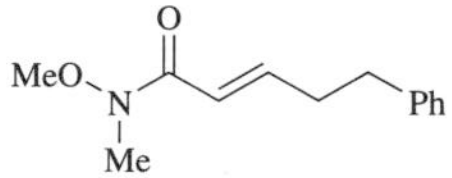	(*S*)-BINOL **17a** (10 mol %), TBHP, Sm(O*i*-Pr)$_3$, TPAO, 4 Å MS, THF, decane, rt	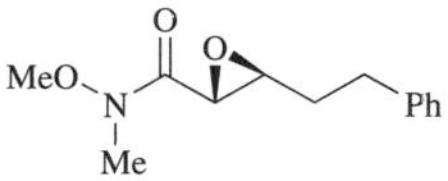 (95), 64	61, 294
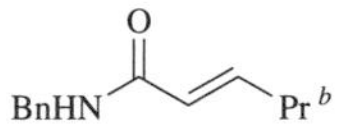	(*S*)-BINOL **17a** (10 mol %), TBHP, Sm(O*i*-Pr)$_3$, TPAO, 4 Å MS, THF, decane, rt, 9 h	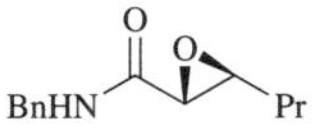 (94), 94	60, 58, 294
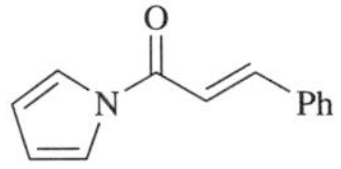	(*R*)-BINOL **15a** (5 mol %), TBHP, Sm(O*i*-Pr)$_3$, TPAO, 4 Å MS, THF, rt, 0.5 h	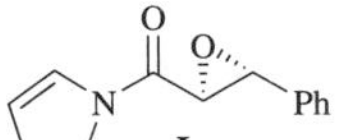 **I** (85), 96	63, 62

Octahydro-(*R*)-BINOL **19** (x mol %), oxidant, $Sm(O\textit{i}\text{-}Pr)_3$, additive, 4 Å MS, solvent, rt — **I**

x	Oxidant	Additive	Solvent	Time (h)		Refs.
5	TBHP	TPAO	THF	0.5	(94), 99	63, 62
5	TBHP	TPPO	THF	0.5	(88), 98	63, 62
5	TBHP	TPPO	THF/toluene	0.2	(97), 99	63, 62
5	CMHP	TPPO	THF/toluene	0.2	(91), >99.5	63, 62
1	TBHP	TPPO	THF/toluene	0.3	(94), 99	63, 62
1	TBHP	TPAO	THF	0.5	(91), 99	62
0.1	TBHP	TPPO	THF/toluene	2	(90), 96	62
0.1	TBHP	TPAO	THF/toluene	0.6	(100), 99	62
0.02	TBHP	TPAO	THF/toluene	1.5	(94), 99	62

Substrate	Conditions	Product(s) and Yield(s) (%), ee		Refs.
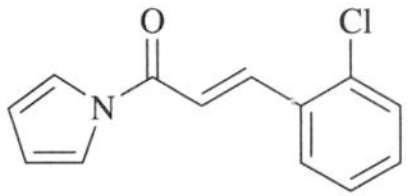	Octahydro-(*R*)-BINOL **19** (5 mol %), TBHP, $Sm(O\textit{i}\text{-}Pr)_3$, TPPO, 4 Å MS, THF, toluene, rt, 0.5 h	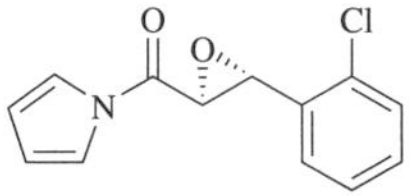	(83), 97[a]	63, 62
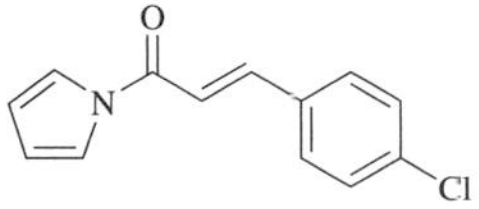	Octahydro-(*R*)-BINOL **19** (5 mol %), CMHP, $Sm(O\textit{i}\text{-}Pr)_3$, TPPO, 4 Å MS, THF, toluene, rt, 0.5 h	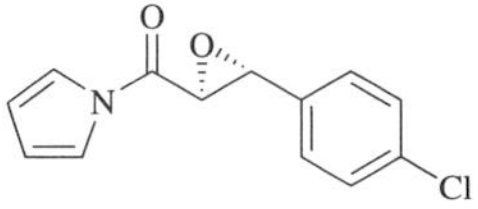	(100), 99[a]	63, 62
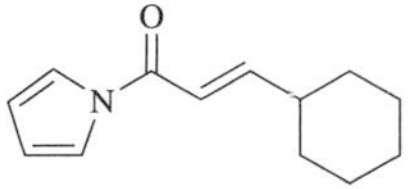	Octahydro-(*R*)-BINOL **19** (5 mol %), CMHP, $Sm(O\textit{i}\text{-}Pr)_3$, TPPO, 4 Å MS, THF, toluene, rt, 2 h	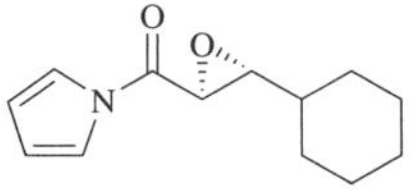	(75), 98[a]	63, 62
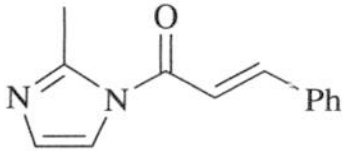	1. (*S*)-BINOL **17a** (20 mol %), TBHP, $La(O\textit{i}\text{-}Pr)_3$, TPAO, 4 Å MS, THF, toluene, rt, 12 h 2. MeOH	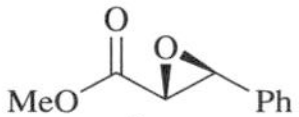 **I**	(70), 77	56, 57, 58
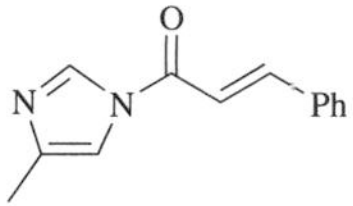	1. (*S*)-BINOL **17a** (20 mol %), TBHP, $La(O\textit{i}\text{-}Pr)_3$, TPAO, 4 Å MS, THF, toluene, rt, 3 h 2. MeOH	**I** (85), 92		56, 57, 58

TABLE 6D. α,β-UNSATURATED AMIDES AND *N*-ACYLATED HETEROCYCLES (*Continued*)

Substrate	Conditions	Product(s), Yield(s) (%), % ee	Refs.
C_{13} 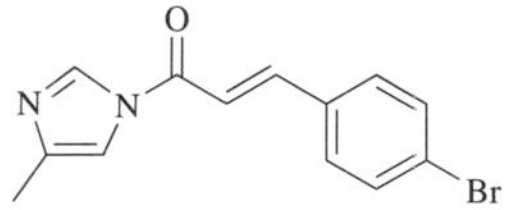	1. (*S*)-BINOL **17a** (10 mol %), TBHP, La(O*i*-Pr)$_3$, TPAO, 4 Å MS, THF, decane, rt, 4 h 2. MeOH	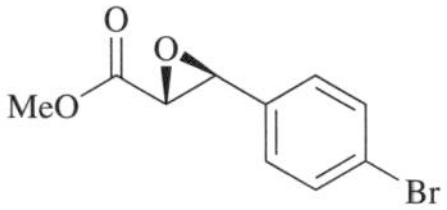 (86), 89	56, 57, 58
	1. (*S*)-BINOL **17a** (20 mol %), TBHP, La(O*i*-Pr)$_3$, TPAO, 4 Å MS, THF, decane, rt, 70 min 2. MeOH, rt, 4 h	**I** (81), 94	265
	1. (*S*)-BINOL **17a** (25 mol %), TBHP, La(O*i*-Pr)$_3$, TPPO, 4 Å MS, THF, decane, 0°, 30 min 2. NaOMe, MeOH, rt, 10 min	**I** (70), 98	265
C_{14} 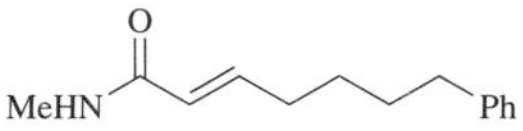	(*S*)-BINOL **17a** (10 mol %), TBHP, Sm(O*i*-Pr)$_3$, TPAO, 4 Å MS, THF, decane, rt, 8 h	 (81), >99	60, 58
	(*R*)-BINOL **15a** (10 mol %), TBHP, Sm(O*i*-Pr)$_3$, TPAO, 4 Å MS, THF, decane, rt	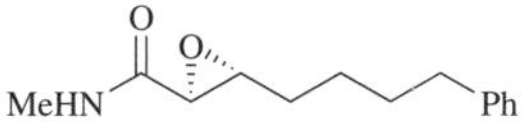 (87), 99	59
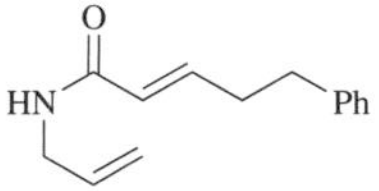	(*S*)-BINOL **17a** (10 mol %), TBHP, Sm(O*i*-Pr)$_3$, TPAO, 4 Å MS, THF, decane, rt, 4 h	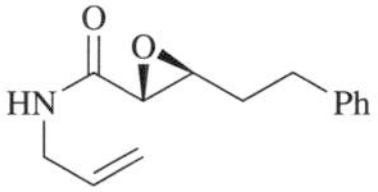 (95), 98	60, 58
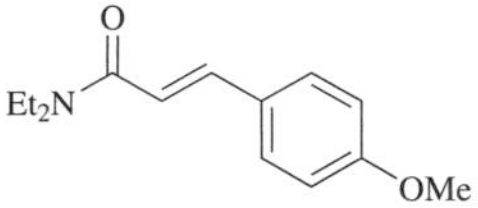	Binaphthyl ketone **41a** (5 mol %), Oxone, $NaHCO_3$, DME, H_2O, 10°, 24 h	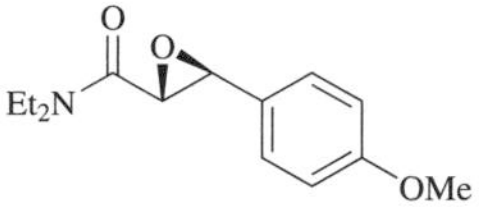 (80), 13	155
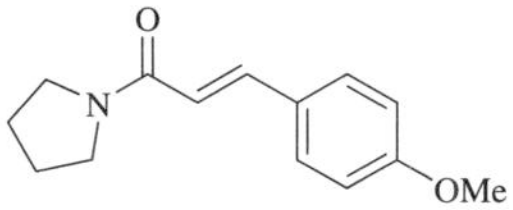	Binaphthyl ketone **41a** (5 mol %), Oxone, $NaHCO_3$, DME, H_2O, 10°, 24 h	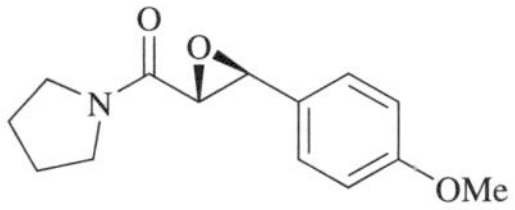 (79), 8	155

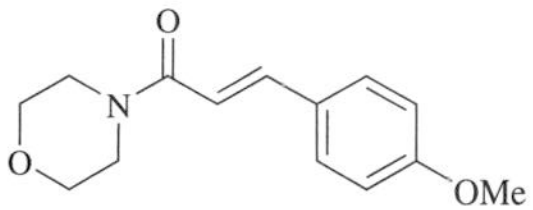	Binaphthyl ketone **41a** (5 mol %), Oxone, $NaHCO_3$, DME, H_2O, 10°, 24 h	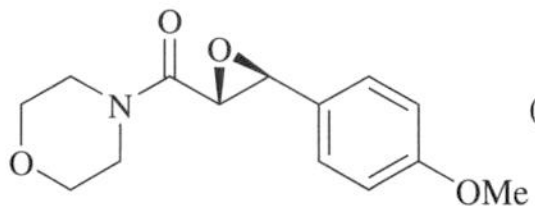	(75), 18	155
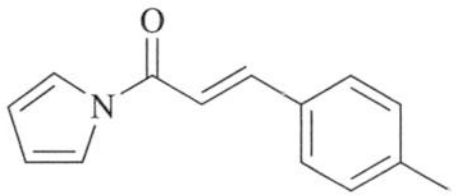	Octahydro-(*R*)-BINOL **19** (5 mol %), CMHP, Sm(O*i*-Pr)$_3$, TFPO, 4 Å MS, THF, toluene, rt, 2.5 h	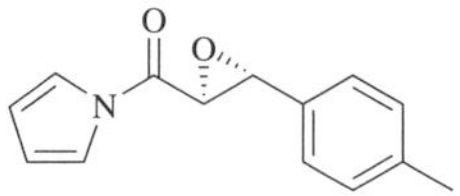	(92), 99[a]	63, 62
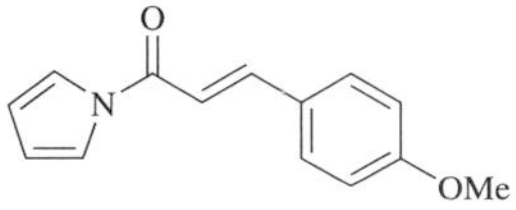	Octahydro-(*R*)-BINOL **19** (5 mol %), TBHP, Sm(O*i*-Pr)$_3$, TFPO, 4 Å MS, THF, toluene, rt, 0.5 h	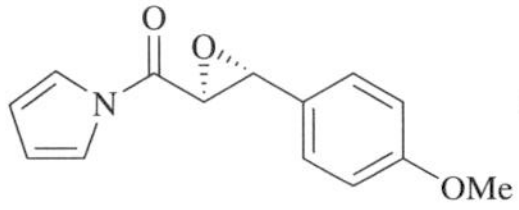	(87), 98[a]	63, 62
C_{15}				
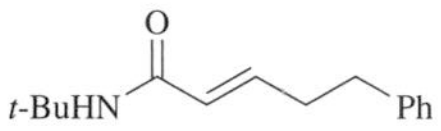	(*S*)-BINOL **17a** (10 mol %), TBHP, Sm(O*i*-Pr)$_3$, TPAO, 4 Å MS, THF, decane, rt, 22 h		(91), 99	60, 58, 294
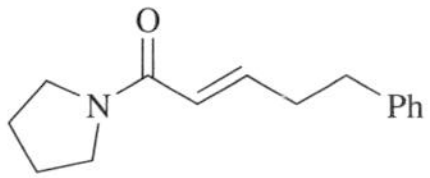	(*S*)-BINOL **17a** (10 mol %), TBHP, Sm(O*i*-Pr)$_3$, TPAO, 4 Å MS, THF, decane, rt, 4 h	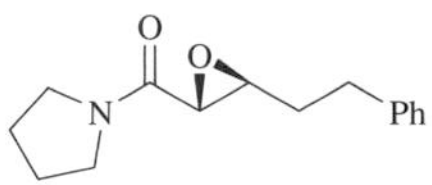	(94), >99	60, 58, 294
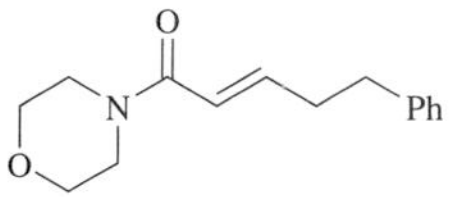	(*S*)-BINOL **17a** (10 mol %), TBHP, Sm(O*i*-Pr)$_3$, TPAO, 4 Å MS, THF, decane, rt	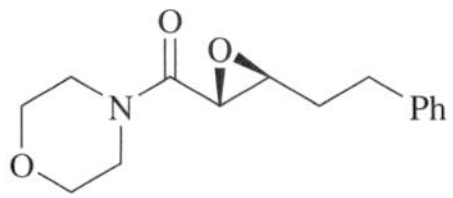	(100), 99	61, 294
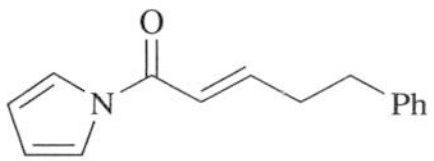	Octahydro-(*R*)-BINOL **19** (5 mol %), CMHP, Sm(O*i*-Pr)$_3$, TFPO, 4 Å MS, THF, toluene, rt, 0.5 h	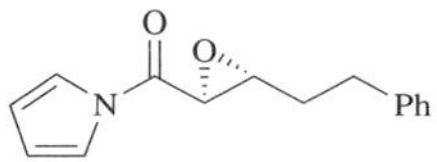	(84), 97[a]	63, 62
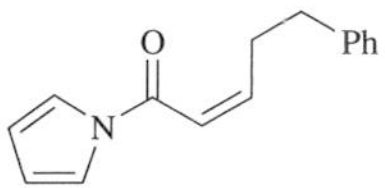	Octahydro-(*R*)-BINOL **19** (5 mol %), TBHP, Sm(O*i*-Pr)$_3$, TFPO, 4 Å MS, THF, toluene, rt, 1 h	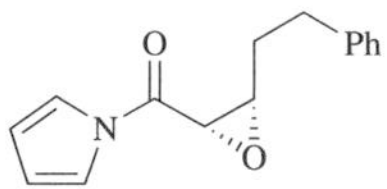	(32), 86	62

TABLE 6D. α,β-UNSATURATED AMIDES AND *N*-ACYLATED HETEROCYCLES (*Continued*)

Substrate	Conditions	Product(s), Yield(s) (%), % ee	Refs.
C_{16}			
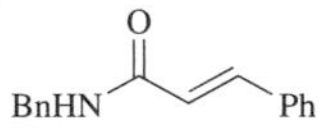	(*S*)-BINOL **17a** (10 mol %), TBHP, Sm(O*i*-Pr)$_3$, TPAO, 4 Å MS, THF, toluene, rt, 18 h	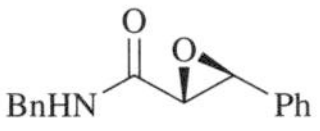 (91), >99	60, 58, 294
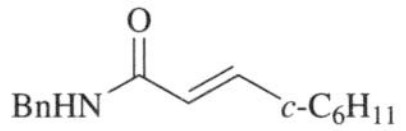	(*S*)-BINOL **17a** (10 mol %), TBHP, Sm(O*i*-Pr)$_3$, TPAO, 4 Å MS, THF, decane, rt, 12 h	 (90), >99	60, 58, 294
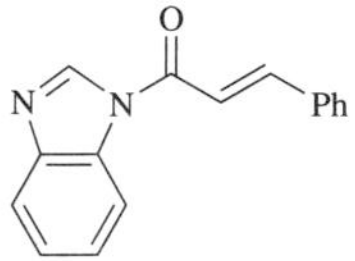	1. (*S*)-BINOL **17a** (20 mol %), TBHP, La(O*i*-Pr)$_3$, TPAO, 4 Å MS, THF, toluene, rt, 24 h 2. MeOH	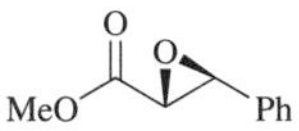 (80), 63	56, 57, 58
C_{17}			
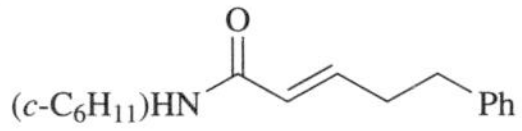	(*S*)-BINOL **17a** (10 mol %), TBHP, Sm(O*i*-Pr)$_3$, TPAO, 4 Å MS, THF, decane, rt, 11 h	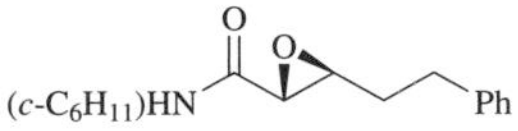 (97), >99	60, 58, 294
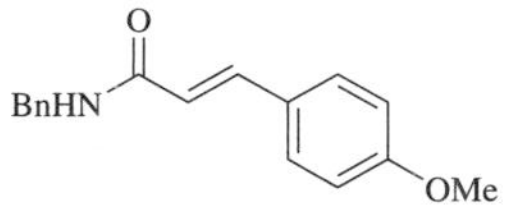	Binaphthyl ketone **41a** (5 mol %), Oxone, $NaHCO_3$, DME, H_2O, 10°, 24 h	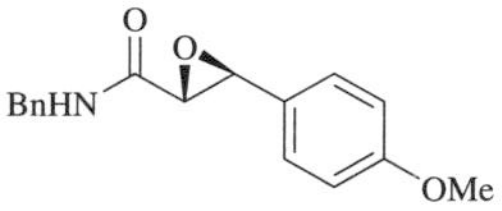 (87), 66	155
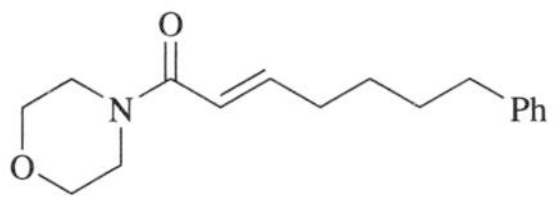	(*R*)-BINOL **15a** (10 mol %), TBHP, Sm(O*i*-Pr)$_3$, TPAO, 4 Å MS, THF, decane, rt	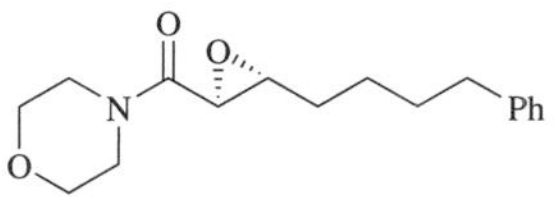 (92), >99	59, 61
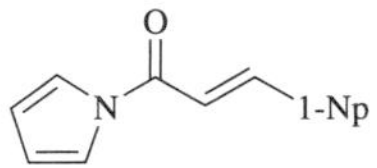	Octahydro-(*R*)-BINOL **19** (5 mol %), CMHP, Sm(O*i*-Pr)$_3$, TPPO, 4 Å MS, THF, toluene, rt, 0.5 h	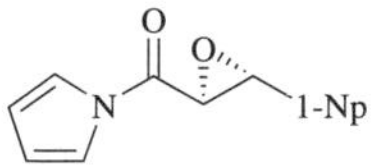 (85), 99[a]	63, 62

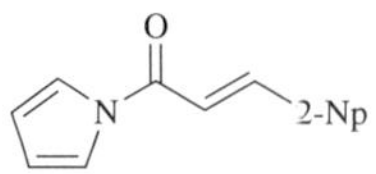	Octahydro-(*R*)-BINOL **19** (5 mol %), CMHP, Sm(O*i*-Pr)$_3$, TPPO, 4 Å MS, THF, toluene, rt, 2 h	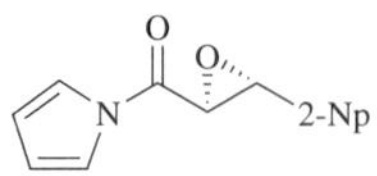	(100), 99[a]	63, 62
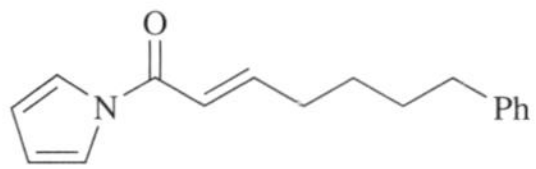	Octahydro-(*S*)-BINOL **20** (5 mol %), CMHP, Sm(O*i*-Pr)$_3$, TPPO, 4 Å MS, THF, toluene, rt, 0.8 h	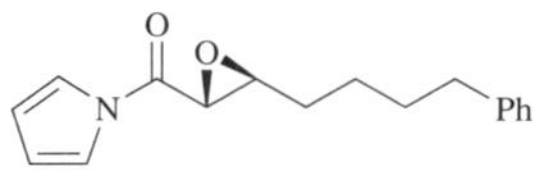	(83), 96[a]	62
	Octahydro-(*R*)-BINOL **19** (5 mol %), CMHP, Sm(O*i*-Pr)$_3$, TPPO, 4 Å MS, THF, toluene, rt, 0.8 h	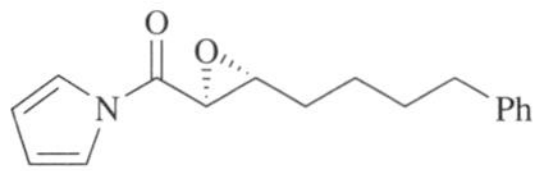	(83), 96[a]	62
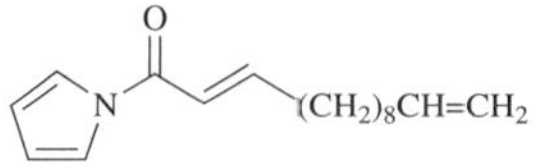	Octahydro-(*R*)-BINOL **19** (5 mol %), CMHP, Sm(O*i*-Pr)$_3$, TPPO, 4 Å MS, THF, toluene, rt, 0.5 h	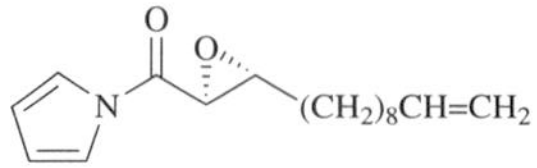	(82), 96[a]	63, 62
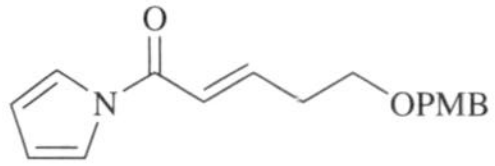	Octahydro-(*R*)-BINOL **19** (5 mol %), CMHP, Sm(O*i*-Pr)$_3$, TPPO, 4 Å MS, THF, toluene, rt, 0.5 h	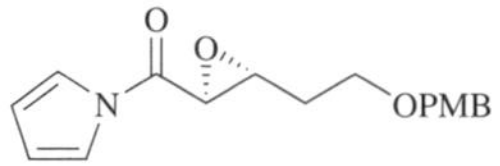	(91), 97[a]	63, 62
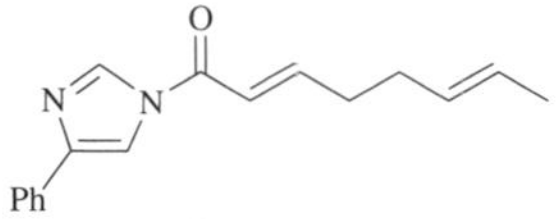	(*S*)-BINOL **17a** (10 mol %), TBHP, La(O*i*-Pr)$_3$, TPAO, 4 Å MS, THF, decane, rt, 1.5 h	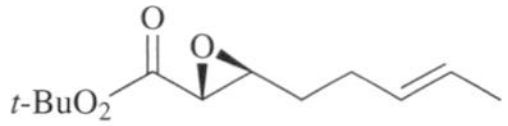	(92), 79	56, 57, 58
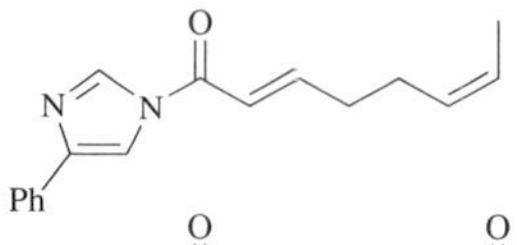	(*S*)-BINOL **17a** (10 mol %), TBHP, La(O*i*-Pr)$_3$, TPAO, 4 Å MS, THF, decane, rt, 2 h		(93), 86	56, 57, 58
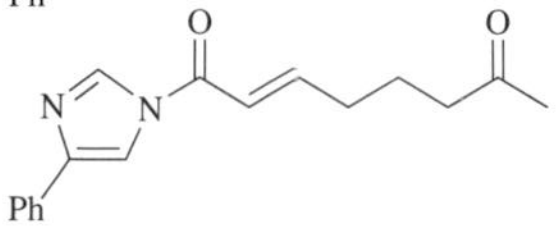	(*S*)-BINOL **17a** (10 mol %), TBHP, La(O*i*-Pr)$_3$, TPAO, 4 Å MS, THF, decane, rt, 4 h	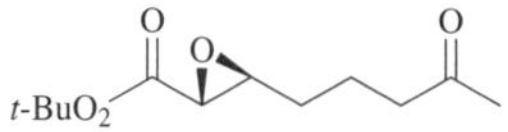	(81), 81	56, 57, 58

TABLE 6D. α,β-UNSATURATED AMIDES AND *N*-ACYLATED HETEROCYCLES (*Continued*)

Substrate	Conditions	Product(s), Yield(s) (%), % ee	Refs.
C_{18}			
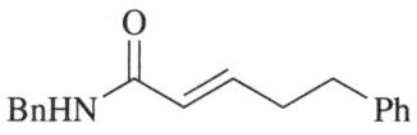	(*S*)-BINOL **17a** (10 mol %), TBHP, Sm(O*i*-Pr)$_3$, TPAO, 4 Å MS, THF, decane, rt, 6 h	 (97), >99	60, 58, 294
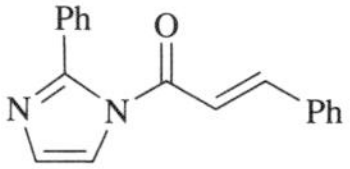	1. (*S*)-BINOL **17a** (20 mol %), TBHP, La(O*i*-Pr)$_3$, TPAO, 4 Å MS, THF, toluene, rt, 12 h 2. MeOH	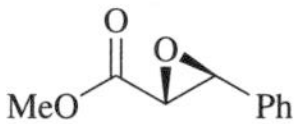 **I** (69), 87	56, 57,[c] 58
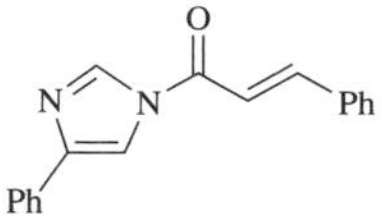	1. (*S*)-BINOL **17a** (x mol %), oxidant, La(O*i*-Pr)$_3$, additive, 4 Å MS, THF, toluene, rt 2. MeOH	**I**	
	x / Oxidant / Additive / Time (h)		
	20 / TBHP / TPAO / 1	(91), 94	56, 57, 58
	10 / TBHP / TPAO / 3.5	(86), 92	56, 57, 58
	5 / TBHP / TPAO / 12	(73), 85	56, 57, 58
	10 / TBHP / TPPO / 3.5	(85), 94	57, 58
	10 / CMHP / TPAO / 18	(47), 94	57, 58
	Octahydro-(*R*)-BINOL **19** (5 mol %), TBHP, Sm(O*i*-Pr)$_3$, TPAO, 4 Å MS, THF, rt, 0.5 h	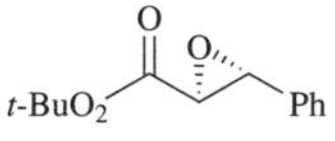 (82), 98	62
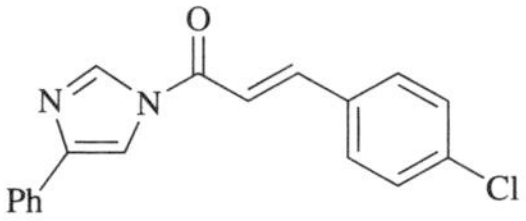	1. (*S*)-BINOL **17a** (10 mol %), TBHP, La(O*i*-Pr)$_3$, TPAO, 4 Å MS, THF, decane, rt, 5 h 2. MeOH	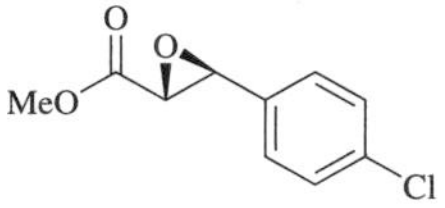 (91), 93	56, 57, 58

	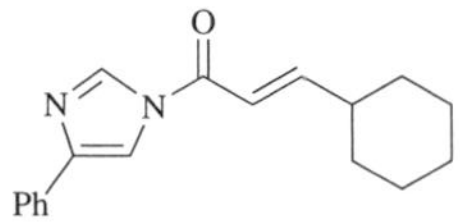	(*S*)-BINOL **17a** (10 mol %), TBHP, Sm(O*i*-Pr)$_3$, TPAO, 4 Å MS, THF, decane, rt, 4 h	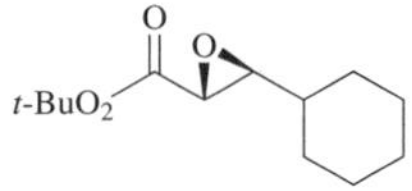	(72), 88	56, 57, 58
C_{19}	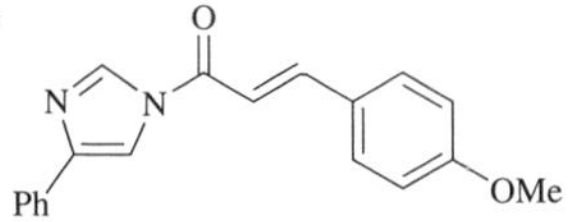	1. (*S*)-BINOL **17a** (10 mol %), TBHP, La(O*i*-Pr)$_3$, TPAO, 4 Å MS, THF, decane, rt, 6 h 2. MeOH	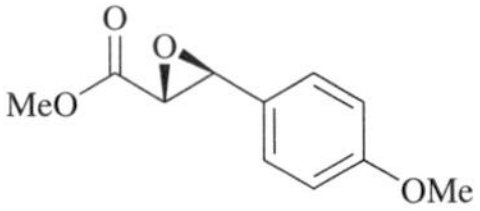	(80), 91	56, 57, 58
	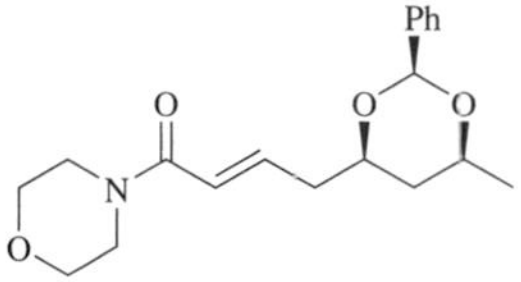	(*S*)-BINOL **17a** (10 mol %), TBHP, Sm(O*i*-Pr)$_3$, TPAO, 4 Å MS, THF, decane, rt	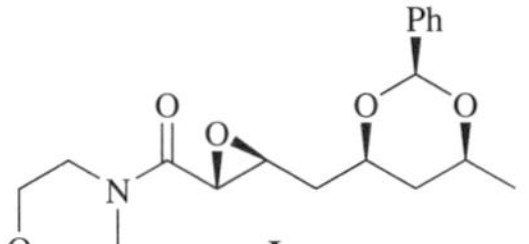 **I** + 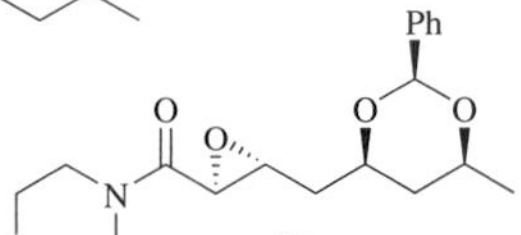 **II** **I** + **II** (90); **I**:**II** > 499:1		61
C_{20}	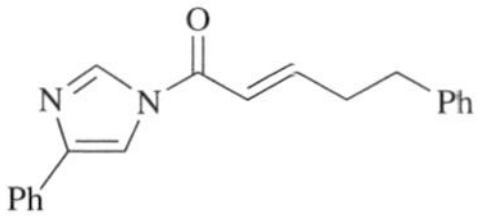	1. (*S*)-BINOL **17a** (10 mol %), oxidant, Ln(O*i*-Pr)$_3$, additive, 4 Å MS, THF, decane, rt, time 2. MeOH			

Oxidant	Ln	Additive	Time (h)		
TBHP	La	TPAO	1	(86), 83	56, 59, 57, 58
TBHP	La	TPPO	1	(84), 87	57, 58
CMHP	La	TPAO	11	(58), 91	57, 58
TBHP	Pr	TPAO	1.5	(87), 85	59, 58
TBHP	Pr	TPPO	4	(82), 86	59, 58
TBHP	Sm	TPAO	6	(81), 79	58
TBHP	Gd	TPAO	3	(81), 67	58
TBHP	Dy	TPAO	3	(79), 77	58
TBHP	Yb	TPAO	20	(78), 68	58

TABLE 6D. α,β-UNSATURATED AMIDES AND *N*-ACYLATED HETEROCYCLES (*Continued*)

Substrate	Conditions	Product(s), Yield(s) (%), % ee	Refs.
C_{21} 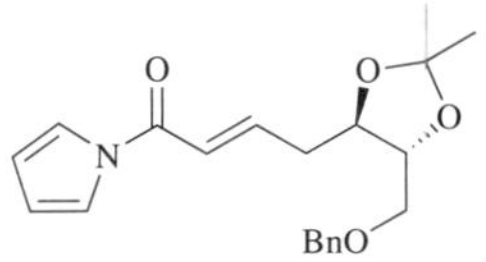	Octahydro-(*R*)-BINOL **19** (5 mol %), CMHP, Sm(O*i*-Pr)$_3$, TPPO, 4 Å MS, THF, toluene, rt, 0.7 h	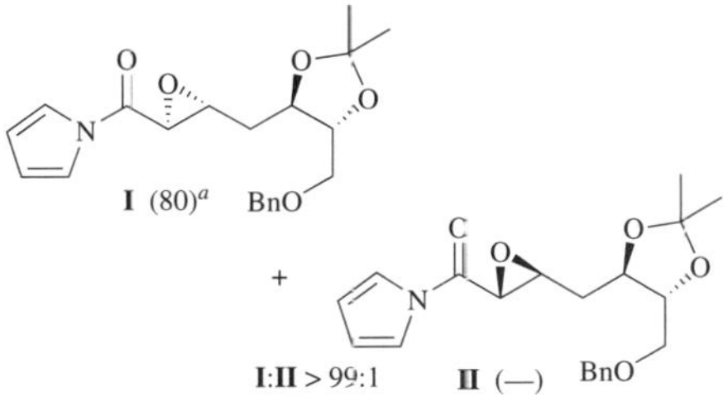 **I** (80)[a] + **II** (—) **I**:**II** > 99:1	63, 62
	Octahydro-(*S*)-BINOL **20** (5 mol %), CMHP, Sm(O*i*-Pr)$_3$, TPPO, 4 Å MS, THF, toluene, rt, 0.9 h	**II** (60)[a]; **I**:**II** = 1:56	63, 62
	Octahydro-(*S*)-BINOL **20** (5 mol %), TBHP, Sm(O*i*-Pr)$_3$, TPPO, 4 Å MS, THF, toluene, rt, 0.7 h	**II** (78)[a]; **I**:**II** = 1:36	63, 62
C_{22} 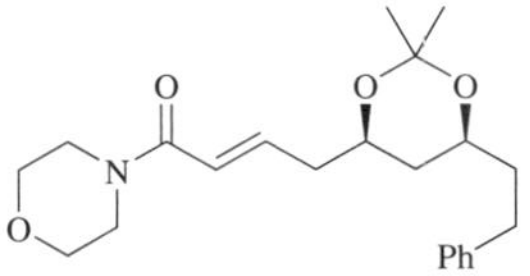	(*S*)-BINOL **17a** (10 mol %), TBHP, Sm(O*i*-Pr)$_3$, TPAO, 4 Å MS, THF, decane, rt	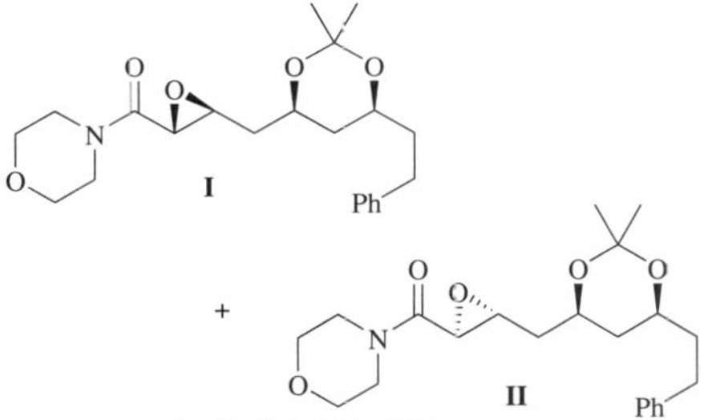 **I** + **II** **I + II** (95); **I**:**II** > 99:1	61, 294
	(*R*)-BINOL **15a** (10 mol %), TBHP, Sm(O*i*-Pr)$_3$, TPAO, 4 Å MS, THF, decane, rt	**I + II** (100); **I**:**II** < 1:99	61, 294

Substrate	Conditions	Product(s) and Yield(s) (%), ee	Refs.
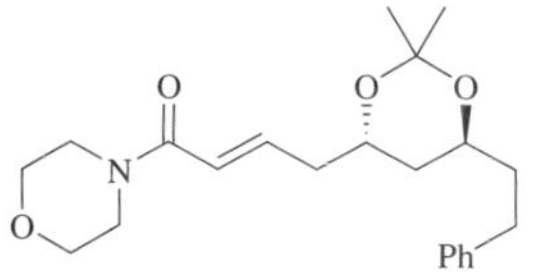	(*S*)-BINOL **17a** (10 mol %), TBHP, Sm(O*i*-Pr)$_3$, TPAO, 4 Å MS, THF, decane, rt	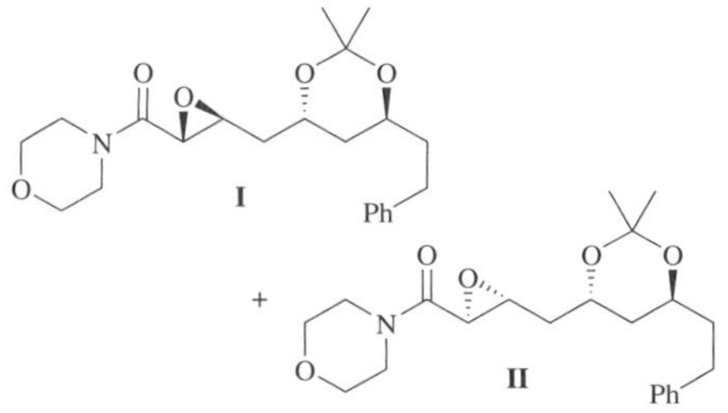 **I** + **II** (99); **I**:**II** > 99:1	61
	(*R*)-BINOL **15a** (10 mol %), TBHP, Sm(O*i*-Pr)$_3$, TPAO, 4 Å MS, THF, decane, rt	**I** + **II** (89); **I**:**II** < 1:99	61
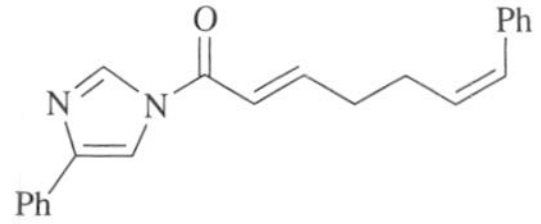	1. (*S*)-BINOL **17a** (10 mol %), TBHP, La(O*i*-Pr)$_3$, TPAO, 4 Å MS, THF, decane, rt, 2 h 2. MeOH	 (85), 82	56, 57, 58
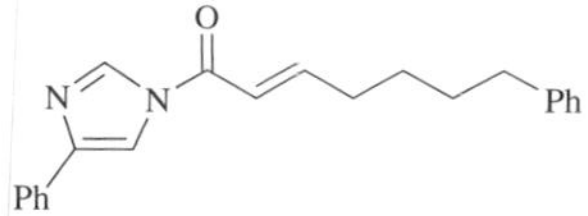	1. (*S*)-BINOL **17a** (10 mol %), TBHP, Ln(O*i*-Pr)$_3$, additive, 4 Å MS, THF, decane, rt 2. MeOH		58

Ln	Additive	Time (h)	Yield (%), ee
La	TPAO	1	(93), 79
Pr	TPAO	1.5	(88), 86
Pr	TPPO	7	(92), 88
Sm	TPAO	5	(88), 65
Gd	TPAO	8	(84), 49
Dy	*TPPO*	*2*	(85), 76
Yb	*TPAO*	*18*	*(71), 56*

TABLE 6D. α,β-UNSATURATED AMIDES AND *N*-ACYLATED HETEROCYCLES (*Continued*)

Substrate	Conditions	Product(s), Yield(s) (%), % ee	Refs.
C_{23} 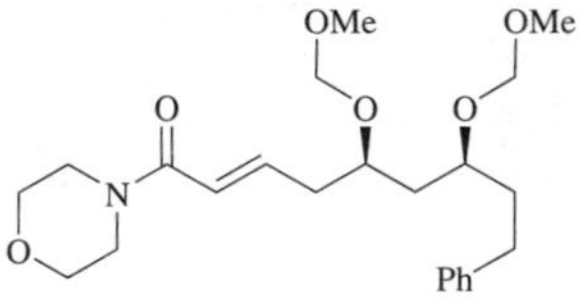	(*S*)-BINOL **17a** (10 mol %), TBHP, Sm(O*i*-Pr)$_3$, TPAO, 4 Å MS, THF, decane, rt	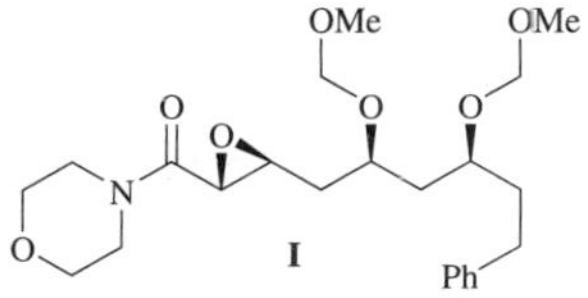 **I** + 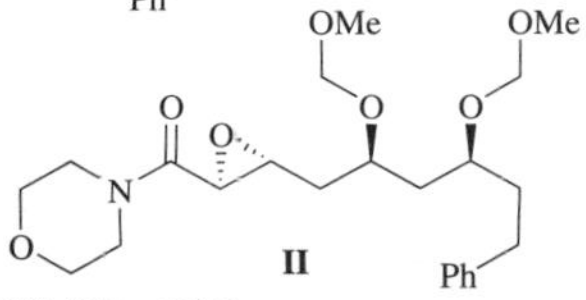 **II** **I** + **II** (90); **I**:**II** > 199:1	61
	(*R*)-BINOL **15a** (10 mol %), TBHP, Sm(O*i*-Pr)$_3$, TPAO, 4 Å MS, THF, decane, rt	**I** + **II** (87); **I**:**II** < 1:199	61
C_{24} 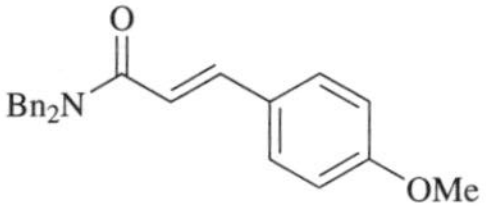	Binaphthyl ketone **41a** (5 mol %), Oxone, NaHCO$_3$, DME, H$_2$O, 10°, 24 h	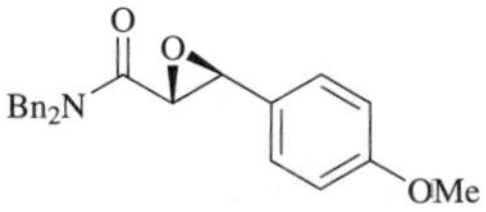 (69), 20	155
C_{27} 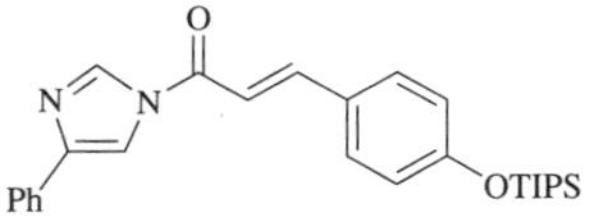	1. (*S*)-BINOL **17a** (10 mol %), TBHP, La(O*i*-Pr)$_3$, TPAO, 4 Å MS, THF, decane, rt, 1.5 h 2. MeOH	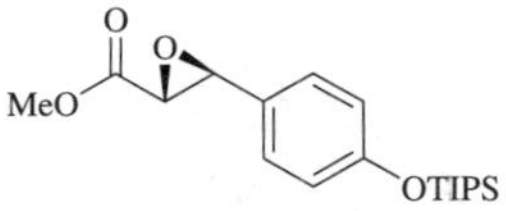 (95), 94	295

C_{31} 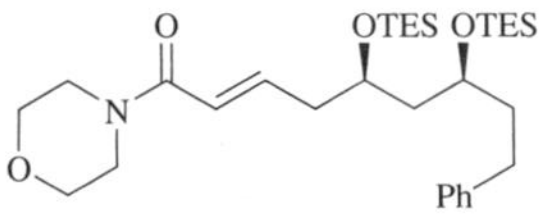	(*S*)-BINOL **17a** (10 mol %), TBHP, Sm(O*i*-Pr)$_3$, TPAO, 4 Å MS, THF, decane, rt	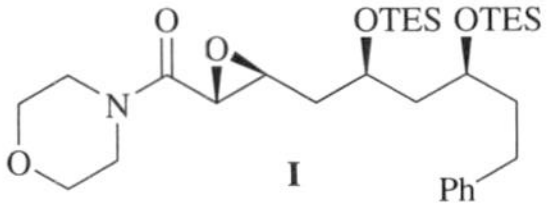 + 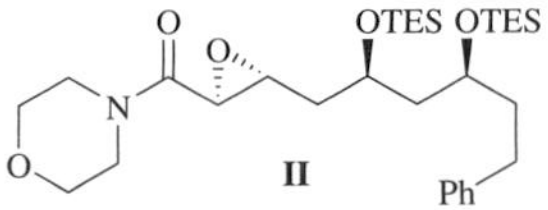 **I** + **II** (81); **I**:**II** > 19:1	61
	(*R*)-BINOL **15a** (10 mol %), TBHP, Sm(O*i*-Pr)$_3$, TPAO, 4 Å MS, THF, decane, rt	**I** + **II** (97); **I**:**II** = 8:92	61

[a] The substrate was prepared in situ through reaction of 1-pyrrol-1-yl-2-(triphenylphosphoranylidene)ethanone with the corresponding aldehyde. The yield refers to the two-step olefination-epoxidation process.

[b] The publication does not specify whether the propyl group is *n*-propyl or *iso*-propyl.

[c] In this publication, the substrate is incorrectly depicted without a phenyl substituent on the imidazole.

TABLE 7A. α,β-UNSATURATED SULFONES

Substrate	Conditions	Product(s), Yield(s) (%), % ee	Refs.
C_9			
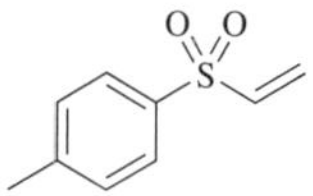	(L-Leu)$_n$ (0.1 mol %), H_2O_2, NaOH, Bu_4NHSO_4, toluene, rt, 18 h	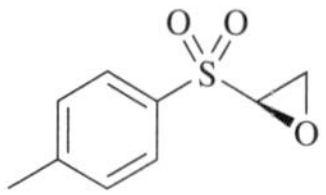 (76), 70	190, 201
C_{13}			
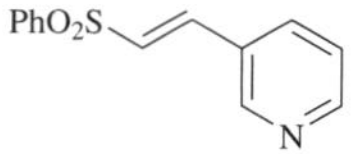	(L-Leu)$_n$ (0.1 mol %), H_2O_2, NaOH, Bu_4NHSO_4, toluene, rt, 6 h	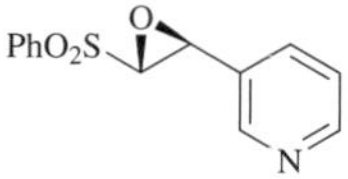 (61), 95	190, 201
C_{14}			
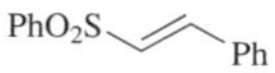	(L-Leu)$_n$ (n_{av} = 33; 11 mol %), H_2O_2, NaOH, Aliquat 336, toluene, H_2O, rt, 30 min	 **I** (95), 64	120, 192
	(L-Leu)$_n$ (n_{av} = 33; 11 mol %), H_2O_2, NaOH, TBAB, toluene, H_2O, rt, 15 min	**I** (82), 68	120
	(L-Leu)$_n$ (0.1 mol %), H_2O_2, NaOH, Bu_4NHSO_4, toluene, rt, 6 h	**I** (66), 91	190, 201
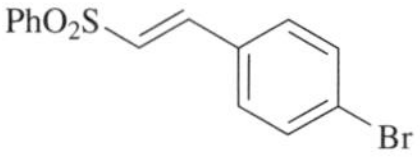	(L-Leu)$_n$ (0.1 mol %), H_2O_2, NaOH, Bu_4NHSO_4, toluene, rt, 6 h	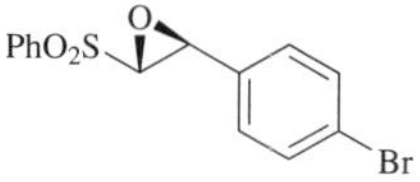 (49), 94	190

TABLE 7B. α,β-UNSATURATED NITRO COMPOUNDS

Substrate	Conditions	Product(s), Yield(s) (%), % ee	Refs.
C_{5-8}			
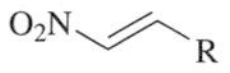	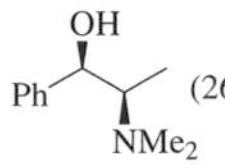 (267 mol %), O_2, Et_2Zn, THF, 0°		42
	R / Time (h)		
	n-Pr / —	(53), 43	
	i-Pr / 3	(53), 42	
	t-Bu / —	(57), 82	
	c-C_6H_{11} / —	(47), 36	
C_9			
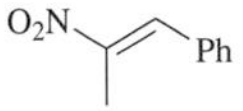	(L-Ala)$_n$ (n_{av} = 10), H_2O_2, NaOH, toluene, H_2O, rt, 24 h	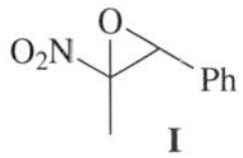 **I** (50), 7	132, 130, 273
	[L-Glu(OBn)]$_n$ (n_{av} =10), H_2O_2, NaOH, toluene, H_2O, rt, 24 h	**I** (67), 4	132, 130
	Benzyl quininium salt **4a** (1.7 mol %), H_2O_2, NaOH, H_2O, toluene, rt, 24 h	**I** (—), —	81
C_{10}			
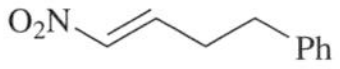	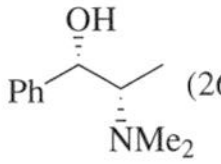 (267 mol %), O_2, Et_2Zn, THF, 0°	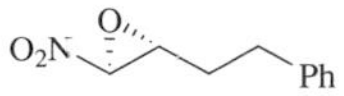 (64), 37	42

TABLE 7C. α,β-UNSATURATED NITRILES

Substrate	Conditions	Product(s), Yield(s) (%), % ee	Refs.
C_{10}			
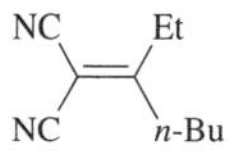	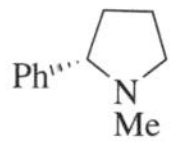 (100 mol %), O_2, DMF, 20°, 65 h	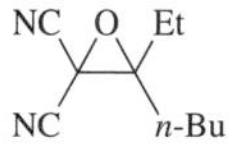 (34), —	202
C_{12}			
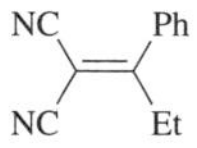	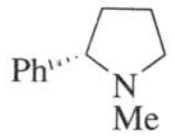 (100 mol %), O_2	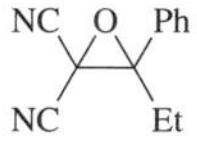 **I** + 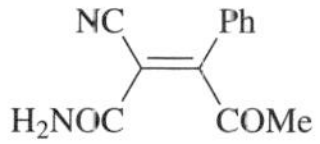 **II**	202
	Solvent / Temp (°) / Time (h)	I / II	
	DMF / rt / 12	(19), — / (23)	
	DMF / 30 / 18	(18), 5.8 / —	
	diglyme / 30 / 76	(18), 7.5 / —	
	anisole / 30 / 47	(8.5), 8.4 / —	
	toluene / 30 / 139	(13), 7.1 / —	
	nitrobenzene / 30 / 67	(15), 8.4 / —	
	DMF / 0 / 17	(13), 7.6 / —	
	DMF / 10 / 18	(17), 6.7 / —	
	DMF / 20 / 20	(19), 6.1 / —	
	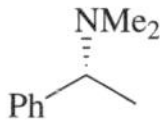 , O_2, DMF, 20°, 7 h	**I** (22), —	202
	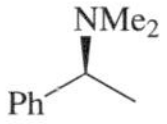 , O_2, DMF, 20°, 7 h	**I** (20), —	202
C_{13}			
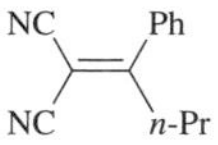	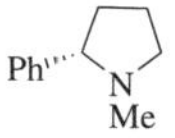 (100 mol %), O_2, DMF, 20°, 104 h	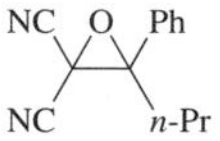 (23), 5.5 + 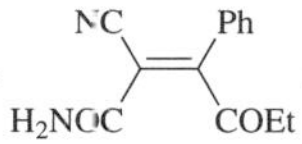 (28)	202

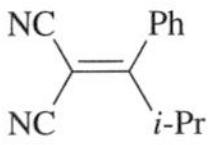	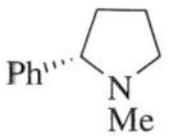 (100 mol %), O_2, DMF, 20°, 45 h	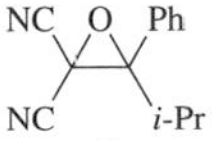 **I** (6), 5.5 + 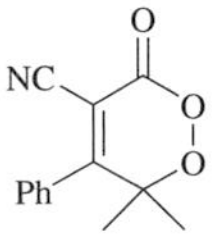 (4)	202
	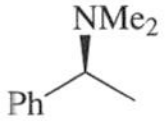, O_2, DMF, 20°, 48 h	**I** (11), —	202
	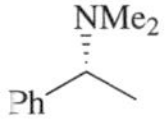, O_2, DMF, 20°, 48 h	**I** (15), —	202
C_{17} 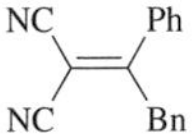	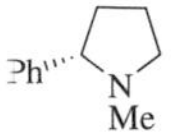 (100 mol %), O_2, DMF, 20°, 93 h	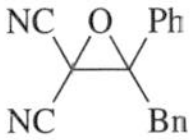 (13), — + 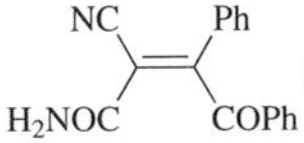 (28)	202

TABLE 7D. α,β-UNSATURATED PHOSPHONATES

Substrate	Conditions	Product(s), Yield(s) (%), % ee	Refs.
C_3 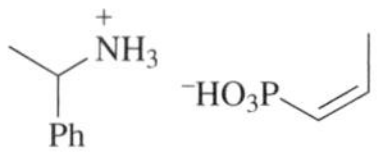	Metal-complex catalyst (5 mol %), H_2O_2, H_2O	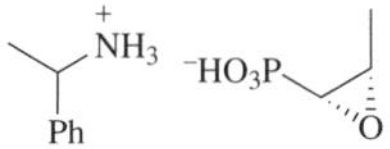	296

Catalyst	Solvent	Temp (°)	Time (h)	Yield (%), % ee
63a	EtOH	rt	24	(100), 45
63a	CH_2Cl_2	50	1	(100), 59
63a	CH_2Cl_2	rt	24	(100), 69
63a	CH_2Cl_2	0	72	(100), 74
63b	EtOH	rt	24	(20), 46
63b	CH_2Cl_2	50	1	(58), 62
63b	CH_2Cl_2	rt	24	(30), 69
63b	CH_2Cl_2	0	72	(24), 75
70a	EtOH	rt	24	(100), 62
70a	CH_2Cl_2	50	1	(100), 67
70a	CH_2Cl_2	rt	24	(100), 74
70a	CH_2Cl_2	0	72	(100), 78
70b	EtOH	rt	24	(100), 63
70b	CH_2Cl_2	50	1	(100), 68
70b	CH_2Cl_2	rt	24	(100), 74
70b	CH_2Cl_2	0	72	(100), 80
71a	EtOH	rt	24	(100), 52
71a	CH_2Cl_2	50	1	(100), 63
71a	CH_2Cl_2	rt	24	(100), 71
71a	CH_2Cl_2	0	72	(100), 76
71b	EtOH	rt	24	(100), 59
71b	CH_2Cl_2	50	1	(100), 64
71b	CH_2Cl_2	rt	24	(100), 72
71b	CH_2Cl_2	0	72	(100), 77

REFERENCES

[1] Katsuki, T.; Martin, V. S. *Org. React.* **1996**, *48*, 1.
[2] Katsuki, T. In *Comprehensive Asymmetric Catalysis*; Jacobsen, E. N., Pfaltz, A., Yamamoto, H., Eds.; Springer-Verlag: Berlin and Heidelberg, 1999; Vol. II, pp 621–648.
[3] Jacobsen, E. N.; Wu, M. H. In *Comprehensive Asymmetric Catalysis*; Jacobsen, E. N., Pfaltz, A., Yamamoto, H., Eds.; Springer-Verlag: Berlin and Heidelberg, 1999; Vol. II, pp 649–678.
[4] Weitz, E.; Scheffer, A. *Chem. Ber.* **1921**, *54*, 2327.
[5] Rosenblatt, D. H.; Broome, G. H. *J. Org. Chem.* **1963**, *28*, 1290.
[6] Yang, N. C.; Finnegan, R. A. *J. Am. Chem. Soc.* **1958**, *80*, 5845.
[7] Bailey, M.; Staton, I.; Ashton, P. R.; Markó, I. E.; Ollis, W. D. *Tetrahedron: Asymmetry* **1991**, *2*, 495.
[8] Bailey, M.; Markó, I. E.; Ollis, W. D. *Tetrahedron Lett.* **1991**, *32*, 2687.
[9] Clark, D. A.; De Riccardis, F.; Nicolaou, K. C. *Tetrahedron* **1994**, *50*, 11391.
[10] Gould, S. J.; Shen, B.; Whittle, Y. G. *J. Am. Chem. Soc.* **1989**, *111*, 7932.
[11] Aisaka, K.; Ohshiro, T.; Uwajima, T. *Appl. Microbiol. Biotechnol.* **1992**, *36*, 431.
[12] Itoh, N.; Kusaka, M.; Hirota, T.; Nomura, A. *Appl. Microbiol. Biotechnol.* **1995**, *43*, 394.
[13] Porter, M. J.; Skidmore, J. *Chem. Commun.* **2000**, 1215.
[14] Nemoto, T.; Ohshima, T.; Shibasaki, M. *J. Synth. Org. Chem. Jpn.* **2002**, *60*, 94.
[15] Vidal, A.; Monroig, J. J.; Rodríguez, S.; González, F. In *Recent Research Developments in Organic Chemistry*; Pandalai, S. G., Ed.; Research Signpost: Kerala, India, 2004; Vol. 8, pp 13–28.
[16] Bonini, C.; Righi, G. *Tetrahedron* **2002**, *58*, 4981.
[17] Lantos, I. In *Asymmetric Oxidation Reactions*; Katsuki, T., Ed.; Oxford University Press: New York, 2001, pp 70–80.
[18] Bunton, C. A.; Minkoff, G. J. *J. Chem. Soc.* **1949**, 665.
[19] Black, W. B.; Lutz, R. E. *J. Am. Chem. Soc.* **1953**, *75*, 5990.
[20] Lutz, R. E.; Weiss, J. O. *J. Am. Chem. Soc.* **1955**, *77*, 1814.
[21] House, H. O.; Ro, R. S. *J. Am. Chem. Soc.* **1958**, *80*, 2428.
[22] Zwanenburg, B.; ter Wiel, J. *Tetrahedron Lett.* **1970**, 935.
[23] Curci, R.; DiFuria, F.; Meneghin, M. *Gazz. Chim. Ital.* **1978**, *108*, 123.
[24] Doherty, J. R.; Keane, D. D.; Marathe, K. G.; O'Sullivan, W. I.; Philbin, E. M.; Simons, R. M.; Teague, P. C. *Tetrahedron Lett.* **1968**, 441.
[25] Curci, R.; DiFuria, F. *Tetrahedron Lett.* **1974**, 4085.
[26] Apeloig, Y.; Karni, M.; Rappoport, Z. *J. Am. Chem. Soc.* **1983**, *105*, 2784.
[27] Adam, W.; Rao, P. B.; Degen, H.-G.; Saha-Möller, C. R. *J. Am. Chem. Soc.* **2000**, *122*, 5654.
[28] Adam, W.; Rao, P. B.; Degen, H.-G.; Saha-Möller, C. R. *Eur. J. Org. Chem.* **2002**, 630.
[29] Adam, W.; Rao, P. B.; Degen, H.-G.; Levai, A.; Patonay, T.; Saha-Möller, C. R. *J. Org. Chem.* **2002**, *67*, 259.
[30] Dwyer, C. L.; Gill, C. D.; Ichihara, O.; Taylor, R. J. K. *Synlett* **2000**, 704.
[31] Aoki, M.; Seebach, D. *Helv. Chim. Acta* **2001**, *84*, 187.
[32] Lattanzi, A.; Cocilova, M.; Iannece, P.; Scettri, A. *Tetrahedron: Asymmetry* **2004**, *15*, 3751.
[33] Kosnik, W.; Stachulski, A. V.; Chmielewski, M. *Tetrahedron: Asymmetry* **2005**, *16*, 1975.
[34] Kosnik, W.; Stachulski, A. V.; Chmielewski, M. *Tetrahedron: Asymmetry* **2006**, *17*, 313.
[35] Terada, M.; Ube, H.; Shimizu, H. WO Patent 2005077921 (2005); *Chem. Abstr.* **2005**, *143*, 248420.
[36] Genski, T.; Macdonald, G.; Wei, X.; Lewis, N.; Taylor, R. J. K. *Synlett* **1999**, 795.
[37] Genski, T.; Macdonald, G.; Wei, X.; Lewis, N.; Taylor, R. J. K. *ARKIVOC* **2000**, *1*, 266.
[38] McManus, J. C.; Carey, J. S.; Taylor, R. J. K. *Synlett* **2003**, 365.
[39] McManus, J. C.; Genski, T.; Carey, J. S.; Taylor, R. J. K. *Synlett* **2003**, 369.
[40] Enders, D.; Zhu, J. Q.; Raabe, G. *Angew. Chem., Int. Ed. Engl.* **1996**, *35*, 1725.
[41] Enders, D.; Zhu, J. Q.; Kramps, L. *Liebigs Ann./Recl.* **1997**, 1101.
[42] Enders, D.; Kramps, L.; Zhu, J. Q. *Tetrahedron: Asymmetry* **1998**, *9*, 3959.
[43] Yu, H. B.; Zheng, X. F.; Lin, Z. M.; Hu, Q. S.; Huang, W. S.; Pu, L. *J. Org. Chem.* **1999**, *64*, 8149.

[44] Minatti, A.; Dötz, K. H. *Synlett* **2004**, 1634.
[45] Minatti, A.; Dötz, K. H. *Tetrahedron: Asymmetry* **2005**, *16*, 3256.
[46] Bougauchi, M.; Watanabe, S.; Arai, T.; Sasai, H.; Shibasaki, M. *J. Am. Chem. Soc.* **1997**, *119*, 2329.
[47] Chen, R. F.; Qian, C. T.; de Vries, J. G. *Tetrahedron Lett.* **2001**, *42*, 6919.
[48] Chen, R. F.; Qian, C. T.; de Vries, J. G. *Tetrahedron* **2001**, *57*, 9837.
[49] Watanabe, S.; Arai, T.; Sasai, H.; Bougauchi, M.; Shibasaki, M. *J. Org. Chem.* **1998**, *63*, 8090.
[50] Watanabe, S.; Kobayashi, Y.; Arai, T.; Sasai, H.; Bougauchi, M.; Shibasaki, M. *Tetrahedron Lett.* **1998**, *39*, 7353.
[51] Daikai, K.; Kamaura, M.; Inanaga, J. *Tetrahedron Lett.* **1998**, *39*, 7321.
[52] Daikai, K.; Inanaga, J. *Kidorui* **2000**, *36*, 284.
[53] Inanaga, J.; Kagawa, T. European Patent 1,127,616 (2001); *Chem. Abstr.* **2001**, *135*, 197209.
[54] Kino, R.; Daikai, K.; Kawanami, T.; Furuno, H.; Inanaga, J. *Org. Biomol. Chem.* **2004**, *2*, 1822.
[55] Nemoto, T.; Ohshima, T.; Yamaguchi, K.; Shibasaki, M. *J. Am. Chem. Soc.* **2001**, *123*, 2725.
[56] Nemoto, T.; Ohshima, T.; Shibasaki, M. *J. Am. Chem. Soc.* **2001**, *123*, 9474.
[57] Nemoto, T.; Tosaki, S.-Y.; Ohshima, T.; Shibasaki, M. *Chirality* **2003**, *15*, 306.
[58] Ohshima, T.; Nemoto, T.; Tosaki, S.-Y.; Kakei, H.; Gnanadesikan, V.; Shibasaki, M. *Tetrahedron* **2003**, *59*, 10485.
[59] Tosaki, S.-Y.; Nemoto, T.; Ohshima, T.; Shibasaki, M. *Org. Lett.* **2003**, *5*, 495.
[60] Nemoto, T.; Kakei, H.; Gnanadesikan, V.; Tosaki, S.-Y.; Ohshima, T.; Shibasaki, M. *J. Am. Chem. Soc.* **2002**, *124*, 14544.
[61] Tosaki, S.-Y.; Horiuchi, Y.; Nemoto, T.; Ohshima, T.; Shibasaki, M. *Chem. Eur. J.* **2004**, *10*, 1527.
[62] Matsunaga, S.; Kinoshita, T.; Okada, S.; Harada, S.; Shibasaki, M. *J. Am. Chem. Soc.* **2004**, *126*, 7559.
[63] Kinoshita, T.; Okada, S.; Park, S. R.; Matsunaga, S.; Shibasaki, M. *Angew. Chem., Int. Ed.* **2003**, *42*, 4680.
[64] Kakei, H.; Tsuji, R.; Ohshima, T.; Shibasaki, M. *J. Am. Chem. Soc.* **2005**, *127*, 8962.
[65] Hayano, T.; Ishida, S.; Furuno, H.; Inanaga, J. *Kidorui* **2003**, *42*, 56.
[66] Wang, X.; Shi, L.; Li, M.; Ding, K. *Angew. Chem., Int. Ed.* **2005**, *44*, 6362.
[67] Inanaga, J.; Furuno, H.; Hayano, T. *Catal. Surv. Jpn.* **2001**, *5*, 37.
[68] Daikai, K.; Hayano, T.; Kino, R.; Furuno, H.; Kagawa, T.; Inanaga, J. *Chirality* **2003**, *15*, 83.
[69] Elston, C. L.; Jackson, R. F. W.; MacDonald, S. J. F.; Murray, P. J. *Angew. Chem., Int. Ed. Engl.* **1997**, *36*, 410.
[70] Jacques, O.; Richards, S. J.; Jackson, R. F. W. *Chem. Commun.* **2001**, 2712.
[71] Tanaka, Y.; Nishimura, K.; Tomioka, K. *Heterocycles* **2002**, *58*, 71.
[72] Tanaka, Y.; Nishimura, K.; Tomioka, K. *Tetrahedron* **2003**, *59*, 4549.
[73] Choudary, B. M.; Kantam, M. L.; Ranganath, K. V. S.; Mahendar, K.; Sreedhar, B. *J. Am. Chem. Soc.* **2004**, *126*, 3396.
[74] Li, C.; Pace, E. A.; Liang, M. C.; Lobkovsky, E.; Gilmore, T. D.; Porco, J. A. *J. Am. Chem. Soc.* **2001**, *123*, 11308.
[75] Lei, X.; Johnson, R. P.; Porco, J. A., Jr. *Angew. Chem., Int. Ed.* **2003**, *42*, 3913.
[76] Li, C.; Johnson, R. P.; Porco, J. A., Jr. *J. Am. Chem. Soc.* **2003**, *125*, 5095.
[77] Kumaraswamy, G.; Jena, N.; Sastry, M. N. V.; Rao, G. V.; Ankamma, K. *J. Mol. Catal. A: Chem.* **2005**, *230*, 59.
[78] Kumaraswamy, G.; Jena, N.; Sastry, M. N. V.; Ramakrishna, G. *ARKIVOC* **2005**, 53.
[79] Kumaraswamy, G.; Sastry, M. N. V.; Jena, N.; Kumar, K. R.; Vairamani, M. *Tetrahedron: Asymmetry* **2003**, *14*, 3797.
[80] Baccin, C.; Gusso, A.; Pinna, F.; Strukul, G. *Organometallics* **1995**, *14*, 1161.
[81] Helder, R.; Hummelen, J. C.; Laane, R. W. P. M.; Wiering, J. S.; Wynberg, H. *Tetrahedron Lett.* **1976**, 1831.
[82] Wynberg, H.; Greijdanus, B. *J. Chem. Soc., Chem. Commun.* **1978**, 427.
[83] Hummelen, J. C.; Wynberg, H. *Tetrahedron Lett.* **1978**, 1089.
[84] Wynberg, H.; Marsman, B. *J. Org. Chem.* **1980**, *45*, 158.

[85] Pluim, H.; Wynberg, H. *J. Org. Chem.* **1980**, *45*, 2498.
[86] Kobayashi, N. *Br. Polym. J.* **1984**, *16*, 205.
[87] Kobayashi, N.; Iwai, K. *Makromol. Chem. Rapid. Commun.* **1981**, *2*, 105.
[88] Dehmlow, E. V.; Düttmann, S.; Neumann, B.; Stammler, H.-G. *Eur. J. Org. Chem.* **2002**, 2087.
[89] Dehmlow, E. V.; Schrader, S. *Pol. J. Chem.* **1994**, *68*, 2199.
[90] Dehmlow, E. V.; Romero, M. S. *J. Chem. Res. (S)* **1992**, 400.
[91] Shi, M.; Kazuta, K.; Satoh, Y.; Masaki, Y. *Chem. Pharm. Bull.* **1994**, *42*, 2625.
[92] Shi, M.; Masaki, Y. *J. Chem. Res. (S)* **1994**, 250.
[93] Takahashi, H.; Kubota, Y.; Miyazaki, H.; Onda, M. *Chem. Pharm. Bull.* **1984**, *32*, 4852.
[94] Baba, N.; Oda, J.; Kawaguchi, M. *Agric. Biol. Chem.* **1986**, *50*, 3113.
[95] Harigaya, Y.; Yamaguchi, H.; Onda, M. *Heterocycles* **1981**, *15*, 183.
[96] Mazaleyrat, J. P. *Tetrahedron Lett.* **1983**, *24*, 1243.
[97] Lygo, B.; To, D. C. M. *Tetrahedron Lett.* **2001**, *42*, 1343.
[98] Corey, E. J.; Zhang, F. Y. *Org. Lett.* **1999**, *1*, 1287.
[99] Lygo, B.; Wainwright, P. G. *Tetrahedron* **1999**, *55*, 6289.
[100] Lygo, B.; Wainwright, P. G. *Tetrahedron Lett.* **1998**, *39*, 1599.
[101] Ye, J.; Wang, Y.; Liu, R.; Zhang, G.; Zhang, Q.; Chen, J.; Liang, X. *Chem. Commun.* **2003**, 2714.
[102] Ye, J.; Wang, Y.; Chen, J.; Liang, X. *Adv. Synth. Catal.* **2004**, *346*, 691.
[103] Arai, S.; Tsuge, H.; Oku, M.; Miura, M.; Shioiri, T. *Tetrahedron* **2002**, *58*, 1623.
[104] Arai, S.; Tsuge, H.; Shioiri, T. *Tetrahedron Lett.* **1998**, *39*, 7563.
[105] Adam, W.; Rao, P. B.; Degen, H.-G.; Saha-Möller, C. R. *Tetrahedron: Asymmetry* **2001**, *12*, 121.
[106] Arai, S.; Oku, M.; Miura, M.; Shioiri, T. *Synlett* **1998**, 1201.
[107] Liu, X. D.; Bai, X. L.; Qiu, X. P.; Gao, L. X. *Chin. Chem. Lett.* **2005**, *16*, 975.
[108] Jew, S.; Lee, J. H.; Jeong, B. S.; Yoo, M. S.; Kim, M. J.; Lee, Y. J.; Lee, J.; Choi, S. H.; Lee, K.; Lah, M. S.; Park, H. *Angew. Chem., Int. Ed.* **2005**, *44*, 1383.
[109] Ooi, T.; Ohara, D.; Tamura, M.; Maruoka, K. *J. Am. Chem. Soc.* **2004**, *126*, 6844.
[110] Barrett, A. G. M.; Blaney, F.; Campbell, A. D.; Hamprecht, D.; Meyer, T.; White, A. J. P.; Witty, D.; Williams, D. J. *J. Org. Chem.* **2002**, *67*, 2735.
[111] Bakó, P.; Makó, A.; Keglevich, G.; Kubinyi, M.; Pál, K. *Tetrahedron: Asymmetry* **2005**, *16*, 1861.
[112] Bakó, P.; Bakó, T.; Mészáros, A.; Keglevich, G.; Szöllösy, A.; Bodor, S.; Makó, A.; Töke, L. *Synlett* **2004**, 643.
[113] Bakó, T.; Bakó, P.; Keglevich, G.; Bombicz, P.; Kubinyi, M.; Pál, K.; Bodor, S.; Makó, A.; Töke, L. *Tetrahedron: Asymmetry* **2004**, *15*, 1589.
[114] Juliá, S.; Masana, J.; Vega, J. C. *Angew. Chem., Int. Ed. Engl.* **1980**, *19*, 929.
[115] Banfi, S.; Colonna, S.; Molinari, H.; Julia, S.; Guixer, J. *Tetrahedron* **1984**, *40*, 5207.
[116] Bentley, P. A.; Bergeron, S.; Cappi, M. W.; Hibbs, D. E.; Hursthouse, M. B.; Nugent, T. C.; Pulido, R.; Roberts, S. M.; Wu, L. E. *Chem. Commun.* **1997**, 739.
[117] Bickley, J. F.; Hauer, B.; Pena, P. C. A.; Roberts, S. M.; Skidmore, J. *J. Chem. Soc., Perkin Trans. 1* **2001**, 1253.
[118] Allen, J. V.; Drauz, K. H.; Flood, R. W.; Roberts, S. M.; Skidmore, J. *Tetrahedron Lett.* **1999**, *40*, 5417.
[119] Da, C. S.; Wei, J.; Dong, S. L.; Xin, Z. Q.; Liu, D. X.; Xu, Z. Q.; Wang, R. *Synth. Commun.* **2003**, *33*, 2787.
[120] Geller, T.; Krüger, C. M.; Militzer, H. C. *Tetrahedron Lett.* **2004**, *45*, 5069.
[121] Geller, T.; Gerlach, A.; Krüger, C. M.; Militzer, H. C. *Tetrahedron Lett.* **2004**, *45*, 5065.
[122] Geller, T.; Gerlach, A.; Krüger, C. M.; Militzer, H. C. *Chim. Oggi* **2003**, *21*, 6.
[123] Itsuno, S.; Sakakura, M.; Ito, K. *J. Org. Chem.* **1990**, *55*, 6047.
[124] Geller, T.; Roberts, S. M. *J. Chem. Soc., Perkin Trans. 1* **1999**, 1397.
[125] Yi, H.; Zou, G.; Li, Q.; Chen, Q.; Tang, J.; He, M.-Y. *Tetrahedron Lett.* **2005**, *46*, 5665.
[126] Flood, R. W.; Geller, T. P.; Petty, S. A.; Roberts, S. M.; Skidmore, J.; Volk, M. *Org. Lett.* **2001**, *3*, 683.
[127] Kelly, D. R.; Bui, T. T. T.; Caroff, E.; Drake, A. F.; Roberts, S. M. *Tetrahedron Lett.* **2004**, *45*, 3885.
[128] Takagi, R.; Shiraki, A.; Manabe, T.; Kojima, S.; Ohkata, K. *Chem. Lett.* **2000**, 366.

[129] Tsogoeva, S. B.; Wöltinger, J.; Jost, C.; Reichert, D.; Kühnle, A.; Krimmer, H.-P.; Drauz, K. *Synlett* **2002**, 707.
[130] Colonna, S.; Molinari, H.; Banfi, S.; Juliá, S.; Masana, J.; Alvarez, A. *Tetrahedron* **1983**, *39*, 1635.
[131] Dhanda, A.; Drauz, K. H.; Geller, T.; Roberts, S. M. *Chirality* **2000**, *12*, 313.
[132] Juliá, S.; Guixer, J.; Masana, J.; Rocas, J.; Colonna, S.; Annuziata, R.; Molinari, H. *J. Chem. Soc., Perkin Trans. 1* **1982**, 1317.
[133] Bentley, P. A.; Flood, R. W.; Roberts, S. M.; Skidmore, J.; Smith, C. B.; Smith, J. A. *Chem. Commun.* **2001**, 1616.
[134] Berkessel, A.; Gasch, N.; Glaubitz, K.; Koch, C. *Org. Lett.* **2001**, *3*, 3839.
[135] Bentley, P. A.; Cappi, M. W.; Flood, R. W.; Roberts, S. M.; Smith, J. A. *Tetrahedron Lett.* **1998**, *39*, 9297.
[136] Kelly, D. R.; Meek, A.; Roberts, S. M. *Chem. Commun.* **2004**, 2021.
[136a] Kelly, D. R.; Roberts, S. M. *Chem. Commun.* **2004**, 2018.
[137] Carrea, G.; Colonna, S.; Meek, A. D.; Ottolina, G.; Roberts, S. M. *Chem. Commun.* **2004**, 1412.
[138] Carrea, G.; Colonna, S.; Meek, A. D.; Ottolina, G.; Roberts, S. M. *Tetrahedron: Asymmetry* **2004**, *15*, 2945.
[139] Mathew, S. P.; Gunathilagan, S.; Roberts, S. M.; Blackmond, D. G. *Org. Lett.* **2005**, *7*, 4847.
[140] Kelly, D. R.; Caroff, E.; Flood, R. W.; Heal, W.; Roberts, S. M. *Chem. Commun.* **2004**, 2016.
[141] Colonna, S.; Manfredi, A. *Tetrahedron Lett.* **1986**, *27*, 387.
[142] Colonna, S.; Manfredi, A.; Annunziata, R.; Spadoni, M. *Tetrahedron* **1987**, *43*, 2157.
[143] Colonna, S.; Manfredi, A.; Spadoni, M. *Tetrahedron Lett.* **1987**, *28*, 1577.
[144] Colonna, S.; Gaggero, N.; Manfredi, A.; Spadoni, M.; Casella, L.; Carrea, G.; Pasta, P. *Tetrahedron* **1988**, *44*, 5169.
[145] Armstrong, A.; Moss, W. O.; Reeves, J. R. *Tetrahedron: Asymmetry* **2001**, *12*, 2779.
[146] Armstrong, A.; Hayter, B. R. *Chem. Commun.* **1998**, 621.
[147] Klein, S.; Roberts, S. M. *J. Chem. Soc., Perkin Trans. 1* **2002**, 2686.
[148] Wang, Y.-C.; Li, C.-L.; Tseng, H.-L.; Chuang, S.-C.; Yan, T.-H. *Tetrahedron: Asymmetry* **1999**, *10*, 3249.
[149] Wang, Z. X.; Shi, Y. *J. Org. Chem.* **1997**, *62*, 8622.
[150] Wu, X.-Y.; She, X.; Shi, Y. *J. Am. Chem. Soc.* **2002**, *124*, 8792.
[151] Shing, T. K. M.; Leung, G. Y. C.; Luk, T. *J. Org. Chem.* **2005**, *70*, 7279.
[152] Bortolini, O.; Fantin, G.; Fogagnolo, M.; Forlani, R.; Maietti, S.; Pedrini, P. *J. Org. Chem.* **2002**, *67*, 5802.
[153] Bortolini, O.; Fogagnolo, M.; Fantin, G.; Maietti, S.; Medici, A. *Tetrahedron: Asymmetry* **2001**, *12*, 1113.
[154] Seki, M.; Furutani, T.; Imashiro, R.; Kuroda, T.; Yamanaka, T.; Harada, N.; Arakawa, H.; Kusama, M.; Hashiyama, T. *Tetrahedron Lett.* **2001**, *42*, 8201.
[155] Imashiro, R.; Seki, M. *J. Org. Chem.* **2004**, *69*, 4216.
[156] Freedman, T. B.; Cao, X.; Nafie, L. A.; Solladié-Cavallo, A.; Jierry, L.; Bouerat, L. *Chirality* **2004**, *16*, 467.
[157] Solladié-Cavallo, A.; Jierry, L.; Klein, A.; Schmitt, M.; Welter, R. *Tetrahedron: Asymmetry* **2004**, *15*, 3891.
[158] Solladié-Cavallo, A.; Jierry, L.; Lupattelli, P.; Bovicelli, P.; Antonioletti, R. *Tetrahedron* **2004**, *60*, 11375.
[159] Solladié-Cavallo, A.; Jierry, L.; Norouzi-Arasi, H.; Tahmassebi, D. *J. Fluorine Chem.* **2004**, *125*, 1371.
[160] Solladié-Cavallo, A.; Jierry, L.; Klein, A. *C. R. Chim.* **2003**, *6*, 603.
[161] Solladié-Cavallo, A.; Bouérat, L.; Jierry, L. *Eur. J. Org. Chem.* **2001**, 4557.
[162] Solladié-Cavallo, A.; Bouérat, L. *Org. Lett.* **2000**, *2*, 3531.
[162a] Baumstark, A. L.; McCloskey, C. J. *Tetrahedron Lett.* **1987**, *28*, 3311.
[163] Wang, Z. X.; Miller, S. M.; Anderson, O. P.; Shi, Y. *J. Org. Chem.* **1999**, *64*, 6443.
[163a] Katsuki, T.; *Coord. Chem. Rev.* **1995**, *140*, 189.
[164] Jacobsen, E. N.; Deng, L.; Furukawa, Y.; Martínez, L. E. *Tetrahedron* **1994**, *50*, 4323.
[165] Chang, S.; Galvin, J. M.; Jacobsen, E. N. *J. Am. Chem. Soc.* **1994**, *116*, 6937.

[166] Zhang, W.; Loebach, J. L.; Wilson, S. R.; Jacobsen, E. N. *J. Am. Chem. Soc.* **1990**, *112*, 2801.
[167] Zhang, R.; Yu, W. Y.; Wong, K. Y.; Che, C. M. *J. Org. Chem.* **2001**, *66*, 8145.
[168] Marigo, M.; Franzen, J.; Poulsen, T. B.; Zhuang, W.; Jørgensen, K. A. *J. Am. Chem. Soc.* **2005**, *127*, 6964.
[169] Zhuang, W.; Marigo, M.; Jørgensen, K. A. *Org. Biomol. Chem.* **2005**, *3*, 3883.
[170] Lattanzi, A. *Org. Lett.* **2005**, *7*, 2579.
[171] Colonna, S.; Manfredi, A.; Annunziata, R.; Gaggero, N.; Casella, L. *J. Org. Chem.* **1990**, *55*, 5862.
[172] Hu, Y.; Harada, A.; Takahashi, S. *Synth. Commun.* **1988**, *18*, 1607.
[173] Sakuraba, H.; Tanaka, Y. *Org. Prep. Proced. Int.* **1998**, *30*, 226.
[174] Kumar, A.; Bhakuni, V. *Tetrahedron Lett.* **1996**, *37*, 4751.
[175] Carde, L.; Davies, D. H.; Roberts, S. M. *J. Chem. Soc., Perkin Trans. 1* **2000**, 2455.
[176] Adger, B. M.; Barkley, J. V.; Bergeron, S.; Cappi, M. W.; Flowerdew, B. E.; Jackson, M. P.; McCague, R.; Nugent, T. C.; Roberts, S. M. *J. Chem. Soc., Perkin Trans. 1* **1997**, 3501.
[177] Dehmlow, E. V.; Sauerbier, C. *Liebigs Ann. Chem.* **1989**, 181.
[178] Kroutil, W.; Mayon, P.; Lasterra-Sánchez, M. E.; Maddrell, S. J.; Roberts, S. M.; Thornton, S. R.; Todd, C. J.; Tüter, M. *Chem. Commun.* **1996**, 845.
[179] Kroutil, W.; Lasterra-Sánchez, M. E.; Maddrell, S. J.; Mayon, P.; Morgan, P.; Roberts, S. M.; Thornton, S. R.; Todd, C. J.; Tüter, M. *J. Chem. Soc., Perkin Trans. 1* **1996**, 2837.
[180] Ray, P. C.; Roberts, S. M. *Tetrahedron Lett.* **1999**, *40*, 1779.
[181] Ray, P. C.; Roberts, S. M. *J. Chem. Soc., Perkin Trans. 1* **2001**, 149.
[182] Augustyn, J. A. N.; Bezuidenhoudt, B. C. B.; Ferreira, D. *Tetrahedron* **1990**, *46*, 2651.
[183] Chen, W. P.; Egar, A. L.; Hursthouse, M. B.; Malik, K. M. A.; Mathews, J. E.; Roberts, S. M. *Tetrahedron Lett.* **1998**, *39*, 8495.
[184] Daikai, K.; Kamaura, M.; Hanamoto, T.; Inanaga, J. U.S. Patent 6,201,123 (2001); *Chem. Abstr.* **2000**, *133*, 177089.
[185] Lasterra-Sánchez, M. E.; Roberts, S. M. *J. Chem. Soc., Perkin Trans. 1* **1995**, 1467.
[186] Lasterra-Sánchez, M. E.; Felfer, U.; Mayon, P.; Roberts, S. M.; Thornton, S. R.; Todd, C. J. *J. Chem. Soc., Perkin Trans. 1* **1996**, 343.
[187] Kroutil, W.; Mayon, P.; Lasterra-Sánchez, M. E.; Maddrell, S. J.; Roberts, S. M.; Thornton, S. R.; Todd, C. J.; Tüter, M. *Chem. Commun.* **1996**, 2495.
[188] Allen, J. V.; Cappi, M. W.; Kary, P. D.; Roberts, S. M.; Williamson, N. M.; Wu, L. E. *J. Chem. Soc., Perkin Trans. 1* **1997**, 3297.
[189] Allen, J. V.; Bergeron, S.; Griffiths, M. J.; Mukherjee, S.; Roberts, S. M.; Williamson, N. M.; Wu, L. E. *J. Chem. Soc., Perkin Trans. 1* **1998**, 3171.
[190] Lopez-Pedrosa, J. M.; Pitts, M. R.; Roberts, S. M.; Saminathan, S.; Whittall, J. *Tetrahedron Lett.* **2004**, *45*, 5073.
[191] Falck, J. R.; Bhatt, R. K.; Reddy, K. M.; Ye, J. H. *Synlett* **1997**, 481.
[192] Bentley, P. A.; Bickley, J. F.; Roberts, S. M.; Steiner, A. *Tetrahedron Lett.* **2001**, *42*, 3741.
[193] Hauer, B.; Bickley, J. F.; Massue, J.; Pena, P. C. A.; Roberts, S. M.; Skidmore, J. *Can. J. Chem.* **2002**, *80*, 546.
[194] Alcaraz, L.; Macdonald, G.; Ragot, J. P.; Lewis, N.; Taylor, R. J. K. *J. Org. Chem.* **1998**, *63*, 3526.
[195] Macdonald, G.; Alcaraz, L.; Lewis, N. J.; Taylor, R. J. K. *Tetrahedron Lett.* **1998**, *39*, 5433.
[196] Lévai, A.; Adam, W.; Fell, R. T.; Gessner, R.; Patonay, T.; Simon, A.; Tóth, G. *Tetrahedron* **1998**, *54*, 13105.
[197] Harigaya, Y.; Yamaguchi, H.; Onda, M. *Chem. Pharm. Bull.* **1981**, *29*, 1321.
[198] Taylor, R. J. K.; Alcaraz, L.; Kapfer-Eyer, I.; Macdonald, G.; Wei, X.; Lewis, N. *Synthesis* **1998**, 775.
[199] Frohn, M.; Dalkiewicz, M.; Tu, Y.; Wang, Z. X.; Shi, Y. *J. Org. Chem.* **1998**, *63*, 2948.
[200] Shi, Y. European Patent 1,021,426 (1998); *Chem. Abstr.* **1998**, *128*, 294999.
[201] Roberts, S. M.; Pitts, M. R. Great Britain Patent 2,399,816 (2004); *Chem. Abstr.* **2004**, *141*, 314142.
[202] Nanjo, K.; Suzuki, K.; Sekiya, M. *Chem. Pharm. Bull.* **1981**, *29*, 336.
[203] Newman, M. S.; Magerlein, B. J. *Org. React.* **1949**, *5*, 413.
[204] Abdel-Magid, A.; Lantos, I.; Pridgen, L. N. *Tetrahedron Lett.* **1984**, *25*, 3273.

[205] Pridgen, L. N.; Abdel-Magid, A.; Lantos, I.; Shilcrat, S.; Eggleston, D. S. *J. Org. Chem.* **1993**, *58*, 5107.
[206] Palomo, C.; Oiarbide, M.; Sharma, A. K.; González-Rego, M. C.; Linden, A.; García, J. M.; González, A. *J. Org. Chem.* **2000**, *65*, 9007.
[207] Ohkata, K.; Kimura, J.; Shinohara, Y.; Takagi, R.; Hiraga, Y. *Chem. Commun.* **1996**, 2411.
[208] Takagi, R.; Kimura, J.; Shinohara, Y.; Ohba, Y.; Takezono, K.; Hiraga, Y.; Kojima, S.; Ohkata, K. *J. Chem. Soc., Perkin Trans. 1* **1998**, 689.
[209] Shinohara, Y.; Ohba, Y.; Takagi, R.; Kojima, S.; Ohkata, K. *Heterocycles* **2001**, *55*, 9.
[210] Schwartz, A.; Madan, P. B.; Mohacsi, E.; O'Brien, J. P.; Todaro, L. J.; Coffen, D. L. *J. Org. Chem.* **1992**, *57*, 851.
[211] Nangia, A.; Rao, P. B.; Madhavi, N. N. L. *J. Chem. Res. (S)* **1996**, 312.
[212] Baldoli, C.; Del Buttero, P.; Licandro, E.; Maiorana, S.; Papagni, A. *J. Chem. Soc., Chem. Commun.* **1987**, 762.
[213] Baldoli, C.; Del Buttero, P.; Maiorana, S. *Tetrahedron* **1990**, *46*, 7823.
[214] Aggarwal, V. K.; Hynd, G.; Picoul, W.; Vasse, J. L. *J. Am. Chem. Soc.* **2002**, *124*, 9964.
[215] Corey, E. J.; Choi, S. *Tetrahedron Lett.* **1991**, *32*, 2857.
[216] Arai, S.; Shioiri, T. *Tetrahedron Lett.* **1998**, *39*, 2145.
[217] Arai, S.; Shirai, Y.; Ishida, T.; Shioiri, T. *Tetrahedron* **1999**, *55*, 6375.
[218] Arai, S.; Shirai, Y.; Ishida, T.; Shioiri, T. *Chem. Commun.* **1999**, 49.
[219] Arai, S.; Ishida, T.; Shioiri, T. *Tetrahedron Lett.* **1998**, *39*, 8299.
[220] Arai, S.; Shioiri, T. *Tetrahedron* **2002**, *58*, 1407.
[221] Bakó, P.; Szöllösy, A.; Bombicz, P.; Töke, L. *Synlett* **1997**, 291.
[222] Bakó, P.; Vízvárdi, K.; Bajor, Z.; Töke, L. *Chem. Commun.* **1998**, 1193.
[223] Bakó, P.; Vízvárdi, K.; Toppet, S.; Van der Eycken, E.; Hoornaert, G. J.; Töke, L. *Tetrahedron* **1998**, *54*, 14975.
[224] Bakó, P.; Czinege, E.; Bakó, T.; Czugler, M.; Töke, L. *Tetrahedron: Asymmetry* **1999**, *10*, 4539.
[225] Annunziata, R.; Banfi, S.; Colonna, S. *Tetrahedron Lett.* **1985**, *26*, 2471.
[226] Takahashi, T.; Muraoka, M.; Capo, M.; Koga, K. *Chem. Pharm. Bull.* **1995**, *43*, 1821.
[227] Meth-Cohn, W.; Williams, D. J.; Chen, Y. *Chem. Commun.* **2000**, 495.
[228] Bhatia, G. S.; Lowe, R. F.; Pritchard, R. G.; Stoodley, R. J. *Chem. Commun.* **1997**, 1981.
[229] Cardani, S.; Gennari, C.; Scolastico, C.; Villa, R. *Tetrahedron* **1989**, *45*, 7397.
[230] Cardani, S.; Bernardi, A.; Colombo, L.; Gennari, C.; Scolastico, C.; Venturini, I. *Tetrahedron* **1988**, *44*, 5563.
[231] Wipf, P.; Kim, Y. T.; Jahn, H. *Synthesis* **1995**, 1549.
[232] Andres, C. J.; Spetseris, N.; Norton, J. R.; Meyers, A. I. *Tetrahedron Lett.* **1995**, *36*, 1613.
[233] Klunder, A. J. H.; Huizinga, W. B.; Sessink, P. J. M.; Zwanenburg, B. *Tetrahedron Lett.* **1987**, *28*, 357.
[234] Pettersson-Fasth, H.; Riesinger, S. W.; Bäckvall, J. E. *J. Org. Chem.* **1995**, *60*, 6091.
[235] De Mico, A.; Margarita, R.; Parlanti, L.; Vescovi, A.; Piancatelli, G. *J. Org. Chem.* **1997**, *62*, 6974.
[236] Schwab, J. M.; Ray, T.; Ho, C. K. *J. Am. Chem. Soc.* **1989**, *111*, 1057.
[237] Schwab, J. M.; Ho, C. K. *J. Chem. Soc., Chem. Commun.* **1986**, 872.
[238] Lin, G.; Xu, H.; Wu, B.; Guo, G.; Zhu, W. *Tetrahedron Lett.* **1985**, *26*, 1233.
[239] Corey, E. J.; Mehrotra, M. M. *Tetrahedron Lett.* **1986**, *27*, 5173.
[240] Kabat, M. M. *Tetrahedron Lett.* **1993**, *34*, 8543.
[241] Molander, G. A.; Hahn, G. *J. Org. Chem.* **1986**, *51*, 2596.
[242] Nicolaou, K. C.; Prasad, C. V. C.; Somers, P. K.; Hwang, C. K. *J. Am. Chem. Soc.* **1989**, *111*, 5335.
[243] Behrens, C. H.; Sharpless, K. B. *J. Org. Chem.* **1985**, *50*, 5696.
[244] Boger, D. L.; Patane, M. A.; Zhou, J. *J. Am. Chem. Soc.* **1994**, *116*, 8544.
[245] Yabe, Y.; Guillaume, D.; Rich, D. H. *J. Am. Chem. Soc.* **1988**, *110*, 4043.
[246] Baldwin, J. E.; Adlington, R. M.; Godfrey, C. R. A.; Patel, V. K. *Tetrahedron* **1993**, *49*, 7837.
[247] Okamoto, S.; Tsujiyama, H.; Yoshino, T.; Sato, F. *Tetrahedron Lett.* **1991**, *32*, 5789.
[248] Mello, R.; Cassidei, L.; Fiorentino, M.; Fusco, C.; Hümmer, W.; Jäger, V.; Curci, R. *J. Am. Chem. Soc.* **1991**, *113*, 2205.

[249] Lindsten, G.; Wennerström, O.; Isaksson, R. *J. Org. Chem.* **1987**, *52*, 547.
[250] Harada, T.; Mai, T.; Tuyet, T.; Oku, A. *Org. Lett.* **2000**, *2*, 1319.
[251] Starks, C.; Liotta, C.; Halpern, M. *Phase-Transfer Catalysis*; Chapman & Hall: New York, 1994.
[252] Baars, S.; Drauz, K. H.; Krimmer, H. P.; Roberts, S. M.; Sander, J.; Skidmore, J.; Zanardi, G. *Org. Proc. Res. Dev.* **2003**, *7*, 509.
[253] Geller, T.; Gerlach, A.; Vidal-Ferran, A.; Militzer, H.-C.; Langer, R. Intl. Patent WO 2003070808 (1-1-2003); *Chem. Abstr.* **2003**, *139*, 214720.
[254] Gerlach, A.; Geller, T. *Adv. Synth. Catal.* **2004**, *346*, 1247.
[255] Marigo, M.; Wabnitz, T. C.; Fielenbach, D.; Jørgensen, K. A. *Angew. Chem., Int. Ed.* **2005**, *44*, 794.
[256] Bickley, J. F.; Gillmore, A. T.; Roberts, S. M.; Skidmore, J.; Steiner, A. *J. Chem. Soc., Perkin Trans. 1* **2001**, 1109.
[257] Yamada, K.; Arai, T.; Sasai, H.; Shibasaki, M. *J. Org. Chem.* **1998**, *63*, 3666.
[258] Jayaprakash, D.; Kobayashi, Y.; Arai, T.; Hu, Q. S.; Zheng, X. F.; Pu, L.; Sasai, H. *J. Mol. Catal. A: Chem.* **2003**, *196*, 145.
[259] Roberts, S. M.; Adger, B. M. U.S. Patent 6,228,955 (2001); *Chem. Abstr.* **1996**, *126*, 18775.
[260] Drauz, K.; Roberts, S. M.; Geller, T.; Dhanda, A. U.S. Patent 6,538,105 (2003); *Chem. Abstr.* **2000**, *133*, 30654.
[261] Gillmore, A.; Lauret, C.; Roberts, S. M. *Tetrahedron* **2003**, *59*, 4363.
[262] Coffey, P. E.; Drauz, K. H.; Roberts, S. M.; Skidmore, J.; Smith, J. A. *Chem. Commun.* **2001**, 2330.
[263] Kagawa, T.; Kambara, T.; Sakka, H.; Yanase, M. U.S. Patent 7,169,944 (2007); *Chem. Abstr.* **2004**, *140*, 303510.
[264] Chen, R. F.; Qian, C. T.; de Vries, J. G.; Sun, P. P.; Wang, L. M. *Chin. J. Chem.* **2001**, *19*, 1225.
[265] Kuboshima, Y.; Matsunaga, S.; Shibasaki, M. U.S. Patent 7,371,578 (2008); *Chem. Abstr.* **2003**, *140*, 28760.
[266] Takagi, R.; Begum, S.; Siraki, A.; Yoneshige, A.; Koyama, K.-i.; Ohkata, K. *Heterocycles* **2004**, *64*, 129.
[267] Baures, P. W.; Eggleston, D. S.; Flisak, J. R.; Gombatz, K.; Lantos, I.; Mendelson, W.; Remich, J. J. *Tetrahedron Lett.* **1990**, *31*, 6501.
[268] Takagi, R.; Manabe, T.; Shiraki, A.; Yoneshige, A.; Hiraga, Y.; Kojima, S.; Ohkata, K. *Bull. Chem. Soc. Jpn.* **2000**, *73*, 2115.
[269] Cappi, M. W.; Chen, W. P.; Flood, R. W.; Liao, Y. W.; Roberts, S. M.; Skidmore, J.; Smith, J. A.; Williamson, N. M. *Chem. Commun.* **1998**, 1159.
[270] Carde, L.; Davies, H.; Geller, T. P.; Roberts, S. M. *Tetrahedron Lett.* **1999**, *40*, 5421.
[271] Savizky, R. M.; Suzuki, N.; Bove, J. L. *Tetrahedron: Asymmetry* **1998**, *9*, 3967.
[272] Bentley, P. A.; Kroutil, W.; Littlechild, J. A.; Roberts, S. M. *Chirality* **1997**, *9*, 198.
[273] Julia, S.; Masana, J.; Rocas, J.; Colonna, S.; Annunziata, R.; Molinari, H. *Anal. Quim. C* **1983**, *79*, 102.
[274] Tanaka, A.; Kagawa, T. European Patent 1,127,885 (2001); *Chem. Abstr.* **2001**, *135*, 195490.
[275] van Rensburg, H.; vanHeerden, P. S.; Bezuidenhoudt, B. C. B.; Ferreira, D. *Chem. Commun.* **1996**, 2747.
[276] van Rensburg, H.; vanHeerden, P. S.; Bezuidenhoudt, B. C. B.; Ferreira, D. *Tetrahedron* **1997**, *53*, 14141.
[277] Nel, R. J. J.; van Heerden, P. S.; van Rensburg, H.; Ferreira, D. *Tetrahedron Lett.* **1998**, *39*, 5623.
[278] Nel, R. J. J.; van Rensburg, H.; van Heerden, P. S.; Coetzee, J.; Ferreira, D. *Tetrahedron* **1999**, *55*, 9727.
[279] Bezuidenhoudt, B. C. B.; Swanepoel, A.; Augustyn, J. A. N.; Ferreira, D. *Tetrahedron Lett.* **1987**, *28*, 4857.
[280] Flisak, J. R.; Gassman, P. G.; Lantos, I.; Mendelson, W. L. European Patent 403252A2 (1990); *Chem. Abstr.* **1991**, *115*, 471370.
[281] Flisak, J. R.; Gombatz, K. J.; Holmes, M. M.; Jarmas, A. A.; Lantos, I.; Mendelson, W. L.; Novack, V. J.; Remich, J. J.; Snyder, L. *J. Org. Chem.* **1993**, *58*, 6247.
[282] Marsman, B.; Wynberg, H. *J. Org. Chem.* **1979**, *44*, 2312.

[283] Chen, W. P.; Roberts, S. M. *J. Chem. Soc., Perkin Trans. 1* **1999**, 103.
[284] Nemoto, T.; Ohshima, T.; Shibasaki, M. *Tetrahedron Lett.* **2000**, *49*, 9569.
[285] Nemoto, T.; Ohshima, T.; Shibasaki, M. *Tetrahedron* **2003**, *59*, 6889.
[286] Fruh, T. *Agro Food Industry Hi-Tech* **1996**, *7*, 31.
[287] Domagala, J. M.; Bach, R. D. *J. Org. Chem.* **1979**, *44*, 3168.
[288] Adam, W.; Fell, R. T.; Levai, A.; Patonay, T.; Peters, K.; Simon, A.; Toth, G. *Tetrahedron: Asymmetry* **1998**, *9*, 1121.
[289] Briggs, J. C.; Hodge, P.; Zhang, Z. P. *React. Polym.* **1993**, *19*, 73.
[290] Shi, Y. Intl. Patent WO 03066614 (2003); *Chem. Abstr.* **2003**, *139*, 179963.
[291] Jacobsen, E. N.; Zhang, W.; Deng, L. U.S. Patent 5,637,739 (1997); *Chem. Abstr.* **1997**, *127*, 108833.
[292] Jacobsen, E. N.; Zhang, W.; Muci, A. R.; Ecker, J. R.; Deng, L. *J. Am. Chem. Soc.* **1991**, *113*, 7063.
[293] Deng, L.; Jacobsen, E. N. *J. Org. Chem.* **1992**, *57*, 4320.
[294] Tosaki, S.-Y.; Nemoto, T.; Kakei, H.; Gnanadesikan, V.; Ohshima, T.; Shibasaki, M. *Kidorui* **2003**, *42*, 62.
[295] Ohshima, T.; Gnanadesikan, V.; Shibuguchi, T.; Fukuta, Y.; Nemoto, T.; Shibasaki, M. *J. Am. Chem. Soc.* **2003**, *125*, 11206.
[296] Wang, X. Y.; Shi, H. C.; Sun, C.; Zhang, Z. G. *Tetrahedron* **2004**, *60*, 10993.

CUMULATIVE CHAPTER TITLES BY VOLUME

Volume 1 (1942)

1. **The Reformatsky Reaction:** Ralph L. Shriner
2. **The Arndt-Eistert Reaction:** W. E. Bachmann and W. S. Struve
3. **Chloromethylation of Aromatic Compounds:** Reynold C. Fuson and C. H. McKeever
4. **The Amination of Heterocyclic Bases by Alkali Amides:** Marlin T. Leffler
5. **The Bucherer Reaction:** Nathan L. Drake
6. **The Elbs Reaction:** Louis F. Fieser
7. **The Clemmensen Reduction:** Elmore L. Martin
8. **The Perkin Reaction and Related Reactions:** John R. Johnson
9. **The Acetoacetic Ester Condensation and Certain Related Reactions:** Charles R. Hauser and Boyd E. Hudson, Jr.
10. **The Mannich Reaction:** F. F. Blickc
11. **The Fries Reaction:** A. H. Blatt
12. **The Jacobson Reaction:** Lee Irvin Smith

Volume 2 (1944)

1. **The Claisen Rearrangement:** D. Stanley Tarbell
2. **The Preparation of Aliphatic Fluorine Compounds:** Albert L. Henne
3. **The Cannizzaro Reaction:** T. A. Geissman
4. **The Formation of Cyclic Ketones by Intramolecular Acylation:** William S. Johnson
5. **Reduction with Aluminum Alkoxides (The Meerwein-Ponndorf-Verley Reduction):** A. L. Wilds
6. **The Preparation of Unsymmetrical Biaryls by the Diazo Reaction and the Nitrosoacetylamine Reaction:** Werner E. Bachmann and Roger A. Hoffman

Volume 15 (1967)

1. **The Dieckmann Condensation:** John P. Schaefer and Jordan J. Bloomfield
2. **The Knoevenagel Condensation:** G. Jones

Volume 16 (1968)

1. **The Aldol Condensation:** Arnold T. Nielsen and William J. Houlihan

Volume 17 (1969)

1. **The Synthesis of Substituted Ferrocenes and Other π-Cyclopentadienyl-Transition Metal Compounds:** Donald E. Bublitz and Kenneth L. Rinehart, Jr.
2. **The γ-Alkylation and γ-Arylation of Dianions of β-Dicarbonyl Compounds:** Thomas M. Harris and Constance M. Harris
3. **The Ritter Reaction:** L. I. Krimen and Donald J. Cota

Volume 18 (1970)

1. **Preparation of Ketones from the Reaction of Organolithium Reagents with Carboxylic Acids:** Margaret J. Jorgenson
2. **The Smiles and Related Rearrangements of Aromatic Systems:** W. E. Truce, Eunice M. Kreider, and William W. Brand
3. **The Reactions of Diazoacetic Esters with Alkenes, Alkynes, Heterocyclic, and Aromatic Compounds:** Vinod Dave and E. W. Warnhoff
4. **The Base-Promoted Rearrangements of Quaternary Ammonium Salts:** Stanley H. Pine

Volume 19 (1972)

1. **Conjugate Addition Reactions of Organocopper Reagents:** Gary H. Posner
2. **Formation of Carbon–Carbon Bonds via π-Allylnickel Compounds:** Martin F. Semmelhack
3. **The Thiele-Winter Acetoxylation of Quinones:** J. F. W. McOmie and J. M. Blatchly
4. **Oxidative Decarboxylation of Acids by Lead Tetraacetate:** Roger A. Sheldon and Jay K. Kochi

Volume 20 (1973)

1. **Cyclopropanes from Unsaturated Compounds, Methylene Iodide, and Zinc-Copper Couple:** H. E. Simmons, T. L. Cairns, Susan A. Vladuchick, and Connie M. Hoiness
2. **Sensitized Photooxygenation of Olefins:** R. W. Denny and A. Nickon
3. **The Synthesis of 5-Hydroxyindoles by the Nenitzescu Reaction:** George R. Allen, Jr.
4. **The Zinin Reaction of Nitroarenes:** H. K. Porter

Volume 21 (1974)

1. **Fluorination with Sulfur Tetrafluoride:** G. A. Boswell, Jr., W. C. Ripka, R. M. Scribner, and C. W. Tullock

2. **Modern Methods to Prepare Monofluoroaliphatic Compounds:** Clay M. Sharts and William A. Sheppard

Volume 22 (1975)

1. **The Claisen and Cope Rearrangements:** Sara Jane Rhoads and N. Rebecca Raulins

2. **Substitution Reactions Using Organocopper Reagents:** Gary H. Posner

3. **Clemmensen Reduction of Ketones in Anhydrous Organic Solvents:** E. Vedejs

4. **The Reformatsky Reaction:** Michael W. Rathke

Volume 23 (1976)

1. **Reduction and Related Reactions of α, β-Unsaturated Compounds with Metals in Liquid Ammonia:** Drury Caine

2. **The Acyloin Condensation:** Jordan J. Bloomfield, Dennis C. Owsley, and Janice M. Nelke

3. **Alkenes from Tosylhydrazones:** Robert H. Shapiro

Volume 24 (1976)

1. **Homogeneous Hydrogenation Catalysts in Organic Solvents:** Arthur J. Birch and David H. Williamson

2. **Ester Cleavages via S_N2-Type Dealkylation:** John E. McMurry

3. **Arylation of Unsaturated Compounds by Diazonium Salts (The Meerwein Arylation Reaction):** Christian S. Rondestvedt, Jr.

4. **Selenium Dioxide Oxidation:** Norman Rabjohn

Volume 25 (1977)

1. **The Ramberg-Bäcklund Rearrangement:** Leo A. Paquette

2. **Synthetic Applications of Phosphoryl-Stabilized Anions:** William S. Wadsworth, Jr.

3. **Hydrocyanation of Conjugated Carbonyl Compounds:** Wataru Nagata and Mitsuru Yoshioka

Volume 26 (1979)

1. **Heteroatom-Facilitated Lithiations:** Heinz W. Gschwend and Herman R. Rodriguez

2. **Intramolecular Reactions of Diazocarbonyl Compounds:** Steven D. Burke and Paul A. Grieco

Volume 27 (1982)

1. **Allylic and Benzylic Carbanions Substituted by Heteroatoms:** Jean-François Biellmann and Jean-Bernard Ducep

2. **Palladium-Catalyzed Vinylation of Organic Halides:** Richard F. Heck

Volume 28 (1982)

1. **The Reimer-Tiemann Reaction:** Hans Wynberg and Egbert W. Meijer

2. **The Friedländer Synthesis of Quinolines:** Chia-Chung Cheng and Shou-Jen Yan

3. **The Directed Aldol Reaction:** Teruaki Mukaiyama

Volume 29 (1983)

1. **Replacement of Alcoholic Hydroxy Groups by Halogens and Other Nucleophiles via Oxyphosphonium Intermediates:** Bertrand R. Castro

2. **Reductive Dehalogenation of Polyhalo Ketones with Low-Valent Metals and Related Reducing Agents:** Ryoji Noyori and Yoshihiro Hayakawa

3. **Base-Promoted Isomerizations of Epoxides:** Jack K. Crandall and Marcel Apparu

Volume 30 (1984)

1. **Photocyclization of Stilbenes and Related Molecules:** Frank B. Mallory and Clelia W. Mallory

2. **Olefin Synthesis via Deoxygenation of Vicinal Diols:** Eric Block

Volume 31 (1984)

1. **Addition and Substitution Reactions of Nitrile-Stabilized Carbanions:** Siméon Arseniyadis, Keith S. Kyler, and David S. Watt

Volume 32 (1984)

1. **The Intramolecular Diels-Alder Reaction:** Engelbert Ciganek

2. **Synthesis Using Alkyne-Derived Alkenyl- and Alkynylaluminum Compounds:** George Zweifel and Joseph A. Miller

Volume 33 (1985)

1. **Formation of Carbon–Carbon and Carbon–Heteroatom Bonds via Organoboranes and Organoborates:** Ei-Ichi Negishi and Michael J. Idacavage

2. **The Vinylcyclopropane-Cyclopentene Rearrangement:** Tomáš Hudlický, Toni M. Kutchan, and Saiyid M. Naqvi

Volume 40 (1991)

1. **The Pauson-Khand Cycloaddition Reaction for Synthesis of Cyclopentenones:** Neil E. Schore
2. **Reduction with Diimide:** Daniel J. Pasto and Richard T. Taylor
3. **The Pummerer Reaction of Sulfinyl Compounds:** Ottorino DeLucchi, Umberto Miotti, and Giorgio Modena
4. **The Catalyzed Nucleophilic Addition of Aldehydes to Electrophilic Double Bonds:** Hermann Stetter and Heinrich Kuhlmann

Volume 41 (1992)

1. **Divinylcyclopropane-Cycloheptadiene Rearrangement:** Tomáš Hudlický, Rulin Fan, Josephine W. Reed, and Kumar G. Gadamasetti
2. **Organocopper Reagents: Substitution, Conjugate Addition, Carbo/Metallo-cupration, and Other Reactions:** Bruce H. Lipshutz and Saumitra Sengupta

Volume 42 (1992)

1. **The Birch Reduction of Aromatic Compounds:** Peter W. Rabideau and Zbigniew Marcinow
2. **The Mitsunobu Reaction:** David L. Hughes

Volume 43 (1993)

1. **Carbonyl Methylenation and Alkylidenation Using Titanium-Based Reagents:** Stanley H. Pine
2. **Anion-Assisted Sigmatropic Rearrangements:** Stephen R. Wilson
3. **The Baeyer-Villiger Oxidation of Ketones and Aldehydes:** Grant R. Krow

Volume 44 (1993)

1. **Preparation of α,β-Unsaturated Carbonyl Compounds and Nitriles by Selenoxide Elimination:** Hans J. Reich and Susan Wollowitz
2. **Enone Olefin [2 + 2] Photochemical Cyclizations:** Michael T. Crimmins and Tracy L. Reinhold

Volume 45 (1994)

1. **The Nazarov Cyclization:** Karl L. Habermas, Scott E. Denmark, and Todd K. Jones
2. **Ketene Cycloadditions:** John Hyatt and Peter W. Raynolds

Volume 46 (1994)

1. **Tin(II) Enolates in the Aldol, Michael, and Related Reactions:** Teruaki Mukaiyama and Shū Kobayashi

2. **Dioxirane Epoxidation of Alkenes:** Waldemar Adam, Chantu R. Saha-Möller, and Cong-Gui Zhao

Volume 62 (2003)

1. **α-Hydroxylation of Enolates and Silyl Enol Ethers:** Bang-Chi Chen, Ping Zhou, Franklin A. Davis, and Engelbert Ciganek
2. **The Ramberg-Bäcklund Reaction:** Richard J. K. Taylor and Guy Casy
3. **The α-Hydroxy Ketone (α-Ketol) and Related Rearrangements:** Leo A. Paquette and John E. Hofferberth
4. **Transformation of Glycals into 2,3-Unsaturated Glycosyl Derivatives:** Robert J. Ferrier and Oleg A. Zubkov

Volume 63 (2004)

1. **The Biginelli Dihydropyrimidine Synthesis:** C. Oliver Kappe and Alexander Stadler
2. **Microbial Arene Oxidations:** Roy A. Johnson
3. **Cu, Ni, and Pd Mediated Homocoupling Reactions in Biaryl Syntheses: The Ullmann Reaction:** Todd D. Nelson and R. David Crouch

Volume 64 (2004)

1. **Additions of Allyl, Allenyl, and Propargylstannanes to Aldehydes and Imines:** Benjamin W. Gung
2. **Glycosylation with Sulfoxides and Sulfinates as Donors or Promoters:** David Crich and Linda B. L. Lim
3. **Addition of Organochromium Reagents to Carbonyl Compounds:** Kazuhiko Takai

Volume 65 (2005)

1. **The Passerini Reaction:** Luca Banfi and Renata Riva
2. **Diels-Alder Reactions of Imino Dienophiles:** Geoffrey R. Heintzelman, Ivona R. Meigh, Yogesh R. Mahajan, and Steven M. Weinreb

Volume 66 (2005)

1. **The Allylic Trihaloacetimidate Rearrangement:** Larry E. Overman and Nancy E. Carpenter
2. **Asymmetric Dihydroxylation of Alkenes:** Mark C. Noe, Michael A. Letavic, Sheri L. Snow, and Stuart McCombie

Volume 67 (2006)

1. **Catalytic Enantioselective Aldol Addition Reactions:** Erick M. Carreira, Alec Fettes, and Christiane Marti
2. **Benzylic Activation and Stereochemical Control in Reactions of Tricarbonyl(Arene)-Chromium Complexes:** Motokazu Uemura

Volume 68 (2007)

1. **Cotrimerizations of Acetylenic Compounds:** Nicolas Agenet, Olivier Buisine, Franck Slowinski, Vincent Gandon, Corinne Aubert, and Max Malacria

2. **Glycosylation on Polymer Supports:** Simone Bufali and Peter H. Seeberger

Volume 69 (2007)

1. **Dioxirane Oxidations of Compounds Other Than Alkenes:** Waldemar Adam, Cong-Gui Zhao, and Kavitha Jakka

2. **Electrophilic Fluorination with N–F Reagents:** Jérôme Baudoux and Dominique Cahard

Volume 70 (2008)

1. **The Catalytic Asymmetric Strecker Reaction:** Masakatsu Shibasaki, Motomu Kanai, and Tsuyoshi Mita

2. **The Synthesis of Phenols and Quinones via Fischer Carbene:** Marcey L. Waters and William D. Wulff

Volume 71 (2008)

1. **Ionic and Organometallic-Catalyzed Organosilane Reductions:** Gerald L. Larson and James L. Fry

Volume 72 (2008)

1. **Electrophilic Amination of Carbanions, Enolates, and Their Surrogates:** Engelbert Ciganek

2. **Desulfonylation Reactions:** Diego A. Alonso and Carmen Nájera

Volume 73 (2008)

1. **Allylboration of Carbonyl Compounds:** Hugo Lachance and Dennis G. Hall

AUTHOR INDEX, VOLUMES 1–74

Volume number only is designated in this index

Organic Reactions, Vol. 74, Edited by Scott E. Denmark et al.

CHAPTER AND TOPIC INDEX, VOLUMES 1–74

Many chapters contain brief discussions of reactions and comparisons of alternative synthetic methods related to the reaction that is the subject of the chapter. These related reactions and alternative methods are not usually listed in this index. In this index, the volume number is in **boldface**, the chapter number is in ordinary type.

Organic Reactions, Vol. 74, Edited by Scott E. Denmark et al.

DATE DUE

CANISIUS COLLEGE LIBRARY
3 5084 00495 8214